EWORK AND EBUSINESS IN ARCHITECTURE, ENGINEERING AND CONSTRUCTION

PROCEEDINGS OF THE 11TH EUROPEAN CONFERENCE ON PRODUCT AND PROCESS MODELLING (ECPPM2016), LIMASSOL, CYPRUS, 7–9 SEPTEMBER 2016

eWork and eBusiness in Architecture, Engineering and Construction

Editors

Symeon E. Christodoulou
University of Cyprus, Nicosia, Cyprus

Raimar Scherer
University of Technology, Dresden, Germany

CRC Press is an imprint of the
Taylor & Francis Group, an **informa** business
A BALKEMA BOOK

CRC Press/Balkema is an imprint of the Taylor & Francis Group, an informa business

Typeset by V Publishing Solutions Pvt Ltd., Chennai, India

Published by: CRC Press/Balkema
P.O. Box 11320, 2301 EH Leiden, The Netherlands
e-mail: Pub.NL@taylorandfrancis.com
www.crcpress.com – www.taylorandfrancis.com

ISBN: 978-1-138-03280-4 (Hbk)
ISBN: 978-1-315-38690-4 (eBook PDF)

eWork and eBusiness in Architecture, Engineering and Construction – Christodoulou & Scherer (Eds)
© 2016 Taylor & Francis Group, ISBN 978-1-138-03280-4

Table of contents

Smart cities

Special session: Energy efficient neighborhoods (ee-Neighborhoods)

eWork and eBusiness in Architecture, Engineering and Construction – Christodoulou & Scherer (Eds)
© 2016 Taylor & Francis Group, ISBN 978-1-138-03280-4

Preface

Dear Reader,

Nowadays, amidst global and prolonged economic slowdown, climatic change, increased urbanization, rapidly evolving technology, and fierce competition in the Architectural, Engineering and Construction (AEC) industry, the development and implementation of scientific and applied knowledge for improved product and process modelling is of paramount importance to not only national agencies and firms but also to transnational networks.

Such scientific knowledge and policies should cover a myriad of issues related to product and process modelling including, but not limited to, Building Information Models (BIM), energy efficiency at the building and urban scales, Information and Communication Technology (ICT) applications in the AEC/FM domains, information and knowledge management, ontologies, data models and interoperability, smart cities, human requirements and human factors.

The following sections discuss most of the aforementioned important components of product and process modelling, as presented at the *Eleventh European Conference on Product and Process Modelling (ECPPM2016)*, held in Limassol, Cyprus (7–9 Sep. 2016). ECPPM is the flagship conference event of the European Association of Product and Process Modelling (EAPPM), with a long standing history of excellence in product and process modelling in the building industry, which is currently known as Building Information Modelling (BIM).

The conference aimed to provide an international forum for the exchange of scientific information and knowledge-sharing on state-of-the-art research efforts and on contemporary product and process modelling issues, covering a large spectrum of topics pertaining to ICT deployment instances in AEC/FM, attracting high quality research papers and providing a platform for the cross fertilization of new ideas and know-how in relation to the special conference themes.

The work presented and included in the conference proceedings constitutes an excellent blending of cutting-edge research, of scientific and applied knowledge, and of case-studies across the globe which should be of great interest to both researchers and practitioners since it offers the European and the international community of product and process modelling professionals a great opportunity to experience the latest achievements in research, science, practice and management related to ECPPM. The proceedings include work from researchers from a total of 25 countries across the globe, and on topics covering the full spectrum of the conference's thematic areas. Further, the conference hosted two special sessions, titled "*Modelling and Simulation for Energy Efficient-Neighbourhoods (ee-Neighborhoods)*" and "*Use Case Data Requirements for Product Models in Building Energy Management Processes*", as well as research outputs from several EU-funded projects.

S.E. Christodoulou, PhD
Conference Host and Chair

eWork and eBusiness in Architecture, Engineering and Construction – Christodoulou & Scherer (Eds)
© 2016 Taylor & Francis Group, ISBN 978-1-138-03280-4

Organization

STEERING COMMITTEE

Chairperson

Raimar J. Scherer, *Technische Universität Dresden, Germany*

Vice Chairpersons

Ziga Turk, *University of Ljubljana, Slovenia*
Symeon Christodoulou, *University of Cyprus, Cyprus*
Ardeshir Mahdavi, *Vienna University of Technology, Austria*

Members

Robert Amor, *University of Auckland, New Zealand*
Ezio Arlati, *Politecnico di Milano, Italy*
Jakob Beetz, *Eindhoven University of Technology, The Netherlands*
Adam Borkowski, *Institute of Fundamental Technological Research, Polish Academy of Sciences, Poland*
Jan Cervenka, *Cervenka Consulting, Czech Republic*
Attila Dikbas, *Istanbul Technical University, Turkey*
Ricardo Gonçalves, *New University of Lisbon, UNINOVA, Portugal*
Gudni Gudnason, *Innovation Centre, Iceland*
Noemi Jimenez Redondo, *CEMOSA, Spain*
Jan Karlshøj, *Technical University of Denmark, Denmark*
Tuomas Laine, *Granlund, Finland*
Karsten Menzel, *University College Cork, Ireland*
Sergio Munoz, *AIDICO, Instituto Technologia de la Construcción, Spain*
Pieter Pauwels, *Ghent University, Belgium*
Byron Protopsaltis, *Sofistik Hellas, Greece*
Svetla Radeva, *College of Telecommunications and Post, Sofia, Bulgaria*
Yacine Rezgui, *Cardiff University, UK*
Dimitrios Rovas, *Technical University of Crete, Greece*
Vitaly Semenov, *Institute for System Programming RAS, Russia*
Ales Siroky, *Nemetschek, Slovakia*
Ian Smith, *EPFL—Ecole Polytechnique Fédérale de Lausanne, Switzerland*
Rasso Steinmann, *Institute for Applied Building Informatics, University of Munich, Germany*
Väino Tarandi, *KTH—Royal Institute of Technology, Sweden*
Alain Zarli, *CSTB, France*

Retired Members

Bo-Christer Björk, *Swedish School of Economics and Business Administration, Finland*
Per Christiansson, *Per Christiansson Ingenjörs Byrå HB, Sweden*
Anders Ekholm, *Lund University, Sweden*
Godfried Augenbroe, *Georgia Institute of Technology, USA*
Matti Hannus, *VTT Technical Research Centre of Finland, Finland*
Ulrich Walder, *Graz University of Technology, Austria*

INTERNATIONAL SCIENTIFIC COMMITTEE

Robert Amor, *University of Auckland, New Zealand*
Chimay Anumba, *Pennsylvania State University, USA*
Ezio Arlati, *Politecnico di Milano, Italy*
Godfried Augenbroe, *Georgia Institute of Technology, USA*
Håvard Bell, *Catenda, Norway*
Michel Bohms, *TNO, The Netherlands*
André Borrmann, *Technische Universität München, Germany*
Tomo Cerovsek, *University of Ljubljana, Slovenia*
Jan Cervenka, *Cervenka Consulting, Czech Republic*
Edwin Dado, *Nederlandse Defensie Academie, The Netherlands*
Nashwan N. Dawood, *Centre for Construction Innovation and Research, University of Teesside, UK*
Attila Dikbas, *Istanbul Technical University, Turkey*
Robin Drogemuller, *Queensland University of Technology UT CSIRO, Australia*
Anders Ekholm, *Lund University, Sweden*
Bruno Fies, *CSTB, France*
Christer Finne, *Building Information Foundation, Finland*
Martin Fischer, *Center for Integrated Facility Engineering, Stanford University, USA*
Thomas Froese, *University of British Columbia, Canada*
Gudni Gudnason, *Innovation Centre, Iceland*
Tarek Hassan, *Loughborough University, UK*
Eilif Hjelseth, *Digitale UMB, Norway*
Wolfgang Huhnt, *Technische Universität Berlin, Germany*
Ricardo Jardim-Goncalves, *Universidade Nova de Lisboa, Portugal*
Peter Katranuschkov, *Technische Universitaet Dresden, Germany*
Abdul Samad (Sami) Kazi, *VTT Technical Research Centre of Finland, Finland*
Arto Kiviniemi, *University of Liverpool, UK*
Bob Martens, *Vienna University of Technology, Austria*
Karsten Menzel, *University College Cork, Ireland*
Sergio Munoz, *AIDICO, Instituto Technologia de la Construcción, Spain*
Svetla Radeva, *College of Telecommunications and Post, Sofia, Bulgaria*
Iñaki Angulo Redondo, *TECNALIA, ICT Division, European Software Institute, Spain*
Yacine Rezgui, *Cardiff University, UK*
Uwe Rueppel, *Technical University of Darmstadt, Germany*
Vitaly Semenov, *Institute for System Programming RAS, Russia*
Miroslaw J. Skibniewski, *University of Maryland, USA*
Ian Smith, *EPFL—Ecole Polytechnique Fédérale de Lausanne, Switzerland*
Rasso Steinmann, *Institute for Applied Building Informatics. University of Munich, Germany*
Väino Tarandi, *KTH—Royal Institute of Technology, Sweden*
Walid Tizani, *University of Nottingham, UK*
Hakan Yaman, *Istanbul Technical University, Turkey*
Pedro Nuno Mêda Magalhães, *Porto University, Portugal*

LOCAL ORGANIZING COMMITEE

Symeon Christodoulou (Chair), *University of Cyprus, Cyprus*
Loukas Dimitriou, *University of Cyprus, Cyprus*
Odysseas Kontovourkis, *University of Cyprus, Cyprus*
Agathoklis Agathokleous, *University of Cyprus, Cyprus*

Keynote speakers

eWork and eBusiness in Architecture, Engineering and Construction – Christodoulou & Scherer (Eds)
© 2016 Taylor & Francis Group, ISBN 978-1-138-03280-4

Data rich design, construction, and operations of sustainable buildings and infrastructure systems

Lucio Soibelman
The Astani Department of Civil and Environmental Engineering, University of Southern California, Los Angeles, USA

ABSTRACT: It is certainly no surprise that construction and operations of buildings and infrastructure systems require a huge amount of information from specifications, plans, construction documents, inventory management, cost estimating, and scheduling, for the design and construction phase and maintenance records, inspections and sensor data for the operations phase. As the AEC industry adopts new computer technologies like laser scanners, sensor networks, RFIDs, digital cameras, among many other data acquisition technologies computerized construction/operations data are becoming more and more available. There exist numerous opportunities to exploit this vast amount of data. Unlike much previous data management research that has been successfully applied in several domains, in the AEC domain, however, the data are of multiple types and from many different sources, some with very low quality.

At the same time that we have increasing access to data, infrastructure systems, broadly defined to include buildings and other facilities, transportation infrastructure, telecommunication networks, the power grid and natural environmental systems will require more and more that engineers provide a continuous state awareness, assessment and proactive decision making for the complete life-cycle of the systems and processes they create. Such continuous state awareness and proactive decision making will allow these systems to be more efficiently and effectively managed in both normal and abnormal conditions.

There are many technological developments and research projects that already support, or begin to support this vision. At this talk professor Soibelman will introduce his vision and work developed within his research group focus on the application and exploration of emerging Information and Communication Technologies (ICT), to a broadly defined set of infrastructure systems and associated processes, such as planning, design, construction, facility/infrastructure management, and environmental monitoring, so as to improve their sustainability, efficiency, maintainability, durability, and overall performance of these systems.

1 ABOUT THE SPEAKER

Professor Soibelman obtained his Bachelor and Masters Degrees from the Civil Engineering Department of the Universidade Federal do Rio Grande do Sul, Brazil. He worked as a construction manager for 10 years before moving in 1993 to the US where he obtained in 1998 his PhD in Civil Engineering Systems from the Civil and Environmental Engineering Department at the Massachusetts Institute of Technology (MIT).

In 1998 he started as an Assistant Professor at the University of Illinois at Urbana Champaign. In 2004 he moved as an Associate Professor to the Civil and Environmental Engineering Department at Carnegie Mellon University (CMU) and in 2008 was promoted to Professor. In January 2012 he joined the University of Southern California as the Chair of the Sonny Astani Department of Civil and Environmental Engineering.

During the last 23 years he focused his research on advanced data acquisition, management, visualization, and mining for construction and operations of advanced infrastructure systems. He published over 100 books, books chapters, journal papers, conference articles, and reports and performed research with funding from NSF (NSF career award and sev-eral other NSF grants), NASA, DOE, US Army, NIST, IBM, Bosch, IDOT, RedZone Robotics among many others funding agencies. He is the former chief editor of the American Society of Civil Engineers Computing in Civil Engineering Journal. In 2010 he received the ASCE Computing in Civil Engineering Award, in 2012 received the 2011 FIATECH Outstanding Researcher Celebration of Engineering & Technology Innovation, or CETI, Award, in 2013 he was elected an ASCE fellow and in 2016 his was appointed a Distinguished 1,000 talent Professor at Tsinghua University.

His areas of interest are: Use of information technology for economic development, information technology support for construction management, process integration during the development of large-scale engineering systems, information logistics, artificial intelligence, data mining, knowledge discovery, image reasoning, text mining, machine learning, advanced infrastructure systems, sensors, streaming data, and Multi-reasoning Mechanisms.

eWork and eBusiness in Architecture, Engineering and Construction – Christodoulou & Scherer (Eds)
© 2016 Taylor & Francis Group, ISBN 978-1-138-03280-4

Virtualizing infrastructure

Ioannis Brilakis
Laing O'Rourke Lecturer of Construction Engineering, University of Cambridge, Cambridge, UK

ABSTRACT: Vertical and horizontal infrastructure is comprised of large assets that need sizable budgets to design, construct and operate/maintain them. Cost reductions throughout their lifecycle can generate significant savings to all involved parties. Such reductions can be derived directly through productivity improvements or indirectly through safety and quality control improvements. Creating and maintaining an up-to-date electronic record of these assets in the form of rich Bridge Information Models (BIM) can help generate such improvements. New research is being conducted at the University of Cambridge on inexpensive methods for generating object-oriented infrastructure geometry, detecting and mapping visible defects on the resulting BIM, automatically extracting defect spatial measurements, and sensor and sensor data modelling. The results of these methods are further exploited through their application in Design for Manufacturing and Assembly (DfMA), augmented-reality-enabled mobile inspection, and proactive asset protection from accidental damage. Virtualization methods can produce a reliable digital record of infrastructure and enable owners to reliably protect, monitor and maintain the condition of their asset.

1 ABOUT THE SPEAKER

Dr. Ioannis Brilakis is a Laing O'Rourke Lecturer of Construction Engineering and the Director of the Construction Information Technology Laboratory at the Division of Civil Engineering of the Department of Engineering at the University of Cambridge. He completed his PhD in Civil Engineering at the University of Illinois, Urbana Champaign. He then worked as an Assistant Professor at the Departments of Civil and Environmental Engineering, University of Michigan, Ann Arbor (2005–2008) and Georgia Institute of Technology, Atlanta (2008–2012), and as a Visiting Associate Professor of Computer Vision at the Department of Computer Science, Stanford University (2014). He is a recipient of the NSF CAREER award, the 2013 ASCE Collingwood Prize, the 2012 Georgia Tech Outreach Award and the 2009 ASCE Associate Editor Award. Dr. Brilakis is an author of over 150 papers in peer-reviewed journals and conference proceedings, an Associate Editor of the ASCE Computing in Civil Engineering, ASCE Construction Engineering and Management, Elsevier Automation in Construction, and Elsevier Advanced Engineering Informatics Journals, and a past-chair and founder of the ASCE TCCIT Data Sensing and Analysis Committee, and the TRB AFH10 (1) Information Systems in Construction Management Subcommittee.

eWork and eBusiness in Architecture, Engineering and Construction – Christodoulou & Scherer (Eds)

What's in a model? Intelligent semantic enrichment of BIM models

Rafael Sacks
Virtual Construction Lab, National Building Research Institute, Faculty of Civil and Environmental Engineering, Technion—Israel Institute of Technology, Haifa, Israel

ABSTRACT: Building Information Modeling is a powerful technology, but transferring information among applications is still limited by the diversity of their internal representation schema. The goal of real 'Open BIM' remains elusive due to the difficulty of the interoperability problem. For similar reasons, model-checking and functional simulations using building models is hampered by the need for careful tailoring of the content of model files exported for these purposes.

Semantic enrichment is a novel approach to this problem. It aims to apply expert system technology to interpret and enrich the semantic content of models so that they can be re-used for multiple purposes with minimal rework. The technique will have application in a wide variety of situations. Among those being developed in current research are precast concrete detailing, cost estimation, compilation of as-built models of bridges for bridge surveys, and acquisition of building data for facility maintenance.

In the context of the SeeBridge project, funded within the EU Infravation Program, the team is using a semantic enrichment prototype named SeeBIM 2.0 to aid in compiling building models of highway bridges from point cloud data that can be used for survey and recording of damage to the bridges. This large scale experimental application has yielded important insights into the ways in which rule sets can be compiled rigorously, to ensure that they can uniquely identify bridge components and the semantic relationships between them. The method, which uses feature vectors and feature relationship matrices, also suggests that the potential exists for an alternative approach using tools developed for computer vision that include machine learning.

Prof. Sacks' talk will explore the need for semantic enrichment, its technology aspects, the ways in which it can contribute to providing interoperability, and the promise of a more advanced approach.

1 ABOUT THE SPEAKER

Rafael Sacks' research interests are focused on the synergies of Building Information Modeling (BIM) and Lean Construction. Prof. Sacks established the BIM and Virtual Construction Laboratories at the Technion. Research in the VC Lab includes development of BIM-enabled lean production control systems; BIM systems for earthquake search, rescue and recovery; innovative approaches to interoperability for BIM; and production system theory in lean construction. The VC Lab's ongoing KanBIM and iKAN research projects are a novel attempt to bring process and product information to the job face.

Prof. Sacks earned his BSc from the University of the Witwatersrand, his MSc from MIT, and his PhD from the Technion – all in Civil Engineering. He served as Head of Structural Engineering and Construction Management in the Faculty of Civil and Environmental Engineering from 2012-2015. He has received numerous awards for research, most recently the ASCE Thomas Fitch Rowland prize. He is a co-author of the "BIM Handbook", some 70 papers in academic journals and numerous conference papers and research reports.

Information & knowledge management (I)—technologies

eWork and eBusiness in Architecture, Engineering and Construction – Christodoulou & Scherer (Eds)
© 2016 Taylor & Francis Group, ISBN 978-1-138-03280-4

SimpleBIM: From full ifcOWL graphs to simplified building graphs

P. Pauwels
Department of Architecture and Urban Planning, Ghent University, Ghent, Belgium

A. Roxin
LE2I Laboratory (UMR CNRS 6306), University of Burgundy, Dijon, France

ABSTRACT: Recent research in semantic web technologies for the built environment has resulted in several proposals to further improve information exchange among stakeholders from the domain. Most notable is the production of several OWL ontologies that allow to capture building data in RDF graphs. For example, an ifcOWL ontology allows to capture IFC data in an RDF graph. As the building data is now available in a semantic graph with an explicit formal basis, it can be restructured and simplified so that it more easily matches the different requirements associated with practical use case scenarios. In this paper, we investigate several proposals and technological approaches to simplify ifcOWL building data, thus addressing the needs of specific industrial use cases.

1 INTRODUCTION

Semantic web technologies have been identified as promising in the field of AEC (Architecture, Engineering, Construction), notably for leveraging the different challenges related to BIM (Building Information Modeling) (Mendes et al. 2015a). Recent research in this domain has resulted in several proposals to further improve information exchange among stakeholders from the domain. Among those efforts, most notable are the production of ifcOWL (Pauwels & Terkaj 2016) and ifcWoD (Mendes de Farias et al. 2015b) ontologies. These ontologies allow the publication of IFC-based building models as directed labelled graphs, represented using the Resource Description Framework (RDF) data model (buildingSMART International 2015a). The resulting RDF graphs can be published as part of a linked building data cloud and can be processed by the latest semantic web technologies, which includes open publication platforms (triple stores), query services (SPARQL endpoints), and inference engines for reasoning.

At the moment, however, the above mentioned ontologies remain close to the original IFC schema as available in the EXPRESS information modelling language. This is for example one of the key criteria outlined in Pauwels & Terkaj (2016) and in Mendes et al. (2014). This criterion is maintained in order to keep the EXPRESS, XSD and OWL schemas as identical as possible, with the IFC EXPRESS schema forming the master schema. This has two key consequences:

1. Many of the EXPRESS-specific semantic constructs (like SELECT data types, LIST data types) are maintained and result in complex and unintuitive constructs in OWL and RDF.
2. The instance graphs (ABox) are at least as large and complex as the original IFC models.
3. Thus, the current ifcOWL ontology does not really simplify handling IFC models as it does not deliver the highly demanded simpler models to AEC practitioners. In other words, the proposed ifcOWL ontology is necessary as an OWL ontology that is nearly identical to the master EXPRESS schema, but in practical engineering use cases, simpler RDF graphs are desirable.

In this article, we therefore look into the generation of simpleBIM models (as RDF graphs) starting from the ifcOWL ontology. Such simpleBIM models in RDF should be more agile than what is available today. In our paper, we first provide an overview of related work (Section 2), which includes a brief overview of the Model View Definitions (MVD) approach and an overview of related suggestions in the Semantic Web domain. Suggestions in the Semantic Web domain typically include simple ontologies to represent buildings without the usage of IFC. We then describe a simple 3D building model (Section 3), for which an RDF graph is generated according to the ifcOWL ontology. Several simplifications of the RDF graph are proposed, resulting in clearly defined approaches to simplify the BIM model. In Section 4, we briefly outline the possible approaches in implementing the transition towards simpleBIM models, which includes (1) straightforward programming using the available semantic web technology APIs (Jena, dotNetRDF), (2) the usage of SPARQL queries, and (3) the usage of logical rules and an OWL-DL based inference engine. We focus mainly on the resulting simplifications that can be made, however. In the concluding section, an overview is given of how the proposed approach can be combined with existing tools and information handling approaches, after which future steps are proposed.

2 RELATED WORK

2.1 *Model View Definitions (MVDs)*

Some proposals have already been made in terms of simplifying building information in an IFC syntax. Within the AEC industry, the most closely related proposal in this regard is the Model View Definition (MVD) approach. This approach is tightly related to the usage of Information Delivery Manuals (IDMs). An IDM aims to methodologically *"capture and specify processes and information flow during the lifecycle of a facility"* (Karlshøj 2011). As such, an IDM supports and assures the efficiency and effectiveness of the AEC business processes described in the IDM. By the combination of an IDM with formally specified MVDs, a structured and well-controlled information exchange can take place among AEC project stakeholders, thus greatly improving the internal cost model of AEC projects.

An MVD in itself formally describes a subset of the full IFC schema, including additional data requirements. This subset is meant to respond to one or more specific Exchange Requirements (ERs) in an Information Delivery Manual (IDM). MVDs can be formally represented using mvdXML files. Not only does an mvdXML formally specify what should be contained in an MVD (*specification*), it can also be used to *validate* to what extent an IFC file actually contains the information specified in an MVD.

The specification of MVDs is currently enabled through the IfcDoc tool (buildingSMART International 2015b). At the moment, the MVD specification relies heavily on modular Concept Templates (Venugopal et al. 2012). These are blocks of IFC schema snippets (e.g. all address information for any IFC object), which can readily be selected when specifying a full MVD in the IfcDoc tool. After composing an MVD in the IfcDoc tool, it can be exported as an mvdXML file, both for future reuse and for inclusion in an IDM. In addition, any IFC file can be validated according to the loaded MVD, which means that the IfcDoc tool visually indicates to what extent the information specified in the MVD is also present in the IFC file.

It is important to point out that an mvdXML file is mainly used to *specify* and *validate* MVDs, and not necessarily to *generate* MVDs. In other words, the main purpose is that the MVD specifies which data needs to be exchanged, after which it is the responsibility of implementers and end users to also generate IFC models that comply with this MVD. So, it could be that the end user generates a full IFC model along with more specific and more limited MVD-compliant IFC models in the STEP Physical File Format (SPFF), but the latter is not necessarily a subset from the former, let alone that the latter is generated from the former. In summary, the mvdXML file is a formal representation schema, as well as an implementation guide as well as a validation tool.

Instead of limiting the role of mvdXML as an implementation guide only, one could also consider an mvdXML file as a *generation* tool, which generates a subset IFC model starting from the full IFC model. As an example, one could in this case implement a simple parser that loads an mvdXML file into an IFC file editor (or BIM authoring environment), after which a small subset IFC file is output that only includes the information as required according to the mvdXML specification.

This is precisely the approach that was proposed in Weise & Pauwels (2015), relying however on semantic web technologies. In Weise & Pauwels (2015), the mvdXML specifications of particular MVDs are translated into logical rules that are combined with an ifcOWL instance graph and ontology, and a semantic inference engine, so that subset IFC graphs are output that are compliant with the mvdXML specifications.

2.2 *Simple BIM ontology*

An entirely distinct strategy to simplify BIM information models, is to entirely disregard the IFC standard and instead consider drastically simplified BIM ontologies. Several authors have suggested such ontologies. For example, Niknam and Karshenas (2015) proposed a 'sumo' shared ontology that only contains the key components of a building model (walls, spaces, elements, floors). Depending on the use case, this shared ontology is then expanded with data following a design ontology, an estimating ontology, and so forth. A much earlier example in which a separate ontology was built from scratch, aiming particularly to represent building knowledge in an ontology-based fashion, was proposed by Lima et al. (2003, 2005) as part of the e-Cognos project. This ontology describes four key elements in construction, namely actors, resources, processes and products. Many of the lessons learnt from the e-COGNOS project are documented in El-Diraby (2013), which presents a domain ontology for construction knowledge (DOCK 1.0) starting from the earlier e-COGNOS work (2005–2013). Similarly, Ruikar et al. (2007) proposed an extensible set of modular ontologies (design-process ontology and team profile ontology) which are then deployed in an ontology-based knowledge-sharing environment (OnToShare) for usage by various stakeholders in construction industry.

Scherer & Schapke (2011) propose a multi-model driven construction management system with a layered ontology framework at the center. The ontology framework includes a Project Collaboration Ontology (PCO), which is composed of 5 sub-ontologies: a Construction Core Ontology (CCO), a multi-media visualization ontology, a software service ontology, an organization ontology, and an information process ontology. Dibley et al. (2012) propose an OntoFM system for real-time building monitoring, which relies on a building ontology based on IFC, a sensors ontology that relies on the OntoSensor ontology, and a general purpose ontology (SUMO) that captures domain independent concepts. Reinisch et al. (2011) and Kofler et al. (2012) developed a ThinkHome OWL ontology, including concepts related to resources (white and brown goods), building (layout, spaces, material), actors (schedules, preferences, context), energy (environmental impact, energy providers), comfort (thermal and visual), and exterior influences (weather, climate). The ThinkHome ontology and project relies heavily on the data coming from household appliances. This is inspired to some extent by the ontology-based household device models in the Domotic OSGi Gateway Ontology (DogOnt) (Bonino & Corno 2008). DogOnt allows only a simplified representation of a building as it focuses mainly on the home automation parts of its ontology.

Several similar proposals are made, their common element being the reliance on a simple building ontology: simpleBIM. In the remainder of this paper, we therefore investigate to what extent simplification strategies can be applied to ifcOWL models so that IFC content can better serve alternative use case that only require a simplified building representation.

3 SAMPLE CASE

For this paper, we have used a simple sample building model, which was modelled in Revit Architecture 2016 and which consists of a simple rectangular building that is located in a flat site. The building is furthermore divided in three spaces, has a floor and a roof and has several windows and floors, as displayed in Figure 1.

The building model was exported to IFC-SPF, in the IFC2X3_TC1 schema, which resulted in a 110kB file. The IFC-SPF was then converted into an RDF graph, using the IFC-to-RDF converter supplied in Pauwels, 2015. The file size of the resulting RDF graph is 767kB. The RDF graph counts 10,173 distinct triples (counted using Jena 3.0.0 and no inferences), and includes 5535 class instances (5580 after running the OWL inference engine). Note that 1313 instances are actually named individuals that are present in the ifcOWL ontology, representing ENUMERATION individuals, so the instance file actually contains only 4222 class instances. All files are made available for reference (Pauwels and Roxin 2016).

The IFC building model in RDF follows the main structure of IFC. The core structure starts with a unique IfcProject instance, any valid IFC file allowing only one instance of this concept. This IfcProject instance can be considered as an aggregation point for the spatial structure of the building model. Figure 2 displays the RDF data structure that represents this data structure. As it can be noticed in Figure 2, a project comprises a number of sites (IfcSite); a site contains a number of buildings (IfcBuilding); a building has a number of building storeys (IfcBuildingStorey); finally a building storey comprises several spaces (IfcSpace). However, in the IFC standard, these relations are represented using intermediate IfcRelAggregates instances. The problem is that these instances are often unneeded in the eventual applications reusing or querying this information, thus their presence in the RDF graph raises its complexity unnecessarily. In the context of a graph approach, direct labelled relations between the different concepts (IfcProject, IfcSite, IfcBuilding, IfcBuildingStorey, and IfcSpace) are preferred by far.

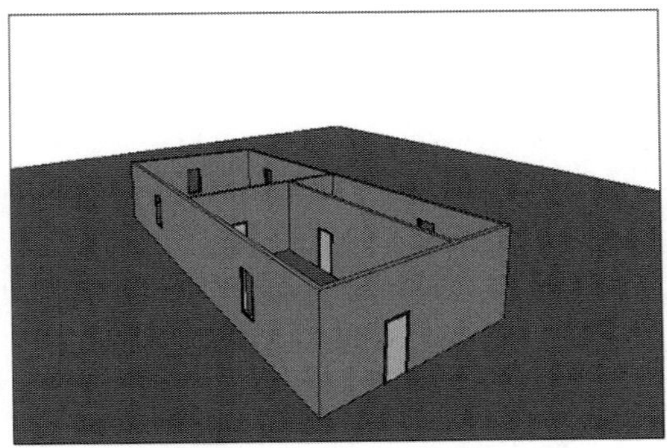

Figure 1. Display of the simple building model that was used here (visualized without roof).

Many similar IfcRelationship instances (which represent the top property for IfcRelAggregates) exist in a typical IFC file. In our sample model, 233 of the 5580 instances are of type IfcRelationship (0 IfcRelAssigns, 19 IfcRelAssociates, 52 IfcRelConnects, 5 IfcRelDecomposes, and 157 IfcRelDefines). Indeed, most of the IfcRelationship instances connect specific building elements to particular IfcPropertySets (see Figure 3). Furthermore the IfcPropertySet, in turn, relates to a set of IfcPropertySingleValue instances, which in turn relate to IfcIdentifier and IfcBoolean instances that finally capture the actual property string and data elements ("isExternal true" in the case of Fig. 3).

Furthermore, 686 out of the 4222 class instances are of type list:OWLList. 417 of these OWLList instances are instances of ifcowl:IfcLengthMeasure_List. In fact, many of these lists relate to geometrical aspects, including IfcDirection and IfcCartesianPoint instances. An example of an IfcCartesianPoint instance and its lists is displayed in Fig. 4, showing three (!) lists to capture that a Cartesian point has coordinates (0,0,0). This clearly becomes even more verbose in the case of IfcPolyline entities and similar geometric objects that are defined by lists of Cartesian points.

As a last element that can be considered for simplification, is the choice to implement EXPRESS simple data types (BINARY, BOOLEAN, INTEGER, LOGICAL, NUMBER, REAL, STRING) using owl:Class wrappers. The choice to implement it as such in ifcOWL is inspired by the choice to do so in most, if not all, EXPRESS-to-OWL conversion efforts (see more detail in Pauwels and Terkaj (2016)). As an example, Figure 4 displays how expr:STRING instances are represented in our sample model. Note that 764 instances out of 4222 are actually such expr:STRING instances.

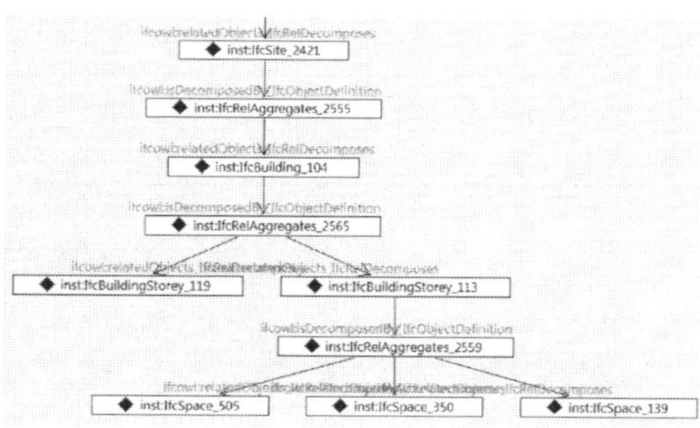

Figure 2. Spatial topology structure as stored in ifcOWL.

4 THE SIMPLIFICATION PROCESS

4.1 *Implementation approaches*

Simplifying the considered graphs can be done in a number of ways. Of course, it is possible to simplify the graphs using procedural programming code. In this case, one could rely on one of the many software libraries available for handling RDF graphs and OWL ontologies. The advantage of this approach is that it can generate *any* kind of simplification and output.

Alternatively, however, one could also rely on the formal basis of the OWL language (Description Logics) and perform the simplification process in a declarative manner. Such an approach typically takes full advantage of query and rule languages commonly available for handling RDF data and OWL ontologies, like SPARQL and SWRL. These languages allow to declare sets of IF-THEN rules, either in SPARQL CONSTRUCT queries or in SWRL rules thus manipulating original considered data. A query or inference engine is then able to match the left hand side IF-parts of these rules and apply them to the available data (ifcOWL graphs), thus deducing the right hand side THEN-part of the rules.

Clearly, this second approach leads to a far more dynamic and on-demand simplification process. The data is natively stored in an ifcOWL graph pattern, and delivered to an end user interface in a simplified manner via such rules. Moreover, diverse different rule sets can be declared, simplifying the

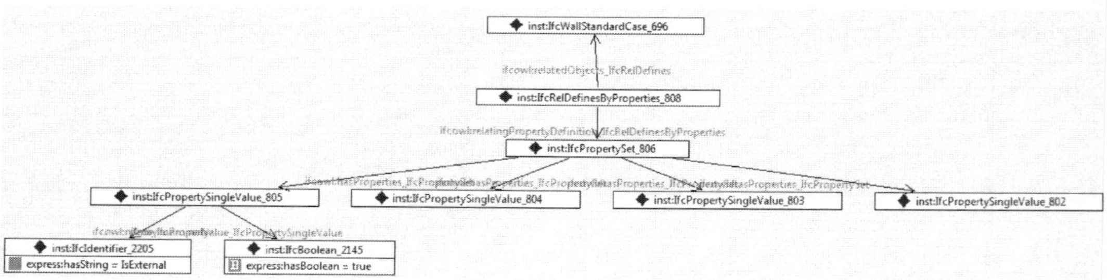

Figure 3. Structure used to relate properties to building elements.

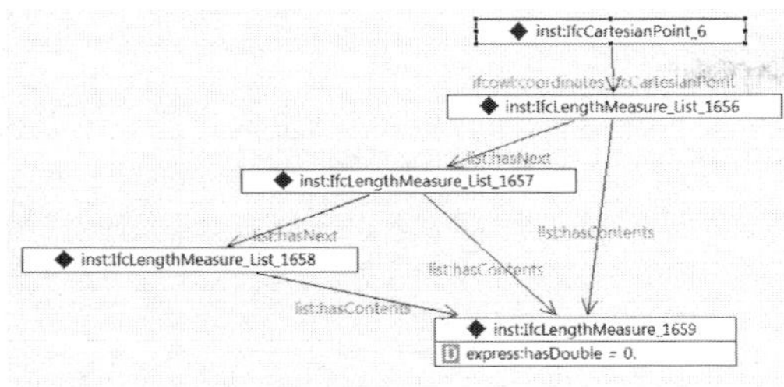

Figure 4. Definition of a Cartesian point.

RDF data depending on the context in which the data is going to be used.

4.2 *Resulting simpleBIM data*

Regardless of the implementation approach taken, a number of relatively straightforward simplification suggestions can be considered here for our use case BIM model. Most of these suggestions relate directly to the presence of EXPRESS relics and features in the ifcOWL ontology. Among those we may cite the usage of data type wrapper classes and the usage of n-ary IfcRelationship instances. In our investigation, we implemented these suggestions in a simple parser and convertor that relies on the Jena software library for handling RDF graphs.

A first simplification that can easily be proposed in the context of a simplified usage of BIM data in a linked data context, is the release of geometrical and (re)presentation data. An IFC-SPF file of a BIM model contains a complete geometrical representation of a building. Such numerical information is seldom used in a linked data context. Clearly such information is far more intuitively handled in a 3D engine or authoring tool, rather than in a pure linked data context. So, it makes sense to remove this data from a simpleBIM building data cloud.

An easy way to make a clear split between parts of the IFC schema, is to use the architecture diagram that is listed for each schema to display resources and domains. In the case of IFC2X3, this diagram can be found in http://www.buildingsmart-tech.org/ifc/IFC2x3/TC1/html/order_by_architecture.htm. In our use case, we removed all instances that were part of the Presentation Resource, the Presentation Definition Resource, the Presentation Appearance Resource, the Profile Resource, the Representation Resource, the Topology Resource, the Geometry Resource, the Geometric Model Resource, and the Geometric Constraint Resource. This includes many of the complex statements, like the one displayed earlier in Fig. 4. After performing this operation, the RDF graph shrank to 6,927 triples (original: 10,173 triples) and 476kb (original: 767kb). This first simplification represents a reduction of the triple number and the file size by 31.9% and respectively 38%.

A second simplification rule relates to Fig. 5 and involves the 'unwrapping' of wrapped data types. Wrapping of data types is performed to allow a safe and uniform conversion of EXPRESS and IFC-SPF constructs into OWL and RDF respectively. In an instance file, however, these wrapped classed can often be unwrapped into explicit datatype properties. The example in Fig. 5 could thus be represented as displayed in Fig 6, using a custom datatype property simpleBIM:globalId that points towards an xsd:string. This can be done for all datatypes, including strings, integers, booleans, and so forth. After performing this operation, the RDF graph shrank to 3,897 triples (original: 10,173 triples) and 279kb (original: 767kb). We thus manage to lower the triple number by 43.74%, while reducing the file size by 41.39%, compared to the figures obtained in the previous step.

A third simplification step involves the rewriting of properties, as it is done in (Mendes et al. 2015b). IfcOWL now handles property values and property sets as displayed in Figure 3, namely using a considerable number of intermediate steps (IfcPropertySet, IfcPropertySingleValue, IfcRelDefinesbyProperties). Although this might make a lot of sense in a BIM authoring tool, which typically uses a relational database with mapping tables between entities and tables of properties, this construct makes little to no sense in a linked data environment. In a linked data environment, a property is ideally immediately attached to an element. By rewriting this information as such, we can obtain a greatly simplified diagram as displayed in Figure 7. After performing this operation in our

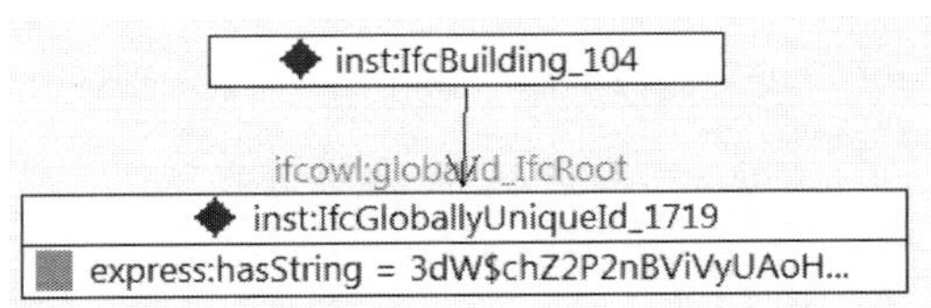

Figure 5. RDF graph representation for an expr:STRING instance related to an IfcBuilding instance.

Figure 6. Simplified RDF graph for the expr:STRING example listed in Fig. 5.

use case model, the RDF graph shrank to 1,630 triples (original: 10,173 triples) and 112kb (original: 767kb). We thus further reduce the number of triples, along with the file size (58.17% and respectively 58.86% less than in the previous step).

A last simplification involves the replacement of the relational instances, which are all subclasses of IfcRelationship, with direct simpleBIM object properties between the applicable instances. This is already implemented in ifcWOD (Mendes et al. 2015b). Indeed, ifcWOD defines IfcRelationship and its related subtypes (e.g. IfcRelDefinesbyProperties) as OWL object properties. In ifcWOD, inverse attributes are also defined in order to link the IFC entities corresponding to related objects to the referring IfcRelationship entities. When applying this to SimpleBIM, we obtain the diagram displayed in Figure 8. After performing this operation in our use case model, the RDF graph shrank to 1,339 triples (original: 10,173 triples) and 83kb (original: 767kb). This represents a reduction of 17.85% for the number of triples and of 25.89% for the file size, when compared to the results achieved with the previous simplification.

4.3 *Overall usability*

The same set of simplifications has also been tested for larger models, to display the significance of the size and complexity improvements that can be made, as well as to show general usability. The results of these tests are briefly outlined in Table 1.

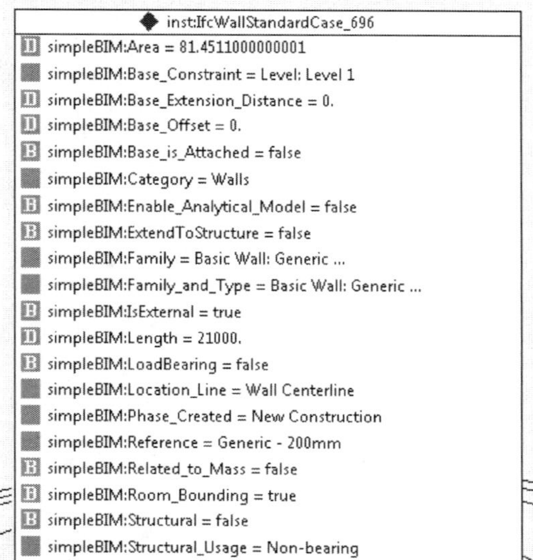

Figure 7. simpleBIM RDF graph with datatype properties directly associated to building elements.

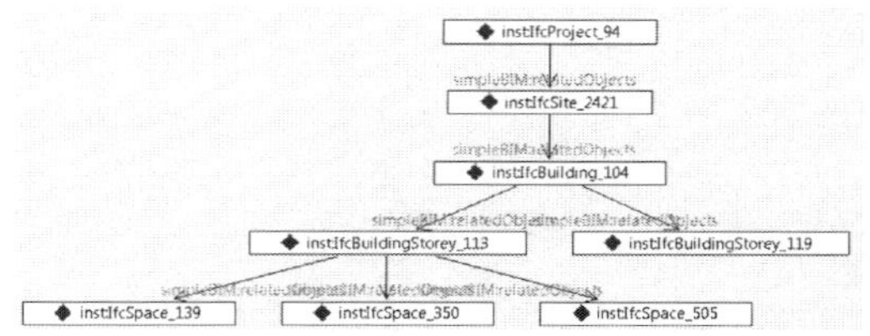

Figure 8. Simplified representation for the spatial structure of a building, as an alternative for the diagram in Fig. 2.

Table 1. Simplification statistics for 2 reference models.

Model	ifcOWL	ifcOWL	simpleBIM	simpleBIM
1	767kb	10,173	83kb	1,339
Impr.	-	-	**89.18%**	**86.84%**
2	16,7 MB	225,135	1.029kb	16,836
Impr.	-	-	**93.98%**	**92.52%**

These results show an average size improvement of 89.68% in terms of triples (91.58% in terms of file size). This in itself shows the dramatic increase in usability of information. Furthermore, we wish to stress here that the actual RDF graphs are considerably simpler, while still representing the information that is typically used in a linked data context. Indeed, the eventually contained information comes close to the information that is typically described in non-IFC ontologies as documented in Section 2.

5 CONCLUSION

In the context of AEC, today's practical use of building information in form of IFC files has arisen several issues notably regarding the simplification of those files. When considering buildingSMART, a first approach for answering these issues comes in the form of IDMs and the related MVDs. Still, the underlying processes (e.g. IDM specification, MVD development) remain complex and time-consuming. Easy and intuitive ways to rapidly browse, query and use BIM information are not often available. The usage of semantic web technologies can remedy this situation, as these technologies allow to more dynamically manipulate the building information in the RDF graphs using web technologies, including query and rule languages.

In this paper, we have defined and exemplified 4 main approaches towards simplification of building models represented in ifcOWL graphs. While the first two of them are totally novel, the last two were adapted from the ifcWOD ontology

(Mendes et al. 2015b). These simplifications can be dynamically applied on existing ifcOWL graphs. Average simplification percentages can be obtained of 89.68% in terms of triple count and 91.58% in terms of file size.

In principle, this method can easily be used to generate MVDs from full ifcOWL graphs. This responds to the idea of *generating* MVDs, in addition to merely specifying and validating Model Views, as outlined in Section 2.1. In this case, an MVD expressed in mvdXML should thus automatically be parsed, so that it informs the simplification process, and the output of the SimpleBIM process is an ifcOWL graph that contains only the information specified in the mvdXML graph. Note that the resulting graph in this case still needs to be valid ifcOWL, because an MVD is also required to result in valid IFC-SPF data. This leave out almost all of the outlined simplification possibilities.

However, additional and alternative simplification approaches can be considered as well, apart from the MVD generation phase, depending on the use case that is considered. For example, some use cases might prefer the usage of IfcPropertySet instances, which are now not considered in the simplification process. Other use cases might focus entirely on the geometry of the ifcOWL graph, a portion of information that is not retained in the presented examples. Hence, the proposed approach has a clear industrial value in the sense that it allows to intuitively supply IFC information in the custom form and custom size that is often required or demanded in the name of interoperability (even much more custom than what is available via MVDs).

The presented technique thus allows to adapt an RDF graph (in ifcOWL) into diverse alternative, less complex graphs, depending on the use case. Indeed, while ifcOWL is intended as a recommended standard, requiring close correspondence with the EXPRESS schema, the graph resulting from this SimpleBIM procedure is more of a usable extension of this standard. The same is true for the ifcWOD extension as proposed in Mendes et al. (2015b) and similar private industrial dynamic simplification mechanisms.

In terms of future work, we plan to investigate how the presented procedure can be used to transform and simplify ifcOWL graphs into the graphs used in the works proposed in Section 2.2: e.g. DogOnt (Bonino & Corno 2008), the OntoFM ontology (Dibley et al. 2012), and the sumo ontology (Niknam and Karshenas 2015). If such a procedure would be available, it means that ifcOWL information can be readily supplied in any of these ontologies, allowing a considerably improved information exchange process.

Furthermore, we intend to perform additional benchmarks, in order to highlight eventual advantages of SimpleBIM for SPARQL query simplification. We have to check that SPARQL queries defined using only SimpleBIM concepts are complete (in other words these queries do return all the expected results). We also want to examine the related query execution time, in order to identify if query simplification results in a sensible reduction of query execution time.

ACKNOWLEDGEMENTS

The authors wish to thank the Special Research Fund (BOF) of Ghent University for their generous support, and the French Agency of Research (ANR) for the financing of the BigSTEP project (funding ANR-16-MRSE-0024-01).

REFERENCES

Bonino, D. & Corno, F. 2008. DogOnt—ontology modeling for intelligent domotic environments. In *Proceedings of the International Semantic Web Conference*. Vol. 5318 of Lecture Notes in Computer Science (LNCS), 790–803.

BuildingSMART International 2015a. Overview page for the Linked Data Working Group. http://buildingsmart-tech.org/future/linked-data.

BuildingSMART International 2015b. ifcDoc Tool Summary. http://www.buildingsmart-tech.org/specifications/specification-tools/ifcdoc-tool.

Dibley, M., Li, H., Rezgui, Y. & Miles, J. 2012. An ontology framework for intelligent sensor-based building monitoring. *Automation in Construction* 28: 1–14.

El-Diraby, T.E. 2013. Domain ontology for construction knowledge. *Journal of Construction Engineering and Management* 139(7): 768–784.

Karlshøj, J. 2011. Information Delivery Manuals. http://iug.buildingsmart.org/idms.

Kofler, M.J., Reinisch, C. & Kastner, W. 2012. A semantic representation of energy-related information in future smart homes. *Energy and Buildings* 47: 169–179.

Lima, C., El-Diraby, T. & Stephens, J. 2005. Ontology-based optimization of knowledge management in construction. *Journal of Information Technology in Construction* 10: 305–327.

Lima, C., Fies, B., Zarli, A., Bourdeau, M., Wetherill, M. & Rezgui, Y. 2002. Towards an IFC-enabled ontology for the Building and Construction Industry: the e-COGNOS approach. In *Proceedings of the eSMRT 2002 Conference*, 254–264.

Mendes de Farias, T., Roxin, A. & Nicolle, C. 2014. A Rule Based System for Semantical Enrichment of Building Information Exchange. In *Proceedings of the RuleML 2014 Challenge and the RuleML 2014 Doctoral Consortium hosted by the 8th International Web Rule Symposium (RuleML 2014)* Prague, Czech Republic, August 18–20, 2014, eds.: Theodore Patkos, Adam Wyner and Adrian Giurca. Aug 2014, Prague,

Czech Republic. Vol-1211, pp.2. Available online at: http://publik.tuwien.ac.at/files/PubDat_231529.pdf

Mendes de Farias, T., Roxin, A. & Nicolle, C. 2015a. Semantic Web Technologies for Implementing Cost-effective and Interoperable Building Information Modeling. In *Proceedings of the 14th International Conference on Informatics in Economy (IE 2015)*.

Mendes de Farias, T., Roxin, A. & Nicolle, C. 2015b. Ifc-WoD, Semantically Adapting IFC Model Relations into OWL Properties. In *Proceedings of the 32nd CIB W78 Conference on Information Technology in Construction*, 175–185.

Niknam, M. & Karshenas, S. 2015. Integrating distributed sources of information for con-struction cost estimating using Semantic Web and Semantic Web Service technologies. *Automation in Construction* 57: 222–238.

Pauwels P. 2015. IFC repository. http://smartlab1.elis.ugent.be:8889/IFC-repo/.

Pauwels, P. & Roxin, A. 2016. Additional files for ECPPM2016 Article. http://users.ugent.be/~pipauwel/ECPPM2016_SimpleBIM/index.html.

Pauwels, P. & Terkaj, W. 2016. EXPRESS to OWL for construction industry: Towards a recommendable and usable ifcOWL ontology. *Automation in Construction* 63: 100–133.

Reinisch, C., Kofler, M., Iglesias, F. & Kastner, W. 2011. ThinkHome Energy Efficiency in Future Smart Homes. *EURASIP Journal on Embedded Systems* 2011: 1–18.

Ruikar, D., Anumba, C., Duke, A., Carrillo, P. & Bouchlaghem, N. 2007. Using the semantic web for project information management. *Facilities* 25: 507–524.

Scherer, R. & Schapke, S.-E. 2011. A distributed multi-model-based Management Information System for simulation and decision making on construction projects, *Advanced Engineering Informatics* 25: 582–599.

Venugopal, M., Eastman, C.M., Sacks, R. & Teizer, J. 2012. Semantics of model views for information exchanges using the industry foundation class schema. *Advanced Engineering Informatics* 26(2): 411–428.

Weise, M. & Pauwels, P. 2015. Best practices for publishing and linking BIM data: scoping of IFC models. *Third International Workshop on Linked Data in Architecture and Construction*.

eWork and eBusiness in Architecture, Engineering and Construction – Christodoulou & Scherer (Eds)
 ISBN 978-1-138-03280-4

IFC model checking based on mvdXML 1.1

M. Weise, N. Nisbet & T. Liebich
AEC3 Ltd., Germany and UK

C. Benghi
Northumbria University, Newcastle Upon Tyne, UK

ABSTRACT: A significant barrier for successful use of BIM is the ability to efficiently and transparently agree on what data should be delivered by the many stakeholders of the supply chain and when. This requires additional agreements and specification work on top of existing standards like IFC. Ideally, these specifications are ready for automatic model checking to ensure the exchange of required BIM data. Based on the IDM/MVD methodology and the mvdXML 1.1 format developed by buildingSMART a web-based requirements management solution called BIM-Q and the mvdXML extension of the XBIM toolkit is discussed that demonstrates how BIM exchange requirements can be configured, managed and used for automatic model checking. All necessary steps are shown using an example from the STREAMER project, namely the Program of Requirements (PoR) and the early design of the room layout for hospitals. Besides presenting preliminary process implementation findings, grounded on data collected from various projects, persisting limitations for managing requirements and in particular for model checking based on mvdXML are discussed. An outlook of potential extensions and improvements of the different tools, mvdXML specification and the whole checking process is presented at the end.

1 INTRODUCTION

1.1 *Framework for BIM information management agreements*

BIM information management and control are fundamental processes in the delivery of the benefits enabled by the use of the BIM technologies. Currently, a significant barrier for beneficial use of BIM is the inability to efficiently and transparently agree on data exchange workflows across the many stakeholders of the supply chain. If expected data is missing, incorrect or misplaced then the project team has to go through additional data correction processes that can be time consuming and lead to critical project delays.

BuildingSMART have developed guidance and standards that help defining a framework for BIM information management agreements; these are based on the production of Information Delivery Manuals (IDM) and the use Model View Definitions (MVD). An IDM is essentially an agreement on the processes and responsibilities of the project partners, whereas an MVD clarifies the data implementation details. Following the prescribed guidelines can result in time-consuming analysis, design and specification work that normally produces descriptive documents that later need to be implemented in software. The adoption of standard computer-interpretable formats, supported by efficient editing and management tools, would help to improve the requirement capturing and implementation processes.

1.2 *Solution approach and structure of the paper*

The work presented in this paper describes how the process of IDM/MVD development for IFC-based data exchange can be efficiently implemented with the use of the latest mvdXML 1.1 specification format through the adoption of a web-based requirements management tool called BIM-Q and the mvdXML extension of the XBIM toolkit.

Chapter 2 will introduce the workflow supported by the developed approach and will give an example defined in the STREAMER research project. All necessary steps are discussed and compared with the state-of-the art technology. The main focus will be on the following steps: (1) capturing data requirements as done by domain experts, (2) linking data requirements to processes, (3) specifying the mapping to IFC by configuring predefined concept templates, (4) the generation of a checkable mvdXML document and (5) the model checking in XBIM and the error reporting using the BIM collaboration format (BCF).

Chapter 3 will give an introduction to mvdXML release 1.1 being published as final version in 2016.

The focus will be on features that are relevant for the configuration of exchange requirements and automatic model checking. It will also clarify the scope of checking exchange requirements in order to avoid misunderstandings about the kind of quality checks that are in focus of the presented scenario and mvdXML.

Chapter 4 will present a solution to capture exchange requirements in a web-based environment, the BIM-Q tool. An important method for defining and managing requirements is the use of templates; available in the BIM-Q database as well as the mvdXML format, templates provide a key feature to reduce the complexity of the requirement definitions. Through templates, technical details can be embedded in preconfigured specifications files that, once refined by skilled specialists, simplify and modularize the usage and understanding of data requirements making them easily accessible by non IT-experts.

The following chapter presents the implementation of mvdXML model checking developed as a plugin for the xBIM Xploerer IFC viewer; it will present implementation objectives and details introducing options in the user interface that has been designed to support a goal oriented interaction with requirement specifications on the foundation provided by the structure of mvdXML.

Usage of the presented solution is shown in chapter 6 where examples from the STREAMER project are discussed to highlight different aspects of the solution provided along with an overview of its limitations. In addition to this, the conclusion in chapter 7 is discussing potential development directions.

2 EXAMPLE—SPACE REQUIREMENTS FOR HOSPITALS

2.1 *Introduction to the STREAMER case studies*

STREAMER is an industry-driven collaborative research project on Energy-efficient Buildings (EeB) that aims to reduce the energy use and carbon emission of new and retrofitted buildings in mixed-use healthcare districts. An important task in that scenario is to achieve unequivocal clarity about the client requirements.

In this particular case, for hospitals, this is achieved starting from the definition of space requirements, which need to be translated to space layouts following given design rules that take into account the constraints of the building site and existing buildings. In case of STREAMER the space layout is generated by an optimization algorithm, the Early Design Configurator, which produces a set of solutions that are evaluated against a set of KPIs, including energy consumption indicators. This simple workflow includes four processes and three data exchanges (see Figure 1).

In order to make sure that each process has a complete set of information the minimum exchange requirements are specified as an MVD using the mvdXML format. This enables to control IFC-based data exchange by checking existence of required information. This will ensure a certain level of quality.

2.2 *IDM/MVD methodology and state-of-the-art*

According to the Information Delivery Manual (IDM, ISO 29481-1) and Model View Definition (MVD) methodology the specification work follows subsequent steps and involves different stakeholders starting with a high-level view on the business processes down to software implementation details. The result of each step is a custom documented agreement or technical specification that forms the basis for further communication and refinements.

The work presented in this paper starts with the definition of Exchange Requirements. Accordingly, relevant processes, involved actors and the data flow as for instance presented in Figure 1 are already available as a reference. One out of the three mentioned data exchanges is the data defined by the Program of Requirements (PoR). Domain experts, in that case mainly the client, have to describe what information is captured in the PoR and ensure there's agreement on terms used, their meaning and the planned arrangement of required data. In case of PoR this results in a set of space

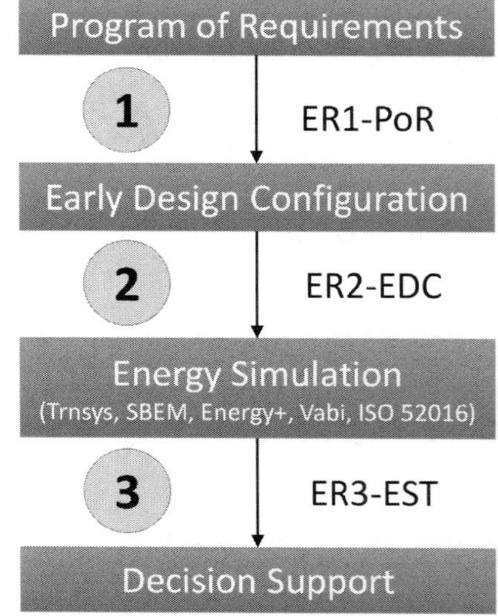

Figure 1. showing the workflow and exchange requirements comprising of (1) client requirement definitions, (2) the space layout, (3) energy simulation using various tools and (4) decision support.

types that are classified by criteria such as comfort, safety, hygiene class, accessibility and others (Di Giulio 2015). For each of those classification criteria allowed ranges have to be defined including terms of parameter constraints, applicable design rules or required technical specifications (a space classified as "A4" means for instance that it should be accessible for staff only). Such classification systems are likely already available for the client and thus need only to be referenced.

The structure of defined requirements may partially fit to other processes. Therefore, it is reasonable to harmonize specifications by reusing them in other processes. If the room type information is needed for space layout but also for energy estimation it should be linked as a requirement to both processes. Traditionally, the main purpose of this step is to prepare implementation of software interfaces, which means to translate the terms of domain experts to a data structure like IFC. This step is done by modelling experts who are familiar with the relevant data structure. For IFC-based MVD developments it means to switch to the ifcDoc tool that enables to work on an mvdXML specification, but will lose the link to the exchange requirements defined by the domain expert.

Today, an MVD even if available as mvdXML is defined mainly for documentation purposes. The ifcDoc tool for instance enables to generate the HTML documentation as known from the IFC4 specification. This is expected to change if mvdXML-based model checking becomes available as presented in this paper. If such MVD specification enables to validate an IFC dataset, then it would not only support software implementation but also the everyday data quality control in real projects. If some requirements are not met the sender could be notified and pointed to the missing data. This can be done via the BIM collaboration format (BCF), which enables to report and visualize identified issues. Ideally, issues are reported using the terms from the original requirements definition and not the attribute or class name of IFC. This would improve communication.

The next chapter will highlight the checking and configuration features of mvdXML, while chapter 4 will detail how it can be generated by our requirements management environment to supports process-specific configurations.

3 MVDXML 1.1

3.1 *Overview and main use cases*

After a two year review period the mvdXML 1.1 specification was published in spring 2016 (Chipman et al.). Besides a couple of minor improvements and simplifications the most notable change is the extended capability for model checking. Although this feature of mvdXML received bigger attention lately the focus is still on MVD documentation purposes, for instance for creating the HTML documentation of the new Design Transfer and Reference View of IFC4. It might also be used for generating an IFC subset schema or data filtering, but both scenarios seem to be less important at the moment.

Although each of those usages has specific requirements the definition of an MVD is always similar. Main elements of each MVD are:

- *ModelView:* one or more of those elements are normally included in an mvdXML file. It is part of the *View* element and is the main container for exchange requirements and root concepts.
- *ExchangeRequirement*: represents the data that is relevant for a use case, either for import, export or both.
- *ConceptRoot:* represents a class of objects for which the same constraints apply. They are normally linked to entities that are derived from IfcRoot, i.e. being a main testable element of an IFC model.
- *Concept:* is part of a root concept and defines a constraint on applicable objects and how it is used in exchange requirements.
- *ConceptTemplate*: defines a unit of functionality that is used and configured by ConceptRoot and Concept elements. It is a selection and basic configuration of IFC definitions that are required to implement a specific functionality such as support of property sets, material layer definition or more complex data like brep geometry.

Each of those elements is able to carry additional meta-data and descriptive text including multilingual support.

3.2 *Concept templates and their configuration*

An important feature in terms of reducing the maintenance effort is the use of configurable concept templates. A concept template defines one or more applicable entities and includes a set of rules that each specifies a sub graph of instantiable attributes. Such sub graph is defined by attribute and entity rules and always starts with an attribute of the applicable entity.

The concept template shown in Figure 2 is defined for all instances of *IfcRoot* entities and contains two rules for the attributes *Name* and *Description*. Both rules define an additional (optional) rule identifier (*RuleID*), which is a unique name used for further configuration. The figure also shows that an *AttributeRule* is followed by (one or more) *EntityRule* that expand the sub graph.

```
<ConceptTemplate
 uuid="c19ec186-9cfd-47fc-a4d4-9fb35008d04a"
 name="User Identity" applicableSchema="IFC4"
 applicableEntity="IfcRoot">
  <Definitions><Definition>
    <Body><![CDATA[Code 020- ...]]></Body>
  </Definition> </Definitions>
  <Rules>
  <AttributeRule RuleID="Name"
                 AttributeName="Name">
    <EntityRules>
      <EntityRule EntityName="IfcLabel"/>
    </EntityRules>
  </AttributeRule>
  <AttributeRule RuleID="Description"
                 AttributeName="Description">
    <EntityRules>
      <EntityRule EntityName="IfcText" />
    </EntityRules>
  </AttributeRule>
  </Rules>
</ConceptTemplate>
```

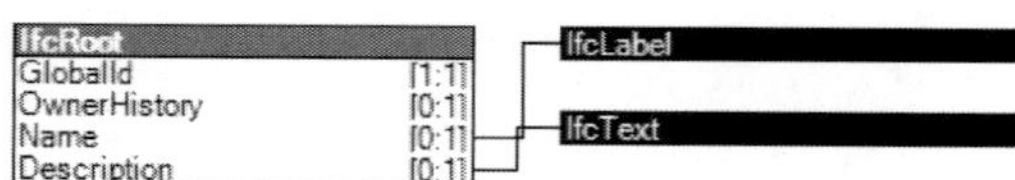

Figure 2. showing the ConceptTemplate "User Identity" and its visual representation as instantiation diagram.

The rule identifier is later used as a parameter in a logical expression to check existence, values, types, size of sets or uniqueness. Accordingly, above shown example enables to configure both attributes, for instance to check for a specific name or existence of a description. However, logical expressions in mvdXML are limited in their expressiveness in order to be as clear as possible both for definition and processing.

3.3 *Checking of exchange requirements*

The principle for defining constraints is based on IF THEN statements. The IF-part is defined in *ConceptRoot* nodes and determines the selection of instances in the IFC model. The THEN-part is defined by *Concept* nodes and defines the constraints that shall be applied to all selected instances. In addition to a "selection by type" (through the *applicableEntity* field) it is possible to define additional constraints. For instance if all load bearing walls shall be checked then all instances of *IfcWall* with a property *Pset_WallCommon.LoadBearing = TRUE* must be selected. Such additional constraints are defined in the *<Applicability>* section of *ConceptRoot*. The mvdXML snippet shown in Figure 3 is selecting instances of IfcBeam with the Name "Beam-206". It is configuring the concept template of Figure 2.

```
<ConceptRoot
 uuid="00000035-0000-0000-2000-000000067001"
 name=" Beam-206"
 applicableRootEntity="IfcBeam">
  <Applicability><Template
  ref="c19ec186-9cfd-47fc-a4d4-9fb35008d04a"/>
  <TemplateRules operator="and">
    <TemplateRule
      Parameters="Name[Value]='Beam-206'"/>
  </TemplateRules>
  </Applicability>
```

Figure 3. showing the configuration of "User Identity" for the selection of an IfcBeam instance.

```
<Concept
 uuid="00000003-0000-0000-0000-000000349910"
 name="Accessibility Labels">
  <Template
   ref="00000000-0000-0000-0001-000000000001"/>
  <Requirements>
  <Requirement applicability="import"
   exchangeRequirement="00000003-0000-0000-
   0000-000000000105" requirement="mandatory"/>
  </Requirements>
  <TemplateRules operator="and">
    <TemplateRules operator="or">
      <TemplateRule Parameters=
      "Set[Value]='STREAMER_Labels_PoR' AND
      Property[Value]='AccessSecurity' AND
      Value[Value]='A1'"/>
      <TemplateRule Parameters=
      "Set[Value]='STREAMER_Labels_PoR' AND
      Property[Value]='AccessSecurity' AND
      Value[Value]='A2'"/>
    </TemplateRules>
  </TemplateRules>
</Concept>
```

Figure 4. showing the constraint for the "Accessibility Labels" defined by the PoR for spaces.

The configuration of constraints works in a similar way; a concept refers to a concept template using its uuid. The *<Requirements>* section then defines the link to exchange requirements and its expected usage. The configuration of rule identifiers starts thereafter, which may be using nested statements logically combined by Boolean operators. Figure 4 shows the configuration of a mandatory space property where only the two values "A1" and "A3" are allowed.

4 CAPTURING REQUIREMENTS WITH BIM-Q

4.1 *Need for a shared, web-enabled requirements management tool*

As outlined in chapter 2.2 exchange requirements are a means for communication and thus need to be agreed and shared between involved participants. Also, many requirements are applicable for several processes so that a lot of definitions can and should be reused.

Today, exchange requirements are typically captured in a spreadsheet format. For each physical or conceptual thing it captures relevant properties, its meaning and use in design processes (IDM). It is simple and straight forward but the more information is captured and shared, the more difficult it is to keep consistency and maintain the content. There are also limitations to evaluate and export requirements, in particular for generating various reports and producing an mvdXML file for checking purposes. Accordingly, there is a need for better tool support which was leading to the development of the presented web-based solution called BIM-Q.

4.2 *Scope related to the IDM/MVD methodology*

Before collecting exchange requirements an initial set-up of the database is necessary. The first step is to define a template guideline that shall group all definitions. This might later be used to configure project requirements. Next to this, the selection of involved stakeholders, covered stages and processes as well as relevant mappings is necessary. Mappings include links to classification systems, translations to other languages and the representation in data structures like IFC. In this initial step it means to set-up the boundaries for the discussed use cases in terms of definitions and standards that becomes relevant to clarify the meaning of terms and to be used for data exchange. Each of those settings can be changed or extended in later stages, but it defines the starting point for defining relevant terms, which is the first main step of capturing domain knowledge.

4.3 *Set-up of reusable concepts*

Definition of exchange requirements follows the object-oriented modelling principle, but with less restrictive rules. Everything is a concept. Each concept can be described, typed, mapped to other definitions and arranged to each other in order to form more complex concept definitions. A concept can for instance represent a class of beam objects whereas another concept represents a simple datatype property for fire rating.

An exchange requirement is typically defined for a property of some object class. A fire safety calculation may requires the fire rating property for all loadbearing building elements. It is a simple and natural way of expressing requirements that can be defined by non-IT experts.

Experiences have shown that a lot of concepts are reused for requirement definitions, in particular in case of generic properties. This is leading to a lot of copied content that is later difficult to maintain. Therefore, the first step is to collect reusable concept definitions that can be arranged in any level of complexity. In that way, a pool of concepts is defined that later can be arranged to any requirement setting that needs to be described. Each reusable concept is linked to default definitions, such as a description or the mapping to IFC, which however can be overridden in a requirement setting if necessary.

The pool of reusable concepts can be organized according to own preferences. Our recommendation based on experiences is to organize similar concepts in groups like classes, properties and geometry. STREAMER is using a labelling approach and thus is using the structure as shown in Figure 5. Further subgroups are recommended, but should be kept as simple as possible. If properly arranged it later helps to find the right concept and to configure the requirement settings.

4.4 *Configuration of exchange requirements*

The next step is to link objects with properties in order to express requirements. This is done by dragging reusable concepts to a new requirements tree as shown in Figure 6. Both trees provide independent search capabilities so that concepts can easily be found and arranged in the requirements tree. In order to speed-up the set-up process it is also possible to drag and drop a concept with all child elements. If reusable concepts are properly arranged it supports an easy and fast set-up process.

Differently to reusable concepts there are some constraints regarding the organization of the requirements tree. Those constraints exist mainly due to the fact that some meaningful reports or an mvdXML file shall be generated out of this tree. By following the idea of having a property of some object class the structure should follow the rule

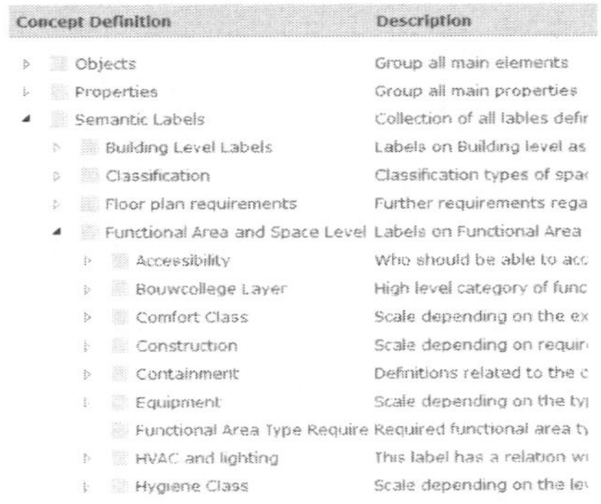

Concept Definition	Description
Objects	Group all main elements
Properties	Group all main properties
Semantic Labels	Collection of all lables defir
Building Level Labels	Labels on Building level as
Classification	Classification types of spa(
Floor plan requirements	Further requirements rega
Functional Area and Space Level	Labels on Functional Area
Accessibility	Who should be able to acc
Bouwcollege Layer	High level category of func
Comfort Class	Scale depending on the ex
Construction	Scale depending on requir
Containment	Definitions related to the c
Equipment	Scale depending on the ty
Functional Area Type Require	Required functional area t
HVAC and lighting	This label has a relation wi
Hygiene Class	Scale depending on the le

Figure 5. Reusable concepts as defined in the STREAMER project.

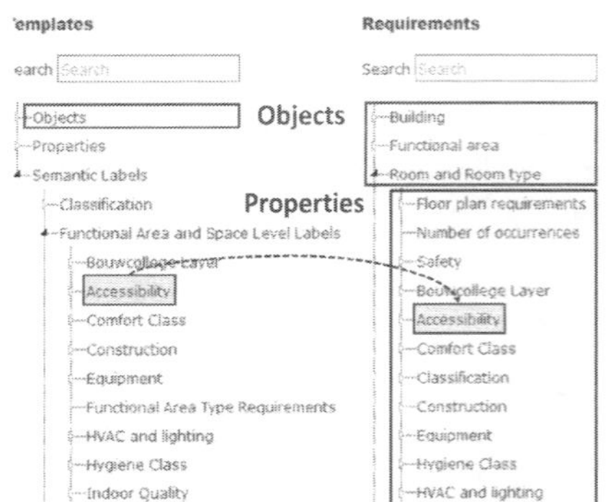

Figure 6. Set-up of the requirements tree by dragging reusable concept from templates (left) to the requirements tree.

Concept Definition	Owner	ER1-PoR	ER2-EDC
Building	Architect	-	MAN
Functional area	-	MAN	MAN
Room and Room type	Architect	MAN	MAN
Accessibility	Building owner	MAN	-
Bouwcollege Layer	Building owner	MAN	-
Classification	Architect	MAN	MAN
Comfort Class	Building owner	MAN	-
Construction	Building owner	MAN	MAN
Containment	Architect	-	MAN
Equipment	Building owner	MAN	MAN
Geometric representation	Architect	-	MAN
HVAC and lighting	Building owner	[illegible]	-
Hygiene Class	Building owner	MAN	MAN
Indoor Quality	Building owner	[illegible]	-
Number of occurrences	Building owner	MAN	NOT

Figure 7. Definition of usage settings and assignment to a concept owner.

of having a property concept, marked as a simple datatype, always as a child element of an object concept. In between there might additional group concepts for better organization of requirements, which are ignored for later model checking. There are special solutions for enumeration datatypes having allowed values as child concepts, which however do not break described general rule. Nevertheless, a risk of configuring a requirements tree that cannot be properly exported to mvdXML checking file remains so that this step should carefully done.

Once the requirements tree is defined the usage settings for the different processes can be configured. It basically means to make a decision what data is required, optional or not allowed. Additionally, an owner of a data concept has to be defined who is responsible to deliver that information (Figure 7).

4.5 *IFC mapping definitions*

Each concept can have any number of mapping definitions to whatever data structure is of interest. In our case the focus is on the open IFC-BIM format that can be formalized by mvdXML definitions.

There are basically two types of mapping definitions:

- *Object concept mappings:* For mvdXML it means to configure a *ConceptRoot* element comprising of the selection of an IFC entity (*applicableRootEntity*) and, optionally, additional *Applicability* settings.
- *Property concept mappings:* This requires the configuration of a *Concept* element, which needs to identify and configure an appropriate *ConceptTemplate*.

The BIM-Q tool supports a simple syntax to easily configure most frequently needed mapping definitions. An object concept for instance maps either 1:1 to an IFC entity, or is additionally restricted by the *PredefinedType* attribute or some property values. The expression *IfcWall.IfcWallTypeEnum.SHEAR* is for instance applicable for all *IfcWall* instances having the *PredefinedType* attribute set to "*Shear*". Similar solutions are available for property concepts, where for instance the configuration of properties and quantities is often needed. Uncommon mapping definitions have to go through a more complex configuration process. This however shouldn't be a problem as this step has to be done by an IFC expert who is familiar with the IFC specification and available mvdXML concept templates.

4.6 *Reporting and mvdXML export*

The final step in the requirements capture process is to produce some sort of evaluable result. This might be a specific PDF report that could act as an contract annex, an mvdXML file for checking purposes or some template documents. In case of mvdXML it is possible to export all settings to a single file. Alternatively, it is also possible to export settings of specific processes or a single owner only.

The export feature itself is translating the used mapping syntax to an mvdXML, which for instance in case of properties expands to a check of *properties on occurrences* and *properties on types*. At the time of this writing there is no consistency check against the IFC specification so that spelling errors are not identified. However, testing a valid file should quickly show wrong mapping definitions.

5 MODEL CHECKING WITH XBIM

5.1 *mvdXML implementation*

In order to test the adoption of mvdXML-based requirement specifications against the model data exchanged between different stakeholders of the STREAMER project, an implementation of the validation features of mvdXML 1.1 has been developed using the infrastructure offered by the open source xBIM toolkit.

The implementation is mainly designed to allow individual stakeholders to independently verify the conformity of received and produced IFC models against the agreed exchange requirements and concept roots in a user friendly visual 3D environment.

To maximize the reusability of the developed components in other validation scenarios the implementation has been divided into two software components:

1. the mvdXML validation library (mvdLib) is a.NET dynamic link library providing validation capabilities that can be consumed in multiple deployment scenarios (e.g. Xplorer UI, web services, cloud environments, command line applications, etc.)
2. the XbimXplorer mvdXML Plugin (mvdUi) is an extension plugin for the pre-existing XbimXplorer IFC viewer that provides the User Interface for interactive validation of models against specification files.

Both modules have further development activities planned in response to feedback from the STREAMER project as well as from scheduled innovations in the underlying xBIM toolkit.

5.2 *User interface development and collaboration workflow*

To enable a complete collaboration workflow between stakeholders of the established IDM processes the mvdUI component has been designed to allow the interactive analysis of models according to arbitrary combinations of exchange requirements, concept roots and IFC classes, the UI allows immediate feedback on the validation status of selected elements as well as whole models; this filtering strategy also helps to improve the responsiveness of the application which can become relevant if thousands of requirements need to be checked for large IFC models. Visual color coding styles have been developed to allow rapid traffic-light model inspection in the 3D viewer of passing and failing requirements.

The development of features for the semi-automatic production of validation reports in the BIM collaboration format (BCF) have required the redesign of the XbimXplorer plugin API in order to allow integration of the MVD plugin with the existing BCF plugin; the designed features allow stakeholders to exchange communication threads on the result of validation tests across different BIM platforms while retaining complete reference of the involved IDM, MVD and IFC background.

6 PROOF OF CONCEPT

6.1 *Preparing client requirements (ER1-PoR)*

Much of the client requirements is shared through informal spreadsheets. In the current case the PoR was prepared using BriefBuilder and the information was shared as a simple CSV file. In order to make this information available for formal checking prior to incorporation into the design process, it is necessary to add the semantic meaning of the individual rows and columns. This was achieved through the use of the AEC3 BimServices Transform1 utility.

The semantic meaning of the rows is by default unknown. The transformation takes a single extra parameter 'topic' which identifies the semantic object represented by the rows. The choices include 'project', 'site', 'building', 'storey', 'zone, 'component', 'system', 'type' or in this case 'space'. The transformation then creates a complete IFC model with the minimum number of other objects necessary to give context for the objects.

Each data field is mapped to a property grouped in a default property set '*Default_SpaceProperties*'. However, the transformation makes use of a global dictionary which contains hints which can add value to the outcome by associating the column headers to specific IFC attributes (Figure 8). The global dictionary can also hold pointers to the expected parent, for example a property set, any synonyms, and any expected values.

6.2 *Checking requirements*

Checking of the resulting IFC model in XBIM is straight forward and shown in Figure 9. After importing both the IFC and the mvdXML file the exchange requirement can be selected and checked (right top view). All constraints that passes or fails are shown with traffic lights in the window below. Each test result is linked with the object in question and can be browsed in the 3d viewer and the properties window. For supporting the communication within the design team a BCF issue can also be generated to point to failures.

Various filter options enable to focus on specific objects or constraints. It is for instance possible to select specific object types or properties of an

```
<concept type="property">
  <term context="BriefBuilder">Room type</term>
  <term context="PoR">RoomType</term>
  <term context="IFC">ObjectType</term>
  <term context="en-GB">
    Space or Component Type</term>
</concept>
```

Figure 8. Example from the global dictionary to control the CSV to IFC mapping.

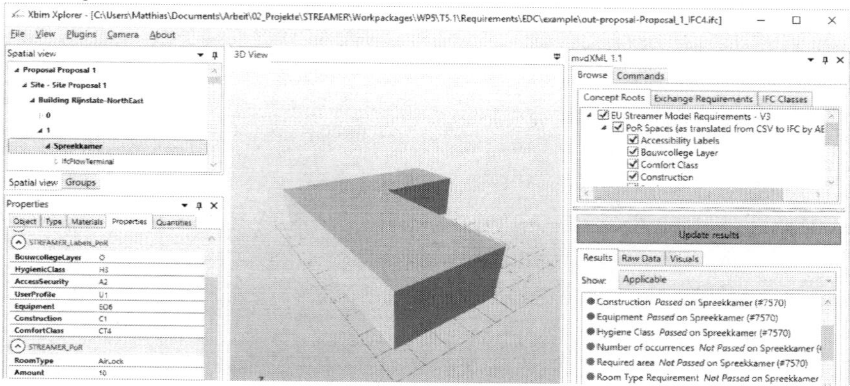

Figure 9. Checking result of an example space layout generated from space requirements

exchange requirement only. It is also possible to select elements in the 3D view that shall be checked by selected requirement definitions. This feature also helps to improve performance in case of very big IFC files and/or constraints.

7 CONCLUSION AND OUTLOOK

An integrated approach for checking exchange requirements based on the mvdXML 1.1 format has been presented. Requirements management is an essential element and is supported by a novel web-based solution called BIM-Q, which not only enables to capture, maintain and easily configure exchange requirements but also allows to specify its mapping to IFC and the generation of an mvdXML file. The mvdXML-based model checking was implemented as a plugin for the open source XBIM viewer and has been validated with examples from the STREAMER project.

Current implementation shows the overall potential of that development. It enables to improve the quality of BIM-based data exchange. However, further research is necessary to develop best practices and more templates in order to reduce the specification effort. Instead of starting from scratch the user can then reuse available requirement definitions and can focus on project specific configurations. Another field of development is to provide consistency checks of configured requirements. And last not least it has to discussed if and how to extend the checking capabilities of mvdXML in order to go beyond yet available fundamental checks.

ACKNOWLEDGMENT

The presented research was done in the frames of the European FP7 Project STREAMER and the buildingSMART Norway BIM Guide developments. We acknowledge the kind support of the European Commission, bS Norway and the project partners.

REFERENCES

BCF. http://www.buildingsmart-tech.org/specifications/bcf-releases.

BriefBuilder. http://www.briefbuilder.nl/.

Chipman, T. Liebich, T. & Weise, M. 2016. mvdXML—Specification of a standardized format to define and exchange Model View Definitions with Exchange Requirements and Validation Rules. *buildingSMART International Ltd. 15.02.2016.*

Di Giulio, R. Quentin, C., van Nederpelt, S. Traversari, R. Nauta, J. & Turillazzi B. 2015. D1.2: Semantic typology model of existing buildings and districts. *Deliverable of the STREAMER project.*

IDM. http://iug.buildingsmart.org/idms/

IFC4 Design Transfer View. HTML documentation: http://www.buildingsmart-tech.org/mvd/IFC4Add1/DTV/1.0/html/

IFC4 Reference View. HTML documentation: http://www.buildingsmart-tech.org/mvd/IFC4Add1/RV/1.0/html/

ifcDoc tool. http://www.buildingsmart-tech.org/specifications/specification-tools/ifcdoc-tool/ifcdoc-beta-summary.

ISO 29481-1:2010. Building information modelling—Information delivery manual—Part 1: Methodology and format. *Published by ISO/TC 59/SC 13.*

MVD. http://www.buildingsmart-tech.org/specifications/mvd-overview/mvdxml-releases/mvdxml-1.1

Nisbet, N. 2010. BimServices—Command-line and Interface utilities for BIM. http://www.aec3.com/en/6/6_04.htm

STREAMER website. http://www.streamer-project.eu/

XBIM websites. http://www.openbim.org/, https://github.com/xBimTeam.

eWork and eBusiness in Architecture, Engineering and Construction – Christodoulou & Scherer (Eds)

Querying linked building data using SPARQL with functional extensions

C. Zhang & J. Beetz
Eindhoven University of Technology, Eindhoven, The Netherlands

ABSTRACT: In this paper, we propose to extend SPARQL functions for querying building models. Building models captured by the IFC data model are the target data sources to develop extended functions. By extending these functions, we aim to 1) simplify writing queries and 2) retrieve useful information implied in building models. These functions are classified into four categories according to required data inputs. A prototype implementation is provided with indicate performance testing. Since this system is compatible with SPARQL and Linked Data environment, building models can be queried with data from other fields. A case study is conducted to query building and regulatory data for code compliance checking purpose. It demonstrates an approach which can be applied and extended for many other use cases.

Keywords: Linked Data, ifcOWL, SPARQL, IFC, use case, vocabulary, function

1 INTRODUCTION

As integrating data in the Architecture, Engineering and Construction (AEC) industry is becoming increasingly important, using the Resource Description Framework (RDF) and Linked Data technologies to represent building data has been proposed time and again (Beetz et al. 2009, Pauwels and Terkaj 2016). Unlike traditional data modeling approaches which are limited by the scope of their underlying schemas, these technologies can provide an open and common environment for sharing, integrating and linking data from different domains and data sources. SPARQL is a generic query language that allows federated queries across heterogeneous Linked Data sources (Harris et al. 2013, Heath & Bizer 2011). In comparison with domain-specific query languages (Mazairac & Beetz 2013, Borrmann & Rank 2009), SPARQL is especially applicable for scenarios when data from multiple sources are needed. Here are some possible scenarios:

- All building objects should be classified according to NL-sfb code.
- The type and thickness of walls can only be modelled according to the right combinations in table X.
- Where are the locations of companies which produce the materials used in walls placed in the hall.

All these scenarios not only need building data as input, but also require data from other fields and documents. They can be more easily implemented with RDF and SPARQL technologies without relying on proprietary systems. As a language for querying generic RDF data, however, SPARQL does not contain specific vocabularies or functions to fulfil requirements common to many use cases in the construction domain. By only using standard SPARQL, many useful relationships and properties e.g. typing, properties, spatial relations etc. that are explicitly defined or implied in building models are either difficult or impossible to be retrieved.

To address this issue, we use SPARQL as a base query language and propose to extend a set of functions specific to the AEC domain. This strategy has been taken by other fields. For example, Open Geospatial Consortium (OGC) has released GeoSPARQL as a set of extended functions for geospatial data. It makes sense that the same approach can be taken for the building industry.

The extended functions in this research are compatible with existing SPARQL environments. They can be used to query building data combined with data from other fields, which in turn may have their own domain specific functions (Fig. 1). It is a generic approach that is reusable in many use cases such as building data query, data integrity checking, code compliance checking, multi-model collaboration etc.. A prototype implementation is described to address related technical issues and demonstrate the effect of the proposed approach. As a W3C standard, SPARQL has been widely implemented by a plethora of RDF APIs and databases, hence many platforms can be used as base environments

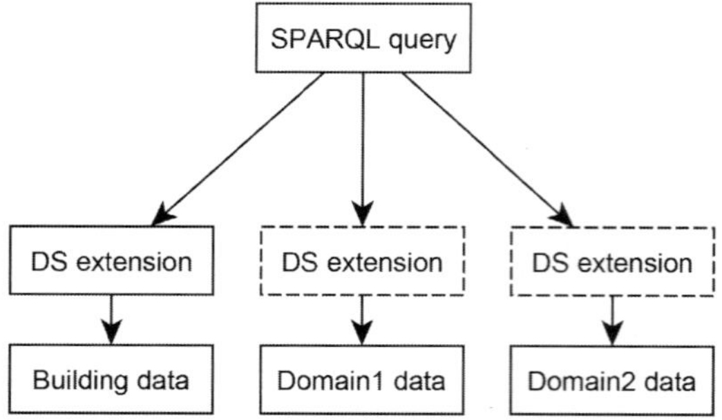

Figure 1. SPARQL query with domain specific functional extensions.

for implementing extended functions (Harris et al. 2013).

This paper is structured as follows. In section 2, SPARQL standard and its extension mechanisms are briefly reviewed. Proposed functions are classified into four groups and described in section 3. In section 4, we present a prototype implementation along with indicative performance tests. In section 5, we provide a case study to evaluate the proposed approach. A discussion about limitations and further work conclude this paper.

2 SPARQL FUNCTIONALITIES

2.1 *SPARQL and official built-in functions*

SPARQL Protocol and RDF Query Language (SPARQL) is a W3C standard for querying RDF data (Harris et al. 2013). It provides an SQL-like syntax and can be used to query RDF triples that are either maintained in local files or triple stores which can expose this data via standardized interfaces using the HTTP protocol as so-called SPARQL endpoints. Traditional relational databases can also provide such services by using e.g. D2R server (Bizer & Cyganiak 2006). SPARQL is versatile and offers full CRUD (create, read, update and delete) functionalities (e.g. SELECT, CONSTRUCT) as well as many logical expressions (Angles & Gutierrez 2008). Additional functions are used in its FILTER expression. In its official specification, these functions mainly include operators to evaluate and manipulate primitive data types and aggregations e.g. numbers, strings and URIs. In the AEC domain, these functionalities are very useful for use cases such as data integrity validation against BIM requirements and MVDs (Zhang et al. 2013). For example, the Dutch Rgd BIM norm requires that all building elements should be associated with appropriate building storey (Rillaer et al. 2012). Provided that the building model is represented in the standardized ifcOWL (Pauwels and Terkaj 2016), the query provided in Listing 1 can check this spatial containment relationship using off-the-shelf SPARQL implementation.

However, since standard SPARQL can only crawl the data graph according to the structure of data models, it cannot query information according to logics of many use cases and cannot retrieve some critical information related to e.g. geometric representations of building objects (Farias et al. 2015, Zhang and Beetz 2015). To achieve this, additional functionality must be provided.

2.2 *Domain-specific functional extensions for SPARQL*

As it is described in section 1, extending SPARQL with additional functions has been proposed and implemented in other fields: The Jena LARQ engine adds a full-text search engine functionality to SPARQL. The stSPARQL in Strabon and GeoSPARQL project from Open Geospatial Consortium (OGC) have specified many topological and geospatial functions (Perry & Herring 2012). They have been implemented by geospatial database systems including Strabon, Parliament and uSeekM (Garbis et al. 2013). Some other RDF APIs and triple stores like the Jena framework, Allegrograph and Virtuoso have also implemented geospatial functions.

In all these extension efforts, there are two ways to extend SPARQL with domain-specific functions:

- The first method is to add operators in the FILTER expression to evaluate data values or RDF nodes.
- The second one is to define a function as an RDF property, which is known as a computed property or property function to generate output based on its bound subject or object (W3C 2014).

The difference is that a property function can also be added into original RDF graphs as additional properties to enrich data, while filter functions are usually only used to evaluate data in the

```
SELECT ?wall
WHERE{
   ?wall a ifc:IfcBuildingElements .
   FILTER NOT EXISTS{
   ?r ifc:relatedElements ?wall .
   ?r a ifc:IfcRelContainedInSpatialStructure .
   ?r ifc:relatingStructure ?storey .
   ?storey a ifc:IfcBuildingStorey .
   }
```

Listing 1. Query to check the spatial containment structure for all building elements not related to a building storey.

query runtime. In the research presented in this paper, all the extended functions are defined as property functions.

3 VOCABULARIES

Building data captured by the Industry Foundation Classes (IFC) data model is the target data sources for developing extended functions in this research. There are many properties and relationships that are repeatedly required in use cases. Functions are wrappers of snippets to derive such information and use them in different scenarios. To formalize them as vocabularies, both real use cases and IFC documentation are reviewed (Rillaer et al. 2012, Statsbygg 2011, IBC 2006, buildingSMART 2013).

Due to the complexity of the building industry, however, it is not possible for us to list needed functions for all scenarios. Instead, we propose to classify them based on required data inputs from IFC building models since they are related to further implementations.

Information in IFC building models can be roughly grouped into a) domain semantics represented by e.g. object types, relationships, and properties, and b) geometric data, which is a low-level technical description captured by geometry objects associated with product instances. The proposed vocabularies are classified into four groups to derive data from these two subsets of IFC models. Section 3.1 and 3.2 describe functions used to derive information from domain semantic subset of models, while section 3.3 and 3.4 describe functions to analyze geometric parts. For each group and subgroup, example functions are provided with a query sample.

Since all these functions are modeled as property functions in SPARQL queries, they are also RDF properties with their respective URIs and namespaces. Due to the flexibility and openness of the RDF technology, additional functions can always be extended.

3.1 *Functions for schema level semantics*

Functions in this group are defined to represent IFC schema level semantics. To identify them we use the prefix qrw:. They are modelled from the fundamental concepts and assumptions specified in the official IFC 4 documentation (buildingSMART 2013). These fundamental concepts describe recommended and commonly used template structures in IFC instances to capture data. Many of them have complex structures to represent semantics. For example, if we need to query the "is placed in" relationship between a window and a wall (Statsbygg 2011), we need to go through a considerably complex model subgraph to retrieve this information as illustrated in Fig. 2. In practices, shortcut relationships are needed to logically adapt with use cases and shorten query writing.

Functions in this category are defined for this purpose. For example, Figure 2 illustrates that direct relationships (in dash line) are defined between concrete building objects. These relationships are extended by functions to be called in a query. Listing 2 shows an example query to check whether a window and its related wall are contained in the same building storey (Statsbygg 2011). This query uses the functions qrw:isPlacedIn and qrw:isContainedIn. Following the same approach, over 40 functions have been defined on top of these fundamental concepts.

3.2 *Functions for instance level semantics*

Functions in this group are defined to represent IFC instance level semantics. The mechanism of property and quantity sets allow the dynamic extension of the semantics of IFC building models beyond the schema. In total, there are more than 2000 properties and more than 200 quantities grouped within 410 property sets and 92 quantity sets in the official IFC 4 documentation. In

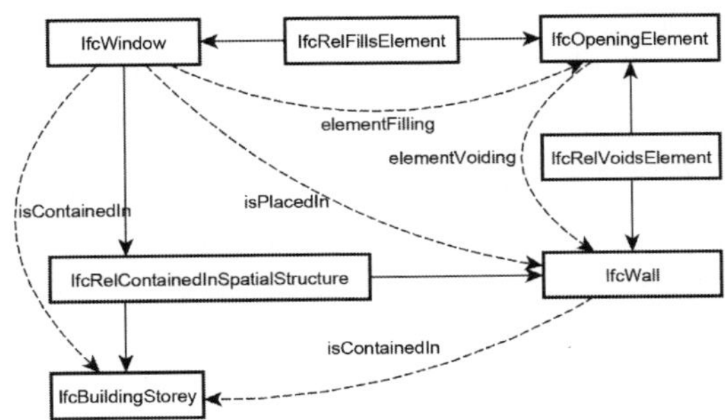

Figure 2. SPARQL query with domain specific functional extensions.

```
SELECT ?window ?wall
WHERE{
   ?window a ifc:IfcWindow .
   ?window qrw:isPlacedIn ?wall .
   ?wall a ifc:IfcWall .
   FILTER NOT EXISTS {
      ?wall qrw:isContainedIn ?storey .
      ?window qrw:isContainedIn ?storey .
      ?storey a ifc:IfcBuildingStorey .
   }
}
```

Listing 2. Query to check the consistency that every window and the wall it is placed in are contained in the same floor. Windows and walls which do not fulfil this requirement will be retrieved.

building models, these properties and quantities are modeled by instances of the IfcProperty and IfcElementQuantity rather than direct properties for building objects. In this vocabulary, all the properties and quantities are defined as RDF properties using the prefixes pset: and qto:.

Listing 3 shows a quantity take off example to count the external walls on each building storey. It uses a function pset:isExternal from this group. These functions provide shortcuts to directly connect building objects and property values instead of using complex structures defined in IFC schema.

3.3 *Functions for product geometry*

Functions in this category are to derive properties based on the geometric representations of a single building product. This vocabulary is identified by the prefix pdt:. There are many properties which can be defined based on product geometry data. These common properties include the following examples:

- pdt:hasOverallHeight is a function to return height of the bounding box of a product.
- pdt:hasVolume is a function to return the volume of a product.
- pdt:hasGrossWallArea is a function to return the area of wall without considering openings.
- pdt:hasWindowArea is a function to return opening area for a window.

Although many of these properties can be represented by property sets and quantity sets (see section 3.2), they are not mandatory and are not always reliable. Directly deriving information from geometric data provides another dimension to enrich data and ensure consistency. Listing 4 shows an example to query heights of all stair flights (IBC 2006).

```
SELECT ?storey (COUNT(?wall) AS ?q)
WHERE{
  ?wall a ifc:IfcWallStandardCase .
  ?wall pset:isExternal true .
  ?wall qrw:isContainedIn ?storey .
  ?storey a ifc:IfcBuildingStorey .
} GROUP BY ?storey
```

Listing 3. Query to retrieve the quantity of all external walls on each storey.

```
SELECT ?sf ?height
WHERE{
  ?sf a ifc:IfcStairFlight .
  ?sf pdt:overallHeight ?height .
}
```

Listing 4. Query to retrieve all stair flights and their heights.

3.4 *Functions for spatial reasoning*

Functions in this group are to derive information related to spatial reasoning, which needs geometric and location data of multiple building objects. This vocabulary is identified by the namespace of spt:.

3.4.1 *Relationships between products*

Functions in this category are used to derive relationships between products. We currently offer general topological relationship functions as follows:

- spt:touches
- spt:intersects
- spt:disjoints
- spt:contains

They can be used to search and evaluate relationships between building objects. Table 5 shows an example to query all walls which touch doors.

3.4.2 *Property for groups of products*

Functions in this category are used to derive properties for a group of products. Distance between products is a typical example. Many building regulations, building codes and BIM requirement manuals constrain the distance of building components,

```
SELECT ?wall
WHERE{
  ?wall a ifc:IfcWallStandardCase .
  ?door a ifc:IfcDoor .
  ?wall spt:touches ?door .
}
```

Listing 5. Query to find all walls (represented by IfcWallStandardCase) which touch doors.

```
SELECT ?column ?window ?d
WHERE {
?column a ifc:IfcColumn .
?column qrw:isContainedIn ?storey .
?window qrw:isContainedIn ?storey .
?window a ifc:IfcWindow .
(?column, ?window) spt:distanceXY ?d .
FILTER (?d<300)
}
```

Listing 6. Query to check the clearance before windows by searching where there are columns too close to a window.

```
SELECT ?wall
WHERE{
    ?wall a ifc:IfcWall .
    ?wall spt:isOutside true .
    FILTER NOT EXISTS{
    ?wall pset:isExternal true .
}
}
```

Listing 6. Query to select all walls which are externally located but do not have "true" value for IsExternal property.

such as clearance before openings, heights between floor slabs etc. The exact semantics of the notion "distances" can between contexts. In our vocabulary we offer the following concepts:

- spt:distance is a function to return the shortest distance between two products in 3D space.
- spt:distanceZ is a function to return the shortest vertical distance between two products.
- spt:distanceXY shortest distance between the projections of two products on XY plane.

An example query is provided to select pairs of column and window which are on the same storey and have in-between distance shorter than 300 millimeters.

3.4.3 *Property for single object based on spatial relationships*

This family contains properties which are related to locations of building objects. We currently only offer one concept in our vocabulary.

- spt:isOutside is similar to the function pset:isExternal (see section 3.2) with the difference that it is based on location and geometric data of all major building objects (wall, door, window etc.) instead of explicitly defined properties that might be provided by the wall type, which in turn might be used wrongly by the modeller. It also returns a boolean value for a specific building object. It also can be used to return building objects according to a boolean value when it is defined.

An example query using this function is shown in Listing 7. It is to check the consistency between the property IsExternal and geometric position for every wall.

4 IMPLEMENTATION

As a proof of concept, these listed functions are implemented based on the open source Jena ARQ query engine. In this implementation, we attempt to minimize hard coding to make defined functions more portable and more transparent for public reviews.

Functions defined in section 3.1 and 3.2 can all be implemented by various rule languages

```
SELECT ?a2
WHERE {
   ?a1 ifc:relatedElements ?arg1 .
   ?a1 ifc:relatingStructure ?a2 .
  ?a1 a
ifc:IfcRelContainedInSpatialStructure .
   }
```

Listing 7. A query as a SPARQL rule to implement the function qrw:isContainedIn.

e.g. SWRL or by SPARQL itself using frameworks like SPARQL Inferencing Notation (SPIN) (Knublauch et al. 2011). We choose SPIN for the implementation, as it is open source and employs the same language. For example, the function of qrw:isContainedIn in Listing 3 is implemented by associating it with the query in Listing 7. It will trigger this query as a subquery when it is called.

Functions in 3.3 and 3.4 are implemented by "black box" coding. In current implementation, geometric representations in ifcOWL instances are preprocessed to triangulated representations to assist computing results. Since developing algorithms for deriving these properties are out of the scope of this research, existing algorithms are used and some simplifications are applied. For example, functions in section 3.4.1 are coded using the same algorithm described in Daum & Borrmann 2014. Functions in section 3.3 only considered bounding box of building products.

Except pre-processing work, all extended functions are backward chaining which are processed during query runtime. The data flow is illustrated in Figure 3. The ifcOWL instances and SPARQL queries are the input of the system. Depending on the size of ifcOWL files, we can choose to load them into memory or disk-based triple stores. In the query runtime when a property function is referred, it will trigger a SPARQL rule as a subquery or a snippet of programming code to retrieve related values. Since a SPARQL rule can also call extended functions, this process recursively continues until no function is called. This process is compatible with other reasoning technologies such as OWL and RDFS reasoners.

4.1 *Performance testing*

According to different sizes of building models, performance testing is conducted. We choose three open IFC models with different sizes for testing

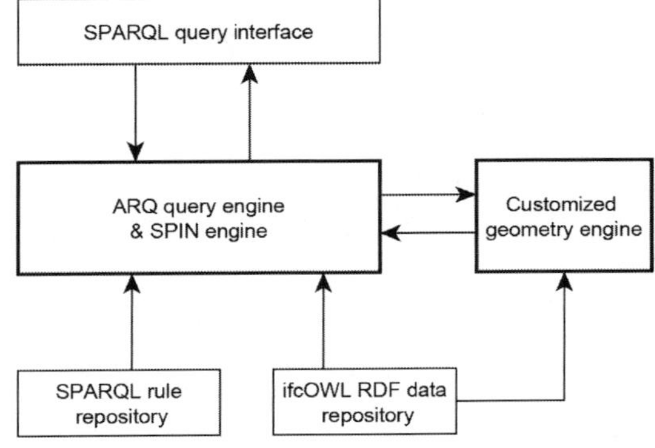

Figure 3. Data flow of the prototype system.

(Table 1). In this test, besides query engine no additional reasoner is used. The Jena TDB triple store is used as query backend. The test environment is a normal laptop with i7 2.0GHz quad-core CPU with OS of Windows 8.1. We show the performance of queries in Listing 3, 4 and 5 on these three models (Table 2).

Q1 contains no function related to geometry data, so its performance only depends on the configuration of ARQ query engine. The performances of Q2 and Q3 are related to both query engine and geometry engine. From Table 2, we can see that Q3 has a considerably long time for query processing. It is expected because this query needs to calculate m*n times to return results, when there are m walls and n doors contained in the model. For the model M3, it needs to calculate 1430*247 times. In further implementations, optimizations are needed and additional 3D geometry libraries might be used.

5 CASE STUDY

The proposed approach inherits the advantages from the technologies of SPARQL and RDF. It can be used to combine data from other sources e.g. classification, product catalogue, sensor network, geospatial etc. to query results (Beetz & de Vries 2009). A use case example is presented in this section in a scenario of regulatory compliance checking.

Table 1. Tested building models and their sizes.

ID	*Name*	*Size*
M1	Duplex_A_20110505.ifc	2.25 MB
M2	Office_A_20110811_optimized.ifc	3.9 MB
M3	091210Med_Dent_Clinic_Arch.ifc	14.7 MB

Table 2. Performance testing results for query 1, 2 and 3.

Query	*ID*	*Query time (ms)*	*Result size*
Q1 (Listing 3)	M1	1062	4
	M2	3250	3
	M3	6984	7
Q2 (Listing 4)	M1	13	2
	M2	14	4
	M3	18	7
Q3 (Listing 5)	M1	437	18
	M2	13387	57
	M3	28152	149

5.1 *Querying building and regulatory data for code compliance checking*

SPARQL CONSTRUCT query can be considered as a rule language to inference and validate RDF data. A case study of building code compliance checking is conducted in this research. This example is from International Building Code 705.8.4 (IBC 2006). This rule is described in Listing 8.

Additionally, the allowable opening areas for both protected and unprotected openings (ap and au) are provided in a table to describe their relations with fire separation distance. Table 3 shows one row of it, which defines that when the fire separation distance is between 15 to 20 feet and the opening is unprotected, the allowed proportion (au) between opening area and external wall area needs is up to 25%. This table is transformed to RDF using the method provided by (W3C 2015). Listing 9 shows a snippet of it, which is equivalent to contents in Table 3.

This transformed RDF data is maintained in a different file, which then is queried with building models in application runtime. The query in Listing 10 is used to check the opening area of external walls on each floor. When a storey violates this rule, this query will generate an issue associates with it.

All the properties starts with ibc: are specifically defined for this example. They are mapped to

705.8.4 Where both unprotected and protected openings are located in the exterior wall in any story of a building, the total area of openings shall be determined in accordance with the following: $(Ap/ap) + (Au/au) \leq 1$

where: Ap = Actual area of protected openings.

ap = Allowable area of protected openings.

Au = Actual area of unprotected openings.

au = Allowable area of unprotected openings.

Listing 8. Rule example.

Table 3. One row of Table 705-8 in International Building Code.

Fire separation distance	Degree of opening protection	Allowable area
15 to less than 20	Unprotected, Nonsprinklered	25%

```
_:table705-8_13
  ibc:minFSDistance "15"^^xsd:double  ;
  ibc:maxFSDistance "20"^^xsd:double  ;
  ibc:openingProtection false ;
  ibc:sprinklerProtection false  ;
  ibc:allowableArea "0.25"^^xsd:double  .
```

Listing 9. Transformed RDF data for Table 3..

vocabularies defined in Section 3 by SPARQL using the method described in Listing 7. For example, the property ibc:protectedOpeningArea is associated with the query in Listing 11, which retrieves the proportion between area of protected openings with area of all external walls for a storey.

A building with 4 floors are used to test this rule. Generated issue report points to the building storey which violated the rules from the table and equation (Fig. 4).

6 CONCLUSION

This research describes using SPARQL with functional extensions to query building models. In comparison with traditional domain specific query languages, it can leverage SPARQL and RDF technologies to facilitate multi-domain query and can accomplish complex use cases as shown in section 5. As more and more data is represented by RDF and Linked Data technologies, this approach will potentially have more effect in the future.

```
CONSTRUCT{
# generate issues for storeies
 _:b0 a spin:ConstraintViolation .
 _:b0 spin:violationRoot ?storey .
 _:b0 spin:violationLevel spin:Warning .

}WHERE {
# derive data from ifcOWL instance
?storey a ifc:IfcBuildingStorey .
?fs a ibc:FireSeparation .
(?storey, ?fs) spt:distanceXY ?d .
?storey pset:sprinklerProtection ?bool .
?storey ibc:protectedOpeningArea ?Ap .
?storey ibc:unprotectedOpeningArea ?Au .

# derive proper ap and au from table
{?b1 ibc:minFSDistance ?min .
 ?b1 ibc:maxFSDistance ?max .
 ?b1 ibc:sprinklerProtection ?bool .
 ?b1 ibc:openingProtection true .
 ?b1 ibc:allowableArea ?ap
 FILTER (?d>=?min && ?d<?max )
}
UNION
{?b2 ibc:minFSDistance ?min .
 ?b2 ibc:maxFSDistance ?max .
 ?b2 ibc:sprinklerProtection ?bool .
 ?b2 ibc:openingProtection false .
 ?b2 ibc:allowableArea ?au
 FILTER (?d>=?min && ?d<?max )
}

 # checking according to the equation
 FILTER (?Ap/?ap+?Au/au>1)
 }
```

Listing 10. Query to check building models.

```
SELECT (SUM(?windowArea)/(SUM(?wallArea)))
WHERE {
   ?wall qrw:isContainedIn ?storey .
   ?wall a ifc:IfcWall .
   ?wall pset:isExternal true .
   ?wall pdt:hasGrossWallArea ?wallArea .
   OPTIONAL
     {?window qrw:isPlacedIn ?wall .
      ?window pset:protected true .
      ?window pdt:hasWindowArea ?windowArea }
   }
```

Listing 11. A query as a SPARQL rule to implement the function ibc:protectedOpeningArea.

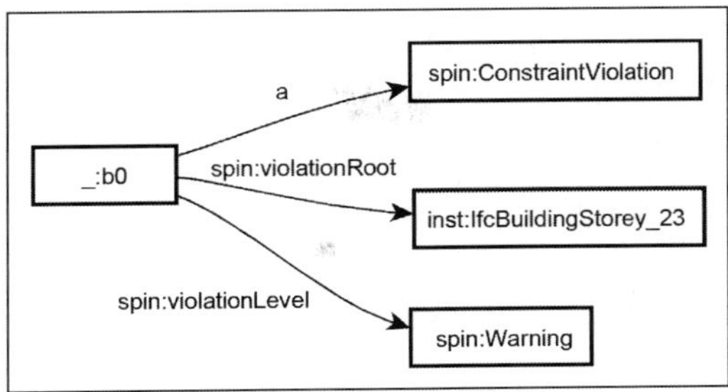

Figure 4. Generated graph data as issue report for a storey inst:IfcBuildingStorey_23.

Current limitations of proposed approach are mainly related to the implementation. The first one is that it sacrifices some portability since extended functions are not implemented in other RDF stores. This can be partly solved by using SPIN framework since it uses SPARQL itself for implementing functions (see section 4). Due to the functional limitations of SPARQL e.g. no fully support for recursive definition, no geometry library, performance etc., however, functions which are related to geometric data can only be implemented by additional coding. The second one is related to algorithms and performance. This is not specifically for this system, but is a common issue no matter using BIMs or Linked Data approaches. Many use cases related to retrieving information related to geometric data are very hard or even impossible to be implemented. More algorithms need to be developed and additional optimizations are needed. In real practices, we also need to consider the balance between preprocessing and query runtime.

For further research, more use cases can be investigated to extend functions for specific domains in building industry and combine more data from different sources. From technical perspective, existing geometric libraries can be integrated into this environment to improve performance.

REFERENCES

Angles, R. & Gutierrez, C., (2008). The expressive power of SPARQL (pp. 114–129). Springer Berlin Heidelberg.

Beetz, J., & de Vries, B. (2009). Building product catalogues on the semantic web. Proc." Managing IT for Tomorrow, 221–226.

Beetz, J., van Leeuwen, J.P. and de Vries, B. (2009). "IfcOWL: A case of transforming EXPRESS schemas into ontologies." Artificial Intelligence for Engineering Design, Analysis and Manufacturing. vol. 23. no. Special Issue 01. 89–101.

Bizer, C., & Cyganiak, R. (2006). D2r server-publishing relational databases on the semantic web. In Poster at the 5th International Semantic Web Conference (pp. 294–309).

Borrmann, A., & Rank, E. (2009). Topological analysis of 3D building models using a spatial query language. Advanced Engineering Informatics, 23(4), 370–385.

Daum, S., & Borrmann, A. (2014). Processing of topological BIM queries using boundary representation based methods. Advanced Engineering Informatics, 28(4), 272–286.

Farias de, T.M., Roxin, A., & Nicolle, C. (2015). IfcWoD, semantically adapting IFC model relations into OWL properties. arXiv preprint arXiv:1511.03897.

Garbis, G., Kyzirakos, K., & Koubarakis, M. (2013). Geographica: A benchmark for geospatial rdf stores (long version). In The Semantic Web–ISWC 2013 (pp. 343–359). Springer Berlin Heidelberg.

Harris, S., Seaborne, A. & Prud'hommeaux, E., (2013). SPARQL 1.1 query language. W3C Recommendation, 21.

Heath, T., & Bizer, C. (2011). Linked data: Evolving the web into a global data space. Synthesis lectures on the semantic web: theory and technology, 1(1), 1–136.

IBC. (2006). International building code. International Code Council, Inc. (formerly BOCA, ICBO and SBCCI), 4051, pp. 60478–5795.

Mazairac, W., & Beetz, J. (2013). BIMQL–An open query language for building information models. Advanced Engineering Informatics, 27(4), 444–456.

Pauwels, P., & Terkaj, W. (2016). EXPRESS to OWL for construction industry: towards a recommendable and usable ifcOWL ontology. Automation in Construction, 63, 100–133.

Perry, M., & Herring, J. (2012). OGC GeoSPARQL-A geographic query language for RDF data. OGC Implementation Standard. Sept.

Rillaer, D. van, Burger, J., Ploegmakers, R., Mitossi, V., 2012. Rgd BIM Standard, version 1.0.1. 1–29.

SPIN. (2011). SPARQL Inferencing Notation. http://spinrdf.org/.

Statsbygg, (2011). Statsbygg Building Information Modelling Manual Version1.2. Available at: http://www.statsbygg.no/bim, accessed January 2014.

W3C (2014). SPARQL/Extensions/Computed Properties. https://www.w3.org/wiki/SPARQL/Extensions/Computed_Properties.

W3C (2015). Generating RDF from Tabular Data on the Web. https://www.w3.org/TR/2015/WD-csv2rdf-20150416.

eWork and eBusiness in Architecture, Engineering and Construction – Christodoulou & Scherer (Eds)
© 2016 Taylor & Francis Group, ISBN 978-1-138-03280-4

Context aware information spaces through adaptive multimodel templates

F. Hilbert
Institute of Applied Computer Science, Dresden University of Technology, Germany

R.J. Scherer
Institute for Construction Informatics, Dresden University of Technology, Germany

ABSTRACT: This paper addresses context aware information supply for information processes in the construction industry. The planning and creating of structures and buildings is based on collaborative construction information processes whose information supply is characterized by a decentralized exchange of heterogeneous application models. Throughout a construction project numerous architecture, engineering and management information models are created and used because of the inherent complexity of the building product and processes as well as the corresponding high number of involved domains. The various specialized applications e.g. CAD, FEM, cost or scheduling software and standards used lead to a weak software integration and inconsistent data handling. Therefore a central challenge in construction projects is still the management of the inhomogeneous distributed information environment where domain models data need to be adequately transformed, exchanged and managed horizontally (between client and contractor and among various discipline specific representations), longitudinally (in their temporal development along the project phases), and vertically (on different levels of abstraction).

To bundle various application models and indicate their dependencies the Multimodel as a kind of an information space was developed. Increased requirements for specialized work processes as well as the growing complexity of construction projects lead to an increase in the amount, scope and complexity of the exchanged information spaces. The corresponding Multimodels quickly become very large and unwieldy. This leads to an inhibitory effect on the construction information processes. The transfer of Multimodels and the orientation in the information spaces will be more complicated. In addition, this effect prevents a mobile use of Multimodel based information spaces. However various construction information processes for planning and creating of structures and buildings usually doesn't need the entire information available in construction projects. When considering the situational information requirements of construction information processes, a context dependence is revealed that determines the quantity and quality as well as the cutouts and linking depth of the required information spaces depending on various aspects of the processing context. The information needs depends on the context. For an efficient information supply of construction information processes by matching information spaces therefore the consideration of the processing situation is necessary. That is why the design of the required Multimodels is determined by various aspects of the process situation. With knowledge of these aspects, context dependent information needs can be anticipated and targeted information supply of construction information processes can be realized.

A prerequisite for such a context aware information logistics is a methodology that enables formal depict dependencies between contextual information and information space elements and evaluate them automatically. For this purpose, in this work an approach is presented to formalize the context dependencies of the information needs of construction information processes. The influence of information logistically relevant context aspects to the design of information needs, the context relations, can be described directly in Multimodel templates. For this a special rule language, called ContextScript was developed that can be annotated instead of static attribute values into the template. In this way, based on Multimodel information spaces, context adaptive Multimodel templates can be defined, which can persisted along with an associated reference process. By evaluating these templates at time of use, situative information needs can be anticipated as MultiModel Template (MMT) and an adequate context oriented Multimodel can be generated. Thus the presented approach allows the realization of a context oriented information supply that allows to anticipate context based information needs and to generate a corresponding situationally information space.

1 INTRODUCTION AND MOTIVATION

1.1 *Interoperability in Construction*

DEY AND ABOWD (2000, S. 304) stated: „By improving the computer's access to context, we increase the richness of communication in human computer interaction and make it possible to produce more useful computational services." Following this main idea, a concept for a context oriented information supply through adaptive information spaces is described in this article.

Throughout a construction project, numerous architecture, engineering and management information models are created and used because of the inherent complexity of the building product and processes as well as the correspondingly high number of involved domains. The various specialized applications (e.g. computer aided design, CAD, finite element method, FEM or scheduling software) and standards used lead to weak software integration and inconsistent data handling. Therefore, a central challenge within construction projects is still the management of the inhomogeneously distributed information environment where domain model data needs to be adequately transformed, exchanged and managed horizontally—between client and contractor as well as among various discipline specific representations, longitudinally—in their temporal development along the project phases, and vertically—at different levels of abstraction.

1.2 *Multimodel based information spaces*

Domain models are rarely processed in an isolated manner. Many construction information processes require the linking of domain model elements. In particular, construction specific nD modeling software applications connect building structures with elements of planning and controlling models. The generated model interdependencies are mapped inside the data model of the application and are stored according to the information process in a proprietary file format or even become discarded. For repeated use this link information must be recreated. One solution is storing interdependencies together with the affected domain models. For such linked information compilation, the term Information Space has been established (Newby, 1996; Van Hoof et al., 2003; Härtwig, 2009). As they address some issues of nD modeling, information spaces are one promising approach to overcoming the disruptive nature of typical construction information environments and improve interoperability in construction information processes.

For the formalization of information spaces, data structures are necessary that suitably reflect the model crosslinking structures. As a recommendation for the realization of such an information space, the generic MultiModel Container (MMC) was developed in the research project Mefisto (www.mefisto-bau.de) which aggregates unmodified domain model instances from different areas of design and construction together with their interdependencies (Scherer and Schapke, 2011; Fuchs et al., 2011). The Multimodel container allows to explicate relations between the domain models by linking their elements in a in a special link model. Figure 1 shows the structural design of a Multimodel container.

The Multimodel approach allows the formalization of information spaces. Their minimum requirements can be summarized as follows:

- An information space has to contain at least one domain model ($fm_i \in FM$).

$$FM_i = (\{fm_1, ..., fm_i\} \mid i \in \mathrm{N}) \tag{1}$$

- The existence of link models is not mandatory ($lm_j \in LM$). Particularly during the first phases of the project this may be absent or it has just been created.

$$LM_j = (\{lm_1, ..., lm_j\} \vee \{\} \mid j \in \mathrm{N}) \tag{2}$$

- An information space has to be described semantically. Therefore, the existence of metadata (*md*) is required.

A Multimodel based information space therefore essentially consists of a tuple of domain models (*FM*), Link Models (*LM*) as well as associated metadata (*md*):

$$InfSpace_{MM} = (FM, LM, md) \tag{3}$$

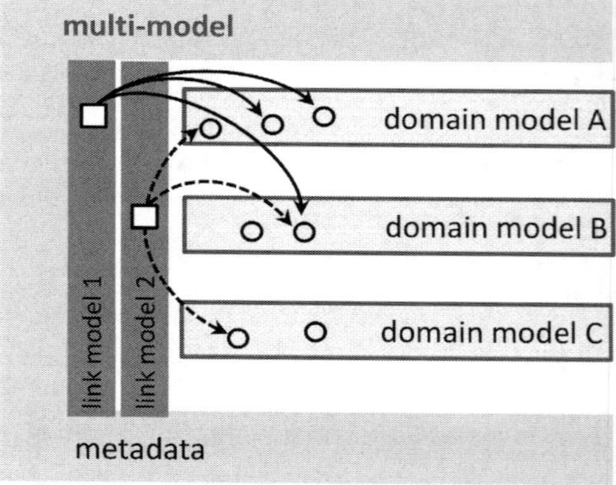

Figure 1. Multimodel structure.

The formal specification of a Multimodel based information space in the focus of this work can thus be represented as follows:

$$InfSpace = (\{\{am_0, ..., am_i\}, md\} \vee \{\{am_0, ..., am_i\}, \{lm_0, ..., lm_j\}, md\} \,|\, i, j \in \mathrm{N}) \quad (4)$$

1.3 *Semantic description of domain models and link models*

For both interdisciplinary collaboration and machine processing, a different interpretation of information involves the risk of incompetent or incorrect use. Therefore, a clear common understanding of project-wide information spaces by concerned actors and applications for crossdomain use has to be ensured (Eigner, 2010, p. 88). The necessary semantic interoperability can be achieved with a homogeneous semantic description of the information spaces as well as the project-wide use of a controlled vocabulary annotation. Enrichment with additional semantic information can be done model internally or externally. Model internal attributions are defined by the respective special applications (or used standards) and are not changeable project-wide. Therefore, to describe the involved domain models and link models in Multimodel based information spaces model external metadata annotations are used (Fuchs, 2015, p. 5). To support a large number of relevant information processes in the construction industry, the semantic description has to cover various specific construction terminology and classifications aspects of the domains involved but still has a sufficiently high generality. For the description of Multimodel based information resources and information needs a marking system is used which is based on an approach by SCHAPKE (2014, S. 156). It consists of a seven part marking system which is divided into four areas. Figure 2 shows the multiple semantic annotations as indices of a domain model.

The top left indices specify the type of the model. Within the different modeling domains (e. g. BIM, CSM, SPM, TSM), there are different types of models, different model grades and contents determined by the domain. The selection of the model types used will depend on the nature of the construction project; not all model types are involved in all projects.

level of detail ***lod***
modeltype ***t*** *model cutout* ***c***

Model Name

project phase ***p*** *model format* ***f***
model status ***s*** *model language* ***l***

Figure 2. Marking system of domain models.

The top right indices represent the model qualities. Here, the level of detail (lod) describes the granularity of the domain model. Different models of the same model type may have different types of model elements. For example, a rough schedule and execution schedule are quite diverse. The same applies to cost models. The determination of the granularity is based on the outline concepts of various specialized model types. They are usually object-related (e.g. CityGML) as well as transaction-based levels of detail.

Alternatively, granularity can also be defined transitively on associated domain models derived (e.g. a cost calculation based on a predetermined construction work model granularity). The model cutout specifies the excerpts from the total models and is based on the outline concepts of various specialized types of models. Here transitive breakdowns by the outline concepts linked trade model types are possible (e.g. a building section which corresponds to the phase of a process model).

The bottom left indices identify the process aspects and describe the integration of the model into collaboration processes. Both, attributes project phase as well processing status, indicate in which phase of the project respectively in which processing status the domain model was created. For the classification of construction projects, different phase models exist and the phase model used has to agree project-wide. Domain models pass different working processes. As the result of which, they receive a different processing status. To document the current processing status, a simple ordinal scale with four processing statuses: request (α), design (β), agreed or released (γ), discarded or obsolete (ε) can be used (Schapke, 2014).

The bottom right attributes describe the formalization aspects. For exchanging domain models, various domain specific and model type dependent file formats exist. The model format attribute describes the used format. As vocabulary for file formats, a model type dependent list of applicable file formats is suitable. The language quality specifies in which language the model was created. Particularly in larger international projects, the use of different languages of the nationalities involved may be caused by communication barriers. This is particularly relevant for domain models with a high amount of semantic content (e. g. bills of quantities). The model name is finally used to identify the domain model. Except for any existing project-wide guidelines for naming, there are no limitations here.

Interdependencies between the model elements are explicitly indicated ID-based in special link models (LM). For different intentions, it is possible to use various link models in a Multimodel (see Fuchs et al., 2011). The description of the link model is mainly based on the specification of the linked domain models and the linked model element types (see Figure 1). In addition, it is possible to provide for the process aspect of both project phase as well as a model state for a link model.

Because a link model connects a lot of application models across domains, no single attribute on quality aspects can be specified. Due to the fixing of the model type, there is also no need to describe other aspects of modeling. The model format is XML, and due to lack of semantics the link model is language neutral. Consequently, no semantic qualities and formalization aspects can be described. Technical qualities of link models can be evaluated on the completeness of the link elements in a considered cutout. For example, it may be required that all the elements of a cost model (e. g. such as GAEB), which have a quantity, even just a part of a building model (e. g. IFC) have to be assigned. Figure 3 shows the semantic annotations for a link model as indices. The bottom left attributes have the same function as with the domain models, but the bottom right indices describe the interdependencies.

In the further work, the presented marking system serves as the basis of the description of Multimodel based information spaces both as an information resource as well as an information need. The vocabulary for the values of metadata has to agree throughout the project. Example vocabulary can be found in (Scherer and Schapke, 2014, Fuchs, 2015).

1.4 *Multimodel templates*

In particular, for the specification of reference processes the semantic descriptions of the Multimodel can also be used in a prescriptive manner. Thus requirements for a Multimodel based information space can be formalized without embedded or referenced domain model instances as a so-called MultiModel Template (MMT).

In this way, the semantic metadata can specify existing information spaces as well as required information. Figure 5 illustrates an example of a Multimodel template and Figure 4 shows a corresponding Multimodel instance. The example focus on an information space for the tendering of construction works. Here three domain models are combined by a link model.

LM Name

·ojectp hase ***p***
ıodelstatus ***s***

model Ref.{***id1***,...,***idi***}
elementtype {$\mathbf{et_1}$,...,$\mathbf{et_i}$}

Figure 3. Specification attributes for link models.

2 CONTEXT AWARE INFORMATION LOGISTIC

2.1 *Context dependent information needs*

The planning and creating of structures and buildings is based on construction information processes. Increased requirements for specialized work processes as well as the growing complexity of construction projects lead to an increase in the amount, scope and complexity of the exchanged information spaces. The corresponding Multimodel s quickly become very large and unwieldy. This leads to an inhibitory effect on the construction information processes. The transfer of Multimodels

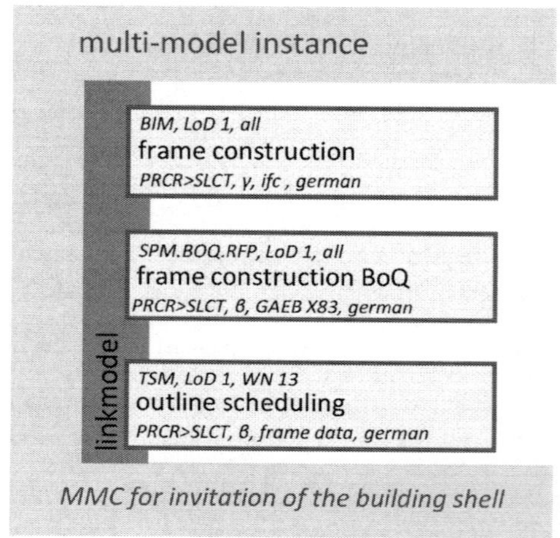

Figure 4. Multimodel instance.

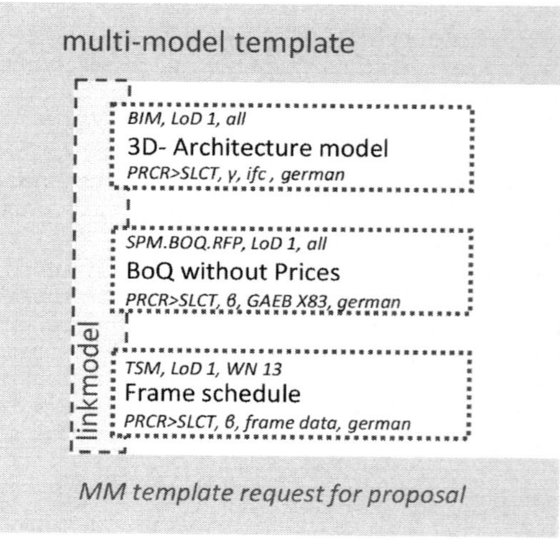

Figure 5. Multimodel template.

and the orientation in the information spaces will become more complicated. In addition, this effect prevents a mobile use of Multimodel based information spaces.

However, various construction information processes for the planning and the creating of structures and buildings usually do not need the entire information available within the construction projects. When considering the situational information requirements of construction information processes, a context dependence is revealed that determines the quantity and quality as well as the cutouts and linking depth of the required information spaces depending on various aspects of the processing context. The information needs depend on the context.

Therefore, it is necessary to consider the processing situation for an efficient information supply of construction information processes by matching information spaces. Figure 6 illustrates this relationship. Conceptually, the collaboration situation determines the need for information, which will ideally be fulfilled by a corresponding adapted information space. In the instance level, the context oriented information supply with Multimodel s is represented. Here the demand for information is specified by a Multimodel template.

However, in a considered information process, not every domain model is equally dependent on context and not every identifiable context aspect of the process context is relevant to the information needs. In order to use the context information for an efficient information supply, in a first step the process specific context dependencies of the information needs must be identified and modeled. In our approach to serialize information spaces, we use Multimodel s and consequently, for representing information needs, Multimodel templates. In the example shown in Figure 7, the context dependent attributes are marked. These templates are innately very static, however. In order to expend the templates for context dependencies, we modify their semantic description. Instead of static attributes, the different influences of various context aspects on the configuration of information needs are annotated as context dependency rules.

Such annotated templates are called context adaptive MultiModel Templates (caMMT).

By evaluating these templates at the time of use, situational information needs can be anticipated as a Template (MMT) and an adequate context oriented Multimodel can be generated. In order to manage these dependencies in the presented approach, a two-step procedure is proposed, as shown in Figure 8. Starting with a context neutral information space (*MMT*), the context dependencies of linked information models are analyzed, formalized and additionally annotated. The result is a context adaptive information space (*caMMT*). At the time of use, the context dependencies are evaluated on the basis of the processing situation and a situation responsive information space is created (*MMT*).

2.2 *Context modeling for construction information processes*

To identify the information logistically relevant context aspects, the processing situation needs to be properly formalized. The factors and dependencies generated by an interpretation and abstraction mental model of a processing situation can be formally specified as context dependent aspects of an intentional context cutout in a hierarchical structure (Schill and Springer, 2011). An appropriate context model allows for complex context requests

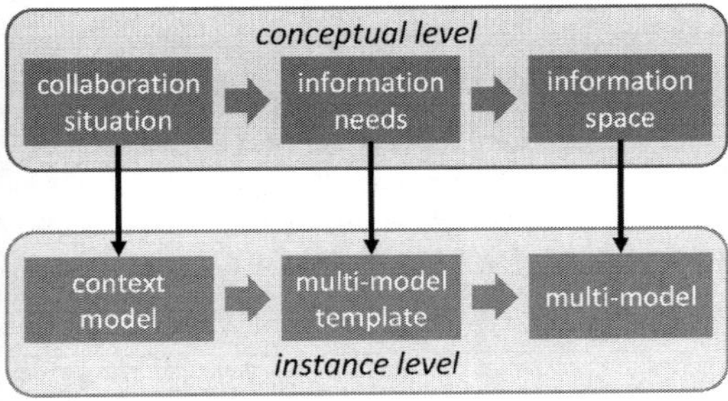

Figure 6. Levels of context-oriented information supply.

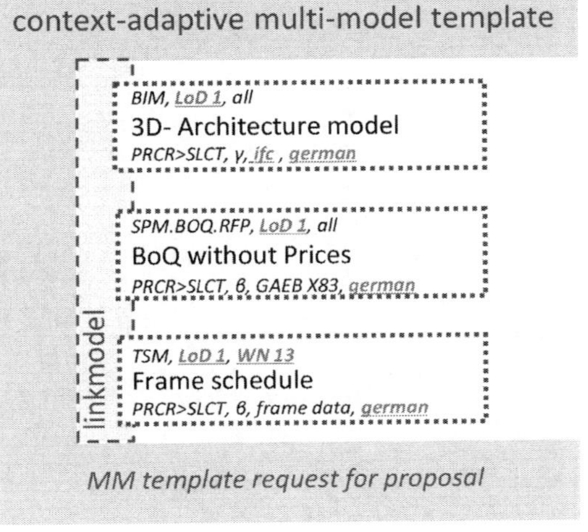

Figure 7. Context dependencies of Multimodel template.

at an abstract level. An overview of the concepts used thereby is expressed in Figure 9.

Because a context is always considered in relation to an entity, there is no universal or entity independent context (Risku, 2004, S. 86). The notion of context always describes an entity dependent section of a theoretically possible overall context. Construction information processes are embedded in a processing context in which three entities operate: actor, task and resource. A context model for mapping processing situation thus ideally comprises the contexts of the entities involved. Figure 10 to 12 shows the entities of the information process and its context areas.

2.3 *Context adaptive multimodel—templates with contextscript*

After a formalization of the Multimodel based information needs as well the context information has been developed, the effects of context aspects to single information space elements can now be formalized. The influence of information logistically relevant context aspects to the design of information needs, so-called context relations, can be described by condition action rules (CA). If at the time of evaluation, a given condition is met, the specified action is executed. Symbolically, a CA-rule can be represented as follows:

$$\{C_1, C_2, \ldots, C_n\} \rightarrow A \tag{5}$$

Based on CA rules, we introduce a rule system to formalize context relations that link individual attributes of the context model with elements of a

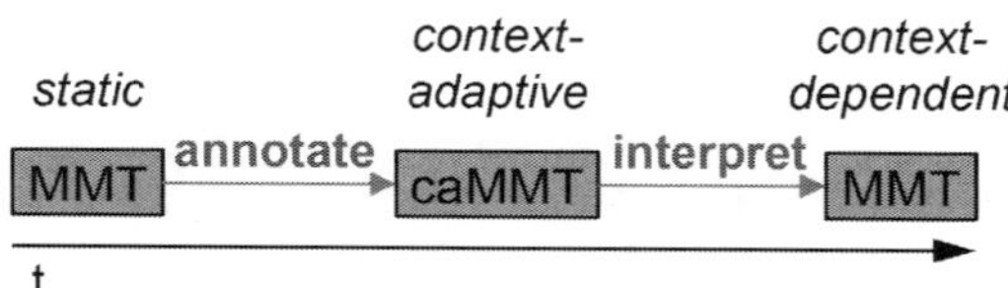

Figure 8. Context usage with multimodel templates.

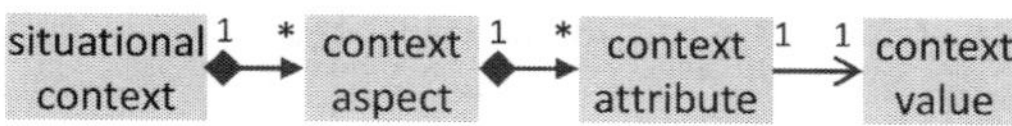

Figure 9. Concepts of context consideration.

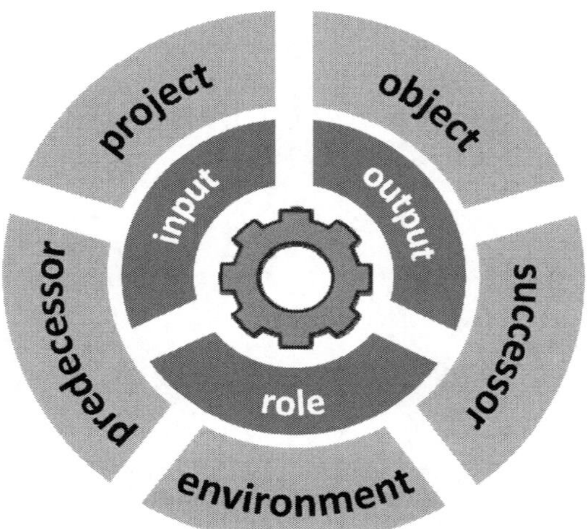

Figure 10. Process context areas.

Figure 11. Actor context areas.

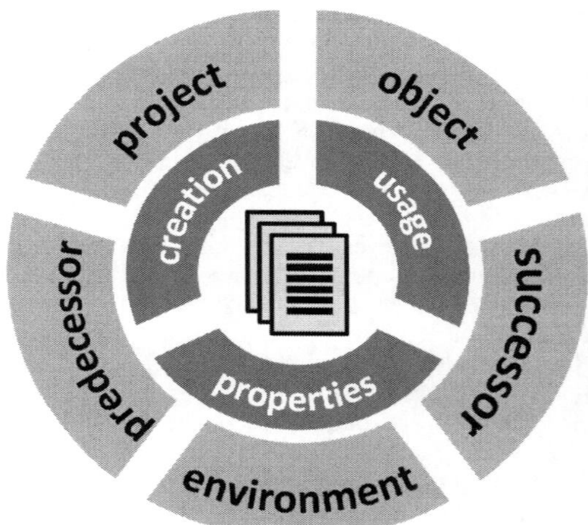

Figure 12. Resource context areas.

Multimodel template using so-called ContextScript rules. ContextScript is a declarative rule language with textual syntax. Each ContextScript usually consists of a rule premise (ContextCondition), which describes the conditions for the rule, and a rule consequence (ValueAssignment), indicating the effect of the rule. Can the rule premise be evaluated as true by a control system, then the rule consequence instruction is performed. Figure 13 (above) illustrates the structure of a ContextScript rule as described. The premise of a ContextScript rule consists of a condition block (ContextCondition), describing the context conditions and a consequence block with attribute values in a "true" and in a "false" branch.

In the rule premise, information logistical context aspects are evaluated and, in the rule consequence, corresponding template attribute values will be generated.

Existence tests and comparisons of content are supported, in particular, in order to use ContextScript rules for the widest possible spectrum of applications and for the formulation of the rule condition linking several contextual conditions. Context attributes and model attributes (of the considered information space template) are both used here as factors. Figure 13 (middle) shows the structure of the Context Condition.

The context-sensitive characteristics of the affected information space parameters are defined in the rule consequence, where constants or variables can also be used (see Figure 13, below).

Since a ContextScript rule only applies to certain information space parameters and is annotated at the relevant location of the Multimodel template, an indication of the target parameter is unnecessary. To identify the context variables used in ContextScript rules, so-called context paths are used that indicate the path from the root of the context model to the context attribute. The segments of the path are divided structurally by dots. Moreover, aliases help in navigating through context model. Corresponding aliases exist for all three collaborating entities (actor, _process and _resource). Through the alias, the context information of the related entity can be determined (actor.<contextattribut>).

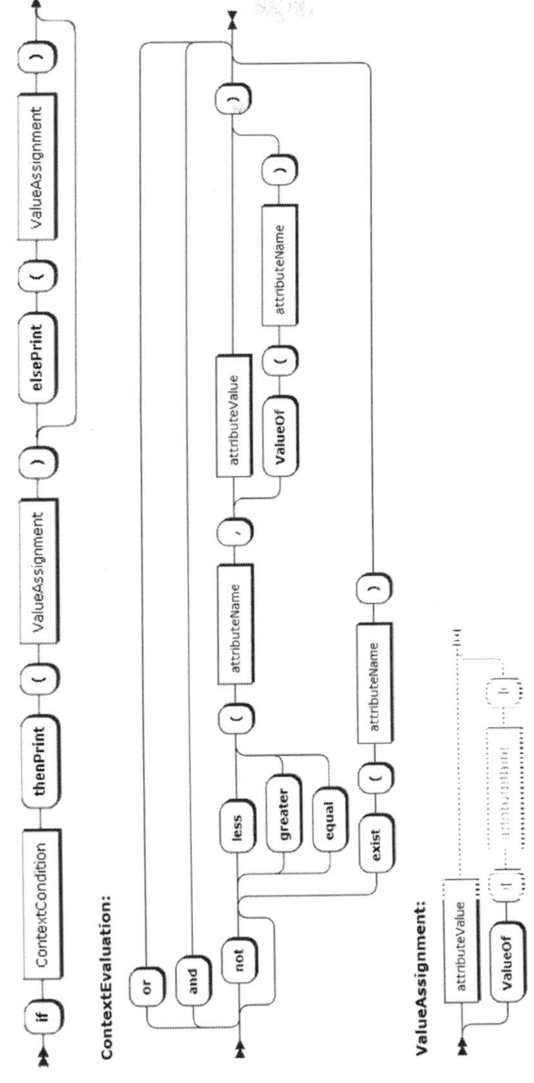

Figure 13. ContextScript.

2.4 *Situational multimodel—templates*

At the time of use the ContextScript rules are evaluated on the basis of current context information and the rule itself is replaced by the determined rule consequence. This evaluation by a control system is done in four steps.

- In a first step, a parser converts the rule premise into an abstract syntax tree, which represents a navigable structure. This syntax tree contains, in its nodes, rule terms that may be combined by logical link operators and modifiers and contain existence or comparison operations. The leaves of the syntax tree contain literals (attribute names and comparative values).
- In the second step of the context integration, the attribute names in the leaves are replaced by the contents of the corresponding context model or template attributes. For this, either the context model or the Multimodel template will be used.
- In the third step, comparison and existence operations of outer nodes are evaluated to boolean values.

- From these results, the propositional link node can now be applied to identify the child nodes. For this, the disjunctive and conjunction expressions are evaluated and replaced by a literal to the determined truth value. This is done iteratively until the root node of the rule premise is evaluated.

The algorithm terminates with the evaluation of the root node, so that the entire rule premise is evaluated into a single logical value. Based on the determined result, the rule consequence can be selected and written to the position of the ContextScript rule in the Multimodel template. Evaluating all ContextScript rules of the context-adaptive Multimodel template results in a situation responsive Multimodel template which represents the situational information needs. This represents an essential prerequisite for the creation of a context-oriented information space.

Two examples illustrate the formulation of ContextScript rules.

- Customizing domain model language to the actor language:
  If exist _context._actor.language thenPrint ValueOf(_context._actor.language)
- Determine usable model formats of specialized applications of the actor and specifying the required formats of the information space:

if exist _context._actor.preferences.application thenPrint(_context._actor.preferences.application.fileFormats)

3 CONCLUSIONS

3.1 *Result*

In this paper, we described the main concepts of a context dependent information supply by Multimodel based information spaces. Through ContextScript a mapping between acting contextual factors and affected information space parameters in Multimodel templates can be represented. For this, the ContextScript rule is annotated directly in the Multimodel template instead of a static attribute value. An annotated Multimodel template with ContextScript rules can be persisted as a so-called context-adaptive Multimodel template along with an associated reference process. This approach enables the implementation of a context-aware information logistics, which accurately provides the information spaces for the project partners in the construction industry, which are needed in a concrete working situation. Due to the short length of the paper, most of the addressed topics could only be briefly highlighted.

3.2 *Outlook*

The approach, however, is not confined to the construction industry. The rule language Con-textScript is a method to describe context relations and form the basis of a context-oriented information supply by Multimodel—based information spaces in all domains. The main intention behind the development of the rule language ContextScript was to build a bridge between the modelers of Multimodel -based reference processes and the developers of context suitable applications. Accordingly, a relatively simple and easy to learn method was created which allows even inexperienced end users to formulate context dependencies for Multimodel -based information spaces. The developed system provides useful guidance for similar further developments as well as clear implementation targets. Beside improved integration, considerable benefits of the approach are expected in the achievement of faster, more accurate and more efficient information processes.

In the future, a steady increase of the information volume can be expected. Therefore, more intensive use of contextual information is becoming increasingly important. Improvements of the described are proposal conceivable rules in the automatic generation of ContextScripts or wider use of user feedback.

ACKNOWLEDGEMENTS

The research described in this paper was enabled by the financial and technical support of the Mefisto project partners and the German Ministry of Education and Research (BMBF). This support is gratefully acknowledged.

REFERENCES

Dey, A.K. & Abowd, G. 2000 Towards a Better Understanding of Context and Context-Awareness. In: *Proceedings of CHI workshop on the What, Who, Where, When and How of Context-Awareness, as part of the 2000 Conference on Human Factors in Computing Systems (CHI 2000)*. The Hague, The Netherlands.

Eigner, R. 2010 KOMODE—A semantic context model for collaborative applications in automotive ad hoc networks *(in german:KOMODE-Ein semantisches Kontextmodell für kollaborative Anwendungen in automobilen Ad-hoc-Netzwerken), Phd,* München: Techn. Univ.

Fuchs, S. 2015 Developing cross-domain information spaces with Multimodel s *(in German: Erschließung domänenübergreifender Informationsräume mit Multimodellen), Phd,* Technische Universität Dresden, Institut für Bauinformatik.

Fuchs, S., Kadolsky, M. & Scherer, R.J. 2011 Formal description of a generic multi-model. In: Enabling Technologies: Infrastructure for Collaborative Enterprises (WETICE), 2011 20th IEEE International Workshops on. IEEE. 205–210.

Härtwig, J. 2009 Konzept, Realisierung und Evaluation des semantischen Informationsraums, *Leipziger Informatik-Verbund (LIV)*, Leipzig.

Newby, G.B. 1996 Metric Multidimensional Information Space. *In Harman, D. (Hrsg.). Proceedings of TREC-5.* Gaithersburg.

Risku, H. 2004 Translationsmanagement, *Narr*, Tübingen.

Schapke S.-E. 2014 Prozessgesteuerter Einsatz von Multimodellen, *In: Scherer, R.J. und Schapke S.-E. (eds.) (2014): Informationssysteme im Baumanagement, Band 1: Modelle und Methoden, Springer Verlag*, Berlin, Heidelberg, 189–205.

Scherer, R.J. & Schapke S.-E. (eds.) 2014 Informationssysteme im Baumanagement, *Band 1: Modelle und Methoden, Springer Verlag, Berlin,* Heidelberg.

Scherer, R.J. & Schapke, S.E. 2011 A distributed Multimodel—based Management Information System for simulation and decision-making on construction projects. *Advanced Engineering Informatics, 25(4), S. 582–599.*

Schill, A. and Springer, T. 2012 Verteilte Systeme. *Springer Berlin* Heidelberg.

van Hoof, A., Fillies, C., & Härtwig, J 2003 Aufgaben- und rollengerechte Informationsversor-gung durch vorgebaute Informationsräume. *in: Fähnrich, K.-P. und Herre, H. (Hrsg.) Content-und Wissensmanagement, Beiträge auf den LIT'03 Content-und Wissensmanagement.* Leipzig, 1–10.

eWork and eBusiness in Architecture, Engineering and Construction – Christodoulou & Scherer (Eds)
© 2016 Taylor & Francis Group, ISBN 978-1-138-03280-4

Access control for web of building data: Challenges and directions

J. Oraskari & S. Törmä
School of Science, Aalto University, Espoo, Finland

ABSTRACT: Web of Building Data enables open publication of building-related data on the Web. However, a large part of building data requires access control for reasons of facility security, resident privacy, competition, and IPR protection. The linking across multiple models at different hosts creates challenges to implement an access control scheme that could avoid repeated and tedious registrations and authentications by users, and enable simple definition of the access control rules at different hosts. In this study the access control challenges of practical use cases are analyzed, objectives for access control are summarized, and possible decentralized access control solutions are explored. The focus is on WebId+TLS and OpenId Connect. The importance of common ontologies for role or property based access control is discussed.

1 INTRODUCTION

Over the lifecycle of a building, a large set of parties produce information related to the building and its environment. The amount and variety of this information is growing due to the ongoing digitalization of society and construction industry. The potentially extremely valuable information remains underutilized: it is fragmented and managed in a disconnected manner, which makes it difficult to access and combine with other relevant information. (Törmä 2014).

Web of Building Data is a set of technologies for publication of building related data on the Web. It enables the online access to an up-to-date version of data in a granular fashion. Each building element has a Web address and can be accessed individually. Using their addresses, building elements and other objects related to them can also be linked with each other across different hosts and models. There are many use cases where these capabilities are valuable, and the increasing openness of data enables efficiency improvements and new innovative applications.

The publication of building data online on the Web creates a very different security environment than the traditional way of exchanging building models between parties, either point-to-point or through a centralized project repository. From the security perspective, Web of Building Data has both advantages and disadvantages. A major advantage is that since building objects can be accessed in a granular manner—one object at a time—it is not necessary in many use cases to transfer complete models to other parties. Transferring a complete model is risky: it requires trust that the receiving party does not misuse the model, will keep it confidential, and store and manage it scrupulously to avoid security breaches. In contrast, granular transfer of data enabled by Web of Building Data can be authenticated, controlled, and monitored to alleviate confidentiality problems. Its disadvantage from the security perspective is the open nature of the Web: all published data is by default accessible to all Web users. If confidentiality is required, the access to data needs to be controlled by proper authentication and authorization mechanisms.

Probably only a minor portion of all building data can be completely open for everyone to access; most of it can only be published to a restricted audience. Confidentiality requirements spring from a variety of sources: the physical security of a building, the privacy of building users, the protection of the intellectual property of designers, the wish to hide operational data from competitors, and the desire to avoid the embarrassment due to the poor quality of data.

The focus of this paper is to study the challenges of access control for Web of Building Data. What are the categories of stakeholders, what are their roles in publishing and accessing data, how sensitive is the published data, and what kinds of threats would the platform be faced with? From what directions could the solutions to the challenges be found?

To answer the above-mentioned questions, we will first analyze a set of use cases for Web of Building Data in the areas of simple model publication, linking of product data, status sharing, and cross-model linking. The stakeholders, their roles in publishing, managing and accessing data, and types and sensitivity of data contents in each case are described, and relevant challenges are identified.

The paper discusses possible approaches to address the security challenges. There is an obvious need to balance the safety concerns with the flexibility and simplicity of the overall security design. As an example, the nature of the platform would suggest properties such as distributed identity management and single sign-on; the combination of these two requirements will constrain the options for platform implementation.

In the remainder of the paper we will first review the concept of Web of Building Data and its implementation in the DRUMBEAT platform, analyze the security challenges with use cases, and then explain the solution approaches. After discussion, we end with concluding remarks.

2 WEB OF BUILDING DATA AND DRUMBEAT PLATFORM

Web of Building Data (WoBD) is a set of technologies for *publication* of building-related data on the Web, instead of *exchanging* it as files between project partners (Törmä 2014). Like the ordinary Web, it is a decentralized and open framework, which can be used by any party as it chooses. WoBD can support information sharing among an extended set of stakeholders over the whole lifecycle of a building.

2.1 *Web of Data*

Web of Data is a term that refers to publication of data on the Web in a way that data objects can be first class citizens: they have a Web address, can be individually accessed, and can be linked with each other. Web of Data consists of basic Web technologies (URI, HTTP), Semantic Web representations and technologies (RDF, Turtle, JSON-LD, RDFS, OWL, etc.), and Linked Data principles (Bizer et al. 2009) that cover the publication and access of data.

The Web of Data approach for data publication can be summarized as following:

- *Identifiers:* URIs (Uniform Resource Identifier) (Berners-Lee 1998) are used as object identifiers.
- *Data:* Data is represented in RDF (Resource Description Framework) (Lassila & Swick 1999), a graph-based representation framework.
- *Concepts:* Conceptual models are ontologies defined typically in OWL (Web Ontology Language) (Dean et al. 2004) or in simple cases with RDFS (Brickley & Guha 2014) or SKOS (Miles & Bechhofer 2009). A same dataset can be partially described by several different ontologies.
- *Serializations:* Data can be serialized in standard formats (e.g., rdfXML, Turtle, JSON-LD) to allow tool chain interoperability and to avoid vendor lock-in (Carroll & Stickler 2004).
- *Access:* Objects are accessed with their URIs over the HTTP protocol (Fielding & Reschke 2014). The description of the immediate RDF graph neighborhood of an object is returned in a suitable RDF serialization. The description can contain further links to other related objects.

2.2 *Web of Building Data*

All major BIM authoring tools can export BIM models in the standard IFC format (Industry Foundation Classes) (ISO 2013). In WoBD approach the exported IFC files can be converted into RDF graphs, so called ifcRDF, which uses the concepts, properties types defined in the ifcOWL ontology (Pauwels et al. 2016).

The ifcRDF models are then published on the Web so that the descriptions of objects are retrievable using the HTTP protocol. The returned descriptions are serializations of the immediate neighborhood of an object in the RDF graph.

ifcOWL is just one ontology in which building related data can be published. Other ontologies based on existing standards could be used—for instance, COBie, CityGML, or InfraGML—and new ontologies can be created for specific use cases, such as status sharing or cross-model linking.

2.3 *DRUMBEAT platform*

DRUMBEAT platform is an implementation of Web of Building Data concept, developed at Aalto University. It contains an IFC to RDF converter called IFC2 LD (Hoang & Törmä 2015), and allows the publication of models on the Web.

In order to make Web of Building Data a practical approach, the DRUMBEAT platform contains several additional mechanisms:

- *Organization of data*: The published data is organized into collections, models, and datasets. These concepts are defined in a metadata ontology LBDHO (Linked Building Data Host Ontology).
- *REST interface:* The creation of the metadata objects (collections, models, datasets) are done through a REST API (Fielding 2000). IFC models that are uploaded into datasets are automatically converted into RDF.
- *Backlinking:* To make the published building data linked into two directions, the platform relies on backlinking functionality: whenever a link to remote object is created a notification to the remote server is sent, and the remote server

can create a backlink, that is, a remote inverse link.
- *Hosting and accessing links:* Each party hosts the *outgoing* links of all the objects that it hosts. When an object is accessed, the outgoing links are included in the returned description. There is thus no need for a separate access to links.

It should be noted that the backlinking is essential function in some of the use cases described below.

2.4 *Deployment*

Web of Building Data is meant to be deployed in a completely decentralized manner. The basic architecture does not require any centralized components, apart from common ontologies. Each party that produces building data, publishes it on its own, and manages the data independently of others. The outgoing links from data are published in a same way as other data. The links can be either embedded in models or published as separate link models, depending on whether the lifecycle of links is the same or different than that of the data.

Using the DRUMBEAT platform software, the options to publish building data are a similar as in the publication of Web pages. A company can have its own Web servers and run the DRUMBEAT software to publish data. If that is too complicated, it can rent a server capacity from a cloud to run the platform, or there could be external service providers that run the platform and provide domains for users to publish their building data.

Applications can access the data at DRUMBEAT platform through a REST API. They can follow the links from one server to another, in a similar manner than Web browsers do with HTML pages. They can also combine data from several different servers according to the needs of a use case.

One advantage of the decentralized publishing is to maintain a clear ownership and control of data. Each party can decide the access control rules for the data as it sees proper.

3 USE CASES

In this section, a set of use cases studied in the DRUMBEAT project will be reviewed. Different use cases are related to various stages of the building lifecycle. Common to all examples is the distributed nature of information production and utilization.

3.1 *Publication of a model*

This is a simple use case where one party publishes a BIM model to allow other parties to access the objects in it or establish links to them. In the DRUMBEAT platform this means that an IFC file is uploaded and the platform converts it into ifcRDF (in which the concepts that property types refer to the ifcOWL ontology), creates URIs for all entities with a GUID, and makes these objects retrievable over the HTTP.

Some obvious requirements arise from this simple scenario: the model author needs to have the rights to add and modify content on the platform, and no authorized parties should not have those rights. To ensure the integrity of data, modification rights cannot be given awarded lightly; the model author can be expected to have the expertise and tools to produce coherent data and be able to take responsibility of it. In the worst case the data would be destroyed by a malicious Internet bot.

The access to a published model is typically given to many users in addition to the publisher. Most of these users cannot have the rights to modify the content but they need to have the right to read it. There will thus be users with different access rights.

Most of the access to building data is instrumental. The need to access data about a building is more based on the role and properties of a user and her relation to the building, than on the preferences or identity of a user. Access rights can be associated to users with the following type of expressions:

- "The site manager of project-12"
- "All employees of the subcontractors of company-23"
- "All employees of company-x working in project-12"
- "The property manager of facility-34"
- "All residents of building-45"

The access control rules therefore refer to relative roles of user (architect, site manager, property manager, employee, resident) or companies (contractor, subcontractor), and certain individuals (project-12, company-23, facility-34, building-45). The identifiers of individuals can naturally be URIs.

In summary, the challenges arising from the model publication use case are:

- The producers and users of data need to be authenticated.

Table 1. Model publication use case.

Stakeholders	Models/datasets
BIM model author BIM model users	BIM model

- Assignment of different access rights for different users must be supported.
- The access control rules should be based on roles or properties of users with respect to certain individuals.

3.2 *Cross-model linking*

An object-level linking can be established between two BIM models. There are several examples:

1. *Structural-Architectural:* The precast elements in a structural model are linked to the walls and slabs in an architectural model.
2. *MEP-Architectural:* The terminals in a MEP model are linked to the spaces they serve in an architectural model.
3. *MEP-Structural:* The MEP objects colliding with structural objects are linked to each other in a form of clashes to be resolved.

Link generation could take place in BIM authoring tool during reference model based design (between the objects in the reference model and the authored model), or after the design work in a cross-model checking system such as Solibri Model Checker.

Once the BIM models have been created and the linking between them generated, links can be used as channels to propagate information across models. For instance, the architect could get information about the status of rooms indirectly through the structural model or MEP model. Status information is important when the possibility of further changes is evaluated.

In a change situation, linking can be used to provide notifications to other parties to indicate how the change would affect them. When an object has changed—when its properties or relations have been modified—and it has a link to an object in another model, the latter object needs to be checked. A notification to the other model will give focused information about the likely impact of the change.

The challenges arising from the access control use case are the following:

- When following links between models, the user may need to be re-authenticated.
- The user should not be burdened by re-authentication requests.

Table 2. Cross-model linking use case.

Stakeholders	Models/datasets
BIM model author	BIM models
Contractors	Link models
BIM model users	

- When following links between models, it should be possible to evaluate the access control rules in a coherent manner.
- When following links between models, if a user does not have read rights to the target object of a link, the link should not be presented as valid.
- It must be possible to publish link models without modification rights to the interlinked models.

3.3 *Linking to product data*

IFC models contain only rudimentary information about the actual products selected as implementations of particular building components. Most of product information is represented in online product catalogs or databases, for instance, at product vendors' sites. Product data is a natural example of enriching IFC models with more detailed data. The building objects in IFC models can be linked to appropriate product data, as long as that product data is accessible on the Web and identified with a URI. This is possible, for instance, with bSDD Data Dictionary and MagiCloud Product Library.

If needed, the linking to product data can be done without touching the original IFC model at all. In the Web of Data approach, the link can be embedded in the model but it can equally well be published in an external link model.

Some product data catalogs are commercial services that can be accessed only by subscription. Also many freely accessible catalogs prefer a user registration. In any sufficiently complex model, there will be links to several different product libraries.

BIM model authors create the models. During the design work, they can select generic or specific products as implementation of components in the model. They associate the product information with components in their model, which means that a linking is created between the objects in the BIM model and the products in the product library. A product selector, typically the contractor, can later change the tentatively selected products, which means the creation of another linking that specializes or overrides the linking created by a designer. The earlier choices of the designer should not

Table 3. Product data use case.

Stakeholders	Models/datasets
BIM model author	BIM models
Product selector	Link models
Product library operators	Product libraries
Product vendors	Product vendors data
BIM model users	

necessarily be accessible to all users, but for some users they need to be preserved for later analyses.

The model users may need to access the BIM model together with detailed product information in a variety or use scenarios: to set up, operate, diagnose, fix, replace, or reuse a product. This presumes that the BIM models will be available to the model users, which naturally is a goal of BIM. However, if the product information is only linked to the BIM model, and the information itself must be accessed from a commercial service, the access right should be delegated, or otherwise all model users need to be customers of that service, either through a subscription or an access fee.

In summary, the additional access control challenges are:

- The access behavior of a user may need to be monitored in order to avoid data hoarding.
- Depending on the business model, it should be possible to delegate the access rights to product data to the users of a model.

3.4 *Status sharing*

During the execution of a construction project, all building objects go through a sequence of status changes as a result of activities performed on them by different parties. For example, a precast element can have the following sequence of states: designed, detailed, accepted to casting, casted, finalized, transported to site, in site storage, installed, inspected.

The different activities causing status changes are generally carried out by different companies. The production of status information is therefore genuinely distributed among all actors participating in the inter-organizational workflow of a building object.

Status information is not very complex or voluminous. However, since status updates are produced by many different companies, and the ICT systems differ from party to party, it has been difficult to organize the status sharing in a manner that all parties interested in the status of the work would be able to see it. The key problem is to configure the complex information flows in an easy and flexible manner.

In the WoBD approach, each party publishes status events at its own domain, and the event contains a link to a building object in a BIM model published elsewhere. The same party can publish multiple status events concerning one object, and different parties can publish status events about same objects.

Through backlinking mechanisms, whenever a new status event is created referring to a building object, a notification is sent, and a corresponding inverse link back to the status event is created at the server hosting the BIM model. Each building object will therefore contain links to all the status events referring to it. That is, it contains its whole status history as links, but the status events themselves remain in the domains of their original publishers.

It would be beneficial to promote open sharing of status information since it can create visibility and transparency to the execution of a project, which again provides opportunities to improve the controllability and efficiency of the project. However, there are also many specific uses for status information. In the workflow of a component, the downstream parties can predict and understand when they need to be ready for their activities, the upstream parties (such as designers) need status information to understand when changes to components are still possible, and the project management wants to monitor the progress of the work and react to deviations.

This use case creates the following challenges to access control:

- If someone wants to access the status history of a construction object, she needs to visit several other sites to access the status events themselves. To make this practical, there needs to be a proper single sign-on solution (Shaer 1995) in place.
- There can be competitors in a same project that produce status events about different construction objects (e.g., two precast concrete fabricators). They may both be ready to publish status events to other parties in a project but not to each other, as they prefer to hide details about their operational efficiency from each other.
- The server that publishes status events referring to an object in a remote server needs to notify that remote server about a new link. The former server needs to authenticate to the latter server.

4 OBJECTIVES FOR ACCESS CONTROL

Common to the use cases is the need to authenticate the user in an effortless way that lessens the burden of memorizing passwords, but there is also the need to identify the communication parties to

Table 4. Status sharing use case.

Stakeholders	Models/datasets
BIM model author	BIM models
Status event provider (detailer, contractor, fabricator, logistics company)	Status datasets
BIM model users	

prevent any bogus activity and to be able to filter the presented content by the business relation that they have.

We presume that co-operation takes place between parties that are equal. They are companies and their employees working for them. To make the solution neutral, if possible, there should not be a central registry or coordinator for the user data or sessions. User management should be decentralized, which requires the standardization of security conventions. In this context, a common ontology for the security roles would be essential.

There may also be the need to be able to refer to projects or facilities in the same way across parties, e.g. an architect could be given read rights for all entities in a certain identified project. All partner firms could have their internal projects but have a way to refer the common grounds of the cooperation.

The autonomous security management can be interpreted so that every data producer owns and manages their data, which would solidify the security arrangements. No external user needs a write access to the system, but she can make change requests or publish her enhanced version of the dataset if some added property, a status event, or an issue would be preferred to be included. The new set could still be linked to the existing data.

The CIA triad (Perrin 2008; Greene 2006) is one of the security models used to assess the safety of systems: C stands for Confidentiality, I for Integrity, and A for Availability. Our focus is on the access control of building data. When that is reflected on the triad, most of the objectives map on the confidentiality. As mentioned in (Rouse et al. 2014), one of the key targets for companies is to assess and classify the severity of the possible damage that could be affected when the confidentiality of a piece of data is broken. The planned access control should enable expressing the shielded entities and make it possible to protect them from any malicious use.

Trespassing is not prohibited on the Internet but a criminal intent is an essential part of any crime (University of Minnesota Libraries 2015). In the legal terms, both cracking and burglary involve unauthorized breaking into something for the purpose felony (Thomson Reuters 2016a; Thomson Reuters 2016b). The goal for the access control is to limit the means the users have to avoid known or unknown threats come true, but if they do, the protections should be valid also in a court. That is, all valuable piece of data should be shielded.

In plain terms, Single Sign-On (Shaer 1995) refers methods that allow end users to authenticate many services from a single point. The authentication is done once and propagated to the associated systems (Parker 1995). In a RESTful implementation, that can be implemented using authorization tokens. However, as RESTful services are stateless by default (Pautasso & Cesare 2013), in a distributed and decentralized network the tokens need to be self-contained in the way that the server can know where to check the validity of the authorization. Instead of using a random number, an address of the authentication server needs to be included.

The solution should reflect the agreements between the companies. That is, in the optimal case, it would suffice that parties agree on the cooperation in a user role level and the responsibility of allocating a specific employer would be delegated to the employer. This increases the flexibility of arranging the authorization but implies that the actual users may not be introduced to the company before the usage of the authentication. Furthermore, if a complete logging is preferred, the login has to contain enough data to identify the user.

In the case of emergency, it may be urgent to give the right information to the right people in right roles as soon as possible. For example, we may not be able to name the firefighters, paramedics, or police officers beforehand, but the roles may be given certain predefined access to the system.

In summary, the main objectives for an access control solution to WoBD are the following:

1. Decentralized implementation.
2. Support for single sign-o.
3. Access rights assigned to roles or properties, instead of individual users.
4. Users should have accessible profile information that determines their roles/properties.
5. Common ontologies for roles/properties.

5 SOLUTION APPROACHES

In the light of the preceding discussion, we review suitable decentralized authentication protocols and Web-oriented authorization ontologies.

5.1 *Authentication*

5.1.1 *OpenID Connect*

OpenID Connect (Sakimura et al. 2014) is a flexible single sign-on authentication protocol for the Internet. It is built on top of OAuth 2.0 (Hardt 2012) as a layer to provide user information in a standardized way. In that, the identification token is used in the role of an ID card. It contains URL of the issuer, but the fields are very limited. However, in principle, the sub claim (Sakimura et al. 2013) of the ID token could contain an address to a WebID profile, or the separately fetched OpenID Connect UserInfo profile. The URL could point to a description of the user role in the cooperation. The solution is only partial, since both profile pages would be needed to be protected separately, and it is left to the content provider to apply limitations.

5.1.2 *WebID-TLS*

The WebID-TLS authentication consists of:

1. A public web accessible user profile that describes the user identity (Sambra et al. 2016),
2. A self-signed X.509 certificate (Housley et al. 2002) that refers to the profile (Story et al. 2016),
3. TLS network protocol (Allen & Dierks 1999) and,
4. the protocol that authenticates the end user when the associated client certificate is introduced (Story et al. 2016).

TLS ensures all communication is encrypted, but it is also used to verify the host of the WebID profile.

The above-described protocol can limit access to the protected resources, but it lacks ways to shield the profile data from malicious data snooping, or attacks from inside like tampering in an untrustworthy server. As proposed in (Wild et al. 2014; Wild et al. 2015), the WebID-TLS should be extended to include signing the WebID profiles and validating the signatures during the login phase, and filtering the access to the profile data.

Instead of testing knowledge like a password, WebID-TLS is based on ownership, i.e. if a user can show that she owns a WebID certificate that gives her rights to a protected resource, she can proceed without a login prompt. It lightens the burden of remembering passwords and thus can serve as a single sign-on system, but care needs to be taken to ensure that nobody can steal the private store. For example, a good protection against computer trojans is recommended. A malicious program that attacked a private key store was described in (Doherty 2011).

In general, WebID gives the users all freedom to create the identification data themselves. That is possible, but in the Web of Data context, added assurance is needed to prove that the identity owner can represent the company that she is working for. A solution is to let companies publish the employee identities as secured WebID profiles. That ensures also that uniform security measures are in use to protect the profiles. Still, at the content provider end, extra care should be taken to check that the associated domains belong to the cooperating companies.

5.1.3 *Authorization*

The ownership based discretionary access control (DAC) is familiar to most of the folder based system users (Ubale & Apte 2014). In that, creators control access rights to their data. These are in line with the objectives of the system, but, in contrast to groups that are typical in a DAC system, we do not have a list of members, but roles assigned by partner organizations. On the other hand, Role-Based Access Control (RBAC) (Gritzalis et al. 1997) systems support user roles. Therefore, if both models can be merged as proposed in (Sandhu & Munawer 1998), the system would fulfill our objectives.

As shown in (Sacco & Passant 2011), WebID profiles can be in unison with additional ontologies like Privacy Preference Ontology (PPO), which enable us to express more precisely not only who can access the data, but also what is shown. This is useful when trying to express restrictions to avoid showing some details to a competitor.

Web Access Control vocabulary (WAC) can describe the access rights for URI referenceable objects in a distributed fashion (Villata et al. 2011). WAC is a discretionary access control schema for the web. That is, access control is given for sets of users that have a WebID identifier. It does not use user profiles, but they can be emulated using the web-accessible lists of groups. If access control lists or parts of them are network-accessible, they need special care in the security considerations.

6 DISCUSSION

Web of Building Data creates a challenging environment for user authentication and access control. Only a small part of building-related data can be published openly. There are obvious needs—security, privacy, competition, and IPRs—for controlling the access to much of the data. In addition, access control should be implemented so that the decentralized nature of WoBD is respected, and no unnecessary centralized mechanisms are adopted.

Web of Building Data approach has the advantage that—as a participant in a decentralized system—every data producer is free to determine the access control policies for the data it owns. However, from the perspective of the user this can be a disadvantage if the access control systems differ among hosts that contain interlinked data. When accessing data from interlinked datasets, a user may need separate authentication to each of the different servers, and for the publishers there can be additional burden to specify access control rules for a heterogeneous set of intermittent users.

The challenges arise as a result of linking. Without linking, the hosts would be separate from each other, and each of them could use traditional access control solutions. The opportunities provided by Web of Building Data to link data across hosts, to follow links and to aggregate data from multiple hosts can create a fragmented environment from the perspective of access control. The data at different ends of a link could have completely different access control policies, and use even different to define those policies. It can become tedious to set up the access rights to each user at each different host, and to users to access interlinked data, if new authentication is required in each host.

A harmonious access control scheme needs the employ some form of a single sign-on solution, and common ontologies for defining access control policies. At the technical level there are existing approaches—such as WebId-TLS or OpenId Connect—that are designed to genuinely distributed environment and that can be adopted in the Web of Building Data platforms. Common ontologies for description of roles and properties of users can be developed, as well as common policies to refer to instances central in access control (projects, facilities, sites). The research on the security solutions for DRUMBEAT platform continues in these directions.

7 CONCLUSIONS AND FUTURE WORK

Although the Web of Building Data creates a challenging framework from the perspective of access control, there is a set of emerging technologies—OpenId Connect, WebId+TLS, WAC, ontologies—that could be used to set up appropriate mechanisms to solve the problems in a reasonable manner.

It should be noted that the current state of the access control of building data is far from perfect. Complete models need to be transferred to some parties—revealing unnecessary information—and other parties may not gain any access to data that they would desperately need. The granular access control to building data would already mean progress with respect to the current situation. The technologies to achieve that already exist.

The work on the DRUMBEAT platform continues with the implementation of and practical experiments with distributed access control discusses above. There is the need for a practical solution that gains the acceptance of different stakeholders by balancing the simplicity and understandability of the design with the level of security achieved.

ACKNOWLEDGEMENTS

This research has been carried out at Aalto University in DRUMBEAT project ("Web-Enabled Construction Lifecycle", 2014–2017) funded by Tekes, Aalto University, and the participating companies.

REFERENCES

Allen, C. & Dierks, T., 1999. The TLS protocol version 1.0.

Berners-Lee, T., 1998. Hypertext Style: Cool URIs don't change. W3. org. Available at: https://www.w3.org/Provider/Style/URI.html.

Berners-Lee, T., Fielding, R. & Masinter, L., 2005. RFC 3986. Uniform Resource Identifier (URI): Generic Syntax.

Bizer, C., Heath, T. & Berners-Lee, T., 2009. Linked data-the story so far. Semantic Services, Interoperability and Web Applications: Emerging Concepts, pp. 205–227.

Brickley, D. & Guha, R.V., 2014. RDF Schema 1.1. W3C Recommendation, 25, pp. 2004–2014.

Carroll, J.J. & Stickler, P., 2004. RDF Triples in XML. In Proceedings of the 13th International World Wide Web Conference on Alternate Track Papers &Amp; Posters. WWW Alt. '04. New York, NY, USA: ACM, pp. 412–413.

Dean, M. et al., 2004. OWL web ontology language reference. W3C Recommendation February, 10.

Doherty, S., 2011. All your Bitcoins are ours.... Symantec Security Response. Available at: http://www.symantec.com/connect/blogs/all-your-bitcoins-are-ours.

Fielding, R. & Reschke, J., 2014. RFC7231: hypertext transfer protocol (HTTP/1.1): semantics and content. Internet Engineering Task Force (IETF).

Fielding, R.T., 2000. Architectural styles and the design of network-based software architectures. University of California, Irvine.

Greene, S.S., 2006. Security policies and procedures, New Jersey: Pearson Education.

Gritzalis, S., Stefanos, G. & Diomidis, S., 1997. Addressing Threats and Security Issues in World Wide Web Technology. In IFIP Advances in Information and Communication Technology. pp. 33–46.

Hardt, D., 2012. The OAuth 2.0 Authorization Framework (RFC6749). Available at: https://tools.ietf.org/html/rfc6749.

Hoang, N.V. & Törmä, S., 2015. Implementation and Experiments with an IFC-to-Linked Data Converter. Available at: http://itc.scix.net/data/works/att/w78–2015-paper–029.pdf.

Housley, R. et al., 2002. Internet X. 509 public key infrastructure certificate and Certificate Revocation List (CRL) profile.

ISO, 2013. Industry Foundation Classes (IFC) for data sharing in the construction and facility management industries, Geneva, Switzerland: International Organization for Standardization.

Kirrane, S. et al., 2013. Secure manipulation of linked data. In The Semantic Web--ISWC 2013. Springer, pp. 248–263.

Lassila, O. & Swick, R.R., 1999. Resource description framework (RDF) model and syntax specification.

Miles, A. & Bechhofer, S., 2009. SKOS simple knowledge organization system reference. W3C recommendation, 18, p.W3C.

Parker, T.A., 1995. Single sign-on systems—the technologies and the products. In European Convention on Security and Detection. Available at: http://dx.doi.org/10.1049/cp:19950488.

Pautasso, C. & Cesare, P., 2013. RESTful Web Services: Principles, Patterns, Emerging Technologies. In Web Services Foundations. pp. 31–51.

Pauwels, P., Pieter, P. & Walter, T., 2016. EXPRESS to OWL for construction industry: Towards a recommendable and usable ifcOWL ontology. Automation in Construction, 63, pp. 100–133.

Perrin, C., 2008. The CIA triad. Dostopno na: http://www.techrepublic. com/blog/security/the-cia-triad/488.

Rouse, M., Haughn, M. & Gibilisco, S., 2014. What is confidentiality, integrity, and availability (CIA triad)?—Definition from WhatIs.com. WhatIs.com. Available at: http://whatis.techtarget.com/definition/Confidentiality-integrity-and-availability-CIA [Accessed April 27, 2016].

Sacco, O. & Passant, A., 2011. A Privacy Preference Ontology (PPO) for Linked Data. In LDOW. Citeseer.

Sakimura, D.N. et al., 2013. OpenID Connect Basic Client Implementer's Guide 1.0-draft 37.

Sakimura, N. et al., 2014. Openid connect core 1.0. The OpenID Foundation, p. S3.

Sambra, A. et al., 2016. WebID 1.0-Web Identification and Discovery (Draft) (April 2016). Dvcs. w3. org. Available at: https://dvcs.w3.org/hg/WebID/raw-file/tip/spec/identity-respec.html.

Sandhu, R. & Munawer, Q., 1998. How to Do Discretionary Access Control Using Roles. In Proceedings of the Third ACM Workshop on Role-based Access Control. RBAC '98. New York, NY, USA: ACM, pp. 47–54.

Shaer, C., 1995. Single sign-on. Network Security, 1995(8), pp. 11–15.

Story, H. et al., 2016. WebID-TLS, WebID Authentication over TLS, W3C Editor's Draft, 05 March 2014. Dvcs. w3. org. Available at: https://www.w3.org/2005/Incubator/webid/spec/tls/.

Thomson Reuters, 2016a. Burglary Overview. Findlaw. Available at: http://criminal.findlaw.com/criminal-charges/burglary-overview.html.

Thomson Reuters, 2016b. Findlaw Legal Dictionary. Available at: http://dictionary.findlaw.com/definition/cracker.html.

Törmä, S., 2014. Web of building data—integrating IFC with the Web of Data. In ECPPM 2014. pp. 141–147.

Ubale, S.A. & Apte, S.S., 2014. Comparison of ACL Based Security Models for securing resources for Windows operating system.

University of Minnesota Libraries, 2015. CRIMINAL INTENT. Criminal Law. Available at: http://open.lib.umn.edu/criminallaw/chapter/4-2-criminal-intent/ [Accessed April 26, 2016].

Villata, S. et al., 2011. An access control model for linked data. In *On the Move to Meaningful Internet Systems: OTM 2011 Workshops*. Springer, pp. 454–463.

Wild, S. et al., 2014. Tamper-Evident User Profiles for WebID-Based Social Networks. In Web Engineering. Springer, pp. 470–479.

Wild, S. et al., 2015. ProProtect3: An Approach for Protecting User Profile Data from Disclosure, Tampering, and Improper Use in the Context of WebID. In Transactions on Large-Scale Data-and Knowledge-Centered Systems XIX. Springer, pp. 87–127.

Collaboration and teamwork

eWork and eBusiness in Architecture, Engineering and Construction – Christodoulou & Scherer (Eds)
© 2016 Taylor & Francis Group, ISBN 978-1-138-03280-4

Integrated buildings and systems design: Approaches, tools, and actors

A. Mahdavi & B. Rader
Department of Building Physics and Building Ecology, TU Wien, Vienna, Austria

ABSTRACT: Buildings involve both predominantly static features such as building construction, fabric, and envelope, as well as dynamically operating environmental control systems. The optimal life-time performance of buildings depends arguably on a well-coordinated design and configuration of these two aspects. However, the communication and collaboration between primary building designers and building service engineers in the design process is frequently sub-optimal. In the present contribution, we address these issues in two ways. First, we report on the results of an inquiry into the mind-set of both primary building designers and building systems specialists regarding building systems integration issues and their mutual roles in the corresponding process. Second, we briefly describe a possible comprehensive workflow process for the systemic generation and assessment of building systems configurations that are tightly integrated with primary (spatial) building designs.

1 INTRODUCTION

Buildings commonly involve two classes of entities. The first class includes mostly static constituents such as building construction, fabric, and envelope, whereas the second class involves dynamically operating environmental control systems. The optimal life-time performance of buildings depends arguably on a well-coordinated design and configuration of these two classes. However, the respective design decisions are frequently made by separate agents (primary designers versus engineering specialists) and in a frequently non-coordinated and asynchronous fashion. Even within the technically focused class of environmental control systems, an insufficient coordination of processes in one domain (e.g. thermal control system) with other domains (e.g. visual control system) can be observed, resulting in considerable inefficiencies. The communication and collaboration between primary building designers and building service engineers in the design process is often sub-optimal. Primary building designers' knowledge of building systems is frequently wanting. On the other hand, building service engineers often work without explicitly reasoned procedures in determination of the type, number, configuration, and placement of technical devices. This can result in arbitrary solutions devoid of a traceable reasoning.

In the present contribution, we address these issues in a twofold manner. First, we report on the results of an inquiry into the mind-set of both primary building designers and building systems specialists regarding building systems integration issues and their mutual roles in the corresponding process. Second, we briefly describe a possible comprehensive workflow process for the systemic generation and assessment of building systems configurations that are tightly integrated with primary (spatial) building design.

2 MIND-SET OF PROFESSIONAL

Recently, we have been conducting interviews with experienced professionals in both architecture and engineering fields to collect insights regarding their educational and professional background (Rader and Mahdavi 2016). So far the interviewed professionals have shared their understanding and views on building systems and their role in the overall planning of energy-efficient buildings. We have been also exploring the professionals' attitudes regarding the feasibility and usefulness of new ideas (i.e., the aforementioned comprehensive workflow process supported by simulation tools and generative methods) toward integrated design of buildings and their systems. Up to now, 14 architects and 8 building service engineers have been interviewed. Basic information about the interviews and a number of respective results are summarized in Table 1.

2.1 *Educational background*

To be successful, the planning of energy-efficient buildings must not only optimise the primary building design (geometry, fabric, envelope, construction) but also integrate the conception and configuration of the necessary environmental

Table 1. Summary of interviewed professionals' background and attitudes (Rader and Mahdavi 2016).

Question	Architects	Building service engineers
Have you learned in your education something about building systems?	Yes 90%; No 10%	
Have you learned in your education something about the professional responsibilities and working habits of architects?		Yes 0%; No 100%
Do you think architects should learn in their education about building systems? (yes, no)	Yes 100%; No 0%	Yes 100%; No 0%
Is the primary design already completed when collaboration between architects and engineers starts?	Yes 40%; No 20%; Partially 40%	Yes 80%; No 20%; Partially 0%
How are decisions regarding the placement of technical made? (experience, simulation/others)	Experience 100%; Simulation/others 0%	Experience 100%; Simulation/ others 0%
Should building systems influence the architectural design process? (yes, no, partially, don`t know, others)	Yes 60%; No 10%; Partially 30%	Yes 100%; No 0%; Partially 0%

control systems. Hence, the latter activity cannot be exclusively left to specialised engineers, but should also involve primary building designers. This implies the importance of building systems as an essential part of the professionals' education.

The educational background of the interviewed architects and engineers are diverse. 93% of the architects attended a university as compared to only 12.5% of the building service engineers. In Austria—as in many other countries—there is a paucity of educational opportunities (specialised professional degrees) for building service engineers. Hence, only a quarter of the interviewed engineers graduated from a "Higher Technical School" (HTL) for building services. 50% of the interviewed engineers completed a vocational training as a plumber or draftsman. According to one of the architects interviewed, plumbing companies frequently extend their services to include those of building service engineers.

93% of the architects interviewed suggested that they had been exposed, in their education, to some material on building systems. However, the majority stated that the building systems content of their university curricula had been overtly compressed and insufficient for professional life. On the other side, only one of the interviewed building service engineers, who had started as a draftsman, was introduced to the professional responsibilities and work habits of architects.

Architects and engineers agree that it is important for architects to learn in their education about building systems. They suggested that basic building systems knowledge is essential for the architects' work. Specifically, architects should be aware of the extent and differences of existing building systems solutions and how different systems could be appropriate for different situations (building type, available primary energy types). It is also highly important that architects correctly estimate the required space for building systems terminals, distribution network (for pipes, ducts, shafts, etc.), and technical rooms.

2.2 *Architects' knowledge of building systems*

We asked architects to evaluate the profession's knowledge about building systems and their planning. 38% suggested that it considerably varies (due to experience and personal interest), whereas 31% characterised it as mediocre. Colleagues' knowledge is rarely attributed as poor (8%) but it is not considered to be very good either. While evaluating their own building systems knowledge, the architects appear to be somewhat more forgiving: half of the interviewed architects classify their knowledge as good and 36% as mediocre. Most service engineers see significant variation in architects' knowledge of building systems.

2.3 *The design process*

For a smooth operation of building systems, the technical elements have to be considered from the onset of the architectural design process. Given the interview results, this does not appear to correspond to reality, at least not according to engineers' judgment, 87.5% of whom reported that the primary building design is already completed when the collaboration with architects starts. The percentage was lower (46%) in case of the architects. This discrepancy may be attributable in part to differences in the personal experiences of the professionals interviewed.

All building service engineers we interviewed are convinced that building systems should influence the architectural design process. But only 64% of the architect stated the same. Some architects argue that strict technical specifications and requirements for building systems would limit their degrees of freedom in the design process. Computational tools can provide support in detailed calculations regarding building systems. However, decisions pertaining to the placement of technical systems' terminals are made preferably based on the professionals' experience and not computational tools. Professionals argue that most computational tools and applications are expensive, time-intensive, and complicated. This specifically represents a challenge for smaller companies.

2.4 *Architects' and engineers' working experience*

Asked about their experiences with service engineers, architects stated that collaboration generally works well if the engineers participate from the beginning of the planning phase. Some architects criticized that engineers often concentrate only on their own concerns and neglect the consideration of the whole building. They also complained that some engineers focus purely on specific brands of products, which may lead to suboptimal selections.

A further problem in collaboration with service engineers results from the relatively small number of sufficiently experienced building service engineers in Austria. Service engineers the collaboration with architects depends on the individuals involved. Some architects deal intensively with building systems whereas others do not. They also criticized the frequently late onset of the collaboration process. It was also noted that primary building designers frequently underestimate the spatial requirements of technical rooms, building systems terminals, and the distribution system. Both professional groups agree, that a basic understanding of building systems and operation of environmental control systems facilitates the working process of the architects and their collaboration with building service engineers.

2.5 *Professionals' attitude toward new design support tools and methods*

The results of in-depth interviews with architects and engineers underline the importance of measures and tools that could effectively support the working process of these professionals and their collaboration. As mentioned earlier, in interviews with professionals, we specifically confronted them with a proposed workflow and associated tools toward a more integrated approach to the design of buildings and their technical systems (see the next section, as well as, Mahdavi et al. 2015, Mahdavi 2004, Mertz and Mahdavi 2003, Mahdavi and Schuss 2013).

Architects' impression of the method was largely positive. The bulk of the interviewed architects think that in case of smaller projects the proposed tools and methods for inclusion of systems issues in preliminary building design could compensate—to a certain degree—for the absence of input from the building service engineers in the very early stages of the design process. Moreover, they could imagine using the method and the respective tools for design support purposes. Toward this end, it would be critical that certain steps in workflow would be taken in a semi-automated fashion. The professionals interviewed also suggested that a user-friendly and graphically supported realisation of the proposed computational environment could facilitate the optimal configuration and placement of indoor environmental control devices, given an underlying spatial layout.

3 A WORKFLOW FOR INTEGRATED BUILDING DESIGN AND SYSTEMS CONFIGURATION

3.1 *Overview*

The starting point for the envisioned workflow (see Figure 1) is a schematic design involving basic characteristics of an indoor space (step i), in which a set of environmental system elements (i.e., device terminals) such as windows, blinds, luminaires, radiators, diffusers is positioned (step ii). Given basic technical properties of these devices, the spatial target (influence) zones of these devices are assessed using simulation-supported methods (step iii). Next, the optimal positions of sensors representing these zones are determined (step iv), again using simulation-based techniques. Given the two entity sets "devices" and "sensors", a general hierarchical and distributed control schema is derived with multiple nodes (placeholders for specific pieces of control code) at multiple layers (step v). The nodes are subsequently filled with fitting algorithmic solutions for the corresponding control tasks (step vi). The resulting structure involving spatial entities, control devices, and control algorithms is virtually enacted within a numeric building and systems simulation environment (step vii), resulting in specific values of relevant performance indicators (e.g., energy use, indoor environmental conditions). Note that the proposed workflow is meant to accommodate iterative interventions at each step.

The above steps cannot be covered in detail here, given the limited scope of the present paper. We thus treat only steps iii to v. The latter step (control nodes schema generation) has been introduced and discussed in previous publications

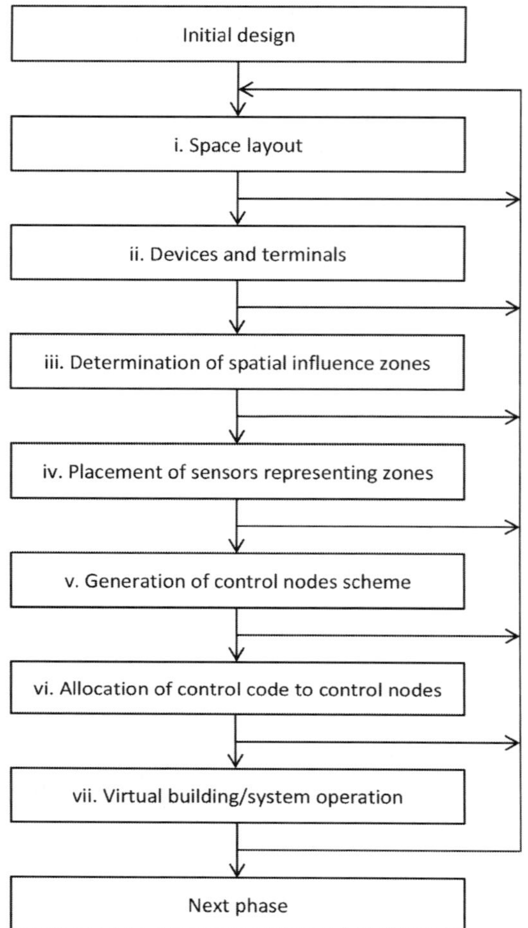

Figure 1. Schematic illustration of the proposed workflow.

(see, for example, Mahdavi et al. 2015). The step iii, however, has been only recently conceived.

3.2 *Determination of spatial influence zones*

The definition and determination of the spatial impact zones of indoor environmental control devices and terminals is not a trivial problem. This perhaps explains the paucity of related formal procedures in the pertinent literature. Practitioners appear to approach the problem using experience and rather simple heuristic tools. In certain high-profile design situations, the performance of a certain configuration of devices (e.g. the lighting system in a large space) may be subjected to simulation-based assessment. Thereby, deviations from the expected performance can trigger iterative changes to the initial design. However, even in such cases, the device-zone relationship is not explicitly formulated: Why were specific devices/terminals selected? What was the reason for their numbers and locations? The lack of clearly documented arguments for such choices makes the reasoning process underlying the design decision making process less transparent and the multi-domain coordination more challenging.

Given these challenges, it may be advisable to approach the problem using the—relatively more straightforward—case of illumination. Provision of sufficient light to a certain location in a room may not be the only purpose of a luminaire, but it is a common one. Moreover, at least in case of electrical lighting, the temporal (dynamic) aspect may be excluded from the initial treatment of the device-zone relationship. In domains such as space heating or ventilation, the device-zone association can be substantially more complex: The thermal and indoor air quality implications of relevant devices (e.g., windows, radiators) and their positioning are not only more complex physically, but also subject to massive dynamic (time-dependent) fluctuations.

Let us thus revert to the simpler device-zone correspondence case of a Luminaire (L) illuminating a part of an architectural space. Let E_p be the horizontal illuminance caused by L in a room at position P at a predefined height above the floor. Furthermore, let E_T be a threshold illuminance level, above which the device may be regarded as having an appreciable impact on the zone. A device's impact zone may be thus defined as the subset of the room's area, to which the following condition applies:

$$E_P > E_T.$$

Note that E_T may be generalised using a reference illuminance level (E_R). This could be, for instance, the illuminance that the luminaire would provide at a certain location (e.g., on a reference plane directly below the luminaire) in a reference space (with standardised dimensions and reflective properties). In that case, we could define E_T as follows:

$$E_T = E_R \,.\, C$$

In this definition, C denotes an expression for the effectiveness fraction: For instance, a value of 0.25 for C would mean that the area in the room where the illuminance caused by L is more that 25% of the E_R value is considered to be within the impact zone of luminaire L.

The above treatment of a device's impact zone is obviously highly simplistic: Horizontal illuminance level is neither the only nor necessarily the most appropriate measure for the determination of a luminaire's impact. Moreover, the height of a reference plane may vary from application to application. Likewise, it may be difficult to define a reference illuminance level (and hence a threshold illuminance level) without making a number of more or less arbitrary assumptions.

An even more formidable challenge must be faced, when trying to apply the threshold-based procedure to the inherently transient processes in daylighting, heating, cooling, and ventilation domains. Whereas we could perhaps simulate—for certain clearly defined boundary conditions—the air or radiant temperature distribution subsequent to the operation of radiator in space, it is difficult to reckon how this could be conveniently done for a window unit or air diffuser.

Despite these challenges, efforts toward formalisation of the correspondence between devices and zones can be highly beneficial. Design procedures already include tacit assumptions regarding device impact zones. It would be thus advisable to explicate the underlying assumptions, not the least for the sake of a rational and retraceable planning process.

3.2.1 *Determination of sensor locations*

Note that a building's environmental control system is "aware" of the impact zones of the devices only to the extent that they are captures via sensors. In our proposed workflow, optimal sensor locations for a specific device impact zone can be inferred in a fairly straightforward manner. For instance, for the aforementioned case of an electrical lighting systems' impact zone, a representative value of the pertinent performance indicator (e.g., average horizontal illuminance level across the identified impact zone) may act as the control parameter. Using simulation, that locus in the space may be identified, at which the local illuminance displays the best correlation with the control parameter value (i.e., mean zone illuminance). Note that the sensor location must not be necessarily confound to the target control zone itself. For instance, in case of electrical lighting system, a location in room's ceiling may be found to provide consistent data with the mean zone illuminance, if modified with a simple multiplicator.

3.3 *Generation of control nodes schema*

3.3.1 *The purpose of the schema*

To add clarity and transparency to the building systems design process, we developed a buildings systems control schema that can be automatically generated based on a limited set of design input data (mainly the associative links between projected device terminals and their intended spatial impact zones). The method to derive this schema has gone through a series of evolutionary steps as documented in previous publications (see, for example, Mahdavi 2014, Mahdavi et al. 2015). The proposed building systems control schema can be generated for an entire building or any part of a building that may be regarded as closed (well bounded) in terms of control actions and their implications.

3.3.2 *Generative rules*

The result of the steps i to iv (Figure 1) and the starting point of the control nodes schema generation is a given space plan with the layout of projected control devices (e.g., windows, luminaires, blinds) and their intended target zones as represented by respective sensors (Figures 2 and 3). If more than one device affects the same sensor, the impact areas of these devices will represented as a single zone. As Figure 3 illustrates, the zones of the two radiators—and the two windows—are merged.

To generate the hierarchical control schema, six rules are deployed as follow:

1. Generate the initial layer (S) of the schema in terms of the intended control zones, as represented by respective sensors.
2. Generate the next layer (D) in terms of the projected devices (or terminals). Every individually controllable device is assumed to have a Device Controller (DC). Multiple devices that are controlled in tandem, require only a single DC.
3. Connect devices to the zones (as represented by sensors), whose states are appreciably influenced by the operation of the devices.
4. Generate the Zone Controllers layer (ZC) as follows: If more than one device influences

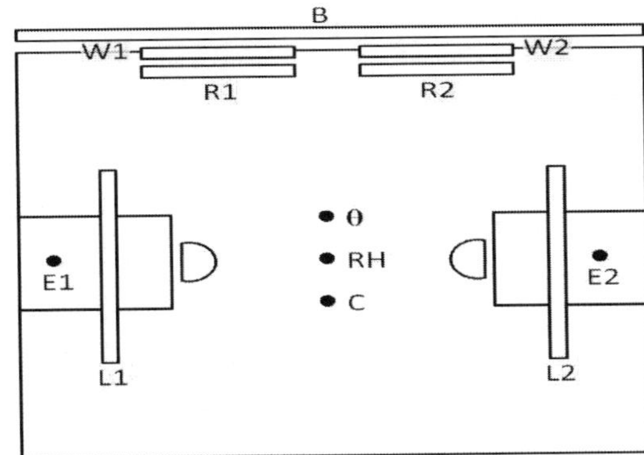

Figure 2. An office space with seven devices (windows W1 and W2, radiators R1 and R2, luminaires L1 and L2, external shade B) and five sensors (illuminance sensors E1 and E2, indoor temperature, relative humidity, and carbon dioxide sensors θ, RH, and C).

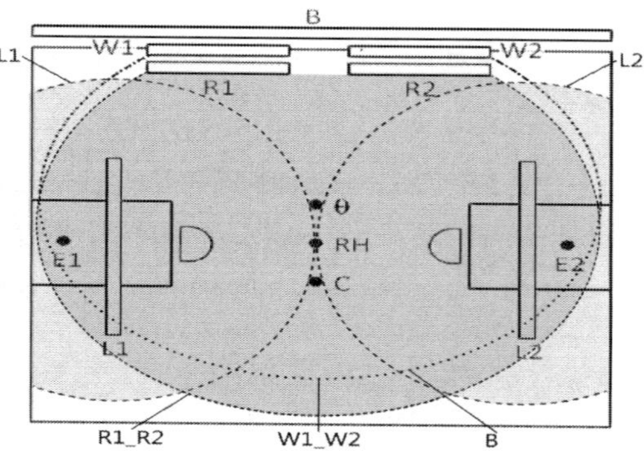

Figure 3. The office space of Figure 2 with illustrative depiction of the spatial impact zones of the devices.

the same zone, a respective zone controller is required to coordinate their operation. This layer accounts thus for the need for zone-specific coordination across multiple devices. Note that in case all devices are individually controllable, the number of resulting zone controllers matches the number of devices.

5. Generate the High-level Controllers (HC) layer as needed: If a DC receives requests from more than one zone controller, a High-level Controller (HC) is generated. This layer accounts thus for the need for device-specific coordination across multiple zones. Note that high-level controllers, which control an identical group of zone controllers can be merged together.
6. If zone controllers receive requests from more than one high-level controller, the schema is terminated with a final layer including a single Meta-Controller (MC).

3.3.3 *Illustration of the schema's application*

Consider the previously treated simple office space of Figures 2 and 3. By following the aforementioned generative steps, the distributed multi-layered control schema of Figure 4 emerges. The control objective is to maintain preferred states (as captures via sensors) in a number of zones (step 1, layer S). These are in this case: air temperature (θ), relative humidity (RH), carbon dioxide concentration (C), and illuminance (E1, E2). The control task is to be accomplished via the operation of windows (W1, W2), a shading device (B), radiators (R1, R2), and luminaires (L1, L2) (step 2, layer D). Application of step 3 results in the connections between devices and zones. Application of step 4 to 6 results in the ZC, HC, and MC layers respectively.

The essential advantage of this schema is the explicit identification of those junctions in a multi-device multi-domain control structure, where

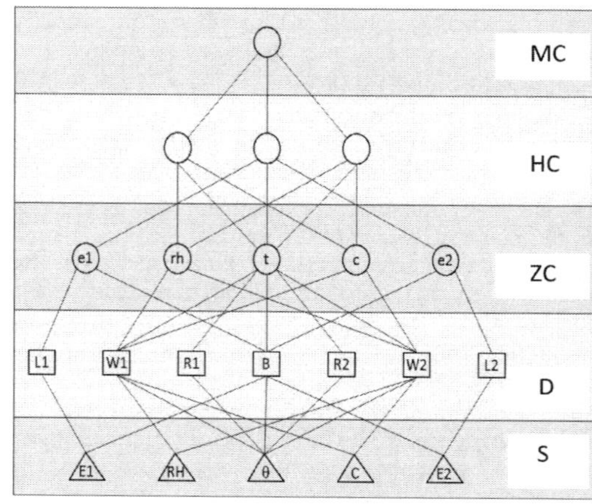

Figure 4. A control logic distribution schema for the office space of Figures 2 and 3.

coordination amongst multiple devices and multiple zones are required. The schema thus does not provide as such semantics (i.e., control code content), but syntax (i.e., nodes for which specific task-oriented control code should be generated). Consequently, the schema does not determine the specific type of algorithmic solution (rule-based, PID, model-predictive. etc.), but rather highlights critical coordination requirement in a given operational scenario.

In a previous application case study involving graduate students, the schema and its usability was tested using concrete architectural instances (Rader and Mahdavi 2014). The participants' impression of the method and its usability was largely positive. The method was found to be effective in supporting the configuration of buildings' technical systems and the communication between architects and engineers.

4 CONCLUSION

Insufficient coordination in the design and implementation of buildings' multiple environmental control systems has been identified as one of the contributors to subpar performance levels in view of energy and indoor environment. In the present contribution, we reported on ongoing inquiry into the insights from both primary building designers and building systems specialists regarding building systems integration challenges. Moreover, we introduced the general structure of a possible comprehensive workflow process for the systemic generation and assessment of building systems configurations that are tightly integrated with primary building design.

Whereas the initial results of this ongoing effort are promising, much additional work is necessary. Specifically, the proposed approach toward the definition of the target zones of environmental control devices must be further developed (in view of the implemented algorithms) and refined (in view of the user interface). Likewise, the feasibility of the introduced methods and tools must be more comprehensively tested based on more detailed and robust implementations and larger experience in practical application situations.

REFERENCES

Mahdavi, A. & Schuss, M. 2013. *Intelligent Zone Controllers: A Scalable Approach to Simulation-Supported Building Systems Control.* Building Simulation 2013—13th International Conference of the International Building Performance Simulation Association. ISBN: 978-2-7466-6294-0; 1498–1505.

Mahdavi, A. 2004. *A combined product-process model for building systems control.* Proceedings of the 5th ECPPM conference. A.A. Balkema Publishers. ISBN 04-1535-938-4. pp. 127–134.

Mahdavi, A., M. Schuss, M. & Rader, B. 2015. *A Multi-Domain Multi-Zonal Schema for Systematic Compartmentalisation of Building Systems Control Logic.* www.itcon.org—Journal of Information Technology in Construction, 20 (2015), Special Issue; pp. 121–131.

Mertz, K. & Mahdavi, A. 2003. *A representational framework for building systems control.* Proceedings of the 8th International IBPSA Conference, Eindhoven, Netherlands; ISBN: 90-386-1566-3; pp. 871–878.

Rader, B. & Mahdavi, A. 2016. *Bridging the gap between Systems Controls and Architectural Design.* Applied Mechanics and Materials. Special Volume: Energy Saving and Environmentally Friendly Technologies—Concepts of Sustainable Buildings (2016), 824; pp. 821–828.

Rader, B. & Mahdavi, A. 2014. *Usability assessment of a generative building control logic distribution system.* Proceedings of the 10th European Conference on Product and Process Modelling (ECPPM2014). Taylor & Francis – Balkema. ISBN: 978-1-138-02710-7; pp. 609–613.

eWork and eBusiness in Architecture, Engineering and Construction – Christodoulou & Scherer (Eds)
 ISBN 978-1-138-03280-4

Ecosystem and platform review for construction information sharing

I. Peltomaa, M. Kiviniemi & J. Väre
VTT Technical Research Centre of Finland, Oulu/Tampere/Espoo, Finland

ABSTRACT: Emerging digitalization is radically changing industrial world. Disruptive technologies give companies great challenges, but also huge opportunities to develop their businesses. One perspective for gaining business benefit from digitalization wave is to utilize ecosystem-based thinking. This paper presents a development of an ecosystem and platform analysis for AEC/FM industry information sharing. The objective of this paper is to identify and classify existing information sharing ecosystems and platforms applicable to AEC/FM industry. Web of Building Data initiative is introduced as one of the construction industry approaches for information sharing. Presented work is a part of a project where the objective is to provide information sharing method and technology for AEC/FM industry. As a part of the project an information sharing platform on the basis of the Web of Building Data is developed. Ecosystem analysis is used as a base for conceptualize an open ecosystem to be formed around platform.

1 INTRODUCTION

Today the wave of digitalization is sweeping over industrial world, including Architecture, Engineering, Construction and Facilities Management (AEC/FM) industry. Mere data digitalization is not enough—the gathered data need to be utilized and make it calve benefit and profit. Emerging digitalization has raised private web-based data repositories and proprietary software platforms; the challenge is to make these islands of data to interoperate and to develop the data into information by combining and inferencing. Globalization and technological development have increased the competition and forced companies to focus on their core business. This has led to demand to increase collaboration between companies and form collaborative networks and ecosystems. Ecosystems are complex instances that support collaboration between organisations; complex because many issues need to be considered in order to ecosystem to be functioning. These issues include economy, technology, humans, laws and regulations to mention few of them. Nevertheless information sharing and information interoperability are the key success factors.

Digitalization and technological development have also leading to transformation from the system of records into a system of engagement. System of engagement is web-based, customer-friendly and consumer-driven Information Technology (IT) system, which is enabled by technology development of cloud, social, mobile, consumer IT and Big Data, and having enormous effect on the way people collaborate and connect in business and personal lives (Moore 2011). System of engagement emphasizes of taking the social aspect in the consideration in enterprise collaboration.

Information exchange in AEC/FM industry has stayed little bit behind from mainstream of implementing emerging technologies for enterprise collaboration. For example in construction process there are still demanding challenges in interoperability between different tasks (design, fabrication, delivery, installation). Even if these tasks would be executed inside one company, there would probably be difficulties in information sharing. Keeping up with competition requires knowledge about and ability to harness these new emerging technologies. The digital world is on now, like for example cloud computing, Internet of Things (IoT), and machine learning are showing as they spread like a wildfire.

This paper presents a development of an ecosystem and platform analysis for AEC/FM industry information sharing. The objective of this paper is to identify and classify existing information sharing ecosystems and platforms applicable to AEC/FM industry. The paper is organized as follows: chapter two presents ecosystem related concepts, chapter three introduces Web of Building Data initiative, chapter four presents the ecosystem and platform classification, and chapters five and six conclude the paper.

2 ESSENTIAL CONCEPTS

In this chapter important concepts around ecosystem unity are introduced. The roots of ecosystems are in ecology: ecosystem is a community of living organisms in conjunction with the non-living components of their environment, interacting as a system, and dynamically acting to adapt constant changes (Chapin et al. 2002). It can also be said that an ecosystem is a loosely coupled, domain clustered environment inhabited by species, each proactive and responsive regarding its own benefit while conserving the environment. Species interact to achieve balance with each other, while the environment supports the needs of its species so they can continue generation after generation. (Boley & Chang 2007) *Human Ecosystem* changes species into actors and the target of interaction into physical and non-physical factors (Bosch 2009).

Digital Ecosystems is an open, loosely coupled, domain clustered, demand-driven, self-organizing agent environment, where each agent of each species is proactive and responsive regarding its own benefit or profit, but is also responsible to its system (Boley & Chang 2007). The definition of Briscoe et al. (2011) is closer to ecology as they suggest that by exploiting the features of biological ecosystem would enable Digital Ecosystems which are robust, scalable, and self-organizing.

Business Ecosystem as a concept was first introduced by Moore already on nineties. Moore suggested that a company should not be viewed as a member of a single industry but as a part of business ecosystem that crosses a variety of industries. Business Ecosystem co-operatively develops a new innovation and new products and services around it according to customer needs. Alongside co-operation there exists competition which at its side drives Business Ecosystem to develop the innovation further. (Moore 1993) Business Ecosystem is dynamic, co-evolving, loosely connected community interconnected to each other, sharing common goal, and having models to control collaboration and competition (Fragidis et al. 2007, Kelly 2015, Rong et al. 2015). Involved companies need to become proactive in developing mutually beneficial (symbiotic) relationships with customers, suppliers, and even competitors (Friess & Riemenschneider 2015). The Business Ecosystem concept examines business enterprise as a whole, considering all stakeholders and relationships with them (Fragidis et al. 2007).

Software ecosystem is a complex web of relationships between different actors during development, selling, and usage of the software in certain context with certain technology (Messerschmitt & Szyperski 2003). Messerschmitt and Szyperski (2003) observe software ecosystem from different points of views covering different stakeholders of the software ecosystem, including users, software creators, managers, industrialists, policy experts and lawyers, and economists. According to Bosch (2009) a software ecosystem consists of the set of software solutions that enable, support and automate the activities and transactions by the actors in the associated social or business ecosystem and the organizations that provide these solutions. A common technological platform is shared factor in many definitions (Jansen et al. 2009, Draxler & Stevens 2011, Manikas & Hansen 2013). Earlier Software Ecosystem was mainly observed from users or software engineering point of view examining only little other views. Later it was noticed, that it is important to take social, legal and economic aspects as integral part of the Software Ecosystem (Draxler & Stevens 2011, Kilamo et al. 2012). From the engineering perspective, a software ecosystem provides the technology for implementation, environment for the overall software project infrastructure and a development methodology (Kilamo et al. 2012).

Innovation Ecosystem is a dynamic, inter-organizational, collaborative, co-creative, economic community sharing complementary technologies and competencies focusing on the development of new value through innovation based on customer preferences (Gobble 2014, Russell et al. 2015, Autio & Thomas 2014, Adner 2006). It is characterized by a continual realignment of synergistic relationships that promote growth of the system (Russell et al. 2015). In innovation ecosystem participants communicate directly to solve challenges instead of intermediary central office; this loosely-coupled structure leads to innovative culture enhancing the flourishment and evolvement of innovative ideas. (Markman 2012).

Platform Economy can be defined as a medium which lets others connect to it. Platform businesses can be found in a growing number of industries including social networking, internet auctions and re-tail, on-line financial and human resource functions, urban transportation, mobile payment, and clean energy. (CGE 2016) Essential properties of a platform economy include direct and indirect network effects giving more value as more users use platform (Seppälä et al. 2015, Evans & Gawer 2016), multidirectional market, product and service components, and cooperative and technical interfaces (Seppälä et al. 2015). Most of today's platforms are digital and they capture, transmit and turn data into money, including personal data, using the Internet (Evans & Gawer 2016). Digital platforms can be defined as information technology systems upon which different actors can carry out valued-adding activities in a multi-sided market environment governed by agreed boundary resources. Typically created, offered and maintained products and services are complementary to one another. (Seppälä et al. 2015) Algorithms, Internet and cloud computing are the building blocks of the platform. Platforms can also grow on platforms: an array of applications forms platform of platforms. (Kenney & Zysman 2015) Platform is typically created and owned by a single business or entity, but deliberately designed to attract the active participation of large numbers of other actors (Kelly 2015). Today the most impressive innovations are not products or services; instead they are platforms on which products and services are built. (Accenture 2016) This leads to fundamental changes in business models. (Evans & Gawer 2016, Accenture 2016, Kenney & Zysman 2015)

In general central components of ecosystem are

- parties participating to ecosystem (actors, species,..),
- interaction between parties in order to affect to other parties, and
- the environment surrounding ecosystem to promote the ecosystem targets.

Interaction and information sharing have essential role in ecosystem, as Moore (1993) highlighted: "Successful businesses are those that evolve rapidly and effectively. Yet innovative businesses can't evolve in a vacuum." From individual company's point of view gaining some kind of benefit from participating in ecosystem is crucial. Benefits can be for example direct revenues or non-monetary, like fame, knowledge, or ideology (Manikas & Hansen 2013).

3 WEB OF BUILDING DATA

Web of Building Data also known as Linked Building Data (Törmä 2013, LDAC 2014) is a new concept for sharing datasets in the AEC/FM sector and have potential to be utilized in platforms and ecosystems in the branch. Web of Building Data aims to combine Building Information Modelling (BIM) and Web of Data technologies to define existing and future use cases and requirements for Linked Data based applications across the life cycle of buildings.

Web of Data is evolvement of World Wide Web (which was created for human readers) making it applicable for machines to understand available data by adding metadata. Web of Data is also known as Se-mantic Web or Linked (Open) Data. This technology enables people to create data stores on the Web, build vocabularies, and write rules for handling data. Linked data are empowered by technologies such as RDF, SPARQL, OWL, and Semantic Web. (W3C 2016) These are described more in standards chapter 4.1.

The motivation behind Web of Building Data is that buildings are becoming increasingly data intensive. They have diverse models or representations like client's requirements, an architectural (design) model, a structural model, a MEP (mechanical, electrical, and plumbing services) model, project plans, and so forth. Furthermore, a lot of external data is available that is to some extent related to the building. These datasets are distributed over various places, in the hands of multiple building stakeholders and it gets complicated to manage data in traditional ways (LDAC 2014).

Using Web of Data technologies for building information management has been in the focus of a number of research and development activities in recent years. In 2014, the W3C Community Group on Linked Building Data was officially started (LBDCG 2016), and in 2015, the buildingSMART Linked Data group was initiated (LDWG 2016).

buildingSMART targets for creating and maintaining a recommendation of an ifcOWL ontology as a derivate of the canonical IFC EXPRESS schema. ifcOWL is a connecting point between Semantic Web technologies and the IFC standard, represented by a Web Ontology Language (OWL) for IFC. Purpose of ifcOWL is to keep on using the long-established IFC standard for representing building data, exploit the enablers of Semantic Web technologies in terms of data distribution, extensibility of the data model, querying, and reasoning, and re-use general purpose software implementations for data storage, consistency checking and knowledge inference. (Pauwels & Terkaj 2016)

Coordinated work to produce single ifcOWL ontology standard started in July 2015. First drafts are available but development is still going on as there are still number of open issues. The ifcOWL ontology is built and maintained by the buildingSMART Linked Data Working Group (ifcOWL 2016). ifcOWL is being standardized also by W3C's Linked Building Data Community Group (ifcOWL 2014). Creation of the current proposed recommendation is described in Pauwels & Terkaj (2016).

The ifcOWL ontology primarily aims at supporting the conversion of IFC instance files into equivalent RDF files. Modelling RDF from scratch using the ifcOWL ontology is only a secondary target. ifcOWL ontology design criteria is set that it must be in OWL2 DL and it should match the original EXPRESS schema as closely as possible (buildingSMART 2015)

The proposed ifcOWL EXPRESS to OWL conversion pattern can be implemented in a number of different algorithms. In tests two parallel converter implementations produced the same results (buildingSMART 2015) proving feasibility of the recommendation.

4 ECOSYSTEM AND PLATFORM ANALYSIS

Objective of the ecosystem and platform analysis is to identify, classify and evaluate existing information sharing ecosystems and platforms applicable to AEC/FM industry. The initial findings are presented in this paper. The perspective of the analysis is to reflect Web of Building Data in selected ecosystems in order to draft an open ecosystem utilizing Web of Building Data concepts.

Ecosystems are emerging way to collaborate, innovate, and do business. Ecosystems can be found from various business areas and the instances are multifold. During research it was noticed that ecosystems often existed on top of a platform (Kelly 2015). It seemed that intents to classify the ecosystem field as a whole were few, and classifying of the identified heterogeneous group of ecosystems proved to be challenging task. A classification of Complex Technical Systems (CTSs), where ecosystems can also be included in, was discovered; Horvath (2015) presents an initial cataloguing of various research domains in CTSs, concentrating on what kinds of foundational research have been and are being done. However this classifying was found unsuitable for our purposes. Instead the ecosystems were grouped according to their perspective and similarity in features.

In connection of platform and ecosystem identifying phase it was noticed that in connection of several ecosystems certain standards or de facto standards were used. Due this also enabling standards were decided to review.

4.1 *Enabling standards*

Standards are important enablers of information sharing and collaboration in ecosystems and in general. It is obvious that many platforms have been evolved without detailed standards but at interoperability level of platforms some common agreements for data structures are needed and will support open ecosystems.

In our analysis standards were classified as presented in figure 1. In the figure standards presented in dark grey boxes are AEC/FM specific standards and others are intended for multiple domain areas. As the digital platforms are operated over the Internet, the *Standardized Communication Protocols* and *Cellular Network Standards* are fundamental standardization for platform economy but those are not examined in this paper even those are ecosystems as such.

Data Exchange Standards are intended to enable human and machine readable information to be transferred from one system to another. One of the oldest standards still in active use is EDI (Electronic

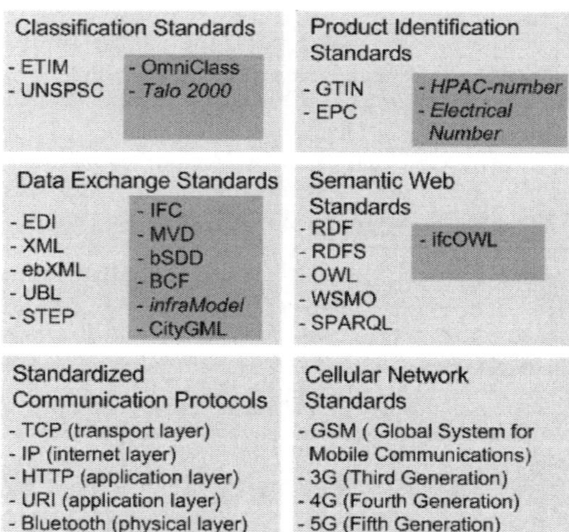

Figure 1. Categories for information sharing enabling standards used in the research. AEC/FM standards are marked with dark background and national concepts with italics font.

Data Interchange), which is an electronic format of business documents for exchanging information between business partners. There are several EDI document standards in different countries for different domains, but they all have same principles (EDI 2016).

Quite many of Data Exchange Standards have origin in XML and XML is one of the most widely-used formats for sharing structured information. For example Electronic Business using eXtensible Markup Language (ebXML) and Universal Business Language (UBL) are both XML-based specifications for business document exchange in electronic business between companies (ebXML 2016, OASIS 2013).

The Standard for the Exchange of Product Model Data (STEP) is a comprehensive ISO standard (ISO 10303) that describes how to represent and exchange digital product information. STEP framework forms of several Application Protocols which all use the same definitions for the same information. STEP standard contains also EXPRESS data specification language for product data definition. (STEP 2016)

In the AEC sector the Industry Foundation Classes (IFC) standards family maintained by buildingSMART is the major standard for exchanging BIM data. buildingSMART have developed several other specifications for enlarging usage of core IFC standard like Model View Definition (MVD), buildingSMART Data Dictionary (bSDD) and BIM Collaboration Format (BCF). MVD defines a subset of the IFC data schema using mvdXML format, which enables automated validation of MVDs. MVD pro-vides a way to add in transformation only needed information as it allows specific data inclusion (buildingSMART 2016). bSDD is a reference library which provides flexible and robust method of linking existing databases with BIM. bSDD defines objects, their attributes, relationships, and properties language independently. BCF is an open file XML format that supports workflow communication in BIM processes. The BCF is implemented as a XML file format and as a RESTful web service, which enables the exchange of BCF data seamlessly in BIM workflows.

Inframodel is open information transfer format for infrastructure building in Finland. Inframodel is based on international LandXML data standard, giving explanation and method how to use LandXML in Finland (Inframodel 2016). CityGML is an open data model and XML-based format for the representation, storage and exchange of virtual 3D city models (OGC 2016)

Semantic Web Standards enhance information integration as they are able to transform also the meaning of the transferred information. During transformation Semantic Web Standards are able to preserve the semantics while transforming the context (Pollock 2001).

One of the most used Semantic Web Standard is W3C developed Resource Description Framework (RDF) for representing information in the web. RDF provides means for annotating information in web with semantic markup to enable machine interpretation. For establishing explicit meaning RDF is not enough but ontologies are required. RDF Schema Language (RDFS) extends RDF by defining terms of a knowledge domain and the relationships between them. (Manola & Miller 2004, Pan & Horrocks 2007, Brickley & Guha 2014, Sikos 2015). Web Ontology Language (OWL) is a knowledge representation language supporting the processing of the content of information instead of presenting of it thus enabling machine interpretability capabilities by providing additional vocabulary along with a formal semantics. OWL is especially designed for creating web ontologies with better properties than XML, RDF, and RDFS. OWL has three increasingly-expressive sublanguages: OWL Lite, OWL DL, and OWL Full. (McGuinness & van Harmelen 2004, Sikos 2015)

Using Web Service Modelling Ontology (WSMO) web services are enriched with semantic markup enabling partially or fully automated discovery, selection, composition, mediation, execution and monitoring. WSMO is a conceptual model for Semantic Web Services in order to model objectives of requester and the capacity of Web Service (Roman et al. 2005, Sikos 2015).

Classification Standards and *Product Identification Standards* are also standards for improving information management and exchange in digital eco-systems.

European Technical Information Model (ETIM) is an open international standardized classification model for technical products. ETIM is two-level classification model consisting product classes and product groups, which are used to organize product classes. Product class describes similar products, and bundles products of different manufacturers or suppliers. Product class can have synonyms and has features which consist of description, feature type, value and/or unit. (ETIM 2016)

The United Nations Standard Products and Services Code (UNSPSC) is an open, global, multi-sector standard for efficient, accurate classification of products and services. It is a five-level hierarchy coded as a ten-digit number, with two digits on each level. (UNSPSC 2016)

The OmniClass Construction Classification System (OmniClass) is a classification system for the construction industry. It provides the means for organizing, sorting, and retrieving information and deriving relational computer applications. OmniClass consists of 15 hierarchical tables, each of which represents a different facet of construction information. (OmniClass 2016). There are several national branch specific classification systems alike, e.g. Talo 2000 in Finland. Similarly there exists a classification system for infrastructure construction.

Product Identification Standards are focused on identifying individual products thus they are serial numbers or object identifiers. GS1 is managing several product identification standards and identifier services. Global Trade Item Number (GTIN) identifies types of trade items and it can be encoded in a barcode or an EPC/RFID tag. Individual trade items can be uniquely identified using a GTIN plus serial number (GS1 2016).

In AEC/FM domain product identification standards seem to be generally national. E.g. in Finland there are national service providers for identification codes of MEP products. The codes are widely used in wholesale trading which has created an ecosystem on them and there are also other types of digital services offered e.g. for comparing prices.

4.2 *Ecosystems and platforms*

This chapter presents classification of ecosystems and platforms in which they are grouped according to their perspective and similarity in features, illustrated in figure 2. Operating systems and related ecosystems are not examined in this paper as they are not in our interest now.

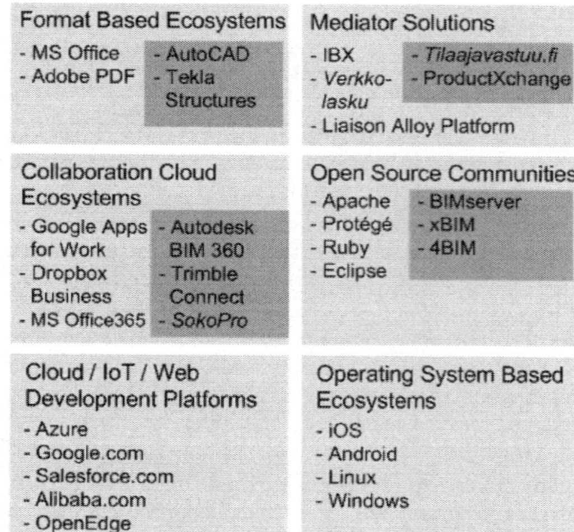

Figure 2. Perspective grouping of Ecosystems and platforms. Entities related to AEC/FM sector are marked with darker background and some national level entities with italics font.

Format Based Ecosystems base on certain (file) format, which are often introduced by a powerful actor in the domain and form a de facto standard. In many cases these de facto standards are later officially standardized. To enable effective information exchange during operation different vendors start to utilize the format in order to improve their product's action as a part of information exchange chain. Programs, applications, software vendors, and users form an ecosystem around used format.

Tekla Structures is a BIM software enabling creation of complex and detailed models supporting BIM. Its features include configurability, modelling of all materials, and support from design to construction. (Tekla 2016)

Mediator Solutions form ecosystem from information systems that are at certain time connected to them. One of the key idea is service platform provider manages different interfaces allowing to form loose couplings between systems, so that the connection and disconnection is easy.

ProductXchange is a communication and BIM product data exchange software for construction projects. It enables automatic data collection and distribution during the whole building life cycle. (ProductXchange 2016)

Open Source Communities have the power in their communities; more value is gained by more users. Achieving the critical user mass is essential in order to success. Open Source Communities usually gather around technical solution, but the starting point for community can be solving a problem. Some of these communities are permanently open source, but for some solutions it is only a phase in their way towards commerciality.

buildingSMART community can be seen as Open Source Community; buildingSMART provides methods for building your own (company-wide, national) data model for building domain and offers means to share your model and to exploit others' models (buildingSMART 2016). Anyway, there are dozens of open source software for BIM but only few are widely used or complete. Most of the open source software is based on IFC family.

BIMServer which could be seen as example of a BIM based ecosystem. It claims to be reliable and ready to use in mission critical applications. BIMServer has lots of interface protocols and libraries and it provides plugin framework for flexibility. BIMServer functionalities are based on plugins but it comes with many preinstalled so that models can be uploaded, browsed and visualized. One can use different serializers, object IDMs, render engines, query engines, model mergers, model compare engines, and model checkers. Projects can be divided into subprojects. There is also user management, access control, and revision management functionalities. The BIMServer can be integrated to other applications using client libraries. To support BIMServer ecosystem opensourcebim.org aims to become a BIM collective for collaboration built around BIMServer and its plugin providers. (BIMServer 2016)

There is also some other notable software. xBIM Tookit is a software development BIM tool that allows developers to read, create and view models in the IFC format (xBIM 2016). Other IFC handling tools are IfcOpenShell, IfcPlusPlus and GeometryGymIFC. Then there are visualisation software like Bimviews and BIM Surfer.

Collaboration Cloud Ecosystems form their ecosystem around cloud-based services. Collaboration clouds enhance formal and unformal information sharing; collaboration solutions enable information transformation from different information system to cloud through APIs.

Autodesk's BIM 360 is an example of private cloud based building information management system enabling collaboration between everyone involved in a construction project and using appropriate tools. BIM 360 core is a cloud service where all data (building models) and metadata (documents, notes etc.) are stored. Autodesk's construction design software has direct access to the cloud for storing and editing models. The BIM 360 cloud is managed mainly via BIM 360 Glue. There is also an open programming APIs by which connectivity to the BIM 360 cloud can be integrated into 3rd party software with most of the functionality available in Autodesk software (Autodesk 2016).

The common type of current web-based collaboration platforms are project repositories for uploading and sharing project documents. Some of those, like Finnish SokoPro (SokoPro 2016), have evolved as value adding service for digital printing.

Cloud/IoT/Web Development Platforms are ecosystems where the base is on platform and usually around which a community is arisen. Vital Development Platforms utilize new emerging technologies rapidly thus they are quickly disposable by the users of Development Platforms. Use of these platforms enhances building of user specific solutions as the foundation of solution already exists.

Beyond these ecosystems and platforms are "*Ecosystemic Totalities*" that are high level concepts in certain area or domain, e.g. Industrial Internet. The key elements of Industrial Internet are intelligent machines, advanced analytics, and people at work (Evans & Annunziata 2012) enhanced with business perspective (IIC 2015).

5 DISCUSSION

The best known part of platform economy is dealing with consumers like mobile software application ecosystems and buying digital content or sharing platforms like Uber and AirBnB or multiservice platforms like Google. Important value adding feature is based on network effect: more users, more value to user (Evans & Gawer 2016) which leads into "winner takes it all" situation in the market. In the B2B markets there are not yet such dominating platforms but e.g. Google has growing services for companies while the users have become familiar with the services in their private life.

In the AEC/FM sector there have been web-based platforms in use due to the needs of comprehensively networked construction projects. The most important feature has been sharing technical documents but also some procurement platforms exist. In the FM there are Software-as-a-Service based applications for e.g. maintenance and for interaction with tenants and those are practically platforms. It is obvious that collaboration of actors is main application area for the platforms in AEC/FM.

In the current situation there can be seen some development trends. Mobile applications for different use cases in AEC/FM are rapidly becoming common. Mobile applications have backend services in the web and those can be considered as platforms. Another trend is development of service bus solutions and this is part of general digitalization of the economy. For example in Finland Tilaajavastuu (Tilaajavastuu 2016) is a service bus provider for contractors to report fulfillment of administrative obligations to the authorities. Third and most interesting trend in this analysis is the transformation of the collaboration platforms. It seems that as the BIM-based design and engineering have become common, the next step is utilizing the information content of BIMs over the Web. The major software vendors that have provided native BIM-software are already offering web-based collaboration platforms for AEC and likely in near future for FM. At the moment it is not clear will these platforms become "private clouds" for maximizing the usage of provider's services or will those be more open for multidimensional offering with complementary services.

6 CONCLUSIONS

This paper presented development of an ecosystem analysis where ecosystems and platforms suitable for AEC/FM industry were identified and classified. The following phase of the analysis process will be the evaluation of the ecosystems and platforms from the Web of Building Data perspective.

The results presented in this paper will act as a base for the evaluation phase of the analysis process and as a starting point for the development of an open building construction ecosystem that will be formed around the Web of Building Data platform. The motivation is to enhance information sharing during construction projects and afterwards in maintenance phase. The classifying of different platforms and ecosystems has been challenging task and refinement of classifying categories needs to be carried on. There are still points of view that need to be considered in order to complete classification categories. The evaluation process needs to be implemented and this requires deeper examination of some selected ecosystems and business cases. The development of Web of Building Data platform has advanced parallel to the presented work.

REFERENCES

Accenture. 2016. Trend 3—Platform Economy: Technology-driven business model innovation from the outside in. Accenture Technology Vision. Available at www.accenture.com/t20160125T111719__w__/us-en/_acnmedia/Accenture/Omobono/TechnologyVision/pdf/Platform-Economy-Technology-Vision-2016.pdf.

Adner, R. 2006. Match Your Innovation Strategy to Your Innovation Ecosystem. *Harvard Business Review* April; 84(4): 98–107.

Autio, E. & Thomas, L.D.W. 2014. Innovation ecosystems: Implications for innovation management? In M. Dodgson, D.M. Gann & N. Phillips (eds) *The Oxford Handbook of Innovation Management*: 204–228. Oxford: Oxford University Press.

Autodesk. 2016. Autodesk. Web-page, available at www.autodesk.com.

BIMServer. 2016. Open source BIMserver | In the heart of your BIM!. Web-page, available at www.bimserver.org/.

Boley, H. & Chang, E. 2007. Digital Ecosystems: Principles and Semantics. In E. Chang & F.K. Hussain (eds), *IEEE Digital Ecosystems and Technologies; Proc. intern. conf., 21–23 February 2007*, pp. 398–403.

Bosch, J. 2009. From software product lines to software ecosystems. In, *Software Product Line Conference; Proc. intern. conf., 24–28 August 2009*. Pittsburgh: Carnegie Mellon University, pp. 111–119.

Brickley, D. & Guha, R.V. 2014. RDF Schema 1.1. W3C Recommendation 25 February 2014, available at www.w3.org/TR/rdf-schema.

Briscoe, G., Sadedin, S. & De Wilde, P. 2011. Digital Ecosystems: Ecosystem-Oriented Architectures. *Natural Computing* 10(3): 1143–1194.

buildingSMART. 2015. buildingSMART Proposed Recommendation EXPRESS-to-OWL conversion routine. Available at www.buildingsmart-tech.org/future/linked-data/ifcowl/express-to-owl-conversion-procedure-proposed-recommendation/view.

buildingSMART. 2016. buildingSMART. Web-page, available at www.buildingsmart.org.

ebXML. 2016. ebXML—Enabling a global electronic market. Web-page, available at www.ebxml.org.

ifcOWL. 2014. ifcOWL ontology file added for IFC4-ADD1. Web-page, available at www.w3.org/community/lbd/ifcOWL.

ifcOWL. 2016. ifcOWL. Web-page, available at www.buildingsmart-tech.org/future/linked-data/ifcowl.

xBIM. 2016. The xBIM Toolkit. Web-page, available at xbim.codeplex.com.org.

CGE. 2016. The Center for Global Enterprises. Web-page, available at www.thecge.net/category/research/the-emerging-platform-economy.

Chapin, F.S., Matson, P.A. & Mooney, H.A. 2002. *Principles of Terrestrial Ecosystem Ecology.* New York: Springer.

Draxler, S. & Stevens, G. 2011. Supporting the collaborative appropriation of an open software ecosystem. *Computer Supported Cooperative Work* 20(4–5): 403–448.

EDI. 2016. EDI Basics. Web-page, available at www.edi-basics.com.

ETIM. 2016. ETIM—Home. Web-page, available at www.etim-international.com.

Evans, P.C. & Annunziata, M. 2012. Industrial Internet—Pushing the Boundaries of Minds and Machines. GE, available at www.ge.com/docs/chapters/Industrial_Internet.pdf.

Evans, P.C. & Gawer, A. 2016. The Rise of the Platform En-terprise—A Global Survey. The Emerging Platform Economy Series No. 1, The Center for Global Enterprise. Available at www.thecge.net/wp-content/uploads/2016/01/PDF-WEB-Platform-Survey_01_12.pdf.

Fragidis, G., Koumpis, A. & Tarabanis, K. 2007. The Impact of Customer Participation on Business Ecosystem. In L. Camarinha-Matos, H. Afsarmanesh, P. Novais & C. Analide (eds), *Establishing the Foundation of Collaborative Networks*. IFIP International Federation for Information Processing 243: 399–406. New York: Springer.

Friess, P. & Riemenschneider, R. 2015. New Horizons for the Internet of Things in Europe. In O. Vermesan, & P. Friess (eds), *Building the Hyperconnected Society—IoT Research and Innovation Value Chains, Ecosystems and Markets* 43: 5–13. River Publishers Series in Communications.

GS1. 2016. GS1-The Global Language of Business. Web-page, available at www.gs1.org.

Gobble, M.M. 2014. Charting the Innovation Ecosystem. *Research-Technology Management* 57(4): 55–59.

Horvath, I. 2015. An initial categorization of foundational research in complex technical systems. *Journal of Zhejiang University SCIENCE A* 16(9): 681–705.

IIC, 2015. Industrial Internet Reference Architecture. Version 1.1. Available at www.iiconsortium.org/IIRA.htm.

Inframodel. 2016. Inframodel. Web-page, available at www.inframodel.fi.

Jansen, S., Finkelstein, A. & Brinkkemper, S. 2009. A sense of community: A research agenda for software ecosystems. In, *Software Engineering; Proc. intern. conf., Vancouver, 16–24 May 2009*. Companion Volume: 187–190.

Kelly, E. 2015. Introduction—Business ecosystems come of age. In M. Canning & E. Kelly (eds), *Business ecosystems come of age*. Part of the Business Trends series: 3–15. Deloitte University Press.

Kenney, M. & Zysman, J. 2015. Choosing a Future in the Platform Economy: The Implications and Consequences of Digital Platforms. Discussion Paper. Prepared for the New Entrepreneurial Growth Conference, Kauffman Foundation.

Kilamo, T., Hammouda, I., Mikkonen, T. & Aaltonen, T. 2012. From proprietary to open source—Growing an open source ecosystem. *Journal of Systems and Softwares* 85(7): 1467–1478.

LBDCG. 2016. Linked Building Data Community Group. Web-page, available at www.w3.org/community/lbd.

LDAC. 2014. Joint Workshop on Linked Data in Architecture and Construction, LDAC 2014. Available at www.linkedbuildingdata.net/LDAC2014.

LDWG. 2016. Linked Data. Web-page, available at www.buildingsmart-tech.org/future/linked-data.

Manikas, K. & Hansen K. 2013. Software ecosystems—A systematic literature review. *International Journal Systems and Softwares* 86(5): 1294–1306.

Manola, F. & Miller, E. 2004. RDF Primer. W3C Recommendation 10 Fegruary 2004, available at www.w3.org/TR/2004/REC-rdf-primer-20040210.

Markman, A. 2012. How to Create an Innovation Ecosystem. *Harvard Business Review* December.

McGuinness, D.L. & van Harmelen, F. 2004. OWL Web On-tology Language Overview. W3C Recommendation 10 February 2004, available at www.w3.org/TR/2004/REC-owl-features-20040210.

Messerschmitt, D.G. & Szyperski, C. 2003. *Software Ecosystem: Understanding an Indispensable Technology and Industry*. MIT Press: Cambridge.

Moore, G. 2011. Systems of Engagement and The Future of Enterprise IT—A Sea Change in Enterprise IT. AIIM White Paper, available at www.aiim.org/~/

media/Files/AIIM%20White%20Papers/Systems-of-Engagement-Future-of-Enterprise-IT.ashx.
Moore, J.F. 1993. Predators and prey: A new ecology of competition. *Harvard Business Review* May/June;71(3): 75–86.
OASIS. 2013. Universal Business Language Version 2.1. OASIS Standard. 04 November 2013. Available at www.docs.oasis-open.org/ubl/os-UBL-2.1/UBL-2.1.html.
OGC. 2016. CityGML, OGC. Web-page, available at www.opengeospatial.org/standards/citygml.
OmniClass. 2016. OmniClass. Web-page, available at www.omniclass.org.
Pan, J.Z., Horrocks, I. 2007. RDFS(FA): Connecting RDF(S) and OWL DL. *IEEE Transactions On Knowledge and Data Engineering* 19(2): 192–206.
Pauwels, P. Terkaj W. 2016. EXPRESS to OWL for construction industry: Towards a recommendable and usable ifcOWL ontology, *Automation in Construction* 63(March): 100–133.
Pollock, J.T. 2001. The Big Issue: Interoperability vs. Integration. *eAI Journal* October: 48–52.
ProductXchange. 2016. Building life cycle collaboration with BIM product data. Web-page, available at www.productxchange.com.
Roman, D., Keller, U., Lausen, H., de Bruijn, J., Lara, R., Stollberg, M., Polleres, A., Feier, C., Bussler, C. & Fensel, D. 2005. Web Services Modeling Ontology. *Applied Ontology* 1(1): 77–106.
Rong, K., Hu, G., Lin, Y., Shi, Y. & Guo, L. 2015, Understanding business ecosystem using a 6C framework in Internet-of-Things-based sectors. *International Journal of Production Economics* 159: 41–55.
Russell, M.G. Huhtamäki, J., Still, K., Rubens, N. & Basole, R.C. 2015. Relational capital for shared vision in innovation ecosystems. *Triple Helix* 2(1).
STEP. 2016. ISO 10303 STEP Standard. Web-page, available at www.steptools.com/library/standard.
Seppälä, T., Halén, M., Juhanko, J., Korhonen, H., Mattila, J., Parviainen, P., Talvitie, J., Ailisto, H., Hyytinen, K.-M., Kääriäinen, J., Mäntylä, M. & Ruutu, S. 2015. The Platform—History, Characteristics, and the Definition. ETLA Reports No 47. Available at www.etla.fi/wp-content/uploads/ETLA-Raportit-Reports-47.pdf, in Finnish.
Sikos, L.F. 2015. *Mastering Structured Data on the Semantic Web: From HTML5 Microdata to Linked Open Data*. Apress.
SokoPro. 2016. SokoPro. Web-page, available at http://www.sokopro.com/english.
Tekla. 2016. Tekla Structures BIM Software. Web-page, available at www. tekla.com/products/tekla-structures.
Tilaajavastuu. 2016. Tilaajavastuu—Contractor's liability. Web-page, available at www.tilaajavastuu.fi.
Törmä, S. 2013. Semantic Linking of Building Information Models. In *Semantic Computing; Proc. intern. conf., Irvine, 16–18 September 2013*. IEEE.
UNSPSC. 2016. UNSPSC. Web-page, available at www.unspsc.org.
W3C. 2016. World Wide Web Consortium. Web-page, available at www.w3.org.

eWork and eBusiness in Architecture, Engineering and Construction – Christodoulou & Scherer (Eds)
© 2016 Taylor & Francis Group, ISBN 978-1-138-03280-4

Building Information Modeling (BIM) for LEED® IEQ category prerequisites and credits calculations

G. Bergonzoni, M. Capelli, G. Drudi, S. Viani & F. Conserva
Open Project S.r.l., Bologna, Italy

ABSTRACT: A good Indoor Environmental Quality is required for any building applying for Leadership in Energy and Environmental Design (LEED®) certification.

The room-by-room calculations required to demonstrate compliance with the Indoor Enviromental Quality (IEQ) prerequisites and credits, starting from the building geometry and use and the HVAC technical data, may be time consuming and error-prone. With Building Information Modeling (BIM) all the required data can be incorporated into the design model, thus speeding up the process, automate the design validation and minimize the errors.

The research explored the use of BIM for the Mechanical design following the LEED IEQ P1 and IEQ C2 qualification, and the automation of calculus for the verification process with the use of Dynamo visual scripting tool for the data exchange between Autodesk Revit and electronic spreadsheets.

Keywords: Interoperability, BIM, LEED, Mechanical Design, Sustainable Buildings

1 PROBLEM STATEMENT

The use of Building Information Modeling (BIM) as a design tool for an energy and environmental design, from the concept stage through the design project and eventually to facility management is a matter of research for the Open Project S.r.l. Company.

The team has a relevant expertise in Leadership in Energy and Environmental Design (LEED), with the design of three LEED Gold certified buildings in its portfolio. Since 2013, when BIM and Autodesk Revit were chosen to be part of the design process, the company is working on the integration of LEED in the Building Information Modeling process. The presented research is included in a larger field of BIM-LEED integration researches. (Azhar et al. 2011) e.g. regarding the Sustainable Sites credits (Chen & Nguyen 2015).

The focus of this study is notably on the application of BIM on the certification program as a tool for both design and verification of the LEED parameters for Indoor Enviromental Quality (IEQ) Prerequisite 1 and IEQ Credit 2 credits.

The research question is whether with the use of BIM is possible to quicken the design process that answers the Enviromental Quality requisites of LEED and speed up the credit verification procedure, and creation of summary schedules.

2 INTRODUCTION

2.1 *LEED references*

The rating sistem “LEED Italia 2009, Nuove costruzioni e ristrutturazioni”, gets reference to EN 15251:2007 for credits regarding Indoor Enviromental Quality Prerequisite 1 and Credit 2 (IEQ P1, IEQ C2).

In zones where mechanical ventilation is used, compliance with IEQ P1 and IEQ C2 can be achieved when in every zone of the building the air flow of outdoor air are granted, with reference to EN 15251 Class II and Class I respectively, and for both credits when the building is considered as in the low polluting category.

The minimum outdoor air flow [l/s] is given by formula:

$$q_{tot} = nq_p + Aq_B \tag{B.1}$$

where
n = zone population [persons];
q_p = minimum outdoor air flow per person [l/s·person]; A = room area [m^2];
q_B = minimum outdoor air flow per room area unit (taking into account for building emissions) [l/s·m^2].

Requirements of the B.1. formula are shown in the table (Figure 1) for the q_p and q_B ventilation rates.

Category	Expected Percentage Dissatisfied	Airflow per person l/s/pers
I	15	10
II	20	7
III	30	
IV	> 30	< 4

Category I:	0,5 l/s, m^2	1,0 l/s, m^2	2,0 l/s, m^2
Category II:	0,35 l/s, m^2	0,7 l/s, m^2	1,4 l/s, m^2
Category III:	0,3 l/s, m^2	0,4 l/s, m^2	0,8 l/s, m^2

Figure 1. EN 15251 Annex B, Table B.1.

2.2 *Without BIM*

Standard procedure for the calculation of IEQ Prerequisite 1 and Credit 2 involves the use of a non-automated iterative process. A cross checking is required between geometrical data, from a CAD file, which shows the design of the air flow of the mechanical systems and its zone location, and the total required air flow and numerical data from a Spreadsheet, the process is adopted room-by-room and iterated every data or design intent change.

2.3 *With BIM*

With the introduction of BIM a lot of the needed parameters can be stored into the BIM model. Required data are from both architectural and Mechanical disciplines; Autodesk Revit is used for modeling and management of data, architectural and mechanical design and its coordination. Calculations according to the B.1. formula requires a spreadsheet. Electronic spreadsheets are widely used in the ASHRAE standards, on which the LEED prerequisites and credits are often based, and are flexible enough to allow the use of any international regulations as allowed by LEED ACPs. LEED Online templates are also often based on electronic spreadsheets. Therefore the research, although limited to a specific case study, could show the potential of a deeper integration between Autodesk Revit and Microsoft Office Excel. The use of spreadsheets also allows to perform calculations using ASHRAE Standard 62.1–2010 as well as EN 13779-2007 standard.

3 METHODOLOGY

3.1 *Data gathering*

The process starts with the collection of the required data, it involves the use of both Architectural and Mechanical disciplines models and their interaction. Air terminals of the HVAC system are placed within rooms and have a designed air flow data. Rooms have the parameter of occupancy and the floor area [m^2] needed to perform the calculation of the B.1 formula. A Room schedule is used to gather Area and Occupancy data, but unfortunately, the Revit schedules do not allow to perform calculations with the B.1. formula units, such as flow per person [l/s·person], thus electronic spreadsheet was used.

3.2 *Interoperability*

A visual script is developed in order to enhance interoperability between the two software, the code is capable of exporting and importing data from and to the BIM model. Room area and occupancy are automatically exported on the spreadsheet where the EN 15251 minimum outdoor air flow is calculated with the B.1. formula. The result of the calculations are automatically imported back to the BIM model where the two numbers can be easily compared in order to perform the verification of the LEED IEQ p1 and IEQ c2 credits (Figure 2).

If compliance is not achieved, following a design change, the script will automatically get the new calculations and credit check, example in the next chapter. Similarly, in any case of the variation of the data, for instance a new owner requirement regarding occupancy, the model updates and remains consistent.

4 CASE STUDY

4.1 *The jewelry factory*

Open Project S.r.l. is an integrated team comprising 40 architects and engineers, working at all of the building scale (from masterplan to interior design) and stages (from concept design to detailed design, project and construction management), with a relevant expertise in LEED, exhibiting three LEED Gold certified buildings in its portfolio. Since 2013, Open Project has been working with BIM.

In 2015 Open Project took advantage of Autodesk Revit to perform the design of the new jewelry manufacturing plant of Bulgari Gioielli S.p.A. which has been registered with the "LEED Italia 2009 Nuove Costruzioni e Ristrutturazioni" rating system.

Building Information Modelling is used to federate 4 models (Figure 3): Architectural (LOD 350), Steel structure (LOD 350), Concrete structure (LOD 300), Mechanical systems (LOD 200) (Figure 4).

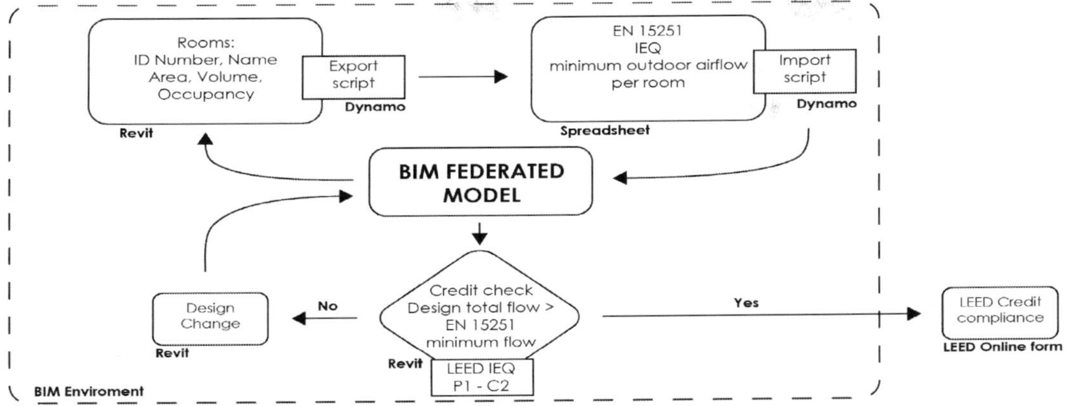

Figure 2. LEED IEQ P1 C2 check methodology flow.

Federated Architectural and Mechanical models are both involved in the process of credit checking.

Rooms are the analytical object used for the Architectural data. A Revit Room schedule is used to gather required parameters: room ID number, room name, room area, room volume, room occupancy (the maximum number of people that can occupy the room, parameter defined by the owner in our case study, or standard reference in EN Standards).

The HVAC system is designed with an high number of room volume related ventilation due to factory production needs, defined by the owner. The design process is able to join both requirements, the LEED compliance to grant IEQ and the defining volume ventilation rate for the factory need. The room volume parameter would not be automated in a 2D CAD workflow. The HVAC system of this case study is designed with a 100% outdoor air but the process can easily be extended to all recirculation air systems.

4.2 *LEED Indoor environmental quality check*

In order to perform the calculations with the B.1 formula a new shared parameter is created "EN 15251 minimum flow" but the formula could not be calculated because the units [l/s·person] or [l/s·m²] are not recognized by Revit in parameter formulas.

The solution is obtained with a script designed on an open source Visual Programming tool, Dynamo, made by Autodesk, an instrument capable of extracting, manipulating and connecting data coming from the model and to the model.

The script is designed to export from the Revit model to an electronic spreadsheet the following parameters of the Room schedule: room ID number, room name, room area, room occupation. The script allows real-time update on data manipulation regarding elements of the project or calculus formulas. In the spreadsheet a column is added with the B.1. formula with the q_p and q_B parameters from the B.1. table (EN 15251, 2007). The second part of the script reads the added q_{tot} column and get back into a Revit parameter the calculated minimum air flow requirement. (Figure 5)

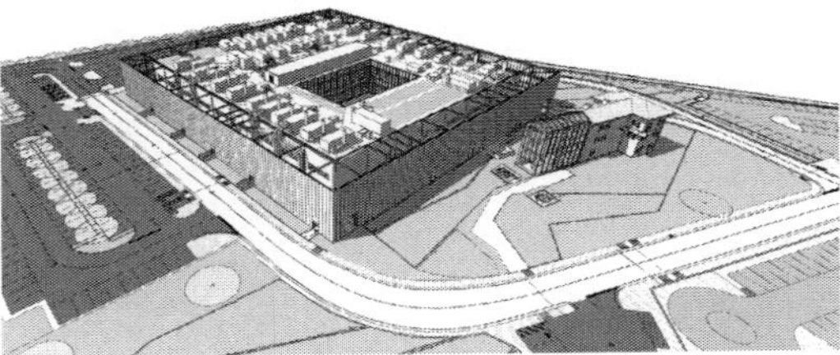

Figure 3. BIM federated model.

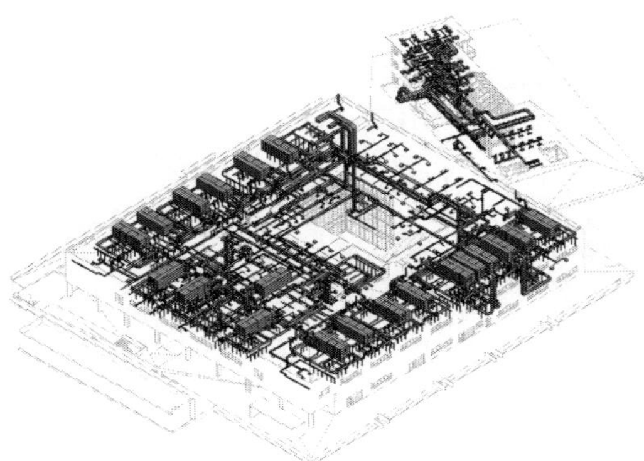

Figure 4. BIM mechanical model.

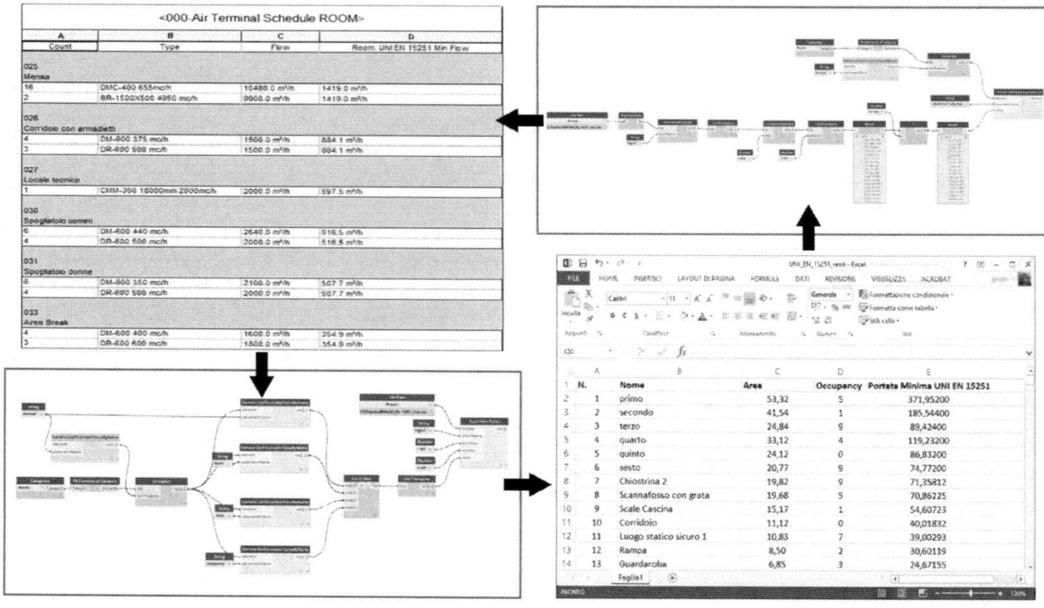

Figure 5. Data iterative flow (Revit-Dynamo-Excel-Dynamo-Revit).

Input data into Revit are placed in the shared parameter "EN 15251 minimum flow", which is assigned to Rooms, but can be recalled into other Revit schedules where also the total design flow per room, otherwise not visible, can be shown.

Thus the Air Terminal schedule, comprising terminals of the mechanical system gathered by room, is chosen to compare the total design flow per room and the shared parameter "EN 15251 minimum flow", in order to verify the LEED credit compliance. La compresenza di due parametri provenienti da sources diverse ci permette di effettuare il confronto in questa schedule. Assignment of the parameter to the respective room also allows to show the minimum flow in the plan in order to facilitate the Testing Adjusted Balance process (TAB).

4.3 *Parameter change process*

The process shown was declined in a 3-step design feedback, an example is given on the Info-point area of the facility, due to an arranged change in occupancy parameter, credits recheck was necessary. (Figure 6)

In accordance with the owner, occupancy of the Info-point area was changed from 15 to 30 people, the 3 steps that followed the change shows the automation granted by the scripted process:

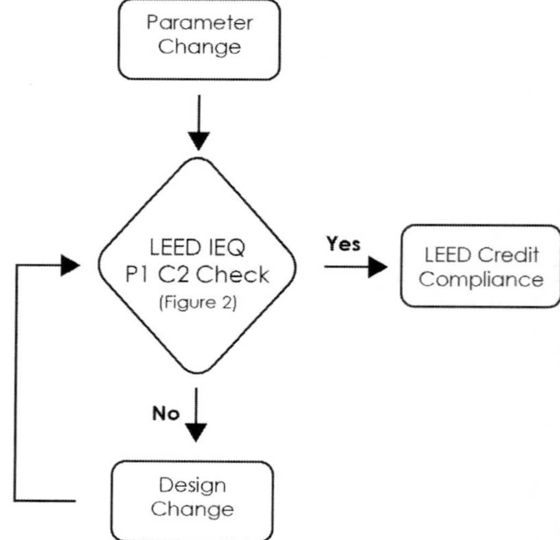

Figure 6. 3-step process.

1. After the data change, the schedule showed that design outdoor air flow did not respond to the IEQ C2 parameters, the check is automated.
2. Mechanical design was consequently changed to increment the air flow and regain the credit. This process involved the work of the MEP Designer thus could not be automated. (Figure 7).

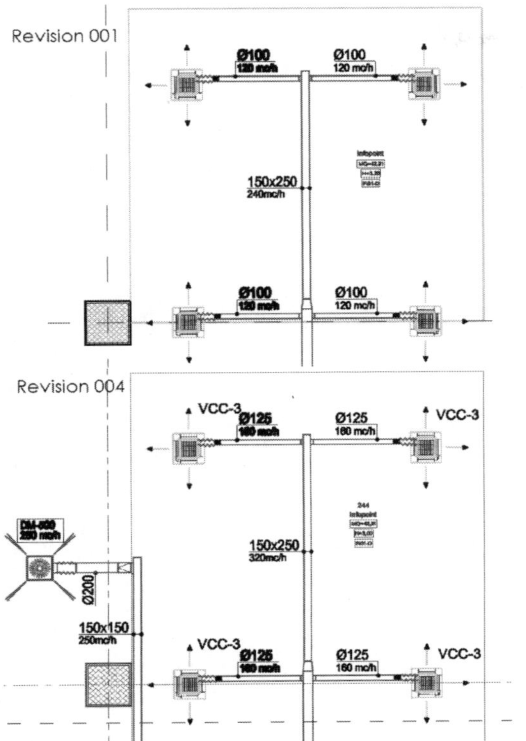

Figure 7. Design change of the mechanical system in the Info-point area.

3. The schedule showed that the new design was in compliance with the LEED IEQ Credit 2, the process iteration ended.

5 CONCLUSIONS

The research shows how to automatically perform the IEQ p1 and IEQ c2 calculations by employing open source software (e.g. Dynamo) for the interoperability between Autodesk Revit and an Electronic spreadsheet.

The potential of such interoperability is very high, thanks to the wide use of electronic spreadsheets in the ASHRAE standards, the EN building codes and the LEED Online templates in general.

The process is also highly reiterative thanks to automated calculations implemented, but presents some disadvantages. It requires data processing knowledge in order to use the script and avoid errors; moreover the workflow needs a complex model, where the input is fulfilled with parameters, it is yet possible to use those data for various purposes.

Further implementation of the proposed script and enhancement of the input data within the BIM model will allow to automate the calculations of other LEED credits and prerequisites.

The increased reliability and the automation of a larger number of calculations will be beneficial for the LEED Integrated Design process, and in a larger perspective will contribute to the delivery of a better and more sustainable design.

REFERENCES

American Society of Heating, Refrigerating and Air-Conditioning Engineers 2010. *ANSI/ASHRAE Addendum h to ANSI/ASHRAE Standard 62.1-2010. Ventilation for acceptable indoor air quality*. Atlanta: ASHRAE.

Azhar, S. Carlton, W.A. Olsen, D. & Ahmad, I. 2011. Building information modeling for sustainable design and LEED® rating analysis. *Automation in construction* 20(2): 217–224.

Chen, P.H. & Nguyen, T.C. 2015. Integration of Building Information Modeling (BIM) and LEED's Location and Transportation Category. *32nd International Symposium on Automation and Robotics in Construction and Mining Proceedings; Oulu, 15–18 June 2015* 32: 347–349.

European Committee for Standardization 2007. EN 13779:2007. *Ventilation for non residential buildings. Performance requirements for ventilation and room-conditioning systems*. Brussels: European Committee for Standardization.

European Committee for Standardization. 2007. *EN 15251:2007. Indoor environmental input parameters for design and assessment of energy performance of buildings addressing indoor air quality, thermal environment, lighting and acoustics*. Brussels: European Committee for Standardization.

Green Building Council Italia 2011. *Manuale LEED 2009 Italia Nuove Costruzioni e Ristrutturazioni*. Trento: Green Building Council Italia.

eWork and eBusiness in Architecture, Engineering and Construction – Christodoulou & Scherer (Eds)
© 2016 Taylor & Francis Group, ISBN 978-1-138-03280-4

Emotional intelligence—improving the performance of big room

O. Alhava
Fira Oy, Vantaa, Finland

E. Laine
European Investment Bank, Luxemburg

A. Kiviniemi
University of Liverpool, Liverpool, UK

ABSTRACT: Given the importance of collaboration in the construction industry, the perspective of emotional intelligence is almost completely absent from construction project management articles and literature. In many other industries collaboration, co-working and co-creation are recognized as common practices to create customer value.

As a crucial soft skill required for efficient collaboration, the role and benefits of emotional intelligence should be investigated and clarified especially when Big Room methods are applied increasingly in construction projects. The purpose of this paper is to test and verify a framework for the assessment of Big Room facilitator's emotional intelligence to improve the performance of Intensive Big Room (IBR) process during the design phase of construction projects.

This study aims to take the performance of the IBR into new level by studying the effects of the emotional intelligence in the co-creation process. The findings indicate that the application of emotional intelligence methods in facilitation immediately improved the performance of the social processes in interdisciplinary collaboration. Furthermore, the participant motivation and stakeholder satisfaction were increased when emotional intelligence methods were included to the facilitation process of IBR process.

Keywords: Emotional intelligence, Big Room, collaboration, Integrated Concurrent Engineering, ICE

1 INTRODUCTION

Even construction projects are emotional. Yet the Emotional Intelligence (EI) has been largely unexplored in construction project management (Zhang & Fan, 2013). Much of the attention in construction project management is directed to the application of tools and techniques and far less attention is given to the role of leadership. However, as the forerunner companies develop and complete relationship based construction projects with far better economical results than the traditional construction industry (Seed, 2014), the role of leadership and relationship management issues have become far more significant for construction industry (Love et al. 2011). Some companies have even come into conclusion that traditionally trained construction Project Manager (PM) is not equipped to deal with the relationship-based nature of modern project delivery (Seed, 2014).

Emotional Intelligence (EI) is a relatively new approach in construction industry, but it has been already demonstrated in large and complex projects that the leadership style and level of team integration predict the success of the project (Zhang and Fan, 2013, Aapaoja et al. 2013). Although a variety of EI theories defined from a different perspective have emerged, each of them suffer from certain limitations when applied in a specific industry (Zhang and Fan, 2013). Most popularized theory of EI, the Goleman's mixed model, has been widely accepted and acknowledged by business (Zhang and Fan, 2013), but it has also received criticism mainly due to populistic statements, which are made in best seller books of Goleman. Especially the Goleman's statement that EI can predict leadership effectiveness has caused multiple academics to stress that EI cannot solely predict the performance of project, but IQ and the "big five" personality traits has to be considered concurrently (Love et al. 2011). However, it is clear, that strictly rational and functionalist approach to the project management has its limitations in effective project management practice (Clarke, 2010).

Another perspective to the improvement of project performance can be taken from the acceleration of the team performance. According to Chachere et al. (2009) an extraordinary rapid design with a high quality has been achieved by using Integrated Concurrent Engineering (ICE) especially in the design phase. Much of the improvement is gained due to reducing the latency time by introducing parallelization of design tasks and enhancing the knowledge and information management by utilizing co-creation and co-working in co-location. Advanced construction companies combine efficiently Virtual Design and Construction (VDC) and ICE, and their approach to the performance acceleration is based on reducing the process equivocality discipline specific modelling and visualization as well as measuring constantly the progress of the team towards clear and congruent goals (Juntunen and Kiviniemi, 2015, Knotten and Svalestuen, 2014).

In the early stages of ICE development, Chachere et al. (2009) identified the importance of leadership as a factor that enables latency reduction in the ICE sessions. Their study concludes that the enablers for acceleration are 1) flat organization hierarchy, 2) culture that enforces the designers' autonomy, 3) high aspiration only to project success, and 4) mutual respect in a high pressure environment. By combining afore-mentioned soft issues to advanced engineering process, utilization of BIM for information management and visualization, and applying Lean tools like Last Planner system and Planned Percent Completed (PPC) metric, very high performance ICE sessions have been achieved with very low response latency time in design phase (Juntunen and Kiviniemi, 2015, Fundli and Drevland 2014).

Simultaneously, the novel project delivery methods, Integrated Project Delivery (IPD) and Alliancing have been introduced and taken to use in the construction industry. Aapaoja et al. (2013) claim that the cornerstones for creating integrated team can be described with 12 characteristics, by which the level of collaboration in teamwork can be increased. Although the framework of their study is IPD, the results can be generalized as the utilization of teamwork can often lead to results that would have remained unrealized with partially integrated or fragmented project team.

As construction management theories is mainly concentrating to the operations paradigm, there should be a switch of lenses to organizational behavior instead (Love et al. 2011). Especially the project leader's emotional awareness and the use of EI seem to have a significant role in how project managers interact with other team members and how relationships between project members are being enacted (Clarke 2010). PM's emotional self-awareness, emotional self-control, empathy, organizational awareness, cultural understanding and communication has been shown to have a strong correlation with project performance (Zhang and Fan, 2013). Yang et al. (2013) have presented a validated model for studying the leadership style of PM and its effects to the project performance. In addition to modelling the leadership style of PM, the proposed model takes into account the team communication, collaboration and cohesiveness as critical components to a successfully implemented project. Finally, the model measures the schedule, cost and quality performance as well as stakeholder satisfaction, and according to study, the PM with transactional and transformational leadership may improve the teamwork performance, which has a strong correlation to the project performance and customer satisfaction.

2 INTENSIVE BIG ROOM (IBR)

Fira has started the development of a Big Room concept and methodology in early 2013. Collaboration and co-creation of customer value has been studied and implemented by using Fira's Verstas-concept from 2009 (Alhava et al. 2014). The very first project, in which the Intensive Big Room (IBR) was tested and further took the shape as a IBR v1.0, was the Rajamäki swimming hall renovation. The challenge for which the IBR was decided to be used up against to, was tight schedule and 20% reduced budget. Even though the scope of the project was radically changed and the length of design phase was unaltered, the project managed to reach the original goals due to use of IBR. The methodology was described partially in Alhava et al. 2015 and the project itself got a special recognition in the Tekla Finland and Baltics BIM Awards 2014 winners for groundbreaking research and development of Big Room methodologies introduced in construction projects (Tekla 2014).

Initially, the IBR method concentrated to the facilitation of ICE sessions, since the coordination of latency was thought to be a unifying performance metric (Chachere et al. 2009), and therefore, reducing it to near-zero was considered to be an ultimate project performance goal. From engineering perspective, reducing waste in design phase was considered to be the success factor in improving the project performance and creating more customer value. Based on internal root-cause analysis and studies made from project performance of Fira's own design and build projects, it was clearly seen that the IBR outperformed the projects in which the design phase was conducted with traditional two weeks' design meeting cycle.

Fira has continued the development of the IBR concept by introducing an application of the Last Planner System (LPS) to coordinate the design management during IBR process (Juntunen and Kiviniemi, 2015). Recently the scope in the development of the IBR concept has changed from operations paradigm to improving the integration level of team, enhancing the facilitation and to the role of emotional intelligence for improving the design process performance.

2.1 *IBR v.1.0 (2013–2014/H1)*

The customer value co-creation and applying service logic into the design phase were in a key role when the development of IBR v.1.0 were started.

The first versions of IBR consisted of weekly sessions each lasting 3 to 4 hours. IBR was a modification of Big Room and the reason for using it was the small size of the projects. In a relatively small project it is not possible to have the participants in the same room for the whole length of the project.

Due restrictions of schedule, the focus in IBR v1.0 was learning in using ICE in reducing the latency and maximizing the customer value. In practice, it was vital for facilitators and team to understand how efficient collaboration in small teams can be facilitated and what changes this would require from participants if compared to traditional siloed design process. The use of BIM and different visualization methods were also tested and adopted into use.

2.2 *IBR v2.0 (2014/H2–2015/H1)*

The second phase in Fira's IBR development started when the use of Lean construction methods, especially the Last Planner System, were taken into use. The increased maturity made it possible to introduce collaborative workflow management and Last Planner System also in IBR environment. The very first PPC-measurements in design process were made in pilot projects and reversed phase scheduling were studied.

Also, the Key Performance Indicators (KPIs) were defined for each consecutive phase of IBR sessions. This underlined the process approach, and therefore technical issues, preparation tasks lists, project goal setting, responsibility matrixes, and IT-tools preceded the leadership issues or human approach. However, the importance of customer's decision process, common organization and training for IBR were also recognized to be relevant for successful implementation of the IBR process.

Simultaneously with the adoption of Lean construction methods, the plus/delta was increasingly used and small-scale stakeholder satisfaction surveys were made. The negative results indicated in some projects that focusing solely to process management and operational efficiency caused negative impact to project performance, which then started the third development phase of IBR.

2.3 *IBR v3.0 (2015/H2-)*

The study in this paper is part of ongoing development of IBR method. By large, the atmosphere has changed emotionally into a more positive direction as Fira has changed over the years from traditional main contractor's role to a customer oriented service company. The change can be seen in the appreciation and encouragement of Fira's personnel show to the other members of the company, partners or customers. The reduced distrust and cynicism inside the company has made it possible to start the study of emotional intelligence and development of the soft skills for IBR facilitators.

3 PROBLEM FORMULATION AND RESEARCH QUESTIONS

The research was conducted in three stages and therefore the research question is tripartite: Firstly, Fira was willing to improve the level of team integration during design phase and provide practical tools for IBR facilitators to be used in IBR sessions or taken into account in preparing to them:

RQ1: Based on the 12 characteristics describing an integrated team (Aapaoja et al. 2013): What are the practical improvements and changes to the leadership and IBR 2.0 model by which the IBR facilitator can improve the project performance?

Secondly, it is vital to understand what are the results in project environment as the developed principles and guidelines are being used in design phase of the project:

RQ2: How effective are the changes made to the leadership and IBR 2.0 model when compared project using improved IBR facilitation model to a project in which conservative IBR method is being used?

Thirdly, is there a correlation between the utilization of the improved IBR facilitation method and key skills and personal characteristics of facilitator as Love et al. (2011) proposes, i.e. is there a correlation of improved use of IE and project performance?

RQ3: Is it possible to identify and further improve the facilitating of IBR sessions by concentrating to the key skills and personal characteristics?

4 METHODOLOGY

The study was based upon literature review, semi-structured interviews with six IBR facilitators and a survey for total of 44 IBR team members from six different projects using IBR.

The six projects were divided into two groups, based on the maturity of the use of IBR method, i.e. the new leadership requirements were used in pilot projects and the reference projects were using IBR 1.0 method. The two projects were chosen to be included into this study based on the highest number of answers to the query. Both pilot project and reference project are at the moment (spring 2016) in the execution phase and there have been more than 25 IBR sessions in both projects.

In order to answer the RQ1, the role of IBR facilitator was to be redefined. This was done based on the literature study and experiences from the previous IBR projects. Also, an organization psychologist and service designer were consulted for improving the IBR process and developing new facilitation methods to be used in IBR sessions. At the same time, the minimum requirements for each IBR session phase were defined and they were later prioritized to help the facilitators to prepare, facilitate, conclude and oversee the sessions and to coordinate the designers individual work or needed co-working in between the sessions in the pilot projects. Group meetings were held, in which all IBR facilitators were invited to discuss of changes in facilitators role and experiences in using the new leadership principles.

During this study, a number of stakeholder questionnaires were proposed and tested for understanding better the facilitators' leadership styles and how successfully they were using the key skills as defined by Love et al. (2011) or increasing the level of integration of the design team. Finally, a decision was made to use the model for measuring the effect of a PM's leadership style as proposed by Yang et al. (2012), and 22 questions were formulated to measure both transactional and transformational leadership, team communication and collaboration, project performance in terms of schedule, cost and quality, and finally stakeholder satisfaction. In model proposed by Yang et al. (2013), the team cohesiveness is also measured, but due to problems in formulating non-ambiguous questions, the cohesiveness (Yang et al. 2013: CH1, CH2 and CH3, respectively) was not included to the study. Similarly, questions which related to the salary and bonuses were excluded since it would have positioned the teams from different companies into an unequal position. Additionally, the question of facilitator's ability to take the needs and feelings of participants into account (Yang et al. 2013: TF3) was divided into two questions. From the rest of the model, 74% of the positions were considered relevant and therefore there were in total 23 questions in the study, whereas the complete model consists of 37 items.

Finally, a survey instrument was used to measure an IBR facilitator's transactional and transformational leadership style, team communication, team collaboration, project performance and stakeholder satisfaction.

5 FRAMEWORKS AND IMPROVEMENTS OF IBR METHOD

From the leadership perspective, the study Aapaoja et al. (2013) points out that level of integration is increased when 1) organizational boundaries are ignored, 2) the team shares mutual focus and objectives, 3) fair and respectful atmosphere is predominant, 4) team members are treated equally, 5) results and innovations are mutually beneficial and 6) team work focus is in solving problems ("no blame" culture). As the collaborative project environment demands members of the team from different organizations work together as an integrated team towards common objectives and mutual benefits, the question will raise who is responsible of increasing the level of integration in the team? During the study, a common conclusion was made that in the design phase of a construction project, the facilitator of the IBR sessions is responsible for increasing the level of integration of the IBR team.

According to Love et al. (2011), there are five key skills of construction managers' - namely instructing, active listening, judgement, decision making, and negotiation—which Love et al. derived from the definition of EI as the "ability to perceive emotions, to access and generate emotions so as to assist thought, to understand emotions and emotional knowledge, and to reflectively regulate emotions so as to promote emotional and intellectual growth".

On the other hand, Love et al. (2011) define also eight key personal characteristics, which suggests the need of EI in construction manager's role. The first three—dependability, leadership and cooperation—require being able to fulfill obligations, willingness to lead and still being pleasant with others, whereas next two—stress tolerance and self-control—require keeping emotions in check, dealing calmly with high stress situations and accepting criticism. Finally, the last three characteristics—initiative, adaptability/flexibility and innovation—are related to the willingness to be open for change, to be creative and to take challenges.

During the development of IBR 3.0, the facilitator's role was enhanced. Firstly, the leadership and management were separated to make sure that managing the issues and leading the people were not to be mixed with each other. Facilitator must have enough management competence from construction design for setting up the design schedule for all disciplines and solid understanding of key components of design for being able to set the goals for design team and for being a supervisor for the team in reaching to goals. Secondly, the leadership role of facilitator was defined by using the 12 characteristics describing the integrated team. A summary of guiding principles for facilitator is depicted in Table 1.

6 CASE STUDY

If the model of Yang et al. (2012) is studied in more detail, there are four main perspectives by which the effect of the PM's leadership style on project performance is studied, namely 1) the leadership style, 2) its subject, the teamwork itself, 3) project performance by terms of schedule, cost and quality, and finally 4) the evidence stakeholder satisfaction. These four perspectives are broke down into more detailed sections as depicted in Figure 1 in which also the results of the conducted survey are presented.

From these four perspectives, the pilot project and reference project seem to be quite similar as how the participants have experienced the team communication and team collaboration. The only significant difference is CL4: "Team members did not show the defensive or mistrustful attitude during group discussions." This might indicate that enhanced leadership skills of facilitator and "no blame" culture reduce first the negative attitudes and atmosphere.

The results how participants experienced the leadership style of the facilitator separates the pilot project from the reference project. There is a significant difference how they perceived the way in which the facilitator praises individuals for exceeding the expectations (TS1), creates a common vision (TF1), encourages for innovation (TF2), expresses emotional awareness (TF3), acts a role model (TF4), and sets up ambitious goals (TF6).

The difference in leadership style might correlate with results, since there is a notable difference in both project quality performance and stakeholder satisfaction in favor for the pilot project. However, the low difference in Net Promoter Score (NPS) does not support this, since it simply measures whether the participant is willing to recommend the IBR to a colleague or not, and therefore one would expect that the difference would be similar as with stakeholder satisfaction.

7 RESEARCH LIMITATIONS

Although the two compared projects were selected from six options, this research compares only two projects and their participants, and therefore the results need to be verified more widely. Especially the fact that there were only two individuals, the facilitators of the pilot project and reference project, whose leadership skills were compared with the survey and no attempts were made to study the impact of their personal characteristics. Therefore, future studies should also take into

Table 1. Practical improvements to the leadership responsibilities for IBR facilitator.

Characteristics	Responsibility for facilitator
1. Organizational boundaries are ignored	Prioritize the customer and Encourage the team to open communication and continuous communication especially during ICE session phase.
2. Team has mutual focus and objectives	Visualize always the goals for the session, goals for each ICE team and their relation to the goals of the customer.
3. Each member is allowed to present their ideas concerning project delivery	Ask secondary opinions during preparation, ICE and conclusion phases for ensuring the collective responsibility for project performance. Finalize the meeting with plus/delta and make required changes to the IBR process accordingly.
4. Each member has an equal opportunity to contribute to the delivery process	Select the members for each parallel ICE team based on the required knowledge and competence. Make sure that the bottleneck resource is identified for each session and his/her time and expertise per each team is shared fairly.
5. Results and innovations are mutually beneficial	In decision making and in case of disputes use as a guideline "best for the project" and "customer decides what is valuable" principles.
6. Focus on solving problems, not finding out who is guilty	Do not accept name blaming, and in case of a mistake find together with the team the root cause and fix it for preventing the same happening again.

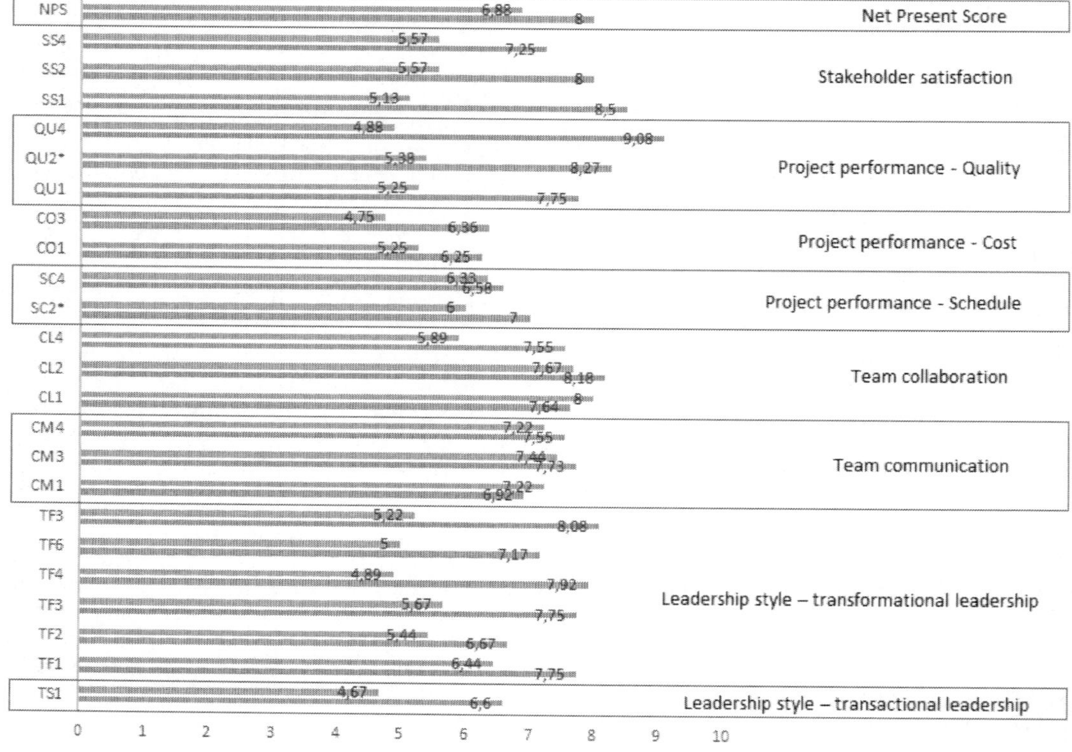

Figure 1. Questionnaire results from pilot and reference projects.

consideration comparing the emotional awareness and personal characteristics in order to rule out the influence of individual facilitators.

Secondly, the research model of Yang et al. (2012), was only partially used due to practical limitations of real life projects: transactional leadership and team cohesiveness might have impact to the results.

8 DISCUSSION

The customer value is co-created in IBR's ICE sessions in parallel teams and at the end of the IBR session in the concluding phase. The paragon of IBR has been the Nasa's Team-X and their extreme collaboration as described by Mark (2002) and Chachere et al. (2009). In final remarks on his study, Mark states that "Almost all current Team X members report being mentally tired at the end of a session." and "One engineer described the experience as exhausting but thrilling, like riding a roller coaster.". These comments are in line with the results from interviews of the IBR facilitators. All six of them reported job stress and a need for social support as the demands and constraints perceived to exceed one's personal resources.

It is hardly a surprise, when the immediate reaction was negative from half of the facilitator groups against defined improvements to the role of facilitator. Facilitators felt that there is an overload of responsibilities and a risk of exhaustion if they have to be responsible for successful implementation of design process and wellbeing of participants as well. Some even declined to take the required actions for increasing the level of integration in their on-going projects and continued with only partially integrated or even fragmented design project team.

According to analyses, there seems to be a strong correlation between leadership style, emotional awareness of the IBR facilitator and the project performance in terms of quality and stakeholder satisfaction. The facilitators of the pilot project had positive attitude towards the responsibilities of their new role and they were interested in learning the transformational leadership skills. During the interviews and internal development meetings they showed interest in understanding the use cases for the key emotional intelligence skills of supervision, which are defined and discussed in Love et al. (2011). As there a no validated method to justify how well each IBR facilitator's personal characteristics correspond to the key personal characteristics of emotionally intelligent PM's profile, it is not possible to argue whether the fairly beneficial results from the survey were result of the improved IBR methodology and facilitators' behavior based on the new role of IBR facilitator, or were the results merely based on innately emotionally intelligent facilitators natural reaction to increased information and understanding of the deep meaning of the EI.

9 CONCLUSIONS

This study is a part of efforts in Fira for taking the performance of the IBR method into a new level by studying the effects of the emotional intelligence in co-creation process and teamwork in ICE sessions. The conducted literature study was successful, because it produced useful 1) framework for studying the integration level of team, 2) definition of the EI to be used in environment of the construction industry, 3) key tasks, skills and personal characteristics for evaluating the EI in supervision tasks of a PM, and 4) validation model for measuring the effects of the leadership style of facilitators.

By the outcomes of the literature study, an answer to the RQ1 was developed by using the 12 characteristics for creating an integrated team. These are summarized in Table 1. Most of the improvements were communicated as guidelines and leading principles. In order to further elaborate those for internal training, further development and studies of use cases during the IBR sessions are needed.

The effectiveness of the changes in the leadership style and practicing the EI in facilitation of IBR were measured with a survey. Results indicate a positive correlation between the emotionally intelligent transformational leadership style and project quality as well as with the stakeholder satisfaction. From the perspective of RQ2 the correlation was not statistically confirmed, but it gives an inspiration to study the EI more deeply and continue the efforts to develop emotionally aware co-creation and collaboration IBR methods for increasing the customer value and at the same time for providing smarter working environment for the IBR teams.

Finally, aspiration of the RQ3 was to improve the project performance by developing the key skills and personal characteristics by means of EI. This question remains open and answering it requires further studies and involvement of competence which does not exist in the domain of engineering. The journey of the construction industry from the restricted emotional atmosphere which is charged with distrust and cynicism to the open and enthusiastic emotional space has just began.

REFERENCES

Alhava, O., Kiviniemi, A., & Laine, E. 2015. Intensive big room process for co-creating value in legacy construction projects. *ITcon, 20*, pp. 146–158. Retrieved from http://www.itcon.org/2015/11

Alhava, O., Laine, E., & Kiviniemi, A. 2014. Interactive Client Centric Design Management Process for Construction Projects. *eWork and eBusiness in Architecture, Engineering and Construction: ECPPM, 2014*. Vienna, Austria. pp. 593–599.

Chachere J., Kunz J. & Levitt R. 2009. The role of reduced latency in integrated concurrent engineering. Working Paper #WP116, Center for Integrated Facility Engineering, Stanford University, Stanford, April.

Clarke N. 2010. Projects are emotional: How project managers' emotional awareness can influence decisions and behaviours in projects. *International Journal of Managing Projects in Business*. Vol. 3 No 4.

Fundli I. & Drevland F. 2014. Collaborative design management—a case study. *Proceedings IGLC-22*, Oslo, Norway, June 2014. pp. 1379–1389.

Juntunen J. & Kiviniemi A. 2015. The Use of Modified PPC Measurement in Design Management. Conference paper of *Lake Constance 5D Conference*, May 4–5, Constance, Germany.

Knotten V. & Svalestuen F. 2014. Implementing Virtual Design and Construction (VDC) in Veidekke—using simple metrics to improve the design management process. *Proceedings IGLC-22*, Oslo, Norway, June 2014. pp. 1379–1389.

Love P., Edwards D. & Wood W. 2011. Loosening the Gordian knot: the role of emotional intelligence in construction, *Engineering, Construction and Architectural Management*, Vol 18, No. 1, 2011 pp. 50–65.

Mark, G. 2002. Extreme Collaboration. *Communications of the ACM*. Vol 45(6), pp. 89–93.

Seed, W. 2014. Integrated Project Delivery Requires a New Project Manager. *Proceedings IGLC-22*, Oslo, Norway, June 2014. pp. 1447–1459.

Tekla Finland and Baltics BIM Awards 2014 winners. 2014- http://www.tekla.com/fi/bim-awards-2014/winners-en.html, last visited 5.4.2016.

Yang L., Wu K. & Huang C. 2013. Validation of a model measuring the effect of a project manager's leadership style in project performance. *Journal of Civil Engineering* Vol. 17 No. 2: pp. 271–280.

Zhang L. & Fan W. 2013. Improving performance of construction projects—A project manager's emotional intelligence approach. *Engineering, Construction and Architectural Management* Vol. 20 No. 2: pp. 195–207.

eWork and eBusiness in Architecture, Engineering and Construction – Christodoulou & Scherer (Eds)
© 2016 Taylor & Francis Group, ISBN 978-1-138-03280-4

A comparative review of systemic innovation in the construction and film industries

S.F. Sujan, G. Aksenova, A. Kiviniemi & S.W. Jones
University of Liverpool, Liverpool, UK

ABSTRACT: Generalising the construction and film industries as a Project Based Inter-Organisational Network (PBION) and Building Information Modelling (BIM) adoption as a systemic change allows phenomena to be compared. The film industry was chosen because of the number of disruptive systemic innovations that have occurred since the late 19th century. Acknowledging that the nature of end products in both industries differ widely, this paper draws fragmentation as common phenomena and is associated with increased transaction costs and lower fixed costs. Due to this research area's highly abstract interdisciplinary nature, two perspectives are investigated for their explanatory and guiding potential. Structuration theory in a macro-perspective associated with the accumulated changes that occur when implementing a form of systemic innovation whereas CHAT is associated with the use of tools in historically developing interacting activity systems and systemic changes.

1 INTRODUCTION

In several countries, the construction industry has recently been forced to undergo changes in the ways it operates, thinks and delivers buildings with the use of Building Information Modelling (BIM). BIM is a process that aims to support new processes of design, procurement, fabrication, construction and maintenance activities with the use of a digital intelligent knowledge-based model (Eastman *et al.* 2011).

The construction industry has already undergone several changes from hand drawings to 2D CAD to 3D CAD, where the innovation process has been mostly incremental (Taylor 2005). In this paper, BIM is viewed as a systemic innovation as it directly affects inter-organisational relationships within construction projects. Despite much hype over the past decade, global BIM adoption has been slow and a vast majority of projects still use traditional methods in design and construction. Researchers discuss various barriers to the adoption of BIM. Some of the problems are related to the organisational nature of the construction industry, which is considered to be highly fragmented and transitory.

The construction industry was described by Taylor (2005) as a Project Based Inter-Organisational Network (PBION). This form of organisation is common in various industries such as the construction, film, and defence industries. This paper focuses on and compares the organisational nature of the construction and film industries.

Since organisational structures of the film and construction industries hold certain similar elements, knowledge developed in the film industry in relation to innovation and organisational structures could shed some light on organisational nature of construction industry and adoption of systemic innovation. Both industries involve the creation of temporary social groups and dissolution of these groups when the project is complete. The transitory nature of these project based industries allows for the application of structuration theory; a theory of the creation and reproduction of social systems that are based on the analysis of both structure and agency; both are dual by nature (Giddens 1986). Since structuration departs from how the relationships are built between stakeholders and organisational structures, (defined by the procurement methods and contractual models), this allows for a valid framework to investigate the aforementioned fragmented nature of the industries.

Cultural Historical Activity Theory (CHAT) sees organisations as networks of activity systems that are distributed, decentred and emergent knowledge systems (Engeström 1987, Blackler *et al.* 2000). These clashing activity systems produce systemic contradictions between use value and exchange value and are developed historically. Understanding the relationship between interacting activity systems may help to reflect on systemic barriers of adoption of BIM since they may occur from project to project at multiple levels and are embedded in organisation of work.

This paper aims to explore the systemic change from organisational perspective in construction projects. The presented knowledge will be conceptualised from the perspectives of structuration theory and CHAT.

2 RESEARCH BACKGROUND

2.1 *Project-Based Inter-Organisational Networks (PBION)*

Jones and Lichtenstein (2009) describe a Project-Based Inter-Organisational Network (PBION) as a temporary alliance of contracted companies linked together by Information and Communications Technologies (ICT) to share skills, information and costs for a common purpose of a specific business task. Taylor (2005) described the construction industry as a PBION, further suggesting that such organisations produce a number of barriers for the diffusion of technological innovation and are not well understood. Existing literature suggests that:

- Inability to learn and manage knowledge on their own adds little value to the entire inter-organisational network (Garcia-Lorenzo 2006);
- The level of fragmentation has increased since 1960 because PBIONs have been put as 'an ideal form to manage markets, cross-functional expertise, customer-focused innovation and technological uncertainty in case of rapidly changing technological and market conditions'. Systemic innovation boosts economic advantages but is inherently weak in coordinating processes, resources and capabilities across the whole project team and organisations (Hobday 2000);
- There are problems with learning closure, communication and feedback loops since multiple ad-hoc actors are involved and they typically change from project to project (Emmitt 2010, Chan and Cooper 2011);
- Lack of *"legitimate organisational authority to arbitrate and resolve disputes that may arise during the exchange"* (Podolny and Page 1998).
- The PBION is inherently weak where functional forms are strong: in executing routine production and engineering tasks, achieving economies of scale and meeting the needs of volume production (Hobday 2000);
- Cross-project integration and management control can also become a problem as any change is costly in integrated organisations, which is contrary to small specialised organisations (Abernathy and Utterback 1978);
- The innovation within the established industry is typically incremental and is driven by efficiency, costs and productivity (Abernathy and Utterback 1978).

2.2 *Systemic and incremental innovation*

This paper builds on the notions put forward by Taylor (2005) that has framed innovations in PBIONs along "a continuum from incremental (implying change inside one company), systemic change (change that affects more than one organisation) to radical (implying a change across the whole project network) change".

Taylor does not refer to BIM as a process that is systemic or incremental. In this paper, we consider the BIM concept as systemic in nature, but has been implemented incrementally. It means that BIM technologies did not trigger the expected change in the processes, but rather were used according to the old ways of work (Miettinen and Paavola 2014), such as the production of 2D drawings out of the Building Information Model.

From Taylor's definition of incremental and systemic innovation, we can see that the difference is in the number of stakeholders that it affects. Furthermore, Taylor finds that there is a lack of theory associated with systemic innovation in PBIONs (Taylor and Levitt 2007).

Therefore, it is suggested that there is need for deeper understanding of the impacts of systemic innovations in the construction industry by bringing perspectives from structuration and CHAT. The theories have been outlined individually in the following sections, followed by a section that explains how the theories would be able to complement one another.

2.2.1 *Structuration theory*

From a sociologist's point of view, procurement of work can be looked at as the creation of social groups using structure and agency. According to Giddens (1986), change is happening naturally but can also be accelerated. In order to assess what can accelerate change, Giddens defined agents as the individuals responsible for change who are controlled by the social structure of their network such as traditions, moral codes or established ways of doing things.

Structuration theory presented by Giddens (1984) focuses on the interaction between agents and structure; this enables us to idealise this process as an iterative one where repeated changes made by individual agents affect organisational structure.

For structuration to apply the micro-level activity needs to be coupled with the macro-level activity; what is changed (through agency) must be with respect to the existing conditions (defined by structure).

Practices within the construction industry stipulate a decentralised way of working where no independent member of the inter-organisational group is an agent, assuming that one change by an agent affects every network member depending on their perspective. It can be argued that various changes by different independent agents' give way to an overall complex change and therefore affect the social structure in a way that is difficult to comprehend.

Broger (2011) explains how structuration theory's integrative perspective is valuable although it is not seen that way by many researchers as they fail to see the explanatory potential of the theory. He also claimed that there is a need for a philosophical companion to structuration theory; for example, CHAT as suggested in this paper.

2.3 *Cultural Historical Activity Theory (CHAT)*

CHAT provides a theory that deals with complex interacting activity systems that are contradictory by nature and where the knowledge and learning are distributed across the elements of the system (Engeström 1987). CHAT gives importance to activities organised around the use of tools. Therefore, in this context, the introduction of BIM tools affected elements of the system in various ways. CHAT gives priority to why networks are functioning in a certain way rather than in other ways in this particular context. CHAT explains that any system is developing historically and only through analysis of historical development of practices it is possible to trace the systemic contradictions that impact slow diffusion of innovations (Engeström 1987).

From the CHAT perspective, the higher the level of fragmentation, the higher the number of interacting activity systems, more tensions and conflicts are manifested in practices. Putham (1986) states that identification of systemic contradictions can help to decipher underlying roots of problems in fragmented and ever-changing organisations as organisations do not develop on their own, people create organisations.

2.4 *Structuration and CHAT*

Structuration theory and CHAT both provide different perspectives that are based on the notions of interacting agency and structure, and that the social and cultural structures produce organisations and limit them at the same time (Blackler *et al.* 2000). However, CHAT suggests that activities are internally contradictory and these contradictions are a source of innovation and trouble (Engeström 1987, Kerosuo *et al.* 2012). This notion of duality is present in structuration theory and CHAT. The transformative nature of agency and structure in CHAT play an important role as compared to structuration theory. While structuration theory explains the relationship between structure and agency where one restricts the other. CHAT offers a theory for a transformation through the opportunities for innovation and the development of collective competencies and learning which is complementary to the descriptive approach of structuration (Miettinen 1999). In other words, CHAT takes into account the individual changes that affect both structure and agency.

It is important to note that what makes synthesis a requirement is structuration's highly general nature, which does not consider the particular role of agents. Therefore, synthesis of the theories would allow for CHAT to address this limitation.

The above-mentioned theories are relevant to the study of systemic changes in the construction and film industries; the transition between the players associated with a project is dependent on the social, cultural and organisational nature of the network relationships. The integrated perspectives of these two theories may help improve understanding of the needs for the systemic change with the increased uptake of BIM. Moreover, these theories have been applied to the studies in the construction and film industries independently to some extent.

3 CRITICAL REVIEW OF SYSTEMIC INNOVATION IN THE CONSTRUCTION AND FILM INDUSTRIES

3.1 *Comparison of the film and construction industries*

The history of innovation in the film industry could provide numerous examples of occurrences where disruptive innovations have altered organisational management.

A key difference between the industries is the economic and social climate. Significant changes occurred in the film industry between the 1950s and 1960s where major studios disintegrated into becoming only financiers and distributors as compared to their original role, which involved every stage of the film making process. These changes in organisational structure allowed for changes in contractual arrangements between organisations where annual contracts became project based contracts. When analysing these two industries, it is critical to keep in mind the economic and social conditions of the environment at the time of innovation occurrence, as historical events such as the Second World War and rise of TV had direct affects how global industries share resources.

A typical film project would be orchestrated by a producer, who would have the knowledge to decide how the network is organised (Windeler and

Sydow 2001). In the construction industry, it seems that no one stakeholder manages change, although the client is often the driving force behind systemic change (Brandon 2006). The client is directly involved in the whole process, however a client is viewed as an agent who does not have sufficient knowledge or incentives for innovative solutions. The producer in the film industry holds a position as a client does in construction, but, contrary to the construction client, the producer is involved directly at each stage (Watson 2004).

One of the critical differences is how projects are funded and earn money. The stakeholders in a construction project are typically funded based on progress towards project completion. The additional income they have is through enforced or intentional change orders. However, in the film industry, the stakeholders are funded depending on the income of the film which depends on the many thousands or even millions of end-users. This is a critical difference as this changes the motivation of individual stakeholders towards delivering a project successfully. In the construction industry, the production process is oriented around just one client, the assigned schedule and the allocated budget; how well the project will generate money to the client is not affecting the reward of the project team.

Organisational practices in the film industry tend to be driven by the need to collaborate, which is a shared interest because of the financial incentives. On the other hand, in the construction industry, the need to collaborate is in contradiction to the dominating contractual practices which do not incentivise collaboration; which may be due to the difference in incentives with respect to the funding structure.

3.2 *Common phenomena in forms of procurement: fragmentation and transaction cost*

Two types of delivery methods from the film industry and construction industry were compared respectively: 1) studio and 2) independently produced films; and 1) traditional Design-Bid-Build (DBB) and 2) the Integrated Project Delivery (IPD). The studio produced film was made within the one studio organisation whereas the independently produced film involved outsourced services. The DBB delivery method is one that involves a highly fragmented environment predominantly due to the separate social entities in design and construction. On the other hand, the IPD is a delivery method that aims to share risks and gains between all the stakeholders (AIA 2007).

Robins (1993) validated the comparison of a studio-produced film to an independently produced film by Warner Brothers. He reiterated the rarity of this form of data as projects in PBIONs are highly subjective and therefore cannot be compared normally. It should be noted that the derivation of phenomena from the film industry is based on this one study.

Fragmentation in the film industry is motivated by cross fertilisation of ideas. Both construction and film are incentivised in eliminating high fixed costs of in-house resource/expertise. Fixed costs are due to standby specialists that have to be paid when there are no projects running.

We can see that the fragmentation of IPD and Studio are a lot less than independently produced films and traditionally delivered projects, however, it should not be mistaken that the studio produced film has the same form of fragmentation as the IPD form of procurement. When reflecting, one should note that the nature of the projects of both industries differ (as described in Section 1) and therefore a solution in one industry cannot be transferred to the other one directly.

Robins (1993) study concluded that the independently produced film cost more to produce than the studio produced film, however, independently produced films are typically more successful. Robins related transaction cost theory to fragmentation to help understand why the more fragmented delivery method costed more. Transaction cost theory from Coase (1960) explains the higher cost of having a service outsourced as one that is due to the increase in social interaction. Therefore, a higher level of fragmentation would stipulate more social groups interacting. This means that there would be a higher transaction cost associated and the magnitude of transaction costs is proportional to the fragmentation and therefore also to social interaction.

3.3 *Historical Development of Tools in the Construction and Film Industries*

The construction and film industries heavily rely on the use of technologies in practices. From the historical perspective it is clear that technologies play an important role in driving innovation.

The film industry is considered to be a constantly changing industry that started with a technological revolution when Thomas Edison introduced the phonograph in 1877 (Encyclopædia Britannica 2016). The following significant systemic change was Edison's introduction of the kinetoscope in 1894. A year later, the Lumière brothers introduced the first commercial projector. On the contrary, the construction industry is considered to be long established. Major innovations, such as moving from hand drawing to 3D CAD, have been introduced to design and construction incrementally meaning that changes were not considered integrally but independently.

As illustrated in Figure 2, the film industry in the 1960s had an urgent need to distinguish itself from the television industry, which allowed people to watch from their homes. Film production was driven by the need to make films that can compete with TV. It had to innovate by adopting new technologies supported collaboration, which made systemic change in how films were produced. This change brought about the need for a more fragmented PBION. Recently, new technologies allow the film production process to become widely distributed, as it is not necessary to have all the people in one single location. Currently film production is tending towards postproduction to minimise the need to use physical objects.

At the dawn of the 21st century, the film industry was one of the fastest growing industries (U.S. National Archives and Records Administration 2001). Recently published statistics from the US Department of Commerce also show this form of growth. Figure 3 below was developed in order to show the cumulative growth from 2008 to 2014 for both industries based on value added to the US economy.

The construction industry has seen several technological changes that have not brought dramatic improvements. The industry was expected to change with the introduction of BIM in the 1990s. Several decades have passed and the industry's productivity was still in decline (Teicholz 2004). This is further supported by Figure 3, which shows the

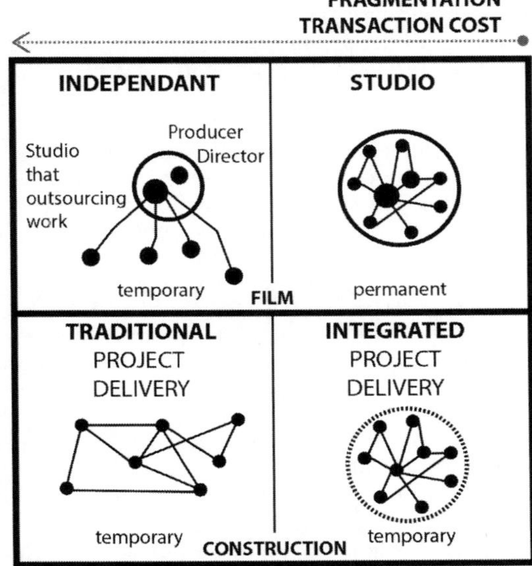

Figure 1. Procurement in Film and Construction.

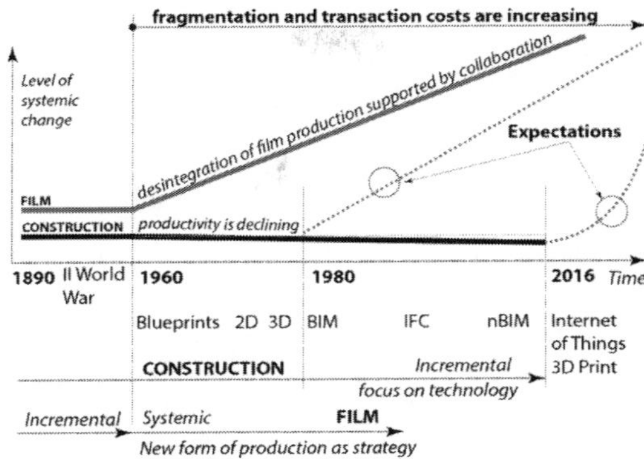

Figure 2. History of technology advance in film and construction industry.

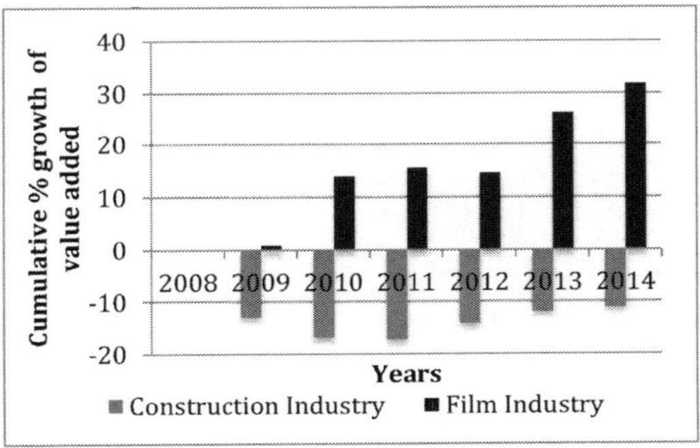

Figure 3. Cumulative % growth of value added to US economy (U.S Department Of Commerce 2016).

negative cumulative growth of construction since 2008. On the other hand, the film industry shows positive cumulative growth.

4 DISCUSSION

Despite all the possible improvements in the industry and multiple efforts such as the introduction of new technologies, research claims that the productivity of construction has been declining since 1968 (Allen 1985). In order to relate this decline to organisational studies it is necessary to look at the history of the network form in both industries. After the Second World War, the film industry disaggregated production, which led to the PBION. In 1960, the construction industry had started to adopt the network form of organisation. Common to both industries was the idea of increased fragmentation, a product of the increasing complexity of technology; due to the increasing number of organisations involved in projects, an increase in stakeholders meant an increase in organisational processes and therefore increased the complexity of processes thereby making it a riskier industry. The network form allowed stakeholders to outsource work which made them less susceptible to liability and reduced fixed costs.

4.1 *Fragmentation and transaction cost*

Figure 4 shows the common elements that motivate PBIONs to be more integrated or fragmented. The film industry in the 1960s disaggregated production (as described in Section 3.3) to support cross fertilisation of ideas to produce more successful movies as small specialist companies are more agile (due to fewer bureaucratic barriers) to the constant changing market needs such as technological changes and generation of new ideas. However, when considering these, we need to consider that motivation towards collaboration could be different as the nature of the product in both industries differ.

While in the construction industry, the process of fragmentation is happening naturally, the industry practices are considered to be conservative and inefficient.

Various researchers discuss fragmentation as a problem; this goes against the low use of the highly un-fragmented procurement method—Integrated Project Delivery (IPD) which is designed to support collaboration and implementation of systemic change. Becerik-Gerber and Kensek (2010) investigated the slow adoption of IPD. Findings suggest that there is a high degree of concern regarding risk in relation to IPD as legal frameworks need to be devised.

Broger (2011) concludes that the more fragmented method (independent) is less cost effective than the less fragmented method (studio). Applying transaction cost theory; the more services run through the market means more transaction cost and since high fragmentation means more separate entities working with one another; this leads to a higher overall transaction cost. In a highly fragmented environment, such as independent filmmaking, there are more transaction costs and since the highly fragmented method is less cost-effective;

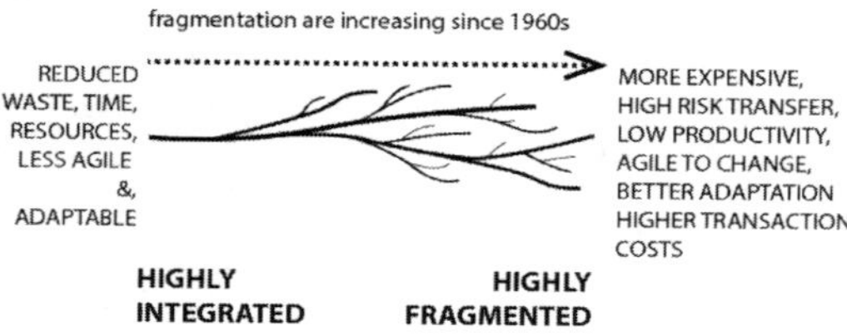

Figure 4. Motivation to fragment or integrate.

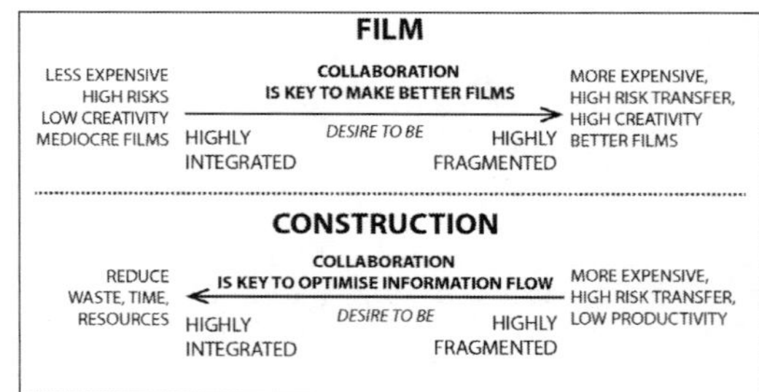

Figure 5. Collaboration in film and construction.

it would be valid to conclude that increased fragmentation would mean increased cost of production due to higher transaction costs. However, it is important to note that fragmentation allows for lower fixed costs and therefore suggests that for a management system to be effective, fragmentation cannot be disregarded.

4.2 *Application of social theory*

Broger (2011) argued that the dualism presented by structuration theory is grounded in the fragmentation of the field concerned. This generalised viewpoint adds to the validity of the application of structuration theory to the highly fragmented film and construction industries.

It is established that in the film industry, the producer is typically one of the agents throughout the project that leads the process. On the other hand, in the construction industry, the agents vary depending on the stage of the project; there is usually no agent that is involved throughout the whole project, for example, in most traditional contracts the contractor (an agent) is not involved in the scope design.

This notion can be related to Taylor's claim in having integrated practices, which would allow companies to establish long term relationships. Long-term relationships would allow for transfer of knowledge and new practices from one project to another that these relationships exist in.

On the level at which stakeholder groups interact, structuration can also be applied. Windeler and Sydow (2001) explained the need for an inter-organisational practice, which has form-specific rules and resources (that control the social structure) produced by knowledgeable agents. It should be noted that the inter-organisational practice is both process and product oriented which means that it considers both process of utilising tools as well as how people collaborate.

Similar phenomena derived from these industries with respect to systemic change are that of fragmentation and transaction cost. These phenomena can be directly linked to the social theory of structuration. However, structuration theory's major limitation in this application is that of a highly generalised concept not taking into account the activity within the system, nor the individuals that are capable of change. Since fragmentation is an important theme in this article, it is not suggested to ignore structuration theory but to fill the gap associated with structuration theory by the use of CHAT. The validity of the synthesis of these two theories is justified in this paper; Giddens (1986) states that structuration theory must be applied both in a micro and macro perspective (considering changes between activities in inter-organisations and accumulated changes that the system becomes familiar too). In this application structuration theory is to be used to explain accumulated changes (macro-perspective) by various agents and how that affects structure. CHAT considers interaction of multiple activity systems and distribution of knowledge and learning between various proponents of the system and tools used, such as implementing a BIM tool, should affect existing structure, for example, rules and division of labour that are devised by both contractual relationships and procurement methods. In other words, the synthesis of the two theories in application to systemic change in project based inter-organisational networks can provide a valuable integrated perspective.

5 CONCLUSION

This paper draws conclusions based on the study produced by Taylor (2005) and has attempted to move the theory forward by bringing perspective from structuration theory and CHAT. It was mentioned by researchers that due to its descriptive nature, structuration theory should have a theoretical companion that might explain the activities that affect the organisational system (Broger 2011). Synthesis of structuration and CHAT can provide a theoretical framework helping to understand why systemic changes diffuse slowly in PBIONs.

Knowledge developed in the film industry on organisational nature and fragmentation was used to reflect on practices in adjacent industries, as by juxtaposing both industries it is possible to see the construction industry from a different angle and potentially bring new understanding. It is obvious that these industries cannot be compared directly and solutions from the film industry cannot be brought directly to construction due to the differences in how both industries are incentivised and motivated. Both industries aim to collaborate to make better products for end-users. However, the key difference is that films are distributed for audience and the revenue is based on the popularity of the product, while in the construction industry the revenue is more or less fixed and based on the location rather than the success of the project.

Although both industries are getting more fragmented, disintegrated and highly specialised, it is obvious that the film industry has found a way to grasp the potential of fragmentation by cross-fertilisation of ideas with the use of cutting-edge technologies, while the construction industry is struggling to find motivation and incentive to adopt new technologies.

Authors of this paper argue that the right balance between fragmentation and integration is required; Integration would allow for easier diffusion of systemic change. The amount of fragmentation and integration required therefore depends on the complexity of the project and its requirements. A piece of evidence to support this notion is Robins (1993) study suggests that independently produced films differ in terms of quality as compared to studio films. However, validation of this claim is limited by the lack of further evidence for Robins' claim due to the lack of non-case study related research methods used to compare projects in PBIONs. Keeping this limitation in mind, how a project is procured affects the final outcome. Since the final outcome in construction is defined from the start, the most efficient way of working (procurement) can be developed from project scope. Suggesting that the most efficient way for a project to be procured depends on the projects' nature/complexity.

From the social science perspective, the slow adoption of BIM is surely related to the business and contractual models that the construction industry adopts. It is suggested by the authors of this paper, that there is need to better understand human relations that make the organisational structure of construction projects in order to propose a better social incentive system that can generate new types of behaviour. This should also be supported by improved methodologies for research (Chan and Cooper 2011). Future research should look beyond transaction costs to a better understanding of how human relations are built at cultural-organisational dimension contributing to the existing knowledge on socio-economic relations.

REFERENCES

Abernathy, W.J. and Utterback, J.M. (1978) 'Patterns of industrial innovation', *Technology review*, 64, 254–228.

AIA (2007) *Integrated Project Delivery: a guide* The American Institute of Architects.

Allen, S.G. (1985) *Why construction industry productivity is declining?*, Internal Report, unpublished.

Becerik-Gerber, B. and Kensek, K. (2010) 'Building Information Modeling in Architecture, Engineering, and Construction: Emerging Research Directions and Trends', *Journal of Professional Issues in Engineering Education and Practice*, 136(3), 139–147.

Blackler, F., Crump, N. and McDonald, S. (2000) 'Organizing Processes in Complex Activity Networks', *Organization Science*, 7(2), 277–300.

Brandon, P. (2006) 'Should Clients Drive Innovation? Mind, Method and Motivation' in Brown, K., Hompson, K. and Brandon, P., eds., *Clients driving construction innovation. Moving ideas into practice*, Cooperative Research Centre for Construction Innovation: Icon.Net Pty Ltd 2006.

Broger, D. (2011) *Structuration theory and organisation research*, unpublished thesis (PhD), the University of St. Gallen.

Chan, p. and Cooper, R. (2011) *Constructing Futures. Industry leaders and futures thinking in construction*, Wiley-Blackwell.

Coase, R.H. (1960) *The problem of social cost*, Springer.

Eastman, C., Teicholz, P., Sacks, R. and Liston, K. (2011) *BIM handbook: A guide to building information modeling for owners, managers, designers, engineers and contractors*, Wiley. com.

Emmitt, S. (2010) *Managing Interdisciplinary Projects: A Primer for Architecture, Engineering and Construction*, Routledge.

Encyclopædia Britannica (2016) 'History of the motion picture'.

Engeström, Y. (1987) *Learning by expanding. An activity-theoretical approach to developmental research*.

Garcia-Lorenzo, L. (2006) 'Networking in organizations: developing a social practice perspective for innovation and knowledge sharing in emerging work contexts', *World Futures*, 62(3), 171–192.

Giddens, A. (1986) *The Constitution of Society: Outline of the Theory of Structure*, Berkeley, CA.: University of California Press.

Hobday, M. (2000) 'The project-based organisation: an ideal form for managing complex products and systems?', *Research Policy*, 29(7), 871–893.

Jones, C. and Lichtenstein, B.B. (2009) 'Temporary Inter-organizational Projects. How Temporal And Social Embeddedness Enhance Coordination And Manage Uncertainty' in *The Oxford Handbook of Inter-Organizational Relations* Oxford Handbooks Online.

Kerosuo, H., Miettinen, R., Paalova, S., Maki, T. and Korpela, J. (2012) 'Challenges of the expansive use of Building Information Modeling (BIM) in Construction Projects', *SPECIAL ISSUE ATWAD—IEA 2012*.

Miettinen, R. (1999) 'The riddle of things: Activity theory and actor-network theory as approaches to studying innovations', *MIND, CULTURE, AND ACTIVITY*, 6(3), 170–195.

Miettinen, R. and Paavola, S. (2014) 'Beyond the BIM utopia: Approaches to the development and implementation of building information modeling', *Automation in Construction*, 43, 84–91.

Podolny, J.M. and Page, K.L. (1998) 'Network forms of organization', *Annual review of sociology*, 57–76.

Putham, L.L. (1986) 'Contradictions and paradoxes in organizations' in Thayer, L., ed., *Organization—communication: emerging perspectives*.

Robins, J.A. (1993) 'Organization as strategy: Restructuring production in the film industry', *Strategic Management Journal*, 14(S1), 103–118.

Taylor, J.E. (2005) *Three perspectives on innovation in interorganizational networks: Systemic innovation, boundary object change, and the alignment of innovations and networks* unpublished thesis Stanford university

Taylor, J.E. and Levitt, R. (2007) 'Innovation Alignment and Project Network Dynamics: An Integrative Model for Change', *Project Management Journal*.

Teicholz, P. (2004) 'Labor Productivity Declines in the Construction Industry: Causes and Remedies', *AECbytes Viewpoint*, 4.

U.S Department Of Commerce (2016) *Interactive Access to Industry Economic Ac-counts Data: GDP by Industry* [online], available: http://www.bea.gov/iTable/iTable.cfm?ReqID = 51&step = 1#reqid = 51&step = 51&isuri = 1&5114 = a&5102 = 10 [accessed April].

U.S. National Archives and Records Administration (2001) 'The migration of U.S. film & television production the impact of 'runaways' on workers and small business in the U.S. film industry.', [online], available: [accessed

Watson, G. (2004) 'Uncertainty and contractual hazard in the film industry: managing adversarial collaboration with dominant suppliers', *Supply Chain Management: An International Journal*, 9(5), 402–409.

Windeler, A. and Sydow, J. (2001) 'Project networks and changing industry practices collaborative content production in the German television industry', *Organization Studies*, 22(6), 1035–1060.

Information & knowledge management (II)—building and urban scale

eWork and eBusiness in Architecture, Engineering and Construction – Christodoulou & Scherer (Eds)
 ISBN 978-1-138-03280-4

An agile process modelling approach for BIM projects

U. Kannengiesser
Metasonic GmbH, Pfaffenhofen, Germany

A. Roxin
Checksem, Laboratory LE2I, University of Burgundy, Dijon, France

ABSTRACT: In the domain of Building Information Modelling (BIM), the open standardisation of methods for product and process modelling is undertaken by the buildingSMART association. Currently, buildingSMART recommends the use of Business Process Model and Notation (BPMN) for creating process models in Information Delivery Manuals (IDMs). This paper argues that BPMN is closely linked to the waterfall nature of today's BIM projects, leading to a number of issues including long project durations, lack of stakeholder involvement, and disconnect of processes and data. Subject-oriented Business Process Management (S-BPM) is introduced as an alternative modelling approach for IDMs, based on its support for agile development that has the potential to address the above issues. Specifically, it is shown that S-BPM supports key concepts of agility, including stakeholder involvement, individual creativity, collaboration, rapid prototyping, and iterative design. The increased agility of S-BPM based IDM development can help make BIM projects faster, more flexible and better adapted to the needs of BIM users.

1 INTRODUCTION

Process modelling has become an increasingly important topic in the domain of AECOO (architecture, engineering, construction, owner and operator). In particular, in the domain of Building Information Modelling (BIM), process models are recognised to provide "a common understanding of the building processes and of the information that is needed for and results from their execution" (buildingSMART 2010). When considering a building's lifecycle, process models provide the scope for defining data ex-changes between stakeholders (Eastman et al. 2010).

For modelling BIM-related processes, the buildingSMART approach relies on Information Delivery Manuals (IDM). Developed by buildingSMART and recognized as an international standard since 2008 (ISO 29481-1:2010), an IDM comprises three main elements: a Process Map (PM), Exchange Requirements (ER) and functional parts (FP). The IDM standard recommends using OMG's (2011) Business Process Model and Notation (BPMN) for defining process maps. While BPMN is a widely accepted approach for modelling well-structured business processes (such as invoicing, order placement, and credit applications), it has significant shortcomings when applied to unstructured or highly dynamic domains such as AECOO. Here, the use of BPMN often leads to long project durations due to a number of issues, including ill-defined execution semantics and high modelling complexity (Zhu et al. 2007; Börger 2012).

This paper argues that "subject-oriented" business process management (S-BPM) (Fleischmann et al. 2012) provides a more suitable approach for process modelling in the built environment. Originally developed in the mid-90s, S-BPM is now receiving growing attention from the academic community as well as from industrial companies worldwide. It differs from traditional process modelling methods such as BPMN in a number of aspects, including executability, simplicity, support for asynchronous communication, interactive tooling support, encapsulation, and integrated process and data perspectives. These concepts provide the basis for agile development of BIM applications, which has the potential to address the aforementioned issues.

This paper is structured as follows: Section 2 outlines the use of process modelling in BIM, including its relation to the IDM approach. A number of issues are identified that can be directly linked to shortcomings of BPMN. Section 3 presents the basics of the concepts and diagrams used in the S-BPM approach. Section 4 shows how S-BPM can provide a better alternative to BPMN, by showing how it supports key concepts of agile development, including stakeholder involvement, individual creativity, collaboration, rapid prototyping, and iterative design. Section 5 concludes the paper by providing a summary and outlook on future work.

2 PROCESS MODELLING IN BIM

2.1 *The IDM approach*

The domain of AECOO involves a considerable number of delivery and operation processes, mainly associated with the management of buildings through their entire lifecycle. Almost every actor from this domain relies on processes: manufacturers need processes for including product data into projects, industry associations depend on process standards' adoption for creating information exchange, while software vendors need to streamline application development processes. Envisioning the delivery of "effective work processes based on collaboration, information technology and standards", the buildingSMART Alliance recognizes in its 2015 strategic plan the importance of the above mentioned needs.

With this objective in mind, process modelling appears as crucial. Today's buildingSMART approach uses IDMs (Karlshøj 2011) for specifying how data can be exchanged among different actors in the context of a building's lifecycle. When two or several actors from the AECOO domain exchange information, the underlying process has to comply with the constraints and requirements defined in the corresponding IDM. In other words, IDMs define what type of information must be contained in a contracted exchange, namely an industry process involving at least two different software applications. Specifications contained in an IDM are then mapped to concepts in the standard model for data exchange, namely the Industry Foundation Classes (IFC) schema (Liebich 2013). This mapping is defined in the form of a Model View Definition (MVD). The concepts of IDM and MVD are further explained in the following paragraphs.

The IDM approach involves producing the following artefacts, as depicted in Figure 1:

- Process Map (PM): is used for defining the industry process to be supported. It should contain a set of activities, roles, and the required data inputs and outputs (notably to ERs).
- Exchange Requirements (ERs): specify the information to be exchanged among the contractors. As a PM identifies several ERs, alignment between PMs and ERs is established using process numbers.
- Functional Parts (FPs): allow mapping IDMs to concepts in Model View Definitions (ISO 29481-3).

IDMs are specified through text-based templates, using the vocabulary and perspective of the professional actor defining the IDM. Until now, several IDMs have been defined, some of which being approved by buildingSMART, others still being at the level of idea or draft. A list of the available IDMs can be found here: http://iug.buildingsmart.org/idms/overview.

When considering information exchange among AECOO actors, the standard information model is the IFC schema (Liebich 2013). This standard comprises hundreds of concepts, as used in real-world building projects and for all building lifecycle phases (e.g. design, construction, management, and demolition). In order to support interoperability across hundreds of software applications and industry domains, the IFC standard allows accommodating several levels of detail and configurations. According to the considered context, a wall can be represented either as a segment between two points, or as a 3D geometry. Also, according to the considered building lifecycle phase, the representation of a wall contains different construction details (e.g. pipe fittings, wirings) along with engineering-, scheduling- and cost-related data.

For addressing these "variables", buildingSMART relies on the definition of MVDs (ISO 29481-3). An MVD is a direct mapping of the exchange requirements expressed textually in an IDM and the corresponding concepts in the IFC schema. It can be considered as a subset of IFC, containing all IFC concepts necessary for the fulfilment of a given ER defined in an IDM. In other words, an MVD specifies a software requirement necessary for implementing a given IFC interface in order to comply with a given ER from a given IDM. In the context of a contracted exchange, one of the parties can simply specify that the data has to be provided according a specific MVD, thus allowing automatic data validation for determining compliance and conformity. MVDs are encoded using the mvdXML format. Mainly used for documenting an MVD, the latest version of the format (mvdXML 1.1) has enhanced the validation of IFC files against ERs defined in a given IDM and the corresponding MVD. Several MVDs exist and have been certified by buildingSMART:

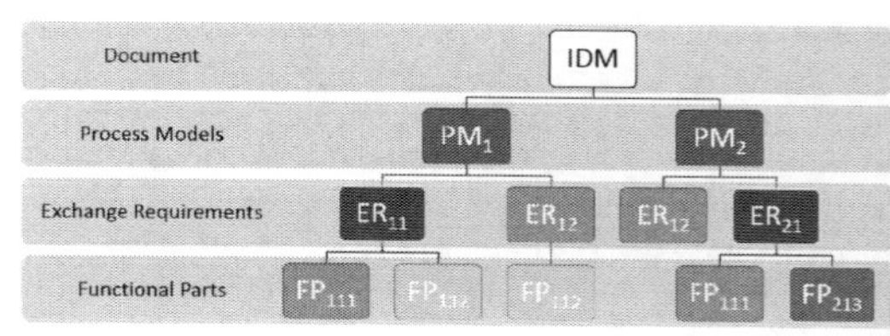

Figure 1. Components of an IDM.

http://www.buildingsmart-tech.org/specifications/ifc-view-definition.

2.2 *Process modelling issues*

For producing PMs, the approach currently recommended by buildingSMART is Business Process Model and Notation (BPMN) (OMG 2011). However, BPMN has a number of known issues that contribute to many BIM projects not meeting expectations in industrial practice. In this Section we point out a few of them.

BIM projects are typically carried out in timeframes of several years (Eastman 2014). As one of the causes one can identify the paper-based definition and exchange of IDM artefacts such as PMs, ERs and FPs. The manual work required for interpreting, maintaining and validating these paper documents is error-prone and time-consuming, especially when many parties are involved as is typical for BIM projects. The method that best describes this way of working is the waterfall approach. In this approach, projects are structured in distinct phases that are separated by stage gates. The typical phases of a BIM project (with their respective timeframes) are shown in Figure 2.

The waterfall approach allows iterations within a phase but discourages iterations across different phases. This is because once a document is produced at the end of one phase, it is regarded as a final agreement that serves as a "contractual" basis for the subsequent phase. Later changes would require not only renegotiation among the parties involved, but also considerable effort to update and validate the consistency of IDM documents.

It is somewhat surprising that BIM development projects still rely almost exclusively on paper-based documents, given that the principal motivation for BIM has been the digitalisation of models and model exchange. Today the benefits of digital information exchange have been realised only for AECOO process applications, but not yet for the development process that generates these applications.

One of the obstacles for digitalising BIM system development is the reliance on a traditional process modelling paradigm that favours producing extensive documentation over rapid design and testing. Exemplary for this paradigm is BPMN,

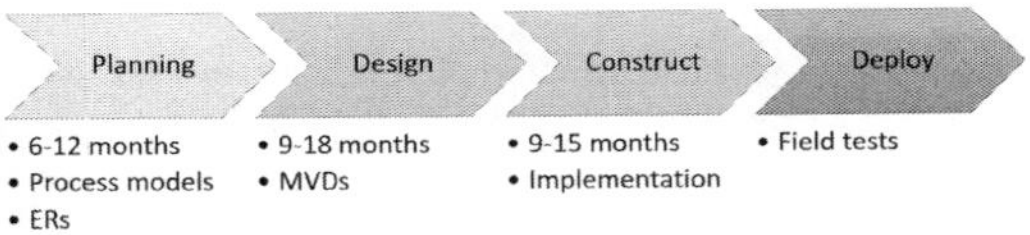

Figure 2. Typical structure of a BIM project (based on Eastman (2014)).

with its focus on graphical rather than computational representations (Börger 2012; Singer and Teller 2012; Sanfilippo et al. 2014). Its main use is in documenting processes for various purposes including analysis, optimisation and compliance (Kocbek et al. 2015). Generating process support software, such as human-centric or automated workflow systems, requires extensive manual work to transform BPMN drawings into computational models. This is because there are considerable gaps in the formal definition of the BPMN notation (Börger 2012; Zhu et al. 2007). In addition, there are fundamental mismatches between BPMN and the Business Process Execution Language (BPEL) (Recker and Mendling 2006) that is commonly used for integrating and executing Web Services in workflow systems.

In addition to its ill-defined execution formalism, BPMN provides inadequate facilities for abstraction (Zhu et al. 2007). This hinders the application of proven concepts for rapid system development that is characterised by frequent cycles of testing and subsequent design changes. These concepts—loose coupling, modularity and architectural design—are all based on the principle of encapsulation that allows separating different functionalities into modules that can be substituted without necessarily affecting other modules. However, the "flat" character of BPMN, where all process parts are exposed using swimlanes, affords producing monolithic rather than modular models. This makes process changes difficult to localise and analyse. As a result, modifying a BPMN model becomes a tedious task. A number of issues inherent in the control-flow paradigm of BPMN (notably regarding the synchronisation of parallel paths) further contribute to difficulties in producing valid, deadlock-free process models (Börger 2012). A recent study (Roy et al. 2014) showed that 81% of a total of 172 BPMN models sampled from industry projects had either syntactic or control-flow related errors.

Despite its declared goal to provide "a notation that is readily understandable by all business users" (OMG 2011), BPMN is very complex (Lee et al. 2013). The complete specification of BPMN defines more than 160 symbols on 532 pages. This often leads to the creation wallpaper-size process maps that are hardly readable or usable. BPMN also requires extensive and costly training. This creates a barrier for many domain experts, such as architects, contractors and engineers, preventing them from being actively involved in the development of a process that adequately reflects their actual work practices. Process acceptance among all process participants is critical for ensuring compliance and preventing ad-hoc workarounds (Muellerleile et al. 2015).

What is needed for BIM projects is a modelling approach that departs from the waterfall nature of process development, by supporting a more agile way of working than is currently possible with BPMN.

3 THE S-BPM APPROACH

Subject-oriented Business Process Management (S-BPM) is a methodology for modelling and executing business processes based on a conceptualisation of processes as a set of interactions between process-centric functionalities that are called "subjects". The subjects in a process encapsulate their individual behaviour specifications (Fleischmann et al. 2012). Subjects exchange messages for coordinating their behaviours. The S-BPM approach is grounded in the Calculus of Communicating Systems by Milner (1999) and Communicating Sequential Processes by Hoare (1978). Abstract State Machines (ASM) (Börger and Stärk 2003) are used as the underlying formalism to allow instant transformation of S-BPM models into executable software. S-BPM mostly targets those applications where a stakeholder-oriented, agile approach to business process management is preferred over traditional methods based on global control flow. An increasing number of field studies demonstrate the use and benefits of S-BPM in practice (Fleischmann et al. 2015). Several open-source and commercial tools are available for modelling and execution support: www.i2pm.net.

S-BPM models consist of two types of diagrams:

- *Subject Interaction Diagram* (SID) specifying a set of subjects and the messages exchanged between them, and
- *Subject Behaviour Diagrams* (SBDs) for every subject, specifying the details of that subject's behaviour as a sequence of states where every state represents an action.

There are three types of states in a SBD:

- *Receive* states for receiving messages from other subjects,
- *Send* states for sending messages to other subjects, and
- *Function* states for performing actions (typically operating on business objects) without involving other subjects.

Examples of a SID and a SBD are shown in Figure 3 and Figure 4, respectively. They represent parts of a material procurement process of a manufacturing company. Here, the SID in Figure 3 includes only the subjects and their interactions via messages. The SBD in Figure 4 shows the internal behaviour specification of the "Inventory" subject; the SBDs of the other two subjects are not shown in this paper. The states in Figure 4 are coloured according to their type: green for receive states, yellow for function states, and red for send states. State transitions are represented as arrows, with labels indicating the outcome of the preceding state.

Communication between subjects may be asynchronous (such as via email) or synchronous (such as via phone). Support for asynchronous communication is provided by the *input pool* concept in S-BPM. An input pool can be viewed as a subject's

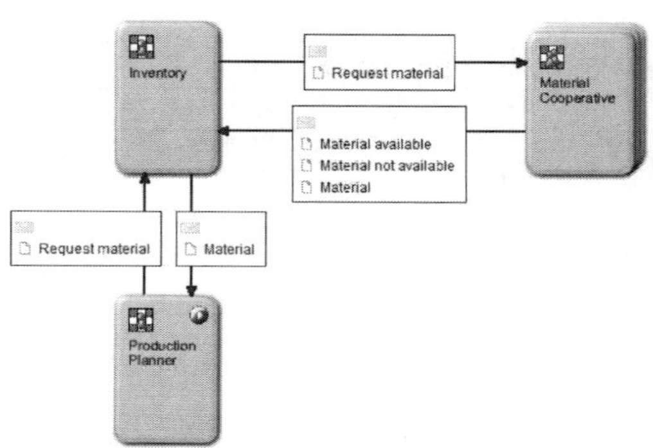

Figure 3. Subject Interaction Diagram (SID) of a material procurement process.

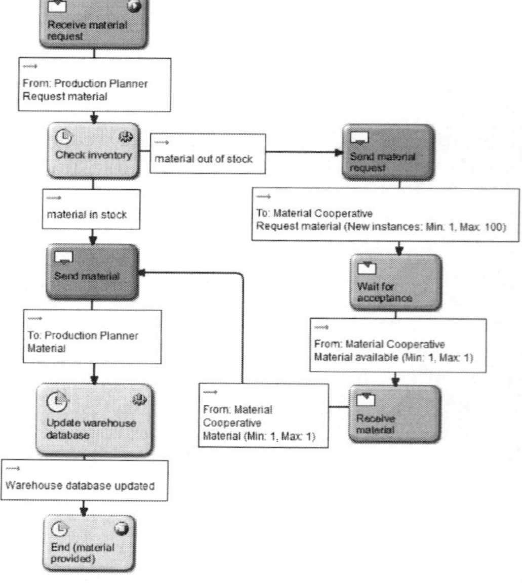

Figure 4. Subject Behaviour Diagram (SBD) of the "Inventory" subject.

inbox for messages. When the subject is in a receive state, it can access its input pool and check for messages. As long as there is no message in the input pool, the subject remains in the receive state. When a message arrives, the subject removes that message from the input pool and follows the transition to the next state as defined in its SBD. The size of the input pool (i.e. the maximum number of messages it can store) can be specified by the modeller. Synchronous communication can be achieved by setting the size of the input pool to zero.

S-BPM allows defining data structures, so-called *business objects*, of arbitrary depth and associating them with some or all of the messages exchanged between subjects and with some or all of the individual states of a subject's behaviour. The business object "Material Information" used in the material procurement process is shown in Figure 5. There is no specific symbol defined in S-BPM for representing business objects diagrammatically.

For every business object an arbitrary number of *views* can be created that define access rights for every field in that business object, including write, read only (or "inactive") and hidden. An example of a view is shown in Figure 6. Views can be associated with individual states in a subject's SBD. Further restrictions (create, display, modify, and delete) can be added for each of these states.

A summary of the principal S-BPM concepts and their definitions is provided in Table 1.

4 S-BPM ENABLING AGILITY IN BIM

Agility has been discussed in a number of domains related to BIM, including manufacturing, project management and software development. Most approaches to agile development share notions of stakeholder involvement, individual creativity, collaboration, rapid prototyping, and iterative design (Fowler and Highsmith 2001). In this Section we will demonstrate how S-BPM has the potential to support these notions.

4.1 *Domain expert involvement*

For involving domain experts in BIM project development, they need to be provided with a quick and easy way to model their processes, without having to conform to overly formal modelling conventions or to struggle with complex tool functionalities. The resulting models may be rough sketches "on

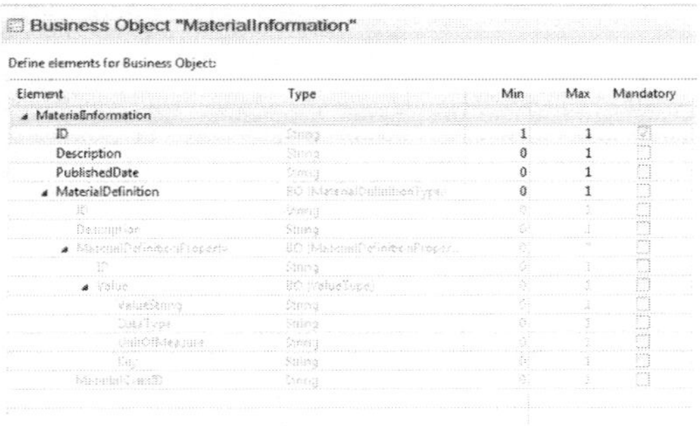

Figure 5. Business object defining a data structure for "material information".

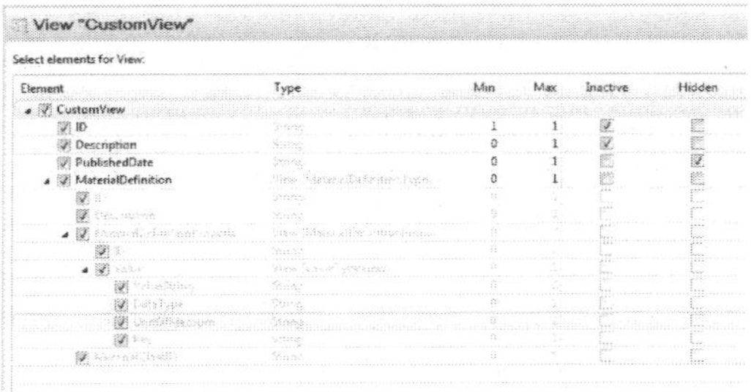

Figure 6. View defined for the business object "Material Information".

Definition of S-BPM concepts.

S-BPM Concept	**Definition**
Subject	Process-centric functionality
Message	Information or material exchanged between subjects
Subject Interaction Diagram (SID)	Diagram showing the subjects and the messages in a process
Subject Behaviour Diagram (SBD)	Diagram showing the behaviour of a subject consisting of states and transitions
Receive state	State in the behaviour of a subject in which it can receive messages
Send state	State in the behaviour of a subject in which it can send a message
Function state	State in the behaviour of a subject in which it can perform actions
Business object	Data structure handled in a process
View	A set of access rights to a business object for a specific subject in a specific context

the back of an envelope" that may even be incomplete or ambiguous. Their main aim is to provide a basis for discussion, reflection and reinterpretation, often among team members in spontaneous, informal meetings. For example, mechanical engineers rarely commence modelling a new product directly using a CAD tool; instead, they commonly produce hand-drawn sketches on paper, whiteboard or other physical surfaces, in order to reflect on their ideas individually or with others. The same applies when people elicit or redesign their work processes. The first time they externalise their view of a process is usually on a piece of paper or whiteboard. Yet, this requires a modelling notation that is simple, and easy to learn and use without necessarily requiring support by a modelling software.

S-BPM provides the notational simplicity required for quick and easy sketching. Since it has only five modelling constructs, their visual syntax can be defined in a way that allows easily distinguishing them from each other. Perceptual discriminability is one of the key factors for the cognitive effectiveness of visual notations (Moody 2009). For example, the syntactic elements used in the example models of the principal reference book on S-BPM (Fleischmann et al. 2012) are perceptually discriminable based on simple shapes. The S-BPM tool Metasonic Suite (www.metasonic.de/en) uses colour as an additional visual variable to further enhance discriminability between the syntactic elements. Producing five basic shapes or colours for modelling can be done using any common sketching tool, such as pen and paper, whiteboards, flipcharts and post-it notes. A number of tangible, interactive modelling tools with embedded object recognition software have been developed to create physical S-BPM models and automatically convert them into computational ones (Oppl 2011; Kannengiesser and Oppl 2015; Fleischmann 2015). An interactive tabletop interface for S-BPM is shown in Figure 7.

The simplicity of the S-BPM method and associated tools democratise process modelling: While traditionally being a craft reserved to only a few highly-trained experts, effective process modelling now becomes accessible to novice modellers after only 20–30 minutes of instruction (Fleischmann 2015). Hence, with S-BPM, all domain experts (including architects, engineers, constructors and facility managers) can be engaged, leading to more accurate process models with fewer errors. In addition, the need for investing in expensive process consultants or trainers is significantly reduced.

4.2 *Multidisciplinary collaboration*

Another benefit of the simple language provided by S-BPM is that it allows various domain experts

Figure 7. Interactive tabletop interface for S-BPM modelling (Source: https://www.metasonic.de/en/touch).

to collaborate, independently of their specific disciplines. This is because S-BPM is based on human communication and organisational theory, providing a common ground of all human interaction. For subjects executed by human actors, the S-BPM concepts appear intuitive as they closely match the individual's perception of organisational reality: One can either *do* something (represented as function states), *send* messages (represented as send states), or *receive* messages (represented as receive states). Even when subjects are executed by software or machines (e.g. an intelligent traffic light in a traffic management process), the cognitive effort needed to conceptualise their interactions in terms of communicative actions can be assumed to be quite low: Using anthropomorphic metaphors is a common human strategy for understanding and predicting the behaviour of computational agents (Dennett 1987).

4.3 *Integrated data perspective*

While BIM defines a clear separation of process models and data structures (IDM and MVD, respectively), there is an increasing interest in linking processes with data to enable reuse and tracking of ERs and FPs (Lee et al. 2013). However, inadequacies of BPMN regarding the explicit representation of data in process models (Meyer et al. 2011) often make it very difficult to establish such links.

S-BPM views processes from a data- and communication-flow perspective rather than solely from a control-flow perspective. It facilitates a clear definition of information exchanges between the stakeholders in a process, which corresponds to the wide-spread view of processes in the engineering community as generating "deliverables"—the outputs of engineering activities to be consumed by other engineering activities (Browning 2001). By treating message exchanges as first-class entities, S-BPM supports common techniques for engineering process optimization, such as value stream analysis (Kannengiesser 2014) and Design Structure Matrix (DSM) (Kannengiesser 2015). As shown in Figure 8, the subjects and messages

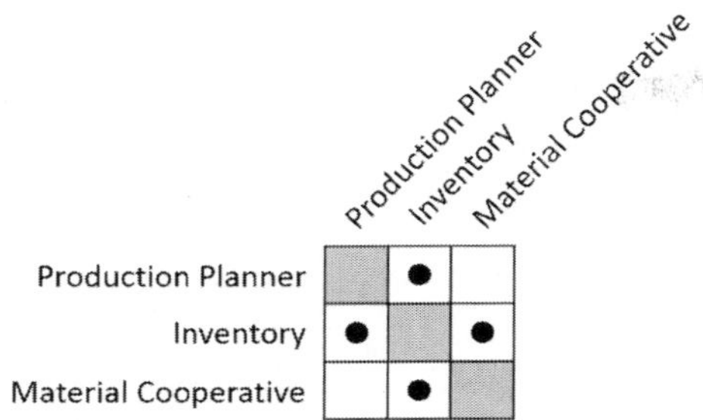

Figure 8. DSM representation of the SID shown in Figure 3.

displayed in a SID can be directly mapped onto a DSM.

The views defined on business objects fit with the notion of MVDs, as they specify restrictions on business object data to form a subset that is tailored to specific subjects in the process. Views in S-BPM allow going even further, in that they can be associated with particular states within a subject behaviour. This results in a more fine-grained, situation-specific definition of views than can be achieved using generic MVDs.

Both views and business objects in S-BPM are always connected to the underlying process model. Once they are specified, they can be reused for further data definitions in the same or other processes, thus reducing modelling time.

4.4 *Process model-driven development*

In most BIM projects, once the ERs, FPs and MVDs are specified, the underlying process models are no longer used or maintained (Zeb and Froese 2014). This has the drawback that any data changes in the later stages of a project cannot be validated in the context of the process in which the data is used. In turn, when there is a late process change (e.g., an additional transaction between two parties), the need to modify the associated data models may remain unnoticed without an up-to-date process model. There are approaches for interlinking process and data models (Lee et al. 2013). However, as they are document-based they are limited to a form of static analysis that does not allow simulating processes and experiencing the effects of design changes at runtime.

The integrated process-and-data perspective of S-BPM, combined with the ASM-based execution formalism, enables process model-driven development of BIM applications. This means that S-BPM process models are not just requirement documents but programs that can be directly executed by workflow engines. This eliminates the model-implementation gap, thus allowing for rapid prototyping of project workflows and their instant assessment by stepping through the specified sequences of subject behaviours (referred to as "validation" in S-BPM; cf. Fleischmann et al. (2012)). As a result, feedback loops back to the planning and modelling stages are minimized, leading to faster development of BIM solutions.

4.5 *Flexibility*

S-BPM supports flexibility of processes based on its use of the well-known principles of encapsulation and loose coupling: Subjects encapsulate different functionalities by exposing only their inputs and outputs (i.e. the messages they receive and send) while hiding their internal behaviour. This allows modifying their behaviour without affecting the rest of the process, as long as the inputs and outputs remain the same. Such loosely-coupled interactions are inadequately supported by BPMN that favours rigidly defined control flows over message passing.

Flexibility is also supported by the clear distinction in S-BPM between subjects (as process-centric functionalities) and agents (as organisational roles that execute subjects) (Fleischmann et al. 2013). As a result, individual agents can be substituted without changing the process model; only the mapping between subjects and agents may need to be modified. For example, the subject "Production Planner" in Figure 3 may be executed by an agent (role) representing an OEM or by an external service provider. This clear separation of process logic from agent structures is not provided by BPMN, where processes are tightly linked to organisational roles.

5 CONCLUSION

This paper has highlighted the need for a new approach to carrying out BIM projects. Among the common issues related to today's approach based on BPMN are long project durations, lack of stakeholder involvement, and insufficient flexibility. Addressing these issues requires a move away from the current waterfall nature of BIM projects, towards a more agile form of system development (Owen et al. 2006). S-BPM can establish such an agile approach. It is not only based on well-known principles of agile software development but also well aligned with existing notions defined in the ISO 29481-1standard (ISO 29481-1: 2010). For example, subject interaction diagrams (SIDs) in S-BPM map onto "interaction maps" in ISO 29481-1, and business objects and views in

S-BPM correspond to "exchange requirements" and "windows of authorisation", respectively, in ISO 29481-1. More generally, S-BPM matches the traditional engineering focus on data and deliverables more closely than the predominantly activity-oriented (or control-flow oriented) focus in the larger BPM community. This has a number of advantages. Particularly, the tight interconnection between processes and data in S-BPM enables the continuous, model-based evolution of data specifications for exchange requirements. This can ease the establishment, validation and maintenance of interoperability in BIM. Finally, the increased agility of S-BPM based development can boost the speed of BIM projects, and make them more flexible and better adapted to the users' needs.

Using S-BPM, with its simple yet well-defined semantics, also facilitates the integration of human-centric and automatic processes. This is useful for supporting end-user centric lifecycle phases of the built environment, notably in the areas of building automation and smart built environments (Zhang et al. 2015). Previous work has already shown that S-BPM can be interfaced with common automation standards such as PLCopen XML (Müller 2012) and OPC UA (Kannengiesser et al. 2015). In addition, the decentralised control of S-BPM based automation systems fits with the growing popularity of agent-based, cyber-physical systems in manufacturing, logistics and construction (Akanmu et al. 2013).

Building on the conceptual foundations laid out in this paper, future work will focus on creating proof-of-concept implementations to demonstrate the benefits of S-BPM. This may occur in several stages. In an initial stage, the ease of modelling of BIM applications using S-BPM can be compared with the traditional BPMN approach. Of particular interest here would be the ease of modelling cross-organisational processes that involves the collaboration of domain experts from multiple disciplines. In a subsequent stage, characteristics of S-BPM related to process model-based development and process flexibility can be assessed for BIM applications. For this purpose, an S-BPM modelling tool may be developed that allows defining and reusing process fragments and associated MVDs, and provides additional support for the validation and simulation of process changes.

ACKNOWLEDGEMENTS

The research leading to these results has received funding from the European Union Seventh Framework Programme FP7-2013-NMP-ICT-FOF(RTD) under grant agreement n° 609190. www.so-pc-pro.eu

REFERENCES

Akanmu, A., Anumba, C. & Messner, J. 2013. Scenarios for cyber-physical systems integration in construction. *Journal of Information Technology in Construction* 18: 240–260.

Börger, E. & Stärk, R. 2003. *Abstract State Machines: A Method for High-Level System Design and Analysis*. Berlin: Springer.

Börger, E. 2012. Approaches to modeling business processes: a critical analysis of BPMN, workflow patterns and YAWL. *Software & Systems Modeling* 11(3): 305–318.

Browning, T.R. 2001. Applying the design structure matrix to systems decomposition and integration problems: A review and new directions. *IEEE Transactions on Engineering Management* 48(3): 292–306.

buildingSMART. 2015. *Alliance Strategic Plan for 2015*, http://c.ymcdn.com/sites/www.nibs.org/resource/resmgr/BSA/bSa2015_StrategicPlan.pdf

Dennett, D.C. 1987. *The Intentional Stance*. Cambridge: MIT Press.

Eastman, C.M. 2014. Business process re-engineering for BIM: New directions in supporting workflow exchanges in IFC. *buildingSMART alliance 2014 National Conference*.

Eastman, C.M., Jeong, Y.S., Sacks, R. & Kaner, I. 2010. Exchange model and exchange object concepts for implementation of national BIM standards. *Journal of Computing in Civil Engineering* 24(1): 25–34.

Fleischmann, A., Kannengiesser, U., Schmidt, W. & Stary, C. 2013. Subject-oriented modeling and execution of multi-agent business processes. *2013 IEEE/WIC/ACM International Conferences on Web Intelligence (WI) and Intelligent Agent Technology (IAT)*. Atlanta, GA, 138–145.

Fleischmann, A., Schmidt, W. & Stary, C. 2015. *S-BPM in the Wild: Practical Value Creation*. Berlin: Springer.

Fleischmann, A., Schmidt, W., Stary, C., Obermeier, S. & Börger, E. 2012. *Subject-Oriented Business Process Management*. Berlin: Springer.

Fleischmann, C. 2015. A tangible modeling interface for subject-oriented business process management. *S-BPM in the Wild: Practical Value Creation*: 135–151. Berlin: Springer.

Fowler, M. & Highsmith, J. 2001. The agile manifesto. *Software Development* 9(8): 28–35.

Hoare, C.A.R. 1978. Communicating sequential processes. *Communications of the ACM*. 21(8): 666–677.

ISO 29481-1:2010. *Building Information modeling—Information delivery manual—Part 1: Methodology and format*. ttp://www.iso.org/iso/iso_catalogue/catalogue_tc/catalogue_detail.htm?csnumber = 45501

ISO 29481-3, Building information modelling—Information delivery manual—Part 3: Model View Definitions. ttp://www.freestd.us/soft4/2085144.htm

Kannengiesser, U. & Oppl, S. 2015. Business processes to touch: Engaging domain experts in process modelling. *BPM 2015*, Innsbruck, Austria, unnumbered.

Kannengiesser, U. 2014. Supporting value stream design using S-BPM. *S-BPM ONE 2014*: 151–160. LNBIP 170, Cham: Springer.

Kannengiesser, U. 2015. Integrating cross-organisational business processes based on a combined S-BPM/

DSM approach. *17th IEEE Conference on Business Informatics*, Lisbon, Portugal, unnumbered.

Kannengiesser, U., Neubauer, M. & Heininger, R. 2015. Subject-oriented BPM as the glue for integrating enterprise processes in smart factories. *OTM 2015 Workshops*: 77–86. LNCS 9416, Springer.

Karlshøj, J. 2011. *Information Delivery Manuals*. http://iug.buildingsmart.org/idms.

Kocbek M., Jošt G., Heričko M. and Polančič G. 2015. Business process model and notation: The current state of affairs. *Computer Science and Information Systems* 12(2): 509–539.

Lee, G., Park, Y.H. & Ham, S. 2013. Extended process to product modeling (xPPM) for integrated and seamless IDM and MVD development. *Advanced Engineering Informatics* 27: 636–651.

Liebich, T., Adachi, Y., Forester, J., Hyvarinen, J., Richter, S., Chipman, T., Weise, M. & Wix, J. 2013. *Industry Foundation Classes, IFC4 Official Release*, buildingSMART, http://www.buildingsmart-tech.org/ifc/IFC4/final/html/

Meyer, A., Smirnov, S. & Weske, M. 2011. Data in business processes. *EMISA Forum*. 31(3): 5–31.

Milner, R. 1999. *Communicating and Mobile Systems: The Pi-Calculus*. Cambridge: Cambridge University Press.

Moody, D.L. 2009. The "physics" of notations: Towards a scientific basis for constructing visual notations in software engineering. *IEEE Transactions on Software Engineering* 35(5): 756–778.

Muellerleile, T., Ritter, S., Englisch, L., Nissen, V. & Joenssen, D.W. 2015. The influence of process acceptance on BPM: An empirical investigation. *2015 IEEE 17th Conference on Business Informatics*, Lisbon, Portugal, unnumbered.

Müller, H. 2012. Using S-BPM for PLC code generation and extension of subject-oriented methodology to all layers of modern control systems. *S-BPM ONE—Scientific Research*: 182–204. LNBIP 104, Springer.

OMG. 2011. *Business Process Model and Notation (BPMN) Version 2.0*, Object Management Group. http://www.omg.org/spec/BPMN/2.0/

Oppl, S. 2011. Subject-oriented elicitation of distributed business-process knowledge. *S-BPM ONE 2011*: 16–33. Berlin: Springer.

Owen, R., Koskela, L., Henrich, G. & Codinhoto, R. 2006. Is agile project management applicable to construction? *14th Annual Conference of the International Group for Lean Construction*. Santiago, Chile, 51–66.

Recker, J. & Mendling, J. 2006. On the translation between BPMN and BPEL: Conceptual mismatch between process modeling languages. *Conference on Advanced Information Systems Engineering*: 521–532, Luxembourg: Namur University Press.

Roy, S., Sajeev, A.S.M., Bihary, S. & Ranjan, A. 2014. An empirical study of error patterns in industrial business process models. *IEEE Transactions on Services Computing* 7(2): 140–153.

Sanfilippo, E.M., Borgo, S. & Masolo, C. 2014. Events and activities: Is there an ontology behind BPMN? *Formal Ontology in Information Systems*: 147–156. Amsterdam: IOS Press.

Singer, R. & Teller, M. 2012. Process algebra and the subject-oriented business process management approach. *S-BPM ONE 2012*: 135–150. CCIS 284, Springer.

Zeb, J. & Froese, T. 2014. Infrastructure management transaction formalism protocol specification: A process development model. *Construction Innovation* 14(1): 69–87.

Zhang, J., Seet, B.-C. & Lie, T.T. 2015. Building information modelling for smart built environments. *Buildings* 5: 100–115.

Zhu, L., Osterweil, L., Staples, M., Kannengiesser, U. & Simidchieva, B.I. 2007. Desiderata for languages to be used in the definition of reference business processes. *International Journal of Software and Informatics* 1(1): 37–65.

eWork and eBusiness in Architecture, Engineering and Construction – Christodoulou & Scherer (Eds)
© 2016 Taylor & Francis Group, ISBN 978-1-138-03280-4

Implications of a BIM-based facility management and operation practice for design-intent models

P. Parsanezhad, V. Tarandi & Ö. Falk
KTH Royal Institute of Technology, Stockholm, Sweden

ABSTRACT: The aim of this paper is to investigate how beneficial contemporary design-intent BIM deliverables could be to FM&O activities and what should be changed with regard to their structure and content for further development of FM&O-intent BIM hand-over. Deliverables of the detailed design phase of a middle-sized educational building were evaluated against requirements derived from a diverse set of resources and directives. The most crucial qualities for the detailed design BIM hand-over documents to be used in FM&O are general concerns about the overall structure of the models as well as classification, attributes and relations of objects. Flexibility and extensibility of the models is also decisive for suitability of the models for FM&O. A number of deficiencies and insufficiencies with the structure and content of the BIM deliverables of the case project have been disclosed. Findings could provide valuable insights for improving the quality of future BIM hand-over documents.

1 INTRODUCTION

1.1 *Background*

Facility Management and Operation (FM&O) constitutes around 5–10 percent of the Gross Domestic Product (GDP) of advanced industrialized countries. The FM&O profession comprises a wide range of activities at strategic, tactical and operational levels from facilities planning, capital budgeting and space optimization to performance evaluation and maintenance. A robust information logistics system for FM&O activities would enable firms to perform activities as such in a more coordinated and resource-effective manner and, thereby, increase their productivity and the quality of their services (Parsanezhad 2014). Devising and implementing such systems has implications for information modelling and management during the earlier life cycle phases of the supply chain of buildings i.e. the planning and design phases.

1.2 *Problem formulation*

Interoperable building information management initiatives based on national and international standards have been widely adopted by design firms in recent decades. Yet, design companies tend to prioritize their short-term needs, economic constraints and contractual considerations over the implications of a total-life approach to buildings. They are often preoccupied by such concerns and constraints when developing building models and their associated databases, also when stipulating their inter- and intradisciplinary information transfer protocols and formats.

Over the past decades, a number of standards such as Industry Foundation Classes (IFC) (Laakso & Kiviniemi 2012), Construction Operations Building information exchange (COBie) (East & Carrasquillo-Mangual 2012) and fi2xml (fi2 2012) have been developed by global and regional actors for supporting and promoting seamless accumulation, evolution and flow of information across the consecutive temporal stages of the building's lifecycle and eventual use of information by FM&O actors. Any further attempt for reconstructing ill-structured models or including missing information later in the process when contracts with the design firms are terminated would be extremely costly, onerous and ineffective,. Yet, no research has been done for evaluating availability and suitability of the required FM&O information within design-intent building models. Currently, the required FM&O information of existing facilities is spread among a mix of digital and paper documents in i diverse set of forms and formats (Parsanezhad 2014). Ideally, an integrated collection of BIM deliverables including all required information in a neutral format should be handed over to the FM&O team.

1.3 *Aim of the paper*

The aim of this paper is to investigate how beneficial contemporary design-intent BIM deliverables could be to FM&O activities and what should be improved with regard to their structure and content so that they could form the basis for prospective FM&O-intent BIM deliverables.

2 THEORETICAL DOMAINS AND CONCEPTS

The concepts addressed and used in this paper have their origins in the two domains of FM&O and BIM. The ontological approach and delimitation of the scope of the paper builds upon two major frameworks in the aforementioned domains, respectively the built environment management model (BEM2) developed by Ebinger & Madritsch (2012) and the BIM Framework developed by Succar (2009). A framework, in this context, could be defined as "the gestalt, the structure, the anatomy or the morphology of a field of knowledge or the links between seemingly disparate fields or sub-disciplines" (Reisman 1994 p.92). The FM&O activities as regarded in this paper span all the four key performance areas (KPIs) of BEM2 at all the three strategic, tactical and operational levels. Among the three fields conceptualized within BIM Framework, this study mainly lies within the technology field. The depth of inquiry, here, corresponds to the microscopic lense as stipulated within BIM Framework. The BIM stage implemented in the case project discussed here is the approximate equivalent of stage 2 according to BIM framework i.e. model-based collaboration. Some actors, however, merely fulfilled the requirements of BIM stages 0 (mere 2D or 3D CAD) or 1 (mere object-based modelling).

Moreover, the two following concept definitions borrowed from COBie specifications are crucial to interpreting parts of the findings:

- Zones "contain groups of spaces that, when connected, provide specific capabilities to the owner" (East & Carrasquillo-Mangual 2012 p.21); and
- Systems are "groups of components that, when connected, provide specific required services" (East & Carrasquillo-Mangual 2012 p.22);

Definitions of other terms and concepts used for describing the findings could be found in IFC2×3 specifications (buildingSMART 2016).

3 METHOD AND RESEARCH DESIGN

Initially, the requirements on building models imposed by needs of the FM&O sector were derived from a variety of data sources i.e. national and international standard specification documents and classification initiatives, earlier research, directives of the client organization for design deliverables as well as BIM guideline documents of the studied project. Next, BIM deliverables of the detailed design phase of the case project (a middle-sized educational building) were comprehensively monitored and evaluated against the requirements mentioned above.

Out of the 12 original BIM deliverables of the case project, building models for major disciplines were selected and analyzed using Solibri Model Checker (Solibri 2016), text-editing software, an application for diagramatic visualization of IFC models called Graphical Instance (Eurostep 2016) and an IFC-explorer providing a simple yet interactive two-dimensional viewer called Floorshow. A Swedish initiative for classification of properties of building components called BIP (BIP 2016) was consulted for this purpose. Figures 1 and 2 demonstrate two snapshots of the procedure mentioned above.

As a complementary step, the possibility of producing COBie deliverables in a low-level format (Excel spreadsheets) for facilitating identification of FM&O information through BIM deliverables was investigated. Solibri Model Checker v9.5 and the COBie extension for Autodesk Revit (Autodesk 2016) were considered for this purpose. Though not all required information was captured in the COBie spreadsheets produced by the two applications.

The findings of analysis of models using the methods and applications mentioned above were complemented with the data collected from the actors participating in the project. The main author was present during BIM coordination meetings of the project and thereby collected the required background data through personal notes and informal talks with the BIM coordinator of the project as well as BIM representatives of the participating firms. The findings were then summarized and presented in tabular form. The resulting table was validated with the BIM-strategist of the project through an in-depth interview session. Figure 3 demonstrates the temporal phase at which the case has been studied as well as the major sources of data used for retrieving information requirements.

4 CASE PROJECT DESCRIPTION

Undervisningshuset (UH) will serve as an educational facility located in the main campus area of the Royal Institute of Technology (KTH) in Stockholm. In the brief document for the project, UH has been envisioned as a flexible and innovative learning environment for students. The building would comprise 7 floors with a total Gross Floor Area (GFA) of 937 m2. The preliminary studies

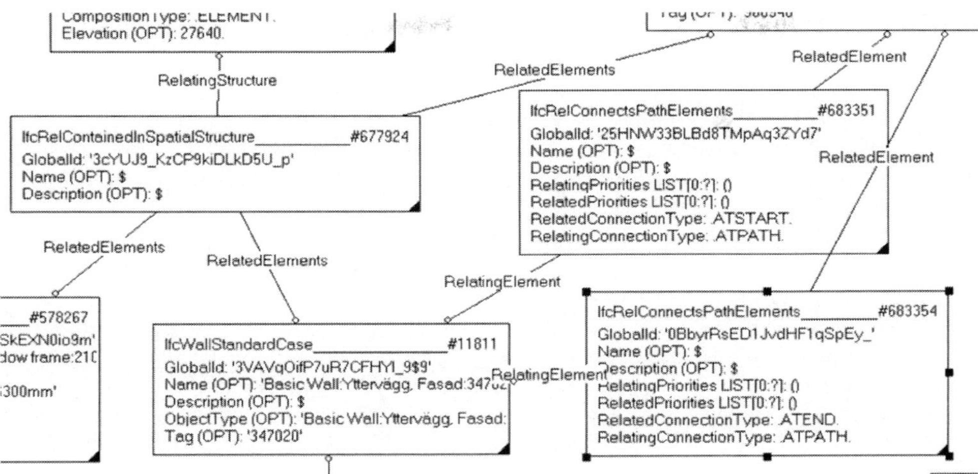

Figure 1. A snapshot demonstrating visual analysis of the data structure of an instance of a 'wall' object using graphical instance.

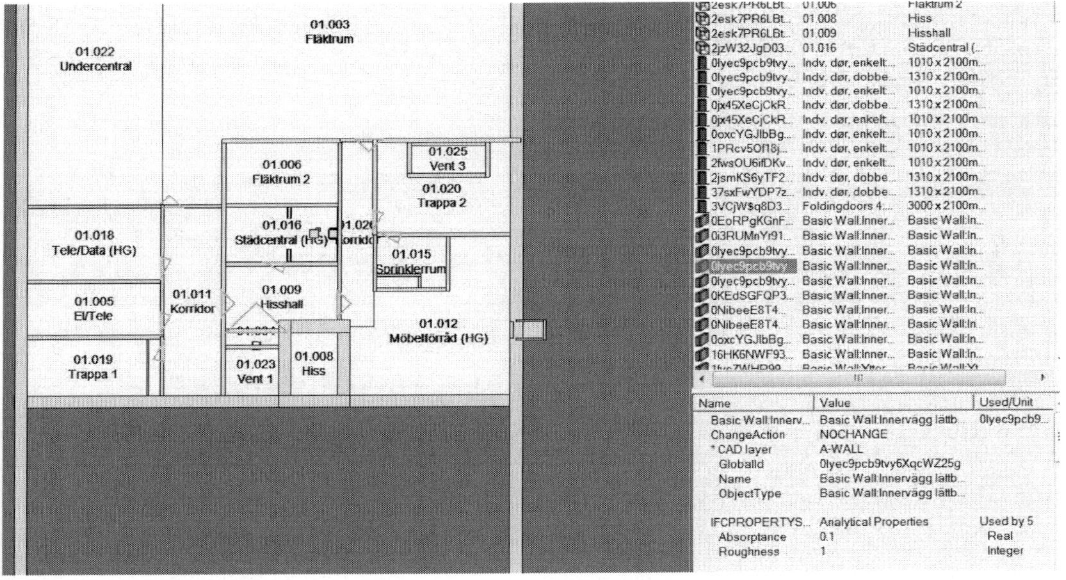

Figure 2. A snapshot demonstrating the process of analyzing properties and propertysets of the BIM-deliverables using floorshow.

for the project were initiated in 2013 and the building is expected to be completed by late 2016. At the time of writing this paper, the building is under construction.

4.1 *Data sources and data description*

The findings of this paper are based on a diverse set of data sources. Of all sources checked initially for identifying FM&O-specific information needs and requirements (see section 3), three sources included the requirements applicable to the detailed design BIM-deliverables: COBie specifications (East & Carrasquillo-Mangual 2012), IFC2×3 specifications (buildingSMART 2016) and the BIM manual of the project (Jongeling et al. 2014). Another source of information was a cloud-based BIM repository where project documents including

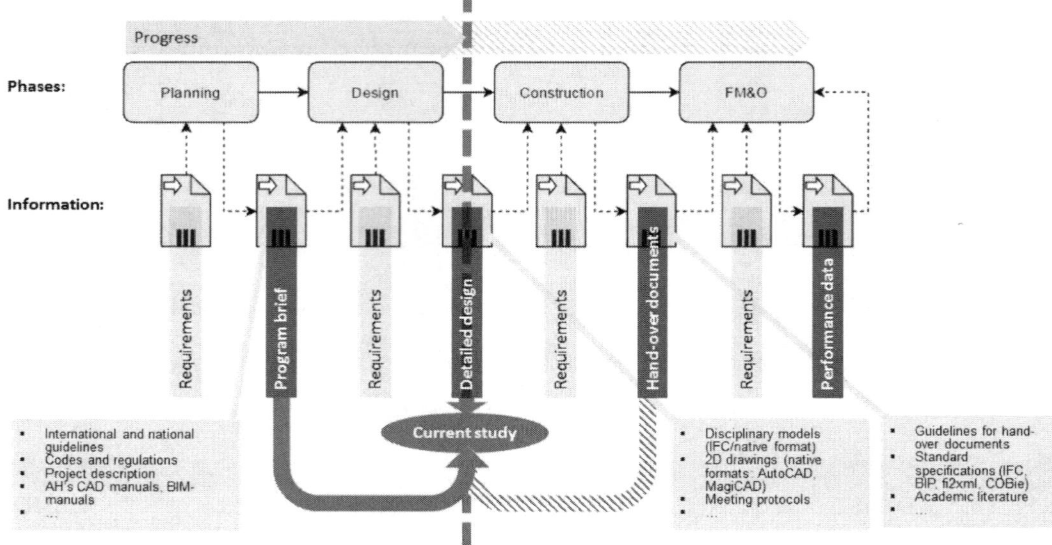

Figure 3. The overall scheme demonstrating the temporal phase of current cross-sectional case study research also disclosing major sources of data used for retrieving information requirements.

disciplinary models were shared and successively updated. Table 1 demonstrates a list of the original BIM deliverables of the detailed design phase together with the BIM-authoring tools originally used by the consultant groups for creating those models. Figure 4 shows an overlaid visualization of all disciplinary models.

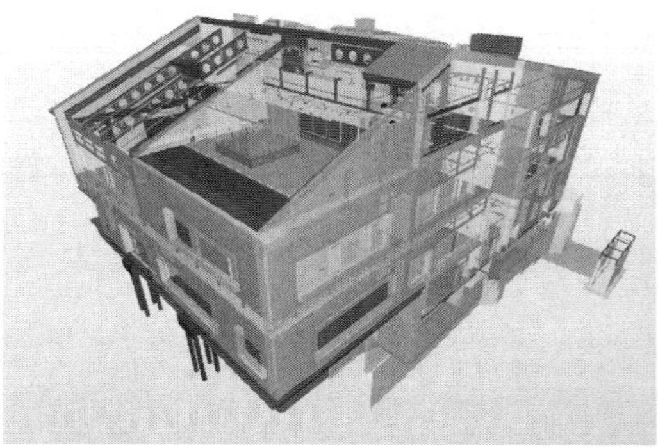

Figure 4. Overlaid visualization of disciplinary models of the project (semi-transparent view).

5 FINDINGS

5.1 *Requirements on BIM deliverables with regard to the FM&O phase*

Prerequisites for an effective use of building models in the FM&O phase could be formulated at two distinct levels: a) requirements on the models' overall structure; and b) requirements on the contents of the models. The latter category could apply to objects, attributes or relations within models. The following section demonstrates the results of evaluation of the detailed design BIM-deliverables of UH against the requirements mentioned above from an FM&O perspective.

5.2 *Results of evaluation of the structure and contents of the detailed design BIM deliverables of UH*

Table 2 demonstrates a summarized account of the findings of this study. Since the studied deliverables are the outcomes of the detailed design phase, requirements pertaining to the ensuing phases i.e. the contents that should be added during procurement and construction were not considered in this evaluation. Below, a summarized explanation of the findings is provided.

Mismatch of the coordination systems used in different models could be partially explained by disparities among the technical approaches taken by different software manufacturers e.g. using local or global coordination systems.

All models fulfill the requirement on unique and consistent floor definitions (as stipulated in Jongeling et al., 2014) except the electrical model

Table 1. Original BIM deliverables of the detailed design phase.

Discipline	Denomination	No. of delivered models	Original modelling software	Version
Architecture	A	1	Revit	2013
Structural engineering	K	1	Tekla	19
MEP engineering	W	2	AutoCAD/MagiCAD	2012/2013.4
Ventilation	V	1	AutoCAD/MagiCAD	2012/2013.4
Electrical engineering	E	3	AutoCAD/MagiCAD	2012/2013.4
Fire suppression	SP	2	AutoCAD/MagiCAD	2012/2013.4
Interior design	I	1	not specified	not specified
Landscape architecture	L	1	AutoCAD	not specified

Table 2. Results of evaluation of the structure and contents of the detailed design BIM deliverables of UH (Part 1).

Requirement	Source	Discipline(s)	Evaluation result
a) The overall structure of the model			
Models should use SWEREF 99 18 00 as coordination system and RH 2010 as the height system with a local origin.	B	All	NF
Floors should be unique and consistent across disciplinary models.	B	A,K,SP,V,W	F
		E	PF
Models should not include duplicate geometries.	C	All	F
Names through deliverables should be unique and consistent.	C	All	NA
GUIDs should be unique and consistent.	I	A,K,E,SP	F
		V,W	NF
Each 3D model should only contain one building.	B	All	F
Divisions of 3D objects should be in accordance with the divisions of the building with regard to levels and spaces i.e. 3D objects should not extend across several floors or several spaces unless in specific cases such as prefabricated components.	B	A,W,K	PF
		E,V,SP	F
3D objects should be modelled as closely as possible to how the building is thought to be built (The aim is to develop a model that is closest to a production model).	B	All	NA
b-1) Objects			
All architectural components that are decisive to the building's design, form and function should be modelled.	B,C	A	F
All structural components should be modeled.	B	K	F
HVAC assets such as chillers, boilers, air handling units, fan coil units, filters, pumps, fans, motors, compressors, Variable Air Volume (VAV) boxes, valves, traps and strainers should be modelled.	C	V,W	F
Plumbing system assets such as water treatment assemblies, valves and plumbing fixtures should be modelled.	C	W	F
Fire suppression system assets such as pumps, valves, sprinkler heads and fire extinguishers should be modelled.	C	SP	F
Electrical system assets such as light fixtures, outlets, switches, distribution panels, switchgear and generators should be modelled.	C	E	F

(*Continued*)

Table 2. (Part 1). (*Continued*)

Requirement	Source	Discipline(s)	Evaluation result
Electrical wires and cables with a diameter more than 40 mm should be modelled.	B	E	F
Food service system assets such as sinks, water disposers, dish-washers, refrigerators, ice-makers, ranges, fryers and freezers should be modelled.	C	A	F
Site assets such as site water distribution system, site fire suppression system, water supply wells and site sanity sewer equipment should be modelled.	C	L	NF
Building components should be modelled with correct modelling tools.	B	K,E,SP,V,W	F
		A	PF
Spaces, systems and zones should be modelled.	C, B	All	PF
Production results such as panels should be modelled when it is required for 3D coordination.	B	All	F
Reserved spaces for maintenance, transportation of equipment and logistics should be modeled as space objects.	B	All	NF
b-2) Attributes			
Project and magnetic north directions should be included in models.	C	All	F
Facility geo-location (longitude, latitude, elevation, rotation) should be specified.	C	All	F
Regional, national and/or client-specific property sets should be included.	C	All	PF
Space, system and zone asset types and attributes should be included.	C	A	F
Room objects should have geometric information, room types and numbers according to project descriptions as well as classification codes according to the national BSAB96 classification system.	B	A	F

that consisted of 20 floors. The multiple floor definitions for different categories of elements i.e. lighting fixtures, cables and under-floor cables could have been merged together. The only occasions where the requirement on unique GUIDs (as stipulated in buildingSMART, 2014) is not fulfilled are the MEP model with 279 duplicated GUIDs for the water supply system elements and 74 duplicated GUIDs through the ventilation model.

In architectural, structural and water supply system models, several occasions of elements associated with wrong floors were observed. Some exterior and interior walls in the architectural model extend across several floors (Fig. 5) which contradicts the requirement of correct spatial association of elements. The structural model includes an additional floor called 'Floor 7' which contains miscellaneous elements such as holes, frames of the skylights and stairs. Divisions of objects within the electrical engineering, ventilation systems and fire safety systems models are in accordance with the divisions of the building with regard to levels and spaces (as stipulated in Jongeling et al., 2014).

Two requirements on the overall structure of models were deemed unnecessary and not checked: building elements are and will be primarily identified by their GUIDs for all future uses including FM&O. Object names could, on the other hand, be discipline-specific or colloquial and need not be unique as required by COBie specifications (East and Carrasquillo-Mangual, 2012). Also, the requirement on models being as close as possible to production models as stated in the BIM manual of the project (Jongeling et al. 2014) was deemed irrelevant for the purpose of this study. In practice, the requirements for enabling future development of the design-intent models to construction-intent models and those of developing FM&O-intent models should be balanced against each other. A practical tool for regulating such a trade-off is providing extensive specifications of the levels of detail/development (LODs) for all major elements in the BIM manual of the project (Jongeling, 2016).

Table 2. Results of evaluation of the structure and contents of the detailed design BIM deliverables of UH (Part 2).

Requirement	Source	Discipline(s)	Evaluation result
b-2) Attributes (cont.)			
Classification (category) codes should be included.	C,B	A,K,V,E,SP	PF
		W	NF
All building components should have tag number-codes and descriptions.	B	All	F
Units of measurement should be included.	C	All	F
Doors and windows should be accompanied with their geometric information (dimensions), material, function (e.g. hinged door, folding door, sliding door or rotating door), fire class, security class and swing direction.	B	A	PF
b-3) Relations			
Relations such as spatial containment, zone association, systems association, spatial placement (e.g. under-floor, above-ceiling, in-wall, on-roof, in-space or on-site) and site spatial containment (e.g. parking lots. loading docks) should be specified.	C	All	PF
Relations of objects should be consistent.	I	A,K,V,E,SP	F
		W	NF

Abbreviations: B = BIM Manual—Undervisningshuset (Jongeling et al. 2014); C = COBie specifications (East & Carrasquillo-Mangual 2012); I = IFC2 × 3 specifications (building SMART 2016)
A = Architecture, E = Electrical engineering, K = Structural engineering; L = Landscape architecture, SP = Fire safety systems, W = Water supply system
V = Ventilation systems, F = Fulfilled, PF = Partially fulfilled, NF = Not fulfilled, NA = Not applicable.

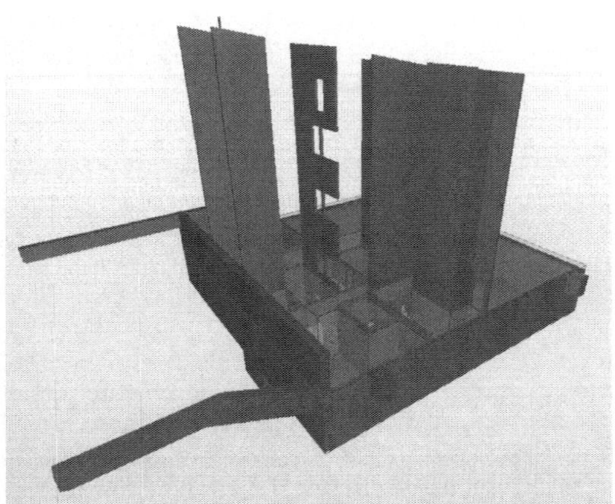
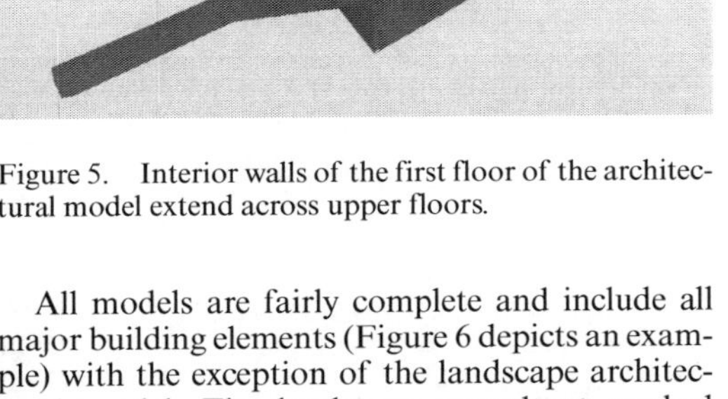

Figure 5. Interior walls of the first floor of the architectural model extend across upper floors.

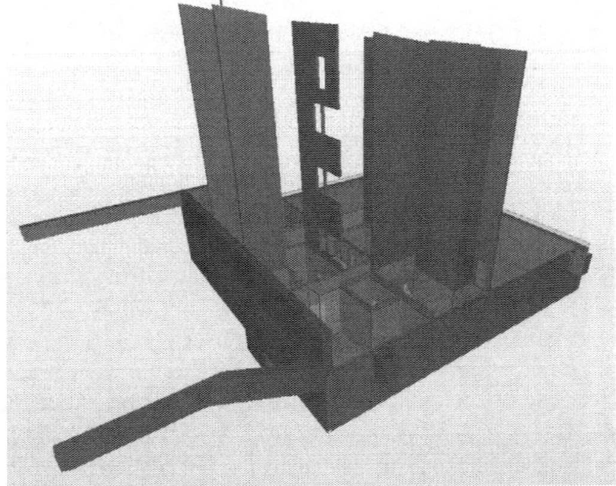

Figure 6. Electrical system assets and wires and cables with a diameter more than 40mm have been modelled.

All models are fairly complete and include all major building elements (Figure 6 depicts an example) with the exception of the landscape architectural model. The landscape consultant worked exclusively with a two-dimensional application (AutoCAD). This was because the BIM software at the time did not include landscape object types. Technologies for creating semantic models of site and infrastructure components in formats that are compatible with BIM deliverables are still in their infancy (Jongeling, 2016).

BIM operators are directly responsible for some of the violations from the modelling rules and requirements such as incorrect semantic definitions of some architectural components. The upper roof has, for example, been modelled with the 'beam' tool. Required spaces (IfcSpace) and systems (IfcSystem) have been modeled. No zone (IfcZone) has been modeled. Fire compartments, for example, are missing. This could be attributed to the relatively lower BIM competency of fire safety consultants. In the case of the studied project, fire safety system components specifications were handed over in two-dimensional drawings which could hamper interoperability when the contents of the deliverables need to be integrated and used for FM&O purposes. Reserved spaces for maintenance, transportation of equipment and logistics are other examples of the missing zone objects that would be required in the FM&O phase. The majority of components are modeled in a sufficiently high level of detail to represent production results.

Most of the requirements on object attributes have been fulfilled. In accordance with the requirements outlined in COBie specifications (East and Carrasquillo-Mangual, 2012), 'room' objects contain attributes representing volume, area, perimeter, room type and room number in compliance with the Swedish classification system, BSAB96.

All major building elements have tag number codes and descriptions. With the exception of the water system components, all objects have BSAB classification codes. The codes, however, appear as different attributes for different objects often depending on the modelling software that has been used. Dimensions of doors and windows are specified through attributes such as OverallHeight and OverallWidth. IfcMaterial holds material types. Function and swing direction have been captured by the enumeration, IfcDoorStyleOperationEnum with a range of selectable values e.g. SINGLE_SWING_RIGHT. Values for fire rating, security rating and sound rating are however missing. Attributes of IfcGeometricRepresentationContext show indications that the north direction could be retrieved from the content of the models. Attributes of IfcSite show indications that the facility geolocation could be retrieved from the content of the models. Units of measurement are often included within the value of quantitative attributes. Some occasions of project-specific property sets were observed n models e.g. KTHU in the structural model.

With the exception of the water supply system elements that have duplicated relations (IfcRelContainedInSpatialStructure), all other relations among objects are consistent. Spatial relations and zone association are not explicitly specified in models, but can be retrieved from names, attributes and attribute values of components.

Of totally 32 requirements considered here, 17 were fully met by the BIM deliverables of the studied case, 10 were partially met, 3 were not fulfilled at all and 2 were deemed non-relevant for this study.

6 CONCLUSIONS

Findings of this study demonstrate that the most important concerns about the detailed design BIM hand-over documents to be used in FM&O are general issues with the overall structure of the models (e.g. mismatch of coordination systems and incorrect spatial association of objects), semantic definition and classification of objects and their attributes (e.g. duplicated names, inconsistent attribute names for different objects, missing values of attribute) and relations (e.g. duplicated relations).

Developed through an in-depth monitoring and examination of BIM deliverables of a case project, this study discloses major deficiencies and insufficiencies with the structure and content of BIM deliverables of the case project with regard to future use of models in FM&O. The findings could, thereby, provide valuable insights for improving the quality of future BIM hand-over documents.

7 FURTHER REMARKS

It should be noted that some of the requirements on BIM models were excluded from this study. Project-specific requirements corresponding to traditional CAD deliverables e.g. specifications of annotations, external references and drawing blocks were, for instance, not considered. Moreover, a substantial amount of the required FM&O information is submitted during the procurement and construction phases and could therefore not be expected to be included within the deliverables of the design phase. Flexibility and extensibility of the models is also a decisive criterion for accommodating the information that would be submitted during consecutive phases and eventually leveraged by FM&O agents. This quality could, however, not be evaluated by the methods used in this study.

Different types of requirements on the commissioned facility are yet another category of information to be archived together with other information and eventually transferred to FM&O agent. Requirements could be retrieved from the design guidelines of the client organization of the case project (Hallen 2011). Such information could be used within KPI 3 (project transaction management) for verification of the building's performance (Ebinger & Madritsch 2012). In the case project studied here, requirements specifications e.g. space program, relationship chart, functional requirements of the building as a whole and individual spaces, accessibility requirements for services and maintenance as well as cost and rent estimations were registered in the cloud-based requirements management system, dRofus (dRofus 2016) and largely implemented during the planning, system design and detailed design phases. Contemporary formats for building information transfer do however not yet possess the capacity for capturing building requirements information. There are some industry standards (e.g. PLCS) that offer such capacities and could thus be also implemented in the AECO industry (Tarandi 2011).

PERSONAL COMMUNICATION

Jongeling, R. (2016). Interview, May 20, 2016

REFERENCES

Autodesk 2016. Autodesk Revit [webpage]. URL https://knowledge.autodesk.com/support/revit-products/learn-explore/caas/simplecontent/content/cobie-extension-for-revit.html.

BIP 2016. BIPkoder [webpage]. URL http://www.bipkoder.se/#/

buildingSMART 2016. Industry Foundation Classes (IFC)—buildingSMART [webpage]. URL http://www.buildingsmart-tech.org/specifications/ifc-overview/ifc-overview-summary (accessed 11.26.14).

dRofus 2016. dRofus [webpage]. URL http://www.drofus.no/en/ (accessed 5.2.14).

East, B., Carrasquillo-Mangual, M. 2012. The COBie Guide: a commentary to the NBIMS-US COBie standard.

Ebinger, M., Madritsch, T. 2012. A classification framework for facilities and real estate management: The Built Environment Management Model (BEM2), *Facilities*, Vol. 30: 185–198. doi:110.1108/02632771211208477

Eurostep 2016. Graphical Instance [webpage]. URL http://www.plcs-resources.org/plcs/dexlib/help/dex/sw_graphinst.htm

fi2 2012. Handbok, fi2xml version 1.3 Del 1 Översikt, fi2 Förvaltningsinformation.

Hallen, T. 2011. Riktlinjer för projektering 31/10 2011, Akademiska Hus, Göteborg.

Jongeling, R., Fransson, E., Buske, E. 2014. IT Handledning/BIM Manual, KTH Undervisningshuset 43:32.

Laakso, M., Kiviniemi, A. 2012. The IFC standard—A review of history, development, and standardization, *Electronic Journal of Information Technology in Construction*, Vol. 17: 134–161.

Parsanezhad, P. 2014. An overview of information logistics for FM&O business processes, in: *Proceedings of the 10th European Conference on Product and Process Modelling (ECPPM 2014)*. Presented at the Proceedings of the 10th European Conference on Product and Process Modelling (ECPPM 2014), CRC Press, Vienna, Austria, pp. 719–725.

Reisman, A. (Ed.) 1994. Creativity in MS-OR: Expanding Knowledge by Consolidating Knowledge, Interfaces: the INFORMS journal on the practice of operations research, INFORMS, Linthicum, Md.

Solibri 2016. Solibri Model Checker [webpage]. URL http://www.solibri.com/products/solibri-model-checker/

Succar, B. 2009. Building information modelling framework: A research and delivery foundation for industry stakeholders, *Automation in Construction*, Vol. 18: 357–375. doi:10.1016/j.autcon.2008.10.003

Tarandi, V. 2011. The BIM Collaboration Hub: a Model Server Based on IFC and PLCS For Virtual Enterprise Collaboration. Presented at the CIB W78-W102 2011: International Conference, Sophia Antipolis, France.

eWork and eBusiness in Architecture, Engineering and Construction – Christodoulou & Scherer (Eds)
© 2016 Taylor & Francis Group, ISBN 978-1-138-03280-4

Seamless integration of common data environment access into BIM authoring applications: The BIM integration framework

C. Preidel & A. Borrmann
Chair of Computational Modeling and Simulation, Technical University of Munich, Munich, Germany

C. Oberender & M. Tretheway
ALLPLAN GmbH, Germany

ABSTRACT: In today's construction industry collaborative processes have received increasing attention due to new digital methods such as Building Information Modeling (BIM). This method bases on the application of digital 3D building models enriched by semantic information. Since a construction project is a composition of several collaborating processes executed by many project participants, a federated model approach has emerged as the most practical solution. It is widely recognized, that for the implementation of this approach and the related collaborative processes digital platforms are required. The Common Data Environment (CDE) is defined as a common digital project space, which provides well-defined access areas for the project stakeholders combined with clear status definitions and a robust workflow description for sharing and approval processes. Since most of today's software solutions lack direct accessibility and integration, we introduce a framework that allows for seamless integration of CDE access and management functionality into standard BIM authoring and analysis applications.

1 INTRODUCTION

Due to the rapid development of digital methods in construction industry during the last years, collaborative processes have become a highly current topic. A substantial part of the technological transformation in construction sector is the Building Information Modeling (BIM) method, which bases on the application of 3D digital building models enriched by semantic information. These models are able to store any information, which is created during the execution of a construction project, and therefore represent a comprehensive digital description of the facility to be constructed. Many of today's work and communication processes can be improved by the help of these structured building models. Therefore, the BIM method can serve as an essential tool for collaborative processes (Borrmann et al. 2015, Young et al. 2009). During the execution of a construction project, the involved engineers and architects cooperate for a long period in various disciplines and different formations. Moreover, a construction project is a composition of several phases starting with the design, following the construction and finally the operating stage. Due to the large quantity of disciplines, project participants and different phases, the requirements for the technical support of this collaboration in terms of consistency and coherence are very high. An essential requirement for a collaborative BIM-based planning is a resilient and technically supported definition of management, data and communication processes. Therefore, technical solutions must be developed in order to implement these. It is widely recognized, that a common data platform can be a solution, since it brings all information together and serves as central data management tool.

The essential requirements and challenges of a digital BIM-based collaboration are presented in this paper. Based on that, current academic as well as commercial approaches for the implementation of such data platforms are discussed. Finally, the authors' own approach towards overcoming the insufficiencies of the presented software products is introduced.

2 BIM-BASED COLLABORATIVE PROCESSES

2.1 *Model-based collaboration*

Starting point for the model-based collaboration is the digital representation of the overall construction process. This representation is created in cooperation of several project participants, who use various BIM-based authoring tools. It has meanwhile become clear, that a direct usage of a single model is not recommended for a number of reasons. Since the working areas of different disciplines may overlap in a single model, the responsibilities cannot be assigned clear and unambiguously. Yet another reason is, that the project stakeholders have usually no interest, to share their intermediate models due to framework conditions such as the contract constellation. Therefore, various guidelines such as the Singapore BIM Guide (BCA Singapore 2013) or the British Publicly Available Specification (PAS) 1192 (British Standards Institution 2013) implement a domain-specific approach, which means, that model authors have full access only to the domain-specific sub-model they responsible for. These subsets describe a certain aspect of the overall model and so they are usually called discipline or domain-specific partial model. According to this resulting federated model approach each domain-specific model is maintained only by an assigned author himself. In this way the responsibilities and authorship of building elements as well as changes during the execution of the construction project are managed unambiguously.

To ensure the integrity and consistency of the overall model, the domain-specific models have to be compared and checked for inconsistencies or collisions at defined regular intervals. For this purpose, the federated model approach provides a central coordination model (Solihin et al. 2016). This model stores the validated and agreed merged information and in turn provides exactly the information, which is required for the collaboration of at least two disciplines. In this way, the coordination model serves not only as a starting point, but also as a checking base for inconsistencies in single domain-specific models. Due to the outstanding importance of the coordination model, its quality must unconditionally comply the agreed project standards at any time. To catch possible errors as early as possible, a further important safety measure is the quality barrier or gate. A domain-specific model has to pass this gate before it can be merged with the coordination model. The quality gate represents specific checking procedures regarding the agreed project standards such as a completeness, clash or suitability check. Depending on the project agreements this first compliance check can be a prerequisite for the merge and therefore has to be executed on side of the discipline model author. As shown in Figure 2, shows the development of the domain-specific due to this intermediate steps until it can finally be merged with the other valid data since it complies the project standards by now.

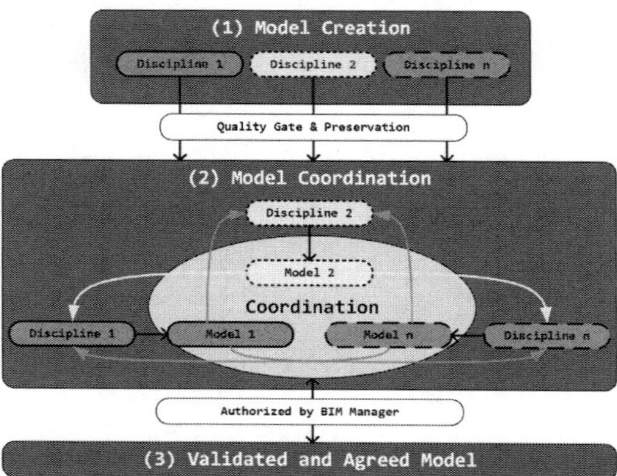

Figure 1. BIM modelling and collaboration procedures according to BCA Singapore (2013).

2.2 *Common data environment*

As discussed in Section 2.1 the management of the digital information and the related processes is a major task during the BIM-based execution of a construction project. It is widely recognized, that for the storage and administration of the different models and to enable the above-mentioned processes, digital data platforms are required. At this point the PAS 1192 (British Standards Institution 2013) specifies a technological foundation, the Common Data Environment (CDE). The CDE basically represents a central space for collecting, managing, evaluating and sharing information. All project participants retrieve the data from the CDE and in turn store their data here. The CDE stores the coordination model, all domain-specific partial models, data bases and documents, which are necessary during the execution of the project. Therefore, it describes the comprehensive BIM-process. The centralization of data storage within the CDE reduces the jeopardy of data redundancy and ensures the availability of up-to-date data at any time. Furthermore, the CDE leads to a higher rate of reusability of

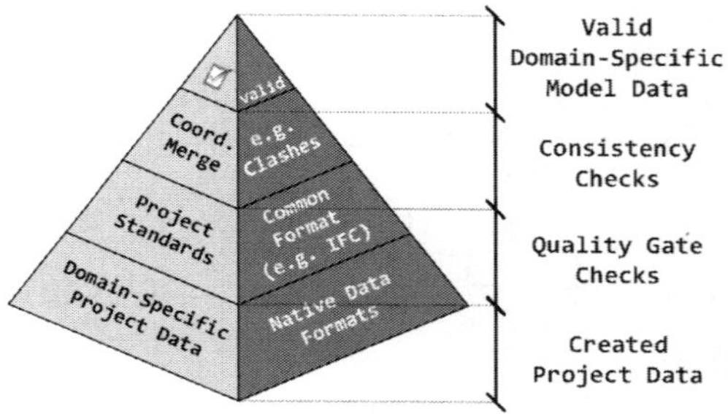

Figure 2. Project data development during collaboration in a BIM-based construction project.

information, simplifies the aggregation of model information and simultaneously serves as a central archive for documentation. Since this environment is accessible for all of the project participants, it can be used as a platform for BIM-based collaborative processes. It should be noted, that the PAS 1192 basically provides recommendations for the technological as well as management-process-based implementation. In this sense, the guideline describes a broad framework for a CDE, but does not set detailed requirements, so that there is room for interpretations and the technical implementation. This setup of a CDE can depend in a variety of ways effected by the application, project volume and participated parties. Nevertheless, a more detailed implementation of such a CDE shall be presented exemplarily. In general, the architecture of a CDE can be described as a layered structure (see Figure 3). On top of the data keeping, the structuring of the stored information is an essential part of the CDE. This structuring has to be agreed on in the beginning of a project and should be updated—if needed contractual—continuously. Next to the actual storage of the information, different processes and workflows, e.g. for the share of information, reviewing, versioning or archiving should be well-defined. Furthermore, the merging process of various discipline models as described in Section 2.1 should be technically defined in here.

In this way the CDE fulfills the requirements, which were set in Section 2.1, and the first basic prerequisites for a cooperative processing of the information is enabled.

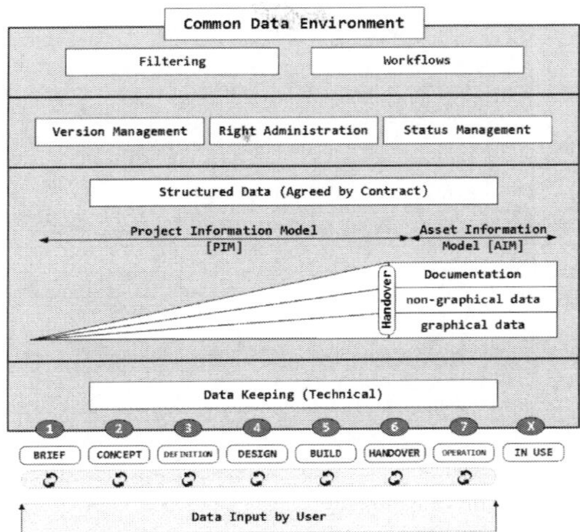

Figure 3. Layer model of a common data environment, inspired by (British Standards Institution 2013).

2.3 *Major challenges*

The presented collaboration processes in the previous sections laid down various requirements and created a framework for technical solutions in order to provide a common environment for BIM-based collaboration processes. A major challenge in this collaboration is the preservation of the overall model quality and the data consistency. An essential tool to keep the model quality continuously on a high level are the agreed project standards, which are valid for any piece of information by merging them with the coordination model at the latest. These standards set legally binding requirements for the way data is stored. Therefore, the created model information has to be checked iteratively regarding these standards. To do so, not only well-defined checking processes are required, but also that any participant is aware of the quality of his created information. Therefore, each author should know and take care about the agreed project standards and must be able to deal with the basic principles of data handling. In order to sensitize the participants for this data handling, the structured information as well as the transportation mechanisms have to be as transparent as possible, so that a user can have an insight and understanding, what the information is and how it is processed. To achieve this, the CDE should be connected smoothly with any kind of BIM-based authoring tool, which is used by the project participants, in order to shorten process paths and to keep the data handling transparent. In order to provide a transparent information transportation, the CDE must provide a full and open access, so that any user is able to have an insight in the processed data at any time. If the data itself or the procedures are black-box processes, the user cannot overtake the responsibility for his data according to his owner- and authorship. This applies in particular for the handling of native data. To prevent this, a smooth implementation of the CDE features directly in the environment of a specific authoring tool reduces significantly the effort for users and therefore represents a step towards breaking down acceptance barriers since unnecessary processing steps are avoided. Furthermore, the communication channels are significantly shortened with this direct implementation and feedback can be directly given to the user.

Another important issue is the consistency of data, which implies that stored information is unified and therefore valid as a whole. Due to the federated model approach, these inconsistencies occur especially, when the domain-specific models are merged with the coordination model. An example for such an inconsistency is shown in Figure 4. In this example user A modifies a model object Y, which only is valid in the context of the

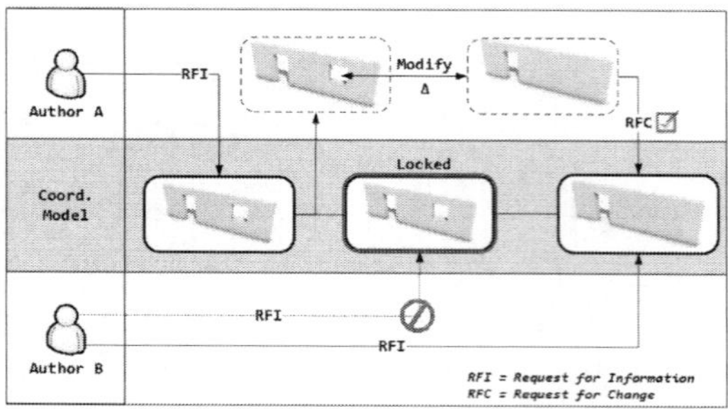

Figure 4. Example for an inconsistency in a BIM-based collaboration process.

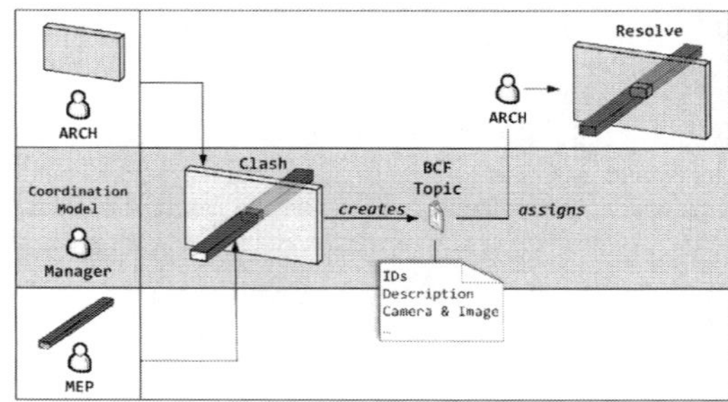

Figure 5. Exemplary locking mechanism for consistency preservation in collaboration processes.

model object X. If user B tries subsequently to access object Y, an inconsistency occurs since the data is not valid anymore. In the given example author A deletes the opening, whose properties are modified by author B at the same time. If the delete operation is executed before the property modification, the handled data object is not valid anymore and so an inconsistency error is raised. To solve these inconsistencies, an administration of data and changes, which we call delta management in the following, can be helpful. The authorship for a subset model is not only related to specific user rights, but also to a certain set of information. The size of this information set can vary considerably from a whole building section to a single granular piece of information. Depending on the discipline and the amount of participants, the granularity of these authorships can be finely organized and even shared with other project participants. To handle the simultaneous data access, a locking mechanism can be used. According to this mechanism a certain amount of data, which is handled by an author, is centrally locked. Other users can have an insight to the last version of the locked data, but are not able to modify this information. A big advantage of this method is the level of granularity of the access, which means, that only small amounts of data (measured in kilobytes) are sent over the network in an editing session. In this way, the amount of data is significantly reduced and the delta management is lean. An exemplary locking mechanism is shown in Figure 5.

A further essential part of the CDE is a robust communication between the collaborating participants. This kind of information can be stored according to the BIM Collaboration Format (BCF), which is an open data schema for BIM-based communication data (buildingSmart 2016). In an BCF-based communication process, the project participants create data objects, called topics, which store several attributes such as a type, a description, a current state and many other kind of information regarding the communication. To connect these details with the digital building model, the topic can be directly linked by storing a certain view position as well as unique identifiers of affected building elements. In this way, the topic is closely related to the building model and helps other project participants to understand the intended meaning of the communication object. In principle this kind of communication replaces the revision cloud as it is used in the 2D-based construction industry. This BIM-based communication plays a special role since it supports not only the assignment of tasks and the exchange of information such as comments but also the documentation of the whole construction process. At this point, the CDE can serve as a central storage for these topics, since it also contains the corresponding model data, which the topics are related with. An exemplary BCF-based communication process is shown in Figure 6.

3 STATE OF THE ART

Several approaches for a technical implementation of BIM-based model servers are provided by software vendors as well as academia. To judge the suitability for an application as CDE, selected representative approaches are shown in the following subsections.

3.1 *BIMserver.org*

The BIMserver.org project was started in 2010 by Beetz et al. (2010) at TU Eindhoven in order to provide a platform for end-users from construction industry to use shared building information models as well as a development environment for fellow academic researchers and commercial software developers. Built-up on a multi-layered software architecture the BIMserver.org is based-on an IFC-based key-value database. In this web-based

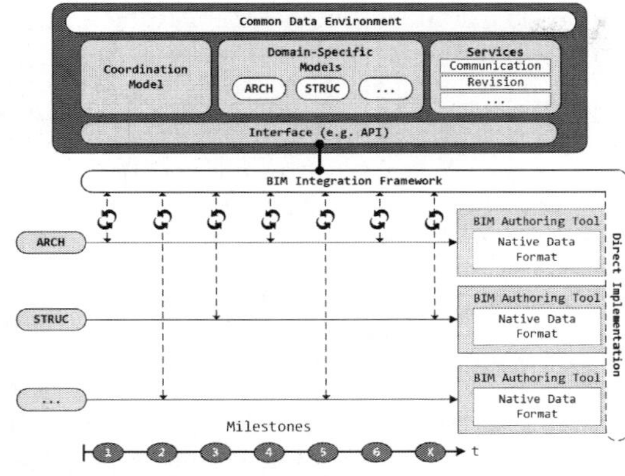

Figure 6. Exemplary BCF-based communication process.

framework all IFC entities are provided as Java classes using an early-binding mechanism, so that the contained information is transformed into a widely accepted data modeling language. In this way the framework simplifies and streamlines the development process. An example for such a development is the Building Information Model Query Language (BIMQL), which is a SQL-based query language for digital building models that can be used to formulate queries on the stored building information (Mazairac 2015, Mazairac & Beetz 2013). A major advantage of the BIMserver.org is the open and full access, so that this platform is technological neutral and therefore can be connected with any kind of BIM-based application. The BIMserver.org project fulfills the general basic prerequisites and therefore is suitable for an application as CDE. However, a direct integration of the provided features in authoring tools are desirable in order to reduce the effort for the user in terms of data handling.

3.2 *A360 & Forge*

A360 is an online project working space for several software products, which are offered by the software vendor Autodesk Inc. (2016a). The platform provides various collaboration features for participants using potentially different Autodesk products in order to share data and use services such as versioning of models or cloud-computing. In most of the Autodesk products the connection to the A360 platform is directly and smoothly integrated, so that all the provided features can be used in a straightforward manner. To realize this, the platform supports the im- and export of various primary native file formats. Furthermore, Autodesk provides a central browser-based application for a cross-platform management of the stored information. Next to this platform, Autodesk Inc. (2016b) offers a development environment for third-party developers, called Forge. Basically it represents a set of cloud services, Advanced Programming Interfaces (API) and Software Development Kits (SDK). With the help of these tool developers can create applications, services and data sets according to their own requirements. However, the A360 services as well as the development environment aims almost exclusively at Autodesk software products and will usually only work with the vendor's BIM tools and proprietary formats. This barrier causes massive problems due to significant data incompatibilities, if A360 services are used as an CDE, and therefore the presented software solution lacks the required accessibility and transparency.

3.3 *BIMcloud*

The Graphisoft (2016a) BIMcloud is a collaboration platform allowing ArchiCAD users to cooperate within a single project of any team size on the basis of native BIM data in real-time. To enable an error-free collaboration the platform provides several technical mechanisms such as a fundamental element reservation system, roles/permission and user logs. The data transactions can be scaled down to any level of granularity so that only very small data packages are transferred. In this way the required bandwidth for collaboration does not need any special network hardware or dedicated fibre lines. In terms of data security BIMcloud can also be set up as a private cloud using a private server as the storage destination. To view and navigate the resulting common model, several applications for desktop and mobile devices—such as Android and iOS—are offered (Graphisoft 2016b, AEC Magazine 2014). The BIMcloud represents a technological mature solution for the BIM-based collaboration. Especially the locking-mechanism and the granular transaction principle build up a solid and resilient base for the collaboration of several project participants. However, the BIMcloud is only available for ArchiCAD users and therefore this software product lacks a major requirement for a CDE—the full interoperability and transparency approach.

3.4 *BIMcollab*

The BIMcollab manager by Kubus (2016) is a web-based platform, which provides communication features for construction projects based on BCF, which was discussed in Section 2.3. For a direct and smooth integration of the communication tools inside of the BIM-based authoring tools, Kubus offers plugins for various tools, such as Autodesk

Revit, Graphisoft ArchiCAD or the Solibri Model Checker. With the help of these plugins the provided features of BIMcollab are uniformly integrated in the user interface of the host applications so that the process paths are shortened. In contrast to the previously shown representatives, BIMcollab does not store any model data but only information concerning the communication within a project. Therefore, this approach represents a very helpful communication feature for a collaboration platform but not a CDE itself. Without a simultaneous administration of communication and model objects, there is high risk of inconsistency between these objects, since there is no mandatory connection. For this reason, the user has to maintain the handled models and their validity on his own, which is cumbersome and error-prone.

3.5 *bim+*

bim+ was founded in 2013 and is by now a part of the Allplan GmbH (2016). Unlike the solutions discussed before, bim+ represents a commercial but open access platform. The core of bim+ are the API services, which provide a full and transparent access to each piece of information stored on the platform. In this way, bim+ provides several services e.g. project, user, model, revision and BCF-based communication services. Since these services are RESTful (Jakl 2006), they are technological neutral, can be used by various programming languages and therefore directly integrable in any authoring tool. For the description of information, bim+ provides a proprietary data model, which is well documented and closely related to the IFC data schema. Based on the API services the platform offers a Web SDK as well as several web-based applications, e.g. for viewing geometric and alpha-numeric information or managing projects, teams, users, rights, models and revisions. Furthermore, the platform can be variously extended by local or server-side services, which reduce expensive computing power on user side. In summary, bim+ provides a resilient technical framework and several features, which comply with the latest technical standards and are suitable for an application as an CDE.

3.6 *Trimble connect*

In a similar fashion as bim+ (see Section 3.5), Trimble connect also represents a BIM-based common environment, but is based on GTeam, which is a collaboration platform developed by Gehry Technologies (Gehry Technologies 2016). Recently a RESTful public API has been released, which can be used to access different kind of data stored on the platform. Trimble connect provides a web-based platform including several features such as management base for teams, users and projects or a model viewer. Due to the technologically neutral API services Trimble connect can also be integrated in arbitrary authoring tools. As an example Trimble connect has been integrated directly in SketchUp 2016. Furthermore, the platform provides several document storage features, which are directly integrated and can also be used for a synchronization with offline files. The same principle is well-known from popular document storage environment, e.g. Dropbox, and provides a direct integration in the user interface of the respective operating system. Since Trimble connect provides similar features as bim+ it brings almost the same advantages and therefore represents also a mature approach for a CDE.

4 BIM INTEGRATION FRAMEWORK

4.1 *Methodology*

As discussed in Section 2.3, the BIM-based collaborative processes imply several requirements, which have to be fulfilled in order to provide a resilient technical foundation. However, even the mature approaches for an CDE presented in Section 3 lack an sufficient, direct integration in BIM-based authoring and analysis tools. In order to overcome these deficiencies, we introduce the BIM Integration Framework (BIF) as a light-weight piece of software, which can be easily integrated into various BIM-based software products. In this way it provides a close and steady connection to a central online data platform for each participant no matter which software he uses. So the primary purpose of the Framework is to provide a seamless integration of CDE access functionality into standard BIM modelling and analysis applications. As a prerequisite, a basic data platform must be given, which fulfills the basic requirements of a CDE in terms of accessibility and transparency.

In this way, the provided features of available CDE approaches can be used, extended and finally integrated in various authoring tools. At the same time several of the technical solutions presented in Section 3 cannot be considered since they do not provide a full open access. By the help of an interface, e.g. an API, the BIF wraps the available features of the database and provides them in the appropriate authoring tool, so that a user is able to directly use these features by means of the User Interface (UI). The framework focusses especially on features that enable a smooth collaboration. In this way, the effort and complexity for various standard processes in the collaborative processes such as the communication or data exchange are

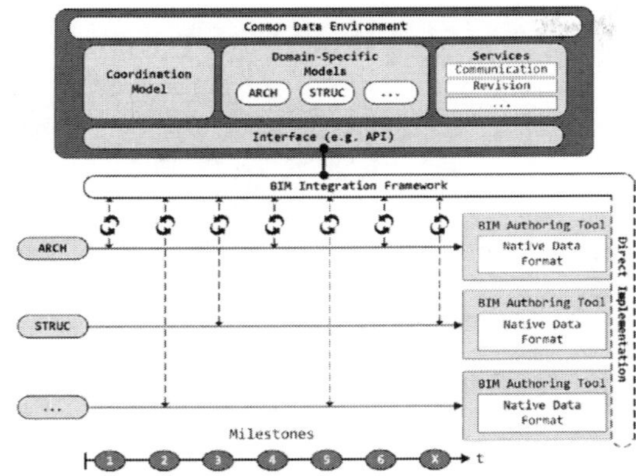

Figure 7. Schematic collaborative process with the help of the BIM integration framework.

significantly reduced. At the same time, the user has access to the stored data all the time and therefore will get a better understanding for processes as well as the processed information. This applies in particular for the communication and the data exchange processes, which are directly available in the UI. In this way, even a locking mechanism as it was discussed in Section 2.3 can be implemented as long as this is supported by the basis data platform. The primary objective of the BIF is not to unify the information regarding a specific native data format, but to provide a direct access to the central platform and the stored data objects according to the agreed project standards. As shown in Figure 7, the project participants can use any authoring tool and therefore also native data format. When specific milestones are reached, the data has to be dropped on the platform. This data drop step contents first of all the quality checking regarding the project standards and the coordination model as described in Section 2.3. At this point, the data has to be converted according to the agreed project standards and checked for compliance. Subsequently the validated information is stored in the common coordination model, the domain-specific model can be stored in a native format instead. In this way unnecessary conversions of the building information data into exchange formats are prevented and data loss due to the transfer processes is avoided. At the same time, the data is always up to date and the validated information is accessible for everyone according to his property rights. As a result, the collaboration processes via the BIF—in comparison to conventional methods—are significantly shortened. Because of the flexible and scalable design, the BIF can be extended by various services, which are subsequently available in all connected applications, so that all of the participants can make use of them. In this way basic services like a project, model or user management as well as further services like a communication or model checking feature can be provided.

4.2 *Proof of concept*

As a proof of concept, the BIF was implemented as a prototype based on bim+ (see Section 3.5), since this platform follows the full access concept, provides a fully open programming interface and therefore represents one of the most advanced representatives of a CDE. As shown in Figure 8, the BIF has been implemented as a light-weight library, which is directly connected with the API of bim+ and therefore has full access to any service, which is provided by the platform. From a technological

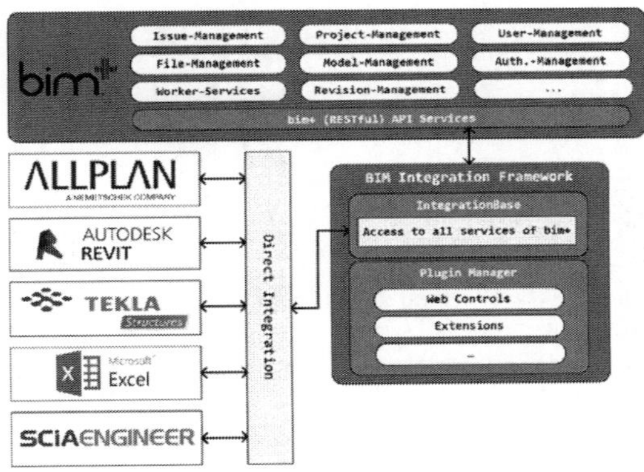

Figure 8. Software architecture of the BIM integration framework based on bim+.

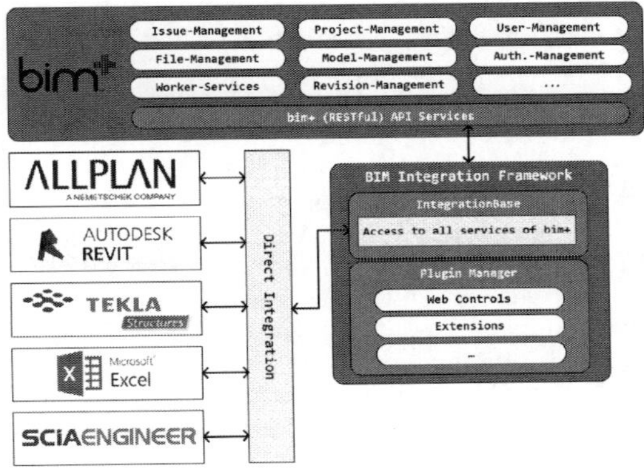

Figure 9. BCF-based communication via the BIM integration framework. Representation of the same topic object in the web application of bim+ (left) and ALLPLAN (right).

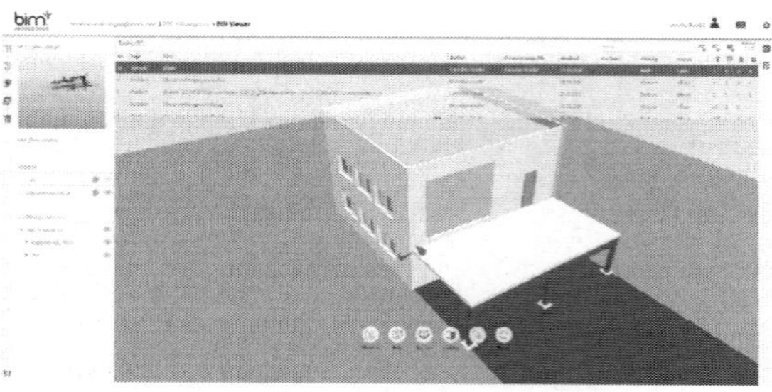

Figure 10. Integration of BIF features in Microsoft Excel.

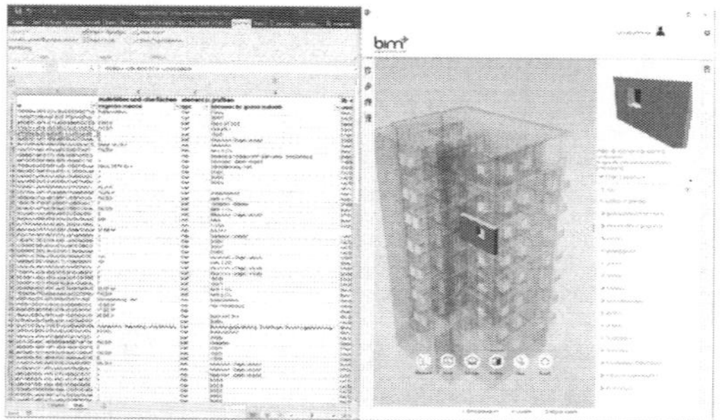

Figure 11. Retrieval of a temporary geometry for a clash detection via the BIF.

point of view, BIF is a dynamic .NET library, which provides various functionalities for the communication with bim+. Since many BIM authoring tools provide a C++ or C# programming interface, BIF can be integrated directly in these products. Of course the basic principle of the BIF can be also implemented for other programming languages, e.g. Java or Python, since the RESTful API services of bim+ are technology neutral. Furthermore, the software architecture of the BIF allows the dynamic extension of the framework by a plug-in mechanism, which is based on the Managed Extensibility Framework (Microsoft Corp. 2016b).

The plugins can be created and customized according to a certain schema at will and are valid for any version of the BIF, so that they can be reused in any BIF-integrated authoring tool. In this way, a unification of all connected authoring tools as well as a flexible scalability can be guaranteed at the same time. As an example for such a plugin, several web controls, which are provided in the bim+ Web SDK, has been implemented in the BIF, so that these can be reused in the local user interface. In this way, unnecessary double development expenses are avoided and the controls are always up-to-date in any version of the BIF and integrated authoring tool.

The BIF-prototype has been successfully implemented in several authoring tools, which are extensively used for BIM-based construction projects. These include ALLPLAN, Autodesk Revit, Tekla Structures as well as Scia at the present time (see Figure 12 & Figure 9). Furthermore, non-BIM-based tools can be connected with the BIF. Microsoft Excel (Microsoft Corp. 2016a) is a spreadsheet tool, which is extensively used in 2D- as well as 3D-based construction industry for various application areas. To show the potential of the CDE-based approach, we implemented the BIF features smoothly in the user interface of Excel to provide direct access for the user. Within this plugin the user is able to connect with bim+, choose a specific project and show any kind of data object stored in this project. Since there is no visualization provided, additional connected web controls are used to support the user. If a specific building element is chosen on the spreadsheet, the web control is automatically instructed to choose the corresponding building element in the viewer and to fly to this object. The presented integration principle of the BIF can be applied for any authoring tool, which provides an appropriate interface to access the user interface. With these prerequisites, further features such as a smart management system for collaboration can be implemented. Due to the direct access to each piece of information, any kind of required information can be retrieved during the modeling process. This feature can be helpful in case of a clash resolving. Currently a BCF-based topic just holds the information, which building elements are affected by the clash, and moreover a screenshot. But for an exact adjustment both clashing models must be transferred, so that the modeler can fix the issues. With the direct access via the BIF, the unique identifier of the clashing object is sufficient, since its geometry can be retrieved from the CDE. In this way, the geometry of the clashing

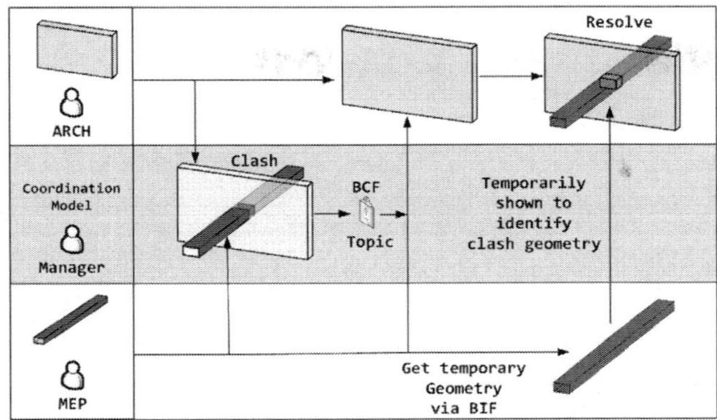

Figure 12. BCF-based communication via the BIM integration framework.
Representation of the same topic object in the web application of bim+ (left) and autodesk revit (right).

object can be shown directly in the authoring tool and just small packages of information are transferred. Furthermore, this transfer of model information could be aligned with the development of the BIM Snippet, which is available in the current BCF by now (buildingSmart 2016).

5 CONCLUSION & OUTLOOK

In the presented paper the importance of BIM-based collaborative processes for today's construction industry is discussed. Due to modern digital methods and the federated model approach there is a sound foundation to enable these. In this context, today's major challenge is to keep the overall consistency and quality during this collaboration. It is widely recognized, that the best way how to achieve this is to use common data platforms. Various technical solutions are already provided by different software vendors in this area. However, most of these approaches focus too much on native data formats and so they lack full access, data transparency and interoperability. Just few of the presented solutions represent a CDE in the proper sense. However, it must be kept in mind, that pure technical solutions are not enough for the successful implementation of collaboration processes. Well-defined workflows must be defined in order to prevent inconsistencies and low model quality. With the implementation of the BIF, the authors show the potential of smooth and direct integrated CDE-features in authoring tools. On the basis of this framework processes can be defined and at the same time the effort is significantly reduced. Due to the scalability and flexibility of the framework, it can be extended at will in order to establish more sophisticated workflows.

As a conclusion it can be noted that, the BIM-based collaborative processes will play an ever more important role in future. Therefore, the basic principles how to work on the basis of a CDE must be implemented in the construction industry. Furthermore, the current technical developments must focus on the implementation of the collaboration processes concerning all related requirements. In this context, more and more academic as well as industrial research projects are started by now. As an example, Project DRUMBEAT (2016) was launched 2015 in Finland. Its major task is to develop new information and communication technology solutions for distributed publication and utilization of building information models on the web. To realize this, the project focusses on the conversion of BIM models from the standardized IFC format to a web-compatible format, such that a unique web address will be assigned to each building part. By this interactive link, the information content of the models can be used on the web and building parts from different models can be linked to each other or to external information systems and data sources.

ACKNOWLEDGEMENTS

The authors gratefully acknowledge the support by the ALLPLAN GmbH and Nemetschek Group for the presented research.

REFERENCES

AEC Magazine. 2014. *BIM in the Cloud* [Online]. Available: http://www.aecmag.com/software-mainmenu-32/623-bim-in-the-cloud [Accessed 08.03. 2016].

Allplan GmbH. 2016. *bim+* [Online]. Available: http://www.bimplus.net/ [Accessed 08.03. 2016].

Autodesk Inc. 2016a. *A360* [Online]. Available: https://a360.autodesk.com/ [Accessed 08.03. 2016].
Autodesk Inc. 2016b. *Forge Platform* [Online]. Available: http://forge.autodesk.com/ [Accessed 08.03. 2016].
BCA Singapore 2013. Singapore BIM Guide—Version 2. Singapore: Building and Construction Authority Singapore.
Beetz, J., de Laat, R., van Berlo, L. & van den Helm, P. 2010. bimserver.org—An Open Source IFC Model Server. *CIB W78*. Cairo.
Borrmann, A., König, M., Koch, C. & Beetz, J. 2015. Building Information Modeling—Technologische Grundlagen und industrielle Praxis, Springer Vieweg.
British Standards Institution 2013. PAS. 1192-2:2013. Specification for information management for the capital/delivery phase of construction rojects using building information modelling.
buildingSmart. 2016. *BCF intro* [Online]. Available: http://www.buildingsmart-tech.org/specifications/bcf-releases [Accessed 14.03. 2016].
DRUMBEAT. 2016. *DRUMBEAT project* [Online]. Available: http://www.drumbeat.fi [Accessed 14.03. 2016].
Gehry Technologies. 2016. Available: http://www.gehry-tech.com/en/ [Accessed 15.03. 2016].
Graphisoft. 2016a. *BIMcloud* [Online]. Available: http://www.graphisoft.com/bimcloud/overview/ [Accessed 08.03. 2016].
Graphisoft. 2016b. *BIMcloud Presentation* [Online]. Available: http://www.graphisoft.com/ftp/marketing/bimcloud/pdf/BIMcloud_Flyer.pdf [Accessed 15.03. 2016].
Jakl, M. 2006. Rest Representational State Tranfer. *Technical report*. University of Technology, Vienna.
Kubus. 2016. *BIMcollab* [Online]. Available: http://www.bimcollab.com/ [Accessed 08.03. 2016].
Mazairac, W. & Beetz, J. 2013. BIMQL—An open query language for building information models. *Advanced Engineering Informatics,* 27, 444–456.
Mazairac, W. 2015. *BimQL* [Online]. Available: http://bimql.org/ [Accessed 14.03. 2016].
Microsoft Corp. 2016a. *Excel 2016* [Online]. Available: https://products.office.com/de-DE/excel [Accessed 15.03. 2016].
Microsoft Corp. 2016b. *Managed Extensibility Framework (MEF)* [Online]. Available: https://msdn.microsoft.com/de-de/library/dd460648 (v = vs.110).aspx [Accessed 15.03. 2016].
Solihin, W., Eastman, C. & Lee, Y.C. 2016. A framework for fully integrated building information models in a federated environment. *Advanced Engineering Informatics,* 30, 168–189.
Young, N., Jones, S., Bernstein, H.M. & Gudgel, J. 2009. The Business Value of BIM. *SmartMarket Report.* Mc Graw Hill Construction.

eWork and eBusiness in Architecture, Engineering and Construction – Christodoulou & Scherer (Eds)
© 2016 Taylor & Francis Group, ISBN 978-1-138-03280-4

Pragmatic use of LOD—a modular approach

N. Treldal
Department of Civil Engineering, Technical University of Denmark, Kgs. Lyngby, Denmark
Rambøll Denmark A/S, Copenhagen, Denmark

F. Vestergaard & J. Karlshøj
Department of Civil Engineering, Technical University of Denmark, Kgs. Lyngby, Denmark

ABSTRACT: The concept of Level of Development (LOD) is a simple approach to specifying the requirements for the content of object-oriented models in a Building Information Modelling process. The concept has been implemented in many national and organization-specific variations and, in recent years, several solutions have been proposed to address the challenge of the LOD concept being either too simple to fully describe the requirements for BIM deliverables or too complex to be operational in practice. This study reviews several existing LOD concepts and concludes that addressing the completeness and reliability of deliveries along with use-case-specific information requirements provides a pragmatic approach for a LOD concept. The proposed solution combines LOD requirement definitions with Information Delivery Manual-based use case requirements to match the specific needs identified for a LOD framework. This framework can act as a basis for future LOD solutions to harmonize the conceptual understanding of LOD definitions.

1 INTRODUCTION

1.1 *Study background*

Level of Development (LOD), Information Levels, and other similar concepts for defining requirements for Building Information Modelling (BIM) deliverables are widely used in the Architectural, Engineering and Construction (AEC) industry. LOD allows for a simple approach for specifying the requirements for the content of object-oriented models in a BIM process, but prior research (Hooper 2015; Berlo et al. 2014; Boton et al. 2015) has established that it is a considerable challenge throughout the AEC industry to define BIM deliveries accurately using existing LOD concepts.

Different design disciplines, project execution models and project organizations require different information to be available at project milestones, so there has to be a granularity within the framework of LOD. For this reason, several organizations have introduced further terms, such as Level of Detail (graphic-oriented), Level of Information (non-graphic-oriented), Level of Accuracy (tolerance-oriented), and Level of Coordination (collaboration-oriented) (BIM Acceleration Committee 2014).

Most solutions differentiate only graphical and non-graphical requirements to limit complexity. For example, the BSI defines *graphical data* as "data conveyed using shape and arrangement in space" and *non-graphical data* as "data conveyed using alphanumeric characters" (BSI 2013).

As the range of options for specifying LOD requirements increases, so does the complexity of defining requirements and the challenge is to achieve actual added project value using such approaches (Hooper 2015). Berlo et al. 2014 also describes a considerable confusion of when a BIM model actually reaches a certain LOD level and it seems that one major challenge is still the misunderstanding of *detailing* as a definition for model progression (NATSPEC 2013).

There is a close correlation between the processes undertaken in AEC projects and the BIM model deliverables (Lee et al. 2007), and this means that any requirements stated will affect how the design is executed. Using LOD can, therefore, make it a complicated matter for clients and others to state requirements that will likely be of value for the entire project. The range of proposed LOD concepts available, however, indicates a need in the AEC industry to have an approach that addresses model deliveries. The challenge seems to be how such a solution can be both unambiguous and operational?

1.1 *Study goals*

The goal of this research was first to compare existing LOD concepts in the AEC industry to clarify their scope and the terminology they use. And then secondly to propose a solution that can harmonize the LOD concept to state unambiguous BIM delivery requirements yet still practical enough to ensure common ground for all the stakeholders in a project.

2 METHODOLOGY

The research started with a review comparing known and widely-used LOD concepts from organizations in several countries. Eight LOD concepts were selected for further analysis in this research based on their individual approach. The main goal was to explore how existing solutions handle the granularity of the LOD concept and to what extent they state unambiguous requirements. Secondly, the findings from work to develop a proposal for a set of new Danish Information Levels were used to define the operational requirements for a LOD concept. The authors have been actively involved in the development of both the prior Danish bips Information Levels and more recently of a new concept by Digital Convergence (DiKon), which is a working group of BIM experts from six of the largest AEC companies in Denmark. The findings from DiKon were identified during multiple workshops within this working group. Based on these findings, a solution is proposed here that builds on top of the DiKon concept in an attempt to harmonize the usage of LOD and also includes recent research on more modular approaches to delivery requirement definitions.

3 REVIEW

3.1 *Development of LOD concepts*

Over the last decade, a number of LOD concepts have been proposed by industry and client organizations. In Denmark, the organization bips based its first proposal for a set of generic Information Levels (bips 2007) on work carried out by the Finish PRO IT organization. The Information levels were later revised (bips 2009) and recently completely reconfigured in a new set called CCS Information Levels (Cuneco 2014). The solutions define high-level and generic descriptions of the Information Levels at model level. There is an intention to define model-element-specific requirements based on the overall levels, but so far this work is still in progress within bips.

In the US, the AIA released their first contracting documents describing LOD requirements in 2008 and revised them in 2013 (AIA 2013a). The documents only cover high-level and short generic descriptions of LOD, but in 2011 the AIA allowed the US organization BIMForum to put further detail into their LOD concept at model element type level. Their latest release (BIMForum 2015) includes more than 140 element-type-specific definitions, and supplementary Element Attribute Tables define requirements per level for non-graphical information for each element type.

Building on top of the work by the AIA, first the U.S. Department of Veterans Affairs and later the Australian NATSPEC organization in 2011 released a BIM Object/Element Matrix (NATSPEC 2013) with requirements for non-graphical object properties for 28 model element types. All the object properties are categorized in groups based on 15 defined use cases and mapped to the buildingSMART IFC specifications (buildingSMART 2007).

In the Netherlands, TNO has developed a proposal for a set of Information Levels focused primarily on the purposes a model can be used for (Berlo et al. 2014). A database is currently under development to define model-element-specific requirements for non-graphical information (TNO et al. 2014). In the UK, the BSI has defined a set of model stages in its PAS standard (BSI 2013) to define requirements at model level for both graphical and non-graphical content based on descriptions of themes and purposes. A BIM Toolkit solution by NBS (NBS 2015) aims to use the PAS model stages to define individual requirements for model element types at different design stages, but this solution also seems to be still under development.

Several other solutions, such as the Finish COBIM, the New Zealand BIM Handbook, and the US Army BIM Minimum Modeling Matrix, have also been introduced, but most other concepts are either limited in their range of model requirements or based on the principles of one of the solutions above.

3.2 *Common Understanding of LOD*

Originally introduced by the AIA as an abbreviation for *Level of Detail*, the term LOD was changed in 2013 to represent *Level of Development* (AIA 2013a) based on conclusions similar to those found elsewhere (NATSPEC 2013; BSI 2013) that LOD represents the combination of requirements for the concretization of both graphical and non-graphical information during a project.

The ambitions for LOD are somewhat multifaceted, ranging from "the degree to which (...) information has been thought through" (BIMForum 2015) to "what the model can be used for" (Berlo et al. 2014).

The AIA defines the term *Model Element* as "a portion of the model representing a component, system or assembly within a building or building site" (AIA 2013a), and most recent LOD concepts define their requirements based on type-specific model element definitions. For the solutions that address model elements, a *LOD* Table like the Model Delivery Specification (DiKon 2015), the Model Element Table (AIA 2013b), or the BIM Object/Element Matrix (NATSPEC 2013) is needed to define delivery requirements for element types at the various project milestones. (Berlo et al. 2014) point out that the complexity of such LOD Tables quickly increases to such an extent that ordinary users lose all track of the relationship between desirable use cases and requirements. Current LOD concepts are therefore challenged by trying to address both a wide range of purposes and the need for simple and operational solutions.

3.3 *Information Delivery Manual (IDM)*

The concept of an IDM is an alternative solution to defining unambiguous exchange requirements for BIM deliveries for specific use cases and has been developed by buildingSMART (See et al. 2012). In an earlier paper, we proposed a solution using IDM Packages—each describing only a single-actor use case per IDM—to allow for a more modular approach to describing information flow in construction (Mondrup et al. 2014). IDM Packages can be rearranged more freely than traditional IDMs, which usually describe large-scale use cases involving several actors. The idea is to have the ability to define unambiguous information requirements at model element level based on specific use cases.

Both the IDM and LOD address delivery requirements, but the origin of the IDM was the need to define object-oriented and property-specific exchange requirements, whereas the origin of LOD was to define generic and high-level requirements. With a use-case-specific and information-intensive LOD concept like the NATSPEC, concepts originating from IDM and LOD get mixed together, illustrating the need for solutions to be both high-level and unambiguous at the same time.

The Norwegian bSN Guiden (buildingSMART Norway 2015) is a solution based on these principles. It provides individual users with a simple database interface for defining use-case-oriented and unambiguous delivery requirements at model element level. However, currently the solution only states limited delivery requirements, none of which relate to graphical information. It therefore needs supplementing with an existing LOD concept to make it fully useful.

3.4 *Aspects of LOD concepts*

To compare the somewhat different LOD concepts, five evaluation aspects were identified during the research:

- Content Aspect—How are completeness and/or detailing of deliveries defined?
- Format Aspect—Is graphical information separated from non-graphical information or are requirements combined?
- Context Aspect—Are levels related to phases and/or related to specific use cases?
- Structural Aspect—Does the concept target overall model requirements or model element requirements?
- Standardization Aspect—Does the concept make use of standardization solutions, like classification systems or exchange formats?

The concepts address these aspects explicitly, implicitly or not at all. The comparison of LOD concepts is shown in Table 1.

3.5 *Review findings*

Notable from the comparison is that although there is still a considerable misunderstanding in the industry that LOD refers to the detailing of deliveries, none of the current LOD concepts address *detailing* explicitly in its definitions. Nevertheless, the concepts that include illustrations of the development stages, e.g. bips, DiKon and BIMForum, do implicitly address detailing to some extent, which can lead to misunderstandings about what is intended. Moreover, the UK BSI concept specifically mentions Level of Detail and includes some illustrations, but all definitions still relate to *completeness*.

No consensus has so far been reached in relation to whether graphical and non-graphical information should be defined separately or combined, whereas most recent solutions agree on defining use-case-related requirements at model element level. The relationship to classification systems and IFC/COBIE is increasing, yet still not implemented throughout the concepts, potentially leading to unclear requirement definitions in some cases.

According to Hooper, there is a lack of research on how useful LOD is in actually benefitting projects as well as a lack of research on the use of IDMs in practice (Hooper 2015). Berlo et al. report that the Dutch General Services Administration has removed all reference to LOD due to uncertainty of deliveries (Berlo et al. 2014), and although the Information Levels from bips are commonly used in public projects in Denmark (bips 2009), five out of the six AEC companies in the DiKon organization have developed supplementary definitions to improve the certainty of agreements. This illustrates the need for further definition of the success criteria for LOD concepts if they are to be unambiguous and operational.

Table 1. Comparison of eight selected LOD concepts based on five defined aspects.

LOD Concepts		Content Aspect		Format Aspect		Context Aspect		Structural Aspect		Standardization Aspect	
Country and Organisation	Denomination	Completeness	Detailing	Combined	Separated	Phase related	Use case related	Focus on model	Focus on model element	Classification	Exchange format
DK bips 2007	Information Levels									DBK	IFC
DK CCS	Information Levels									CCS, intended	IFC, intended
DK DiKon	Information Levels										
US AIA 2013	Level of Development										
UK BSI	Level of Definition										
AUS NATSPEC	Level of Development									Uniformat	IFC + COBIE
US BIMForum	Level of Development									Uniformat	Partly IFC
NL TNO	Information Levels									NISfb	IFC, intended

Explicitly addressed Implicitly addressed Not included

4 FINDINGS FROM DANISH DEVELOPMENT EFFORT

4.1 *Initial work by bips*

During the development of the first set of Information Levels (bips 2007), it was concluded that the levels must be detached from phases because different project constellations require information to be utilized at different stages. However, the solution should 1) have levels representing deliveries in all main phases, and 2) allow for different parts of a delivery to be represented by different levels.

In the latest version from bips (Cuneco 2014) the levels have lost explicit connection to phases and are now defined as generic steps in the concretization of building projects ranging from 1 to 7. The challenge with this approach is that the levels are now defined so generically that it can be complicated to relate the levels to desired use cases in an unambiguous way.

4.2 *DiKon Information Levels*

So in 2015, the Danish organization DiKon decided to expand the latest set of Information Levels with a range of model element type-specific definitions and create a LOD table to link requirements to phases. The solution includes specific descriptions for 22 commonly used model element types (DiKon 2015).

More than 15 workshops were conducted by DiKon first to define the scope and then to review the content of the proposed model element definitions. The following findings summarize the conclusions from the workshops and define the scope of the proposed solution:

- The LOD levels from AIA/BIMForum do not match the delivery requirements common in the Danish AEC industry.
- The main goal is a tool for agreeing on the scope of deliverables that must be operational for clients and project managers with limited BIM experience
- The solution must make it possible to state unambiguous delivery requirements throughout a project without obstructing the processes and being too workload intensive.
- A LOD Table is required to allow for individual element types to be assigned different LODs at specific deliveries depending on 1) their type (prefab/build-on-site, etc.) and 2) their location in the building (e.g. differentiating HVAC components in plant rooms and shafts from similar components in other room types).
- Requirements for graphical and non-graphical information must be defined separately in the LOD table.
- The requirements for non-graphical information must be based only on high-level use cases and must be part of the information currently available in BIM models.

Based on the above findings, a solution was developed (DiKon 2015), as illustrated in Table 3. Only graphical requirements are illustrated in the table. Additional requirements for non-graphical information are also part of the solution, defined per level for each selected model element type.

The graphical requirements are defined based on three criteria:

- The reliability of the elements, ranging from *Expected*, *Specified* to *Final*.
- The shape of elements related to reliability, e.g. the max. outer contours for the expected level of reliability or contours reflecting the final dimensions for the final level of reliability.
- The completeness of the elements, referring to the element representation (e.g. generic, assembly or element-divided) and the scope

Table 2. BIMForum 2015 LOD definition for D2010.20—domestic water equipment supplemented with an interpretation of corresponding detailing, Level of Reliability (LOR) and Level of Completeness (LOC).

LOD 100	LOD 200	LOD 300	LOD 350	LOD 400
Diagrammatic or **schematic** model elements; **conceptual** and/or **schematic** layout/flow diagram; design performance parameters as defined in the BIMXP to be associated with model elements as non-graphic information.	**Schematic** layout with **approximate** size, shape, and location of equipment; **approximate** access/code clearance requirements modeled; design performance parameters as defined in the BIMXP to be associated with model elements as non-graphic infomation.	Modeled as **design-specified** size, shape, spacing, and location of equipment; **approximate** allowances for spacing and clearances required for all specified anchors, supports, vibration and seismic control that are utilized in the layout of equipment; actual access/code clearance requirements modeled.	Modeled as **actual** construction elements size, shape, spacing, and location/connections of equipment; **actual** size, shape, spacing, and clearances required for all specified anchors, supports, vibration and seismic control that are utilized in the layout of equipment.	Supplementary components added to the model **required for fabrication** and field installation.
Interpretation: Detailing = Diagrammatic LOR = Conceptual LOC = Diagrammatic	**Interpretation:** Detailing = Medium LOR = Approximate LOC = Generic Level	**Interpretation:** Detailing = Fine LOR = Design-specified LOC =Type Level	**Interpretation:** Detailing = Fine LOR = Actual LOC = Component Level, Design	**Interpretation:** Detailing = Fine LOR = Actual LOC = Component Level, Fabrication

Table 3. DiKon 2015 information level definition for heating and sanitation components supplemented with an interpretation of corresponding detailing, Level of Reliability (LOR) and Level of Completeness (LOC).

Information Level 2	Information Level 3	Information Level 4	Information Level 5	Information Level 6
Not defined				
	Components are modelled as **generic** volume objects in **expected** max. outer contour. **Expected** location and orientation of components.	Components are modelled in **specified** max. outer dimensions incl. **Specified** location and orientation of components.	Components are modelled in **final** outer dimensions. **Final** location and orientation of components.	Components are modelled in **final** dimensions based on actual choice of product. **Final** location and orientation of components.
	Interpretation: Detailing = Coarse LOR = Expected LOC = Generic Level	**Interpretation:** Detailing = Coarse LOR = Specified LOC = Type Level	**Interpretation:** Detailing = Medium LOR = Final LOC = Component Level, Design	**Interpretation:** Detailing = Fine LOR = Final LOC = Component Level, Fabrication

Table 4. Framework for a generic LOD solution—must be detailed on model element level based on national or organizational needs.

	LOD 0	**LOD 1**	**LOD 2**	**LOD 3**	**LOD 4**	**LOD 5**	**LOD 6**
Scope	Specification	Idea	Outline	Proposal	Design	Construction	Handover
Level of Reliability	Final (requirements)	Expected	Expected	Specified	Final	Final	As-build
Level of Completeness	Descriptive Level	Volume Level	Generic Level	Type Level	Component Level, Design	Component Level, Fabrication	Component Level, Handover
Level of Information (based on the four high-level use cases)	- Identification - Scope	- Identification - Size	- Identification - Type - Size	- Identification - Type - Size - Material - Performance requirements	- Identification - Type - Size - Material - Performance requirements	- Identification - Type - Size - Product-specific values	- Identification - Type - Size - Product-specific values

of components to include along with the main element.

5 DESIRED LOD TERMINOLOGY

5.1 *Comparison of DiKon and BIMForum*

The DiKon concept is very similar to the BIMForum concept because they share common goals. In Table 2 and Table 3, a comparison is made of definitions for graphical requirements of comparable building services based on the DiKon and BIMForum concepts. Three interpretations of each definition described below have been added to the tables to make it possible to compare the similarities and differences of the two concepts.

5.2 *Level of Completeness (LOC)*

The review in section 3 concluded that LOD definitions describe completeness and not detailing, so we introduce the concept of Level of Completeness (LOC) to address this need directly. LOC is defined on the basis of the concretization of the model element and the scope of included components. The combination of a description and an illustration defines each LOC. The comparison in Tables 2 and 3 illustrates why the BIMForum solution is not directly applicable in a Danish context: as similar levels, LOD 350 and Information Level 5, might both be at component level, but whereas e.g. hangers are included in LOD 350, this is not the case in Information Level 5 because this is not Danish practice. This indicates that while the concept of LOC could be used to harmonize the definitions of graphical requirements at a generic level, national or organisation-specific definitions are needed to match the content required for local practices or needs.

5.3 *Level of Reliability (LOR)*

The DiKon workshops concluded that reliability is a useful factor to include so that the concept can be used as part of a contractual agreement. It is clear that BIMForum reached similar conclusions because the terms *Conceptual*, *Approximate*, *Design-specified* and *Actual* are used throughout their definitions. A harmonization of such terms would further add to the common ground on expectations for deliverables.

5.4 *Detailing*

Believed to be of less relevance to the deliverables is the detailing or coarseness of the model elements. Detailing describes how objects are presented visually, but does not address the content. BIM authoring tools like Autodesk Revit have a functionality for easily changing the detailing of objects from *Coarse* to *Medium* or *Fine*. However, this does not necessarily imply that the LOD level has increased. The above tables indicate the interpreted detailing level on the basis of the illustrations available and this clearly shows that only limited consensus has been reached because the detailing is not aligned in the concepts. A review of illustrations for other model element types—particularly in the BIMForum concept—further adds to the conclusion that there is limited consensus about the detailing level at different LOD levels.

This partly explains the concern expressed by Berlo et al. that if people are asked to review different models, they reach very limited common agreement about what LOD level a particular model has reached (Berlo et al. 2014). Most likely this is because they focus on the detailing level of the model elements as opposed to the completeness and reliability of the elements.

5.5 *Focus on use cases*

Berlo et al. conclude that this confusion should lead to LOD levels focusing purely on use cases. The more use cases included to define a LOD level, the narrower the reuse of similar LOD levels can be in different project constellations. The review and findings in this paper indicate that there is a need to be able to define the level of concretization of model elements in a generic, simple and unambiguous way and this is why the use of LOC and LOR seems more valid as the foundation of a LOD framework. BIMForum acknowledges the need to link requirements to use cases and select 1) Quantity take-off, 2) 3D coordination, and 3) 3D control and planning

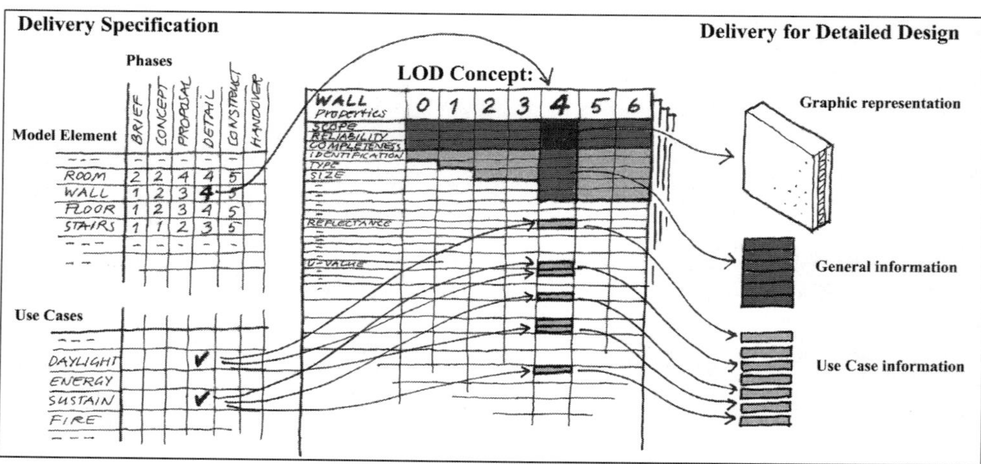

Figure 1. Modular approach of defining delivery requirements based on LOD selection for model elements and supplementing requirements with desired use cases.

as the high-level use cases which graphical and non-graphical delivery requirements should address as a minimum. Supplementing the above with drawing production as a use case still seems necessary, but leaving out requirements for additional use cases keeps the concept generic and still unambiguous.

5.6 *Level of Information (LOI)*

To define the requirements for non-graphical information accurately, we make use of the Level of Information (LOI). Such requirements are stated explicitly as object properties in some LOD concepts and in others described implicitly just as information needed to fulfil specific use cases, such as energy simulation or cost calculation. Koskela et al. conclude that the use of relevant software tools has limited impact if the AEC process is confused at the outset, so unambiguous information requirements are needed (Koskela et al. 2002). For this reason, LOI requirements should be defined explicitly and a solution like IDM Packages could be included to define requirements for additional use cases, adding a more modular approach to the LOD concept.

6 PROPOSAL FOR A PRAGMATIC LOD APPROACH

6.1 *Generic Framework*

Based on the above conclusions, we propose a solution for a generic set of LOD levels as shown in Table 2. This framework is intended to act as a basis for future LOD solutions to harmonize the conceptual understanding of LOD content. As previously indicated, there is a need to customise the LOC definitions to match local practices, while the framework is still seen as generic.

6.2 *Scope*

BIMForum uses LOD 100 to define requirements for diagrammatic or schematic layouts of model elements. As argued also by NATSPEC, 2D drawings and other informational representations could just as well be defined by different LOD levels (NATSPEC 2013). Accordingly we argue, based on the original findings from DK bips 2007, that LOD levels should be available to represent all the main phases of the AEC industry, focusing on delivery milestones. This is why we propose a total of seven levels spanning from LOD 0 to 6. The hundred-concept used by BIMForum to allow for custom LODs like 120 or 340 has also been dropped because we argue that such in-between levels are not desirable. Instead, local variations of the LOC of the seven levels must be accepted. Some LOD concepts assign *Operation* as the last level, but we argue that operation, maintenance, renovation, etc. are all use cases which use data from the milestone *Handover*.

6.3 *LOD table and use case connection*

To make it possible for additional use cases to be addressed, we propose to use the concept of IDM Packages to define use-case-specific information requirements. Including use cases in a Delivery Specification (LOD Table) based on such IDM Packages would allow for a configuration system to point directly to the information required by the additional use cases, as illustrated in Figure 1. The sum of information required as standard based on LOD 4 and the additional information required by the two selected use cases constitute the total requirements of what is to be delivered at (in this case) the Detailed Design phase.

6.4 *Pragmatic and modular*

The proposed framework and practical solution will allow clients and project managers with limited BIM knowledge to use the Delivery Specification to agree on the scope of deliveries.

Since the Delivery Specification is backed up by unambiguous information requirements, the BIM modellers will know what content should be included in the BIM models later, how it should be modelled, and what non-graphical information should be included.

The concept is only tied to a few high-level use cases and the modular approach allows for any additional requirements to be included as long as they are derived from a use case.

7 CONCLUSIONS

In this study, we review several existing LOD concepts and conclude that addressing the completeness and reliability of deliveries along with use-case-specific requirements can provide a pragmatic approach for a LOD framework. This framework can act as a basis for future LOD solutions to harmonize the conceptual understanding of LOD definitions and, because it combines LOD definition requirements with IDM-based use case requirements, the solution is also highly modular.

The use of LOD is linked to the need for a pragmatic approach to agreeing on model deliveries, but the concept still requires human interpretation of graphical requirements to be translated into individual model-specific requirements. As the AEC industry matures further in relation to BIM modelling, it would be appropriate to focus more on

IDM requirements—potentially fully integrating the LOD concept into IDM.

ACKNOWLEDGEMENTS

The authors would like to thank the bips organization and the members of DiKon for their fruitful and intense discussions on the development of Information Levels for the Danish AEC industry.

REFERENCES

AIA 2013a. Document E203™—2013—Building Information Modeling and Digital Data Exhibit. US: The American Institute of Architects.

AIA 2013b. Document G202™—2013—Project Building Information Modeling Protocol Form. US: The American Institute of Architects.

Berlo, L. van, Bomhof, F. & Korpershoek, G. 2014. Creating the Dutch National BIM Levels of Development (extended). Proceedings of the 2014 International Conference on Computing in Civil and Building Engineering, 129–136.

BIM Acceleration Committee. 2014. New Zealand BIM Handbook—Appendix C—Levels of Development definitions. NZ: The Building and Construction Productivity Partnership.

BIMForum 2015. Level of Development Specification (Version 20). US: BIMForum.

bips 2007. 3D Working Method. Translated into English in 2007. Ballerup, Denmark: bips.

bips 2009. C102 e—CAD manual 2008, Instructions. Translated into English in 2009. Ballerup, Denmark: bips.

Boton, C., Kubicki, S. & Halin, G. 2015. The Challenge of Level of Development in 4D/BIM Simulation Across AEC Project Lifecyle. A Case Study. Procedia Engineering, 123, 59–67.

BSI 2013. PAS 1192–2:2013 Specification for information management for the capital/delivery phase of construction projects using building information modelling (March). UK: The British Standards Institution.

buildingSMART 2007. IFC2x Edition 3 Technical Corrigendum 1. Retrieved April 10, 2016, from http://www.buildingsmart-tech.org/ifc/IFC2x3/TC1/html/index.htm

buildingSMART Norway 2015. bSN Guiden—Prosessveiledninger. Retrieved October 19, 2015, from http://buildingsmart.no/bs-guiden/prosessveiledninger

Cuneco 2014. CCS Informationsniveauer (R0 ed.). Herlev, Denmark: bips.

DiKon 2015. Level of Information for Building Models—for selected building parts. DK: Digital Convergence. Retrieved from http://www.digitalkonvergens.dk/building-standards/model-delivery-specification/

Hooper, M. 2015. BIM Anatomy II—Standardisation needs and support systems. Lund University.

Koskela, L., Huovila, P. & Leinonen, J. 2002. Design Management In Building Construction: From Theory To Practice. Journal of Construction Research, 03(01), 1–16.

Lee, G., Eastman, C.M. & Sacks, R. 2007. Eliciting information for product modeling using process modeling. Data & Knowledge Engineering, 62(2), 292–307.

Mondrup, T.F., Treldal, N., Karlshøj, J. & Vestergaard, F. 2014. Introducing a new framework for using generic information delivery manuals. In Proceedings of the 10th European Conference On Product And Process Modelling In The Building Industry (ECPPM) (pp. 295–301).

NATSPEC 2013 BIM Paper NBP 001. AUS: Construction Information Systems Limited.

NBS 2015. BIM Toolkit. Retrieved March 20, 2016, from https://toolkit.thenbs.com/

See, R., Karlshøj, J. & Davis, D. 2012. An Integrated Process for Delivering IFC Based Data Exchange. Retrieved March 22, 2016 from http://iug.buildingsmart.org/idms/methods-and-guides/Integrated_IDM-MVD_ProcessFormats_14.pdf/view

TNO, BuildingSMART, Benelux & Stumico. 2014. BIM informatieniveaus. Retrieved April 10, 2016, from http://niveaus201402.nationaalbimhandboek.nl/

eWork and eBusiness in Architecture, Engineering and Construction – Christodoulou & Scherer (Eds)
© 2016 Taylor & Francis Group, ISBN 978-1-138-03280-4

Identifying and addressing multi-source database inconsistencies: Evidences from global road safety information

L. Dimitriou & P. Nikolaou
Department of Civil and Environmental Engineering, University of Cyprus, Nicosia, Cyprus

ABSTRACT: The data collection and the creation of extensive databases for the investigation of different global phenomenon, such as road traffic fatalities, inherent risks of information inconsistencies. The current paper is presenting a novel approach on the efficiently detection of potential data anomalies (inconsistencies), using wide in range socio-economic factors from different 'instances' years (2010 and 2013), for the investigation of the phenomenon of road traffic fatalities concerning 121 UN countries (restricted to UN countries with significant population). Unfortunately, collecting information from different, even reliable, sources (global organizations) raises speculations of uncertain, implausible, inconsistent and unstable information, which can be transparent with different data-model analysis likewise Principal Component Analysis, Negative Binomial regression analysis and Structural Equation Modeling.

Keywords: Global Statistics, Road Fatalities, Inconsistent Data, Structural Equation Modeling, Principal Component Analysis, Negative Binomial Regression.

1 INTRODUCTION

The internet being a major source of collecting data for processing a large number of studies, however, collecting data, even from reliable sources (global organizations) inherent speculations for uncertain, implausible, inconsistent and unstable information. Usually, these speculations are observed on the models' behavior and on the results they obtain.

The objective of the current paper is to investigate the phenomenon of road traffic fatalities in two 'in-stances' namely years 2010 and 2013, while the sample concerns 121 UN countries (restricted to UN countries with significant population) and to shed light on the models' anomalies due possible information inconsistencies. For analyzing the traffic accident factors effectively an extensive traffic accident historical database is needed. In particular, in the current paper 25 socio-economic and risk exposure data were collected from four global organizations (e.g., World Health Organization, World Bank, Natural Earth and World Atlas). Form the initial historical database two separate databases were arose. The two databases had similar forms, i.e. they included the same variables but referring on different years. In particular, database 1 was referring for the year 2010 and database 2 for the year 2013.

At first, a parsimonious description of the data was provided explaining the variance of the variables, by Principal Component Analysis (PCA) implementation. Subsequently, two different methodologies were developed for investigating the phenomenon of road traffic fatalities in a macro level and across the globe and also for detecting and addressing information inconsistencies, towards data management. Furthermore, a series of models (16 models) were developed for studying the contribution of socio-economic factors (based on relevant literature) on the phenomenon, implementing Negative Binomial (NB) regression analysis and Structural Equation Modeling (SEM). The resulting models from the current application present a robust and a good-of-fit form by making them reliable in use for estimating the phenomenon. Notwithstanding this fact, the final form of the 2010 models was different from the 2013 models. This finding leads to the conclusion that this kind of information must be consciously used for road traffic fatalities estimation and not for extrapolation, due to data inconsistency.

Smart Cities use 'Big Data' that have been collected from several sources (smart meters, RFID tags, street security cameras, Foursquare, public transportation organizations, smart phones, logistics companies, sensor networks, etc.) for spotting patterns, predicting potential crisis situations and in general making smart decisions based on real facts (Hinssen, 2012). Consequently, Smart Cities can provide real information that can be processed and so meaningful and robust results can be provided for the benefit of many research fields, one of which is road safety.

The rest of the paper is organized as follows. Section 2 provides previews studies similar to the current paper. Section 3 presents the data collection procedure. Section 4 describes the methodological approach that was followed. Section 5 offer some conclusions, remarks and highlights some points of further research.

2 BACKGROUND REVIEW

Suspicions of information inconsistencies in macro-level data inherent risks, risks that must be taken under consideration especially when these data are used towards policy/decision making.

In details, socio-economic information was used (based on relevant literature) for investigating the phenomenon of road traffic fatalities. Yu (2014), explored differences in crash frequency across neighborhoods with different economic statuses and ethnic compositions, and further tested the potential moderator effect of socio-demographic characteristics on the built environment-traffic safety association.

For identifying the information inconsistencies NB regression and SEM, models were developed. In general, both methodologies have a wide implementation in the field of road safety. For instance, Coruh et al. (2015) analyzed the factors that affecting the frequency of accidents in 81 cities over a three-year period (2008–2010) with monthly data using random-parameters negative binomial panel count data models. Furthermore, Hassan et al. (2013) implemented SEM for identifying and quantifying the impacts of significant variables influencing crash size.

There are many who have implemented various methods for detecting and addressing data inconsistencies. Ma et al. (2009), developed a method for information inconsistencies detection for real-time information in dynamic decision-making. Fomina et al. (2014), presented some methods and approaches to deal with inconsistent and noisy databases used for the inductive notion formation. Deb & Liew (2016), developed a methodology for imputing missing data of numerical or categorical values in a traffic accident historical database.

The current paper is proposing a new approach on detecting data inconsistencies and offers hints on how to avoid inconsistent information.

3 DATA COLLECTION

For the investigation of the phenomenon of road traffic fatalities in 121 UN countries, a time series, macro-level socio-economic information was collected. In particular 25 variables were collected from four global organizations (e.g., World Health Organization, World Bank, Natural Earth and World Atlas). From this initial time series database, only two years were chosen for investigating the phenomenon of road traffic fatalities, which are the years 2010 and 2013. Thus, two databases were created including the same 25 variables with each database referring to the above 'instances' years. The reason for selecting those many variables was to cover a wide range of socio-economic factors but also for treating endogeneity.

At the beginning the sample was concerned 193 UN countries, however the 72 out of the 193 UN countries were omitted from the database, because their population is below the 1 000 000 people (meaningless studying the countries with such a small number of population).

Figure 1, presents the data visualization of both databases and also the difference between the databases. From Figure 1 (a) and (b) it is not easy to observe significant differences between the two databases, however Figure 1 (c) depicts the differences between the two databases and it is obvious that with the yellow color are presented the vari-

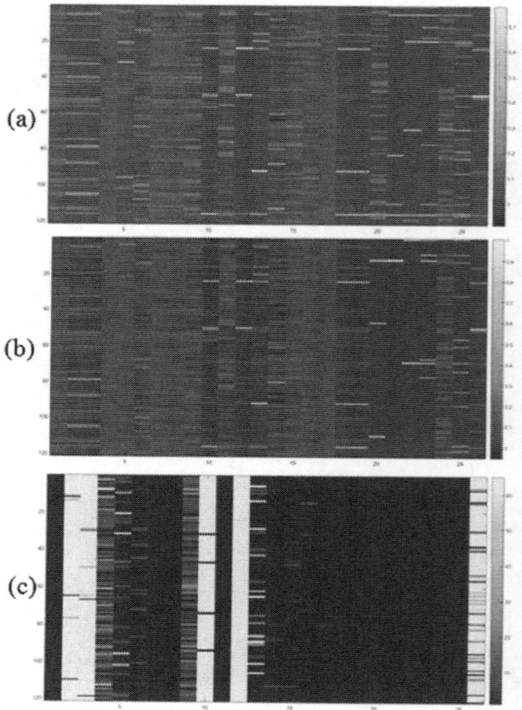

Figure 1. Data visualization (a) 2010 database; (b) 2013 database; (c) differences between both databases.

ables with the most significant differences. These variables are, Gross National Income (GNI), Gross Domestic Product (GDP), number of registered cars, population, reported number of road traffic fatalities.

From each database, four data-sets were created with all containing a different number of socio-economic sectors. Utilizing the data, which were collected from open sources without any preprocessing, inherent risks of information inflations. The next section provides information of treating these inflations and implementing PCA, NB regression and SEM.

4 METHODOLOGICAL APPROACH

This section presents the methodological approach, which was followed, and provides information about treating information inflation and implementing PCA, NB regression analysis and SEM for identifying data inconsistencies by comparing the obtained results from the different in year of study models.

4.1 *Information Inflation (collinearity) treatment*

After the data-set's creation the next step was to check the data for multicollinearity, so the results to be unbiased. Thus correlation tables were created for all data-sets, for the investigation of the collinear variables. In every correlation table the high correlated variables (pairs) were investigated as possible collinearity variables.

High correlated pairs were those who belonged in the range above 0.7 and below –0.7. The criterion for choosing which variable was the collinear from the high correlated pairs was the correlation that every of those two variables had with the dependent variable, i.e., the variable with the smallest correlation with the dependent variable was the collinear.

4.2 *Principal component analysis*

After the omission of the collinear variables from the data-sets the next step was to implement PCA.

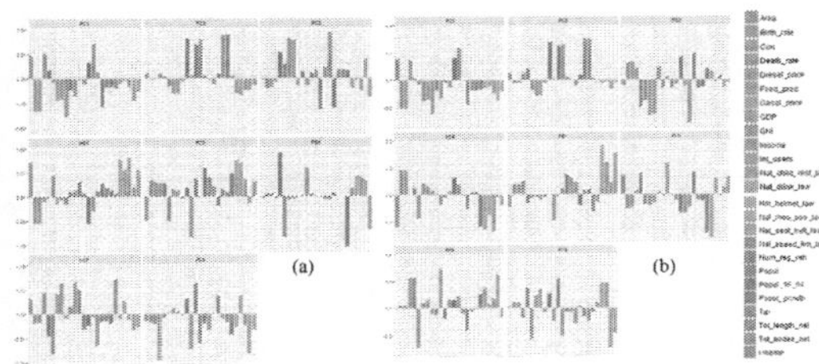

Figure 2. Principal components (a) 2010 database; (b) 2013 database.

One of the primary objectives PCA has is to interpret data. This object is accomplished by explaining the variance-covariance structure using a few linear combinations of the originally measured variables. Through this process a more parsimonious description of the data is provided-reducing or explaining the variance of many variables with fewer well-chosen combinations of variables (Washington et al., 2011). In the current paper, PCA was implemented for estimating the proportion of variance that the data of each database has and to correlate to each other.

From both databases, 8 principal components were selected since the cumulative proportion was above 80%. Figure 2, presents the variance of the variables in each principal component. Additionally, from Figure 2 it can be observed that each database, regardless the fact that they include the same variables from different year of study, however the variables concluded to have, in large scale, different principal component coefficients. Thus, the PCA implementation assisted the detection of data inconsistencies. Once again the claim that these data cannot be used for extrapolation is strengthened.

4.3 *Negative binomial regression analysis*

In this step the Negative Binomial (NB) regression analysis was implemented for investigating the relationship between the independent variables with the road traffic fatalities by estimating the coefficients of the variables in each model and comparing the resulted statistically significant models that were obtained from the different databases.

In particular 8 models were developed for estimating the road traffic fatalities; 4 concerning 2010; and 4 concerning 2013. From Table 1 it can be observed that the 2013 models have a better fit (according AIC and R-squared) instead of the 2010 models. Additionally, the final form of each different year's model concluded to include, more or less, the same number of variables. From this table it is not easy to detect any data inconsistencies, thus some researchers are falling into the trap of using these data for extrapolation, which is very dangerous considering the damage that might be occurred towards policy/decision making.

For space saving reasons, only two equations will be presented. Equation 1 presents the final form of model 1 for the year 2010 and equation 2 presents the model 1 for the year 2013.

$$\ln(\text{report\_acc}) = 7{,}54 + \text{Income(Low)} + \text{Income(Middle)} - 0{,}81 \times \text{Diesel\_price} + 3{,}77 \times \text{Num\_reg\_veh} \quad (1)$$

Table 1. The resulted models after implementing NB regression analysis.

Models/ Sectors	Economy (*10 variables*)	Demograpic/ Geographic (*7 variables*)	Network (*2 variables*)	Enforcements (*6 variables*)	*Total*	R^2	AIC	Null deviance	Residual deviance
2010 Models									
Model 1	3	–	–	–	3	0.407	2101	384.2	137.1
Model 2	1	4	–	–	5	0.413	2114	376.8	137.4
Model 3	1	3	1	–	5	0.408	2107	393.5	136.7
Model 4	1	3	1	1	6	0.400	2104	414	120
2013 Models									
Model 1	3	–	–	–	3	0.495	2078	393.3	137.2
Model 2	2	4	–	–	6	0.430	2094	389	137.4
Model 3	3	1	1	–	5	0.462	2090	374.6	137.9
Model 4	3	1	1	3	8	0.458	2088	393.5	137.2

$$\ln(\text{report\_acc}) = 8{,}36 - 1{,}86 \times \text{GDP} - 0{,}76 \times \text{Diesel\_price} + 3{,}25 \times \text{Num\_reg\_veh} \quad (2)$$

where, "report_acc" = reported road traffic fatalities

Num_reg_veh = number of registered vehicles

Income (Low) = 0,55 presenting the coefficient for the low level of income for the variable "Income"

Income (Middle) = 0,82 presenting the coefficient for the middle level of income for the variable "Income".

Despite the fact that both models included the same variables but with the only difference to be the year of study, however they concluded to have a different form with different intercepts. Equation (1) shows that the fluctuation of the road traffic fatalities in 2010 for the 121 UN countries was according the income levels, diesel price and the number of registered cars. Equation (2) shows that the road traffic fatalities in 2013 are correlated with the GDP, diesel price and the number of registered cars. Both models included the variables diesel price and number of registered cars and they have almost the same coefficients which means that these two variables are good to be used for future studies. As has been mentioned above implementing NB information inconsistencies are not easily observed. Further investigation of the latent information of the data will be offered in the next subsection.

4.4 *Structural equation modeling*

The current study did not confine only to the implementation of NB, SEM was also implemented, incorporating observed and latent structures in a seamless manner.

SEM model can accommodate a latent variable as a dependent variable something that cannot be done in a standard regression analysis. Furthermore, SEMs provide a way to check the entire structure of data assumptions not just whether the dependent variable predictions fit observations well. Also the complexity of variables relationships accommodated in the SEM framework translates to a significant increase in the potential for data mining. In a SEM, there is a risk that the number of model parameters sought will exceed the number of model equations needed to solve them (Washington et al., 2011).

In the current implementation 8 models were developed, 4 concerning 2010 and 4 concerning 2013. Table 2 shows the statistical information of the 8 models' final form. Subsequently, four good-of-fit indices (Akaike Information Criterion (AIC), Bayesian Information Criterion (BIC), Goodness of Fit Statistic (GIF), Root Mean Square of Approximation (RMSEA)) were used for selecting models with good fit.

Notwithstanding, the fact that in NB all the models for 2013 had a better good-of-fit (according AIC) from 2013 models, in SEM implementation it seems that the opposite is true according RMSEA and AIC, i.e. 2010 models have a better fit than 2013 models (always according RMSEA and AIC indices). Furthermore, it seems that the variables from the sector "Network" were omitted in every case they been used. Also the variables from the sector "Economy" were omitted, when they been used in the model 3 for both 'instances' years.

For space saving reasons two models were selected to be depicted. Figure 3, shows the initial and the final form of the model 1 for the year 2010 and Figure 4 shows the initial and the final form of the model 1 for the year 2013.

From Figure 3, it can be observed that the model started with 10 variables and concluded to have 4

Table 2. The resulted models after implementing SEM.

Models/ Sectors	Economy (*10 variables*)	Demograpic/ Geographic (*7 variables*)	Network (*2 variables*)	Enforcements (*6 variables*)	*Total*	AIC	BIC	GFI	RMSEA
2010 Models									
Model 1	4	–	–	–	4	26.44	54.39	0.98	0.05
Model 2	4	2	–	–	6	66.73	105.87	0.92	0.12
Model 3	0	1	0	–	1	6.00	14.39	1.00	1.33
Model 4	4	1	0	4	9	100.67	159.38	0.91	0.08
2013 Models									
Model 1	5	–	–	–	5	54.34	87.89	0.92	0.14
Model 2	5	1	–	–	6	66.85	105.99	0.92	0.12
Model 3	0	1	0	–	1	6.00	14.39	1.00	0.21
Model 4	5	1	0	0	6	66.85	105.99	0.92	0.12

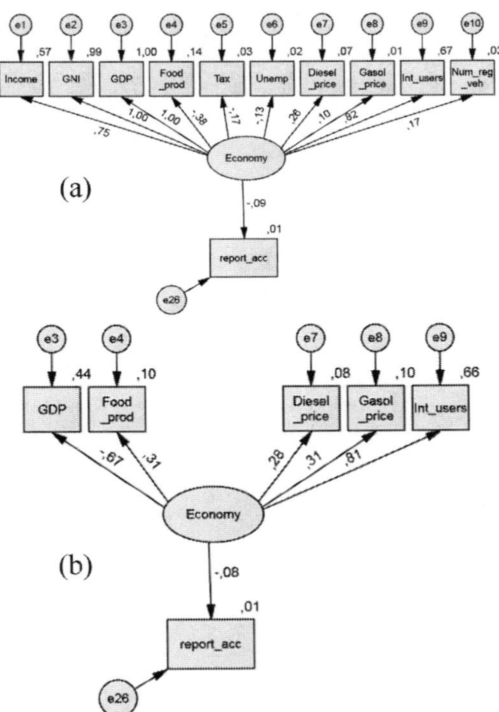

Figure 3. SEM model 1, year of study 2010 (a) initial form; (b) final form.

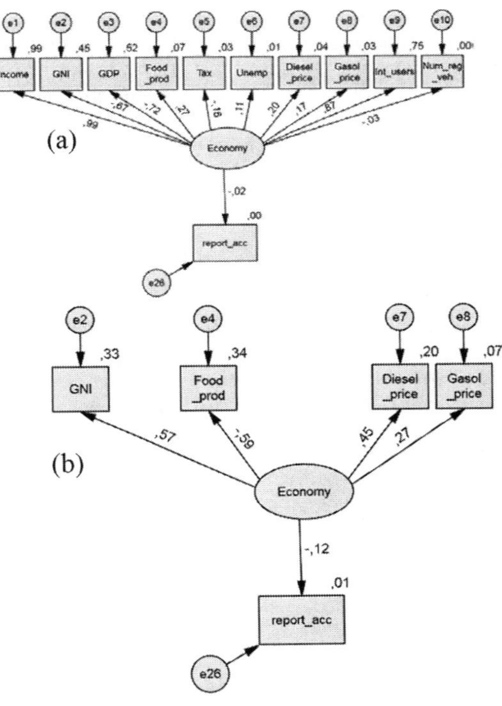

Figure 4. SEM model 1, year of study 2013 (a) initial form; (b) final form.

variables. The variables that are correlated with the road traffic fatalities in 2010 according model 1 are, GNI, food production, diesel price and gasoline price.

Only the variable "diesel price" seems to be as a common variable as the NB model 1 for estimating road traffic fatalities in 2010. The variables GNI, diesel price and gasoline price tend to increase the latent variable "Economy", instead of the variable food production which seems to decrease it. In general, "Economy" seems to decrease the road traffic fatalities.

Despite the fact that model 1, which is estimating the road traffic fatalities in 2013 had the same initial form as model 1 for estimating the road traffic fatalities in 2010, although it concluded with the variables GDP, food production, diesel price, gasoline price and internet users. This SEM model included the variable GDP and diesel price as the NB model. The variables food production, diesel price, gasoline price and internet users seem to increase the latent variable "Economy" instead of the variable GDP which tend to decrease it. In this model the variable food production is increasing the latent variable "Economy" in contrast with the variable food production from 2010 which seems to decrease it. In general, for both models the latent variable "Economy" appears to have a negative 'influence' on road traffic fatalities' increment.

5 CONCLUSIONS AND OUTLOOK

The current paper offers a novel methodological approach on detecting data inconsistencies and hints on which socio-economic data can be trusted for investigating the phenomenon of road traffic fatalities in a macro-level across the globe.

First of all both databases were thoroughly investigated for possible information inflations, omitting the collinear variables and so creating robust data-sets. Continuously, the PCA was implemented for the explanation of the variables' variance in each database. PCA was able to show dissimilarities that the databases have between them. As a first point, these dissimilarities can be explain as data inconsistencies, because the time gap is that small (3 years) and the variables diverge that much. However, before any hasty conclusions the methodology continued by implementing NB regression analysis and SEM.

By implementing NB regression analysis no great differences were detected between the models of estimating road traffic fatalities in 2010 and 2013. Although, some small differences were observed in which variables are interpret the phenomenon in each time 'instance'. The variables diesel price and number of registered cars in the model 1 of 2010 and 2013 appeared to be correlated with the road traffic fatalities, indicating that these two variables can boldly be used for investigating the road traffic fatalities.

Moreover, SEM models helped to detect the data inconsistencies between those two databases. Despite the fact that SEM models, more or less, concluded to the same form (same included variables), though the goodness-of-fit indices were those which stressed the differences between the databases (look Table 2, model 1 and 4).

Collecting data from several sources, even global organizations, inherent risks of data inconsistencies. Particular attention should be given by the researchers when using macro-level data toward decision/policy making. Furthermore, the current databases can be used for investigating and estimating the phenomenon of road traffic fatalities for the year that the data are referring to. All these precautions must be taken under the procedure of data collection before further methodological implementations.

These estimations can be used on short term policy making. Additionally, for avoiding using inconsistent information meaningless data (in terms of statistics) must be omitted from the databases.

Transport observatories can assist on the in attempt of collecting real information from monitors. Also, within Smart Cities data warehouses can be stored, so significant and open information will be available for researchers, and not only, to be exploited. Systems like smart meters, RFID tags, street security cameras, Foursquare, public transportation organizations, smart phones, logistics companies, sensor networks, etc., will installed in the Smart Cities and will offer valid information.

REFERENCES

Coruh, E., Bilgic, A., Tortum, A., 2015. Accident analysis with aggregated data: The random parameters negative binomial panel count data model. Analytic Methods in Accident Research 7: 37–49.

Deb, R., Liew, A.W.C., 2016. Missing value imputation-for the analysis of incomplete traffic accident data. Information Sciences 339: 274–289.

Fomina, M., Morosin, O., Vagin, V., 2014. Argumentation Approach and Learning Methods in Intelligent Decision Support Systems in the Presence of Inconsistent Data. Procedia Computer ScienceVol. 29: 1569–1579.

Hassan, H.M., Dimitriou, L., Abdel-Aty, M.A., Al-Ghamdi, A.S., 2013. Analysis of Risk Factors Affecting the Size and Severity of Traffic Crashes in Riyadh. Transportation Research Board Compendium.

Hinssen, P., 2012. Open Data Power Smart Cities. Across Technology.

Ma, J., Lu, J., Zhang, G., 2009. Information inconsistencies detection using a rule-map technique. Expert Systems with Applications 36: 12510–12519.

Washington, S.P., Karlaftis, M.G., Mannering, F.L., 2011. Statistical and Econometric Methods for Transportation Data Analysis Second Edition. CRC Press.

Yu, C.Y., 2015. Disparity in Traffic Safety across Neighborhoods with Different Economic Statuses and Ethnic Compositions. Transportation Research Board Compendium.

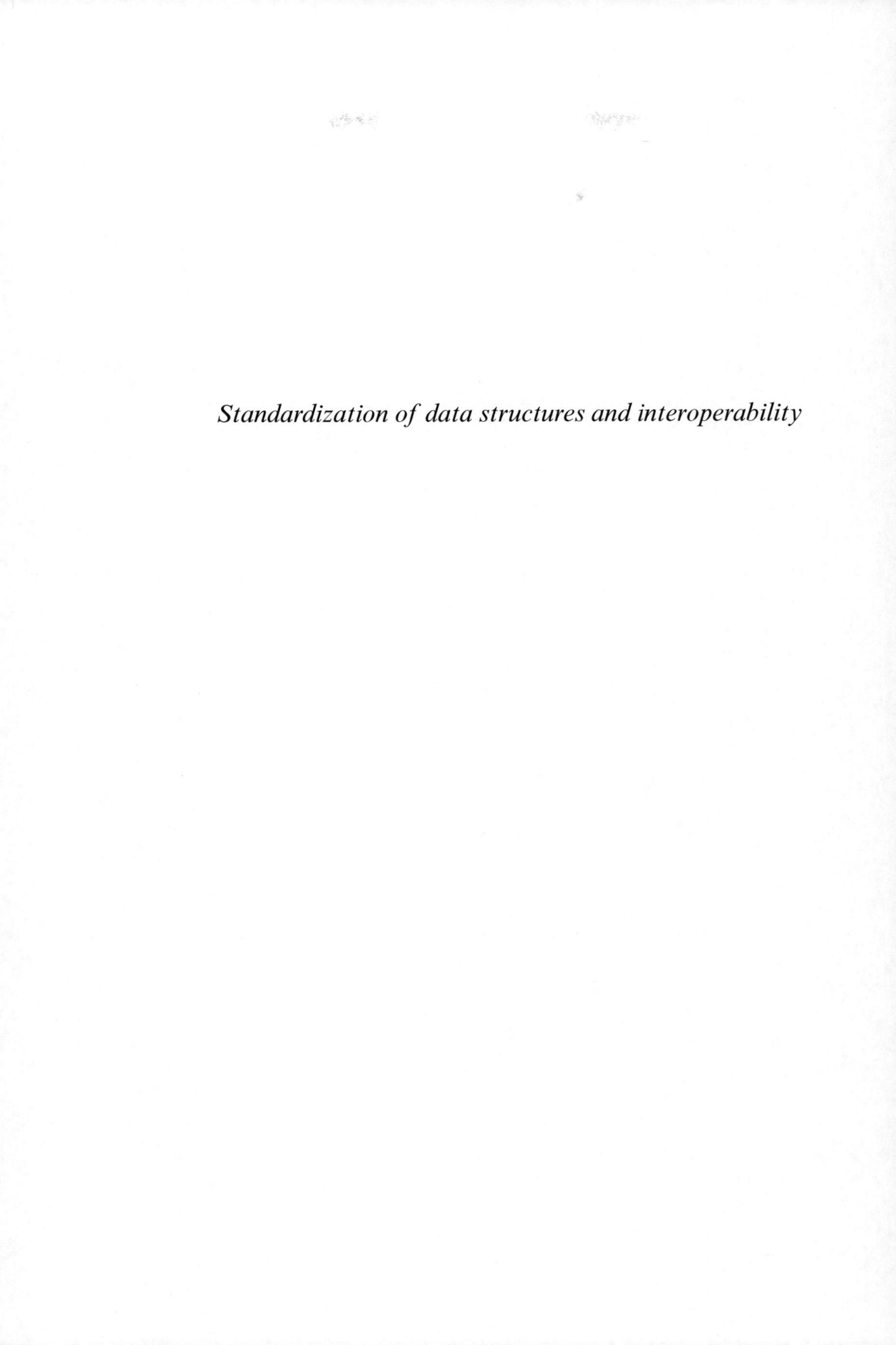

Standardization of data structures and interoperability

eWork and eBusiness in Architecture, Engineering and Construction – Christodoulou & Scherer (Eds)
 ISBN 978-1-138-03280-4

Facilitating the BIM coordinator and empowering the suppliers with automated data compliance checking

L.A.H.M. van Berlo
Netherlands Organisation for Applied Scientific Research TNO, Delft, The Netherlands

E. Papadonikolaki
Faculty of Architecture and the Built Environment, Delft University of Technology, The Netherlands

ABSTRACT: In projects with Building Information Modelling (BIM), the collaboration among the various actors is a very intricate and intensive process. The various suppliers and engineers provide their input in Industry Foundation Classes (IFC), which in turn is used for design coordination. However, the IFCs have to undergo an intermediate checking process to ensure compliance with various client-set and technical requirements. The paper focuses on the potential of (semi-)automatic IFC compliance checks and discusses a relevant recent initiative in the Netherlands, according to which several IFC compliance criteria were agreed among 14 contractors. This study aims to unravel the changes induced by this development not only as to the IFC compliance checks, but also as to potentially balancing the roles between the BIM coordinator and the suppliers.

1 INTRODUCTION

The use of Building Information Modelling (BIM) technology in Design and Construction has proliferated the last years. The benefits of BIM have contributed to the increasing number of adopters in Architectural, Engineering and Construction (AEC) industry. The information exchange among the various disciplines has been made possible via Industry Foundation Classes (IFC), which is the main open data standard (Berlo et al., 2015). Although it is argued that IFC still faces semantic challenges (Amor, 2015), particularly due to its evolving structure, it offers satisfactory and consistent information flows.

Currently, BIM implementation has been synonymous with IFC exchange among the various multi-disciplinary actors. Recently, the periodic—usually weekly—control sessions of IFC have become increasingly important and complex. These periodic controls involve various actors of the AEC supply chain, such as engineers, consultants and suppliers, who generate their own version of the building project. An emerging role in this process is the role of the BIM coordinator, who is in charge of the process of IFC exchange and federation of the various aspects of BIM models from the different disciplines respectively, under the concept of 'aspect models' (Berlo et al., 2012). However, this process is continuously undergoing change, as there are many efforts taken to automate the process of receiving, checking and federating the IFC aspect models. Subsequently, such changes induce transformations in the roles of the BIM coordinator and the involved suppliers.

The contribution of this paper is to provide new insights into the process of automated data compliance checking and the respective changes in the roles of the BIM coordinator and the suppliers. The study draws data and presents recent efforts from both 'top down' mandates and 'ground-up' industry initiatives for compliance checking criteria to ensure that every actor could always access, handle and reuse information in a consistent manner.

The paper is organised as follows. First, the background, related work and research gap pertinent to automated data (IFC) compliance checks and the associated emerging roles are presented. Second, the selected methodology to present and analyse the data is described. Then, the data from the semi-automatic improvements to periodic controls are presented, and their inter-organisational implications are discussed. Finally, the paper concludes with a summary of the main benefits and repercussions of the phenomenon under study and sets points for future research.

2 BACKGROUND, RELATED WORK AND RESEARCH GAP

2.1 *Industry Foundation Classes (IFC)*

Previous research has underlined the potential of BIM for reliable collaboration through the combination of IFC and model-checking software, such as Solibri Model Checker and Tekla BIMSight. The IFC is currently the main open data standard for practitioners in AEC. The IFC was initially introduced in the 1990s from an international consortium of software vendors and researchers, who formed the International Alliance for Interoperability (IAI), now called buildingSMART. The development of the IFCs has undergone various versions and additions the past decades, which have become a burden for the software developers and vendors (Amor, 2015). Despite those challenges, the IFCs provide freedom to the end-users to use the proprietary BIM authoring application of their preference to design their part.

Accordingly, as the most readily used open data structure, the IFC model has continued gaining traction among the AEC professionals, and in particular in combination with model checking software. Solibri Model Checker, BIMserver.org, and Tekla BIMSight are popular model checking applications that rely heavily on the use of IFC data. Most of the model-checking applications support the data exchange about specific issues that arise in a project in the BIM Collaboration Format (BCF). By aggregating and viewing simultaneously the IFC data from various disciplines in a project, e.g. engineers and suppliers, these applications report on coordination issues among the input of the various involved actors, using built-in or custom-made checking rules.

2.2 *Collaborative Engineering with IFC*

Various experiments with IFC and model checking software have been conducted the last years. These efforts were triggered by the curiosity to understand and explain the collaboration with BIM (Berlo et al., 2012). These research efforts have taught us that the dogma of a 'central BIM repository' itself does not bring (much) added value in a team collaboration. Unfortunately, the misconception that when everyone uses the same data structure, collaboration will immediately happen still exists. This idea probably adds to the confusion about how to collaborate efficiently and effectively with BIM, given that it could essentially be defined as *'a multifunctional set of instrumentalities for specific purposes that will increasingly be integrated, but to what extent is an open question'* (Miettinen and Paavola, 2014). Thus, the processes and functions to collaborate within the concept of BIM are still under development and rely heavily on the technological advancements.

In the Netherlands, there are numerous studies on the emerging BIM-based collaboration processes via IFC and model-checking software, following the concept of 'reference models' (Berlo et al., 2012, Berlo et al., 2015). After experimenting with the use of BIM in a central data repository it was concluded that the concept of exchanging 'reference models' (or 'aspect models') was a stable way to exchange data produced by the various disciplines, in their preferred BIM software, in an asynchronous manner, e.g. weekly (Berlo et al., 2012). Following-up experiments in the Netherlands further confirmed the use of IFC and model checking software as a common engineering practice in a BIM environment and underscored the challenge of BIM to align with the existing project phases (Berlo et al., 2015). From the previous studies, one could conclude that the periodic controls of the IFCs consist of three inter-connected steps:

- Compliance check with the requirements (data level);
- Aggregation of the IFC files (information and/or data level);
- Coordination of the design (information level).

2.3 *Data compliance checks with IFC*

To perform high-quality coordination and model checking, the data has to be of high quality as well. The data requirements are usually stored in a BIM Protocol or BIM Execution plan. In the recent years, a commonly accepted standard has emerged for IFC data requirements. These requirements are heavily influenced by the Dutch Rijksvastgoedbedrijf BIM norm. Rijksvastgoedbedrijf (2012) started mandating IFC data requirements with a norm that in turn seemed to have been influenced by the Norwegian equivalent authority of Rijksvastgoedbedrijf, called Statsbygg, which previously published similar BIM requirements (Statsbygg, 2011).

The IFC data requirements have formed a baseline for checking the IFC data requirements in the industry for a long time. In the recent years, the industry has extended and additionally fine-tuned the original IFC requirements set by the Rijksvastgoedbedrijf. From anecdotal sources and informal interviews with practitioners, the following set of requirements has now emerged as a common requirement set used by most AEC contractors:

- There should be only one IfcProject object per file (no more, no less);
- There should be only one IfcSite object per file;
- All objects should be linked to an IfcBuildingStorey object;
- There should be at least one IfcBuilding object in the dataset;
- There should be at least one IfcBuildingStorey in the dataset;
- The naming of the building storeys should be consistent and in order, i.e. floor-numbers;

- The length unit should be millimeters;
- The area unit should be square meters;
- The volume unit should be cubic meters;
- The objects should be 'close' to the origin point (0,0,0) of the dataset;
- Objects found across multiple 'aspect/reference models' should have the same position and orientation point;
- There should be no intersections in the individual 'aspect models';
- There should be no duplicate objects in the entire dataset.

The clients' firms have mandated the same requirements and further added the following requests:

- The dataset should have the true North set;
- The site elevation should be set;
- The site latitude and longitude coordinates should be set;
- The site cadaster ID should be available.

Most of the above requirements come from a combination of the Statsbygg requirements, the Rijksvastgoedbedrijf BIM norm, and practical insights. On April 2016, a new initiative in the Netherlands reached an agreement about the criteria for the compliance check to ensure that every party could always access, handle and reuse information uniformly. This initiative was initiated by BuildingSMART Benelux and was executed with a core team of fourteen large and medium size Dutch contractors. This action could simplify, improve and reduce the conflicts during the periodic controls of the IFC within and across construction projects. The goal of this initiative, called 'Information Delivery Specification' was to align the BIM data requirements that contractors mandate to their project partners. The agreed criteria for the compliance check are (National BIM Guidelines, 2016):

1 Uniform file naming of the various reference models from the difference disciplines;
2 Same position and orientation point;
3 Consistent naming and appending of the various building levels;
4 Correct generation and structure of IFC objects;
5 Correct names in the IFC entities;
6 Consistent classification of the objects under the NL/SfB system;
7 Correct attribution of the materials' description;
8 Elimination of duplicated entities and internal clashes per aspect of the federated model.

From the above, there is an apparently extended overlap between the 'Information Delivery Specification' and the requirements previously described at the beginning of this sub-section, derived from literature and anecdotal evidence from interviews.

2.4 *Emerging functions during collaborative engineering with IFC*

The above means, i.e. IFC, and processes, i.e. periodic controls of the IFC, affect the functions and the responsibilities of the various involved professionals during the information exchange in a BIM environment. The process of the—usually weekly—periodic controls of the IFC from the various disciplines is governed by BIM coordinator, who is responsible for the check, aggregation, and coordination of the data and information. However, given that the concept of collaborating in a BIM environment is currently under development, there is a lot of ambiguity about the appropriate functions to support a BIM project.

Gathercole and Thurairajah (2014) provided evidence of this ambiguity by mapping how the various BIM functions are described in BIM-related job advertisements in the United Kingdom (UK). From their analysis, three main BIM-related roles emerged: 'BIM manager', 'BIM coordinator' and 'BIM technician'. Whereas the BIM managers had mostly BIM-related project administrative functions and the BIM technician had mostly modeling duties, the BIM coordinator's role was inconclusive and assumed responsibilities from both administrative and modeling domains (Gathercole and Thurairajah, 2014). Thus, there is a gap of knowledge in the function and responsibilities of the BIM coordinator.

Another aspect to be considered about the BIM-related functions is apart from the technical and administrative skills, the soft competences that are mobilised in a multi-disciplinary setting. Davies et al. (2015) underscored that skills such as communication, conflict management, negotiation, teamwork, and leadership were also deemed essential for a BIM environment. This paper distinguishes the firm-based BIM manager who is responsible for BIM adoption in a firm including BIM strategy and training, from the project-based BIM coordinator, who is responsible for the periodic IFC controls and operational issues, and focuses only on the latter.

As presented through the various efforts to coordinate the periodic IFC controls across firms, there are still opportunities to further improve these controls and particularly regarding the compliance check to the requirements. The changes and improvements in IFC compliance checking could probably lead to a reconsideration of the function and the role of not only the BIM coordinator but also the various other parties involved in the periodic controls with IFC, i.e. the engineers and suppliers. Thus, there is a lack of understanding about the changes that the semi-automated IFC compliance checking entails for the (a) periodic IFC controls and (b) the function of the BIM coordinator and the various involved actors, such as engineers and suppliers. The study sets out to explore the following research question: *How does the pre-processed automated IFC compliance checking affect the process and the functions required for the periodic controls of IFC?*

3 RESEARCH APPROACH

3.1 *Setting of the study*

The Netherlands has been selected for this study because it has been displaying a variety of examples on self-regulation in the AEC market. Also, this study could be considered a follow-up of other relevant previous studies about collaborative engineering with IFC in the Netherlands, e.g. Berlo et al. (2012) and Berlo et al. (2015). Moreover, compared to other countries who are usually described as forerunners in BIM adoption, in the Netherlands there is a balanced distribution of mandatory and suggestive publicly available publications such as guides, protocols, and mandates (Kassem et al., 2015). The Dutch BIM policy-making authorities have been so far reluctant to publish mandatory documents and have been mainly relied on a 'ground-up' rather than 'bottom-up' BIM diffusion approaches. After all, Winch (2002:25) has described the Dutch construction industry as a *Corporatist type System* according to which, the "social partners" are more keen to negotiate and coordinate to control the market. Accordingly, the Dutch AEC firms are proactive in adopting new technologies with a consensus-seeking culture.

3.2 *Hypothesis*

A common iteration in the Dutch industry between contractors and suppliers is described by Berlo et al. (2015) and shown in the top part of Figure 1(a). The suppliers send their design input as IFC data to the contractor during the week. The contractor checks and aggregates the data (usually) on a Friday and performs several checks of the model as part of their project coordination role.

Unfortunately, not all IFC requirements are met by suppliers, according to testimonials from practice. This in turn induces a great burden on the BIM coordinator, who would have to rectify the situation, during the coordination sessions. The BIM Coordinator has less time available to spend on actual project coordination among the various disciplines, because he would spend much

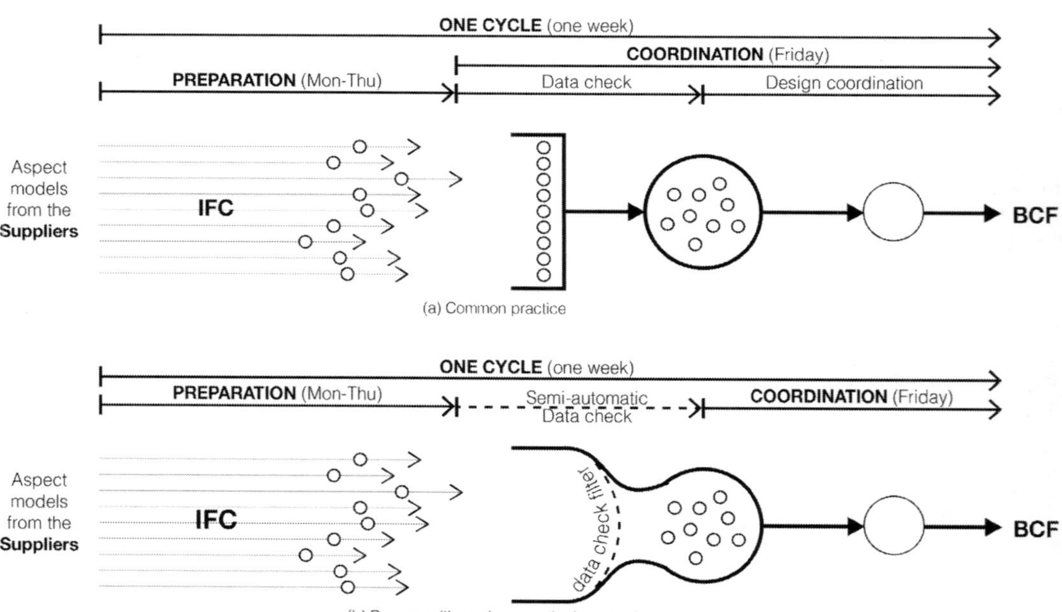

Figure 1. (a) The common practice of coordinating IFC data where the data are checked on compliance by the BIM coordinator and (b) the new process featuring semi-automatic data checks.

time on fixing the non-compliant models instead, given that the sub-process between the submission of the data and the compliance check behaves as a bottleneck. The research hypothesis is formed as follows:

The BIM Coordinator (project coordinator) could spend more valuable time on actual coordination tasks when the data from the suppliers is meeting the requirements.

The hypothesis would be tested by comparing how the checking tasks could be reduced in a new process, shown at the lower part of Figure 1(b). The data from the suppliers that is usually sent via e-mail or put in a shared folder, e.g. Sharepoint, Dropbox, and many others, could be instead uploaded to an online model server. During the uploading of the data, the dataset is checked against basic requirements. When the data does not meet the requirements, the upload function is rejected by the server. This process ensures that only qualified data reaches the BIM coordinator. The suppliers are 'obliged' to improve and send valid data for the coordination.

3.3 *Research methods*

The main research method was an exploratory case study, using reports on IFC data from three projects in the Netherlands. Out of the fourteen contractors that participated in the previous BuildingSMART initiative, three were selected by their long-lasting experience with using an (online) checking platform. The three studied projects used recent data, not older than a year. The IFC and model-checking processes were described, analysed and evaluated as to following aspects:

- The rules used for checking the BIM models;
- The BIM protocols as ground for setting requirements for the engineers and suppliers;
- The tasks and the role of the BIM coordinator;
- The software infrastructure for the controls;
- The BIM readiness and capability of the contractors, designers, engineers, and suppliers.

This study analysed several datasets from suppliers and checked them against the requirements in an automated manner. The following checks were performed:

1. Is there one (and only one) IfcProject object?
2. Is there one (and only one) IfcSite object?
3. Are all objects linked to an IfcBuildingStorey object?
4. Is there at least one IfcBuilding object?
5. Is there at least one IfcBuildingStorey?
6. Is the naming of the building storeys in order (numbers)?
7. Is the length unit in millimeters?
8. Is the area unit in square meters?
9. Is the volume unit in cubic meters?
10. Does the dataset have the true North set?
11. Is the site elevation set?
12. Are the site latitude and longitude coordinates set?
13. Is the site cadaster ID available?

For one case (case A) also additional checks were performed using the semi-automated approach:

- Are the objects 'close' to the origin point (0,0,0) of the dataset?
- Are all objects from multiple 'aspect models' on the same position and origin point?
- Are there no intersections in the aspect models?
- Are there no duplicate objects in the entire dataset?

4 DATA ANALYSIS

The study compared the IFC periodic controls of three cases. The cases featured the common practice of iterative collaboration based on IFC, during which once a week, the segregate models from the various disciplines were checked, aggregated and co-ordinated via model-checking software by the BIM coordinator as described by Berlo et al. (2015). The cases followed the experimental process for automated IFC compliance checks. The various involved suppliers used an online platform to upload and automatically check their IFCs mid-week. In total 88 files from suppliers were analysed in the three cases (A, B, and C). The results are shown in Table 1. The rows contain the various suppliers per case, and the columns contain the different checks performed in the datasets. The checks are compared using the symbols "v" and "-" when the file was or not compliant respectively. Three requirements were met in all the files: only one IfcProject object (requirement 1); at least one IfcBuilding (requirement 4) and the length unit (requirement 7). Therefore, these results are not included in Table 1.

For case A, an extra semi-automated analysis was conducted using four additional requirements. These analyses could also be supported by the automated proposed process, however, to gain more insight into the actual issues and challenges arising, Solibri was used instead. Table 2 presents the data from these additional checks. Again, the checks are compared using the symbols "v" and "-" when the file was or not compliant respectively. Also, where indicated, the total number of errors is calculated accordingly.

The checks that were performed focus on the geometric compliance of data. All 49 IFC files of the case A were analysed. All files were close to the origin point. 53% of the files had internal clashes.

Table 1. Results from the data compliance checks per case.

Supplier per case	No. of IFC files	Check #2*	Check #3	Check #5	Check #6	Check #8	Check #9	Check #10	Check #11	Check #12	Check #13
A1	1	–	–	–	v	v	v	–	–	–	–
A2	1	v	–	v	–	v	v	–	–	–	–
A3	3	–	–	–	v	v	v	–	–	–	–
A4	4	v	–	v	v	v	v	–	–	–	–
A5	1	v	v	v	v	v	v	v	v	v	–
A6	1	v	v	v	–	v	v	v	v	v	–
A7	1	v	–	v	–	v	v	–	v	–	–
A8	3	v	–	v	–	v	v	–	v	–	–
A9	13	v	v	v	–	v	v	v	v	v	–
A10	1	v	–	v	–	v	v	v	v	v	–
A11	1	v	v	v	–	v	v	–	v	–	–
A12	4	v	–	–	v	v	v	–	v	–	–
A13	4	v	–	v	–	–	v	–	–	–	–
A14	11	v	–	v	v	–	v	–	–	–	–
B1	1	v	v	v	–	–	–	v	v	v	–
B2	2	v	v	v	v	v	v	v	v	v	–
B3	1	v	v	v	–	v	v	–	v	–	–
B4	1	v	v	v	v	–	–	v	v	v	–
B5	1	v	v	v	v	–	–	–	v	v	–
B6	1	v	–	–	v	v	v	–	v	–	–
B7	1	v	v	v	–	–	–	v	v	v	–
B8	1	v	–	v	–	–	–	–	–	–	–
B9	4	–	–	–	v	v	v	–	–	–	–
B10	1	v	v	v	–	–	–	v	v	v	–
B11	3	v	–	v	–	v	v	–	v	–	–
B12	1	v	–	v	–	v	v	–	v	–	–
B13	1	v	v	v	–	v	v	–	v	–	–
B14	1	v	–	v	–	–	–	v	v	v	–
C1	3	–	–	–	v	v	v	–	–	–	–
C2	3	–	–	v	–	v	v	–	–	–	–
C3	3	–	–	–	v	v	v	–	–	–	–
C4	1	v	–	v	–	v	v	–	v	–	–
C5	3	v	v	v	–	v	v	v	v	v	–
C6	4	v	–	v	–	v	v	v	v	v	–
C7	2	v	v	v	–	v	v	v	v	v	–
Sum	88	29	14	28	13	26	28	13	24	14	0
%	–	83	40	80	37	74	80	37	69	40	0

*The checks' numbers refer to the numbered list in section 3.3.

These are clashes in the individual files, not in the coordination among the various disciplines' input. The BIM/project coordinator would come across these internal clashes during the coordination sessions and would turn his attention away of the actual design coordination of the project. When clashes were found, the number of clashes (intersections) is listed. Only 76% of the files did not have duplicate objects. The number of duplicate objects per file is also shown in Table 2. However, it strongly depends on the type of coordination (and the phase of the project) if the work from the BIM/project coordinator is affected by these duplicate objects.

The compliance overall percentage of the cases is:

- Case A: 64%
- Case B: 65%
- Case C: 64%

These numbers are quite similar. This is due to the high overlap of requirements that are met, e.g. of the 1st requirement about having only one ifc-Project, and not met, e.g. of the 13th requirement

Table 2. Additional criteria check for the IFC files of Case A.

No. IFC files	Origin point	Intersections	Errors intersecting components	Duplicates	Errors duplicate components
26	v	v	N/A	v	N/A
1	v	–	16	–	1
1	v	–	16	v	N/A
1	v	–	56	v	N/A
1	v	–	208	–	10
1	v	–	136	–	6
1	v	–	121	–	6
1	v	–	103	–	5
1	v	–	6	v	N/A
1	v	–	N/A	v	N/A
1	v	–	1039	v	N/A
1	v	–	38	–	9
1	v	–	4	–	10
1	v	–	36	–	15
1	v	–	429	–	5
1	v	–	3	v	N/A
1	v	–	7	v	N/A
1	v	–	1	v	N/A
1	v	–	395	–	11
1	v	–	178	–	15
1	v	–	79	v	N/A
1	v	–	68	–	2
1	v	–	16	v	N/A
1	v	–	6	v	N/A
(Sum:) 49	100%	53%	–	76%	–

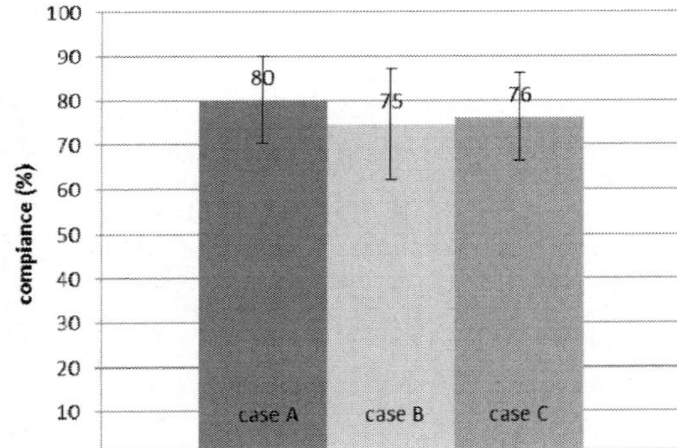

Figure 2. Average compliance to the requirements per case (and standard deviation).

about the cadaster information (see again Table 1).

Only two IFC files completely met the requirements: one in case A and one in case B. When we do not take the requirements 10 to 13 into account still only two models met 100% of the requirements (1 in case A and 1 in B). The average percentages of compliance, for only requirements 1 to 9, and the standard deviation, are shown in Figure 2.

5 DISCUSSION

5.1 *Inter-organisational implications*

After discussing and validating the research findings with the case participants, it was concluded that the automated compliance-checking process initially carries inter-organisational implications for the two main categories of project actors, i.e. the suppliers and the BIM coordinators, and afterwards, technical repercussions about the data compliance checks. Concerning the inter-organisational implications, the workload of the BIM coordinator and the suppliers was interdependent. The IFCs that were properly checked and prepared by the suppliers did not need additional coordination and communication back to the BIM coordinator to the suppliers.

First, the various engineers and suppliers were able to check beforehand their data for compliance with the requirements of the BIM protocols, without waiting for the approval of the BIM coordinator during the periodic controls. The suppliers were again responsible for the (data) compliance of their work to the requirements, as it was originally in traditional non-BIM-based projects. Such increased responsibility of the suppliers corroborates with the discussions of Nederveen et al. (2010) that with integrated design process across all tiers, the suppliers would assume a more dominant role in the design and thus, the construction project design process would be more 'ground-up.'

Second, the project-based BIM coordinator was able to evaluate the information provided by the various partners only as to its technical and engineering feasibility and not as to the syntactical conformance to the rules of the BIM protocol. The BIM coordinator's role became less related to BIM knowledge and more related to their domain expertise, e.g. architecture, engineering or quantity surveying. Simultaneously, the BIM coordinators would have the room to develop more "soft" capabilities from collaborating with the various partners, which could, in turn, corroborate the discussions of Davies et al. (2015) about an increased demand for additional soft skills to support BIM-based projects. One might claim that the role of the BIM coordinator might also be aligned or concur with the traditional role of the 'design coordinator,' who used to be responsible for the coordination of the multi-disciplinary input, within the contractors' firms.

5.2 *Benefits and challenges of the automated data compliance check*

The benefit of this process lies in avoiding the uncertainties of data compliance during the scheduled periodic controls for the alignment of the multi-disciplinary information flows. Looking back to Table 1, there are some interesting observations about the various checks performed, apart from the requirements checks #1, #4, and #7, which are not included in Table 1. For example, it seems difficult to link all the IFC objects to an ifcBuildingStorey (requirement 3), given that only 40% of the suppliers met that requirement.

Regarding the 6th requirement, i.e. the naming convention and numbering of building storeys, surprisingly, it also presented a lot of errors in the automated check. This is probably because no specific naming convention is given, and therefore the automated check can only work partially.

The requirements that are specific to client requests also scored relatively low in the analysis, e.g. the 10th and 13th requirements about the site elevation, true north, the latitude and longitude of the site, and the Cadaster information. Experiments have shown that it is very simple to inherit these properties from the original design model and have the BIMserver add them automatically to the supplier models. These checks are almost never mandated by contractors to the suppliers, through the BIM protocols, and therefore, probably they do not a significant influence on the coordination work of the BIM project coordinator

The requirements for the length, area, and volume unit (7th, 8th and 9th requirement respectively), are quite questionable as to their influence to the work of the BIM coordinator during the coordination. Most software tools recalculate different units during the coordination so the end user might not be hindered by the mismatches in these checks. Subsequently, the most essential requirements for the design coordination seem to be met most of the time.

In all cases and suppliers' files, the 13th requirement, about cadaster information, was never found compliant using the automated check. To investigate this issue further, some complementary checks with the automated checking tool in other datasets proved that the feature was completely functional.

5.3 *Research limitations and applicability*

Whereas the study reviewed only three cases, the findings could be generalised in other settings. Given that the IFC compliance process is an inseparable process of the federations of the aspect (or reference) models from various suppliers, the above implications and challenges could apply to more projects. At the same time, the presented recent initiative among the fourteen contractors in the Netherlands, who agreed to specific requirements for the IFCs, resembles a 'middle-out' BIM diffusion mechanism, previously described by Succar and Kassem (2015), which could in turn influence more firms and disciplines in the Dutch AEC sector.

At the same time, the above findings regarding the transforming roles of the BIM coordinator and the suppliers could be transferred in other countries, where similar checks have been defined 'top-down', such as in Norway. Regardless the content of the checks, which could differ from country to country and from client to client, the balance between the BIM coordinator and the contributing actors to the project design, would undoubtedly continue to shift due to the increasingly stricter and probably more sophisticated data compliance checks for the IFCs.

6 CONCLUSIONS AND FUTURE WORK

Some recent developments in the Netherlands included joint agreements on behalf of fourteen contractors for ensuring consistent IFC compliance checks throughout their projects. The impact of this initiative extends not only to these specific contractors, their projects, and the local Dutch AEC market but also aligns with the vision for the use of open data standards in AEC. At the same time, given that these contractors usually have long-term relations and form partnerships with several sub-contractors and suppliers, this initiative and way of establishing joint agreements in the Dutch AEC sector would potentially transform the smaller enterprises in the market. The study identified the conditions to popularise the adoption of automated compliance checks in AEC.

The automated compliance check process presented the potential for a more balanced division of tasks in BIM-based projects, which could not only improve the quality of the building product but also increase the satisfaction among the involved actors. Specifically, the study identified two tendencies across the roles of the involved parties in the exchange and federation of IFC aspect models. On one hand, the study sheds new light on the existing and vividly discussed, role of the BIM coordinator. In particular, it differentiated it to the various BIM-related roles, such as BIM managers and BIM technicians. Simultaneously, by providing evidence from automated data compliance checks, this study highlighted the potential of the BIM coordinator to resume a more design-related rather than a BIM-routinised role. On the other hand, the study highlighted the changing role of the engineers and the suppliers who would assume higher responsibility from checking, preparing, and ensuring that their aspect models would comply with the jointly agreed criteria for IFC-based project delivery.

Whereas this study has shown the potential of online automated compliance checking, there is no clear evidence that the quality of models will improve in the process. Further research would be required across multiple projects to validate this hypothesis. Online checking in this research was done with standardised requirements that were implemented in a custom-made checker tool. To fully reap the benefits of online model checking, a more dynamic approach should be additionally developed. Specific checks per supplier or per type of supplier would then be an option that potentially would contribute to an even higher level of data quality. Several technologies like Model View Definitions, Concept-libraries and BIM Query languages need further research and development to be used in a practical use case as described in this paper.

ACKNOWLEDGEMENTS

This study would not be possible without the help of the data providers. The open dataset Schependomlaan (2016) was used as case A. We thank all the people involved in this research for their openness and new insights to this topic.

REFERENCES

Amor, R. 2015. Analysis of the evolving IFC schema. *In:* Beetz, J., van Berlo, L., Hartmann, T. & Amor, R. (eds.) *CIB W78*. Eindhoven, Netherlands.

Berlo, L., van, Beetz, J., Bos, P., Hendriks, H. & Tongeren, R., van 2012. Collaborative engineering with IFC: new insights and technology. *9th European Conference on Product and Process Modelling, Iceland.*

Berlo, L., van, Derks, G., Pennavaire, C. & Bos, P. 2015. Collaborative Engineering with IFC: common practice in the Netherlands. *In:* Beetz, J., van Berlo, L., Hartmann, T. & Amor, R. (eds.) *Proc. of the 32nd CIB W78 Conference 2015*. Eindhoven, The Netherlands.

Davies, K., McMeel, D. & Wilkinson, S. 2015. Soft skill requirements in a BIM project team. *In:* Beetz, J., van Berlo, L., Hartmann, T. & Amor, R. (eds.) *CIB W78*. Eindhoven, Netherlands.

Gathercole, M. & Thurairajah, N. 2014. The influence of BIM on the responsibilities and skills of a project delivery team. *In:* Amaratunga, D., Richard Haigh, R., Ruddock, L., Keraminiyage, K., Kulatunga, U. & Pathirage, C. (eds.) *International Conference on Construction in a Changing World.* Heritance Kandalama, Sri Lanka.

Guidelines, N. B. 2016. *BIM Basis* [Online]. Available: http://nationaalbimhandboek.nl/onderwerpen/bim-geboden/.

Kassem, M., Succar, B. & Dawood, N. 2015. Building Information Modeling: analyzing noteworthy publications of eight countries using a knowledge content taxonomy. *In:* Issa, R. & Olbina, S. (eds.) *Building Information Modeling: Applications and practices in the AEC industry.* University of Florida: ASCE Press.

Miettinen, R. & Paavola, S. 2014. Beyond the BIM utopia: Approaches to the development and implementation of building information modeling. *Automation in construction,* 43, 84–91.

Nederveen, S., van, Beheshti, R. & Ridder, H., de 2010. Supplier-driven integrated design. *Architectural Engineering and Design Management,* 6, 241–253.

Rijksgebouwendienst. 2012. *Rgd BIM Standard, v. 1.0.1* [Online]. The Hague, The Netherlands: Rijksgebouwendienst. Available: http://www.rijksvastgoedbedrijf.nl/english/documents/publication/2014/07/08/rgd-bim-standard-v1.0.1-en-v1.0_2.

Statsbygg. 2011. *Statsbygg Building Information Modelling Manual Version1.2 (SBM1.2)* [Online]. Available: http://www.statsbygg.no/Files/publikasjoner/manualer/StatsbyggBIMmanualV1-2Eng2011-10-24.pdf [Accessed 29 April 2016 2016].

Succar, B. & Kassem, M. 2015. Macro-BIM adoption: Conceptual structures. *Automation in Construction,* 57, 64–79.

Winch, G. M. 2002. *Managing construction projects,* Oxford, Blackwell Science Ltd.

eWork and eBusiness in Architecture, Engineering and Construction – Christodoulou & Scherer (Eds)
 ISBN 978-1-138-03280-4

DRUMBEAT platform—a web of building data implementation with backlinking

N. Vu Hoang & S. Törmä
Aalto University, Espoo, Finland

ABSTRACT: The purpose of Web of Building Data (WoBD) is to support the sharing of building data among the diverse stakeholders over a construction lifecycle. DRUMBEAT platform is an open-source, proof-of-concept implementation of WoBD. Many practical issues have been addressed during its development, concerning the organization and description of published data, URI design for objects, publication and management of links, and the users' access to links. The paper describes in detail the architecture used to publish multiple manifestations of the same model: versions, views, and partial exports. Different ways to publish links are supported but the object access interface, and a special backlinking protocol ensure that links can always be included in object descriptions without complicating the interface with explicit link access methods.

1 INTRODUCTION

Over the lifecycle of a building, many different parties are involved in the production and utilization of building-related data. They include regulatory authorities, urban planners, emergency services, architects, designers, contractors, fabricators, product vendors, facility managers, maintenance persons, owners, residents, and visitors. For simplicity, in the remainder of this paper we will use the term *building data* to refer to the variety of data in the process ranging from requirements, specifications, urban plans, BIM (building information modelling) models, infrastructure models, project plans, and product data to all data produced in fabrication, construction, and operations.

As of now, the valuable building data produced by different parties is managed in a fragmented manner with fragments disintegrated with each other. It is underutilized because it is difficult to access and disconnected from potential users. (Törmä 2014)

Web of Building Data (WoBD) is a concept developed for information sharing between all stakeholders over the lifecycle of buildings (Törmä 2013, Törmä 2014). It is based on the Web of Data technologies of W3C, including basic Web standards (URI, HTTP), the Semantic Web representations (RDF, OWL, SPARQL), and Linked Data principles (Berners-Lee 2006). Web of Building Data means the application of these technologies to building data domain, taking into account the existing standards, for instance, the IFC (Industry Foundation Classes) format of BIM models. The IFC schema has been translated into an OWL ontology called ifcOWL and IFC files can be converted into RDF graphs (Beetz et al 2009, Vu Hoang 2015, Pauwels 2016).

Web of Building Data supplements the existing approaches for building information exchange with new capabilities. It supports the decentralized publication of building data and granular object-level access to data. The published data is online, up-to-date and addressable, and can easily be linked to and from other data on the Web. The definition of the semantics of data with common ontologies will in the future enable greater automation of processes.

When attempting to publish building data on the Web using ifcOWL and ifcRDF to implement practical use case, there are many additional questions that need to be solved:

1. How to organize the published data on a host?
2. How to describe the published data (metadata)?
3. How to define the identifiers (URIs) of objects?
4. How to publish and manage links?
5. How to provide the access to links to users?

This study addresses these questions. It describes how they are solved in the DRUMBEAT platform, an open-source proof-of-concept implementation of the Web of Building Data framework produced in the DRUMBEAT research project.

As described below, these questions are not as simple to answer as they may initially appear. For instance, the publication of data needs to address the problem of how to deal with different versions, views, and partial exports of a same model.

Or linking of data needs to address the question, how can all relevant parties become aware and get an access to published links in a decentralized system?

We present a set of interrelated solutions for these questions, and describe their implementation in DRUMBEAT platform. It should be noted that topics concerning security and access control are outside of the scope of this paper, and they will be discussed in Oraskari & Törmä (2016).

In the remainder of this paper we first review the world of building data (Section 2), technologies of Web of Building Data (Section 3), and the implementation of the DRUMBEAT platform (Section 4). After discussion (Section 5), we end with conclusions and identification of topics for future research (Section 6).

2 BUILDING DATA

During the recent two decades, advanced BIM tools for building design have emerged. As a result, a lot of structured information is created during the design stage of modern construction projects. However, the design work is split into different disciplines (such as requirements engineering, architectural design, structural engineering, MEP engineering, and sometimes subdivisions of these) and the resulting models are mostly disconnected from each other. There are multiple BIM models that describe different aspects of the same physical entities, yet the connections between the models are almost completely based on geometrical information—which objects appear to be at overlapping locations—that has to be repeatedly reconstructed for coordination purposes.

The production of the different models progresses partly sequentially and partly in parallel. Changes are ubiquitous throughout the design stage and when inconsistencies across designs appear, the models need to be brought into agreement with each other. Otherwise, inconsistencies will materialize during physical construction activities, when bits are turned into atoms. The ensuing redesign and rework will result in cost overruns and delayed schedules; a resort to shortcuts can cause long-term quality problems.

IFC (Industry Foundation Classes, ISO 16739:2013) is a standard format for sharing BIM models, currently supported by all major BIM tools. Each tool has its own native format in which models are edited, but they can all export the models as IFC files. In the IFC world, there is also a standard way of defining views to the models, called Model View Definitions (MVD). IFC and MVD standards have been designed with the underlying assumption that models are exchanged with other parties using file transfer. Nowadays file transfer can be mediated by project repository systems that typically are cloud-based systems containing shared folders.

The amount of data generation just from one BIM model during a project is large. As the work on a BIM model progresses, many manifestations of its content will be created (Fig. 1). Typically, several versions of the model are produced in order to coordinate work with other design disciplines and with project management. Sometimes the versions are grouped into major phases, called level of developments (when the amount of information significantly increases) or level of details (when the design progresses to a different level of granularity). Moreover, sometimes only a part of the model (e.g., the precast elements of the second floor) will be produced as a partial export from a model. Furthermore, there can be different views of the model, produced with MVDs for purposes of coordination or reference design, that are produced to simplify the model for a given purpose and the check whether all data needed for that purpose has been given.

The information quantities generated in the design stage are thus large, but they are only a part of what is produced. There is much information related to project and process management, issue management, specifications, status monitoring, and so on.

The information produced at the design and fabrication stages connects to information existing in the broader environment of the building: urban plans, building codes, building permits, infrastructure models, transportation and communication network information, and product and material catalogs.

In the operational stage of a building lifecycle, the information related to the physical manifestation of the building starts to accumulate.

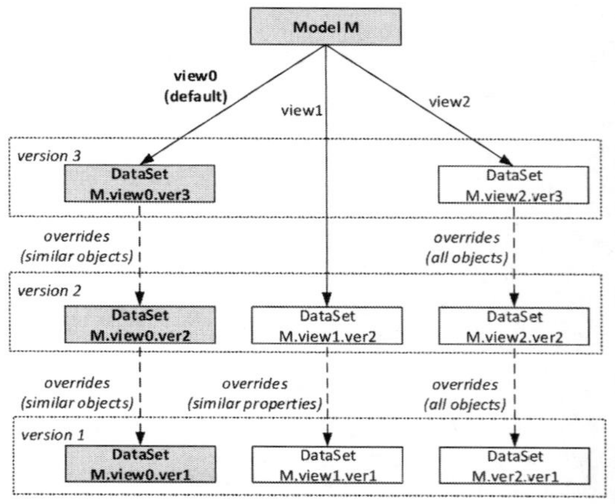

Figure 1. Datasets of one model belong to different views or versions.

There are sensor measurements, consumption readouts, work orders, maintenance reports, and asset management data.

A major problem with building data is fragmentation (Eastman 2011). The data is produced with different tools and in different formats. It is managed in isolated manner in separate systems. Objects represented in different models have obvious relationships with each other, but these relations are not represented in an explicit manner anywhere. The users of the information must reconstruct the relationships in their cognition every time they engage in problem solving or coordination activities. Some relations can be periodically recomputed in collision detection or cross-model checking tools. These kinds of repeated reconstruction tasks, when performed by humans, are tedious and error-prone, and even when aided by computer, are resource-consuming. Many relations could be computed statically and objects in different models could be linked together to enable a much smoother and fine-grained workflows, for instance, in change management.

3 WEB OF BUILDING DATA

3.1 *Motivation*

The main purpose of the development of WoBD is to facilitate information sharing among different stakeholders over the whole lifecycles of buildings. Obviously, this calls for a decentralized and loosely coupled approach because there are public and private parties whose data cannot be centralized into any single place. In addition, the approach should enable the controlled opening of data to a larger set of users than what is currently possible.

The role of WoBD should not be confused with that of the currently emerging BIM collaboration platforms, such as Autodesk A360, Trimble Connect, Nemetschek's bim+, or OpenSource BIMServer. The aim of BIM collaboration platforms is to support project collaboration during the project phases within a project lifecycle. They are logically centralized (often cloud-based), their scope is within the time-frame and consortium of one project, and they are focused on managing data produced within the project. Moreover, some of them are dependent on the use of concrete BIM tools of a single vendor.

WoBD aims at a genuinely distributed setting, to support the information sharing over the lifecycle of a building and to link together the information both produced in projects and external to them (for instance, related to infrastructure, regulations, urban plans, or building products). The approach is based on standard, vendor-independent representations.

WoBD is therefore not competing with BIM collaboration platforms. Rather, there might be possibilities to combine the capabilities of former for interoperability reasons.

3.2 *Characteristics of web of data*

Web was born as a decentralized platform for publication of documents—HTML pages—for human consumption. Web of Data is based on same protocols as the ordinary Web but it provides representation and techniques to publish data objects, in addition to documents, which allow applications to consume the data without human intervention.

The identifiers of data objects are the same as Web pages: URIs (Uniform Resource Identifiers). However, while documents are represented in HTML, for data there are number of other specifications to address different aspects of data representation:

- *Primitive datatypes*: XSD (XML Schema Definition) Datatypes
- *Structural data*: RDF (Resource Description Framework), a graph structured language that consist of directed relations between objects
- *Semantics of data*: OWL (Web Ontology Language) or RDFS (RDF Schema) for the definition of the types of objects and links in RDF graphs
- *Serialization of data*: Various standards (rdfXML, JSON-LD, Turtle) to represent RDF graphs as strings.
- *Query and rule languages*: RDF models can be queried and manipulated in a declarative manner with SPARQL query language and there are several experimental rule languages available.

In addition, Web of Data contains so-called Linked Data principles that in effect say that published data objects should be accessible over the HTTP protocol, and that the return result of the access is a serialization of the RDF representation of the object (in the desired serialization format), possibly including URIs of related objects.

The basic principles of Web of Data:

- *Decentralization*: Data is published and accessed in a decentralized manner, and there are no centralized node in the architecture.
- *Publication*: Data is not exchanged across parties, but published and accessed on demand. The models stay at their home domain.
- *Object home*: Each object has a home domain from which its description can be accessed. There is a unique home for each object but the home domain of different objects is generally different. The state of an object can be followed by subscribing to the changes in its home domain.

- *Linkability*: Published objects have a URI and can be individually accessed and linked to. The granular access to information provides more fine-grained access control and monitoring since complete models need not be exchanged. Moreover, the linkability of objects allows modular enrichment of data (e.g., association of properties, groupings, issues, or status events without touching the original data). Furthermore, linkability is important to enable a growing and evolving ecosystem of linked building data.

The main advantages of Web of Data are the following:

- *Online*: Data is online and accessible for anyone with network connection.
- *Up-to-date*: Different parties access the newest version of the data.
- *Clear ownership*: The ownership and control of data is with the party that published it.
- *Loosely-coupled*: Using linking, data from different sources can be integrated in a loosely-coupled manner. There need not be any large-scale upfront integration efforts, but integration can evolve according to needs of new applications.

3.3 *Mapping building information to web of data*

The basic idea of Web of Building Data is to map building-related data to Web of Data representations. Most of the attention in the field has been on BIM models represented in the standard IFC format.

Table 1 shows different levels of the mapping. The conceptual schema of the IFC is converted to OWL, and the resulting ontology is called ifcOWL. Using the concepts defined in ifcOWL for object types, properties and relations, BIM models are converted to RDF (often called ifcRDF). In the process the GUIDs of IFC objects are converted to URIs taking into account the domain where the RDF graph is published. The RDF graph is stored in a RDF database, and a HTTP interface for URI lookup using the RDF database is provided.

4 DESIGN OF DRUMBEAT PLATFORM

This section explains how DRUMBEAT platform (for short, DRUMBEAT) solves the typical issues faced when implementing the WoBD approach in a practical setting. Most of these issues are already described in Sections 2 & 3.

4.1 *Basic concepts and requirements*

In order to simplify and unify the data organization, DRUMBEAT takes into account only the basic concepts shown in the domain model (Fig. 2).

There are four levels of data containers (Fig. 3):

- *Host* is a DRUMBEAT platform running as a Web application. Thanks to information technologies, one Host can be deployed in many physical servers with the same base URI. This is needed, for instance, for load balancing and data backup.
- *Collection* is simply a group of Models. It can be created, for instance, for a site, a facility, a project, a customer order, or a product category. DRUMBEAT does not support multilevel

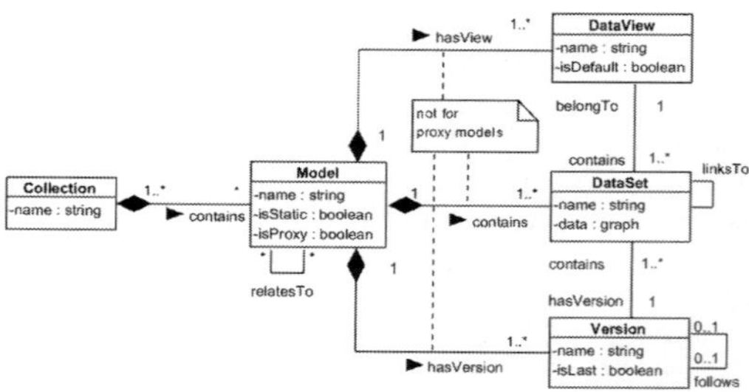

Figure 2. Domain model of DRUMBEAT platform.

Table 1. Representation of IFC models as Web of Data.

	IFC	Web of Data
Reasoning	–	SWRL, SPIN, RIF, …
Dataset description	–	VOID
Linking	–	URI, RDF
Query	–	SPARQL
Access	–	HTTP / REST
Publication	Files	RDF stores, Files
Ontology/schema	EXPRESS	OWL
Data representation	IFC data	RDF graph
Serialization formats	STEP File, ifcXML	JSON-LD, Turtle, RDF/XML,
Object identifiers	GUID	URI

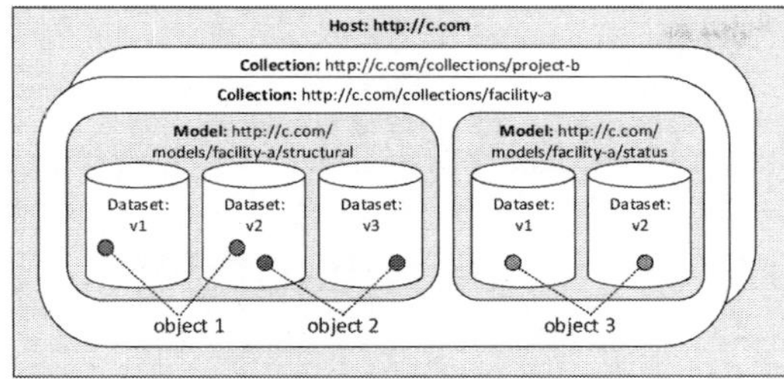

Figure 3. Four levels of data containers.

collection hierarchies. Each collection is published in one Host.

- *Model* represents one building-related-data model, which has an identified data origin (often discipline-specific, for instance, a structural engineering company), certain data type and known relationships with other Models. Examples are different kinds of BIM models, property sets, status event containers, and so on.
- *Dataset* contains a set of Model's *Objects*, which are nodes in RDF graphs. One Model can contain many Datasets that represent different subsets of different versions of the Model. One Object can belong to many Datasets, but only one Model. A Dataset is usually generated from an external representation such as an IFC file.

Each Model has one or more *Versions* ordered chronically. Each Model also has several *DataViews*, one of which is set as *default*. Each Dataset belongs exactly to one Model's Version and one DataView. For instance, in Model *M* in Figure 1 there are three DataViews (*view0*, *view1* and *view2*) and three Versions (*ver1*, *ver2* and *ver3*). Dataset *M.view0.ver3* belongs to Version *ver3* and DataView *view0*. It is also the default dataset returned when Objects from a Model are requested without additional filters.

Inside one DataView, each DataSet can *override* another fully or partially. There are there types of overriding relationships between two Datasets:

- *graph-level* when all objects of the older Dataset are replaced;
- *object-level* when all similar objects (with same URIs) of the older Dataset are replaced;
- *property-level* when all similar properties of similar objects (with same subject URIs and property names) are replaced.

In case of object- or property-level overriding, part of data from an older Version also belongs to the newer one. In means that in order to get the most recent data for the *view2* of Model *M* in Figure 1, data processors must go through three DataSets: *M.view0.ver3*, *M.view0.ver2*, and *M.view0.ver1*, where each DataSet overrides similar objects of another, older than itself.

The same Model can belong to many Collections. In this case, *proxy-models* are used. They contain no data, but the reference to the real Model.

A Model can be *static* or *dynamic*. The contents of DataSets of a static Model are non-changeable. While BIM models are often static, status event containers are dynamic.

DRUMBEAT maintains model- and object-level links, which are directed links between two Models or two Objects. Model-level links imply that there may be object-level links between their Objects.

4.2 *Architectural model*

Figure 4 shows the architecture of the DRUMBEAT platform. There are two major layers in the system:

- *Data (storage) layer* stores RDF data in an RDF database (triple store) and other data as files in the server file system.
- *Middleware layer* implements business logic for data management, publishing and accessing, including the security and access control mechanisms, and provides a REST API to client applications.

The different parts of the architecture are described more in detail below, except security functionalities that are discussed in (Oraskari 2016).

4.3 *Data organization*

DRUMBEAT stores all RDF-data in an RDF database. Contents of every Dataset is saved in a separate *named graph*.

Besides, there is one named graph used for metadata about Datasets. The metadata schema is

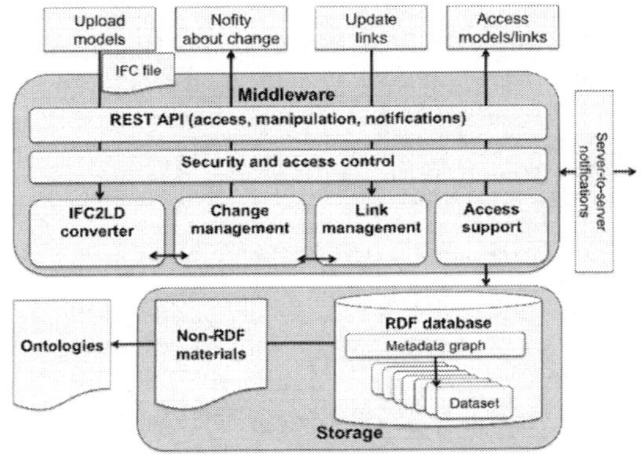

Figure 4. The architecture of DRUMBEAT platform.

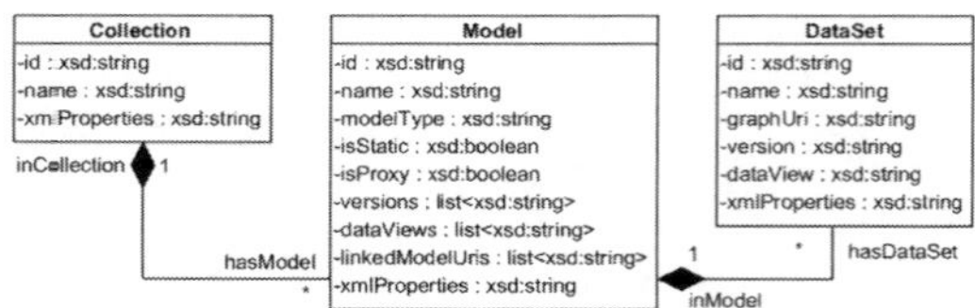

Figure 5. Linked building data host ontology.

specified in Linked Building Data Host Ontology that is derived from the class model illustrated in Figure 5, which in its turn is based on the domain model in Figure 2. In this ontology, Versions and DataViews are represented in string properties of classes Model and Dataset. A Model has an list of Versions, ordered from the newest to the oldest. A Model also has a list of DataViews, first of which is the default one. Model-level links are defined in property *linkedModelUris*. In case of proxy models, this property is used for referring to the real model.

DRUMBEAT stores non-RDF data as files in the file system. Usually they are files uploaded by clients to the platform. However, depending on the file type, DRUMBEAT can convert it to RDF format and then store both RDF-data and the compressed original file as a reserve copy.

4.4 *Naming RDF-resources with URIs*

URIs are used as identifiers of resources in Web of Data. They are not just globally unique identifiers but they also serve for retrieving and managing resource descriptions over network. To facilitate use cases of WoBD, URIs should be designed carefully:

1. URIs need to be *unique*: One URI can only denote one object.
2. URIs should to be *stable* over time. They should not change during the evolution of data, as long as the objects stay semantically the same.
3. Each object should preferably have just a *single* URI. Otherwise, to maintain the semantic integrity of data, it would be necessary either to duplicate links or manage *sameAs*-relations between the names. The latter solution is more preferred in the Semantic Web as it reduces data redundancy.

DRUMBEAT names its resources with the following URI format:
<objectUri>=<baseUri>/<resourceType>/<localId>
<baseUri>=<protocol>://<domain>[:*<port>*] [/*<appName>*]
<resourceType>= collections | *models* | *datasets* | *objects*
<localId>= <collectionId>[/*<modelId>*[/*<datasetId>*]][/*<objectId>*]]

where

- *baseUri* is the Host's base URI;
- *resourceType* is the type of resources;
- *localId* consists of one or more identifiers depending on the *resourceType* (Fig. 3).

In IFC models, all significant objects have a GUID as an identifier. DRUMBEAT also uses these GUIDs in URIs of these objects. If the same GUIDs appear in different versions or subsets of an IFC model then they identify semantically same objects. Similarly that, objects with the same GUIDs in two Datasets of one Model in DRUMBEAT must have a single URI. Thus, the object URI must not include the Dataset IDs:
<localId>=<collectionId>/<modelId>/<GUID>,
for example:
https://construct.com:1234/objects/site1/structural/F445F4F2-4D02-4B2A-B612-5E456BEF9137

In the same time, URIs of objects with identifiers, which are unique in a Dataset, but not globally unique, must include Dataset ID, otherwise unrelated objects may have the same URI. It is recommended to use additional prefix (e.g., "/B") to *objectId* for facilitate filtering objects in a Dataset:
<localId>= <collectionId>/<modelId>/<datasetId>/B/<objectId>.

4.5 *RESTful APIs and data manipulation*

Most of standard Linked Data solutions are based on the idea of publishing data in RDF format on the Web with proper URIs and offering a SPARQL endpoint through which users can issue queries on the RDF data. They support well basic functions like reading, writing and querying RDF data with primitive URI lookup algorithm. However, these functionalities are insufficient for a practical industry-oriented platform, such as DRUMBEAT, because they turned out to be too far from the toolset used in the mainstream software development.

DRUMBEAT combines URIs of RDF-resources (defined in Chapter 4.4) with HTTP methods, such as *GET*, *PUT*, or *DELETE*, to create REST APIs for resource management in Web of Data. Conceptually, REST APIs suit well with Web of Data because they both are based on primacy of resources identified with URIs, one is about HTTP-based mechanism of manipulating data, and the other is about graph-based data representation. However, it all DRUMBEAT's REST APIs must response with data in RDF format so that they could be used from from Web of Data tools. Depending on the HTTP-request, the response can be in any RDF serialization format (e.g., JSON-LD, Turtle, NTriples, etc.) or it can be a HTML document to show the RDF data in a Web browser.

Currently, DRUMBEAT supports the following types of data manipulation:

1. *Metadata management* for creating, reading, updating and deleting metadata about data containers, such as Collections, Models, and Datasets.
2. *Converting and loading data to datasets* for adding new objects and properties to the named graphs of Datasets, which are already defined in the metadata. The data source can be an URL-resource, a server file, a client file, or a string content. The data can be uncompressed or compressed with GZIP or ZIP format to reduce the network traffic. The data model can be either *RDF* (with any RDF serialization format), or *IFC 2 × 3*, or *IFC 4* (with STEP physical file format or ifcXML format). IFC data is converted to ifcRDF by the configurable *IFC2LD* tool described in (Vu Hoang 2015) and integrated into DRUMBEAT.

4.6 *Accessing to objects of datasets*

All object descriptions can be requested just by using REST API with the URI is the object URI and HTTP method is *GET*. However, as mentioned in Section 2 and Chapter 4.1, a Model can contain many Datasets, each of which can represent a view to the data or a version (Fig. 1). In order to filter data from a certain data view and/or version, the optional parameters *dataView* and *version* can be added to the object URIs. Their default values are: the default *dataView* and the newest Version of the Model. If no *objectId* is indicated then all objects with GUIDs of that Model will be returned. For example, the response of the following request:

GET http://c.com/objects/facility-a/structural/?dataView=view1&version=ver3

will be all objects from the dataset of *view1* and *ver3*. If this dataset overrides similar objects of an older one then some objects from the older dataset may be also returned. In this case, additional operations, such as comparing and merging graphs, can reduce the query performance than when objects are selected from a single RDF graph. However, it is the cost we must pay in order to enable versions, views and partial models of BIM models in WoBD.

4.7 *Linking management*

There are three interrelated questions concerning link management:

1. *Publication of links*—Should links be embedded in existing models or published in separate link models?
2. *Access to links*—Should links be accessed separately or is the link access be integrated to object access (that is, links returned within object descriptions)?
3. *Inverse links*—How can the target object of a link get to know about the existence of the link?

RDF graphs consist of triples, that is, named relations composed of three parts: <subject, predicate, object>. All triples are like links, but within the context of Linked Data, links typically mean a subset of triples where subject and object belong to a different models and often in different domains. Since links, in this more restricted meaning, are a subset of triples, they can be managed in two different ways:

1. *embedded* in an existing RDF graph, or
2. *separately published* as a new RDF graph.

The first question is whether links should be managed in an embedded or separate fashion. The answer is that it depends on practical considerations. The most important issue is whether links are an integral part of existing data or whether they are generated separately from all other data. In the first case their lifecycle and authorship is the same as that of the data; in the latter the lifecycle is different because links are created afterwards and possibly by a different author. Both approaches can be supported in the platform since they are not mutually exclusive. Some links can be embedded in existing data and some published separately.

The second question is whether the access to links should be separate from object access. That is, when an object is accessed, its description is returned without external links, and external links of an object could be accessed as a separate action. Alternatively, all the links could by default be included in the object description.

In the DRUMBEAT platform the latter approach is adopted. The links can be published either as embedded in data or in external link models. However, when links are stored separately, the access to objects needs to be implemented in a way that all links of the object from an external link models are also included in the object description (unless user specifically requests otherwise).

The third questions is whether the target of the link should be able to use link information. Since the object descriptions will only include outgoing links from the object, in order for the target of the link to include the link information, the inverse links need to be computed. As an example, consider a graph shown in Figure 6 that consists of three different kinds of triples:

1	type	http://s.com/123, rdf:type, ifc:IfcWallStandardCase>
2	property	http://s.com/123, rdfs:label, "Element 123">
3	relation	http://s.com/123, blo:implements, http://a.com/456>

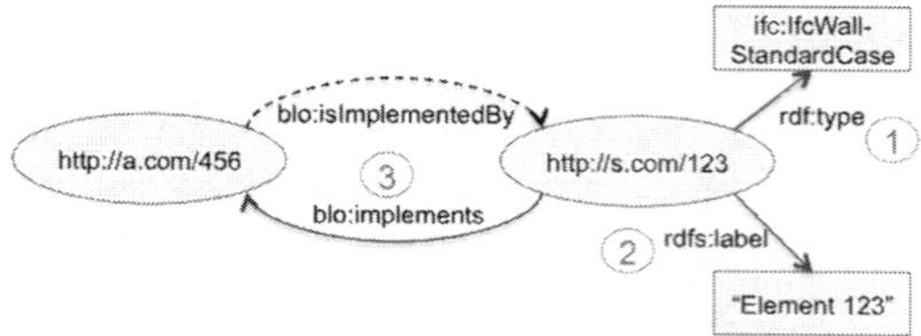

Figure 6. RDF graph with an inverse link (dashed line).

The natural place to host the inverse links is at the domain of the target object as shown in Figure 7. Backlink, a remote inverse link (dashed line).

The target domain (a.com in the example) can utilize the inverse links because they are outgoing links from the objects it hosts.

In DRUMBEAT platform these kind of *remote inverse links* are called *backlinks* (Stefanidakis 2011, 2012). The platform implements a notification protocol to enable their creation and maintenance. It works as follows:

1. When a link is established that refer to an object at another domain, a notification is sent to the remote server, including the information about the source object and the type of the link.
2. Using the information provided in the notification, the target domain can access the ontology to determine the inverse property of the original link.
3. The target domain creates an inverse link and stores it locally.
4. The target domain can do additional things such as access the referring object and cache it locally.

The creation of backlinks means that all link information is accessible either by accessing the source (subject) of a link or the target (object) of the link. Since these are the only places affected by the link, the links themselves need not anymore be independently accessible, and they do not need to have URIs (which a simple link does not have since it is just a single triple).

Backlinking behaviorally decouples the two domains in the sense that neither domain needs to access any link models from the other domain because it already has all the link relevant to its own behavior. The backlinking increases the the number of cross-model links and makes the data more connected. In addition, it enables the implementation of many interesting use cases, for instance, related to status sharing and issue management.

5 DISCUSSION

DRUMBEAT platform is a proof-of-concept implementation of the Web of Building Data

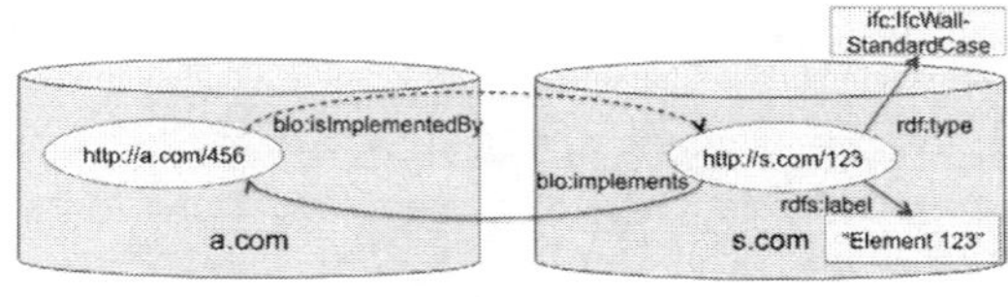

Figure 7. Backlink, a remote inverse link (dashed line).

concept. The development work has made visible many practical issues that need to be addressed to make the concept a reality.

Two central issues are the dealing of multiple manifestations of the same model—versions, views, and partial exports—in a the platform, and the management of links—with backlinking and integrated access—so that their existence does not make the interaction of applications with the platform more complex.

Many of the issues could be solved also in other mechanisms. The overall philosophy in DRUMBEAT platform is to hide unnecessary complexities from the user. For instance, a straightforward implementation of model publication and linking may lead into situation in which the users is forced to specify explicit version information in every access or to access linksets with separate explicit calls. If these kinds of complications are revealed in the structure of the interface, the result may become unusable in its complexity.

It is likely that the concept of publishing building data on the Web in a granular and retrievable manner will gain alternative implementations in the future. Consequently, there will be needs for standardization of the interfaces and representation formats used. DRUMBEAT platform offers one proposal for such client program interfaces. It also addresses the question of server-to-server communication.

It is possible to implement the DRUMBEAT interfaces to existing BIM collaboration platforms. They should in the minimum implement (1) a mapping from their local identifiers to URIs in a unique and stable manner, (2) a URI lookup protocol to retrieve a description of an object in some RDF serialization format, and (3) the sending and receiving of server-to-server notifications.

6 CONCLUSIONS AND FUTURE WORK

DRUMBEAT platform is a proof-of-concept implementation of WoBD that addresses the issues concerning the organization of the published data on a host, the description of published data (metadata), URI design for objects, publication and management of links, and the provision of the access to links to users. The problems are solved by (1) implementing a platform with three-layer architecture, (2) creating four-level metadata structure host-collection-model-dataset, (3) providing a REST API and a URI design scheme, and (4) providing mechanisms for uploading, converting and storing data files; for versioning and filtering partial modes, and creating links and backlinks.

The future research consists of methods for automatic link generation, new data converters (e.g., from COBie and LandInfra), change management, change notification, integration with security and access control mechanisms, and on-the-fly generation of views.

ACKNOWLEDGEMENTS

This research has been carried out at Aalto University in DRUMBEAT "Web-Enabled Construction Lifecycle" (2014–2017)—funded by Tekes, Aalto University, and the participating companies.

REFERENCES

Beetz, J., J. Van Leeuwen, & B. De Vries, 2009. IfcOWL: A case of transforming express schemas into ontologies, *Artificial Intelligence for Engineering Design, Analysis and Manufacturing*, vol. 23, no. 01, pp. 89–101, 2009.

Berners-Lee, T., 2006. Linked Data—Design Issues, *W3C*, vol. 2009, no. 09/20.

Kevin R. Page, David C. De Roure, & Kirk Martinez. 2011. REST and Linked Data: a match made for domain driven development? In *Proceedings of the Second International Workshop on RESTful Design* (WS-REST '11), Cesare Pautasso and Erik Wilde (Eds.). ACM, New York, USA, 22–25.

Oraskari, J. 2016. Access Control for Web of Building Data: Challenges and Directions, ECPPM 2016.

Pauwels, P., & Walter, T., 2016. EXPRESS to OWL for construction industry: Towards a recommendable and usable ifcOWL ontology. Automation in Construction, 63, pp. 100–133.

Stefanidakis, M., & Papadakis, I. (2011, May). Linking the (un) linked data through backlinks. In *Proceedings of the International Conference on Web Intelligence, Mining and Semantics* (p. 61). ACM.

Stefanidakis, M., & Papadakis, I. (2012, May). A decentralized infrastructure for the efficient management of resources in the web of data. In *Proceedings of the 4th International Workshop on Semantic Web Information Management* (p. 7). ACM.

Törmä, S. 2013. Semantic linking of building information models. In *IEEE Seventh International Conference on Semantic Computing (ICSC)*: 412–419. IEEE.

Törmä, S. 2014. Web of Building Data—integrating IFC with the Web of Data. *eWork and eBusiness in Architecture, Engineering and Construction: ECPPM 2014*: 141.

Vu Hoang, N. 2015. IFC-to-Linked Data Conversion: Multilayering Approach. In *the Third International Workshop on Linked Data in Architecture and Construction, Eindhoven, The Netherlands.*

Vu Hoang, N. & Törmä, S. 2015. Implementation and Experiments with an IFC-to-Linked Data Converter. In *the 32nd International CIB W78 Conference on Information Technology in Construction, Eindhoven, The Netherlands.*

eWork and eBusiness in Architecture, Engineering and Construction – Christodoulou & Scherer (Eds)
 ISBN 978-1-138-03280-4

A versatile and extensible solution to the integration of BIM and energy simulation

D. Mazza, E. El Asmi & S. Robert
CEA, LIST, Gif-sur-Yvette, France

K. Zreik
CiTu Paragraphe, Université Paris 8, Paris, France

B. Hilaire
CSTB Sophia Antipolis, France

ABSTRACT: Energy simulation represents now a critical step of the building design process, but the integration between BIM authoring tools and simulation tools is still a source of issues and problems. Most simulation engines still require the definition of a specific input model, thus making impossible to directly use the BIM model information to configure and run simulations. The lack of agreements on standards for simulation input files makes software integration between BIM and simulation tools insufficient. With this aim, this paper presents a software framework for BIM-Simulation integration aiming at both effectiveness and extensibility. An intermediary conversion step for the generation of simulation-specific yet tool-agnostic building design model (here called Building Simulation Model) allows to tackle the problem through a two-step generation scheme (BIM-to-BSM and BSM-to-simulation tool) leveraging the principles of the Model-Driven Engineering methodology. Targeted engines have been EnergyPlus and COMETH (the French regulatory thermal engine) in the scope of an on-going European project (FP7 HOLISTEEC).

Keywords: BIM, Energy simulation, integration, model conversion, MDD, MDE

1 INTRODUCTION

Energy simulations play nowadays an important phase of the building design process. Energy consumption is one of the main aspects tackled in current times, and it has really become a sensible issue for clients towards the sustainability of the living practices of our and future society. For these reasons, the current building design process cannot avoid to take this aspect and all the related issues into consideration in the proper and detailed way.

Evaluation of energy consumption is usually performed starting from a CAD-authored building model and given as input to energy simulation tools that compute the energy consumption of building different areas (thermic zones) over a specified period of time (e.g. a year).

Integration between authoring tools and simulations engines is still source of issues and problems, due to the lack of widely-adopted standards for the specification of building information or (wherever standards are present) to the lack of direct integration through adopted data format between authoring tools and simulation engines.

In the following BIM models and related context will be referenced by the BIM acronym, while the simulation engines specific formats and context will be referenced by the SIM acronym.

This article presents a framework for the BIM-SIM integration, which exploits the concepts of the model-driven methodology for the conversion of a BIM model into a SIM one. A model-based approach is straightforward, given the need of converting a BIM model into a SIM. Indeed simulation engines do not accept BIM models as direct inputs, and an engine-specific model should be created starting from the BIM one. BM-SIM conversion can be represented as a model-to-model conversion, thus the concepts developed by the model-driven branch of software engineering can help providing best practices and well-developed procedures for conversion.

2 BIM AND ENERGY SIMULATIONS

BIM-SIM integration is complex because of the different points of view adopted by BIM and SIM models, and by the different information needed by simulation engines to perform computations. In particular:

- simulation engines generally do not directly accept BIM models as input. Thus the need of model conversion to SIM model
- additional data not originally provided with BIM models are needed for SIM models (e.g., physical properties of materials). BIM models usually represent geometric information while simulation engines need different information to perform calculations. Thus the need of integrating the inputted SIM model
- simulation engines usually adopt proprietary formats for their models in input. Thus a once-for-all conversion is not possible due to the need to be separately addressed each simulation engine.

These issues likely clarify the needs for an environment that could manage the input of BIM building projects to simulation engines. Such framework should be able to abstract the specific features and the technicalities of each single simulation engine and should be able to provide a procedure that, by focusing on the main aspects of the conversion of a model from a BIM format to a simulation engine-specific format, could allow to a generalization of the conversion procedure in order to possibly target different simulation engines.

This the purpose of the work here presented and of the framework described in the following.

3 A FRAMEWORK FOR THE BIM-SIM CONVERSION

The main concern in interfacing BIM and SIM worlds is to guarantee that the same building project and its characteristics are considered and conserved in both domains. Equivalence of the considered models is thus the driving criterion and rationale. A simulation engine is used indeed to verify specific energy-efficiency or consumption constraints, thus considering the same building under design is of utmost importance.

Model conversion is at the core of the BIM-SIM integration here proposed. This is anyway a non-trivial task:

- conversion cannot be as easy as a 1-to-1 correspondence between BIM—and SIM-model elements. Indeed for one BIM object more than one SIM objects need typically to be specified
- not all BIM elements have a corresponding SIM element. Different realms tackle different concepts
- specific parameters or object properties cannot be directly mapped (e.g., value treatment during conversion is required) or are not defined in the target model
- origin and target model could propose a way to model real building objects from different points of view, thus leading to a lack of information and data to be used during the conversion, which should be integrated somehow.

In order to specifically tackle the above issues, this work proposes an approach to BIM-SIM model conversion based on an extensible architecture, where those general issues common to all BIM-to-SIM conversions can be managed once-for-all, without further customization or intervention, and when those conversion logics specific for a targeted simulation engine can be fine-tuned by the user and adapt to the specific needs of each case.

This work stemmed from an original approach devoted to the specification of conversion procedure addressing *EnergyPlus*[1] as target simulation engine, and later extended to another proprietary simulation engine called *COMETH (COre for Modelling Energy and THermal comfort)*, developed and sold by CSTB (the French Centre for Building Science and Technology). The design of the solution for the latter engine highlights the existence of common needs and logics to generalize and apply to the conversion process, which led to a refinement of the original solution for more structured software architecture. A *model-driven design* approach has been adopted, which is explained in the following.

3.1 *The model-driven methodology*

The model-driven methodology is an approach of designing solutions to problems where models play a key role. The idea behind such approach is to obtain a solution to a problem through the *transformation* of an initial model into another one. In the software engineering domain, models are used to abstract objects and their features for the need of a specific model to be able to describe a general solution regardless its final delivery (i.e., technology or platform on which this solution has to run or execute). This *Platform-Independent Model (PIM)* contains all the relevant information and concepts to completely describe the solution from a general point of view.

The idea underlying model transformation is then the possibility of converting the initial PIM into a more specific model that represents a

[1]EnergyPlus is one of the most known and used simulation engines in building design and AEC domain. Further details can be found on the official web site: http://apps1.eere.energy.gov/buildings/energyplus/

solution for the problem on a specific platform, leading to what is called *Platform-Specific Model (PSM)*. Model-to-model transformation therefore involves mainly a transformation of the concepts (model objects) information (model data, objects features) from the initial, abstract model to a low-level, technology-specific model, thus "rearranging" the original information and objects in order to build up a solution-specific model. Obviously, the advantage of such an approach is that different PSMs, one for each specific platform, can stem from the same PIM, as shown in Figure 1.

The model-driven approach has therefore been applied in our case: by considering the definition of a general simulation model, that we called *Building Simulation Model (BSM)*, we represent a way to model the building under study adopting an energy-simulation point of view. This way it is possible to model all the concepts involved in energy simulation at a higher and abstract level: this model is our PIM and it is a simulation engine-independent way of describing a building. Through a model transformation procedure we can then obtain a PSM for the simulation engine of interest (EnergyPlus, COMETH, others at need).

The PIM model contains the necessary information in order to transform the model to the specific PSM. Information exchange between the PIM and PSM is thus an important specification step. The use of a standardized framework as *Information Delivery Manual (IDM)* describes the process of data exchange for the BIM domain. It is developed by the BuildingSmart® Association, and it allows specifying the data to exchange among the involved parties through the use of a standardized format. The IDM can thus be generally meant as formalism to model the process of information exchange for the BIM domain. Involved concepts are related to a specific target domain (e.g., energy simulation in the particular case of this work). IDM will then be translated by IFC experts and leads to a second document named MVD (Model View Definition), defining a subset of the IFC schema focusing on the information to exchange. Figure 2 sketches the whole process.

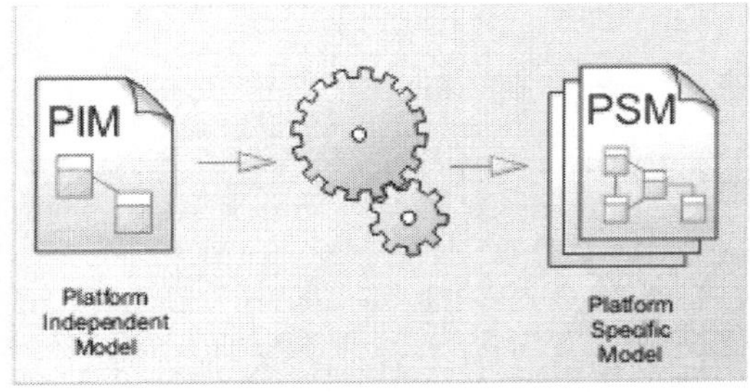

Figure 1. The model transformation process.

For energy simulation, we are defining a SIM-MVD describing the IFC elements we need to exchange to move from a BIM to a SIM model. In parallel, missing IFC entities will be specified and developed by IFC experts to fully describe the SIM model; the combination of the SIM-MVD with the missing IFC entities, will corresponds to what could be called a Building Simulation Model (BSM) (cfr. Figure 3).

To sum up, in energy simulation domain and regarding the model-driven methodology, a specific PIM model can be defined and considered as a BSM. And by using IDM-MVD methodology, we are able to define its content.

3.2 *Solution design and implementation*

In order to adopt the explained model-driven approach, there was the need of a model able to catch all the aspects covered by simulation models, thus a model that could play the role of common denominator for all the engine-specific simulation models.

The introduced Building Simulation Model (BSM) plays the role of the PIM in the context of the model-driven approach explained before. In our case, the platform-independency has to be

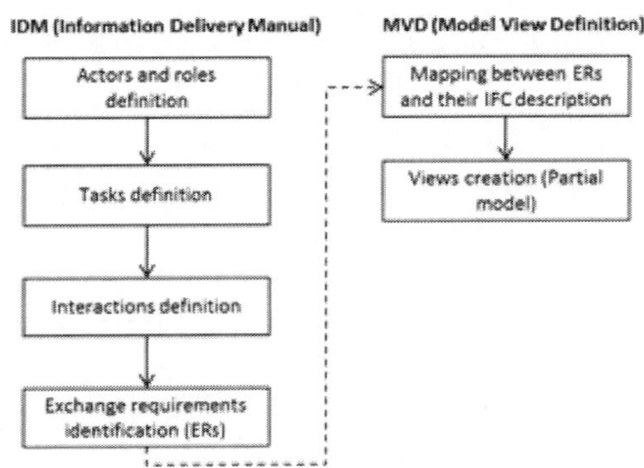

Figure 2. The transformation process of IDM to MVD.

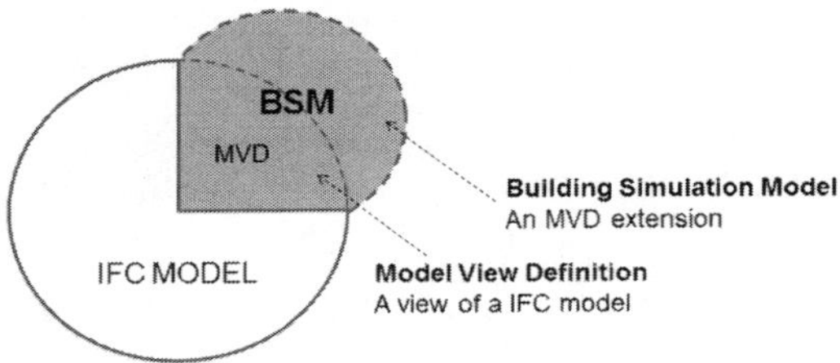

Figure 3. BSM as an extension of IFC concepts and data.

meant as simulation engine-independency: in order to target multiple simulation engines, we need an abstract model that allows us to catch those concepts related to the simulations, and not linked to a specific engine. BSM plays this PIM role, while in our case the role of PSMs are played by the engine-specific models required by the targeted simulation engines.

BSM model contains all the objects for the geometric definition of a building (storey, ceiling, floor), its detailed elements (door, window, wall), and detailed composition of elements (material).

Each object has a set of associated properties that represent the peculiarities of each instance (i.e., a specific door, wall or window) of a single object type. Properties describe the different aspects of an object, that can be again related to its geometry (e.g., width, height), its properties related to its object type (e.g., the color) or related to its nature (e.g., the composition material, usually a reference to a material meant as another object).

The BSM model has been defined incrementally: an initial version of BSM has been defined with the purpose to abstract from the specific input format accepted by EnergyPlus, the first targeted simulation engine; at a later stage of the work, the BSM has been adapted and extended in order to cover and abstract the additional modeling peculiarities needed by the input format of COMETH, the second targeted simulation engine. During this adaptation process for the latter simulation engines, some aspects were already covered by the first BSM version defined for targeting EnergyPlus, thus highlighting indeed the possibility of identifying a minimum set of simulation-specific concepts common to all the simulation engines.

Extension to the first initial BSM with additional elements of features has been necessary to cover simulation aspects that were not needed for the definition of the EnergyPlus simulation model; anyway only those non-COMETH specific details have been added, in order keep the BSM as an engine-independent simulation model, as detailed in the following.

3.2.1 *energyplus vs. COMETH models*

EnergyPlus and COMETH models differ in terms of the concepts used for modeling building and its thermal characteristics.

While the former adopts mainly a geometry-based building modeling, with the further specification of the thermal properties of the involved objects, the latter adopts a thermal engineering point of view, thus defining for defining the building through its separation in thermal zone, and the specification of the characteristics of these elements in terms of heat exchange.

EnergyPlus model stresses more on the geometry of the building, and then decorates the elements with the specification of those physical properties necessary for energy simulation computations; the COMETH model consists instead of *project* composed of a set of *zones* (i.e. thermal zones) defined according to their *use* and considered in *groups*, each representing a homogeneous thermal unit. This unit then contains a set of *walls*, *windows* and *thermal bridges* that model the heat exchange between adjacent zones.

The differences in building modeling capabilities between the two simulation engines have raised issues in the specification of the common BSM model, that had to be adapted to manage properly those information specific of given model. It is worth mention how the EnergyPlus model better link with the IFC building model, given their common geometric approach in building specification. For the sake of generalization, adaptation of BSM model due to the peculiarities of COMETH model has thus been necessarily introduced.

3.3 *Model-to-model conversion*

The initial building model for the overall process is a BIM model. As such model, BuildingSmart® IFC (Industry Foundation Classes) has been considered as the building design format of reference; it is therefore the input format of the procedure presented in this work. This choice has been mainly driven by the role of standard that IFC would like to play in the BIM world and, despite the still rather limited number of commercial products and authoring tools currently supporting IFC as storage format, its use and adoption among practitioners have been increasing in recent years.

For this work, IFC4 (the latest one) is considered as the reference version, and all conversion procedures have been designed starting from this version.

The design procedure consists therefore in two model-to-model transformation:

- the first step is the conversion from IFC to BSM. Indeed IFC contains all the necessary information to model a building from the geometric point of view, but it is not a format developed to support energy simulations. This first conversion is therefore necessary to adapt the IFC model to a simulation-oriented model, like BSM wants to be
- as further step, the conversion of BSM to a simulation engine-specific format is then performed.

Each model transformation is performed by dedicated converter in the architecture of the framework that it is explained in Section 3.4.

Figure 4 sketches the overall conversion process. With respect to the first conversion, from IFC to

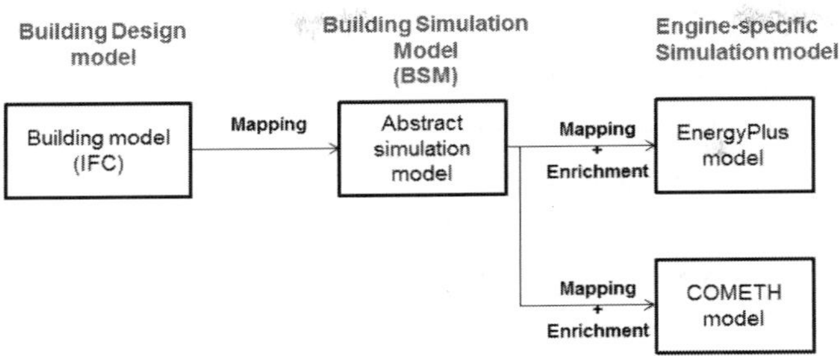

Figure 4. The overall conversion process.

BSM, it is worth remarking here that IFC projects mainly describe buildings as a set of interconnected objects; it is mainly a geometric, structural description of buildings. Of course, this information are useful for the definition of simulation models, which have the need as well to describe buildings from a structural point of view; anyway, simulation engines need also other information to perform their task, e.g., material properties or other physical attributes, which can allow them to build an inner energy model to perform computations.

These additional data cannot be retrieved from IFC projects, and must be separately supplied during the conversion process in order to build up a complete simulation model.

Materials and related data provide a good example for information missing in IFC projects to be integrated to the generated BSM model. Indeed, in a BSM model materials need to have physical and energy-related properties specified, all information that a IFC project does not originally contain; usually IFC files contains material names or other properties for general reference, but no more specific information about materials. In order to create a complete BSM, material information have been provided and integrated to the building project through the aid of an external database that, containing the additional information related to material and relevant for energy simulation purposes (e.g. physical properties such as material conductivity, roughness; etc.), has allowed us to add such information in the IFC project file, specifying these data as additional properties to be associated with specific materials. Thanks to the addition of these extended data, a complete BSM could be defined, thus easing the further conversion step towards the engine-specific simulation model.

With respect to the second conversion, from BSM to engine-specific format, it is worth notice here that additional data could be needed here as well, but in this case these data are specific settings expected by the targeted simulation engine. These data can be fixed parameters setting, that could cover most of the simulation engine uses, or other parameters that can be fine-tuned by the user and therefore set before the execution of the converter.

3.4 *Architecture of the conversion framework*

In order to perform the BIM-SIM conversion, software architecture for the converter has been put in place. In particular, during the design of the software architecture the attention has been put into the specification of *converter* and *engine-specific* elements. Specifically:

- *converter* elements are those software elements that perform the conversion procedure and that do not change with respect to the target simulation engine
- on the other side, *engine-specific* elements are those elements that are responsible of the conversion and of the construction of destination model specific of a target simulation engine. This way, these components are supposed to implement those conversion logics that define the specific way in which the model-to-model transformation has to occur according to the targeted simulation engine, and these are therefore the components that are likely to be redefined (or the logics implemented by them) modified in order to perform the conversion for the destination simulation engine.

Figure 5 explains the schema of the converter architecture, highlighting fixed and variable parts.

The *converter* parts have the purpose of analyzing and scanning the input file in order to identify the composing elements, and to perform *pre-* or *post-processing* operations when necessary. These elements coordinate the overall conversion procedure.

The *engine-specific* parts are devoted to the conversion of specific objects: according the type of the recognized object, a specific set of rule is applied for the conversion. This *type-based conversion* allows the user to define the conversion logics

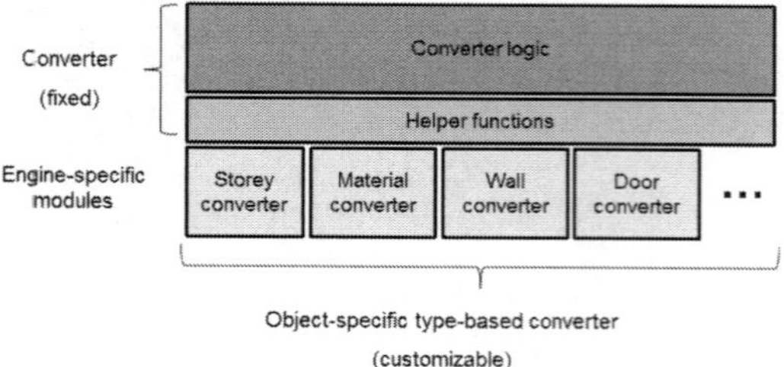

Figure 5. The conversion framework architecture.

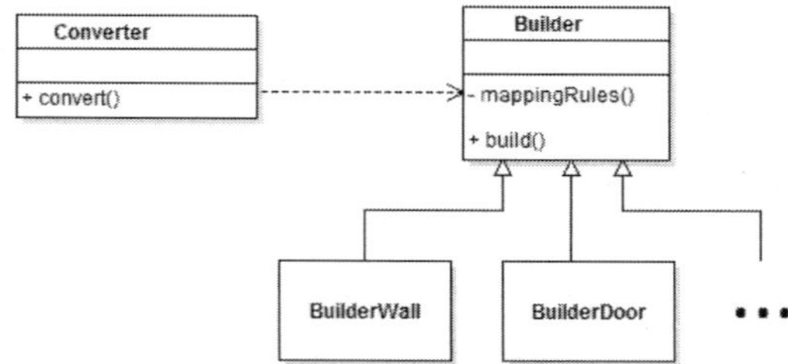

Figure 6. The UML diagram of the foundation objects of the conversion framework.

by re-defining and *customizing* the rules implemented by those modules devoted to the conversion of a specific object type. This way, a general conversion procedure can be put in place, by leveraging on two main axes: a) the type of the object to convert b) the targeted simulation engine.

3.5 *Modularization and customization of the conversion software procedure*

All the conversion software architecture is implemented according in an *object-oriented* fashion through specific C++ objects devoted to the performance of specific conversion tasks. Figure 6 shows the objects that compose the framework; specific methods exposed by these objects are expected to be redefined and overridden according to the targeted simulation engine for which the model conversion procedure has to be implemented:

- *Converter* is the object that performs the actual conversion procedure, through the execution of its *convert()* method
- *Builder* is the abstract object devoted to the conversion of specific model object; builder objects for different objects types are expected to the derived from this abstract object.

Customization of the conversion procedure is thus implemented through the *overriding* of the methods of these specific classes; in particular, the overriding of method *convert()* of *Converter* object allows to define the order and the prioritization of the conversion for each specific object type, while the overriding of the method *mappingRules()* of the *Builder* object allows to specify the conversion rules according to the object type.

3.6 *Evaluation of the approach*

The approach for the evaluation of a model transformation has mainly to verify that the output model contains those concepts of the input model that can be specified in the former, representing according the format allowed by the latter.

In order to verify this, the typical technique adopted in the model-driven is *formal checking* (Buettner et al. 2012). This approach consists thus in the definition of a set of constraints (or conditions) supposed to be conserved during the model transformation and that the output model has to possess and present. In particular, it can be easily understood how part of the information of the output model or the way in which data are structured and related each other can depend a) on the information and data provided by the input model b) can be additional parameters/elements not present in the input model or derived from the information of the input model; in this case, the values of this additional parameters can be constant values or data derived (directly, or indirectly thus through further elaboration/computation) from input model data.

Formal checking techniques have thus been applied by testing the model transformation approach on the test case described in the following for both the simulation engines (EnergyPlus and COMETH) considered in this work. The test has consisted in executing the conversion procedure starting from the IFC4 project file as input and converting it for the use by a simulation engine. Given the differences in the conversion procedure, and the different concepts (point of view) of the two simulation engines, the approaches to define the goodness of the conversion have to be different between the two engines, because some measures or aspects taken into consideration by the EnergyPlus model are not described in the COMETH one, and vice versa.

Preliminary step for evaluation is the generation by the conversion procedure that guarantees the executability of the generated model by the targeted simulation engine. Although this step does not say anything particular about the goodness of the quality of the model conversion, it represents anyway an obvious but necessary condition for the evaluation of the approach.

3.7 *Test case*

The solution has been tested on a real project of a family house, designed and then constructed in Southern France. For this project, the original IFC4 files created by the engineering company that managed the project were available.

The house of the test case has 1 floor, 4 main rooms and 1 other service space, and the project defines also the position, dimension and materials of doors and windows.

3.7.1 *Test with EnergyPlus*

Given the specific features of EnergyPlus and its input model file format (IDF), in order to check the goodness of the results of the conversion, the outcome model has been check on the following aspects:

- *geometry*: it has been checked whether the geometry of the output modelled building has been the same of the input model. This has been done by checking:
- *number of elements with respect to their types*: this is a *sanity check*, in order to be sure that the number of geometric elements (doors, windows, walls, floors, etc.) composing the building has been kept during the conversion process
- *building (elements) topology*: adjacency of rooms (spaces) and walls, as well as belonging of openings (doors and windows) to walls have been checked in the output model and compared with the adjacency and relationship between the same objects in the input model
- *coordinates position*: actual coordinates of the positioning for doors, walls and windows have been checked in the output model with respect to the input model. For this check and the previous one, the *EPDrawGUI* tool[2] have helped a lot, thanks to its functionality to open the generated IDF files and to convert it as DXF file, thus giving the possibility to visually check the topology of the generated IDF model and to compare it with the one of the original IFC model
- *properties' definition*: those attributes that could tracked and assigned to a specific object of the input model have been checked for existence and correct assignment to the same object in the output model. Check is done on:
- *number of elements with respect to their types*: here again, this is a sanity check with the focus on those objects defining the property of elements: a typical example here is the materials definition, which are defined each as specific objects in IDF files, then applied as properties to a geometry element.

[2]The *EPDrawGUI* tool is supplied with EnergyPlus with the purpose to provide a DXF format file out of an IDF given in input. This way geometry analysis of IDF model can be done more easily, given that few tools are able to natively read IDF models and provide a graphical representation of them.

3.7.2 *Test with COMETH*

COMETH takes as input an XML document to set up the building simulation. The characteristics of these XML input files were specified in an XSD. We were able to implement COMETH data model based on this XSD file. As part of validation of the developed mapping process, it has been verified that the generated XML conforms to the original XSD schema.

The analysis of the generated file has been performed manually at first, by verifying that all the necessary fields were present and fit for the expected / required values. We have noted that the level of detail of the information present in the input COMETH file was significantly less than the one of the IFC source file.

We have been then able to calculate the B_{bio} coefficient (i.e. a factor, according to French regulations, representing the thermal characteristics of the building in passive simulation mode). The B_{bio} coefficient indicates the energy efficiency of a building without considering HVAC, taking into considerations those parameters like building location, surface, shape, orientation, exposition to daylight, thermal transmittance, openings and walls properties.

3.7.3 *Considerations about the validation approach*

Unfortunately, given the nature of the addressed conversion procedure, a way to automatize the comparison of output models vs. input models has been hardly to find. Indeed, the design of the conversion procedure has been mainly driven keeping in mind that output models had to be equivalent to input models. This meaning, even considering the peculiarities and technicalities of the addressed simulation engine, we wanted mainly to assure that the model on which the simulation had to be conducted was equivalent to the input one.

For this reason, the above-mentioned aspects targeted to validate the outcome of the procedure for the two simulation engines, strive to check this equivalence between input and output model, but such an equivalence check is hard to automatized or generalized in an algorithmic procedure. This is the main reason for which tests have been mainly done visually or anyway by hand.

4 RELATED WORKS

Connection and integration between BIM and energy models is regarded as a critical research challenge in literature. Various approaches have been adopted during the years and a definitive solution has not yet been found, mainly due to the peculiarities or the lack of simulations support by BIM-specific formats (namely, IFC) on one hand and the need of detailed and specific data by simulation engines to execute computations on the other hand.

Interoperability with BIM has been addressed in several research publications. The work of (Bazjanac 2007) is a reference pillar for the integration of BIM and SIM worlds: it presents the importance of the use of the *National Building Information Model Standard (NBIMS)* for energy performance simulations and, in particular, an IFC HVAC interface for EnergyPlus relying on IFC property sets is described, allowing the mapping of EnergyPlus input data with IFC-compliant BIM authoring tools.

(Moon et al. 2011) gives an evaluation of the integration between a BIM model and different energy simulation tools (e.g., EnergyPlus is among those considered). The focus is here on the *gbXML* exchange format which allows describing building geometry, space composition, building construction, internal load, operation schedule and HVAC systems. It shows the compatibility of all these tools with BIM models, although at a different level of interoperability.

Other adopted approaches in literature involve *Modelica Libraries* and link existing building performance simulation tools with such libraries through the Functional Mockup Interface standard. The advantage of using Modelica is that it has a growing research community and that it is becoming a *de facto* standard for the simulation domain thanks to the development of several modules supported different simulation engines. The works of (Cao et al. 2014) (Wimmer et al. 2014) are worth mentioning due to the model transformation from IFC based BIM to Modelica they propose. The focus is on HVAC system conversion from SimModel to Modelica, where SimModel is meant as a general placeholder model for IFC.

Other works propose a more direct approach through the development of devoted libraries for the BIM-SIM integration: this is the case of (Kim et al. 2015) or (Robert et al. 2014). The work presented here, in particular, stemmed from the latter and represent and extension and a generalization of the approach there adopted. Most approaches recognize the need of IFC models to be enriched in order to be suitable for the conversion into complete energy simulation models (Hitchcock & Wong 2011).

Further research works have aimed at assessing the capabilities of the IFC format, including the latest released version (IFC4). They show that cannot be the locus for the full specification of the energy simulation information (Robert et al. 2014). For instance, in our specific scope, it does not include the elements required for energy simulation in the COMETH simulation engine (Haas & Corrales 2013).

In general, with respect to above-mentioned works and to literature, the approach here presented can be more regarded as trying to merge the operational-specific aspects involved in the BIM-to-SIM models conversion with the possibility of the conceptual specification of this conversion, in order to tackle this task in a general way through a framework which represents a customizable conversion software architecture to address different energy simulation engines.

5 CONCLUSION AND FINAL REMARKS

This article presents an approach for the integration between BIM projects and energy simulation engines. Energy simulation is now playing a relevant role in the design of buildings and energy consumption and efficiency are fundamental aspects to be taken into consideration during the design of new building solutions. Unfortunately, the obstacles now present for the easy integration of BIM projects and simulation engines makes the energy evaluation process rather heavy, requiring users to re-define the models of the designed buildings for the simulation engines to use, given that the latter usually do not accept BIM projects as direct input and mainly rely on proprietary formats for modeling the buildings to evaluate.

This work proposes an approach to the BIM-SIM integration issue based on the conversion between models, from the BIM model to the simulation model, and designs and implements a solution providing a conversion framework with capabilities of adaptation to different simulation engines. The proposed way and technology has been demonstrated to be rather effective allowing concentrating on the crucial conversion aspects, by focusing only the data transformation from BIM model to SIM model really needed and providing a software architecture able to tackle those aspects common to all BIM-SIM conversion as overhead and thus managed once-for-all.

The application of the implemented solution to a real test case shows that this approach represents a viable solution for the BIM-SIM integration, still is an open problem according to the survey literature. The obtained results, especially from the point of view of the model-to-model conversion, shows that conversion is effective and feasible and preserving model equivalence during transformation, although initial models are not self-contained and additional information need to be supplied to obtain an actual simulation model.

REFERENCES

Bazjanac, B. 2007. Impact of the U.S. national building information model standard (nbims) on building energy performance simulation. In *International Building Simulation Conference proceedings*, Beijing, China.

Buettner, F., Egea M., Cabot J. & Gogolla M. 2012. Verification of ATL transformations using transformation models and model finders. In Aoki T., Taguchi K. (eds), *Formal Methods and Software Engineering*, Vol. 7635 of Lecture Notes in Computer Science, Springer Berlin Heidelberg, pp. 198–213.

Cao, J., Maile, T., O'Donnell, J., Wimmer, R. & van Treeck, C. 2014. Model transformation from simmodel to modelica for building energy performance simulation. In Proceeding of BauSIM2014 Conference (eds), *Proceedings of BauSIM 2014*, Aachen, Germany.

Haas, B. & Corrales, P. 2013. Solution pour l'interoperabilité avec COMETH. In *Proceedings of Building Simulation 2013: 13th Conference of International Building Performance Simulation Association*, Chambery, France.

Hitchcock, R.J. & Wong, J. 2011. Transforming IFC architectural view BIMs for energy simulation. In *Proceedings of Building Simulation 2011: 12th Conference of International Building Performance Simulation Association*, Sydney, Australia.

Kim, J.B., Jeong, W., Clayton, M.J., Haberl, J.S. & Yan, W. 2015. Developing a physical BIM library for building thermal energy simulation. In *Automation In Construction,* vol. 50, no. 1628

Moon, H.J., Choi, M.S., Kim, S.K. & Ryu, S.H. 2011. Case studies for the evaluation of interoperability between a BIM-based architectural model and building performance. In *Proceedings of Building Simulation 2011: 12th Conference of International Building Performance Simulation Association*, Sydney, Australia.

Robert, S., Mazza, D., Hilaire, B., Sette, P. & Vinot B. 2014. An approach to enhancing the connection between bim models and building energy simulation HVAC systems in the loop. In *eWork and eBusiness in Architecture, Engineering and Construction ECPPM 2014*, Vienna, Austria.

Wimmer, R., Maile, T., O'Donnell, J. & van Treeck, C. 2014. Data-requirements specification to support BIM-based HVAC definitions in Modelica. In Proceeding of BauSIM2014 Conference (Ed.), *Proceedings of BauSIM 2014*, Aachen, Germany.

eWork and eBusiness in Architecture, Engineering and Construction – Christodoulou & Scherer (Eds)
 ISBN 978-1-138-03280-4

IfcTunnel—a proposal for a multi-scale extension of the IFC data model for shield tunnels under consideration of downward compatibility aspects

S. Vilgertshofer, J.R. Jubierre & A. Borrmann
Chair of Computational Modeling and Simulation, Leonhard Obermeyer Center, Technical University of Munich, Munich, Germany

ABSTRACT: The Industry Foundation Classes (IFC) provide a comprehensive, standardized and neutral data format to enable the exchange of digital building models. However, the current version of IFC lacks the ability to comprehensively describe infrastructure facilities such as roads, bridges, railways or tunnels in detail. This paper shows the general concept for a space oriented approach to describe shield tunnel models by extending the IFC and the integration of multiple levels of detail into the IFC standard in the scope of considering downward compatibility aspects. The proposal therefore introduces three consecutive levels of extension. Thus, we enable any IFC-viewer supporting IFC4 to visualize the exemplary instance files created in the first level by using proxy objects. The higher levels extend the standard IFC4 schema by tunnel-specific semantic elements. They also integrate an approach aiming at the representation of multi-scale models by integrating multiple levels-of-detail.

1 INTRODUCTION

The Industry Foundation Classes (IFC) provide a comprehensive, standardized and neutral data model to enable the exchange of digital building models.

Although IFC models are able to represent a wide variety of buildings, they are not explicitly well suited for exchanging product data models of infrastructure constructions such as roads, bridges or tunnels. Additionally, IFC does not support the exchange of models implementing the concept of different Levels of Detail (LoD), which is widely used in the GIS domain and particularly useful for the display of a product model's geometry in varying scales. This multi-scale representation is particularly important in the context of planning infrastructure projects like tunnels, as they typically extend over long distances. For the routing of the alignment, a kilometer scale is required, while a centimeter scale needs to be considered for the detailed design of connection points and to avoid spatial conflicts.

This paper presents a proposal for a shield tunnel product model based on the IFC under consideration of preliminary work by Yabuki (Yabuki et al., 2007; Yabuki et al., 2013). It is based on a formerly published conceptual proposal (Borrmann & Jubierre, 2013; Borrmann et al., 2014a, 2014b) and further enhances this approach in the scope of considering downward compatibility aspects. It shows the general concept for this space oriented approach to describe shield tunnel models by extending the IFC and the integration of multiple levels of detail into the IFC standard. Additionally, we introduce three consecutive levels of extension. Thus, we enable any IFC-viewer supporting IFC4 to visualize the exemplary instance files created in the lowest level. The higher levels then extend the standard IFC4 schema by tunnel-specific semantic elements (Figure 1).

The presented product model for shield tunnels fulfills the requirements on data exchange in the context of the design and engineering of large infrastructure projects. As demanded by IFC modeling guidelines, our proposal provides a clear separation between semantic objects and the corresponding geometry. The concept also implements the association of semantic entities with a particular LoD in order to enable LoD-dependent model views.

In order to maintain downward compatibility with the current IFC standard, we make use of the space structure concept provided by the IFC to model refinement relationships across the different LoDs. In the IFC standard, this concept is applied to provide a hierarchical aggregation structure for buildings (using *IfcSite*, *IfcBuilding*, etc.) and to organize them by means of the relationship *IfcRelAggregates*. In the proposed data model, this space structure concept is used to introduce corresponding spatial containers which describe

Level of Extension		Tunnel Model	Levels of Detail
	1	tunnel entities modeled as proxy objects	not available
	2	integration of designated tunnel entities	not available
	3	integration of designated tunnel entities	integration of aggregation entities to include LoDs

Figure 1.The three levels of extension. Models generated in Level 1 can be imported by any IFC4-capable viewer.

different spaces of a tunnel. Additionally, we make use of this space structure concept for modeling cross-LoD refinement relationships.

A key aspect of our approach is that the refinement hierarchy is created with the help of space objects, while physical objects are modeled on the finest level only. It allows us to use spaces as placeholders on coarser levels which prevents overlapping of physical objects. It differs from the multi-scale concept of CityGML, which allows the description of physical objects on every LoD (Borrmann et al., 2014b; Kolbe, 2009). Additionally, the explicit dependencies defined by the refinement relations allow the execution of cross-LoD consistency checks.

The presented approach introduces five levels of detail. In LoD 1 the tunnel is geometrically represented by a curve describing the main axis. For the levels 2 to 4 a strict containment hierarchy is employed: The spaces on a finer level are fully included in a space provided by the coarser level. In the 5th LoD, each physical object is placed in one of the spaces of the coarser LoDs. Figure 2 depicts the main space objects and physical objects described by this extension schema.

The feasibility of our approach is demonstrated by a collection of example files, describing an exemplary tunnel in different LoDs by using various methods of geometric representations that the IFC supports. The example files are available online and can be downloaded at: *www.cms.bgu.tum.de/ifctunnel*.

In the first level of extension we create these models by employing the standard IFC4 schema and use *IfcProxy* objects to model the tunnel spaces. The models created this way can be displayed by any IFC viewer capable to read IFC4 files. In a second and third level of integration we introduce the specific tunnel entities and LOD-related information as described in the proposed tunnel extension. These models can only be displayed in an IFC viewer that implements the proposed extension, though.

This paper is structured as follows: In Section 2 previous and related works as well as the theoretical background of this proposal are outlined. Section 3 gives a description of the shield tunnel product model in terms of semantics and geometry. Section 4 shows the three levels of extension, which ensure downward compatibility. The final Section concludes the paper and summarizes its main findings.

2 RELATED WORK AND THEORETICAL BACKGROUND

The Industry Foundation Classes (IFC) is a comprehensive data model for the exchange of information in the domain of building design, engineering and construction (Liebich et al., 2013), which has been developed over the last decade by the international organization buildingSMART. IFC aims at being a vendor-independent open data format to support the interoperability in the AEC industry. The IFC data model is defined by the data modeling language EXPRESS, which is a part of the ISO standard 10303 "STEP—Standard for the exchange of product model data"(Laakso & Kiviniemi, 2012). An important feature of the IFC model is the use of objectified relationships and inverse attributes. The IFC data model consists of several hundred entities, which enable the description of the semantics and geometry of buildings and building parts.

There is only little research that addresses the extension of the IFC model to enable the description of tunnel facilities. The first contributions date back almost 10 years to a fundamental study on the development of a shield tunnel product model (Yabuki et al. 2007). This proposal was further developed (Yabuki et al. 2013) but has not been integrated into the IFC standard as of yet. Another IFC-based date scheme for the representation of NATM tunnels has been developed by Lee et al. (2015). In Stascheit et al. (2013) a holistic IFC-based tunnel product model is discussed. It includes a ground data model and a tunnel boring machine model as well as a tunnel product model. This tunnel product model is based on the following approach by Borrmann et al.

Under consideration of the preliminary work by Yabuki et al., a concept for a shield tunnel product model has been developed (Borrmann & Jubierre, 2013; Borrmann et al., 2014b). This concept aims to fulfill the demands of data exchange in the context of the design and engineering of large infrastructure projects and forms the basis of the product model that is described in Section 3 in detail. As it

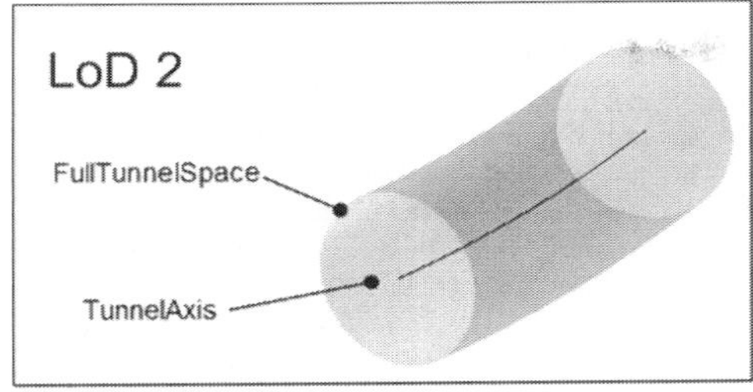

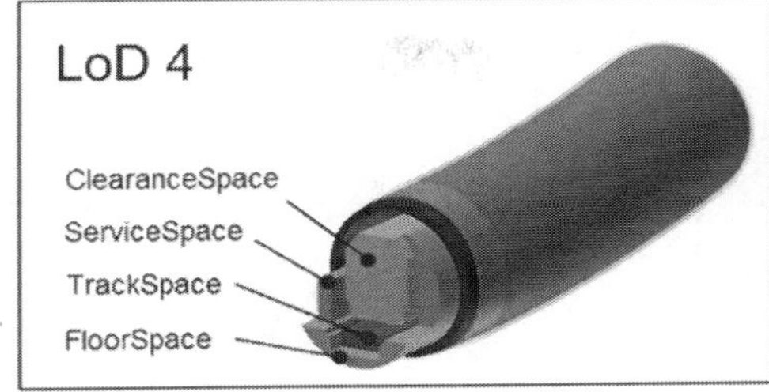

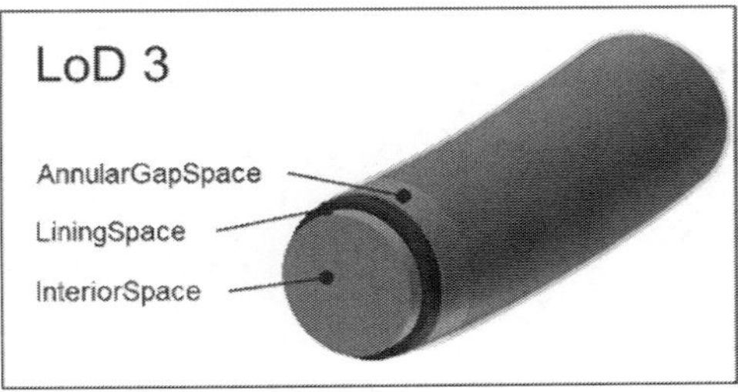

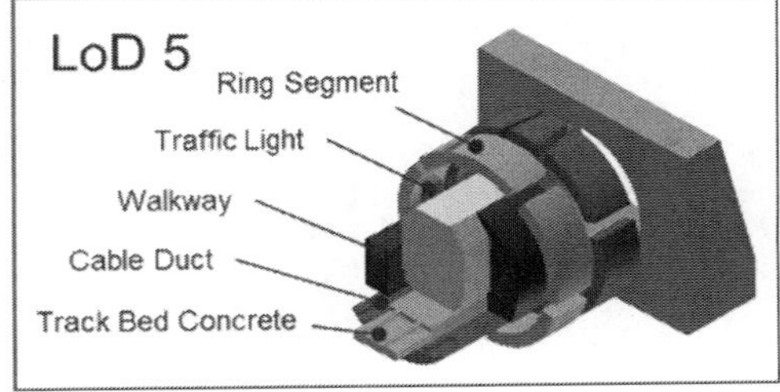

Figure 2. A 3D representation of the different LoDs of the multi-scale tunnel product model.

is not yet part of the IFC this paper demonstrates its applicability by introducing a low level extension that can be used with current IFC viewers.

An important aspect of the proposal by Borrmann et al. is the integration of multi-scale modeling into an infrastructure product model, which is a concept well established in the GIS domain. An important example is CityGML, an XML-based data model for the representation of 3D city models, which comprises five LoDs (Kolbe, 2009).

Multi-scale models provide different geometric representations of a semantic object in each LoD. These representations are then used to visualize the modeled buildings or infrastructure facilities in different scales.

3 DESIGN OF THE SEMANTIC SHIELD TUNNEL PRODUCT MODEL

The following section gives an overview of the design of the semantic and geometric parts of the proposed shield tunnel product model that was introduced by Borrmann et al. It is the basis for the three level extension approach that is shown in Section 4.

The presented product model for shield tunnels aims to comply with the requirements of data exchange necessary for the design and engineering of large linear infrastructure projects, i.e. shield tunnels.

In compliance with the IFC standard, the tunnel product model is based on a clear separation between semantic objects and the geometric representation of those objects. A key point is the association of each semantic entity with a particular LoD to achieve semantic-geometric coherence of the overall model.

The IFC standard provides a space structure concept to provide a hierarchical aggregation structure for buildings (Figure 3). It uses *IfcSite*, *IfcBuilding* and *IfcBuildingStorey* objects and orders them by employing the relationship *IfcRelAggregates*. This space structure concept is used to model refinement relationships across the different LoDs in the proposed product model. Doing so, downward compatibility with the current IFC standard is maintained.

Figure 2 depicts the defined LoDs 2–5 of the tunnel model as 3D views. The extensions that need to be added to the IFC data model to describe shield tunnel specific elements and enable multi-scale representations are listed in Figure 5.

The extensions necessary to implement the proposed tunnel model consist of space objects and physical objects in compliance with the IFC model. To group and to select all elements belonging to a certain level of detail, a new relationship entity is used: *IfcLoD*. Instances of this entity aggregate all spatial and physical objects at a certain LoD. The LoD is specified by the entity's attribute *Level*. To explicitly model the refinement hierarchy, aggregation relationships across the different LoDs are maintained. This is realized by the newly introduced relationship entity *IfcIsRefinedBy*, which inherits from *IfcRelAggregates*.

The definition of the refinement hierarchy is realized by the use of space objects as placeholders.

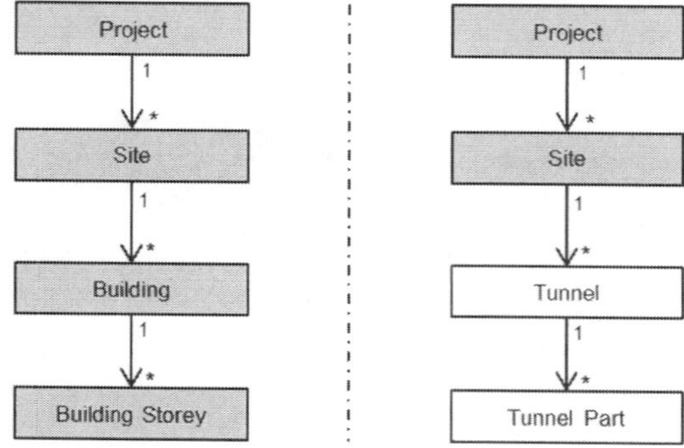

Figure 3. Left: Hierarchical space structure in the IFC standard. Right: Space aggregation structure in the proposed extension. The prefix 'Ifc' is omitted.

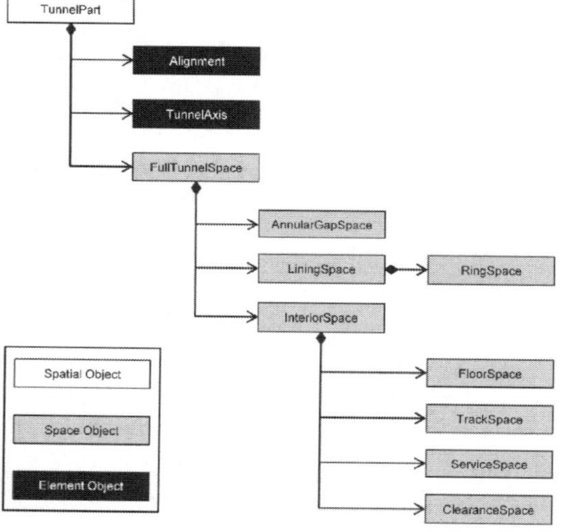

Figure 4. Space aggregation structure of a TunnelPart in the proposed extension. The prefix 'Ifc' is omitted.

The physical objects are modeled only on the highest level (Figure 2). This use of spaces as placeholders on coarser levels prevents an overlapping of physical objects, which could be wrongly interpreted as clashes, thus being fully compliant with the general IFC approach. On the contrary, the LoD concept of CityGML allows the description of physical objects on any LoD (Kolbe, 2009).

On LoD 1, the tunnel is represented geometrically by a curve representing the alignment. To this end, the respective *IfcTunnelPart* object is associated with an *IfcTunnelAxis*. This object then refers to its geometric representation in form of a curve representing the underlying alignment. The alignment is of major importance in the design and engineering of infrastructure facilities such as tunnels and it is therefore essential for a product model to include alignment objects e.g. lines, arc segments and clothoids (Amann et al. 2013).

The containment hierarchy of the levels 2 to 5 defines that spaces on a finer level are completely included in a space of the coarser level. For example, all the spaces defined in LoD 4 are refining the *InteriorSpace* of LoD3. The relations between these semantic objects is realized by the space structure concept and thereby creates the containment hierarchy (Figure 4).

In the following listing the spaces of the LoDs 2–4 are presented:

- LoD 2: The *FullTunnelSpace* is used to provide an object that represents the complete outer bounding of the tunnel.
- LoD 3: The three non-overlapping space objects *AnnularGapSpace*, *LiningSpace* and *InteriorSpace* refine the *FullTunnelSpace*.
- LoD 4: The *InteriorSpace* is refined by the space objects *ClearanceSpace, FloorSpace, TrackSpace* and *ServiceSpace*.

In LoD 5 any physical object of the tunnel can be modeled and is also assigned to a certain space by the *IfcContainedInSpatialStructure* relationship.

The proposed model makes use of very general entities and provides them with an attribute to declare the specific type of an object based on a predefined enumeration. It is thereby avoided to define a particular entity to represents every element component, which is the case in the approaches described by Yabuki et al. and Lee et al. This general concept follows the principles of object-oriented modeling and the IFC modeling guidelines and allows easy maintenance and extendibility (Borrmann et al., 2014b).

By this paradigm the diverse tunnel spaces are not modeled as individual entities. Instead they are combined in the entity *IfcTunnelSpace*, which provides a type attribute that explicitly defines the space types of the object (FullTunnelSpace, InteriorSpace, etc.). The same applies to the physical tunnel objects. The only objects that are modeled by a dedicated entity due to its importance and particular characteristics are the RingSegments. That means most objects have to be interpreted as instances of TunnelSpace or TunnelElement, instead of as instances of a specific entity.

The level of detail concept is introduced into the class model with the help of a dedicated relationship entity *IfcLoD*, which is a subclass of the existing relationship entity *IfcRelAggregates*. This relationship is used to assign instances of subclasses of *IfcProduct* to a given level of detail. Additionally, the relationship entity *IfcIsRefinedBy* has been integrated for modeling the refinement relationships.

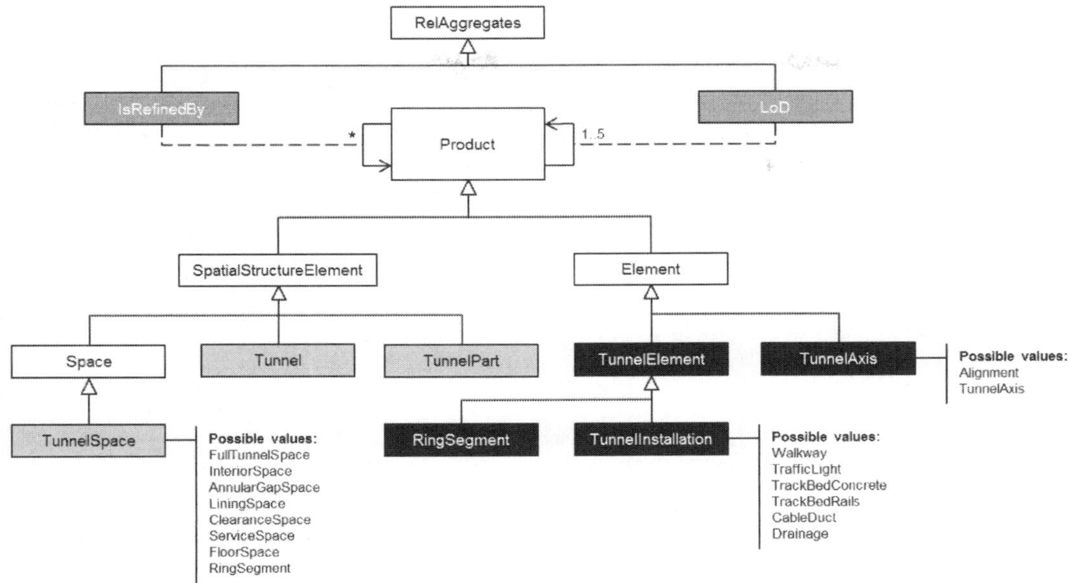

Figure 5. UML class diagram depicting the introduced relationship and object classes.

4 ENABLING DOWNWARD COMPATIBILITY

The concept described in Section 3 and the other approaches of creating a shield tunnel product model, which were listed in Section 2, are not yet implemented in the IFC standard. Even up-to-date IFC-viewers are not able to interpret the described models, as neither the extension of the IFC data model by semantic shield tunnel objects nor the introduction of multi-LoD concepts into IFC are close to standardization. Thus, the possibilities to demonstrate the use of the developed data models are severely limited.

Based on the product model presented in Section 3, we created a set of schemas and corresponding instance files, to show the applicability of this approach. These schemas and instance files are provided in three different consecutive levels of extension and therefore enable downward compatibility with the current IFC4 standard (Figure 6).

At *www.cms.bgu.tum.de/ifctunnel* we present example shield tunnel models in different levels of detail, which were created with the proposed product model extension. As only models of the first extension level can be displayed by IFC-viewers, we can only present images of these models. The IFC files are imported and displayed with the FZK Viewer developed at the Karlsruhe Institute of Technology ("FZKViewer," 2016).

4.1 *Overview*

On the lowest level of extension, we use the standard IFC4 schema without any tunnel-specific extensions. Instead, we make use of the *IfcProxy* entity for representing tunnel spaces and objects. The semantic description is provided by the *Name* attribute of the *IfcProxy* objects. The instance files created in this level of extension can be interpreted and visualized by any IFC4-capable viewer.

On the second level the schema extends the standard IFC4 schema by tunnel-specific semantic elements, e.g. *IfcTunnel*, *IfcTunnelPart* and *IfcTunnelSpace*. This extension (or a variation) is proposed to form part of a concept used in a future IFC for infrastructure data model.

The third level of extension expands the second level schema by entities that allow the explicit representation of the levels of detail. On this level we introduce the entities *IfcLevelOfDetail* and *IfcRelIsRefinedBy*. The entity *IfcLOD* defines the level of detail that a tunnel object belongs to *IfcRelIsRefinedBy* is used to describe the refinement relationships among the objects belonging to different LoDs.

For each level, we generated examples with different geometry representations. Which type of representation is used when a model is generated should always depends on the geometric properties of an object. This means that as far as possible implicit representations should be preferred, while

IfcProxy: schema		LoD1	LoD2	LoD3	LoD4	LoD5	All LoDs
IfcFacetedBrep			📷 ❁	📷 ❁	📷 ❁	Work in progress	❁
IfcAdvancedBrep (NURBS)			Work in progress	Work in progress	Work in progress	Work in progress	
IfcExtrudedAreaSolid + IfcArbitraryClosedProfileDefinition + IfcCircle + IfcBooleanResult			📷 ❁	📷 ❁	📷 ❁	Work in progress	
IfcSweptDiskSolid	IfcCompositeCurve	📷 ❁	📷 ❁	📷 ❁			
	IfcBSplineCurve	❁	❁	❁			
IfcFixedReferenceSweptAreaSolid	IfcCompositeCurve	📷 ❁	❁	❁	❁	Work in progress	
	IfcBSplineCurve	❁	❁	❁	❁	Work in progress	

IfcTunnel without LoD: schema		LoD1	LoD2	LoD3	LoD4	LoD5
IfcFacetedBrep			Work in progress	Work in progress	Work in progress	Work in progress
IfcAdvancedBrep (NURBS)			Work in progress	Work in progress	Work in progress	Work in progress
IfcExtrudedAreaSolid + IfcArbitraryClosedProfileDefinition + IfcCircle + IfcBooleanResult			❁	❁	❁	Work in progress
IfcSweptDiskSolid	IfcCompositeCurve	❁	❁	❁		
	IfcBSplineCurve	❁	❁	❁		
IfcFixedReferenceSweptAreaSolid	IfcCompositeCurve	❁	❁	❁	❁	Work in progress
	IfcBSplineCurve	❁	❁	❁	❁	Work in progress

Figure 6. Example IFC-files available at *www.cms.bgu.tum.de/ifctunnel.*

explicit representations should only be used when no implicit representation is possible or available.

The entities that are used for geometry representation purposes are listed as follows:

- *IfcFacetedBrep*: A triangle-based explicit representation of the elements' geometry.
- *IfcAdvancedBrep*: A NURBS-based explicit representation of the elements' geometry. We make use of the respective geometry entities introduced in IFC4. NURBS representations are particularly advantageous in the context of tunnels, as their elements possess a high number of curved surfaces.
- *IfcExtrudedAreaSolid*: An extrusion of the tunnel profile along a straight axis. By definition of the entity, the extrusion must be a straight path. Therefore, this geometry representation can only be used as an approximation, if the real geometry is based on a curved axis (linear approximation by segmentation).
- *IfcSweptDiskSolid*: The geometry is created by sweeping a circular disk along a given axis. We created different examples, which use either an *IfcCompositeCurve* (a composition of linear and arc segments) or an *IfcBSplineCurve* as sweeping axis. As this representation supports only the sweeping of a circular disc, models on LoD 4 and LoD 5 cannot be modeled.
- *IfcFixedReferenceSweptAreaSolid*: The geometry is created by sweeping an arbitrary closed profile along a given path. For the definition of the path the same methods as listed in the previous paragraph are used. As this representation supports the definition of arbitrary geometry, all levels of detail are modeled.

4.2 *Level 1*

We use the standard IFC4 schema without any tunnel-specific extensions for the first level of extension. Accordingly, we model the *Tunnel* and *TunnelPart* objects by using the *IfcProxy* entity. The tunnel spaces (*IfcFullTunnelSpace*, *IfcLiningSpace*, etc.) as well as the physical objects (*IfcTunnelElement*) are also modeled as instances of the *IfcProxy* entity. The relations between those objects is realized with *IfcRelAggregates*.

As the schema employed on the Level 1 is the standard IFC4 schema, any IFC viewer capable to read IFC4 files is able to display the model correctly.

However, the tunnel-specific semantic information can only be represented in a reduced manner, as the *IfcProxy* entity is applied. In order to associate the tunnel specific semantic information with the objects, we make use of the attribute *Name* of *IfcProxy*, which labels whether the object is a Tunnel, TunnelPart, a certain type of tunnel space or a tunnel element. The attribute *ProxyType* is left as *Notdefined*.

On the first level, there is not yet an explicit representation of the different levels of detail, as an integration of this concept is not supported by the current IFC4 schema.

4.3 *Level 2*

For the second level of integration we introduce the tunnel entities described in the proposed model extension. Hence, our examples start with the compulsory *IfcProject* and *IfcSite* objects. Then we incorporate the new *IfcTunnel* and *IfcTunnelPart* objects as is shown in Figure 3.

We do not use the *IfcProxy* object to represent these objects and the different spaces, but the *IfcTunnel*-spaces as defined in the extension model. Thus, we are able to model the complete tunnel-specific semantic information by containing it in the attributes of the introduced entities. Although we do not introduce the LoD concept yet, we structure the spaces under the same hierarchy we introduced in Figure 3 and 4 by means of *IfcRelAggregate*.

The examples on the second level are based on a customary extension of the IFC product model, which is not yet part of the standard. Therefore, the resulting examples cannot be interpreted by any of the currently available IFC-viewers.

4.4 *Level 3*

Only on the third and highest level of extension, we introduce the aggregation entities *IfcLod* and *IfcRelIsRefinedBy*, which substitute *IfcRelAggregates* used in the previous levels. The aggregation *IfcLod* is used to connect the different spaces and elements with *IfcTunnelPart*. This way, a capable viewer can filter the model based on the levels of detail and thereby show only relevant information. *IfcIsRefinedBy* is used to reproduce the hierarchical structure of spaces and physical elements.

Moreover, when the aggregation is done between a space and an element, the aggregation *IfcRelContainedInSpatialStructure* is maintained. This allows the standard IFC viewers to recognize the relation between the spatial structure and the element containment independently of the level of detail.

The examples generated within this extension level are also based on an extension of the IFC product model. Therefore, they cannot be interpreted by any of the currently available IFC viewers.

5 CONCLUSION

The current version of the IFC standard is not well suited for representing and exchanging product models of infrastructure facilities.

This paper gives an overview of the concept for a future extension of the IFC standard in order to enable the detailed modeling and exchange of shield tunnels product models. The planning of such large infrastructure facilities requires the consideration of differing scales. Therefore, the proposed extension introduces the concept of multi-scale modeling by enabling the representation of objects in different levels of detail.

Based on the presentation of the product model, the paper focuses on the issue of compatibility between the proposed model extension and current applications for interpreting IFC-files. As the definition is not yet included in the IFC standard, there are no IFC-viewers capable of importing IFC-files that use shield tunnel specific entities.

Therefore, we present the shield tunnel product model in three consecutive levels of extension. This gradual approach provides a low level implementation that does not require an extension of the existing IFC 4 standard, but introduces the concept of the presented product model by using the *IfcProxy* entity. This downward compatibility allows existing IFC-viewers to interpret example files generated in scope of this research. Only on the second and third level of extension, new IFC entities are defined to model tunnel—or LoD-specific objects and relations.

By the introduction of the shield tunnel product model in these three levels of extension we aim at demonstrating the use of the developed data model. Thereby, we present the advantages of the product model and show how the application of the proposed approach can be applied.

ACKNOWLEDGEMENTS

We gratefully acknowledge the support of the German Research Foundation (DFG) for funding the project under grant FOR 1546.

REFERENCES

Borrmann, A., Flurl, M., Jubierre, J.R., Mundani, R.-P., & Rank, E., 2014a. Synchronous collaborative tunnel design based on consistency-preserving multi-scale models. Advanced Engineering Informatics 28, 499–517.

Borrmann, A., & Jubierre, J.R., 2013. A multi-scale tunnel product model providing coherent geometry and semantics. In: Proc. of the 2013 ASCE International Workshop on Computing in Civil Engineering. Los Angeles, pp. 291–298.

Borrmann, A., Kolbe, T.H., Donaubauer, A., Steuer, H., Jubierre, J.R., & Flurl, M., 2014b. Multi-Scale Geometric-Semantic Modeling of Shield Tunnels for GIS and BIM Applications. Computer-Aided Civil and Infrastructure Engineering 30, 263–281.

FZKViewer [WWW Document], 2016. URL http://www.iai.fzk.de/www-extern/index.php?id = 1931 (accessed 4.14.16).

Kolbe, T.H., 2009. Representing and Exchanging 3D City Models with CityGML. In: Lee, J., Zlatanova, S. (Eds.), Proceedings of the 3rd International Workshop on 3D Geo-Information, Seoul, Korea, Lecture Notes in Geoinformation and Cartography. Springer Berlin Heidelberg.

Laakso, M., & Kiviniemi, A., 2012. The IFC Standard—A Review of History, Development and Standardization. ITcon Journal of Information Technology in Construction 17, 134–161.

Lee, S.H., Park, S.I., & Park, J., 2015. Development of an IFC-Based data schema for the design information representation of the NATM tunnel. KSCE Journal of Civil Engineering 00, 1–12.

Liebich, T., Adachi, Y., Forester, J., Hyvarinen, J., Richter, S., Chipman, T., Weise, M., & Wix, J., 2013. Industry Foundation Classes: Version 4. BuildingSmart International (Model SupportGroup).

Stascheit, J., Meschke, G., Koch, C., Hegeman, F., & König, M., 2013. Processoriented numerical simulation of mechanized tunneling using an IFC-based tunnel product model. In: Proceedings of the 13th International Conference on Construction Applications of Virtual Reality. London, UK.

Yabuki, N., Aruga, T., & Furuya, H., 2013. Development and application of a product model for shield tunnels. In: Proceedings of the 30th ISARC. Montréal.

Yabuki, N., Azumaya, Y., Akiyama, M., Kawanai, Y., & Miya, T., 2007. Fundamental Study on Development of a Shield Tunnel Product Model. Journal of Civil Engineering Information Application Technology 16, 261–268.

eWork and eBusiness in Architecture, Engineering and Construction – Christodoulou & Scherer (Eds)
ISBN 978-1-138-03280-4

Delivering of COBie data—focus on curtain walls and building envelopes

J. Karlshøj
Technical University of Denmark (DTU), Kongens Lyngby, Denmark

P. Borin, M. Carradori, M. Scotton & C. Zanchetta
University of Padova, Padova, Italy

ABSTRACT: COBie is a standard data framework whose main purpose is to transmit useful, reliable and usable information collected throughout the whole building process and to be consumed in order to properly maintain the facility. Focusing on Facility Management information exchanges and considering the UK BIM policies and requirements, this paper shows the results obtained applying COBie to complex products such as curtain walls. Two Information Delivery Manuals (IDMs) were also developed, in order to provide a commonly known and standardized framework, which can regulate the COBie-based information exchanges. Future developments of this study could concern the application of the developed IDMs to different case studies in order to overtake that specificity characterizing each single project and verify the validity of the proposal.

1 INTRODUCTION

The way through which information is exchanged between the different stakeholders during the building process is a theme on which international organizations and software developers are deeply involved. In the last years, specifications and applications had been developed in order to make information exchange as efficient as possible.

This research looked at this topic focusing on that part of the entire set of information exchanged during the project realization needed for Facility Management. Considering the Industry Foundation Classes schema (IFC—ISO16739:2013), the open data format developed by buildingSMART International, this subset of information requires to be defined by a specific Model View Definition (MVD).

The Construction Operation Building information exchange (COBie) is a standard developed by the North American chapter of buildingSMART, buildingSMART alliance, that aims to define a MVD able to collect all that information needed to properly maintain the facility. In this study, the exchange of information following the COBie standard was applied to specific products, namely curtain walls and building envelopes, in order to understand if this existing standard could be a proper solution also for a complex product in terms of geometry, description and classification, such as those considered ones.

2 STATE OF THE ART

COBie provides a structured framework to collect specific data across the different project stages, ensuring an effective data transmission between the main actors involved in the building process: the Design and Construction teams, the Owner and the Facility Manager. This set of information aims to provide a comprehensive knowledge in order to properly run the maintenance operations.

COBie was developed starting from the assumption that entrusting the delivery of FM information to paper documents and PDFs, the current information vehicles, was not and will never be the best solution. Nowadays, in fact, this behavior has the only result to produce hundreds of paper or file that are virtually useless to facility managers and often these pieces of information are not available at the time of occupancy during the building process, but months or even years after the actual beginning of the 'in use' phase, they are stored in a room and never used (East et al., 2013). Moreover, a lot of time could be wasted by the builder delivering handover documents every time he has to recreate information already specified by architects and manufacturers, but not in a way that allows its inclusion in such documents.

Before COBie, two different approaches have been noticed regarding electronical capture of facility management information:

- the owner has maintenance staff involved in retyping information from the handover documents to the chosen maintenance management software;
- the owner requires the contractors to provide information directly in the chosen FM application or in a specific data format ready to be imported in the FM system (East et al., 2013).

COBie aims to provide a solution to the problems related to the methods described above, defining a unique container for this pieces of information, delivered in an electronic format with a standard, open and reusable structure based on the IFC schema (ISO 16739:2013) that allows facility managers to handle a concretely useful and usable set of information. COBie is defined, in fact, as a MVD, since it represents a specific subset of the building information model, and it does not handle any geometric information, as it is "a performance-based specification for facility asset information delivery" (East, 2009, p.18).

The first COBie release was published in 2007 by the Construction Engineering Research Laboratory of US Army Corps of Engineer. Today the definition of COBie is jointly maintained by buildingSMART alliance, and the British and Irish chapter, buildingSMART UKI. The technical definition of the standard can be found in the National BIM Standard-United States version 3 (NBIMS-US v.3), lastly updated in July 2015 and in the buildingSMART alliance website, where COBie was officially published as a MVD in October 2013. Moreover, since COBie is expected to become mandatory for public commissions in UK from April 2016, the British Standard BS1192-4:2014 provides a code of practice that should be followed in the UK scenario. COBie-compliant information can be delivered in three formats: IFC, IFCXML and the XML format, a simple spreadsheet file editable with Microsoft® Excel that is the format chosen by the British Government.

As a part of the IFC schema and BIM literature (ISO/TS 12911:2012) COBie captures information for both spatial and physical assets in the facility. A unique COBie file is created for each building in a project and once the required information is collected, it is organized following the diagram reported below (fig. 1). Using a standard for FM information exchange requires the relations between the information and the different stakeholders involved in the project to be precisely defined. The owner, in particular, is then responsible to specify: which assets are going to be managed, which is the level of detail of the information to be provided and when, during the building process, information has to be transmitted. These specifications make COBie's content 'project-specific'. Regarding the physical assets, all the equipment that needs preventive maintenance plan, has consumable parts and requires management and regular inspections, are supposed to be included in COBie. COBie could also handle architectural and structural elements if they are subjected to maintenance operations (East, 2013). As stated above, the conducted research focused on curtain walls and building envelopes (from now the term curtain wall will describe both the products considered by this research). Two main reasons drive this paper focusing on the UK scenario:

- the BIM Task Group work and the UK Government's initiative could become a framework for other states of the European Union, which does not yet submit a BIM policy.
- the UK construction business shows how notable the impact of new private housing and private commercial is (UK Office for Nation Statistics, 2016). Since curtain walls are commonly used in both these fields, it becomes an important research discipline.

From a construction point of view, the research faces the complexity within curtain walls' definition and description.

1. Product. A 'curtain wall' can be simplified as a unique object, the whole envelope system, but it can be also seen as an assembly composed by any subsystems that constitute the envelope (Herzog et al., 2004). Therefore, the whole system can be further subdivided in the several elements that make up the envelope.
2. Sub components variability. In buildings like skyscrapers, for example, the hundreds of panels that cover the whole building could seem at a first sight the same panel type repeated for hundreds of times, but, actually, they usually differ, for instance, for dimensions. Namely building envelopes imply a high number of different objects belonging to different types.
3. Integration. Curtain walls' information overlap among building based information, such as design collaboration, construction and FM, and manufacturing based information, which have different requirements and procedures.
4. Interferences. Curtain walls are related to other discipline components, such objects from structure and MEP systems.

By the COBie point of view, this complexity requires specific considerations about the exportation of Type, Component, Assembly and System worksheets, in order to align the COBie schema and requirements to the particular features of curtain walls.

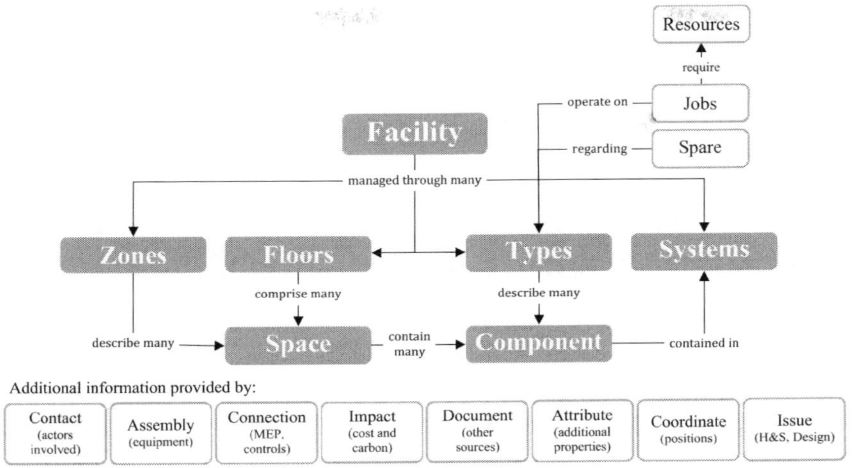

Figure 1. COBie's content structure.

3 METHODOLOGY

3.1 *The problem of the manageable assets*

As previously said the content of COBie in a specific project should be specified by the employer, who firstly has to decide which facility assets are going to be managed during the facility life-cycle. However, the definition of 'manageable assets' could potentially include a wide set of different objects.

The COBie specification reported in NBIMS-US v.3 provides a list of the IFC entities which are not part of the IFC subset expected by the COBie MVD and so which are not supposed to be handled by COBie. Regarding the specific purpose of the conducted research, it must be underlined that entities such as IfcCurtainWall, IfcPlate, IfcMember and the respective IFC Type entities, are part of this exclusion list. However, the COBie specification itself specifies that software vendors may also allow the user to apply different exclusion lists, in order to satisfy specific owner's requirements or regional directions (NBIMS-US v.3). Since building envelopes are products that typically require to be maintained during the facility life-cycle, in this research it was chosen to loosen the constraints imposed by the COBie specification and include the considered product in the outgoing COBie file.

One of the main purpose of this report, pursued through the development of two Information Delivery Manuals (IDM) on the delivery of COBie data, was to identify the best way to manage curtain walls within COBie, defining what can be considered a manageable asset in this particular case and how the template expected by COBie should be compiled.

3.2 *Producing COBie data*

To understand how to produce COBIE-compliant information, an experimentation phase was also included in this study. Since the XML format of COBie is the one which is going to be required in UK, the tests aimed to create COBIE files in spreadsheet format.

Apart from filling the COBie template manually with the relevant information, which is an extremely time consuming procedure and easily related with compiling errors, two different methods were followed: creating COBie data directly from the chosen BIM authoring tool, using a specific exporter; creating COBie data exporting an IFC model from the native one and using external applications able to transform the IFC file in COBie XML format.

Considered the complexity of the analyzed product and its relationships, the use of the first method aims to demonstrate the information exchange's maximum quality from the point of view of a unique stakeholder. However, it is clear that it requires each actor in the process to produce his single COBie file from his own model. In this case, at the end of the process, there could be several issues in aligning the information deriving from different professionals, with manual copy and paste operations and corrections. An OpenBIM scenario in which every contribute is merged in a unique model, necessarily through the IFC format, and then the required information is extracted (following the COBie MVD or a specific FM MVD) seems to be the best solution. The second procedure tries to verify the same quality standards used before.

3.3 *FM MVD and IDM procedure*

It is not possible to deal with the COBie standard without referring to IFC: as already highlighted,

COBie is not just a spreadsheet; it is intended to be a standard way to provide a defined information content of an IFC model. The IFC format should enable the different stakeholders to exchange information, thanks to its interoperability and standardization supported by software systems implemented for different fields (See et al., 2012).

However, at the time of writing, it has been noticed that there is not any Model View Definition named 'COBie' supported by the used software application. The current procedure asks the user to export data through the so called 'Extended FM Handover View', a specific MVD made to cover the FM information exchanges between the design and operational phase, but still not officially approved by buildingSMART International and, therefore, not stable. In other words an MVD enabling the different stakeholders to exchange usable and functional information for the Facility Management during the whole building process is still needed. As a consequence it is not possible to exchange just COBie-related information through the IFC format so far, since the available Model View Definitions include a greater amount of data.

In addition to these considerations, it must be pointed out that the IFC format is responsible to carry the informative content of a model, but it does not provide any information related to the process. To gain the best results from interoperability, the quality of communication must be as high as possible and the process should be standardized and commonly recognized (Wix & Karlshøj, 2010). The IDM is intended to define in detail a specific business process where at least one information provider and one receiver are involved: 'who', 'when', 'what' and 'why' represent the content of the IDM in relation to an information exchange. The main purpose of IDM is to apply a standardized methodology to describe and define already existing or new processes, in order to assist the different stakeholders in their tasks throughout the entire lifecycle of a facility. It must be noted that there is not a direct correspondence between the content of an IDM and the IFC data: the MVD is responsible to establish a link between the technical specifications of the IFC schema and the human-readable information defined by the IDM.

The development of an IDM is one answer to the AEC industry need of a structured framework for its business processes, throughout all their complexity. For this reason, the cited IDM methodology has as a primary focus the definition of the flow of information, besides the information content itself: once identified the specific need associated to a business case and the purpose subtended by the information exchange, the development process of an IDM begins. The attention could be focused on the different building systems, with the main purpose of managing their peculiarity and specific characteristics.

4 DEVELOPMENT

4.1 *Brief analysis of the used model*

The research consequently moved to the experimentation phase, where the two possible ways to deliver COBie data from a project during its different stages were tested. The tests were conducted using Autodesk Revit 2014 as BIM authoring tool; the typical curtain walls' breakdown made of panels and mullions, also proposed by the used software, was not implemented in the analyzed model, where the curtain systems were modeled as an association of panels only. Consequently, each panel actually represents a container (from now named as 'cell', intended as the main component of a unitized curtain wall systems) of typologically different elements.

4.2 *COBie Extension*

The COBie Extension for Revit is an internal add-in of the used authoring software that allows the export of a COBie XML file directly from Revit. In particular it allows exporting the following worksheets: Contact, Facility, Floor, Space, Zone, Type, Component, System, Attribute and Coordinate. For the specific scope of this study, the attention was mostly directed to those worksheets that expect information for which a specialist subcontractor involved in the field of interest is responsible, namely the ones from Type to Coordinate. It must be said that once the information is inserted in the model, the COBie Extension for Revit requires the user just to define some settings to export the COBie data. For how concern the classification system, that is an information to be included in COBie, it was possible to force Revit reading Uniclass2015 values, instead of OmniClass ones, even if the followed procedure is not sufficient to immediately assign the expected value to the relevant properties, as it requires the user to select it between a range of possible choices.

4.3 *COBie via IFC*

Since the COBie spreadsheet format is just a mapping of the COBie MVD developed to show the information content in a human-readable way, the route of getting COBie via IFC is the natural way of doing and the one which can, referring to the whole building process, return a complete and representative COBie file.

The delivery of COBie data via IFC is a process that has to be subdivided in two phases:

1. the exportation of the native model in IFC format;
2. the transformation of the IFC model in COBie spreadsheet format.

Of course, the way the chosen BIM authoring tool translates the native model into IFC format depends on the specific exporter application used by the software.

It must be underlined that even if the used model did not include Curtain Mullions, some tests conducted on a trial model showed some issues in the exportation of mullions from Revit. In particular, each mullion instance is correctly defined by the IFC entity IfcMember, but no type entity (IfcMemberType) is assigned to the mullion instances. Furthermore, mullions are not correctly named in the IFC file; it means that the name associated to mullions in the Revit model, is not the same through which mullions are identified in the IFC file, while it occurs for all the other elements in the model.

It is also possible to specify the MVD to be applied as a filter in the model exportation. The standard MVD that Revit uses is the Coordination View 2.0 based on IFC2 × 3, but the exporter allows to choose other MVDs like the mentioned IFC2 × 3 Extended FM Handover View, used in these tests.

The properties associated to each object in the Revit model are exported in IFC through specific data records called property sets: it is possible to map data defined within the used BIM authoring tool to the desired parameters in the outgoing IFC file.

To realize the second step in the process and obtain the outgoing COBie spreadsheet file from the IFC model, four different external applications were tested: BIMserver, COBie Toolkit (based on BIMserver and issued before the inclusion of a specific application within BIMserver itself), and the first and the last version of BimServices, a software developed by AEC3 since the first COBie release. However only COBie Toolkit and the last version of BimServices allowed the generation of COBie data for curtain walls, since they are the only ones that allow the user to modify the exclusion list the COBie specification proposes.

4.4 *Brief analysis of the resulting content of the COBie worksheets*

The results obtained through the export of the relevant COBie data of the entire model were evaluated in relation to the product and the specific stakeholder considered and as a consequence those COBie worksheets whose compilation he is responsible for.

About the exportation of the Type and Component worksheets, that can represent the basic content of COBie in the considered case, both the analyzed procedures (namely the COBie Extension and the IFC method) showed the expected results. In particular the Type worksheets was filled with what is defined as type in Revit; as a consequence Types were exported regardless dimensions, since this is the meaning of 'type' in Revit, even if the COBie schema expects Types to be defined also in relation to dimensional parameters (*NominalLength*, *NominalWidth* and *NominalHeight*). However, it must be underlined that the COBie Extension do not allow the exportation of the curtain system Type, so the entire curtain system cannot be described by a type of product in the COBie file; instead, following the IFC method, it is not possible to export mullion Types, because of the explained issues about the exportation of curtain mullions from Revit to IFC.

For both Types and Components the relevant properties were correctly exported in the outgoing COBie file, using the values introduced by the user in the specific COBie fields generated in the type and instance property menu by the COBie Extension (if this was the chosen procedure), or using the values still introduced by the user in Revit, but included in the property sets of the IFC file. About the association of the relevant Space to each Component, the undertaken tests did not show any value in the relative column since no information about the spatial subdivision of the facility was included in the used model. This fact reinforces again the use of IFC models for information exchange between different stakeholders: in this way the considered subcontractor, which is not supposed to deal with the spatial organization of the building, can work on a model that already contains this information, producing consistent COBie data.

About the other COBie worksheets, many of them are not supported by the COBie Extension; these are: Assembly, Connection, Spare, Resource, Job, Impact, Document, Issue. Considering these worksheet, the IFC method showed a partial exportation of some of the Assembly one, describing each Curtain System as an assembly of the several panels and mullions (each one included in COBie as a row in the worksheet) that compose the entire system. The fact that both the analyzed procedures do not allow the exportation of some information is because Revit, as a design software, is not able to handle information that are typically exchanged during the construction phase, for example that one related to Spare, Job and Resource worksheets.

Finally, all the methods allow the user to export in the Attribute worksheet all those properties not relevant to the columns of the other ones in the COBie template.

It must be underlined that the existing literature is still not clear enough to define the content of many COBie worksheets (i.e. Impact, Connection) or their content could be strongly different from project to project in relation to specific requirements (i.e. System, Coordinate). For these reasons, and considering those worksheets whose compilation is defined as 'Optional' by BS1192-4 (Assembly, Connection, Coordinate, Issue), the results had been evaluated without considering the mentioned worksheets.

4.5 *IDM development*

The development of the IDM, needed for the definition of the information to be exchanged, followed the specific methodology proposed by (Mondrup et al., 2014), based on a modular approach in the organization of the workflow and the management of data. As a consequence it is possible to define reusable IDM packages that specify the information exchanges that occur at a specific life cycle stage involving specific actors of the process.

With this perspective, for the purpose of this research, two IDMs were defined in relation to the specific actor involved and to the phases of the process during which the exchanges of COBie data occur, the so called COBie Data Drops. The two IDM packages were then named 'COBie Data Drop 3—Specialist Designer' (fig.2) and 'COBie Data Drop 4—Subcontractor', considering the roles proposed by the British regulation BS 1192:2007 and the COBie Data Drops specified by the BIM Task Group (BrydenWood, 2012), both specifically referred to the British scenario. Each IDM package should be composed by four key documents: Business Use Case, Process Map, Exchange Requirements and Exchange Requirements Model. For the sake of simplicity, the latter IDM deliverable was not considered, since it exceeded the boundaries of this study. Moreover, the Business Rules must be considered as a constituent part of the IDM, even if they are just external references used to define constraints, indications, regulations to be observed in the management of specific data, within the framework outlined by the IDM.

The expected result deriving from the use of this approach should be a more reliable exchange of data during the lifecycle of a facility between the different actors developing BIM data. The proposal suggested by (Wix & Karlshøj, 2010) was taken as a reference concerning the practical development path to be followed to deliver a complete IDM. The *Business Rule Localization* was the followed method, since the COBie IDM already existed, as well as the Exchange Requirements associated to the information exchange of interest. The Business Rules that drove the writing of the IDMs were the BS1192:-4:2014 and the Uniclass2015, in order to get aligned to the UK scenario. What the research aimed to do was the re-definition of the already existing IDM in relation to the COBie UK standard, in relation to a specific actor, involved in the field of curtain walls and building envelopes, during specific stages of the building process. The problem of the 'manageable asset', namely the definition of which object and related information should be included in COBie, concretely affected this phase.

The deep analysis of the COBie 2.4 specification and the available COBie spreadsheet examples taken from the BIM Task Group website led to a final consideration concerning the actual practice related to COBie: where precise indications from the employer are not given, as long as a wide and

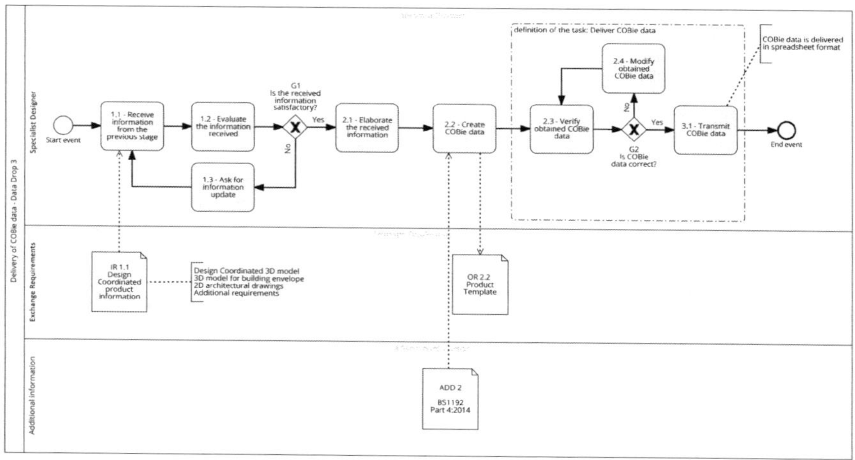

Figure 2. COBie data drop 3—specialist designer, process map.

well-known practice on COBie is not established, the project team has to define which the best way to model COBie data is, according to the specific knowledge each member is able to provide on its particular field of work.

To fulfill the objective of the research, the COBie 2.4 specification (NBIMS-US v.3) and the COBie Responsibility Matrix were taken as references; in particular, the first was used to establish a connection between the Exchange Requirements (ERs) expected by COBie and the COBie Data Drops, the second allowed to precisely define when each cell of the COBie template should be compiled during the project life-cycle. Regarding the definition of what to include in COBie, the leading consideration was the following: every element of the building envelopes which needs maintenance operations, is provided with replacement parts or requires to be specified by additional documents, is described as a specific Type in COBie. Moreover, all those objects that need to be specified with those properties expected by the Type worksheet (for example those objects for which warranty information is required) are exported as Types too. It must be clearly pointed out that the criteria used to define the object Types is not dimension-related. This way of doing was chosen in order to mitigate the complexity of the considered product and to simplify the exportation which, otherwise, will lead to the definition of hundreds of different Types and an unusable COBie file.

Concerning this matter, only the cells are exported as Components, even if this procedure would not be allowed by BS1192:2014, since *"every Type should apply to at least one Component"*. This simplification, however, seems to be necessary for managing the complexity of the specific case, keeping into account that the relation between different objects can be showed by the Assembly worksheet. Finally, in the general case where a subdivision in cells is not needed or useless, all the elements of the building envelope should be exported as Components. Panels and mullions, in fact, represent in the most general and diffuse practice the composing elements of curtain walls: therefore each one of them must be seen as a single component, needing specific maintenance, or being part of a specific system, for example.

Through all these considerations the expected standardization and specificity were pursued.

5 CONCLUSIONS AND FUTURE DEVELOPMENTS

5.1 *Considerations upon the way of modeling*

Regarding how the model should be produced, the process and the standard through which information is exchanged need to be known and considered. Starting from the objective, namely the delivery of COBie data, its requirements, and considering the IFC schema, its structure and the possible relations to define the elements in their complexity, the user should find the best way of modeling in order to ensure the best result in the final COBie file. In this sense and referring to the curtain wall product, the definition of IfcPlate given by buildingSMART seems to be enlightening, enhance different types of relations with the spatial structure and its subcomponents. Of course, the results highly depend on the specific BIM authoring tool used and the ability of its IFC exporter to translate the native model in IFC format.

5.2 *Consideration about the two methods*

A first observation regarded the purpose of the generation of COBie data: if the objective is simply delivering COBie in its spreadsheet format, in order only to fulfil a contractual requirement, the COBie Extension for Revit can be seen as the best solution (fig. 3).

The reason supporting this statement is the ease of using and customizing the Extension within the Revit environment, without the necessity of additional IT capabilities, besides the fact that it allows to define all the same user-defined properties that could be exported to COBie via the IFC file obtained from Revit. The manual compilation of the several blank COBie fields is required in any case and it is comparable between the two proposed methods.

Still, if the expected output is a COBie spreadsheet, the method used to obtain it does not affect its final appearance: an XML file is just a table, in any case, either it is generated through the COBie Extension or transforming an IFC file. However, a wider perspective leads to reconsider the method

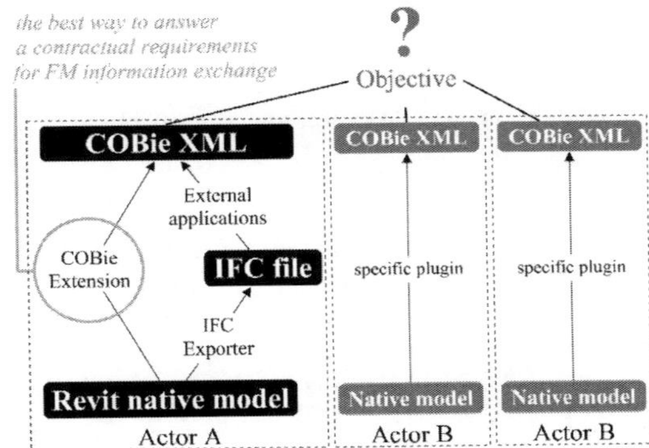

Figure 3. Issue in delivering COBie data using a single stakeholder perspective.

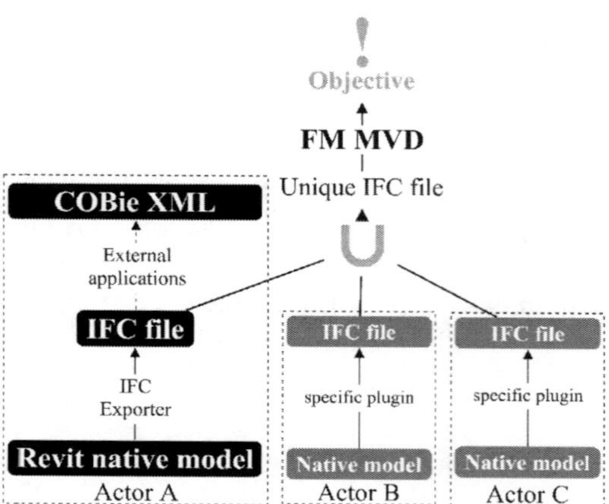

Figure 4. Delivering COBie data using a comprehensive perspective.

based on the translation of an IFC model into the XML format. It must be kept in mind that IFC is the open format designated for enabling interoperability between the different actors of the building process. In this light, delivering COBie data through IFC could be considered as just one aspect of the entire building process based on IFC in order to perform interoperable exchanges where functional and usable information is transmitted (fig. 4).

Moreover this aspect is of primary interest also in the UK scenario, where the IFC standard is one of the essential requisite to reach BIM maturity Level 3. For this reason, gaining familiarity with the IFC environment could represent a wise decision for any stakeholder in the AEC industry. These considerations make clear that an MVD supporting this kind of information exchange is strongly needed: the different actors of the building process need to manage information in a collaborative way, where the content of the exchange is clearly defined and the data flow is structured in a standardized way. IFC seems to be the natural answer to these requirements.

As said, panels and mullions represent the standard way through which curtain system are organized. Regardless the encountered issue related to mullions, this is also the organization used within the IFC schema. However, IFC allows also the definition of curtain systems made by cells that contain sub-elements, but it was discovered the used BIM authoring tool do not give the possibility to map this kind of relation between the different components.

With this regard, future consideration upon the adequacy of how BIM tools translate curtain wall product's complex relations, from the literature review point of view, within IFC schema could be made. Future developments of this study could also concern the application to different case studies of the developed IDMs, in order to overtake that specificity characterizing each single project and verify the validity of the proposal. With this wider perspective, the IDMs themselves could be adjusted in relation to the actual requests of the AEC industry, in order to reach an as high as possible optimization of the building process.

REFERENCES

BrydenWood 2012. *Cobie Data Drops—Structure, uses & examples*. BrydenWood Limited, London (UK).

BSI 2007. *BS 1192:2007—Collaborative production of architectural, engineering and construction information—Code of practice*. British Standard Institution, London (UK).

BSI 2014. *BS1192-4:2014—Collaborative production of information—Part 4: Fulfilling employer's information exchange requirements using COBie—Code of practice*. British Standard Institution, London (UK).

East, W. 2009. Performance Specification for Building Information Exchange. In: Journal of Building Information Modeling, Fall 2009, buildingSMART alliance, pp. 18–20.

East, W. 2013. Using COBie. In: *BIM for Facility Managers* (P. Teicholz, Ed.). John Wiley & Sons Inc., Hoboken, New Jersey (USA), pp. 107–143.

East, W., Liebich, T., Nisbet, N. 2013. Facility Management Handover Model View. In: Journal of Computing in Civil Engineering (27, Issue 1). American Society of Civil Engineers, pp. 61–67.

Herzog, T., Krippner, R., Lang, W. 2004. *Façade Construction Manual*. Basel: Birkhäuser.

ISO16739:2013—Industry Foundation Classes (IFC) for data sharing in the construction and facility management industries. ISO, Geneva (Switzerland).

Mondrup, T.F., Treldal, N., Karlshøj, J., Vestergaard, F. 2014. Introducing a new framework for using generic Information Delivery Manuals. In*: eWork and eBusiness in Architecture, Engineering and Construction. Proceedings of the 10th European Conference on Product & Process Modelling (ECPPM 2014)*. Taylor & Francis Group, London, pp. 295–302.

NBIMS-US 2015. Construction Operation Building information exchange (COBie)—Version 2.4. In: *National BIM Standard—United States Version 3*. National Institute of Building Science, Washington, DC (USA).

Office for National Statistics 2016. *Output in the construction industry: January 2016 and new orders Quarter 4 2015*.

See, R., Karlshøj, J., Davis, D. 2012. *An Integrated Process for Delivering IFC Based Data Exchange*. buildingSMART International.

Wix, J., Karlshøj, J. 2010. *Information Delivery Manual—Guide to Components and Development Methods*. buildingSMART International.

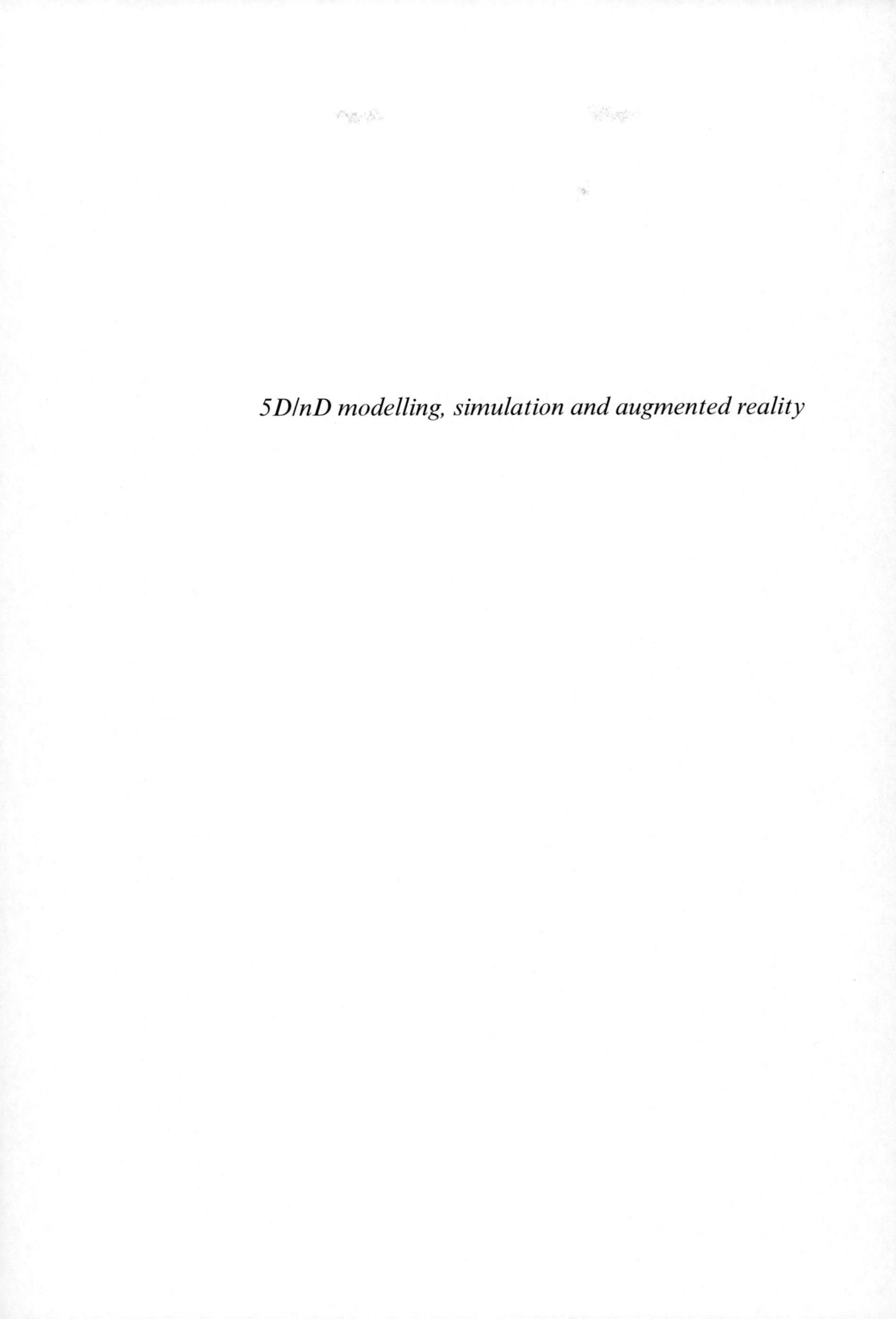

5D/nD modelling, simulation and augmented reality

eWork and eBusiness in Architecture, Engineering and Construction – Christodoulou & Scherer (Eds)
 ISBN 978-1-138-03280-4

Software library for spatial-temporal modeling and reasoning

V. Semenov, K. Kazakov, K. Petrishchev & V. Zolotov
Institute for System Programming RAS, Moscow, Russia

ABSTRACT: Visual 4D modeling and planning technologies have recently begun to play a crucial role in the realization of complex construction projects and programs. They enable to improve communication and coordination among stakeholders and, thereby, to reduce risks and waste during the project implementation by means of simulation and visualization of project activities in space dimensions and across time. Modern project modeling systems possess some underlying functions for such purposes, but are still limited in reasoning capabilities necessary for automated validation of the project schedules in terms of their feasibility and absence of spatial-temporal conflicts. Basically, these are collisions and interferences of construction objects, missing collision-free paths to deliver the objects to destination positions in planned time, and invalid schedules leading to the objects hanging without any reliance. A software library providing advanced capabilities for spatial-temporal modeling and reasoning is presented in the paper. The library called Constructivity4D supports a representation of all of the core concepts in the scene modeling and possesses a wide range of operators and validation functions. The modeled dynamic scenes are suggested to be hierarchically organized with the objects obeying particular behavioral patterns. Being compliant with popular qualitative reasoning formalisms and relation algebras, the supported operators and functions can be combined to identify non-trivial spatial-temporal conflicts originated from the construction project validation problems. The paper discusses the design considerations of the library, its functionality, interfaces. Particular attention is also paid to applied computational methods and implementation details.

Keywords: 4D modeling and planning, collision detection, spatial-temporal reasoning, software engineering

1 INTRODUCTION

In recent years, visual 4D modeling and planning technologies have begun to play a crucial role in the realization of complex construction projects and programs. They enable to improve communication and coordination among stakeholders and, thereby, to reduce risks and waste during the project implementation by means of simulation and visualization of project activities in space dimensions and across time. Popular commercial 4D modeling systems like Autodesk Navisworks, Synchro, Bentley Schedule Simulator, Intergraph Schedule Review, Trimble Vico, Rib iTwo provide some underlying functions for such purposes, but are still limited in reasoning capabilities necessary for automated validation of the project schedules in terms of their feasibility and absence of spatial-temporal conflicts.

In particular, the mentioned above systems are capable of detecting simple conflicts caused by collisions or interferences of construction objects, but are useless in identifying non-trivial situations such as missing collision-free paths to deliver the objects to destination positions in planned time, invalid schedules leading to objects hanging without any reliance or to unsteady object configurations, unavailability of required workspaces for performing planned activities. Automated analysis of such situations would be another step towards comprehensive and trustworthy project plans and schedules.

It is noteworthy, that modern 3D geometric kernels (ACIS, Parasolid, C3D, OpenCascade), popular game engines (Bullet, Havok, MuJoCo, ODE, PhysX, Ogre, Unity, Jmonkey, Panda3D) (Erez, et al., 2015) and various tools for qualitative spatial calculus (SparQ, GQR, QAT, CLP(QS)) (Freksa, 2013) do not fulfill this functional gap because their features are basically designed to meet the specific requirements of advanced geometry modeling, realistic rendering and retrieving spatial knowledge correspondingly.

The aim of this paper is to present a software library providing advanced capabilities for spatial-temporal modeling and reasoning. The library called Constructivity4D supports a representation of all of the core concepts in scene modeling and provides a wide range of operators and functions for identifying spatial-temporal conflicts originated from the construction project validation problems.

The paper discusses the design considerations of the library, its functionality, interfaces. Particular attention is also paid to applied computational methods and implementation details. The rest of the paper is organized as follows: we consider the library organization in Section 2 putting emphasis on underlying modeling concepts and software interfaces to access the model data. Section 3 introduces formal definitions and meaningful illustrations of the operators and functions supported by the library. Computational methods used for their implementation are discussed in Section 4. The benefits of the library and perspectives of its introduction to industry are shortly summarized in Conclusions 5.

2 4D MODELING

Constructivity4D is an object-oriented library that defines an extensive set of classes and associated functions to support dynamic scene modeling, reasoning in such scenes and identifying spatial-temporal conflicts.

The library has been implemented in C++ language for a number of reasons. As a middle-level language comprising both high-level and low-level features, it enables to develop high-performance computation applications. Being a statically typed, compiled, general-purpose, case-sensitive, free-form programming language that supports procedural, object-oriented, and generic programming, it allows a systematic approach to development of reusable software components and complex systems.

In this Section we consider only those library components which are responsible for the representation of underlying modeling concepts and for accessing of corresponding data.

Model is a key concept associated with the dynamic scene being a collection of physical objects with prescribed behaviors. It can be thought of as corresponding to a construction project consisting of building elements with scheduled activities to install the elements at a construction site, to move them to other locations or to remove them from the site at all. An instantiated *Model* object has its own GUID, name, comment, and timeframe. It also contains an arbitrary number of the *GeometryObject* instances organised into a scene tree with the *CompoundObject* root. Reusable objects can be placed separately as library resources and be referenced directly from the scene representation using the *ReferenceObject* instances. Each geometry object has its own OID and name, as well as optional *Behavior* association in case it reveals any dynamics.

In general, the geometrical objects are subdivided in *CompoundObject*, *ReferenceObject* and *SimpleObject* categories. The *CompoundObject* allows the construction of complex scenes and object assemblies. The *ReferenceObject* enables to avoid redundant representations of the same assembly parts by referencing to available library resources and by applying transforms preserving relative locations in parent coordinate systems. There is only one restriction: cyclic dependencies of compound objects and referenced objects are not allowed.

The *SimpleObject* concept allows to further categorize the following components in the well-known concepts of geometric modeling such as points, curves, surfaces and solids. The latter may in turn be categorized as solid primitives, polyhedra, booleans, bends, direct edits, fillet and chamfers, louvers, reinforcing ribs, sections and cuts, shells with drafted faces, stamping, and thin-walled solids. A more detailed consideration of the geometry modeling concepts is beyond the paper subject.

The *Behavior* concept allows to describe the dynamic deterministic behavior of a geometrical object by means of defining and sequencing *Event* instances. Each event may be either appearance or disappearance of an object or a movement to another location which occurs in discrete time points. An optional transformation enables to localise the place where the event is happening and the position the object is moving to. Pairs of subsequent events and the life-circle periods over which the object events occur are represented as the *Interval* instances.

Being specified in EXPRESS language formally the discussed concepts drive model data access interfaces in implementation languages in the same way as it is regulated by the STEP family standards. The specified concepts are implemented as C++ classes with provided methods to manage model data within the library representation. These classes enable third-party modeling and planning applications to access the model data, read/write from/to standard STEP physical exchange files and to check its compliance with the original specification.

3 REASONING

In addition to the model data access interfaces the Constructivity4D library provides validation interfaces through which the third-party applications can invoke spatial-temporal operators and functions. For brevity we confine ourselves to the operator semantics omitting implementation details. Formal definitions of the operators and simplified illustrations have been compiled in the Tables 1–4.

The supported operators are typical of qualitative reasoning formalisms which are usually based on common-sense abstractions of quantitative

Operator	Formalization and illustration
Before (E1,E2) *E1 < E2*	*E1.time < E2.time*
AtOnce (E1,E2) *E1 = E2*	*E1.time = E2.time*
After (E1,E2) *E1 > E2*	*E1.time > E2.time*
Next (E1,E2)	*neither E1.time ≤ E2.time* *nor E3 E1.time > E3.time > E2.time*
Previous (E1,E2)	*neither E1.time ≥ E2.time* *nor E3 E1.time < E3.time < E2.time*
Fork (E1,E2)	*E3 E3.time < E1.time ∧* *E3.time < E2.time*
Independent (E1,E2)	*neither E1.time ∧ E2.time* *nor E3 E1.time < E3.time < E2.time*
Before (I1,I2)	*I1.start < I1.end < I2.start < I2.end*
Meets (I1,I2)	*I1.start < I1.end = I2.start < I2.end*
Overlaps (I1,I2)	*I1.start < I2.start < I1.end < I2.end*
Starts (I1,I2)	*I1.start = I2.start < I1.end < I2.end*
During (I1,I2)	*I2.start < I1.start < I1.end < I2.end*
Finishes (I1,I2)	*I2.start < I1.start < I1.end = I2.end*
AtOnce (I1,I2)	*I1.start = I2.start < I1.end = I2.end*

Table 1. Temporal operators.

Operator. 9IM representation	Relation formalization and illustration
Equal (A, B) $\begin{bmatrix} * & F & F \\ F & * & F \\ F & F & * \end{bmatrix}$	$\sim A^0 \partial B \wedge \sim A^0 B^- \wedge \sim \partial A B^0 \wedge \sim \partial A B^- \wedge \sim A^- B^0 \wedge A^- \partial B$
Disjoint (A, B) $\begin{bmatrix} F & F & * \\ F & F & * \\ * & * & * \end{bmatrix}$	$\sim A^0 B^0 \wedge \sim \partial A B^0 \wedge \sim A^0 \partial B \wedge \sim \partial A \partial B$
Touches (A, B) $\begin{bmatrix} F & F & * \\ F & T & * \\ * & * & * \end{bmatrix}$	$\sim A^0 B^0 \wedge \sim \partial A B^0 \wedge \sim A^0 \partial B \wedge \partial A \partial B$
Within (A, B) $\begin{bmatrix} * & * & F \\ * & F & F \\ * & * & * \end{bmatrix}$	$\sim A^0 B^- \wedge \sim \partial A \partial B \wedge \sim \partial A B^-$
Contains (A, B) $\begin{bmatrix} * & * & * \\ * & F & * \\ F & F & * \end{bmatrix}$	$\sim \partial A \partial B \wedge \sim A^- B^0 \wedge \sim A^- \partial B$ *Within (B, A)*
Overlaps(A, B) $\begin{bmatrix} 3 & * & T \\ * & * & * \\ T & * & * \end{bmatrix}$	$A^0 B^0 \wedge A^- B^0 \wedge A^0 B^-$
Intersects (A, B)	$A^0 B^0 \vee \partial A B^0 \vee A^0 \partial B \vee \partial A \partial B$ *Not Disjoint (A, B)*
Clashes (A, B)	$A^0 B^0 \vee \partial A B^0 \vee A^0 \partial B$

Table 2. Topological operators.

temporal and spatial relations peculiar to the physical reality world. Numerous formalisms have been pointed out over the past 30 years. The well-known examples of temporal calculi include the interval algebra introduced by Allen, the temporal modal logics, and the point algebra by Vilain and Kautz. On the spatial side, the popular mereotopological calculi, Frank's and Freksa's relative orientation calculi, Cardinal Direction Calculi (CDC), Oriented Point Relation Algebra (OPRA), Egenhofer and Franzosa's 4- and 9-intersection calculi, various Region Connection Calculi (RCC), Occlusion Calculi (OCC) are worth mentioning. These formalisms have been utilized by many researchers for their simplicity and the ease of implementation. (Cohn & Renz, 2008) (Egenhofer, et al., 1993) (Mossakowski & Moratz, 2012) (Eloe, 2015).

The Constructivity4D provides advanced operator sets which can be related to temporal, topological, metrical and directional formalisms. The operators are based on quantitative estimations of the object relations, at the same time not preventing the further qualitative analysis. The operators can be called in combination to return a verdict about non-trivial spatial-temporal conflicts.

3.1 *Temporal operators*

The temporal operators are used to determine how the model events are related to one another. For

Table 3. Directional operators.

Operator	Relation formalization
InDirectionOf (A, B, D)	$\forall b \in B \exists a \in A : (\bar{a} \cdot \bar{D}) > (\bar{b} \cdot \bar{D})$
StrictlyInDirection Of (A, B, D)	$\forall a \in A, b \in B : (\bar{a} \cdot \bar{D}) > (\bar{b} \cdot \bar{D})$
SupportsIn Direction (A, B, D)	$\partial A \cap \partial^D B \neq \phi \wedge A^0 \cap B^0 = \phi$ where $b \in \partial^D B \Leftrightarrow b \in \partial B \wedge \begin{pmatrix} \forall \varepsilon > 0 \exists b^0 \in B^0 : (\overline{b^0} - \bar{b}) \times \bar{D} \wedge \\ \left(0 < (\overline{b^0} \cdot \bar{D}) - (\bar{b} \cdot \bar{D}) < \varepsilon\right) \end{pmatrix}$

example, it can be ascertained whether the event *E1* happened *Before* the event *E2*, *AtOnce* with it or *After* it. Additional operators *Next*, *Previous* help to refine whether the events *E1*, *E2* occurred immediately one after the other or any other events could have happened between them. The operators *Fork*, *Independent* are designed to determine whether there is a common predecessor *E3* of the events *E1*, *E2* or they should be considered as independent over time.

The next operators enable to establish seven basic relations of the Allen's algebra that can be held between two intervals *I1*, *I2*. These are *Before, Meets, Overlaps, Starts, During, Finishes, and AtOnce*. The main difference is that the intervals are reinterpreted in terms of the events as it is defined by the model specification. Formal definitions and graphic illustrations of the temporal operators can be found in Table 1. The supported operator suite is sufficient to solve typical temporal reasoning problems.

3.2 *Measurement functions and metric operators*

The Constructivity4D performs calculations of geometry object properties necessary for many purposes. Diameter, area, volume, centre of mass and planar projections of a solid object can be calculated when it has been preliminary tessellated and represented as polyhedron. The library also provides functions to compute the distance between the objects and the penetration depth when the objects overlap each other. Special metric operators enable to determine which of the two given objects is closer to the reference object and which object is located farther. These are basically aimed at those cases in which a verdict may be quickly deduced using qualitative assessments.

All measurements are performed to a specified model focus time $t \in Model.timeframe$ since the geometrical representations and object positions in dynamic scenes are allowed to be changed throughout the model timeframe. The focus time is implicitly applied to all spatial operators and functions discussed below. This parameter has been removed from the signatures for simplicity.

Table 4. Validation functions.

Validation functions	Formalization of conditions
ClashCheck (M)	$\forall t \in M.timeframe$ $\forall A, B \in M.objects$ *Not Clashes(A(t), B(t))*
GravityCheck (M)	$\forall t \in M.timeframe$ $\forall B \in M.objects$ $B.fixed \vee \exists A \in M.objects$ *Supports (A(t), B(t))*
PathCheck (M)	$\forall A \in M.object$ $\forall E \in A.events(INSTALL)$ $\exists Q(\tau), \tau \in [0,1]$ $Q(0) = Q_\infty$ $Q(1) = E.transform$ $\forall B \in M.objects, \forall \tau \in [0,1]$ *Not Clashes(A(Q(τ)), B(E.time))*

3.3 *Topological operators*

Although numerous topological relation models have been proposed during the last decades, we have had to define our own set of topological operators to fulfil the contradictory requirements of the computational feasibility and usefulness. Table 3 provides formal definitions of the introduced operators and the necessary illustrations.

The introduced operator set is similar to some extent to the well-known Dimensionally Extended nine-Intersection Model (DE-9IM). This model is intended to provide qualitative reasoning about geometries with the identified interior, boundary and exterior. In the notation of topological space operators, these regions are denoted as $A^0, \partial A$, and A^- correspondingly. The model abstractly describes geometries by their possible relations to each other using a 3×3 intersection matrix. For objects A and B its elements store the maximum numbers of dimensions of the intersection of the geometric regions:

$$\begin{bmatrix} \dim A^0 \cap B^0 & \dim A^0 \cap \partial B & \dim A^0 \cap B^- \\ \dim \partial A \cap B^0 & \dim \partial A \cap \partial B & \dim \partial A \cap B^- \\ \dim A^- \cap B^0 & \dim A^- \cap \partial B & \dim A^- \cap B^- \end{bmatrix}$$

The dimension of an empty set is denoted as F (false). The dimension of a non-empty set is denoted as T (true) or specifically 0 for points, 1 for lines, 2 for areas and 3 for volumes.

The main differences between the introduced operator set and DE-9IM are the following:

- the concepts of points, lines, and areas are interpreted in terms of general topology, i.e. as having boundaries, but not having interiors;
- the geometries are assumed to be closed, regular and bounded, and may consist of multiple disjoint parts;
- the operators are intended for both two- and three-dimensional geometries;
- the operators and corresponding relations are formalized using original intersection matrices;
- the operators are applicable for spatial-temporal analysis of the geometries reconstructed to some model focus time.

3.4 *Directional operators*

Direction is a binary relation between an ordered pair of objects *A* and *B* in a given reference frame, where *A* is the reference object and *B* is the target object. For point objects Frank's cardinal direction model and start-like model are typically applied for qualitative reasoning (Frank, 1996). The models have two principal shortcomings. In the three-dimensional case the number of operators grows significantly. Another drawback relates to the uncertainties when geometries have extended borders.

Figure 1 provides a few meaningful cases when an intuitive identification of the relations between object *A* and other objects is difficult and needs an additional formalization. The previously made attempts to handle such cases are worth mentioning (Borrmann & Rank, 2009). We also distinguish a rigorous case when the related objects can be separated by a plane perpendicular to a given direction *D* and, thereby, can be recognized as being strictly located in this direction. We also consider a relaxed interpretation according to which the related objects cannot be entirely separated, but it is possible to determine the relative location of their separate parts.

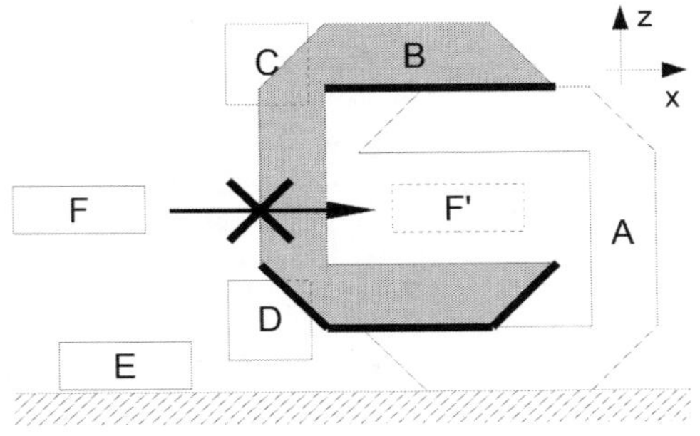

Figure 1. A scene with uncertain relations among objects.

The Constructivity4D provides three underlying directional operators with an additional operand *D* denoting the direction in which the relations are to be checked. Various sorts of directional relations can be covered by choosing different operand values. Formal definitions and signatures of the directional relations are given in Table 3. Scalar and cross products of two vectors formed by the point $a \in A$ and the direction *D* are denoted there as $(\bar{a} \cdot \bar{D}), \bar{a} \times \bar{D}$.

The first relation *InDirectionOf* is a relaxed interpretation with the property of transitivity. The second relation *StrictlyInDirectionOf* assuming rigorous interpretation is transitive, irreflexive, and asymmetric. The applied formalization can be easily understood if the operand *D* is directed vertically. The implied relations take the trivial definitions:

$$Above(A,B) \Leftrightarrow \forall b \in B\, \exists a \in A : a_z > b_z$$

$$StrictlyAbove(A,B) \Leftrightarrow \forall a \in A, b \in B : a_z > b_z$$

The third operator *SupportsInDirection* is intended to qualify whether the object *A* does support the object *B* in a given direction *D*. It can be formalized by defining the supporting boundary $\partial^D B$ as a subset of the boundary of the object *B* with neighboring interior points located along the given direction *D*. In Figure 1 the supporting boundary of the object *B* is highlighted with bold lines on the suggestion that the direction points up.

The relation becomes transparent if it is required to determine whether the object *B* can be placed on top of the object *A* without any gaps violating the gravity laws. Let $\underline{\partial} B$ be a supporting bottom boundary of the object *B* so that

$$b \in \underline{\partial} B \Leftrightarrow b \in \partial B \wedge$$
$$\left(\forall \varepsilon > 0\, \exists b^0 \in B^0 : (b_x^0 = b_x) \wedge (b_y^0 = b_y) \wedge (0 < b_z^0 - b_z < \varepsilon)\right)$$

Then the relation can be formally defined as follows:

$$Supports(A,B) \Leftrightarrow \partial A \cap \underline{\partial} B \neq \phi \wedge A^0 \cap B^0 = \phi$$

Obviously, this contiguity relation is not equivalent to any topological or directional relation considered above. Moreover, it cannot be expressed in terms of the spatial algebras known to authors today.

3.5 *Model validation functions*

Model validation is intended to detect collisions and other sorts of spatial-temporal conflicts. Three complementary sorts of conflicts can be identified and properly reported using the library functions presented in Table 4. Notations $A(t)$ and $A(Q)$ mean object A positioned in a scene with the transform Q computed on model focus time t. Each function takes entire model data M as an input parameter to return a verdict on the availability of conflicts among the model objects *M.objects* within the model timeframe *M.timeframe*.

The function *ClashCheck* enables to detect explicit collisions between the objects using the topological operator *Clash* that reveals interferences of the object interiors but ignores touches of the boundaries. The function generates a report with detected pairs of clashed objects and time intervals when it is happening.

The function *GravityCheck* helps to identify error situations typical of the construction project schedules: the objects are placed without any support or removed before the top objects are fixed. Before checking some objects such as ground or priory known static infrastructures must be qualified as already fixed, otherwise the validation process becomes meaningless. The function reports about the model events resulting in such violations.

The function *PathCheck* allows to detect situations associated with the inability to deliver the objects in a given place at a given time in accordance with the prescribed model events. For example, object F in Figure 1 cannot be placed inside the internal region formed by preliminary installed objects A and B. Formally, for any installed object A we require the existence of a continuous collision-free path $Q(\tau)$ starting at some point at infinity Q_∞ and ending in the destination position defined by the event transform *E.transform*. We suppose that the replacement occurs instantaneously at the event time *E.time*. It is noteworthy that if the object replacement has been preliminary defined in the model as an event with a detailed delivery path, there is no the need in validating the model against path conflicts. However, specifying such paths becomes extremely difficult for large-scale construction projects and appears less productive than comprehensive validation.

4 IMPLEMENTATION

Due to spatial-temporal modeling and reasoning problems being computationally hard, most basic operators and functions of the library have been implemented using special data structures such as spatial-temporal indexes, caches of geometric representations and constraint networks. Such a combined strategy provides high efficiency of the supported operators and functions when proceeding on preliminary indexed, cached and qualified data rather than on original model data. The implementations based on naive algorithms and direct access to the model data would result in high computational complexity and, ultimately, in the inability to solve the problems in a reasonable time. The general organization of the library as well as the original and derived model data involved in the main computational processes are presented in Figure 2.

4.1 *Spatial-temporal indexes*

An event tree plays the role of primary temporal index which can be computed once and then be updated only if the model events have happened. It has been implemented in the Constructivity4D as an AVL binary search tree ordered by the event timestamps. It allows fast lookup, efficient retrieval of events in a given time interval and quick updates when registering new events in the model. All these operations take time proportional to the logarithm of the number of model events and this fact makes possible the use of the event tree for various computational tasks. The event tree is used in the library for the implementation of the temporal operators discussed above as well as for the reconstruction of a model scene at a given focus time.

Spatial decomposition trees are secondary indexes which are computed or updated whenever a model focus time is changed. They depend on the entire model data, and that is why their

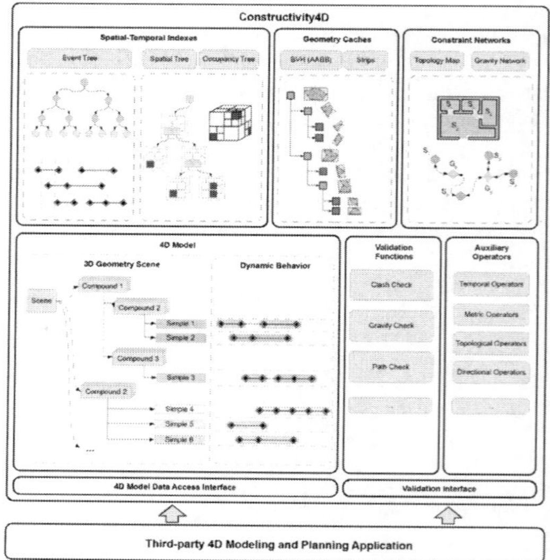

Figure 2. A general organization of the Constructivity4D library and its spatial-temporal data representation.

updates should be organized effectively. Our previous research has proved that regular octrees being used as spatial indexes exhibit high performance for such queries as frustum culling, nearest neighbor search, collision localization, hidden surface removal meanwhile being applicable to objects with extended borders. At the same time non-expensive incremental updates are admitted to bring the indexes to consistent states.

The traditional regular octrees use recursive decomposition of the scene space into eight equal octants and their successive association with the scene objects. This procedure is ineffective if the scene consists of a huge number of the objects and the requests are addressed to both simple and compound objects.

Nested regular octrees have been adopted in the Constructivity4D for such purposes. Partial octrees are deployed for those compound objects which contain a relatively large number of children. We have reported uniformly high efficiency of such indexes for scenes of varied complexity and hierarchical organization (Semenov, et al., 2015).

An occupancy tree is convenient to think of as a spatial decomposition tree which cells are provided with the occupancy status and the distance to the nearest object. The status points out whether the cell is partially occupied, entirely full, or entirely empty. The occupancy tree is utilized in the library to generate topology maps and reasoning about path conflicts. A detailed description of the computational methods and industrial applications can be found in (Semenov, et al., 2012).

4.2 *Geometry caches*

The library caches alternative geometric representations of the individual objects if requested repeatedly. It helps to avoid redundant computations and to minimize CPU resources.

Within the accepted conceptualization each simple object can be provided with alternative geometric representations. These are an Axis-Aligned Bounding Box (AABB), an Oriented Bounding Box (OBB), tessellated representations at different Levels Of Details (LOD), Bounding Volume Hierarchies (BVH), triangle strips and vertex buffers. The requested geometric representations are automatically stored on a hard disk in expectation of further use. When required the available representations can be obtained directly from one of the deployed caches. If an object is changed, then the caches are updated in a lazy manner assuming the computations only when the object representation is really requested. The representations are always computed in local coordinate systems and do not depend on positions of the objects in a scene, therefore such updates can be performed locally and handled trivially.

The following representations are intensively used in the library. Bounding boxes are applied in the operations with spatial decomposition trees and in localization of object collisions. Tessellated boundary representations are used in the measurement functions and in the implementations of the topological and directional operators for solid objects. BVH structures are applied for the optimization of the underlying operators and collision detection methods (Semenov, et al., 2010). Triangle strips, vertex buffers, and other display structures are used for rendering purposes.

4.3 *Constraint networks*

The library also keeps specially pre-computed constraint networks which might help to solve qualitative spatial-temporal reasoning problems. These problems are usually stated as follows: given the possibly indefinite knowledge of the relations between some objects we need to deduce the strongest assertions possible about the relations between some of the objects or all of them. By applying the relation composition tables and constraint networks, new knowledge can obtained in a formal way.

As an example, the library supports topology maps which can be considered as specific constraint networks with vertices corresponding to free regions of the model scene and with edges corresponding to the connectivity relations between them. Establishing the relations between individual regions, it becomes possible to pave the object route between two given positions using qualitative reasoning methods. It is interesting that the same problem can be interpreted and resolved in terms of graph theory.

A gravity network consists of vertices being the scene objects and directed edges represent the vertical contiguity relations between them. Having the network, a verdict on the gravity conflicts in a scene can be quickly made by analyzing vertices with missing incoming edges. To sure the model validness throughout its timeframe such analysis is repeated every time when the network is updated.

Since the validation functions call different operators, the indexes, caches and networks are deployed and maintained optionally within the library sessions. If the library is not configured properly, the validation procedures are based on naive algorithms with expected restrictions on the model size.

5 CONCLUSIONS

The Constructivity4D software library has been introduced in the paper. By supporting a representation of all of the core concepts in the scene modeling and possessing a wide range of temporal, metric, topological, and directional operators it provides advanced capabilities for spatial-temporal modeling and reasoning. In particular, it can be effectively applied to identify non-trivial conflicts originated from the construction project validation problems.

Some components of the library have been previously approved and successively employed in a few industrial projects (Semenov, et al., 2011). Currently, the software library is being implemented in full in accordance with the declared features and design decisions described in the paper.

REFERENCES

Borrmann, A. & Rank, E., 2009. Query Support for BIMs Using Semantic and Spatial Conditions. In: J. Underwood & U. Isikdag, eds. *Handbook of Research on Building Information Modeling and Construction Informatics: Concepts and Technologies.* Hershey: IGI Global, pp. 405–450.

Cohn, A. G. & Renz, J., 2008. Qualitative Spatial Representation and Reasoning. In: *Handbook of Knowledge Representation.* Amsterdam: Elsevier, pp. 551–596.

Egenhofer, M. J., Sharma, J. & Mark, D. M., 1993. *A Critical Comparison of the 4-Intersection and 9-Intersection Models for Spatial Relations: Formal Analysis*. Minneapolis, Autocarto 11.

Eloe, N., 2015. *VRCC-3D+: Qualitative spatial and temporal reasoning in 3 dimensions.* Missouri University of Science and Technology: Missouri.

Erez, T., Tassa, Y. & Todorov, E., 2015. *Simulation Tools for Model-Based Robotics: Comparison of Bullet, Havok, MuJoCo, ODE and PhysX.* Seattle, IEEE, pp. 4397–4404.

Frank, A., 1996. *Qualitative Spatial Reasoning: Cardinal Directions as an Example*. Vienna: Springer Berlin Heidelberg.

Freksa, C., 2013. Spatial computing. In: *Cognitive and Linguistic Aspects of Geographic Space.* Heidelberg: Springer Berlin, pp. 23–42.

Mossakowski, T. & Moratz, R., 2012. Qualitative Reasoning about Relative Direction of Oriented Points. *Artificial Intelligence*, Volume 180, pp. 34–45.

Semenov, V. A., Kazakov, K. A. & Zolotov, V. A., 2012. Global Path Planning in 4D Environments Using Topological Mapping. In: G. Gudnason & R. Scherer, eds. *eWork and eBusiness in Architecture, Engineering and Construction.* London: CRC Press, Taylor & Francis Group, pp. 263–269.

Semenov, V. A., Kazakov, K. A. & Zolotov, V. A., 2015. Effective spatial reasoning in complex 4D modeling environments. In: A. Mahdavi, B. Martens & R. Scherer, eds. *eWork and eBusiness in Architecture, Engineering and Construction.* London: CRC Press, Taylor & Francis Group, pp. 181–186.

Semenov, V. A., Kazakov, K. A., Zolotov, V. A. & Dengenis, T., 2011. Virtual Construction: 4D Planning and Validation. In: H. Bargstädt & K. Ailland, eds. *Proceedings of the International Conference on Construction Applications of Virtual Reality* . Weimar: Bauhaus-Universität Weimar, pp. 135–142.

Semenov, V. A., Kazakov, K. A., Zolotov, V. A. & Dengenis, T., 2011. *Virtual Construction: 4D Planning and Validation.* Weimar, s.n.

Semenov, V. A., Kazakov, K. A., Zolotov, V. A. & Jones, H., 2010. Combined strategy for efficient collision detection in 4D planning applications. In: W. Tizani, ed. *Proceedings of the International Conference on Computing in Civil and Building Engineering.* Nottingham: Nottingham University Press, pp. 31–39.

eWork and eBusiness in Architecture, Engineering and Construction – Christodoulou & Scherer (Eds)
© 2016 Taylor & Francis Group, ISBN 978-1-138-03280-4

BIM registration methods for mobile augmented reality-based inspection

M. Kopsida & I. Brilakis
University of Cambridge, Cambridge, UK

ABSTRACT: On-site construction inspection for progress monitoring is a manual, time consuming and labour intensive process consumed by exhaustive manual extraction of data from drawings and databases. Efforts have been made to facilitate the inspection process by using emerging technologies such as Augmented Reality (AR). AR based systems can simplify and reduce the time of inspection by providing the inspector with instantaneous access to the information stored in the Building Information Modelling (BIM). However, precise alignment between the BIM model and the real world scene is still a challenge. For estimating the position and orientation of the user, methods have been proposed that either use markers or confine the user to a specific location, or use Global Positioning System (GPS) which cannot operate efficiently in an indoor environment. This paper presents an evaluation of different methods that could potentially be used for a marker-less BIM registration in AR. We implemented and tested line, edge, and contour detection algorithms using images, data from LSD and ORB Simultaneous Localisation And Mapping (SLAM) methods and 3D and positioning data from Kinect sensor and Google Project Tango. The results indicate that sparse 3D data is the input dataset that leads to the most robust results when combined with XYZ method.

1 INTRODUCTION

Time and cost overrun is a common challenge for many construction projects. Only 16% of construction projects are able to a) finish on time b) within budget and c) meet required quality standards (Frame 1997). Approximately 60% of British construction project organisations encounter time and cost overrun on more than 10% of their projects (Olawale & Sun 2010).

The problem of time and cost overrun is mainly caused by poor construction project monitoring and control (Memon et al. 2012). Although it has been proven that successful progress monitoring can lead to a 15% reduction in execution slip (IBC 2000), more than 10% improvement in cost (CEC 1999) and reduces the cost of reworks, disputes and claims (Yates & Epstein 2006), current practice remains mainly manual (Zavadskas et al. 2014) and labour intensive (Navon 2007). Project monitoring is conducted by visual inspections and the inspector needs to fill several forms, write reports and perform extensive information extraction from drawings. Interior inspections can be even more complex (e.g. installation of Mechanical, Electrical, and Plumbing, etc.), The quality of the collected progress data, can be subjective, and can be influenced by the inspector's level of education, training (Elazouni & Abdel-Wahhab 2009) and experience (Golparvar-Fard et al. 2009).

In order to facilitate the process of inspection, inspectors have now started to use web-based technologies to improve their onsite efficiency and management of tasks and data handling. However, inspection reports remain manual and subjective. Automated systems are now becoming available and new commercial inspection software packages such as LATISTA, Autodesk BIM 360 Field, Field 3D, xBIM, etc. are now available to aid the inspection process. These software packages replace the use of paper drawings and documents by providing the inspector with on-site access to a Building Information Model (BIM) via a mobile device. Although this improves information management, navigation within the BIM model during inspection is still manual. BIManywhere tries to automate navigation in the BIM model for facilities management applications by using QR codes. The inspector simply scans the QR codes with a mobile device in order to access the required BIM view. This system is not so efficient for on site use since it is difficult to install and maintain the QR codes in a dynamic construction site environment.

A survey that was conducted for facility management purposes, has shown a preference amongst inspectors for a mobile-based augmented reality solution. (Gheisari et al. 2014) indicating a shift towards augmented reality solutions for the automation and efficiency of construction progress monitoring inspection.

2 BACKGROUND

There has been a lot of interest in the application of AR to many aspects of the construction cycle including design, building and maintenance. However, one of the remaining challenges for the implementation of such technology is the accurate estimation of the user's pose in order to align accurately the virtual and real world data (Azuma 1997, Koller et al. 1997).

Current research in construction has attempted to apply AR technology in order to improve the performance of many activities (Shin & Dunston 2008). Researchers have also investigated the implementation of AR for progress monitoring purposes (Lee & Peña-Mora 2006, Golparvar-Fard & Peña-Mora 2007) in order to visualise the progress status by aligning the as-built data with an as-planned 3D model. The alignment of the virtual and real data has been performed manually (Memon et al. 2005, Zhang & Arditi 2013) or in a semi-automated way (Golparvar-Fard et al. 2009, 2011, 2012, Bosché 2010, 2012) where the user first picks points, for the initial coarse registration, between the as-planned 3D model and the point cloud. In this case the point cloud has been acquired either from laser scanners, or from images and reconstruction algorithms, and then fine registration is achieved by using iterative optimisation methods. An automated method has also been presented (Kim at al. 2011) but is successful only in a few specific cases. Although these methods are static and do not perform in real time on site, they demonstrate the potential implementation of AR to facilitate and automate Construction Progress Monitoring.

Researchers have also tried to develop mobile based AR methods for inspection. For example, Cote et al. (2013) and Shin & Dunston (2009, 2010) have developed a system that achieves accurate registration between the virtual model and the real data, but the equipment used is tripod mounted and cumbersome to use on site as it needs to be located at fixed static points. There are also AR systems to provide onsite activity information (Wang et al. 2014) or to manage defects of reinforced concrete (Kwon et al. 2014) using fiducial markers. The use of the markers requires additional time, cost and effort for their installation and maintenance. In order to eliminate these constraints, Irizarry et al. (2013) presented a mobile AR system for facility management purposes that uses the BIM2MAR application (Williams et al. 2014a, 2014b) for the visualisation of the Building Information Modelling (BIM) information and a system that consists of a three-axis gyroscope, accelerometer, Wi-Fi, and a digital compass to estimate the pose of the user. This system, however, restricts the user's location to a specific location and requires the use of Wi-Fi which is not always available on site. Other mobile AR systems either use Global Positioning Systems (GPS) or compasses (Woodward et al. 2010) to estimate the pose of the user, but the accuracy is insufficient and cannot be used for indoor applications (Wing et al. 2005).

In computer vision domain, additional methods for marker-less AR have been introduced but have not been tested with BIM models and on construction environment. The geometric primitives that are mostly used are points (Dementhon & Davis 1995, Lu et al. 2000, Ansar & Daniilidis 2003), segments (Dambreville et al. 2008), lines (Ansar & Daniilidis 2003), contours or points on the contours (Drummond & Cipolla 2002), conics (Neumann et al. 1999, Tarel et al. 2000), cylindrical objects (Yoon et al. 2003) or a combination of these different features (Marchant & Chaumette 2002, Comport et al. 2003). To overcome the registration problem, geometric (Yoon et al. 2003) or numerical and iterative (Dementhon & Davis 1995, Marchant & Chaumette 2002) approaches can be used. Linear approaches use a least-squares method to estimate the pose and non-linear techniques aim at minimising the error between the observation and the forward-projection of the model using numerical iterative algorithms such as Newton-Raphson or Levenberg-Marquardt. Non-linear approaches provide accuracy but they are subjected to local minima and divergence. Additional augmented reality approaches rely on relative camera motion (Chia et al. 2002), planar homography estimation (Simon & Berger 2002) or optical flow based techniques (Brox et al. 2006) for pose estimation.

In order to overcome problems such as occlusion, changes in illumination and errors in tracking, statistical methods for pose estimation can be considered (Comport et al. 2003). Marchant & Chaumette (2002), Comport et al. (2003) and Comport et al. (2006) proposed a framework for a real time marker-less tracking for augmented reality based on the visual servoing approach known in robotics which uses only one monocular camera. The principal of their algorithm is to iteratively modify the parameters of the camera using the visual servoing paradigm to register the back projection of the model with the data extracted from the images. The features used are circles, lines and cylinders and the tracking of the features is performed using moving edge algorithm at video rate. This is achieved by minimising the error between the desired state of the image features and the current state. The algorithm can be performed in real time, however, detection of lines in images is a classic challenge in computer vision and image processing.

In the literature, there are also monocular Simultaneous Localisation And Mapping (SLAM) methods for constructing a dense or a semi-dense map of an environment and tracking the user's location and pose in real time. Some of the most representative monocular SLAM methods are PTAM (Klein & Murray 2007), LSD (Engel et al. 2014) and ORB (Mur-Artal et al. 2015). The reconstruction and pose tracking of some of these methods have also been tested for mobile AR applications.

In the recent past new systems have been that are able to provide both colour and depth images (e.g Kinect Sensor, Project Tango, Microsoft Hololens, etc.) and which facilitate the development of AR applications by using the 3D perception of the world to estimate the pose of the user. However, these RGBD systems have been designed mostly for interior applications and thus they are, in practice, restricted to indoor applications. In addition to this, the problem of drift error has not been satisfactorily solved and error increases when there is no depth variation (Newcombe et al. 2011).

3 ANALYSIS

The objective of this paper is to present an evaluation of different methods that could be implemented for a marker-less mobile BIM based AR solution for inspections. We compared 3 groups of methods. The first group uses 2D images taken from mobile devices and the model based AR framework as presented from Comport et al. (2006), the second group uses 3D and pose estimation data acquired from 2D images using the LSD and ORB SLAM methods, and finally, the third group uses 3D and positioning data acquired from Microsoft Kinect and Google Project Tango which can provide RGBD data directly.

3.1 *Model based Augmented Reality using 2D Features and Virtual Visual Servoing*

For the first group of the 2D imaging methods, common line, edge and contour detection algorithms were tested. These features are in abundant in buildings since they consist mainly of surfaces (e.g. walls, columns, floor, ceiling, etc.). These features can then be used along with the BIM model for applying the servoing method of the marker-less model based AR of Marchant & Chaumette (2002), Comport et al. (2003) and Comport et al. (2006) as depicted in Figure 1. This frame illustrates the potential for real time AR application on regular mobile devices. For testing these methods, we first visited the Dyson construction site in Cambridge, UK for data collection (Figure 2). A Samsung S6 Edge device was used for capturing videos and images. Then, OPENCV library was used for line (Figure 2 and 3), edge (Figure 4) and contour (Figure 5) detection.

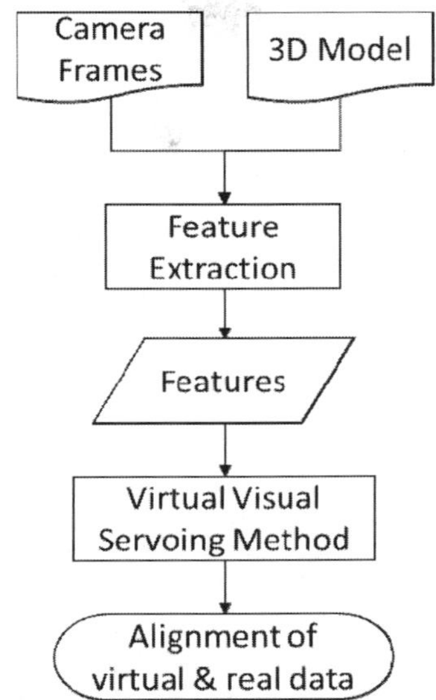

Figure 1. Implementation steps of Group 1 methods using 3D model-based AR.

Figure 2. Dyson Construction Site, Cambridge, UK.

Figure 3. Line detection using OPENCV.

As illustrated in Figures 2–6, the scene of a building under construction is complex and the common algorithms for extracting features such as line, edges and contours do not perform well in order to proceed to the next step which is the registration of the back projection of the model with the data extracted from the construction site images. This restricts the purpose of a robust marker-less mobile AR solution for inspection.

3.2 *Marker-less Augmented Reality using monocular SLAM*

In order to find a more robust solution for marker-less mobile AR inspection, we tried two of the state-of-the-art SLAM methods, the LSD and ORB SLAM. These use 2D data as an input and provide as an output not only the estimation of the pose of the camera, but also a 3D point cloud of the scene which we could leverage and refine the registration of the BIM model by using iterative optimisation methods such as the Iterative Closest Point algorithm as illustrated in Figure 7.

Figure 4. Line segment detection using OPENCV.

Figure 5. Edge detection using OPENCV.

Figure 6. Contour detection using OPENCV.

In order to test these methods, we again used data from the Dyson construction site in Cambridge, UK (Figure 8). The device used was a Samsung S6 Edge. Both of the methods provided noisy and sparse 3D point cloud of the scene (Figure 9). Tracking was more robust when we used the ORB SLAM compared to the LSD SLAM. However, in both cases the device's autofocus proved to be problematic and detrimental to the results. In order to avoid this issue, we again tested this method at the Construction Information Technology (CIT) Lab, Engineering Department, Cambridge, UK in a more controlled environment using a variety of web-cameras with different frame rates and selected optimum settings provided by the developers in order to improve performance. Again, the results were similar. ORB SLAM performed better in tracking, and although the point cloud was of better quality, it was still very noisy and too sparse to be used for refining the registration between the 3D model and the as-built data (Figure 10 and 11).

3.3 *Marker-less Augmented Reality using RGBD devices*

The third group of methods use RGBD devices (Microsoft Kinect and Google Project Tango) for the marker-less mobile AR solution for inspections. Both Kinect and Project Tango provide the user with the 6 Degrees Of Freedom (DOF) camera pose estimation and a 3D reconstruction of the as-built scene. Similar to the group 2 methods, for the AR application both the pose data and the 3D reconstructed point cloud can be used along with iterative optimisation methods for refining the alignment between the as-planned and as-built data (Figure 12).

In the case of the Kinect, as presented in previous work of the authors (Kopsida & Brilakis 2016) in more detail, Kinect v2 was used and an AR platform was built using Windows 8.1, Visual Studio 2013, and.NET Framework 4.5 based on XbimXplorer which is an open source BIM viewer written in C# and WPF. For the Project Tango research, Windows 8.1 and Unity Platform were used for the AR implementation. Similar to the group 2 methods, experiments held at the CIT Lab and within scenes that have sufficient depth variation (same as Group 2 methods).

The experiments showed that Project Tango offers a more robust solution in terms of tracking the pose. As a complete mobile device it also uses other motion sensors such as an Inertial

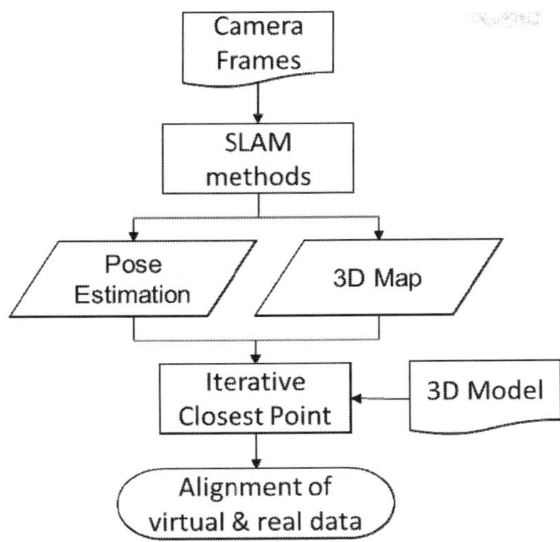

Figure 7. Implementation steps of Group 2 methods using SLAM methods.

Figure 8. Dyson Construction Site, Cambridge, UK.

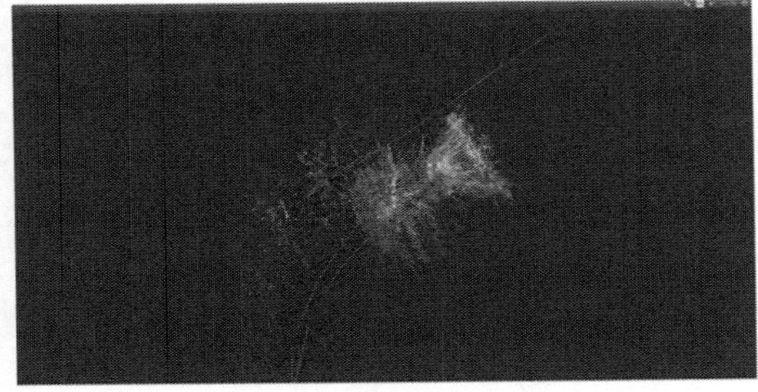

Figure 9. LSD SLAM using data from Dyson Construction Site, Cambridge, UK.

Measurement Unit (IMU) which in combination with the embedded depth sensor and the point cloud, significantly improves the tracking performance. Although the 3D reconstruction of

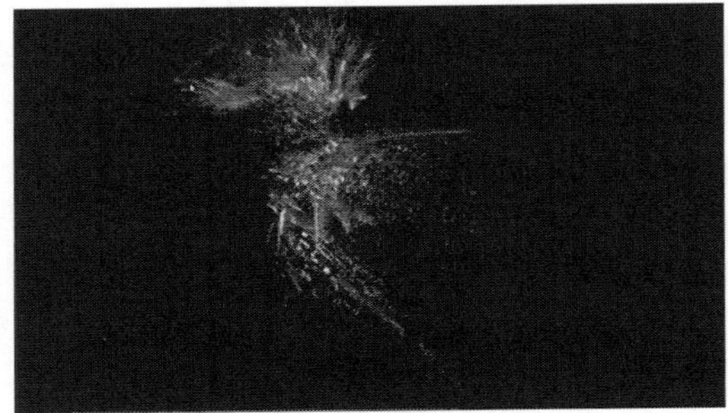

Figure 10. LSD SLAM using data from CIT Lab.

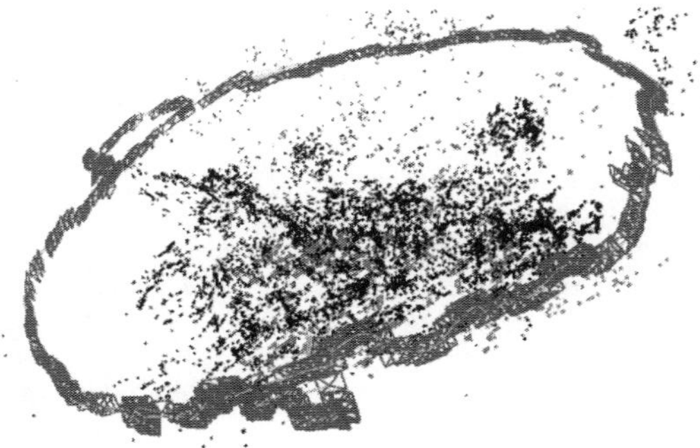

Figure 11. ORB SLAM using data from CIT Lab.

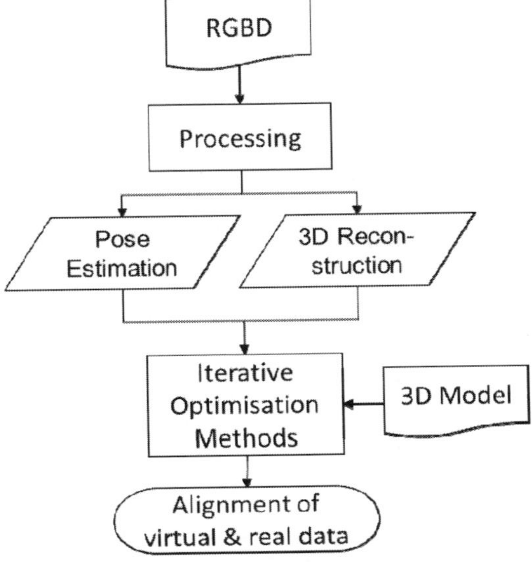

Figure 12. Implementation steps of Group 3 methods using RGBD devices.

the Kinect, using Kinect fusion, is more detailed and accurate, it only provides local meshes of the surroundings and it is slower than Project Tango. Project Tango's 3D reconstruction is noisier but it performs in near real-time which is very important for AR applications and offers a complete map of the as-built scene. For the further refinement of the alignment between the as-built real data and as-planned virtual data, further optimisation methods could be implemented that would leverage the 3D reconstruction of the as-built scene.

4 CONCLUSION

As described, currently, there are no efficient mobile AR solutions for on-site inspections. Similar systems in literature present limitations as they either use markers, or they restrict the user to operate from specific locations. Other methods for marker-less AR have been introduced in computer vision literature but have not yet been tested with and/or on construction sites. This paper presents an evaluation of the different methods that could be, potentially, used for a marker-less BIM based mobile AR solution for on-site inspections. The methods evaluated can be divided into three groups; first the group that uses only 2D data, second the group that uses monocular SLAM methods, in which they process 2D data and extract 3D and motion data and finally, the group that uses RGBD devices. Experiments have shown that neither the first group nor the second group of methods could efficiently provide a robust AR solution for on-site inspections. However, the use of RGBD devices shows significant potential for AR applications in the construction arena. Compared to the Kinect sensor, Project Tango offers a more robust motion tracking and although the 3D reconstruction is noisier than Kinect, it can capture larger scenes and operates more quickly, providing real time advantages for AR inspection implementations required on busy construction sites.

ACKNOWLEDGEMENTS

The research leading to these results has received funding from the European Community's Seventh Framework Programme (FP7/2007–2013) under grant agreements n°247586 ("BIMAutoGen") and n°334241 ("INFRASTRUCTUREMODELS"). This Publication reflects only the author's views and the European Community is not liable for any use that may be made of the information contained herein.

REFERENCES

Ansar, A., & Daniilidis, K. (2003). Linear pose estimation from points or lines. Pattern *Analysis and Machine Intelligence, IEEE Transactions on*, 25(5), 578–589.

Azuma, R.T. (1997). A Survey of Augmented Reality. *Presence-Teleoperators and Virtual Environments* 6, 4(August), 355–385.

Bosché, F. (2010). Automated recognition of 3D CAD model objects in laser scans and calculation of as-built dimensions for dimensional compliance control in construction. *Advanced Engineering Informatics*, 24(1), 107–118.

Bosché F. (2012). Plane-based registration of construction laser scans with 3D/4D building models. *Advanced Engineering Informatics*. Jan;26(1): 90–102. Available from: 10.1016/j.aei.2011.08.009

Brox, T., Rosenhahn, B., Cremers, D., & Seidel, H.P. (2006). High accuracy optical flow serves 3-D pose tracking: exploiting contour and flow based constraints. *In Computer Vision–ECCV 2006* (pp. 98–111). Springer Berlin Heidelberg.

Chia, K.W., Cheok, A.D., & Prince, S.J. (2002). Online 6 dof augmented reality registration from natural features. *In Mixed and Augmented Reality, 2002. ISMAR 2002. Proceedings. International Symposium on* (pp. 305–313). IEEE.

Comport, A.I., Marchand, E., & Chaumette, F. (2003). A real-time tracker for markerless augmented reality. *In The Second IEEE and ACM International Symposium on Mixed and Augmented Reality, 2003*. Proceedings. (Vol. 03, pp. 36–45). IEEE Comput. Soc. doi:10.1109/ISMAR.2003.1240686

Comport, A.I., Marchand, E., Pressigout, M., & Chaumette, F. (2006). Real-time markerless tracking for augmented reality: the virtual visual servoing framework. *IEEE Transactions on Visualization and Computer Graphics*, 12(4):615–628.

Côté, S., Trudel, P., Desbiens, M., Giguère, M., & Snyder, R. (2013). Live mobile panoramic high accuracy augmented reality for engineering and construction. *Proceedings of the Construction Applications of Virtual Reality (CONVR)*, London, England.

Dambreville, S., Sandhu, R., Yezzi, A., & Tannenbaum, A. (2008). Robust 3d pose estimation and efficient 2d region-based segmentation from a 3d shape prior. *In Computer Vision–ECCV 2008* (pp. 169–182). Springer Berlin Heidelberg.

Dementhon, D.F., & Davis, L.S. (1995). Model-based object pose in 25 lines of code. *International journal of computer vision*, 15(1–2), 123–141.

Drummond, T., & Cipolla, R. (2002). Real-time visual tracking of complex structures. *Pattern Analysis and Machine Intelligence, IEEE Transactions on*, 24(7), 932–946.

Elazouni, A., & Abdel-Wahhab, O. (2009). Progress monitoring of construction projects using pattern recognition techniques. *In Construction Research Congress 2009* (Vol. 0082, pp. 1068–1078).

Engel, J., Schops, T., & Cremers, D. (2014). LSD-SLAM: Large-scale direct monocular SLAM. *In European Conference on Computer Vision (ECCV)*, Zurich, Switzerland, September 2014, pp. 834–849.

Frame, J.D. (1997). Establishing project risk assessment teams. *Managing risks in projects*, K. Kahkonen and K.A. Artto, Eds.: E & FN Spon, London.

Gheisari, M., Williams, G., Walker, B.N., & Irizarry, J. (2014). Locating Building Components in a Facility Using Augmented Reality Vs. Paper-based Methods: A User-centered Experimental Comparison. *Proceedings of International Conference on Computing in Civil and Building Engineering* (pp. 850–857).

Golparvar-Fard, M., & Peña-Mora, F. (2007). Application of Visualization Techniques for Construction Progress Monitoring. *Computing in Civil Engineering* (pp. 216–223). Reston, VA: American Society of Civil Engineers.

Golparvar-Fard, M., Peña-Mora, F., & Savarese, S. (2009). Monitoring of construction performance using daily progress photograph logs and 4d as-planned models. *In ASCE International Workshop on Computing in Civil Engineering*.

Golparvar-fard, M., Peña-mora, F., & Savarese, S. (2011). Integrated Sequential As-Built and As-Planned Representation with D 4 AR Tools in Support of Decision-Making Tasks in the *AEC/FM Industry*, (December), 1099–1116. doi:10.1061/(ASCE)CO.1943-7862.0000371.

Golparvar-Fard, M., Peña-Mora, F., & Savarese, S. (2012). Automated Progress Monitoring Using Unordered Daily Construction Photographs and IFC-Based Building Information Models. *Journal of Computing in Civil Engineering*, 04014025. doi:10.1061/(ASCE) CP.1943-5487.0000205

IBC 2000 Project Control Best Practice Study by IPA.

IBC Cost Engineering Committee (CEC).

Irizarry, J., Gheisari, M., Williams, G., & Walker, B.N. (2013). InfoSPOT: A mobile Augmented Reality method for accessing building information through a situation awareness approach. *Automation in Construction*, 33, 11–23.

Kim, C., Lee, J., Cho, M., & Kim, C., (2011). Fully automated registration of 3D CAD model with point cloud from construction site. *Proc. 28th International Symposium on Automation and Robotics in Construction*, Seoul, Korea, pp. 917–922.

Klein, G., & Murray, D., (2007). Parallel tracking and mapping for small AR workspaces. *In IEEE and ACM International Symposium on Mixed and Augmented Reality (ISMAR)*, Nara, Japan, November 2007, pp. 225–234.

Koller, D., Klinker, G., Rose, E., Breen, D., Whitaker, R., & Tuceryan, M. (1997). Real-time vision-based camera tracking for augmented reality applications. *Proceedings of the ACM symposium on Virtual reality software and technology—VRST '97* (pp. 87–94). NY, USA.

Koo, B., & Fischer, M. (2000). Feasibility study of 4D CAD in commercial construction. *Journal of construction engineering and management*, 126(4), 251–260.

Kopsida, M., & Brilakis, I., (2016). Markerless BIM Registration for Mobile Augmented Reality Based Inspection. *16th International Conference on Computing in Civil and Building Engineering (ICCCBE2016)*, Osaka, Japan.

Kwon, O.-S., Park, C.-S., & Lim, C.-R. (2014). A defect management system for reinforced concrete work utilizing BIM, image-matching and augmented reality. *Automation in Construction*.

Lee, S., & Pena-Mora, F. (2006). Visualization of Construction Progress Monitoring. *In Joint International Conference on Computing and Decision Making in Civil and Building Engineering* (pp. 2527–2533).

Lu, C.P., Hager, G.D., & Mjolsness, E. (2000). Fast and globally convergent pose estimation from video images. *Pattern Analysis and Machine Intelligence, IEEE Transactions on*, 22(6), 610–622.

Marchand, E. & Chaumette, F. (2002). Virtual visual servoing: a framework for realtime augmented reality, *Proceedings of the EUROGRAPHICS Conference*, Vol. 21 (3 of Computer Graphics Forum), p. 289–298, Saarbrücken, Germany.

Memon, Z.A., M.Z. Abd.Majid, et al. (2005). An Automatic Project Progress Monitoring Model by Integrating Auto CAD and Digital Photos. *Proceedings of the 2005 ASCE International Conference on Computing in Civil Engineering*, Cancun, Mexico.

Memon, A.H., Rahman, I.A., & Aziz, A.A.A. (2012). The cause factors of large project's cost overrun: a survey in the southern part of peninsular Malaysia. *International Journal of Real Estate Studies*. 7(2).

Meža, S., Turk, Ž., & Dolenc, M. (2014). Component based engineering of a mobile BIM-based augmented reality system. *Automation in Construction*, 42, 1–12.

Mur-Artal, Raul, Montiel, J.M.M., Tardos, & Juan D. (2015). ORB-SLAM: a versatile and accurate monocular SLAM system. *IEEE Transactions on Robotics*, 2015.

Navon, R. (2005). Automated project performance control of construction projects. *Automation in Construction*, 14(4), 467–476.

Navon, R. (2007). Research in automated measurement of project performance indicators. *Automation in Construction*, 16(2), 176–188.

Newcombe, R.A., Izadi, S., Hilliges, O., Molyneaux, D., Kim, D., Davison, A.J., Kohli, P., Shotton, J., Hodges, S., & Fitzgibbon, A. (2011). KinectFusion: Real-time dense surface mapping and tracking. *Proceedings of the 2011 10th IEEE International Symposium on Mixed and Augmented Reality*, pp. 127–136, October 26–29, doi 10.1109/ISMAR.2011.6092378.

Neumann, U., You, S., Cho, Y., Lee, J., & Park, J. (1999, January). Augmented reality tracking in natural environments. *In International Symposium on Mixed Realities* (Vol. 24). Tokyo: Ohmsha Ltd and Springer-Verlag.

Olawale, Y. & Sun M. (2010). Cost and time control of construction projects: Inhibiting factors and mitigating measures in practice. *Construction Management and Economics*. 28(5), 509–526.

Shin, D.H., & Dunston, P.S. (2008). Identification of application areas for augmented reality in industrial construction based on technology suitability. *Automation in Construction*. 17(7), 882–894.

Shin, D.H., & Dunston, P.S. (2009). Evaluation of augmented reality in steel column inspection. *Automation in Construction*, 18(2), 118–129.

Shin, D.H., & Dunston, P.S. (2010). Technology development needs for advancing Augmented Reality-based inspection. *Automation in Construction*, 19(2), 169–182.

Simon, G., & Berger, M.O. (2002, September). Reconstructing while registering: a novel approach for markerless augmented reality. *In 2013 IEEE International Symposium on Mixed and Augmented Reality (ISMAR)* (pp. 285–285). IEEE Computer Society.

Tarel, J.P., & Cooper, D.B. (2000). The complex representation of algebraic curves and its simple exploitation for pose estimation and invariant recognition. *Pattern Analysis and Machine Intelligence, IEEE Transactions on*, 22(7), 663–674.

Wang, X., Truijens, M., Hou, L., Wang, Y., & Zhou, Y. (2014). Integrating Augmented Reality with Building Information Modeling: Onsite construction process controlling for liquefied natural gas industry. *Automation in Construction*, 40, 96–105.

Williams, G., Gheisari, M., Chen, P., & Irizarry, J. (2014a). BIM2MAR: An Efficient BIM Translation to Mobile Augmented Reality Applications. *Journal of Management in Engineering*, 10.1061/(ASCE) ME.1943-5479.0000315, A4014009.

Williams, G., Gheisari, M., & Irizarry, J. (2014b). Issues of Translating BIM for Mobile Augmented Reality (MAR) Environments. *Construction Research Congress 2014*: pp. 100–109. doi: 10.1061/9780784413517.011

Wing, M.G., Eklund, A., & Kellogg, L.D. (2005). Consumer-grade Global Positioning System (GPS) accuracy and reliability. *Journal of Forestry*, 103(4), 169–173.

Woodward, C., Hakkarainen, M., & Rainio, K. (2010). Mobile augmented reality for building and construction. *VTT Technical Research Centre of Finland. Mobile AR Summit MWC 2010*.

Yates, J.K., & Epstein, A. (2006). Avoiding and minimizing construction delay claim disputes in relational contracting. *Journal of Professional Issues in Engineering Education and Practice*, 132(2), 168–179.

Yoon, Y., DeSouza, G.N., & Kak, A.C. (2003, September). Real-time tracking and pose estimation for industrial objects using geometric features. *In Robotics and Automation, 2003. Proceedings. ICRA'03. IEEE International Conference on* (Vol. 3, pp. 3473–3478). IEEE.

Zavadskas, E.K., Vilutienė, T., Turskis, Z., & Šaparauskas, J. (2014). Multi-criteria analysis of Projects' performance in construction. *Archives of Civil and Mechanical Engineering*, 14(1), 114–121.

Zhang, C., & Arditi, D. (2013). Automated progress control using laser scanning technology. *Automation in Construction*, 36, 108–116.

eWork and eBusiness in Architecture, Engineering and Construction – Christodoulou & Scherer (Eds)
© 2016 Taylor & Francis Group, ISBN 978-1-138-03280-4

Generation of serious games environments from BIM for a virtual reality crisis-management system

A. Wagner & U. Rüppel
TU Darmstadt, Darmstadt, Germany

ABSTRACT: The presentation of building information as game content enjoys an increasing interest from various fields throughout the civil-engineering sector. The application of serious games or virtual reality visualisations for clients are just two examples.

With this work, we present a concept for a translation from building models to a game environment. For this purpose, city and building models are merged and the resulting geometries retrieved. The geometry then is used to define a map in the game environment. To achieve realistic results, an ontology maps defined building materials to textures existing in the game content.

The demonstrator can be modified to work with other game engines and offers an easy and forward way of automatically generating a gaming map from building information models.

1 INTRODUCTION

1.1 *Motivation*

With recent events like the ones in Paris or Brussels in mind, the protection against threats to public safety in cities is gaining importance. In addition to more conventional dangers, such as fire, earthquakes or floods, new dangers are occurring, terrorism and shooting sprees for example. First responders have a wide variety of tools ready to hands for managing their task forces or getting an overview of information on the current crisis. However, a combination of those tools in a virtual reality environment does not exist yet.

To facilitate the rendering of simulation results into serious game content, we are developing a tool that generates a map for game engines based on a city's infrastructure and building information. The examined file formats for this purpose are CityGML and Industry Foundation Classes (IFC). The former stores various information on the infrastructure as streets and building geometries as well as the surrounding terrain. The latter describes a more detailed digital model of singular buildings. The goal is to allow an easy and fast forward approach for creating the base of a virtual reality crisis-management system independent of the chosen game engine. As a demonstrator module, the tool was applied to the Unity Game Engine.

In the following chapters of this paper, we first describe the used files and environments including the IFC and CityGML file formats, as well as the Unity Game Engine chosen for the demonstrator. Chapter 2 will give an overview of the adopted concept, in detail the database and conversion of the file formats. The following chapter will show problems that occurred during the implementation of the tool. Finally, a conclusion is given in chapter 5.

1.2 *Industry Foundation Classes (IFC)*

The IFC are a standard based on the EXPRESS notation and are used for the exchange of building information models. They are developed and promoted by the BuildingSmart Foundation and can be imported and exported by various modelling software applications (buildingSMART International Ltd. 2016a). The file structure is STEP-based, though an alternate IFC file format exists, basing on XML. The current release is the IFC4 Add1.

They do not only store geometric information but also semantic. It is possible to memorize multiple buildings in one file, however, infrastructure as streets and railroads are not yet considered. Although, these domains are announced to be included in the next release IFC5 (buildingSMART International Ltd. 2016c). Still, the surrounding terrain of the building as well as trees or plants in general cannot be pictured by the IFC and it is not yet estimated if and when these aspects will be regarded in a new release.

1.3 *CityGML*

CityGML is an XML-based information model for 3D city and landscape models. It can store information regarding the geometric, topologic, visual,

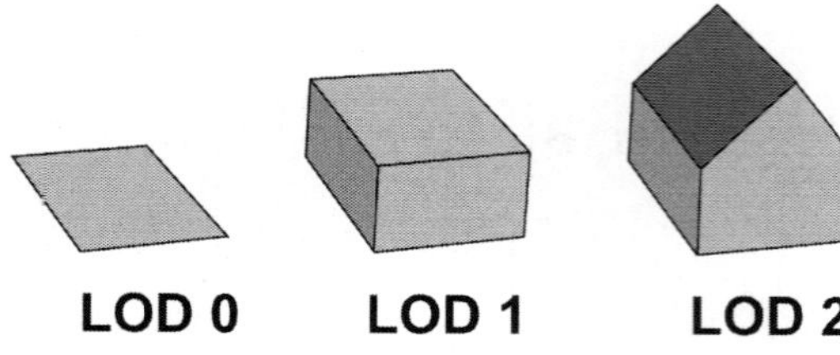

Figure 1. Definition of LOD 0 to 2.

and semantic features of the model in various detail levels (Kolbe 2012).

Five Levels Of Detail (LOD) describing the needed information of the model's elements are defined. Figure 1 shows an illustration of the LOD 0 to 2 as shown by (Biljecki 2015). The first two levels, 0 and 1, depict the basic geometry of the element. LOD 0 possesses the layout of the elements, with LOD 1 the overall height of the building is added. Each LOD includes more information like the roof shape, balconies, windows, and semantic properties. LOD 4 is the highest quality that can be achieved using CityGML models and delineates not only the exterior but also the interior of buildings. Since it is probable to have an uneven quality of information within an entire city model, it is possible to combine elements of various LODs in one model.

For the further work, it is relevant to know the position of the singular elements, e.g. described by their address. This information is stored within the element's description in the model and can be retrieved. Besides the postal address, the geographic coordinates of the buildings can be stored.

1.4 *Game engine*

As we expect the result of this research to be a universally applicable approach for the generation of game content, the chosen engine for the demonstrator will not be explained in detail.

However, it was important that the engine also supports virtual reality and mobile games, since the field of application of a universal translator is not clearly defined. With this in mind, we decided to utilise the Unity Game Engine for the demonstrator module (Unity Technologies 2016).

The presented approach is applicable to other game engines as long as they have a code based editor or an Application Program Interface (API).

1.5 *Related work*

Serious games are already being used in the field of civil engineering. The scope of recent and current researches varies.

(Schatz 2014) evaluated the suitability of serious games based on building models concerning fire safety. The main focus was set on interactive evacuation analysis.

Apart from fire safety, serious games are also used in the field of crisis management. Among research projects currently funded by the EU, for example the project "eVACUATE" can be found (Vassiliadis and Petrantonakis 2016). The project's result serves as a teaching tool in crisis situations as the evacuation of football stadiums or tube stations.

Another field of application is represented by (VIMtrek 2016). They offer a software application based on the Unity Game Engine that renders building models to game content for visualisation. The user can then explore the building in game on PC or in virtual reality.

Also, the conversion of the model standards IFC and CityGML has been examined. While most researchers focussed on the unidirectional conversion from IFC to CityGML (El-Mekawy, Östman, and Hijazi 2012), the Karlsruhe Institute of Technology (KIT) developed the FZKViewer allowing users to export CityGML files to IFC standard (KIT 2015).

2 CONCEPT

This chapter will present the concept for merging the different standards and translating them to game content. First, the database and ontology storing building information and relating materials and textures defined in the building models and game engine are discussed. Then, the conversion will be in focus. For the conversion, geometries and properties from all files and their elements have to be combined in one singular file. That final file subsequently needs to be loaded into the game engine to be processed to gaming content.

2.1 *Database*

The Database for coordinating the models contains two tables. The first one stores information regarding the existing and more detailed IFC files. The second handles the building elements of the city model.

To identify the buildings stored in IFC files, a data-table called "detailedModel" containing their addresses, memory location and Global Unique Identifiers (GUID) is used. Both properties can be retrieved from the building model, but the handling in an external database causes less effort while searching for an explicit model.

The "cityModel" table stores the geographic location of the elements centre, memory location and GUID of each building element of the city model. Additionally, the GUID of a more detailed model, in case one exists, can be saved for each building.

2.2 *Ontology*

The ontology connects textures of materials as they are stored in the IFC or CityGML standard. This will facilitate the conversion between those formats. Also, a conjunction with the available textures of the game engine takes place.

By using an ontology for this purpose, the universal approach for generating game content from building information is promoted. In case the user wants to apply our demonstrator to another game engine, the ontology can be expanded by the textures of the new engine. Since multiple classification is allowed within an ontology, the number of regarded game engines is limitless.

2.3 *Merging of the standards*

Before merging the two standards, a decision had to be made which standard should be the one they are merged to. Then, the concept for the merging itself had to be developed.

2.3.1 *IFC or CityGML*

Libraries for both standards exist and can be used to read from the files. The structure of the files is diverse, but to simplify this, IFC files could be stored in an XML-based format. On the downside, this format needs approx. 300–400% more memory space (buildingSMART International Ltd. 2016b).

Another aspect is that the IFC are still being developed. Further fields of the civil engineering applications will be added during the coming releases. It is likely that the surrounding terrain will be considered in the future releases of the IFC as well. BuildingSmart is also ambiguous to promote the IFC to the general exchange standard for building projects and the IFC are involved in the formulation of new national and international standards for exchange data in the civil engineering domain (NHRD and Heating and Ventilation Technology Standards Committee 2015).

The CityGML standard on the other hand has not been updated since 2012 (Kolbe 2012). This originates from the Infrastructure for Spatial Information in the European Community (INSPIRE) initiative founded in 2007 (European Parliament 2007). Its goal is to develop a new standard that is consistent throughout the European Union to ensure international interoperability. The new standard is supposed to be presented in 2019. Therefore, this research will still focus on CityGML.

Since the conversion in general is possible in both directions, we decided in favour of the IFC standard. This is mainly substantiated by the potential of the IFC in the future.

The using of ifcXML files is thereby no longer of interest, because the file containing all building and elements of a city will presumably be large and a further enlargement is not desirable. Also, existing libraries for both standards facilitate the translation process which makes the ifcXML standard dispensable.

2.3.2 *CityGML to IFC*

The developed process of converting a CityGML model to an IFC model is pictured in figure 2. On the left side, the tasks for the existing IFC files are presented. It is assumed that each building model is saved in its own file. If this is not the case, the file will be separated into as many files as the original file contained. The files are then read out and each building is registered in a database containing its GUID, memory location and postal address of the building.

On the right hand of figure 2, the working steps of the CityGML model are shown. First the initial file will be converted to IFC. The newly generated model then is divided into as many IFC files as buildings exist in the model. For each model a

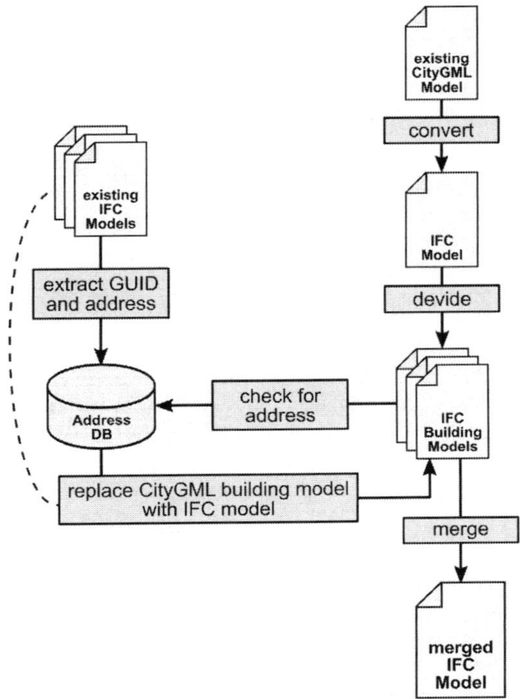

Figure 2. Process overview for merging a CityGML file with multiple IFC files to a singular IFC file.

query for the database is performed checking if an entry exists for its postal address. If this is the case, the memory location and GUID of the appropriate building model is retrieved from the database and the corresponding city model's IFC file will be replaced by the existing IFC file or a part of it in case the file holds more than one building.

After the substitution of building models, the files will be merged into a singular building model. To secure the realistic structure of the city model, the geographic position of each building has to be considered.

2.4 *Conversion to game content*

The resulting IFC file is then read out and rendered into the game engine as content. For this, the geometrical and semantic information is extracted. Based on this information meshes are defined and collision boxes determining the physical size of elements in the game engine are created. These elements are then loaded into the game's map.

An ontology relates surface textures of the model's materials with textures from the game engine. The meshes of the model will be equipped with the texture matching the described material of the model. This allows a realistic look of the building elements. If no materials are provided within a building model, previously defined default materials will be used.

3 IMPLEMENTATION

In this chapter the implementation of the demonstrator module will be discussed. This includes the presentation of used libraries and occurred problems. The discussion begins with the implementation of the standard's merging, then the generation of game content is focussed.

The ontology of textures and materials has already been implemented and will not be described further at this point. Since the integration of materials and game engines only have to happen once, this process is conducted manually until now.

3.1 *Merging of the standards*

First, the user references all existing IFC building models as well as the CityGML model he wishes to translate to game content. Then, each IFC file is examined, if it contains more than one building, the second resp. third building is retrieved and saved in its own file. Afterwards the excess buildings are deleted. The postal address and GUID of every building as well as the memory location of the file storing the model are extracted and saved in the "detailedModel" table of the database.

For the translation of the CityGML model to an IFC model KIT's FZKViewer was used (KIT 2015). But the viewer can only convert CityGML elements of LOD 2 or higher. In figure 3 the problem is depicted using the "Waldbruecke" CityGML model (Karlsruhe University of Applied Science 2012). The dark buildings are LOD 2 and can thereby be exported to an IFC file. The grey blocks symbolise buildings of LOD 1 meaning only the layout and height of the buildings are given. Also, the white areas cannot be saved in IFC since the standard does not consider infrastructure elements or surrounding areas.

To depict the complete model in IFC, the viewer was expanded to export LOD 0 and LOD 1 elements as well. The element's height of LOD 0 elements is determined by an analysis of surrounding elements and equals their mean. For now, the roof shape is considered to be a flat roof, but the shape could also be determined by finding the most common shape of the surrounding roofs. (NOTE: We are currently working on implementing this feature, it is likely to be included in the final paper.)

Besides the extraction of elements with low LODs, infrastructure elements as well as the surrounding terrain are reproduced as individual buildings. This step is taken to depict these elements in an IFC4 Add 1 file, which does not allow a portrayal of these kinds of elements yet.

In the next step, the IFC file is separated into as many files as building elements are stored in it. The "cityModel" table in the database stores the geographic and memory location as well as the GUID of these files. Then, each building's postal address is extracted and compared with the entities of the "detailedModel" table. If a more detailed IFC model exists for the building, the memory location in the "cityModel" table will be updated with the detailed file's location.

After all buildings have been checked, the entities of the "cityModel" table are merged into a singular IFC model and the IFC files resulting from the division of the city model are deleted. To avoid

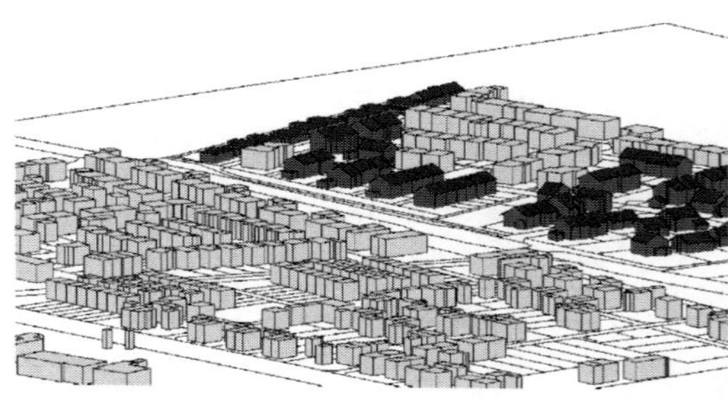

Figure 3. Visualisation of a CityGML model with highlighted LODs.

overlapping of models and a realistic array of the elements, the geographic location is used. Since the location stored in the "cityModel" table pertains for the elements centre, the placement of them can easily be realised. A local coordinate system with its origin at the first elements geographic location is generated. The local coordinates of the remaining elements can then be calculated.

3.2 *Generation of game content*

To facilitate the implementation of the translator of building information to game content, the IfcLibrary was used (RDF ltd. 2016). A wrapper class then retrieves geometric and semantic information for each building element stored in the IFC file. The geometric information is then converted to meshes and bounding boxes that can be displayed by the engine, as shown in figure 4.

After the meshes are generated, they will be linked to the texture associated to the elements material. E.g. figure 5 shows the texture associated to a brick wall.

Figure 6 shows the resulting map of the example case. As depicted, the translator considers solid elements such as walls, ceilings and stairs only. Dynamic elements like doors or windows can be translated to game content, too, but their dynamic functions will be lost during the process.

The generated map is solid and allows players to walk around the building. It does not allow movement through walls and prevents the player to fall through ceilings or floors, as is demonstrated in figure 7.

4 CONCLUSION AND OUTLOOK

The presented game content generator is a promising approach for a general-purpose transformation from building information to game content.

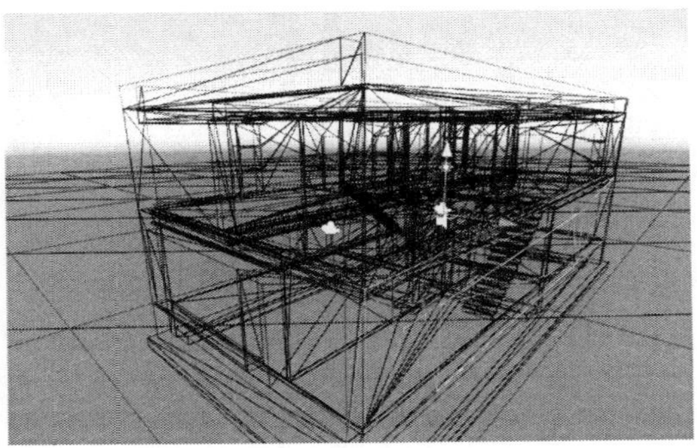

Figure 4. Unity Game Engine's depiction of meshes generated from building information.

Figure 5. Rendering of the "brick wall" texture in the Unity Game Engine.

Figure 6. Screenshot of the example map as it is shown in-game.

Figure 7. Screenshot of a player walking through the example map.

We were able to show the applicability of the presented methods.

This tool may serve as a fundament of a universal description of crisis scenarios. Thereby, the scenarios can easily be translated to game engines for visualisation, teaching or research. As mentioned before, this approach can be applied to any game engine.

For now, terrain and infrastructure elements cannot directly be depicted with IFC. Further developments of this standard have to be awaited. The same problem applies to plants. Since there is no easy workaround as with terrain and infrastructure elements, the ontology could be expanded to link custom IFC elements to trees or other plants. This would make it possible to place plants as game content where the city model describes their location.

Besides the consideration of plants by a connection in the ontology, dynamic objects could be integrated this way as well. (Schatz and Rüppel 2012) already examined the aptitude of ontologies to describe dynamic building information for an integration in game environments.

In general, the ontology used within this research was developed as an example. Additional researches need to be conducted in this field.

To achieve a more realistic mapping of city models, a connection to online sources such as published pictures of buildings or parks could be established. Matching images could then be rendered to textures and fitted to the generated meshes. This could also be exerted with pictures recorded in the database.

REFERENCES

Biljecki, Filip. 2015. 'Filip Biljecki—PhD Research'. *Filip Biljecki—PhD Research*. May 6. http://filipbiljecki.com/research/phd.html.

buildingSMART International Ltd. 2016a. 'buildingSmart'. *buildingSMART-Tech.org*. http://www.buildingsmart-tech.org/.

———. 2016b. 'IFC Overview'. *IFC Overview Summary*. http://www.buildingsmart-tech.org/specifications/ifc-overview/ifc-overview-summary.

———. 2016c. 'IFC4 Add1 Release'. *IFC4 Add1 Release*. April 12. http://www.buildingsmart-tech.org/specifications/ifc-releases/ifc4-add1-release.

El-Mekawy, Mohamed, Anders Östman, and Ihab Hijazi. 2012. 'An Evaluation of IFC-CityGML Unidirectional Conversion'. *International Journal of Advanced Computer Science and Applications(IJACSA)*. http://dx.doi.org/10.14569/IJACSA.2012.030525.

European Parliament, Council of the European Union. 2007. Directive 2007/2/EC of the European Parliament and of the Council of 14 March 2007 Establishing an Infrastructure for Spatial Information in the European Community (INSPIRE). http://data.europa.eu/eli/dir/2007/2/oj.

Karlsruhe University of Applied Science, media::lab. 2012. 'Waldbrücke'. CityGML. Waldbrücke. http://www.citygml.org/fileadmin/count.php?f=fileadmin%2Fcitygml%2Fdocs%2Fwaldbruecke_v1.0.0.zip.

KIT, Institute for Applied Computer Science. 2015. *FZKViewer* (version 4.5). FZKViewer. KIT. http://www.iai.fzk.de/www-extern/index.php?id=2315&L=1.

Kolbe, Thomas H. 2012. 'CityGML'. *Homepage of CityGML*. February 24. http://www.citygml.org/.

NHRD, DIN-Normenausschuss Heiz—und Raumlufttechnik sowie deren Sicherheit, and Heating and Ventilation Technology Standards Committee. 2015. 'ISO167575-1'. Beuth.

RDF ltd. 2016. *IFC Engine DLL*. IFC Engine DLL. http://www.ifcengine.com/.

Schatz, Kristian. 2014. *Bauwerksmodellbasierte Serious Games als Ingenieurmethoden im Brandschutz am Beispiel interaktiver Entfluchtungsanalysen*. Vol. 1. Berichte des Instituts für Numerische Methoden und Informatik im Bauwesen 2014. Aachen: Shaker Verlag.

Schatz, Kristian, and Uwe Rüppel. 2012. 'Ontology-Based Approach for Integrating Building Information Modeling into Serious Gaming Environments'. In *Proceedings of the 14th International Conference on Computing in Civil and Building Engineering*. http://www.icccbe.ru/paper_long/0111paper_long.pdf.

Unity Technologies. 2016. 'UnityEngine'. *Unity—Game Engine*. https://unity3d.com/.

Vassiliadis, Dimitris, and Dimitris Petrantonakis. 2016. 'eVACUATE'. *eVACUATE*. http://www.evacuate.eu/.

VIMtrek. 2016. 'VIMtrek'. *WIMtrek*. vimtrek.com.

eWork and eBusiness in Architecture, Engineering and Construction – Christodoulou & Scherer (Eds)
© 2016 Taylor & Francis Group, ISBN 978-1-138-03280-4

Simulation model generation combining IFC and CityGML data

G.N. Lilis, G.I. Giannakis & K. Katsigarakis
Department of Production Engineering and Management, Technical University of Crete, Chania, Greece

G. Costa & Á. Sicilia
ARC, La Salle Engineering and Architecture, Ramon Llull University, Barcelona, Spain

M.Á. Garcia-Fuentes
Department of Energy, CARTIF Foundation, Valladolid, Spain

D.V. Rovas
Institute for Environmental Design and Engineering, University College London, London, UK

ABSTRACT: The energy efficiency requirements at district scale revealed the need for detailed building energy simulations, with which the overall district energy demand can be estimated with an acceptable degree of accuracy. In order to meet this need, an automated simulation model generation process is introduced at the context of the European project OptEEmAL, which includes: a query stage where data are gathered from IFC, CityGML files, and a transformation stage where a single IDF file is generated for a building in a district environment, suitable for EnergyPlus simulations. The queried data are assumed to conform to certain correctness, completeness and consistency conditions across district and building scales. As a demonstration example, a simulation model is generated for a specific building. Future improvements of this work are discussed related to the integration of all the data requirements of the proposed process, in a District Data Model under an ontological framework.

1 INTRODUCTION

The recent building energy footprint reduction requirements highlighted the value and promoted the use of detailed building thermal energy simulations. Furthermore, since the thermal performance of buildings in a district environment is affected by phenomena caused by neighbor building topologies, treating buildings individually and neglecting their neighbor building topologies impact (e.g. shading effects and microclimate), results to reduced accuracy in their simulation results. The need for accurate thermal energy simulations is met by a plethora of simulation engines, developed for either building-scale simulations (e.g. EnergyPlus), where detailed building geometry information is required, or district-level simulations (e.g CitySim), where the geometry details of buildings are omitted. The input data to the above programs are formatted properly according to the specific program requirements.

From the input data availability perspective, two popular data schemas have been developed for different purposes: the first is a BIM scheme called IFC (ISO 16739, 2013), designed for building-scale data and promoting interoperability among AEC industry programs; the other is a GIS schema, called CityGML (Kolbe et al., 2005), structured in order to render city scale data. These data sources cannot be used directly as inputs to thermal energy simulation programs as they require further processing related to the generation of the second-level space boundary geometric topology (Bazjanac, 2010).

Incorporating the districts environmental impact in building-scale simulation programs improves the quality of their results. Attempts towards this direction have been reported via the use of co-simulation (Thomas et al., 2014). In order to perform detailed building thermal energy simulations including the buildings district environmental impact, a simulation model generation process is introduced, as the main subject of the present work. Parts of this process will be used at the context of the European project OptEEmAL[1], which aims at automating the selection of the best energy conservation refurbishment scenario of a district, according to specific performance indicators. In order to include the districts environmental

[1]https://www.opteemal-project.eu

impact to the generated simulation models, input data from GIS and BIM sources are integrated. This idea of integrating data across building and district scales has appeared in past research efforts: SEMANCO project (Sicilia et al., 2014), GeoBIM extension (de Laat & Van Berlo, 2011), ontology-based Unified Building Model (El-Mekawy & Östman, 2012), virtual 3D city model (Döllner & Hagedorn, 2007).

The rest of the paper is organized into two parts. In the first part, the current simulation model generation process is described which uses data from an IFC file of a building, combined with district data of surrounding buildings extracted from a CityGML file, in order to generate inputs suitable for the energy simulation program EnergyPlus. Initially, the data requirements of the process are highlighted, followed by the definition of data quality rules, these data have to satisfy, in order to be suitable for simulation model generation. Then, the algorithmic parts of the process are described, followed by an application example, on specific district data.

In the second part of the paper, it is discussed how the current simulation model generation process can be improved by moving some of its functional components and organizing its input data, in a single district data model, using an ontological framework. Such extensions will introduce interoperability possibilities to the process by establishing semantic connections to other simulation tools.

2 DATA REQUIREMENTS

Building simulation data models require a variety of data, ranging from pure geometric descriptions, to operation characteristics of micro-climate control devices installed in building interiors. In order to organize such versatile data across building and district scales, an initial classification is attempted. According to this classification, the required data for a simulation model generation in a district environment can be classified into three categories (as illustrated in figure 2): BIM data, GIS data, and Contextual data. The characteristics of each category are described in the following sections.

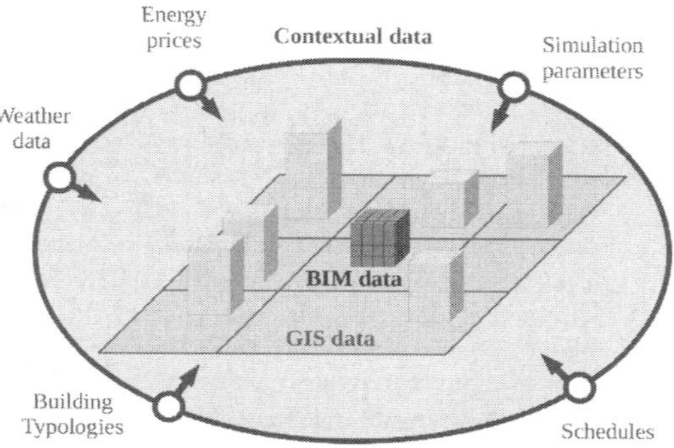

Figure 1. Data requirements of the simulation model generation process.

2.1 *BIM data*

The required BIM data for simulation model generation are related to building geometric descriptions, material thermal properties, and the characteristics of the systems installed in the buildings. The required geometric BIM data contain either: the solid geometric representations of architectural elements (walls, slabs, roofs, coverings, openings and others, illustrated in figure 2A), or their respective boundary surface topology (illustrated in figure 2B).

The boundary surface topology is the only geometric prerequisite of a simulation model generation process. If it is missing, it can be obtained from the solid geometric representations of the architectural elements via geometric operations as illustrated in figure 2 (from a set of architectural element solid representations (part A), the respective boundary surface topology (part B) is produced). The boundary surface topology consists of data belonging to three categories: (a) second-level space boundaries, (b) virtual space partitions and (c) shading surfaces. The second-level space boundaries (Bajzanac, 2010), are boundary surface pairs, through which thermal energy flows either among building spaces (internal space boundaries, green surfaces of figure 2B) or between a building space and its environment air/ground (external space boundaries, orange surfaces of figure 2B). Virtual space partitions are also boundary surface pairs, which separate virtually internal building spaces, without the use of a building construction (blue surfaces of figure 2B). Finally, shading surfaces are boundary surfaces which play an indirect role in a thermal simulation by blocking sunlight.

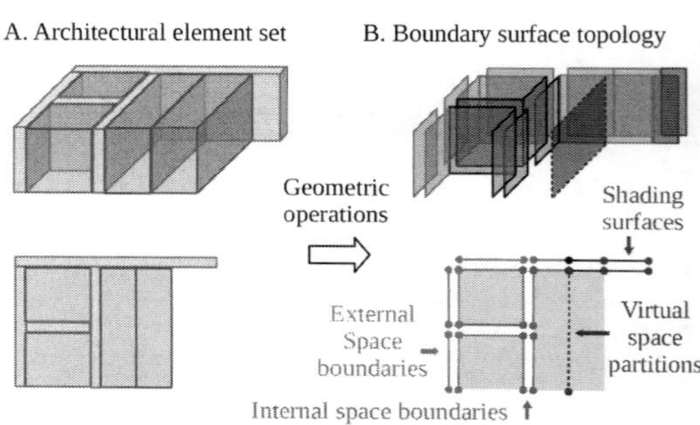

Figure 2. Architectural element set and its second-level space boundary topology obtained from CBIP.

The thermal properties of the material layers used in the building constructions are also required for a simulation model generation, since they provides necessary information in order to determine thermal energy exchange among building spaces.

The properties of opaque constructions refer to values of quantities related to every layer in the construction, such as: thickness, density, thermal conductivity and specific heat and others. If layer bedding information is not available, values of quantities referring to the whole construction, such as: total thermal mass and thermal resistance, have to be specified. For transparent constructions only quantity values referring to the overall construction, such as: the U-value and the solar transmission coefficient, are required to be specified as material properties.

Finally, the operation characteristics of energy consuming, or self-sufficient devices installed in buildings which alter their internal thermal conditions, defined in a broad sense as systems, are BIM data which are also required.

2.2 *GIS data*

Building simulation models at district scale require data referring to the overall district, which are not included in the BIMs and are defined as GIS data. Similarly to the BIM data types, there are three GIS data types, characterized as geometric, material property and system data types. GIS geometric data contain descriptions of the district building envelopes as polygon surface sets. These polygons are used in neighbor building shading calculations. GIS material property data contain the reflectivity coefficients of neighbor building surfaces, used for solar calculations. Finally, GIS system data refer to characteristics of district-scale systems which may be present in the district servicing multiple buildings.

2.3 *Contextual data*

Finally simulation models require data, which cannot be classified as BIM-related or GIS-related data and are defined as Contextual data. These data can be classified into the following five data types:

- *Weather data* include the values of weather quantities required for building thermal simulations, such as: Outside dry-bulb temperature and pressure and others, which are contained in weather files.
- *Schedules* are vectors formed by a time series which contains values describing the presence of users (occupancy) and the operation of passive and active devices installed in the buildings.
- *Simulation parameters* are values assigned to specific parameters required in order for a simulation to be initiated, such as: warm up time, starting and ending time instances, time step and others.
- *Energy prices* refer to the cost in monetary units of the energy use in the district.
- *Building typologies* contain data values which are fixed for all buildings and are based on the geographical location, use and other classification parameters (Vimmr et al., 2013). In case building data are not available for specific buildings, they can be inferred using building typologies.

3 DATA QUALITY

The required data, classified as BIM, District or Contextual data, are obtained either from CityGML and IFC data sources, or they are inserted manually. These data pass three stages of data quality checking operations in order to be suitable for a simulation model generation. These operations include consistency, correctness and completeness checks and are explained analytically in the following sections.

3.1 *Data consistency*

The first checking operation ensures that an inserted IFC model is consistent with the underlying CityGML model. Although BIM geometric data obtained from an IFC file, might be visually correct, they may be inconsistent with the CityGML geometric data, which are described in world coordinates. Such inconsistencies occur when the geometric definition of a building in IFC model appear slightly rotated or translated with respect to the CityGML shell geometric definition of the same building. The inserted IFCmodel is considered CityGML-consistent if all IFC architectural elements are located inside a single CityGML shell. In any other case an inconsistency is declared and is communicated back for correction.

3.2 *Data correctness*

Both IFC and CityGML data should be checked for correctness, before being used as inputs to the simulation model generation process. Incorrect data have many causes and the respective errors have different characteristics, as discussed next.

3.2.1 *Error causes*

There are three different sources of the errors appearing in IFC and CityGML data files, which can be ordered, depending on their causes, as follows:

- *Scanning errors*. Some of the CityGML geometric data are generated from point clouds obtained from terrestrial or airborne scanning devices, which contain errors related to malefaction of these devices or incorrect geo-referencing of the obtained points.

- *Design errors*. Oftentimes IFC and CityGML files contain errors caused by incorrect design, where the designer specifies incorrectly an architectural element, or material property or system.
- *Exporter errors*. Finally there are cases where either the IFC or the CityGML exporters generate errors by populating incorrectly the data classes in the respective IFC and CityGML files.

3.2.2 *Error classification*

Errors appearing in IFC and CityGML files can be classified, with respect to their characteristics, into the following two categories:

- *Missing data*. There are cases where the data in IFC or CityGML files are not complete. For example, certain data might be omitted from the specification of the material layer properties of a construction. These errors are characterized as missing data errors.
- *Incorrect data*. Apart from the missing data errors, there are cases where IFC or CityGML data are incorrect. For example a solid geometric representations of architectural element, might be misplaced with respect to other element representations.

A more detailed investigation of the geometric errors encountered in IFC files, and correction techniques, can be found in (Lilis et al., 2015).

3.2.3 *Data completeness*

Finally, the inserted IFC models are checked for completeness. More specifically, IFC data have to satisfy certain minimum data requirements expressed as a set of conditions in order to be suitable for simulation model generation, which are:

- *Boundary conditions*. The boundary surfaces, at which the buildings of the district are attached to the outside air and ground, should be explicitly defined for every building.
- *Conditioned spaces*. The inserted IFC file should have at least one conditioned space volume, i.e. a building space which is going to be studied thermally.
- *Material properties and space boundaries*. Every second level space boundary surface pair of the boundary surface topology of a building, related to either a thermal or an opening element, should be linked semantically to a building construction, characterized by a set of thermal properties.

Provided that the above conditions are satisfied, the simulation model generation process can be performed, as described in the following section.

4 SIMULATION MODEL GENERATION

The purpose of the simulation model generation process is the creation of an Energy Plus compatible input data file <*.idf$_b$>, referring to a building b, in a district setting, from its IFC file (IFC$_b$) and a CityGML file describing the geometry of the district. Both IFC$_b$ and CityGML data files are assumed to conform to the data quality rules described in previous sections.

Algorithmically, the simulation model generation process is divided into three stages: (a) Data query stage, where data from IFC$_b$ and CityGML files are extracted, (b) Boundary Surface topology generation stage, where the boundary surface topology of building b, is generated and augmented with shading surfaces from neighbor buildings and (c) Input Data File generation stage, where the augmented space boundary topology and the construction material property data, are used in order to produce a single input data file suitable for Energy Plus simulations. These stages are illustrated in figure 3 and are explained in detail in the following sections.

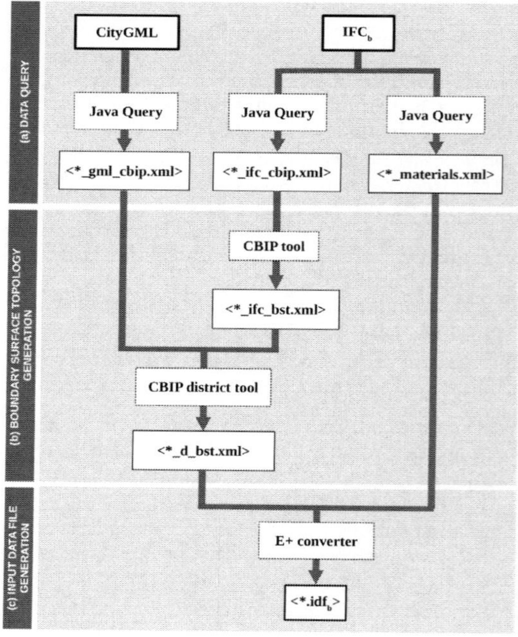

Figure 3. Overview of the simulation model generation process.

4.1 *Data query*

The required data for the simulation model generation, of a single building in a district, are obtained from two different meta-models, which both are defined in a textual format, are usually object-oriented and tend to be large. The most popular BIM meta-model is the IFC, which is based on the EXPRESS data schema, which is available as a STEP file. Additionally, the most common GIS meta-model is the CityGML, which is based on a XML data model language and its schema is defined by a set of XSD files. Because the proposed query components have been written in Java, which is an object-oriented language, it is useful to have on-memory the populated Java objects. In order to achieve this functionality, third-party frameworks have been chosen: for the conversion of the IFC schema the Eclipse Modeling Framework (EMF) was used and for the CityGML schema, the Java Architecture for XML Binding (JAXB) was chosen. Specifically, in the case of IFC schema the BuildingSMART library, which is a part of the BIM Server project, has been used for the automated generation of certain Java classes. On the other hand, in the case of CityGML schema the Reference Implementation (RI) of JAXB, which is a part of the GlassFish project, has been used. As mentioned before, these classes are used to store on-memory BIM and GIS models and pass them to the proposed components for further manipulation. Both IFC query and CityGML data queries are described next.

As far as the CityGML query process is concerned, only geometric surface descriptions contained in the CityGML file, which refer to the district building envelopes, are queried while any appearance related data structures are omitted. Additionally, elements not related to building envelope surfaces (transportation objects, vegetation objects and others), were not taken into account. The coordinates of the points of the building envelope surfaces, were gathered and used to populate appropriate data structures in the <*_gml_cbip.xml> data file (figure 3). Apart from the geometric content of the envelope surfaces, the semantics of these surfaces classifying them as either: wall, ground or roof surface, were gathered as well in order to be used for future reference.

Regarding IFC data, two data types are queried: geometric data and material property data. The geometric data refer to characteristics of geometric solid representations of architectural elements contained in the IFC file, such as the extruded area solid, the manifold and the faceted boundary representation. Each architectural element and the characteristics of its solid geometric representation, are written in the output <*_ifc_cbip.xml> file, which is used as an input file to the boundary surface topology generation process described in section 4.2. Regarding the material property data, certain thermal property values contained in the IFC file are queried, referring to building constructions and their respective layer bedding, as mentioned in section 2.1. These properties are written in file <*_materials.xml>, which according to the process diagram of figure 3, is used as input, to the input data file generation process described in section 4.3.

4.2 *Boundary surface topology generation*

Sometimes, although IFC files contain the necessary structures to support the geometric data requirements of a simulation model generation process, the geometric data of IFC files referring to the boundary surface topology of buildings, are incorrect, or they are missing. In such cases the generation of the boundary surface topology of building is required, and is performed using the Common Boundary Intersection Projection Algorithm (CBIP) (Lilis et al., 2014). In a nutshell, CBIP receives as input the geometric solid descriptions of architectural building elements and performs the necessary geometric operations, illustrated in figure 2, in order to generate the boundary surface topology of a building described in section 2.1. Additionally, apart from the boundary surface topology geometric data requirements referring to a single building, shading surfaces of neighbor buildings must also be determined. CBIP process is extended in order to include such surfaces, as described next.

Algorithmically, the boundary surface topology generation process in a district environment (illustrated by the process block (b) in figure 3), includes two stages. In the first stage CBIP is used in order to transform the queried IFC architectural geometric data of the building, contained in xml file <*_ifc_cbip.xml>, into an intermediate xml data file containing the boundary surface topology data of the building <*_ifc_bst.xml>. In the second stage, the CBIP district tool augments the generated <*_ifc_bst.xml> file from the first step, by including the neighbor shading building surfaces obtained from the queried CityGML data, contained in <*_gml_cbip.xml>, and generates a new xml data file containing the boundary surface topology of the building in the district environment <*_d_bst.xml>.

4.3 *Input data file generation*

In EnergyPlus, input data are defined by two ASCII (text) files: the Input Data Dictionary (IDD) and the Input Data File (IDF). The IDD file contains all possible EnergyPlus classes and descriptions of their properties. Each version of EnergyPlus has a different IDD file. The IDF file consists of all the necessary data to properly define a thermal simulation model of a certain building, described using appropriate IDD classes.

The input data file generation process aims at creating an IDF file from boundary surface topology data of a building in a district setting, contained in <*_d_bst.xml> file and material thermal property data, contained in <*_materials.xml> file, as illustrated in the process block (c) in figure 3. This process is developed using the MATLAB programing environment and includes two stages. In the first stage, a MATLAB script has been developed that identifies the version of the IDF file (<*.idf> file), parses the appropriate IDD file (<*.idd> file) and creates a library (MatlabIDDxx, where xx is the EnergyPlus version) of MATLAB classes, corresponding to EnergyPlus classes. In the second stage, the generated MATLAB classes are populated based on the geometric and material data contained in <*_d_bst.xml> and <*_materials.xml> files, and exported into a final input data file (<*.idf> file).

This input data file generation process, referring to a single building, conforms to certain transformation rules described thoroughly in (Giannakis et al., 2015). In summary, certain IDD classes are populated from data obtained from: (a) CBIPs xml geometric output and contained in <*_ifc_bst.xml> file (figure 3) and (b) material properties queried from the IFC file contained in <*_materials.xml> file (figure 3). From CBIPs geometric output the classes Zone, BuildingSurface:Detailed and FenestrationSurface:Detailed are populated, while the classes Construction, Material and WindowMaterial:SimpleGlazing System are populated from the material property file.

Geometric data referring to neighbor buildings, which are queried from the CityGML file and added to the <*_ifc_bst.xml> file, via the CBIPs district tool, generating the final boundary surface topology file <*_d_bst.xml> of the building ina district environment (illustrated in figure 3). These geometric data, are used to populate the IDD class Shading:Building:Detailed in the final IDF file.

5 EXAMPLE

The proposed simulation model generation process is demonstrated on a residential district, which is a part of Santiago de Compostela city in Spain, displayed in Figure 4A. The chosen LoD2 CityGML model of the district consists of 65 building envelopes out of which one was selected (indicated in Figure 4B), in order to be replaced by a more detailed IFC building model geometric representation.

A single IDF file was obtained following the stages of the described simulation model generation process. The geometric content of the generated IDF file is displayed in figure 4C, where different colors are used, in order to highlight the characteristics of the different surfaces contained in the IDF file (blue color is used to display external space boundaries; green color is used for internal space boundaries; purple color for neighbor building shading surfaces; and yellow color for site boundaries). Although all neighbor building shading surfaces were considered here as a proof of concept, algorithms selecting the ones which have considerable impact to the building simulation, using proximity criteria, will be developed in the future.

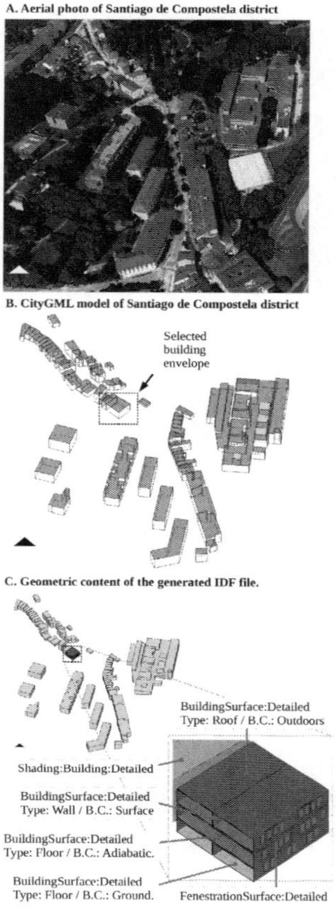

Figure 4. (A) Aerial photo of Santiago de Compostela district (B) CityGML model geometric representation, (C) Geometric content of the generated IDF file (B.C.: Boundary Condition).

6 FUTURE WORK

6.1 *Energy data model*

Although the simulation generation process mentioned earlier is automated for geometry and material thermal property input data, building system characteristics have not yet included in this process as input data. The versatility of such input data highlights the need for integrating all these diverse data sources under a common data model.

A suitable data model towards this direction, which provides the necessary input data structures for the Energy Plus simulation program, is the SimModel (O'Donnell, 2013). SimModel is an XML-based data scheme, which is designed to support building-scale simulation models and does not contain district-related data structures.

Consequently, in order to integrate the input data of a district environment, an extension to the current SimModel schema is viewed as a potential future work path. This extension will be used to populate a new Energy Data Model, as illustrated in figure 5 with a dashed rectangle. Furthermore, according to figure 5, the Energy Data Model (EDM), is visualized as a subset of a District Data Model (DDM), which will be a central functional component of OptEEmAL platform (Izkara et al., 2016), designed to provide the necessary data to support, one (for the whole district), or multiple (for every building) simulations. Additionally, in order to provide interoperability links to other simulation programs, an ontology based structure of the DDM is visualized, as described in the following section.

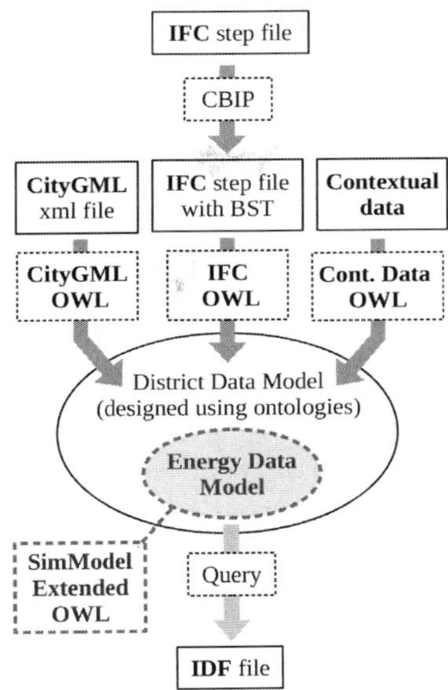

Figure 5. Illustration of the inclusion of an ontology-based District Data Model into the current simulation generation process (BST: Boundary Surface Topology).

6.2 *Ontology-based district data model design*

Data integration based on ontologies can be useful to effectively combine data from multiple heterogeneous data sources (Wache et al., 2001). This way, by means of ontologies it is possible to integrate data from different sources when there is an ontology that represent them. When these data sources are based on open standards, such as CityGML and IFC, is easier to carry out this integration since there are ontologies already implemented that can be reused. In this context, ontologies are useful to facilitate data linking between different data models.

The use of ontologies to integrate information to an energy domain is not new. For example, this approach was applied in SEMANCO project at the urban level, multi-scale analysis of carbon reduction problems and integration of GIS (Sicilia, et. al., 2014). In the case of OptEEmAL platform, input data from IFC, CityGML files and contextual data can be transformed into RDF according to existing domain ontologies such as ifcOWL. Thereby, the District Data Model can be redefined as an ontology-based framework where the role of ontologies is to facilitate the integration of these three input data sources (Costa et al., 2016). In this framework, the EDM is based on a version of the SimModel in OWL (SimModelOWL). The EDM is populated through an ETL process taking data from the different ontologies of the DDM, as illustrated in Figure 5.

To carry out this ETL process, the DDM ontologies has to be aligned to SimModelOWL using ontology matching methods. For example, ifcOWL is aligned with SimModelOWL. Then, the RDF data derived from the input sources can be transformed according to the EDM structure. Since the data in the DDM is already in RDF, it becomes a problem of RDF reshaping.

This ontological approach also facilitates that integration of new simulation tools. The generation of energy simulation models for such tools is a matter of aligning the EDM which contains all the parameters needed to carry out a detailed energy simulation with the input model of such tool. This approach would also be extrapolated to other data model for other kinds of simulations such as cost analysis.

6.3 *CBIP as an IFC data completeness tool*

The ontological structure of the energy data model, described earlier, will enable CBIP to be used as an IFC data completeness tool and not as a part of the simulation model generation process. Such use is supported by the fact that CBIPs output, the boundary surface topology (second-level space boundaries, virtual space partitions, shading surfaces), can be adequately described by the *IfcRelSpaceBoundary2ndLevel* and *IfcShadingDevice*, data structures. Therefore CBIPs operation can be a part of the IFC data completeness process, before DDM insertion. During this process, IFC data can be enriched by the boundary surface topology, obtained by CBIP. After the IFC data being enriched, they can be transformed to RDF using the ifcOWL ontologies and become a part of the DDM. A visualization of CBIPs independent operation as a data completeness tool, is illustrated in figure 5.

7 CONCLUSIONS

A semi-automated simulation model generation process capable of forming input data files suitable for Energy Plus calculations in a district environment, combining data from IFC and CityGML files, was presented. This process is a part of the European research project OptEEmAL, which aims at selecting the best according to certain performance measures, district retrofitting solution. Consistency, correctness and completeness data quality rules, for both IFC and CityGML input data files, were also discussed. Provided that these data quality conditions are met, the three stages of the proposed simulation model generation process, were described in detail. The overall process was demonstrated successfully, on a selected building defined by an IFC file, in a demo district described by a CityGML file.

Future improvements of the process related to: the integration of its input data using ontologies into an energy data model, and the use of the process' boundary surface topology generation tool, as an IFC data completeness tool, were also discussed.

Conclusively, there are certain key concepts and challenges which have to be addressed, arising from the proposed process, which include: the need for establishing data quality validation processes, which will ensure consistency, correctness and completeness of the available input data; the use of mechanisms for automating the query of input data from other sources (apart from BIM and GIS), such as weather data should also be included; and finally the possibility of creating links of CBIPs geometric output to other simulation tools, should also be examined.

ACKNOWLEDGEMENTS

Part of the work presented in this paper is based on research conducted within the project "Optimised Energy Efficient Design Platform for Refurbishment at District Level", which has received funding from the European Union Horizon 2020 Framework Programme (H2020/2014–2020) under grant agreement n° 680676.

REFERENCES

Bazjanac, V., 2010. Space boundary requirements for modeling of building geometry for energy and other performance simulation. In *proc. of CIB W78*, Cairo, Egypt.

Costa, G., Sicilia, A., Lilis, G.N., Rovas, D.V., & Izkara, J.L., 2016. A comprehensive ontologies-based frame-work to support retrofitting design of energy-efficient districts. In *proc. of European Conference on Product & Process Modelling*, Limassol, Cyprus.

de Laat, R. & Van Berlo, L., 2011. Integration of BIM and GIS: The development of the CityGML GeoBIM extension. In *Advances in 3D geo-information sciences*, pp. 211–225. Springer.

Döllner, J. & Hagedorn, B., 2007. Integrating urban GIS, CAD, and BIM data by service based virtual 3d city models. *R. e. al.(Ed.), Urban and Regional Data Management-Annual*, pp. 157–160.

El-Mekawy, M. & Ostman, A., 2012. Ontology engineering method for integrating building models: The case of IFC and CityGML. *Universal Ontology of Geographic Space: Semantic Enrichment for Spatial Data. IGI Global*, pp. 151–185.

Giannakis, G.I., Lilis, G.N., Kontes, G., Valmaceda, C., Garcia-Fuentes, M.Á., & Rovas, D.V., 2015. A methodology to automatically generate geometry in-puts for energy performance simulation from IFC BIM models. In *proc. of Building Simulation IBPSA conference*, pp. 504–511, Hyderabad, India.

ISO 16739, 2013. Industry Foundation Classes (IFC) for data sharing in the construction and facility management industries. <https://www.iso.org/>.

Izkara, J., Prieto, I., Sicilia, A., Costa, G., Madrazo, L., Bayili, S., & Katsigarakis, K., 2016. Requirements and Specification of the District Data Model.

Kolbe, T.H., Gröger, G., & Plümer, L., 2005. CityGML: Interoperable access to 3d city models. In *Geo-information for disaster management*, pp. 883–899. Springer.

Lilis, G.N., Giannakis, G.I., Kontes, G., & Rovas, D.V., 2014. Semi-automatic thermal simulation model generation from IFC data. In *proc. of European Conference on Product & Process Modelling*, pp. 503–510, Vienna, Austria.

Lilis, G.N., Giannakis, G.I., & Rovas, D.V., 2015. Detection and semi-automatic correction of geometric in-accuracies in IFC files. In *proc. of Building Simulation IBPSA conference*, pp. 504–511, Hyderabad, India.

O'Donnell, J., 2013. SimModel: A domain data model for whole building energy simulation. In *proc. of Sim-Build*, pp. 382–389, Sydney, Australia.

Sicilia, A., Madrazo, L., & Pleguezuelos, J., 2014. Integrating multiple data sources, domains and tools in urban energy models using semantic technologies. In *proc. of European Conference on Product & Process Modelling*, pp. 837–844, Vienna, Austria.

Thomas, D., Miller, C., Kämpf, J., & Schlueter, A., 2014. Multiscale co-simulation of EnergyPlus and CitySim models derived from a building information model. In *proc. of BauSIM IBPSA conference*, pp. 469–476.

Vimmr, T., Loga, T., Diefenbach, N., Stein, B., & Bachová, L., 2013. Tabula—Residential Building Typologies in 12 European Countries—Good practice example from the Czech Republic. In *proc. of CESB13*, Prague, Czech Republic.

Wache, H., Voegele, T., Visser, U., Stuckenschmidt, H., Schuster, G., Neumann, H., & Hübner, S., 2001. Ontology-based integration of information—A survey of existing approaches. In *proc. of IJCAI-01 workshop: Ontologies and Information Sharing*, pp. 108–117, Seattle, USA.

Information & knowledge management (III)—
life cycle operations and energy efficiency

eWork and eBusiness in Architecture, Engineering and Construction – Christodoulou & Scherer (Eds)
© 2016 Taylor & Francis Group, ISBN 978-1-138-03280-4

Information requirement definition for BIM: A life cycle perspective

G.F. Schneider
Fraunhofer Institute for Building Physics IBP, Valley, Germany
Energie Campus Nürnberg, Nürnberg, Germany

A. Bougain
Fraunhofer Institute for Building Physics IBP, Valley, Germany
Energie Campus Nürnberg, Nürnberg, Germany
Vienna University of Technology, Vienna, Austria

P.S. Noisten & M. Mitterhofer
Fraunhofer Institute for Building Physics IBP, Valley, Germany

ABSTRACT: Adopting the Building Information Modeling (BIM) methodology assists in fulfilling stringent cost and time objectives in architecture, engineering, construction and facility management industry. To enable further adoption of the technology a life cycle oriented approach needs to be pursued. In this work, information requirements for the whole life cycle of a building are defined and the capabilities of existing BIM models to assist the identified information areas are evaluated. The results indicate missing support on information about equipment efficiency curves, description of decentralized energy conversion devices, occupant behavior, thermal and visual indoor comfort and information on the preservation status. Furthermore, to support future BIM model development, the novel information management method GASCEeIiL is presented which allows tracing the use of a piece of information by each stakeholder along the information life cycle.

1 INTRODUCTION

The introduction of digital tools in Architecture, Engineering, Construction and Facility Management (AEC/FM) industry has led to a general change in the way buildings are designed, constructed and operated. Following the Building Information Modeling (BIM) paradigm assists in fulfilling time and cost requirements of complex facility development projects (Eastman et al. 2011).

Implementing and utilizing the BIM methodology is an active field of development. BIM adoption and deployment in recent years mainly focused on early phases of the building life cycle. To further encourage the adoption of BIM technologies in the AEC/FM industry, there is a need to also consider later stages of the life cycle, e.g. deployment of BIM technologies in existing buildings (Volk et al. 2014).

A key component of the BIM method is the use of an encompassing digital model of the building (BIM model) to exchange information required in the process. BIM model development and standardization has therefore been an active area of research in the past (Cerovsek 2011). However, issues related to the interoperability of digital tools based on these models remain unsolved (Cerovsek 2011). One reason for this is that existing BIM models do not satisfy the information needs of stakeholders especially in later phases of the building's life cycle.

In literature limitations of existing BIM models and information requirements for BIM models are defined. In particular, for the purpose of Building Energy Performance Simulation (BEPS) during the design phase, additional information to current state-of-the-art BIM models must be added, such as energy efficiency curves of technical equipment and specific material properties, e.g. thermal conductivity (O'Donnell et al. 2011, Wimmer et al. 2014). To investigate before construction the performance of decentralized energy conversion devices in buildings such as photovoltaic solar panels, additional technical specifications of these devices need to be included in a BIM model (Gupta et al. 2014). Becerik-Gerber et al. (2012) identify information needs in facility management especially for locating building equipment and gathering their documentation. Specific information for decontamination planning or deconstruction tracking is demanded by Volk et al. (2014).

Efforts on the definition of whole life cycle information requirements are also reported. Wong & Zhou (2015) focus on buildings' sustainability and energy issues and the application of BIM in this domain. According to the authors, information to support the reduction, reuse and recycling of building components in the refurbishment and deconstruction phases is required. For the purpose of an integrated energy efficiency and life cycle management of public use facilities the HESMOS project defines information requirements across the building's life cycle with a view on energy related information (Liebich et al. 2011).

Thus, a reasonable amount of information requirements has been collected in literature in the past. This is often done focusing solely on single phases of the building's life cycle with limited discussion on information exchange to adjacent phases. Alternatively, whole life cycle definitions are reported with specific focus on energy relevant information, which consequently puts limitations for a generic approach.

The contribution aspect of this study resides in providing an information requirement definition for each phase of the life cycle of a building in a generic manner. By this, missing information in existing BIM models is identified. Furthermore, to support future BIM model development, a novel information management method is proposed which allows tracing the use of a piece of information by each stakeholder along its life cycle.

2 METHODOLOGY

The first step encompasses the definition of information requirements of a BIM model for each respective life cycle phase (section 3). The information requirement definition is based on the authors' own experience and literature review (Liebich et al. 2011, O'Donnell et al. 2011, Becerik-Gerber et al. 2012, Volk et al. 2014, Wimmer et al. 2014, Wong & Zhou 2015). Next, existing BIM models are evaluated against the defined requirements to identify missing areas of information not covered so far (section 4).

To enhance future BIM model development a novel framework is proposed which integrates the stakeholders' views on information at different phases of the life cycle and the status of this information. The framework includes a suggestion on implementing a modular BIM model structure to support BIM use during the whole building life cycle (section 5).

3 LIFE CYCLE CENTRIC INFORMATION REQUIREMENT DEFINITION

Prior to data modeling, the development of building life cycle oriented descriptions necessitates the definition of information requirements of all stakeholders during each phase of the life cycle. For the purpose of this discussion the building life cycle phases are considered as depicted in Figure 1. It is assumed in this study that each phase comprises a closed procedure for each discipline. Also, a project may start with any of the described phases, yet may not necessarily follow the sequence in a timely order.

3.1 *Definition of the life cycle phases*

The *planning phase* encompasses all actions and tasks done prior to any real design. The tasks involved are mainly to identify the needs by consulting the building owner and users, to define a rough construction schedule and collect data about the construction site.

During the *design phase* the full building is defined in an increasing level of detail and all information required for the start of the building construction is prepared, including specifications for operation and maintenance purposes. On top of defining the shape of the building itself, the design phase also embraces the structural design, the design of the technical service system such as Heating, Ventilation and Air-Conditioning (HVAC), lighting, safety and others. During the *construction phase* the building itself is built and all necessary service equipment is installed. Applied changes are documented and, depending on completion, there may be overlaps with the following commissioning phase. In the course of *commissioning*, the building and its systems are set into operation for the first time. Systems are configured and all operation parameters are tuned according to the users' needs. *Operation and maintenance* has the longest duration compared to the other phases. During this phase the building and its systems are used on a regular basis. The facility processes are adapted to suit best the user's primary activities (British Standards Institution 2015). Repair and maintenance are necessary due to "natural degradation, disrepair, or poor operational practices" (Bailey 2002). The *refurbishment* of a building may be required, when a building and/or its systems

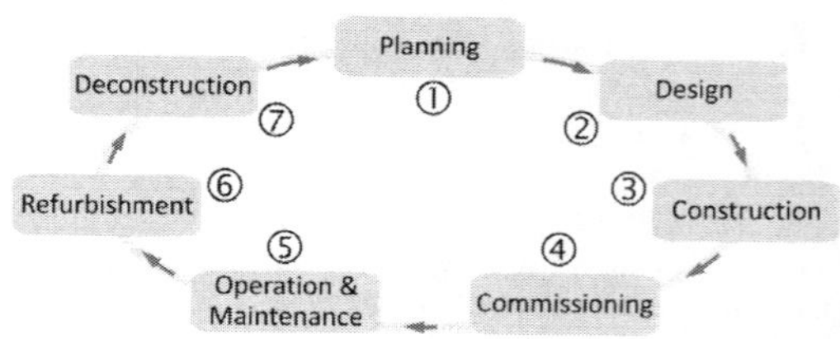

Figure 1. Life cycle phases of a building as considered in this analysis. Sequence not necessarily in timely order. (adapted from Kotaji 2003).

are prone to repeated overhaul or when major functional requirements change as defined by new users and/or owners. This requires the refurbishment of the building with probably some reiterations of the prior phases of the life cycle. At the end of its use, the building and all its elements are removed from the site during the *deconstruction phase* to be recycled, reused or disposed.

In the following, the areas of information identified are firstly described before defining their requirements over a building life cycle.

3.2 *Identified information areas*

Information for the description of three-dimensional geometry (3D geometry) forms a basis for subsequent activities such as quantity take-off calculations and needs to be apparent in BIM models. It summarizes information necessary to describe the appearance and position of building elements and equipment, as well as their aggregation in the entire building. This information is provided on different levels of detail for example to create 3D mock-ups for presentation and/or for automatically generating quantity take-offs for procurement.

Process and Project Management encompasses information needed at a respective phase for management purposes. For process management this includes information describing and defining the properties of digital tools to be used and related to the work flow. Project management information focusses on information about the related stakeholders, e.g. contract and contact data, information on the time schedule of the project, regulatory and legal information and information on the budget and costs.

Various information about *Elements and Equipment* installed needs to be included in BIM information. In particular this comprises:

- Information about elements of the enclosing surfaces such as walls, windows, slabs, doors etc.;
- Technical equipment, e.g. HVAC and automation equipment;
- Equipment for decentralized energy conversion, e.g. photovoltaic or ground heat collectors;
- Interior fit-out and furnishings, e.g. furniture, floors and electrical appliances;
- Topological information on the location in the building;
- Manufacture information like year of manufacture, model, manufacturer, material composition;
- Maintenance information e.g. contact of the service provider;
- Technical documentation with maintenance guidelines, mounting and dismounting procedure, as well as sequence of operation for technical equipment.

Global coordinates of the project, climate conditions, intended country of construction as well as the terrain topology and geology define information abbreviated in *Site Information*.

Material Property summarizes information on the materials utilized in the construction process for building elements and equipment. This may include abstract trade names as well as specifications based on physical properties. Important is the specification of toxicity and emissions created by the materials as well as whether they are recyclable and reusable for sustainable refurbishment and deconstruction.

The ultimate purpose of a building is to fulfil the needs of its *Users*. Information on requirements set by users, e.g. indoor comfort and air quality, user behavior, e.g. occupancy schedule, visual comfort and user feedback are of key interest, especially as there may be frequent changes.

Documentation of the project and processes such as energy audits and as-built status is critical, especially in case of warranty-related issues.

Sequences of operation relates to information on the intended or actual operation of the building and its systems. An example is the regularly planned rise of the hot water system's temperature above a threshold to prevail bacterial contamination.

Preservation covers specific information related to the protective status of a building and its elements and equipment. To preserve the conservational status of cultural heritage buildings, a complete documentation of the performed changes as well as their reversibility must be provided to the authorities. This is of significance considering that the BIM paradigm should also be deployed in the context of existing buildings (Volk et al. 2014).

Simulation-assisted methods are utilized throughout the whole life cycle of a building, such as for calculations of energy performance, for model-based predictive control during operation or for process simulations of the deconstruction. *Simulation Specific* includes specific information required for applying these methods. In particular for energy-relevant simulations, information on the modelling assumptions, boundary conditions and second level space boundaries (Bazjanac 2010) are included. Also, information on internal loads and specific material properties is needed, such as thermal conductivity or emissivity.

3.3 *Information requirement definition*

Table 1 presents the areas of information (column 1) required at the respective phases of the building life cycle. While some areas appear to be required for all seven phases, it does not mean that the same pieces of information are continuously required all along the life cycle. For instance, the Documentation almost only includes phase-specific contents.

Table 1. Areas of information identified for each life cycle phase (1: Planning, 2: Design, 3: Construction, 4: Commissioning, 5: Operation & Maintenance, 6: Refurbishment, 7: Deconstruction.) •: Information required for the respective phase.

	Phases						
Information requirement	1	2	3	4	5	6	7
3D Geometry	(•)	•	•	•	•	•	•
Process and Project Management	•	•	•	•	•	•	•
Elements & Equipment		•	•	•	•	•	•
Site Information	•	•	•	•	•	•	•
Material Property		•	•		•	•	•
User	•	•			•		
Documentation	•	•	•	•	•	•	•
Sequences of Operation		•		•	•		
Preservation	(•)	(•)				•	•
Simulation Specific		•			•	•	•

3.3.1 *Planning*

3D information is required only to a limited extent in this phase as the design has barely started.

The emphasis is rather on information for process and project management. The intended digital tools are also determined. For the planning, site information results in requirements for the further development process (ISO 1994).

3.3.2 *Design*

Three-dimensional geometric information needs to be covered as well as process and project information to keep timelines and goals while fulfilling initial requirements set by owners, users and regulations. The documentation of compliance to these requirements is needed. Properties of used materials are defined at this stage. Detailed information on the site is an essential input for the design. For the application of simulation-based design, simulation-based information must be defined. Information on interior fittings and furnishing as well as maintenance process is necessary to allow integrated design including all stakeholders. As an outcome of the design process, detailed information on all building elements and equipment is generated and stored as well as their sequences of operations. As the design phase may also relate to a complete redesign of a building possibly with preservation status, this information also needs to be available.

3.3.3 *Construction*

Detailed geometrical information is needed in this phase, e.g. for correct assembly, as well as information on the workflow for process management. Installed equipment needs to be technically documented and detailed information on the terrain topology and geology is needed for correct construction. Information on the installed materials as well as the construction progress needs to be provided.

3.3.4 *Commissioning*

For the commissioning phase, process and project management information is needed and the as-built geometry of the building and its equipment is required. Storing the documentation of audits and acceptance tests helps to ensure traceability in case of warranty claims. The commissioning process relies on site information and on requirements defined by its users in earlier phases.

3.3.5 *Operation and maintenance*

Necessary information involves geometry of the building and its elements and equipment, site information and the behavior and needs of the occupants. Managerial information on the process is needed to enable for example continuous improvement of the operational performance.

In particular for maintaining and operating the installed technical equipment information on its location, technical documentation and sequences of operation are required. Storing documentation of audits and compliance to regulations is needed.

3.3.6 *Refurbishment*

Detailed information on process management during refurbishment is needed as well as the as-built three-dimensional information. Specific information is necessary for the case of refurbishing cultural heritage buildings. Also, as defined above, reiterations of the previous phases may be required depending on the size of the refurbishing operation. Subsequently affiliated information may be included as well.

3.3.7 *Deconstruction*

Here, information needs are primarily related to geometry for estimating waste quantities as well as the location of critical elements and equipment. Information on the materials is also required to clarify its toxicity and the possibility to recycle or reuse. Information on the installed technical systems allows for instance to decide whether a unit may be reused.

4 EVALUATION OF EXISTING BIM MODEL STANDARDS

Existing standardized BIM models are evaluated on their capabilities to describe the aforementioned information areas (see Table 2 and textual description). It should be noted that non-standardized extensions of the models are considered insufficient

Table 2. Evaluation of existing BIM models. (+) information area covered, (•) information is covered with some limitations, (–) information is not covered.

Information requirement	IFC 4 Add 1	gbXML
3D Geometry	+	•
Process and Project Management	•	–
Building Elements and Equipment	•	•
Site Information	+	–
Material Properties	•	+
User	–	–
Documentation	–	–
Operation Schedule	–	–
Preservation	–	–
Simulation Specific	•	+

as these would create new interoperability issues. Despite that numerous BIM model standards exist, the Industry Foundation Classes version 4 Addendum 1 (IFC) (buildingSMART International Limited 2015) and the Green Building XML Schema (gbXML) (Green Building XML 2016) are evaluated here as two prominent implementations widely accepted in the field.

4.1 *Industry Foundation Classes (IFC)*

The IFC data model supports a sound basis of possibilities to describe three-dimensional geometry of objects either based on boundary representation or constructive solid modeling (Borrmann et al. 2015). The data model specifies classes to model process and project management information such as time schedules and constraints, cost information and sub-contracting (buildingSMART International Limited 2015). However, most applications using IFC do not focus on these capabilities.

A wide variety of classes for the description of building elements and equipment is provided. Limitations exist in particular for the descriptions of decentralized energy conversion devices.

Elaborate description of site relevant information is possible but weather information is missing. Material properties may be added to each object covering a wide range of applications. IFC offers some support for life cycle assessment of components e.g. Pset_EnvironmentalImpactIndicators. However, for supporting waste treatment during refurbishment and deconstruction more detailed descriptions is required on reusability and recyclability. Information describing the user behavior or comfort preferences on indoor climate is not supported up to the latest release.

Information on sequences of operation and schedules in the realm of technical equipment is not backed yet by the standard. Nonetheless, concepts are provided to model workflow processes.

An area of information not covered so far is the description of information related to the preservation status of a building. Some simulation specific information may be stored using the IFC model such as secondary space boundaries, however, information to specify the dynamic behavior is not covered, e.g. efficiency curves of boilers or pumps (O'Donnell et al. 2011).

4.2 *Green Building XML schema (gbXML)*

gbXML (Green Building XML 2016) was introduced in 2000 in order to facilitate the exchange of data between BIM software and building energy simulations. Instead of IFC, gbXML can be exported from a BIM tool and thanks to its specific features improves the connection of building energy simulation programs to the original building model. Contrary to the IFC, geometries are reduced to rectangular shapes, thereby possibly changing the original geometry. With its limitation to energy simulation related data and a less complex data scheme, it offers lower file size and is easier to implement in software (Dong et al. 2007).

Interfaces to an increasing number of tools have already been realized. The import and export process, however, is still error-prone, leading to the need for manual revision in building energy simulation programs which feature no or low visualization capabilities. HVAC information can be transferred but scheme extension is necessary to include full specifications and detailed schedules (Maile et al. 2007).

As a domain specific format, the gbXML aligns with the concept of IFC model definition for energy performance simulation. However, it is implemented in a different, more manageable format. Its schema is not intended to serve for holistically capturing a building's informational content over several life cycle phases. gbXML is preferably used for connecting simulation software to the BIM data pool.

5 INFORMATION MANAGEMENT METHOD FOR LIFE CYCLE AND STAKEHOLDER CENTRIC BIM MODEL DEFINITION

To spread BIM adoption in the building industry, the novel method "GASCEeIiL" is proposed, aiming to support a building life cycle and stakeholder centric handling of building information. The method is developed by first analysing the handling of information along its life cycle and subsequently developing the GASCEeIiL matrix. Finally, an interconnected modular BIM framework for utilizing the BIM methodology along the life cycle of a building is proposed. To differentiate clearly between the life cycle of a building and the life cycle of a piece of information, the latter is distinguished as "Information Life Cycle" (ILC) in the present section.

5.1 *Information life cycle requirement definition*

The analysis of the information handling has led to an exhaustive list of states of a piece of information along its life cycle. The ILC requirement definition comprises three exclusive phases: generation, modification and archiving. During its ILC, a piece of information is handed over between several stakeholders and may be used by different people at the same time for different purposes. For optimal handling, the purpose and the later users must be known when the data is produced, in order to deliver it in the most appropriate form.

Once available, the data is used by the stakeholders differently at different times. Five exclusive modes can be distinguished: disregarded, read only, supplemented, edited and transformed. The latest modes are three forms of modification. For supplementation new information is added to existing without changing previous contents. In editing, existing information is deleted and replaced with new contents. On the contrary, information may also remain the same in terms of contents while it undergoes a transformation to be presented under another form, better adapted to the next user.

To support data quality with unambiguous descriptions in the data entries, information structures must be defined explicitly, i.e. with a sufficiently detailed breakdown that partial data entries are avoided. This breakdown can be flexible as to handle the fluctuating levels of information occurring within a phase. Structure definitions may for instance be associated with the Levels Of Details (LODs), a measure to define the accuracy of a BIM model in five levels, termed LOD100 to LOD500 (AIA 2013). LODs help to support the increased detailing happening during the planning phase. Other phases may require adaptations such as a reduced LOD500 for operation. Using such explicit structures, supplementation would not occur anymore and could be disregarded. However, it is not realistic to do this for the moment as validated explicit data structures are not available yet.

As a result, the analysis characterized the information on its state within its own ILC and on its future use. For transparent information management, the connection between the deliverables and the team members must be further specified. This is solved in project management using responsibility assignment and design responsibility matrices. In the first one, activities are allocated in a chronological order to the corresponding stakeholders, so that at least and only one party is assigned as accountable for each activity (Project Management Institute 2008). Additionally, the design responsibility matrix increases transparency related to specific aspects of the design. With indications on the level of detail and the level of information exchanged, it defines "precise requirements as to what individual, from which relevant firm, will be responsible for the production of specific data" (East & Carrasquillo-Mangual 2013, British Standards Institution 2015). One common responsibility assignment matrix used is the RASCI method: the statuses Responsible, Accountable, Support, Counsel and Informed are used to clarify the types of involvement expected for each task (Hightower 2008, Project Management Institute 2008). The RASCI's application in the building sector provides proof of the engagement of the numerous members of the team to follow a collaborative approach in the project: while prepared by the design and construction team for approval by the owner's representative, it should involve all main stakeholders of the building life cycle, including facility managers (British Standards Institution 2015).

In conclusion, the analysis showed that four dimensions characterize a piece of information at any stage of the building's life cycle:

1. Its specific definition to the present phase i.e. its current level of detail;
2. Its connection to each stakeholder with responsibility assignments;
3. Its current status within its own life cycle, i.e. generation, modification, archiving;
4. Its intended future use by the stakeholders.

The GASCEeIiL method is suggested as a novel comprehensive information management method with a combined view of these four dimensions.

5.2 *A novel information management method*

The GASCEeIiL method may be used to depict the evolution of a particular piece of information along its ILC, as exemplarily shown in Table 3 with a matrix excerpt for a door's geometrical characteristics. The first dimension represents the ILC-specific activities on the piece of information and appears in rows. The second dimension relates to the involved stakeholders and is modeled in columns. The present state of the piece of information and its intended use are described by the capital letters *GASCEIL* (third dimension) and the lower case letters *ei* (fourth dimension), respectively.

While the duties of the *Accountable*, *Support* and *Counsel* remain unmodified, other RASCI roles are adapted to better respect the dimensions

Table 3. The GASCEeIiL information management matrix displays the relation between the stakeholders and the data along its life cycle. Example for geometrical characteristics of a door. The relation is given as a combination of the respective characters: G—Generation, A—Accountable, S—Support, C—Counsel, E—Editing, e—editable, I—Information preparation, i—informational, L—Long-term storage.

Information's ILC-specific definition	Level of Detail LOD	Architect	Structural eng.	BIM Manager	Contractor	Manufacturer	FM	Owner/ End-user
(Design) Simple description of requirements		Ge	Si	A	–	–	C	C
(Design) Specified overall performance	100	Ee	Si	A	–	–	–	–
(Design) prescribed generic products performance	200	Ee	Si	A	–	–	–	–
(Design) design characteristics as specifications for tender	200	I	–	A	i	i	–	–
(Design) Tenderers' information: producer, door type,...	300	Le	i	–	GAi	Ci	–	–
(Design) Acceptance of tenderer's information	300	ELi	Si	A	Se	Ci	–	–
(Construction) prescribed manufacturer products for construction plans	400	Si	Si	A	Ee	Si	–	–
(Construction) Changed characteristics (depending on procurement)	500	Li	Ci	A	Ee	Si	–	–
(Commissioning) key properties saved into an asset database	500^{-}	Si	i	A	I	Ci	e	L
(O&M) updated characteristics	500^{-}	–	–	A	Si	Si	Ee	L

described. The RASCI's "Responsible" still applies to the unique stakeholder in charge of the current ILC-activity. It is however detailed to differentiate between the data-related specific activity for which the responsibility apply: the *Generation*—for the data's very first creation or its supplementation for non-explicit data structures—the *Editing*—for the data's modification from an existing state to an over-written state—the *Information preparation*—for preparing the data to suit a new output format without modification of its contents—and the *Long-term storage*—for the data archiving.

As it is not known for sure to which point in time the data will be retrieved again and by whom, the only specification on the archive's output format at the time of archiving is to ensure that the data can be accessed again and stays readable over time. This may be facilitated by using open standards, such as IFC.

To ensure that the data is produced in a form facilitating its future use, the GASCEeIiL indicates whether the data is required in an *informational* status—read-only form—or as *editable* data—expected to be modified in a later step by the assigned stakeholder. The Responsible should then get into direct contact with the future stakeholder needing the data in a later phase. Since the Accountable's task also implies validating the quality of the data entered, it is assumed that s/he always gets informed of the data evolution at a later stage; however s/he is not responsible for performing the necessary changes.

It must be noted that the method only depicts the responsibilities towards the operative data handling. Therefore, although the owner remains accountable for the data contents and its adequacy to the contracted requirements, this responsibility does not fall into the scope of the GASCEeIiL. In this way Table 3 depicts a case where the owner has enrolled a BIM-Manager to manage the information quality, for the data archiving.

5.3 *Interconnected modular BIM model framework*

The study has highlighted that essential information is not yet available in existing BIM models. To cope with this, a framework is proposed relying on the following process. First, before starting each phase, the information requirements must be defined. Then, each information requirement identified must be associated to a GASCEeIiL matrix to clarify the exchanges and the stakeholders' responsibilities towards it—which highly depend on the chosen project team structure. The GASCEeIiL therefore gives the necessary information for the data transfer at the interface. Finally, the information handover is managed using the interconnected modular BIM Model used as follows.

To manage the information in the most adapted manner for the specific needs of each building life cycle phase, a modular BIM model may be defined as a set of interconnected BIM models. Building life cycle phase-specific BIM models are created, including only information directly relevant to this phase or subject to revision because of their dependency on the stated data. Within each building life cycle phase, the information is managed by the stakeholders considering the *GASCEI* statuses defined earlier. At the interface between two phases, one part of the information contained in the phase-BIM model is transferred to the next phase considering the requirements on the future handling mode (statuses *e* and *i*), while the other part is archived for use at a later stage (status *L*), as depicted on Figure 2. All data archived is stored on a common repository, accessible by all stakeholders involved in the project, at any phase of the building life cycle. The archived data should be directly reusable in later phases by uploading predefined sets of information e.g. Model View Definitions (MVDs) (buildingSMART 2016) into the corresponding phase-BIM model. A set may either be a snapshot of a complete BIM model or only constitute of a part of the model. For data safety, it is advised to store a snapshot of the complete model at each phase end, on top of the specifically archived sets.

There are two main reasons for *long-term storage*. On one side, it may be wished to make a snapshot of the current state of the information. The predefined set is then used as comparison reference or as template in a new cycle of the same phase. On the other hand, it may not be expected

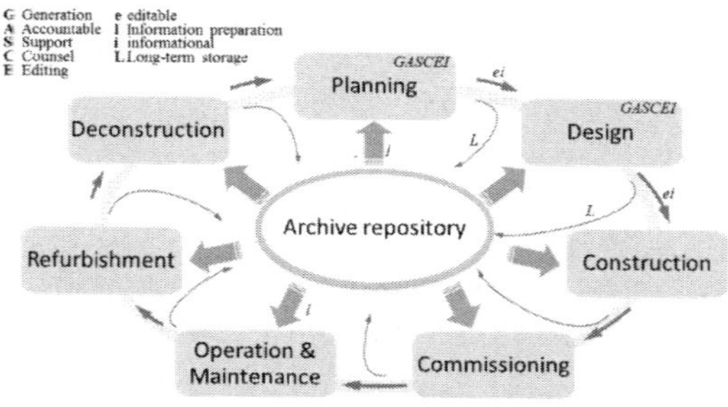

Figure 2. Schema of the interconnected modular BIM model framework: information handled in the phase-specific sub-models is transferred to other sub-models and to the archive repository according to the GASCEeIiL statuses.

to use the information for a long time. For instance documentation of the construction progress may only be of interest in case of later liability and warranty claims. To limit the size of information exchanged, such unused data is archived as a predefined set. It is later uploaded to complement the current phase BIM model. In this case, inconsistent data between the repository version and the current BIM model must be highlighted for the user to decide on which is the up-to-date information.

6 DISCUSSION

BIM model definition is never complete and information requirements change over time as new technologies evolve. The authors understand the definitions made in this work as a non-permanent and future iterations are required.

Some problems appear when implementing a modular BIM framework. From the necessity of providing a piece of information to various stakeholders during different life cycle phases, different storage and exchange methods are required. Flexible data bases with the capability to grow over time, or adaptable exchange and conversion formats can be useful remedies in this context. Implementation issues of BIM models such as versioning, exchange scenarios and model repositories are important questions but beyond the scope of this work.

An easier information management over the whole life cycle is expected as a benefit of applying the GASCEeIiL methodology. The methodology reveals in a transparent manner how and when a piece of information is needed by a stakeholder. Its capabilities help structuring exchange requirements and can assist in the definition of MVDs.

Even if information in a BIM model is assumed to be complete issues related to faulty design remain. This is because these depend on external factors. For instance, the requirements on the BIM model cannot ensure that the HVAC systems are oversized by the design engineer.

7 CONCLUSION

In this work the information requirements for a building over the whole life cycle are defined. Information areas of interest are identified and the capabilities of the latest release of the Industry Foundation Classes (buildingSMART International Limited 2015) and of Green Building XML schema (Green Building XML 2016) are evaluated on their backing of the identified information areas.

Missing areas of information identified are technical details on building equipment such as efficiency curves and support of decentralized energy conversion devices such as photovoltaic panels. With regards to the building users missing support of describing user behavior, thermal and visual comfort and air quality is found. In many cases site specific information needs to be supplemented as no backing for weather and climate data can be provided. For further adoption of the BIM methodology throughout the industry, information about the preservation status needs to be included.

In addition to the information requirement definition a novel information management method termed "GASCEeIiL" is proposed which allows tracing the use of a piece of information by each stakeholder along its life cycle and supports future BIM model development by providing a life cycle centric view on information.

Future work is intended to extend the number of BIM models evaluated and the application of the GASCEeIiL methodology to a practical test case.

An interesting development in pursuing a multi-model approach is the connection of established vocabularies following the linked building data paradigm as proposed by Pauwels & Terkaj (2016).

The results of the present study show that a multi-model approach should be developed and followed, which is promising in fulfilling the varying information requirements of each stage of the buildings life cycle.

ACKNOWLEDGEMENTS

Parts of this research have been performed as part of the Energie Campus Nürnberg and supported by funding through the "Aufbruch Bayern (Bavaria on the move)" initiative of the state of Bavaria.

REFERENCES

AIA 2013: *AIA Contract Document G202-2013: Building Information Modeling Protocol Form*: American Institute of Architects.

Bailey, R. 2002: *Unsustainable Promises: Not-So-Green Architecture*.

Bazjanac, V. 2010. Space boundary requirements for modeling of building geometry for energy and other performance simulation. In: *CIB W78: 27th International Conference*. Cairo, Egypt.

Becerik-Gerber, B., Jazizadeh, F., Li, N. & Calis, G. 2012. Application Areas and Data Requirements for BIM-Enabled Facilities Management. *Journal of Construction Engineering and Management* 138 (3): 431–442.

Borrmann, A., König, M., Koch, C. & Beetz, J. (eds) 2015: *Building Information Modeling: Technologische Grundlagen und industrielle Praxis*. Wiesbaden: Springer Fachmedien Wiesbaden.

British Standards Institution 2015 *BS 8536-1: Briefing for design and construction—Part 1: Code of practice for facilities management (Buildings infrastructure)*: BSI Standards Limited.

buildingSMART 2016: *Model View Definitions* [online]. Available from: http://www.buildingsmart-tech.org/specifications/ifc-view-definition/summary [Accessed 23 Mar 2016].

buildingSMART International Limited 2015: *Industry Foundation Classes (IFC) 4 Add 1* [online]. Available from: http://www.buildingsmart-tech.org/ifc/IFC4/Add1 [Accessed 23 Mar 2016].

Cerovsek, T. 2011. A review and outlook for a 'Building Information Model' (BIM): A multi-standpoint framework for technological development. *Advanced Engineering Informatics* 25 (2): 224–244.

Dong, B., Lam, K.P., Huang, Y.C. & Dobby, G.M. 2007. A comparative study of the IFC and gbXML informational infrastructures for data exchange in computational design support environments. *Proceedings: Building Simulation*: 1530–1537.

East, B. & Carrasquillo-Mangual, M. 2013: *The COBie Guide: a commentary to the NBIMS-US COBie standard* (Release 3).

Eastman, C.M., Teicholz, P., Sacks, R. & Liston, K. 2011: *BIM handbook: A guide to building information modeling for owners, managers, designers, engineers and contractors*. 2nd ed. Hoboken, N.J.: Wiley.

Green Building XML 2016: *gbXML: Current Schema* [online]. Available from: http://gbxml.org/Schema_Current_GreenBuildingXML_gbXML.

Gupta, A., Cemesova, A., Hopfe, C.J., Rezgui, Y. & Sweet, T. 2014. A conceptual framework to support solar PV simulation using an open-BIM data exchange standard. *Automation in Construction* 37: 166–181.

Hightower, R. 2008: *Internal Controls Policies and Procedures*. New York, NY: John Wiley & Sons.

International Organization for Standardization 1994 *ISO 9699: Performance standards in building—Checklist for briefing—Contents of brief building design*. Switzerland, Genève.

Kotaji, S. 2003: *Life-cycle assessment in building and construction: A state-of-the-art report*. Pensacola, FL: Society of Environmental Toxicology and Chemistry (SETAC).

Liebich, T., Stuhlmacher, K., Katranuschkov, P., Guruz, R., Nisbert, N., Kaiser, J., Hensel, B., Zellner, R., Laine, T. & Geißler, M.C. 2011: *HESMOS Deliverable D2.1: BIM Enhancement Specification*. Brussels, Belgium: © HESMOS Consortium.

Maile, T., Fischer, M. & Bazjanac, V. 2007: *Building Energy Performance Simulation Tools—a Life-Cycle and Interoperable Perspective*. CIFE Working Paper. Stanford University.

O'Donnell, J., See, R., Rose, C., Maile, T., Bazjanac, V. & Haves, P. 2011. SimModel: A domain data model for whole building energy simulation. In: *Proceedings of Building Simulation 2011*.

Pauwels, P. & Terkaj, W. 2016. EXPRESS to OWL for construction industry: Towards a recommendable and usable ifcOWL ontology. *Automation in Construction* 63: 100–133.

Project Management Institute 2008: *A guide to the project management body of knowledge: (PMBOK guide)*. 4th ed. Newton Square, Pa.: Project Management Institute.

Volk, R., Stengel, J. & Schultmann, F. 2014. Building Information Modeling (BIM) for existing buildings—Literature review and future needs. *Automation in Construction* 38: 109–127.

Wimmer, R., Maile, T., O'Donnell, J.T., Cao, J. & van Treeck, C. 2014. Data-Requirements Specification to Support BIM-Based HVAC-Definitions in Modelica. In: C. van Treeck & D. Müller (eds), *Proceedings of BauSIM 2014: 5th German-Austrian Conference of the International Building Performance Simulation Association*. Human-centred building(s): 99–107.

Wong, J.K.W. & Zhou, J. 2015. Enhancing environmental sustainability over building life cycles through green BIM: A review. *Automation in Construction* 57: 156–165.

eWork and eBusiness in Architecture, Engineering and Construction – Christodoulou & Scherer (Eds)
 ISBN 978-1-138-03280-4

A flexible and scalable approach to building monitoring and diagnostics

M. Schuss, S. Glawischnig & A. Mahdavi
TU Wien, Austria

ABSTRACT: This paper presents a flexible and easily adaptable building monitoring and diagnostics approach that combines an IoT-influenced hardware setup with a web based application design. Small data concentrators and logger modules implemented on Arduino YUN development boards are used to collect data in a distributed way over the Internet. This data could be synced with a data repository and analyzed in a modular web application. The typically used hardware and a general system setup is presented in detailed, followed by a description of developed modules used in the web based diagnostics application. Finally, we demonstrate the flexibility of the approach in use cases of real project related implementations.

Keywords: building monitoring, building performance, building diagnostic, data analysis, embedded systems

1 INTRODUCTION

Systematic monitoring of buildings plays an important role for optimized operation of the existing building stock. The evaluation of monitored data leads to a better understanding of the building and the resulting actual performance. A detailed analysis of all relevant physical parameters could be beneficial for optimizing the operation of the respective building. Resulting information can be used to plan hardware related changes or could target software related control system adaptions. All improvements toward optimized building management and operation require efficient and timely data monitoring from multiple sources. Various scientific projects focus on manufacturer independent solutions for data access and the latest developments of IoT solutions give them new options. For instance, the open source MOST project (MOST 2016, Zach and Mahdavi 2012) provides a platform for data from different sources and enables the user to analyze the data or export it after preprocessing. In parallel, a development focuses on combined control of systems with programming possibilities for users. Those features are available in different software applications such as openHAB (openHAB 2016) and Fhem (Fhem 2016).

Experiences from previous research projects underline the need for an easily adaptable, low-cost, and easy to set up monitoring infrastructure that could provide pertinent data for the system modeling and performance evaluation. This is not exhaustively covered by the aforementioned approaches. Especially in the case of MOST and openHAB, the adaption of the JAVA code was not practical. This led to the development of a new monitoring concept, which targeted fast adaption possibilities related to the project needs.

The increasing availability of small powerful embedded system development boards (e.g., Arduino) and mini computers (e.g., raspberry or beagleboard) facilitates the conception and implementation of a cost-efficient infrastructure for data collection and building diagnostics within a reasonable framework in terms of time and effort.

For the purposes of the present approach, the Arduino Yún (Arduino 2016) was selected as a general platform to create a data logger that obtains data from wireless sensors, stores it locally, and is able to sync it with a central data repository.

2 MONITORING SYSTEM SETUP

Our concept for an easily adaptable, low-cost. and easy to set up monitoring infrastructure is based on a cloud solution with a number of data collectors and loggers connected to the Internet. This cloud-based solution is illustrated in Figure 1. Communication between the different hardware components, as well the software components is based on a concept that uses RESTful data interfaces/web services.

2.1 *Flexible hardware for data collection and logging*

The Arduino Yún (Arduino 2016) was selected as a platform to create an embedded system based

Figure 1. The monitoring system combines data from data collectors and web services in the cloud.

data collector and data logger. In contrast to other embedded developer boards, the Yún is a combination of two embedded systems that are internally bridged. One is a classical, very stable microcontroller system based on an Atmel ATmega32U4 processer running with 16MHz that offers implementation possibilities with precise real time behavior. The other one is an Atheros AR9331 based system with 400MHz that uses OpenWRT (OpenWRT 2016), a small scale Linux operating system. This Linux system could also be used in a number of commercial Linux based Ethernet/WiFI Routers. This enables us to adapt the system with additional packages for hardware or applications depending on the use cases.

Figure 2 shows the board together with the attached Enocean USB300DB-transmitter.

As it is visible in Figures 2 and 3, a classical Ethernet connection as well as a WiFi Interface are included in the Arduino Yún and thus offer easy web integration.

The classic Arduino microcontroller component of the YUN could be programmed in C with the related Arduino IDE-software. This enables us to integrate analogue and digital low cost sensors directly via the on board inputs as well as data access in our logging routines/scripts. The board offers in general 12 analogue inputs and up to 20 digital inputs, depending on the number of used analogue inputs. For example, up to 12 analogue inputs can be used for a non-intrusive electric energy monitoring with CT-Sensors similar to the solution of open energy monitoring project.

Until now we used the developed data concentrator and logger for sensors that are communicating wired or wireless and are located in different areas of buildings. A general implementation setup with commonly applied sensors is illustrated in Figure 3. For indoor parameters such as temperature, humidity, CO_2-concentration, window/door positions, and local electricity metering, standards of the shelf Enocean sensors are used together with the Enocean USB-300DB—transmitter connected to the USB of the YUN. A detailed list of those sensors is included in Table 1.

Figure 2. Data concentrator and logger using an Arduino Yún developer board and Enocean USB Stick.

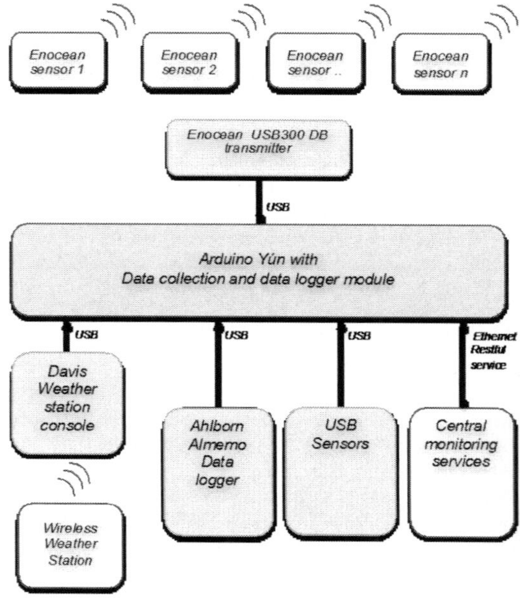

Figure 3. Typical system setup of an Arduino Yún based data collection and local data logging implementation.

For precise long term thermal comfort measurements regarding ISO 7730 standard a desk measurement station (Figure 4) was developed. The unit uses a precise anemometer (SWEMA 03) together with two accurate digital temperature/humidity sensors (Sensirion SHT 75). One of the sensors is embedded in a black bulb for glob temperature measurement. The station uses a similar setup such

Table 1. Standard Enocean sensors used for the flexible monitoring approach.

Enocean wireless sensor	
PRESSAC—Enocean CO2: solar powered room sensor measuring CO_2 levels, temperature, and humidity	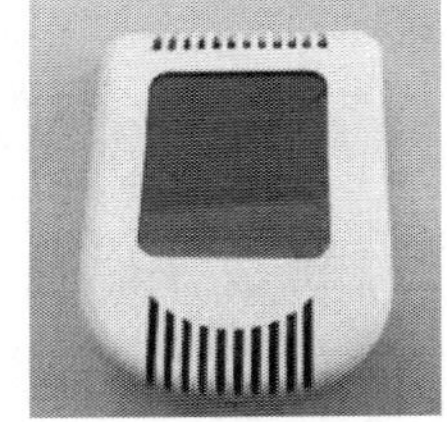
Thermokon—SR04 CO2 rH: room temperature, humidity and CO_2 sensor	
Thermokon—SR-MDS Solar Motion/occupancy and brightness sensor	
Thermokon—SRW01: window contact sensor	
Thermokon—SR65 TF: contact temperature sensor	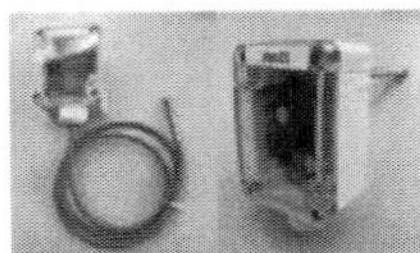
Thermokon—SR65 TF: Adapted for glob temperature measurement	
PRESSAC—Enocean single current clamp v1: single phase current measurement	
Eltako—FWZ61: single phase energy meter	
Enocean—STM110: Enocean standard transmission module—is used to integrate sensor with classical 0 to 10 V or 0 to 20 mA outputs.	

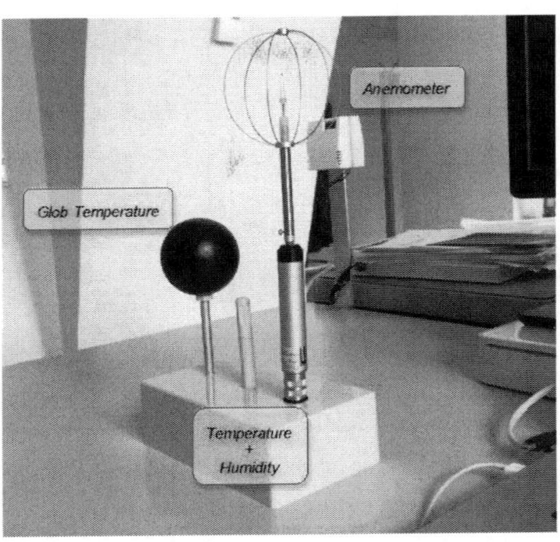

Figure 4. Thermal comfort station for long-term measurement on desk level.

as our other data loggers—an Arduino YUN—and is connected to the Internet.

2.2 *Software modules for data collection and logging*

Sensed data is collected and locally stored with device related connection modules. The developed modules support different storage possibilities such as CVS files, SQLite, and MySQL databases.

Figure 5 illustrates the system related parts of the Arduino YUN and the developed software components. Data collection modules for Enocean, Ahlborn, Swema and Davis have been developed and implemented. Enocean is a common wireless protocol for sensors and actuators in building automation at field level in commercial setups. The development of semi-professional low-cost do-it-yourself home automation solutions makes it convenient to additionally integrate other wireless protocols with a 433/868 MHz RF connector in the future. As a general connection to existing building management systems, an integration of a BACnet connector that uses the Python Library BACpypes (Bacpypes 2016) can be seen as an option for the future.

In addition to data collection and storing modules a modular web based user interface is developed for the devices.

3 WEB BASED USER INTERFACE

For administration, visualization, data analysis, and data export a flexible graphical user interface was

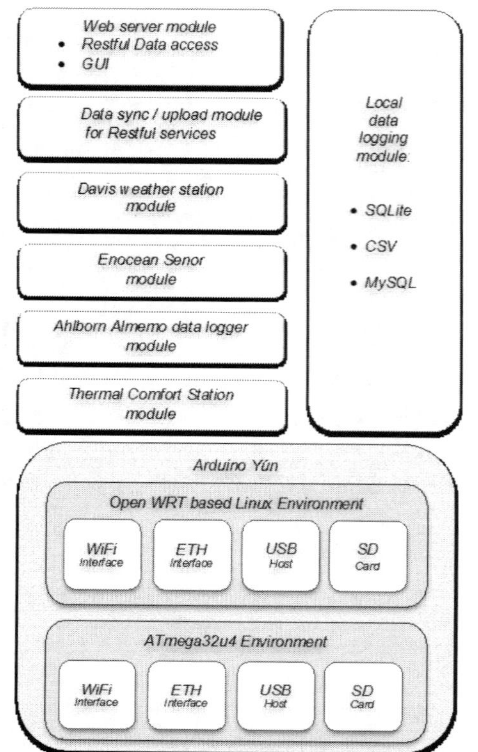

Figure 5. System components of the Arduino Yún based data collector and logger.

developed. Main components are programmed in PHP and executed on the sever side. Resource intensive calculations for data analysis are implemented as java scripts and are executed on client side to minimize the resulting loads on the server side. This concept makes it is possible to install the interface on different systems in terms of processing power.

Currently our developed user interface for monitoring applications is applied to two different types of hardware. It is practically used in actual projects and the related monitoring use cases. The first one is a classic web installation. It uses an apache web server on a powerful machine and integrates the data from different data repositories and sources into one central application. Necessary monitoring data requests use a typical RESTful communication. The requirement for integration of a new data source is the accessibility via web with a RESTful data interface that returns the data in CSV-format. Specifically, a simple module that could be implemented as a PHP script has to be programmed. This script has to convert raw data from different streams into a uniform data representation for all other modules.

The second installation case of our developed web based interface was the integration into our data concentrator and logger units. For this purpose we used the integrated webserver of the Arduino Yún with additional packages for PHP and a session management. This enabled us to provide a similar interface for all our monitoring system parts. It is possible to reach the local user interface via Ethernet or directly connect to a wireless network that is operated by the logger.

An overview of main user interface components together with related components from the RESTful data service is shown in Figure 6.

A responsive webpage design was used for the graphical interface and makes the usage on different devices easier. The navigation menu block reduces to a drop down menu on small devices. Figure 7 shows an example with the expanded navigation block on the right side of the frame.

A very simplified content management system approach was integrated in the interface. In detail we decided to use a concept that is based on a folder structure and related PHP-files with page content instead of a classic resource intensive concept with a database. Figure 8 shows a typical folder structure on the left side and the dynamically generated navigation block on the right side.

First modules for data illustration and visualization are implemented in PHP and java scripts. These modules can be used in the related PHP-files of the content management system. Figure 9 shows a typical dashboard block that is generated

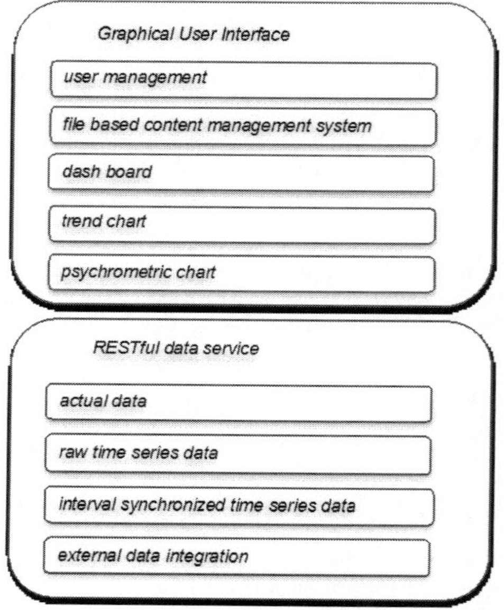

Figure 6. Components of the Graphical User Interface and the RESTful data service.

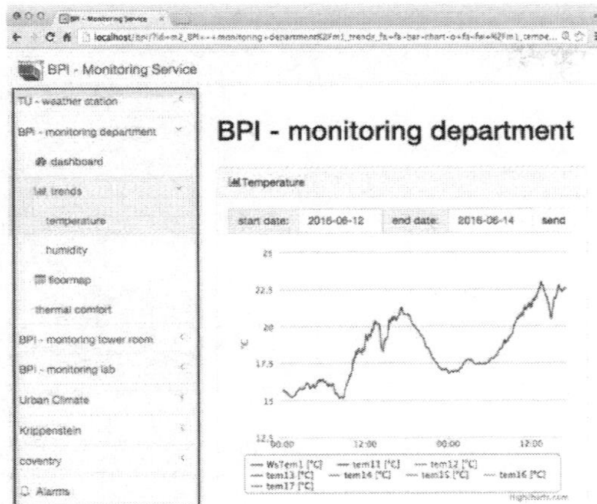

Figure 7. Graphical User Interface—example with expanded navigation (red box) on the left side.

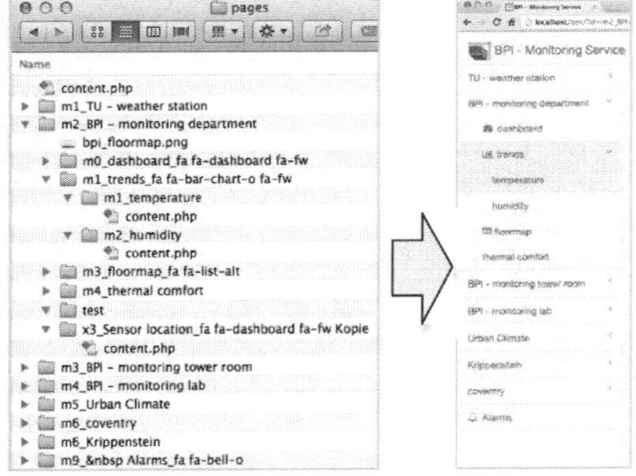

Figure 8. Graphical User Interface with dynamic navigation menu based on folder structure/names.

temperature VIP RS

T_room1-local:	29.9 °C (07 mins ago)
T_room2-local:	26.5 °C (07 mins ago)
T_VIP_RS_1-local:	19.7 °C (07 mins ago)
T_VIP_RS_2-local:	19.6 °C (07 mins ago)
T_VIP_RS_C-local:	22.6 °C (07 mins ago)
T_VF-local:	22 °C (07 mins ago)
T_out-local:	27.3 °C (07 mins ago)

Figure 9. Graphical User Interface—dashboard example.

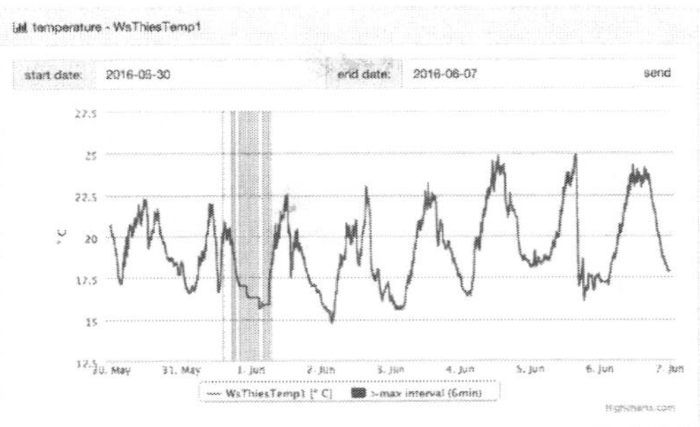

Figure 10. Graphical User Interface—trend plot example with marked gaps.

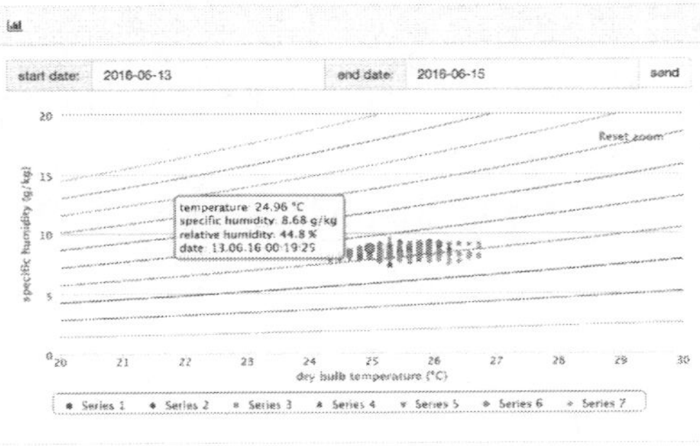

Figure 11. Graphical User Interface—interactive psychrometric chart.

by a function based on a list of sensor-IDs. These blocks also include direct links to trend chart plots (see Figure 10). The Highchart (Highcharts 2016) java script library was used for the interactive integration of different plot types.

The gap-marking option of the trend can be used to identify monitoring gaps as shown in Figure 10.

It is possible to generate classic psychometric charts as shown in Figure 11. The deployed interactive plot library enables the user to zoom in and also check the data by selection of a specific measurement point and the generated information box.

A combination of a standard html image-map and sensor related info boxes enable the user to generate interactive floor maps or sections for the web pages as presented in Figure 12.

4 CASE STUDIES AND MONTITORING INSTALATIONS

The following monitoring implementations illustrate the flexibility of the described system structure and the web based approach.

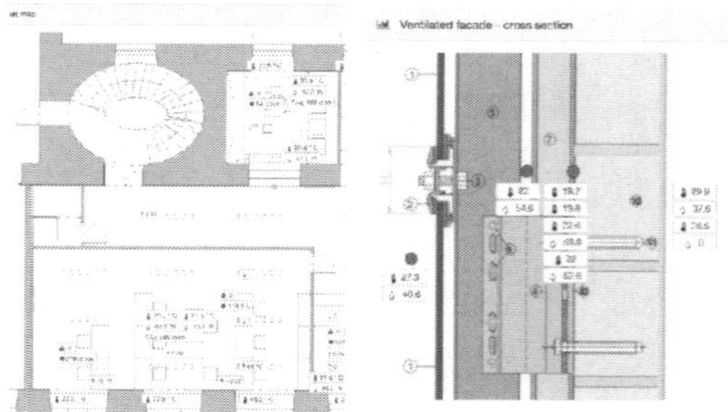

Figure 12. Graphical User Interface—interactive floor map and section plot with sensor icons and actual values.

Figure 13. External view of Krippenstein cable car top station.

4.1 *Dachstein–krippenstein*

The Dachstein–Krippenstein case study focuses on indoor thermal comfort and the evaluation of a load balancing approach together with the discussion of possible indoor climate fluctuations caused by the load balancing influence of the electric heating system. The main target of this long-term monitoring campaign is the evaluation of the thermal comfort independent of the building automation system. The external view of the building is show in Figure 13. Figure 14 shows an internal view of the monitored restaurant area.

Wireless multi sensors for temperature, humidity, and carbon dioxide concentration were installed in a restaurant and seminar area located on the ground and first floor of the building (see Figure 15).

It is obvious that this is not a typical building. It has very specific characteristic and faces extreme outside conditions. However, latest adaptions with a new electric infrared heating and a load balancing render it an interesting use case. Changes related to the usage of the building made it necessary to refurbish and to adapt the building accordingly. The original building design included a large hotel area that is no longer in use. The new concept involved an electric heating system as the old water based heating system was no longer viable.

Parallel to the monitoring of the restaurant, we also started (in February 2016) with the monitoring of the ventilation system.

Figure 14. Internal view of Krippenstein cable car top station (restaurant area).

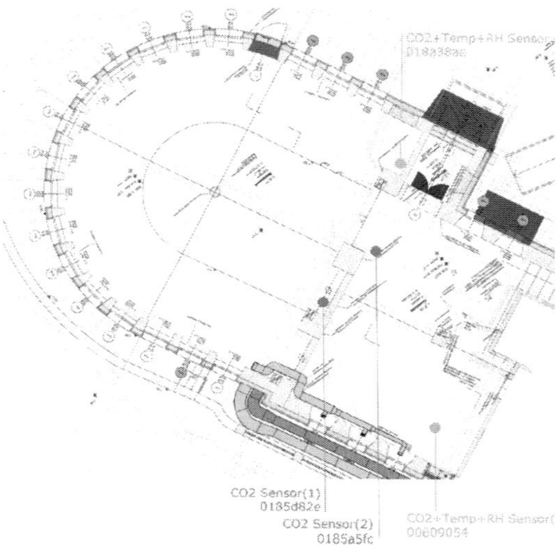

Figure 15. Restaurant area floor map with locations of the multi sensors.

First results will be available after the next heating period.

4.2 *FP7 Project: campus 21- ERI BUILDING*

The EU-FP7 project "Control & Automation Management of Buildings & Public Spaces in the 21st Century" or CAMPUS 21 explored the potential of integrated security, control, and building management software systems for improving the energy efficiency of buildings (CAMPUS 2011). The project involved participants from both academic institutions and industry and was supported by the European Union (EU 2012).

Within this project, a detailed monitoring for a selected open office was applied in one of the demonstration buildings at the University in Cork,

Figure 16. Internal view of the monitored test room 1.23 at the ERI building.

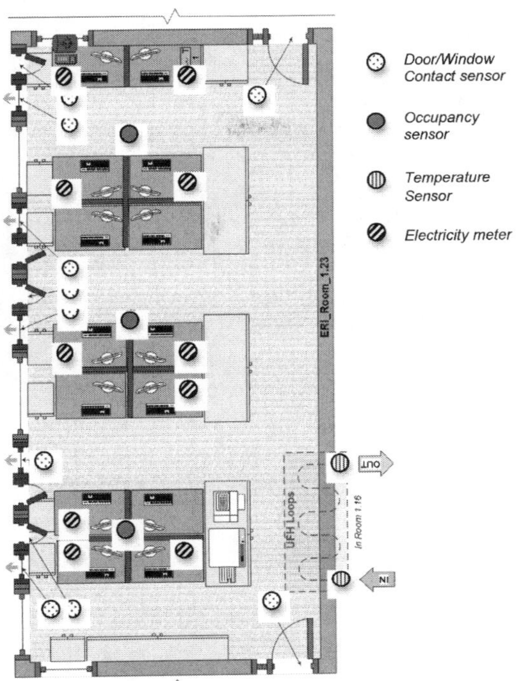

Figure 17. Layout and sensor locations of the living laboratory room 1.23 (ERI building).

Irland (Mahdavi et al. 2014). The Environmental Research Building (Figure 16) is situated in the west of Cork city, on the Lee Road. The ERI is not part of the main campus but provides a demonstration site with renewable energy sources such as solar thermal and geothermal systems. In addition, it is supplied with gas for hot water boilers and electricity from the national grid for lighting and other electro-mechanical systems. The building was inaugurated in 2006.

For the development of model based predictive control approach, an open office room (Figure 16) was adapted and used as a living laboratory (Mahdavi et al. 2014). The room was intensively monitored for more than a year to develop the needed room and occupancy prediction modes. Therefore, additional sensors were installed as illustrated in Figure 17.

A Yun based data collector and local logger was used to collect additional data from the wireless sensors and to insert them live via Internet into a central project data warehouse. This data was then hourly interpolated and used in the context of a predictive control approach. The resulting control commands were communicated via Internet to the building. The project demonstrated that a real implementation that focuses on optimized building operation is realistic and could help to reduce the building energy demand by adapting predefined building control schedules.

5 CONCLUSION

This contribution showed, how low-cost hardware components can be used to develop efficient, distributed, real-time building monitoring systems. The locally stored data can be synchronized via standard and custom interfaces (e.g. RESTful implementations) with a monitoring cloud. The data is thus easily accessible for various users and can be embedded into a wide variety of building optimization use cases. Furthermore, a flexible content management system can be used to administrate the connected devices and to analyze the recorded data.

ACKNOWLEDGEMENT

The research presented in this paper was supported in part by a fund from Seventh Framework Programme—ICT "CAMPUS 21" (Project-Nr: 285729).

REFERENCES

Arduino 2016, Arduino Yún, http://www.arduino.cc/en/Main/ArduinoBoardYun, accessed Mai 2016.

Bacpypes 2016, BACpypes library for building BACnet applications using Python, http://bacpypes.sourceforge.net/, accessed Mai 2015.

CAMPUS 2011. Control and Automation Management of Buildings and Public Spaces in the 21st Century. http://zuse.ucc.ie/campus21/.

EU 2012. FP7—ICT—Information and Communication Technologies. http://cordis.europa.eu/fp7 /ict/.

Fhem 2016, Fhem—A single user interface to all your systems, http://fhem.de/fhem.html, accessed Mai 2016.

Highcharts 2016, Highcharts: Interactive JavaScript charts for your webpage, http://www.highcharts.com accessed: Mai.2016.

Mahdavi A., Schuss M., Menzel K., & Browne D. Realization of ICT potential in improving the energy efficiency of buildings: The CAMPUS 21 project. In Architecture, Engineering and Construction, G. Gudnason, R. Scherer et al.; Taylor & Francis, (2012), ISBN: 978-0-415-62128-1; S. 131–137.

OpenHAB 2016, OpenHAB—A single user interface to all your systems, http://www.openhab.org/, accessed Mai 2016.

Zach R. & Mahdavi A. 2012. MOST-designing a vendor and technology independent toolkit for building monitoring, data preprocessing, and visualization. In *Proceedings—First International Conference on Architecture and Urban Design.* Vol. 1.

eWork and eBusiness in Architecture, Engineering and Construction – Christodoulou & Scherer (Eds)

Acquisition and processing of input data for building certification: An approach to increase the reproducibility of energy certificates

U. Pont, O. Proskurnina, M. Taheri & A. Mahdavi
Department of Building Physics and Building Ecology, TU Wien, Vienna, Austria

B. Sommer, G. Nawara & G. Adam
Energy Design, University of Applied Arts Vienna, Vienna, Austria

ABSTRACT: One of the key aspects of the European Building Performance Directives (EPBD) of 2002 and 2010 was the definition of the obligation to issue building energy certificates for the majority of existing and all to-be-constructed buildings. EU-member states were required to reflect this obligation into their national legal frameworks and derivative guidelines. Thereby, the primary intentions were to: i. define calculation rules for certain key energy and thermal performance indicators (KPI) for buildings (in Austria a normative monthly method was formulated to derive buildings' heating demand (HWB) and related indicators); ii. facilitate an easy-to-understand comparison between different buildings for tenants and owners via ranking of such KPIs (illustrated as energy-performance-class figures); iii. provide clear and well-defined evaluation schemes for minimum requirements of the thermal quality of buildings and for public subsidy schemes (for instance grants for realizing efficient retrofit strategies). However, since the legal implementation of energy certificates in Austria several problems have emerged: First, the calculation guidelines and corresponding standards changed several times, including a change in the heating-demand threshold values underlying different KPI classes. This renders comparison between energy certificates issued in different years difficult. Furthermore, guidelines entail a number of vague formulations and miss some relevant definitions. Thus, Issuers of energy certificates "enjoy" a high degree of interpretative freedom, as the resulting energy certificates do not necessarily reveal the underlying assumptions. This obviously renders the targeted comparability of buildings rather difficult. Moreover, the expected quality assurance regarding thermal building performance levels cannot be satisfactorily achieved given such uncertainties. This contribution reports on recent efforts and progress of an ongoing research project (EDEN), which addresses the above issues. The project aims to define rigorous procedures for comprehensive documentation of input data used in energy certificates. The primary objective is to achieve, thus, more transparent and reproducible energy certificates. The paper documents the major steps involved in the implementation of the project as follows: i. Identification of uncertainties concerning the data derived from buildings' geometry and topological circumstances (e.g., neighboring buildings, surrounding obstructions); ii. Identification of uncertainties concerning physical attributes of the buildings' envelope as well as other input assumptions concerning building usage and operation; iii. Extensive test series (and comparative outcome analyses) with professional and semi-professional participants, who were requested to independently perform energy certificate calculations for the same set of buildings; iv. Sensitivity analysis based on comprehensive test series to identify and document input assumptions with high and low impact on the values of the final KPIs.

1 INTRODUCTION

It is a well-known fact that the built environment contributes significantly to both global energy demand and emissions. As a reaction, the European Union started an incentive to establish building energy certification about 15 years ago (EBPD 2002, EBPD 2010): Each European state was obliged to implement building energy certification in their national legislation. Correspondingly, energy certificates are stipulated by law in the majority of EU-member states since 2006. In Austria this was realized via a law (the so called "Energieausweisvorlagegesetz", EAVG 2006), which was later adapted and amended in 2012 (EAVG 2012). Additionally, a set of rules and guidelines was developed, and efforts were started to harmonize the legislation of the different Austrian federal states (OIB 2015a, 2015b). Furthermore, national standardization was adapted to provide a calculation method for energy certificates that is in line with European standards (ASI 2016, ISO 2008). The guidelines encompass rules for calculation, threshold values for thermal quality of building components and building-related key performance indicators. Moreover, they try to bridge gaps in the input data by offering default values for existing building constructions that are not properly documented. However, the existing guidelines do not comprehensively describe the input data collection process and the derivation of substantial key performance indicators. Moreover, practitioners do not share a common understanding of the approach to energy certificates, resulting in different practical approaches and different interpretation of the calculation guidelines. This leads to prevailing uncertainties regarding input data and corresponding assumptions. The explanatory power of building energy certification is thus severely reduced by these obstacles. It can be regularly observed that different energy certificate issuers obtain in different results for the same building, even when using of the same building documentation. Under the assumption that professional issuers do not deliberately make mistakes, this circumstance can be regarded highly critical. Considering that building permits and subsidy grants often depend on the results of building energy certification, the current practice has to be critically questioned. Furthermore, there is an urgent need to study and understand the differences in results, to identify critical input data, and to develop methods that help in ensuring reproducibility of energy certificates.

The EDEN-project addresses the abovementioned issues. Thereby a set of objectives were defined which are pursued:

- Exploring the sensitivity of different input data assumptions and their impact on the resulting key performance indicators of the Austrian Energy Certification systematics.
- Exploring and categorizing input data that is prone to errors or difficult to clearly capture.
- Performing case studies on a set of buildings (different in in size, morphology, usage, input data availability and age).
- Derivation of a standardized and accurate documentation of input data for energy certificates.

All of the mentioned objectives target an increased degree of reproducibility and legal currency of energy certificates, while reducing errors and manipulation risk.

Related current and past efforts that influence the research design of this study include a number of contributions; Pont et al. (2011) examined a new detached residential building consisting of 8 different units on the impact of varied input data assumptions. For instance, they separated the building in its constituent units and compared their heating demand results with the heating demand (HWB) of the overall building, and assessed different assumptions regarding the conditioning of the basement of the units. Kaiser (2009) suggested that more than 50% of the issued certificates include severe errors. Currently, an EU-wide incentive tries to increase the quality of building energy certification. This incentive is named Qualicheck (2016), and discusses the different procedures of energy certification in Europe. Moreover, it tries to identify advantages and disadvantages of the different methods and to formulate recommendations for the further development of building energy certification.

This contribution reports on a set of efforts in the framework of the project. We present a first categorization of uncertainties based on the analysis of typical energy certificates, and the results of a study conducted with graduate students, in which we discuss the differences in results and input data assumptions for a limited set of buildings. In this context we present two methodological approaches in evaluation of differences. Moreover, we present a sample building database, which acts as guide for further investigation.

2 METHODOLOGY

2.1 *Building database*

In order to pursue the goals of the EDEN project a building data base was developed including circa 150 buildings different in size, morphology, usage, input data availability and age. The structure of the database facilitates the selection of buildings for specific analysis. For each building in the database, the data was organized as illustrated in Table 1.

For the purposes of the study with graduate students, 16 buildings were selected from the database. The selection was mainly based on availability of the specific required information for that study. Key data of these buildings is described in Table 2.

2.2 *Categorization of uncertainties*

To be able to assess the different uncertainties regarding building energy certification and related input data, we categorized the different input data streams. Hereby, we followed a "bottom-up" approach: We analyzed the process of issuing energy certificates and thereby identified aspects that pertain to different uncertainty categories. These categories are described in table 3. In this paper, we focus on uncertainty categories (i)–(iv).

2.3 *Study with graduate students*

Two independent groups of graduate students were asked to calculate energy certificates for 16 predefined buildings from the database. The two groups were provided with the same materials

Table 1. Required information to structure the building database.

Building properties	Plausible options
Building usage profile	– WE: Residential single-family house – WM: Residential multi-family house – B: Office building – HS: School – KP: Kindergarten
Construction year	{year of construction}
Building morphology/ General Form	(R), (I), (L), (O), (U), (T)
Urban situation:	– F: Freestanding structure – S: In a street front – E: In a corner location
Number of floors	{Integer}
Building height	{Height in metres}
Shading information	– Opposite building height and distance – Self-shading
Usage of each building floor (zones):	– W: Residential space – B: Office space – HS: Educational space – KP: Kindergarten – V: Retail – L: Restaurant – T: Technical room
Decorative façade	{Yes/No}
Orientation of the main façade(s)	{0–360° azimuth angle}
Roof shape	{Flat, Sloped, Mansard, …}
Building systems	{name, description, and key performance data of existing HVAC systems}
Building address	{address}
Zoning plan (GIS data)	{information from GIS-systems, if available}
Preserved façade / Listed building	{Yes/No}

including the existing building plans, architectural documents and pictures. In case of missing information, making assumptions and using standard default values were allowed. The groups were asked to describe potential uncertainties and their input data assumptions in detail via spreadsheets. Following this method, we facilitated a later identification of differences both in results and input data assumptions. In a later step, a detailed analysis of the impact of input data variation of uncertainty categories (i)–(iv) on the results of the energy certificates was conducted. The 16 buildings were pre-selected from the database to fit to this later

Table 2. List of the selected buildings and some of their key parameters.

Building	Construction year	Usage	Urban situation	Morphology	Main façade orientation
1	1828	V, W	E	O	ENE-NNW
2	1889	V, L, W	S	T	N
3	1896	W	S	L	WSW
4	1914	B, W	E	U	SSW
5	2005	HS	S	U	NW
6	1953	W	S	R	ENE
7	1912	V, W	S	I	NNW
8	1973	B	S	I	NNW
9	1946	L,V, W	E	L	W
10	1953	V, B, W	F	R	SW
11	1870	L, V, W	S	T	SW
12	1960	KP, W	S	R	NW
13	1820	L, V, T, W	E	U	NNW
14	1996	W	S	I	WNW
15	1992	W	S	T	S
16	1990	W, V, W	S	L	N

Table 3. Categorization of uncertainties of input data for building energy certificates.

(i) Urban context of the building (insolation, adiabatic components)
(ii) Usage of the building (heated/unheated, zoning, internal gains)
(iii) Thermal properties of building envelope (Detailed Layered constructions vs assumptions vs default values)
(iv) Glazing attributes and solar gains (frame/glazing ratio, g-values of windows)
(v) HVAC systems of the building
(vi) Insufficiently considered parts and components of the building envelope (geometry)

evaluation step. Students were asked to use tools they had previously learned (Archiphysik 13 (A-Null 2016) for issueing energy certification; SketchUp (2015) and a spreadsheet for modelling of the building's geometry).

To assess the results of this study with the graduate students, we applied a rather simple evaluation method: We analyzed the absolute values of resulting Key Performance Indicators (KPIs) and a set of input settings, and contrasted between the two groups. This evaluation method addresses potential uncertainties in input data in practice.

2.4 *Sensitivity and elasticity analysis of input data variation*

In a further effort we examined the impact of uncertainties regarding input data differently:

Table 4. Examined parameters.

Category	Independent variables	Dependent variables
(i) Urban context	– Percentage of adiabatic firewalls	– Overall Heating demand – Transmission losses
(ii) Usage	– Minimum zoning – Intermediate zoning – Maximum zoning	– overall heating demand – Heating demand per zone
(iii) Thermal properties	– U-value variation of external wall component – Variation of weighted U-value of all building elements	– Transmission Losses – Overall Heating Demand
(iv) Glazing properties	– Glazing/frame ratio of windows	– Overall heating demand – Solar gains

We varied, for a set of input parameters, the assumptions, and evaluated the absolute and relative impact on key performance indicators. Table 4 provides an overview about the examined parameters ("independent variable") and KPIs ("dependent variable").

To study the response of the independent variables to changes in the dependent variables, we used the elasticity analysis. This method, originally used in economics to evaluate demand and price relation for goods, provides a single number indicator η. This value states, if the reaction of the independent variable is strong ("elastic") or weak ("inelastic"). Equation (1) illustrates one expression of elasticity (Graham 2013, Hofstrand 2007).

$$\eta = \frac{(Q_1 - Q_2) \; / \; (Q_1 + Q_2)}{(P_1 - P_2) \; / \; (P_1 + P_2)} \tag{1}$$

Hereby, η is the elasticity, Q_0 is the initial demand (dependent variable) when the price is P_0 (independent variable) and Q_1 represents the new demand when the price changes to P_1 (Graham 2013). The same method can be used in other practices with similar relationships between the variables. The demand is *elastic* if η is greater than 1 which means that the dependent variable changes faster than the independent variable. If the number is smaller than 1, it is *inelastic* and the dependent variable changes slower than the independent variable. The change rates are equal if η is equal to 1.

In the following subsections we explain the elasticity scenarios examined in this paper.

2.4.1.1 *Urban situation*

This category addresses the impact of urban surroundings on overall heating demand of the building. The former includes such parameters as percentage of adiabatic and non-adiabatic walls in a building envelope, obstruction caused by neighboring buildings etc. Here, we focused on the impact of adiabatic firewalls in a building on its overall heating demand. The percentage varied from 0 to 100% with 25% iteration steps.

The calculation case with 100% of adiabatic walls was taken as the base case for the calculation of sensitivity analysis and elasticity values, assuming that all fire walls are adjacent to conditioned volumes of neighboring buildings. Steps 1, 2, 3, and 4 reduce each the ratio by 25%

2.4.1.2 *Usage*

During energy certification process professionals must divide the building into thermal zones (respectively zones of different usages). This may impact indicators such as the overall heating demand, air flow rates and heat transfer between thermal zones. For the purposes of this study, zoning was defined only by the usage profile of examined spaces. Therefore, the following cases were considered:

- Minimum zoning, where building is modelled as a single thermal zone, defined by its main usage profile.
- Intermediate zoning, where the building is modelled with 2 to 3 thermal zones; attic spaces are included into the main thermal zone.
- Maximum zoning, where the building is modelled with 3 to 4 zones. Attic and basement spaces are modelled as separate thermal zones.

The case of maximum zoning was taken as a base case for the sensitivity and elasticity calculation of the overall heating demand in response to the considered cases.

2.4.1.3 *Thermal properties*

In order to examine the sensitivity of the overall building heating demand to the thermal properties of building elements, two calculation cases were considered: case 1—variation of U-values of a single building component (in this contribution—external walls due to their high surface area ratio); case 2—variation of the weighted U-value of all building components according to defined ranges. Variation steps are presented in Table 5, where step1 is a default U-value for a building component constructed around 1900s, step 3 is a required heat transfer coefficient for a newly constructed building, steps 2 and 4 were assumed as intermediate U-values. The relative change for both cases, as well as the elasticity was calculated in relation to the default U-value (step 1).

Table 5. U-value variation for both calculation cases, [$W.m^{-2}.K^{-1}$].

	Step 1	Step 2	Step 3	Step 4
Case 1	1.55	0.70	0.35	0.17
Case 2	1.53	0.86	0.55	0.38

2.4.1.4 *Glazing properties*
The overall building heating demand and solar gains are highly influenced by properties such as glazing/frame ratio of a window, its g-value, shading etc. In this paper we focus on glazing/frame window ratio and its influence on overall building heating demand. A range of configurations was examined: from 55/45 to 90/10 in 5% step.

Sensitivity and elasticity calculations were carried out in relation to configuration 70/30 as a base value. This is due to the fact that this configuration is considered by default as glazing/frame for windows in many cases in practice.

3 RESULTS AND DISCUSSION

3.1 *Absolute differences in KPIs and input data assumptions*

Figures 1 to 3 show the relation between the results of the student groups. These figures reveal that the calculated area, volume and glazing percentages provided by the two groups present a better linear relationship compared to the averaged weighted U-values.

Figure 4 presents the deviations in the results of the two groups on heating demand, total area, total volume, average weighted U-values and glazing percentage for each building. Based on these results, deviations in heating demand mostly relate to the deviations in average U-value and glazing percentage.

3.2 *Sensitivity and elasticity analysis*

To perform a profound analysis of the impact of variation of input data we assigned the 16 examined buildings to the 4 different categories of uncertainties (compare table 3). Energy certificates were created for each building and for all of the above-mentioned input data variations.

3.2.1 *Urban situation*
Figure 5 presents the relative change of the overall heating demand in response to percentage variation of the adiabatic firewalls in a building.

Elasticity estimate in this case presents only small changes in heating demand (Table 6). In other words, in the selected case studies the heating demand is not very responsive to the variation of adiabatic wall area assumptions. Note that the calculated elasticities are negative: Smaller adiabatic wall ratios increase the calculated heating demand.

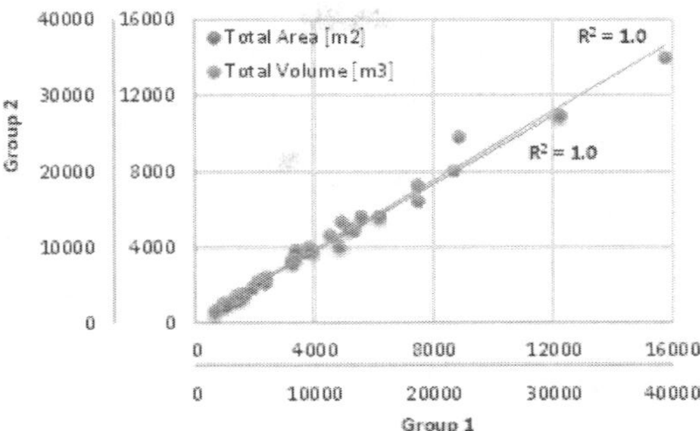

Figure 1. Total area and volume of 16 buildings calculated by group 1 vs. group 2.

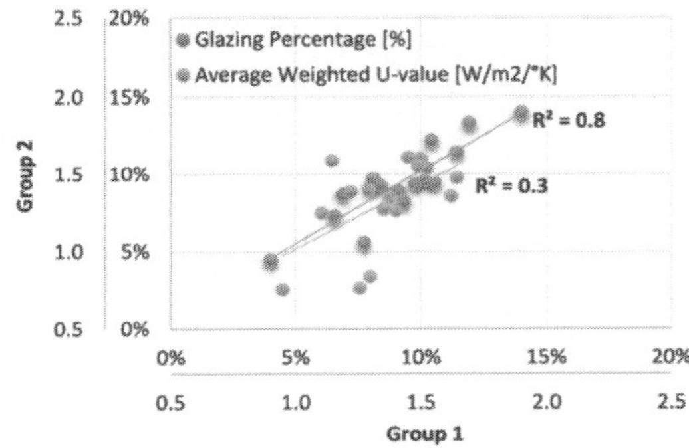

Figure 2. Glazing percentage and average weighted U-value of 16 buildings calculated by group 1 vs. group 2.

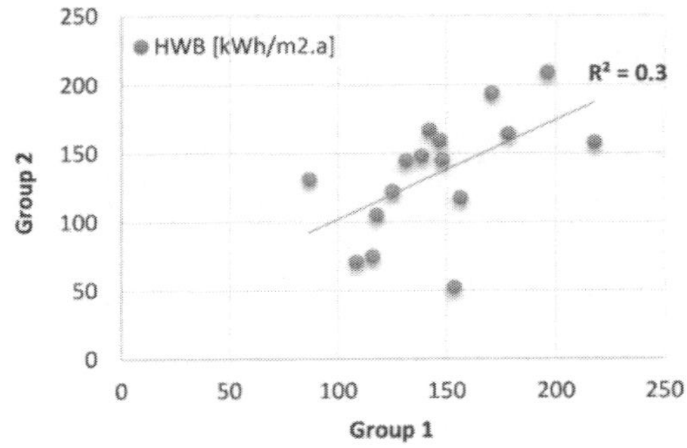

Figure 3. Annual heating demand of 16 buildings calculated by group 1 vs. group 2.

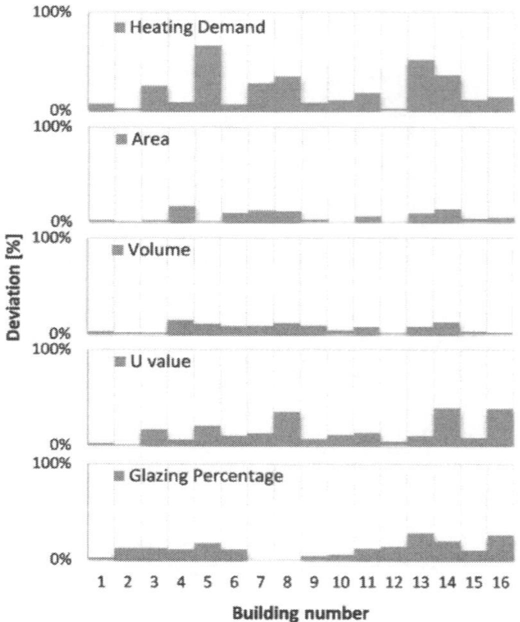

Figure 4. Relative deviation in the results for each building.

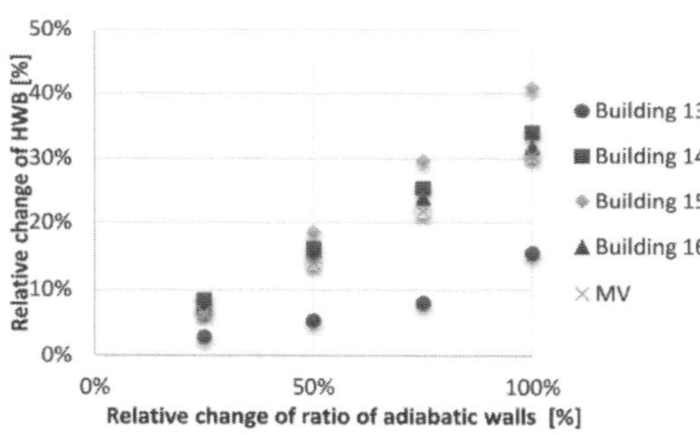

Figure 5. Relative change of overall heating demand in response to percentage variation of adiabatic walls.

Table 6. Elasticity of heating demand to the change of urban properties in respective buildings.

Building number	Elasticity (η)				
	Step 1	Step 2	Step 3	Step 4	Average
13	−0.1	−0.1	0.0	−0.1	−0.1
14	−0.2	−0.2	−0.2	−0.2	−0.2
15	−0.1	−0.1	−0.1	−0.1	−0.1
16	−0.2	−0.2	−0.2	−0.2	−0.2

3.2.2 *Usage*

Based on the results, the application of various zoning scenarios resulted in up to 9% relative change in overall heating building demand (intermediate zoning). The case with the maximum number of zones demonstrated a relative change in HWB of up to 27%. Such a radical deviation of the overall heating demand underlines the importance of a careful decision making regarding the zoning of evaluated buildings, and, moreover, the need for of a thorough documentation of these efforts.

3.2.3 *Thermal properties*

Figure 6 illustrates the relative change of the overall heating demand of the four buildings in the category of thermal properties, in response to U-value variation of external wall component (case 1). The relative change of the overall heating demand in response to the weighted U-value variation of all building components is shown in Figure 7 (case 2).

The results reveal that buildings 3 and 4 with higher percentage of external walls showed a higher response to U-value variation of the external wall

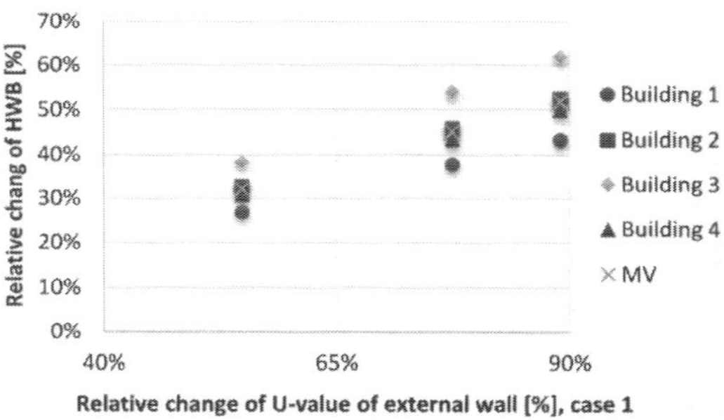

Figure 6. Relative change of heating demand and mean relative change in all buildings in response to U-value variation of external wall component.

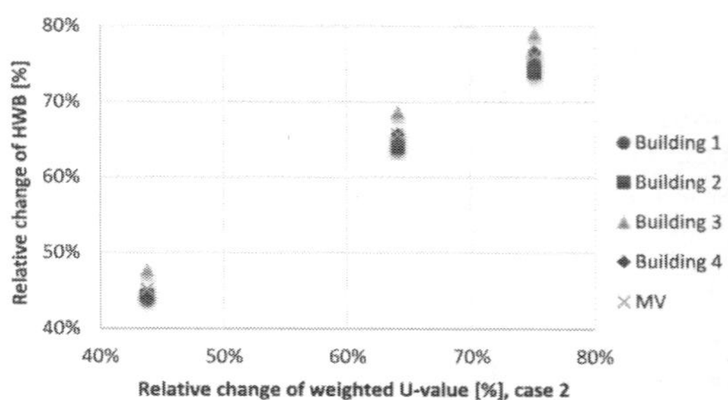

Figure 7. Relative change of heating demand and mean relative change in all buildings in relation to weighted U-value variation.

in all calculated ranges (see Fig.6). The response of the heating demand in these buildings was higher in Step 3 and 4, where the lowest U-values were applied. In case 2, a similar tendency of relative changes can be identified in all examined buildings, though the response of the HWB stayed constant in all steps and did not vary according to applied U-value.

Table 7 and 8 present the elasticity of the overall heating demand to the change of thermal properties in respective scenarios for the selected buildings. Based on the results, changes in the independent variable (i.e. U-value), resulted in a similar rate of changes for the dependent variable (i.e. heating demand) for all the four case studies in each scenario. Elasticities equal to 1 presented in Table 4 indicate that the change rates in heating demand and weighted U-value are equal. Obviously, the 4 examined buildings all show a strong connection between overall heating demand and average U-value. This emphasizes the importance to accurately define the thermal properties of the building envelope. Note that the positive values indicate the expected similar relationship between the change of U-value and heating demand.

3.2.4 *Glazing properties*

Figure 8 illustrates the relative change of HWB as a function of the glazing/frame ratio. According to the obtained results, the overall heating demand increases in the variation steps with higher percentage of glazing area.

Buildings 6 and 7, which have the lowest percentage of transparent elements in the sample, show the lowest response to the applied changes (2%). Contrary to this, buildings 5 and 8, which feature a higher glazing percentage, show a higher response of 6% in the results of the overall heating demand.

Table 7. Elasticity of heating demand to the change of thermal properties in respective buildings.

Building number	Elasticity (η) Step 1	Step 2	Step 3	Average
1	0.4	0.4	0.3	0.4
2	0.5	0.5	0.4	0.5
3	0.6	0.6	0.6	0.6
4	0.5	0.4	0.4	0.4

Table 8. Elasticity of heating demand to the change of weighted U-value in 4 examined buildings.

Building number	Elasticity (η) Step 1	Step 2	Step 3	Average
1	1.01	1.01	1.00	1.0
2	1.02	1.00	0.97	1.0
3	1.12	1.11	1.09	1.1
4	1.02	1.04	1.03	1.0

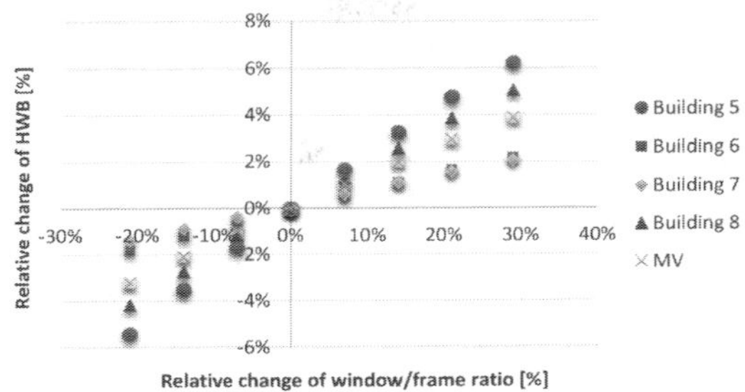

Figure 8. Relative change of heating demand in response to the variation of glazing/framing ratio of the window.

Table 9 presents the elasticity of the overall heating demand to the change of glazing to frame ratio. The applied changes in the independent variable (i.e. glazing ratio) resulted in only small changes in the dependent variable (i.e. the heating demand). The negative values are due to the inverse relations between the increase/decrease of the glazing ratio and the heating demand: As the glazing to frame ratio decreases, solar gains decrease as well, and (given the same component U-value of the overall window) the heating demand increases.

It is important to add that the glazing/frame ratio has to be considered critically. More detailed calculation schemes (numeric thermal simulation) would consider the impact of the glazing-frame-joint as thermal bridge and thus show different results (which might be closer to the reality than a ration-based U-value calculation and solar gain adaptation).

4 CONCLUSION AND FUTURE RESEARCH

In this contribution we presented the recent efforts and progress of the ongoing research project EDEN. We examined the consistency of energy certificates issued by different individuals for the same buildings. Two groups of graduate students ("semi-professionals") were asked to generate energy certificates for 16 case studies. Typically the documentations of existing buildings provide sufficient geometry information. However the data on the thermal quality of the building envelope is often insufficient and inconsistent. Accordingly, participants had to frequently use default values based on the year of the construction or make educated

Table 9. Elasticity of heating demand to the change of glazing/frame ratio in respective buildings.

Building number	Elasticity (η)							
	Step 1	Step 2	Step 3	Step 4	Step 5	Step 6	Step 7	Average
5	–0.2	–0.2	–0.2	–0.2	–0.2	–0.3	–0.3	–0.2
6	–0.1	–0.1	–0.1	–0.1	–0.1	–0.1	–0.1	–0.1
7	–0.1	–0.1	–0.1	–0.1	–0.1	–0.1	–0.1	–0.1
8	–0.2	–0.2	–0.2	–0.2	–0.2	–0.2	–0.2	–0.2

assumptions. We compared the building geometry information and thermal properties of the building envelope as well as the resulted annual heating demand. The results suggest that the deviations in the calculated energy certificates were mainly affected by the assumptions and simplifications relating to the thermal properties of the building.

Moreover we applied sensitivity and elasticity analysis to a set of building properties to identify input assumptions with large (or little) impact on the outcome of energy certificates. Four categories of uncertainty parameters were studied. The results suggest that thermal properties and assumptions about adiabatic firewalls (urban situation) have considerably higher impact on the overall heating demand than the examined glazing/framing ratio.

As such, this contribution can be considered as a interim step in the ongoing efforts, as the discussion of other attributes, such as the g-value or shading effects due to urban surroundings were not included. Additional efforts are currently undertaken to perform in-depth test series of energy certificate calculations with professional users taking such parameters under consideration.

The sensitivity and elasticity approach, as presented in this contribution, is being currently extended to the rest of the building sample, which will increase the sample size from 4 to 16 buildings. Furthermore, topological circumstances, physical attributes of the building envelope as well as other input assumptions concerning building usage and operation will be considered in future more thoroughly. Moreover, we currently evaluate other statistical/mathematical methods for evaluation of uncertainty impact on energy certificates.

ACKNOWLEDGEMENT

The research project presented in this paper is supported by the Austrian Research Promotion Agency FFG (Grant-No.: 850101). The authors also gratefully acknowledge the contribution of a number of students toward conducting the study presented in this paper. These include Alexandra Balmus, Milica Dukic, Triin Liis Palm, and Jakub Rozumek.

REFERENCES

A-Null 2016. "Archiphysik 13", www.archiphysik.com (last accessed, Mar 2016)

ASI 2016. Complete Standard ÖNORM B 8110 (Part 1–part 7, different years of publication). Austrian Standardization Institute.

EAVG 2006. Österreichisches Energieausweisvorlagegesetz 2006. Available via www.ris.gv.at

EAVG 2012. Österreichisches Energieausweisvorlagegesetz 2012. Available via www.ris.gv.at

EBPD 2002. European Directive on Buildings, 2002. Directive 2002/91/EC (available via http://eur-lex.europa.eu/)

EPBD 2010. European Directive on Buildings, 2010. Directive 2010/31/EU (available via http://eur-lex.europa.eu/)

Graham, R.J. 2013. Managerial Economics. John Wiley & Sons, Inc., Hoboken, New Jersey, USA.

Hofstrand, D. 2007. Elasticity of demand. IOWA State University, USA. Available online: https://www.extension.iastate.edu/agdm/wholefarm/pdf/c5–207.pdf. Accessed 14 April 2016.

ISO 2008. Energy performance of buildings—Calculation of energy use for space heating and cooling. International Standardization Organization, Geneve, Switzerland.

Kaiser, J. 2009. Tickende Zeitbomben, Immo-Report 03/2009.

OIB 2015a. OIB-Richtlinie Energieeinsparung und Wärmeschutz. Vienna, Österreisches Institute für Bautechnik (available via www.oib.or.at)

OIB 2016b. Leitfaden zur OIB-Richtlinie 6. Energietechnisches Verhalten von Gebäuden. Vienna, Österreisches Institute für Bautechnik (available via www.oib.or.at)

Pont, U., Sommer, B., & Mahdavi, A. 2011. Sources of uncertainty in compilation of energy certifcates.; Speech and published contribution in "Buildings & Environments 2011—Visions, Common Practice, Legislation" Bratislava, 20.10.2011, R. Rabenseifer et al. (Ed.), STU Bratislava, Paper-No 36.

Qualicheck 2016. QualiCheck-Platform (Towards better quality and compliance); http://qualicheck-platform.eu (last accessed, Mar 2016)

SketchUp, 2016. "SketchUp Make 2016", http://www.sketchup.com/ (last accessed Apr 2016).

eWork and eBusiness in Architecture, Engineering and Construction – Christodoulou & Scherer (Eds)
© 2016 Taylor & Francis Group, ISBN 978-1-138-03280-4

Utilization of GIS data for urban-scale energy inquiries: A sampling approach

N. Ghiassi & A. Mahdavi
Department of Building Physics and Building Ecology, TU Vienna, Vienna, Austria

ABSTRACT: Over the past years, new energy supply and management paradigms, such as distributed power and heat generation, have highlighted the significance of urban-scale energy assessments. The present contribution briefly presents an ongoing research effort towards development of a bottom-up simulation supported urban energy model for the hourly estimation of heating demand in the city of Vienna, Austria. The presented research project adopts a sampling approach towards high-resolution urban energy modeling and employs a well-known data mining method, Multivariate Cluster Analysis, to select representative buildings based on energy-related building characteristics. The selected sample is subjected to detailed performance assessments, the results of which are up-scaled to obtain the overall energy profile of the neighborhood. Focusing on the data-related challenges of urban energy modeling, the paper describes the informational requirements for the adopted approach, and elaborates on the underlying data structure and the data processing methods developed to overcome the encountered challenges.

1 INTRODUCTION AND BACKGROUND

Over the past years, new energy supply and management paradigms such as distributed power and heat generation, smart or responsive grid concepts, and grid-independent urban clusters have highlighted the significance of urban-scale energy assessments. Such energy paradigms perceive the urban stock as an interconnected network of energy demand and supply nodes, rather than a sum of isolated and independent entities. This integrated perception allows for more efficient allocation of resources and shared use of infrastructure, suited to the energy use patterns and sustainable energy harvest potentials of an assembly of buildings. Towards this end, an overall understanding of the temporal and spatial distribution of energy demand and supply potential is essential. Although smart metering and large scale monitoring activities are the most reliable methods of building energy data acquisition, extensive investment costs and associated data privacy issues hinder the realization of such schemes in the near future. In the meantime, development of computational methods to estimate the energy behavior of the urban building stock based on available information seems to be a feasible option to bridge the information gap.

In this regard, bottom-up engineering models, EMs, (Swan & Ugursal 2009) are considered to be highly effective due to their independence from historical data and their capability to capture unprecedented climatic, technological or behavioral changes. The modeling capabilities, informational requirements, and the precision and resolution of the outcome of EMs, are proportional to those of their underlying computational methods. In order to arrive at a comprehensive modeling environment, which not only enables detailed analysis of physical interventions and boundary condition changes, but also accommodates elaborate occupant modeling and investigation of demographic and behavioral changes, Building Performance Simulation (BPS) tools represent a suitable option. However, these tools require substantial amounts of high resolution data on buildings and their operation (extremely difficult to acquire at urban scale) and are computationally resource-intensive. To overcome the high informational and computational demand of BPS tools, and enable their utilization in urban-scale inquiries, the present research effort adopts a sampling approach towards reducing the computational domain (Fig. 1). The project is concerned with the development of a bottom-up simulation supported urban energy model for the hourly estimation of heating demand in the city of Vienna, Austria. Although stock sampling and archetyping have been the basis of many former urban energy modeling efforts, persistent lack of implicit reasoning for the selection of sampling criteria (see Ghiassi et al. 2015) has led the authors to attempt at developing a systematic approach towards automated GIS-based stock sampling.

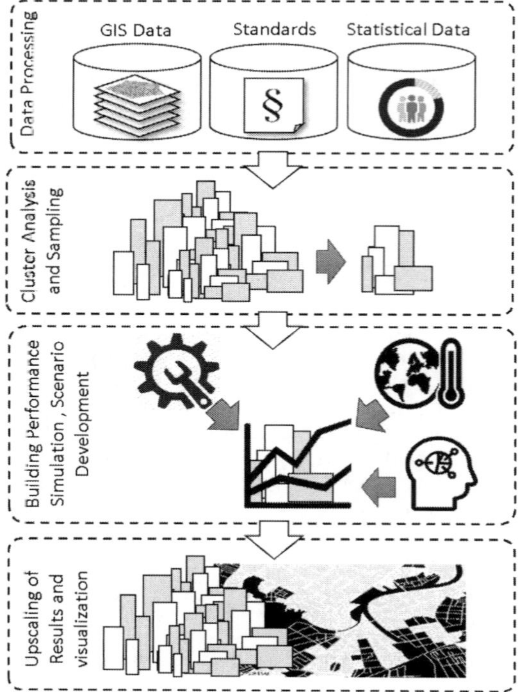

Figure 1. Schematic illustration of the suggested approach.

The developed method incorporates available large-scale data, including official and crowd-sourced GIS data, as well as building energy assessment codes and standards to generate energy based representations of individual buildings, in an automated process. It then employs a well-known data mining technique, Multivariate Cluster Analysis, MCA (Hair et al. 2010), to select a sample of buildings representative of the heating demand diversity of the entire building assembly. The selected representatives are subjected to detailed performance assessments, the results of which are up-scaled to obtain the overall energy use patterns of the neighborhood.

Focusing on the data-related challenges of urban energy modeling, the present contribution describes the informational requirements of the adopted approach, gives an account of the recurring data inadequacies and inconsistencies (in the context of the Viennese building stock), and elaborates on the underlying data structure and data processing methods developed to overcome the encountered challenges. The developed data processing routine is a versatile block that not only can support the above-mentioned cluster sampling approach, but also caters for a variety of logical or computational processes aimed at urban energy analysis. It establishes an interface between the largely available GIS data, building standards, and the explicit requirements of building energy computation.

2 APPROACH

2.1 *Study area and associated data*

The project is carried out on a neighborhood in Vienna city center, which includes over 750 buildings of various construction periods, usages, and morphologies (Fig. 2). Special buildings such as a church, subway stations, and kiosks in a permanent market place are excluded from the study. Available large-scale data, as well as relevant standards pertaining to the area were collected. The available data sources and the associated information are summarized in Table 1.

2.2 *Energy-based building classification*

The sampling procedure is intended to reduce the computational domain down to a manageable number of buildings, while reflecting the energy diversity of the neighborhood as much as possible. For this purpose, buildings have to be partitioned into groups with similar energy behavior. Representative buildings can then be selected from the partitioned data domain to insure that various energy behavior "types" are represented by the sample. The present research effort proposes the utilization of Multivariate Cluster Analysis, MCA, for data partitioning.

Multivariate Cluster Analysis is a fundamental technique in exploratory data-mining concerned with the identification of homogeneous groups of objects in a data set. It has been widely employed in various fields of science including medical studies and market research. Its potential towards

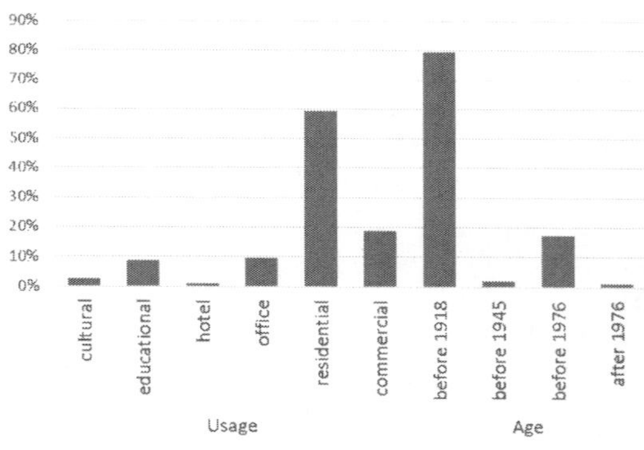

Figure 2. Distribution of buildings by usage and age.

Table 1. Available large scale data.

	Source	Data type	Contained data
GIS DATA	ViennaGIS (2015)	Land Use Plan	Building Footprint Polygons Construction type (main/annex) Relative height on the perimeter Elevation from ground
		Digital Elevation Model	Relative height of each pixel
		Building Inventory	Building construction period/year Main building usage Number of Floors
	Hammerberg (2014)	Sky View Factor Map	Sky View Factor of each pixel
	OpenStreetMap (2015)	Land Use Plan	Building Footprint Outline Building usage
STANDARDS	OeNorm B8110-5 (OeNorm 2011)	Thermal insulation in building construction: Model of climate and user profiles	Usage-based internal gains Usage-based infiltration rate Usage-based use hours
	OeNorm B8110-6 (OeNorm2014)	Thermal insulation in building construction: Principles and verification methods, Heating demand and cooling demand	Average window to wall ratio Average frame to window ratio Average net to gross ratio
	OIB-RL 6 (OIB 2015)	Guidelines: Energy-technical behavior of buildings	Age-based component U-values Age-based window solar transmittance

building stock classification and archetyping however, has not been sufficiently explored. To be able to subject the building stock to MCA methods, buildings have to be represented in terms of multi-dimensional data points. Since an energy-based clustering is desired, the multiple dimensions of these data points should collectively cover building characteristics, which influence the energy performance of a building. The MCA algorithms adopted by this research require all variables to be of numerical type.

Any demand estimation procedure relying on building heat balance, regardless of the resolution and precision of the adopted computational method, requires a description of physical, contextual, and operational characteristics of a building. Figure 3 displays the building properties required to assess various factors contributing to a building's heating demand. These building properties being influential in determining the energy behavior of a building, similarity among buildings with regard to these properties, is expected to result in similar performance. MCA variables were defined in such a way as to involve the main aspects of the above-mentioned building properties. Table 2 displays the adopted MCA variables, as well as the information involved in the computation of these variables. For more information on the selected variables see Ghiassi et al. 2015, and Ghiassi & Mahdavi 2016a.

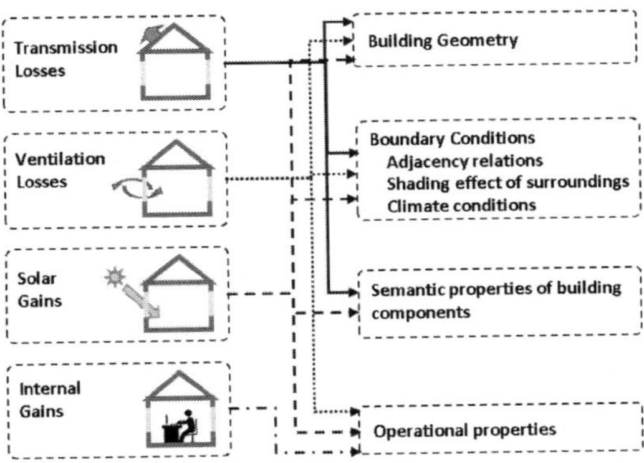

Figure 3. Building properties determining heating demand.

In the next step, available data was examined for consistency and accuracy and methods were developed to extract the values of the pertinent building parameters, through an automated process.

2.3 *Utilized tools*

QGIS is a user friendly Open Source Geographic Information System (GIS) licensed under the GNU General Public License (QGIS 2016). It was adopted as the analysis and visualization

Table 2. Multivariate Cluster Analysis variables.

Variable	Involved building parameters
Net Volume	Building volume Net to gross volume ratio
Thermal compactness	Building volume Area of building enclosures Adjacency relations of enclosures
Effective floor height	Building volume Number of floors Area of each floor
Effective glazing ratio	Area of external walls Orientation of external walls Window to wall ratio Extent of shading received on walls Glazing to window ratio Solar transmittance of windows
Effective average envelope U-value	Area of building enclosures Adjacency relations of enclosures U-values of building components
Daily area related internal gains	Area related internal gains Operation hours
Daily air-change rate	Air-change rate Operation hours
Ratio of daytime to total use hours	Annual operation hours Daytime operation hours

environment of the project. This GIS interface is developed in and therefore compatible with the general-purpose high-level Python programming language (Python 2015). Thus all computations and analyses leading to the selection of samples could be performed through a QGIS plug-in generated by the authors in Python. Computation of the Sky View Factor map was facilitated by the existing DEMTOOLS plug-in for QGIS (Hammerberg 2014).

Cluster analysis and advanced statistical investigations were carried out by several packages of the R Project for Statistical Computing (2015). Also a GNU project, R can be run and operated from Python. All utilized tools are openly and freely accessible. The sampling plug-in developed in the framework of the current project will be available on the QGIS plug-in repository upon finalization.

2.4 *Data verification and preprocessing*

The official land use data of the city of Vienna identifies all polygons belonging to a single building with a reference number. An investigation of the data revealed that in some cases, multiple buildings forming a single legal entity (several buildings situated on one parcel) were associated with a single reference number. The data was manually edited so as to differentiate between buildings.

The same reference number is used to link the building inventory information (information on building age and usage) with the geometric data. The two data layers were examined to make sure all buildings were associated with a single inventory data point. In cases were several data points were linked to a single building (e.g., due to multiple surveys), the information was merged into a single entry.

No information concerning the date of latest refurbishment activities or construction of roof-top extensions was available in our dataset. As such, buildings were assumed to be in the original conditions. Any missing information was filled in with average values through the data extraction routine.

To save computation time, average height of every polygon is computed based on the digital elevation model and stored in a separate column in a copy of the land use layer, as a property of the feature. For this purpose, the relative height of pixels that fall within the boundaries of a feature in the land use layer are averaged to arrive at the mean height of the feature.

2.5 *Data extraction*

Data extraction was facilitated through an object-oriented framework. Figure 4 demonstrates the simple data model adopted to generate the necessary building representation. Below is a description of how various data is incorporated to generate the desired representation.

2.5.1 *Envelope geometry and adjacency relations*

Land use layer features (polygons with additional attributes) constituting a building's footprint are identified by the reference number. Features might be associated with secondary (annex) buildings. Spaces in such buildings are assumed to be unconditioned. Figure 5 and Table 3 illustrate an example involving four buildings. Building 1's footprint (B1) is composed of three polygons (P1, P2, P3) all associated with a main building. Building 2 (B2) is associated with a single polygon (P4), which is declared to represent an annex building. B2 is therefore assumed to be unconditioned.

Based on the elevation from ground level (provided by the land use plans), the implemented code can determine if a lowermost building enclosure is earth adjacent or if it is elevated from ground, therefore exposed to outside air. For instance, in the above example P1 represents the floor element of a protruding built volume, exposed to outside air, whereas P2 and P3 represent ground adjacent building floors (Fig. 6).

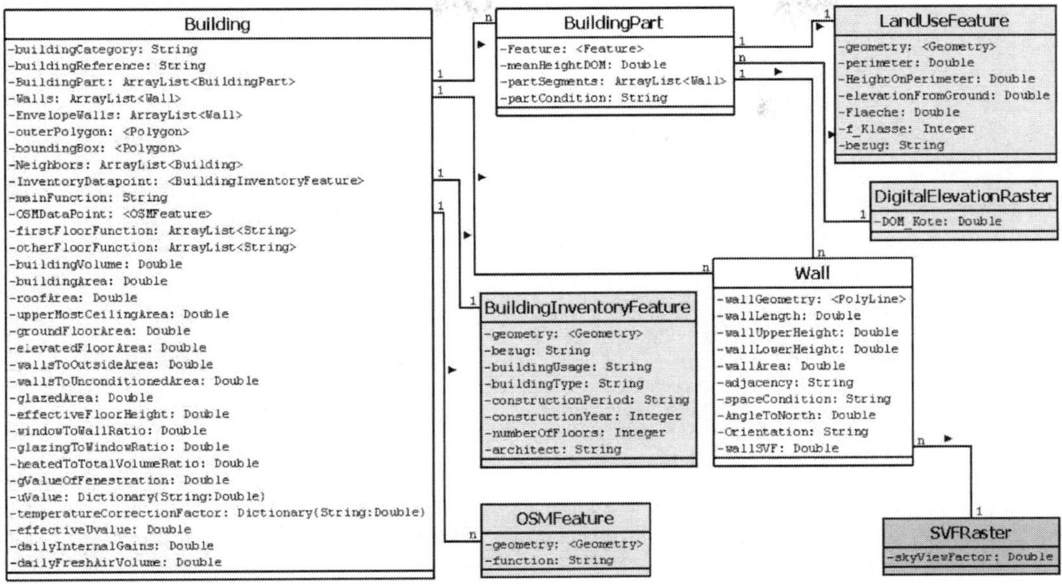

Figure 4. Underlying building data representation.

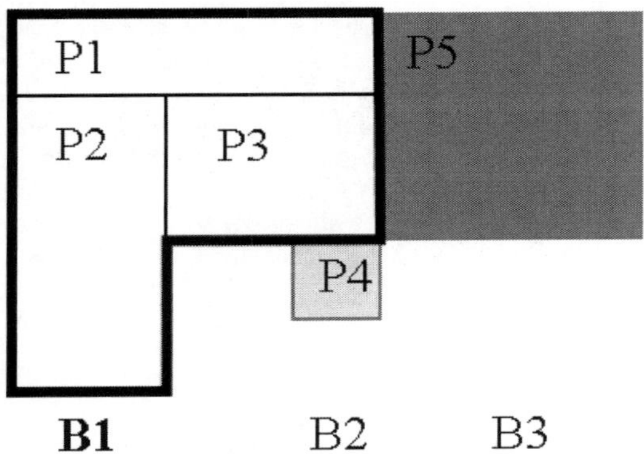

Figure 5. An illustrative example featuring four buildings.

Table 3. Feature properties in the illustrative example.

Building	Polygons	Type	Ground elevation	Relative height	Average height
B1	P1	main	3.00	9.00	9.00
	P2	main	0.00	9.00	10.80
	P3	main	0.00	9.00	12.50
B2	P4	annex	0.00	3.00	3.00
B3	P5	main	0.00	7.50	7.50

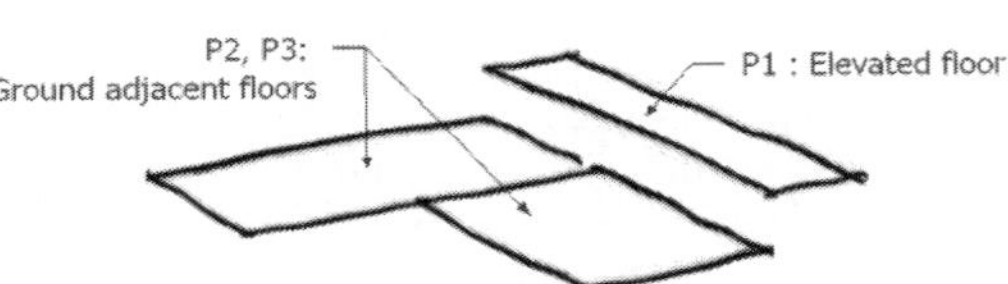

Figure 6. Elevation of the Building 1's lowermost enclosures from ground.

The relative height associated with each polygon (perimeter height) is compared with the average height computed based on the digital surface model. If no significant difference is observed, the represented building part is assumed to have a flat roof (e.g., P1). If the average height is more than the height given in the land use data—but no more than 3 meters—the polygon is assumed to represent the base of a sloped roof, i.e., the uppermost enclosure is a ceiling below an unheated attic space (e.g., P2). If the height difference exceeds 3 meters, the volume below the roof is assumed to be inhabited and therefore conditioned (e.g., P3). The area of the polygon is in this case multiplied by 1.5 to approximate the area of the roof, as the uppermost envelope enclosure. In this case, the attic volume is also considered in the computation of the overall volume of the envelope. (Fig. 7).

Polygons constituting the building are examined against one another. Heights of polygons with colliding edges are compared to determine the vertical elements of the building envelope (represented as Wall objects).

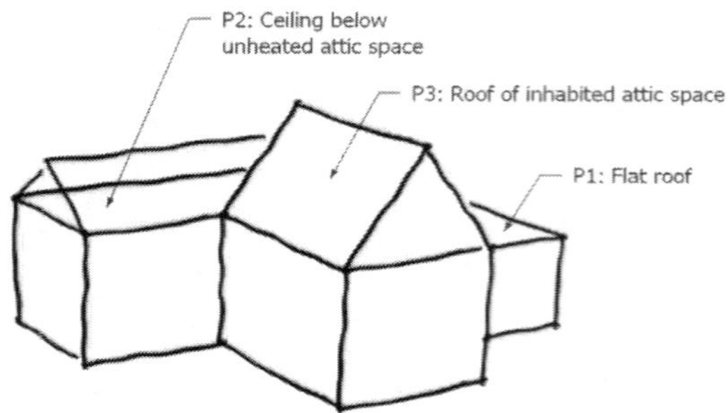

Figure 7. Determining the uppermost building enclosures as represented by the features.

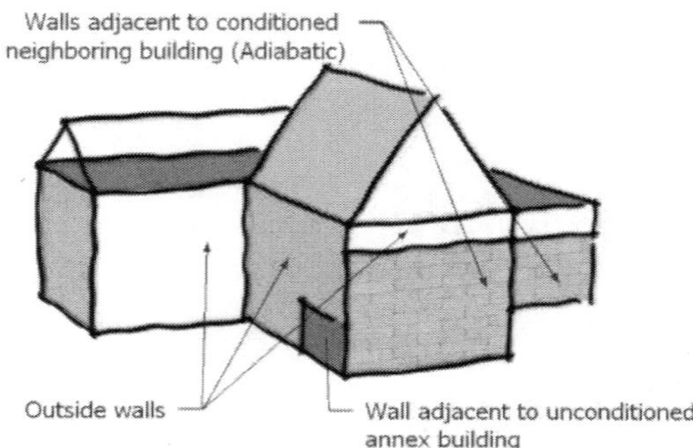

Figure 8. Adjacency relations of walls.

For each building, a neighbor search is done (via the bounding box geometries). The perimeter walls of the building are examined against those of the neighbors to enhance the Wall objects with adjacency information. The thermal condition of the adjacent building is considered (Fig. 8). The orientation of the Wall is geometrically derived.

The number of floors is given by the building inventory data. This information is however not entirely reliable in case of larger buildings, where the number of floors in various parts of the building may vary. The number of floors is used to derive the effective floor height as an indicator of the ratio of volume to useful floor area. In cases where the computed effective floor height falls outside of a plausible range, the number is replaced by the average effective floor height in the data set.

For the approximation of transparent building enclosure elements, information pertaining to average window to wall ratio and glazing to window ratio (derived from standards), as well as the extracted wall descriptions (wall area and orientation) are used.

2.5.2 *Semantic properties of the components*

From the building inventory data layer, the construction year or, in its absence, the construction period of the building is extracted. Each construction period is associated with a set of building component U-values derived from the standards for the energy assessment of historical buildings in Austria. If no information on the age of the building is available (incomplete or missing inventory data), the building is assumed to have been built before 1900. In cases where the period of construction of the building overlaps with two different historical periods mentioned in the standard, the older period is considered for the determination of component thermal properties.

2.5.3 *Operational parameters*

The building inventory data layer also provides information on the primary function of buildings (residential, educational, office, etc.). In order to verify and enrich the building use information of multi-purpose buildings, the OpenStreetMap GIS data of the neighborhood is used. Every data point contained within the perimeters of a building is linked with that building. A rule-based system is developed in order to associate fractions of the building volume with various usages present in a building. For instance some usages are assumed to be associated with the entire building (e.g., school, hotel, etc.). In these cases, if the official GIS data contradicts the OSM data, the latter is relied on, since crowd-sourced data repositories tend to be more frequently updated. The ground floor is divided among usages that are usually located at the street level such as shops, banks, restaurants, etc. Similar rules are foreseen for other functions.

Use profiles provided by the standards include usage-based descriptions of various operational parameters.

2.5.4 *Shading effect of surrounding buildings*

The exact computation of the solar radiation incident on building enclosures is theoretically possible, based on the information contained in the GIS layers. However, for urban scale inquiries, the required computational effort and time may be unreasonable. As such, the shading effect of the surrounding buildings is approximated by the Sky View Factor (SVF) on a point on the ground in the vicinity of the building's walls. Accordingly, if a larger portion of the sky is seen from a point close to the wall, that wall will receive less shading from its surrounding elements.

For every Wall object, the SVF of a point in the middle of the wall footprint, towards the outside of the building is extracted from the SVF raster data. If the footprint of the wall is longer than 5 meters, three points along the wall are selected, and the corresponding SVF values are averaged to better represent the shading conditions.

All information on the glazed enclosures including area, orientation, solar transmittance and SVF, as well as the aggregate area of outside walls are used to determine the effective glazing ratio of the building.

3 RESULTS AND DISCUSSION

Once an energy assessment compliant model of every building is constructed based on the available data, the MCA variables can be computed according to the generated data. For more details on the computation of MCA variables see Ghiassi et al. 2015, and Ghiassi & Mahdavi 2016a.

The matrix containing values of the variables for the building assembly is standardized prior to the cluster analysis. The standardization is necessary to prevent the difference in the magnitude of variables from affecting the clustering process as an unwanted weighting effect. In the absence of standardization, the clustering process will be dominated by variables such as building's net volume, which are represented by larger numbers than variables like effective envelope U-value.

The standardized data matrix can then be subjected to cluster analysis methods. The details of the implemented algorithms and some preliminary results have been presented in Ghiassi & Mahdavi 2016b.

Figures 9 and 10 display the initial GIS data layers and the results of one of the implemented cluster analysis algorithms (hierarchical agglomerative method) as visualized by the developed plug-in. The resulting building clusters are stored in a copy of the land use data layer enhanced with a column containing the number of the cluster, with which every building and its corresponding features are associated. The generated map offers visual clues and a more intuitive understanding of the resulting building classifications.

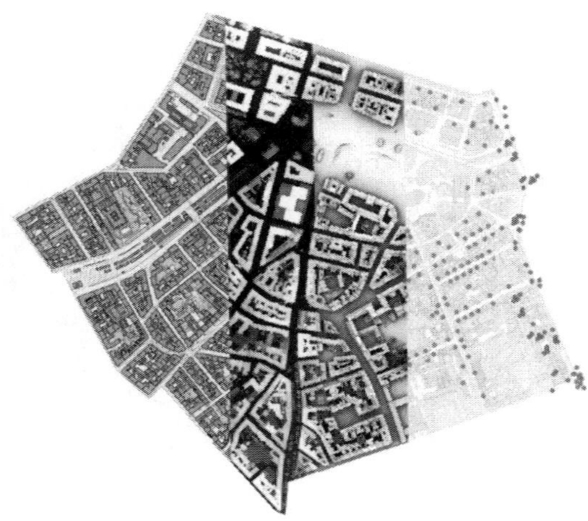

Figure 9. GIS data layers incorporated by the developed approach. Left to right: Land use plan, Digital elevation model, Sky View Factor raster, Building Inventory data points, and Open Street Maps data points.

Figure 10. Building clusters represented by different colors.

Currently, the described building representations are generated on the run, and only the necessary values for the MCA process are stored in csv format. It is however theoretically possible to store the created building objects, their properties and associations as an energy assessment compliant urban stock representation.

Data processing, extraction of the MCA variables, implementation of three MCA algorithms and cluster performance assessment procedures (see Ghiassi & Mahdavi 2016b), as well as the visualization of results for the adopted case study require approximately 12 to 14 minutes on a typical laptop (Quad core processor, 8GB RAM).

The developed code can be optimized through implementation of GIS data processing techniques, such as introduction of grids, to reduce computational time for larger data sets.

4 CONCLUSION

The present contribution reports on the development of an automated data processing method, which bridges the gap between the available large scale building data and the requirements of building energy assessment. The suggested data processing approach aims to facilitate the utilization of urban big data towards urban energy modeling. The ongoing research project employs the developed method towards automated energy-based sampling of the Viennese building stock for simulation supported urban energy modeling.

Implemented as a plug-in for an open source GIS environment, the data processing routine employs several layers of official, computed, and crowd-sourced GIS data, as well as building standards to generate a comprehensive energy-oriented model of every building within an urban neighborhood. The generated model is then used to arrive at a summarized building representation for Multivariate Cluster Analysis purposes, leading to the performance-oriented partitioning of the data space and selection of samples from the generated clusters. The resolution of the developed building representation is of course bound by the limitations of the available data. However, a superposition of various data sources can help refine the generated representation.

In any modeling effort, the quality of the generated model is highly dependent on the quality of input data. Reliable GIS data is indispensable to any bottom-up urban energy modeling environment. Unfortunately, GIS data accumulation, surveying activities, and public data distribution have not been performed with energy aspects of buildings in perspective. As such, essential data with regard to the state of the building envelope and potential extensions are not provided. Also, information on building usage may be outdated and inaccurate, given the paucity of periodical data verification.

The utility of the developed toolbox is not limited to energy-based sampling or archetyping efforts. The generated building representation can also cater for other approaches towards urban-scale energy inquiry. For instance, Glawischnig (2016) employs a similar (yet less detailed) building representation for web-based computation of heating demand.

The proposed framework can facilitate the generation of energy-oriented urban stock representations through establishment of mapping procedures between the internal data model of the plug-in and standard urban representations such as CityGML (OGC 2016). Since the link to the GIS data is preserved, such a mapping schema can be easily developed.

Future research intentions include the evaluation of the performance of the integrated MCA methods towards identification of appropriate building clusters; design and implementation of detailed upscaling methods, which can provide more realistic approximations of aggregate load profiles; development and comparative analysis of various realistic change and intervention scenarios, and comparison of upscaling-based projections with available monitored data.

REFERENCES

Ghiassi, N., Hammerberg, K., Taheri, M., Pont, U., Sunanta, O., & Mahdavi, M. 2015. An enhanced sampling-based approach to urban energy modelling, Proceedings of the 14th IBPSA Conference, Hyderabad India.

Ghiassi, N., & Mahdavi, A. 2016a. A GIS-based framework for semi-automated urban-scale energy simulation, CESB Conference, Prague Czech (accepted).

Ghiassi, N., & Mahdavi, A. 2016b. Urban energy modeling using Multivariate Cluster Analysis, BAUSIM Conference, Dresden, Germany (under review).

Glawischnig, S. 2016. An urban monitoring system for large-scale building energy assessment. Diss. TU Vienna, Austria.

Hair, J.F., Black, W.C., Babin, B.J., & Anderson, R.E. 2010. Multivariate Data Analysis—a Global Perspective, Pearson Global Editions, New Jersey, USA.

Hammerberg, K. 2014. DEMTools. QGIS plugins repository<https://plugins.qgis.org/plugins/DEMTools> (cit. 12.03.2016)

OeNorm. 2014. B 8110-6 Thermal insulation in building construction, Part6: Principles of verification methods—Heating demand and cooling demand—National application, national specifications and national supplements to OeNoem EN ISO 13790, Austrian Standards Institute Vienna Austria.

OeNorm. 2011. B 8110-5 Thermal insulation in building construction, Part5: Model of climate and user profiles, Austrian Standards Institute Vienna Austria.

OGC. 2016. CityGML <http://www.opengeospatial.org/standards/citygml> (cit. 07.04.2016)

OIB. 2015. RL-6: Energy behavior of buildings. Austrian Institute of Construction Technology, Vienna Austria.

OpenStreetMap. 2015. <www.openstreetmap.org> (cit. 18.03.2015).

Python. 2015. <www.python.org> (cit. 11.18.2015).

QGIS. 2016. <www.qgis.org> (cit. 10.03.2016).

R Project for Statistical Computing. 2015. <www.r-project.org> (cit. 11.18.2015).

Swan, L. G., & Ugursal, V. I. 2009. Modeling of end-use energy consumption in the residential sector: A review of modeling techniques. Renewable and Sustainable Energy Reviews, 13(8), pp. 1819–1835.

Vienna GIS. 2015. <https://www.wien.gv.at> (cit. 18.03.2015)

eWork and eBusiness in Architecture, Engineering and Construction – Christodoulou & Scherer (Eds)

Semantic interoperability for holonic energy optimization of connected smart homes and distributed energy resources

S. Howell, Y. Rezgui, J.-L. Hippolyte & M. Mourshed
BRE Trust Centre for Sustainable Engineering, Cardiff School of Engineering, Cardiff, UK

ABSTRACT: Recent work has attempted to deliver optimized distributed energy resource management, including the use of demand side management through smart homes. This aims to reduce power transmission losses, increase the generation share of renewable energy sources and create new markets through peak shaving and flexibility markets. Further, this leverages the development of product models at the device, building, and network level within the operational lifecycle stage, beyond the conventional role of BIM between design and construction stages. However, the management of heterogeneous software entities, incompatible data models and domain perspectives, across systems of systems of significant complexity, represent critical barriers to sustainable urban energy solutions and leads to a highly challenging problem space. The presented work describes a systemic approach based on the concept of holonic systems, which exemplify the role of autonomy, belonging, connectivity, diversity and emergence across entities. This reduces the decision complexity of the problem and facilitates the implementation of optimized solutions in real power systems in a scalable and robust manner. Further, the concept of a flexibility market is introduced, whereby smart appliance owners are able to sell load curtailment and deferment to a local aggregator, which interfaces between a small number of homes and a distribution system operator. Artificial intelligence is present at each of the entities in order to express constraints, trade energy and flexibility, and optimize the network management decisions within that entity's scope. Specifically, this paper focuses on enabling interoperability between system entities such as smart homes, local load aggregators, and last mile network operators. This interoperability is achieved through ontological modelling of the domain, based on the existing standards of CIM, OpenADR, and energy@home. The produced ontology utilizes description logic to formalize the concepts, relationships and properties of the domain. A use case is presented of applying the ontology within a multi-agent system, which enables the optimization of day-ahead markets, load balancing, and stochastic renewable generation, and closely aligns with the holonic approach to deliver a holonic multi-agent system. The use case assumes a scenario in line with the emerging energy landscape of a district of domestic prosumers, with a high penetration of micro-generation, energy storage and electric vehicles. Initial results demonstrate interoperability between heterogeneous agents through ontological modelling based on an integration and extension of existing standards, which acts as a proof of concept for the approach.

Keywords: Information management, interoperability, MAS, OWL, ontology, energy management, DER, prosumers, smart grid

1 INTRODUCTION

With varied and mounting challenges facing the urban environment and its energy supply, urban energy systems have undergone increasingly rapid change from centralized systems to the distributed energy systems currently deployed and reported in research. This is due in part to the growth of Smart Grids (SGs) (El-Hawary 2014, Giordano et al. 2011, Ahat et al. 2013, Werbos 2011), Distributed Energy Resources (DERs) (Jiayi 2008, Manfren 2010), and Demand Side Management (DSM) (Palensky et al. 2011, Wang et al. 2015). The growing interest in these areas embodies an underlying shift in the energy landscape towards sustainability (Baños et al. 2008, Dominguez-Garcia et al. 2012) and resilience (Dominguez-Garcia et al. 2012, Ghosn et al. 2010) through distributed resources and intelligence. However, this highly distributed energy landscape brings many challenges, including the interoperability of the software artefacts managing the power network, and the Home Energy Management System (HEMS) software which implements such DSM schemes.

This paper therefore presents a means to overcome the semantic heterogeneity amongst the highly distributed and varied agents in the domain. Specifically, the paper presents an ontology, and accompanying implementation, towards the management of domestic flexibility markets. These allow consumers to sell the deferment and curtailment of their loads to a distribution service operator, through a flexible hierarchy of aggregation, and close integration with smart appliances and DERs. Further, the concept of a holonic architecture is utilized to optimize this aggregation hierarchy through an adaptable system of systems approach.

First the background of the study is presented to introduce the evolution of energy systems and the requirement for semantic interoperability. Next, the target system and knowledge management approach is described, before the ontology is discussed. A primary use case is given, and a benefit of the semantic approach is highlighted through another use case, before a discussion and conclusion.

2 BACKGROUND

2.1 *Evolution of the urban energy landscape*

The energy landscape has evolved at an accelerating pace since the creation of the first generators by Michael Faraday, through centralization and subsequently towards increasing penetration of renewable generation and decentralization (REN21 2014). More recently, the addition of monitoring and intelligent management through software has led to the paradigm of cyber-physical systems, where ICT and physical entities are tightly coupled (El-Hawary 2014, Giordano et al. 2011, Ahat et al. 2013, Werbos 2011).

Centralized electricity systems consist of a small number of large power plants which generate higher voltage power, a high voltage transmission network, transformers which produce lower voltage power for a distribution network or subtransmission, local transformers which produce low voltage power, and a low voltage grid connected to energy consumers. In this system architecture, power flows in one direction only, consumers aren't aware of the state of the network or of others' consumption, and producers view consumers as passive entities. This lack of predictability and load control results in the need to overproduce energy and store it as reserve capacity in case of load spikes (Martin 2009). Increasingly, this has resulted in wasted energy and is one reason why this approach is being recognized as inefficient, alongside losses over long transmission distances and the waste of thermal energy produced during electricity generation.

Given the environmental imperative of decarbonization, and following several key international events and policies (UNEP n.d., UN 1995, UN 1998), global reliance on Renewable Energy Sources (RESs) doubled between 2004 and 2014 (REN21 2014). It is important to note that whilst coal and oil reliance has been reducing, the contribution of gas to the generation mix has been increasing globally alongside renewables due to its low emissions, and arguably, its simplicity to integrate with existing markets and business models.

Global RES policies initially caused change without disrupting the centralized paradigm, through large plants which each contributed significantly to the national grid. Large wind farms (Department of energy n.d.) and solar farms (Go Solar California n.d.) as well as geothermal and biofuel plants dominated the non-hydro renewable generation domain and have gradually supplemented conventional energy sources (REN21 2014). The management and underlying architecture of energy systems has remained centralized until recently (IEA-RETD 2014), with little microgeneration (Harper 2008) or energy storage occurring and consumers only acting as passive agents. Arguably this was due to a lack of technology and incentives, causing barriers for prosumers and small scale renewables to enter the energy landscape.

The end of this generation of gradual shift to renewables without dramatic system or paradigm reform has been brought about in the research community during the popularization of the smart grid concept (Hommelberg et al. 2007). Whilst inherently related to intelligent management, the smart grid concept is now specifically associated with the integration of modern technologies such as microgrids, distributed energy resources and virtual power plants. This has somewhat coincided with the 'smart city' and 'smart planet' revolution which has manifested throughout national policies and has contributed to investment in DER technologies and management systems (IEEE 2013, Rey-Robert 2009). Further, recent state-backed financial incentives to invest in microgeneration technology have contributed to its accelerating uptake within the energy sector (Ofgem, n.d., Energy Saving Trust n.d., Jones et al. 2010).

Another key smart grid concept, Demand Side Management (DSM), refers to the systemic interaction with consumers and active loads to directly or indirectly affect demand profiles. This includes concepts such as active loads, demand management, and demand response, which extends the problem space to that of a complex socio-technical system. For example, (Mohsenian-Rad et al. 2010) considered automated load scheduling at the smart meter level, by using dynamic pricing to incentivize

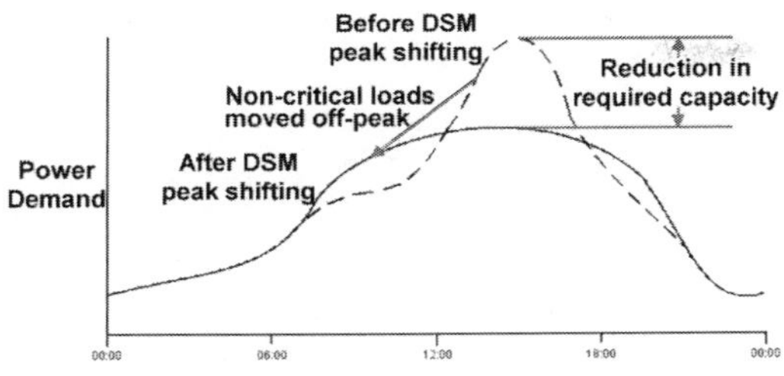

Figure 1. Illustration of DSM peak-shifting.

consumers whilst optimizing locally and globally, and successfully stabilized the demand profile and reduced overall system costs. One key intended outcome of DSM is load shifting to reduce the total generation capacity required in a system, as shown in figure 1. Within such objectives, it is critical that a consumer-focused approach is adopted to improve acceptance by consumers in competitive market-driven environments.

2.2 *Semantic interoperability in energy systems*

The challenge of interoperability in energy management is becoming increasingly important (IEEE 2011, Baclawski et al. 2014, Von Dollen 2009, a significant factor of which is semantic heterogeneity between the vocabularies and data representations used by the numerous software and hardware artefacts penetrating the domain. In line with this concern, the IEEE standards committee identified the challenge of interoperable protocols, data formats and meaning (IEEE 2011). Alongside the requirement for a secure framework, and the benefits of a service oriented architecture (IEEE 2011), the use of a common vocabulary and data model mitigates the effort required for software artefacts to communicate effectively with others in the energy management system (IEEE 2011, Van Dam and Keirstead 2009). These common models must standardize the concept descriptions and data representations within the shared domain of the software artefacts.

One manifestation of these semantic models, taken from semantic web technologies, is that of a domain ontology; a machine-interpretable description of the concepts in a domain, the relationships between these concepts and domain data and the restrictions on how these concepts can manifest (W3C 2009). This domain ontology can then be instantiated with the specific objects, relationships and data present in an instance of the domain, such as an individual energy system, to create a comprehensive knowledge base. Further, each domain ontology can re-use parts of other ontologies which map upper domains or related domains, to further facilitate interoperability and reduce development time.

Within related fields, the benefits of common data models are being widely recognized. For example, within the building construction and lifecycle management domain, open 'Building Information Modeling' (openBIM) has recently boomed through the use of BuildingSMART's common data model for representing building level information (buildingSMART 2013), which significantly improves data utilization through its handover between the design, construction and management stages of a building's lifecycle. In the smart city field, the British Standards Institute has recently published a standard towards developing smart city data models for interoperability (BSI 2014), and the Open Geospatial Consortium has released CityGML; an internationally standardized model for describing spatial and semantic information in the urban environment (Gröger et al. 2012). Finally, an understanding of the benefits of a shared ontology within multi-agent systems is commonplace and forms a central aspect of the international MAS modelling standards of FIPA Poslad 2007). Given the recent trend towards agent-oriented energy management systems, the development and use of shared domain ontologies is then a natural progression.

2.3 *Smart grid, building and appliance modelling*

Based on the observed requirement for semantic interoperability, significant steps have been taken towards semantic modelling of the energy domain, including the domestic and DSM sub-domains. Arguably, the most widely noted example is the Common Information Mode (CIM) (IEC 2011), and IEC 61968 (IEC 2003) although the OpenADR (OpenADR 2013) and energy@home (Energy@home 2015) models are also highly relevant, and these are all now discussed briefly.

The Common Information Model (CIM) is a broad domain model based on the UN electronic exchange standards, and is comprised of 3 layered parts. It aims to facilitate power management processes such as outage management, asset management and customer information management, amongst many others. This is expressed as a UML class model, and includes concepts such as topology, asset descriptions and component descriptions.

Whilst the CIM takes a significant step towards interoperability in power systems, it is arguably not well suited to DSM due to its lack of modelling at the last mile of the supply chain. Further, it lacks the expressivity of OWL, and so is not suitable for semantic web applications.

The OpenADR (Open Automated Demand Response) standard models concepts regarding demand response, which is a subset of DSM, through a data model and communication specification. As such, it formalizes concepts between the consumer and the supplier of energy, such as market context, dynamic pricing and event descriptions. Again, this represents a critical aspect of modelling the domain, but still fails to integrate data and commands at the device level. To this end, the energy@home data model is highly beneficial.

The energy@home data model specifies a representation model for home area networks, based on the CIM approach (through its evolution into the SEP2 model), and it is broadly aligned with the OpenADR schema. The specification includes concepts regarding smart appliances, power profiles, renewable energy generation, smart meters and smart user interfaces.

Arguably, the union of these three standards represents the scope required for the delivery of DSM across the supply chain to consume appliances, but their lack of harmonization, and semantic expressivity, greatly limits actual interoperation in use cases which span across their individual scopes. The detailed analysis of these three artefacts led to an understanding of their potential alignments. Further, this highlighted that concepts regarding flexible loads, flexibility markets, and aggregation of flexibility, were currently not evident, but were deemed essential towards intelligent DSM.

3 HOLONIC ENERGY MANAGEMENT PLATFORM

3.1 *Holonic multi-agent architecture*

The semantic artefacts developed herein were utilized within a multi-agent system, structured so as to implement the concept of holonic energy systems. The holonic system approach is based on the concept of a dynamic hierarchy of holons, where each holon represents an autonomous and self-contained system but can contain or be contained within other holons, and the topology of this hierarchy is flexible towards the overall benefit of the super-system, or holarchy. Further, each sub-holon can change which super-holon it is a part of or become a part of multiple super holons. In this way the holonic approach is a hybrid between the distributed approach where autonomous subsystems adapt within a static framework, and the centralized approach where subsystem behavior is prescribed by a supervisory controller. Further, as each holon can be composed of sub-holons, these sub-holons interact through cooperation and competition to produce the emergent behavior which characterizes their super-holon; in this way the approach is similar to multi-agent systems. This can be reflected by adopting a holonic system

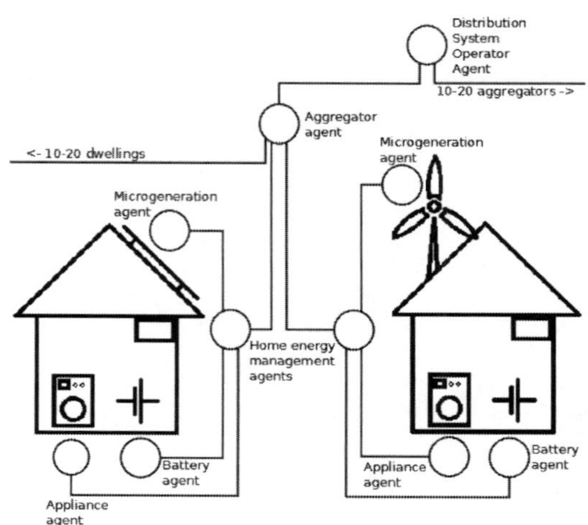

Figure 2. Illustration of the target HMAS.

approach within a multi-agent system, to produce a Holonic Multi-Agent System (HMAS) (Pahwa et al. 2015), as shown in figure 2. The holonic system approach lends itself well to the management of energy systems (Pahwa et al. 2015, Vašek et al. 2014) as they are typically conceptualized in a hierarchical manner whereby components form a network, such as a microgrid, which in turn forms part of a larger network, and so on. However, the holonic approach also exploits the potential autonomy in each system and the aggregation flexibility afforded by virtualization.

3.2 *Agent and service integration through semantics*

The integration of the agents with web services is achieved through a semantic web approach which utilizes a domain ontology, described in the next section, along with instance data, within a triple store. This data can then be accessed through an ontology service wrapper around a SPARQL endpoint, based on the JENA ARQ API. The entities within the domain ontology also then comprise the content ontology for FIPA-ACL message payloads, as illustrated in figure 3. This allows a coherent representation of the domain within the triple store, as well as the integration of both web services and traditional agents.

4 PROSUMER DEMAND SIDE MANAGEMENT ONTOLOGY

The developed domain ontology was based primarily on an OWL representation of the previously

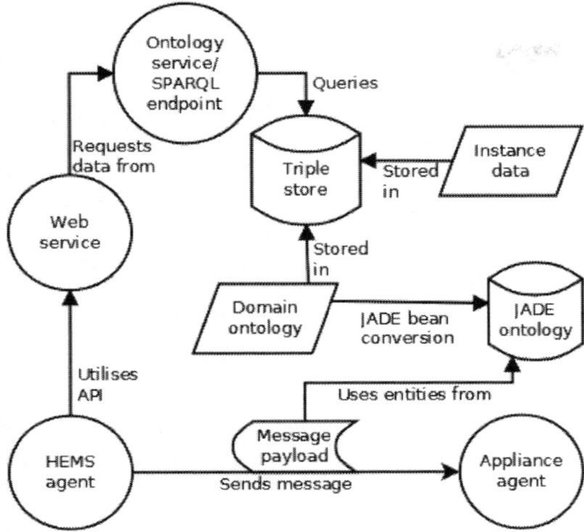

Figure 3. Knowledge management approach for web service and agent integration.

identified semantic resources; the CIM, IEC 61968, OpenADR, and the energy@home data model, by reusing and extending concepts from these. Firstly, classes and slots were elicited from the described resources, to produce a coherent model across the domains of smart appliances, demand response and smart grid, whilst still respecting the scope of the ontology prescribed by the use case defined in section 5. The ontology also formalized the concept of domestic load flexibility, and included concepts related to the trading and aggregation of this flexibility. The ontology is now presented briefly, followed by the formalisation of the flexibility concept, and finally the alignment with existing models.

4.1 *Candidate domain ontology*

Following the elicitation of domain knowledge through a use case and standards analysis driven process, the resultant knowledge was formalised in description logic using basic OWL (OWL 2009) constructs so as to produce a candidate ontology which reused the existing standards identified. This was conducted in the Protégé software (Stanford 2016), and serves to produce an ontology with value outside of its MAS application, as it is also suitable for web service deployment, or direct use in a C++ or Java program through a relevant OWL library.

Figure 4 presents an excerpt of the concepts and relationships formalised; the physical system taxonomy and the energy scheduling ontology. The energy scheduling ontology formalises energy scheduling classes, as specified by the energy@home data model (Energy@home 2015), which expresses that a power profile is made up of an ordered set of modes, which are then composed of energy phases. Further, the concept of flexibility was formalised, subject to the intended modelling pattern described in the next subsection.

4.2 *Domain perspective of energy flexibility*

Load flexibility is here defined as a market commodity of utilised peak load reduction through optional deferment and/or curtailment of consumer demand, expressed as a unit of energy. Deferment is the shifting of a load to a time more favourable to the network operator, where the amount of flexibility is equal to the amount of energy shifted. In this way the extent of the shift is independent to the flexibility, as the consumer sets a deadline for the task completion. This is represented in figure 5 below, where Qtot is the total energy consumption of the task, Qf is the flexibility utilised, t0 is the earliest start time of the task, t1 is the task completion deadline, and Tmin is the minimum amount of time the task requires to be completed.

Curtailment of load is then the supply of a quantity of energy over time which is less than the desired quantity. The flexibility is then the difference between the desired quantity and the supplied quantity, again expressed as an amount of energy. This is shown in figure 5, where t0 is the earliest start time of the task, t1 is the non-negotiable deadline of the task, Qf is the amount of flexibility utilised, and Pmin is the minimum amount of energy to be supplied (such as when a heating device must meet a minimum room temperature).

4.3 *Alignment with existing ontologies*

Alignments with the existing standards were formalised as OWL annotations, an excerpt of which is shown in figure 6 below. These could be used to trivially perform a schema conversion through a SPARQL CONSTRUCT query, or to produce an 'owl:equivalentClass' instantiation. Whilst it would be incorrect to state that this represents full compliance or alignment with the standard, it demonstrates broad coherence with the domain perspectives of the existing standards, and paves the way for genuine compliance if the existing standards are developed into full semantic models in the future.

4.4 *Ontology conversion to JADE*

In order to utilise the ontology to formalise the semantics of FIPA-ACL encoded payloads (the most prevalent MAS implementation language), the OWL candidate ontology was converted into a set of JADE concept and predicate java bean classes. This process was automated by using Apache Jena to interpret the OWL file expressing

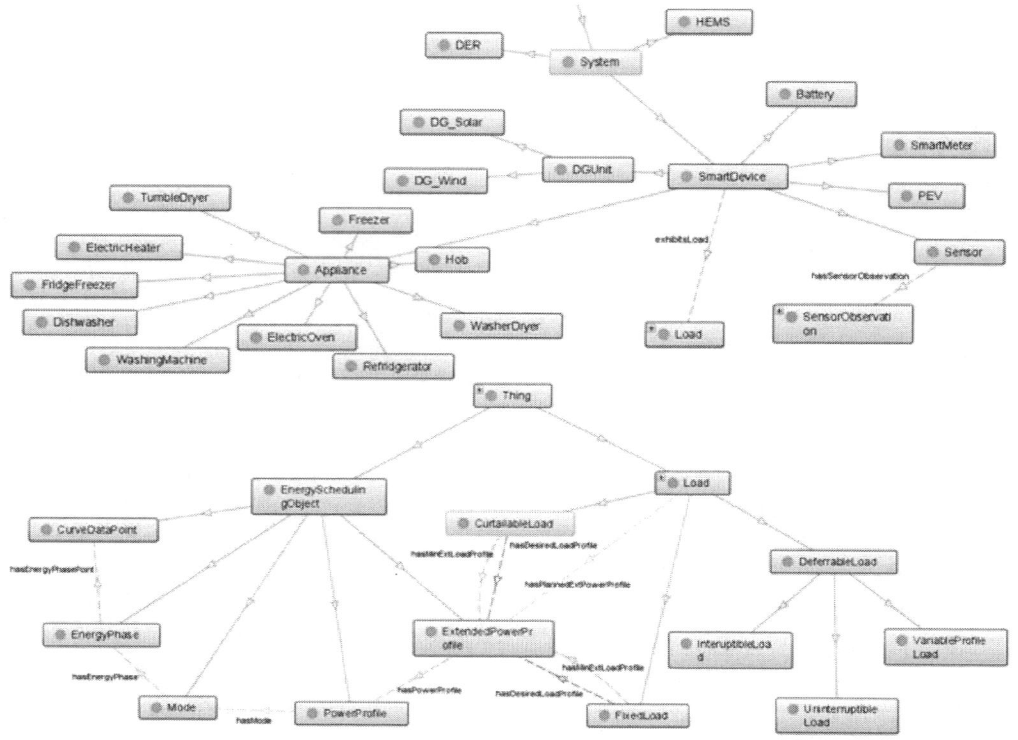

Figure 4. Excerpt of the ontology showing the system taxonomy and load scheduling concepts.

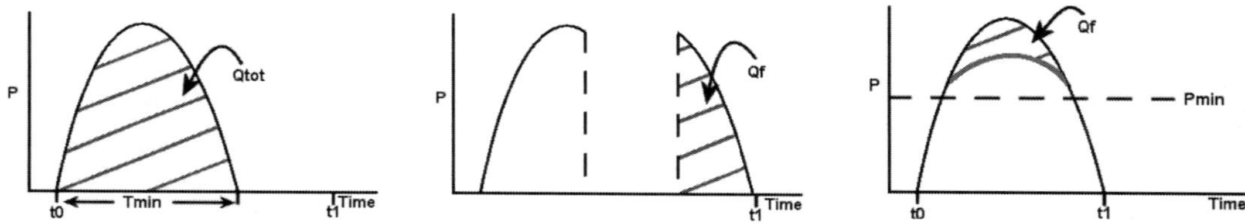

Figure 5. Definition of deferment and curtailment: left—desired load, middle- deferred load, right—load curtailment.

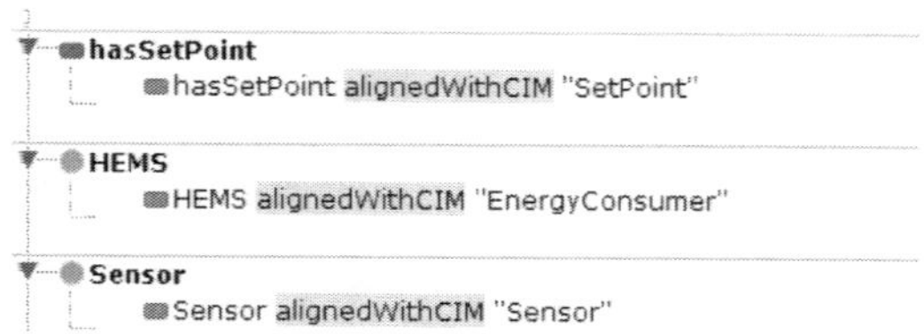

Figure 6. Alignment of ontological concepts with CIM.

```
public class DeferrableLoad extends DomesticLoad {

    public void setHasTotalEnergyDemand(java.lang.Float HasTotalEnergyDemand) {
        this.HasTotalEnergyDemand = HasTotalEnergyDemand ;
    }
    public java.lang.Float getHasTotalEnergyDemand(){
        return this.HasTotalEnergyDemand;
    }
```

Figure 7. JADE ACL ontology class excerpt.

the candidate ontology, and Eclipse JDT to manipulate Java source code.

Through this conversion process, the ontology's axioms were formalised using JADE constructs, as shown in figure 7, which shows the partial formalisation of a class (including inheritance and data properties).

5 USE CASE: FLEXIBILITY TRADING AND APPLIANCE AUTOMATION

The primary intended use case of the semantic modelling conducted was prosumer demand side management; specifically with a close coupling of home appliance automation and the pea k shaving goals of the smart grid. This was to be achieved through the emergent properties of the considered holonic multi-agent system. As a prerequisite to achieving this, effective interoperability would be required between the domestic smart appliances, DERs, and the home gateway (home energy management system), and between the home gateway and network management agents. These agents could include an aggregator agent, similar to a microgrid controller, but with the added flexibility of being able to change the homes which the agent virtually interfaces with. This benefit is derived from the notion of holonic systems, as described previously. The use case was further focused through the requirement for a flexibility market to emerge, whereby prosumers sell flexibility to their DSO, which the DSO can then intelligently manage and trade amongst intermediate aggregator agents, and hence present a more stable demand profile to the higher voltage grid.

6 USE CASE: INTEROPERATING SMART HOME APPLICATIONS

As well as the primary intended use case of intelligent prosumer DSM, a key benefit of the semantic web approach is the potential for re-use and integration of knowledge through ontological extension, re-use, and alignment processes. One potential use case which demonstrates the value of this is shown in figure 8, which illustrates the hypothetical case of a consumer with both a water feedback app and a HEMS interacting with their devices. Figure 8 shows the objects (physical and otherwise) which are relevant to the repositories

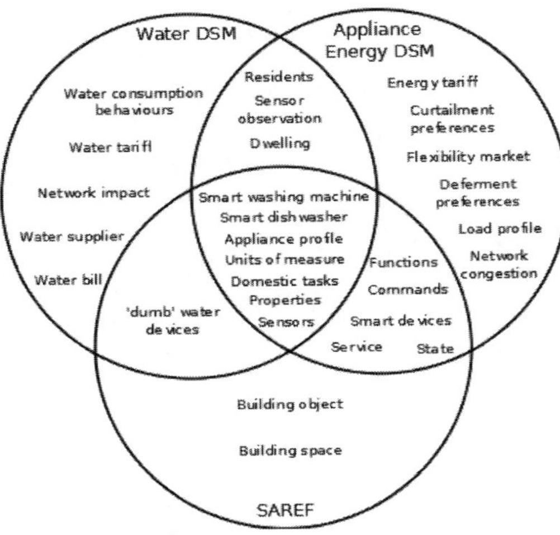

Figure 8. Reuse across smart home applications.

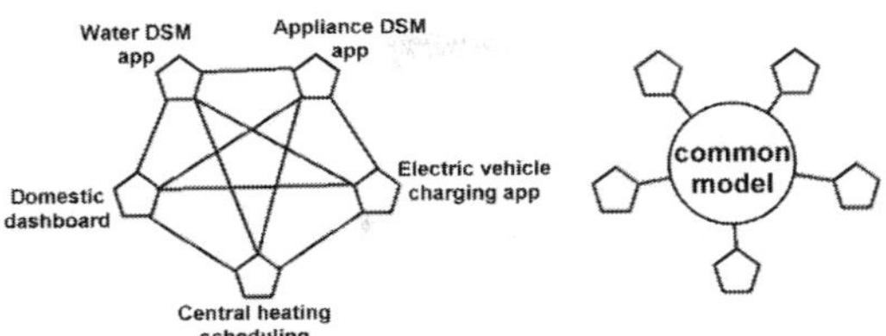

Figure 9. Mapping effort reduction through common model.

of both applications, including those which can be reused from the HEMS ontology by aligning it with a reference ontology such as SAREF (Daniele 2015). If a third application is introduced, and the previous mappings to SAREF have already been completed, this would mean that only one mapping was required to integrate the application, as opposed to mapping to both of the other applications. This is illustrated further in figure 9 as a means to avoid exponential mapping task growth in the likely future case of many integrated software artefacts.

7 DISCUSSION & CONCLUSION

This paper has described the evolution of the energy landscape towards a highly distributed system of systems, with a close coupling between the physical artefacts involved in energy generation, distribution, and consumption, and the cyber artefacts which optimize their management. From this evolution, the requirement for highly effective interoperability between agents in the domain was observed, which has been expressed by authoritative sources as an issue of protocol and semantic heterogeneity, the latter of which was explored in some detail. This included a contribution towards overcoming the semantic heterogeneity present in the domain, through a semantic web approach within a holonic multi-agent system environment.

The holonic system approach was introduced, then applied to the problem space through the a multi-agent architecture intended to optimize demand in a network by intelligently trading flexibility, whilst respecting individual desires and beliefs. A knowledge management architecture was presented which allowed the integration of both traditional agents, and web services, which are entities that do not exhibit traits such as autonomy, but which have the advantage of integrating traditional agents with modern web technologies. Such a scenario could be beneficial if weather prediction knowledge was produced by a web service which applied artificial intelligence to historical and real time weather data in the cloud.

The paper then presented the main contribution; a domain ontology suitable for message payload specification and semantic web deployment. The ontology's class hierarchy and object properties were presented, and then the concept of flexibility was formalized to facilitate this key aspect of the intended use case. Its alignment with existing models was then indicated, before a description of its conversion to a format suitable for use in the JADE platform. The primary use case of the ontology and system was then described, and the key benefits of the semantic web approach of re-use and integration of knowledge were illustrated through another use case.

Further work will develop the surrounding eco-system of value-added services required to deliver the described use cases. Also, further validation of the ontology, methodology, and system design, will be conducted by analyzing its performance against functional requirements, and comparing it to existing approaches.

The potential for semantic models to facilitate a common domain conceptualization and hence the interoperability of virtual artifacts represents a critical enabling step towards highly distributed energy systems. The diversity and prevalence of interoperating components is increasing and is leveraged as an opportunity in holonic systems, but the ability to share and fully utilize data between these components is critical to their intelligent management.

REFERENCES

Ahat M, Ben Amor S, & Bui A. Agent Based Modeling of Ecodistricts with Smart Grid. Adv Comput Methods Knowl Eng 2013;479:307–18.

Baclawski K. Semantic Interoperability for Big Data. Northeastern University 2014.

Baños R, Manzano-Agugliaro F, Montoya FG, Gil C, Alcayde A, & Gómez J. Optimization methods applied to renewable and sustainable energy: A review. Renew Sustain Energy Rev 2011;15:1753–66. doi:10.1016/j.rser.2010.12.008.

BSI. Smart city concept model- guide to establishing a model for data interoperability. PAS 182:2014. 2014.

buildingSMART. Industry Foundation Classes Version 4 2013.

Daniele, L. Smart Appliance Reference Ontology. 2015.

Department of Energy. History of Wind Energy | Department of Energy n.d. http://energy.gov/eere/wind/history-wind-energy (accessed August 14, 2015).

Dominguez-Garcia AD, Hadjicostis CN, & Vaidya NH. Resilient Networked Control of Distributed Energy Resources. IEEE J Sel Areas Commun 2012;30: 1137–48. doi:10.1109/JSAC.2012.120711.

El-Hawary ME. The smart grid—State-of-the-art and future trends. Electr Power Compon Syst 2014;42: 239–50.

Energy Saving Trust. Feed-in Tariff scheme | Energy Saving Trust n.d. http://www.energysavingtrust.org.uk/domestic/feed-tariff-scheme (accessed August 14, 2015).

Energy@home. Enegy@home data model 2.1. 2015.

Ghosn SB, Ranganathan P, Salem S, Tang J, Loegering D, & Nygard KE. Agent-oriented designs for a self healing smart grid. Smart Grid Commun. SmartGridComm 2010 First IEEE Int. Conf. On, IEEE; 2010, p. 461–6.

Giordano V, Gangale F, Fulli G, Jiménez MS, Onyeji I, & Colta A, et al. Smart Grid projects in Europe. JRC Ref Rep Sy 2011;8.

Go Solar California. History of Solar Energy in California—Go Solar California n.d. http://www.gosolarcalifornia.ca.gov/about/gosolar/california.php (accessed August 14, 2015).

Gröger G, & Plümer L. CityGML—Interoperable semantic 3D city models. ISPRS J Photogramm Remote Sens 2012;71:12–33. doi:10.1016/j.isprsjprs.2012.04.004.

Harper R. UK Trends in Microgeneration Adoption. RWE 2008.

Hommelberg MPF, Warmer CJ, Kamphuis IG, Kok JK, & Schaeffer GJ. Distributed Control Concepts using Multi-Agent technology and Automatic Markets: An indispensable feature of smart power grids, IEEE; 2007, p. 1–7. doi:10.1109/PES.2007.385969.

IEA-RETD. Residential Prosumers—Drivers and Policy Options 2014. http://iea-retd.org/wp-content/uploads/2014/09/RE-PROSUMERS_IEA-RETD_2014.pdf (accessed September 3, 2015).

IEC, "Energy management system application program interface (EMS-API)—Part 301: Common information model (CIM) base (IEC 61970–301:2011)," British Standards Institution, Nov. 2011.

IEC. IEC 61968: Common Information Model (CIM) / Distribution Management. 2003.

IEEE Standards Committee, IEEE Standards Coordinating Committee 21 on Fuel Cells P Dispersed Generation and Energy Storage, Institute of Electrical and Electronics Engineers, IEEE-SA Standards Board. IEEE guide for smart grid interoperability of energy technology and information technology operation with the Electric Power System (EPS), end-use applications and loads. New York, N.Y.: Institute of Electrical and Electronics Engineers; 2011.

IEEE. The Relationship Between Smart Grids and Smart Cities—IEEE Smart Grid n.d. http://smartgrid.ieee.org/may-2013/869-the-relationship-between-smart-grids-and-smart-cities (accessed August 14, 2015).

Jiayi H, Chuanwen J, & Rong X. A review on distributed energy resources and MicroGrid. Renew Sustain Energy Rev 2008;12:2472–83. doi:10.1016/j.rser.2007.06.004.

Jones AE, Irwin M, & Izadian A. Incentives for Microgeneration Development in the US and Europe. IECON 2010-36th Annu. Conf. IEEE Ind. Electron. Soc., IEEE; 2010, p. 3018–21.

Manfren M, Caputo P, & Costa G. Paradigm shift in urban energy systems through distributed generation: Methods and models. Appl Energy 2011;88:1032–48. doi:10.1016/j.apenergy.2010.10.018.

Martin J. Distributed vs. centralized electricity generation: are we witnessing a change of paradigm. Introd Distrib Gener Paris HEC 2009.

Mohsenian-Rad A-H, Wong VWS, Jatskevich J, & Schober R. Optimal and autonomous incentive-based energy consumption scheduling algorithm for smart grid. Innov. Smart Grid Technol. ISGT 2010, 2010, pp. 1–6. doi:10.1109/ISGT.2010.5434752.

Ofgem. About the Domestic Renewable Heat Incentive | Ofgem n.d.

OpenADR. OpenADR 2.0 Profile Specification. 2013.

Pahwa A, DeLoach SA, Natarajan B, Das S, Malekpour AR, Shafiul Alam SMS, et al. Goal-Based Holonic Multiagent System for Operation of Power Distribution Systems. IEEE Trans Smart Grid 2015:1–1. doi:10.1109/TSG.2015.2404334.

Palensky P, & Dietrich D. Demand Side Management: Demand Response, Intelligent Energy Systems, and Smart Loads. Ind Inform IEEE Trans On 2011;7: 381–8. doi:10.1109/TII.2011.2158841.

Poslad S. Specifying protocols for multi-agent systems interaction. ACM Trans Auton Adapt Syst 2007;2:15–es. doi:10.1145/1293731.1293735.

REN21. 10 years of renewable energy progress 2014.

Rey-Robert X. Smarter Cities—Dublin event. IBM 2009.

Stanford University. Protégé 5.0. 2016.

UNEP. Agenda 21—United Nations Environment Programme (UNEP) n.d. http://www.unep.org/Documents.Multilingual/Default.asp?documentid=52 (accessed August 26, 2015).

United Nations framework convention on climate change 1995. http://www.official-documents.gov.uk/document/cm28/2833/2833.pdf (accessed November 6, 2015).

United Nations. Kyoto Protocol to the United Nations Framework Convention on Climate Change 1998.

Van Dam KH, & Keirstead J. Re-use of an ontology for modelling urban energy systems. 3rd Int. Conf. Gener. Infrastruct. Syst. Eco-Cities INFRA 2010—Conf. Proc., 2010. doi:10.1109/INFRA.2010.5679232.

Vašek L, Dolinay V, & Sysala T. Heat production and distribution control system based on holonic concept. WSEAS Trans Heat Mass Transf 2014.

Von Dollen D. Report to NIST on the smart grid interoperability standards roadmap. Prep Electr Power Res Inst NIST June 2009 2009.

W3C. OWL2 Web Ontology Language Document Overview. 2009.

Wang Q, Zhang C, Ding Y, Xydis G, Wang J, & Østergaard J. Review of real-time electricity markets for integrating Distributed Energy Resources and Demand Response. Appl Energy 2015;138:695–706.

Werbos PJ. Computational Intelligence for the Smart Grid-History, Challenges, and Opportunities. IEEE Comput Intell Mag 2011;6:14–21. doi:10.1109/MCI.2011.941587.

Smart cities

eWork and eBusiness in Architecture, Engineering and Construction – Christodoulou & Scherer (Eds)
© 2016 Taylor & Francis Group, ISBN 978-1-138-03280-4

Energy matching and trading within green building neighborhoods based on stochastic approach considering uncertainty

S.S. Ghazimirsaeid & T. Fernando
University of Salford, Salford, UK

M. Marzband
The University of Manchester, Manchester, UK
Department of Electrical Engineering, Lahijan Branch, Islamic Azad University, Lahijan, Iran

ABSTRACT: Non-dispatchable generation resources can be installed as small scale generation units that environmentally and economically could be competitive with conventional power generation. To reach this aim, a hybrid system including several types of non-dispatchable generation, dispatchable generation resources incorporated with energy storage assets can provide a sustainable necessary electricity/thermal/water pumping power during a green building's daily operation. The objective of this paper is to model a dynamic system for a single green building considering several generation resources for feeding of some electrical and thermal specific load demands needed in a sustainable way. The proposed model based on a dynamic decision process is implemented to manage and monitor a complex hybrid system encompassing several generation resources and load demands by considering various uncertainties. In order to handle the uncertainties, scenario generation approach is utilized. The model is developed in The General Algebraic Modeling System (GAMS) environment in order to determine the optimal solution with scheduling resources by setting up the optimal power set-points for them. The optimization model is applied to a case study where the produced power is also used to supply water pumping for domestic consumption. Furthermore, other capabilities such as extendibility, reliability, and flexibility are examined about the proposed approach.

Keywords: Green buildings, energy management system, mixed integer linear programming, distributed energy resources

1 INTRODUCTION

A continuous and optimal supply of energy demand throughout a day in green building is an incredibly complex issue which could be addressed by future research. Due to the intermittency nature of the renewable energy resources and the load demand, some alternative cooling/heating systems with integrated energy storage are poised for success. To accomplish this, some strategies can be also profitably used to manage the energy storage. It can be adjusted to improve efficiency, to maintaining stability under increasingly environment robustness. The most important feature of this may be its ability to reduce carbon dioxide emissions as well as to provide an optimal mode scheduling in hybrid systems which will be installed on a green building [1]–[3]. Under this direction, renewable energy technologies that utilize renewable energy resources such as the sun, the wind, biomass, geothermal, water, and so on can be used in the green buildings [4]. These energy generation resources are naturally available and environmental friendly. Since the energy supplied by the renewables is unpredictable and intermittent, uncertainty analysis is essential tool to obtain a robust representation of model predictions consistent with the state-of-knowledge [5]. Furthermore, Energy Storage (ES) devices integrated with these generation resources based on renewable energy sources can be a great way to increase power reliability, improve power quality, decrease electricity bills, reduce carbon footprint, or even realize energy autonomy [3]. During daily operation in these systems, the power exchange between household and upstream grid can be treated if there is insufficiency or excess generations in either side. It can help to verify the possibility of a power sharing optimization between these parties. In this regard, a proper Energy Management System (EMS) is crucial to avoid any mismatch in power [5]. Moreover, EMS in the top-level management should be able to supply all load demands including large residential/commercial buildings and industrial consumers without any interruption [4]. It means that if the total generated power by renewable resources at each time interval in the green building is bigger than the load demand, this excess generation can be applied to supply ES, to feed responsive load demand as well as to sell to the upstream grid. In addition, it is possible that this green building may have a generation shortage during daily operation system. All these cases, and others, can be controlled and monitored by the developed EMS.

The contributions of this paper are as follows:

1. To propose a comprehensive model based on Mixed Integer Non-Linear Programming (MINLP) within a green building toward scheduling for the day-ahead operation plan of a grid-connected green building. In this context, a comprehensive model is presented and illustrated that it is flexible and could be used for different configurations of green building systems;
2. To formulate an uncertainty mechanism based on scenario analysis approach.

2 PROBLEM STATEMENT

A green building connected to the upstream grid with several energy generation facilities and energy storage technology and load demand is depicted in Figure 1. The proposed system is developed for a household where electrical, thermal and water pumping is needed at the same time by the consumers, so that the distributed energy resources (DER) have to be the optimal choice to determine the supply and load demand. The important goal of this structure is the maximum use of renewable and existing resources in green buildings and also optimum management of existing loads in it while satisfying pay-off function for both sides. In the proposed structure, the decision making variables are the quantities (the production and consumption resources offers). The performance of green building is studied in a typical day on an hourly basis, under the assumption that a similar operation could be done for each day. The proposed problem in this paper is multi-period and thought to be applied for planning and scheduling hourly day-ahead purposes of green building and DER, that is, not in real time. However, it can be easily expanded and adopted to the real-time applications. As seen in this figure, the system has several DERs which can easily deliver the desired output power within a board range of operating parameters and some given constraints.

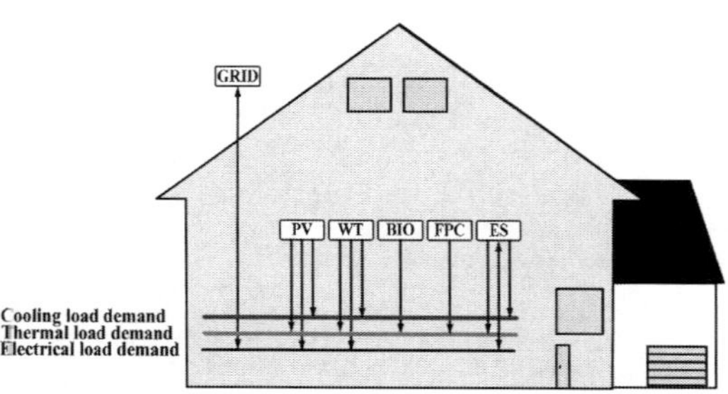

Figure 1. Energy flows between the hybrid system and the green household.

3 EMS-MINLP ARCHITECTURE

The proposed framework in this paper to develop an optimal decision support system is shown in Figure 2. This framework for energy management system based on mixed integer non-linear programming (termed EMS-MINLP herein) is composed to two sub-units, namely scenario generation unit, and Energy Management System (EMS) unit. As seen in this figure, first, the predicted values of electrical/thermal and water pumping load demand, wind power generation and solar irradiation should be sent to scenario generation unit in order to integrate uncertainties inherent to the evaluations. The relevant scenarios of each parameter with the related probabilities can be generated in this unit. This unit is modelled using the scenario decision tree method. After calculation of weighted values of each uncertainty parameters in scenario generation unit, these values will be sent to EMS unit. In this unit makes decisions, for each period, about the quantities to electricity and thermal generate and related procedures for electricity supply, to buy/sell from/to the upstream grid, considering technical and economic constraints.

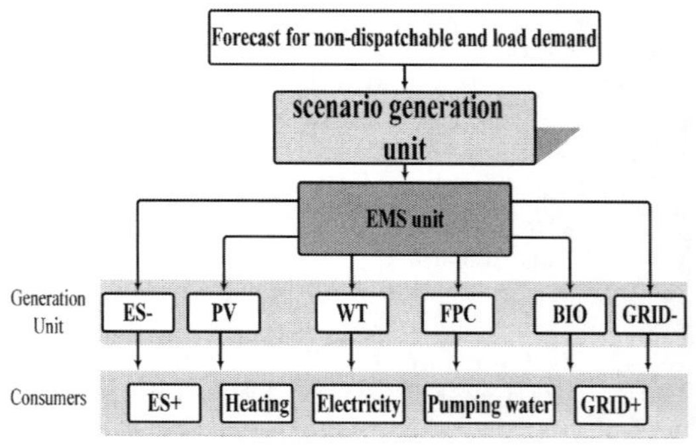

Figure 2. EMS-MINLP architecture.

The EMS unit assign priority to overcome the shortage of electricity/thermal, offer to decrease production or increase consumption in situations when there is a surplus of electricity in the system based on offer prices associated with generation units or the defined objective function. In addition, it can choice best possible energy alternative to trade energy between green building and upstream grid aiming to optimize a certain objective function.

The optimization problem proposed in EMS-MINLP unit is a non-linear programming in the general case and a linear programming in some particular cases. This optimization problem consists of an objective function and a set of constraints. A general description of this optimization problem is provided in what follows, and they will be described in detail in the problem formulation section.

4 PROBLEM FORMULATION

The system under study is considered a grid-connected green building including non-dispatchable generation resources (wind turbine-WT, photovoltaic-PV and flat plate collector-FPC) and dispatchable generation resources (biomass-BIO) and energy storage-ES supplying some non-responsive load demand (electrical, thermal and water pumping). The aim of the proposed EMS is to minimize thermal and electrical losses, to maximize the energy sold to the upstream grid, to increase the energy stored in the ES or to improve state-of-charge (SOC). The optimization problem is defined as the following objective function:

$$Z=\sum_{t=1}^{T}\left(\begin{array}{l}\alpha\cdot\left(\begin{array}{l}E_{(t,s)}^{WT,th}+E_{(t,s)}^{PV,th}+E_{(t,s)}^{FPC,th}+\\E_{(t,s)}^{BIO,th}+E_{(t,s)}^{GRID,th}+\\E_{(t,s)}^{ES,th}-E_{(t,s)}^{D,th}\end{array}\right)^{2}+\\\beta\cdot\left(\begin{array}{l}E_{(t,s)}^{WT,ele}+E_{(t,s)}^{PV,ele}+\\E_{(t,s)}^{GRID,ele}+E_{(t,s)}^{ES,ele}-E_{(t,s)}^{D,ele}\end{array}\right)^{2}+\\\theta\cdot\left(\begin{array}{l}\dfrac{E_{(t,s)}^{WT,ele}+E_{(t,s)}^{PV,ele}+E_{(t,s)}^{GRID,ele}+E_{(t,s)}^{ES,ele}}{\rho gh}\cdot\eta_{ps}-\\E_{(t,s)}^{D,wp}\end{array}\right)^{2}+\\\varphi\cdot E_{(t,s)}^{GRID,ele}+\gamma\cdot E_{(t,s)}^{ES,ele}-\rho\cdot SOC_{(t,s)}^{ES}\end{array}\right)\tag{1}$$

$\forall t\in\{0,\cdots,T\}$

where T is the number of simulation periods in time interval t. The objective function in (1) allows autonomous or grid connected decision making to determine the hourly optimal dispatch of generators depending on system technical and economic constraints. The three first six items in (1) represent the energy losses due to the transport of electrical/thermal and water pumping power energy, related to the characteristics of the distribution system, the supplied demand and types of equipment in use. The forth item is the net electricity power between upstream grid and green building. The two last items are included in the objective function to improve ES performance in order to supply the load demand as economically as possible within pre-defined continuity, quality and security patterns. $\alpha,\beta,\varphi,\gamma$ and ρ are weighting factors. $E_{(t,s)}^{A,th}$ is the generated energy by A resources which can be estimated by

$$E_{(t,s)}^{A,th}=P_{(t,s)}^{A,th},\forall A\in\left\{\begin{array}{l}WT,PV,FPC,BIO,GRID+,\\GRID-,ES+,ES-\end{array}\right\}\tag{2}$$

The decision variables of the optimization problem are the power produced by generation resources and the consumed power by consumers, and SOC. The minimization of the objective function is subject to the following constraints.

A *Wind turbine (WT)*

The following model is used to simulate the electrical power generated by WT [6]:

$$P_{(t,s)}^{WT}=\begin{cases}0 & V_{(t,s)}\le V_{ci}\\P_r^{WT}\times\dfrac{\left(V_{(t,s)}\right)^2-V_{ci}^2}{V_r^2-V_{ci}^2} & V_{ci}\le V_{(t,s)}\le V_r\\P_r^{WT} & V_r\le V_{(t,s)}\le V_{co}\\0 & V_{(t,s)}>V_{co}\end{cases}\tag{3}$$

where $V_{(t,s)}$ is the wind speed in time interval t under scenario s (m/s). It is worth to mention that the wind speed is predicted by some meteorological model and hence these predictions are retained as realistic ones. P_r^{WT} represent the rated electrical power, V_{ci}, V_{co} and V_r are the cut in, the cut off and the rated wind speed, respectively. In general, the wind speed measurements are given at a height different than the hub height of the wind turbine which can be expressed by

$$V_{(t,s)}=\tilde{V}_t\times\frac{\ln(H_{hub}/z)}{\ln(H_{meas}/z)},\quad\forall t\tag{4}$$

where H_{hub} and H_{meas} are the hub height and the height of the measurement, respectively. z is the surface roughness length and $\tilde{V}_t$ is the forecasted wind speed at the height of the measurement.

B *Photovoltaic (PV) [7]*

The power generated from PV modules can be calculated using the following formula [8]:

$$P_{(t,s)}^{PV} = S^{PV} \cdot \eta^{PV} \cdot p^{f} \cdot \eta^{PV} \cdot G_{(t,s)} \quad (5)$$

where S^{PV} is the solar cell array area, η^{PV} is the module reference efficiency, p^{f} is the packing factor, η^{PV} is the power conditioning efficiency and $G_{(t,s)}$ is the forecasted hourly irradiation.

C *Flat Plate Collector (FPC)*

The useful thermal energy extracted from the water collector depends on the instantaneous incident solar irradiation, the plate area, and its efficiency [9]. It can be formulated as:

$$E_{(t,s)}^{FPC} = \eta^{FPC} \cdot A^{FPC} \cdot G_{(t,s)} \cdot \Delta t \quad (6)$$

where η^{FPC}, A^{FPC} and $G_{(t,s)}$ are the efficiency of the solar FPC, the area and the forecasted hourly irradiation, respectively. Δt is the energy management time step. During summer period when normal heat supply to the district heating is required less, water for heating passing through the FPC may be stopped. This constraint can be formulized as follows:

$$E_{(t,s)}^{FPC} \geq E_{(t,s)}^{FPC,th} \quad (7)$$

D *Biomass (BIO)*

Energy provided by the biomass heating plant depends on the used biomass quantity $u_{(t,s)}$, the biomass Volumetric Mass (VM) (i.e. the ratio between the dry mass and the volume), the Lower Heating Value (LHV).

The LHV assumes that the latent heat of vaporization of water in the fuel and the reaction products is not recovered. Then, it can be calculated once Higher Heating Value (HHV) and Moisture Content (MC) are known. The HHV is the total energy release in the combustion with all of the products at 273 K in their natural state when water has released its latent heat of condensation. In the present work, the HHV is evaluated from the basic data analysis of biomass. The biomass MC represents the water amount present in the biomass and it can be expressed as a percentage of the dry weight. As regards production plant, the plant is supposed to operate at the maximum productivity level. The following equation provides the plant developed energy [10]:

$$E_{(t,s)}^{BIO} = f \cdot \eta^{BIO} \cdot LHV \cdot u_{(t,s)} \cdot VM \quad (8)$$

Biomass may not be used (especially during summer, heating is not necessary and cannot be sent to the network), as a result this constraint can be defined as follows:

$$E_{(t,s)}^{BIO} \geq E_{(t,s)}^{BIO,th} \quad (9)$$

E *Energy storage (ES) [3], [11]–[13]*

$$SOC_{(t,s)}^{ES} \leq C_{Tot,(t,s)}^{ES} \quad (10)$$

$$E_{(t,s)}^{ES} = E_{(t,s)}^{ES,ele} + E_{(t,s)}^{ES,th} + E_{(t,s)}^{ES,wp} \quad (11)$$

$$E_{(t+1,s)}^{ES} = E_{(t,s)}^{ES} + E_{(t+1,s)}^{WT,ele} + E_{(t+1,s)}^{PV,ele} + E_{(t+1,s)}^{GRID,ele} \quad (12)$$

F *Upstream grid (GRID)*

$$E_{(t,s)}^{GRID,ele} \leq E_{(t,s)}^{WT,ele} + E_{(t,s)}^{PV,ele} \quad (13)$$

G *Equations*

$$E_{(t,s)}^{th} = \begin{bmatrix} E_{(t,s)}^{WT,th} + E_{(t,s)}^{PV,th} + E_{(t,s)}^{FPC,th} \\ + E_{(t,s)}^{BIO,th} + E_{(t,s)}^{GRID,th} + E_{(t,s)}^{ES,th} \end{bmatrix} \quad (14)$$

$$E_{(t,s)}^{ele} = E_{(t,s)}^{WT,ele} + E_{(t,s)}^{PV,ele} + E_{(t,s)}^{GRID,ele} + E_{(t,s)}^{ES,ele} \quad (15)$$

$$E_{(t,s)}^{wp} = E_{(t,s)}^{WT,wp} + E_{(t,s)}^{PV,wp} + E_{(t,s)}^{GRID,wp} + E_{(t,s)}^{ES,wp} \quad (16)$$

The amount of pumped water is proportional to the energy used for this purpose, that is: [10]

$$Q_{(t,s)}^{wp} = \frac{E_{(t,s)}^{wp}}{\rho g h} \cdot \eta_{ps} \quad (17)$$

5 RESULT AND DISCUSSION

The operation of the case study is optimized based on available day-ahead hourly non-dispatchable and dispatchable generation resources for given electrical, thermal and water pumping demands. The uncertainty parameters are the electricity and thermal load demand in green building, wind speed, solar radiation. The input of optimization model can be classified as follows:

1. Electrical/thermal and water pumping demands by the consumers;
2. Data about locally available energy resources including solar irradiation data (w/m^2) and wind speed (m/s) as shown in Figure 3;

3. Technical and economic performance of non-dispatchable and dispatchable DERs. These characteristic include, for example, rated power for PV, power curve for WT, the capacity of ES and the initial value of SOC.

To begin, EMS-MINLP algorithm receives data including the generated power by non-dispatchable DERs and load demands provided by the scenario generation unit, the ES SOC. Then, all the optimal power set-points of other generation resources will be dispatched to them at each time interval based on the EMS algorithm in this unit. The electrical/thermal and water pumping load demand profiles are shown in Figure 4. The real life experimental data carried out from [3], [10] are used to simulate WT, PV and load demand profiles.

The power generated by non-dispatchable and dispachable DER resources (BIO, ES during discharging operating mode, FPC, PV, and WT) and the electricity bought from upstream grid are presented in Figure 5. The obtained results, indicated in Figure 5, show that the most of the necessary

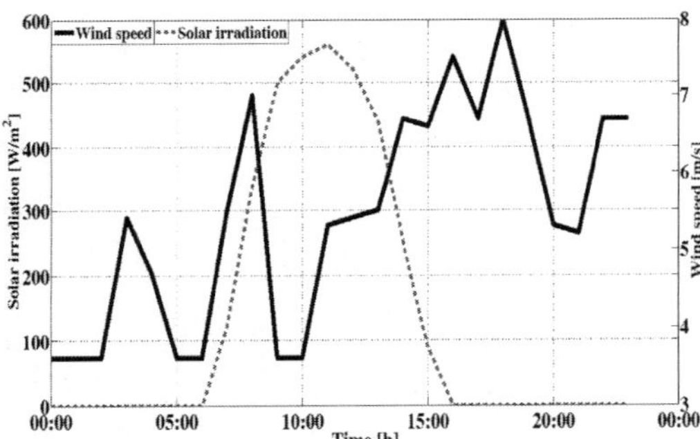

Figure 3. Input wind speed and solar irradiation as used in the model.

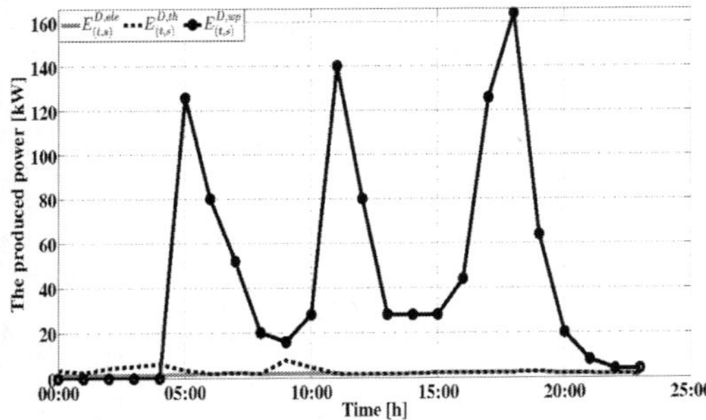

Figure 4. Electrical/thermal and water pump load demand profiles.

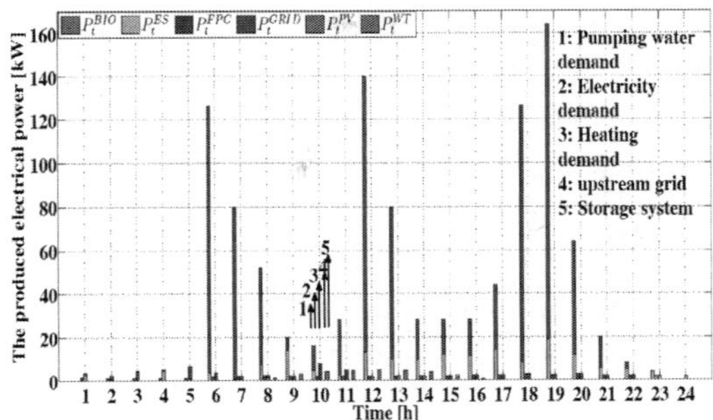

Figure 5. The bar-graph related to the DER resources during system performance.

power for supplying the water pumping and electricity load demands (around 85% and 93%, respectively) is provided by electricity purchased from the upstream grid and the rest of power is supplied by the ES if it is possible according to the present value of SOC and WT generation unit (about 14.5% and 3.5%, respectively). This is while all generated power by PV is only used to supply ES during charging mode. Upstream grid has still a significant role to play for supplying the thermal demand (almost 47%). FPC, BIO and ES are available generation resources that are identified to produce the rest of thermal power (28%, 9.5%, 16%, respectively).

6 CONCLUSION

Since renewable resources have intermittent characteristics, approaches to analyse the energy management in green buildings would be stochastic rather than deterministic. To take the uncertainties into account, scenario generation method is implemented. Decision making model base on stochastic algorithm is presented for energy management within the residential buildings. The principal benefits of the proposed algorithm can be summarized as follows:

1. maximize usage of non-dispatchable based on renewable generation;
2. prioritization for the charging/discharging of the ES devices inside green building as a result reliability enhancement;
3. To better manage, leverage and utilize energy resources and sustained economic growth.

This model also has ability to add a new generation resource which can be usually utilized to install in various hybrid systems. Furthermore, it can identify possible capability in the distributed economic dispatch strategy, where multiple green building with independent EMS taking into account load sharing function can be exploited without accordant modification in design/requirement model. The obtained simulation results show the significant reduction of the total electrical/thermal loses (about 10%) and the significant improvement in ES operation in each time interval. Furthermore, the proposed model can also be applied for the real-time energy management online application.

NOMENCLATURE

The main notation used throughout the report is stated below for quick reference. Other symbols are defined as required.

Acronyms

BIO	Biomass.
DER	distributed energy resources.
ES	energy storage.
ES+, ES-ES	during charging/discharging modes.
FPC	Flat plate collector.
MINLP	Mixed-integer nonlinear programming.
PV	Photovoltaic.
GRID	Upstream grid.
SOC	State-of-charge.
WT	Wind turbine.

Variables

$E^{A,th/ele}_{(t,s)}$	: the produced electrical/thermal energy by A at each time t under scenario s (kWh)
$E^{D,th/ele}_{(t,s)}$	: the electrical/thermal load demand during time interval t under scenario s (kWh)
$E^{D,wp}_{(t,s)}$	: the amount of load demand for water pumping (kWh)
$A \in$	: {WT, PV, FPC, BIO, GRID+, GRID-, ES+, ES–}
$Q^{wp}_{(t,s)}$	: the amount of pumped water in time t under scenario s (m^3/h)
η_{ps}	: pumping water efficiency (%)
ρgh	: density of pumping water (kW/m^3)

REFERENCES

[1] Marzband, M. & Ghadimi, M., Sumper, A. & Domínguez-García, J.L. 2014. Experimental validation of a real-time energy management system using multi-period gravitational search algorithm for microgrids in islanded mode, *Appl. Energy*,128: 164–174.

[2] Marzband, M. & Sumper, A. & Ruiz-álvarez, A. & J.L. Domínguez-García, & Tomoiagâ, B. 2013. Experimental evaluation of a real time energy management system for stand-alone microgrids in day-ahead markets, *Appl. Energy*, 106: 365–376.

[3] Marzband, M. & Sumper, A. & J.L. Domínguez-García, & R. Gumara-Ferret, 2013. Experimental validation of a real time energy management system for microgrids in islanded mode using a local day-ahead electricity market and MINLP, *Energy Convers. Manag.*, 76: 314–322.

[4] Marzband, M. Experimental validation of optimal real-time energy management system for Microgrids, 2013. Ph.D. dissertation, Dept. d'Enginyeria Elèctrica, EU d'Enginyeria Tècnica Industrial de Barcelona, Universitat Politècnica de Catalunya.

[5] Marzband, M. & Azarinejadian, F. Savaghebi. M & J.G. Guerrero, 2015. An Optimal Energy Management System for Islanded Microgrids Based on Multiperiod Artificial Bee Colony Combined With Markov Chain, *IEEE Syst. J.*, 1–11.

[6] Mohamed. F.A. & Koivo, H.N. Multiobjective, 2012. Optimization using Mesh Adaptive Direct Search for power dispatch problem of microgrid, *Int. J. Electr. Power Energy Syst.*, 42(1): 728–735.

[7] Marzband, M. & Sumper, A. 2014. Implementation of an Optimal Energy Management within Islanded Microgrid, *presented at the Int. Conf. Renewable Energies Power Quality*, Cordoba, Spain.

[8] Hocaoğlu, F.O. & Gerek, Ö.N. & M. Kurban, 2009. A novel hybrid (wind-photovoltaic) system sizing procedure, *Sol. Energy*, 83(11): 2019–2028.

[9] Fadar, A. El & Mimet, A. & M. Pérez-García, 2009. Modelling and performance study of a continuous adsorption refrigeration system driven by parabolic trough solar collector, *Sol. Energy*, 83(6): 850–861.

[10] Dagdougui, H. & Minciardi, R. & A. Ouammi, M. Robba, & R. Sacile, 2012. Modeling and optimization of a hybrid system for the energy supply of a 'green' building, in *Energy Conversion and Management*, 64: 351–363.

[11] M. Marzband, N. Parhizi, & J. Adabi, 2015. Optimal energy management for stand-alone microgrids based on multi-period imperialist competition algorithm considering uncertainties: experimental validation, *Int. Trans. Electr. ENERGY Syst.*,30(1): 122–131, Mar.

[12] M. Marzband, & E. Yousefnejad, & A. Sumper, & J.L. Dominguez-Garcia, 2016. Real Time Experimental Implementation of Optimum Energy Management System in Standalone Microgrid by Using Multi-layer Ant Colony Optimization, *Int. J. Electr. Power Energy Syst.*, 75: 265–274.

[13] M. Marzband, & A. Sumper, & M. Chindris, & B. Tomoiaga, *Energy Management System of hybrid MicroGrid with Energy Storage, presented at the Int. Word Energy System Conf.*, Suceava, Romania, 1–3 Jun 2012.

eWork and eBusiness in Architecture, Engineering and Construction – Christodoulou & Scherer (Eds)
© 2016 Taylor & Francis Group, ISBN 978-1-138-03280-4

Using a mobile application to assess building accessibility in smart cities

N. Forcada, M. Macarulla & R. Bortolini
Department of Construction Engineering, Universitat Politècnica de Catalunya, Barcelona, Spain

ABSTRACT: Mobile applications and well-implemented technology can help governments increase the efficiency and effectiveness of information analysis and exchange. This paper presents a case study on the implementation of a mobile application to record accessibility information in the commercial buildings of Terrassa. To record and analyze the accessibility of the commercial buildings a taxonomy including general information and accessibility data was created. The analysis results confirmed the only 1% of the recorded commercial buildings fulfilled all accessible parameters. The main problems were in the access steps or ramps and also in the interior furniture. This data can then be implemented in other applications to be used by people with disabilities. Results also demonstrate that mobile and well-implemented technology can help governments save money and be more efficient. The use of this application simplified and reduced the time needed to record accessibility information while the standardized information helped them obtaining consistent data.

1 INTRODUCTION

Smart cities apply innovative solutions to managing its services and resources to improve citizens' quality of life, whether it's improving the timing of traffic lights or creating useful applications, which becomes more powerful as smartphone penetration continues to increase (Perez et al. 2015).

Some applications enable citizens to obtain real-time information about events in the city finding free parking spaces (Albakour et al. 2014) or report local problems such as illegal trash dumping, faulty street lights, broken tiles on sidewalks, etc. which allow local government agencies to take action to improve the city. Others help government act and implement improvement measures. This is the case of the evaluation of the air quality, building inspections, energy efficiency audits, etc. (Beltran et al. 2015, Kukka et al. 2013).

Within this type of applications, the recording of public building characteristics to evaluate their level of accessibility is rarely taken into consideration. Although there are some applications to inform about accessible restaurants, shops or bars, their data source normally comes from manual analysis and studies. Nowadays, many municipalities still rely on manual processes and paper-based techniques for their building and city inspections and when they incorporate mobile computing applications they are not normally integrated in all municipal systems (Beltran et al. 2015). This drawback can lead to time and information loss.

Mobile applications and well-implemented technology can help governments increase the efficiency and effectiveness of information analysis and exchange (EPRS 2015).

This paper presents a case study on the implementation of a mobile application to record accessibility information in the commercial buildings of Terrassa.

2 METHOD

Firstly, an analysis of the regulation on accessibility in buildings was carried out with the aim to define the parameters to check the fulfilment of the accessibility in buildings.

Secondly, the area where to carry out the study and the methodology and technologies to use to collect the data was determined. For the specific technology a taxonomy on accessibility was required.

Then, field data was registered and analyzed and improvement proposals presented.

Finally an evaluation of the usefulness of the application was carried out.

3 ACCESSIBILITY OF THE BUILT ENVIRONMENT

In almost all countries there exists a general obligation to make public buildings accessible, with specific requirements for certain categories of buildings often extending to both public and private providers. These general obligations are primarily grounded in national non-discrimination legislation with subsidiary national building regulations and codes.

Accessibility of the built environment in Spain is covered by the Código Técnico de la Edificación' (CTE) (Ministerio de Vivienda 2006). The CTE Act provides grounds for eliminating physical barriers to disabled people both for new and existing buildings. CTE is divided in different sections.

3.1 *Approach and access to the building*

The Act (Ministerio de Vivienda 2006) states that the approach to the entrance from the adjacent road, carpark or other area accessible to motor vehicles should be level. Where a level approach is not possible, a step lower than 2 cm or a sloped approach with as gentle a gradient as possible should be provided. Where the slope is provided: the surface should be suitable for wheelchair traffic and reduce the risk of slipping; slopes and landings should have a clear unobstructed width of at least 1.0 m; the gradient should not be higher than 10% in 3.0 m length slopes; a landing of 1.5 m long should be incorporated at the end of the slope to facilitate wheelchair turning; and a handrail on each side of slopes and landings should be incorporated.

The access doorway should provide a minimum clear opening width of not less than 0.8 m and 2 m height. Double doors should include at least one leaf which provides a minimum clear opening width of not less than 0.8 m.

3.2 *Circulation within a building*

Internal doors should be so designed and constructed that provide a minimum clear opening width of not less than 0.8 m.

A corridor or passageway accessible to wheelchair users should have a clear unobstructed width of at least 0.8 m. Where a change of level within a store of a building is necessary because of site constraints or design considerations, either— suitable means of access should be provided between the levels by graded or sloped access or by means of a passenger lift or platform lift, or the same range of services and facilities which is available to able-bodied users of the store should be available and accessible on the level to which independent access for people with disabilities is available.

Toilets should be designed to be able to allow a wheelchair move inside (minimum a circumference of 1.5 m diameter). All the accessories should be adapted.

The furniture should also be designed to allow disabled people access the desk (lower than 0.85 m) and other furniture elements.

4 PICK@GO APPLICATION

Pick & Go is a mobile application developed by the GRIC (Group of Research and Innovation in Construction) in close cooperation with the Municipality of Terrassa. It has been already successfully implemented in other fields such as defect tracking in construction site (Macarulla et al. 2012) and simplifies the information recording process using images and standardized tags.

Pick & Go uses images as a unique entry point to capture field information to spend a limited time by adding tags from an accessibility information classification system to contextualize the image. The use of a classification system (standardized tags) contributes to the standardization and integration of data (Park et al. 2013), makes it easier to analyze the information, and prevents ineffectiveness such as that caused by arbitrary descriptions of observed issues or inconsistences in the fields form (Lin et al. 2014, Abudayyeh et al. 2006).

Tags can be automatically added using the implemented functionalities in the smartphone, such as GPS location, date and hour. This type of tags do not increase the time required to capture field information.

The other tags to characterize the image should be defined by end users. These tags can be numerical or categorical. In this context, an accessibility information classification system is developed based on the existing regulation (See section 5).

Pick & Go also uses voice annotations to transcribe comments using speech recognition software, and graphical annotations such as arrows or rectangles. See Figure 1.

Pick & Go also allows textual annotations to be added, but the aim is for end users to avoid this functionality, because typing a large amount of text in a smartphone is a weighty, onerous task.

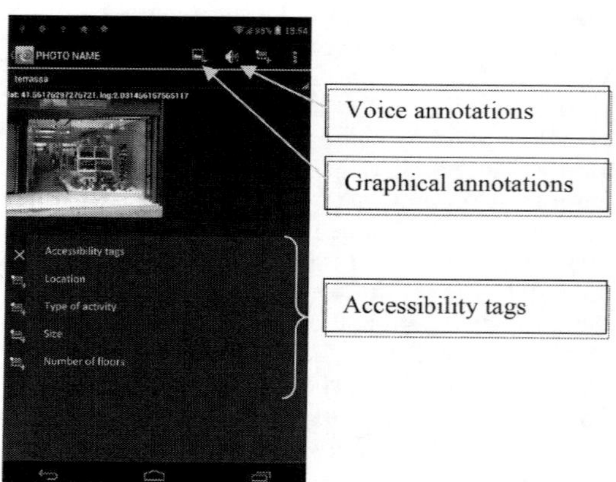

Figure 1. Pick & Go screenshot.

Each image has an associated XML file, containing all the attached information such as the tags. The information is sent to a dropbox account.

Finally an MS Excel file is created to read the information, and add to or modify it. The MS Excel allows basic statistics to be prepared.

5 ACCESSIBILITY INFORMATION CLASSIFICATION SYSTEM

To record and analyze the accessibility of the commercial buildings a taxonomy including general information and accessibility data was created.

General information included:

- Location (address and GPS location).
- Size, divided in to four big groups:
  - Less than 50 m2
  - Between 50 and 100 m2
  - Between 100 and 150 m2
  - Higher than 150 m2.
- Number of floors.
- Type of activity classified by:
  - Bars and restaurants.
  - General services: including travel agencies, banks, casinos and lotteries, toys, games and hobbies, books, music and multimedia, leisure and entertainment, gifts, services, telephony and electronic cleaners and arrangements clothing;
  - Health and aesthetics services: including pharmacies and health and beauty;
  - Food: including markets, supermarkets and hypermarkets;
  - Clothing and accessories: including jewelry, watches and accessories, lingerie and underwear, fashion (clothing, footwear and sports) and optics;
  - Home accessories: including housekeeping, gardening and decorating and home.
  - Others

Accessibility information was oriented to qualify the fulfillment of the accessibility parameters in the entrance, the itinerary, the lifts, the toilets and the furniture. These are classified according to whether they hinder the access to the building or the mobility inside the building:

Access to the building

- Existence of a step higher than 2 cm.
- Existence of a ramp with a slope greater than 10%.
- Entrance door width less than 0.80 m.

Interior of the building

- Existence of a step higher than 2 cm.
- Existence of a ramp with a slope greater than 10%.
- Hallway width less than 0.80 m.
- Toilets adapted for disabled people.
- Desk adapted for disabled people.

6 DATA COLLECTION

Data was collected from 600 commercial buildings in Terrassa.

Data collection started manually by measuring the height of the steps, length and height of the slopes, width of the entrance door, etc. and using paper-based support. This information was then complemented by the use of notes on pictures or drawings. Then, this information was transferred into an excel file. Once 15 commercial buildings were analyzed, the time required to collect and transfer the information into a databased was recorded (8.4 min/premise). This method was considered to be very time consuming and the management of collected information was an arduous task which lead to missing information. During this process, problems such as loss of information, misunderstandings, and unclear information were also common due to the manual data collection and transcription.

Then, Pick & Go application was tested so as to analyze the feasibility to use it for this task. Data from 16 commercial buildings were collected using Pick & Go. Taking into account that data transfer is automatic and data collection is simpler, only 3.1 min per premise were needed (Table 1). Therefore, Pick & Go was used for the collection of accessibility data in the rest of the commercial buildings.

31% of the analyzed commercial buildings were clothing and accessories followed by general

Figure 2. Terrassa's streets where analyzed commercial buildings were located.

services (18%), bars and restaurants (14%) and health and aesthetics services (13%) (Table 2).

7 DATA ANALYSIS

Only 1% of the recorded commercial buildings fulfilled all accessible parameters.

7.1 *Access to the building*

Only 10.5% of the commercial buildings had an accessible access (Table 3). Examples of not accessibility fulfilment include access ramp with dimensions outside the limits, ramps that although within the slope limits include a step at the beginning, etc.

Terrassa is a city located in a very sloppy area. Then, many streets have longitudinal slope. During the data capture, 46% (275 commercial buildings) of the analyzed commercial buildings were located in sloppy streets (Fig. 3). In these cases, the height in the midpoint of the step was collected.

78% of the analyzed commercial buildings included a step to access the premise and the other 32% an access ramp.

From those commercial buildings that did not meet the accessibility parameters in the entrance, 80% had an entrance step higher than permitted.

Only 4.8% of the commercial buildings with access ramps met the requirements for disabled access.

Table 1. Analysis of the time required to collect accessibility information manually and using Pick & Go.

Method	Manually	Pick & Go
Nº of analyzed commercial buildings	15	16
Time to collect data (min.)	76	50
Time to transfer data (min.)	50	0
Total time (min.)	126	50
Total time / premise (min.)	8.4	3.1

Table 2. Distribution of commercial buildings by type.

Type of premise	N	%
Bars and restaurants	83	14
General services	108	18
Health and aesthetics services	76	13
Food	54	9
Clothing and accessories	186	31
Home accessories	38	6
Others	55	9
Total	600	100

Table 3. Number of commercial buildings fulfilling accessibility requirements.

Accessibility elements		N	%
Access		63	10.5
	Step	20	3.3
	Ramp	29	4.8
	Doors width	556	92.7
Interior		468	78.0
	Steps	0	0.0
	Ramps	5	0.8
	Corridors' width	531	88.5
	Furniture	197	32.8

Figure 3. Example of non-accessible entrance due to a sloppy access.

The vast majority of the commercial buildings (92.7%) fulfil the width of the access doors (wider than 0.8 m).

7.2 *Interior of the building*

78.0% of the commercial buildings fulfilled accessibility regulations within the premise, either, because they only had one floor or because there were mechanisms or ramps to access the different floors.

Only 20% of the analyzed commercial buildings had steps inside the premise but if so, they were normally wrongly dimensioned due to space restrictions.

Only 6% of the analyzed commercial buildings had ramps within the premise. From these commercial buildings, 86% exceed the maximum slope permitted by the accessibility regulations.

The vast majority of the commercial buildings (88.5%) fulfilled the width of the interior corridors.

Figure 4. Example of non-accessible entrance due to a sloppy ramp.

Only 32.8% of the commercial buildings fulfilled with the furniture requirements in terms of height, dimensions, etc.

Generally, the main problems were in the access steps or ramps and also in the interior furniture.

A proposal for the adaption of the access of these commercial buildings to fulfil accessibility regulations was carried out. The main proposals were: incorporation fixed or mobile ramps both with variable height and slope. Based on the results of the analysis, the estimated costs for each action was determined. The total cost came to 592,080.70 €. Terrassa municipality used this information to determine grants for refurbishing commercial buildings and adapting at least the entrance to these buildings.

8 PICK & GO UTILITY ANALYSIS

For the purpose of this study Pic&Go supported the collection of accessibility data in a consistent and standardized way and reduced the time needed to capture the information.

The application included all necessary functionalities to transfer the data from the device to the cloud. No voice annotations were used because it was not needed. However, for other purposes it might be an interesting feature.

Speech recognition feature was a way to overcome situations in which it is impossible to type textual annotations, for example when wearing gloves. For this study, speech recognition was neither used but also found to be important in other potential studies in colder locations.

The availability of structured information and the automatic generation of excel files with all the structured tracked information facilitated the analysis of data.

This case study suggests that this application can be used for other purposes because it provides efficiency and effectiveness of the field information tracking process.

The major characteristic is the simplicity and the collection of data based on images. For that purpose, the most important and necessary task is to create predefined classification systems for each case study.

Statistical analyses were easily done thanks to the standardized data. Other analysis such as correlations, analysis of variance and contingency analysis can also be undertaken.

Pick & Go was already successfully used for construction follow up, quality incidences in construction works and defect management in urban works. This study confirms the usefulness of this application in different fields that require data capture.

9 CONCLUSIONS

This paper presented a case study on the implementation of a mobile application to record accessibility information in the commercial buildings of Terrassa city center.

The application was developed by the GRIC in close cooperation with the Municipality of Terrassa. It has been already successfully implemented in other fields such as defect tracking in construction site and simplifies the information recording process using images and standardized tags. The use of tags prevents ineffectiveness such as that caused by arbitrary descriptions of observed issues, inconsistent references to the same information, or inconsistences in the fields form. To record and analyze the accessibility of the commercial buildings a taxonomy including general information and accessibility data was created. General information included: location, type of activity, size and number of floors. Accessibility information was oriented to qualify the fulfillment of the accessibility parameters in the entrance, the itinerary, the lifts, the toilets and the furniture. To do so, the entrance width, the height of the steps, the slope of ramps, the dimensions of toilets, etc. were collected. The analysis results confirmed the municipality's initial fear. Only 1% of the recorded commercial buildings fulfilled all accessible parameters. The main problems were in the access steps or ramps and also in the interior furniture.

Results demonstrate that mobile and well-implemented technology can help governments be more efficient. The use of this application simplified and reduced the time needed to record accessibility information while the standardized information helped them obtaining consistent data.

The results of this study allowed Terrassa's City council create an accessibility map of all commercial buildings in the city center and implement improvement actions or sanctions to owners. It also allowed them to determine grants for refurbishing commercial buildings.

These results can be put at the disposal of citizens to know which commercial buildings are adapted and which are the features of each premise related to adapted systems. Information that previously was only accessible at certain points and to certain people can be put at the disposal of any citizen, helping improve their relationship with the city managers, as well as to the work crews that operate in the city.

There is a long list of domains that can benefit from services built around accessibility information. This includes tourism, old people, etc.

On the other hand, the same analysis can be done in other areas such as transport systems, urban areas and housing for renting or selling to enlarge the scope of the results and integrate all elements related to accessibility and mobility in only one system.

However, information is alive so new commercial buildings might be adapted, other might be constructed, demolished or regulations might change. Then, this database should be updated continuously. To do it in the most efficient way, Pic & Go might be put at disposal to the citizens to refine existing data and to collect other accessibility information from other fields such as transportation or town public spaces. In general, it aims to improve transparency on all areas of a city. This data might be integrated in other smart city applications for the integral management of the city, including areas such as energy efficiency, environment, eGovernment, etc.

With the aim to provide a global framework and put all city information at the disposal of the public citizens the system should be connected through different terminals, such as smartphones, PCs, tablets or even TVs.

REFERENCES

Abudayyeh, O., Fredericks, T.K., Butt, S.E. & Shaar, A. 2006. An investigation of management's commitment to construction safety, Int. J. Proj. Manag. 24, 167–174. doi:10.1016/j.ijproman.2005.07.005

Albakour, M.-D., Macdonald, C., Ounis, I., Clarke, C.L.A. & Bicer, V. 2014. Information Access in Smart Cities (i-ASC), Advances in Information Retrieval. *36th European Conference on IR Research, ECIR 2014. Proceedings: LNCS* 8416, 810-14, 2014; 10.1007/978-3-319-06028-6_102.

Beltrán-Ramírez, R., Maciel-Arellano, R., Gómez-Barba, L. Stokes L. & Gonzalez-Sandoval, C. 2015. Mobile applications utilized for the prevention of potential epidemics in smart cities, *Smart Cities Conference (ISC2), 2015 IEEE First International*, Guadalajara, 2015, pp. 1–4. doi: 10.1109/ISC2.2015.7366158

European Parliamentary Research Service EPRS (2015) eGovernment Using technology to improve public services and democratic participation 10.2861/150280

Kukka, H., Kostakos, V., Ojala, T., Ylipulli, J., Suopajrvi, T., Jurmu, M. & Hosio, S. 2013. This is not classified: everyday information seeking and encountering in smart urban spaces. *Personal and Ubiquitous Computing*. 17(1), 15–27.

Lin, K.-Y., Tsai, M.-H., Gatti, U.C., Je-Chian Lin, J., Lee, C.-H. & Kang, S.-C. 2014. A user-centered information and communication technology (ICT) tool to improve safety inspections, Autom. Constr. 48 53–63. doi:10.1016/j.autcon.2014.08.012

Macarulla, M., Forcada, N., Casals, M. & Kubicki, S. 2012. Tracking construction defects based on images, *eWork Ebus. Archit. Eng. Constr.—Proc. Eur. Conf. Prod. Process Model. 2012, ECPPM 723–729.*

Macarulla, M., Forcada, N., Casals, M., Gangolells, M., Fuertes, A. & Roca, X. 2013. Standardizing Housing Defects: Classification, Validation, and Benefits, *J. Constr. Eng. Manag*. 139, 968–976. doi:10.1061/(ASCE)CO.1943-7862.0000669

Ministerio de Vivienda. 2006. Código Técnico de la Edificación. 314/2006.

Park, C.-S., Lee, D.-Y., Kwon, O.-S. & Wang, X. 2013. A framework for proactive construction defect management using BIM, augmented reality and ontology-based data collection template, Autom. Constr. 33 61–71. doi:10.1016/j.autcon.2012.09.010

Perez, I, Poncela, J, Moreno-Roldan, J.M. & Memon, M.S. 2015. IntelCity, Multiplatform Development of Information Access Platform for Smart Cities, *Wireless Personal Communications*, 85(2) 463-81, 10.1007/s11277-015-2749-8

eWork and eBusiness in Architecture, Engineering and Construction – Christodoulou & Scherer (Eds)
© 2016 Taylor & Francis Group, ISBN 978-1-138-03280-4

Prediction of traffic characteristics in smart cities based on deep learning mechanisms

V. Gkania & L. Dimitriou
Laboratory of Transportation Engineering, Department of Civil and Environmental Engineering, University of Cyprus, Nicosia, Cyprus

ABSTRACT: Cities worldwide face rapid growth and huge transportation challenges. By monitoring traffic performance and patterns over time, cities can ensure they operate at full capacity. The prediction of traffic characteristics such as traffic flow and travel time stands for an important feature in Advanced Traveler Information Systems (ATIS). In the era of data availability, the dissemination of accurate traffic information to travelers could have a huge impact on their trip choices and thus in systems' performance. The scope of this paper is traffic modeling and prediction based on artificial neural networks namely deep learning mechanisms. The proposed framework enables the estimation of traffic characteristics between a predefined set of Origin-Destinations (O-D) locations, by taking into account available disaggregate traffic data. The proposed application is tested on a realistic road system, namely that of Cyprus. The aim of the study is to provide reliable travel information to users in order to improve significantly the use of the existing transportation networks.

1 INTRODUCTION

Traffic prediction methods applied both for urban and interurban areas draw the attention of researchers and the industry for the past decades. The scope of these methods was to forecast traffic states of the road network (both for short-term as well as for longer prediction horizon), by estimating variables such as traffic flow or travel time between predefined locations. This type of information is valuable in Advanced Traffic Management Systems (ATMS) for surveillance, managerial, or control purposes (e.g. in ATIS applications).

Typical predictions approaches are based on the traffic models, enhanced with statistical methods and time series analysis for estimations. Moreover, in the applications that have been tested up to now, prediction accuracy or application value are limited to small part of larger systems, especially within the urban environment where the road networks are extremely extended and the traffic conditions are changing rabidly. With the last decades evolution of computational intelligence, data-driven methods gained ground in predictive traffic analytics. Many methods belonging to the Artificial Intelligence class of models (e.g. Artificial Neural Networks-ANN, Support Vector Machines, etc.), able to capture interrelations among very complex traffic flow phenomena even under unstable traffic conditions have been tested with notable performance in location specific test-cases.

The proposed study investigates different ANN configurations in order to model the physical process that generates the observed traffic flow. Different combinations of input information (mainly flow and average speed) are used. The output results are evaluated using statistical terms. As a result, an overall view of the loading aspects is achieved, with network-wide coverage.

The paper is organized into 6 Sections. In Section 2 related work and motivation are presented while in Section 3 the analysis area is introduced. In Section 4 the experimental setup is analyzed while Section 5 details the results of the particular framework. Finally Section 6 summarizes and concludes the contribution of the study as well as further work remaining.

2 BACKGROUND AND MOTIVATION

One of the most critical aspects of ATIS success is the provision of accurate real-time information and short-term predictions of traffic parameters such as traffic volumes, travel speeds and occupancies. Thus considerable research efforts have been focused on short-term traffic forecasting. In Vlahogianni et al. (2004) a detailed overview of objectives and methods for short-term traffic forecasting is presented.

Clustering short-term forecasting based on the area of implementation, the large part concerns freeways and highways. Relatively work can be found in van Lint et al. (2002) where the authors predicted freeway travel time using State-Space Neural Networks. In cases of urban areas, forecasting problem becomes more complex. Some examples of forecasting methods in urban areas can be found in Stathopoulos and Karlaftis (2003) where a multivariate state space approach was used for urban traffic flow modelling and prediction. In Dimitriou et al. (2008) an adaptive hybrid fuzzy rule-based system approach was used for modelling and prediction of urban traffic flow. In Vlahogianni et al. (2005) an optimized and meta-optimized neural networks for short-term traffic flow prediction was attempted in an urban arterial. In Ghosh et al. (2009) traffic flow forecasting methodology including time series analysis, was applied at a network of ten intersections in the city of Dublin. In Csikos et al. (2015) traffic speed prediction was attempted using simulated data and ANN, applied at a specific intersection in Budapest.

Interestingly, most studies usually focus on urban arterials or urban intersections. In this study the prediction of traffic characteristic in urban environment is attempted for a whole traffic system. Thus, traffic data are used from several inductive loop detectors and Bluetooth sensors located in the real network of the city of Nicosia, in Cyprus. The motive of this study is to highlight the complexity of developing urban traffic patterns for a whole system compared to small scale prediction applications (intersection or urban arterials) and to achieve an overall view of the loading aspects.

3 ANALYSIS AREA

3.1 *Study area*

In order to predict traffic variables in urban traffic systems, a suitable testing area was selected. The study area is located in the network of Cyprus, namely in the city of Nicosia and includes 15 inductive loop detectors that measure flow, occupancy and spot speed. Bluetooth sensors are also located in order to achieve travel time estimation between specific locations in the network. The locations both of the loop detectors labeled with numbers and the Bluetooth sensors are depicted in Figure 1.

3.2 *Data*

For the particular framework a representative data set of spatial data, flow and speed for both direction of each loop detector was collected for the last week of November 2015. As prediction interval, a 15-min interval was selected due to strong

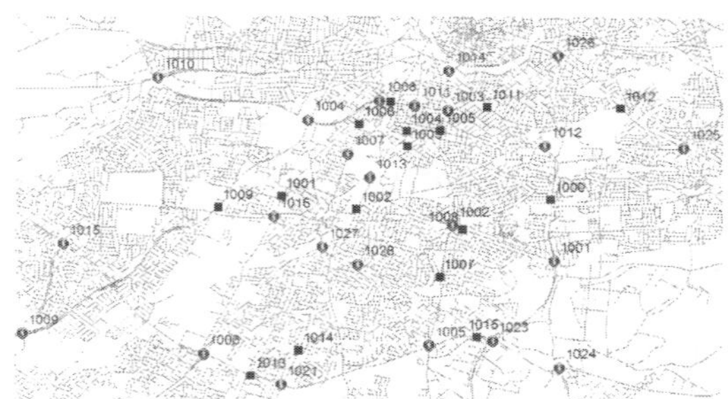

Figure 1. The road network of Nicosia city in Cyprus and the locations of the loop detectors and the Bluetooth sensors.

fluctuations that traffic flow exhibits at shorter intervals. The original dataset contains 60 different variables of flow and speed as both direction of the 15 loop detectors were used and 672 observations for each variable using 15-mins interval for the selected week. Figure 2 shows the loading and speed patterns for the selected week from all the loop detectors. As it can be seen from the same figure the loading patterns present differences between weekdays and weekend, as it was expected. As far as speed patterns is concerned great variability is observed depending on the location of the loop detector and the time of the day.

4 EXPERIMENTAL SETUP

After the first visualization of the data set, the authors continued with five different experiments in order to investigate the possible extensions (or computational burden) of the well-documented data-driven network loading projection problem, from a limited (or single point) number of locations to a system-wide coverage. The analysis was based on the promising (and with valuable extensions) methodological framework of time series while the results were limited to the use of alternative ANN configurations, since they are powerful and flexible. The scope of training a neural network is not only to obtain an exact representation of the training data, but also to build a model that represents the physical process that generated that data.

Many different ANN models have been proposed the last three decades. Perhaps the most influential models are the Multi-Layer Perceptrons (MLP). An MLP is typically composed of several layers of nodes. The lowest layer is an input where independent or predictor variables is received. The highest layer is an output layer where the problem solution is obtained. The input layer and output layer are separated by one or more inter—mediate layers called the hidden layers. The nodes in

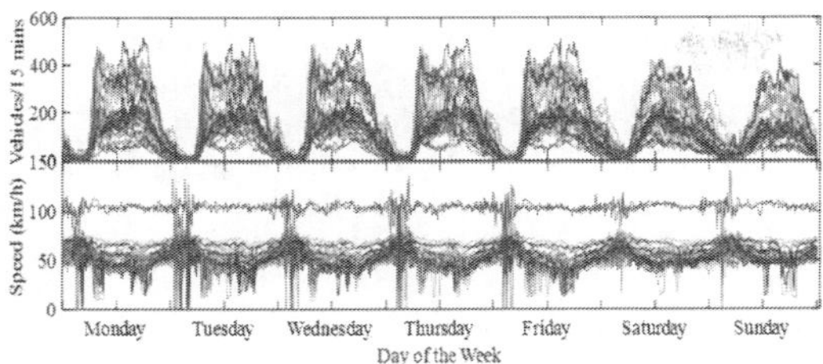

Figure 2. Loading and speed patterns using a 15-min time interval for the last week of November of 2015 from 15 loop detectors located in the city of Nicosia.

adjacent layers are usually fully connected by acyclic arcs from a lower layer to a higher layer, Zhang et al. (1997).

The functional relationship estimated by the ANN in the case of a Nonlinear Autoregressive with External (Exogenous) Input (NARX) can be written as the following Equation 1:

$$y(t)=f(x(t-1),x(t-2),x(t-d),y(t-1), y(t-2),y(t-d)) \quad (1)$$

Where d is the number of delays, and as Equation 2 in case of a Nonlinear Autoregressive (NAR).

$$y(t)=f(y(t-1),y(t-2),y(t-d)) \quad (2)$$

The total available data is usually divided into a training set (in-sample data), a validation set and a test set (out-of-sample or hold-out sample). The training set is used for estimating the arc weights while the test set is used for measuring the generalization ability of the network. The training algorithm is used to find the weights that minimize some overall error measure such as Mean Squared Errors (MSE) which is the average squared difference between outputs and targets. Usually, Regression (R) which measures the correlation between outputs and targets and MSE are used to measure the performance of the ANN.

The proposed study includes five different experiments using NARX or NAR models in order to predict traffic characteristics of the road network of Nicosia. The first two experiments investigate traffic flow patterns for specific point locations. The last three experiments investigate travel time patterns between predefined Origin-Destination (OD) locations, namely between Bluetooth detectors. In the first experiment all the initial variables of flow and speed were used as inputs in order to predict the flow variables, using a NARX model. The number of hidden neurons was 50 in all five experiments while the number of delays was 2. The train algorithm was Levenberg-Marquardt and two trains were executed for all the experiments. The selected percentages for the target timesteps were 70% for training, 15% for validation and 15% for testing. In the second experiment, only the variables from the specific loop detectors that PCA indicated having higher loadings, were used as inputs. In order to predict travel time between OD pairs, all the available variables from nearby detectors were used at the third experiment. Although the fifth experiment that used only travel time as input information using an NAR model appeared better performance. Last in the sixth experiment all the available data was used as input information in order to predict travel time for 16 urban OD but also traffic flow and speed at specific location. The input or activation function that was used was tanh while the output function was linear for all the experiments. A summarized table which contains all the general information of the experiments, follows.

As it can be observed higher number of input variables gave higher MSE, while the choice of particular input variables, and gave lower MSE in the second experiment. As a result a dynamic selection of input variables depending on the time of the day is essential according to the MSE results. The last three experiments appeared lower values of R as travel time prediction is more complex compare to traffic flow. As far as the MSE values for the last three experiments the results differ significantly from the rest as travel time is measured in minutes and the errors between target and output can be equal to seconds. A detail analysis of the results is given in the next section.

5 RESULTS AND DISCUSSION

5.1 *Traffic flow prediction models*

In order to compare the performance of the different experiments regarding traffic flow loading patterns, the regression diagrams from the ANN were evaluated. Specifically for the second experiment that appeared higher value of correlation R overall and lower Mean Square Error (MSE), compared to the second experiment, an overview is given in the following Figure 3, showing the linear relationship between output and target variables for each target timestep.

Although correlation between observed and predicted (output) values of flow vary depending on the loop detector as it can be observed in Figure 4. Many loop detectors such as 1000 with southbound direction appeared higher R values. In order to investigate these differences between each loop detector and for all the first two experiments, the value of R and MSE were calculated for each

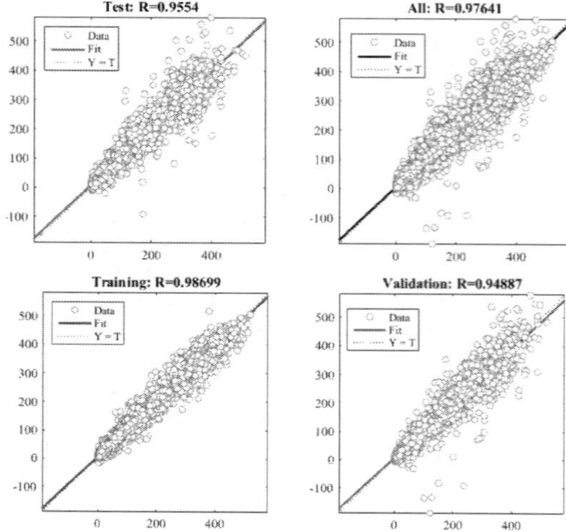

Figure 3. Regression results for the second experiment for each target timestep.

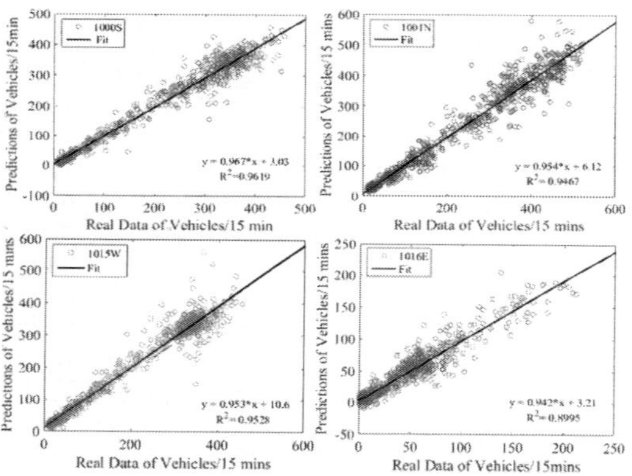

Figure 4. Linear Regression between observed and predicted flow for specific loop detectors.

case. Figure 5 illustrates the values for R for each experiment. The fourth experiment appeared lower values of R and higher MSE for many loop detectors and as a result greater number of delays failed to give better predictions. Another import conclusion derived from Figure 6, is that in some loop detectors, the MSE appeared the same low range of values for all the experiments. These loop detectors are located in the center of the city and they have in common smaller number of lanes 2 or 3 for both directions that indicates a more stable loading pattern.

5.2 *Travel time prediction models*

In order to compare the performance of the different experiments regarding travel time prediction, the regression diagrams from the ANN were evaluated. The fifth experiment, an NAR model using all the available variables as inputs appeared better performance in travel time prediction according Table 1. For the fifth experiment an overview is given in the following Figure 7, showing the linear relationship between target and output variables for each target timestep.

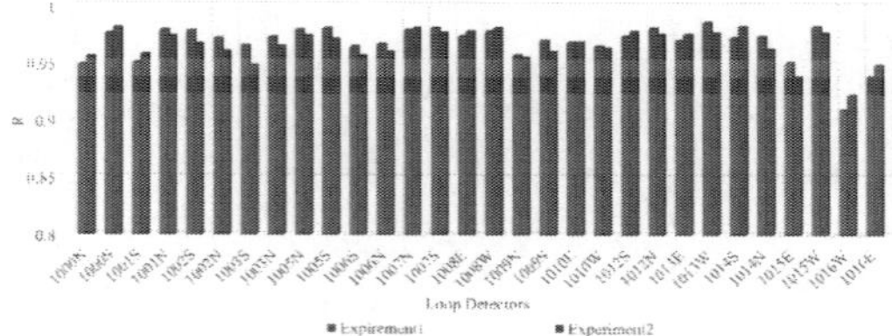

Figure 5. Correlation values between observed and output values of flow from each loop detector and for the first two experiments.

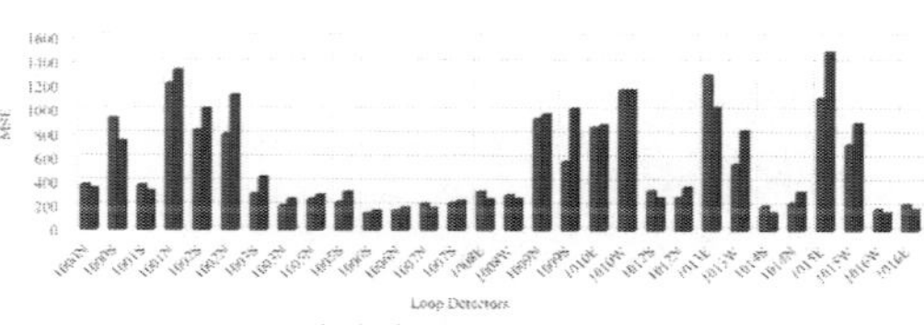

Figure 6. MSE values of flow predictions for each loop detector and for the first two experiments.

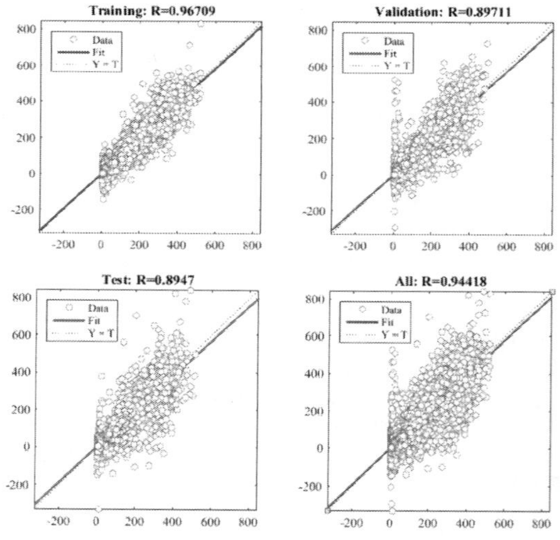

Figure 7. Regression results for the fifth experiment for each target timestep.

Figure 8 depicts the sixteen urban selected OD that was used in order to predict travel time loading

Table 1. Experiments details.

Target	Flow		Travel Time		
Experiments	1	2	3	4	5
Model	NARX	NARX	NARX	NAR	NAR
Input Variables (Number)	Flow Speed (60)	Flow Speed (9)	Flow Speed T.Time (38)	T.Time (6)	Flow Speed T.Time (76)
Output Variables (Number)	Flow (30)	Flow (30)	T.Time (6)	T.Time (6)	Flow Speed T.Time (76)
Input Dataset	40320	6048	25536	4032	51072
Output Dataset	20100	2010	4020	4020	51040
R	0.949	0.955	0.827	0.868	0.895
MSE	1329	1107	0.630	0.546	1718

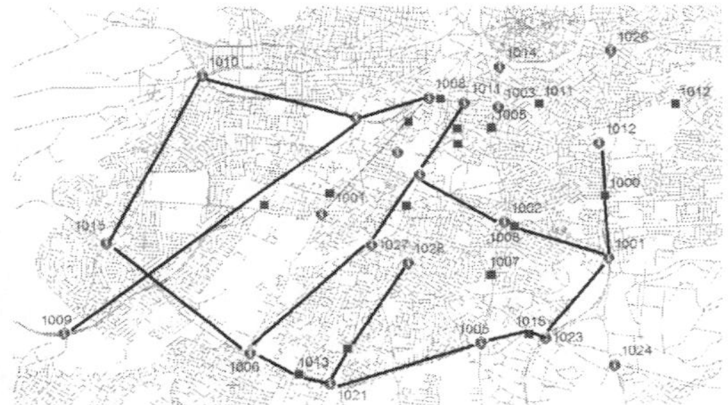

Figure 8. Location of the traffic counters and OD pairs used in the travel time prediction models.

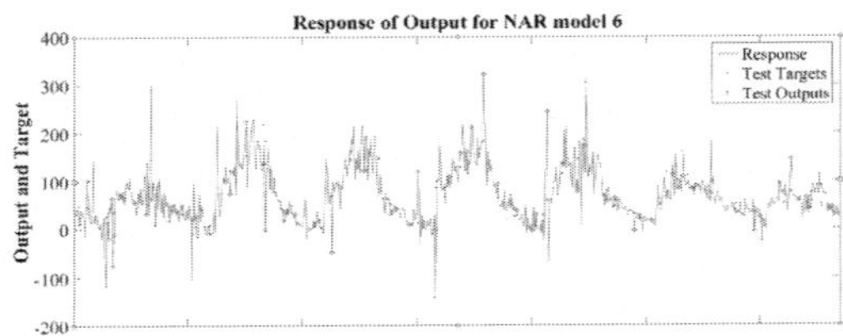

Figure 9. Time series response for the fifth experiment.

patterns and the also the traffic counters locations. In Figure 9 a comparison between target and output values for the fifth experiment is presented. The errors between target and output values are smaller than a minute in most cases, according to the time series response.

6 CONCLUSION

To conclude the proposed study attempts predictions of traffic characteristics in smart city, from a limited number of locations to a system-wide coverage using alternative ANN configurations. At first visualization of the data and the relationship of the variables were explored, in order to gain an insight of the traffic characteristics of the system overall. Reasonable experiments were conducted that led to the following:

- Higher number of input variables gave higher MSE for all the experiments.
- The choice of particular input variables gave lower MSE in the second experiment.
- Dynamic selection of input variables depending on the time of the day is essential according to the MSE results.
- Correlation between observed and predicted (output) values of flow vary depending on the loop detector.
- Loop detector with high value of R presented high values of MSE and vice versa.
- Travel time prediction models appeared lower values of R compared to traffic flow prediction models
- Travel time prediction models appeared lower values of MSE as travel time is measured in minutes and the errors between target and output can be equal to seconds.

Future work remains in order to optimize the architecture of these models taking into account seasonality and time—space diversity of the traffic system for the whole city.

REFERENCES

Csikós, A., Viharos, Z.J., Kis, K.B., Tettamanti, T. & Varga, I., 2015, June. Traffic speed prediction method for urban networks—an ANN approach. In Models and Technologies for Intelligent Transportation Systems (MT-ITS), 2015 International Conference on (pp. 102–108). IEEE.

Dimitriou, L., Tsekeris, T. & Stathopoulos, A., 2008. Adaptive hybrid fuzzy rule-based system approach for modeling and predicting urban traffic flow. Transportation Research Part C: Emerging Technologies, 16(5), pp. 554–573.

Ghosh, B., Basu, B. & O'Mahony, M., 2009. Multivariate short-term traffic flow forecasting using time-series analysis. Intelligent Transportation Systems, IEEE Transactions on, 10(2), pp. 246–254.

Stathopoulos, A. & Karlaftis, M.G., 2003. A multivariate state space approach for urban traffic flow modeling and prediction. Transportation Research Part C: Emerging Technologies, 11(2), pp. 121–135.

Van Lint, J., Hoogendoorn, S. & Van Zuylen, H., 2002. Freeway travel time prediction with state-space neural networks: modeling state-space dynamics with recurrent neural networks. Transportation Research Record: Journal of the Transportation Research Board, (1811), pp. 30–39.

Vlahogianni, E.I., Golias, J.C. & Karlaftis, M.G., 2004. Short-term traffic forecasting: Overview of objectives and methods. Transport reviews, 24(5), pp. 533–557.

Vlahogianni, E.I., Karlaftis, M.G. & Golias, J.C., 2005. Optimized and meta-optimized neural networks for short-term traffic flow prediction: a genetic approach. Transportation Research Part C: Emerging Technologies, 13(3), pp. 211–234.

Zhang, G., Patuwo, B.E. & Hu, M.Y., 1998. Forecasting with artificial neural networks: The state of the art. International journal of forecasting, 14(1), pp. 35–62.

eWork and eBusiness in Architecture, Engineering and Construction – Christodoulou & Scherer (Eds)

Monitoring drivers' perception of risk within a smart city environment

K. Stylianou & L. Dimitriou
Laboratory of Transportation Engineering, Department of Civil and Environmental Engineering, University of Cyprus, Nicosia, Cyprus

ABSTRACT: An important aspect of a Smart City framework is its transportation system and the existence of an efficient traffic management system. Efficiency in a traffic management system is characterized—among others—by the network's level of road safety. This paper studies a road safety oriented contribution to Smart Cities by quantifying driver's risk perception in relation to vehicle to vehicle interaction. The objective of this study is to propose an automatic driving behavior monitoring mechanism for an urban environment, which identifies near-crash phenomena by capturing rear-end potentials at a microscopic level, while furthermore to induce driving behavioral aspects, valuable for understanding drivers' perception on rear-end collision risk. The disaggregated data utilized in the study were obtained by inductive loop detectors in the urban network of Nicosia, Cyprus. The data gathered from the loop detectors was post-processed and a risk index based on rear-end potential was derived, which was used to classify drivers into four risk levels describing whether given their individual characteristics, drivers would engage in a potential rear-end collision. The proposed risk index results showed that 65% of the car-following events were considered as potentially unsafe. It was also shown that when engaged in car-following situations with Heavy Goods Vehicles-HGVs mean speeds of the following vehicles are lower. The proposed methodology enables the identification of potential near-crash events in an urban environment in real time. The information and knowledge collected by real-time data processing are key aspects of an efficient traffic management system and consequently a Smart City as a whole.

1 INTRODUCTION

The accelerating urban growth of population and rapid urbanization, require upgraded infrastructure and application services that are able to accommodate the needs of the population. Urban performance currently, depends not only on physical infrastructure but also, on the availability and quality of knowledge and social infrastructure (Caragliu et al., 2011). Within this concept, smart cities are introduced as a mechanism of clever solutions through qualitative advances within an urban environment. The label of a smart city therefore, leads to an increase in data by several orders of magnitude (Hashem et al., 2016). At present, data is available from various sources such as, sensors, cameras, smartphones and social networks and thus, analyzing data based on user needs and choices makes cities smarter (Rathore et al., 2016).

The concept of smart cities contributes in different areas of everyday life including the transport sector among others. The improvement of the urban transportation system and especially in the aspect of traffic efficiency is of high importance for the urban population. The effective analysis and utilization of the data provided by traffic monitoring systems within a smart city, is a key factor for the successful operation of an urban network.

Traffic performance defined by efficiency and safety is targeted through understanding and dealing with complex transportation problems by identifying the dynamic nature of traffic conditions. The recent technological advances in Intelligent Transport Systems (ITS) and the increased use of dynamic traffic control systems has turned research attention into investigating the real-time relationship of traffic and crash prediction. Traffic parameters such as volume and speed can be collected in real-time through various traffic surveillance systems and a variety of studies have focused on developing crash prediction models based on real-time traffic data in the scope of dynamic safety management systems.

An important issue in road safety, within traffic monitoring, stands for the quantification of driving style, especially related to car-following behaviour. In car-following situations vehicles interact with adjacent vehicles and it has been identified that crashes occur when interactions between vehicles become unstable (Oh and Kim, 2010). Car-following behaviour is defined by evaluating the driving behaviour of the following vehicle with respect to the lead vehicle in the same lane, by usually inspecting safe distance. It is important for drivers to be 'trained' such as to maintain a proper distance from the leading vehicle, especially in an urban environment where traffic conditions vary rapidly. It has been repeatedly reported that rear-end collisions is one of the most frequent among all road crash types. Under the dynamic conditions of driving, maintaining a safe distance from the lead vehicle becomes a demanding task, as distance estimation should change accordingly to the traffic ahead. Safety distance models hypothesise that drivers try to maintain a proper distance by controling their speed, so that there is sufficient distance to stop if the leading vehicle suddenly breaks (Duan et al., 2013).

In car-following events, when the following distance decreases then the risk of potential rear-end collision increases. According to Summala (1988), with experience driving becomes a largely automatized activity in which risk control is based on maintaining safety margins. Risk homeostasis theory (Wilde, 1982; Wilde, 1998) hypothesises that people do not attempt to minimize risk, but rather optimise it; it is assumed that at any moment drivers perceive a certain level of subjective risk they will compare it to their acceptable risk level and adjust their behaviour accordingly in an attempt to eliminate discrepancies. According to standard assumptions, in car following situations drivers will adjust their speed or headway in order to maintain a safe distance.

The objective of this study is to propose a driving behaviour control mechanism in an urban network, which identifies hazardous events, based on driver choices when engaged in car-following events. The focus is to identify near-crash phenomena by capturing rear-end potentials at a microscopic level. This study introduces a method of an induced monitoring of driving behaviour following a thorough investigation of disaggregated loop detector data. Data used in this study was collected for the urban network of Nicosia in Cyprus. The real-time traffic parameters (volume, speed) were obtained from the inductive loop detectors, along with individual vehicle information such as time stamp, vehicle type and lane of travel. To examine risk level exposure a risk index is derived in order to classify driver behaviour based on rear-end potential. The effects of leading vehicle, and lane of travel on individual driver choices are also examined. In the following sections, a brief presentation of the research background is provided, followed by the methodological approach and the experimental results.

2 BACKGROUND

As mentioned before, the concept of a smart city relies on the quantity and quality of information. The key concept within this concept is to obtain the right information at the right time and place in order to make a city-related decision with ease (Rathore et al., 2016). Regarding the transport sector, the electronic equipment which smart cities utilize, such as cameras and sensors, provide real-time information of traffic and its characteristics. This type of information is used by analysts to explore real-world traffic data. The objective of this paper is to investigate car following behavior as described by disaggregated loop detector data in order to identify rear-end potentials. Below, a brief background of car following models is presented.

Car following models are widely used to capture driving behaviour between consecutive vehicles in a single lane. In car-following situations drivers devote their attention to the lead vehicle, and accordingly adjust their driving style so as to decrease risk of possible collision. According to Olstam and Tapani (2004), four main classes of car-following models can be distinguished based on utilized logic: the Gazis-Herman-Rothery (GHR) family of models; safety distance or collision avoidance models; psycho-physical or action point models; and fuzzy-logic models. The GHR models assume that the following vehicle's acceleration is proportional to the speed difference and the space headway of the lead vehicle (Gazis et al., 1961). Safety distance models assume that the driver of the following vehicle tries to maintain a safe distance from the lead vehicle, in order to avoid collision (Gipps, 1981; Pipes, 1953). Gipps model is widely adopted in microsimulation to date. Contrary to GHR models which hypothesise that the following vehicle reacts to the action of the leading vehicle even if the distance between the two is large, psycho-physical models use thresholds (or action points) to determine where drivers change their behaviour. Fuzzy-logic has also been utilised in car-following. In these cases it assumed that the car following parameters (speed, distance etc.) are not known as such, but instead drivers are assumed to only conclude other vehicles characteristics (e.g. Al-Shihabi and Mourant, 2003).

The relationship between car-following behaviour as described by traffic flow characteristics and crash occurrence is of high interest among the research community. Real-time crash risk prediction models have been developed in order to estimate the likelihood of crash occurrence, with recent studies showing that driving behaviour is different under different traffic states (e.g. Christoforou et al., 2011; Xu et al., 2015). The technological advances in Intelligent Transport Systems has contributed to the availability of traffic parameters, leading to the development of more advanced car-following models and real-time crash prediction models. However, these models are mainly focused on freeways (Xu et al, 2012, Xu et al, 2013; Abdel-Aty et al., 2012; Ahmed and Abdel-Aty, 2013; Kwak and Kho, 2016) where traffic flow is unobstructed from signalization or other features presented in an urban network. The influence of different traffic environment and oncoming vehicle interaction is ignored. In fact, driving in an urban environment is a much more demanding task in comparison to freeway driving, in the sense that drivers are more frequently required to adapt their driving style to cope with traffic changes. Driving adaptation is characterised by the individual's choices while driving. Here data are used in order to capture risk perception and in particular car following behavior related to drivers' perception of risk for rear-end collisions.

3 METHODOLOGICAL APPROACH

In order to develop a methodology for evaluating risk prone behaviours and quantifying perceived risk in urban networks, this study utilises individual vehicle information and car following techniques. Individual vehicle information can be collected through loop detector technology which provides detailed information for each vehicle passing over a station, in real time. A variety of useful information can be collected through loop detectors, depending on their technology, including vehicle characteristics such as timestamp, count, vehicle type and speed which aggregated provide fundamental traffic characteristics such as volume, flow and mean speed.

A risk index is derived using the disaggregated data based on individual driving behaviour as given by drivers' choices of speed and time headway, in order to identify near-crash phenomena in real time. The disaggregated data collected is used in developing a safety mechanism which evaluates drivers' risk perception as obtained from the calculation of safe stopping distance in multiple occasions. With respect to the leading vehicle, when following a vehicle steadily, according to the Homeostasis theory proposed by Wilde (1982), drivers try to maintain a constant level of risk exposure by adjusting their choices, e.g. speed and headway. The risk of engaging in a rear-end collision increases as the following distance decreases; in order to avoid a rear-end collision, the safe stopping distance of the following vehicle should be smaller of the stopping distance of the leading vehicle. Individual data of each vehicle are used to derive the proposed risk index, and to identify the drivers who engage in a risk prone behaviour. Further on, the effect of the leading vehicle is explored, in order to identify how drivers control risk exposure in mixed traffic. In car-following situations drivers focus their attention on the leading vehicle while using peripheral view to scan the rest of the environment (Duan et al., 2013). Leading vehicle size obviously poses different effects on following behaviour, as larger vehicles are generally considered to be more dangerous than private vehicles due to the fact that large vehicle accidents usually present higher injury severities.

3.1 *Vehicle monitoring and risk index derivation*

To estimate the relationship between risk perception and near crash phenomena, the proposed methodology utilises inductive loop detectors, collecting spot volume, occupancy, vehicle speed, type, length and weight in real-time. Additionally, individual vehicle information can include lane of travelling and timestamp of a vehicle passing over the detector. This study uses the disaggregated data obtained from the loop detectors in order to derive the risk index based on safe stopping distance.

In order to prevent a collision the distance between two consecutive vehicles should be such as the following vehicle will have the appropriate distance to safely stop in the case of a sudden break of the leading vehicle or if any other hazardous situation occurs in the network. Rear end collisions are significantly featured in road-crash statistics; it has been reported that around 28% of all road accidents are rear-end collisions. Research focused on driving behaviour in different states of traffic, has shown that a major contributor to rear-end collisions is a short headway, which does not allow for the following vehicle to react in an appropriate manner in the case of a sudden break of the leading vehicle (Taieb-Maimon and Shinar, 2001). As such, monitoring the distance between consecutive vehicles in real-time situations could be valuable in the identification of near-crash phenomena and crash prediction.

The development of vehicle monitoring in this study is based on car-following behaviour in the urban network. In this study a cardinal assumption is made, that vehicle i follows vehicle $i+1$ as presented in Figure 1. Vehicle $_{i+1}$ is leading with a speed V_{i+1} and vehicle i is following with a speed V_i, and between the two vehicles exists a time headway, HW. Given this setup in a single lane and the data collected from the loop detectors (namely speed and time headway) the proposed risk index, based on stopping distances, can be derived. The explanatory variables of this study include speed, time headway, reaction time, vehicle deceleration rate, vehicle break time and vehicle length. The risk index is derived as follows:

where RI is the risk index for each pair of consecutive vehicles, HW: Headway (s), L: vehicle length (4.5 & 8 m) RT: Driver's Reaction Time

$$
RI = \begin{cases} \text{Class\_0, case } HW - L_{i+1} - RT - BT_i > 0 & (1a) \\ \text{Class\_1, case } HW - L_{i+1} - RT - BT_i < 0 & (1b) \\ \text{Class\_2, case } HW - L_{i+1} + BT_{i+1} - RT - BT_i < 0 & (1c) \\ \text{Class\_3, case } S_i - S_{i+1} - S_{HW} < 0 & (1d) \end{cases}
$$

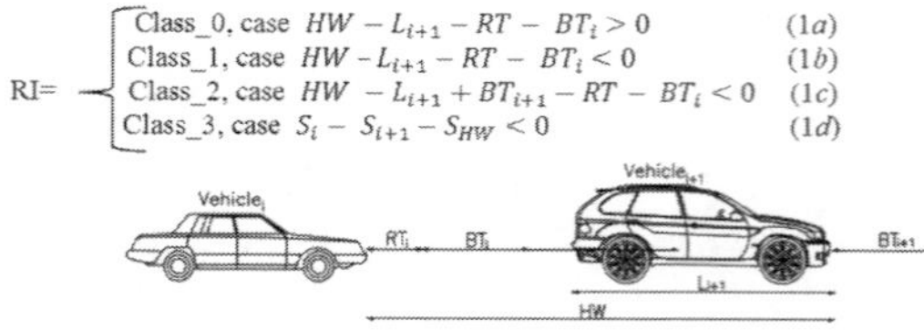

Figure 1. Car following setup.

(=3 s), BT: Break Time (s) calculated as $\frac{V}{dr'}$, V: Speed (km/hr), dr: Deceleration Rate of Vehicle (=0.5 g m/s^2), S: Distance Travelled in time t (km) calculated as Vt, t = 10 s.

The above Risk Index represents four risk levels that drivers come under while driving in the urban network, derived by whether they would engage in a potential rear-end collision under different circumstances. As such, RI=0 indicates a risk averse/safe driving behaviour while keeping a safe distance from the leading car; RI=1 indicates a risk prone behaviour as the distance kept from the leading vehicle is not sufficient in the case of a sudden incident in the road ahead; RI=2 indicates a risky behaviour as there is not sufficient distance for the following vehicle to stop when the leading vehicle emerges in a break; RI=3 indicates a highly risky behaviour as that the distance kept is not sufficient to avoid collision with the leading vehicle in the next 10 seconds if no action is taken (over-speeding of the following vehicle). The above analysis provides a classification of the individual observations based on risk willingness, which will later on be further analysed by specifying the lane of travel and the type of vehicle following and leading respectively.

4 APPLICATION AND RESULTS

4.1 *Data collection and preparation*

To accomplish the research objective, data was obtained from the urban network of Nicosia, Cyprus. In detail, 16 inductive loop detectors in northbound, southbound, eastbound and westbound directions in the network are located sporadically in the city. This study used data from 12 inductive loop detectors with a northbound direction toward the city centre, as presented in Figure 2. Each loop detector would cover either one or two lanes in each location (depending on the carriageway design), totaling in 18 lanes. The total volume collected for one day was equal to 150,327 vehicles, corresponding to a large-scale and representative database.

The data collected for each individual vehicle were timestamp of vehicle passing over the loop detector, speed and vehicle type. In addition, other information including loop detector location and lane of travel were also collected. Apart from the individual vehicle information which provides an insight of driving behaviour for each driver separately, in order to associate traffic state with safety performance, traffic flow and average speed are good indicators of traffic conditions in the network (Xu et al., 2012). Following the collection of the individual vehicle information, the average speed and flow in 5-minute aggregation was calculated through the disaggregated data. Research on crash prediction has shown that 5-minute aggregation of the data provides accuracy in crash prediction (Abdel-Aty et al., 2012) and also that crash occurrence is mostly related to the six 5-minute intervals prior to the crash (Xu et al., 2012). This aggregation of traffic data has been adopted in previous studies in developing real-time crash prediction models (Xu et al., 2013, Abdel-Aty et al., 2012, Oh et al., 2001).

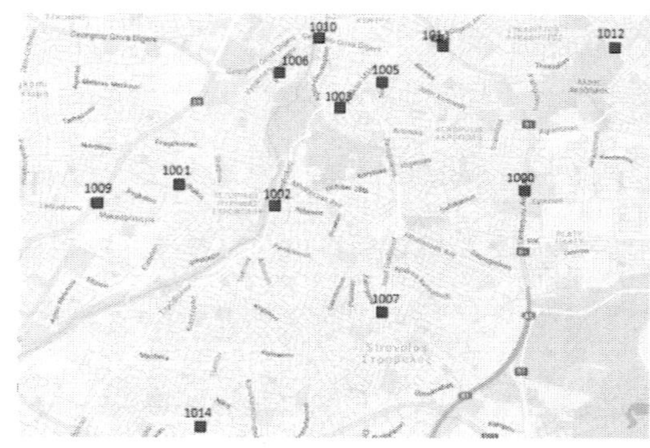

Figure 2. Loop detector locations.

As mentioned above, the traffic data used in the current study were collected from loop detectors, and the individual vehicle information collected (timestamp, speed, location, lane, vehicle type) was used to form the original databases of the study. The traffic variables used in the analysis to derive the Risk Index as described above, were individual vehicle speed and time headway between two consecutive vehicles. The vehicle speed was directly given from the loop detectors whereas time headway was calculated as the difference in timestamps between two consecutive vehicles presenting a car-following behaviour. Finally, in order to associate traffic state with safety performance, average flow and speed were calculated in 5-minute intervals.

4.2 *Data analysis*

Figure 3 presents the descriptive statistics for the sample explanatory variables of flow, speed and speed standard deviation. The traffic flow variables depicted consist of 5-minute observations of flow, speed and speed standard deviation. The traffic data was extracted in a 5-minute interval to identify the possible relationship of traffic conditions and near-crash phenomena. The statistical results show that the mean speed of the sample was near 46 km/h with a standard deviation equal to 14.05. It should be noted that in the study area there are two distinct speed limits; 50 km/h and 65 km/h, however at the locations of the loop detectors the speed limit is 50 km/h. Taking into account the fact that this is an urban network the effect of traffic congestion was clear through the data; nearly 5% of the observed speeds were below 20 km/h and these were presented in the morning peak hours (an expected result as the current study explores only one driving direction; toward city centre) as shown in Figure 2. Another observation was that even though the overall mean speed of the sample was below the speed limit, a little over 6% of the observed speeds exceeded a speed equal to 65 km/h. These results are similar to a self-reported survey in which responders were asked to state their normal operating speed on interstates in the United States under distinct speed limits, where it was reported that drivers' choice of speed was around 13 km/h over the speed limit (Anastasopoulos and Mannering, 2016).

Other than speed observations, headway is also considered in this study as an explanatory variable of driver choices, in car-following situations. Through the headway data it was shown that mean values were influenced by some high values of outliers. However, 77% of the headway observations regarded values below 10 seconds with the average time headway kept between two consecutive vehicles equal to 10.30 s.

For each location where loop detectors covered a two lane carriageway the nonparametric test of equality, Kolmogorov-Smirnov (two-sample) test, was employed between the two lanes examining speed and headway, in order to identify the sample's distribution similarity. At each location the KS-tests for both the explanatory variables (namely speed and headway) showed that the two samples are not from the same continuous distribution, implying that driving behaviour differs between lanes. This result provides evidence that drivers engage into different choices regarding their lane of travel. Cyprus is a left side driving country, and as such the left lane is considered as the 'slow lane' whereas the right lane is the 'overtaking lane'. The speed observations confirmed this statement, as mean speed on the left lane (43.99 km/h) was lower compared to the mean speed on the right lane (46.16 km/h). Mean headway was slightly higher (12.39 s) in the left lane compared to the right lane (12.02 s). These results suggest that drivers with a more conservative driving style (lower speeds and higher headways) choose to drive on the left lane, whereas drivers who exhibit a risk prone behaviour choose the right lane. Even though speed observations provide information on driving style and as consequently risk perception, based on traffic flow theory speed is linearly correlated to network density, with higher densities leading to low speed values. Therefore, as an alternative the variation of

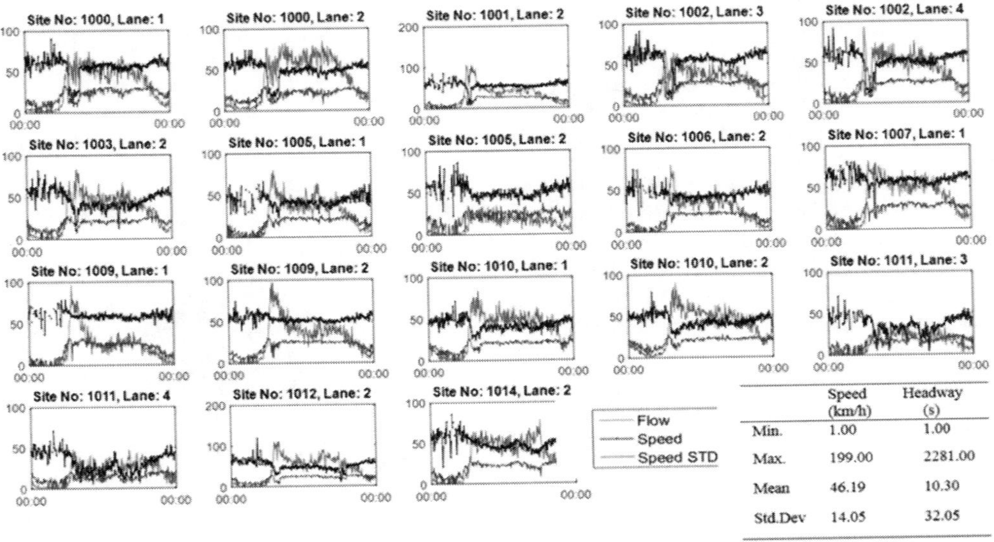

	Speed (km/h)	Headway (s)
Min.	1.00	1.00
Max.	199.00	2281.00
Mean	46.19	10.30
Std.Dev	14.05	32.05

Figure 3. Speed, flow and speed standard deviation time series and database descriptive statistics.

speed is also considered as an explanatory variable in order to analyse the relationship of traffic state and risk perception. Studies have shown that speed variance is a contributing factor in crash risks. A study by Oh et al., (2001), showed that the standard deviation of speed aggregated in 5-minute intervals was the best indicator of 'disruptive' traffic flow leading to a crash as opposed to 'normal' traffic flow. Disruptive traffic conditions are represented by high variations in traffic parameters. Figure 4 presents the relationship between flow and speed standard deviation aggregated in the 5-minute interval, for two selected locations in the network. The parabolic shaped flow-speed standard deviation curve shows that speed variation increases with flow increase until it reaches a critical point where then the negative slope of the curve implies traffic coordination.

4.2.1 *Risk level classification*

Individual vehicle speed and headway between two consecutive vehicles were used to derive the RI, which is used as a classification indicator of the represented risk level. According to equations 1–4 each observation was appointed with a risk level from 0 (risk averse) to 3 (highly risky behaviour). Following this classification method, it was shown that 35% of the observed car-following events are considered to be safe (RI=0), 22% are considered as risk prone (RI=1), 38% are considered risky and the remaining 5% are considered as highly risky. These values represent that about 65% of the observed car following events passing the loop detectors would be identified as unsafe driving. In a similar study Oh et al (2006) reported that 35% of car-following events in a freeway in California would be identified as unsafe conditions.

Through the RI classification each location examined in the network, was characterized according to its risk level. Locations which had a percentage of drivers over 5% come under class1_RI=4 (eq.1d) were characterized as highly risky, locations with a percentage over 30% come under class3_RI=3 (eq.1c) were characterized as risky, locations with a percentage over 20% in class2_RI=2 (eq.1b) were characterized as risk prone and the remaining were characterized as safe. Figure 5 presents an instantaneous representation of the risk level in the network.

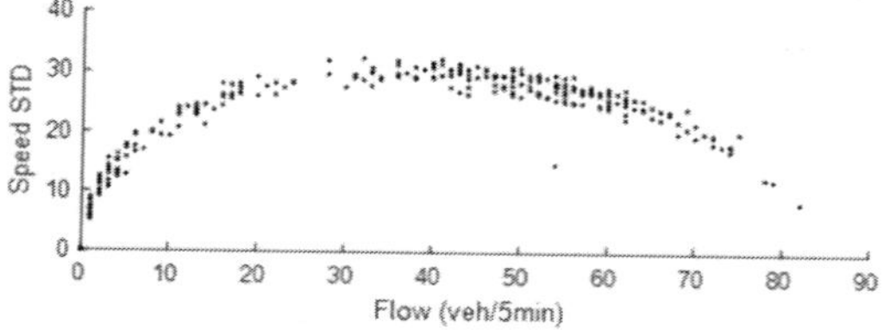

Figure 4. Flow speed STD relationship.

4.2.2 *Leading vehicle effect*

Loop detector technology allows for the identification of vehicle type in real time, based on individual vehicle characteristics such as weight and axles. This allocation of identification to each observation provides valuable information for identifying drivers' risk perception in mixed traffic conditions. Considering the fact that this study analyses an urban network and focuses on only unidirectional (towards city centre), 98% of the vehicles were found to be private vehicles. However, the cases where Heavy Goods Vehicles (HGV) were presented in car following behaviour were distinguished and analysed separately. The effects of leading vehicle type on speed and headway of the following vehicle in this study were found to differ according to vehicle size. Figure 6 shows that the mean speed was slightly lower and the time headway was higher when the lead vehicle was a HGV (44.56 km/h and 17.46 s respectively), compared to when the lead vehicle was a private vehicle (45.85 km/h and 11.30 s respectively).

Figure 6 presents the total means of the sample in black, whereas the mean values of speed and headway observed in each location are presented in light gray color.

The headway results are similar to the results of a driving simulator experiment, by (Duan et al., 2013) in which the distance headway increased as the vehicle size increased at the speed of 45 km/h in a 4 km single carriageway scenario.

5 CONCLUSIONS AND OUTLOOK

This study proposed a driving behaviour monitoring mechanism aimed to identify hazardous events in an urban environment, based on driver

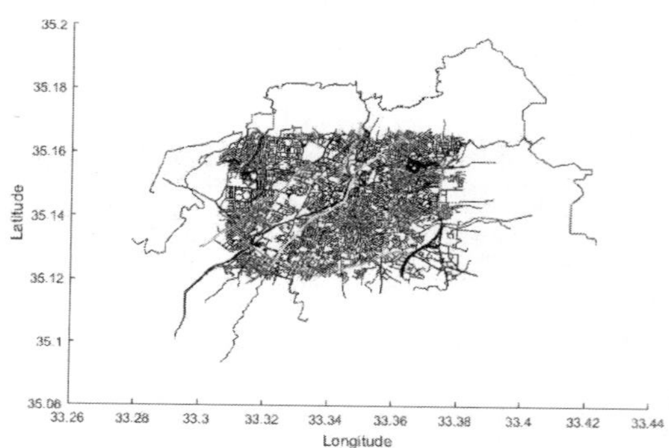

Figure 5. Network risk level.

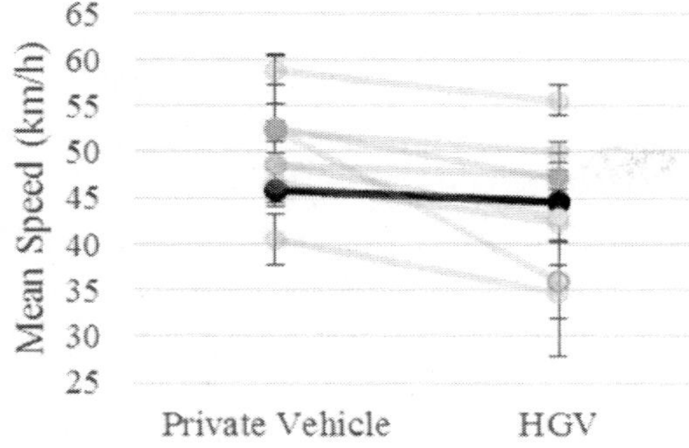

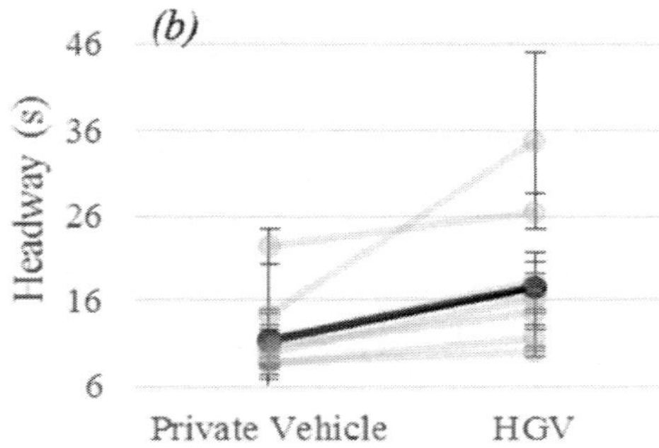

Figure 6. Leading vehicle effect.

choices when engaged in car-following situations. The objective of the study is to identify near-crash events by capturing rear-end potentials at a microscopic level, utilising disaggregated data from inductive loop detectors. The effective analysis and utilization of the data provided by traffic monitoring systems within a smart city, is a key factor for the successful operation of an urban network.

The availability of individual vehicle information, through inductive loop detectors in the urban network, such as speed, enabled the detection of potential rear-end conflicts by the derivation of a Risk Index (RI) based on distance keeping between two consecutive vehicles. Through an extensive real-time database, consisting over 150,000 observations in the urban network of Nicosia Cyprus, it was shown that around 65% of drivers presented a risk prone behaviour which could potentially lead to a rear-end collision. A statistical analysis was also employed, in order to investigate the effects of the leading vehicle type and lane of travel. Through this analysis it was shown that mean speed is lower and headway is higher when the leading vehicle is a HGV. Also, the Kolmogorov-Smirnov test was employed to explore the relationship of drivers' characteristics within the different lanes of the network, which provided evidence that a different driving style is observed between the slow lane and the overpassing lane.

To establish a reliable model a larger dataset with mixed traffic should be investigated, and more analytical factors should be taken account of in the risk index derivation, such as different deceleration rates based on vehicle type. Also car-following behaviour could be investigated in relation to further traffic, and not restricted to the leading vehicle. Vehicle trajectories, from video capture should also be used as inputs of the models. Collecting individual vehicle information and identifying unsafe events on a microscopic level with high accuracy can be of great importance in collision control mechanisms. Potential collision identification mechanisms can be a valuable support tool in urban traffic management. Finally, other dynamic safety management systems can also benefit from a real-time detection of risky events and aim transport professionals in developing effective crash prevention strategies. Within this concept, of traffic efficiency, the performance of smart cities will benefit and a more sustainable urban environment can be achieved.

REFERENCES

Abdel-Aty, M., Hassan, H., Ahmed, M. & Al-Ghamdi, A. 2012. Real-time prediction of visibility related crashes. *Transportation Research Part C: Emerging Technologies*, 24: 288–298.

Ahmed, M. & Abdel-Aty, M. 2013. A data fusion framework for real-time risk assessment on freeways. *Transportation Research Part C: Emerging Technologies*, 26: 203–213.

Al-Shihabi, T. & Mourant, R. 2003. Toward More Realistic Driving Behavior Models for Autonomous Vehicles in Driving Simulators. *Transportation Research Record: Journal of the Transportation Research Board*, 1843: 41–49.

Anastasopoulos, P. & Mannering, F. 2016. The effect of speed limits on drivers' choice of speed: A random parameters seemingly unrelated equations approach. *Analytic Methods in Accident Research*, 10: 1–11.

Ben-Akiva, M. and Lerman, S. 1985. *Discrete choice analysis*. Cambridge, Mass.: MIT Press.

Caragliu, A., Del Bo, C. & Nijkamp, P. 2011. Smart Cities in Europe. *Journal of Urban Technology*, 18(2): 65–82.

Christoforou, Z., Cohen, S. & Karlaftis, M. (2011). Identifying crash type propensity using real-time traffic data on freeways. *Journal of Safety Research*, 42(1): 43–50.

Duan, J., Li, Z. & Salvendy, G. 2013. Risk illusions in car following: Is a smaller headway always perceived as more dangerous?. *Safety Science*, 53: 25–33.

Gazis, D., Herman, R. and Rothery, R. 1961. Nonlinear Follow-the-Leader Models of Traffic Flow. *Operations Research*, 9(4): 545–567.

Gipps, P. 1981. A behavioural car-following model for computer simulation. *Transportation Research Part B: Methodological*, 15(2): 105–111.

Hashem, I., Chang, V., Anuar, N., Adewole, K., Yaqoob, I., Gani, A., Ahmed, E. & Chiroma, H. 2016. The role of big data in smart city. *International Journal of Information Management*, 36(5): 748–758.

Kwak, H. and Kho, S. 2016. Predicting crash risk and identifying crash precursors on Korean expressways using loop detector data. *Accident Analysis & Prevention*, 88: 9–19.

Oh, C. & Kim, T. 2010. Estimation of rear-end crash potential using vehicle trajectory data. *Accident Analysis & Prevention*, 42(6): 1888–1893.

Oh, C., Oh, J. & Ritchie, S. 2001. Real-time estimation of freeway accident likelihood. *Presented at the 80th Annual Meeting of the Transportation Research Board, Washington DC.*

Oh, C., Park, S. & Ritchie, S. 2006. A method for identifying rear-end collision risks using inductive loop detectors. Accident Analysis & Prevention, 38(2): 295–301.

Olstam, J. & Tapani, A. 2004. *Comparison of Car-Following Models*. Swedish National Road and Transport Research Institute, 13–14.

Pipes, L. 1953. An Operational Analysis of Traffic Dynamics. *J. Appl. Phys.*, 24(3): 274.

Rathore, M., Ahmad, A., Paul, A. & Rho, S. 2016. Urban planning and building smart cities based on the Internet of Things using Big Data analytics. *Computer Networks*, 101: 63–80.

Summala, H. 1988. Risk control is not risk adjustment: the zero-risk theory of driver behaviour and its implications. *Ergonomics*, 31(4): 491–506.

Taieb-Maimon, M. & Shinar, D. 2001. Minimum and Comfortable Driving Headways: Reality versus Perception. *hum factors*, 43(1): 159–172.

Wilde, G. 1982. The Theory of Risk Homeostasis: Implications for Safety and Health. *Risk Analysis*, 2(4): 209–225.

Wilde, G. 1998. Risk homeostasis theory: An overview. *Injury Prevention*, 4(2): 89–91.

Xu, C., Liu, P., Wang, W. & Li, Z. 2012. Evaluation of the impacts of traffic states on crash risks on freeways. *Accident Analysis & Prevention*, 47: 162–171.

Xu, C., Liu, P., Wang, W. & Li, Z. 2015. Safety performance of traffic phases and phase transitions in three phase traffic theory. *Accident Analysis & Prevention*, 85: 45–57.

Xu, C., Tarko, A., Wang, W. & Liu, P. 2013. Predicting crash likelihood and severity on freeways with real-time loop detector data. *Accident Analysis & Prevention*, 57: 30–39.

Special session: Energy efficient neighborhoods (ee-Neighborhoods)

eWork and eBusiness in Architecture, Engineering and Construction – Christodoulou & Scherer (Eds)
© 2016 Taylor & Francis Group, ISBN 978-1-138-03280-4

A collaborative environment for energy-efficient buildings within the context of a neighborhood

M. Bassanino, T. Fernando, K. Wu & S. Ghazimirsaeid
University of Salford, Salford, UK

K. Klobut, T. Mäkeläinen & M. Hukkalainen
VTT Technical Research Centre of Finland, Helsinki, Finland

ABSTRACT: This positioning paper explains our approach to creating a collaborative environment to assist multi-disciplinary teams in designing better energy efficient buildings with consideration of their neighborhoods. A scenario-driven approach was used here to define the energy related activities, the actors involved and the tools required to perform these activities. More specifically, the paper has a focus on energy matching activities throughout the whole product's life cycle. The paper will go on to suggest an appropriate User Interface to allow multi-disciplinary teams to collectively explore various solutions as they visualize BIM models and data models in a 3D interactive workspace to achieve optimum energy efficient buildings at neighborhood level.

1 INTRODUCTION

The multidisciplinary nature of design teams in construction requires a high level of collaboration from various stakeholders to produce the final product (Wikforss & Lofgren, 2007). Advanced technologies such as BIM (Building Information Modelling) has already been utilized as an enabler to meet these challenges by bringing data from a range of design disciplines to identify potential clashes during the design process (Eastman et al, 2011). However, the demand for considering energy efficiency has certainly added another challenge for the multidisciplinary design team. Despite various attempts to integrate building energy simulations within the design process, issues still arise in regard to fully integrating these simulations extending from the analysis of the building's indoor climate to the outdoor surroundings and incorporating energy production models within the BIM design models with consideration of the surroundings/neighborhood. Currently, buildings are still designed as individual components with minimal consideration of their surroundings (IREEN, 2013). Even when design tools are used to calculate energy performance, they are largely focused on the building level rather than on the neighborhood level (Mäkeläinen et al. 2014).

Therefore, in order to offer a holistic approach for designing energy efficient buildings with consideration of their surroundings, the Design4Energy project selected three scenarios to illustrate design activities. The first scenario concerns a neighborhood energy matching context in building design, which is the focus of this paper. The second scenario is about holistic energy design optimization during the early design stage (Bassanino et al. 2014a). The third scenario concerns the use of operational and maintenance data in retrofit design (Fouchal et al. 2015). This paper will focus on the collaboration aspect of the Design4Energy platform, taking the first scenario as the context with an emphasis only on those stages where energy matching activities are usually performed, as illustrated in Figure 1.

The proposed paper will address the following research questions:

- RQ1: What are the collaborative energy related activities that should be performed during the product's life cycle in order to design energy efficient buildings within the context of a neighbourhood?
- RQ2: What types of tools are required to support these activities to allow design teams to collectively make better informed decisions in designing energy efficient buildings within the context of a neighbourhood?

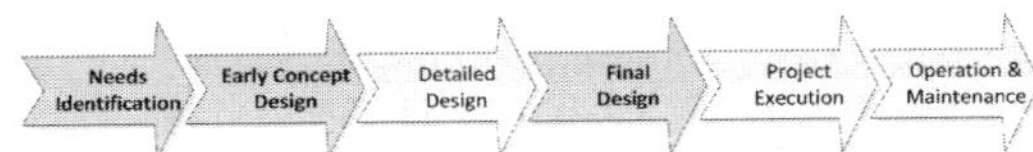

Figure 1. A project's life cycle with highlighted stages where energy matching activities take place.

The paper will suggest a new design methodology to improve current practice, which utilizes energy simulation tools integrated within a 3D virtual workspace. This will allow various actors to explore what-if-scenarios in a new building design with a consideration of the neighborhood and to make better informed decisions in optimizing energy performance.

2 RELATED WORK

The development of collaborative working environments has been explored in various European projects over the last two decades, such as DIVERCITY (Arayici & Sarshar, 2002) and CoSpaces (Fernando et al. 2013). Although these projects had a focus on multi-user interaction and the management of data models for collaboration by providing design teams with tools to enhance communication and explore various design options (Bassanino et al. 2014b), they did not explore collaboration in the context of energy efficiency.

A number of recently completed EU projects have demonstrated different approaches to create a sustainable built environment through an efficient usage of building information and data by integrating innovative software tools. Both the HESMOS (2013) and ISES (2014) projects developed an energy-extended BIM platform with enhanced data exchange capabilities between various stakeholders, using advanced energy analysis, simulation and tools from an early design phase throughout operation and use.

On-going EU projects such as eeEmbedded, HOLISTEEC and STREAMER together with the Design4Energy project have extended the work of the HESMOS and ISES projects to the neighborhood of surrounding buildings and energy systems. Although these projects share a similar aim in developing an open BIM-based holistic collaborative design and simulation platform to design energy efficient buildings in the context of their neighborhood, various approaches have been considered in relation to the neighborhood. The HOLISTEEC project utilizes a Neighborhood Information Model (NIM) which contains the data about the surrounding environment where the building is located (Mazza et al. 2015). This way, the HOLISTEEC approach considers the requirements needed at both building and district level for various physical simulations including energy, acoustics, light, environment and business services (Delponte et al. 2014). In the STREAMER project, a semantic approach to design starting from the combination of BIM and GIS is used (Iadanza et al. 2014). In the case of the eeEmbedded project, urban design is considered as defining the building envelope geometry and the district energy system (Geißler et al. 2014).

The above various approaches clearly highlight the differences in defining the neighborhood. In the Design4Energy project the definition of a neighborhood consists of a group of buildings with the same geographical location, served by common energy networks. For this reason, the focus of this paper is on energy matching activities to balance the energy supply from local renewable sources with the energy demand of a neighborhood. The Design4Energy project is also in the process of exploring the concept of Green Buildings in Neighborhood Systems (GBNS), providing a more realistic approach by looking beyond a single building and considering energy usage at a neighborhood level. Against this background, members of the design team need to consider a cluster of buildings as an energy system where buildings can optimally interact with each other through smart grid technologies, for example, in order to reach more energy efficient buildings and neighborhoods (Marzban et al. 2013).

3 METHODOLOGY

The Design4Energy methodology was described in detail by (Mäkeläinen et al. 2014) explaining the combined approaches of top down and bottom up, applied to define a set of usage scenarios. A number of road maps were analyzed and studied in detail to provide a vision for designing energy efficient buildings in the context of their neighborhoods.

The REEB roadmap (Hannus et al. 2010) identified five categories to be considered as priorities for ICT-enabled energy efficiency in buildings. Energy management and trading is one of these categories that corresponds to the construction industry's priority of a life cycle support service. This category directs its long term vision towards providing a holistic optimization of the built environment with consideration of the energy performance of buildings and the energy balance between buildings within a district responding to the local grid.

The IREEN roadmap's vision emphasizes that of REEB in regard to its concept of achieving a holistic design as: 'energy efficiency at the heart of holistic planning of both new and existing neighborhoods'. In this way, buildings should be designed to be interconnected within their neighborhoods, instead of being considered as a single entity. These challenges have been taken into account in this research to define a holistic approach for designing energy efficient buildings with a consideration of their neighborhoods.

The bottom up approach was carried out by analyzing the stakeholders, their roles and responsibilities, in addition to their activities, related to designing energy efficient buildings in the context of their neighborhoods. The main actors involved, as illustrated in Table 1, need to collaborate (Huifen et al. 2003) to produce energy efficient buildings with a consideration of their neighborhood.

During the design process, the client/developer's role is to ensure that energy efficiency is appropriately considered as part of the project brief. The client responsibility here is to attend design review meetings to assess design alternatives and ensure that the brief's objectives are properly met. The architect's role is to make sure the architectural design is energy efficient and thus he/she needs to collaborate with other team members to reach an energy efficient design. All the engineering disciplines (mechanical, electrical, and so on) need to work with other team members and to check the impact of their design on the building design. The energy expert's role is to advise on energy efficient solutions; this involves assisting the client to define the project's KPIs (Key Performance Indicators) as well as assisting the architect in conducting energy simulation and matching activities.

4 VISIONARY SCENARIO

The scenario starts at the needs' identification stage, when the client assisted by the energy expert defines the project's KPIs and sets key target levels accordingly. During this early stage of the project, the energy expert collects boundary conditions about the site such as city plans, terrain/soil model, climate data and energy prices before exploring a number of possible renewable energy options to consider as part of the feasibility studies. The first part of the concept design stage has a focus on a single building design and so it starts with the architect sketching a design of the energy efficient building using a BIM model as he/she drags certain components from his/her personalized component catalogue. In doing so, the architect takes into consideration a number of parameters such as the local weather profile and lighting to decide the building orientation as well as the use of raw materials relating to carbon emissions before the conceptual design is completed. The architect then defines the physical appearance of the building including internal spaces, external openings, use of materials and so on. Next, the energy expert imports the BIM model into the energy simulation software to simulate and assess the energy profile of the architect's design. In doing so, what-if-scenarios are explored for energy demand, solar gain and natural light and discusses several design options with the client during a design review meeting to decide the most suitable one.

Table 1. Main actors with their roles and responsibilities.

Stakeholder	Role/responsibility
Developer (Client)	Ensure that energy efficiency is considered as part of the project brief for both design and operation
Architect	Ensure that architectural design is energy efficient
Energy Efficient Expert	Advise on energy efficient solutions
Mechanical Engineer	Ensure that HVAC design complies with energy indicators
Electrical Engineer	Ensure that electrical design is energy efficient

The second part of the early design stage has a focus on energy related activities with consideration of the neighborhood to establish if there is any mismatch between production and consumption.

At this point the energy expert performs energy matching and energy trading activities. For uncomplicated designs, the use of interactive tools to conduct energy matching at a neighborhood level (Klobut at al. 2016) could be adequate. This enables the energy expert to make an early assessment of energy estimations from the architect's conceptual sketches by integrating a single renewable energy source into the building's system. However, when renewable energy is introduced to a building, the energy matching process could become a challenging issue since there could be a mismatch between the energy produced by the non-dispatchable generation units (such as wind turbines and photovoltaics) and the non-responsive load demand (energy to be utilized instantly on request for lighting, heating and appliances) at a given time for a building (Ghazimirsaeid at al. 2016). In this case, more complex modelling will be performed by the energy expert to conduct energy matching and trading in order to add a profit. When complex modelling is performed, energy storage capabilities are usually used by the energy expert to establish additional demand response in a holistic, integrated way (REEB roadmap).

The architect then compares the energy matching results with the indicators previously set up and adjusts accordingly. The architect then improves the BIM models to discuss with the client in a collaborative review meeting.

Next, the design is passed on to the mechanical and electrical engineers as a BIM model. Once each engineer proposes his/her own design, a simulation is run to enable each member to analyze the design, to compare the results with the expected energy indicators and improve the design accordingly.

Following this, at the final design stage, and once the architect has performed the integrated design and performance simulation activities to bring together all the design specialisms that were considered separately, a more refined set of energy matching and energy trading activities will be carried out by the energy expert. Once the client has reviewed these results to ensure that the project brief with its objectives are properly met, the project execution starts after the final selection and approval.

5 ACTIVITIES

Although the visionary scenario covered the whole product life cycle to design energy efficient buildings with a consideration of their neighborhoods, this section will only focus on energy matching and trading activities and mapping them to a set of KPIs, since other energy related activities of a building design were previously discussed by (Bassanino, et al. 2014a). As a result, three use cases were identified within the whole project's life cycle where energy matching activities take place: namely, needs' identification, early concept design and final design stages. These use cases serve to illustrate the detailed set of activities, the actors performing these activities and the tools required to carry out such activities.

The key activities for use case 1: needs' identification (Figure 2) can be summarized in the following:

- Based on the client's design criteria and requirements, the energy expert sets key target levels through searching and defining benchmark data based on parameters either from his/her own database or external sources. The possibility of using renewable energy as part of energy matching is considered here as part of the target setting activity.
- The energy expert collects boundary conditions about the site such as city plans, terrain/soil model, climate data and energy prices to define renewable alternatives to study.
- The energy expert runs feasibility studies and publishes the results on the Design4Energy collaborative workspace. Again, energy matching activity is considered at this initial stage by conducting a simple analysis of the renewable alternatives by the energy expert.

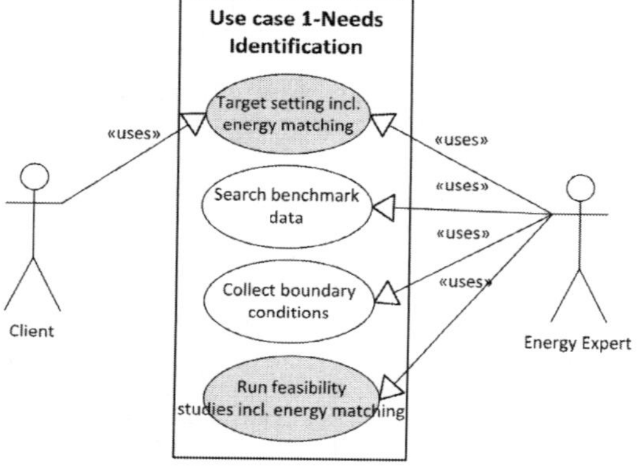

Figure 2. Use case 1 diagram: needs' identification stage.

The key activities for use case 2: early design concept can be summarized in the following:

- The architect and the energy expert execute the initial building design and simulation respectively.
- The energy expert runs a simulation to calculate the energy performance at the neighbourhood level (energy matching and trading) with a reference to the energy price model, renewable energy and energy production.
- Based on the energy matching results, the client refines the project's target settings.
- The architect then improves the design according to the energy matching results making a reference to the indicators prior to publishing the design as a BIM model onto the Design4Energy collaborative workspace.
- The client conducts a design review with the architect to assess design alternatives as a BIM model according to the simulation results and indicators before final approval.

The key activities for use case 3: final design can be summarized in the following:

- The architect conducts an integrated design and simulation by bringing together all the design specialisms' models that were considered separately.
- The energy expert performs another energy matching and energy trading to refine the previous results.
- Based on the energy matching and the energy trading results, the client refines the target settings and checks the building design's impact on the neighbourhood.
- The architect together with the client have a face-to-face or remote design review meeting to assess/discuss the previous results. The final selection is then made by the client.

The previous three use cases provided a detailed set of activities of those stages where energy matching and trading are performed as summarized in Figure 5.

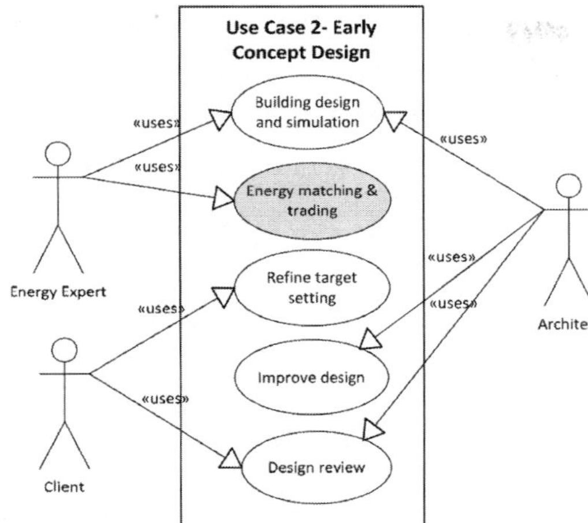

Figure 3. Use case 2 diagram: early concept design.

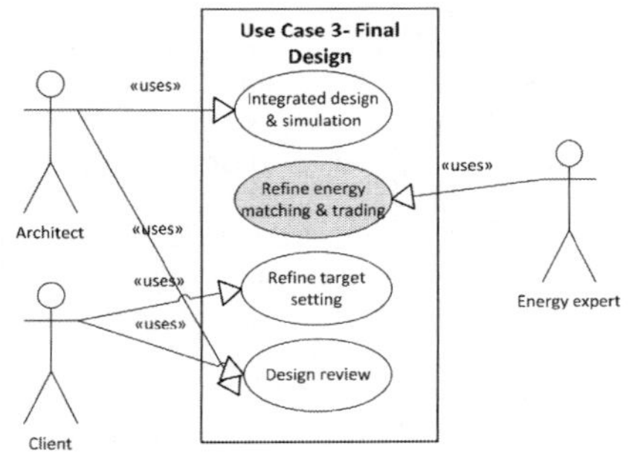

Figure 4. Use case 3 diagram: final design.

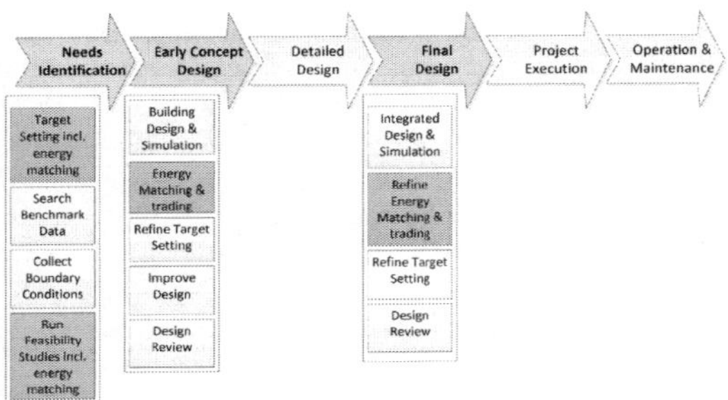

Figure 5. Summary of energy matching and trading activities in use cases 1–3.

The main activities defined in relation to the neighborhood comprise: target setting, energy matching and energy trading. Table 2 maps these activities to the set of tools required to perform such activities.

Table 2. Mapping the activities to the required tools.

Activities	Tools
Target setting	Target setting & assessment tool
Energy matching for uncomplicated designs at a neighborhood level (with single renewable energy source)	Interactive energy matching tool
Energy matching and trading for a complex dynamic system (with multiple renewable energy sources)	Complex systems' modelling tools
Visualization of models and data	Design review tool

6 TOOLS

Following on from the discussion of the detailed set of activities performed in the above use cases, this section describes in detail the previously defined tools required to support multi-disciplinary teams in designing better energy efficient buildings with consideration to the neighborhood.

6.1 *Target setting and assessment tool*

Since this tool combines two separate features, this section will first explain the target setting feature then the assessment feature of the tool.

The frame of the criteria areas structures the indicators in the target setting tool (Figure 6) which is used at the needs' identification stage to describe the client's value profile of a particular project. This is achieved by first defining a set of Key Performance Indicators (KPIs) (Ala-Juusela, Sepponen, & Crosbie, 2014) with a value for each (A-E) and assessing later on if the design alternatives achieved these indicators or not.

The value profile template of the target setting tool comprises four value areas (known as areas of concern and factors of influence) where a set of indicators and sub-indicators are defined for each area (Fraunhofer et al. 2015). These value areas are: economic performance, performance in use, environmental performance and impact of neighborhood performance. The criteria required to explore the building's impact on neighborhood performance for example involves two indicators: energy and quality. The energy indicator consists of three sub-indicators: energy balancing, peak power need and storage capacity while the quality indicator doesn't have any sub-indicators.

During the target setting stage, the client decides the level of performance required to achieve each indicator. At this time, the energy expert starts by assigning a weight to each area as well as to the area's indicators and sub-indicators. The template

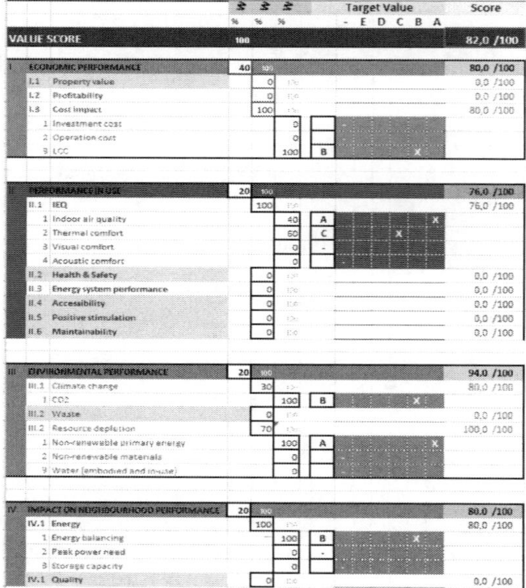

		%	%	%	Target Value (- E D C B A)	Score
	VALUE SCORE	100				82,0 /100
I	ECONOMIC PERFORMANCE	40	100			80,0 /100
I.1	Property value		0			0,0 /100
I.2	Profitability		0			0,0 /100
I.3	Cost impact		100			80,0 /100
1	Investment cost			0		
2	Operation cost			0		
3	LCC			100	B	
II	PERFORMANCE IN USE	20	100			76,0 /100
II.1	IEQ		100			76,0 /100
1	Indoor air quality			40	A	
2	Thermal comfort			60	C	
3	Visual comfort			0	-	
4	Acoustic comfort			0		
II.2	Health & Safety		0			0,0 /100
II.3	Energy system performance		0			0,0 /100
II.4	Accesaibility		0			0,0 /100
II.5	Positive stimulation		0			0,0 /100
II.6	Maintainability		0			0,0 /100
III	ENVIRONMENTAL PERFORMANCE	20	100			94,0 /100
III.1	Climate change		30			80,0 /100
1	CO2			100	B	
III.2	Waste		0			0,0 /100
III.2	Resource depletion		70			100,0 /100
1	Non-renewable primary energy			100	A	
2	Non-renewable materials			0		
3	Water (embodied and in-use)			0		
IV	IMPACT ON NEIGHBOURHOOD PERFORMANCE	20	100			80,0 /100
IV.1	Energy		100			80,0 /100
1	Energy balancing			100	B	
2	Peak power need			0	-	
3	Storage capacity			0		
IV.1	Quality		0			0,0 /100

Figure 6. Indicator framework template explaining value profile of the client for a specific project.

provides three levels of weights; the first level is for the value area, the second level is for the indicators within each area, while the third level is for the sub-indicators. It should be noted that the weights of each level should add up to 100.

Following the weighting process, the user then sets his/her desired target values (A-E) with A being the best and E the worst for each criterion and indicator of the value profile. The energy expert starts by defining the initial/estimated target levels that could be achieved for the KPIs based on the benchmark data. Later on during the feasibility studies, he/she adjusts these values and proposes more accurately defined target values for the project KPIs. This is achieved by collecting existing and predicted data on the building site and the neighborhood before exploring a number of possible renewable energy solutions as part of the feasibility studies that match the project's objectives. Based on a review of the feasibility studies, the client makes a decision on the KPIs' target values for the initial design brief. As soon as the client approves the value profile, it becomes part of the project's brief.

For the assessment feature of the tool, the above frame of indicators is also used as a multi-criteria assessment tool to evaluate the output of the simulation/analyses and the calculations behind each indicator. The assessment of design alternatives takes place during the feasibility studies to assess the building's geometry (based on best practice of building's volumes) or towards the end of the early design phase to assess whether the design alternatives have achieved these target values or not.

6.2 *Energy matching tool*

The energy matching tool is designed to achieve a satisfying target level of OER (On-site Energy Ratio) by matching the building's energy performance and local renewable supply within the neighborhood. The OER calculates from hourly values the balance between the annual energy demand (of energy types altogether) and the annual energy supply from local renewable sources (respectively, of energy types altogether) (Ala-Juusela et al. 2014).

The energy matching tool is first used during the feasibility study to calculate OER as part of the indicator framework template. During this stage, simple analyses' models are generated using the energy expert's templates to identify a target value for the energy matching.

The tool is then used for the second time during the early concept design stage to conduct numerous energy simulations carried out at both building and neighborhood levels (Figure 7). During this process, the OER is first calculated on a building level (see the top part of Figure 7), before repeating the process for a number of buildings within the pre-defined neighborhood. At this point, the neighborhood is comprised by selecting the right set of pre-calculated buildings to achieve a target OER value for the whole district. This procedure is explained in detail in (Klobut et al. 2016).

The set of indicators is used here to measure the energy positivity of a neighborhood. This set includes the yearly OER and the energy mismatch indicators of each type (heating, cooling and electricity) (Fraunhofer et al. 2015). The mismatch indicators include: Annual Mismatch Ratio (AMRs), Maximum Hourly Surplus (MHSx), Maximum Hourly Deficit (MHDx) and the monthly ratio of peak hourly demand to lowest hourly demand (RPLx). Here x is replaced by an indicator for the different energy types respectively (h for heating, c for cooling and e for electricity) (Ala-Juusela et al., 2014).

The numerous energy simulations potentially produce extensive amounts of results. The graph in the second part of Figure 7 provides an example of a possible visualization that could be achieved using the tool developed by (Laine et al. 2013) in the ISES project. This kind of tool enables not only a useful visualization of a large amount of calculated results, (including a link with their input data) but it can also be used for the what-if-scenario analysis. In such an analysis, the elimination of undesired design options could be carried out by limiting the

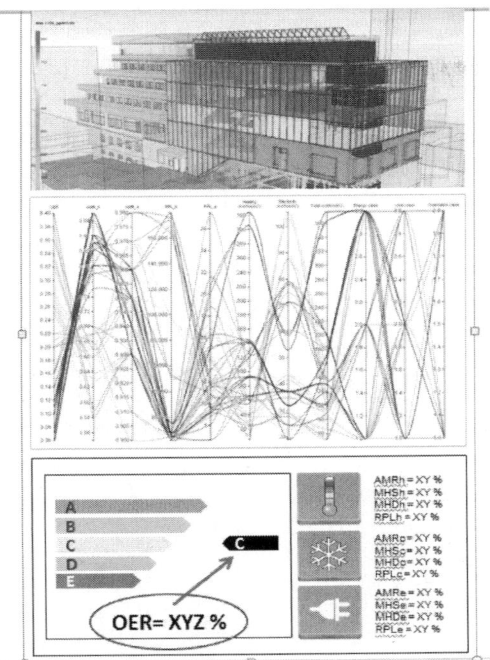

Figure 7. Use of energy matching tool to perform numerous energy simulations at a building and neighbourhood level.

considered cases to those within acceptable ranges of selected parameters, such as external wall transmittance (U-value), window type, building orientation, etc.

The above results will indicate whether the building has positive (OER>1), zero (OER = 1) or negative energy (OER<1) which will then be compared to the target values that were previously set up with the client during a design review meeting.

To support the communication between the architect and the client, the last part of Figure 7 represents an alternative way of visualizing the target level of OER using labels.

The above tool could perhaps be adequate for conducting energy matching for uncomplicated designs with a single renewable energy source. However, with complicated designs when multiple generation resources are taken into consideration, a more complex modelling based on a dynamic decision process is required to conduct energy matching and trading activities (Ghazimirsaeid et al. 2016).

Such complex modelling is usually used for the day ahead to schedule an operational plan for a grid-connected green building. In Green Buildings' Neighborhood Systems (GBNS), the integration of on-site energy production systems, the energy storage systems, as well as the communication between the building and the upstream grid, poses great challenges for the energy systems' efficient control. A multiple GB system usually includes producers based on renewable energy resources, non-renewable energy resources, consumers, and a Neighborhood System Operator (NSO), as illustrated in Figure 8.

For such complicated designs, a Multi-Agent System (MAS) could perhaps be used as an approach to integrate several types of generation and consumer units within a GB. Examples may include PV (Photovoltaic), WT (Wind Turbine), MT (Diesel generators and Micro-Turbines), ES (Energy Storage) and response load demand and non-response load demand. Such a system is described in detail by (Ghazimirsaeid et al. 2016).

6.3 *Design review tool*

Results from performing the set of activities described earlier on using the above mentioned tools could perhaps be visualized using a design review tool. This tool is a collaborative platform designed to organize meetings and manage project team resources. This includes providing a space for collaborative discussions to take place between team members as they visualize various design alternatives and resources.

Moreover, the visualization of IFC models will allow the users to see various views of the BIM model, while mark up and annotation tools allow users to draw simple shapes to annotate and highlight any particular aspect of the object (3D model or document). In addition, video/audio conferencing facilities are provided to offer support to organize synchronous distributed design review meetings (Huifen et al. 2003) when needed throughout the project's life cycle. The advanced visualization and interaction techniques provided in the design review tool will enable project team members to better communicate their views and reach a shared understanding about the design

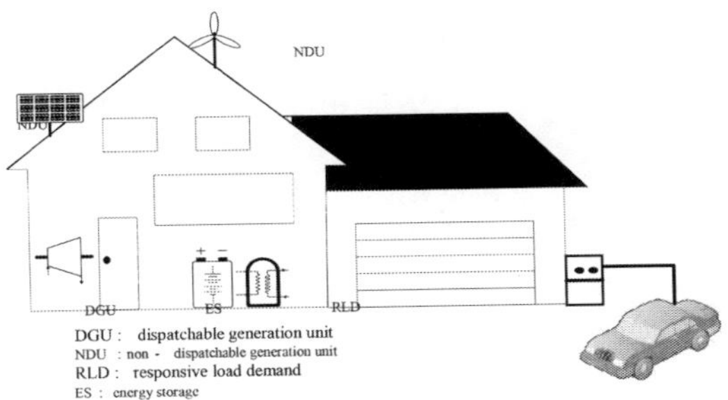

Figure 8. A typical green building with various producers and prosumers.

during design review meetings (Bassanino, et al. 2014b).

7 THE DESIGN4ENERGY PORTAL

This paper described the authors' approach to designing a collaborative environment where architects are provided with tools to assist them in designing better energy efficient buildings with a consideration of their neighborhoods. This section discusses the visualization and the use of the portal from the end-users' point of view.

This Design4Energy collaborative environment is a process driven system. The users are presented with the workflow of the whole product's life cycle described earlier on. Each stage comprises a number of activities and a set of tools to support each one of these activities.

The structure of the portal's user interface, as illustrated in Figure 10, comprises a project workflow navigation bar at the top for users to access different stages of the workflow, and a project configuration and tools' menu on the left hand side. This setting will allow users to configure the project settings and activate the tools based on that particular stage of the workflow. The selected tools will be presented in the tools' user interface. A set of collaboration tools with audio/video functionalities are also provided on the right hand side for supporting remote meetings.

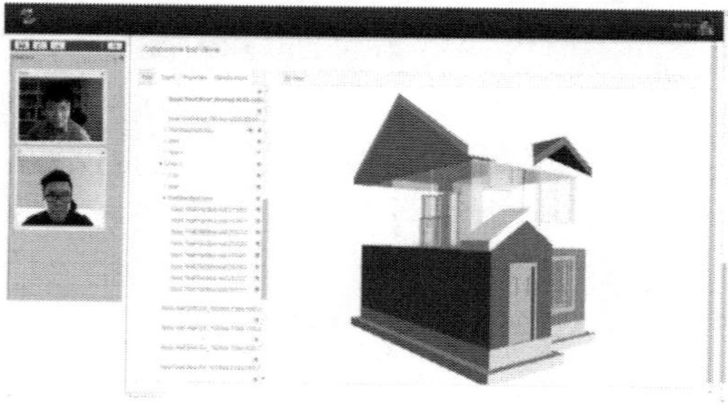

Figure 9. The design review tool used to support collaborative design review meetings.

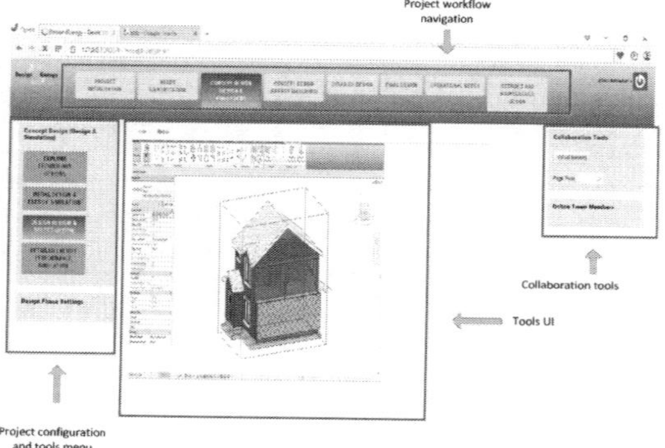

Figure 10. The D4E user interface.

8 CONCLUSIONS

This paper presented the Design4Energy approach to creating a collaborative environment to support multi-functional teams in designing energy efficient buildings with a consideration of their neighborhoods. A scenario approach was utilized to identify the activities the project team performs together with a list of required tools to assist them to carry out these activities.

ACKNOWLEDGEMENT

This research is funded by the European Commission under contract FP7–2013-NMP-ENV-EeB through the Design4Energy project (Grant agreement no: 609380). We would like to acknowledge all the project's partners and, in particular, the project's architects, GSM (Spain), 3L (Germany) and TPF (Poland), for their valuable contribution.

REFERENCES

Ala-Juusela, M., Sepponen, M. & Crosbie, T. 2014. Defining the concept of an energy positive neighbourhood and related KPIs. *Sustainable Places Conference*, Nice, 1–3 October.

Arayici, Y. & Sarshar, M. 2002. DIVERCITY: A Virtual Construction Design & Briefing Environment. *The 3rd International Conference on Decision Making in Urban and Civil Engineering Proc.*, London, 6–8 November.

Bassanino, M., Fernando, T. & Wu, K.-C. 2014b. Can virtual workspaces enhance team communication and collaboration in design review meetings? *Architectural Engineering and Design Management, 10*(3–4): 200–217.

Bassanino, M., Fernando, T., Masior, J., Kadolsky, M., Scherer, R.J., Fouchal, F., Hassan, T.M., Firth, S., Mäkeläinen, T. & Klobut, K. 2014a. Collaborative environment for energy-efficient buildings at an early design stage. In A. Mahdavi, B. Martens & R. J. Scherer (ed.), *eWork and eBusiness in Architecture, Energineering and Construction Proc.*, Vienna, 17–19 September. CRC Press.

Delponte, E., Ferrando, C., Di Franco, M., Hakkinen, T., Rekola, M., Abdalla, G., Casaldaliga, P., Ortiz, C.P., Vega, A.L. & Shih, S.-G. 2014. Holistic and Optimized Life-cycle Integrated Support for Energy-Efficient Building Design and Construction: HOLISTEEC methodology. In A. Mahdavi, B. Martens & R. J. Scherer (ed.), *eWork and eBusiness in Architecture,*

Energineering and Construction Proc., Vienna, 17–19 September. CRC Press.

Eastman, C., Teicholz, P., Sacks, R. & Liston, K. 2011. *BIM Handbook: A Guide to Building Information Modelling for Owners, Managers, Designers, Engineers and Contractors (Second Edition ed.)*. John Wiley & Sons.

Fernando, T., Wu, K.-C. & Bassanino, M. 2013. Designing a Novel Virtual Collaborative Environment to Support Collaboration in Design Review Meetings. *ITCon*, 18: 372–396.

Fouchal, F., Masior, J., Wei, S., Hassan, T. & Firth, S. 2015. Decision support to enable energy efficient building design for optimised retrofit and maintenance. The 32nd CIB W78 Conference Proc., Eindhoven, 27–29 October.

Fraunhofer Team, Mäkeläinen, T., Klobut, K., Sepponen, M., Shemeikka, J., Hassan, T., Fouchal, F. & Firth, S. 2015. Design4Energy Deliverable D2.1: Indicators and success factors for holistic energy matching.

GeiBler, M.-C., Woudenberg, W. & Guruz, R. 2014. Processes and requirements for an eeEmbeded Virtual Design Laboratory. In A. Mahdavi, B. Martens & R. J. Scherer (ed.), *eWork and eBusiness in Architecture, Energineering and Construction Proc.*, Vienna, 17–19 September. CRC Press.

Ghazimirsaeid, S., Fernando, T. & Marzband, M. (in press) 2016. A multi-objective decision optimization model for green buildings based on stochastic approach considering uncertainties. *11th European Conference on Product and Process Modelling Proc.*, 7–9 September. Limassol.

Hannus, M., Kazi, A. S. & Zarli, A. 2010. ICT Supported Energy Efficiency in Construction; Strategic Research Roadmap and Implementation Recommendations.

Huifen, W., Youliang, Z., Jian, C., Lee, S. F. & Kwong, W. C. 2003. Feature-based collaborative design. *Journal of Materials Processing Technology, 139*(1–3): 613–618.

IREEN—The ICT roadmap for energy-efficient neighbourhoods. 2013, available from http://cordis.europa.eu/project/rcn/100736_en.html

Iadanza, E., Truillazzi, B., Terzaghi, F., Marzi, L., Giuntini, A. & Sebastian, R. 2014. The STREAMER European Project. Case Study: Careggi Hospotal in Florence. *The 6th European Conference of the International Federation for Medical and Biological Engineering Proc.*, Dubrovnik, 7–14 September. Springer.

Klobut, K., Hukkalainen, M. & Ala-Juusela, M. (in press) 2016. New Potential Indicators for Energy Matching at Neighbourhood Level. *The CIB World Building Congress Proc.*, Tampere, 30 May–3 June.

Laine, T., Idman, T., Rekola, M., Kukkonen, V., Forns-Samso, F., Karola, A., Kukkonen, E., Kavcic, M. & Katranuschkov, P. (2013). ISES Deliverable D6.2: Prototype of the simulation synthesis and version management service.

Marzban, M., Sumper, A., Dominguez-Garcia, J. L. & Gumara-Ferret, R. 2013. Experimental validation of a real time energy management system for microgrids in islanded mode using a local day-ahead electricity market and MINLP. *Energy Conversion and Management, 76*: 314–322.

Mazza, D., Linhard, K., Mediavilla, A., Michaelis, E., Pruvost, H. & Rekkola, M. 2015. An integrated platform for collaborative performance efficient building design: the case of HOLISTEEC project. *Sustainable Places Conference*, 16–18 September, Savona.

Mäkeläinen, T., Klobut, K., Fernando, T., Hannus, M., Masior, J., Fouchal, F., Hassan, T., Firth, S. & Bassanino, M. 2014. New methodology for designing energy efficient buildings in neighbourhoods. In A. Mahdavi, B. Martens & R. J. Scherer (ed.), *eWork and eBusiness in Architecture, Energineering and Construction Proc.*, Vienna, 17–19 September. CRC Press.

Wikforss, O. & Lofgren, A. (2007). Rethinking Communication in Construction. *ITCon*, 12: 337–345.

eWork and eBusiness in Architecture, Engineering and Construction – Christodoulou & Scherer (Eds)
© 2016 Taylor & Francis Group, ISBN 978-1-138-03280-4

KPI framework for energy efficient buildings and neighbourhoods

K. Klobut, T. Mäkeläinen, A. Huovila, J. Hyvärinen & J. Shemeikka
VTT Technology Research Centre of Finland, Espoo, Finland

ABSTRACT: Design4Energy (D4E) project aims at developing a design methodology that is able to create energy-efficient buildings within the context of their neighbourhoods. An indicator framework for managing project strategic objectives has been set up, with Key Performance Indicators (KPIs) to assess buildings' performance in use. Selected performance, economic and environmental indicators, enriched with neighbourhood energy efficiency indicators are defined with assessment criteria and metrics. The D4E usage scenarios are formalised following Information Delivery Manual (IDM) methodology: capturing the KPI target setting and their assessment processes in project definition and early design phases in Business Process Modelling Notation (BPMN) process maps, and specifying the content of each of those exchanges as Exchange Requirements (ER). This paper will demonstrate how D4E design methodology with a set of KPIs can be used to improve building design process, including neighbourhood context.

1 INTRODUCTION

At present, buildings are often designed as individual units seldom considering their energy exchange with the surrounding neighbourhood. Current practices do not conduct enough reliable energy performance predictions at the design stage due to both methodological and technological issues (Sepponen et al., 2013).

The aim of the D4E project is to create new methodology and tools to move on from this current position to allow design teams to take a holistic view in designing energy efficient buildings within the context of neighbourhood.

2 VISION OF NEIGHBOURHOOD PLANNING

Sepponen et al. (2013) formulates the vision for neighbourhood planning as: "Energy efficiency at the heart of holistic planning of new and existing neighbourhoods", and underlines the availability aspect of relevant data and ICT tools to support holistic decision making in neighbourhoods. Relevant, timely and linked data accessible via open platforms supports transparency and close cooperation in the planning and management of neighbourhoods, involving a range of partners in the realisation of energy efficient neighbourhoods from public, private and civil society sectors. The aim is that neighbourhood planning incorporates all aspects of energy conservation relating to the built environment and associated socio-economic activities, such as working, commuting as well as access to services and leisure. ICT systems enable robust planning decisions based on optimising energy efficiency and economic performance. The purpose of ICT in this context is to provide access to methodologies, tools and data that facilitate accurate simulations of a neighbourhood's energy characteristics based on interdependencies as part of the wider ecosystem. As part of the on-going maintenance and management of energy efficient neighbourhoods, a range of energy services will offer ICT-enabled automated responses to conserve energy in relation to utilities, transport and passive energy design.

2.1 *Neighbourhood definition*

Energy demand of a neighbourhood comprises the energy demand of buildings (and depending on the scope considered, also other urban infrastructures, such as waste and water management, parks, open spaces and public lighting, as well as the energy demand from transport). Renewable energy includes solar, wind and hydro power, as well as other forms of solar energy, biofuels and heat pumps (ground, rock or water), with the supply facilities placed where it is most efficient and sustainable. (Ala-Juusela, 2014).

Originated from the concept of Nearly Zero Energy Buildings in the context of EPBD directive, the broader context of "neighbourhood" was proposed for consideration by the European Commission in the FP7 ICT calls.

The neighbourhood definition within D4E project is currently evolving through discussions among the project members on what should be included in the concept of neighbourhood in this case, the appropriate time scales and the energy variants to be included in the demand and as renewable supply. A contemporary working definition of a neighbourhood at this stage is limited to the proposal that neighbourhood should be considered as a group of buildings:

- with a common geographical location,
- served by common energy networks,
- with energetic quality to be quantifiable by indicator metrics.

3 KEY PERFORMANCE INDICATORS IN D4E PROJECT

The performance based approach focuses on the required end performance of a building instead of prescribing the technical solutions to achieve that performance (Gibson, 1982). Compared to the prescriptive way to govern the construction process, the advantage is in shifting the focus to user requirements leaving room for innovation in finding the best technical solutions. Performance indicators are quantifiable metrics (values that can be measured) that describe the performance in use. They are useful to set strategic targets for a project. Those can be formulated as client's minimum requirements for a project and then the experts (architect, designer, energy engineer) can specify the range of possible solutions to fulfil those requirements, and by comparing those, find the optimal solution.

Performance indicators have been developed from different perspectives in earlier European projects such as Perfection (Loomans et al., 2011), SuPerBuildings (Häkkinen et al., 2012), ValPro (Hyvärinen et al., 2012) and Ecobim (Huovila et al., 2014). Those are also used in numerous sustainability rating systems. In the D4E methodology, presented in this paper, the aim is not to develop another rating system but rather select Key Performance Indicators (KPIs) that can be used to set strategic project objectives by the client and facilitate the exchange with the experts (architect, energy engineer) in early design of energy efficient buildings including their impact on neighbourhood. Since the most important decisions affecting the building performance through its life-cycle are made in this early design phase, it is important to define the targets already at this stage by using a holistic range of measurable performance indicators. However, those cannot be too specific and technical at so early phase and a rather limited number of KPIs covering the most important aspects is recommended.

A long list of D4E indicators was presented by (Fouchal et al., 2014). The next step was to select the most important KPIs that are relevant in the early energy design phase through D4E usage scenarios (presented in chapter 4.1) and concrete cases. In D4E the framework for the KPIs was built around the most important performance categories: 1) Economic performance, 2) Performance in use, 3) Environmental performance, and 4) Impact on neighbourhood performance. The three first categories assess the performance of a building and the third category examines its impact on the surrounding neighbourhood. The framework with selected main categories, sub-categories and KPIs, as implemented in D4E KPI tool, is presented in Figure 1. Each KPI can be assessed on a five level scale ranging from A (best) to E (worst). This uniform assessment scale makes the KPIs comparable and allows seeing on which aspects a building is performing well and on which aspects worse.

In the first scenarios analysed six most important KPIs were selected: 1) Life Cycle Costs (EUR/m[2]), 2) Indoor Air Quality (CO_2 concentration ppm), 3) Thermal comfort (temperature range in °C), 4) Climate change (CO_2 eq. emissions), 5) Energy consumption (kWh/m^{2}a) and 6) Onsite energy ratio (%). Numerical target values have been developed for those on the previously mentioned five-level assessment scale. The tool presented in Figure 1 then enables to calculate a score based on the performance levels of the KPIs. Also weights can be used to indicate the relative importance of different aspects. As such, this tool will be useful in client's requirement setting and exchange with the architect to easily analyse the importance of different aspects and see what minimum performance

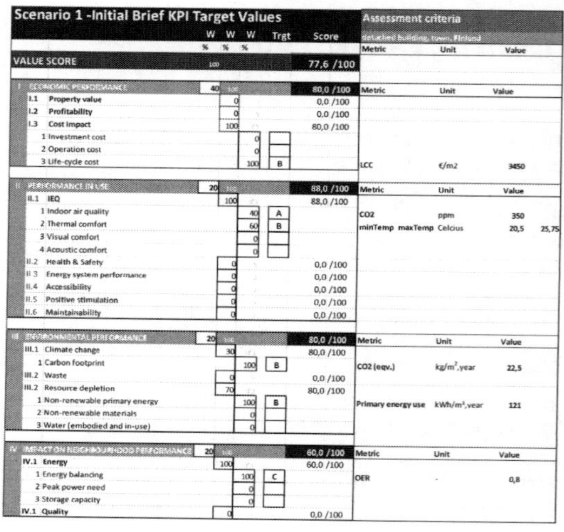

Scenario 1 -Initial Brief KPI Target Values	W %	W %	W %	Trgt	Score	Assessment criteria (detached building, town, Finland) Metric	Unit	Value	
VALUE SCORE	100				77,6 /100				
I ECONOMIC PERFORMANCE	40	100			80,0 /100	Metric	Unit	Value	
I.1 Property value		0			0,0 /100				
I.2 Profitability		0			0,0 /100				
I.3 Cost impact		100			80,0 /100				
1 Investment cost			0						
2 Operation cost			0						
3 Life-cycle cost			100	B		LCC	€/m2	3450	
II PERFORMANCE IN USE	20	100			88,0 /100	Metric	Unit	Value	
II.1 IEQ		100			88,0 /100				
1 Indoor air quality			40	A		CO2	ppm	350	
2 Thermal comfort			60	B		minTemp maxTemp	Celcius	20,5	25,75
3 Visual comfort			0						
4 Acoustic comfort			0						
II.2 Health & Safety		0			0,0 /100				
II.3 Energy system performance		0			0,0 /100				
II.4 Accessibility		0			0,0 /100				
II.5 Positive stimulation		0			0,0 /100				
II.6 Maintainability		0			0,0 /100				
III ENVIRONMENTAL PERFORMANCE	20	100			80,0 /100	Metric	Unit	Value	
III.1 Climate change		30			80,0 /100				
1 Carbon footprint			100	B		CO2 (eqv.)	kg/m^2,year	22,5	
III.2 Waste		0			0,0 /100				
III.2 Resource depletion		70			80,0 /100				
1 Non-renewable primary energy			100	B		Primary energy use	kWh/m^2,year	121	
2 Non-renewable materials			0						
3 Water (embodied and in-use)			0						
IV IMPACT ON NEIGHBOURHOOD PERFORMANCE	20	100			60,0 /100	Metric	Unit	Value	
IV.1 Energy		100			60,0 /100				
1 Energy balancing			100	C		OER	-	0,8	
2 Peak power need			0						
3 Storage capacity			0						
IV.1 Quality		0			0,0 /100				

Figure 1. Setting target levels with KPI tool.

is required on a certain aspect, how low level is allowed on another one and how different solutions can fulfil such requirements.

4 NEW METHODOLOGY TO CAPTURE NEIGHBOURHOOD ENERGY CONTEXT

This section describes a usage scenario, developed in D4E project, for building energy design in the context of the district level energy system, where multi production/consumer environment and open district heating network are considered. The formal description of the usage scenario follows the Information Delivery Manual (IDM) methodology (ISO, 2010), by capturing the activities and information flows in process maps (in Business Process Modeling Notation; OMG, 2011), where exchanges between project partners are identified and specified as exchange requirements.

Design principle is: "First minimize energy demand of the building, and then choose solutions for needed energy production for the building."

Business model principle is: "Energy matching is usually profitable if energy can be produced and used in same system without selling it to the energy operator of municipality (when feed-in tariff not used)."

Also we believe that the future energy efficient buildings using fluctuating or unpredictable renewable energy sources are complex, meaning high uncertainties for the design selections. We assume that by running the building energy simulations as early as possible in the design phase, we will have a true possibility to have an effect on the overall energy design of a future energy efficient building. Similar kind of early stage approach has been stated for example in (Attia et al., 2012) for individual building design. Our methodology goes beyond and widens the scope to consider also the energy impact of a single building on the neighbourhood level.

4.1 *Usage scenario story*

Activities in this scenario belong to the Preparation and brief stage, and Concept design stage in RIBA plan of work 2013 (RIBA, 2013).

In the Briefing stage, architect and/or energy designer start performance planning together with the client. This includes first selecting key criteria to govern the project, reviewing benchmarks and conducting feasibility studies, which is followed by detailed target setting for each KPI: Life Cycle Costs, Indoor Air Quality, Thermal comfort, Climate change, Energy consumption and Onsite energy ratio. A KPI tool has been developed in D4E project to make this process easy, see Figure 1.

The communication between experts and client is eased by the normative indicators, meaning that discussion can be conducted using only value levels (A-B-C-D-E), not getting too deep into measurable target values of each indicator, nor any technical details of solutions that could fulfil the requirements. At this stage client has an opportunity to declare his preferences on two levels: individual indicators and indicator categories. This is managed by weighting—first valuing the four categories against each other and then valuing individual indicators within each category against each other (Figure 1).

Upon receiving an initial brief with KPI target levels set, architect starts the first concept design by reviewing the basic parameters for the energy efficiency of design solutions, and producing different building massing alternatives with minimising the energy demand of the building in mind as an aim.

Cost estimates are based on space types at this stage. Proposed architectural concept solutions are reviewed by the client (and building control authorities).

Energy designer then conducts numerous energy simulations to obtain the energy performance of buildings with different design alternatives (energy demand of the building, on-site energy production). Alternatives are reviewed by the client, and the best ones are selected for the next step.

Value score provided by the tool as a composed outcome of assessment of indicators helps to assess the maturity of each design solution and its variations. It enables the design management and client to focus on target and requirement management as part of their decision making during the review sessions. Indicator level information outcome from assessment supports the whole design team in value configuration.

Above mentioned process tasks are part of the design process described and supported in D4E. The other tasks connected to holistic energy efficiency design include consideration of local energy trading and urban design executed by stakeholders not normally active in building design. Holistic energy trading aspects are analysed by energy designer and verified with help of models and simulations on neighbourhood level. Urban designers maintain dynamic district plan model, where new buildings are placed, and existing ones updated as retrofitted. In future we foresee that these processes can be managed as part of building level design.

4.2 *Process maps*

In the Building energy concept design, yet considering neighbourhood energy system, the client together with key experts (architect and energy designer) is looking for optimal solutions to satisfy the project key objectives (generally concerning economic, environmental, social and cultural values, etc.). In this scenario the focus is on financial and energy objectives.

This process has four sub-processes ("mini scenarios"), as depicted in overall process map in Figure 2:

- Set target levels,
- Sketch design alternatives,
- Analyse energy efficiency (Figure 3),
- Analyse energy matching alternatives (Figure 4).

Information exchanges, identified as data objects in the usage scenario process maps, are listed and defined in the deliverable (Hyvärinen et al., 2016).

Here an example is shown regarding results of energy matching analysis, i.e. exchange requirement *Matching results*, as executed using XML message, according to schema definition developed for D4E (Figure 5).

5 NEIGHBOURHOOD ENERGY EFFICIENCY INDICATORS

Ala-Juusela et al. (2014 and 2015) have developed a set of key performance indicators (KPIs) for assessing the matching of energy demand and supply of a neighbourhood. In addition to meeting the overall annual energy balance, it is important that different types of energy are taken into account

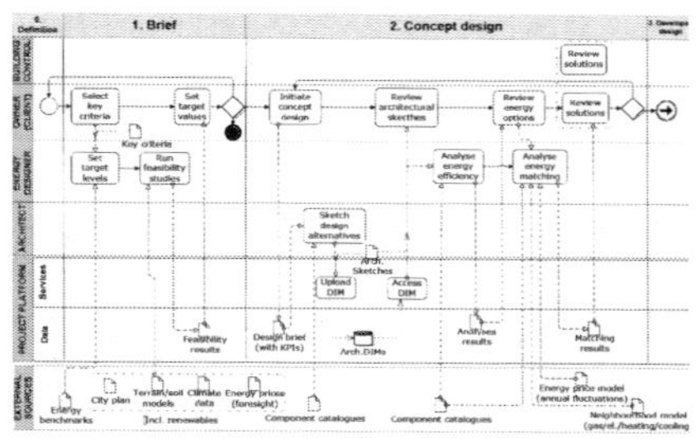

Figure 2. Top level process map of the usage scenario.

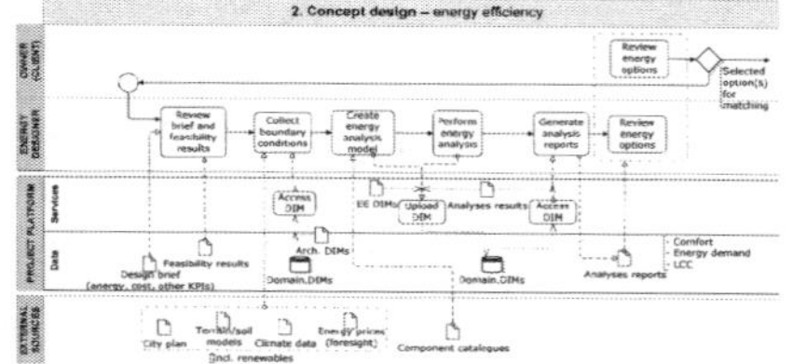

Figure 3. Energy efficiency analysis on the process map.

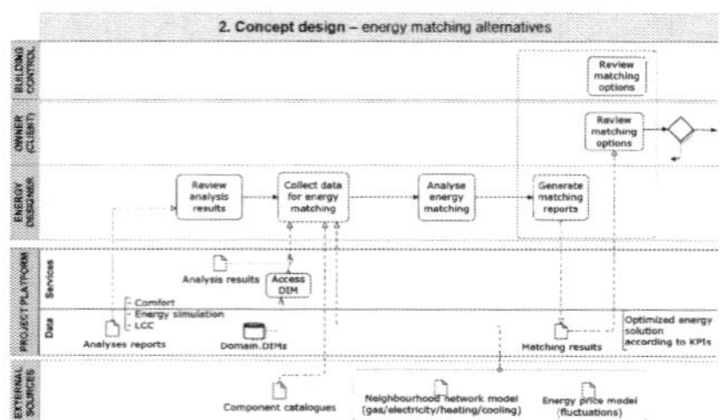

Figure 4. Energy matching analysis on the process map.

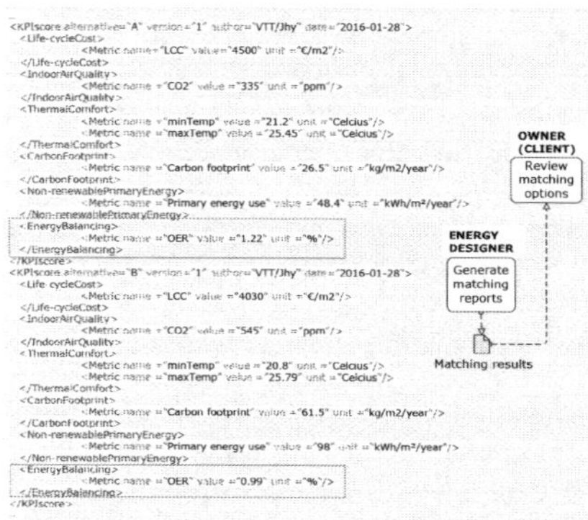


```
<KPIscore alternative="A" version="1" author="VTT/Jhy" date="2016-01-28">
  <Life-cycleCost>
        <Metric name="LCC" value="4500" unit="€/m2"/>
  </Life-cycleCost>
  <IndoorAirQuality>
        <Metric name="CO2" value="335" unit="ppm"/>
  </IndoorAirQuality>
  <ThermalComfort>
        <Metric name="minTemp" value="21.2" unit="Celcius"/>
        <Metric name="maxTemp" value="25.45" unit="Celcius"/>
  </ThermalComfort>
  <CarbonFootprint>
        <Metric name="Carbon footprint" value="26.5" unit="kg/m2/year"/>
  </CarbonFootprint>
  <Non-renewablePrimaryEnergy>
        <Metric name="Primary energy use" value="48.4" unit="kWh/m²/year"/>
  </Non-renewablePrimaryEnergy>
  <EnergyBalancing>
        <Metric name="OER" value="1.22" unit="%"/>
  </EnergyBalancing>
</KPIscore>
<KPIscore alternative="B" version="1" author="VTT/Jhy" date="2016-01-28">
  <Life-cycleCost>
        <Metric name="LCC" value="4030" unit="€/m2"/>
  </Life-cycleCost>
  <IndoorAirQuality>
        <Metric name="CO2" value="545" unit="ppm"/>
  </IndoorAirQuality>
  <ThermalComfort>
        <Metric name="minTemp" value="20.8" unit="Celcius"/>
        <Metric name="maxTemp" value="25.79" unit="Celcius"/>
  </ThermalComfort>
  <CarbonFootprint>
        <Metric name="Carbon footprint" value="61.5" unit="kg/m2/year"/>
  </CarbonFootprint>
  <Non-renewablePrimaryEnergy>
        <Metric name="Primary energy use" value="98" unit="kWh/m²/year"/>
  </Non-renewablePrimaryEnergy>
  <EnergyBalancing>
        <Metric name="OER" value="0.99" unit="%"/>
  </EnergyBalancing>
</KPIscore>
```


Figure 5. XML exchange of energy matching results.

and the timing of the supply and demand of these different types of energy is matched.

The KPIs developed to measure the energy positivity level of a neighbourhood include yearly On-site Energy Ratio (OER) and energy mismatch indicators for each energy type (heating, cooling and electricity). The mismatch indicators include Annual Mismatch Ratio (AMRx), Maximum Hourly Surplus (MHSx), Maximum Hourly Deficit (MHDx) and monthly ratio of peak hourly demand to lowest hourly demand (RPLx), where x is replaced by an indicator for the different energy types respectively (h for heating, c for cooling, e for electricity). (Ala-Juusela et al. 2014 and 2015)

The overall balance between annual energy demand and local renewable supply is indicated with the On-site Energy Ratio (OER), which is the ratio of these two:

- Annual energy supply from local renewable sources (all energy types together)
- Annual energy demand (all energy types together).

The short term imbalances are indicated with:

- Annual Mismatch Ratio (AMR), which indicates how much energy needs to be imported into the area for each energy type on average. It is the annual average ratio of these two, for those

hours when the local demand exceeds the local renewable supply:
 - o hourly difference between demand and local renewable supply (by energy type)
 - o hourly demand (by energy type) during that same hour
- Maximum Hourly Surplus (MHS), which is the maximum yearly value of how much the hourly local renewable supply overrides the demand during one single hour (by energy type)
- Maximum Hourly Deficit (MHD), which is the maximum yearly value of how much the hourly local demand overrides the local renewable supply during one single hour (by energy type)
- Monthly Ratio of Peak hourly demand to Lowest hourly demand (RPL) indicates the magnitude of the peak power demand, and it is calculated as the ratio of these two (by energy type):
 - o The highest value for hourly demand over the month
 - o The lowest value of hourly demand over the month (0-values are ignored)

It is worth noticing that OER = 1 means zero energy building or neighbourhood and OER>1 means energy positive building or neighbourhood. OER<1 indicates that a building or a neighbourhood requires imported energy, i.e. it represents a typical situation today.

AMR indicator can have values between 0 (meaning perfect match) and 1 (no match at all), i.e. the smaller value AMR has, the better the local renewable supply matches with the demand.

5.1 *Energy analysis in practice*

Preparation for the analysis starts by architect and client defining together the building variables and the range of their changes. Typically such variables could be: external wall U-value, type and thermal transmittance of windows, building orientation, efficiency of heat recovery system, characteristic of building usage. For example, Klobut et al. (2016) describes a study where single family houses and office buildings were used in 60 energy simulations.

For the office building, three different types of construction and building technology were simulated: A building with normal insulation level and heat recovery, a low energy office building and an old house with high energy use and no ventilation heat recovery. The second variable parameter was the orientation of the building, either to South or to East.

For the simulation of office buildings, also two types of window shading options were used: no shading, or external shading, lights and schedule control. All different building technologies, orientations and window shading options were combined,

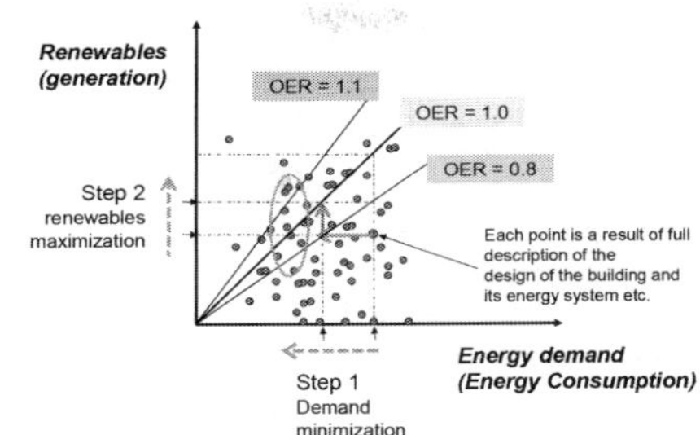

Figure 6. Energy matching tool used to compose neighbourhood satisfying the target level of OER.

resulting in total of 12 different configurations for the office building (3 construction types x 2 orientations x 2 window shading options). Similar analysis was done with single family buildings, summing up to 60 simulations all together.

Generally, the amount of simulation to be carried out equals the number of all possible combinations of considered variables values, i.e. grows very quickly. However, the methodological approach in D4E proposes a relatively high amount of energy simulations to be carried out already in the very preliminary stage of the project. Subsequently, in two steps, following the principle stated in section 4, unfavourable solutions are gradually eliminated based on the newly developed indicators (Figure 6).

6 CONCLUSIONS

The goal of the D4E project is to develop a Building life-cycle evolutionary design methodology able to create Energy-efficient Buildings which are flexibly connected with neighbourhood systems. D4E contributes to bridging barriers in the implementation of the Integrated Energy Efficient Project Delivery (IEEPD) concept.

The methodological approach of D4E comprises using information management tools and methods in the process that is modularly defined, and digitally supported tasks within the phases. The whole process is guided with a collaborative virtual workspace managed by the chief architect (Fernando et al., 2015).

KPI framework for managing lifecycle performance is a useful instrument for reaching ambitious target values during design, construction and retrofit projects. In D4E methodology it is seen as a core instrument of requirement management and can be used alongside with risk management. Additionally, KPI framework enables the following up of the target values during the lifecycle of the building. It can be used alongside with quality management if expanded to cover process related criteria indicators of work-flow or data-flow.

KPI framework includes indicators relevant to D4E project targets. Several indicator listings exist in the area of building-level energy efficiency (Fouchal et al., 2014). However, they do not cover energy issues of buildings linked to the context of their neighbourhood. The developed indicator of energy balancing, measured with On-Site Energy Ratio, demonstrates that this dependence can be quantified and considered already during building design.

KPI framework of D4E presents one relevant set of indicators clustered to 3 performance categories and 1 impact category. This set will support the design team to achieve high level of sustainability in the neighbourhood, when actively used during design and project management.

In an ideal case the categories and indicators of KPI framework should be selectable on the ground of project type and other circumstances, in relation to the specific focus of the project in question.

According to our knowledge, there is not available a specific integrated software for architects and energy designers that combines neighbourhood KPI-framework, energy simulation and KPI-optimisation. A collaborative environment to support multi-functional team in designing energy efficient buildings with consideration of their neighbourhoods, being developed in D4E project (Bassanino et al., 2016), can be considered as a step in this direction. The work presented in this paper supports such development. The planning process of architect needs this kind of seamless support tool in an early stage to tackle the future energy efficient building design challenge.

ACKNOWLEDGEMENT

This research is funded by the European Commission under contract FP7-2013-NMP-ENV-EeB through the Design4Energy project (Grant agreement no: 609380).

REFERENCES

Ala-Juusela, M., Sepponen, M. & Crosbie, T. 2014. Defining the concept of an Energy Positive Neighbourhood and related KPIs. Proceedings of Sustainable Places Conference, October 2014, Nice, France.

Ala-Juusela, M., Sepponen, M. & Crosbie, T. 2015. Defining and Operationalising the Concept of an Energy Positive Neighbourhood. Proceedings of the 10th SDEWES Conference on Sustainable Development of Energy Water and Environmental Systems. September 27 - October 2, 2015, Dubrovnik, Croatia.

Attia, S., Gratia, E., De Herde, A. & Hensen, J. 2012. Simulation-based decision support tool for early stages of zero-energy building design. *Energy and Buildings, 49 (2012), pp. 2–15.*

Bassanino M., Fernando T., Wu K., Ghazimirsaeid S., Klobut K., Mäkeläinen T. & Hukkalainen M. 2016. Approach to design energy-efficient buildings within the context of neighbourhood. Manuscript proposed for Conference on Product and Process Modelling (ECPPM), 7–9 September 2016, Limassol, Cyprus.

Fernando, T. & Bassanino, M. et al. 2016. Design and implementation of the virtual workspace to explore various design options. Deliverable D7.3, Design4Energy project. September 2015.

Fouchal, F., Hassan, T., Firth, S., Klobut, K., Sepponen, M., Heimonen, I., Mäkeläinen, T., Malo, P. & Almeida, B. 2014. Indicators and success factors for holistic energy matching. Deliverable D2.1, Design4Energy project. September 2014.

Gibson, E. 1982. Working with the Performance Approach in Building. International Council for Research and Innovation in Building and Construction (CIB), CIB report, publication no. 64. Rotterdam, Netherlands.

Huovila, A., Corredor Ochoa, A., Huovila, P., Antuña Rozado, C. & Lommi, J. 2014. ecobim guidelines for sustainable eco-innovative construction business models. ecobim project deliverable 2.3. http://www.vtt.fi/sites/ecobim/en/what-is-ecobim

Hyvärinen, J. et al. 2016. Definition of usage scenarios with economic and environmental criteria. Deliverable D2.2b of Design4Energy project (in prep.)

Hyvärinen, J., Huovila, p. & Porkka, J. 2012. Value procurement approach and tools for real estate sector. In: Issa, R. & Issa, C. (eds.) CIBW078 2012 Conference Proceedings, Beirut, Lebanon, October 17–19, 2012. Lebanese American University (2012), 523p., pp. 203–212.

Häkkinen, T. et al. 2012. Sustainability and performance assessment and benchmarking of buildings. VTT Final report. Available at http://cic.vtt.fi/superbuildings/

ISO 29481–1:2010 Building information modelling—Information delivery manual—Part 1: Methodology and format.

Klobut, K., Hukkalainen, M. & Ala-Juusela, M. 2016. New Potential Indicators For Energy Matching At Neighbourhood Level. Manuscript accepted for the CIB World Building Congress 2016, May 30–June 3, 2016, Tampere, Finland

Loomans M., Huovila A., Lefebvre P.-H., Porkka J, Huovila P., Desmyter J. & Vaturi A. 2011. Key Performance Indicators for the Indoor Environment. Proceedings of the World Sustainable Building Conference 2011, Helsinki, 18–21 October 2011. Theme 4. Sustainable processes and eco-efficient technologies, pp. 1666–1675.

Mäkeläinen, T., Klobut, K., Hannus, M., Sepponen, M., Fernando, T., Bassanino, M., Masior, J., Fouchal, F., Hassan, T.M. & Firth, S. 2014. A new methodology for designing energy efficient buildings in neighbourhoods. Proc. European Conference on Product and Process Modelling (ECPPM), 17–19 September 2014, Vienna, Austria.

OMG Document Number: formal/2011–01-03. Business Pro-cess Model and Notation (BPMN) Version 2.0. January 2011.

RIBA 2013. Editor: Dale Sinclair. RIBA Plan of Work 2013 Overview. ISBN 978 1 85946 519 6.

Sepponen, M., Hannus, M., Hedman, Å., Tommis, M., Bour-deau, M., Decorme, R., Zarli, A., Huerva, A., Wagner, M., Blanco, A., Carlos, J., Hernandez, R., Martinez, J., Vlasveld, J., Beurden, H., Bouricius, M., Chricchio, F., Mastrodonato, C. & Blöchle, M. 2013. Roadmap for European-scale innovation and take-up. Deliverable 3.3.2 of the EU-project The ICT Roadmap for Energy-Efficient Neighbourhoods. 17.10.2013.

eWork and eBusiness in Architecture, Engineering and Construction – Christodoulou & Scherer (Eds)

Multiscale building modelling and energy simulation support tools

A. Romero, J.L. Izkara, A. Mediavilla, I. Prieto & J. Pérez
Sustainable Construction Division, TECNALIA, Derio, Spain

ABSTRACT: Building and district modelling (BIM, CityGML...) are key technologies for the deployment of energy efficiency strategies at building and district level, from the initial stages of planning and design to the operation and maintenance ones. These technologies allow satisfying the interoperability requirements that facilitate the cooperation among the multiple stakeholders and provide the framework to develop more intelligent tools. This paper introduces five complementary European R&D projects in which TECNALIA is collaborating, very good examples of innovative systems based on these concepts. MOEEBIUS enhances passive and active building elements modelling approaches enabling improved building energy performance simulations. HOLISTEEC focuses on building multi-physical simulations considering the neighborhood context. FASUDIR exploits the high potential of GIS tools for urban sustainability analysis and accurate building energy performance evaluation. EFFESUS integrates district and building scales in historic districts. OPTEEMAL develops a platform at district level, based on an IPD approach.

1 INTRODUCTION

Increasingly there is a need for interactive and user-friendly decision support tools that enable analysis of the impact of the building energy oriented projects on the sustainability of the urban district in a holistic way, and facilitate the necessary communication mechanisms between the multiple stakeholders that are involved in the process (Egusquiza et al. 2014).

The five projects presented in this paper aim to develop the knowledge, strategies, decision support models and tools to meet the challenges of linking the strategic urban level to the executive building level from a multiscale perspective considering innovative energy simulation approaches.

At building level the main research focus of the MOEEBIUS (Modelling Optimization of Energy Efficiency in Buildings for Urban Sustainability) project is on developing tools to monitor and assess actual building energy performance, considering relevant factors such as user behaviour, complex energy systems performance and weather forecast, and to be able to predict accurately building energy loads and consumption along the whole lifecycle.

The HOLISTEEC (Holistic and Optimized Life-cycle Integrated SupporT for Energy-Efficient building design and Construction) project aims to provide a collaboration platform for performance-based building design during the entire life-cycle, based on multi-physical simulations (energy, acoustics, lighting and environment) from early design phases and considering interactions with the neighbourhood, built upon IFC (Liebich et al. 2015) and CityGML (Gröger et al. 2012) standards, and using BCF (BIM Collaboration Format) mechanism.

The FASUDIR (Friendly and Affordable Sustainable Urban Districts Retrofitting) and EFFESUS (Energy Efficiency for EU Historic Districts' Sustainability) projects propose a seamless integration of district and building scales through a unique data model based on CityGML standard, combining the high potential of GIS tools for urban sustainability analysis with an accurate energy performance evaluation at building level to allow the selection of the most suitable strategies on energy retrofitting interventions in districts.

The OPTEEMAL (Optimised Energy Efficient Design Platform for Refurbishment at District Lev-el) project aims to devélop a design platform, based on an IPD (Integrated Project Delivery) approach and supported by the utilization of BIM models, for an integrated, optimized and systemic energy oriented refurbishment at district level.

2 MOEEBIUS PROJECT

MOEEBIUS project (www.moeebius.eu) aims to reduce the gap between energy prediction and real/measured energy performance of buildings to values below 10%, by addressing occupants' behaviour, real HVAC (Heating, Ventilation and Air Conditioning) performance and real weather conditions both at the building energy performance simulation (commissioning), as well as during the operation phase (real-time optimization on the basis of fine-grained control and automation).

MOEEBIUS solutions will be validated in real-life conditions over an extensive 20-month pilot roll-out period (that includes equipment and systems installations, base-lining activities, models and systems calibration and actual validation of MOEEBIUS in real-life situations) in a variety of buildings (office, retail, educational, sports, residential, hotel) and building blocks (considering their interactions in energy performance optimization) under different environmental, social and cultural contexts in three dispersed geographical areas (London in UK, Mafra in Portugal and Belgrade in Serbia).

2.1 *Modelling approach*

MOEEBIUS introduces a holistic modelling approach that focuses on appropriately addressing and accurately understanding all sources of uncertainty and inaccuracy in building performance assessment.

MOEEBIUS adopts a hybrid approach that com-bines white-box modelling techniques (at the level of BIM) with black-box modelling approaches (focusing on occupants' behaviour) to deliver an innovative system that captures the real complexities of actual buildings and allows for the correct understanding of user behaviour's impact.

Enhanced, accurate and dynamic behavioural (individual and/or group) profiles complement improved static BIM models (with reduced simplifications and able to accommodate life-cycle assessment and life-cycle cost parameters to enable advanced and optimized predictions through, the appropriately configured, MOEEBIUS building energy performance simulation engine.

2.2 *Simulation approach*

MOEEBIUS uses a two-step calibration process of dynamic simulation tools that considers as-built characteristics, set-points, local real weather data, real occupancy (first step) and sub-metering HVAC/lighting data, real indoor temperature (second step).

The MOEEBIUS performance optimization mechanisms are based on an enhanced version of the already available open-source and widely used EnergyPlus simulation engine. It accommodates enhanced algorithmic concepts for bringing together improved BIM models, semantically improved with DER (Distributed Energy Resources) models and dynamically updated with occupant behaviour profiles, schedules and weather forecasts and utilizing them in building performance simulation iterations towards offering optimized performance predictions of high accuracy.

Even though modelling comprises a focal point of MOEEBIUS, the core outcome and main innovation introduced in the project lies upon the MOEEBIUS dynamic assessment engine. At building level, the dynamic assessment engine serves two distinct, but also interrelated functions: (i) fault detection and diagnosis and (ii) distributed fuzzy model predictive control.

Simulation outputs and real-time measurements from BEMS (Building Energy Management Systems), the MOEEBIUS wireless sensor network and external sources (weather data), are fed to the dynamic assessment engine and comparatively assessed for the identification of performance deviations and their root causes. Through fault detection and diagnosis, the dynamic assessment engine is able to recognize whether the building is beginning to operate sub-optimally; and proactively identify specific performance trends at different spatio-temporal granularity (e.g. abnormal HVAC consumption increase in a specific room) that could progressively lead to significant performance deviations. Subsequently, it is able to drill-in and analyze parameters affecting the deviating metrics (e.g. ambient/behavioural trends) to define the root cause of the evolving deviation.

The definition of the root cause triggers the activation of an innovative distributed fuzzy model predictive control engine. The engine allows for short-term prediction of the building performance outcome (every few minutes) under alternative automated control strategies that aim at mitigating the identified deviation. This is achieved by adapting the operation (self-adaptation) of the building to performance targets, while preserving occupants' comfort and health at acceptable levels. Optimization performed through the dynamic assessment engine is an iterative and continuous process that allows for the prompt identification of deviations (continuous assessment through the fault detection/ diagnosis) and execution of short-loop, few-minutes long simulations for the definition of optimal automation strategies.

In case no relevant root cause is identified, the problem is passed to the predictive maintenance and retrofitting advisor modules, for further investigation and definition of alternative maintenance and retrofitting actions, respectively, to effectively mitigate the identified deviation. To enhance user experience in this area, a virtual reality environment using BIM information (3D BIM mapping) and mobile device sensors (for location tracking) effectively complements the operation of the maintenance subsystem with a highly intuitive, friendly and useful user interface.

The multiscale energy simulation focuses on real time optimization of energy demand and supply with the objective of reducing the difference between peak power demand and minimum night time demand. Low-level information and intelligence (building) will be uplifted to the high-level (district) towards assessing (in real-time) the energy performance and demand flexibility at the level of blocks of buildings. This is achieved through an innovative forecasting, aggregation and flexibility module that enables real-time dynamic virtual power plants formulation with enhanced aggregated flexibility capabilities to participate in event-, time-, location- or price-based demand side management strategies.

2.3 *Innovation*

The main innovation introduced in MOEEBIUS is the holistic modelling approach that focuses on appropriately addressing and accurately understanding all sources of uncertainty and inaccuracy in building performance assessment. Dynamic models reflecting user comfort and overall behaviour in the built environment, enhanced DER and district heating models and short-term weather forecasts feed the MOEEBIUS building energy performance simulation system, allowing for predictions of high accuracy.

Novel end-user applications and decision support tools developed in the project are oriented to (i) real-time optimization through automated control, (ii) predictive maintenance diagnostics and decision making, including sanitary maintenance of HVAC systems to preserve high indoor environmental quality and (iii) advanced retrofitting advising through automated criteria-based configuration, evaluation and selection of optimal integrated retrofitting interventions.

2.4 *Acknowledgements*

This project has received funding from the European Union's Horizon 2020 research and innovation program under grant agreement No 680517.

3 HOLISTEEC PROJECT

The central aspect of HOLISTEEC (www.holisteecproject.eu) is to enable multi-physical simulations and optimization during all building life-cycle design stages, relying on a collaborative and loop-based design methodology (Delponte et al. 2014, Mazza et al. 2015).

In order to achieve this goal it is essential that (i) simulation tools are adapted to models in different LOD (Level of Development), (ii) there is a proper interoperability between BIM and simulation that can handle model changes and versions, (iii) collaboration mechanisms are implemented to support open workflows and loop-based design between designers and simulation experts in different disciplines (energy, acoustic, lighting and environment) so that conflicts can be evaluated and different decisions can be tracked back and (iv) KPIs (Key Performance Indicators) definitions and requirements which cover all disciplines are accordingly adapted to the design phases and can be continuously monitored.

HOLISTEEC will be validated in four real building projects provided by end-users in the consortium. Targeted projects cover different countries and climates (Turkey, Finland, the Netherlands and Belgium), different stages (conceptual designs, detailed designs, retrofitting projects, etc.) as well as different uses and typologies (residential and holiday complexes, office buildings, student dormitories, etc.).

3.1 *Modelling approach*

In order to achieve an interoperability mechanism between spatial domains and technical disciplines a neutral simulation model based on XML (SIM model) is defined. Algorithms have been developed to extract and adapt the geometry and topology from the BIM model (IFC) and the neighbourhood model (CityGML) and combine into a single source which accommodates different representation details.

In addition, the model adapts to the building LOD in terms of building products and systems. Thus, in early phases the SIM model matches the basic composition information provided in the BIM (perhaps only a name or tag representing the typology) with performance information provided by generic e-catalogue products. In detailed phases (higher LOD), the details of product physical properties can be directly taken from IFC material layers. An e-catalogue system adapted to various LOD and supporting both generic and vendor-specific products is developed as part of the HOLISTEEC platform.

The generation of simulation models is a three-step process: (i) an automatic generation of the aforementioned neutral or "raw" SIM model with geometry/topology suitable for various disciples, (ii) A loop process, where the SIM model is linked to e-catalogue products creating a set of SIM model variants or configurations. It is a user-driven process but supported by HOLISTEEC services and user interfaces and (iii) transformation of those simulation variants into different input files for each of the provided simulation engines. In some cases tools provided by project partners and tailored to HOLISTEEC are used, in other cases existing third-party engines.

3.2 *Simulation approach*

In relation to energy simulation, the multiscale issue is targeted by providing two approaches, each one focused in different users, different project phases and different decisions. For early stages a neighbourhood level simulation is offered, intended for early decisions by the architect (e.g. building shape type, orientation, glazing ratios). Influences of the surrounding environment are considered in terms of shadowing, reflections, etc. but also impacts of the target building into the rest in a bidirectional way.

The main innovation for this approach is to extend the capabilities of an existing building scope simulation engine like EnergyPlus for urban scale simulation. Thus, the CityGML model is transformed into the aforementioned SIM model. For the target building envelope is created (external wall, windows, roof...) and when there is not yet information about windows rules for glazing ratios creation are defined. Storey-level thermal zones are then automatically generated. Other buildings are considered as shadowing surfaces. To reduce the computational resources of a complex urban model filtering mechanisms are implemented to consider only relevant shadowing surfaces depending on distance and height. The process is iteratively repeated, considering a different target building in each iteration, yielding from one city model a set of EnergyPlus simulations (one per target building), whose results are then combined and visualized in 3D. Different design variants can be tried and compared.

The second approach for energy simulation focuses on a single building (the target one) and relies on a detailed representation of the building geometry, internal zone partitioning and details about HVAC systems, as well as more detailed occupancy and behavior models. In this approach, the core of the information to the SIM model comes from IFC and is suited for more advanced LOD, oriented to consultants or simulation experts with a deeper insight, since further manual interaction and enhancing of the SIM model could be required. Existing software implementations are adapted to the platform and data models.

3.3 *Innovation*

Energy simulations are considered for a variable range of levels of detail through an intermediate neutral SIM model which can combine several building and several representation modes for the same building, merging information coming from either IFC or CityGML, which then is converted to various simulation engines targeted to various end users and project stages.

Neutral SIM models are transformed to different existing building simulation engines through custom adaptations. In the case of the neighbourhood, the main innovation consists of the extension of building scope engines (EnergyPlus) for urban extent.

3.4 *Acknowledgements*

This project has been funded by the European Union Seventh Framework Programme under Grant Agreement No 609138.

4 FASUDIR PROJECT

FASUDIR (www.fasudir.eu) main result is the Integrated Decision Support Tool (IDST), developed to help decision makers to select the best energy retrofitting strategy to increase the sustainability of the whole district (Mittermeier et al. 2014). The IDST features a 3D graphical user interface in order to facilitate the interaction between the multiple stakeholders involved in the decision making process (Romero et al. 2014).

FASUDIR IDST is being validated in three different European urban developments that are representative of different district typologies that are common in Europe, and especially in need of energy retrofitting initiatives (Santiago de Compostela in Spain, Frankfurt in Germany and Budapest in Hungary).

4.1 *Modelling approach*

In FASUDIR a unique building/district energy model, which enables the documentation of all the information required for the development of the FASUDIR IDST, is designed and developed.

The data model structure is based on the CityGML standard extended with domain specific information. Data model compiles the required information for KPI calculation at both levels (building and district). Data model is divided into CM (City Model) and SM (Simulation model). The CM, which is based on CityGML, contains all the data representing the current state of the district. All the parameters required for triggering the required simulations are stored into the SM.

CM is completed with the indicators information (KPIs) after their calculation to determine the current state of the district. The information collected into the FASUDIR CM is grouped into the different city elements represented into the CityGML standard (building, transportation, vegetation, city furniture, etc.). CM extends CityGML with required ADE (Application Domain Extensions) in order to represent all the required information. The way to access the FASUDIR CM is through the use of standard web services defined by OGC (Prieto et al. 2013).

4.2 *Simulation approach*

Once the building/district energy model is created, the simulation process can start through the simulation server. Instructions and data are sent to the simulation server through the user interface and record data from the CM database are received. SM gets the information from the CM and adapts it to the needs of the simulation tool to be used (IES <VE>). The results of the simulation of different variants are represented by the results of the KPIs and are stored in the CM. The SM is generated on the fly for the simulation.

The simulation server creates/maintains the simulation job queue through the simulation scheduler for the calculation of KPIs, creates batch instructions for each job (e.g. solar simulations + thermal simulations + value) and returns data to the CM database once simulations are completed to be stored in the database.

4.3 *Innovation*

The software enables modelling the district and building with an adequate level of definition, in such a way that evaluation results are precise enough, but the input data to define the retrofitting project is easily supplied. FASUDIR model is based on CityGML and is supported by GIS capabilities and accurate energy performance evaluation at building level.

4.4 *Acknowledgements*

This project has been funded by the European Union Seventh Framework Programme under Grant Agreement No 609222.

5 EFFESUS PROJECT

The main output of the EFFESUS project (www.effesus.eu) is a DSS (Decision Support System), a software tool, which includes all the parameters needed to select suitable energy efficiency interventions for historic districts (Eriksson et al. 2014). The main software modules developed in EFFESUS are: a multiscale data model, a categorization tool, a repository on technologies and the DSS

EFFESUS DSS is being validated in two different European urban developments that are representative of different district typologies that are common in Europe, which are Santiago de Compostela (Spain) and Visby (Sweden).

5.1 *Modelling approach*

The multiscale data model defined in EFFESUS is a virtual 3D city model based on the CityGML standard. The multiscale data model provides a representation of the urban information at different levels (from the city level to the building component level). It represents graphical appearance of the city as well as semantic or thematic properties.

A four step methodology has been defined for the 3D geometry generation and storage. Generation methodology is based on available data sources (footprints, cadaster, LiDAR, images, etc.). As a result of the generation process terrain, roads, green areas and buildings at different LoD (Level of Detail) are obtained.

The categorization tool provides the user an easy and intuitive way for the identification of building typologies in an urban district. The categorization tool is based on data included in the EFFESUS data model. Algorithms for building categorization have been implemented based on the building stock categorization methodology defined within the project. Geometry of the multiscale data model has been used for the visualization of the building typologies as well as the most representative building of each typology. The user interacts with the tool editing properties of representative buildings as well as editing parameters and thresholds for categorization. EFFESUS DSS is based on the identification of different typologies of building stock.

5.2 *Simulation approach*

The DSS software tool is based on a holistic methodological framework for the assessment of energy-related interventions in built cultural heritage. This methodological framework for decision-making aims to identify and classify actions according to: compatibility with the cultural significance, energy saving, habitability and economical, technical and legislative feasibility. The methodology has been developed at two scales: the urban scale and the building level. The methodology defines the specifications for its implementation in the expert system of the DSS.

The required information for the decision making process is based on the "simple hourly calculation procedure" included in the EN ISO 13790:2008 (ISO 2008). This procedure has been selected because it has been proved to generate satisfactory results with a limited input data requirement, which is crucial for EFFESUS taking into account the scale in the decisions.

An EPDE (Energy Performance Domain Extension) has been designed. It has been implemented as an ADE for the CityGML standard. This ADE has the purpose to storage the information regarding the energy related parameters that will allow proper decision making. The calculation procedure and the re-quired input data has to provide enough accuracy and flexibility to model the building stock of any district, aiming to accomplish this calculation with-out relying heavily on a large amount of input data.

The EPDE includes in-formation at different scales: district, building, building envelope (wall, roof, ground, etc.) and building installation (demand and generation installations). Information related with climate is referenced at district level, information related with geometry, occupancy and use is set at building level, while material properties, type and size of the windows and relation between opaque and opening areas must be identified at the level of the envelope elements of the building. Other element relevant for the identification of the energy efficiency of the building is related with the energy installations: type, efficiency, etc.

5.3 *Innovation*

The EFFESUS data model has been generated from different available data sources (geometric and semantic). The categorization tool provides the user an easy and intuitive way for the identification of building typologies in an urban district.

A CityGML ADE has been designed and implemented in order to structure and storage all the required information that is not included in the CityGML core. Information is structure according to the representative-ness of the elements of the historic district for the energy assessment and management.

5.4 *Acknowledgements*

This project has been funded by the European Union Seventh Framework Programme under Grant Agreement No 314678.

6 OPTEEMAL PROJECT

The objective of OPTEEMAL project (www.opteemal-project.eu) is to develop an optimized energy efficient design platform for refurbishment at district level. The platform delivers an optmised, integrated and systematic design based on an IPD approach for building and district retrofitting projects. This is achieved through development of holistic and effective services platform that involves stakeholders at various stages of the design while assuring interoperability through an integrated ontology-based DDM (District Data Model).

OPTEEMAL will be validated in three different urban districts in three European cities, which are San Sebastian (Spain), Lund (Sweden) and Trento (Italy).

6.1 *Modelling approach*

The DDM plays a key role to ensure the interoperability between different standard data models. The proposed DDM is a comprehensive semantic framework which facilitates the intertwining of standard data models with domain specific ontologies.

The DDM will be implemented as a set of interoperable data repositories. A data repository is a DDM component whose goal is to manage the information required to carry out the platform's processes. Furthermore, the outputs of these processes as well as the users' inputs are stored in the data repository.

At this stage of the project five repositories have been envisaged: BIM repository, city repository, contextual repository, ECM (Energy Conservation Measures) catalogue and platform database.

BIM repository stores the models of the buildings of the case studies. The final enhanced BIM models generated by the platform will also be stored in this repository. The number of BIM models to be included in the repository will depend much on the availability of those models. An optimum scenario will include one BIM model for each building, however currently it is not very common to have such models, even less probable for existing buildings. In such case the minimum number of BIM models will be one and will be desirable to have at least a BIM model for each building typology.

City repository stores a district model. The buildings represented in this model are linked to the BIM models stored in BIM repository.

Contextual repository stores the contextual data of the case studies such as weather data, economic indicators, social data, and environmental data, among others. These data are linked to the data stored in BIM and city repositories.

ECM catalogue stores the energy conservation measures used to generate the refurbishment scenarios to be optimized by the OPTEEMAL platform.

Platform database stores the data generated within the platform such as DPIs (District Performance Indicators), platform users, scenarios, user's inputs (e.g. barriers, targets, boundaries, priorities) and simulation models (energy, economic, environmental…) automatically generated.

Assuring the interoperability between heterogeneous information is a major requirement of the

OPTEEMAL platform. In addition to the use of standards like IFC and CityGML, it is necessary to create links between different data models to perform the functionalities foreseen in the platform. Three different types of link can be identified at this stage: (i) links between different data models in different repositories describing the same district, (ii) links between objects into different data models (e.g. a building into the IFC and CityGML) and (iii) links between alternative retrofitting scenarios describing the same building or district.

6.2 *Simulation approach*

In the OPTEEMAL platform, the current scenario is represented by the district data (CityGML model), the building data (BIM model) and the contextual data (urban data, climatic data, energy and environment data, social data, etc.). To complete the description of the current situation, according to the targets, boundaries, barriers and the prioritization criteria, for the refurbishment of the case study pro-vided by the user, the platform calculates the set of DPIs applicable to the current situation of the district.

IFC data model is defined for being used in a broad range of applications and domains. Due to the vast extent of IFC specialized domain application, models shall be defined as an intermediate step be-tween the BIM and the simulation engine. This intermediate step is represented in OPTEEMAL by the simulation models. It will be necessary to generate different simulation models for each simulation tool (EnergyPlus, CitySim, NEST and OPTEEMAL tools).

The approach in OPTEEMAL is to use IFC and CityGML as common data models for input and output of the district and building information, which will be further transformed into simulation models tailored for each domain and simulation engine, keeping the traceability and mapping between elements and concepts with the original models. According to the list of DPIs and the way they are grouped into categories, 6 domain models are identified: energy model, environmental model, comfort model, economic model, social model and urban model.

6.3 *Innovation*

In OPTEEMAL, the DDM will be modeled using IFC and CityGML standards linked with domain specific ontologies. Based on the DDM, different simulation models for the simulation tools identified in the project will be generated in the process to DPI calculation. As a result of the optimization process an enhanced BIM model including most suitable ECM will be obtained.

6.4 *Acknowledgements*

This project has received funding from the European Union's Horizon 2020 research and innovation programme under grant agreement No 680676.

7 CONCLUSIONS

The five projects described in this paper provide complementary approaches to building modelling and energy simulation with multiscale perspective.

MOEEBIUS addresses the challenges and factors that hinder the capabilities of current simulation and control frameworks to provide highly accurate predictions and fine-grained optimization considering the complexities induced during buildings' and districts' real time operation.

HOLISTEEC addresses the issue of enabling multi-physical simulations in all design stages, flexibly adapting to the available information in each stage and covering different spatial extents. This is achieved through a neutral SIM model, which integrates information from IFC and CityGML by providing automatic geometry and topology transformation routines common to all disciplines. Additionally, an intelligent e-Catalogue system is provided for providing different performance simulation scenarios in each phase. Finally exporters to specific simulation engine input files are developed. All the actors involved collaborate using BCF standard.

Integrated approaches presented in FASUDIR and EFFESUS with intuitive user-friendly software represent an innovative alternative for decision-making to prioritize the action to be taken and to improve the sustainability of urban districts and their subsequent management. The strategic management of the information generated by a city should be a key part of this process. The development of data models based on the international CityGML standard allows GIS and BIM concepts to be integrated within the same model. The information contained in the model is unique and can be used to develop various applications that the different agents (city managers, technicians and members of the public) employ.

OPTEEMAL provides a holistic platform to design efficient refurbishment projects at building at district level supported by a comprehensive ontologies-based framework for district information representation based on the relation of existing semantically enriched data models (CityGML, IFC) with existing ontologies/data models in the main fields for urban sustainable regeneration. The interoperability among inputs and outputs of the platform with external tools is assured by the definition of simulation models.

Consequently, this collection of projects visualizes the revolution in building and cities design and management procedures that is already starting. The growing complexity of buildings and cities and the urgency to make them more sustainable requires the development of knowledge based design and management tools supported by powerful data models. The combination of the IFC/BIM (building dimension) and CityGML (district and city dimension) has demonstrated that is the most promising approach, providing the modeling detail that is requested by each stakeholder and seamless integration between building and city.

Finally, one of the main challenges encountered that needs further research and improvements is to reach a successful integration and interoperability between different approaches: (i) physical models (BIM/IFC and CityGML) vs simulation models (which rely on a conceptual zone-based geometry definition) and (ii) adaptation of simulation engines usually focused on design aspects to accommodate real-time data.

REFERENCES

Delponte, E., Ferrando, C., Di Franco, M., Hakkinen, T., Rekola, M., Abdalla, G., Casaldàliga, P., Pujols, C., Lopez Vega, A. & Shih, S.G. 2014. Holistic and Optimized Life-cycle Integrated Support for Energy-Efficient Building Design and Construction: HOLISTEEC methodology. *European Conference on Product and Process Modelling (ECPPM), Vienna. Conference Proceedings. ISBN: 978-1-138-02710-7, pp. 899–906.*

Egusquiza, A., Gandini, A., Izkara, JL. & Prieto, I. 2014. Management and decision-making tools for the sustainable re-furbishment of historic cities. *V Congreso Latinoamericano REHABEND 2014 sobre Patología de la Construcción, Tecnología de la Rehabilitación y Gestión del Patrimonio.*

Eriksson, P., Hermann, C., Hrabovszky-Horváth, S. & Rodwell, D. 2014. EFFESUS methodology for assessing the impacts of energy-related retrofit measures on heritage significance. *The Historic Environment: Policy & Practice, 5(2), pp.132–149.*

Gröger, G., Kolbe, T.H., Nagel, C. & Hafele, K.H. 2014. OpenGIS City Geography Markup Language (CityGML) Encoding Standard (OGC 12-019). Version 2.0.0. OGC 12-019. *Open Geospatial Consortium.*

ISO, E., 2008. 13790: 2008. Energy performance of buildings—Calculation of energy use for space heating and cooling.

Mazza, D., Linhard, K., Mediavilla, A., Michaelis, E., Pruvost, H. & Rekola, M. 2015. An integrated platform for collaborative performance efficient building design: the case of HOLISTEEC project. *Sustainable Places—Conference Proceedings ISBN13:979-10-95345-00-8, pp. 151–160.*

Mittermeier, P., Essig, N. & Romero-Amorrortu, A. 2014. Evaluation and Development of Indicators for Sustainability Assessments of Urban Neighbourhood Renovation Projects. *World Sustainable Building 2014 Barcelona Conference—Conference Proceedings—Volume 3, ISBN: 978-84-697-1815-5.*

Prieto, I., Izkara, J.L. & Egusquiza, A. 2013. Architectural heritage 3D and semantic information visualization based on open standards. *Virtual Archaeology Review, 4(9), pp.70–75.*

Romero, A., Egusquiza, A. & Izkara, J.L. 2014. Integrated decision support tool in energy retrofitting projects for sustainable urban districts. *World Sustainable Building 2014 Barcelona Conference—Conference Proceedings—Volume 3, ISBN: 978-84-697-1815-5.*

eWork and eBusiness in Architecture, Engineering and Construction – Christodoulou & Scherer (Eds)
© 2016 Taylor & Francis Group, ISBN 978-1-138-03280-4

From District Information Model (DIM) to Energy Analysis Model (EAM) via interoperability

N. Rapetti, M. Del Giudice & A. Osello
Politecnico di Torino, Turin, Italy

ABSTRACT: Energy saving for existing buildings is considered one of the most important issues in the last years for the development of a smart city. The DIMMER (District Information Modeling and Measuring for Energy Reduction) project starts from the development of a 3D parametric model, able to collect heterogeneous data at different scale. It aims at improve the energy consumption optimization, monitoring real-time data and simulating all energy flows.

Starting from Building Information Models (BIMs), based on the development of DIM models, it is possible generate an Energy Analysis Models (EAM) able to simulate building energy usage compared with indoor temperature coming from real time temperature sensor.

Finally, the DIMMER project is focused on management, modelling and visualization of different data that describes the district, connecting different data-sources with different level of information. In order to achieve these goals, interoperability is considered a crucial step for sharing data between different environments, across various software and different professionals.

1 INTRODUCTION

According to Horizon 2020 policy, the main aim of public administration is to reduce energy consumption. For this reason, the main challenge, for a public administrator, is to face the complexity of an urban energy system. As well as planning optimization for urban utilities requires an appropriate urban modelling and simulation calculation methods address these following key issues. (Melia et al. 2015)

In this panorama, the DIMMER project proposes an interdisciplinary approach, that involves different disciplines, called domains, able to manage heterogeneous information at different scale. Taking into account the complexity to manage the information at different scale, it was necessary to set up a standardization of components and material library, in order to facilitate the data sharing between different service provider, reducing the redundancy of data.

In this field, many researchers are investigating the possibility to collect heterogeneous data about an urban district into a unique model. It can be useful for many purposes such as i) the design or refurbishment of buildings; ii) maintenance and monitoring of energy consumption iii) data visualization for increasing user awareness.

In order to create a District Information Model (DIM), the main challenge of this approach concerns the ability to manage a big amount of data, coming from different domains. In this field the interoperability plays a key role for data exchange among different domains. Adding to this, an important step is the definition of the right Level Of Development and Detail (LODs), to facilitate the querying of the information, based on the stakeholder's profile, that are: i) public administrator; ii) energy utility manager; iii) estate manager.

Following the idea to link different domain, the DIMMER project aims at develop a web-service oriented platform with capabilities of real-time district level data processing and visualization. (Del Giudice et al. 2014)

The case study chosen for this work is "the Politecnico district" situated in Turin (Italy) that is composed by public and private buildings. For public buildings like schools, university campuses or municipal buildings.

2 RELATED WORKS

Nowadays, one of the main challenges in the Information and Communication Technology (ICT) field is the integration among different systems providing new tools to optimize data management at district level. About this topic, several projects are investigating this issues, proposing different approaches:

SEMANCO (Semantic Tools for Carbon Reduction in Urban Planning) approach proposes the use of semantic technologies to create a model able to manage energy information coming from different data sources at different scale and from different domains. (Corrado et al. 2015)

INDICATE (Indicator-based Interactive Decision Support and Information Exchange Platform for Smart Cities) project aims to improve a master-planning tools that should support decision maker and stakeholders to facilitate the transition from cities to Smart Cities. The goal of the project is to optimize the interaction among building, energy network and renewable technologies using dynamic simulation models and algorithms. (Turner et al. 2014)

SUNSHINE (Smart Urban Services for Higher Energy Efficiency) project proposes the development of smart platform. It should provide, for planners and public administrator, analytical indicators useful to define energy saving policies for the existing buildings and pre-certification mechanism. Moreover, SUNSHINE delivers a smart platform accessible from web-based application. (Schrenk et al. 2013) (Sebastian et al. 2013)

STREAMER project aims to reduce the energy use and carbon emission of healthcare districts optimizing the design methodologies for energy efficient building. Thanks to development of a methodology based on an interactive system composed by different datasource as Information Communication Technology, BIM and GIS. (Iadanza et al. 2015)

SEEMpubs (Smart Energy Efficient Middleware for Public Spaces) project exploits the potentiality of ICT and control services to reduce energy usage and CO2 footprint in existing building. One of the main goals of the project is to manage the interaction among different system through the implementation of LinkSmart Middleware for the integration of heterogeneous networked devices. (Osello et al. 2011), (Osello et al. 2012), (Osello et al. 2013)

Basing on this background, the DIMMER approach aims at develop a District Information Modelling (DIM) that, compared with a Unified Building Model (UBM) (El-Mekawi et al. 2012) or a Virtual City Model (Melia et al. 2015), integrates several information collected into different data sources, such as BIM service, Geographic Information System (GIS) service, EAM service and Building Management System service, described with geometric and alphanumeric data. Hence, to implement data sharing between different systems the development of a multi-service platform based on Middleware service is necessary (Krilovskiy et al. 2015), where data came both from the real-time monitoring and digital data.

Furthermore, taking into account the complexity to manage the data flow coming from each domains. The setting of many standard formats exchange such as Industry Foundation Classes (IFC) and green building eXtesible Markup Language (gbXML), Intermediate Data Format (IDF), JavaScript Object Notation (JSON) was fundamental to facilitate the exchange process, avoiding loss of data that is produced by the interoperability process that currently it is not yet error free.

3 METHODOLOGY

The DIM model, that it is a simplify representation of real urban district, composed by several domains, each domain corresponds at service provider that collect information and they are divided into: i) Building Information Modelling (BIM); ii) Geographic Information System (GIS); Energy Analysis Model (EAM); Building Energy Analysis Measuring (BEAM).

The connection among different service providers it is possible through the middleware based on microservices architecture. In this way each datasource share data via interoperability, thank to data exchange format. (Brundu et al. 2015)

3.1 *From building information modeling at district information modeling*

Starting from the parametric models as visible in the Figure 1 below, it is necessary understood how

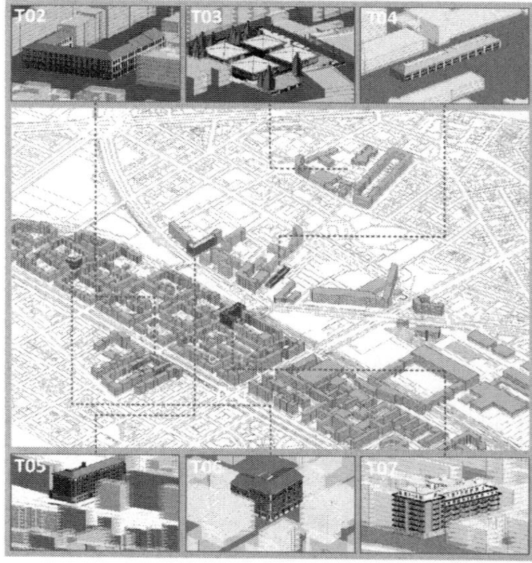

Figure 1. The Turin pilots district available on revit environment.

different BIM models can co-exist in a unique environment, without exceed the critical dimension, around 150 MB. For this reason, it has been tested the use of different approach to collect the models: i) link; ii) copy/past; iii) workset.

The first way concerns the use of links method. This approach allows a horizontal interoperability between model created with the same interface software. Moreover, the peculiarity of link consists in the fact that it is possible considerate the model linked as an "object" added in the project, allowing to move, to rotate, to query the object. Nevertheless, the connected model through the link method doesn't allow any type of change, because these have to be done in the original file. With this approach the models are imported into a district mass model and thank to the use of share coordinates they are located into the model in the right position. Regarding the critical dimension of the file, the use of link to a mass model, has under lighted a very small changes respect the original dimension.

The second approach involves the use of the copy/past strategy that is the easiest way to merge different model into a unique file project. Once picked out all the elements that compose the BIM model to copy and paste the entire model into another file is enough. Also in this case, the position of the model is defined from sharing coordinates, set up in the source model. This approach, is not error free, in fact, it is possible observe that several errors in the import phase such as the disjunction of elements, the lack of material properties, the separation of construction elements and etc. are produced. Although, this is the easiest way compared with the others, the district model file grown up reaching critical dimensions.

The merge of more models into a unique one caused a higher growth of the file dimensions than the first approach.

The last way concerns the use of workset method. It was developed to allow the worksharing between different actors and discipline. Starting from an empty central file, it is possible to generate a local file for each model. The central file is able to contain all the models, while the local model is a copy of central model the other models can be displayed or hided.

Comparing the different approaches, as visible in Figure 2, the link method represents the best choice in terms of file dimensions, whereas the workset method presents the biggest dimension.

Nevertheless, the workset approach is the best solution and this is due to the fact that copy/past method generates error and loss of data. Even if the link method maintains the smallest dimension than the other ways, this strategy doesn't allow to modify and to query the connected models.

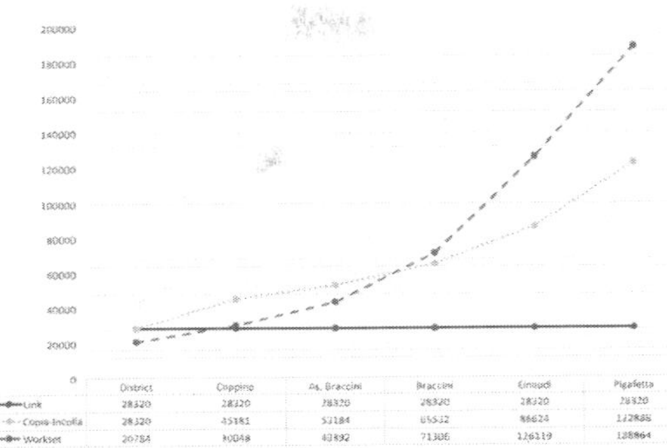

Figure 2. District information modelling file dimension trend.

3.2 *Set up of standards libraries and objects*

The main difficulty working on existing buildings, compared to new constructions, consists in having all information available about the components stratigraphy, the thermal characteristics, etc. For this reason, the DIMMER project proposes a typological approach, based on TABULA experience. The TABULA European project provides a typological description for different residential building typologies, indicating the U value for several building typologies, such as roofs, slabs, walls, etc. Basing on this data, the BIM model are reached correctly it is assigning the right U value, to facilitate the interoperable process with the energy analysis software.

Moreover, how it will be said in the paragraph 3.4, thermal properties that are possible to share and to manage into BIM model. In fact, for this reason a material library was created, sharable through the.adsklib exchange format, based on Tabula U values. Starting from this data, as visible in Figure 3, for each material component a material is assigned, in order to obtain the U value indicated into TABULA. (Ballarini et al. 2014)

In this way it is possible draft a standard for energy design based on components and material libraries, to optimize the interoperable process between both BIM utilities and Energy utilities.

3.3 *Level of development detail for Energy Analysis Model (EAM)*

Considering that the BIM model is a digital representation of the physical and function characteristics of a facility, this representation is composed by digital objects that correspond to real world components such as doors, walls, and windows with associated relationships, attributes and properties.

In this field, the concept about Level Of Development and Detail (LODs) plays a key role, in order to communicate the right level of information

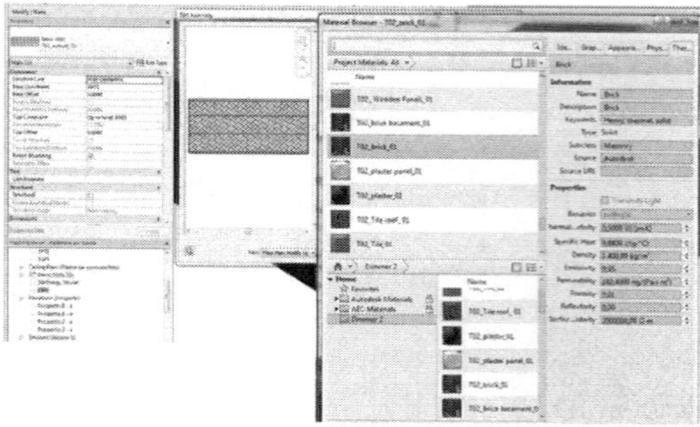

Figure 3. An example of DIMMER objects/materials library.

according to the final goal of each project. Starting from the definition of PAS 1192–2:2013 (PAS et al. 2013) which provides a specific guidance for the information management requirements associated with projects delivered using BIM. The scale used is composed by three main LODs as:

- LOD 2 is developed using masses (that can be visualized in wireframe or surface/solids), containing data needed to estimate per square meter rates and other similar metrics;
- LOD 3 is developed using generic components (for both objects and systems) with detailed form and function, defining all components in terms of overall size, typical detail, performance and outline specification (if available, working on existing/historical buildings);
- LOD 4 is developed using specific components (for both objects and systems) detailing assemblies accurate in terms of specification, size, form, function and location (if available, working on existing/historical buildings).

Moreover, as visible in the Figure 4 below, in order to reduce interoperability errors among energy software, the model was necessarily simplified. In fact, how it will say later, the interoperability process is not error free and simplification involves different parts of the model as: i) to attach roofs with walls; ii) internal partitions; iii) fixed and operable windows; vi) absence of all HVAC systems and equipment.

3.4 *Data exchange format and simulation software test*

This paper aims to validate the DIMMER approach testing interoperability through the use of different simulation engines such as Design Builder, IES environments and Open Studio. So, as visible in the Figure 5, starting from a 3D parametric model, developed until LOD 200, it was possible to set

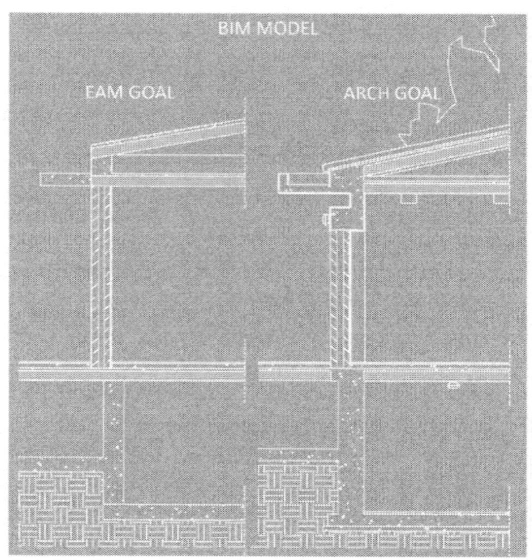

Figure 4. Comparison between EAM and architectural model.

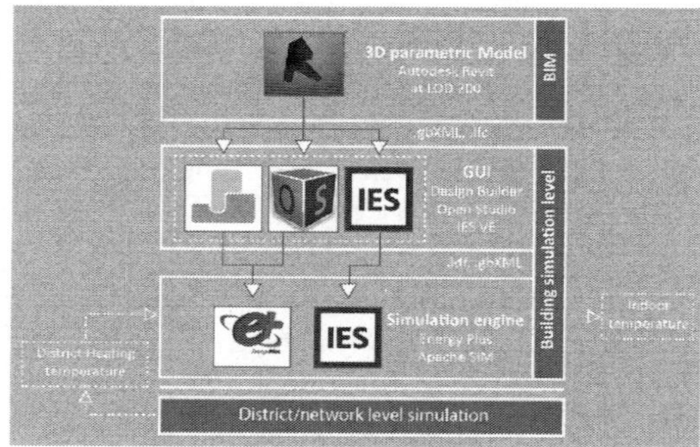

Figure 5. Interoperability methodology for building simulation level using design builder, open studio and IES VE.

up a data flow, generating an EAM model, where geometrical information has been converted into analytical surfaces. So, the energy model simplifies the architectural components into planar surfaces, defining the dispersion surfaces, preserving thermal properties like U value, thermal mass, permeability, density, emissivity and so on. Once, created the EAM model, it is possible to continue with the data flow, exporting geometrical and alphanumerical information into different energy software. In this phase data was filtered in order to preserve physical and thermal characteristics. This process may be through the use of exchange formats such as Industry Foundation Classes (IFC) standard, green buildings eXtensible Markup Language (gbXML), etc. After the import/export process, the EAM model has been then completed considering the specific use of the building, adding the occupancy and activity profiles of the users and editing the real characteristics of the heating system.

Therefore, the interoperability process is not error free as is visible in Figure 6, each exchange format does not exchange all information. The first tested approach concerned the use of.ifc format, even if it is not supported by the majority of energy simulation software. For this reason, Open Studio has been chosen because it supports both.gbXML and.ifc formats. Although, the.ifc format seems to maintain the objects geometry like the source model, in reality it is an empty box without the useful information for energy simulation. In fact, the model imported with.ifc format doesn't recognize:

- Material properties.
- Rooms and thermal zone.
- Heating System.

The second test was carried out exporting the energy model, through the use of.gbXML format into Design Builder and Open Studio. Both the Graphic User Interfaces (GUIs) have as a simulation engine the Energy Plus open-source software, developed from U.S Department of Energy Building Technologies Office.

Unlike the.ifc format. the.gbXML format preserves several information useful for energy simulations and both software support this format. However, the process is still not error free and although, the attention to simplify the model, the interoperability process, several errors has been generated such as a lack of geometrical information like:

- Room is incorrect due to heterogeneous connection between wall and curtain wall.
- Roof is considered as a shading.
- Windows framing are missing.
- Thermal Properties are not imported.
- Curved surfaces are not imported.
- Adjacent Rooms are incorrect due to the different of level.
- Windows or openings below ground floor are not imported.
- High Number of surface for room are not supported.
- High Complexity of the rooms are not supported.

All these problems have to be correct with the simplified model as it has been said in paragraph 3.3, in order to reduce the errors and to allow the next steps.

The last energy software tested is IES. The export process also in this case, was provided through the use of.gbXML format or a plug-in, that it generates in any case a.gbXML file.

Compared with the other software, IES presents a different simulation engine, based on ApacheSim Calculation.

Also in this case it is possible to observe several interoperability errors as:

- Voids in proximity of stairs.
- Problems to recognize curtain walls, columns and roof.
- Generation of air gaps.
- Impossibility to associate more types of construction elements for each external wall.

The EAM aims at define the energy demand for heating and cooling, Domestic Hot Water (DHW), lighting, and other different appliances. Independently of tool used, the simulation model is an approximate imitation of the real-world systems that necessarily introduce some simplifications and thus some behavioral differences. Simulated data is used to verify that the model describes accurately the real system with respect to the metrics (kWh, CO_2) that represent the final DIMMER outputs. The model validation is thus performed in parallel with its calibration, which usually implies an iterative and step-by-step process aiming at tuning its behavior.

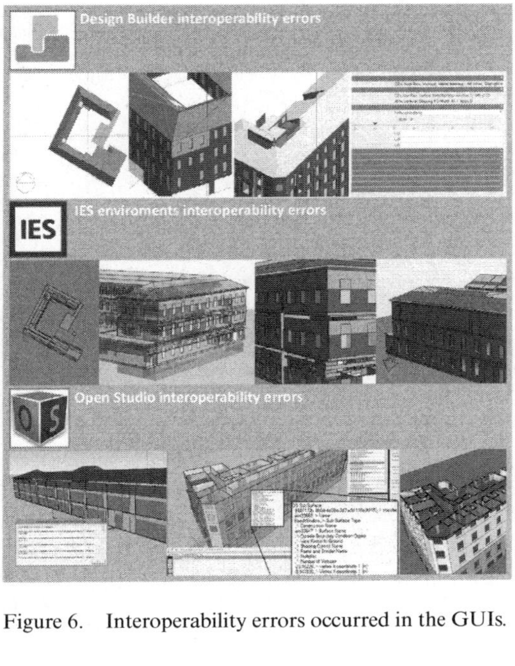

Design Builder interoperability errors

IES

IES enviroments interoperability errors

Open Studio interoperability errors

Figure 6. Interoperability errors occurred in the GUIs.

3.5 *Energy simulation with design builder*

Taking into account the previous considerations about modelling and interoperability errors, a rating was established, as visible in Table 1, to discover whose is the best software on the basis of several parameters such as preservation of geometric and material data, thermal zone identification and data re-association velocity.

Although Design Builder (DB) and IES environment have achieved both good scores, the choice fell on DB thanks to its simulation energy engine, Energy Plus. IES and Open studio were discarded, because the first software uses ApacheSim as simulation engine; while Open Studio was discarded on the basis of the worst score.

Once imported the Energy Model into DB, it was necessary to set up the activities template. It manages the other settings about occupancy, heating profile, natural and mechanical ventilation, metabolic heat, DHW, lighting, and so on.

The occupancy represents the time using of the building and it changes based on building use. Moreover, this field influences the heating profile, concerning the switching on/off the heating system. This is one of the information that is missed during the interoperability process causing the duplication of information in the GUI interface, imposing a supply thermal flow at a fixed flow rate and a fixed temperature. The temperatures of the thermal flow, exiting from the heat generator and reaching the final thermal units installed in the rooms, can be imposed in the energy model by setting a schedule. Therefore, in a second step, it will be possible to compare the indoor simulated temperatures with the monitored ones to validate the strategy of the energy saving at district level. The heat flow produced by the heat source described above, reaches the radiators modelled in each thermal zone. These terminals have a fixed nominal power and a presetted flow rate of the hot water.

As in the reality thermostats are not present, in the energy model they are not filled, allowing the indoor the freedom to fluctuate not influenced neither by the internal gains, nor by the solar gains and or by the fluctuation of the outdoor temperature.

Table 1. Score table for the energy software ranking.

	Geometrical data	Material data	Thermal zone identificatio	data re-association velocit	Total score
Design Builder - gbxml	2	2	2	2	8
IES VE - gbxml	2	2	2	2	8
Open studio - gbxml	2	0	2	0	4
Open studio - ifc	1	1	0	0	2

Once finished the above phases the simulation can be run obtaining several information such as indoor temperature, PMV, etc. Figure 7.

Finally, as is visible in the Figure 8, simulated data can be compared with the measured sensor data. This step is necessary to calibrate the simulated model with the reality, obtaining a starting point to develop other different scenarios of refurbishment. Obviously the model validation step is already in beta phase and it will be improved by the end of the project.

4 DATA VISUALIZATION

Data visualization in the DIMMER project plays a key role to communicate large amount of data,

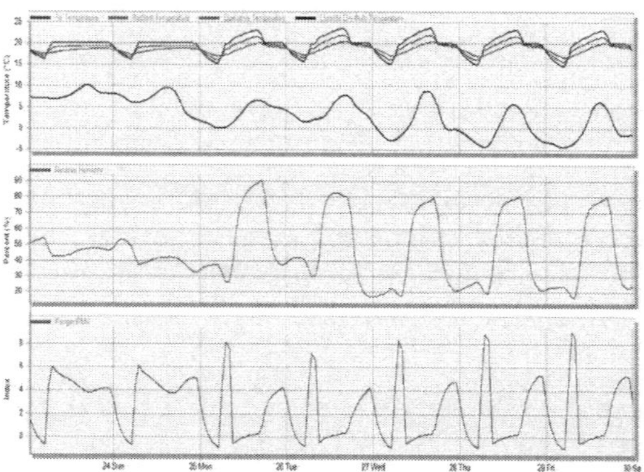

Figure 7. Example of simulated data with design builder.

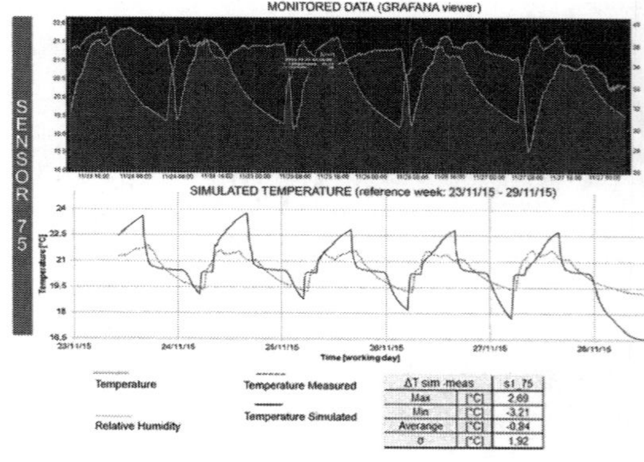

Figure 8. Comparison between simulated data and real time sensor data.

coming from several data sources. The challenge is to collect different kind of data into a unique platform able to querying information coming from different domains.

For this reason, one strategy to goal this objective consists in the use of Autodesk Infraworks. In fact, as visible in the Figure 9 below, this tool provides the overlay of different datasets, merging BIM (.rvt,.fbx), GIS (.shp,.geotiff,.sda, Citygml) and SIM (.shp,.dxf) information. Thanks to the use of Infraworks a DIM model was created and was able to reproduce a simplified model of the reality. Obviously, this process is based again on the interoperability and on data exchange formats.

The advantages of this methodology is due to the fact that each component is connected as a proper link in the platform, allowing the refresh of information.

In this way, in accordance with the DIMMER approach, the information is filled as layers that can stratify the DIM model, in order to simplify a complex organism like a district. Unfortunately, at present only geometrical information can be visualized, because the alphanumeric information is not imported. Adding to this, real time data sensors, and building attributes coming from BIM and GIS domain are not imported.

5 RESULTS

One of the most important results of the DIMMER project is the creation of standards and guidelines for data management at building/urban scale. This phase can be considered a mile stone for the development of smart cities.

Data sharing, service providers, real time data sensors, interoperability among software and platform carry out the DIMMER project in a new dimension where the interdisciplinary and collaborative process allows many professionals to manage data about a complex system as an urban district.

Figure 9. An example of DIM model with Autodesk Infraworks.

6 CONCLUSION AND FUTURE WORKS

In the last years, interoperability is considered the right way for sharing information in the building process among different users. The use cases are realizing that the way to conceive the building process is changing. In order to achieve the energy saving policies, professionals should consider the building characteristics and the geographical information about the context where they are located. Creating a system that share data between different datasources at different scales should be a possible way to develop smart cities and smart districts merging the needs of different users.

At present the DIMMER platform is in a beta phase and for this reason needs to be improved, because interoperability is not error free and the software infrastructure for district data management is not ready yet.

Finally, data visualization will be investigated using Virtual and Augmented Reality(V/AR) in order to improve the awareness of citizens about energy consumption, which is another goal of the DIMMER project.

ACKNOWLEDGEMENT

The research is funded by DIMMER, District Information Modelling and Management for Energy Reduction, an EU FP7 SMARTCITIES 2013 funded project. Moreover, the authors thank to the students Daniela De Luca, Arianna Fonsati and Michela Giori, Federico Magnea for them works of master thesis.

REFERENCES

Ballarini, I., Corgnati, S.P., Corrado, V. 2014. Use of reference buildings to assess the energy saving potentials of the residential building stock: The experience of TABULA project. *Energy Policy*, *68*(MAY), 273–284.

Brundu, F.G., Patti, E., Del Giudice, M., Osello, A., Macii, E., & Aquaviva, A. 2015. DIMCloud: A distributed framework for district energy simulation and management. *Lecture Notes of the Institute for Computer Sciences, Social-Informatics and Telecommunications Engineering, LNICST*, *150*, 331–338. http://doi.org/10.1007/978-3-319-19656-5_45

Corrado, V., Ballarini, I., Madrazo, L., & Nemirovskij, G. 2015. Data structuring for the ontological modelling of urban energy systems: The experience of the SEMANCO project. *Sustainable Cities and Society*, *14*(1), 223–235.

Del Giudice, M., Osello, A., & Patti, E. 2014. BIM and GIS for district modeling. In eWork and eBusiness in Architecture, Engineering and Construction-Proceedings of the 10th European Conference on Product and Process Modelling (ECPPM), vol. 31, pp. 851–855.

El-Mekawy, & M. Östman, A. & Shahzad, K.2012: A unified building model for 3D urban GIS. ISPRS International Journal of Geo-Information 1, 120–145.

Iadanza, E., Turillazzi, B., Terzaghi, F., Marzi, L., Giuntini, A., & Sebasian, R. 2015. 6th European Conference of the International Federation for Medical and Biological Engineering. *IFMBE Proceedings*, *45*, 162–165. doi:10.1007/978-3-319-11128-5

Krylovskiy, A. & Jahn, M. & Patti, E. 2015: Designing a smart city internet of things platform with microservice architecture. 3rd International Conference on Future Internet of Things and Cloud (FiCloud 2015).

Melia, A.M.A., Nolan, E.N.E., & Kerrigan, R. 2015. Indicate: Towards the Development of a Virtual City Model, Using a 3D Model of Dundalk City, 925–930.

Osello, A., 2012. Il futuro del disegno con il BIM per Ingegneri e Architetti/The Future of Drawing with BIM for Engineers and Architects. DARIO FLACCOVIO EDITORE, Palermo, pp. 1–323.

Osello, A., Acquaviva, A., Aghemo, C., Blaso, L., Dalmasso, D., Erba, D., Virgone, J. 2013. Energy saving in existing buildings by an intelligent use of interoperable ICTs. *Energy Efficiency*, *6*(4), 707–723. doi:10.1007/s12053-013-9211-0

Osello, A., Cangialosi, G., Dalmasso, D., Paolo, A. Di, Turco, M. Lo, Piumatti, P., & Vozzola, M. 2011. Architecture data and energy efficiency simulations: BIM and interoperabiliyt standards. Politecnico di Torino-DISET, Turin, Italy. *Proceedings of Building Simulation 2011: 12th Conference of International Building Performance Simulation Association, Sydney, 14–16 November.*, 14–16.

PAS 1192-2:2013, by British Standards Institution 2013, BSI Standards Limited 2013. URL:http://shop.bsigroup.com/Navigate-by/PAS/PAS-1192-22013/.

Schrenk, M., Wasserburger, W.W., Č, B.M., & Dörrzapf, L. (n.d.). SUNSHINE: Smart UrbaN ServIces for Higher eNergy Efficiency, (2013), 18–24.

Sebastian, R., Böhms, H.M., Bonsma, P., & Helm, P.W. Van Den. 2013. Semantic Bim and Gis Modelling for Energy-Efficient Buildings Integrated in a Healthcare District, *II*(November), 27–29.

Turner, W.J.N., Kinnane, O., & Basu, B. 2014. Demand-side characterization of the Smart City for energy modelling. *Energy Procedia, 62*(0), 160–169.

eWork and eBusiness in Architecture, Engineering and Construction – Christodoulou & Scherer (Eds)

Energy modelling of existing facilities

N. Nisbet
AEC3 UK Ltd., Bucks, UK

J. Cartwright & M. Aizlewood
Rotherham Hospital, Rotherham, UK

ABSTRACT: The emphasis of both building and energy modelling has often been on new-build with existing buildings having minimal attention due to incomplete information or the lack of any immediate prospect of validation. This paper is about discussing a new approach of energy modelling for existing facilities using UK Rotherham Hospital (TRF, 2013)[1] as an existing facility with no BIM models and few available records to be explored during the EU STREAMER, 2013[2] project.

This paper will examine in detail the methods used to capture sufficient information relating to the existing building stock, from written text, poorly reproduced drawings and on-site monitoring. The paper will also explain the tools used to capture that information including textual analysis and mark-up, mapping of semi-structured information to IFC, non-interactive simulation and the deployment of gaming strategies to identify optimum strategies. Multiple strategies were tested against two departments of similar size and construction but with differing operational needs.

1 BACKGROUND

1.1 *Client*

Rotherham Hospital (TRF) is a large general hospital serving an industrial town in the north of England. Like many public sector services, it is under unremitting pressure to save running costs and avoid capital expenditure wherever possible.

The Facilities Management services have a need to identify economic strategies for energy saving. In practice this means funding system improvements out of the operational budget with the prospect of re-investing savings into future upgrades. To develop such strategies needs a clear understanding of how potential upgrades will impact capital expenditure and annual savings, without necessarily having as much information as conventional thermal modelling requires.

1.2 *Project*

EU Streamer is an industry-driven collaborative research project on Energy Efficient Buildings (EeB) in the healthcare district, funded by the European Union. It is a European initiative with a 4 year duration commencing in September 2013, and is aimed at reducing the energy use and carbon emissions of both new and retrofitted buildings in healthcare districts of the EU by 50% in the next 10 years. As part of the EU STREAMER, four European hospitals are being taken as case studies for the improvement of the energy performance of districts. Hospital campus sites are ideal for such studies as they typically come under a single administration but have many different operational regimes housed in a building stock typically acquired over many decades and in some cases over many generations. Whilst the Dutch, French and Italian case studies are predominantly concerned with new-build projects in a semi-urban district context, the Rotherham case-study is focused primarily on existing facilities. Within the estate, TRF identified two operational departments for detailed investigations.

1.3 *Approach*

Initially it was intended to adopt the conventional approach of fully modelling the existing spaces, equipment and systems of the two departments using conventional BIM design authoring applications and using gbXML to transfer the fabric into a high-end dynamic energy modelling application. It would then have been necessary to add details relating to the equipment, occupancy and energy systems and controls. Much of this detailed information was not readily available: This was a potentially intensive and risky process in that mistakes made in early stages of the modelling activity might prove difficult to remedy, especially as the fabric model had to be absolutely integral if the transfer was to be useful. Modelling of different fabric and system options could have compounded the resources required. Lastly the time required to perform the simulations was anticipated to be unsustainable if more than a few options were to be examined.

AEC3 (AEC3 BimServices, 2010)[3] therefore proposed to examine an alternative strategy, based on exploiting any existing information resources and the maximum amount of automated processes. Whilst this approach was acknowledged as being innovative and risky, the process could be trialed repeatedly: it had only to succeed once and then be refined. It was hoped that the process would be more appropriate to the strategic questions that were being asked about the two departments, and that it would be scalable to as many departments as were of interest.

1.4 *Resources*

A number of existing resources relevant to the strategic goals were identified. Each was assessed for the most efficient and repeatable means of extracting information. The target for each information extraction was selected as IFC. This was chosen over gbXML as AEC3 had access to considerable libraries of methods for processing IFC.

Resource 1 was schematic ground plans of the five levels of the main hospital building, along with mapping information from Open Street Map and Google Maps. A prototype block modelling add-on called Quarter1, (Quarter1 Limited, 2015)[4] within Sketch-Up (Trimble Navigation, 2012)[5] was used to capture the block model (Fig. 1). Whilst this block model played little part in the eventual analysis, it provided a visual representation for the idea of a 'building information model', even if in the event, most of the relevant information proved to be non-geometric. This modelling exercise was conducted once, and then corrected after a site visit when the vertical sequence of levels was found to be A-E descending rather than our initial assumption of A-E ascending.

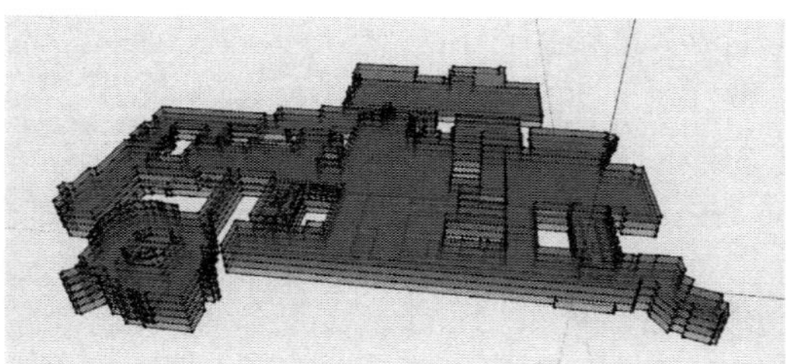

Figure 1. Block model.

Resource 2 was a schedule of these five levels with the correctly sequenced datum and storey heights (Table 1). A similar resource was created to correctly locate the building geospatially and as an address. These were transformed into IFC.

Resource 3 was the gbXML model of one of the two departments, which was a legacy of the earlier approach. This was converted to IFC primarily so as to have the specification of the major fabric elements, ceilings, external walls and openings available (Table 2). In the event these were found to have been left as default values and so no useful information was obtained.

Resource 4 was a schedule of all the departments and out buildings in the hospital campus (district). The main departments were associated to a named floor (Table 3). This was provided as a spreadsheet and was transformed into an abstract (non-geometric) spatial structure.

Resource 5 was the written and photographic description of the two departments which had been prepared as part of the original proposals (Fig. 2). As well as detailing the major fabric elements, it also contained summaries of the operational regime. Crucially it also described a number of potential fabric and system upgrades.

Table 1. Extract from schedule of levels.

Name	Description	Elevation m
R	Roof level	6.90
A	Top level	3.27
B	Ward level	–0.36
C	Entry level	–3.99
D	Admin level	–7.62
E	Lowest level	–11.25

Table 2. Materials found in gbXML model.

Location	ObjectType
Ceiling	ASHIF5
ExteriorWall	ASHWL-66
InteriorWall	ASHIW23
NonSlidingDoor	MDOOR
OperableWindow	DGL-R-I
RaisedFloor	con-c23

Table 3. Extract from schedule of departments.

Description	Level	Area sq.m
Outpatient's Department (OPC)	C	500
Ward B6—Ophthalmology	B	623

Resource 6 was a manually maintained spreadsheet of the readings for the main heat and electrical meters associated to the two departments, amongst others (Table 4). These presented as dates with meter readings.

Resource 7 was the weekly reports generated by the detailed monitoring of the sub-circuits of the main distribution boards (Table 5). This wireless sub-circuit monitoring was installed in the early months of the project and reports were received weekly from then on, so that by the time of writing over a year's data was available for over forty individual sub-circuits.

2 CAPTURE TOOLS

The resources described above can be divided into three main categories. Each category required a different processing solution. In each case, the outcome was a single coherent IFC project model. For clarity these are referred to as sub-models, similar to but smaller than 'discipline' models used in federated BIM.

1 OPD

1.2 Construction

1.2.1 Windows. The windows are full height with the bottom two panels being opaque, insulated and block work up to sill height and UPVC double glazed

Figure 2. Extract from marked up text.

Table 4. Extract from meter reporting.

Description	Name	Heat Annual Estimate kw.h
Ward B6 Heat Meter	B6 Heat Meter	111983

Table 5. Extract from sub-meter report.

Export of Custom Report	RH1-AEC3-B-WardB6-IE-L-0001
Created by	Bob Wakelam
Data display interval	Daily
Unit	kVAh
EntryDate	Lighting—DB40—Ward B6 Mains Cupboard
2015-04-23	0.546599998
2015-04-24	3.370399887

2.1 *Written material*

The first category includes the written and photographic material. This was marked up using the RASE (Hjelseth & Nisbet, 2011)[6] (requirements applicability selection exception) methodology, where the four different types of descriptive text and paragraphs are identified (Fig 2 above). Phrases that described the attributes of the zones and systems were marked as Requirements, though in this case they were definitive descriptions of the hospital as built. Phrases that refined the relevance of these descriptions were marked as Applicability. Any phrases that expanded the relevance of the descriptions were marked as Selections. In the event, there were no phrases found that generated Exceptions, which was not surprising given the specific nature of the document. Using the standard transformations of RASE mark-up documents, it was then possible to automatically generate schedules of attributes of the two Zones (Table 6) and of the Systems (Table 7).

2.2 *Tables*

This processing (2.1) meant that Resource 5 was now effectively moved to the second category which covered the collection of single sheet spreadsheets. Each spreadsheet was classified to identify the (single) type of object being described in each row. This allowed a single transformation to pass over the spreadsheet generating a coherent IFC BIM, with the appropriate number of the chosen object types present, one per row found.

2.3 *Exceptional tables*

The third category included the two exceptional spreadsheets, the manual meter readings and the automated sub-circuit meter readings (Table 4

Table 6. Extract from schedule of Zones.

Space ID	STREAMER Use Labelling 2015
OPC	U1: Office Use
B6	U4: Hotel Use

Table 7. Extract from schedule of Systems.

Name	Description	Classification
OPCHT0	Frenger heated ceilings with a small proportion of wet heating systems	Ss_60_40_37: Heating Systems
OPCHT1	Underfloor heating system	Ss_60_40_37: Heating Systems

and 5 above). Both were exceptional in that each column represented an instance of a specific type of object, rather than each row. A special transformation was developed that processed the columns systematically, and by taking the first and last readings and date, an estimated annual consumption was calculated. As with the other spreadsheets, these calculated values were then associated into multiple elements of the same type within a generic BIM model.

3 VOCABULARY

3.1 *Problem*

Whilst the resultant sub-models were valid, they embodied a significant latent issue. Each sub-model used the names of the entities exactly as found in the source material, and properties were named from the column (or row) headings exactly as found in the source material. The former problem was anticipated to cause problems when the sub-models were merged together and all the variations of naming (Ward B6, Ward-B6, and B6 for example) became critical. The latter problem was anticipated to cause problems when the energy analysis was run, as significant values (Area, Floor Area, Floor Area Planned for example) might not be found.

3.2 *Approach*

A decision was made to prepare two resources, a global dictionary to contain generally reusable terms such as attributes and object types (Fig. 3), and a local dictionary to contain specific terms relating to the names of the hospital project, site, facility, systems and zones (Fig. 4). In both dictionaries, the names and descriptions found were grouped with more preferable names and descriptions. All the existing sub-models were processed to seed populate the dictionaries.

Both dictionaries were accessed by the transformations described above: this meant that the entire capture process could be re-run at any time, typically taking 30–40 minutes, and thereafter the sub-models contained IFC preferred attribute names and entities and TRF preferred object names. It should be emphasized that the need for evolving global and local dictionaries arises primarily from the novelty of the process and should be expected to reduce in future.

```
<concept type="property">
  <term context="BriefBuilder">
        7. Types of space</term>
  <term context="STREAMER">
        BouwcollegeLayer</term>
  <term context="IFC">
        BouwcollegeLayer</term>
  <term context="en-GB">
        Four-way classification of hospital spaces by activity</term>
</concept>
```

Figure 3. Extract from global dictionary (XML).

```
<concept type="object">
  <term context="IFC">IfcBuildingStorey</term>
  <term context="local">Level-C</term>
  <term context="local">Level C</term>
  <term context="local">C</term>
  <term context="global">
        2xxdpcjVrU85py4317b61l</term>
  <term context="en-GB">Level C</term>
  <term context="en-GB">Entry level</term>
</concept>
```

Figure 4. Extract from local dictionary (XML).

3.3 *Sub-model development*

In parallel with the development of the model capture process, other work-packages within the EU STREAMER project were developing a design methodology to simplify and make manageable the design of hospitals and their energy profile. The recommendations that emerged included the use of Uniclass 2015 for the classification of Systems, and the classification of Zones and Spaces according to a number of criteria. An example categorisation was to assign all Spaces or Zones to be either 'Hot' for medical and emergency procedures, 'Hotel' for wards and residential accommodation, 'Office' for interview, out-patients and administration and 'Factory' for all industrial and laundry activities. These additional attributes were added to the spreadsheets as they were announced.

3.4 *Association of costs*

An independent consultant was asked to provide indicative cost rates for the System upgrades. These figures were not verified and are unsupported. Each Zone was given values for each characteristic measure (Table 8), each potential upgrade was given a named characteristic measure and a rate (Table 9).

4 MODEL COMPOSITION

It was realized that the sub-models individually were of limited value in energy analysis, but if taken together they held a comprehensive high level model. To make sure that this detail was available to the energy analysis, it was necessary to merge the sub-models. This was

Table 8. Extract from schedule of Zones with characteristic measures.

Name Description	Floor Area sq.m
OPC Out-patients Level C	500
B6 Ward 6 level B	623

Table 9. Extract from schedule of Systems with characteristic measures.

Name	Construction Area	Cost Rate £ GBP/sq.m
OPCHT0	Floor Area	0.00
OPCHT1	Floor Area	325.00

expected to highlight and help eliminate any inconsistencies.

4.1 *Federation and integration*

Many applications and processes support the use of federated models, for example Solibri and XBIM both allow multiple sub-models to be viewed simultaneously and for an extraction to COBie to be performed. However, the merging of multiple IFC files is not commonplace.

4.2 *Approaches*

Two approaches were trialed. Firstly, the DataCubist SimpleBIM (DataCubist, 2009)[7] application has facilities to request the merging of multiple IFC files. There was no documentation available on any assumptions or pre-requisites. The process itself was reasonably efficient using a graphical user interface. A later version of the application offered some automation of the repetitive process.

The second approach developed tools previously used in the US Army ERDC LCie (East, E William & Nisbet, Nicholas. 2010)[8] project which sought to re-enact many information exchanges from the project life cycle. A merge transformation (Fig. 5) took the current model and a second model and effected the following three rules.

A. Keeping old objects but updating object attributes from matching upgrade objects.
B. Adding upgrade objects except those matching old objects but re-pointing all references to these omissions to the matching old objects.
C. Merging any matched singular relationship (for example, the assignment of a type to a list of multiple instances).

BimServices Transform64 licensed to AEC3 UK Ltd
Old:
RH1-AEC3-Z-Z-3D-A-2002_OPCEG3.ifc
Upgrade:
RH1-AEC3-Z-Z-3D-A-2002_OPCEG3_asfromSBEM.ifc
Keeping old objects but updating object attributes from matching upgrade objects:
Project Description was RH1 Project now RH1 Refurbishment
Adding upgrade objects except those matching old objects but repointing all references to these omissions to the matching old objects.
IfcProject old i1kept, upgrade ji100011 not copied
Created:
RH1-AEC3-Z-Z-3D-A-2002_OPCEG3_Merge.ifc
Finished.

Figure 5. Extract from merge log report.

4.3 *Correspondences*

This ruleset still leaves the question of what constitutes a correspondence of an entity in the first sub-model and an entity of the second sub-model. In ideal circumstances the presence of GUID (globally unique identifiers) on each entity would provide a robust method of matching. However, in practice the entities in the sub-models had unique and separate provenance, and so it was inevitable that no such correspondence would be found. The same situation could arise in new build design, if a surveyor and an architect both commenced work simultaneously. The surveyor's IfcSite will have a different GUID from the architect's IfcSite, even though it is useful to bring these models together. This meant that correspondence was established by name, and in the special cases of the singular IfcProject, IfcSite and IfcBuilding by definition, though this last assumption would not necessarily be valid if a more complex district or city model was being created.

4.4 *Policy*

Of the two methods the second was taken forward because of the direct control on the algorithm, and because of the ability to effect the merge process without user intervention. A complete merge process of 8 sub-models was taking 25–35 minutes.

5 OPTION SELECTION

The fully merged model was exceptional in one respect: it contained system definitions for all the proposed upgrade options. Before running an energy simulation, it was necessary to decide which systems would be included and more particularly which systems would be removed prior to the analysis.

5.1 *Approaches*

It would have been possible to create a repetitive process that extracted every possible combination of systems, and the analysis run on every such model.

5.2 *Selected approach*

However, one of the objectives of the EU STREAMER project was to disseminate the energy strategies and implications to as wide an audience as possible. It was therefore decided to create a gaming scenario where different attendees at the dissemination workshops would be able to choose a selection of system upgrades which they intuited would give a best cost-benefit ratio. The necessary web-page was automatically generated from the model content (Fig. 6). This web page offered a number of radio and toggle buttons, so that a team could enter their name, their selection and then submit their choices for immediate simulation. Each team's results (as described in the following section) will be plotted on a dynamic results page, so that the best performing teams can be commended.

6 OPTION SIMULATION

Having been notified of the desired options, the next step is to perform an energy simulation.

6.1 *Preparation*

The web-page submitted a command line with the chosen options listed. A filtering transformation then removed the unwanted options and passed the new model to a transformation which processed the IFC file and prepared a simulation input file for the UK NCM SBEM (UK NCM, 1998)[9] application. This application is sup-ported by the relevant UK regulatory authority (DCLG) and maintained on their behalf by the BRE. A key feature of the tool is that whilst it is delivered with a user interface for preparing the input files, the simulation run can be initiated automatically on a command line. Initially the mapping of the model into the SBEM input file was able to use only the most basic information: the area of the department, and the occupancy schedules suggested by the activity classification. Gradually more sophisticated interpretation of the zone and system descriptions are making the SBEM file more complete and accurate and the simulation less dependent on default and typical values.

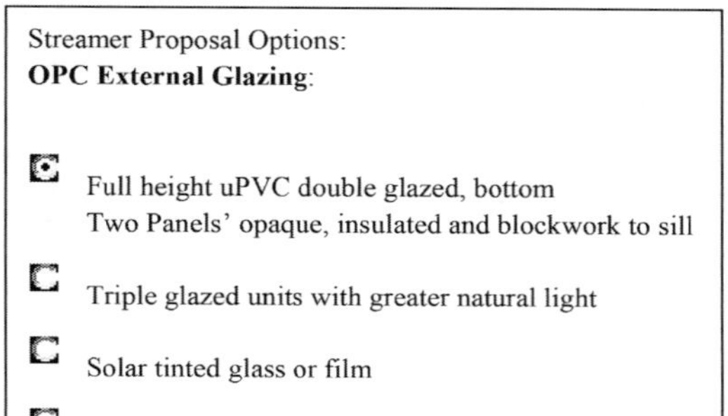

Streamer Proposal Options:
OPC External Glazing:

- (selected) Full height uPVC double glazed, bottom Two Panels' opaque, insulated and blockwork to sill
- Triple glazed units with greater natural light
- Solar tinted glass or film
- Solar shading

Figure 6. Extract from interactive web page form.

6.2 *Output*

A summary of the input and the results of the simulation are generated as a CSV file, and again this was transformed into an IFC model comprising just a building with the associated energy demand (by purpose) and consumption (by fuel) associated to it. This small sub-model was then merged with the in-put model to create a final representation of the option and its outcomes in terms of cost and energy. The model was then processed to generate an HTML report page summarizing the cost and energy savings (Table 10). In the context of EU STREAMER, the model is also delivered to a project options dashboard which is being deployed for the final EU STREAMER demonstration late

Table 10. Extract from Outcome web page.

Name	Value kw.h / sq.m / £ GBP
Building	Rotherham Hospital
Annual Energy Demand	1752056.31
Annual Energy Consumption	1332397.95
Zone	Outpatient's Department (OPC) - ADMIN
Gross Area Planned	500.00
Zone	Ward B6 – Ophthalmology Level B—HOTEL
Gross Area Planned	623.00
System	Solar Shading
Capital Cost	1485.00

in 2016. This dash-board will include additional life cycle costing estimates.

7 CONCLUSIONS

This project has explored the potential for energy strategies for existing facilities to be guided by information that is already available. A continuous process has been created that exploits these information resources and, pausing only to choose the options of interest, immediately generates energy and cost results. The tools used to facilitate this improved process are all relatively simple: data transformations that operate on free text, on simple spreadsheets and on simple IFC models. Within the EU STREAMER project, each step has then been improved and enhanced to the point where meaningful and relevant results can be obtained so that a possible energy upgrade strategy has started to emerge for TRF to consider.

Many of the methods presented here could be applied to new build, as soon as there is some knowledge of the shape, size and specification of the departmental zones in a hospital campus, or indeed any other building type. The results should be available to inform and critique the design much earlier. This is in contrast to the process currently typically applied to new build, where only one energy simulation is conducted, long after the major formal, fabric and system choices have been made.

ACKNOWLEDGEMENT

The presented research was done in the frames of the European FP7 Project STREAMER developments. We acknowledge the kind support of the European Commission and the project partners.

REFERENCES

All accessed 2016-04-12.

1. TRF. 2013. The Rotherham NHS Foundation Trust—Project Streamer. retrieved April 2016 from http://www.rotherhamhospital.nhs.uk/ProjectStreamer/
2. EU Streamer. 2013. Streamer. retrieved April 2013 from http://www.streamer-project.eu/
3. AEC3 BimServices. Nisbet, Nicholas, 2010. BimServices—Command-line and Interface utilities for BIM. Retrieved April 2016 from http://www.aec3.com/en/6/6_04.htm/
4. Quarter1. Quarter1 Limited, 2015. retrieved April 2016 from http://www.quarter1.com/
5. SketchUp. Trimble Navigation, 2012. retrieved April 2016 from http://www.sketchup.com/
6. RASE. Eilif, Hjelseth & Nisbet, Nicholas 2011. Capturing normative constraints by use of the se-mantic mark-up (RASE) methodology. In CIB W78 2011 28th International Conference—Applications of IT in the AEC Industry.
7. SimpleBIM. DataCubist, 2009. retrieved April 2016 from http://www.datacubist.com/
8. LCie. East, E William & Nisbet, Nicholas. 2010. Analysis of life-cycle information exchange. Retrieved April 2016 from http://projects.buildingsmartalliance.org/files/?artifact_id = 3864
9. SBEM. UK NCM, 1998. retrieved April 2016 from http://www.uk-ncm.org.uk/

eWork and eBusiness in Architecture, Engineering and Construction – Christodoulou & Scherer (Eds)

A collaborative platform integrating multi-physical and neighborhood-aware building performance analysis driven by the optimized HOLISTEEC building design methodology

H. Pruvost & R.J. Scherer
Institute of Construction Informatics, Technische Universität Dresden, Germany

K. Linhard & G. Dangl
Institute of Applied Building Informatics, Munich, Germany

S. Robert & D. Mazza
CEA, LIST, Gif-sur-Yvette, France

A. Mediavilla Intxausti
TECNALIA, Derio, Spain

D. Van Maercke
Centre Scientifique et Technique du Bâtiment, Grenoble, France

E. Michaelis & G. Kira
GEM Team Solutions, Neustadt, Germany

T. Häkkinen
VTT—Technical Research Centre of Finland, Helsinki, Finland

E. Delponte & C. Ferrando
D'Appolonia, Genoa, Italy

ABSTRACT: The paper presents the technical developments made in the course of the HOLISTEEC project. HOLISTEEC is an EU FP7 project whose objective is to first, formalize a new optimized building design methodology, and second, to develop the software platform and modules that implement that design methodology. The paper focuses on the developments made on that software infrastructure that shall at the end of the project provide a BIM-based collaborative platform supporting the optimization of the building design in terms of performances and also enhancing the building design process itself. As an important driver for that optimized methodology the project sets a focus on the consideration of interactions between the building and its neighborhood as well as multi-physical aspects including energy, acoustics, lighting and environmental impact.

1 INTRODUCTION

Currently, BIM technologies encompass not only data model specifications such as building and neighborhood information models but also the elaboration, use and iteration of standardized workflows and codes of conducts for building design. For several years, software applications have begun supporting BIM modeling and are now quite sophisticated. However, interactions and information exchange between systems are still in a primary stage and users do frequently encounter problems when they want to collaborate with other actors from different companies who use other systems. Additionally, there is still a lack of tools that do support a collaborative workflow in performance-based design and exploit all the capabilities that BIM grants.

Taking these gaps into consideration and in order to implement the innovative HOLISTEEC building design methodology, the collaborative platform integrates different software modules that belong to different functional layers. The central layer is constituted by a collaborative web-based software environment that implements the BIM Collaboration Format (BCF) which is a BuildingSMART open standard (BuildingSMART, 2016). This web environment allows for different users to register to the same project and exchange information. The capabilities provided by BCF enable for establishing, executing and monitoring predefined sequences of tasks that are performed by end user tools or internal software modules of the platform. Besides performance management and workflow management this web environment provides different functionalities such as messaging, project information setup and browsing as well as design variants management.

Complementary to this central web environment, the HOLISTEEC platform disposes of additional components and services that are essential for the execution of the whole design workflow. On a ground layer several data servers are used and provide all the required data. For the Building and Neighborhood Information Models (BIM and NIM models) two servers support respectively the IFC and CityGML standards in their latest release. Moreover a specific KPI server is newly developed as part of a KPI management system that is for itself composed of a KPI requirement setting tool and a scoreboard for comparison of simulated values against requirements. On a core services layer generic modules for data access and manipulation are provided. On this layer, besides a common interface enabling the HOLISTEEC platform for accessing data on the different servers, generic services for BIM data filtering and NIM data integration are provided. These core services are meant to retrieve and preliminary prepare in a task-independent manner the data for use by specific task-dependent business services. On this business services layer, modules for data mapping are used for among others transforming BIM and NIM data into a simulation-ready data format. In order to allow for simulation of numerous design alternatives, a specific variant model is introduced. This variant model enables the definition and configuration of several variants of simulation parameters. In this way a single variant model can specify and drive numerous simulation runs. Once the data is ready for use in a specific application, it can then be provided to the last layer composed of all end user tools. Those include classical CAD software but also simulation engines and an expert system providing for design recommendations.

2 SUMMARY ABOUT THE HOLISTEEC BUILDING DESIGN METHODOLOGY

HOLISTEEC building design methodology is a BIM-based design methodology that connects ICT tools into the design process. The aim is to help implementing integrated design and performance-based design principles, taking a step forward in comparison to traditional practices that tend to consider separately design and assessment activities. The focus of the methodology and tools in this project is in design phases. It is also possible to use simulation tools in the construction and operation phases, but the need is greater in design phases.

Both AS-IS and TO-BE processes are presented below. As shown, the use of simulation tools in traditional processes is mainly related to the design phases, while in the TO-BE process a widespread use of simulation tools is foreseen along the whole process implementation. The HOLISTEEC BIM-based collaborative platform will include an evaluation framework common to all project phases so that to try to optimize and harmonize as much as possible the outcome of each phase.

The proposed methodology is based on the following key elements:

- **Collaborative approach:** at this purpose meetings are of fundamental importance, since the very beginning of the building design process and throughout the whole process. This approach is foreseen in order to involve in the decision-making process all the strategic stakeholders, increasing the harmonization among the design of the different building design components.
- **Performance-based methodology** with reference to HOLISTEEC evaluation framework, in different domains: acoustics, lighting, energy and environmental impact.
- **BIM use since the early design stages** taking into account an increasing Level of model Development (LOD).
- **Use of simulation and multi-criteria tools** to assess the impact of the design decisions taken at

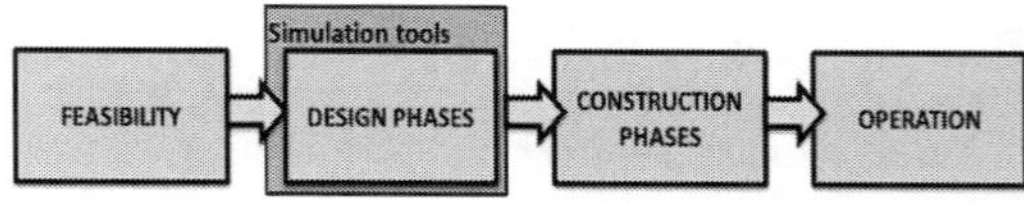

Figure 1. AS-IS building design process.

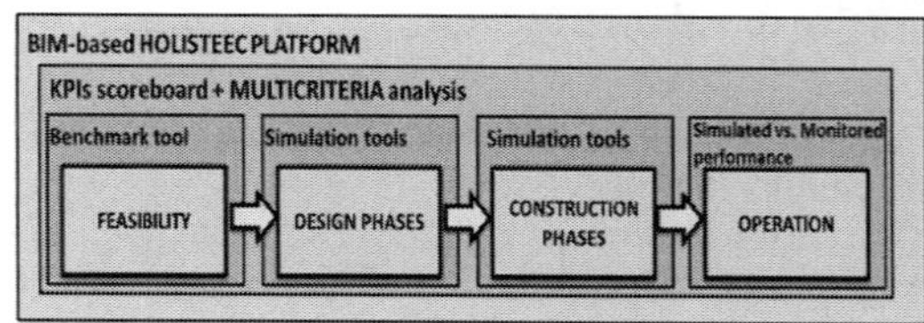

Figure 2. HOLISTEEC TO-BE building design process (emphasis on early project phase, design and assessment activities combined, collaboration enhanced by the BIM-based platform).

the different design stages, using as reference the information included in the BIM model. Energy, acoustic, lighting and environmental impact calculation engines will be connected to the HOLISTEEC BIM-based collaborative platform.

3 BCF-DRIVEN HOLISTEEC BUILDING DESIGN WORKFLOW

The HOLISTEEC project is offering a comprehensive design approach to the European AEC/FM industry. It takes the whole building life cycle into account with the objective of providing a basis for an energy efficient building design process. The project introduces a new methodology for design and discusses it with relevant stakeholders involved in the energy efficient construction value chain. It develops and demonstrates also a collaborative BIM-based building design software platform in the cloud, featuring advanced design support for multiple building optimizations considering external influences from the neighborhood. The central role herby is the role of the BIM Manager. Besides managing processes and procedures for information exchange, he/she should have a more complex array of responsibilities.

3.1 *BIM functions*

Based on the process mapping components that can be used in a BIM integration process (Bjork, 1992), a procedure has been identified to create process maps expressed in the Business Process Model and Notation (OMG, 2016) for the main design stages within a construction project. The activities in the BPMN notation are characterized as BIM functions which provide following functionalities:

- add and optimize the set of typical steps/activities executed during the design stages;
- identify the actors involved;
- identify what resources will be referred to or used for the completion of the task;
- organize the steps in each design phase in a reasonable sequence;
- define activity types for an accumulative tracking and assessment of activity performance quality;
- BIM Functions dependencies—one function can have more than a single input, respectively can distribute its output to many successors (Figure 3).

3.2 *HOLISTEEC BCF-extension-functions*

To support the HOLISTEEC workflow, BCF has to be extended by some data objects and related creation functions, that allow to map complex HOLISTEEC BIM functions to a sequence of BCF (Extension) functions. The following list enumerates objects that are all implemented by defining fixed topics, each with an individual set of parameters:

- create_Milestone
- create_Repository
- create_Requirement
- create_Snapshot
- create_Variant
- increment_Phase
- increment_Version
- markout_Phase
- request_Assessment
- request_Model
- upload_Assesment
- upload_Model

These topics will be created and accessed via standard BCF commands. The BCF client's implementation of the HOLISTEEC BCF extension should guarantee the HOLISTEEC data model integrity of these topics and their relationships. Since BCF does not support a transaction mechanism that would allow to group a sequence of BCF commands into a single HOLISTEEC command transaction, the client's implementations will have to take this fact into special consideration.

3.3 *Definition of sets of data models with BCF and URIs*

Within the process maps different data exchanges occur which are formalized as Exchange Requirements (ER). According to this, model links represent the first step in the implementation of the exchange requirements and more specifically, serve the purpose of associating and locating the different data models involved in one or more ERs. In

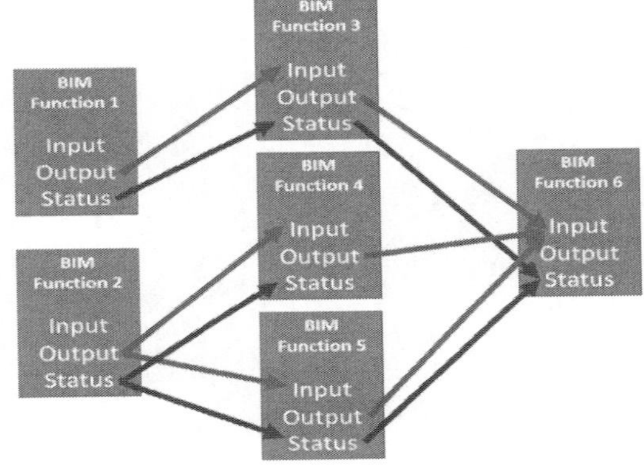

Figure 3. BIM function input output dependencies.

the BCF specification the fundamental element for making model links is the attribute "Related Topic" which is used to define relationships between two or more BCF topics. Additionally, BCF topics can provide, through the attribute "ReferenceLink", references to the data models of interest using their URIs. Those references are made by following a 1–1 relationship between BCF topics and data model URIs. The following figure represents how data models are linked by applying URI references and combining topics together. More specifically the data models are referenced by their URLs pointing to the servers they are located on like the BIM, the NIM server, the KPI server or external databases which are all accessible through the web. As illustrated, in order to execute the design workflow, the BCF topics referencing the data models involved in an ER are related to another BCF topic which is used for communication and data exchange between two machines or two actors involved in two distinct tasks.

4 HOLISTEEC COLLABORATIVE ENVIRONMENT

4.1 *BIM—it*

BIM—it (iabi, 2016) is a cloud-based collaboration platform that is accessible by any client with a web browser and an internet connection. It is fully compatible with BCF and supports the BCF API as well as the BIM Model API that are both described in chapter 6. For the HOLISTEEC platform, it is the central entry point to setup a new project.

The BIM Manager section offers distinct functionalities, the first one being the configuration of the project extensions which means for HOLISTEEC projects, loading the BCF-extension-functions. Another topic that is managed by the BIM Manager is the fine tuning of who is able to access a

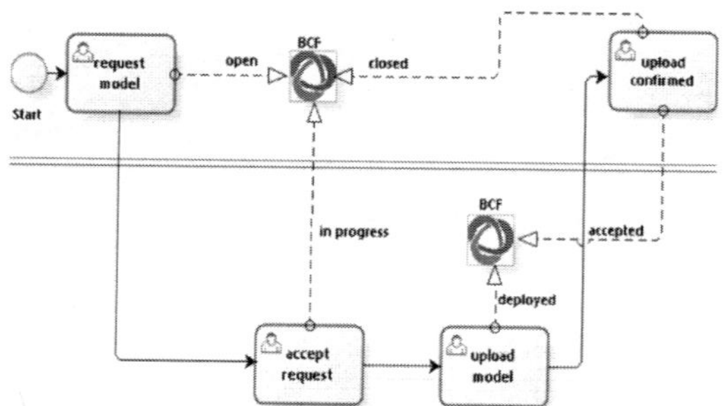

Figure 4. Extension of BPMN actions with BCF-extension-functions.

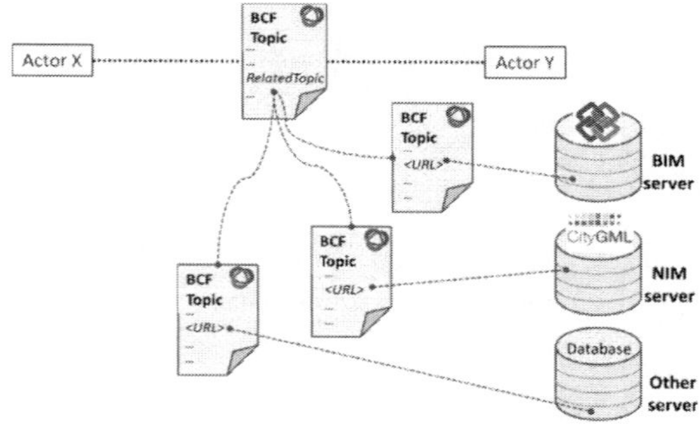

Figure 5. Use of BCF topics relationships and data model URLs for making model links.

project. In general, users are added or removed from a project via the "users" property of the project extensions. If users are not yet registered on the platform, BIM Managers are presented a choice to send an invitation e-mail to the user. The user will then be added to the project after having registered on the platform. Projects that call for a more granular access matrix can so be configured by the "User Access Rights" section. Basically, the BIM Manager may choose which labels respectively domains a single user can access. Repositories (BIM, NIM, KPI) supporting the BIM Model API can then be set up and related to specific labels (domains).

4.2 *Dashboard*

Since the HOLISTEEC methodology introduces a specialized communication and design workflow, a general HOLISTEEC user interface has to support the key concepts of this workflow. This user interface is provided by the dashboard which represents the main communication interface.

Based on the BCF communication technology described in the previous chapter, the dashboard structures the communication and data topics into context dependent user views. On one side the project scope tree (design project, design variant, design version) is used to delimit the view focus of the topics. On the other side the HOLISTEEC topic type (Milestones, Files, Requirements ...) and inter-topic relationships are used to filter away items that are not relevant within a specific context. In this sense the dashboard handles HOLISTEEC specific data types with specialized user interfaces as partially shown in Figure 6.

5 KPI MANAGEMENT SYSTEM

One of the main advantages of the HOLISTEEC design platform is its support for performance-based

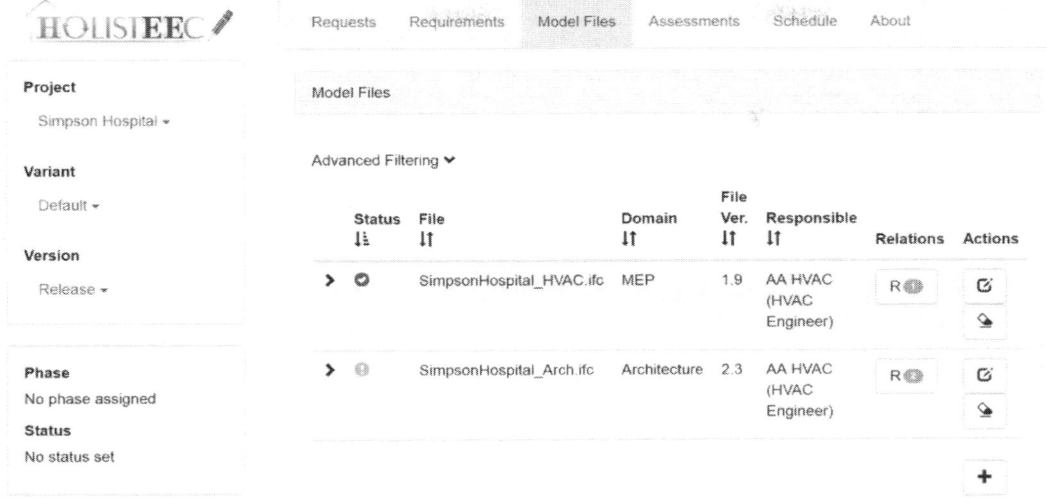

Figure 6. Screenshot of the HOLISTEEC dashboard showing a list of two topics from type model.

design. This is realized on the basis of the HOLISTEEC KPI management system and simulation engines. More specifically, KPI management is enabled by several software modules and simplified by a classification system.

By defining and using a model classification structure the complexity of mapping KPI requirements and performance attributes to building model elements is greatly reduced. The classification enables the user to treat groups of model elements like having the same virtual type and therefore reduces the amount of mapping work.

The modules which support the KPI management features are the following:

- dashboard for communication and workflow management
- the KPI requirement setting tool for defining requirements
- an intra platform tool or external CAD plugins to help making the classification into the CAD model
- a KPI server to store the KPI related data and present an interface access to the other modules that have to deal with the KPI data like simulation engines
- the scoreboard to present the results.

5.1 *KPI requirement setting tool*

The KPI requirement setting tool supports the definition of performance requirements with the help of indicators. The tool enables the definition or selection of any number of performance indicators, to define the desired levels or values for these indicators, to allocate requirements for different spaces or space types, and to define the steps for the follow-up during the design process.

One part of the indicator data enables to link each indicator to the modeled content ("model linking") for visualization purposes in the scoreboard. Therefore, the indicators can be considered early in process, and when the design plans are available (e.g. as an IFC file), the set of requirements and/or results can be visualized with certain components, like spaces of the building.

The tool can be used repeatedly in such a way that the required values are first defined on general level (like excellent quality) and later interpreted into specific classes or numerical values. The different steps that are performed during KPI requirement setting are presented in Figure 7.

The present HOLISTEEC platform supports the design under consideration of energy, environmental, lighting and acoustical performance, thus by providing simulation engines for each of these performance aspects. These simulation engines work with specific key performance indicators. The requirement setting tool supports the definition of required values for these indicators, but additionally other indicators can also be defined.

5.2 *KPI server*

As represented in Figure 8, the KPI Server is the central repository for all performance related data. The server stores KPI definitions, KPI requirements, KPI targets and results, as well as the classification trees that are project, version and phase specific. Additionally the KPI server provides a

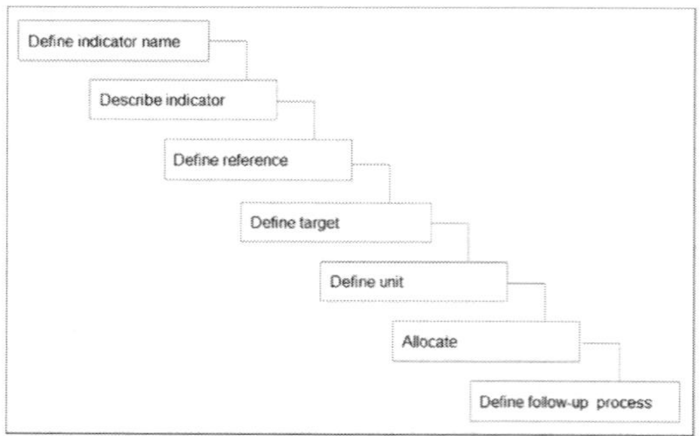

Figure 7. Performed steps during KPI requirement setting.

RESTful web service that allows all components like simulation engines, scoreboard, requirement setting tool, CAD systems or the HOLISTEEC dashboard to access and eventually modify the performance related data.

6 CORE SERVICES

6.1 *HOLISTEEC API*

The **BCF REST API** enables software applications to exchange BCF data seamlessly within BIM workflows by fully supporting the BCFv2 schema, using the HTTP protocol as well as simple JSON data formats. For HOLISTEEC, it is especially viable since there exists an open API specification that allows BCF implementations to utilize centralized storage and server features, thus enabling working in the cloud throughout different environments.

The **Model REST API** was developed in the eeEmbedded (eeEmbedded, 2013) research project and adopted from it. It offers services that support CRUD (Create, Read, Update and Delete) operations for managing structured data, for example IFC model files with attached metadata. The API was developed to have a vendor neutral and open service specification available and is offering basic functionality that was identified as being necessary for interfacing with repositories with regard to storing and retrieving building model data and related information.

The **KPI REST API** enables all HOLISTEEC software components and services to exchange KPI related data seamlessly within the HOLISTEEC workflow. The REST service itself does include creation, modification and reading access to all the categories of HOLISTEEC KPI related exchange data. Those latter are composed of KPI definitions specified and setup independently of a design project, KPI targets set as design project or design

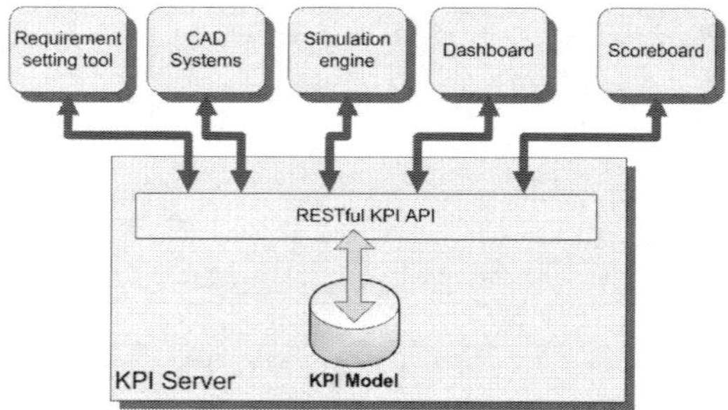

Figure 8. Principal architecture of the KPI management system.

variant related requirements, and finally individual KPI instances related to a specific design model context together with their computed or otherwise determined performance values. Filter parameters allow to access KPI data distinguished by projects, variants, versions and snapshot models (an explicit list of all the model files that form the base of a model assessment), also by KPI definition types and individual model context.

6.2 *BIM core services*

In order to provide the different end-user tools with the required data, the HOLISTEEC platform offers a BIM-based filtering service used to extract selective information from the Building Information Model. This information is meant to be provided to any kind of end-user tool and mainly to simulation engines in the context of multi-physical simulation. In that sense the provided information has to correspond to the data exchange requirements set by such specific tools. The Exchange Requirements in HOLISTEEC are specified with the help of the IDM methodology established by BuildingSMART (ISO 29481:1, 2010). In order to automate the data filtering process these Exchange Requirements are implemented in so called Model View Definitions (MVD) using the mvdXML format developed by BuildingSMART (Chipman et al. 2012). A MVD is a resource that defines what kind of information is required from an IFC model for a certain simulation. On that basis a filter tool, as shown in Figure 9, can retrieve the needed information from the data model and produce a so called model view (Katranuschkov et al., 2010). This model view represents a reduced building model that contains only the required BIM data.

The filter tool is composed of a generic filter layer reading mvdXML and a data access layer accessing data on a server. It then produces a BIM view which is a reduced BIM model containing only the necessary information for a simulation.

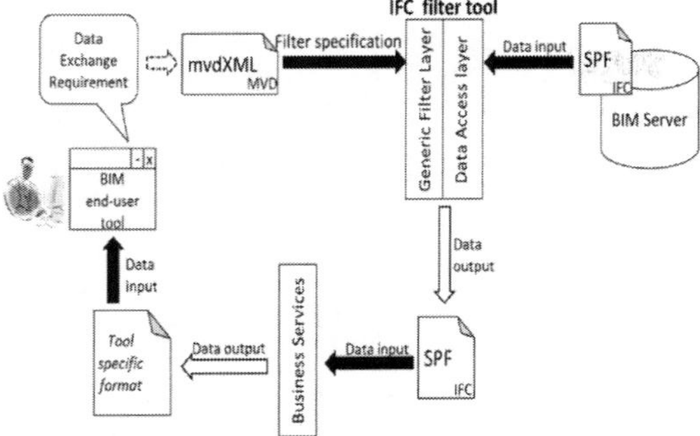

Figure 9. BIM-based filetring workflow.

The BIM-based filtering services are part of the core services as they can access the data, filter it and provide it in the same structure as the initial data model namely IFC. Afterwards the filtered information is passed to and manipulated by the business services in order to give information to end-user tools with respect to their specific proprietary implementation.

6.3 *NIM core services*

The fundamental component of the NIM Server is the 3D City DB repository (3DCityDB, 2016), which is a free implementation of the CityGML model into a relational database structure, based on PostgreSQL. Utilities for converting the database schema to a CityGML file and vice versa are provided, as well as utilities for deploying services following the Open Geospatial Consortium (OGC) specifications.

From the basis of these utilities a set of low-level services has been developed called the NIM core services. They are deployed as web services, using RESTful technologies, in accordance with the overall HOLISTEEC approach. These core services perform a set of basic functionalities which can be applied and extended to multiple vertical domains. In the context of HOLISTEEC they are used by business services built on top of them in the field of performance-based design and simulation.

The first step done is the adaptation of the OGC services to the project. In particular, a Web 3D visualization service based on W3DS has been deployed. It allows generating a 3D model directly embeddable into web applications, with navigation and interaction capabilities.

Secondly a set of REST services for querying buildings by address has been deployed:

- Get streets: it returns all the streets of the model in XML format
- Get portals by streets: it gets all the portals of a given street
- Get building by street and portal: it returns a building ID reference matching the query.

Finally, a set of REST services is provided for enabling basic operations with CityGML files, as follows:

- File upload/download services
- File merge services: it merges two files passed by reference. A typical use case is to add a new building (file 2) into an existing model (file 1) which still does not contain the new building
- Replace service: it allows replacing a given building inside one file with another building.

The above mentioned file-based utilities are further exploited in the variant generation process during the design workflows, e.g. different new building proposals which are merged with the existing built environment in order to create different city layouts to be simulated and compared among them.

7 BUSINESS-FOCUSED SERVICES

The business-focused services are the top-most layer of HOLISTEEC architecture, the closest one to the end user. They make use of the different data sources provided by the platform (BIM, NIM and KPI models), as well as the core services layer and the BCF communication mechanisms in order to provide functional modules in the performance-based design workflows.

One category is focused on the model auto-generation processes, mainly applicable to NIM. In the case of BIM, models will be created by designers in their authoring tools, but since there is not a NIM/CityGML authoring tool, algorithms have been developed for transforming 2D GIS data into 3D semantic models, combined with height data provided by LiDAR scans. This is dependent of the geo-graphical location, but in general there is a wide range availability of GIS layer (e.g. from public ad-ministrations or OpenStreetMap) and LiDAR flights across Europe.

For new building designs (coming from BIM and not available in GIS) a simplification process has been deployed to convert conceptual IFC files into CityGML and merge into the original CityGML model (see section 6.3).

Special care has been taken to create proper links and synchronization mechanism between the BIM/IFC model and the corresponding CityGML representation of the same building, preserving the data coherence.

Both the IFC and CityGML models are stored in a project structure which permits to keep track of the different design versions/variants properly linked.

Since one of the biggest gaps in current design practices is a reliable data transfer from design to simulation tools, another main category of business services is meant to ease the generation of simulation (SIM) models from BIM and NIM models, combined into a single conceptual neutral model which accounts for building and neighborhood interactions and is usable from many domains (energy, acoustics…). These transformations mainly target geometrical and topological relationships. This model will be kept updated in subsequent BIM and NIM model updates.

It must be noticed that, after this neutral transformations more specific transformations will occur, possibly locally and requiring user interaction, in order to properly link the models with external e-Catalogues and to transform them to specific input formats for each of the simulation tools considered.

In fact, a key component in the HOLISTEEC architecture corresponds to the e-Catalogue of products, covering both generic and vendor-specific products focused on building envelope and HVAC systems. The transition from generic to specific product will be smoothly done depending on project phase and LOD. It supports standard properties, compatible with IFC.

8 INTEGRATION OF A MULTI-CRITERIA DECISION MAKING SYSTEM

Among the services offered by HOLISTEEC platform there is the possibility of performing an analysis of the current building project. The service allows designers and platform users in general to get an assessment of the current designed solution, and to obtain as result a set of recommendations, i.e. warnings or advices about risks or better design choices for an improvement of the currently proposed solution.

This Multi-Criteria Analysis tool (MCA) is based on the expert system technology that allows the storage of the knowledge of a domain expert and then to apply this "codified knowledge" on the analyzed solution. Thus it performs an evaluation of it and suggestions for improvement are made exploiting the expert know-how, as if a human expert were at work.

Outcome of such a process can lead to alerts for foreseen risks in some of the building phases, to suggestions for improvements towards the achievement of better performances, to the assessment of the compliance of the proposed solution to regulatory constraints or other external requirements.

The power of this tool resides at first in the level of accuracy and precision of the knowledge expressed, thus the degree of quality of it in representing the domain knowledge to apply. In addition, the fuzziness capabilities associated with employed technology allow the system to analyze and evaluate the submitted solution in a way similar to that of a human actor, thus trying to replicate the reasoning process for assessing the solution as followed by a real expert having the same domain knowledge.

Collection of the knowledge that led then to the codification of the same know-how inside the expert system has been done through a set of interviews with real domain experts, in order to get a broader and extensive classification of the points of view and best practices of a field.

The system offers the possibility to choose among the domains along which the analysis can be conducted. Domain is thus one of the axes for the analysis, and the possibility to evaluate and assess the solution on different domains make the use of the tool flexible and powerful at the same time. For the HOLISTEEC project, the basic domain set has been identified in line with those of reference for simulations (i.e. energy, acoustics, environment, lighting), which thus represent the axes along which the multi-criteria analysis can be conducted.

9 CONCLUSION

The HOLISTEEC project has a very challenging character with regard to the functionalities and tools that are planned to be introduced and all integrated into a common platform. Much implementation has been already done so that at the moment a first functional prototype of the platform is available. This first prototype disposes for the moment only of some basic functionality and many tools are still under integration. The next steps of the implementation look rather optimistic and all services should be able to be tested in the context of pilot projects.

Next to the challenging development activities, one other and major challenging aspect will be the transfer of the new HOLISTEEC design methodology and the dedicated platform into the daily industrial practice. For that purpose the project aims at solving real issues occurring in traditional collaborative design. The main strategic issue is about interoperability which should be considerably enhanced through the core and business-focused services, which together with common interfaces significantly increase the information space around the building information model as well as its management. Furthermore key features are introduced that shall support engineering work consequently. Among them, the very innovative KPI management system, its combination with multi-physical simulation at neighborhood scale and with assisted design through expert knowledge, and facilitated design and testing of variants, shall bring substantial time and cost savings within the design process.

REFERENCES

3DCityDB, 2016. 3D City DB (The CityGML database). Retrieved from http://www.3dcitydb.org/3dcitydb/welcome/. Last accessed April 2016.

Bjork, B.-C. 1992. "A unified approach for modeling construction information." Building and environment, 27(2), 173–194.

BuildingSMART. 2016. BCF releases. Retrieved from http://www.buildingsmart-tech.org/specifications/bcf-releases.

Chipman, T., Liebich, T. & Weise, M. 2012. mvdXML—specification of a standardized format to define and exchange model view definitions with exchange requirements and validation rules, © buildingSMART International.

eeEmbedded 2013–17, EU FP7 Project No. 609349. eeEmbedded [online], http://www.eeEmbedded.eu.

ISO 29481–1. 2010. Building information modelling—Information delivery manual—Part 1: Methodology and format.

iabi. 2016. Welcome to BIM—it.net. Retrieved from https://bim--it.net/.

Katranuschkov, P., Weise, M., Windisch, R., Fuchs, S. & Scherer, R. J. 2010. BIM-based generation of multi-model views. In: Proc. 27th CIB W78 International Conference "Applications of IT in the AEC Industry & Accelerating BIM Research Workshop", 16–19 Nov. 2010, Cairo, Egypt.

Liebich, T. 2012. mvdXML 1.0 released. http://www.buildingsmart-tech.org/blogs/msg-log/mvdxml-1.0-released, accessed December 2012.

OMG. 2016. Business Process Model and Notation. Retrieved from http://www.bpmn.org/.

Scherer, R. J., Weise, M. & Katranuschkov, P. 2006. Adaptable views supporting long transactions in concurrent engineering. In: Proc. Joint International Conf. on Computing and Decision Making in Civil and Building Engineering, Montreal, Canada, June 14–16 2006, pp. 3677–3686.

Weise, M., Katranuschkov, P. & Scherer, R. J. 2003. Generalised Model Subset Definition Schema. In: Amor R. (ed.) "Construction IT: Bridging the Distance", Proc. CIB-W78 Workshop, NZ, 16 p.

eWork and eBusiness in Architecture, Engineering and Construction – Christodoulou & Scherer (Eds)
© 2016 Taylor & Francis Group, ISBN 978-1-138-03280-4

Collaboration requirements and interoperability fundamentals in BIM based multi-disciplinary building design processes

Gloria Calleja-Rodriguez
Centro de Estudios de Materiales y Control de Obra S.A., Spain

Romy Guruz
Technische Universität Dresden, Germany

Marie-Christine Geißler
IBAM Deutschland AG, Germany

R. Steinmann, K. Linhard & G. Dangl
Institute of Applied Building Informatics, Munich, Germany

ABSTRACT: This paper is reflecting the concepts of the buildingSMART standards IDM (Information Delivery Manual) and MVD (Model View Definition) based on specific energy efficient BIM-information levels within design processes. Taking into account the needs of different LOD (Level of Development) stages, the impact of LoD (Level of Detail), and LoA (Level of Approximation) an extended definition of LoI (Level of Information) is being proposed. The paper also reflects the potential to derive well defined purposes for BIM-information exchange on the basis of so called BIM-Functions. The research that this paper is based on was conducted within the eeEmbedded project. eeEmbedded is funded by the European Commission within the Seventh Framework Programme.

1 INTRODUCTION

With the ever increasing usage of BIM in projects within the construction and design industry, interoperability of software systems gains more and more significance. Currently, BIM processes are often only loosely defined. While ever more regulations do provide guidelines, such as RIBA or CoBIM, they, too, are mostly targeting higher levels of collaboration without specifying many details of a software exchange. Additionally, new technologies lead to a rising demand in both quantity and quality of construction and design processes.

Identification and description of collaboration scenarios was a key part in the eeEmbedded project. Significant use cases, i.e. functionalities that are not restricted to a single project but have importance in different building projects, were investigated. Common characteristics were extracted and analyzed so to be able to distill concepts and methodologies that are generally shared in BIM processes. Further research has been conducted into traits such as involved actors and tools as well as input and output requirements. This resulted in a basic eeEmbedded library of BIM use cases that are applicable to building projects.

This paper describes the ways that make specifications of data exchange scenarios between different software systems possible and introduces the concepts and technologies behind them.

The foundation for functional interoperability is the definition of requirements. On a high level, Information Delivery Manuals are introduced that represent a way of collecting all information regarding how data has to be shared. Furthermore, software data schemas are used as methods to qualify information contained in models and set requirements for necessary scenarios.

The whole BIM process is then divided into small functional parts. Each of these can be described via exchange requirements, allowing consistent quality checking in small, manageable steps as well as reusability of projects parts that have significance in not only a single project.

2 COLLABORATION PRINCIPLE CONCEPTS

Structured methods of collaboration encourage introspection of behavior and communication. These methods specifically aim to increase the success of teams as they engage in collaborative problem solving. Forms, rubrics, charts and graphs are useful in these situations to objectively document personal traits with the goal of improving performance in current and future projects.

Information Delivery Manual (IDM)

Information Delivery Manual is a methodology which was developed within buildingSMART International as the standard for processes and later published as ISO 29481-1. IDM's aim is to offer standardized methods to answer the following questions:

- Who needs the information extracted from the building information model?
- At which point in time this information is needed?
- Which minimal amount of data has to be exchanged?

The essential parts of this information delivery manual are:

- Defining "who" and "when" by means of a general process map
- Defining "what", thus the required data as exchange requirements listed in a tabular form.

The steps of the IDM-methodology (Liebich et al., 2013) are as follows:

IDM step 1: Identify Processes and Actors;
Requirement: Clear definition of roles, processes and functions
Concept: Specification of use cases defined in BPMN
IDM step 2: Identify Exchanges;
IDM step 3: Create Exchange Requirements.

After that the following steps of the MVD-methodology or in other words the technical implementation can be added:

MVD step 1: Extend to Exchange requirements Models;
MVD step 2: Unify to Model View Definition.

3 INTEROPERABILITY PRINCIPLE CONCEPTS

A wide variety of software systems are used during the entire life cycle of a building. These software systems support different processes of different users with different data models. So, the interoperability problem in construction industry is inherent: if one user wants to exchange his information with another user using another software system, both software systems should be interoperable to ensure an efficient information exchange. If software systems are not interoperable, then they cause loss of time and they are more prone to errors, so that they raise the cost of a project [Gallaher et al., 2004].

1. Interoperability is not necessary if only one software would exist covering all use cases in construction industry, but this is idealistic.
2. Another way to guarantee high interoperability is to use software from same vendor. Software vendors are interested in making their software products interoperable to each other because they want to make their customers to use as many software products as possible. And because software vendors know their own software structure and data formats, interoperability of their own software products can be done easily. But this way is also not practical in construction industry, because no software vendor is offering all tools and functionalities, which are required in construction projects.
3. In case software from different vendors needs to communicate, then communication rules need to be agreed. This can be done if one vendor has documented its exchange file format and another vendor follows the rules of this document.
4. Another method of interoperability is the use of open standard for exchanging information from one software system to another software system. This is quite common in IT industry.
5. The final method of interoperability is that actually non-interoperable software systems don't communicate directly with each other instead they are using a service. The software systems are transferring their data to the service in their proprietary data format, where the data are filtered and mapped into an internal data format of the service, and another software system can demand the data in its also proprietary data format [Katranuschkov et al. 2011]. Creating such an interoperability service needs knowledge of all proprietary data format of all participant software systems.

For using interoperability method (4), in construction industry there are open data standards available for transferring data between software systems, such as: Green Building XML (gbXML) and Industry Foundation Classes (IFC). IFC was developed within buildingSMART. Additionally, to IFC buildingSMART developed other open standards to enhance information exchange: Information Delivery Manual (IDM) including Model View Definition (MVD), BuildingSMART Data Dictionary (bSDD) based on International Framework for Dictionaries (IFD) and BIM Collaboration Format (BCF).

IFC as data standard in construction industry covers a high number of domains, so that different actors in construction process can work with the same data standard and share their information via one common data schema. But in some cases IFC has its limits and it cannot be used to store or to transfer information of special issues. In such a case an extension of IFC data schema would be necessary or additionally data models are used to cover up such specific issues. Changes in standardized data schema have to pass through standardization committees which takes time, and some software tools supporting such standards have a hard rime catching-up changes in the data schema. Therefore, the usage of data schemata in addition to IFC will lead to faster results.

The following requirements of interoperability are basically structured in 4 levels:

Level 1: Technical interoperability

Requirement: Achieving technical interoperability

Concept: Distributed service-oriented web architecture

Defines the communication protocol for exchanging data between the participating systems; on this level, a communication infrastructure is established allowing systems to exchange bits and bytes, and the underlying networks and protocols are unambiguously defined.

Level 2: Syntactic interoperability

Requirement: Syntactic interoperability

Concept: Common data formats

Defines a common structure to exchange information, i.e., a common data format is applied. On this level, the format of the information exchange is unambiguously defined.

Level 3: Semantic interoperability

Requirement: Semantic interoperability

Concepts: Agreed common model schemas, Model View Definitions, Common ontology and Link Model

Defines common reference model(s) that enable common understanding of the data, i.e. the meaning of the data is shared. On this level, the content of the information exchange requests is unambiguously defined.

Level 4: Functional interoperability

Requirement: Functional interoperability

Concepts: Common Multi-Model Services—from MMC to Multi-Model Service Common ontology and ontology management methods

Defines common methods of interpreting the shared information, i.e. requesting and responding system have a common understanding how the data is to be processes. On this level, the management of the information exchange requests is unambiguously defined.

4 SPECIFICATION OF THE EEEBIM INFORMATION LEVELS WITHIN DESIGN PROCESSES

4.1 *State of the art*

The Level of Detail and Level of Development concepts have been used in a number of references. VICO software created a Level of Detail framework in 2004 to deal with the growing potential of BIM as a process and as a virtual 3D model. This framework was adopted and augmented into specific Levels of Development (LOD 100, LOD 200, LOD 300, LOD 400 and LOD 500) by the American Institute of Architects (AIA) and turned into the contract document AIAE202-2008 BIM Protocol Exhibit, released in 2008. In 2011, the [BIMForum] initiated the development of the LOD Specification and formed a working group comprising contributors from both the design and construction sides of the major disciplines. The working group first revisited the AIA's basic LOD definitions for each building system, and then compiled examples to illustrate the interpretations. In August 2013, BIMForum published their « Level of Development Specifications, Version 2013 », [BIMForum, 2013]. It rapidly became one of the most referenced BIM documents, independent of the user's geographic location. The AIA integrated the work of the BIMForum into an updated version of its protocols published in 2013 as Document E203™—2013—Building Information Modelling and Digital Data Exhibit.

4.2 *Objectives*

We have identified as a requirement the development of a framework based on levels of information that can be used for defining where an element is along its design process as well as for defining the information that should be in a model to perform a specific task. In this paper, the term Level of information can be defined as a concept or group of concepts which are used for describing the information contained in a model or object. We use the term **Levels of Information** (LOI) to refer to the group of three concepts:

- Level of Development (LOD)
- Level of Detail (LoD)
- Level of Approximation (LoA).

Therefore, our purpose is to improve and enrich the existing approaches using the three concepts—Level of Development, Level of Approximation

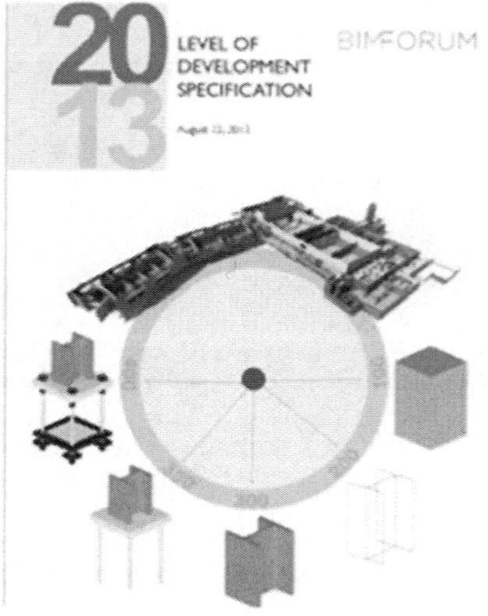

Figure 1. Document E203™—2013—building information modelling and digital data exhibit.

and Level of Detail—and linked them with the design process [Calleja-Rodriguez et al., 2014].

4.3 *Concepts and definitions*

In general, a model element is a digital representation of the physical and functional characteristics of an actual building component to be used in the project. Level of Detail and Level of Development are both about Model Elements and its attached information, but they stand for different concepts.

Level of Development (LOD) is the degree to which the element's geometry and attached information has been thought through (has been effectively designed and decided upon). As such, its intention is to provide a measure of the degree to which project team members may rely on the information when using the model. The reliable information is defined for the end of each LOD-phases "Data Drop Point—DDP" in an "Object Element Matrix—OEM". This matrix is a spreadsheet used for identifying and tracking BIM information during the project. It's important to note that not all elements/disciplines, depending on their domain, may be in the same LOD.

Level of Detail (LoD) is essentially how much detail/information is included in the model or object. It is the specific information needed to do a design task (BIM function, see section 4.3) or provided after the design task (BIM function). This is the input and output information for a task. LoD is not defined by numbers such as 100, 250 but by exchange requirements.

Figure 2. BIM functions as the driving aspect of LOI.

Level of Approximation (LoA) indicates how accurate are the analysis results, means how far we are from the reality. The accuracy in the estimate of the various physical parameters is refined in each new LoA by devoting (more precise information and) some additional time to the analyses, which leads to better (more accurate) estimates.

The concept **"BIM Function"** has been developed through the years. In the beginning, it was used to refer to the simple visualization. Later, it was enriched and used for calculations and simulations. In this paper, we have found out that the BIM Functions must be the purpose of the model and the driving aspect of the Levels of Information. Therefore, required information levels will be setup on the basis of BIM Functions.

4.4 *Modeling perspectives*

This paper proposes to use the Object Element Matrix (OEM) to define every LOD-DDP (Level of Development—Data Drop Point). The Object Element Matrix can be represented via a Model View Definition (MVD) for a specific design phase.

Another modelling perspective proposed in this paper consists of:

Defining the eeE-BIM functions to specify for which purposes we want to use BIM at the beginning of the project (e.g. define building cubature, define HVAC system, calculate costs, etc.).

Defining the Level of Detail that is required to perform a BIM function (input) as well as the Level of Detail that will be reached after the BIM Function (output).

BIM Function inputs and outputs are subsets of the MVD and can be defined as exchange requirements.

The main strength of the resulting approach is that LoD and LOD are defined as MVD and on the basis of BIM functions which are reusable and include the end-user wishes. Therefore, the standardization of eeE-BIM functions deployment is possible and required within borders of an extended enterprise (company & preferred partners) because the information level specification per project is too labour intensive.

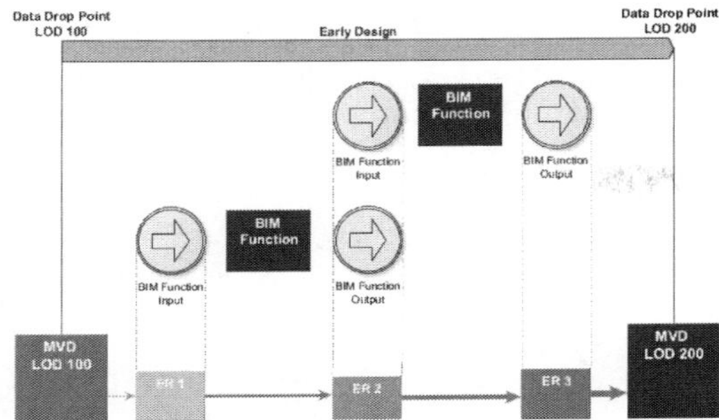

Figure 3. LOD-Data Drops as MVD.

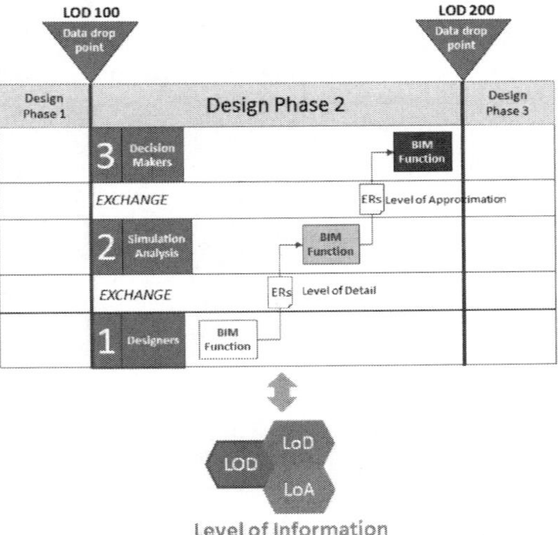

Figure 4. LoD and LoA based on exchange requirements.

Figure 3 shows the integration of LoD, LOD and LoA approach within design processes The Level of Development (LOD)—the degree to which model elements and attached information has been effectively designed and decided upon—will be defined at the end of each LOD "Data Drop Point—DDP" in an "Object Element Matrix—OEM" to increase interoperability between the project phases. Once the LOD framework is set up by specifying the reliable output of each phase, the Level of Detail (LoD) of the model elements needed can be defined, based on the Exchange Requirements (ERs) to perform a certain eeE-BIM function. The LoA must be specified in the outputs of the BIM functions that involve analysis, calculations or simulations. That way design alternatives can be easily analysed based on simulation/computation results with the needed accuracy (Level of Approximation—LoA) to choose the one with the highest value for the client.

The advantages expected with the new and enhanced framework of Levels of Information are:

- It will ensure that the client and the collaborative partners get the models they need to support the task and objective they want.
- It will avoid coordination errors and rework because the necessary information is available when it´s needed.
- Modelling efforts will be reduced, the design team will save time and the companies will save money.

5 CONCLUSION AND OUTLOOK

Within the increasing experience in using IFC for BIM-based information exchange the essential role of MVDs becomes very obvious. Very often, when end-users claim that "IFC doesn't work" the actual reason behind this experience is, that IFC is not being used for a clearly defined purpose, in order to meet the objectives of specific Exchange Requirements. However, the appropriate mean to technically specify these needs are MVDs.

Current discussions in buildingSMART tend to demand a huge number of MVDs, in order to meet the various requirements in BIM-based information exchange. This however will lead to a confusing long list of MVDs which is difficult to understand and to handle in daily practice.

Experts in buildingSMART are proposing to agree on a limited number of MVDs, which are covering general needs of information exchange, like coordination, design transfer or hand-over. Such general MVDs should contain very focused Exchange Requirements, which meet the specific needs of information exchange within the workflows of BIM-processes. While applications don't necessarily need to support a whole MVD, they need to fully support specific Exchange Requirements of a MVD. This would also be a solid foundation for a rigorous software certification.

This paper is offering valuable input for the future improvement of the MVD-concept and with this for improved implementation and certification and finally an improved and seamless way of IFC-based BIM-information exchange.

ACKNOWLEDGEMENTS

This paper is part of the research developed in eeEmbedded project. We kindly thank the support of the European Commission to the eeEmbedded project, Grant Agreement No. 609349, http://eeEmbedded.eu.

REFERENCES

Aurelio Muttoni, The Levels of Approximation approach in MC 2010, Encontro Nacional BETAO ESTRUTURAL-BE2012 FEUP, 24–26th October 2012.

BIMForum, https://bimforum.org/, Last access: September 2014.

BIMForum, Level of Development Specification, 2013, http://bimforum.org/lod/

Calleja-Rodriguez, G. & Guruz, R., (2014); eeEmbedded Deliverable D1.3: Interoperability and collaboration requirements, © eeEmbedded Consortium, Brussels.

Chipman, T. & Liebich, T.: IFC Documentation Guide, 2012.

Chipman, T., Liebich, T. & Weise, M.: mvdXML—Specification of a standardized format to define and exchange Model View Definitions with Exchange Requirements and Validation Rules, 2012.

Construction Information Systems Limited (2011), NATSPEC BIM Object/Element Matrix, http://bim.natspec.org/index.php/natspec-bim-docu-ments/national-bim-guide

eeEmbedded 2013–17, EU FP7 Project No. 609349.

Gallaher, M.P., O'Connor, A.C., Dettbarn, J.L., & Gilday, L.T. 2004. Cost Analysis of Inadequate Interoperability in the U.S. Capital Facilities Industry. National Institute of Standards and Technology Publication, NIST GCR 04–867.

Geißler, M.C., Guruz, R. & van Woudenberg, W. (2014); eeEmbedded Deliverable D1.1: Vision and requirements for a KPI-based holistic multi-disciplinary design, © eeEmbedded Consortium, Brussels.

Katranuschkov, P., Fuchs, S., Muntzinger, H.-D., Hienz, R., Liebich, T., Hauber, T. & Zeller, G. (2011) Mefisto Bericht B–4.1: Interoperabilitätsmethoden und Filtern von Informationen und Modellen—Stand der Forschung und Technik und Anforderungsspezifikation.

Liebich, T., Stuhlmacher, K., Weise, M., Guruz, R., Katranuschkov, P. & Scherer, R.J. (2013): HESMOS D+ Additional Deliverable Information Delivery Manual Work within HESMOS, © HESMOS Consortium, Brussels.

Ralph G. Kreider & John I. Messner, Penn State, The Uses of BIM—Classifying and Selecting BIM Uses, September 2013.

Robert L., Lloyd H. & Kurt Knight, Renée Tietjen, The VA BIM Guide, U.S. Department of Veterans Affairs, Office of Construction & Facilities Management, April 2010.

eWork and eBusiness in Architecture, Engineering and Construction – Christodoulou & Scherer (Eds)
 ISBN 978-1-138-03280-4

Technical challenges and approaches to transfer building information models to building energy

F. Noack, P. Katranuschkov & R. Scherer
Institute for Construction Informatics, TU Dresden, Germany

V. Dimitriou, S.K. Firth & T.M. Hassan
Loughborough University, UK

N. Ramos, P. Pereira & P. Malό
UNINOVA, Portugal

T. Fernando
Salford University, UK

ABSTRACT: The complex data exchange between architectural design and building energy simulation constitutes the main challenge in the use of energy performance analyses in the early design stage. The enhancement of BIM model data with additional specific energy-related information and the subsequent mapping to the input of an energy analysis or simulation tool is yet an open issue. This paper examines three approaches for the data transfer from 3D CAD applications to building performance simulations using BIM as central data repository and points out their current and envisaged use in practice. The first approach addresses design scenarios. It focuses on the supporting tools needed to achieve interoperability given a wide-spread commercial BIM model (Autodesk Revit) and a dedicated pre-processing tool (DesignBuilder) for EnergyPlus. The second approach is similar but addresses retrofitting scenarios. In both workflows gbXML is used as the transformation format. In the third approach a standard BIM model, IFC is used as basis for the transfer process for any relevant lifecycle phase.

1 INTRODUCTION

Building Information Modelling (BIM) is used as a technology to model and manage the digital representation of a building over its entire lifecycle. BIM acts as a bridge between the industry and information technology (Eastman et al. 2011), which makes the entire building lifecycle more efficient and effective. Energy modeling gives the designers an outlook of expected energy consumption regarding various designs prior to construction. To achieve energy-efficient building lifecycle performance it is important to consider the overall processes together with the information requirements and the ways these requirements can be efficiently dealt with whereby suitable decisions in the very early stage of the design process usually have the strongest influence on the energy consumption.

With the recent advance of BIM in AEC practice a sound basis for collaboration using the same model data by all actors in the process is becoming possible. However, the enhancement of the model data with additional specific energy-related information and the subsequent mapping to the input of an energy analysis or simulation tool is yet an open issue. Interoperability in this regard is still on relatively low level, the related workflows are not studied in detail and a lot of preparatory error-prone work is still necessary to obtain the desired results. Reasons for that are various gaps in the BIM export from CAD authoring tools, the heterogeneity of the data exchange formats, and the requirements of the simulation tools which use their own data structures, optimized for the particular focused physical problem but generally not well aligned with BIM.

Whole building simulation requires the exact preparation of the needed input data. The original data about the building is not generated by the software tools for building energy performance like e.g. EnergyPlus. This data represents a different view of the building using a different data structure. Building geometry in a CAD file characterizes the architect's view of the building, not its thermal view, so that it has to be transformed appropriately to represent the thermal view. Similarly, definitions of construction materials defined in a CAD tool have to be extended with data that define their respective thermal properties.

In the energy modelling process, object relationships and semantics of architectural models are often abstracted when such model information does not substantially affect simulation results. For example, building components such as walls, roofs, and floors are frequently simplified as surfaces and then heat transfer through the surfaces is simulated in the energy models. When designers modify building design according to the abstracted simulation results, energy performance of each surface needs to be mapped back to corresponding building components. Such abstractions can delay result interpretation.

Currently, the Industry Foundation Classes, IFC (buildingSMART, 2016) and Green Building XML, (gbXML.org, 2014) present the two most widely acknowledged information infrastructures in the AEC industry. They are both used for common data exchange between AEC applications such as CAD and building simulation tools. The goal of IFC is to provide a universal basis for process improvement and information sharing in the construction and facilities management industries. Compared to IFC, the gbXML schema is simpler and easier to understand. It can be used to represent energy-related building information, such as geometry and material properties of building elements (floors, walls, windows, etc.) using the Extensible Markup Language (XML) format. Its main purpose is to facilitate the transfer of the building information between current building design tools and a variety of engineering analysis software applications. The "bottom-up" approach adopted from the outset in the gbXML schema makes it relatively flexible and straight-forward (Dong et al. 2007).

One critical point lies already in the use of the CAD software at the beginning of each process. In order to export usable models either as gbXML or IFC the energy-relevant aspects have to be considered already in the CAD system itself and equally in its export functionalities. In the end, it is necessary to create the input files for the specific executing engine using the information included either in a gbXML or in an IFC file. To achieve that, appropriate transformations are indispensable. These transformations have to be automated as far as possible in terms of user-friendliness, time saving and error-free performance.

In practice, different IT workflow approaches may be suitable regarding the BIM model in different design situations. This paper examines three such approaches investigated in the Design4Energy project (www.design4energy.eu) pointing out their respective advantages and disadvantages and their envisaged use in practice. However, while distinctly different, all developed approaches are grounded on a common overall concept which is outlined in the following chapter.

2 OVERALL CONCEPT

The Dynamic Energy Efficient oriented Building Information Platform (DEEBIP) developed in the Design4Energy project represents an open platform that will allow different stakeholders to integrate their energy related service modules to further enhance energy efficiency of future buildings.

At the outset, the architect defines the physical appearance of the building including internal spaces, external openings, use of material and so on. He takes into consideration a number of parameters such as the local weather profile and lighting to decide the house orientation as well as the use of raw material related to carbon emissions before the conceptual design is completed. The design is then exported as a BIM model and passed to energy performance simulation. The collaborative environment runs the simulation enabling team members to compare the simulation analysis result with the expected energy indicators. It provides team members the possibility to explore what-if-scenarios and investigate the impact of various parameters on their design to estimate energy performance through the project lifecycle already from an early design stage. The team members can simulate the impact of weather, occupants' behavior, component deterioration, material running cost and propose future maintenance plans. The design can be modified by using components from the Design4Energy Component Database by drag & drop. Once the design is optimized, validated and approved by all stakeholders, the solution is agreed.

For consistency of the following discussion in all studied cases the same target energy application is used which is the well-known EnergyPlus simulation software (EnergyPlus, 2003–2016). Among the various existing energy consumption simulation engines, EnergyPlus developed by the Lawrence Berkeley National Laboratory is widely acknowledged as the mainstream engine. The key features of EnergyPlus include:

- heat and moisture balance,
- integrated building loads/systems solution,
- sub-hourly time steps,
- configurable HVAC systems,
- broad limits regarding surfaces, zones or systems,
- links to other tools.

EnergyPlus provides a huge amount of simulation options that require specific input including simulation control parameters (surface convection and heat balance algorithm options, equipment and system sizing options), daylighting analyses, room air and airflow analysis models, economics calculations, and an enormous array of output possibilities. Many features require domain expertise for input specification and output assessment but not each detail is necessary for all use cases.

Using EnergyPlus we have three major parts: the definition of building geometry and data related to it, the definition of HVAC equipment and data related to them, and the definition of internal loads as well as use and operating schedules for the building. Regardless of the specific approach the resulting input file for the calculation with EnergyPlus must be created fitting these criteria. At this point it is possible to perform several calculations in parallel using an appropriate grid or cloud environment.

The results of the simulation with EnergyPlus flow directly in the calculation of energy key performance indicators (eKPIs) which summarize the building's energetic and thermal behavior for a certain period and furthermore in consideration of the energetic interrelationships within the neighborhood. That way it is possible to compare the impact of varied properties of a building on its thermal and energetic behavior. These eKPIs which represent for example energy costs or CO_2 emissions need to be optimized. This optimization for a certain user-purpose is the aim of the results evaluation. To achieve it, the decision maker uses a number of customable visualizations to compare different design solutions. Such visualizations must be capable to consider the effects of the variables taken into account for the different analyzed eKPIs and the effect on the different alternatives. Moreover, visualizations should be easy to understand and should enable intuitive reasoning. Hence, the graphic representation of solutions and decision making visualizations requires the use of advanced techniques. The Design4Energy development used for decision making analysis enables multiple user-driven ways to visualize main results from different KPI and parameter viewpoints. This may include scatter diagrams, hyper-radial visualization via parallel coordinate plots as well as summarized radar charts providing fast overview and result comparisons.

Energy performance of individual elements can be applied to the BIM model using color-coded visualization in the collaborative virtual workspace. A similar approach is applied to a neighborhood area. Each property is given a traffic light color (e.g. green yellow and red) to show the energy in high, medium and low performance, respectively. The properties can be represented both in icons for sorting/filtering purpose and blocks for displaying on the geographical map (Fernando et al. 2015).

3 POSSIBLE SOLUTIONS REGARDING THE OUTLINED GENERAL CONCEPT

3.1 *First approach—focusing on the early design phase*

The first investigated approach addresses early design scenarios. It focuses on the supporting tools needed to achieve interoperability given a wide-spread commercial BIM model (Autodesk Revit) and a dedicated pre-processing tool (DesignBuilder) for EnergyPlus. Its generalized workflow is shown schematically in Figure 1.

EnergyPlus simulation software was chosen due to its capability to estimate the energy consumption of the building, taking into account heating and cooling systems, as well as lighting, and plug and process loads, but also to show the most straight-forward approach to create simulation input from BIM.

As shown in Figure 1, the very first step of the proposed approach starts with a complete definition of the building under analysis, in a design tool such as Archicad or Autodesk Revit. The Industry Foundation Classes (IFC) model is widely used to describe buildings and construction data, aiming to facilitate the interoperability between different architecture, engineering and construction software due to its open specification. Due to the fact of being an open format, it has become an international AEC standard (ISO 16739). Consequently, Revit software also offers such type of output files. However, when exporting a file to perform energy simulations, the IFC model obtained by Revit does not contain all the information needed even though it is available in the internal Revit model. To overcome this drawback, the gbXML output file, that contains all of the heating and cooling information, is adopted in this approach. The gbXML file format is quickly gaining popularity as a standard

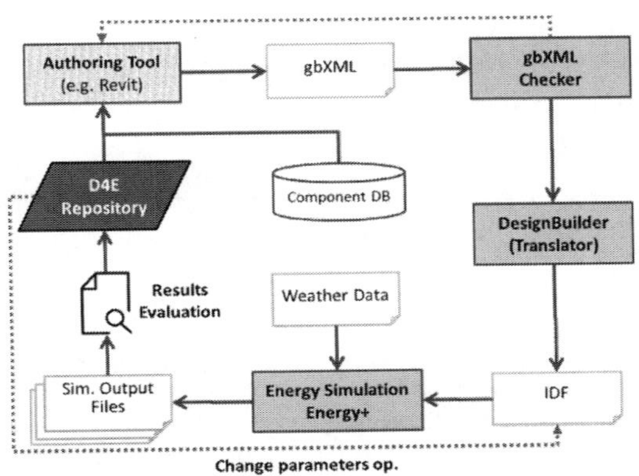

Figure 1. Flowchart of the proposed approach.

file for exchange of information for the purposes of energy analysis. GreenBuildingStudio, DesignBuilder, Arup's Energy2, IES VE, DoE's energy-10 and NREL's OpenStudio are some of the examples of gbXML compatible energy software (gbXML.org, 2014). The main advantage is that the gbXML schema enables easy manipulation and incorporation of additional information. One such example can be found in the work of Dong et al. (2007) on adding sensor data to the gbXML file.

Aiming to perform an energy simulation of the building under study, several steps must be executed where the information has to be exchanged accordingly. In the first step, after obtaining the gbXML output file from Revit, a validation process is executed to guarantee that the extracted model includes all the necessary information to go through the energy simulation. If some information is missing, the checker procedure is capable to recover it from Revit. In a second step, DesignBuilder is used which provides interoperability with the BIM models through its gbXML import capability. This allows users to import 3-D architectural models created by other drawing systems that support gbXML data exchange, such as Revit. In this approach DesignBuilder is used as black box to translate the gbXML file to an IDF file, as this is the type of file supported by EnergyPlus. The third step refers to the energy simulation itself, where the IDF file previously obtained is uploaded to EnergyPlus software. At this point, minor changes to the building model (IDF file) are allowed, in order to improve the energy efficiency of the building. Moreover, weather data (if available) can be included in EnergyPlus software with the purpose of obtaining more realistic results. Finally, the simulations results are stored in a repository and offered to the end user.

3.2 *Second approach—focusing on the retrofitting phase*

The second approach is especially targeted on the retrofitting phase. Due to the specific requirements of the retrofitting phase, using the same tools as in the first approach is less efficient here. Therefore a different workflow with different intermediate stages for the transformation from the BIM model to EnergyPlus is proposed.

In order to perform retrofit analysis of existing buildings a calibrated whole-building base-case model needs to be developed and used to apply and assess alternative retrofit scenarios (Miller et al., 2014). BIM for sustainable design of existing buildings often uses estimated values of the building's thermal parameters that are not always representative of the real building characteristics. Since the retrofit decision-making in buildings is not supported by appropriate data, and the retrofit decision analysis is prone to mistakes. To avoid some of the uncertainties introduced by the estimated parameter values, operational data from the building can be used to inform the model building (Motawa & Carter, 2013). At the same time a significant number of options need to be set by the user to perform energy analysis using EnergyPlus. These can vary in how significantly they impact on the energy results. Invariably all the options set will have some effect on the final results. Hence, the user resorts often to default values to resolve this issue in order to perform the analysis. Therefore, a more transparent workflow is needed that provides more control over the default options set.

Figure 2 presents the proposed workflow focusing on the tools needed to achieve interoperability between BIM and EnergyPlus with increased transparency and potential for operational data feedback. The selected BIM authoring tool is again Autocad's Revit and the gbXML export file format is again used. However, in most cases the gbXML export from Revit needs to be edited to include additional information or to implement changes to the building's thermal properties.

To achieve that, the gbXML file is imported in the GbXML Editing Tool which is linked to the Design4Energy Component Database. The Editing Tool can be used to implement the required changes to the building characteristics and to embed additional information to the gbXML file when needed. The output of the Editing Tool is the edited gbXML file containing all the information describing the building's thermal performance. The edited gbXML file is then imported into the developed gbXML to idf Conversion Tool. This Conversion Tool uses as much information as possible from the gbXML file and additional parameter values that are specified

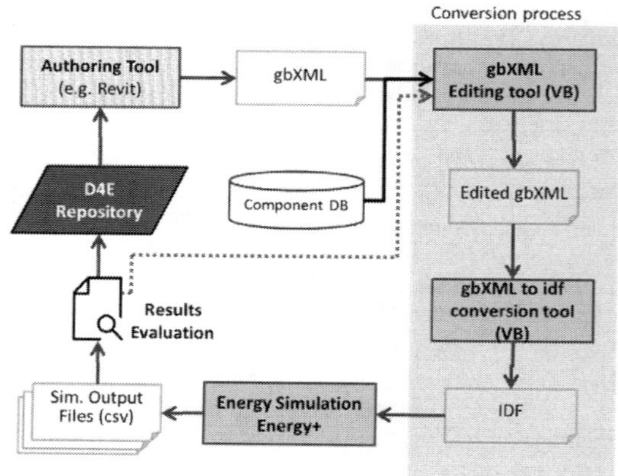

Figure 2. Proposed BIM to EnergyPlus.

by the user to create the idf file. These two tools constitute the main part of the conversion process. EnergyPlus, the selected Building Energy Modelling software, uses idf as its input file to perform energy analysis and produces CSV files with the results. CSV files can easily be used for calibration purposes or for retrofit decision support provision. In existing buildings, the modelled results may deviate significantly from the measurements hence calibration of the model is needed. This can be performed by applying changes to the thermal parameters using information from the Design4Energy Component Database to increase the model's adequacy in representing the real building. Multiple iterations of the calibration process using the gbXML Editing Tool might be required to achieve good agreement between the modelled and the actual building energy performance. The calibrated and verified model can then be used as a base-case to model alternatives for retrofit decision support. Again, the Editing Tool and the Component Database can be used to create gbXML files of retrofit alternatives.

Figure 3 shows a flowchart focusing on the conversion process of the proposed BIM to EnergyPlus workflow. The Revit gbXML export is imported into the Editing Tool. The Editing Tool is linked to the Component Database in which thermal properties for different building elements can be found. The Editing Tool also offers the possibility to add additional information and data to the gbXML file to achieve a more comprehensive gbXML representation of the building under study.

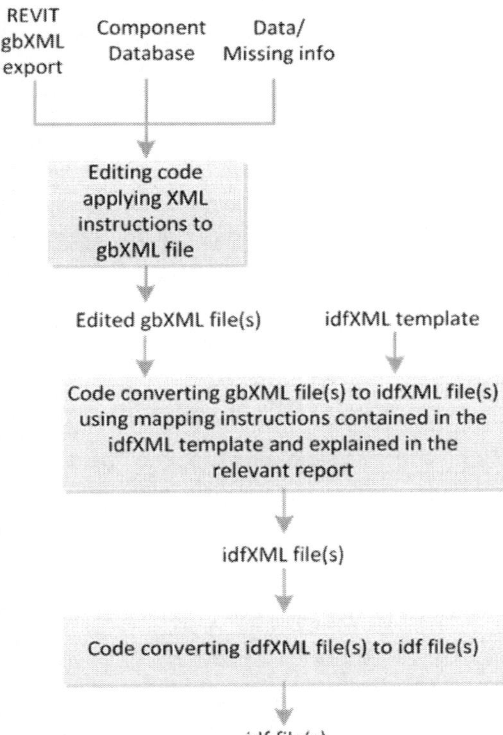

Figure 3. gbXML to idf conversion process using the Editing and Conversion tools developed.

The Editing Tool is able to apply user defined XML instructions to the gbXML file to add to the gbXML file or modify it as required. The edited gbXML file is the input file to the conversion tool. The first step is the conversion of the gbXML file in a XML file containing all the idf objects required. This file is named the idfXML file and serves as a facilitator for better control over the defaults assigned to the idf parameters and improved transparency of the conversion process. A template idfXML file contains the mapping instructions of the idf objects to the relevant nodes of the gbXML file. The idfXML file is then converted into the native idf format. Throughout this conversion process multiple files can be handled at the same time. Further explanation of the process, with more detailed description of each step and verification using a case study is available in (Dimitrou et al. 2016).

3.3 *Third approach—focusing on the use of the universal IFC open standard*

Figure 4 shows an approach for the data transfer from 3D CAD applications to building performance simulations using the standard BIM schema IFC (ISO 16739) as basis for the BIM workflow. IFC is an open and comprehensive data format used to exchange data within the BIM context. It is the most mature open standard to represent building information and BIM collaboration tools including the BiMserver and the EDM-modelServer support the storage, maintenance and query of IFC-based BIM models.

Typically, the entire BIM will contain several domain models such as architecture, HVAC, BACS etc. Most of them are already considered in the IFC but external (non-BIM) data has to be included too. Such data encompass climate files, occupancy profiles, detailed material characteristics, detailed equipment performance specifications and so on. Consequently, in the IFC-based approach, the model enrichment and model transformation to EnergyPlus is achieved via an additional energy-extended BIM data structure (eeBIM) and a set of model manipulation tools (model validator, model combiner, model filtering, eeBIM2SIM converter).

To establish the eeBIM model the IFC schema is firstly reduced to a BIM schema containing

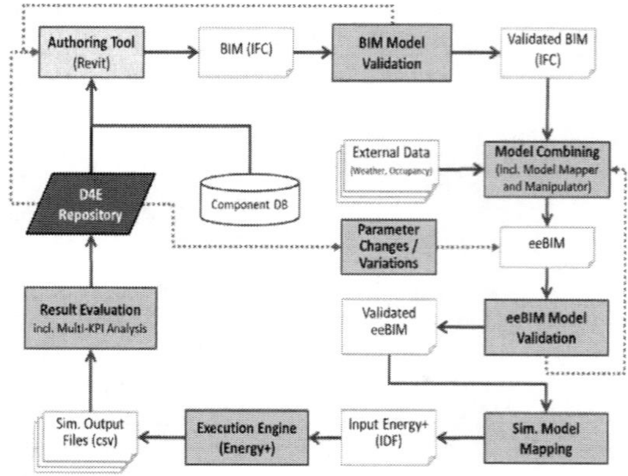

Figure 4. Simplified general procedure using IFC.

only the IFC entities required for the applications according to a Model View Definition (MVD). Filtering operations represent a key aspect at this point. In Design4Energy the filtering is realized by using and extending the filter functionality of the Java-based filter toolbox BIMFit (Wülfing et al. 2012) and a batch support tool providing automated server-side processing of the data (Pirnbaum, 2015). These tools enable the filtering of building model subsets as well as the filtering of certain elements which fulfil domain or task specific information requirements.

Within the eeBIM framework we use the possibilities of direct integration into the IFC via appropriate IFC objects as well as the approach of linking of external data with the IFC model. A so-called Link Model connects elements of different elementary models via unique identifiers (IDs) but is held completely outside the elementary models offering a set of tools for model querying, object selection and user defined constraint checking.

The specification of the eeBIM goes hand in hand with the development of operational methods bringing the original data into the required form as well as methods working on the eeBIM model and preparing it to serve as input for the envisaged applications. Central functions are:

- to support the checking of the IFC model to ensure that it contains the necessary thermal performance data,
- to support mechanisms to add thermal performance data for various components if this data is missing,
- to complete the basic BIM model with different design components, not included in the CAD output
- to support the interface to the simulation execution engine.

Energy simulation models essentially employ the same data as already contained in the eeBIM, but structured completely differently. Generally two possible approaches can be envisaged for the following mapping process:

1. customary one step mappings to each solver integrated in the eeBIM-based virtual laboratory platform, and
2. the more difficult but also more promising two step approach involving development of a harmonised simulation model and first mapping eeBIM to it and only then to one of various possible simulation tools, thereby achieving higher level of interoperability on medium term.

As in the previous approaches EnergyPlus is used as the simulation software at the end of the process chain. Since the input data format for EnergyPlus is IDF, the transformation from IFC data to IDF data constitutes the key point of directly using the IFC-based design result in the energy simulations with EnergyPlus. Its essence is to generate information from one model to another model by simplification, translation and interpretation (Bazjanac & Kiviniemi 2007; Bazjanac, 2008). To establish the mapping between the IFC entities and the IDF entities in the corresponding models the relationships between the IFC-based information model and IDF-based information model for energy-efficient design must therefore be analyzed. Based on these relationships, appropriate algorithms for the transformation from IFC data structures to appropriate definitions in the IDF files are established (see Figure 5).

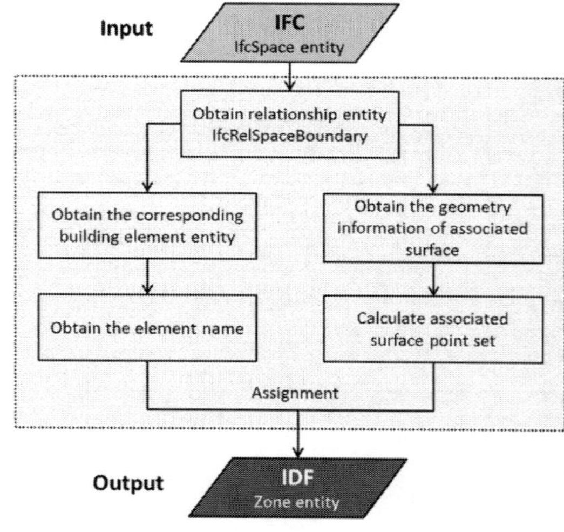

Figure 5. Example of the transformation algorithm for the IfcSpace entity.

The transformation of the geometry information is the most important and simultaneously the main challenge. In IFC data, the geometry information of the building elements and thermal zones can be expressed by various geometric modeling methods, e.g. solid model or surface model, whereas only surface model is used in the IDF data. The model transition has to include the generation of a 3D surface model from the IFC-based 3D model, the configuration of that model with the specific requirements of IDF data and finally the calculation of other IDF data like the internal or external attribute of boundaries. Some specific tasks are:

- the verification of the normal vectors (direction) for space boundaries that represent surfaces of walls, slabs, doors etc.,
- the determination of thermal zones,
- the association of space boundaries with the construction materials of the object
- the correct arrangement of material layers for interior walls, etc.

The transformation methodology should avoid arbitrary and manual data definition to preserve the integrity of the original data, and assure their transformation by unambiguous rules embedded in software (Bazjanac & Kiviniemi 2007). In addition to the details regarding the geometry, other information like occupancy schedules, HVAC data as well as runtime performance control specifications have to be included to constitute the comprehensive input of the simulation using EnergyPlus.

The results of the simulations with different input parameters and construction variants can be used afterwards to figure out the most suitable solution with regard to the energy efficiency of the building. This decision making process is similar to the previous two described approaches. It is supported by the calculation and visualization of selected representtative eKPIs.

Overall, the IFC-based approach is more complex than the other two approaches. However, it is also much more flexible as it allows integration of different specialized tools in the same manner and using the same platform tools, whereas the first approach and to some extent the second are more restricted to the use of a specific simulation tool, EnergyPlus.

4 DISCUSSION AND CONCLUSIONS

This paper examined three approaches developed in the frames of the Design4Energy project for the solution of the technical challenges to transfer building information models to building energy simulation and analysis tools.

In the simulation data preparation process, the fragmented connection between BIM design authoring tools and simulation tools causes the difficulty for the domain experts to obtain the original building data in the design phase. In order to meet the resultant requirements we evaluated different workflows, which we connect to an integral tool chain. Based on the proposed methods architects, engineers and energy consultants are enabled to consider aspects of energy performance simulation that are not typically available from the design documents, but are critical in performing realistic energy simulations for design optimisation. Such aspects are related to the physical environment in which the building is located and how it influences the energy consumption as well as to the actual functional use of the building and its individual spaces. With the proposed approaches, multi-disciplinary teams are enabled to investigate distinct energy solutions already in the early design phase of energy-efficient buildings up to the later retrofitting phase of their life cycle.

Autodesk Revit, which is applied in this research, provides an opportunity to develop APIs that can be used to create custom tools that plug directly to it, but this does not provide a general approach. Firstly, it limits the design team to the use of Revit as authoring tool which reduces the possibilities for cross-company cooperation, and secondly, it makes developers of sophisticated specialized software vulnerable to future changes of the proprietary non-standard system and model schema of Revit. Therefore, in the presented solutions acknowledged model schemas, gbXML and IFC, are used for the data exchange between CAD and simulation software. Both data models are able to potentially encapsulate almost all information related to a building in comprehensive manner and facilitate the sharing of pertinent information from design (for evaluating the energy building performance) to construction (for evaluating cost and schedules) and to operation (occupancy and environmental sensing and system controls).

The aim of gbXML is to facilitate the exchange of data among CAD and energy analysis software. Hence, the gbXML schema is simpler and easier to understand than IFC, which facilitates quicker implementation of the schema and its extension for different design purposes. On the other hand, the goal of IFC is to provide a universal basis for process improvement and cross-company information sharing in the construction and facilities management industries and above all it represents an open standard, ISO 16739. Furthermore the IFC focuses on the representation of building descriptions throughout all building life cycle phases.

In general, each of the presented approaches is suitable to reach the preset objective. Compared to the two workflows using gbXML, the approach using IFC as basis shows greater generality and flexibility with regard both to the input authoring tool in which the BIM is created and the targeted simulation/analysis tools. However, it is also more difficult to implement and due to yet existing shortcomings in IFC export, as shown for example in (Katranuschkov et al. 2015), it is also the one which is still least prepared for practice.

Due to the complexity of the task and the intermediate decisions that have to be taken at several sub-steps of the overall process, the conversion from BIM to an energy simulation can hardly be a fully automatic or a single-click solution. However, for further improvement of energy efficiency, more work towards automating the process from BIM to simulation analyses is necessary. Currently, the extraction of geometrical information from the architectural models for energy simulations is considerably hampered by inconsistent geometrical and other related definitions in the planning phase which can be avoided in many cases. The current necessity to fix and complete geometry models hinders a smooth conversion process into thermal models. Up to now not all possibilities provided by IFC or gbXML are used in the export functionality of CAD tools. Here, especially, the development of usable standards for the verification of building models will be of vital importance for the acceptance of BIM in practice. In any case further tests with predefined usage scenarios and broadening the scenarios to cover more use cases and building types are indispensable.

ACKNOWLEDGEMENT

This research is funded with the support of the European Commission to the FP7 Design4Energy project (Grant agreement no: 609380) and partially the FP7 eeEmbedded project (Grant agreement no: 609349). We would like to acknowledge the Commission's support and all the project's partners, in particular the architects, for their valuable contribution to this work.

REFERENCES

Bazjanac, V. & Kiviniemi, A. 2007. Reduction, simplification, translation and interpretation in the exchange of model data. In Proc. CIB W78, 24th Int. Conf. on Bringing ICT Knowledge to Work, Maribor, Slovenia, pp. 163–168.

Bazjanac, V. 2008. IFC BIM-Based Methodology for Semi-Automated Building Energy Performance Simulation. Lawrence Berkeley National Laboratory. Retrieved from: http://escholarship.org/uc/item/0 m8238pj.

BiMServer. 2014–16. BiMServer Documentation, Available at: http://bimserver.org/documentation/.

buildingSMART 2008–2016. IFC Overview, Available at: http://www.buildingsmart-tech.org/specifications.

DesignBuilder 2005. DesignBuilder—building simulation made easy, DesignBuilder Software Ltd., Available at: http://www.designbuilder.co.uk/

Dimitriou, V., Hassan, T.M. & Fouchal, F. 2016. BIM enabled building energy modelling: development and verification of a gbXML to idf conversion method. In Proc. BSO2016.

Dong, B., Lam, K.P., Huang, Y.C. & Dobbs, G.M. 2007. A comparative study of the IFC and gbXML informational infrastructures for data exchange in computational design support environments. In 10th Int. IBPSA Conf., pp. 1530–1537.

Eastman, C., Teichholz, P., Sacks, R. & Liston, K. 2011. BIM Handbook. Wiley & Sons.

EDMmodelServer 2016. EDMmodelServer for BIM and VDC using IFC, http://www.epmtech.jotne.com/solutions/

EnergyPlus. 2003–2016. EnergyPlus, US Department of Energy (US-DOE), Available at: https://energyplus.net.

Fernando, T., Bassanino, M., Tan, Z., FHR Team, Scherer, R., Kadolsky, M., Hassan, T., Fouchal, F., Firth, S., Wei, S., Nguyen, V.K. & Hasani K. 2015. Design4Energy Deliverable D7.2: 1st Phase -3D Online Gaming Environment for Learning How to Build Green Buildings, Available at: http://www.design4energy.eu/Download.html.

gbXML.org. 2014. gbXML—Software green building XML schema, Available at: http://www.gbxml.org/soft ware.php.

ISO 16739 (2013). Industry Foundation Classes (IFC) for data sharing in the construction and facility management industries. International Organization for Standardization, Geneva, Switzerland.

Katranuschkov, P., Scherer, R.J., Baumgärtel, K., Hoch, R., Laine, T., Protopsaltis, B., Gudnason, G., Balaras, C., Dolenc, M., Mansperger, T., Leskovsek, U., Semenov, V. & Christodoulou, S. 2015. ISES Final Report (Updated Version), EU FP7 Project 288819 ISES, © ISES Consortium, Brussels.

Miller, C., Thomas, D., Irigoyen, S.D., Hersberger, C., Nagy, Z., Rossi, D. & Schlueter, A. 2014. BIM-extracted EnergyPlus model calibration for retrofit analysis of a historically listed building in Switzerland. In Proc. Sim-Build.

Motawa, I. & Carter, K. 2013. Sustainable BIM-based evaluation of buildings. Procedia Social and Behavioral Sciences, 74, 419–428.

Pirnbaum, S. 2015. The XMLToFilterConverter Tool, Int. Report, Institute of Construction Informatics, TU Dresden, Germany.

Wülfing, A., Baumgärtel, K. & Windisch, R. 2012. BIMfit—A Modular Software Tool for Querying and Filtering of building Models (in German), 24th Conference 'Forum Bauinformatik', Bochum, Germany.

eWork and eBusiness in Architecture, Engineering and Construction – Christodoulou & Scherer (Eds)
© 2016 Taylor & Francis Group, ISBN 978-1-138-03280-4

Task-specific linking for generating an eeBIM model based on an ontology framework

M. Kadolsky & R.J. Scherer
Institute of Construction Informatics, Technische Universität Dresden, Dresden, Germany

ABSTRACT: In the last decades scarcity of resources and global warming have led to a more and more efficient building design and usage aimed to reduce energy consumption and CO_2 emission. For increasing energy efficiency of buildings over the whole life cycle analyzing and simulation tools became an important technology. For such engineering analyses the use of one domain model is mostly not enough. Often, additional information coming from other domains are required and have to be combined creating an overall information basis and providing the input information for the envisaged external simulation tools and their related simulation models. These link models are task specific and in the most cases the link models differ only slightly regarding the usage of different simulation software. So, embedding link models and relating quality checks in a context and making them available for a restricted group or even for unrestricted use could lead to more reliability.

In this paper an approach will be presented describing a generic framework for efficiently using of link models and quality checks based on a certain context definition. As description method for this framework an ontology approach is considered linking and consolidating the different input sources and creating the base for the quality checks.

1 PROBLEMS AND CHALLENGES

The focused energy performance analysis of buildings needs next to the BIM information describing the architecture of the building, information about the climate, the user behavior, etc. Next to harmonized information basis building energy performance analyses require a certain quality of the given domain models serving as input for the envisaged analyzing tools. The corresponding quality check of input models is mostly done by the analyzing tool itself. Cause of the reason, that simulations need to be configured before starting pre-checking of a simulation model can cost much time. These link models and pre-checks are task specific and in the most cases the link models and checks differ only slightly regarding the usage of different simulation software. So, embedding link models and checks in a context and making them available for a restricted group or even for unrestricted use could lead to more reliability.

In the whole integration process the linking and the pre-checking represent only two steps. The different steps starting with the preparation of a multi model appropriate for the envisaged simulation and ending with a quality check of the simulation results filtering out results with less quality can be listed as follows:

1. Multi Model Generation: In this step the different domain model instances will be linked. The linking can be done manually supported by template usage. The template provides different linking types for different domain models on different level of detail (Linking between classes, linking between object types, etc.). The envisaged ontology can store the templates and select the templates regarding the task and the models to be linked.
2. Multi Model Filtering: The multi model filtering follows directly the Exchange Requirements (ER) specifications. This means the task related ERs are stored as model views and queries will be applied on the previous generated multi model to get the partial model required to execute the task. The ontology can store the queries and select them for each task.
3. Multi Model ER Checking: Since in the multi model generation step link types would be offered to support the user by finding the correct linking or finding missing links in this step the mandatory links and the mandatory attributes will be checked based on the ER specifications. Here, it is not checked if the attribute values make sense, but only the existence of values is checked. The existence check can be realized as query or as ontology rule. The queries and rules are attached to the tasks.
4. Deriving extended semantic concepts: For rule defining and checking and for easy filtering it

is necessary to use the explicit given information for deriving new information existing previously only in implicit form. The new concepts could be new building elements, which usage is very specific and can be described in ontology rules. Furthermore, these new concepts defining new filter conditions the model could be filtered for.

5. Link quality control/ pre-condition Check: In the Multi Model Check step the mandatory links and the mandatory attributes were checked based on the ER specifications. Since this was just an existence check in this step the quality of the links and the attributes will be checked based on regulations and stored internal experience. So, it will be checked if the linked U-value range of coming from a material library fits to the related building element.
6. Partial model filtering: Since the Multi Model Filtering describes a filtering on the origin models, the partial model filtering operates on the semantically extended Multi Model. This means the temporary generated output model contains also the new derived concepts.
7. Process Input Transformation: In this step the certain calculation process will be invoked. This could be an external simulation triggered by the workflow manager or an internal simplified analysis method encoded in ontology rules also triggered by the workflow manager.
8. Process Output Transformation: The results will be written back after the calculation. In general this means simulation results will be returned and a new link model will be generated with the results linked to the building elements or spatial elements of the building model.
9. Quality Check: Here, the simulation results will be checked regarding their quality. Similarly to the ER check the returned As-Is values will be compared with the stored To-Be values representing the quality range.
10. Quality Check Results Transformation: The results of the comparison are generated and the workflow manger will be informed, so that the results could be displayed and visually evaluated.
11. Result Evaluation: Based on the underlying evaluation methods alternatives could be deleted (KO Selection) and only the accepted results will be stored and provided for the next step. A list of the best alternatives could be generated (List Selection). This list has a predefined certain length. A Mixed Selection is selected, if only a certain number of the accepted alternatives is stored and provided for the next step. Furthermore these alternatives are also ordered by their quality.
12. Evaluation Results Transformation: The results of the evaluation are generated and the workflow manger will be informed, so that the results could be displayed and visually evaluated.
13. Error Handling: It exists the possibility, that design alternatives not fulfilled the result check could be correc\ted. This could be done semi-automatically by ontology based correction methods or in an additional iterative step.

In this paper the focus will be on step 1 Multi Model generation and step 5 Link quality control/ pre-condition Check.

2 LINKING TABLES AND MATCHING TABLES

In this paper a link describes the relation between different models each represented in a different format. This means no internal relationships will be captured by the following concepts, but only external relationships. For capturing the external relationships an external model is preferred in contrast to storing the links internally in an existing model. The advantage is, that an external link model can specify links between different models format independent especially on schema level. This is for the envisaged task specific approach and the aim of increasing the reliability of linking an important point.

2.1 *Linking on instances level using linking tables*

The simplest way to describe a link between two elements of two different models is to merge their IDs, so that a new unique ID is created representing the link. Pre-condition for this is, that each element, which should be linked got an unique ID. The link itself represents an element opens up the possibility to realize n-ary relationships.

Based on the idea making linking practicable and easy to manage so called linking tables are introduced representing the linking between the elements of two models in table form.

The concepts comprised by this table are:

- Link Model
  - ID
  - Process/Task
- Domain Models
  - ID
  - Process/Task
  - Format
  - LoD
- Domain Model Elements
  - ID
  - Instantiated Class/Type

Model *Domain Format*	Class/ Type	ID	Link *ID Type*	ID	Class/ Type	Model *Domain Format*
BIM *IFC-Step*		fhzru34	95vbnf	dfrsg5		Costs *GAEB XML*
BIM *IFC-Step*	Wall	sdgtu4sdt	36trsd	sd354szt	Wall Costs	Costs *GAEB XML*
BIM *IFC-Step*	Steel Column	asderf4345	56tlkg	jfhdtu490t	Welding Costs	Costs *GAEB XML*

Figure 1. Example of a linking table.

Model *Domain Format*	Class/ Type	Link *Type*	Class/ Type	Model *Domain Format*
BIM *IFC-Step*	Wall	Inner Domain Link	Wall Costs	Costs *GAEB XML*
BIM *IFC-Step*	Steel Column	Inner Domain Link	Welding Costs	Costs *GAEB XML*

Figure 2. Example of a matching table.

Model *Domain Format*	Class/ Type	ID	Link *ID Type*	ID	Class/ Type	Model *Domain Format*
BIM *IFC-Step*	Wall	sdgtu4sdt	36trsd Inner Domain Link	sd354szt	Wall Costs	Costs *GAEB XML*
BIM *IFC-Step*	Steel Column	asderf4345	56tlkg Inner Domain Link	sdgtu4sdt	Formwork Costs	Costs *GAEB XML*

Figure 3. Wrong linking within a linking table.

- Links
  - ID
  - Link Type

The table form allows three kinds of linking:

- Linking of two models: In this case only the first and the last column and the link column containing values. If elements of two models are linked this kind of linking is automatically derived.
- Linking between elements and models: This is the case for example if building elements are not directly related to a certain material, but to the catalogue the material is specified. Here, the element side is completely filled and on the model side only the model column contains values.
- Linking between elements: Linking between elements requires all columns of the table filled with values.

The specification of the instanced classes allows the search for specific relationships and views like the search for all walls related to costs. Next to this it allows to describe linking on a higher level required for supporting reliability.

2.2 *Linking on schema level using matching tables*

Matching tables describe the linking between models on schema level and therefore format independent. From the technical point of view matching tables are linking tables without the instances columns. So, the meaning of a row is that a class of the one model matches to a class of the other model. Thereby, both classes could be connected by a certain linking type.

Similarly to a lexicon this stored information can be used to check an existing linking table regarding correctness. This will be done by checking if the instantiated classes in the linking table also exist in the matching table applied for the check.

The matching tables can be also used during the linking and generating a linking table. In this case all possible candidates for an element of a model will be listed based on the used matching table.

Cause of the fact that links are also elements and can be also linked in separate linking table with other elements, the type of the error by a mismatch can be additionally linked.

2.3 *Task-specific linking*

Linking is a task specific issue. So, for an office building the energy requirements and the related material could be fulfilled, but for a hospital the quality for the related material is not enough. Furthermore, for a specific task not all possible checks should be applied and in this sense only an appropriate matching table should be provided.

Link models can be described by an explicit model or by a functional model. Using an explicit model the links are already created before the runtime. However, using a functional approach, the links will be generated during the runtime. On the one hand it doesn't make sense to store an explicit link model for each task, on the other hand it is not efficient to generate each link model reiteratively during runtime. The envisaged approach is based on an ontology framework, which opens up the possibility to (de)select links and reducing the link model as well as extending the link model by using deriving rules. In this sense the ontology approach follows the idea of identifying similar tasks, which could be grouped and whose link models can be developed by selection and extension based on an

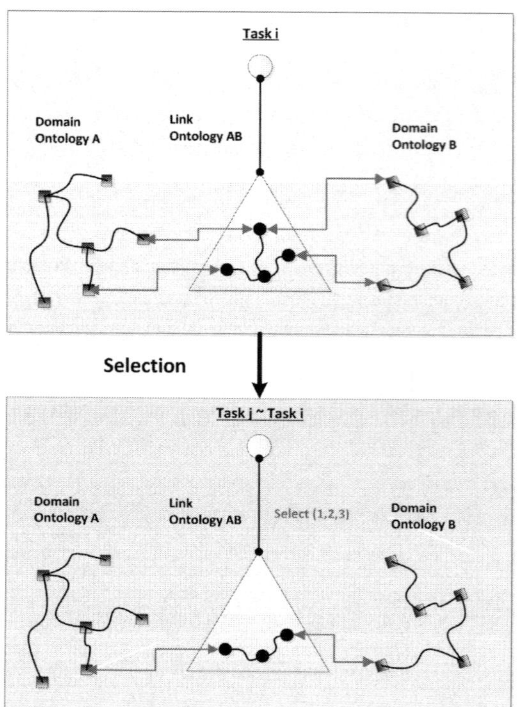

Figure 4. Task-specific selection of link concepts.

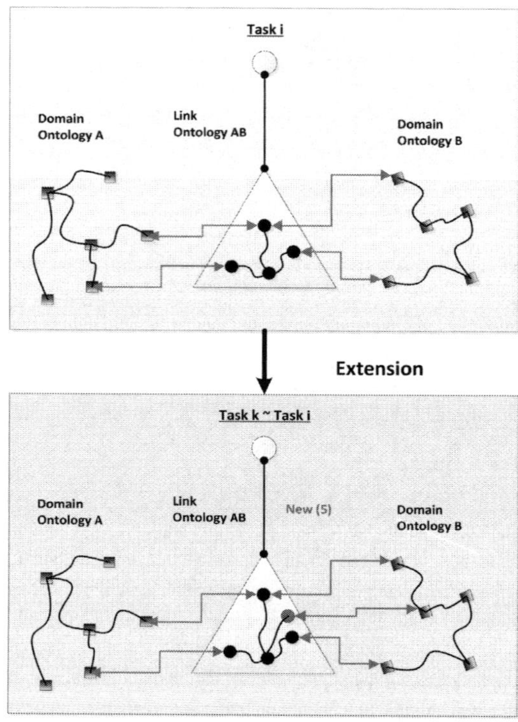

Figure 5. Task-specific extension of link concepts.

explicit link model as starting point. So, the implementation result for such task groups is a core link model explicitly represented and selection queries and deriving rules attached to this core model and related to the corresponding tasks.

3 ADVANCED CHECKING OF LINKS

The previous description of the linking tables and matching tables shows a method for a simplified check of correctness. This check is a simple syntax check, which could be done in a spreadsheet application. For a more complex check Link Types will be introduced. Here, a very generic approach will be presented, which can be easily extended.

3.1 *Link types*

In this paper the generic approach of link types is based on idea, that a model can be adapted, detailed, be linked with another model and changed based on a new version:

- Reduction Link Type
  - Abstraction Link Type
  - Selection Link Type

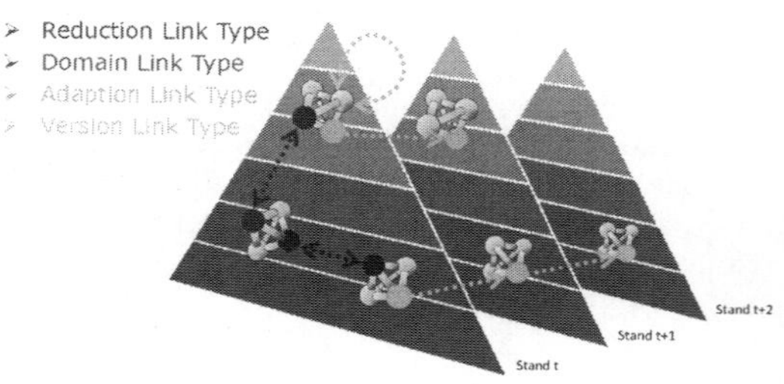

Figure 6. Link type concepts.

- Domain Link Type
  - Outer Domain Link Type
  - Inner Domain Link Type
- Adaption Link Type
  - Correction Link Type
  - Variation Link Type
  - Version Link Type

First, this definition of types gives the user of the linking tables new information. So, an inner link type indicates overlapping section and a simulator could react in an appropriate way. He could delete overlapping section or create variants based

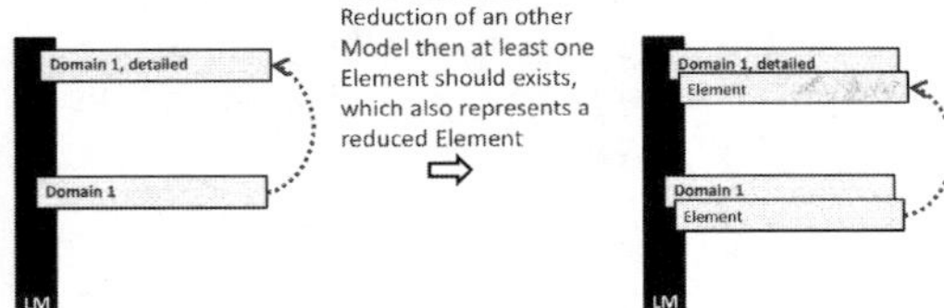

Figure 7. Example for a transitivity check.

on these. Next, the link types open up new possibilities for more complex checking methods.

3.2 *Transitivity checks*

Since the previous link checks considered only one relation by checking two classes the link checks based on link types aims to check link chains and transitivity of links.

4 SPECIALIZATION OF CLASSES

Matching tables and the checking of corresponding classes are more meaningful, if the classes are more specialized. A class can only be specialized, if there is information at least indirectly existing. The specialization can then be done automatically by triggering appropriate deriving methods or manually.

4.1 *Classification of Matching Tables*

The idea of matching tables is to use them not once, but to use them several times. However, the use of these tables depends on the identification of the classes specified in the matching tables and their counterpart in the linking tables. This requires that specialized classes could be derived from the existing model. Matching Tables (MT) allows more checks and more not even obvious checks by a more specialized Class/Type specification. If a detailed class specification is not in the origin schema deriving methods could provide such a detailing. So, depending on the underlying Model Schemas different kinds of MTs could be defined:

- **First Order MTs** are MTs including **no class/type**, which could not directly mapped from the origin schema to the MT.
- **Second Order MTs** are MTs including at least **one class/type**, which could not directly mapped from the origin schema to the MT, but requires a deriving method for specializing a given class using **one additional information** like an attribute (e.g. Material: Steel).
- **Third Order MTs** … using **two additional information** … (e.g. Material: Steel + Structural Properties: Loadbearing -> Loadbearing Steel Column).

4.2 *Specialization of classes*

The specialization of classes and their representation in the table aims to serialize a complex structure comprising different hierarchies to a simple string, so that the comparison between linking tables and matching tables is reduced to a string comparison. Deriving methods are hidden and the complexity of the linking problem divided into several steps. The disadvantage is that hierarchy in the table is flat and inheritance structures are not considered, so that additional work could arise.

5 UNDERLAYING ONTOLOGY CONCEPT

The linking tables and matching tables show a practicable approach for linking and checking. As mentioned before using a simple spreadsheet application for the complete concept of these tables would lead to limitations. In this sense the tables should be more seen as application layer based on an implementation layer, which can realize the advanced checking and the specialization of classes. In this paper an ontology framework represents the implementation layer.

The ontology framework forms the core approach for realizing the interoperability between the different domain models and external data. Thereby, this approach comprises two level of data integration. On the lower level only the linking information between the different models is harmonized and represented in a uniform semantic, the origin models remain as they are. This approach is very simplified and mostly made for data exchange, this means for linking tables containing no specialized classes. For a controlled data exchange the semantic information of the origin models is mapped into ontology format offering the possibility for deriving information, additional filtering methods, rule based calculation methods and link checks.

5.1 *BIMOnto*

The BIM- ontology (BIMOnto) forms the center model of the ontology framework. Following the core idea of ontology use as a method for semantic integration geometrical information as not meaningful information and unsuitable for pre-checking were not mapped, but referenced to the ontology.

The integration step comprises the specification of new BIM-elements extending the existing ele-

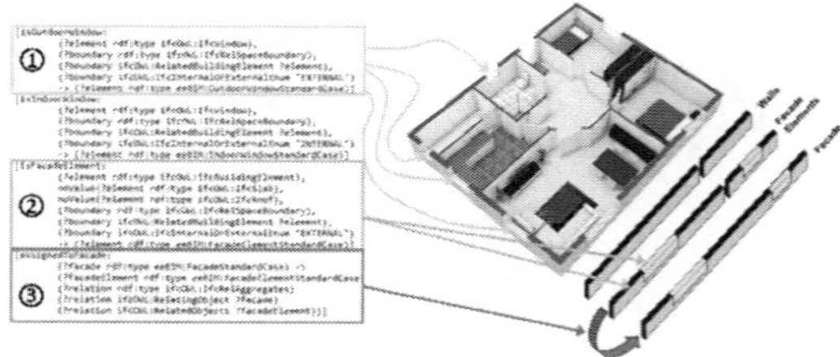

Figure 8. Rule definition for generating eBIM.

ments to serve as connection points for the additional non-BIM information. In order to come to energy interpretable BIM we defined a step-wise data extension so that energy-related entities are explicitly given and forming the connection points to non-BIM data. For doing so, we check the LoD the information is structured first. As an absolute requirement we suppose that space boundaries are given where associated building elements can be considered geometrically as external or internal. With such information we can consider if e.g. a wall is an outdoor or indoor wall etc. Figure 8 presents an excerpt of our rule set for bringing the BIM entities to an extended BIM (eBIM) representation which is more type-safe and tool interpretable. Rule number one checks if at least one space boundary is described as "EXTERNAL" and if this is the case the whole window is inferred as outdoor window. Rule number two checks if a building element is a façade element by analysing the space boundaries in the same manner as in the orange rule. Rule number three expresses that all façade elements are part of the building façade.

In another rule set we make a similar analysis where IfcSpace entities are evaluated. For example, based on the name of the room and the definition given by a room book an IfcSpace is assigned to be an office room (defined as OWL class "WorkRoom") or a technical room ("TechnicalRoom") which both extends IfcSpace. With such an association heated rooms (e.g. office, living area etc.) can be isolated from unheated rooms (e.g. cable funnel, elevator etc.) in a next step.

After the extended elements were generated, a check is performed if all individuals were inferred correctly. When rule sets which bring the BIM to an eBIM cannot be applied (and validated), it indicates that there is a problem regarding the possibility to perform an energy analysis. The checking against constraints and reasoning with rules are big advantages when working with BIM. While there are many existing specifications how a well-modelled building should be provided (e.g. through LoD), it is not explicitly defined in the model schema and therefore not mandatory. The way of expressing model requirements through rules leads to a consistent workflow later and saves time in the case, that an energy simulation fails or produces wrong results.

5.2 *Interface ontology*

As mentioned before, the focused energy performance analysis of buildings needs next to the BIM information describing the architecture of the building, information about the climate, the user behaviour, etc. To combine these different domain information and to come up to an energy enhanced BIM (eeBIM) two core strategies can be distinguished: Centralized information integration and decentralized information integration. By using the centralized approach only one domain model representing the centre model is related to the other domain models. The decentralized approach is characterized by domain models, which can be combined arbitrarily with each other. Caused by the fact, that for building design issues the building and the corresponding BIM model is focused, the ontology approach follows the centralized integration with the BIM model as centre model. BIM model implementations like the IFC aim to provide general concepts to cover common building design scenarios, but for specific engineer tasks additional BIM information are required to extend the architecture model. This extended BIM (eBIM) (Kadolsky, Baumgärtel, & Scherer, An Ontology Framework for Rule-Based Inspection of eeBIM-Systems, 2014) has to offer the connection points needed to link the information of the non-BIM domain models and to complete the overall energy enhanced BIM model (eeBIM).

In the ontology approach the BIM model is represented by the BIM ontology, the extension of the BIM model is represented by the Interface ontology and for realizing the eeBIM model the non-BIM ontologies have to be linked (Figure 9). To transform the raw data into ontology information they have to be mapped and linked:

1. 1 to 1 mapping just transforms the data format into ontology format. In the most cases it is a XML to OWL or IFC to OWL transformation.
2. The reduced mapping does not map the complete data into the ontology. To achieve a lean ontology model not all data are integrated (e.g. geometry data are not considered). Instead of the whole geometry model only derived abstract data are included (e.g. the width or height of a room instead of the whole room geometry).
3. N to 1 mapping combines several data to get one ontology element. Vice versa 1 to N mapping splits one data element into several ontology elements.

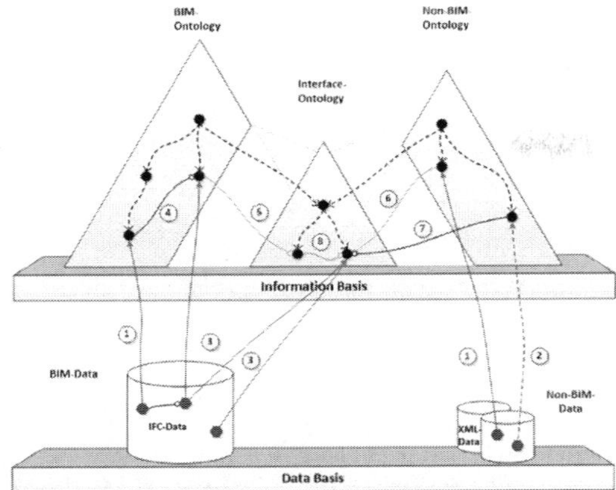

Figure 9. Interface ontology concept.

4. Direct linking is done in the course of the element mapping operations or manually, if corresponding elements cannot be identified by algorithms.
5. Implicit linking is based on ontology rules used to interpret implicit information for generating explicit links. This is mainly done for linking the BIM- with the Interface-ontology.
6. External linking describes the linking between an ontology and a data element (mapping ops/ manually).

In summary, the ontology allows the integration of data on three abstraction levels. On the lowest level the external data are only linked by a certain URL. The content of the linked data file is unknown. This corresponds to Element-Model Link in the linking tables. On the next level additionally to the URL link-address Meta data are integrated. This corresponds to Element-Element Link in the linking tables, in which one model was reduced to a Meta model. On the highest level the external data are completely integrated. This corresponds to the Element-Element Link in the linking tables. The integration of the data is next to its representation in different abstraction levels dependent on the data set, which is selected to be integrated in the ontology.

6 MANAGEMENT OF LINKING OPERATIONS

The idea of the linking tables, matching tables and the ontology is to increase the reliability. To allow a widespread use of them an appropriate open platform is required.

6.1 *General concept of a document store*

For a widespread use the linking, checking and integration operations should be presented in a way, that it could be applied by common tools. Based on this the operations are not represented in a programming language, but in a specific language. So, mapping rules are represented in Xquery, linking and checking operations in SPARQL or SWRL and underlying model concepts in OWL. In this sense the open platform is planned as eDocument Store.

6.2 *Specification of a document framework*

At the top of all documents is the specification of the tasks. Connected to this task are the different steps. Thereby, different steps could be combined to one. For reaching a complete integration each specification has to be embedded in a frame containing the connection points to the previous and following step.

6.3 *Realization of the document store*

The realization of such an eDocument Store could be done by using the ResearchGate sever. It allows uploading documents and generating unique DOIs (Digital Object Identifier). The DOIs can then be used for establishing the connection points.

7 CONCLUSION

In this paper an ontology approach was presented forming the base for a semantic integration of different domain models for supporting the design of an energy efficient building. Based on this ontology approach an application layer was set providing a practicable specification of links on instances level and schema level. Here, the advantage of ontology-based description was used to consolidate the heterogeneous resources and to formulate linking rules with regard to the envisaged energy performance analysis. So, on the one side the ontology provides the description of the domain models by specifying certain concepts and on the other side the underlying mathematical basis enables the definition of logical rules and in this way the interpretation of the made specifications and their interoperable use. Cause of the reason, that simulations costs much time and before simulations can be done the configuration can be very huge and the pre-processing erroneous, the ontology approach aims to pre-check the given input data by defining and applying appropriate logical rules. This kind of inspection allows the identification of modeling mistakes in a very early stage of the design process by using only common semantic tools. The ontology and the corresponding ontology platform were developed and will be further developed in the European eeEmbedded project.

REFERENCES

[1] Baumgärtel, K. 2013. Multimodellintegration in einem virtuellen Labor. Forum Bauinformatik, 25, S. 25–39.

[2] Baumgärtel, K., Katranuschkov, P. und Scherer, R.J. 2012. Design and Software Architecture of a Cloud-based Virtual Energy Laboratory for Energy-Efficient Design and Life Cycle Simulation. In: European Conference on Product and Process Modelling (ECPPM).

[3] Bazjanac, V. 2010. Space boundary requirements for modeling of building geometry for energy and other performance simulation. In: 27th International Council for Research and Inovation in Building Construction (CIB W78).

[4] Beetz, J., van Leeuwen, J. und de Vries, B. 2008. IfcOWL: A case of transforming EXPRESS schemas into ontologies. In: Artificial Intelligence for Engineering Design, Analysis and Manufacturing.

[5] Curry, E., et al. 2012. Linking building data in the cloud: Integrating cross-domain building data using linked data. In: Advanced Engineering Informatics 27 (2), S. 206–219.

[6] Fuchs, S., Katranuschkov, P. und & Scherer, R.J. 2010. A Framework for Multi-Model Collaboration and Visualisation. In: 8th European Conference on Product and Process Modelling (ECPPM).

[7] Hietanen, J. 2006. IFC Model View Definition Format. International Alliance for Interoperability.

[8] Kadolsky, M., et al. 2014. ISES Deliverable D3.1: Ontology specification, ISES Consortium.

[9] Windisch, R., Katranuschkov, P. und Scherer, R. 2012. A Generic Filter Framework for Consistent Generation of BIM-based Model Views. In: 2012 Eg-Ice Workshop.

eWork and eBusiness in Architecture, Engineering and Construction – Christodoulou & Scherer (Eds)
 ISBN 978-1-138-03280-4

Visual support for multi-criteria decision making

T. Laine, F. Forns-Samso & V. Kukkonen
Granlund Oy, Helsinki, Finland

ABSTRACT: Successful building project needs sufficient feedback for the decision making. Currently only a few alternative solutions are analyzed and decision making is not able to support multi-discipline collaboration or multi-criteria view. By introducing the new Key Point methodology and visual KPA tool for decision making the different sustainability aspects can be effectively analyzed and optimized in a totally new way.

1 INTRODUCTION

Successful building project needs sufficient feedback for the decision making. To find optimal solution(s) a large number of alternatives should be produced and analyzed from many different eeKPI (energy enhanced key performance indicator) point of view. Currently only a few alternative solutions are analyzed in building or district design and decision making is not able to support multi-discipline collaboration or multi-criteria view. The use of multi-criteria KPIs and decision making have shown positive results in previous studies (Saniuk and Jakabova, 2012). The use of KPIs allowed the re-planning of objectives and the decision-making process to be improved.

2 BACKGROUND

Decision making is a collection of different activities that include gathering, interpreting, and exchanging information; creating and identifying alternative courses of action; choosing among alternatives by integrating the often differing perspectives and opinions of team members; and implementing a choice and monitoring its consequences (Guzzo, 1995). We acknowledge the multi-disciplinary nature of a construction project; it employs several experts performing different types of simulation analysis and creating a vast number of results. Therefore, a general case of decision-making can be expressed as an end decision that is the product of a whole series of domain decisions (i.e. energy, thermal comfort, lifecycle cost, and CO_2 emissions) and the interaction of many interested parties.

Decision making is not the only activity of relevance to teams, but it is a key activity contributing to team effectiveness (Guzzo, 1995). Decision making in a team/group is quite distinct from individual decision making and groups are more effective and offer more advantages than individuals (Rosenfeld and Wilson, 1999). There have been a number of techniques introduced to make groups work together although all have had limited success. In construction, no project is free from group decision making and the decision often needs close collaborative working methods. Usually, every discipline in a construction project works independently of another but inevitable makes decision that affects others. Hence, there is an increase need for effective collaborative decision making between all the key participants in any construction project. Even though, in practice it is not common that construction teams use formal collaborative approaches, latest trends indicate the opposite. For instance, the use of new project delivery methods such as Integrated Project Delivery (IPD) or Alliance contracting along with the use of Building Information Modeling (BIM) are main drivers for enhanced collaboration between stakeholders in recent years. However, many times these collaborative processes do not employ a structured way of collaborating. In this context we intend to cover that gap by using methodologies such as the Key Point methodology, which allows teams and experts from different disciplines to work in such collaborative environment.

The methodology is supported by the use of software applications. The Key Point Analysis (KPA) tool is a software component within the Key Point Framework responsible to provide appropriate cross-domain visualization and decision support for the computed Key Performance Indicators (KPIs) and Decision Values (DVs). The next section explains further the Key Point methodology and KPA tool.

3 METHODS

This paper studies the involvement of many different disciplines during the design phase of a construction project supported by the use of building information modeling. BIM tools are capable of simulating and predicting different parameters that can be utilized to support collaborative way of decision making. In such way the use of BIM forms the basis for the development of the Key Point methodology and KPA tool.

3.1 *Key Point Framework*

The Key Point Framework is defined as a set of standardized concepts, procedures and software solutions whose relations are defined and which constitute the fundamental structure and basis of a work environment to support end-users to evaluate alternatives and select the best one in a cost-effective way.

The overall concept of the Key Point Methodology consists of translating the requirements into verifiable Key Points in order to evaluate the design alternatives and discard them progressively based firstly on Key Design Parameters (KDPs) in the design domains, secondly on Key Performance Indicators (KPIs) in simulation/calculation domains and thirdly on Decision Values (DVs) in Decision Making Domain.

The Key Points are energy related verifiable design check points, which are providing domain related requirements in form of target values, which can be checked after common design steps. The Key Points are hierarchically categorized into:

- Decision Value (DV): Represents the preferences of the decision makers related to the project goals. This allows prioritizing KPIs by means of a weighting factor.
- Key Performance Indicators (KPIs): Numeric metrics of energy usage, cost and building performance. They are influenced by Key Design Parameters and are additionally the basis for evaluation via Decision Values.
- Key Design Parameters (KDPs): Represent the mandatory building properties and usually have a limited or a range value.

The Key Point Methodology is an integrated holistic design system to guide through the numerous design alternatives and choose the best ones for further evaluation. Therefore, the aim of this methodology is to keep and improve those alternatives which are potentially optimal solutions and discard those alternatives which are far from being potentially optimal as soon as possible in order to save effort time and computational resources.

The explained overall concept can be split into 3 sub-concepts illustrated in the Figure 1: Requirements decomposition, (2) Domain Task and (3) Results aggregation.

- *Requirements decomposition* refers to the translation of key requirements determined by the client, regulations, site, etc. into the Key Points starting from the Decision Values, then the Key Performance Indicators and finally the Key Design Parameters. The outputs will be the target values of the Key Points: DVs TO-BE, KPIs TO-BE, KDPs TO-BE.
- *Domain task* is the development of the design process tasks specified in D1.2 (Geißler et al, 2014b). The outputs will be the AS-IS values of the Key Points: DVs AS-IS, KPIs AS-IS, KDPs AS-IS.

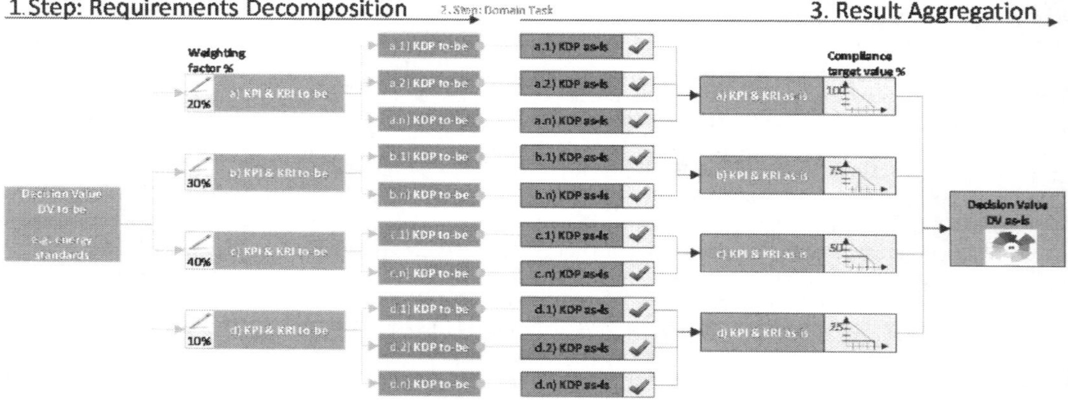

Figure 1. Key Point methodology decomposition.

- *Result aggregation* is the comparison of Key Points AS-IS against Key Points TO-BE which will be used as basis to rank, select and discard design alternatives. It will start comparing KDPs, then KPIs and finally DVs.

3.2 *Key Point Analysis tool*

In addition to the Key Point Methodology, this paper presents the Key Point Analysis (KPA) tool, which enables visual support for multi-criteria based decision making by combining and weighting of simulation results related to energy, emissions, indoor climate and cost. The KPA tool is an extension of what has been previously developed in the EU Project ISES (https://ises.eu-project.info). The new features of the KPA tool include the Decision Value analysis, which will form the main focus of this paper.

The development of the KPA tool is intended to support multidisciplinary work and enhance collaboration between project teams. Also, it should support the experts and the decision maker in analyzing the impacts of key performance indicators. In addition, it is a central component of the Key Point Framework described in the previous section.

The KPA tool helps teams to work in a more structured way, but enabling flexibility and interaction in the analysis of alternatives facilitating the decision making process. The KPA should function as a critical tool for the rapid comparison of different evaluation criteria. Therefore, KPA tool should help the complex decision making process, assist in the evaluation of alternative options or scenarios, deal with complexity and have a clear, reproducible procedure.

The KPA tool is developed as a web-based application. It uses a graphical multi-attribute utility analysis to evaluate and compare alternatives based on key performance indicators. In addition, it contains a decision value analysis, which represents the preferences of the decision makers related to the project goals. This allows prioritizing KPIs by means of a weighting factor.

3.3 *Functional requirements for the-KPA tool*

The KPA tool is developed to support multidisciplinary decision making and enhance collaboration with project teams. The functional requirements are developed based on research on multidisciplinary approaches, the Key Point Framework, end-users input and our professional industry experience. The main requirements are listed below.

Communication of the decision process based on a Decision Value (DV): Decision makers must be able to understand and visualize their preferences in the decision making process. This is important for prioritizing the more important parameters for the decision makers as well as to establish clear objectives for the project. A decision value as way to translate the quality requirements and summarize the decision making process with a single determining value.

Visualization of simulated results and interaction of professional expertise: Applying advanced graphical tools that enable multi-dimensional data visualization such as pareto graphics, hyper radial visualization, parallel coordinate plots and radar charts have proven useful for understanding general performance trends as well as variable sensitivities in order to make informed decisions in guiding the optimization process. The design objectives and requirements often change during the design process based on analysis and interpretations of results. The interaction of professional expertise and computer-based exploration therefore is essential for the process to be successful.

Generation and analysis of design alternatives: The literature has proven that successful designs are achieved by investigating a large set of design alternatives. (Akin 2002). The generation of design alternatives in the KPA tool are explained in more detail in Laine et al., (2014). Once an alternative is generated practitioners should be able to assess the performance of the alternative across a wide range of criteria or also known as KPIs in order to make an informed decision.

Integrate different simulation results: KPA tool should have the capability to be interoperable and integrate different simulation results from different software engines. This would truly allow multi— criteria view and support multi-disciplinary decision making. Simulation results can be related for instance to energy efficiency, life cycle cost and emissions.

Customize optimization strategies: Optimization strategies are based on specific requirements of the design problem. The *user* should be able to select a*n* optimization strategy including the targets, number of variables, maximizing, minimizing or simply complying *to* design requirements.

4 RESULTS

4.1 *Decision Value*

The Decision Value is a representation of the preferences of the decision makers related to the project objectives. This allows prioritizing KPIs by means of weighting factors. The next section explains the decision value calculation and the optimization strategies implemented.

4.1.1 *Decision value calculation*

Decision Values (DVs) are used for a quick comparison of different design solutions. To facilitate intuitive comparisons, each design solution is rated with a single number from 0 to 100, calculated as a weighted sum of decision values assigned to each KPI of a design solution. The calculation of a decision value for a solution follows the following four-step process:

1. Configure decision value calculation parameters. This includes grouping of KPIs, setting target ranges and optimization methods for each KPI, and determining weights for each KPI group and individual KPI.
2. Calculate the decision value for each KPI in a given design solution by using target ranges and optimization strategy.
3. Calculate the decision value for each KPI group as a weighted sum of the decision values in that group.
4. Calculate the final decision value as a weighted sum of the different KPI groups' decision values.

To calculate a decision value for a single KPI in a design solution, two parameters are required: target range, which roughly translates to accepted range of variance, and an optimization method, which defines how to use the target range to determine a score of 0 to 100. Three types of optimization methods are defined: "minimize", which interpolates values inversely, yielding a decision value of 100 on the lower bound of the target range and below, and conversely 0 on the upper bound and above; "maximize", which is the exact opposite of minimize; "fit", which yields a decision value of 100 to any KPI within the target range.

Figure 2 shows the current user interface for configuring parameters; effectively, this is where step 1 of the four steps above is done by the user. As the parameters are updated, the decision values are recalculated when necessary.

Figure 3 shows a view of the decision values; a chart for highlighting design solutions, and a breakdown of the decision value for a highlighted solution. Hovering over a slice of the aster plot will show names and values of KPIs and DVs.

4.1.2 *Optimization strategies*

As explained in the functional requirements. The selection of the appropriate optimization strategies depend upon the formulation of the optimiza-

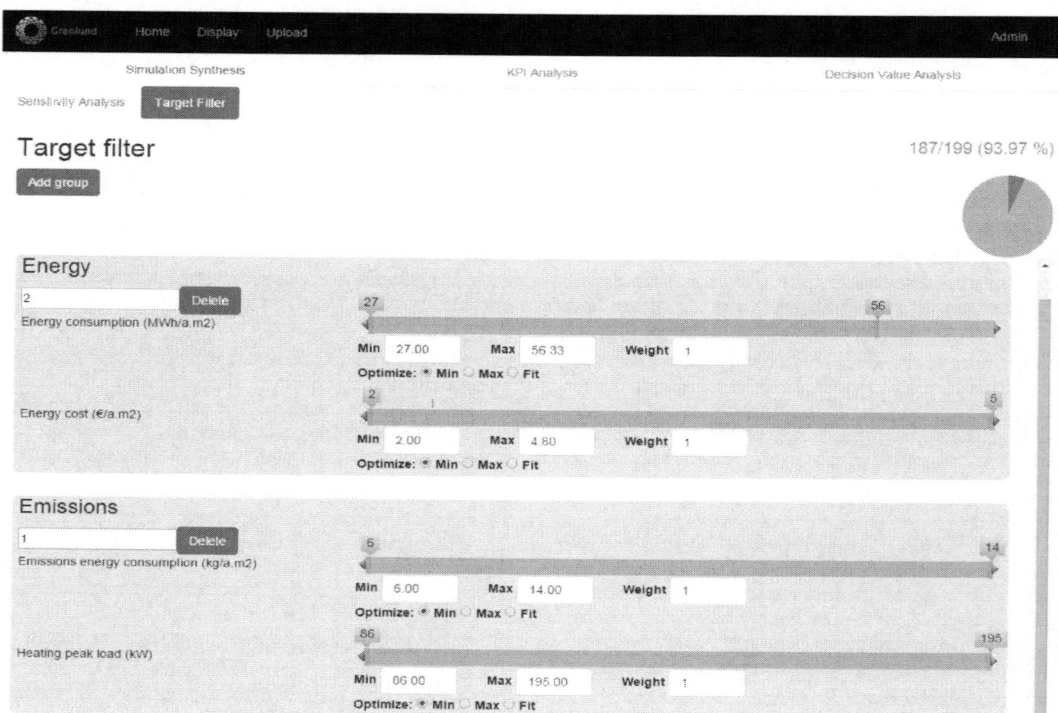

Figure 2. User interface for setting decision value parameters.

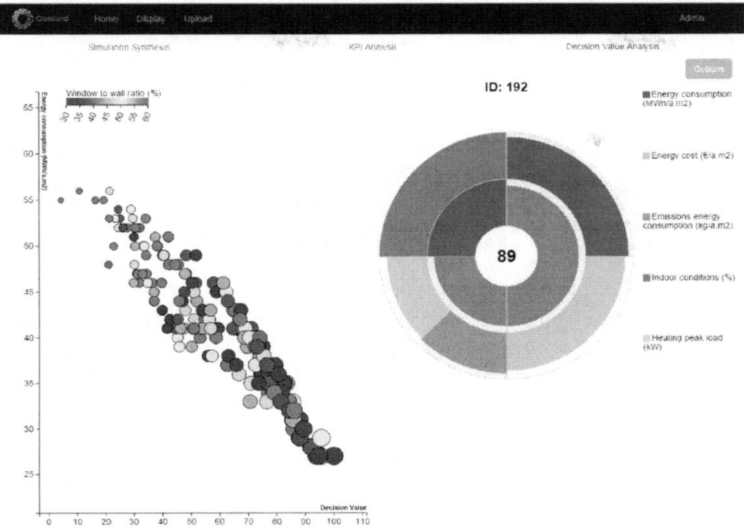

Figure 3. Decision value analysis view.

tion problem including the targets, constraints and variables. In this context, in our development practitioners can customize the optimization strategies based on the particular design problem.

Our current implementation uses a target range instead of a single target value. By default, if no other input has been given, the target range is the minimum and maximum of the given data. Three strategies are implemented for calculating the decision value: minimization, maximization and fitting. These are explained below.

KPI_{MIN}: Minimum value of the target range
KPI_{MAX}: Maximum value of the target range

Minimization aims to assign greater decision values to lesser KPI values. All KPI values less than the minimum value of the range are assigned the decision value of 1, and all values above the maximum are given a value of 1.

$$KPI_{NORMALIZED} = \begin{cases} 1, \textit{if } KPI_{AS\_IS} \leq KPI_{MIN} \\ \dfrac{KPI_{MAX} - KPI_{AS-IS}}{KPI_{MAX} - KPI_{MIN}}, \textit{if } KPI_{MIN} < KPI_{AS\_IS} < KPI_{MAX} \\ 1, \textit{if } KPI_{AS\_IS} \geq KPI_{MAX} \end{cases}$$

Maximization aims to assign greater decision values to greater KPI values. All KPI values less than the minimum value of the range are assigned the decision value of 0, and all values above the maximum are given a value of 1.

$$KPI_{NORMALIZED} = \begin{cases} 0, \textit{if } KPI_{AS\_IS} \leq KPI_{MIN} \\ \dfrac{KPI_{AS\_IS} - KPI_{MIN}}{KPI_{MAX} - KPI_{MIN}}, \textit{if } KPI_{MIN} < KPI_{AS\_IS} < KPI_{MAX} \\ 1, \textit{if } KPI_{AS\_IS} \geq KPI_{MAX} \end{cases}$$

Fitting assigns the maximum decision value to all KPI values within the range, and 0 for all other values.

$$KPI_{NORMALIZED} = \begin{cases} 0, \textit{if } KPI_{AS\_IS} < KPI_{MIN} \\ 1, \textit{if } KPI_{MIN} \leq KPI_{AS\_IS} \leq KPI_{MAX} \\ 0, \textit{if } KPI_{AS\_IS} > KPI_{MAX} \end{cases}$$

5 CONCLUSION

The developed KPA tool utilizes the Key Point methodology. Key Points are energy related verifiable design check points, which are providing domain related requirements in form of target values, which can be checked after common design steps.

These Key Points will also allow planners to easily structure the design process in individual evaluable parts and will thus help them to concentrate on high-level strategic decision making tasks. The eeE Key Point driven design process is expected to lead to greater efficiency in the planning procedure to get final design results of higher quality. At the same time, it will provide an opportunity of weighting up many more alternatives than currently possible.

By utilizing visual KPA tool for decision making in design the different sustainability aspects can be effectively analyzed and optimized in a totally new way. This supports also collaboration in projects, especially with new contract models such as Alliance or IPD (Integrated Project Delivery).

ACKNOWLEDGEMENTS

The work was done in the eeEmbedded project by FP7 European Union Funding for Research & Innovation.

REFERENCES

Akin Ö. (2002). Variants in Design Cognition. Design Knowing and Learning Cognition in Design Education. C. Eastman, M. McCracken and W. Newstetter. Amsterdam, Elsevier: 105–124.

Geißler, M.C., Guruz, R., van Woudenberg, W. (2014); eeEmbedded Deliverable D1.1: Vision and requirements for a KPI-based holistic multi-disciplinary design, © eeEmbedded Consortium, Brussels.

Guzzo, R.A., & Salas, E. (1995). Team effectiveness and decision making in organizations (Vol. 22). Pfeiffer.

Laine T., Forns-Samso, F., Katranuschkov P.„ Hoch R.„ and Freudenberg P., Application of multi-step simulation and multi-eKPI sensitivity analysis in building energy design optimization. eWork and eBusiness in Architecture, Engineering and Construction. Aug 2014, 799–804.

Katranuschkov, P., Scherer, R.J., Baumgärtel, K., Hoch, R., Laine, T., Protopsaltis, B.,... & Leskovsek, U. (2015). ISES Final Report (Updated Version),© ISES Consortium.

Rosenfeld, R.H., & Wilson, D.C. (1999). Managing Organizations: Text, Readings, and Cases. McGraw-Hill.

Saniuk, A., & Jakábová, M. (2012). Key Performance Indicators for Supporting Decision-Making Process in Make-To-Order Manufacturing. *Research Papers Faculty of Materials Science and Technology Slovak University of Technology in Trnava*, *20*, 182–190.

eWork and eBusiness in Architecture, Engineering and Construction – Christodoulou & Scherer (Eds)
© 2016 Taylor & Francis Group, ISBN 978-1-138-03280-4

An IT-based holistic methodology for analyzing and managing building life cycle risk

H. Pruvost, T. Grille & R.J. Scherer
Institute of Construction Informatics, Technische Universität Dresden, Dresden, Germany

ABSTRACT: By nature every building project is affected by uncertainties that can result at some point in the future in building performance deviations. In order to avoid or at least minimize that unwanted effect, uncertainty should be taken into consideration as early as possible, and in particular since the design phase of a building. With the aim of analyzing and keeping a certain control over uncertainty several techniques and standards have emerged in the field of risk management. Unfortunately the AEC industry still makes poor use of such techniques and risk analysis is still often done manually and sporadically. Moreover, often only one specific category of uncertainty is analyzed, namely cost risk. Nevertheless, even if cost is one of the most important decision criterions, risk can have effect on all building performance aspects. In order to close those gaps this research proposes an approach that relies on IT-methods for systematizing as well as automating as much as possible the analysis and management of risk in the building life cycle. The overall aim is then to provide building project stakeholders with a holistic view of risk as basis for decision making.

1 INTRODUCTION

Nowadays risk has become a common and quite important concept in engineering but it is still difficult to define, model and assess it in a proper manner. In general risk is associated with negative events and related consequences. However there exist different definitions of risk that depend on the domain in which it is considered. For example, risk can be defined as a possible error in a process, a failure in a system, a probable financial lost or even a catastrophe. As stated by (Smith, Merna, & Jobling, 2006) all those definitions can be referred to as risk for the domain they are applied in. With the aim of harmonizing the existing different views of risk, the international standard (ISO/IEC 31000:2009) about risk management has assembled several previous standards and guidelines that have been developed in the last two decades. This has made it an internationally recognized reference that can be applied to any organization regardless of its size, activity or sector.

In parallel to the development of standards, relative young fields have emerged, like among others Operational Risk Management (ORM), Enterprise Risk Management (ERM) and Project Risk Management (PRM). Their initial application area has been corporate and market finance which are precursor domains in the development of risk management techniques. With the time such techniques and related concepts spread to the industries of manufacturing and services, and especially in activities commonly associated with high risk levels like in chemical industries. Nevertheless, the AEC industry either neglects risk management techniques, or applies methods that examine risk as a rather isolated purpose in a specific domain like project scheduling or cost planning. In that case, risk analysis is done only occasionally during a project, as it should begin in a very early project state and be accomplished continuously throughout all project phases. In order to bridge these gaps, it is of interest to take benefit of the latest risk management concepts applied in precursor fields. Additionally, the principles and methods from Building Information Modeling (BIM) can be of great support for integrating concepts from the risk management field into the building design process.

2 BACKGROUND ASSUMPTIONS ABOUT RISK

2.1 *Main outcomes of the risk standards*

As cornerstone the risk standards establish a general concept about risk in which it is defined as the effect of uncertainty on objectives. The standards also describe four tasks that are now widely recognized as the essential steps of the risk management cycle. This cycle comprises risk identification, risk analysis, risk treatment and risk monitoring. This risk management cycle is meant to be performed repetitively and in a dynamic way all along a project. Risk identification performs an exhaustive listing of all possible threats, their categorization, and the recognition of the affected items. Risk analysis enables the evaluation and prioritization of risk, mostly with help of established assessment techniques. Risk treatment consists of determining preventive measures and reaction strategies to reduce threats to the project objectives and to enhance reactivity. The risk monitoring task controls and re-evaluates the suitability and effectiveness of the implemented treatment measures. It also supports identifying new emerging risks. This overall procedure is quite universal and followed as it is in several application fields with sometimes some minor singularities.

2.2 *Existing approaches for risk modeling*

Several efforts have been already made to classify risk in building projects. Traditional perceptions of risk define it as an event that can occur in a project. From this paradigm Perry & Hayes (1985) drew up a list of such events that could then be used for making risk catalogues. From such risk catalogues typical risks and measures can be retrieved and selected for a particular project. Under the same paradigm Tah & Carr (2001) used a Hierarchical Risk-Beakdown Structure (HRBS) in order to classify risk in categories and use them as basis for defining a risk class diagram. Other works have also been done trying to overcome the lack of formalism in risk management for building projects (Cooper & Chapman,1987, Wirba et al.,1996). On a more institutional level the ICE & FIA (1998) have established the Risk Analysis and Management for Projects (RAMP) framework. This latter introduces a risk rating system based on the traditional probability-impact matrix tool. Besides the use of matrices, other methods are available for risk evaluation (Smith, 2003). Consultants can use decision trees and fault/event trees, especially in reliability analysis of technical systems. Some other techniques consist of spider diagram, influence diagram, neural networks, sensitivity analysis or Monte Carlo simulation.

All the classification efforts made to categorize risk and formalize a terminology are quite useful for understanding what can happen during a project. Nevertheless such formalisms are quite difficult to apply and are often neglected by stakeholders. Indeed by trying to make comprehensive lists of threats and breaking down all their causes and consequences, one can lose visibility in what effect risks concretely have and which of them are in fact really worth to be considered. In such classifications, focus is set more on the origin or cause of risk. The presented approach rather tackles the problem the other way around and focuses more on the impact of risk on targeted performances about the building. Another limitation by applying risk concepts comes from its former definition as event that can be easily applied in some cases, but is less adapted when considering overall building performance. The presented approach to risk modeling is meant to go beyond these limitations. Subsequently it also forsakes old definitions of risk by applying the concept of uncertainty that has a more universal character as recognized by the ISO.

2.3 *Uncertainty in the building life cycle*

In practice there exists a distinction between two types of uncertainty, the aleatoric and epistemic uncertainty. A description about their general influence in engineering and corresponding modeling approaches is provided by Der Kiureghian & Ditlevsen (2009).

Aleatoric uncertainty is also known as statistic uncertainty. Uncertainties of that kind include a random character and thus are not directly controllable by the designer. Some typical examples are among others the climatic conditions at the building location, the fluctuation of energy prices, building occupancy and occupant behavior or intrinsic material properties.

Epistemic uncertainty is also called systematic uncertainty and describes uncertainty that is introduced by people being not able or not willing to dispose of information exactly. Often, an epistemic uncertainty relates to information that could be known in theory but is not known in practice. For example, the U-value of a wall by itself is not uncertain, but in practice we will use an estimated or expected value and thus an inaccurate value.

The performance-based approach to risk presented in this work aims mostly at exploring aleatoric uncertainty and their consequences in terms of performance deviations. Following a BIM approach to model that uncertainty is the most suited. This implies to interlink uncertainties with related BIM objects, thus extending the classical design parameter space with uncertain parameters.

3 PERFORMANCE-BASED RISK MODELING APPROACH

The purpose of a performance-based approach for managing risk is less to try to inventory all problems that can occur within the building life cycle, but rather to focus on unwanted deviations in targeted performances while sorting out the causes of such deviations. Nevertheless, inventories of possible threats can be of additional support in specific building life cycle phases and for specialized actors like facility managers or construction foremen.

The intended approach relies first on the identification of the main objectives and related important performance criteria. Then main threats can be identified and accordingly analyzed. In this context, concepts from the risk management field can become very useful. For example, Beasley et al. (2011) say about ERM that it provides the opportunity for organizational leaders to achieve a robust and holistic enterprise-wide view of potential risks that may affect the achievement of the organization's objectives. As in ERM a building project is an organized and collaborative activity that aims at the achievement of a final product with regard to precise requirements expressed as performances. In view of that, a good overview of it risks is necessary to support design and investment decisions that minimize the probability of performance mismatches. In order to enable a holistic view and efficient control of risk, the concept of Key Risk Indicator (KRI) is introduced.

KRIs are used in ERM and ORM for supporting the different steps of risk identification, assessment, treatment and monitoring. According to the Institute of Operational Risk (IOR, 2010), a KRI is a metric that provides information on the level of exposure to a given risk at a particular point in time. More specifically a KRI can be used to measure:

- the amount of exposure to a given risk or set of risks,
- changes in the risk exposure level giving early warning of increasing risk exposure
- the effectiveness of any action made to reduce or mitigate a given risk exposure,
- how well the risk exposure is being managed and if it fits in a predefined risk appetite.

Moreover, latest management techniques typically make use of three different types of indicator: Key Performance Indicator (KPI), key risk indicator (KRI) and Key Control Indicator (KCI). A detailed description of such indicators is provided by Smart & Creelman (2013). As shown in Figure 1, the three kinds of indicators are closely interrelated. The first ones relate the organization performance goals, the second ones give the risk profile with regard to goals achievement, and the third ones assess the effectiveness of controls which refer to decisions and actions made to mitigate risk. Often a distinction is made between leading and lagging indicators. The first ones provide an early signal/warning that targets will or not be achieved. The second ones provide a signal that the targets have or not been achieved. Principally,

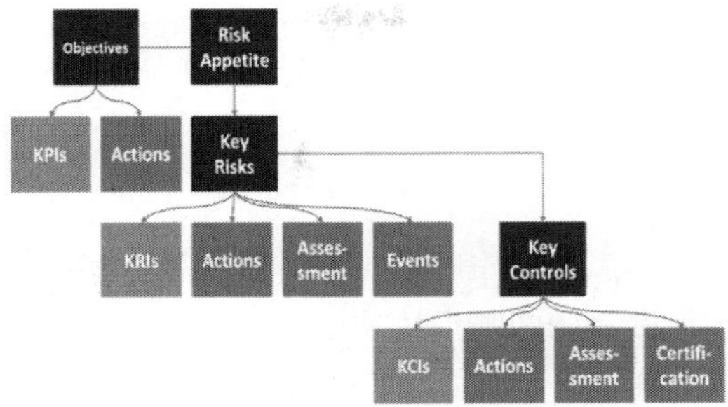

Figure 1. Interrelationships between KPIs, KRIs and KCIs (StratexSystems, 2012).

when applying such concepts to the building life cycle, the first category will be preferably used in the design phase. The construction and operation phase will rather use both categories.

KRIs often serve their most practical purpose in conjunction with a system of thresholds (Immaneni et al., 2004) which can be formalized under the terms of risk appetite or risk tolerance. When a KRI breaches its associated threshold, it triggers a management action that can be a design alternative in terms of building design. Per definition, the risk tolerance reflects the maximal uncertainty accepted by a stakeholder. In view of that the decision making process will support taking investment decisions and design actions that bring the risk level back to an acceptable range.

To summarize, the difference between a KPI and a KRI is that the first is meant as a measure of how well something is being done while the second is an indicator of the possibility of adverse impact. In a sense it also reflects the volatility of performances, while the first rather indicate the fulfillment of goals and reflect less the notion of deviation. Even if both concepts have the same roots, in the BIM community KPI is today a well-established concept but there is still no application or even mention of KRI.

4 GENERAL METHODOLOGY

Backbone of the methodology is the integration of KRIs into the building design process with the support of BIM principles. KRIs can be used in any building life cycle phase. In the operation phase the sensor technologies and building monitoring techniques can provide useful information about the level of risk related to the fulfillment of expected building functions. The same remark can be made for the construction phase in which from the real work progress indicators can tell if the building planning is correctly followed. By contrast, during the design phase the assessment of risk, as of building performances, rather relies on computational methods.

4.1 *KRI-driven and BIM-based risk management cycle for building design*

To compute KRIs existing software tools can be used in combination with stochastic methods. In this context there are two possibilities: 1) the analysis model consists of stochastic variables, e.g. Stochastic Differential Equations (SDE), or 2) the analysis model does not use stochastic variables directly but a deterministic paradigm. In the second case the risk analysis can rely on a statistical experiment consisting of additional stochastic pre-processing and post-processing. As almost all building performance analysis tools available today rely on a deterministic model e.g. energy calculation tools, this latter approach is very useful. Indeed, there is a lack of tools that provide building performance calculation on the basis of stochastic models. Nevertheless, for some aspects like cost or energy system reliability the first approach can be easily applied by entering simple analysis models into a classical statistic and numerical simulation software.

The pre-processing performs a stochastic sampling after which parameter samples are stored in a dedicated simulation data model. In this context, the simulation algorithm of the used building performance analysis tool remains unchanged and several simulation runs are performed in order to simulate all samples. Then the risk analysis is achieved by a post-processing that provides the KRIs after statistical treatment of the different simulation results. Like the KRIs the stochastic variables specified in a building project and used as input for risk analysis are part of the risk model. As shown in Figure 2, the integration of the risk model into the project data resource follows a BIM approach in which the risk model is interlinked with the building information model as well as with other related data models. The links specify to which object uncertainty applies and can for example link a probability distribution to a cost item, a failure rate to an energy system component and a KRI to a building. The overall data interlinking approach follows the method defined by Scherer & Schapke (2011).

The overall purpose of the previously described procedure is then to support making design changes and investments which reduce the level of risk reflected by the KRIs. To complement this approach a knowledge base is introduced and built upon an ontology. It provides the decision maker with best and expert knowledge about building components and techniques that reduce the exposure to certain risks. For further assisting risk treatment decisions, the risk model contains target KRIs that can be predefined by the decision maker to express his risk tolerance.

4.2 *Key risk indicators for energy-efficient building design*

Per definition, a KRI can be specified as any metric that indicates if a KPI target value range is not reached with regard to the simulated performance values. It can also for example quantify how much deviation exists between a KPI target and simu-

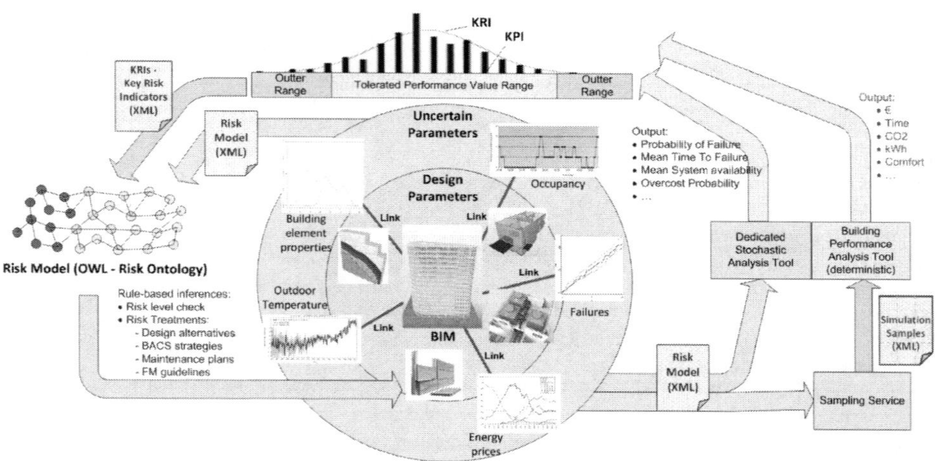

Figure 2. Illustration of the overall performance-based risk analysis methodology and of the integration of uncertainty in the BIM information space.

Building Performance Criteria	Key Risk Indicators
heating demand – [kWh/(m².a)]	σ, mode, mean, P(KPI$_i$> X$_i$)
cooling demand – [kWh/(m².a)]	σ, mode, mean, P(KPI$_i$> X$_i$)
electrical demand – [kWh/(m².a)]	σ, mode, mean, P(KPI$_i$> X$_i$)
primary energy use – [kWh/(m².a)]	σ, mode, mean, P(KPI$_i$> X$_i$)
total energy demand – [kWh/(m².a)]	σ, mode, mean, P(KPI$_i$> X$_i$)
CO_2 emission – [kWh/(m².a)]	σ, mode, mean, P(KPI$_i$> X$_i$)
Energy System Reliability	Failure rate – [y^{-1}]
	Mean Time To Failure (MTTF) – [h]
Energy System Maintainability	Mean Time To Repair (MTTR) – [h]
Energy System Availability	Average system unavailability - [h/y]
investment costs [€]	σ, mode, mean, P(KPI$_i$> X$_i$)
operation costs [€]	σ, mode, mean, P(KPI$_i$> X$_i$)
maintenance costs [€]	σ, mode, mean, P(KPI$_i$> X$_i$)
construction costs [€]	σ, mode, mean, P(KPI$_i$> X$_i$)
demolition costs [€]	σ, mode, mean, P(KPI$_i$> X$_i$)
revenue [€]	σ, mode, mean, P(KPI$_i$< X$_i$)

Figure 3. Examples of Key Risk Indicators (KRI) applicable for risk analysis during the building design phase.

lated value. Because such KRIs are rather from a primary nature and do not really quantify uncertainty, we introduce more elaborated KRIs that reflect the degree of uncertainty.

In the context of energy-efficient design KRIs can either provide a measure of uncertainty around certain KPIs or they can specifically express certain underperformances and critical states of the building life cycle. A KRI can then be defined on the basis of the value distribution of a KPI resulting from uncertainty. In other words it can be defined as a statistical indicator deduced from the probability distribution of a KPI. This case is illustrated in figure 2. For that purpose indicators like standard deviation (σ), mean and mode (most probable value) are used. Other interesting indicators are also the probability of being over or beneath the KPI target value or even being outside a certain value range. Such indicators are also easily deduced from the statistical value distribution of a KPI.

For reliability aspects other kinds of indicators have to be introduced that express how building energy system performance can be threatened by intrinsic uncertainties within the system. Most of them can also be defined in term of probabilistic value. In view of that a set of KRIs for energy-efficient building design has been specified. As example, a partial list of such KRIs is provided in Figure 3.

5 UNCERTAIN PARAMETERS

The following chapter introduces uncertain parameters considered of importance in energy-efficient design. They describe stochastic influences on building performance one has no direct control on. In this paper, the focus lies on four categories: climate, occupancy, cost and energy system reliability.

Because those parameters shall be modeled in a reliable manner it is of interest to use parameters that dispose of historical data. Climate and cost parameters are best examples of such values, as external temperature and market prices are monitored all the time. For such information there exist a lot of sources. For occupancy there is also the possibility to rely on observed data. On one hand there already exists some available information e.g. results of household surveys (Ipsos-RSL, Office for National Statistics, 2003). On the other hand it can be assumed that such information will be more and more available in the future thanks to sensor technologies. The same remark can be done for reliability aspects in the energy system as BACS systems can provide much monitoring and statistical data.

5.1 *Climate data*

When performing simulations to predict the energy consumption of a building, the climatic conditions play a decisive role. To predict these climatic trends one has different types of weather data at his disposal. The classical example is Test Reference Year (TRY) that has been further developed to Typical Meteorological Year 2 (TMY 2) and Weather Year for Energy Calculations 2 (WYEC 2) (Chan et al., 2006). These are synthetic compilations of weather relevant values like dry bulb temperature, wind direction, direct normal illuminance and many more, belonging to indicated locations.

When the simulation is planned to span a whole building life cycle, the effects of climatic change can also be considered. There are two common methods to achieve this. One is the so-called analogue scenario, which means the usage of present day weather data from a location whose current climate complies with the target location predicted climate. The second method is the use of global circulation models, which are thermodynamic climate models, relying on differential equations. As these models have a very coarse resolution, a downscaling technique is needed to gain the fine time resolution needed for energy simulation.

The use of those different data sources and the historical climate profiles make it possible to select different samples of climate data for use in simulation, as it has been performed for example in (Yang et al., 2008). At the simplest, the samples can be set as worse, most probable or best climate conditions. A bigger set of samples can be built upon the different climate profiles provided among the several years of historical data.

There also exist stochastic models, mainly based on stochastic differential equations (SDE) that represent climatic sub-processes. Examples of such techniques are described in (Majda et al., 2001). Their application however is mainly useful and restricted to climatologists.

5.2 *Occupancy modeling*

The simulation of occupancy has two important reasons. Firstly, a present person is a source of heat and moisture and thus directly affects energy and computational fluid dynamics simulations (Ainsworth et al., 1993). Secondly, there are numerous energy consuming devices that need manual interaction for their activation, e.g. home appliances or manual lighting, so occupancy directly impacts the energy performance.

There are several distinct ways to model occupancy. The simplest way is deterministic modeling by setting average values like occupant density and occupant activity level. This approach is very simple to apply but does not assess the uncertain nature of occupants' behavior and its impacts on KPIs. Thus the following presented models use a stochastic approach.

A very detailed and computationally expensive method is agent-based modeling (Liao et al., 2010). Hereby each occupant is modeled and simulated individually. That means that each occupant gets a personal arriving time distribution or probability of using e.g. the coffee machine assigned. Obviously the possibilities of detailing are limitless. On the other hand, in many cases this amount of detailed information will either not be available or the modeler is interested in a faster and more accessible way of modeling. Then the next approach is advisable.

A more computational efficient approach, that is also easy to implement, is the inhomogeneous first-order Markov chain model as described by (Richardson et al., 2008). In this model the number of occupants is simulated directly. Figure 4 shows an example of two samples generated with such a model.

First-order Markov chain implies that the number of occupants at a time $t - 1$ only depends on the number of occupants at time $t - 1$ and the probability of transition between states. The term inhomogeneous means that the probabilities of transition depend on the point of time. To apply this model one needs several transition matrices, one for every time step and for every different zone. These matrices contain the probabilities of a change of state, e.g. from two occupants to zero. They can be computed from monitored occupancy data or be estimated by experts.

When the raw occupancy data can be generated, the next step can be the modeling of user interac-

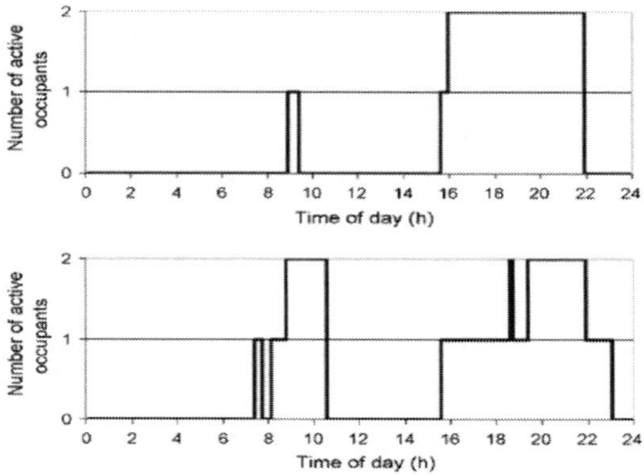

Figure 4. Example runs for a residential zone with max. two occupants on a weekday (Richardson et al., 2008).

tion. Especially energy relevant examples for such interactions are the opening of windows and the manual operation of the lighting system. An example for a model that simulates manual control of lighting and blinds is called Lightswitch-2002 and can be found in (Reinhart et al., 2003).

5.3 *Cost*

One of the most important considerations when carrying out a building project is the expected cost. While the calculation of the construction costs can be done reliable to a certain degree, the prediction of the operation costs and the monetary returns over the building life cycle is highly uncertain. On the cost side, a major influence is the development of energy prices over the next decades. On the income side, the discounted cash flow method helps to estimate the profitability of a project.

5.3.1 *Energy prices*

The observation and prognosis of the energy price can be divided into two perspectives: (1) a long-term perspective with relevance for investment and design decisions until around 30 years and (2) a short-term perspective covering the following hours up to several days.

When modeling the future development of energy prices, two distinct approaches are available. First, there are so-called fundamental models. These are numerical non-stochastic models, which are based on the technical characteristics of the energy market, like among others plant capacities, restrictions in transmission, local laws, type as well as localization of power plants and storages. Examples are MARKAL, TIMES and ELMOD (ETSAP, 2015). Based on the rich information imported into these models, they are able to give very accurate predictions, but on the other hand they need huge initial investments, capital-wise as well as knowledge-wise. To bypass these restrictions, it is possible to order predictions created on the basis of such models. As fundamental models are non-stochastic, each set of inputs delivers the same outputs.

Even if such models provide one unique predicted data set used as input for cost calculation, they give an insight into possible additional expenses and respectively cost risk due to market price changes. Additionally, uncertainties can be inserted in the inputs of fundamental models.

The second approach relies on stochastic models built upon historical data. An overview of such approaches can be found in (Möst & Keles, 2010). Especially time series models like ARIMA models and GARCH models were developed to predict the development of market prices. Theoretically, these can be used for predictions over any time scale, but because the margin of error increases with the time period they are mostly used for short-term prediction.

Another downside is that the classical approach uses one stochastic process per energy source. So in its basic form, it assumes that the prices of the different energy sources are independent. As the studies in (Bencivenga & Sargenti, 2010) and (De Jong & Schneider, 2009) show, this is not the case. So an advanced time series model needs to take the correlations between the energy source prices into consideration. If one relies on stochastic models, these need to be supplied with relevant data to calibrate their predictions. As an example, European data sources can be found at (Eurostat, 2016) and (EC, 2016).

To sum up, for short term prediction, like one-day ahead, stochastic time series models are very well established and their functionality is proven. For long term forecasts however, more intelligent approaches are needed, as technological improvement and unforeseeable structural changes need to be taken into consideration. Fundamental models may serve this purpose.

5.3.2 *Discounted Cash Flow (DCF) method*

To estimate the financial return a building project will yield, real estate valuation techniques such as the Discounted Cash Flow method (DCF) are used. In DCF, the Net Present Value (NPV) or Internal Rate of Return (IRR) are determined by looking at the projects expenses and the project incomes on a yearly basis. Hereby the investment costs, the income and the operating expenses are estimated based on assumptions for financial key values like e.g. rents and periodic rental growth. A discount rate is used to discount future cash flows.

This calculation can be done completely deterministic. When calculating the NPV of a real estate, one assumes that the cash flow for every year is known. This cash flow depends on the different cost models, as well as on the income generated by the property. All these values can be estimated, but can never be determined precisely. This uncertainty can be incorporated into the calculations by using random variables instead of fix values for the assumed calculation inputs. These can either be estimated from historical values or, if no historical values are available, a normal distribution with the former assumed value as mean and a reasonable variance can be used. To perform the NPV calculation, it is no longer sufficient to do a single calculation, but a Monte Carlo simulation has to be performed. This yields a distribution of NPV values. Their mean provide the Expected Net Present Value (ENPV) as final outcome of the probabilistic NPV calculation.

In the former approach, the input variables are static, which means they maintain their value over the complete time span. As NPV calculations are executed over a long period of time, it is to expect that the economic conditions change during this time. To incorporate this behavior in the calculations, instead of static variables, a stochastic process, called random walk can be used (Leung, 2014). In a random walk, the change of a value between the years is modeled by a random variable, e.g. a normal distribution with zero mean, if an increase and a decrease are seen as equally likely. This process may either be estimated based on historical values or judged by common knowledge.

5.4 *Energy system reliability*

Reliability defines the ability of a system to perform its function over time without disturbances. In order to assess energy system reliability, the impacts and probabilities of component failures are analyzed. For that purpose, the system must be modeled in a qualitative as well as a quantitative manner.

On the one hand, the qualitative model of the energy system is required to describe cause/effect relationships between components. They can model how failures of some components spread into the overall system. An ontology model, as the one mentioned in chapter 4, can support modeling such relationships and resulting failure scenarios or modes. Such scenarios can then be stored at instance level in the ontology used as knowledge base. Treatment measures can then be associated to failure modes and expressed as preventive or corrective actions in a maintenance plan or as guidelines for the energy system designer.

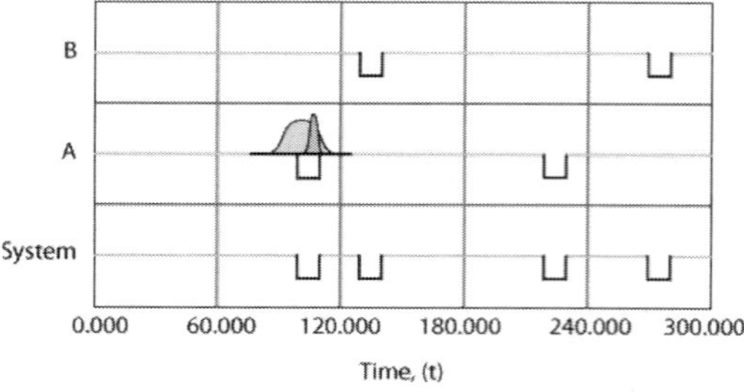

Figure 5. Up-down time curves for a system with two components with random failure and repair time (ReliaSoft, 2015).

On the other hand, a quantitative model is required that attributes reliability parameters to the system components. Such attributes are for example the hazard rate, component life time or the Mean Time To Failure (MTTF). On the basis of both qualitative and quantitative models a reliability analysis can be performed. For that purpose there exist already numerous techniques that can be applied according to the system characteristics. A comprehensive overview about those techniques can be found in (Rausand & Høyland, 2004). Principally, the purpose of reliability analysis is to aggregate components characteristics into whole system characteristics. Many of these techniques rely on stochastic models including for example Poisson process or Weibull distribution as most widely used ones. As a result availability diagrams also called up-down curves can be simulated as shown in Figure 5.

From these, relevant KRIs such as the average system unavailability time or the system mean time to failure can be deduced. Those indicators can then be used as basis for design decisions and optimizations trying to reduce the level of reliability risk.

6 CONCLUSION

The uncertainties presented in this work do not represent an exhaustive list for building design. The choice of relevant parameters for risk analysis is conditioned by the building performance requirements as well as the reliability of the available data itself. Moreover it is up to the designer or project manager himself to select what uncertainty he wants to analyze the effect of. In the presented methodology uncertain parameters are part of a risk model. They can then be used as input for performing specific risk analyses that can rely on existing tools and methods. The overall purpose of the methodology is then to provide the decision maker with relevant KRIs storing them into the risk model.

The presented KRIs have been defined for the design phase, but the methodology and its components are meant to be continuously enriched within the whole building life cycle. For the next phases, extension possibilities could involve disturbances in construction processes, fire safety aspects, maintenance and retrofitting aspects. For that purpose, monitoring technologies can provide much support in measuring KRIs. Moreover, following a BIM process, the risk model shall be enriched with new data and remain part of the stakeholder's project data resource over the building life cycle.

REFERENCES

Ainsworth, B. E., Haskell, W. L., Leon, A. S., Jacobs Jr, D. R., Montoye, H. J., Sallis, J. F., & Paffenbarger Jr, R. S. 1993. Compendium of physical activities: classification of energy costs of human physical activities. Medicine and science in sports and exercise, 25(1), 71–80.

Beasley, M.S., Branson B.C. & Hancock, B. 2001. Developing Key Risk Indicators to Strengthen Enterprise Risk Management. Published by the Committee of Sponsoring Organizations of the Treadway Commission (COSO), New York, NY, January 2011.

Bencivenga, C., & Sargenti, G. 2010. Crucial relationship among energy commodity prices (No. 5).

Cooper, D. F. & Chapman, C. B. 1987. Risk analysis for large projects, Wiley, Chichester, U.K.

De Jong, C., & Schneider, S. 2009. Cointegration between gas and power spot prices. The Journal of Energy Markets, 2(3), 27–46.

Der Kiureghian, A., & Ditlevsen, O. 2009. Aleatory or epistemic? Does it matter?. Structural Safety, 31(2), 105–112.

EC: European Commission. 2016. Energy Statistics. Retrieved from http://ec.europa.eu/energy/en/statistics.

Energy Technology Systems Analysis Programme [ETSAP]. 2015. MARKAL. http://www.iea-etsap.org/web/Markal.asp [Status 2015–06-08].

Eurostat. 2016. Your Key to European statistics. Retrieved from http://ec.europa.eu/eurostat/web/energy.

Immaneni, A., Mastro, C. & Haubenstock, M. 2004. A structured approach to building predictive indicators. The RMA Journal, Special Edition: Operational Risk, May 2004.

IOR: Institute of Operational Risk. 2010. Sound Practice Guidance: Key Risk indicators. Nov. 2010.

ISO/IEC 31000:2009. Risk Management—Guidelines for principles and implementation of risk management (15 November 2009).

Institution of Civil Engineers (ICE) & Faculty and Institute of Actuaries (FIA). 1998. RAMP: Risk Analysis and Management for Projects, Thomas Telford, London.

Ipsos-RSL, Office for National Statistics. 2003. United Kingdom Time Use Survey, 2000. [data collection]. 3rd Edition. UK Data Service. SN: 4504, http://dx.doi.org/10.5255/UKDA-SN-4504-1.

Leung, K. C. K. 2014. Beyond DCF analysis in real estate financial modeling: probabilistic evaluation of real estate ventures (Doctoral dissertation, Massachusetts Institute of Technology).

Liao, C., & Barooah, P. 2010. An integrated approach to occupancy modeling and estimation in commercial buildings. In American Control Conference (ACC), 2010 (pp. 3130–3135). IEEE.

Majda, A.J., Timofeyev, I. & Eijnden, E.V. 2001. A mathematical framework for stochastic climate models. Communications on Pure and Applied Mathematics, Vol. LIV, 0891–0974 (2001) John Wiley & Sons, Inc.

Möst, D., & Keles, D. 2010. A survey of stochastic modelling approaches for liberalised electricity markets. European Journal of Operational Research, 207(2), 543–556.

Perry, J.G. & Hayes, R.W. 1985. Risk and its management in construction projects. In: Proc., Instn. Civ. Engrs., London, 1(78), 499–521.

Rausand, M., & Høyland, A. 2004. System reliability theory: models, statistical methods, and applications (Vol. 396). John Wiley & Sons.

Reinhart, C. F., Bourgeois, D., & Dubrous, F. 2003. Lightswitch: a model for manual control of lighting and blinds.

ReliaSoft. 2015. System analysis reference: reliability, availability & optimization. Tucson, AZ: ReliaSoft Publishing.

Richardson, I., Thomson, M., & Infield, D. 2008. A high-resolution domestic building occupancy model for energy demand simulations. Energy and buildings, 40(8), 1560–1566.

Scherer, R.J. & Schapke, S.-E. 2011. A distributed multi-model-based management information system for simulation and decision-making on construction projects. Advanced Engineering Informatics 25, Elsevier, pp. 582–599.

Smart, A., Creelman, J. 2013. Risk-Based Performance Management: Integrating Strategy and Risk Management. Palgrave Macmillan, Sept. 2013.

Smith, N.J. 2003. Appraisal, Risk and Uncertainty (Construction Management Series), London: Thomas Telford Ltd, UK.

Smith, N., Merna, T. & Jobling, P. 2006. Managing risk in construction projects. Blackwell Publishing.

StratexSystems. 2012. Managing with KPIs and KRIs, prepared for StratexSystems Webinar Series, 1 November 2012.

Tah, J. & Carr, V. 2001. Knowledge-Based Approach to Construction Project Risk Management. Journal Of Computing in Civil Engineering.

Wirba, E. N., Tah, J. H. M., and Howes, R. 1996. “Risk interdependencies and natural language computations.” J. Engrg., Constr. And Arch. Mgmt., 3(4), 215–269.

Yang, L., Lam, J. C., Liu, J., & Tsang, C. L. 2008. Building energy simulation using multi-years and typical meteorological years in different climates. Energy Conversion and Management, 49(1), 113–124.

eWork and eBusiness in Architecture, Engineering and Construction – Christodoulou & Scherer (Eds)
© 2016 Taylor & Francis Group, ISBN 978-1-138-03280-4

Open eeBIM platform for energy-efficient building design

R.J. Scherer, P. Katranuschkov & K. Baumgärtel
Institute for Construction Informatics, Technische Universität Dresden, Dresden, Germany

ABSTRACT: The application of BIM as new working methodology in AEC has many facets ranging from team collaboration tasks to the support of individual engineering work and enhanced tool interoperability. Great progress has been made in the last decade with regard to the BIM-based support for design communication, data exchange and data sharing but the use of BIM for creative design work performed with the help of specialized engineering applications is still limited to partial solutions with considerable interoperability constraints. The idea of an open, configurable and vendor-independent BIM-based design platform acting as a virtual lab for design practitioners is thus yet to be realized. In this paper we describe the concept and a first reference implementation of such a platform, which was initially developed by the TU Dresden for the domain of energy-efficient building design, and discuss options for possible further development and use of the platform in research and practice. Drawing upon results from the EU projects HESMOS and ISES the paper presents the current achievements accomplished in the frames of the on-going open eeBIM initiative supported by the EU projects eeEmbedded, Streamer, HOLISTEEC and Design4Energy.

1 INTRODUCTION

Use of Building Information Modelling (BIM) in building design and construction has many aspects ranging from a common repository for cooperative work up to support for the interoperability of individual independently developed specialized applications. However, while for the first type of tasks extensive experience has been already gathered and a number of tools and services such as BIM servers, BIM viewers, model checkers etc. do already exist, for the second—addressing a variety of problems in various AEC subdomains—a unifying approach is still unavailable. Interoperability with regard to the use of BIM for creative design work performed with the help of specialized applications is still limited to partial proprietary solutions or at best some fixed "interoperable chains" such as e.g. Revit—gbXML—DesignBuilder—energyPlus in the energy domain. A structured platform where third-party software can be flexibly chosen using standard BIM (IFC) and common engineering middleware services is missed. Such a platform would enable using in much the same way e.g. EnergyPlus, Riuska or TRNSYS for energy simulations or ANSYS, SOFiSTiK or Dlubal for structural analysis with only minimal re-configuration.

In this paper we describe the architecture and a first reference implementation of such a platform, conceptualized as an open and extensible PaaS system. Its essence is in using standardized BIM data, consistently aligned with a collaboration approach based on the use of a common BIM repository, and extend that to an energy-enhanced BIM (eeBIM) taking in consideration various additional information resources such as climate data, occupancy data, building services equipment and automation data, and so on. These resources are integrated with the BIM data using a flexible multi-model method. Integration of analysis and simulation tools is achieved by a set of general-purpose model filtering, model transformation and analysis/simulation management services. The capability for parallel computations using a compute cloud or grid environment is thereby also given.

The developed concept targets an *open environment* where a set of free kernel services can be flexibly configured with specialized free or commercial engineering services, tools and applications. The paper describes the first prototype realization of the envisaged open platform which was initially developed by the TU Dresden for the domain of energy-efficient building design in the EU FP7 projects HESMOS and ISES and is currently substantially extended in the FP7 IP eeEmbedded. Further development is under consideration by an initiative of related large EU projects (Streamer, HOLISTEEC, Design4Energy).

2 PROBLEM STATEMENT AND OBJECTIVES

The main problem regarding the use of BIM in AEC practice is that the same conceptual model attempts to satisfy distinctly different goals, i.e.:

1. Use of BIM as design resource (CAD)
2. Use of BIM as team coordination and cooperation resource (data exchange and sharing, common information repository)
3. Use of BIM as process coordination resource (4D modelling)
4. Use of BIM as basis for simulations, analyses and decision support (virtual BIM Lab).

All these goals could be theoretically achieved if BIM would encompass all needed building data and their relationships for all domains of design and construction and if all software tools would understand and support one and the same BIM schema. Unfortunately, these requirements are commonly seen as impossible to achieve on full scale. Therefore for the practical use of BIM it is necessary to apply different BIM management tools and schema extensions as well as various workarounds and external (non-BIM) data. However, while for the first two of the abovementioned goals a number of tools are already available and for the third extensive development work up to market introduction is currently taking place, for the fourth aspect, virtual BIM Lab, there is still no clear strategic concept and suitable BIM tools are largely missing. For that reason BIM can yet be mainly applied as common Reference Model (IFC), as basis for a proprietary integrated platform on top of a CAD system like Allplan, ArchiCAD or Revit, or as basic architectural model with limited interoperability features. This provides good but not yet sufficient basis for the goals of complex domains like energy-efficient building design. Hence, in practice, the support of specialized tools for analysis and simulation of various domain aspects happens largely manually, with the help of onboard tools like Excel, or by means of individually contracted software developments, which is mainly possible for large construction projects. Achievement of general interoperability, a goal defined already at the outset of BIM/IFC development, is still on relatively low level. This is especially noticeable when sophisticated engineering applications are involved.

Thus, without the development of flexible and efficient BIM Labs the expected benefits in terms of time-saving, design quality and error reduction cannot be fully reached, which may in turn influence the overall acceptance of the BIM method in AEC practice. However, as Table 1 shows, a BIM Lab requires much more than all other BIM-related goals. Along with the obvious need for a rich set of appropriate simulation/analysis tools, a central issue for the realization of the virtual lab idea is the achievement of seamless information interoperability. This can be done by means of an energy extended BIM (eeBIM) concept as suggested in the EU project HESMOS (Katranuschkov et al., 2011). Accordingly, our approach is to build the virtual lab as an open cloud platform based on eeBIM and extended by the development of missing functionalities and services for intelligent access to ICT control systems and advanced energy analysis and simulation tools.

Table 1. Needed concepts and tools for the achievement of various BIM goals.

BIM Goal	Concepts	Needed tools (except CAD)
Team Communication	- BIM Schema - MVDs	- BIM Export, Import - MVD support (optional)
Team Cooperation	- BIM Schema - MVDs	- BIM Server- BIM Export, Import - BIM Management (read, write, edit, filter) - Access Management (Security, Authentication, Access Control) - MVD Support (optional) - Model Merging
Process Coordination	- BIM Schema - Schema extensions or external process schema	- BIM Export, Import - Process integration (BPMN or similar) - Process simulation - Cost estimation (optional)
Virtual Lab	- BIM Schema - MVDs - External models - Multi-model concept - Rules	- As in Team Cooperation, but also: - Linking and management of external non-BIM resources - Model Transformation (Filtering, Mapping) - Model validation (Ontology, Rules) - Pre- and Post-Processors - Platform configuration tools - Process and Model Navigator

The objective of the open eeBIM platform initiative is to provide modular software services which are dedicated to support BIM working. The idea is that many of these services can be generic or general-purpose developments that can be combined in different configurations with analysis and simulation services on the market on pay-per-use basis. Such open services should help to downscale and expedite the realization of BIM software systems by avoiding repetitive development of the same kind of basic services.

3 STATE OF THE ART

Today, there exist many state-of-the-art software tools for component and building design, for cost analysis, energy analysis, facility management and life cycle analysis. These tools have proven their efficiency and reliability in their particular domains, like CAD systems in building design, Multizonal Building Energy Solvers (MBES) in energy consumption analysis and simulation, Building Envelope Solvers (BES) in the analysis and simulation of heat and moisture transportation, facility management systems for the management of building operation and maintenance, cost calculation and cost estimation tools for life cycle assessment and so on. However, as already mentioned, a common model and comprehensive model interoperability methods to integrate all such tools on a common platform are still missing. Hence, an integrative holistic approach is not easy to achieve, making data gathering for any energy study a very tedious and to a large extent manual effort.

A critical issue for running energy simulations is the data that is available about the component product and/or the facility to be built or renewed. In the last years BIM has indeed become a key technology for collecting data about products in the AEC industry (Eastman et al. 2011; Mahdjoubi et al. 2015). It consolidates and manages available product data from different sources to provide high quality up-to-date information about the buildings. It thus acts as main point of information that shall be used by energy simulation services to avoid time consuming and costly re-entering of differently structured component product and building data.

Being a powerful collaboration concept, BIM needs to be implemented in software and data models. In this regard the international IFC Standard (ISO 16739, 2013) developed by the non-profit buildingSMART initiative is taking a leading role and is meanwhile supported by all major software vendors in the AEC market.

However, the IFC is not qualified for storing all data of involved actors in a building life cycle because (1) it does not provide all entity types of all involved domains needed, and (2) it is not laid out to incorporate extensions for the full (and large) amount of all unspecified information. Other data models such as the open Green Building XML Schema (gbXML, 2015), or legacy models from software vendors may thus be interesting as well and cannot be excluded from the required data access specifications.

A solution in this regard has been established by the IDM approach (ISO 29481–1, 2010) that first concentrates on specifying business needs, which are independent from any particular data model. On that level an IDM (Information Delivery Manual) defines processes and exchange requirements that formally clarify the interaction with other participants such as architects, building services engineers or facility managers. The second, ICT-related step is to provide mappings to data models such as the new IFC4 release (buildingSMART, 2016) including appropriate implementation agreements, the so called MVDs. Both steps are necessary to improve the interoperability of BIM-based AEC tools. However, MVD developments are still in an early stage.

To calculate and predict energy consumption appropriate simulation models are additionally needed. Such models are already in development, most notably the SimModel (O'Donnell et al. 2012; Pauwels et al. 2015), but their use in conjunction with BIM/IFC is not yet consolidated. For example, SimModel requires a number of specialized dedicated adapters and the idea of an eeBIM is not supported. Moreover, to serve all users adequately, such models should be reusable and interoperable in a distributed heterogeneous environment. The simulations themselves must be flexible and hence should be provided as a service (SaaS). The integration of such services in a Service Oriented Architecture (SOA) warranting the efficient collaboration of the involved users and their applications is considered to be one of the most promising approaches today (Baumgärtel et al., 2012).

In (Stack et al. 2009) a standard set of interfaces based on SOA and defining extensible and reusable segments offered to all other components in an integrated system is described. The developed and used components provide services for energy simulation, maintenance, monitoring, sensors and actuators and the building product model. The core of the system is a data warehouse, which uses extracted data from different sources on three layers: a network layer that senses and communicates performance data, a data layer that stores the data, and a tool layer that involves end-user tools and graphical user interfaces. However, even though many energy and facility management functions are covered, the whole life cycle and the interaction with architects and other designers are not addressed.

OpenStudio (https://www.openstudio.net/) provides a collection of software tools to support whole building energy modeling using EnergyPlus and advanced daylight analysis using Radiance. It is an open source (LGPL) development to facilitate community development, extension, and private sector adoption. The graphical applications include the OpenStudio SketchUp Plug-in, a ResultsViewer and a Parametric Analysis Tool. OpenStudio supports import of gbXML and IFC for geometry creation but the latter is still in development.

Simergy (http://simergy.lbl.gov/) provides a comprehensive GUI for EnergyPlus that has been developed by a public and private partnership effort. It gives users the ability to import geometry from BIM or 2-D CAD, as well as to create geometry directly. Translation of geometry from BIM to Building Energy Models (BEM) has always been a challenge, and Simergy presents a developing solution for this including the use of IFC. However, as in all other approaches, conception, construction and use of an explicit eeBIM platform is not provided.

There are also many EU projects developing ICT support for energy-efficient building regarding design, redesign, management, control etc. These projects have delivered and are currently developing a number of useful ICT tools, components and services applying also various re-usable components such as the open BiM Server (http://bimserver.org). However, although these developments feature similar central management software components, they are proprietary with regard to software interoperability. This fact hinders that ICT results of the EU projects can be combined, integrated and built upon.

Hence, whilst much research work has been done in recent years a consistent data management for an open and flexibly configurable virtual lab platform for life cycle building energy management is not yet available. All this shows the need for a common ICT foundation for the eeB market.

4 PRINCIPAL CONCEPT

Starting in 2010, the Institute of Construction Informatics at TU Dresden, Germany, is in accumulating way continuously researching and developing BIM management and middleware services for AEC domain models, combined multi-models and related interoperability issues on semantic, functional and technical level. In this context, the concept of an open eeBIM platform as technical expression of the virtual lab paradigm emerged.

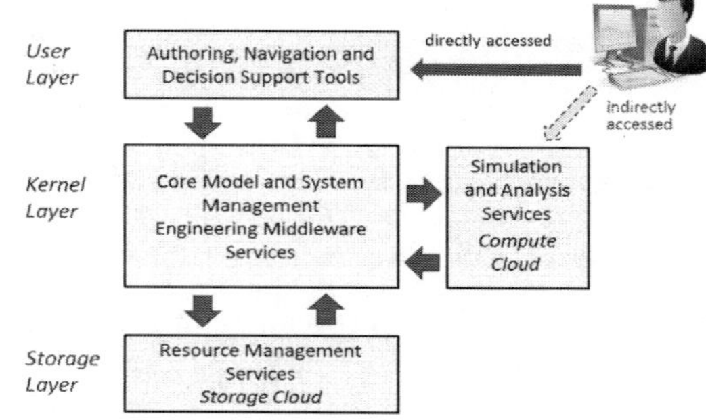

Figure 1. Principal architecture of the eeBIM platform.

The overall idea of the platform is exposed by its principal service-oriented architecture shown on Figure 1. According to that, the platform is organized into four distinct blocks.

The user accesses all platform services through the Authoring, Navigation and Decision Support tools on the upper User Layer. Via the core model and system management services on the Kernel Layer indirect access to all simulation and analysis services located on a compute cloud and the resource management services located on a storage cloud are provided. Thus, although services and information resources can be highly distributed they appear as one coherent system to the designer and can be uniformly used by the whole design team. However, cloud and distributed resource use are not mandatory. If required, the platform can easily be configured for local or LAN use as well.

Figure 2 presents an expanded view of the platform architecture, showing the envisaged tools and services in each block from Figure 1.

They are divided in seven categories, thereby sub-dividing the kernel into three distinct sub-categories as follows:

1. Modelers (CAD and other authoring tools)
2. BIM Navigator, for product and optional process navigation
3. Collaboration Management Services
4. BIM Management Services
5. Simulation Management Services
6. Information repositories and the related resource management and access services
7. Computational analysis and simulation services.

For the envisaged open eeBIM platform the services highlighted in Figure 2 are already provided as reference implementation. They are described in more detail below.

The BIM Navigator (2) is conceived as primary access point to the platform. It is an end user

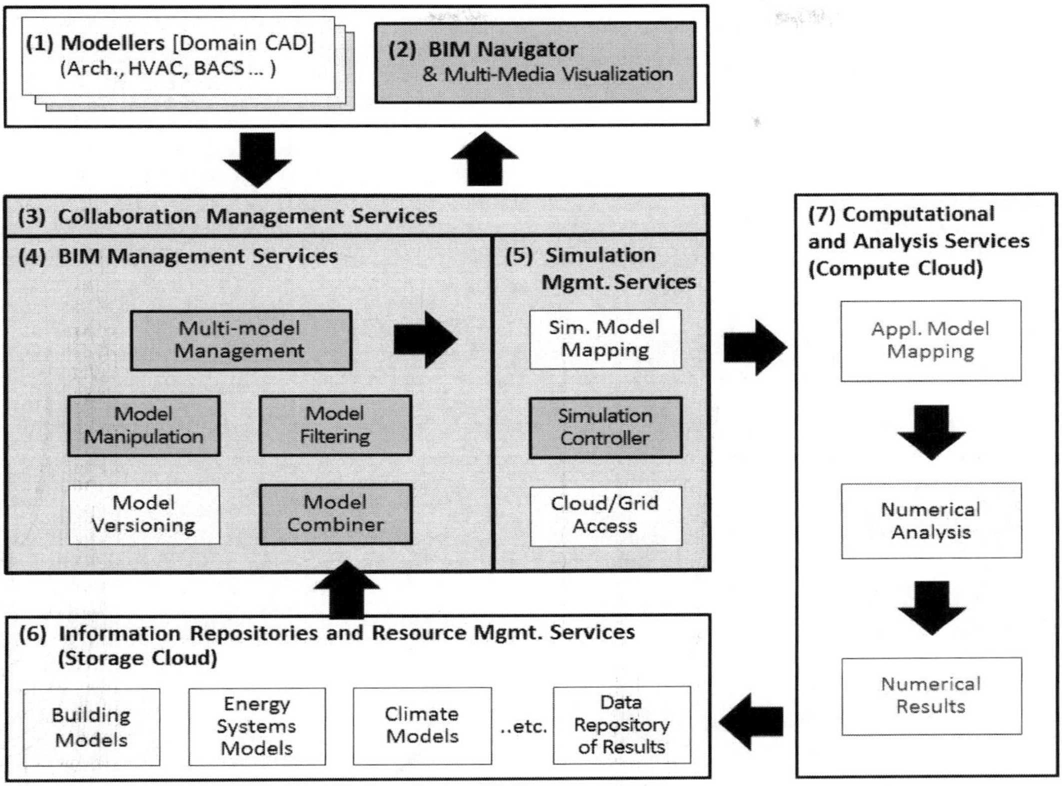

Figure 2. Expanded view of the eeBIM platform architecture showing the principal modules and highlighting currently available open software developments (grey boxes).

application which enables navigating, visualizing and managing big datasets of BIM information in design process context, utilizing several core BIM management services to select, filter or edit the relevant data. Users and their files for different projects can be serviced and multiple domain models can be viewed and manipulated separately or as multi-models, together with their link information.

The BIM Collaboration Management Services (3) are responsible for user management, file management and internal (Kernel) and external service management. They provide for workflow configuration, service execution orders, managing user access rights for files and projects and validating the input data against task requirements in the configured workflows.

The BIM Management Services (4) are a collection of four dedicated modules for the management of BIM information. They provide the access to various data models and enable selection, filtering, manipulation and combination of the model data (Figure 3).

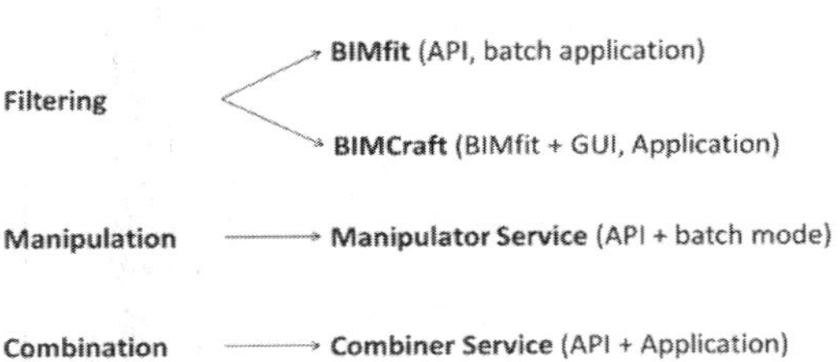

Figure 3. BIM Management Services on the eeBIM platform.

The concept is based on a multi-model approach where the BIM data are distributed in separate domain models and are inter-linked via an explicit Link Model (Fuchs et al., 2010). This provides for structured data access and management without scaling problems and without changes to the domain models. The architectural BIM-IFC model is thereby used as basis and inter-linked with other resources like climate data, material data, occupancy data, HVAC model data etc.

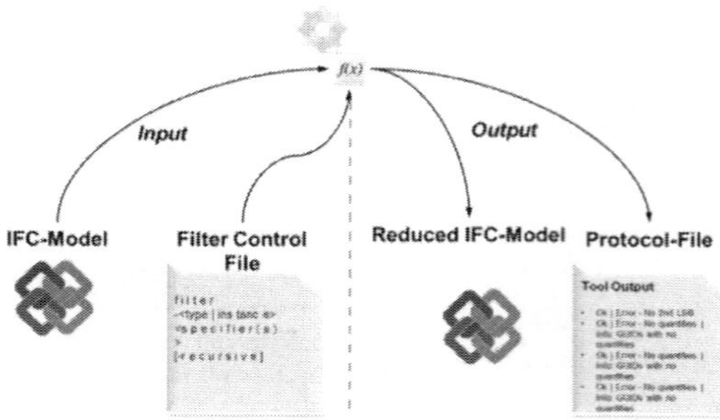

Figure 4. Schematic presentation of the inputs and outputs of the BIMfit Toolbox.

Specifically for checking and filtering purposes the BIMfit toolbox has been developed. It can be used both as a software library for other applications and as an independent filtering service. In the latter case, an IFC file and a filter control file are sent as input, and a respectively reduced IFC model and a protocol file are obtained as output (Figure 4).

BIMfit supports both descriptive model filtering, via a formal language which lays out how the model schema should be reduced to a sub-schema and how the model content should be reduced accordingly, and prescriptive filtering, via a set of functions that can be put together as logical procedures executed at runtime. For more comfortable user interaction, an additional end user tool BIMCraft has been also developed. It provides a GUI that enables visual input of filter commands by the user. These commands are then automatically mapped to BIM functions (Wülfing et al. 2014).

The BIM Manipulator Service enriches the eeBIM model with more semantics. It can change attributive information of model objects, create certain new objects, add/remove property sets and attach more information to IFC attributes (Name, Description etc.) in order to satisfy the information demands of the analysis models. An important part of that service is the conversion of first to second level space boundaries by using the BSPro library (http://www.granlund.fi/en/software/design-tools/). The result is a new IFC file with the second level space boundary objects that can then be used for thermal energy analyses.

The BIM Combiner Service integrates several domain models to a multi-model. By using the filter services the domain models can be filtered to sub-sets before their combination and hence task-specific multi-model views can be generated. In addition, the BIM Combiner Service also provides functions to generate the external, explicit link models needed for the definition of multi-models. In the HESMOS and ISES projects we have used the eeBIM approach to connect IFC models with external data like material files, occupancy schedules, sensor readings and climate data (Katranuschkov et al. 2011; 2015). The first version of the BIM Combiner Service developed in HESMOS was later on extended to use ontologies where all linked objects of the domain are specified in the ontology and the link model is exported as RDF file. Thus, it is possible to use a query language like SPARQL to search in all domains and to apply rules.

The Simulation Management Services (5) control simulation pre-processing, execution and post-processing. They comprise three sub-services (see Figure 1). The first analyzes incoming user requests to create the appropriate input data format for the energy solvers, the second controls simulation execution for one or more (up to hundreds) parallel simulation runs on the cloud, and the third provides for the actual cloud access. To enable variant studies a new model called Simulation Matrix has been specified. It enables various parameter variations as e.g. changing the building orientation, material transmittance values, or occupancy information, links these to the respective BIM objects (spaces, walls, floors, windows, storeys etc.) and defines the valid variant combinations to be run on the cloud. On the basis of these combinations the Simulation Management Services know how many energy simulations must be started and in which order, and how should result post-processing take place.

5 CURRENT IMPLEMENTATIONS

5.1 *Developed services*

The following of the services outlined in the previous chapter are currently provided as reference implementations:

- Collaboration Management Services
- BIMfit Toolbox
- BIMCraft
- Manipulator Service (Space Boundary Converter)
- Combiner Service
- Simulation Management Service

Contributors, service type and license type are summarized in the following table. All developed services are in prototype status and may be updated in future. Nevertheless they are all fully functional and demonstrable.

5.2 *Reference platform implementation*

The services outlined in the previous section are integrated in a first reference implementation of

Table 2. Currently implemented services for the open eeBIM platform.

No. in Concept & Name	Provider	Technology	License
(3) Collaboration Services	TU Dresden, Germany, IABI, Germany	Java and C#	Open Source
	Services for user management, file storage and SOA management based on the new BCF 2.0 specification of buildingSMART and including a reference server and client with a GUI. The client allows Drag&Drop of files and packing these in the BIM Snippet of a BCF message. A BPMN-based workflow can also be supported based on project and user needs.		
(4) BIMfit library and service	TU Dresden, Germany	Java	Open Source
	Filter tool box for IFC2 × 3 & IFC4 that allows filtering on class (type) and instance (object) level		
(4) BIMCraft	TU Dresden, Germany	Java	Free for non-commercial use
	Graphical user interface for BIMfit using a dedicated visual query language (VBQL)		
(4) Manipulator Service (Space Boundary Conversion)	TU Dresden, Germany, Granlund, Finland	Java and C#	Free for non-commercial use
	Converts first level space boundaries in an IFC file to second level space boundaries as needed for thermal energy analysis. The service exports the resulting higher level of detail as new IFC file. The BSPro library of Granlund (compatible to IFC2 × 3 and IFC4 is used as basis.		
(4) Model Combiner Service	TU Dresden, Germany National Res. Centre, Iceland	Java, Jelly	Open Source
	Combines IFC with external resources defined in an ontology and exports a link model with all defined links in RDF. Supported are SPARQL queries and the application of rules.		
(5) Simulation Management	TU Dresden, Germany	Java	Open Source
	Provides for the execution of simulation variants by using the Simulation Matrix model which defines building targets, parameter variations and their combinations in a XML schema over a basic BIM model, thereby enabling multiple parallel simulation runs. The explicit open XML schema specification can be used for several simulation solvers like EnergyPlus, TRNSYS, Riuska etc. The EnergyPlus mapping is currently still under development.		

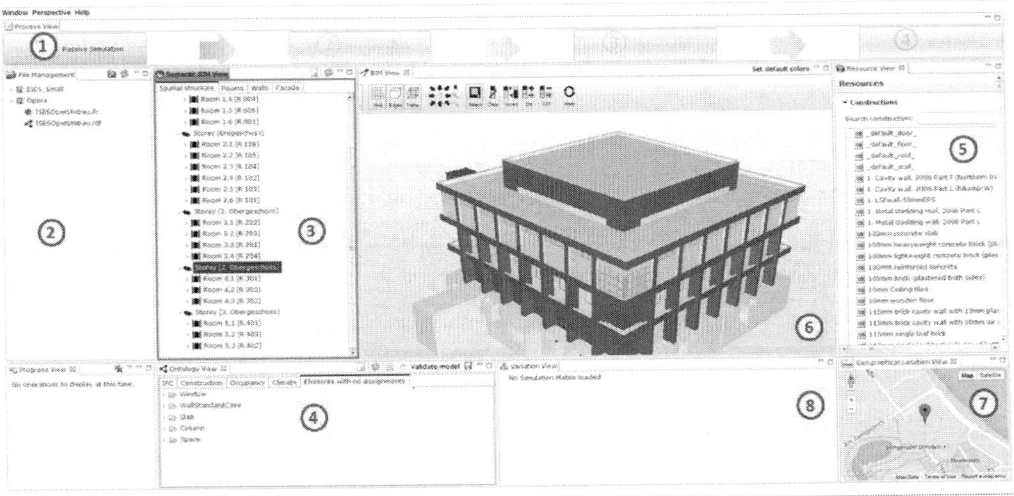

Figure 5. Screenshot of the developed Multimodel Navigator.

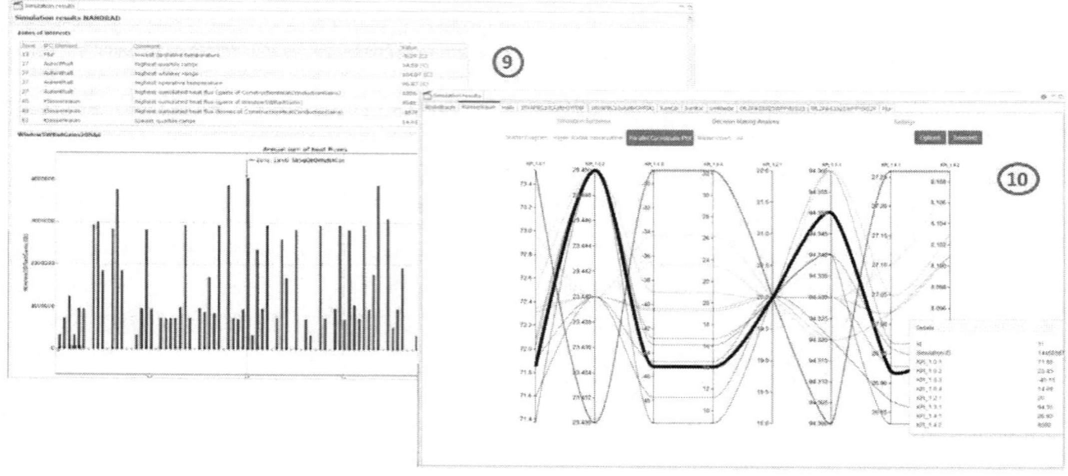

Figure 6. Simulation result view and Decision maker view integrated in the Multimodel Navigator.

the eeBIM platform which can be used to upload BIM files and inter-link them with external data provided as templates to execute thermal energy simulations. The prototype is divided in back end and front end services and is highly configurable. Figure 5 below shows the BIM perspective of the prototype. Several views are provided that can be flexibly and easily adjusted. The Collaboration Services are realized by two views—the process view (1) and the project view (2). The BIM Management Services are provided via the BIM structure and filtering view (3), and the multi-model services are enabled via the explicit link model view (4), the template view (5), the BIM visualization view, utilizing a dedicated version of Open IFC Tools of Bauhaus Universität Weimar (6), and the GIS view (7). Simulation Management is provided by the simulation configuration view (8), which is empty in the shown screenshot, as well as the energy result view (9) and the decision maker view (10) which are shown on the next Figure 6. The latter includes several presentation options such as scatter diagrams, radar charts, hyper-radial visualization, parallel coordinate plots etc. This enables decision makers to focus on different relevant aspects to find the best solution alternative.

More information on the concept and its implementation provides (Katranuschkov, 2015).

6 CONCLUSION

Building information modeling introduces a new work paradigm that helps designing more efficient and better quality buildings. Via buildingSMART a growing number of valuable free or open source BIM applications like Open IFC Tools or BiM Server become available which can be (and are) used to develop more complex BIM-based frameworks and systems. However, there is still no common concept to integrate such developments in an open, configurable and extensible platform.

The approach presented in this paper attempts to fill in this gap. The outlined general platform concept is lean and easily configurable to various scenarios and project needs. It suggests a number of re-usable services that can be applied or adapted for different platform layouts due to the clear structuring of the functionality of the software modules and the standardized modeling basis they are built upon. Such a platform can be used in a variety of ways. Firstly, it can provide a home for better coordinated and compatible tools and service developments, tightly aligned with the modeling standards of buildingSMART. Secondly, it gives software vendors the opportunity to build specialized applications upon common re-usable libraries, services and data models, thereby improving their interoperability and competitiveness. Thirdly, software vendors can join efforts more easily by sharing the benefits of an interoperable platform on pay-per-use basis. Last but not least, AEC practitioners can benefit in improving teamwork and cross-company cooperation.

The described platform in the paper is specifically related to the requirements and needs of energy-efficient building design. On that basis the joint non-profit and voluntary open eeBIM inititative was inaugurated in autumn 2014 (www.openeebim.org). However, we believe that using the proposed concept various dedicated engineering labs can be established to serve much broader AEC needs. In that regard, buildingSMART is hoped to provide valuable support.

ACKNOWLEDGEMENTS

This research was made possible through funding support of the European Commission to the FP7 projects HESMOS (Grant agreement no. 260088) ISES (Grant agreement no: 288819) and the IP project eeEmbedded (Grant agreement no: 609349). This support is herewith gratefully acknowledged.

REFERENCES

Baumgärtel, K., Katranuschkov, P. & Scherer, R. J. 2012. Design and software architecture of a cloud-based virtual energy laboratory for energy-efficient design and life cycle simulation. In: Gudnason, G. & Scherer, R. J. (eds.): eWork and eBusiness in Architecture, Engineering and Construction (ECPPM), Reykjavik, Iceland, ISBN 978-0-415621281, CRC Press/Balkema.

buildingSMART 2008-2016. IFC Overview, Available at: http://www.buildingsmart-tech.org/specifications.

Eastman, C., Teichholz, P., Sacks, R. & Liston, K. 2011. BIM Handbook, ISBN 978-0-470-54137-1, John Wiley & Sons.

Fuchs, S., Katranuschkov, P. & Scherer, R. J. 2010. A framework for multi-model collaboration and visualisation. In: Menzel K., Scherer R. J.: eWork and eBusiness in Architecture, Engineering and Construction (ECPPM), Cork, Ireland, ISBN 978-0-415-60507-6. CRC Press/Balkema.

gbXML 2015. gbXML – Current Schema (Version 6.01 – Nov. 2015), Available at: http://www.gbxml.org.

ISO 16739:2013. Industry Foundation Classes (IFC) for data sharing in the construction and facility management industries, ISO, Geneva, Switzerland.

ISO 29481-1:2010 Building information modelling – Information delivery manual - Part 1: Methodology and format, ISO, Geneva, Switzerland.

Katranuschkov, P., Guruz, R., Liebich, T. & Bort, B. 2011. Requirements and gap analysis for bim extension to an energy-efficient bim framework. In: Zarli, A. (ed.): Proc. 2nd Workshop on eeBuilding Data Models (CIB W078 – W102), Nice, France.

Katranuschkov, P. (ed.) 2015. ISES Final Report (Updated Version), EU FP7 Project 288819 ISES, © ISES Consortium, Brussels, Belgium.

Mahdjoubi, L., Brebbia, C. A. & Laing, R. (eds.) 2015. Building Information Modelling (BIM) in Design, Construction and Operations, WITpress, ISBN 978-1-84564-914-2, UK.

O'Donnell, J., See, R., Rose, C. Maile, T., Bazjanac, V. & Haves, P. 2012. SimModel: a domain data model for whole building energy simulation, Report LBNL-5566E, Lawrence Berkley National Laboratory, Berkeley, USA.

Pauwels, P., Corry, E. & O'Donnell, J. 2015. Making SimModel available as RDF graph. In: Mahdavi A., Martens B., SchererR. J. (eds.): eWork and eBusiness in Architecture, Engineering and Construction (ECPPM), Vienna, Austria, ISBN 978-1-138-027107, CRC Press/Balkema.

Stack, P., Manzoor, F., Menzel, K. & Cahill, B. 2009. A service oriented architecture for building performance monitoring. In: Proc. 18th Int. Conf. on the Application of Computer Sceince and Mathematics in Architecture and Civil Engineering, Weimar, Germany.

Wülfing, A., Windisch, R. & Scherer, R. J. 2014. A visual BIM query language. In: Mahdavi, A., Martens, B. & Scherer R. J. (eds.): eWork and eBusiness in Architecture, Engineering and Construction (ECPPM), Vienna, Austria, ISBN 978-1-138-02710-7, CRC Press/Balkema.

ONLINE TOOLS AND PROJECT REFERENCES

BiMServer: http://bimserver.org
BSPro: http://www.granlund.fi/en/software/design-tools/
Dlubal: http://www.dlubal.com
EnergyPlus: https://energyplus.net
EU project Design4Energy: www.design4energy.eu/
EU project eeEmbedded: http://www.eeembedded.eu
EU project HESMOS: http://www.hesmos.eu
EU project HOLISTEEC: www.holisteecproject.eu/
EU project ISES: http://ises.eu-project.info
EU project Streamer: www.streamer-project.eu/
Open eeBIM platform: https://www.openeebim.org
Open IFC Tools: http://www.openifctools.com/Open_IFC_Tools
OpenStudio: https://www.openstudio.net/
Riuska: http://www.granlund.fi/en/software/design-tools
Simergy: http://simergy.lbl.gov/
SOFiSTiK: http://www.sofistik.com
TRNSYS: http://www.trnsys.com

BIM implementation and deployment (I)—principles and case studies

eWork and eBusiness in Architecture, Engineering and Construction – Christodoulou & Scherer (Eds)
 ISBN 978-1-138-03280-4

Is BIM-based product documentation based on applicable principles?—Practical use in Norway and Portugal

E. Hjelseth
Department of Civil Engineering and Energy Technology, Faculty of Technology, Art and Design, Oslo and Akershus University College of Applied Sciences, Oslo, Norway

P. Mêda
Construction Management Division, Faculty of Engineering, Porto University, Porto, Portugal

ABSTRACT: This study explored implementation solutions for BIM-based product documentation in Norway and Portugal, two countries with different BIM maturity. In this respect, significant differences in BIM-based product documentations are expected. The work was based on a selection of small, medium and large companies (type of company: architects, engineers, contractor, owner and Facility Management (FM), both private and public sector) used as case study of the two countries. Integrated Design and Delivery Solution (IDDS) is the framework for structuring findings and analysis. The study confirmed that BIM-based product documentation is limited compared to traditional document-based solution on paper and/or PDF documents. There were no significant differences between Norway and Portugal, but the expectations were higher in Norway. However, companies with high implementation of digital solutions understood BIM-based product documentation in a different and more integrated way than as just an extension of properties in BIM-based objects. The impact of this contributes to increase understating of product data as extension of BIM or need for development of Asset Information Modelling (AIM) -based solutions.

1 INTRODUCTION

Product documentation is mandatory in trading and, choice of relevant products, purchase, logistics, production in addition to facility management. In this respect, product documentation should be regarded as an embedded part of every construction product. The use of BIM-based processes and tools should therefore be expected to be highly integrated within solutions for product documentation. Based on these preconditions, should one expect that countries with high maturity in BIM have a significantly higher implementation of BIM-based solutions for product documentation. This study will explore the relation between BIM maturity and the use of BIM-based product documentation, which can be useful to predict use of product-documentation solutions. A better understanding of these relations can contribute to an improved perception of what must be included in solutions for utilization of product documentation in the entire life cycle of the building process. Even if this study only uses Norway and Portugal as references, the principles should be relevant for other countries.

BIM-based solutions are emphasized by multiple sources (McHill, 2014) as one of the major contributions to developing the information flow in the AECOO industry (architects, engineers, contractors, operators and owners). There are many definitions of BIM, and this study will not go into discussions about BIM definitions. Countries and companies with high maturity in use of BIM should be expected to have a significantly higher implementation of digital solutions for product documentations. It is therefore more relevant to focus on integrated solutions at the national level. To explore implementation of BIM-based or digital-based solutions, the researchers have chosen to look into one country that is considered to have a high level of BIM implementation, in the study represented by Norway, and one country that is in an early stage of BIM implementation, in the study represented by Portugal (Venâncio, 2015 and Ribeiro, 2012). The expected outcome is that the AECOO-industry in Norway should have a significantly higher implementation of BIM-based solutions for product documentation. However, how digitalization and BIM are understood can have a significant impact on implementation. The study intends to identify different approaches for development and implementation of digital solutions for product data.

The research question is therefore: *How is the relation between digital solutions of product documentation and use of BIM based on observations from Norway and Portugal.*

2 MATERIALS AND METHODS

2.1 *Choice of methodology*

This study explores the practice related to product documentation in general and BIM-based documentation with highlight in Norway and Portugal. This is based on a combination of exploring BIM-based software, literature review and semi-structured interviews with relevant stakeholders in the AECOO industry. The use of a wide-distribution questionnaire to question a large number of stakeholders was considered, but this method would likely have prompted an underrepresentation of "non-users." Interviews would also give a better understanding of the situation and problems.

The analysis is performed using a multi-disciplinary approach in which the three aspects of IDDS—integrated process, collaborating people and interoperable technology—are used to analyse aspects that can contribute to understanding relevant factors for BIM-based/digital solutions for product documentation. The technology aspect is based on exploring digital solutions for specification, distribution and the acquiring of product documentation.

2.2 *Overview of interview with stakeholders in Norway and Portugal*

The aspects of integrated processes and collaborating people are completed by semi-structured interviews with stakeholders that have a high level of knowledge and understanding regarding BIM methodologies, as presented in Table 1.

2.3 *Integrated Design and Delivery Solutions (IDDS)*

The analysis of the interviews is based on IDDS as a theoretical framework. IDDS focuses on the integration of collaboration between people, integrated processes and interoperable technology (IDDS, 2013), as illustrated in Figure 1. These three aspects are used to structure findings and discussions.

The technology aspect is based on exploring digital solutions for specification, processing and distribution of product documentation. The aspects of integrated processes are related with regulations, including legislation, standards and market-related processes for product documentation. Assessing the collaboration of people includes focusing on the awareness of how people collaborate to utilize processes and technology. It is important to be aware that the integration of all three aspects can result in something new and different from the sum of all "elements."

Product documentation calculations can be regarded as integrated deliverables. The aspects of IDDS can be embedded into product documentation through the following ways:

- Integrated Processes -> use of standards and methods for reliable processing of product documentation.
- Collaborative People -> multidisciplinary collaboration ensuring validity of input in product documentation.
- Interoperable Technology -> software technology based on IFC/openBIM input for product documentation.

IDDS was a special theme initiated by CIB in 2009 (CIB, 2016). This indicates that IDDS is a relevant framework for analysis of AECOO-related problems. IDDS can also be regarded as a simplification of the sociotechnical theory (Bostrom & Heinen, 1977) adapted to the AECOO industry.

Table 1. Interviews conducted.

Reference	Type of company	Role in company/organization	Country	Interview technique
N-Public	Public client	Facility manager	Norway	Face-to-face
N-Design	Design company	Consulting engineer	Norway	Face-to-face
N-Contractor	Contractor	Tendering office	Norway	Face-to-face
N-Hospital	Hospital owner	Facility manager	Norway	Face-to-face
N-Soft	BIM software developer	Management	Norway	Face-to-face
N-Legal	Public authority	Professional consultant	Norway	Face-to-face
P-Design	Design company	Architect	Portugal	Skype
P-Design	Design company	Structural engineer	Portugal	Face-to-face
P-Design	Design company	Project manager	Portugal	Face-to-face
P-Contractor	Contractor	Tendering office	Portugal	Skype
P-Assoc.	Manufacturers association	Secretary-general	Portugal	Face-to-face

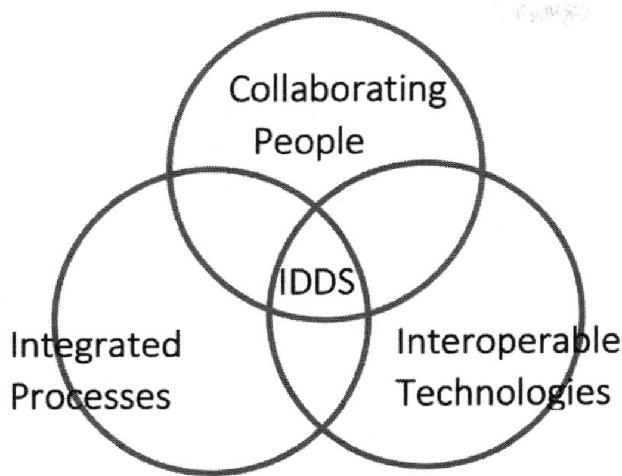

Figure 1. The three aspects of IDDS (2013); figure simplified by the authors.

2.4 *Overview of product documentation*

The characterization of products through organized information/parameters is fundamental, and it is an enabler for the evaluation and selection of a certain product over other that is similar. The default rule is that products shall be documented (DiBK, 2016). Figure 2 presents an overview of different types of regulations that contribute to different types of documentation of products.

2.4.1 *European level*

At the European level, the European Parliament has laid down harmonized conditions for construction products trading (EU Regulation, 2011). This regulation has an impact on a large number of other regulations at the European or national level. REACH is a regulation of the European Union that requires unambiguous substance identification as a pre-requisite. Actors in the supply chain must have sufficient information to identify their substances (REACH, 2016).

However, it is important to be aware of the difference between documentation for marketing and use of building materials in a specific project. This implies that not all building materials that are traded on the European market may be allowed for use in Norwegian or Portuguese buildings.

CE-labelling/Declaration of Performance (DoP): This mark proves that a construction product complies with Construction Products Regulation, either by confirming that the construction product is in conformity with a harmonized product standard or a European technical assessment. This means that CE-labelled construction products shall be freely trade throughout the EEA. However, CE labelling does not provide complete product information or constitute a quality mark—conformance must not be mixed with performance to the product.

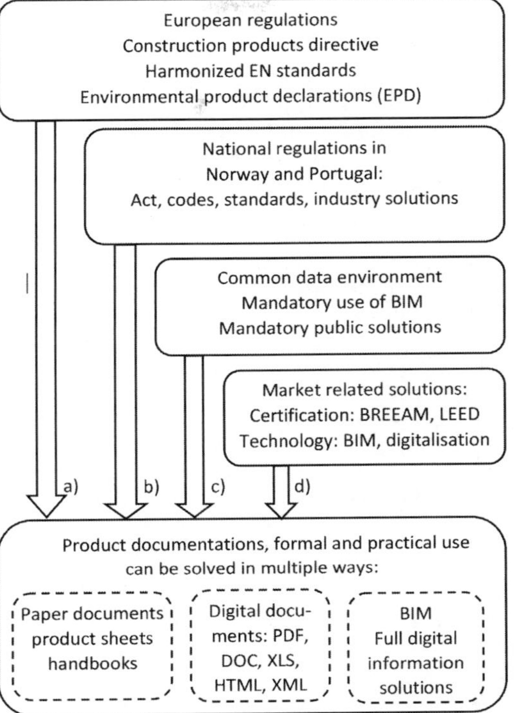

Figure 2. Overview of public and commercial regulations.

2.4.2 *National requirements*

In Norway there are the technical requirements for product types specified in technical code called TEK10 (supplemented with a guideline approx. 300 pagers), while the specifications for each product shall be documented according to a special code called Code of Documentation (DOC) (DiBK, 2016). The DOC code state that relevant information shall be submitted to the building owner by hand-over. However, the code do not give help to the important issue in specifying relevant information.

In Portugal, the technical requirements applicable for product types are spread in different regulations. These are applicable by construction types, such as buildings and urban areas, roads or others. In terms of building construction, there can be assumed two main regulations, (RGEU,

2008 and RJUE, 2014), and others that derive from these, geared for specific requirements/ design disciplines. Fire and acoustic performance or electric systems regulations are some examples. The Technical Housing Sheet is a mandatory document with product information that needs to be delivered to the owner before inhabiting a house (Portaria, 2004). Usually, it is provided on paper format. In terms of civil-engineering construction types, there are requirements for products on the public owner guidelines. On a local level, there are complementary provisions that can add other requirements.

2.4.3 *Common Data Environment (CDO)*

Establishing a Common Data Environments (CDO) is essential for exchange of information. In this study, BIM-based solutions are used with a wide interpretation to covers software solutions for exchanging specified information. This scope covers more than file-based exchange and use of IFC format, such as the Enterprise Resource Planning (ERP) system. Refvik (2014) indicates a global trend going towards model-server-based solutions rather than single-file exchange of information.

2.4.4 *Market related*

Even if product data traditionally has been regarded as a legal issue with a focus on minimum specifications, we observe that BREEAM, LEED and other environmental or energy certification or labelling solutions results in increased focus on digital product documentation to support the assessment process.

2.4.5 *Formal and practical product documentation*

The overview of regulations in Figure 1 is not related to defined phases, role or purposes in the building life cycles. Legal documentation require only a limited number of properties. DIBK (2016) says that only one single property is enough to declare for allow trading of a product. The use of Product Data Template (PDT) will include documentation of more criteria.

Product information can assume different visions. From the product regulation point of view, the product types must present specific properties according to the harmonized standard. These properties result from tests and define the product characteristics. The regulations set information or characteristics that need to be declared for specific applications and in specific parameters as fire resistance, thermal or acoustic isolation, among others.

These two visions of information must be compatible for all products, as the product's Declaration of Performance (DoP) must be confronted with the minimum requirements from the regulations in order to obtain product compliance for a specific use/application.

Environmental Product Declarations (EPD) are additional documents that provide complementary product properties. These documents are not yet mandatory, but they can provide a set of complete product information. PDTs are structures geared for the systematization of all the product properties, from the ones mentioned to product dimensions or their classification. These structures fit and resume global data in a way that can be used by information systems.

3 FINDINGS AND ANALYSIS (RESULTS)

3.1 *Criteria for analysis*

The findings from interviewees and literature review are structured after the following three aspects in the IDDS framework: Integrated Processes, Collaborative People and Interoperable Technology. Findings from Norway and Portugal are briefly mentioned, as well as similarities and differences presented in the analysis part.

3.2 *Integrated process*

3.2.1 *General introduction*

The integrated process aspect includes procedures for how product documentation is processed, presented and distributed to relevant stakeholders. Legislation and standards do therefore have a central role in structuring processes, supported by collaborative people and interoperable technology.

3.2.2 *Findings from Norway*

Act and code are mainly function based. They require documentation of "as-built", but the required information is only specified as: "*documentation should be adequate to secure operation and maintenance of the building*" (DiBK, 2016). The perspective is illustrated by the following: "*The purpose of public legislation of product documentation is securing health and security with minimum level of documentation*" (N-Legal). This is further confirmed at the European level: "*European regulations have priority to documentation for reducing trade barriers. References to harmonized standards are one important element*" (N-Legal).

Even if hand-over of documentation is included in the contract, this is often very limited "*There is a challenged that the information we need in design is not identical with the owner's need for operation. The clients must therefore be more specific about their needs. We deliver documentation in PDF format, and from our point of view, it is much easier to deliver a complete product catalogue than identifying relevant parts*" (N-Design).

3.3 *Findings from Portugal*

Traditional design (CAD + documents) has a low level of integration. BIM adoption case studies are focused on the introduction of product information to get outputs for the construction process stage. For example, a structural engineer places the relevant product information on the BIM model to support the dimensioning task. A contractor who develops a BIM model during the tender action has the major concern of quantity take-off, and therefore it works out so that only the product properties can help support this task.

The situation regarding product documentation can be illustrated as follows: "*Product documentation constitutes a challenge for the coming years. Some manufacturers are starting to provide BIM objects. This is important from the marketing point of view, and it can foster their competitiveness. The challenge of providing complete information for the supply chain is not easy. Improved awareness of the benefits and importance of product documentation is important for the manufacturer's work as a driver of change*" (P-Assoc.).

An interesting action toward process integration was developed by a Public Owner that adopted a Government tool to set lists of requirements according to standards and regulations for the products. These need to be fulfilled by the design team for the production of the Bill of Quantities. The design teams that participated in these projects comply with the specification requirements, and good documents were produced to support the construction stage (Mêda, 2014).

The situation can be illustrated by following quote: "*On the school refurbishment process, product documentation was pushed from traditional design processes with the use of ProNIC—Construction Standardization Protocol. Through this methodology standard bills of quantities for design delivery were produced, and specifications were set in digital format. The drawings were produced following the "traditional CAD tools." Communication of this application with modelling tools would provide interesting capabilities for the different agents*" (P-Design-Arch).

The trends point to the ability of this tool to also produce documentation to link with other applications such as modelling software or provide outputs for asset management.

3.3.1 *Analysis of integrated processes*

There is a lack of focus on integrated processes, due to a narrow focus on minimum requirements for specific tasks such as trade, rather than information to support life cycle or the building.

3.4 *Collaborating people*

3.4.1 *General introduction*

This aspect focuses on who are working with product documentation and with whom they collaborate. This perspective includes the priority to collaborate (which also intersects with processes).

3.4.2 *Findings from Norway*

Product documentation in general is a low priority and something that is done after the busy construction phase is over. An example of this situation can be illustrated by the following: "*As operator of the building, we face the problem with lack of product documentation. It can take up to two years to get all necessary information. A challenge is that we can get a lot of documentation, both as physical documents (product-sheets/catalogues) and as PDF files. However, much documentation does not imply that we receive relevant information*" (N-Public). This priority is also confirmed by the contractor's experience: "*We work in an industry with hard competition—and deliver what the client wants—and are paid for it. We mostly forward information that comes from our suppliers and sub-contractors. So far, this has been good enough. We have never lost contracts due to product documentation, and the specifications are often limited*" (N-Contractor).

3.4.3 *Findings from Portugal*

The roles are similar to those of Norway and can be illustrated by following: "*The scope of developing a BIM model from our point of view is the ability to get quantity take-off in order to help the tendering process. Therefore, the introduction of product documentation is low. If the process goes further, to construction, some systems can be more developed in order to support the construction planning*" (P-Contractor).

The focus on collaboration and level of skills is illustrated by the following quote: "*The supply-chain behaviour in terms of design delivery can be highly varied. In terms of product documentation, it's worth mentioning that the experience proves that architects and structural engineers can provide interesting BIM models in terms of visualization and embedded product documentation. MEP engineers are a little behind. Yet, BIM objects for electrical products are in accelerated development. Much information that comes from the models is not useful for the process. The introduction of information does not lead directly to good information*" (P-Design-Pro).

3.4.4 *Analysis of collaborating people*

People/professionals working with product documentation are different from the professionals working with design and project management.

Understanding each other's needs for relevant information as a challenge and an underlying reason for more focus on the "push in" of all information, preparing digital solutions for the user helps him/her to "pull out" relevant information when needed.

3.5 *Interoperable technology*

3.5.1 *General introduction*

Interoperable technology is in this study defined as technology where information can be captured, processed, presented and interchange between multiple software solutions. BIM-based software for Facility Management (FM) is generally at an early stage in development. FM software is in general profiled to administrative solutions, not construction (building) solutions.

3.5.2 *Findings from Norway*

Due to high focus on openBIM and a very active buildingSMART community, interoperability is in focus, and IFC import/export is given priority.

Even if digitalization in general has a high focus, this is not reflected in legislation as the following quote expresses: "*Development of regulations has so far not been adapted to digital solutions; however, they should not be regarded as barriers. Please note that they require documentation, not necessary documents*" (N-Legal). See Figure 4 for difference between "documentation" and "documents".

The situation for practical use can be illustrated by the following statement: "*Even if we demand BIM-based solution, our experience is that product documentation and facility management is a poorly developed digital solution compared with BIM in the design phase, even if the FM phase is much longer and needs solutions that can be continually updated*"(N-Public).

Norwegian software developers for the AECOO industry have high priority in IFC-based exchange of information. Examples of software developers are:

CoBuilder (www.cobuilder.com) is a pioneer in product documentation. Its solutions include collection and processing of information to document legislation and support for FM.

Areo software for FM (http://areo.io/) is a new company with a focus on IFC-based exchange of facility management information.

Byggtjeneste (www.byggtjeneste.no) has a long history of product documentation for the distribution, logistics and retail market. They have recently developed a solution together with My Building Folder (www.boligmappa.no) for distribution of digital documents directly from selected retailers or electrical and plumber contractors (which have specific codes that demand documentation to private clients).

Catenda (www.catendea.no) offers a customizable solution based on BIM server. They have started developing solutions for FM linked to import and visualizations of IFC models.

The situation for software developers can be illustrated by the following quote: "*Developing simple solutions for distribution of documents (PDF) is quite easy. Developing digital and integrated solutions for specified elements of information requires adaptations to the client's needs. Chemical documentation is an example of a specialized adaptation to a group of clients. However, having quality, secured information with correct references is critical for correct decisions*" (N-Software).

The need for integrated solutions for information exchange can be expressed by the following quote: "*Based on my background in cybernetics, I find it hard to see that the current file-based exchange on one single format (IFC) can solve the variety in sources and use cases of information*" (N-Hospital). The hospital uses model-server-based solutions for facility management. This way of thinking indicates that asset information is a part of the corporate information management system.

3.5.3 *Findings from Portugal*

The Portuguese construction sector is behind schedule in this issue. There are few works and communities perched on this subject. Most of them are part of universities. There is still no involvement on the part of the buildingSMART community.

The situation regarding use of BIM is illustrated by the following quote: "*BIM is not mandatory for public works. However, some private owners are already demanding it*" (P-Design-Arch).

3.5.4 *Analysis of interoperable technology*

There were significant differences in expectations, but regarding practical implementation, both counties were in the early stage of digitals solutions. In Norway, there was an expectation related to extensions of BIM bases by use of IFC files with extended property sets, including the FM phase. Examples for the Norwegian software industry indicate opportunities for new types of solutions based on multiple interlinking sources. This solution was only observed in pilot studies in large organizations, and they were supported by manual effort in order to be applicable. These "expectations" are not enough to conclude that BIM-based solutions are practical for use.

4 DISCUSSION

4.1 *Direct and indirect findings*

The outcome from the interviewees and literature review in Norway and Portugal did not provide support for the identification of significant differences in practical use of BIM-based product documentation, based on the expectations that a high degree of BIM implementation in general should result in BIM-based product documentation.

The AECOO industry seems to be based on a traditional way of thinking about digitalization, where paper documents just become replaced with digital documents in PDF or HTML format. Lack of understanding of digitalization and information exchange appears to be an underlying factor in the industry. A study by Codinhoto et al. (2013) supports the by following statement: *"Results indicate a lack of awareness related to the benefits that BIM can offer to FM processes."* Further studies by Bew and Underwood (2010) and discussions by Eastman et al. (2011) confirm that it is rare to find evidence that demonstrates benefits within FM.

BIM is an enabler, but not a driver for change. Due to the defragmentation in the industry, it is found that public regulations or targets like environmental assessment (BREEAM, LEED, etc.) will have a significant impact on collecting and structuring product data.

4.2 *Analysis by use of integrated perspectives*

Through use of IDDS, this study intends to present an explanation where on can not continue with "more BIM is better". Studies by Succar (2009) point out a general lack of understanding of BIM capabilities and maturity levels for understanding of BIM related to development and solutions. In this respect, IDDS can be a supplement to increased understanding. The aspects of IDDS were presented in Figure 1 as three integrated circles of identical size. This ideal solution is not observed in this study. The results identified a lack of integration between: 1) integrated processes, 2) collaborative people and 3) interoperable technology. This lack of integration is the most important reason for the limited implementation of BIM-based LCC calculations. A visualization of status is illustrated in Figure 3.

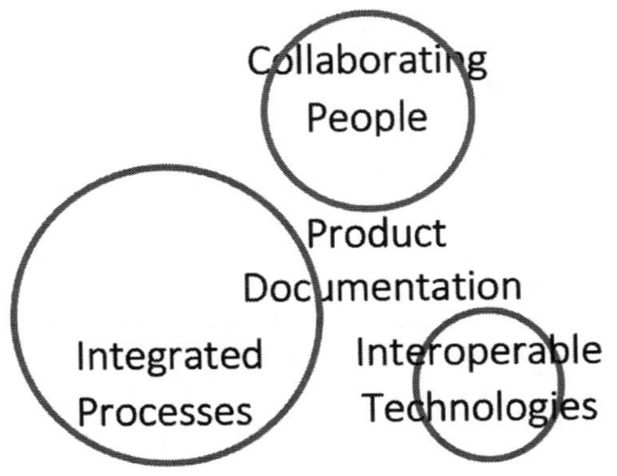

Figure 3. The current aspects of BIM-based product documentation.

4.3 *Difference between document and documentation*

The impact of this study indicates that it is not enough only to focus on traditional BIM implementation. BIM-based product data, (despite the naming) should not be static related to the BIM object, but it should have a dynamic relation to phase, role and purpose. Product documentation can be regarded as Asset Information Modelling (AIM). A more dynamic model that enables AIM to support BIM is needed. This can be expressed as: **AIM** + **BIM** = > **CIM** to be adapted to phase, role and purpose. The C has an inclusive interpretation and can be an abbreviation for: communication, construction, commercial, client requirements and collaboration. Figure 4 illustrates this in the following way:

Results identify an increasing focus to include product documentation as part of the building process in general and hand-over specifically. This development is more related to change of business model supported by digitalization—and not the utilization of traditional file-based use of BIM.

Product in the CIM-based context is related to information about the product, which can be a virtual object with varying degrees of relevant information.

Use of systems such as levels of model detail (LOD), which relates to the graphical content of models, and levels of model information (LOI), which relates to the non-graphical content of models (PAS 1192-2:2013) is still missing in practical use of AIM. The current focus has a static way of thinking, The technical solution is file-based with information embedded as property sets, resulting in heavy and static models in IFC format, or as separate file in PDF format. Information as services related to users current need is missing.

Virke (2015) presents this as the dominating perspective in the BIM community. The report highlights implantation on sematic-based solutions based on buildingSMART DataDictionary as enablers. Global Trade Item Number (GTIN)

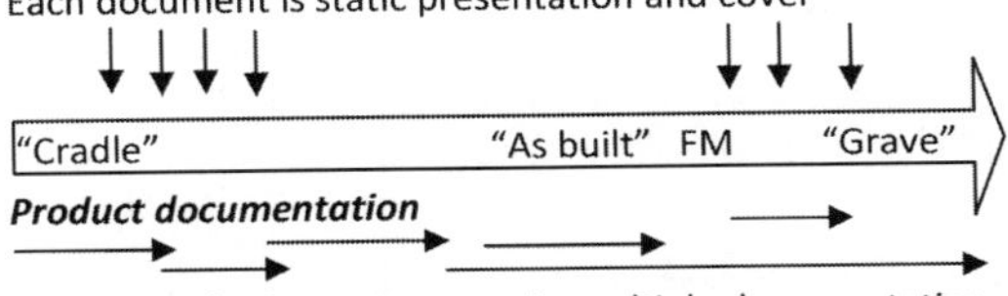

Figure 4. Product document vs Product documentation.

can be another enabler. GTIN describes a family of global data structures that employ 14 digits, and can be encoded into various types of data carriers. Currently, GTIN is used exclusively within bar codes (GTIN, 2016), but it could also be used in other data carriers such as radio frequency identification (RFID). Change of paradigm about BIM as a solitary solution for the AECOO industry and integration of GTIN-based solution within the entire value chain, not only trade, can have a significantly impact on the value or relevant product data in supporting good solutions.

4.3.1 *Understanding innovations*

Although the current use is limited, the expectations for BIM-based improvements are relatively high, especially in Norway. However, this view regards increased use of BIM-based technology to play the major role in change of solution. This can be expressed as follows: "BIM revolution has started with architects—and other actors are following—and FM is expected to be then next step." The interviewees from the AECOO industry, especially in Norway, have high expectations for future BIM-based solutions. This is representing a techno-centric perspective where BIM in general and IFC (open-BIM) specifically act as drivers for change. Use of IDDS in this study will instead classify BIM as an enabler. In the fragmented AECOO industry, two of the most significant types of drivers for change will be: 1) use of regulation with new requirements for distribution and access to documentation; and 2) manufacturers can provide relevant information as an embedded part of the product. The information delivery can be customized to the clients need, at be exchanged in relevant formats (IFC, XML, XLS etc.) and media (e.g. file by e-mail, commercial service solutions or different cloud solutions).

Regarding different maturity of BIM in Norway and Portugal, some systematic differences should be expected. The general level of BIM implementation was not identified as a driver for change, and other reasons such as demand from public authorities or market forces have significantly more influence on use of product documentation. Products with multiple pieces of public legislation, such as chemical products or products related to safety issues, are in general covered. However, this type of information does not include the building (maintenance, performance) within its scope. This type of information will not contribute to, e.g., LCC calculation or maintenance schedules.

4.4 *Directions for further studies*

Further studies are needed in two directions: The first direction is to explore countries with different contexts (regulations and BIM focus). In this respect, the UK is interesting due to its pervasive focus and "mandatory" use of BIM in building projects. The second direction is depth studies within each country that uses the IDDS framework to identify: a) drivers for change, b) enablers, c) barriers and d) opportunities in the short term, medium term and long term related to the use of BIM-based product documentation.

5 CONCLUSIONS

The use of product documentation is very fragmented in the AECOO industry and plays a limited degree of support to the life cycle of buildings. There is no joint understanding of the concept and of content in product documentation. Use of IDDS as theoretical framework has contributed to the discoverer aspect of cross-disciplinary understanding.

The study confirmed that paper-based documentation, such as PDF files or printed documents, is dominating in both Norway and Portugal. It has not been possible to identify general differences between the countries in implementation of BIM-based solution that have affected the national industry in general. Traditional solution and manual processes are dominating. However, based on higher BIM maturity, the Norwegian industry, especially public builders and large companies, have a different attitude/approach, which includes product documentation as part of the building process in general and hand-over specifically. This development is more related to the changes in the business model supported by digitalization and less related to utilization of traditional file-based use of BIM.

The impact of this study indicates that it is not enough only to focus on BIM-based solution as extension of information into static objects (property sets) with file-based exchange, even if open format like IFC is used. Product documentation is the key element in asset information must be related to real needs in the different phases, roles and purposes to enable an "***AIM + BIM => CIM***" way of thinking.

The C is dynamic and can be replaced with; Communication, construction, commercial, collaboration, other related to users current need. This way of thinking implies that digital solutions for product documentation should not only fulfil minimum public requirements, but must be regarded as value adding services to information management throughout the entire life cycle of the building.

REFERENCES

Bew, M., & Underwood, J. 2010. Delivering BIM to the UK market. Handbook of research on building infor-

mation modeling and construction informatics: *Concepts and technologies* (pp. 30–64).
Bostrom, R.P. & Heinen, J.S. 1977. MIS problems and failures: A socio-technical perspective, *MIS Quarterly, Vol. 1, No. 3*, pp. 17–32. http://misq.org/cat-articles/mis-problems-and-failiures-a-socio-technical-perspective-part-i-the-causes.html
CIB. 2016. Priority Theme, Integrated Design and Delivery Solutions" (IDDS), *International Council for Building (CIB)*. http://www.cibworld.nl/site/programme/priority_themes/integrated_design_and_delivery_solutions.html.
Codinhoto, R., Kiviniemi, A., Kemmer, S. & Gravina da Rocha, C. 2013. BIM-FM Implementation: An Exploratory Investigation. International, *Journal of 3-D Information Modeling*, 2 (2). pp. 1–15.. http://repository.liv.ac.uk/id/eprint/2007718
DiBK. 2016. Building Acts and Regulations, Norwegian Building Authority. https://www.dibk.no/no/byggeregler/Gjeldende-byggeregler/Building-Regulations-in-English/
Eastman, C., Teicholz, P., Sacks, R., & Liston, K. 2011. BIM handbook: A guide to building information modelling for owners, managers, designers, engineers and contractors (2nd ed.). Wiley. ISBN: 978-0-470-54137-1
EU-regulation. 2011. EU-regulation No 305/2011 of the European Parliament and of the Council of 9 March 2011 laying down harmonised conditions for the marketing of construction products http://eur-lex.europa.eu/legal-content/EN/TXT/?uri = celex%3A32011R0305
GTIN. 2016. Global Trade Item Number (GTIN), *The Global Language of Business*. http://www.gs1.org/gtin
IDDS. 2013. Integrated Design and Delivery Solutions (IDDS), CIB Publication 370. Editors: Owen, R., Amor, A., Dickinson, J., Matthjis, P. & Kiviniemi, A. *International Council for Building*. ISBN 978-90-6363-072-0. http://site.cibworld.nl/dl/publications/pub_370.pdf
McGrawHill. 2014. Business Value Of BIM In Global Markets 2014, Smart Market Report, Design and construction Intelligence. *Mc GrawHill Construction*. https://synchroltd.com/newsletters/Business%20Value%20Of%20BIM%20In%20Global%20Markets%202014.pdf
Mêda, P. 2014. *Integrated Construction Organization - Contributions to the Portuguese Framework*. Master Thesis in Civil Engineering, Faculty of Engineering of the University of Porto, Portugal. https://sigarra.up.pt/flup/pt//pub_geral.show_file?pi_gdoc_id = 327084
PAS 1192-2:2013 Specification for information management for the capital/delivery phase of construction projects using building information modelling, *British Standards Institution*. http://shop.bsigroup.com/navigate-by/pas/pas-1192-22013/
Portaria. 2004. Portaria n.º 817/2004 de 16 de Julho, Housing Technical Data Sheet. Presidency of the Council of Ministers and Ministries of Economy and of Public Works, Transports and Housing. *Diário da República, I série-B—N.º 166*.
REACH. 2016. The European Chemicals Agency (ECHA) http://echa.europa.eu/web/guest/regulations/reach/
Refvik, R. 2014. ByggNett – Status survey of solutions and issues relevant to the development of ByggNett. *Norwegian Building Authority report*. dibk.no/globalassets/byggnett/byggnett_rapporter/byggnett-status-survey.pdf
RGEU. 2008. General Regulation of Urban Buildings. Decreto-Lei n.º 50/2008, de 19 de Março, *Diário da República, 1.ª série—N.º 56*, Ministry of Public Works, Transports and Communications, Portugal.
Ribeiro, D.C. 2012. *Avaliação da aplicabilidade do IPD em Portugal*. Master Thesis, Porto University - Faculty of Engineering. https://paginas.fe.up.pt/~gequaltec/w/images/Tese_DavidRibeiro.pdf
RJUE. 2014. Legal Regime for Building and Urban Development. Decreto-Lei n.º 136/2014, de 09 de Setembro, Ministry of Environment, Planning, Territories and Energy, *Diário da República, 1.ª série—N.º 173*, Portugal.
Succar, B. 2009a. The difference between BIM capability and BIM maturity. BIM ThinkSpace Blog. Retrieved September 22, 2011, from http://www.bimthinkspace.com/2009/06/index.html
Venâncio, M.J. 2015. *Avaliação da Implementação de BIM—Building Information Modeling em Portugal*. Master Thesis, Porto University - Faculty of Engineering. https://paginas.fe.up.pt/~gequaltec/w/images/Dissertacao_VersaoFinal.pdf
Virke. 2015. Verdikjede (Value chain). Byggevarehandelens rolle i dagens og fremtidens BIM-modeller, *Virke report series*, http://www.virke.no/bransjer/bransjeartikler/Documents/ Byggevarehandelens%20rolle%20i%20dagens%20og%20fremtdens%20BIM%20modeller,%20Verdikjede%20juni%202015.pdf

eWork and eBusiness in Architecture, Engineering and Construction – Christodoulou & Scherer (Eds)
© 2016 Taylor & Francis Group, ISBN 978-1-138-03280-4

Necessary conditions for the accountable inclusion of dynamic representations of inhabitants in building information models

A. Mahdavi & F. Tahmasebi
Department of Building Physics and Building Ecology, TU Wien, Vienna, Austria

ABSTRACT: To cater for the informational requirements of building assessment applications, building information models need to include representations of inhabitants. Thereby, the representation of people as passive and static entities is unlikely to yield reliable building performance assessment and building operation planning. Rather, adequate representations of building inhabitants should account for user-initiated actions (e.g., interactions with buildings indoor environmental control devices and systems). To address these requirements, many recent model development efforts have explored the potential of sophisticated mathematical formalisms. However, the resulting occupancy-related behavioural models have rarely gone through a rigorous evaluation process. The present contribution is indeed motivated primarily by the lack of general procedures and guidelines for the evaluation of proposed user-related behavioural models. Specifically, we formulate a number of conditions that are necessary for systematic and dependable enrichment of building information models with representations of buildings' inhabitants. Toward this end, we discuss both general model evaluation requirements as well as specific circumstances pertaining to models of building inhabitants. Moreover, we present, as a case in point, a model evaluation study involving a number of recently proposed window operation models. Thereby, our main objective is to promote a rigorous process toward quality assurance while considering and integrating behavioural representations in building information models that are meant to meet sound scientific requirements as well as professional accountability criteria.

1 MOTIVATION AND BACKGROUND

Building Information Models (BIM) have the potential to support building performance assessment applications via efficient and structured provision of required input information. Specifically, building performance simulation tools require information on context (climate), building geometry, construction, and internal processes. Whereas inclusion of information on physical building components and properties (pertaining, for example, to buildings' fabric and construction) in BIM is fairly well advanced, representations of inhabitants (presence, movement, behaviour, perception, and evaluation) are frequently rudimentary. In fact, assumptions regarding user presence and behaviour in building are not explicitly specified in BIM, but rather injected quasi downstream, i.e. in the course of simulation model generation. To accommodate a number of salient applications—particularly in building performance simulation and building systems control domains—building information models need to include representations of inhabitants. Moreover, such representations should ideally reflect the nature of building occupancy processes: Specifically, the representation of people as passive, static, and exchangeable entities would not cater for reliable building performance assessment and building operation planning. Rather, adequate representations of building inhabitants should address not only passive presence, but the multiplicity of active user-initiated actions (e.g., interactions with buildings indoor environmental control devices and systems). Moreover, such representations should reflect the dynamic nature of such actions. A related phenomenon that needs to be considered in any model development activity is the inhabitants' behavioural diversity (inter-individual differences amongst attitudes, preferences, and habits).

Conventional representations of buildings' inhabitants in performance simulation models mostly consist of fixed schedules (so-called diversity profiles) and rule-based action models. As such, these kinds of representations do not realistically reflect the inherent temporal fluctuations of occupancy-related processes and events (e.g. entering, leaving, and moving in buildings, operation of devices such as windows, blinds, luminaires, manipulation of control set-points). There has been thus recently a number of efforts—for instance, by the professionals in the building performance simulation community—to develop more sophisticated dynamic models of people's presence and actions in buildings (e.g., stochastic algorithms, agent-based representations).

A significant number of such efforts have focused on the potential of probabilistic methods and associated formalisms. Thereby, a stated objective has been to replace fix schedules and rule-based actions models in performance simulation with stochastically based models. A number of such models have been and are being incorporated in building performance simulation applications. However, this process has not been too immune to a number of unwarranted claims, misconceptions, and fallacies (Mahdavi 2011, Mahdavi 2015, Mahdavi & Tahmasebi, 2016). Models have been at times prematurely promoted as valid and reliable, despite wanting empirical evidence and despite ignorance regarding the down-stream deployment scenarios. It is conceivable that these kinds of insufficiently tested representations could migrate up stream to BIM products—implemented, for instance, in terms of software agents representing inhabitants. But while the inclusion of sophisticated and realistic behavioural models in BIM is desirable as such, it must be done in a careful and systematic manner, lest confusion and poor decision making may result.

Given this background, the present contribution is primarily motivated by the lack of general procedures and guidelines for the evaluation of proposed user-related behavioural models. Specifically, we formulate a number of conditions that are necessary for systematic and dependable enrichment of building information models with representations of buildings' inhabitants. Toward this end, we discuss both general model evaluation requirements as well as specific circumstances pertaining to models of building inhabitants. Moreover, we present, as a case in point, a potentially paradigmatic model evaluation process using a comparison of a number of recently proposed window operation models. Thereby, our main objective is to promote a rigorous process toward quality assurance while considering and integrating behavioural representations in building information models that are meant to meet sound scientific requirements as well as professional accountability criteria.

2 ABOUT MODEL VALIDATION

A central trust of scientific activity is the development of models that are used to describe phenomena and predict events. Given the persistence and historical evolution of model development activity across a variety of scientific disciplines, one would expect that there would be no need to revisit the question of model validation in the rather narrow context of the occupancy-related models. However, at least a brief treatment would be in order, given the aforementioned shortcomings in the building inhabitants model development domain. Note that a number of such shortcomings could be the consequence of the following three circumstances:

- Firstly, occupancy-related studies in the context of the built environment belong to a relatively young field of inquiry. The strength of research standards in a domain typically results from expected utility and a critical mass of projects and researchers. As compared to many other areas of scientific inquiry (such as medical sciences or information technology), occupancy-related research is significantly underdeveloped.
- Secondly, a perilous problem for both model development and model evaluation lies in the rather limited availability of large-scale observational data. Consequently, the demographic basis of most proposed models is often very small. The coverage and representativeness of behavioural models of buildings' inhabitants depends on the availability and fidelity of observational data. As such data is still hard to come by, models are often proposed with insufficient empirical backing.
- Thirdly, behavioural models require—in principle—the concurrent consideration of multiple operative parameters of physical, physiological, psychological, and socio-cultural nature. To conduct field or controlled studies addressing this complex pattern of potential causal factors is indeed anything but trivial.

Obviously a number of above-mentioned challenges in behavioural model development and evaluation cannot be met in the short run. Collection of vast amount of reliable observational data in the course of field studies is laborious, time-consuming, and costly. Likewise, conducting experimental behavioural studies is exceedingly difficult and the corresponding results cannot be readily generalised. These observations, however, do not absolve the invested community from trying to do better, and an indispensable precondition for doing better is a self-critical assessment of the past efforts in model development and application. Specifically, avoiding certain common mistakes and fallacies would help to further the behavioural modelling discourse in a more reasoned manner (Mahdavi 2015). Specifically:

- We should not confuse simulation (computational, typically dynamic representation of a system's behaviour) with prediction;
- We should not generally claim that the mismatch (the so-called performance gap) between simulation-based predictions and observations of

energy use is necessarily due to behavioural factors. Long-term predictions of building performance indicators are difficult (if not impossible) to make due to an extensive list of uncertainties, pertaining not only to internal (occupancy-related) processes, but also to building fabric, building systems, and especially boundary conditions (climate);

- We should not use the expression "deterministic", which has a weighty philosophical baggage, while meaning to refer to fixed diversity profiles (e.g., assumed fixed schedules of occupants' presence) and rule-based behavioural models;
- We should not claim building performance simulation results would be necessarily more "accurate" if we simply replace occupancy-related diversity profiles and rule-based assumptions with probabilistic ones (see for example Tahmasebi & Mahdavi 2015, 2016);
- We should not confuse code-based benchmarking with energy use prediction. Specifically, we should not assume that a specific modelling approach or technique can be appropriately applied to all kinds of use cases (see Gaetani et al. 2016, Mahdavi & Tahmasebi, 2016).
- We should properly and meticulously document the model development and evaluation procedures (research design, empirical basis, hypotheses and assumed causal factors, limitations, etc.), such that others could independently retrace, comprehend, and reappraise them;
- We should not claim we have "validated" an occupancy model without (or with just a "quick and dirty") comparison of calculations and a limited set of observations;
- We should not conflate data sets for model development and model evaluation. Testing a model based on the same data set, which was used for its development, is not sound methodologically and hence entirely unconvincing;
- We should not extrapolate from a single limited behavioural study to all kinds of populations, building types, locations, and climates. Specifically, it is hard to see why black-box models—devoid of first explicit principles based causal explanations—should be generally applicable;
- We should safe-guard against bias in model evaluation. As such, internal evaluation by model developers does not provide conclusive evidence for a model's general reliability. External evaluation procedures, double blind studies, and round robin tests are in a better position to convincingly support the evaluation of a model's credibility;
- We should be careful while incorporating insufficiently documented and rudimentarily tested behavioural models in simulation tools lest potential users are misled into assuming such models necessarily capture "reality".

In the next section of this paper, we address some of these considerations through a specific case study of behavioural models pertaining to inhabitants' operation of windows for natural ventilation in buildings. Note that the case study itself has a number of key limitations (small set of reference empirical data from only one location, small number of models considered, etc.). However, the structure and embedded procedure of this external evaluation exercise of a number of window operation models provides a useful context to specifically address a number of the aforementioned model evaluation challenges.

3 CASE STUDY: EXTERNAL EVALUATION OF WINDOW OPERATION MODELS

3.1 *Empirical data for model calibration and evaluation*

An office area at TU Wien (Vienna, Austria) including an open space with multiple workstations and a single-occupancy closed office acted as the data source for external model assessment. We specifically focused on seven workstations, in which each occupant has access to one manually operable casement window. The occupants' presence, state of windows and a number of indoor environment variables (including air temperature, humidity, and CO_2 concentration) are monitored on a continuous basis. Outdoor environmental parameters (including air temperature and precipitation) are also continuously monitored via building's weather station. For the present study, we used 15-minute interval data from a calendar year (referred to as calibration period) to calibrate the coefficients of stochastic window operation models. As such, this option is only of interest, if the model deployment scenario involves already existing buildings (e.g., model use for optimisation of building operation). A separate set of data obtained from another calendar year (referred to as validation period) was used to evaluate the predictive performance of the models.

3.2 *Selected window operation models*

We studied three existing stochastic and three simple non-stochastic window operation models. The stochastic models (referred here as A, B, and C) are derived based of occupant behaviour at office buildings and are widely referenced in the building performance simulation community. They are all Markov chain based logistic regression models that estimate the probability of window opening and closing actions based on the previous window state and a number of occupancy-related and environmental independent variables. To our knowledge, at least two of these models are implemented within well-known building performance simulation tools (model A in ESP-r and model C in IDA ICE), despite the rather limited underlying empirical basis and despite the lack of conclusive evidence for their conclusive general validity and applicability.

The non-stochastic models (referred as D, E, and F) are defined based on simple rules according to the common practice in use of building performance simulation tools without integration of stochastic models (models D and F are, for example, integrated in EnergyPlus).

In our study, we also included new variations of models A and C (denoted as A* and C*), as the original models did not capture a key behavioural feature in the building under study where the inhabitants are requested not to leave the windows open when they leave the office due to storm damage risk. In addition, we considered two benchmark pseudo-models (denoted as G and H), whose purpose is to put the performance of the selected models into perspective. A brief description of the aforementioned models is provided below:

- Model A, developed by Rijal et al. (2007), estimates the probability of opening and closing windows based on outdoor and operative temperature, when operative temperature is outside a dead-band (Comfort temperature ± 2°C). This model is derived based on data obtained from 15 office buildings in UK between March 1996 and September 1997.
- Model A*, a variation of Model A, always returns a closing action upon each occupant's last departure.
- Model B, developed by Yun and Steemers (2008), is derived based on summer data (from 13 June to 15 September 2006) obtained from a naturally ventilated office building in UK without night time ventilation. It estimates the probability of opening windows upon first arrival and the probability of window opening and closing actions within intermediate occupancy interval (i.e. after first arrival and before last departure) based on indoor temperature.
- Model C, developed by Haldi and Robinson (2009), estimates the probability of opening and closing actions at arrival times (first and intermediate ones), intermediate occupancy intervals, and the departure times (intermediate and last ones) based on a number of occupancy-related and environmental independent variables (see Tahmasebi & Mahdavi 2016, for the list of independent variables, and the original and adjusted estimates of the coefficients used in this study). This model has been developed based on data obtained from 14 south-facing cellular offices in a building located in the suburb of Lausanne, Switzerland for a period covering 19 December 2001–15 November 2008.
- Model C*, a variation of Model C, always returns a closing action upon each occupant's last departure.
- Model D, a non-stochastic model, operates as follows: windows are opened if indoor temperature is greater than outdoor temperature and indoor temperature is greater than 26°C. Otherwise the windows are closed.
- Model E, a non-stochastic model, is formulated as follows: windows are opened if indoor temperature is greater than outdoor temperature and indoor temperature is greater than 26°C. Windows are closed if the indoor temperature is less than 22°C.
- Model F, a non-stochastic model, operates as follows: windows are opened if the operative temperature is greater than the comfort temperature calculated from the EN15251 adaptive comfort model. Following the definition of comfort temperature for free-running period in EN15251, the windows can be opened only if weighted running average of the previous 7 daily average outdoor air temperatures is above 10°C and below 30°C.
- Model G, a benchmark pseudo-model that "predicts" windows are always open.
- Model H, a benchmark pseudo-model that "predicts" windows are always closed.

In case of the stochastic window operation models, to conduct the evaluation in a comprehensive manner, we used both original and adjusted coefficients of the logit functions. Whereas the original coefficients are published by model developers, the adjusted coefficients are obtained from re-fitting the models to a separate set of data obtained from the building under study in the calibration period. We specify the models with original coefficients with a subscript "O" and the ones with calibrated coefficients with a subscript "C". As mentioned before, the latter option (adjusting model coefficients based on observations in actual buildings) has no relevance to model deployment scenarios pertaining to building design support, but may be of some interest in operation scenarios of existing buildings.

3.3 *Office area calibrated simulation model*

The previous studies on evaluation of stochastic window operation models (Schweiker et al. 2012, Fabi et al. 2015) neglected the models feedback. This circumstance represents a simplification regarding window operation model validation. Therefore, we suggest the use of a building calibrated simulation model as a test bed for evaluation of window operation models, which includes the models feedback, i.e. the impact of models' output (window states) on models' input (indoor temperature). To fulfil this purpose, we needed a model that could represent the building performance in validation period with high accuracy. Therefore, firstly, the building model was subjected to an optimisation based calibration to adjust the fixed parameters governing the multi zine air flow simulations (For details on calibration procedure, see Tahmasebi & Mahdavi 2012). Secondly, we incorporated the monitored data pertaining to occupancy, plug loads, use of lights, and operation of heating system into the calibrated building model as a set of full-year data streams with a resolution of 15-minute intervals. This data set was obtained in the validation period. The resulting model, when fed with actual window operation data as the benchmark model, predicts the hourly indoor temperatures in validation year with a Normalized Mean Bias Error of 2.8% and a Coefficient of Variation of Root-Mean-Square Error of 4.8% (Tahmasebi & Mahdavi 2016).

The described building simulation model served as a platform, into which the selected window operation models were integrated, such that in each variation of the building model, the occupants' interactions with windows are represented using one of the selected window models. For each occupant in the building, individual occupancy data and zone-level indoor environmental factors are provided for the window operation model. That is, at each simulation time-step, the window model is executed separately for each occupant. We also built a benchmark model, which contained the actual operation of windows based on the monitored data obtained in the validation period.

The described building model was exposed to the outdoor environmental conditions in the validation period, using a weather data file generated from the on-site weather station measurements. The measured dataset included outdoor air temperature, air humidity, atmospheric pressure, global horizontal radiation, diffuse radiation, wind speed, and wind direction.

3.4 *Evaluation statistics*

One of the fundamental challenges of evaluation procedures pertaining to behavioural models of building inhabitants pertains to the paucity of systematically classified model performance metrics. For the purpose of the current study, we used the following indicators to evaluate the predictive performance of window operation models:

- Fraction of correct open state predictions [%]: This is the number of correctly predicted open state intervals divided by the total number of open state intervals.
- Fraction of correct closed state predictions [%]: This is the number of correctly predicted closed state intervals divided by the total number of closed state intervals.
- Fraction of correct state predictions [%]: This is the number of correctly predicted interval states divided by total number of intervals.
- Fraction of open state [%]: This is the total window opening time divided by the observation time.
- Mean number of actions per day [d^{-1}] averaged over the observation time.
- Open state durations' median and interquartile range [hour].
- Closed state durations' median and interquartile range [hour].

From the above indictors, the fraction of correct open state predictions (as "true positive rate"), fraction of open state, mean number of actions per day, median open state duration, and median closed state duration have been suggested in previous studies to evaluate the predictive performance of window operation models. We added three indictors to the previous work, namely fraction of correct closed state predictions to express models' state prediction performance, and the interquartile range of open state and closed state durations to capture the spread of window states' durations.

To ensure the robustness, transparency, and integrity of model evaluation procedures, the selection of reliable, expressive, and consistent model performance metrics is indispensable. Related future efforts in this direction are thus of utmost importance.

3.5 *Results*

The values of evaluation indicators for different window operation models are given in Table 1. These values are obtained from model executions in the whole validation period (a full calendar year). To better illustrate the performance of models in terms of different evaluation indicators, Figure 1 to Figure 3 show the models' prediction errors under consideration of their feedback. Note that in these Figures, models' relative error percentages are displayed in a logarithmic scale: For instance, a value of 1 read from the y-axis denotes a relative error of 10% in the evaluation indicator with reference to the benchmark. This mode of representation facilitates a better visibility of the differences in models' behaviour.

3.6 *Discussion*

A fundamental question with regard to the application of behavioural models concerns their

Table 1. The values of evaluation statistics obtained from model executions with feedback.

Models	Fraction of correct open state [%]	Fraction of correct closed state [%]	Fraction of correct states [%]	Fraction of open state [%]	Actions per day [d^{-1}]	Opening duration [hour]		Closing Duration [hour]	
						Median	IQR	Median	IQR
Observed	100.0	100.0	100.0	4.1	0.28	1.8	5.3	23.5	55.3
A_o	44.0	85.2	83.5	16.0	0.05	18.6	59.0	152.2	308.8
A_o*	47.2	96.9	94.9	4.9	0.21	5.7	5.3	22.4	66.0
B_o	41.8	88.4	86.5	12.9	5.2	0.5	0.5	0.5	0.8
C_o	54.2	78.2	77.2	23.1	0.07	37.1	91.2	133.7	313.2
C_o*	30.9	97.5	94.7	3.7	0.18	4.5	4.9	56.4	120.9
A_c	41.3	86.0	84.2	15.1	0.04	19.8	93.1	172.5	408.2
A_c*	44.4	97.5	95.3	4.2	0.18	5.4	5.4	23.6	76.2
B_c	44.6	96.4	94.3	5.3	0.31	2.8	5.9	38.3	76.3
C_c	47.9	83.9	82.5	17.4	0.16	3.7	22.8	63.0	128.5
C_c*	35.4	97.2	94.7	4.1	0.24	3.2	5.8	45.8	97.6
D	36.0	97.6	95.1	3.8	1.25	0.3	0.3	0.5	2.5
E	54.3	95.8	94.1	6.3	0.23	6.8	6.0	18.8	47.9
F	44.1	94.8	92.8	6.8	1.78	0.3	0.5	0.5	1.3
G	100.0	0.0	4.1	100.0	0.0	8760.0	0.0	–	–
H	0.0	100.0	95.9	0.0	0.0	–	–	8760.0	0.0

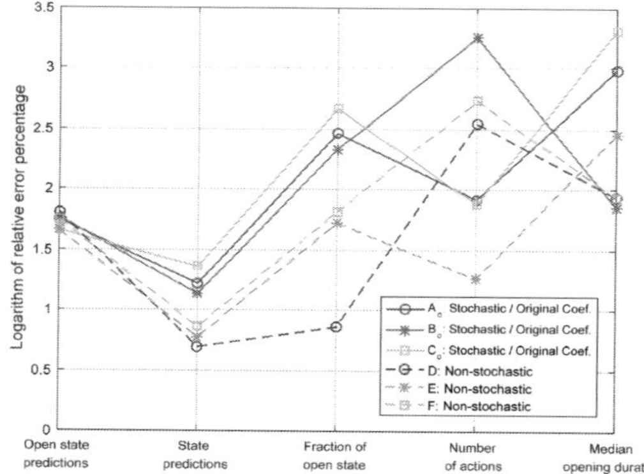

Figure 1. Errors of stochastic window operation models with original coefficients and no adjustment (A_o, B_o, and C_o) as well as non-stochastic models D, E, and F in terms of 5 evaluation statistics.

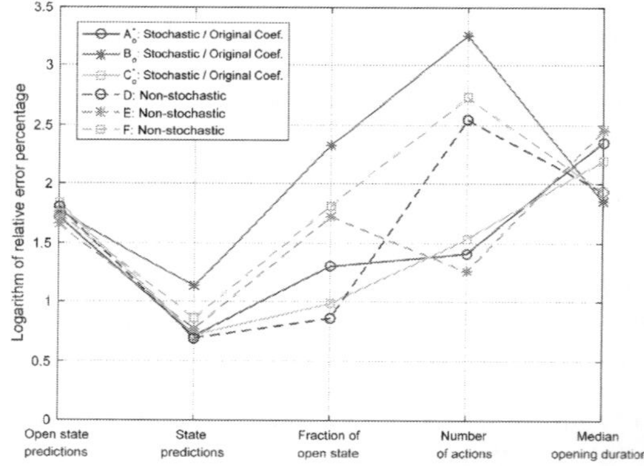

Figure 2. Errors of stochastic window operation models with original coefficients and adjusted to buildings without night time ventilation (A_o*, B_o, and C_o*) as well as non-stochastic models D, E, and F in terms of 5 evaluation statistics.

capability in reproducing empirical observations. We may thus first ask if the models could, in the present case, provide acceptable approximations of the observations. Assuming a threshold of ± 20% for the relative error of model predictions as a reasonable benchmark, we must conclude that without adjustments (night-time ventilation, calibrated coefficients), none of the studied models performs satisfactorily (see Table 1 and as well as Figure 1). Only regarding the indicator "fraction of correct state predictions" do the non-stochastic models meet this criterion. However, the night-time ventilation adjustment markedly improves the performance of the stochastic models A_o* and C_o* (see Figure 2). Furthermore, calibrating the coefficients of stochastic models via observational data results in a significant improvement of their predictive performance. Specifically, for indicators "fraction of correct state predictions", "predicted fraction of open state", and "the number of daily actions", these models' relative errors remain roughly under 30% (see Figure 3).

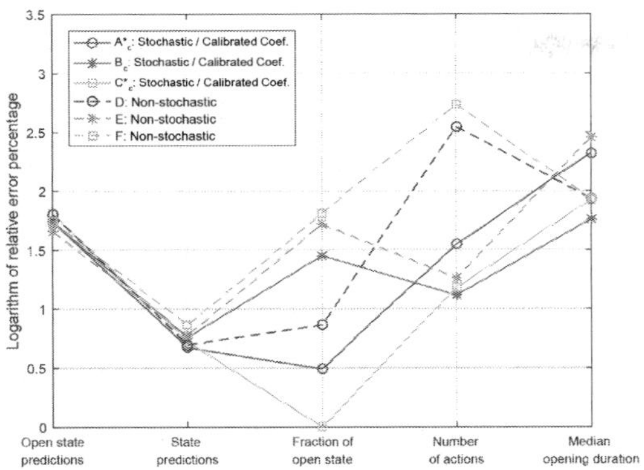

Figure 3. Errors of stochastic window operation models with calibrated coefficients and adjusted to buildings without night time ventilation (Ac*, Bc, and Cc*) as well as non-stochastic models D, E, and F in terms of 5 evaluation statistics.

4 CONCLUSIONS

Building information models can be significantly improved in their coverage and applicability if they are enriched with high-resolution representations of inhabitants. Many recent model development efforts have explored the potential of sophisticated mathematical formalisms for such representations. However, rigorous evaluation processes are needed to ensure the usability and reliability of occupancy-related models in BIM. Given the lack of general procedures and guidelines for such models, we formulate a number of relevant conditions and requirements. Furthermore, we presented a demonstrative model evaluation study involving a number of recently proposed window operation models. Thereby, the observed large deviations from reality underlines the need for clear documentation of associated uncertainties with existing behavioural models in different deployment scenarios as well as development of more generally applicable occupancy-related models. Definition and pursuit of rigorous model validation procedures in the behavioural modelling field may be seen as work in progress. As a consequence, both model developers and potential users would be well-advised to be careful with regard to introduction and application of behavioural models pertaining to inhabitants' actions in buildings. Specifically, statements concerning models' validity and overall applicability in the building delivery process would be of little credibility without comprehensive empirical backing and careful model testing procedures.

ACKNOWLEDGEMENTS

The research presented in this paper benefited from the authors' participation in the ongoing efforts of the IEA-EBC Annex 66 (Definition and Simulation of Occupant Behaviour in Buildings) and the associated discussions.

REFERENCES

Fabi V., Andersen R.K. & Corgnati S., 2015. Verification of stochastic behavioural models of occupants' interactions with windows in residential buildings, Building and Environment, 94(1), pp 371–383, doi:10.1016/j.buildenv. 2015.08.016.

Gaetani I., Hoes P. & Hensen J.L.M. 2016. Occupant behavior in building energy simulation: Towards a fit-for-purpose modeling strategy, Energy and Buildings doi:10.1016/ j.enbuild.2016.03.038.

Haldi F. & Robinson D., 2009. Interactions with window openings by office occupants. Building and Environment, 44(2009), pp 2378–2395, doi:10.1016/ j.buildenv.2009.03.025.

Jun G.U. & Steemers K., 2008. Time-dependent occupant behaviour models of window control in summer. Building and Environment, 43(2008), pp 1471–1482, doi:10.1016/j.buildenv.2007.08.001.

Mahdavi, A. 2011. The Human Dimension of Building Performance Simulation, Proceedings of the 12th Conference of the International Building Performance Simulation Association, K16 - K33, ISBN: 978-0-646-56510-1.

Mahdavi, A. 2015. Common fallacies in representation of occupants in building performance simulation, Proceedings of Building Simulation Applications 2015 - 2nd IBPSA-Italy Conference, pp 1–7, Bozen-Bolzano University Press, ISBN: 978-88-6046-074-5.

Mahdavi, A. & Tahmasebi, F. 2016. The deployment-dependence of occupancy-related models in building performance simulation, Energy and Buildings 117 (2016) 313–320, doi:10.1016/j.enbuild.2015.09.065.

Rijal H.B, Tuohy P., Humphreys M.A., Nicol J.F., Samuel A. & Clarke J., 2007. Using results from field surveys to predict the effect of open windows on thermal comfort and energy use in buildings. Energy and Buildings, 39(2007), pp 823–836, doi:10.1016/j. enbuild.2007.02.003.

Schweiker M., Haldi F., Shukuya M. & Robinson D., 2012. Verification of stochastic models of window opening behaviour for residential buildings, Journal of Building Performance Simulation, 5(1), pp 55–74, doi:10.1080/19401493. 2011.567422.

Tahmasebi F. & Mahdavi A. 2012. Optimization-based simulation model calibration using sensitivity analysis, 7th Conference of IBPSA-CZ, Brno, Czech Republic.

Tahmasebi F. & Mahdavi A. 2015. The sensitivity of building performance simulation results to the choice of occupants' presence models: a case study. Journal of Building Performance Simulation, (2015), doi:10.1080 /19401493.2015.1117528.

Tahmasebi F. & Mahdavi A. 2016. An inquiry into the reliability of window operation models in building performance simulation, under review in Building & Environment.

eWork and eBusiness in Architecture, Engineering and Construction – Christodoulou & Scherer (Eds)
 ISBN 978-1-138-03280-4

Methodology for tracking BIM benefits on project level

T. Mäkeläinen & J. Hyvärinen
VTT Technical Research Centre of Finland, Espoo, Finland

M. Rekola
Senate Properties, Helsinki, Finland

ABSTRACT: A BIM Benefit Matrix is introduced as a tool to clarify advantages and to identify benefits from BIM. Methodology is based on indicators, related to a BIM functionalities and n characteristics of BIM process. The methodology was developed and tested with four follow-up indicators, one of them for design performance, during the design of a large infrastructure development project in Finland. The client made remarkable investment in implementing BIM and was interested to show the return of the investment. Objective of the study was to measure and make visible the benefits from using BIM in the studied project, and to add understanding about potential benefits and how they can be achieved. The findings suggest that BIM advantages are relatively clear and easy to formulate, but measuring benefits is challenging and time consuming effort. This paper clarifies the discussion of benefits with a framework that separates BIM advantages from measured benefits.

1 INTRODUCTION

Building Information Modelling (BIM) as a technology for construction project delivery has been growing in use in house building industry during the last decade and is now also extensively being implemented in civil engineering and construction of infrastructure (e.g. Borrmann et al, 2015). Building information Model (BIM) has been defined as a digital representation of physical and functional characteristics of a facility (NBIMS, 2015). A BIM is a shared knowledge resource for information about a facility forming a reliable basis for decisions during its life-cycle; which defined as existing from earliest conception to demolition (NBIMS, 2015). Building Information Modelling refers to a process focused on the development, use, and transfer of a digital information model of a building project to improve the design, construction, and operations of a project or portfolio of facilities (NBIMS, 2015).

In civil engineering the term BIM is also used, although often understood to refer only to buildings. In general, it can be used for the overall process of creating, sharing and utilising models in construction. Also terms civil information modelling, CIM (e.g. Guo et al, 2014), and BIM for infrastructure (BIM Task Group, 2016) are used. In this paper we use the term BIM (for infrastructure).

Soon after implementation of BIM started, first in house building, the interest to measure the benefits of BIM raised (e.g. Fox 2008; Leicht & Messner 2008). BIM is a tool that is changing the ways of design and construction industry; thus, implementation is not straight forward thing, but needs learning and adaptation. When considering investing in new technologies and competence development, stakeholders raise the question of Return Of Investment (ROI). There have been a lot of research about the benefits and ROI of BIM during the years but objective measures still remain debated. In this paper we define general potentials to result in measured benefits as "BIM advantages". For example, one advantage of BIM is its visualization power that can be used to advance understanding and communication in several use cases; in one particular case, the benefits of this would be earlier detection/prevention of clashes in design (i.e. fewer iterations) and their more efficient resolution (decreased delivery time).

This paper focuses on the findings in the planning phase of the Pisara railway project in the city centre of Helsinki, Finland. It is a large and complex project where the client wants to utilize BIM in managing the complexity and the risks of the project. The project has a wide diversity of design and engineering disciplines, requested to start modelling their plans right from the start of the project. Main use of modelling is to support design development, communication and design coordination. The client is making a remarkable investment in implementing BIM to the full in the project and is keen to show that the investment is worthwhile. This was the motivation to develop the tool presented in this paper.

Both standard-based and non-standard-based data exchange are considered in the paper when addressing BIM. The aim of the project was to enable interoperability using open BIM standards (IFC, LandXML), but because of the nature of the project, including a large variety both civil and building design and planning disciplines, there are no standards to cover the total addressed scope.

2 RESEARCH METHOD AND OBJECTIVES

2.1 *Objectives*

Objective of the study was to measure and/or make visible the benefits of using BIM in the studied infrastructure development project. Since the study was carried out during planning phase (mainly concept design of the railway and the stations) it was clear that some possibilities, for example for measuring overall cost savings were not realistic. This study presents a framework to identify potential benefits, and proposes some quantifiable measures to be reviewed in the project later on.

Also the aim was to help the client to form a total picture of what benefits can be achieved and what kind of steering mechanisms could be used to achieve them. Taking into account that in civil engineering the BIM based projects are in their early days, the study also served for learning and longer term planning of strategies in implementing BIM in Finnish transport infrastructure sector.

2.2 *Research method*

The paper consists of a literature study and a study of the case project (described in chapter 4.1). First, a literature study was carried out on the methods presented for measuring and studying benefits of implementing BIM, and about benefits reported so far (chapter 3). Based on the literature findings and conclusions from earlier research, a framework (presented in chapter 4.2) was developed to review BIM benefits in the case project.

Data about the case project was gained by observation (Cooper et al, 2006) and complementary interviews. The authors have participated for two years in the work of BIM steering group and BIM management and development group of the project, observing and collecting information about the BIM based working during planning and early design.

Eight semi-structured interviews were made with the key representatives of the project partners to deepen the understanding of the total process and organization of the project, since the direct observations were made only on BIM related activities of the project.

Assessment of the benefits was made by the researchers and a number of project participants using the presented tool.

3 LITERATURE REVIEW OF BIM BENEFITS

This chapter reviews literature about benefits and added value reported on implementing BIM. Because so far there are less BIM projects in civil engineering than in house building industry, there is also less research available related to BIM in infrastructure projects. Because many benefits of BIM are project management benefits (Allison, 2010), benefits of BIM in civil engineering and transport infrastructure development are believed to be very similar to those experienced in house building industry. Therefore, reference is made also to earlier research in house building industry.

3.1 *Benefits of BIM*

Many studies emphasize the benefits of BIM in promoting interaction and collaboration of team members from different design disciplines (Saini & Mhaske 2013; Eadie et al. 2013; Porwal & Hewage 2013). Allison (2010) points out that many benefits of BIM are different kinds of benefits for project management.

A similar survey of BIM implementation has been carried out in Finland and in the UK. In the surveys the respondents were asked to answer how much they agree to the given statements of BIM benefits. The results are summarized in Table 1 based on (Finne et al. 2013) and (Anon 2014).

Barlish and Sullivan (2012) summarised a key list of the top mentioned benefits of BIM reported in earlier studies. Top mentioned benefits of BIM that 4 or more of the references mention

Table 1. Percentage of respondents agreeing with stated benefits in Finnish and UK survey of BIM use according to (Finne et al. 2013) and (Anon 2014).

Benefit of BIM	Finland	UK
Improves visualization	85%	83%
Improves data management	77%	
Improves the coordination of construction documents	39%	77%
Improves profitability	27%	45%
Brings cost savings	24%	61%
Shortens the time of delivery	22%	52%

were summarised by (Barlish & Sullivan 2012) as following:

- schedule (11)
- sequencing coordination (7)
- rework (5)
- visualization (5)
- productivity (5)
- project cost (5)
- communication (4)
- design/engineering (4)
- and physical conflicts (4).

Bryde et al (2013) reviewed 35 published BIM case studies to make a secondary analysis of which of these projects benefitted from BIM with regard to set success criteria. The study did not show any dependence of experienced benefits on the size of the project or type of the building. It could also happen that different analysed projects experienced both positive and negative benefits with regard to one success criterion (Table 2).

Kreider et al. (2010) made a survey of BIM utilization frequency and the benefits. 3D modelling, design reviews and design authoring were assessed very positively both in terms of perceived benefits and frequency. They outlined several BIM use cases, and their results showed that in many of these, BIM was used only by some, although the perceived value was positive. BIM was assessed to bring benefits for structural analysis, energy analysis, cost estimation, sustainability analysis, mechanical analysis, and lighting analysis but the use frequency was moderately low.

Ganah and John (2013) claim that using BIM can be justified by benefits in areas like buildability, quality assurance, cost and scheduling, but it is not so obvious how the utilisation of BIM in visualisation and knowledge embedment will enhance the management of the long term benefit especially during post-construction phase. Also, Malleson (2014) state that although the level of BIM use is moderately good in Britain, there is still much to do before the overall building performance is managed with the help of BIM, and the long-term benefits can be achieved based on that.

Table 2. Experienced positive and negative benefits in BIM case studies (Bryde et al 2013; Table 5).

	Number of projects +	Number of projects
Success criterion	Positive benefit	Negative benefit
Cost reduction or control	21	2
Time reduction or control	12	3
Communication improvement	13	0
Coordination improvement	12	3
Quality increase or control	12	0
Negative risk reduction	6	1
Scope clarification	3	0
Organisation improvement	2	2
Software issues	0	7

3.2 *Measurement and analysis methods*

Barlish and Sullivan (2012) reviewed 21 literature references about BIM benefits and found only 4 giving quantifiable results. Whether reported measures are qualitative or quantitative, most of the data about the benefits is based on surveys asking respondents' opinions and experiences, either open ended or with pre-set multiple choices (e.g. Kreider et al 2010; Finne 2013; Anon 2014; Gerbov 2014). Some measurable indicators have been proposed for mainly construction phase to measure either time or cost savings resulting from assumed higher quality of design documents (e.g. Li et al 2014; Barlish & Sullivan 2012).

Barlish and Sullivan (2012) aimed at assessing the benefits of BIM on the basis of cost analysis by considering not only returns but also the investments required for BIM. They compared non-BIM and BIM projects and made preliminary estimates of overall savings. They propose as metrics to be tracked:

- change orders (% of standard cost),
- avoidance log and associated cost,
- RFI quantities non-BIM vs. BIM,
- Offsite prefabrication man-hours,
- OCIP insurance cost savings,
- reconciliations of savings from designers and contractors using BIM,
- actual durations as % of standard duration.

Often the sources that give examples of saved costs due to BIM use the reasoning where first a lump sum is assumed as cost of an average error found out on site during construction. Then, all the issues found out during BIM coordination are assumed as errors that would have been found out only during construction and the saved sum is declared based on all resolved issues during design. (Or at least no method is described to evaluate which errors have been counted as not resolved without BIM.) However, it can be argued that some of the errors may have been detected before construction also with traditional coordination. Other ways of looking at benefits of better quality of design documents are the number of reported errors (e.g. TU, 2013) or number of requests for additional information (e.g. Leicht and Messner, 2008).

Barlish & Sullivan (2012) criticize that the indicators suggested in the literature are not precise enough and result in an overload of subjective measurements. Gerbov (2014) stated that it was impossible to measure because organisations did not even collect the kind of information that would make possible to measure and compare multiple projects.

A basic problem in most of the studies on benefits is that the case projects utilizing BIM are of type piloting or demonstration. There the needed practices for efficient BIM use have not yet become routine. Becerik-Gerber and Rice (2010) present that the lack of proper experience in the use of BIM hinders the determination of the value of BIM.

Project success is a multi-dimensional concept. According to Shenhar et al (2001), ironically, some benefits may be overlooked because the success or benefit is seen too narrow as cost and schedule benefits. Li et al (2014) bring up the point of view of limited natural resources. BIM could have beneficial impact on materials management in projects but it is never measured.

4 BIM BENEFIT MATRIX METHOD

4.1 *Project environment for the BIM benefit matrix development*

The project under study was an underground railway loop being planned in the city centre of Helsinki, Finland. The length of the railway loop is 8000 meters, including 6000 meters of tunnel and three underground stations. Also traffic and landscape arrangements connecting the new structure to the existing infrastructure were included. The project had wide diversity of design and engineering disciplines; altogether over 20 separate contracts were issued.

According to their current BIM implementation strategies the client organisations (Finnish Transport Agency and City of Helsinki) decided to use BIM in the project and start as early as possible. The consultants were requested to start modelling their plans right from the start of the project. Modelling should be implemented as fully as possible, including integration of building modelling (underground stations) and transport infrastructure modelling. The biggest challenges were clear from beginning: steering a large number of different design disciplines and wide range of stakeholder groups from the city departments to external stakeholders like facility owners. The location in Helsinki city centre which is filled with existing underground structures making the task even more demanding.

The city of Helsinki was updating the city plan as a parallel process due to the planned railway.

4.1.1 *Targeted benefits in project's BIM strategy*

The project was intended to be carried out as fully BIM based as possible. This was a bold decision for a large complex project and a clear statement for implementing BIM in infrastructure and city planning. Five high level objectives for using BIM in the project were defined in the project's BIM strategy:

1. Communication and sharing information in the project team and with stakeholders and interest groups;
2. Quality assurance of design and construction;
3. Procurement and contracting of construction works based on BIM;
4. Production design, management and control;
5. Taking into consideration Life cycle information management.

During the planning phase study only the first two were followed up. The expected benefits of using BIM for these objectives were listed as:

- Ensuring financing for construction investment
- Supporting coordination with city planning by fluent communication and information exchange
- Observing potential design alternatives
- Identifying best design solutions
- Managing costs better
- Improving design quality
- Allowing for improved productivity in construction.

To enforce BIM based working and information management in the project the client had procured services for acquiring and quality checking initial data of present state, as well as project portal with file management and BIM support services. Data portal consultant was tasked to create a 3D model of the initial state. Also during the design there were milestones defined to create a combined model of all the design information and the initial data in order to enforce the understanding of the total project.

Uncertainties of the process and contents of early phase modelling and interoperability between BIM tools used in building design, rock engineering and railroad engineering had to be solved. To support innovation during modelling process a BIM development group was established for the project. All the BIM coordinators of different design and engineering contracts were invited to the group discussing and solving problems faced during the design. In addition a group of client organisations' representatives, representatives of project management consultant and information management consultant (portal support), and external advisers did regularly meet to ensure quick decision making and steering the process.

4.2 *BIM benefit matrix development*

Based on the literature study it was clear that straight forward measurements of BIM benefits were not possible, especially not yet during the early design phase. However, the client was interested to collect evidence of benefits by examples of good practices or beneficial impacts identified during the design. A BIM benefit matrix (Figure 1) was developed as a systematic tool to identify benefit scenarios: how general BIM advantages have potential to result in benefits if employed in appropriate use cases as BIM functionalities. The matrix supports finding follow up indicators which could better track, and when possible, quantify the benefits for process and end product (and other benefits).

First, a list of beneficial characteristics was formulated for BIM process (Figure 2). Sacks et al (2010) have identified how characteristics of BIM and principles of Lean construction can be matched and their analysing frame served as reference for the process characteristics. Also, some characteristics were formulated to describe the desired outcome of the planning phase. Owners defined their expectations for a well-managed modelling process and for advanced information management during the life-cycle of the end product. In the matrix, BIM benefits can then be grouped under three categories: ones for the process, for the final end-product, and other benefits.

Second, a list of BIM functionalities utilized in Pisara project use cases was listed as rows of the matrix (Figure 3). The cross section of process characteristic and BIM functionality was formulated as a benefit scenario (Figure 1).

The benefit scenarios were formulated as sentences like:

- Correct and exact initial data decreases changes in design.
- High quality in modelling work and data specifications decreases uncertified from design. This is basis for cost assurance.

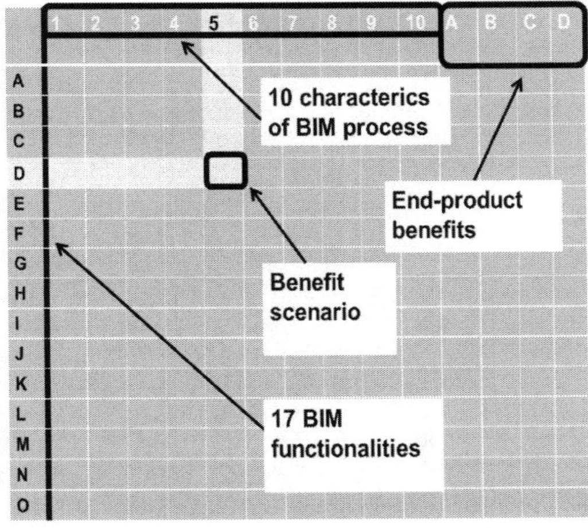

Figure 1. BIM benefit matrix outlined for Pisara case, for row and column headings, see Figures 2 and 3.

Consideration of maintenance process
Anticipation of production methods and processes
Target and requirement management
Supporting collaboration, team work and common decision making with reasoning
Supporting collaboration an communication with interest groups an stakeholders

Figure 2. BIM process characteristics (columns of BIM benefit matrix). Pisara case focus marked in grey.

Analysing and structuring initial data for the design
Information transfer and collaboration (communication) during design work flow
Technical visualization 1 (coordination model of whole project)
Technical visualization (viewing design models in design meetings)
Visualisations (public presentations, work with interest groups)
Rapid production of alternative design solutions
Compliance with usability /constructability
Assessement of investment cost (model based quantity take off)
Compliance with the requirements (structural requirements, Impact to environment, ..)
Use of models fro information maintenance and managements of information after project time
Coordination models / merged model (Class detections)
Semi automated production of design documents

Figure 3. BIM functionalities (rows in BIM Benefit matrix). Pisara case focus marked in grey.

- Reachable, frequently presented model based information enable transparency of design process. There is no extra waiting for information. Parallel design activities are well coordinated.

Next, the follow up indicator able to define or measure each benefit was defined and the indicator was named. For example, "Number of design changes due to initial data" and "Number of requests for information" were specified follow-up indicators.

During the conceptualising, the methodology was tested with four defined follow-up indicators and using one of them in assessing design performance. "Number of design changes due to initial data" indicator was co-developed and discussed in the group of BIM coordinators with several examples from each design domain.

Since the study was carried out during planning phase (mainly concept design of the railway and the stations), the benefits from BIM functions in use cases of later project phases could not be assessed.

4.3 *BIM advantages and benefits in Pisara project*

4.3.1 *Observed advantages and benefits*

During project work flow several more BIM advantages with potential benefits were identified by the BIM coordinators, such as

- Coordination within the design groups of three underground stations in defining model information presentation levels
- Studying constructability by merging models when all disciplines have started modelling
- Quick finding of initial data and reference design data (initial data model)
- Certainty that the data is accurate and detailed enough (initial data model)
- Understanding geometrical boundaries in underground design (present state model and coordination model)
- Coordination model as a handy interface for updated data collation relevant for each design domain.
- Fluent information manipulation and filtering for certain BIM functionalities e.g. cost estimation).

Also the general advantage of BIM based working was highly appreciated. For example, it enables to solve conflicts and issues in time, when design solutions are drafted and boundary conditions tackled in earlier design phases than in traditional process. These conditions will arise because the designers have to study design aspects in more detailed way in creating a whole information model. Some design solutions were only drafted as they will be studied in more detail in the next detailed design stage.

Earlier literature address that benefits of BIM based design is closely linked to collaborative working methods and skilled teams and individuals. If the advantages are made clear in the mind-sets of project team members, the project has a high probability to reach benefits as well. Discussion and exercises on advantages of BIM was facilitated in Pisara project at very early stage within the BIM coordinators. Both trust and open collaboration took place with the BIM development group meetings creating basis for good team spirit and knowledge sharing.

4.3.2 *Gaps in planned and realized benefits*

Our finding is that the above listed of observed BIM benefits were on more concreate level than the targeted benefits mentioned in project strategy (chapter 4.1.1). In achieving some of the benefits and ensuring smooth process with reliable data transfer between different planning and design disciplines the guidelines provided by the BIM development group play an important role. In Pisara project there were resources to establish such a group, but it is not the case in all infrastructure or building design projects.

With the BIM benefit matrix method we were able to make visible the underlying mechanisms by which BIM based working creates benefits, and help to identify these benefits. This analysis also clarifies which characteristics of the process should be focused on in order to maximize the probability to realize potential benefits, and which functionalities fit best to the specific project and should be supported.

4.4 *Role of planning and steering the BIM process*

One aspect in gaining benefits is the importance of creating the right support for BIM based working on process level. The role as observers gave researchers a good outlook for elements which were used in holistic building information management. These elements were structured as a framework to be used in subsequent projects (see Figure 4).

BIM planning and steering can be roughly captured as (1) planning and setting up basic resources, (2) defining and steering of process and BIM functionalities, (3) focusing on leadership and collaboration. (4) Deep know-how of the usage of BIM tools (Figure 4). Further best practice examples showing specific methods or guidelines used during Pisara-project in BIM steering and coordination were collected.

Findings underline the importance of planning and preparations for BIM process work. In the integrated design work it is fundamental that BIM functionalities are planned properly, process maps defined with integrated information flow and usage with roles and tasks. Any gaps of the interoperability of the design tools need to be identified in advance. Also the process tasks need to be scheduled realistically taking into consideration that design work flow is a chain from designer to the other.

4.5 *BIM benefit steps*

BIM based benefits for projects are well known in general (see chapter 3.1) and there exits categorizing models for benefit areas (see chapter 3.2). In Pisara case the BIM coordinators were well aware of the BIM advantages in general and the technical know-how needed for smooth inter-disciplinary collaboration and reliable data transfer etc. was on high level. Based on their many years' experience they could also formulate deeper views of advantages and see benefits from each stakeholder's viewpoint. They saw that the client/ owner will be the one getting the final benefits. Also BIM benefits connected to the

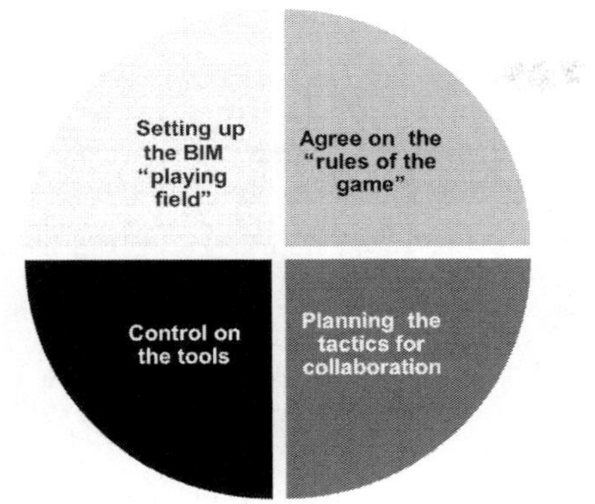

Figure 4. Main areas to steer BIM based integrated working.

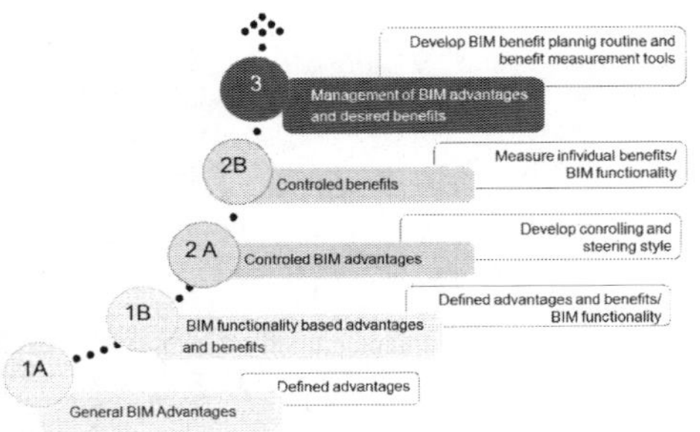

Figure 5. BIM benefit steps support the maturity of managing of BIM advantages and desired benefits.

most mature BIM functionalities like quantity take-off and clash detection with merged models were easy to formulate by the BIM coordinators.

However, to measure and analyse BIM benefits by BIM functionality, or in general on project level, is still uncommon in Finland, though BIM project environments exist. Therefore, also common knowledge on real benefits is rare.

To get one step ahead in controlling the advantages and measured benefits, they were structured in a "benefit steps" (Figure 5). This structure can hopefully serve as a part of a framework to clarify the discussion on BIM benefits and help to focus on areas which need further development.

5 DISCUSSION

Pisara project served as a development pilot environment for the BIM Benefit Matrix. It was not implemented directly, but only for gathering BIM functionalities into same framework. Also, the project environment enabled to define and prioritize the benefits important specifically in the Pisara project's process (not in general). The tested follow-up indicators triggered an open discussion on different problems from delays to design changes. The discussion helped the BIM coordinators to overcome challenges related to BIM and maturity of BIM tools by learning from other design disciplines.

We believe that follow-up indicators can be developed towards a support action to steer BIM projects. They give signals to the project management if BIM functionality level advantages and benefits are exploited fully. This does not eliminate the need to study BIM benefits further. There is a lack of evidence of quantified benefits of BIM, though interesting results in this area has been introduced by many research teams (chapter 3.2).

One method to quantify benefits is to estimate the risks which were avoided by using BIM based working. Clash detection is a very promising BIM functionality to study further in the benefit analyses/calculations. When design team is studying merged models and finding solutions for constructability, maintainability and usability, the cost of the risk avoided can be estimated. The design team can also contribute directly to risk management monitoring if any risk factor decreases with optimised design solution.

The matrix is developed for process level planning and following up the BIM advantages and/or potential benefits defined for the project. It does not focus on benefits of BIM based working in general. These kinds of frameworks are valid in organizational BIM transformation processes. Here for example Conceptual frameworks of BIM assessment and maturity evaluations may serve a better ground (Abdirad, H, 2014), (Cerovsek, 2011).

5.1 *Further development*

The BIM benefit matrix introduced as a concept in this paper needs further development. The benefit scenarios in each cross section of the benefit matrix need to be analysed and more relevant follow-up indicators formulated to clarify the whole picture.

Follow-up indicators could be chosen wisely if benefits target were also specified as measurable as possible. For example:

- The target of the project team is 50% timesaving in examining of initial project information.
- Decreasing of design faults in comparison to 2D design: within design collaboration 80% and within contractor collaboration 50%.

Large design, construction and maintenance projects including both building and transport infrastructure domains can serve as excellent pilot developments environment. Fruitful innovativeness and learning enable rapid adaptation of digitalisation.

6 CONCLUSION

The purpose of this paper was to explore ways how expected BIM benefits can be followed up during design process. The primary aim was to develop an approach to define BIM benefits in a novel way based on BIM benefit scenarios. Secondary aim was to develop a method to master expected BIM benefit scenarios during project management with follow-up indicators.

The findings of this research suggest that advantages of using BIM on project level are relatively clear and easy to formulate. Also, research exists widely on this area. However, measuring the gained benefit is a challenging and time consuming effort, and many research teams are contributing to development of measuring formulas.

Our study shows that the BIM benefit matrix can help in planning a firm BIM strategy, and choose the follow-up indicators to ensure the targeted benefits already as part of project planning. The target setting in briefing is the natural phase to define BIM benefits and BIM benefit scenarios. The scenarios can be traced in BIM benefit matrix and ensure that BIM functionalities are chosen in an optimal way and BIM process characteristics supported accordingly.

The studied Pisara case showed that innovative process level activities occur when the project team is motivated and eager to "making the most out of BIM". This observation suggest that we need to analyse more deeply the integrated BIM process task by task and activate research studies in ongoing BIM pilot project environments based on most interesting ways of working.

From the traffic infrastructure owners' point of view the Pisara project can be seen as an important project to manage the transformation towards BIM based ways of working in design collaboration. Management relies on know-how to adapt BIM functionalities and leadership, and courage to adopt and implement them fully. Challenges lie in giving up the old way of producing "design that has been modelled" and move to the new way of "producing model based design".

ACKNOWLEDGEMENTS

The authors wish to thank the Finnish Transport Agency for their financial support. Also, all the input received from those involved in the Pisara project at the Finnish Transport Agency and City of Helsinki as well as in the design teams is gratefully acknowledged.

REFERENCES

Abdirad, H. and Pishdad-Bozorgi, P. 2014. Trends of Assessing BIM Implementation in Construction Research. Computing in Civil and Building Engineering (2014): pp. 496–503.

Allison, H. 2010. Reasons Why Project Managers should Champion 5D BIM software. Vico Software Blog, Nov 23, 2010. Available at http://www.vicosoftware.com/vico-blogs/guest-blogger/tabid/88454/bid/27701/10-Reasons-Why-Project-Managers-Should-Champion-5D-BIM-Software.aspx [Accessed 1.12.2015].

Barlish, K. & Sullivan, K. 2012. How to measure the benefits of BIM - A case study approach. *Automation in Construction*, 24, pp.149–159.

Becerik-Gerber, B. & Rice, S. 2010 the Perceived Value of Building Information Modeling in the U. S. Building Industry. *Journal of information technology in construction ITCon*, 15(February), pp.185–201.

BIM Task Group. 2016. UK BIM Task group web site, http://www.bimtaskgroup.org/bim-4-infrastructure-uk/.

Borrmann, A., Flurl, M., Jubierre, J., Mundani, R., and Rank, E. (2014) Synchronous collaborative tunnel design based on consistency-preserving multi-scale models. *Advanced Engineering Informatics*, vol: 28 (4), pp: 499–517.

Bryde, D., Broquetas, M. & Volm, J.M. 2013. The project benefits of building information modelling (BIM). *International Journal of Project Management*, 31(7), pp.971–980. Available at: http://dx.doi.org/10.1016/j.ijproman.2012.12.001. [Accessed November 19, 2015].

Cooper, D.R., Schindler, P.S. & Sun, J. 2006. *Business research methods*. Vol. 9. New York: McGraw-Hill.

Damian, P., Walters, D. (2014) The advantages of information management through building information modelling, Construction Management and Economics, Volume 32, Issue 12, 2014.

Eadie, R. et al, 2013. BIM implementation throughout the UK construction project lifecycle: An analysis. *Automation in Construction*, 36, pp.145–151. Available at: http://dx.doi.org/10.1016/j.autcon.2013.09.001.

Finne, C., Hakkarainen, M. & Malleson, A. 2013. Finnish BIM Survey 2013. Available at: https://www.rakennustieto.fi/material/attachments/tutkimus_ja_kehittamistoimita/6JKJM353s/BIM_Survey_Finland_Tulokset.pdf.

Fox, S. 2008. Evaluating potential investments in new technologies: Balancing assessments of potential benefits with assessments of potential disbenefits, reliability and utilization. *Critical Perspectives on Accounting* ***19***(8), 1197–1218.

Ganah, A. & John, G.A., 2013. Suitability of BIM for enhancing value on PPP projects for the benefit of the public sector, PPP. In *International Conference 2013 Body of Knowledge, 18–19 March 2013, Preston, UK*. pp. 347–356. Available at: http://clok.uclan.ac.uk/7751/1/Ganah%26 John 2013 PPP 2013.pdf.

Gerbov, A. 2014. *Process improvement and BIM in infrastructure design projects – findings from 4 case studies in Finland*. Master's Thesis. Espoo: Aalto University,

School of Science, Degree Programme in Industrial Engineering and Management.

Guo, F., Turkan, Y., Jahren, C., and David Jeong, H. 2014. "Civil Information Modeling Adoption by Iowa and Missouri DOT". Computing in Civil and Building Engineering (2014): pp. 463–471.

Kreider, R., Messner, J. and Dubler, C., 2010. Determining the frequence and impact of apllying BIM for different purposes on projects. Available at: http://bim.psu.edu/Uses/Freq-Benefit/BIM_Use-2010_Innovation_in_AEC-Kreider_Messner_Dubler.pdf [Accessed November 9, 2015].

Leicht R. & Messner J. 2008. Moving toward an 'intelligent' shop modeling process, ITcon Vol. 13, Special Issue Case studies of BIM use, pg. 286–302

Li, J., Hou, L., Wang, X., Wang, J., Guo, J., Zhang, S., & Jiao, Y. 2014. A project-based quantification of BIM benefits. *International Journal of Advanced Robotic Systems*, *11*.

Malleson, A., 2014. BIM Survey: Summary of findings. *National BIM Report 2014*, pp.12–20.

NBIMS. 2015. *National BIM Standard—United States® Version 3, 5 Practice documents.* National Institute of Building Sciences buildingSMART alliance®.

Porwal, A. & Hewage, K.N., 2013. Building Information Modeling (BIM) partnering framework for public construction projects. *Automation in Construction*, 31, pp. 204–214.

Sacks, R., Koskela, L., Dave, B. & Owen, R. 2010. The interaction of lean and building information modeling in construction. *Journal of Construction Engineering and Management*, **136** (9), pp. 968–980.

Saini, V. & Mhaske, S., 2013. BIM an emerging technology in AEC industry for time optimization. *International Journal of Structural and Civil Engineering Research*, 2(4), pp.196–200.

Shenhar, A.J., Dvir, D., Levy, O., & Maltz, A.C. 2001. Project success: a multidimensional strategic concept. *Long range planning*, *34*(6), 699–725.

TU. 2013. 3D-modellering av veiprosjekter reduserer ekstrakostnadene fra 20 til 5 prosent. 6. nov. 2013 http://www.tu.no/artikler/3d-modellering-av-veiprosjekter-reduserer-ekstrakostnadene-fra-20-til-5-prosent/233934 Teknisk Ukeblad Media AS [Last Accessed 12.4.2016].

eWork and eBusiness in Architecture, Engineering and Construction – Christodoulou & Scherer (Eds)

BIM for the integration of building maintenance management: A case study of a university campus

R. Bortolini, N. Forcada & M. Macarulla
Department of Construction Engineering, Universitat Politècnica de Catalunya, Terrassa, Barcelona, Spain

ABSTRACT: In the Operation and Maintenance (O&M) phase of an existing building, frequently, different systems are available to manage data about building maintenance. However, current practices do not integrate these systems and still manually process dispersed and unformatted data. The purpose of this paper is to investigate the potential benefits of the integration of data of Building Maintenance Management (BMM) in Building Information Modelling (BIM) in existing buildings. To do so, a questionnaire to evaluate the Facility Management practices in buildings is presented. This questionnaire is then used in the case study of Terrassa Campus at Universitat Politècnica de Catalunya (UPC). Conclusion demonstrated that the integration of BMM in BIM can enable support for maintenance decisions and increase the efficiency of the maintenance process.

1 INTRODUCTION

Throughout the life cycle of a building, the largest fraction of expenses is incurred during O&M phase, which consist approximately 60% of the total cost (Akcamete et al., 2010). According to Mostafa et al. (2015), most of the maintenance activities in this phase are corrective actions executed in emergency conditions. However, this attitude is no longer acceptable and maintenance role is recognized as a strategic element for any organization (Mostafa et al., 2015). Rather than reacting to failures, there is a need for supporting more planned maintenance approaches (preventive, predictive) (Akcamete et al., 2010).

An effective Building Maintenance Management (BMM) must integrate and manage information such as inspections, drawings, manuals, contracts (Chen et al., 2013) checklists, maintenance reports and maintenance records (Lin & Su, 2013). Generally, the information includes 2D drawings, spreadsheets, bar charts and field reports which are typically handed over from the building design and the construction phase (Koch et al., 2014) and available as text based and maintained as handwritten record papers (Chen et al., 2013; Motamedi et al., 2014). Additionally, information are distributed to different locations and the maintenance process must deal with Big Data managing information and multiple variables that are available from disparate systems (e.g. Building Monitoring Systems—BMS, Computerized Maintenance Management Systems—CMMS). Traditionally, these datasets have been stored and analyzed independently with one another. The lack of integration of information makes it impossible for managers to make optimum maintenance management decisions (Chen et al., 2013) and creates the process of correlating the information time consuming and less intuitive (Motamedi et al., 2014).

There is a significant waste of time and money for search information or to try to make decisions with limited information (Akinci, 2014). The efficiency of the maintenance process is expected to be highly increased with the availability of an integrated digital database (Akcamete et al., 2011). There is an opportunity for facility managers to improve the current practice and use Building Information Modelling (BIM) as a decision making tool (Carbonari et al., 2015). BIM applications on maintenance process can provide an integrated database of historical record of all equipment, providing information about the actual equipment performance within the building and provide instant access to facility information through an easy interface (Oskouie et al. 2012).

This paper reports on an ongoing research project within the maintenance department of UPC Campus, Spain. This project is developing a framework to integrate BMM data with BIM. In the following sections, the current problem of BMM is described in more detail. The development of a questionnaire for data collection is described and used in the case study of UPC. Finally, a discussion of the potential benefits for utilizing BIM for maintenance management is provided.

2 BACKGROUND

2.1 *Facility Management*

Facility Management (FM) can be defined as a profession that encompasses multiple disciplines to ensure functionality of the built environment by integrating people, place, process and technology (IFMA, 2015). FM constitutes an extensive field involving multidisciplinary disciplines such as: maintenance, space, asset, catering, cleaning, among other. A key focus point of FM is maintenance management (Lewis et al., 2011).

2.2 *Building Maintenance Management*

All building structures, material, finishes, and services deteriorate over time through an inevitable process of the effects of climate and usage. This process of decay can be controlled and the physical life of the buildings extended if they are properly maintained (Chew et al., 2004).

BMM can be classified in three types: corrective, preventive and predictive. The corrective maintenance concerns about a reactive maintenance in response to a cause of failure or breakdown (Motawa & Almarshad, 2013). Preventive maintenance is carried out by periodically undertaking routine tasks necessary to maintain component or system in a safe and efficient operating condition on a regularly schedule. More recently, the advance in technology made the development of another maintenance category called predictive maintenance (Yam et al., 2001). This approach detect the system degradation and conduce maintenance on the actual condition of the facility (Sullivan et al., 2010).

Currently, most of the maintenance work is reactive (Sullivan et al., 2010). This fact pointed that there is a need for supporting more planned maintenance work (Akcamete et al., 2010) since reactive maintenance involves more costs (Sullivan et al., 2010). To reduce the number of reactive maintenance it is necessary effective planning strategies. Initial steps are integrating information about maintenance data and also capturing information about how a building deteriorates.

2.3 *BIM and BMM*

In order to improve the identification of buildings deteriorisation during the BMM, it is necessary to integrate data about building context and investigating the performance history of components to be able to identifying reoccurring problems, abnormalities and find causes of problems (Akcamete et al., 2011). The problem is that the available data about BMM is dispersed through different kind of data sources. The potential of connect these data using open standards greatly increase the data that is available for such analysis. Some authors stress that BIM can be used as 3D visual database to store, organize and exchange information (Lavy & Jawadekar, 2014; Akcamete et al., 2011) about how buildings have been behaving and deteriorating over time to improve the current practice and use BIM as a decision making tool to plan maintenance work (Akcamete et al., 2010).

Although the recognition of the importance of BIM in FM and the initiatives by industry in terms of standardization (Industry Foundation Classes—IFC and Construction Operations Building Information Exchange—COBie) (Patacas et al., 2015), the benefits of BIM in O&M phase is still relatively a new topic (Eastman et al. 2011; Kassem et al., 2015). There is a vastly number of studies about the application of BIM during design, planning and construction phases developed both by practioners and academics. Dong et al. (2014) pointed that specifically the use of BIM for FM during the O&M phase is still limited.

Some recent research studies proposed the use of BIM to integrate maintenance information (Lin & Su, 2013; Motawa & Almarshad, 2013; Irizarry et al., 2014), however, not all necessary information is currently available in a digitally integrated and standardized model (Koch et al., 2014). Existing buildings often face outdated information due to omitted updating in as-built documentation which configures a challenge to the BIM use in this context (Volk et al., 2014). Interoperability between BIM technologies and current FM technologies is still an issue in the handover of information and data to O&M stage (Akcamete et al., 2011).

3 METHOD

A case study approach was conducted in order to analyze real problems and to be able to describe potential benefits of the integration of maintenance data with BIM.

The justification for the selected case study was the willingness of the maintenance staff to share their experiences in improving the maintenance management processes at the university campus.

Data collection used by this research was obtained through a questionnaire, field observation, maintenance plans and other existing databases.

Interviews with key maintenance personnel were conducted to understand how the building is routinely managed and operated. To obtain a structured collection of information, a questionnaire was constructed. The aim of this questionnaire was to evaluate in a standardized way the FM practices in buildings. The questionnaire includes a mix of open ended questions (no options or predefined categories are suggested) and closed ended questions (respondents' answers are limited to a fixed set of responses) (Annex A).

The questionnaire was structured based on a literature review on FM (Talamo & Bonanomi, 2015; Coenen & Felten, 2014; Akcamete et al., 2010; Chotipanich & Nutt, 2008). To validate the questionnaire, some applications were set with maintenance staff of UPC and FM experts. These meetings had the objective to review the questionnaire considering the understanding of the questions by the respondents and the time of application of the questionnaire.

The first part includes questions about the analyzed company and the role and experience of the respondent. The second part includes questions to obtain a broad view of FM practices. This section includes the description of the different disciplines of FM that are managed with the type of contractual arrangement (internal, external or a mix of them). Also, some questions were included to describe the different systems used to capture and to manage information and how the information is structured. Taking into account the use of BIM for FM, the question includes the structured of information based on COBie to analyze how similar is the structure of information of the company analyzed to that standard.

The third section aims at characterizing maintenance management practices for being a key point of FM (Lewis et al., 2011). It includes questions about how the maintenance planning of the buildings is organized, the frequency of meetings, what is considered in the prioritization of activities (based on Preiser & Nasar, 2008), how the plans is the structured (based on Duffuaa et al., 2000) and which systems to manage maintenance work are used. It also includes some specific questions about corrective and preventive maintenance (Motamedi et al., 2014). Finally, information about the parameters to take into account to prevent breakdowns, improve building performance and reduce cost of maintenance work are also requested. These parameters were established through an initial literature review on factors that affect the performance of building elements and systems: age of element or system, type of material, solar orientation, weather condition, surrounding environment, cleaning frequency (Wu et al., 2010; Olanrewaju & Abdul-Aziz, 2015; Flores-Colen et al., 2010; Teo et al., 2005).

4 CASE STUDY

4.1 *Campus description*

The description of the case study is structured by the previously presented questionnaire. The maintenance and space management were the focus in this description due to the importance of these activities that are being carried out by the campus.

The campus is located in a small urban area (less than 0.15 km^2) in the city of Terrassa (Barcelona), involving facilities, blending educational, research and office. Terrassa campus, manage the maintenance of 25 buildings. The typology of UPC buildings of Terrassa campus is quite varied. The age of the buildings are significantly quite disparate, there are buildings with more than 100 years and buildings with less than 10 years. About 98% of the buildings have the energy consumption monitored. 60% of the buildings have automated HVAC systems and only 20% of the buildings have automated lighting.

4.2 *BMM at UPC campus*

The maintenance activity is characterized by a mixed of internal and external contractual arrangement. The major part of maintenance activities is subcontracted. The internal operators are only responsible for painting and minor electric repairs activities. The structure of the department is composed by the head of the department, one assistant, two technical, two supervisors, two internal operators and five external operators.

The data used for maintenance management can be distinguished in three types: Building characteristics data, Building monitoring and control data and Maintenance management data.

4.2.1 *Building characteristics data*

Building characteristics are related to the data of the existing building such as: age of construction, location, type of materials, technology of the equipments, etc. The majority of this kind of data about the buildings of the university campus is paper-based and stored in an intranet platform. UPC campus utilizes a database (Somdoc) to store documents and provide an easy access to relevant information through the component of interest itself. The platform organizes contracts, building projects, certifications, etc. The new buildings have the drawings available in the native CAD format. However, not all data is available in digital form. Mainly the data about the design (drawings, materials) of old building are hand-writing paper-based. Therefore, a physical stock of this documentation in the maintenance department exists. Maintenance department updates the projects if there is some refurbishment or extension on the buildings. This activity require manual update in all the systems used, creating time loss.

4.2.2 *Building monitoring and control data*

Building monitoring and control data is automatically captured by devices such as sensor and includes temperature, humidity, energy consumption, etc. At UPC, this information is obtained from devices that send information to a Building Management System (BMS) and visualized in an intranet application. The BMS (TAC VISTA Schneider) has the capability to monitor and control the pressure, temperature, and humidity of the HVAC system, compressed air for specific laboratories, switch on/off boilers, etc. There are alerts when dysfunctions occur. Temperature set points are defined for winter and summer and also based on the schedule. End users can monitor the room offset in 0,5°C. Cooling and heating systems are switched on in reference to the exterior temperature threshold.

The system to monitor and control water and energy consumption of UPC buildings is Power Studio® while the information is visualized on Sirena (System of Information on Consumption of Energy Resources and Water) which is an intranet platform (Mata et al., 2009). Clear graphs and comprehensive reports improve the visualization of the monitored consumptions. Although Power Studio® and TAC are databases related to energy, they are not integrated. Therefore, the potential of integrating this data to improve the maintenance management is not yet explored.

Regarding the weather condition, data is available from the meteorological station of National Meteorological Institute located in Terrassa. This station (Informet) registers data about the external temperature, humidity, wind speed, and historic external temperatures of the city. However, that data is also not related to the BMS system.

4.2.3 *Maintenance management data*

Regarding the maintenance management data, UPC campus adopted a standardized classification of elements. All building elements and systems are classified with a predefined code based on COBie. The structure of information has a standard classification and it is organized by: campus, building, floor, room and type of element.

The campus adopts a preventive and corrective maintenance approach. Regular meetings are set to discuss maintenance planning of corrective actions. Those meeting are organized with the presence of one technical and one supervisor once a week. The prioritization of activities is done considering the available resources and the risk involved in the maintenance request actions. There are some standard maintenance procedures to conduct the maintenance activities that are paper-based.

For preventive maintenance, the university campus use Archibus® system, which consist of a Computer Automated Facility Management (CAFM) and Integrated Workplace Management System (IWMS) for FM. The visualization of information is text based and includes 2D drawings of the buildings. The system generates an automatic long-term schedule for preventive maintenance of the campus. The schedule contains activities for a year and the system send automatic notifications to the maintenance staff about the required planned maintenance work of each system or component of a building. The frequency of inspections comes from ITeC software—DicPla (2016) which is based on regulations.

The corrective maintenance is also conducted through the Archibus® system. An intranet platform is connected to that system and it is available for users to make maintenance complaints. However, it is only available for regular users, mainly administrative sector and professors are available to request a corrective action. This platform enable the standardization of data of corrective maintenance complaints. There are forms to fill with pre-existing standard information to select regarding the type of maintenance required, the building and the specific room of the solicitude. The maintenance staff receives that information and analyzes the requests each day. The maintenance planning is not connected with the preventive maintenance and is done in a short-term by day. The system can send notifications to the person that made the request as a feedback about the current situation. However, the operator who executes the work order still use paper, with hinders an automatic updated of the CAFM system. This situation results in time loss to processing this paper-based data.

During the year 2014, the maintenance department received 2.342 requests for corrective actions. The majority of the corrective work orders were obtained from building B01 (with more than 100 years), buildings B05 (between 50 and 70 years) and B25 (between 10 and 50 years) (Fig. 1).

These data shows that the number of requests is not directly related to the age of the buildings. Buildings with more than 100 years (B01) and with 10 years (B25) have a great number of corrective maintenance requests. The major corrective actions are related to: electricity, air conditioning and plumbing.

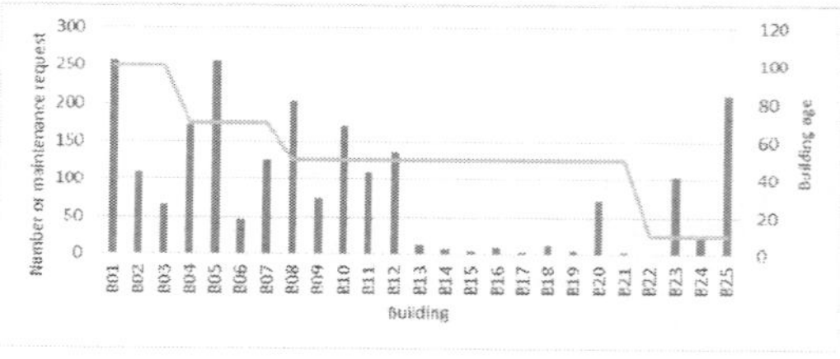

Figure 1. Building age and number of requests.

4.3 *Space management at UPC campus*

Some activities related to space management are planned together with the maintenance planning in Archibus® system. During summer period, with no academic activities, there is a planned activity to check the conditions of all classrooms. In this activity, the actual condition of equipment (e.g. chairs, projectors) is checked. Another activity related to space management is the transference of the workplace of teachers. The work in this case is related to relocate the equipment (e.g. table, chair, computer) to another room.

The information available about the use of spaces in the campus is a timetable of the classes during the year and the rooms occupied for researches, professors and administrative people. To request a room for any kind of activity (e.g. for a meeting), there is an intranet platform with the available list of rooms for each specific data and hour.

4.4 *Summary of data and systems used*

Maintenance management at university campus revealed a number of shortcomings as it still use paper-based form and unsystematic database to manage a complex and huge amounts of data including monitored data and tracking of complaints. Table 1 provides the overview of the different kind of data with an example and where the data is located with the commercial name of the systems that are managed from the university maintenance department.

Table 1. Kind of data and where is located.

Kind of data	Example	Located
Building characteristics data	Type of material, age of equipment	Paper-based and Somdoc
Building monitoring and control data	Exterior Temperature, humidity	Informet
	Interior Temperature, boilers switch on/off	TAC
	Electrical, gas, water consumptions	Power Studio
Maintenance management data	Inventory, schedules, equipment's lifespan	Paper-based and Archibus
Space management data	Relocation of equipment	Archibus
	Day room reservation	School intranet

5 BIM FOR THE INTEGRATION OF BMM

Within the understanding of all managed data at the university campus, it was proposed an integration conceptual framework among the various sources of buildings' data into a BIM model. Figure 2 shows an example of the relationship among the existing data and databases that were categorized in the previous description: building characteristics, building monitoring and control and maintenance management data. Through the definition of relationships between these data and considering BIM as a mandated, the data integration can be obtained.

Figure 3 shows the generic framework proposed to be a decision support tool for maintenance management. An activity to digitalize the paper-based information was proposed. Moreover, a categorization of this data is necessary in order to access whether the information is geometric or nongeometric. The geometric information is incorporated in BIM, the nongeometric information is integrated in an external Maintenance Database to the parametric model.

The integration of data through BIM and the external Maintenance Database is possible through the definition of a Unique Identifier (ID) for each object and space. The unique identifier is implemented as a GUID (globally unique identifier) for generate a unique number for each object.

In the proposed conceptual framework, keeping these different kinds of sources integrated has a number of advantages:

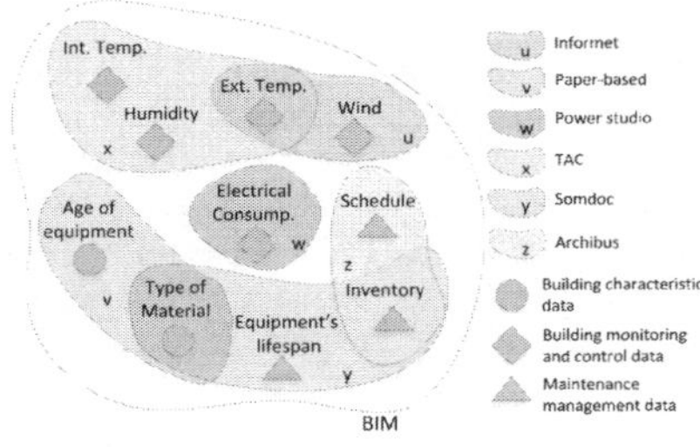

Figure 2. Data integration.

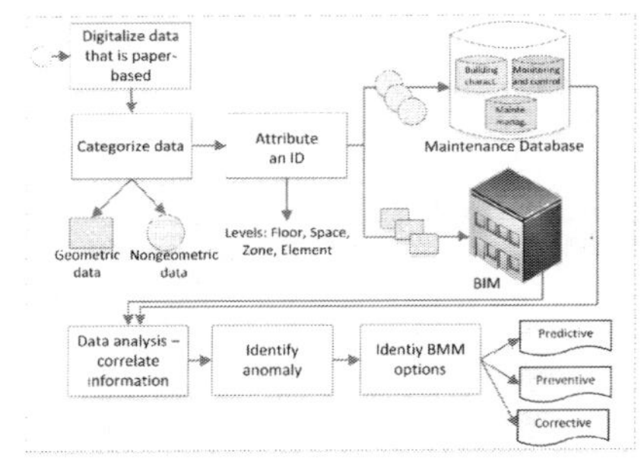

Figure 3. Systems integration.

- A set of information can be frequently updated and revised automatically through the definition of BIM as a database for storing and retrieving data (Motawa & Almarshad, 2013).
- The use of a unique identifier for each entity in all maintenance management software application permits connect the databases of these applications, thereby opening up query capabilities across different datasets (Ballesty et al., 2007).
- The integration of databases enable maintenance staff analyze different kind of data, identify abnormalities and decide for a predictive, preventive and corrective maintenance management options.
- The possibility to integrate information about many data sources increases the possibility of predicting the performance of elements and systems of a building instead of reacting to failures or breakdowns. Therefore, such integration facilitates the maintenance planning process, and consequently, increases the building performance through the predicting of failures. For example, considering the energy consumption, within all integrated systems, maintenance staff can obtain information necessary using BIM model and directly accessing many variables such as external and internal temperatures, maintenance records, type of equipments, manufactures, location of equipments, to be able to predict possible abnormalities and identify if there is any equipment that is not performing correctly. Therefore, maintenance staff can plan actions to fix the equipment and save energy.
- After digitalizing of all campus in BIM, the maintenance department will be able to automatic update the data related to all systems and resolve the problem of the drawbacks of traditional paper-based information management. The result is the improvement of the manual process of information handover that have not been possible to achieve with the current processes utilized in the university. This will result in a huge improvement in the accuracy of maintenance data.
- The integration of information related to corrective actions can provide an easily updated and reported information. Such integration improve the effectiveness of information flow in the maintenance process (Lin & Su, 2013) and lead to reduction in response times to the person that did the request.
- The use of BIM models on the inspections would improve facilities visualization through the provision of 3D views (Akcamete et al., 2010).

6 CONCLUSIONS AND FUTURE STEPS

A BMM of an existing building can implement many sophisticated sensors and computerized systems capable of delivering data about the status and performance of elements and systems of the building. However, there is little or no practical use regarding the most of this data. The case study of the university campus confirms that there is no practice of a continuous and seamless flow of information throughout maintenance processes. One cause of this problem is due to the understandable format available related to this data, which makes impossible to correlate different kind of data sources. The capabilities if each system used by the university are different, so there is a lack of standardization among software platforms used for maintenance management. And another cause is configured by the use of some paper-based data.

To structure the collection of information, a questionnaire was developed. The interview questions mainly focused on extracting insight as to how the buildings are maintained and used on a day-to-day basis. This questionnaire will be used in future case studies.

The case study illustrates the huge potential for increasing efficiency in maintenance process. By allowing the monitoring and analyzing of the performance of multiple systems (air-conditioning, lights, water usage) BIM can lead to more thoughtful energy use, lowered expenses, and increased asset value.

It is concluded that BIM is an appropriate beneficial technology enabling storage and retrieval of integrated building, maintenance and management data for existing buildings. Using this approach yielded several advantages such as consistency in the data, intelligence in the model, multiple reports generation, integrated source of information and integrated views across all existing facility systems. The standardized building model acted as main data structure which could be extended with other data sources as each element of the model such as a wall, furniture, a room, or a grouping of elements had a unique identifier.

The old buildings of the university will face more problems with the implementation of BIM due to the obsolete information available. However, the university shows a great interest in connecting maintenance plans to BIM so as to adapt and improve FMM practices. The standard codification of buildings and elements will facilitate this task.

Additional work is needed in order to better link dispersed data and also to comprehend how the availability of data integrated affect the maintenance planning. Some steps for further investigation and discussion are presented in the following:

- Develop of a methodology for modeling existing buildings in BIM.
- Integration of BMM systems into BIM models.
- Integration of maintenance management approaches: corrective, preventive and predictive.

REFERENCES

Akcamete, A. et al. (2010). Potential utilization of building information models for planning maintenance activities. *Proceedings of the International Conference on Computing in Civil and Building Engineering*, (October 2015), 151–157.

Akcamete, A. et al. (2011). Integrating and Visualizing Maintenance and Repair Work Orders In BIM: Lessons Learned From A Prototype. *CONVR 2011, International Conference On Construction Applications of Virtual Reality.*

Akinci, B. (2014). Situational Awareness in Construction and Facility Management. *Frontiers of Engineering Management*, 1(3), 283.

Ballesty, S. et al. (2007). Adopting BIM for facilities management: Solutions for managing the Sydney Opera House. *Cooperative Research Centre (CRC) for Construction Innovation*, Brisbane, Australia.

Carbonari, G. et al. (2015). Building information model implementation for existing buildings for facilities management: a framework and two case studies. *Building Information Modelling (BIM) in Design, Construction and Operations*, 149, 395.

Chen, H.M. et al. (2013). A 3D visualized expert system for maintenance and management of existing building facilities using reliability-based method. *Expert Systems with Applications*, 40(1), 287–299.

Chew, M.Y.L. et al. (2004). Building Maintainability — Review of State of the Art. *Journal of Architechtural Engineering*, 10(September), 80–87.

Chotipanich, S., & Nutt, B. (2008). Positioning and repositioning FM. *Facilities*, 26(9/10), 374–388.

Coenen, C., & von Felten, D. (2014). A service-oriented perspective of facility management. *Facilities*, 32(9/10), 554–564.

Dong, B. et al. (2014). A BIM-enabled information infrastructure for building energy Fault Detection and Diagnostics. *Automation in Construction*, 44, 197–211.

Duffuaa, S.O., Raouf, A., & Campbell, J.D. (2000). Planning and control of maintenance systems. *Willey and Sons*, 31–32.

Eastman, C. et al. (2011). BIM Hanbook: A guide to building information modeling for owners, managers, designers, engineers and contractors. John Wiley & Sons.

Flores-Colen, I. et al. (2009). Discussion of criteria for prioritization of predictive maintenance of building façades: Survey of 30 experts. Journal of *Performance of Constructed Facilities*, 24(4), 337–344.

International Facility Management Association (2015), available at: www.ifma.org

Irizarry, J. et al. (2014). Ambient intelligence environments for accessing building information. *Facilities*, 32(3/4), 120–138.

Instituto de Tecnología de la Construcción (ITeC) (2016) www.itec.es/programas/dicpla/libro-edificio-plan-mantenimiento-estandar/.

Kassem, M. et al. (2015). BIM in facilities management applications: a case study of a large university complex. *Built Environment Project and Asset Management*, 5(3), 261–277.

Koch, C. et al. (2014). Natural markers for augmented reality-based indoor navigation and facility maintenance. *Automation in Construction*, 48, 18–30.

Lavy, S., & Jawadekar, S. (2014). A Case Study of Using BIM and COBie for Facility Management. *International Journal of Facility Management*, v. 5, n. 2.

Lewis, A. et al. (2011). Linking energy and maintenance management for sustainability through three American case studies. *Facilities*, 29(5/6), 243–254.

Lin, Y.-C., & Su, Y.-C. (2013). Developing mobile- and BIM-based integrated visual facility maintenance management system. *The Scientific World Journal*, 2013, 124249.

Mata, É. et al. (2009). Optimization of the management of building stocks: An example of the application of managing heating systems in university buildings in Spain. *Energy and Buildings*, 41(12), 1334–1346.

Mostafa, S. et al. (2015). Lean thinking for a maintenance process. *Production & Manufacturing Research*, 3(1), 236–272.

Motamedi, A. et al. (2014). Knowledge-assisted BIM-based visual analytics for failure root cause detection in facilities management. *Automation in Construction*, 43, 73–83.

Motawa, I., & Almarshad, A. (2013). A knowledge-based BIM system for building maintenance. *Automation in Construction*, 29, 173–182.

Olanrewaju, A.L., & Abdul-Aziz, A.R. (2015). *Building Maintenance Processes and Practices*. Springer.

Oskouie, P. et al. (2012). Extending the interaction of building information modeling and lean, 1(617).

Patacas, J. et al. (2015). Bim for Facilities Management : Evaluating Bim Standards in Asset Register Creation and Service Life Planning,. *Journal of Information Technology in Construction*, 20(January), 313–331.

Preiser, W.F., & Nasar, J.L. (2008). Assessing building performance: Its evolution from post-occupancy evaluation. *International Journal of Architectural Research*, 2(1), 84–99.

Sullivan, G.P. et al. (2010). *Operations & Maintenance Best Practices*. U.S. Departament of Energy, Federal Energy Management Program, (August), 321.

Talamo, C., & Bonanomi, M. (2015). Knowledge Management and Information Tools for Building Maintenance and Facility Management. Springer.

Teo, E.A.L. et al. (2005). An assessment of factors affecting the service life of external paint finish on plastered facades. *In The proceedings of 10th international conference on durability of building material and components.*

Volk, R. et al. (2014). Building Information Modeling (BIM) for existing buildings — Literature review and future needs. *Automation in Construction, 38*, 109–127.

Wu, S. et al. (2010). Research opportunities in maintenance of office building services systems. *Journal of Quality in Maintenance Engineering,* 16(1), 23–33.

Yam, R.C.M. et al.. (2001). Intelligent Predictive Decision Support System for Condition-Based Maintenance. *The International Journal of Advanced Manufacturing Technology*, 17(5), 383–391.

Annex A

1. Data of the Company

1.1. Name of the company:	**1.4.** Position of the respondent in the company:
1.2. Type of activity of the company:	**1.5.** Years of experience of the respondent in FM:
1.3. Dimension of the company (number of employees):	

2. Facility Management

2.1. Which facility management areas are there in the building or company?
[ ] Maintenance [ ] Space [ ] Asset [ ] Cleaning [ ] Gardening [ ] Vending [ ] Other____________________

2.2. Select the type of contractual arrangement of each FM area (if the type is mixed, indicate the percentage of internal and external contracts)

	Area: Maintenance	Area: Space	Area:	Area:	Area:	Area:
1. Internal (I)	1.[ ]	1.[ ]	1.[ ]	1.[ ]	1.[ ]	1.[ ]
2. External (E)	2.[ ]	2.[ ]	2.[ ]	2.[ ]	2.[ ]	2.[ ]
3. Mixed (I and E)	3.[ ]I_%_E%	3.[ ]I_%_E%	3.[ ]I_%_E%	3.[ ]I_%_E%	3.[ ]I_%_E%	3.[ ]I_%_E%

2.3. If the contractual arrangement is internal, how is the structure of each FM area?
1. [ ] All FM areas are manage integrated by one team 2. [ ] By department 3. [ ] By sector 4. [ ] Other:____________

2.4. Select the number of employees in each FM area.
1. [ ] Between 1 and 5 2. [ ] Between 6 and 10 3. [ ] More than 10

2.5. Select if the systems used to capture information about FM is manual or automatic. Write the name of the system for each FM area. (e.g. information about maintenance inspections)

	Area: Maintenance	Area: Space	Area:	Area:	Area:	Area:
1. Manual (paper-based checklists)	1.[ ]_____	1.[ ]_____	1.[ ]_____	1.[ ]_____	1.[ ]_____	1.[ ]_____
2. Automatic (intranet)	2.[ ]_____	2.[ ]_____	2.[ ]_____	2.[ ]_____	2.[ ]_____	2.[ ]_____

2.6. Select if the systems used to manage information about FM is manual or automatic. Write the name of the system for each FM area.

	Area: Maintenance	Area: Space	Area:	Area:	Area:	Area:
1. Manual (paper-based plans)	1.[ ]_____	1.[ ]_____	1.[ ]_____	1.[ ]_____	1.[ ]_____	1.[ ]_____
2. Automatic (CMMS, BMS, CAFM)	2.[ ]_____	2.[ ]_____	2.[ ]_____	2.[ ]_____	2.[ ]_____	2.[ ]_____

2.7. Select the type of visualization of information.
1. [ ] Text 4. [ ] BIM
2. [ ] 2D drawing 5. [ ] Other:______________
3. [ ] 3D drawing

2.8. Select the type of structure of information (based on COBie)
1. [ ] Facility 5. [ ] Type
2. [ ] Floor 6. [ ] Component
3. [ ] Space 7. [ ] System
4. [ ] Zone 8. [ ] Other:______

3. Maintenance Management

3.1. Which maintenance approach is carried out in the building?

[ ] Preventive Maintenance
(The preventive maintenance is the process of periodically undertaking routine tasks necessary to maintain component or system in a safe and efficient operating condition.)

[ ] Corrective Maintenance
(The corrective maintenance concerns about a reactive action in response to a cause of failure or break down.)

[ ] Predictive Maintenance
(The predictive maintenance detect the system degradation and conduce maintenance on the actual condition of the machine.)

3.2. Are there regular meetings conducted to discuss the maintenance plan?
1. [ ] Yes - Which frequency? ([ ] by day, [ ] by week, [ ] by month, [ ] by year) 2. [ ] No

3.3. Who participate of the meetings?
1. [ ] Head of the department 2. [ ] Outsourcing firms 3. [ ] Other:____________

3.4. Which management parameters are taken into account to plan the maintenance? (prioritization)

Preventive Maintenance
1. [ ] Time available 5. [ ] Occupant satisfaction
2. [ ] Resources available 6. [ ] Normative
3. [ ] Costs involved 7. [ ] Manufacturers recommendations
4. [ ] Risk involved 8. [ ] Other:______________

Corrective Maintenance
1. [ ] Time available 4. [ ] Risk involved
2. [ ] Resources available 5. [ ] Occupant satisfaction
3. [ ] Costs involved 6. [ ] Other:______________

3.5. How is the structure of the maintenance plan?

Preventive Maintenance
1. [ ] Long-term ([ ] week, [ ] month, [ ] year)
2. [ ] Mid-term ([ ] week, [ ] month, [ ] year)
3. [ ] Short-term ([ ] day, [ ] week)

Corrective Maintenance
1. [ ] Long-term ([ ] week, [ ] month, [ ] year)
2. [ ] Mid-term ([ ] week, [ ] month, [ ] year)
3. [ ] Short-term ([ ] day, [ ] week)

3.6. What is the planning technique used?

Preventive Maintenance
1. [ ] Critical Path Method
2. [ ] Line of balance
3. [ ] Program Evaluation and Review Technique
4. [ ] Other:____________

Corrective Maintenance
1. [ ] Critical Path Method
2. [ ] Line of balance
3. [ ] Program Evaluation and Review Technique
4. [ ] Other:____________

3.7. Are there any standard maintenance procedure to conduce the maintenance activities?
1. [ ] Yes Which kind?______________________ 2. [ ] No

3.8. Are there any measurement and control of maintenance activities? (indicators for the control and improvement of maintenance management)
1. [ ] Yes Which kind?______________________ 2. [ ] No

3.9. Who can request a work order?
1. [ ] End user 2. [ ] Maintenance staff 3. [ ] Monitored system alarm 4. [ ] Other:__________

3.10. Which system is used to request a work order?
1. [ ] Internet platform 2. [ ] Mobile application 3. [ ] Paper request 4. [ ] Other:__________

3.11. Which parameters are taken into account to plan the preventive maintenance and to define the frequency of inspection?
1. [ ] Age of system/component
2. [ ] Type of material
3. [ ] Equipment technology
4. [ ] Solar orientation
5. [ ] Weather condition (temperature, UV, moisture, wind)
6. [ ] Surrounding environment (industrial, seaside, vegetation, etc)
7. [ ] Type of use (classroom, teacher room, reception, etc)
8. [ ] Cleaning frequency
9. [ ] Other: ______________

eWork and eBusiness in Architecture, Engineering and Construction – Christodoulou & Scherer (Eds)

A comparative case study of coordination mechanisms in design and build BIM-based projects in the Netherlands

A.A. Aibinu
Faculty of Architecture Building and Planning, The University of Melbourne, Melbourne, Australia

E. Papadonikolaki
Faculty of Architecture and the Built Environment, Delft University of Technology, Delft, The Netherlands

ABSTRACT: BIM implementation can affect the project coordination mechanisms in unexpected ways, even in widely-applied project procurement structures. Apart from the chosen procurement approach, the BIM technology and the distribution of roles in the project team influence and shape the project coordination. This paper aims to explore the emerging coordination structures and processes from BIM implementation in design-build procurement. An exploratory comparative case study has been undertaken. The findings included two main coordination mechanisms: a centralized and decentralized structure and a hierarchical versus participative decision-making processes. These two patterns subsequently open a debate about the relationship between BIM implementation and business models in AEC and particularly the emergence of specialized all-around BIM firms versus BIM-knowledgeable engineering firms.

1 INTRODUCTION

The Architecture, Engineering and Construction (AEC) industry is usually described as highly fragmented. The conventional design and construction process of a building project involves multiple interactions among various domain experts responsible for the design as well as multiple sub-contractors and suppliers on site, arranged by a contractor on site. The project team of a construction project is usually a temporal network (Winch, 2002), which is believed to be responsible for fragmented information flows between design and construction. Accordingly, the design and construction processes are clearly separated and the project information generated and shared across these two phases is often unreliable and difficult to access due to poor k coordination among the work of the various domain experts and the those responsible for the executing the work on site. This interface between design and construction is managed by project managers. With the advent of the digital technologies in AEC, and particularly of Building Information Modeling (BIM), the chasm between design and construction is deemed to be closer to being bridged.

In past decade, BIM has been considered a solution to that fragmentation, poor project coordination and information management problems (Eastman et al., 2008). The promise is that BIM and its associated technologies and processes, can facilitate simultaneous work by multiple design disciplines. However, the BIM collaboration process is often asynchronous under most circumstances (Cerovsek, 2011). Also, despite the popular and utopic belief that BIM could enable a centrally controlled flow of information—and thus centralized collaboration—this is not possible due to computational limitations (Miettinen and Paavola, 2014). Howard and Björk (2008) claim that "the single BIM (model) has been a holy grail but it is doubtful whether there is the will to achieve it" and thus directly defying the claims for centrally controlled BIM. However, BIM sufficiently supports a centrally performed federation of multidisciplinary information from the various actors (Berlo et al., 2012). Accordingly, BIM challenges the traditional coordination mechanisms, roles and workflows in construction. On one hand, many BIM-specialized firms have emerged to offer all-inclusive BIM-related services to AEC firms and projects. These services sometimes encapsulate the traditional project management as well as technology and information management-related services. On the other hand, various in-house roles pertinent to BIM have emerged within existing firms.

Various coordination mechanisms could be applicable for BIM implementation. To investigate the emerging BIM coordination mechanisms, this paper focuses on Design-Build (DB) procurement, within which according to Eastman et al. (2008) "the use of BIM (...) is clearly advisable". This paper aims to showcase coordination structures from BIM implementation in two cases in the Netherlands. It would examine and compare the emerging project coordination from BIM implementation in DB projects, and the various actors' roles. It will also attempt to shed light on the impact that these mechanisms had and the challenges and the outcomes of the cases. Thereafter, the findings would attempt to inform and assist AEC practitioners to improve their BIM adoption processes and reap its acclaimed benefits.

2 BACKGROUND, RELATED PREVIOUS WORK AND GAP

2.1 *The interactions of project procurement and project coordination with BIM*

Building Information Modelling (BIM) has been defined as tools, processes, and technologies that are facilitated by digital, machine-readable, documentation about a building, its performance, its planning, its construction, and later its operation' (Eastman et al., 2008). BIM entails the use of many tools, processes and technologies to produce a building information model. In a BIM-based project delivery process, input from the various design disciplines, contractor, suppliers and subcontractors can be sought early in the design process, be visualized and the potential disciplinary coordination problems could be detected and resolved. This process requires close and ongoing collaboration among the project team members. Eastman et al. (2008) advice that DB procurement "may provide an excellent opportunity to exploit BIM technology, because a single entity is responsible for design and construction", as it is more cost-efficient and shorter than the Design-Bid-Build (DBB) approach.

In general, the procurement methods, BIM technology and the distribution of responsibilities have a major impact on the coordination process and project success. Whilst the procurement governs 'design, construction and commissioning of projects' (Holzer, 2015), the coordination is the underlying abstract pattern of decision-making and communication among the project team. The coordination plays a crucial role in every project procurement method and is needed for managing the tasks interdependences (Malone and Smith, 1988). Thus, the project procurement method would interact with project coordination structure and thereafter influence the success of BIM implementation. The DB procurement approach could support BIM coordination, by creating an environment that fosters concurrent interactions among team members, and especially in the interface of design and construction.

2.2 *Project procurement and BIM*

Procurement can be defined as 'the organizational structure adopted by the client for the management of the design and construction of a building project" (Masterman, 1992). Uher and Davenport (2009) describe it as 'the process by which the client seeks to satisfy his [or her] building requirement, characterized by a particular organizational form, distribution of responsibility, tasks and risk allocation'. Turner (1997) identified two essential decisions in procurement (a) the organization for the overall project management, and (b) the organization for design and construction. The organization for the overall management of project involves client's decisions for either using an in-house project manager or an external project management or a combination of the two.

In the AEC industry, various procurement methods have been used before BIM. Turner (1997) classify them into (a) design-led (b) designer-led, and (c) management-led. Others include Public Private Partnerships (PPP), alliancing, and Integrated Project Delivery (IPD). There is agreement in the literature that the Design-led procurement is not an arena for realizing the full benefit of BIM (Loke, 2012, Sebastian, 2011b). Holzer (2015) conducted an analysis of the opportunities and challenges of BIM under the contract procurement methods as applied in Australia and deduced that IPD is the closest fit, contractually speaking, for full BIM implementation, although it is not applicable to all local markets (Sebastian, 2011a, Holzer, 2015). In DB procurement, some potential opportunities for BIM use and issues identified by Holzer's (2015) analysis are:

- BIM facilitates increased transparency in setting up and pricing tender packages,
- The stakeholders can set up their models up with Construction BIM requirement in mind,
- BIM increases the potential for interfacing information between consultants and trade-contractors in construction documentation,
- The risk lies with the contractor to maximize BIM knowledge transfer,
- It requires skilled contractors who understand BIM workflows and
- The input from client to help define operational requirements is not automatically guaranteed.

Holzer's (2015) work is theoretical. Sebastian (2011a) reaches to comparable conclusions as to the fit of DB for BIM when he compared various procurement approaches using a single case study. Whereas, all procurement routes could support BIM, the DBB would add to the fragmentation of information between design and construction, and on the other hand, the DB discourage a potential involvement of the client in design and construction phases (Sebastian, 2011a). Therefore, there is always a trade-off between the project scope and the extent of the client's involvement and the coordination of the information flow from design to construction.

2.3 *Project coordination structure and BIM*

The coordination structure is regarded as the pattern of decision-making and communication among a set of actors who complete tasks to achieve project goals (Malone and Smith, 1988). It is the underlying abstract decision-making that characterizes every project procurement method and is needed for managing the tasks dependences. The task interdependences in a construction project require that: for each party to complete their task, they must receive information needed from another party. In order to fulfil the client's needs, there is need for coordinated teams, dynamic information flow, and efficient communication and interaction among actors and tasks. The success of coordination would depend on interactions among parties and the communication paths that could enabled by digital technologies. Thus, coordination in BIM-based projects involves technology and human interactions. Dabbish et al. (2010) distinguish between formal and informal coordination. For early organizational theorists, formal coordination is needed where uncertainties are low, e.g. where the tasks are clear and based on routine and involving 'a priori definition of organizational structures and processes for managing dependences including supervision, rules, routines, standardization, scheduling, pre-planning, and division of labor into minimally dependent units' (March and Simon, 1958). Informal coordination is interpersonal coordination, better suited for managing highly interdependent and complex tasks where the actors interact directly to exchange task information and negotiate the tasks' dependences (Malone and Crowston, 1994).

In the context of product development in organizations, Olson et al (1995) classified formal coordination structure into seven structures ranging from the most mechanistic, e.g. bureaucracy, to the most organic and participative structure, e.g. design centers. These are characterized by varying degrees of complexity (simple to complex), distribution of authority (centralized to decentralized), formalization (formal to less formal), autonomy (low to high). Such mechanisms are also characterized by processes that affect decision-making/conflict resolution (hierarchical to participative) and information flow (formal to informal). Figure 1 illustrates Olson et al's (1995) types of coordination mechanisms.

According to Malone (1987), the costs of coordination structures include production cost, coordination cost and vulnerability cost. Production cost include the cost of delays in finishing tasks; coordination cost are the cost of the maintaining the communication links among the parties as well as cost of exchanging 'messages', e.g. information, whereas the vulnerability cost is the cost of failure of parties to perform their tasks or failure to make decisions. Drawing on Williamson's (1975) transaction cost economics concept, there are two means of coordination for tasks: (a) internal coordination for tasks using in-house capacity and (b) market coordination for the same task, based on outsourcing. According to Williamson (1975), the choice between in-house and market coordination is that of differences in transaction costs of the two means of coordination. Coordination of

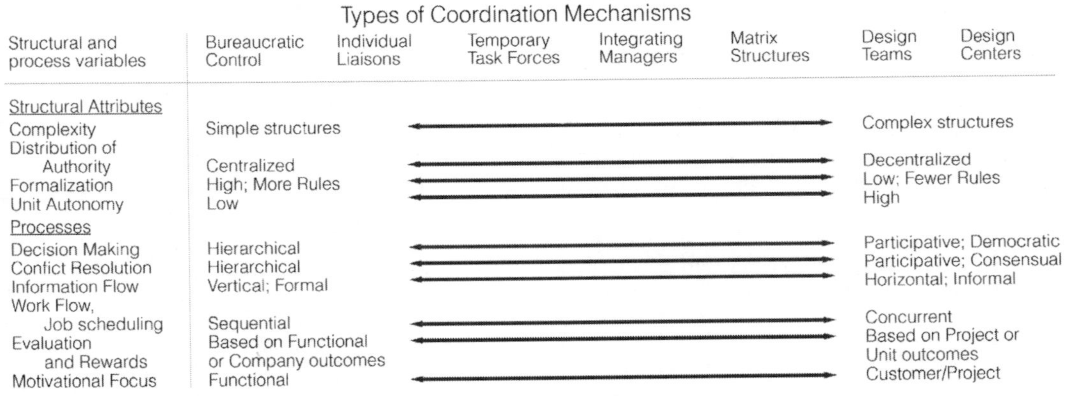

Types of Coordination Mechanisms

Structural and process variables | Bureaucratic Control | Individual Liaisons | Temporary Task Forces | Integrating Managers | Matrix Structures | Design Teams | Design Centers

Structural and process variables	Bureaucratic Control ⟵	⟶ Design Centers
Structural Attributes		
Complexity	Simple structures	Complex structures
Distribution of Authority	Centralized	Decentralized
Formalization	High; More Rules	Low; Fewer Rules
Unit Autonomy	Low	High
Processes		
Decision Making	Hierarchical	Participative; Democratic
Conflict Resolution	Hierarchical	Participative; Consensual
Information Flow	Vertical; Formal	Horizontal; Informal
Work Flow, Job scheduling	Sequential	Concurrent
Evaluation and Rewards	Based on Functional or Company outcomes	Based on Project or Unit outcomes
Motivational Focus	Functional	Customer/Project

Figure 1. A continuum of interfunctional coordination mechanisms: Attributes and processes (Adapted from Olson et al. 1995).

production in construction is often achieved by the use of the market. The overall cost of market coordination structure can vary according to two types of coordination namely—centralized and decentralized (Malone (1987). In centralized coordination, there is a centralized manager who coordinates the activities of the various actors. The manager has a communication link to each actor and is responsible for ensuring that all the tasks are performed appropriately and on time and are brought together with other tasks to fulfil the goals (client needs), whereas, in decentralized coordination, the actors interact with each other and there are communication links among all actors. Thus, the communication links is denser than in centralized structure. According to Malone (1987) the coordination cost is proportional to the number of connections between the actors. In centralized coordination, the failure of the manager to act, make decision or perform can delay the overall production. However, in decentralized coordination, the failure of an actor to perform could result into termination of the contract of the actor and selection of another actor of similar expertise. Thus, vulnerability cost is lower in decentralized structure than in centralized structure whereas coordination cost is higher in decentralized than centralized structures. Nevertheless, Stank et al. (1994) anticipated that the centralized firms would have better information support than decentralized firms, whereas there was no significant differences between the level of information support for the two structures. They also hypothesized that the sophisticated information systems would handle information requirements regardless of the organizational structure." (Stank et al., 1994).

Project coordination in AEC is highly dynamic and entails complex interdependent tasks often targeting new solutions and involving frequent changes. Using design artefacts, such as BIM models and web platforms, such as Common Data Environment (CDE) to connect the actors and integrate design work is then crucial to support information exchange, and coordination. It can help the actors to understand each other's view, negotiate and resolve conflicts in an ongoing basis. Based on the above theorizations, it would be useful to explore the emerging coordination mechanisms arising from BIM and the disturbances in the traditional project phases and roles dictated by BIM implementation. This study will highlight how these opportunities and issues play out in real world Design-Build (DB) projects. It will contribute to the discourse on BIM and its practical implementation by showcasing lessons learned from BIM implementation in two DB cases in the Netherlands by examining the project coordination mechanisms that emerged from BIM.

3 RESEARCH APPROACH

The paper used a case study methodology. The BIM-based projects were analyzed as to the, (a) BIM management structure, i.e. distribution of roles, responsibilities and tasks, (b) BIM-related activities and processes, (c) outcomes. Two cases in the Netherlands, cases A and B, were analyzed and compared. The Dutch AEC was selected for this study, because BIM adoption in the Netherlands presents a balanced mix between policy-driven BIM roadmaps and emerging BIM practices (Kassem et al., 2015). Whereas, the BIM-related policies are not very advanced in the Netherlands, the construction firms have been quite proactive in adopting BIM technologies. Both cases had a DB procurement method, but used opposite approaches for managing and coordinating the BIM implementation process. In case A, a specialist BIM consulting company was hired for BIM implementation, whereas in case B the various relevant BIM functions were performed by in-house BIM-knowledgeable employees from the various firms. Case A was studied during early 2013 and case B during late 2015.

The case studies were exploratory cases. The exploration involved interviews with the project actors, analysis of project documentation, and live observations of BIM clash and engineering sessions. The interviews were semi-structured and addressed to various actors, e.g. contractor, client, engineers and the BIM consultancy firm (only in case A). The first set of questions was about the firms' BIM adoption history, challenges and outcome. The next set of questions were about BIM implementation at a project level, e.g. motivation for BIM, BIM workflow, contractual strategies, BIM roles and responsibilities and technical challenges from BIM. The case description is presented in text, and the responses to the questions about BIM activities and roles are presented in tables, to facilitate the case comparison.

4 COMPARATIVE CASE ANALYSIS

4.1 *Case description*

Case A (2013) is a housing project of 40 rental apartments with five apartments per floor for single and two-person households, using industrialized building systems. The client is a housing association in partnership with a property developer. For the project, BIM was not a contractual requirement. The use of BIM was part of the contractor's tender proposal to the client with the goal of using BIM to achieve 'a better building delivered at the lowest possible cost'. It was envisioned that BIM and VDC methodology will be used for reducing design errors and clashes and deliver the project faster (time), cheaper (cost) and better (quality).

In case B (2015), whereas the client did not require BIM, the contractor and his partners decided to use BIM to increase project quality. The project was a housing tower with 12 stories and 83 housing units of two to four bedroom apartments, to buy or rent. BIM also supported the technical challenges in the site logistics. It was a tower in a small plot, adjacent to a shopping center, whose operation could not be disturbed. This project also used industrialized building systems and dry construction, which is very common in the Netherlands. The architect stated that they did not dare to do this project without BIM. The motives for using BIM in case B was also strategic, because the contractor and their partners wished to deliver "as-built" drawings and potentially master the use of BIM for their future projects.

The procurement of case B was a less complex DB than that of case A. The client hired the contractor to deliver design and construction and gave them complete power over next actions. The contractor had long-term partnerships with the architects and the structural engineer (Str. Eng.). Also, the contractor had long-term partnerships with a MEP firm, sub-contractors and suppliers. Most firms involved in case B adopted BIM as a means to control the information flows in the project. The adoption of BIM was triggered from either internal or external reasons. On one hand, for the architect, the structural engineer and the contractor, BIM adoption was a natural decision to improve their businesses. On the other hand, the concrete sub-contractor, the suppliers and the MEP engineering firm adopted BIM to comply with customer demand and because it was requested from the market.

4.2 *Project procurement*

Figure 2a illustrates the project procurement structures of the two DB projects. In case A, the designers were engaged by the client to define the client's requirement and the scope of the project via development of conceptual design (Level of Development (LOD100) to schematic design (LOD200). The contractors tender was based on the LOD200 documents and model, then the contractor was selected and thereafter the architect was novated to the contractor. The contractor afterwards hired the BIM firm, whereas the designers worked under the BIM manager's leadership. Various suppliers and subcontractors were also selected by the BIM managers, after consulting with the contractor, on the basis of their experience with BIM.

4.3 *BIM implementation and coordination*

4.3.1 *Overall management structure*

In case A, after the project award to the contractor, the BIM consulting firm i.e. BIM managers, was hired by the contractor. The BIM managers were responsible for not only the overall management of the project, as project managers, but also for the generation of the BIM models based on models produced by the designers and several subcontractors, as coordinators. To ensure the success of BIM implementation, an initial project workshop was conducted i.e. a BIM "kick-off" meeting. The purpose was to ensure that all the parties understood the project and agreed about the way of working and BIM use. All parties had to sign the BIM execution document as a part of their contract.

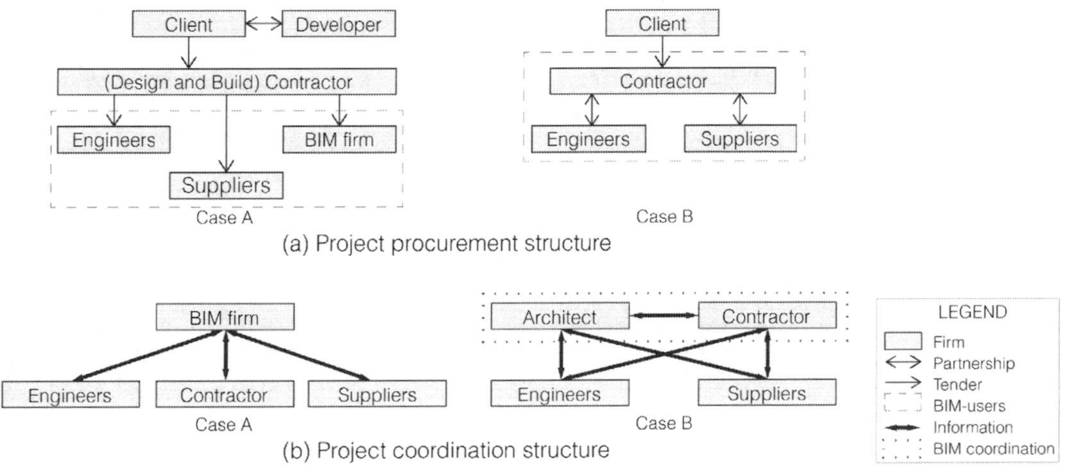

Figure 2. Project procurement structure (a) and project coordination structure (b) in cases A and B.

Table 1. BIM implementation, coordination structure and BIM roles in cases A and B.

Phase	Activity	Responsible party	
		Case A	Case B
BIM "Kick-off"	Transfer of documents to all parties	All parties	Contractor
	Presentation of BIM methodology	BIM firm	Architect
	Preparation of the online platform (CDE)	BIM firm	Contractor
	Providing the BIM execution plan to all parties	BIM firm	Architect
	Verifying & agreeing on the execution plan	All parties above	All parties above
BIM drafting until LOD300	Set-up of the architectural model	BIM firm	Architect
	Set-up of the structural model	BIM firm	Str. Eng.
	Clash detection: architectural and structural model	BIM firm, Contractor	Architect
	Revising the previous models	BIM firm	Architect, Str.Eng
	Transferring the revised models to MEP engineers	BIM firm	Architect
	Preparation of MEP models	MEP engineers	MEP engineers
	Clash detection: architectural, structural and MEP model	BIM firm	Architect
	Revising the models	BIM firm, MEP engineers	Architect, and engineers
	Sharing working model with the subcontractors/suppliers	N/A	Architect
	Verifying & agreeing on models	All parties above	All parties above
BIM drafting until LOD400	Sharing LOD300 model with the subcontractors/suppliers	BIM firm	Architect
	Identifying key constraints based on subcontractors/suppliers input	Subcontractors/suppliers	Contractor
	Preparation of subcontractors/ suppliers models	Subcontractors/suppliers	Suppliers
	Clash detection of the models	BIM firm, Contractor	Architect
	Revising the models	Subcontractors/suppliers	Suppliers, engineers
	Verifying & agreeing on models	Contractor and all parties above	All parties above
BIM until Pre-construction phase	Identifying required information for working drawings	Contractor	Architect
	The list of required working drawings	Contractor	Contractor
	Clash detection and Processing design changes of models from the suppliers to LOD400 model	BIM firm	Contractor
	Preparation working drawings out of LOD400 model	BIM firm	Suppliers, engineers
	Control of the working drawings	Contractor, Architect	Contractor
	Adjusting/revising the working drawings	BIM firm	Engineers, suppliers
	Verifying & agreeing on technical drawings	All parties above	All parties above
General activities in all phases	Consultation with the contractor	All parties	N/A
	Consultation with the client	Contractor	Contractor
	Process management	BIM firm	Contractor
	Consultation with other parties e.g. authorities	Contractor	Architect
	Specific explanation of methodologies	BIM firm	Architect
	Collocations	N/A	All parties
	Maintenance of the CDE	BIM firm	Contractor
	Evaluation of the project	All parties	All parties

The BIM process was supported by BIM protocols and management plan from the early stage of the workshop and the project.

In case B, BIM was applied from various roles within the involved firms. The main project management function was held by the contractor. The architect, structural engineer and MEP firms had at least one BIM-savvy engineer, alongside the project engineer. A "kick-off" session and a BIM protocol took place from the start, to coordinate the BIM scope. The BIM process was supported also from frequent collocations. The architect was the BIM coordinator until the pre-construction phase. Thereafter, a site engineer from the contractor's firm coordinated the BIM process. Figure 2b illustrates the BIM coordination structures in case A and B.

4.3.2 *Processes and activities*

In case A, BIM was used from Definitive Design, with LOD200 until the Construction Preparation phase with LOD400. BIM was used for the following activities: design coordination, clash detection, design visualization, quantities take-off, cost estimation, preparation of working (shop) drawings and information exchange (Table 1). The project schedule was prepared in different software and was not linked to BIM tools. The authoring tools employed include: Revit Architecture, Revit MEP, Navisworks, ArchiCAD, HiCAD, Tekla, and BIM-ID (for cost calculation). The BIM manager was responsible for modelling, cost calculation and clash detection. The federated model formed a basis for the subcontractors. There were a lot of formal and informal coordination activities with the various subcontractors to produce jointly a working model for construction. Interestingly, some of the suppliers were collocated in the same office building with the BIM managers. According the BIM managers, this greatly influenced the team collaboration. A project website hosted on the servers of the BIM managers was used as a Common Data Environment (CDE) to share project information using Industry Foundation Classes (IFC).

In case B, BIM was used from the Initiation phase, i.e. LOD100 until the Hand-over (as-built BIM). It was used for design exploration, visualization, design coordination, cost estimation, clash detection, quantity take-off, information exchange and site resource management. The authoring BIM tools used were primarily Revit and Tekla Structures and the BIM checking tool used was Solibri Model checker. Similar to case A, the information exchange took place in a CDE, where all parties uploaded their IFC files. Afterwards, the various reference models were federated to perform clash detections as described in Berlo et al. (2012). The contractor used preliminary input from the architectural and structural models to perform the budget estimation, and early informal discussions with the suppliers. The suppliers were involved early in the process after the LOD300 phase and provided preliminary input.

4.3.3 *Outcomes of the cases*

Case A was delivered ahead of schedule. The client was satisfied with the quality. All parties had better understanding of the BIM process but some challenges included time pressure because of the contractual obligations and late completion of tasks by some parties. The contractor's expectations were too high because it was their first BIM project, which also put work and time pressure on the other parties. The BIM managers had to work overtime to meet the BIM management function. Case B is an ongoing project and so far no time delays have been reported. Time pressure was reported by various project actors, but according to them it was not due to the BIM implementation, but rather due to the strategic decisions of the contractor's commercial managers. However, some coordination issues surfaced regarding the role of the BIM coordinator. In the beginning, the architect performed this function, but later, after request from the partners, a site engineer was trained to become a BIM coordinator, so as to combine technical expertise from the site to technical BIM expertise. Also, frequent collocations of the partners increased the understanding and knowledge about BIM process. Table 1 contains the case comparison across the processes and roles of BIM implementation.

5 DISCUSSION AND IMPLICATIONS

5.1 *Structural attributes of coordination*

In Case A, the BIM coordination structure and the project management were highly centralized. The BIM managers were responsible for BIM modelling and coordination, project and cost management. This made the coordination structure more simple according to Olson et al (1995), and more cost-efficient according to Malone (1987). The BIM managers also exerted control over the MEP, sub-contractors' and suppliers' models. The BIM managers send their staff to support the other BIM users whenever issues arose. The BIM managers performed an 'integrating manager' role (Olson et al., 1995). They also exerted informal influence from their central position (see Figure 2b). However, this structure, whereas very controlled, would potentially have a greater vulnerability cost (Malone, 1987), as the BIM implementation would solely depend on one actor in the chain.

Whereas Case B was also DB procurement, had an opposite BIM coordination structure to case A, because of the multiple partnering relations among the firms. The contractor executed the project management activities. All engineers and suppliers were then responsible for their BIM input to the federated model. The paradox in case B was that although the project management was centralized, the BIM coordination structure was decentralized and more complex. Both the contractor and the architect were BIM coordinators and this lead to a decentralized BIM structure (see Figure 2b and Table 1), providing evidence of highly autonomous and less formal coordination structures from Olson et al. (1995). According to Malone's (1987) categorization, this BIM coordination structure would induce a highly costly BIM coordination, but also less vulnerable to failure.

5.2 *Attributes of the coordination process*

Surprisingly, whereas the control in case A was centralized, the decision-making was not strictly hierarchical. This was possible because the CDE ensured participative structure and a quasi-concurrent workflow. Most of the interactions were between the BIM managers and the suppliers and subcontractors, and were facilitated by the CDE. There were also a lot of informal interactions. The CDE was critical for the interaction of the BIM users. Case A also shows that the designer's and contractor's roles were less visible due to the power of the BIM management firm.

In case B, the role of the BIM coordinator, included the tasks of distributing the information about the BIM process among the partners, assigning tasks, model federation and model checking. The design process of the engineers and the suppliers was more participative and consensus-seeking than in case A, as they were responsible for creating and revising of their own models, and also ensuring that their models were in the correct form for the federation. The engineers and suppliers were empowered to apply BIM and responsible for their work. Given that not all actors had the same BIM capabilities, frequent collocations, informal communication and shared learning took place. However, the decentralization of coordination in case B means that failure to maintain the density of communication would result in poor coordination among the engineers, and suppliers, and thus higher production cost. Future research would be useful to further investigate the vulnerability and coordination costs of various coordination structures in BIM-based projects.

5.3 *Comparison of emerging BIM business models*

The cases carry implications for the business models in AEC firms. The BIM management firm of case A was originally cost managers, who reinvented themselves into an all-round BIM firm that provided information management, cost and project management services. This could lead to rise in mergers, consortiums, and acquisitions of firms that previously provided auxiliary services. From case B, the contractor seems to have incorporated the information management services. Also, in case B, there was an increase in the engineers' and supplier's empowerment and responsibilities to provide their services using BIM standards and agreements. This could be a sign that the future AEC business models AEC would offer integrated BIM and discipline-related services. Accordingly, it would be interesting to explore the clients' preferred approach for reducing the risks of BIM adoption, i.e. choosing between specialized or integrated BIM and engineering firms.

6 CONCLUSIONS

Various procurement routes have been discussed as to their suitability to support BIM implementation. Given the promise of BIM for consistent information flows, it is considered that BIM benefits can be better realised when combined with integrated delivery processes, e.g. IPD. However, this paper provided evidence that even simpler procurement routes, such as Design-Build could provide integrated processes for BIM implementation. To this end, it was revealed that not only the procurement, but also the selection of various firms to work together, affect the project coordination mechanisms and in particular the structures and processes. In particular the two cases presented two structures of BIM coordination: centralized and decentralized supported by hierarchical and participatory decision-making processes respectively. These BIM coordination mechanisms subsequently carry various implications for future business models in the AEC sector.

Due to the increasing adoption of BIM, the various firms would gain experience from BIM-projects and become increasingly aware of its potential. The two DB cases presented two opposite approaches to BIM implementation. The cases used either specialized BIM consulting firms or integrated BIM solutions within existing firms, e.g. hiring BIM-savvy engineers or training their in-house personnel, to reduce the cost of outsourcing BIM. There are lessons to be learned from both cases: a centralized and inclusive approach towards BIM (from the BIM consulting firm) sets high-quality standards that challenge any ad-hoc BIM approaches. On the other hand, a decentralized approach to BIM coordination, might soon gain more traction, given that the use of BIM technology gradually becomes an industry requirement, which could be partially or wholly supported by BIM-savvy professionals, thus making full outsourcing of BIM work unnecessary. Nevertheless, the engagement of firms in both 'centralized' and 'decentralized' BIM coordination structures could potentially contribute to greater development of BIM knowledge and higher BIM maturity across AEC firms.

REFERENCES

Berlo, L., van, Beetz, J., Bos, P., Hendriks, H. & Tongeren, R., van 2012. Collaborative engineering with IFC: new insights and technology. *9th European Conference on Product and Process Modelling, Iceland.*

Cerovsek, T. 2011. A review and outlook for a 'Building Information Model'(BIM): A multi-standpoint framework for technological development. *Advanced engineering informatics,* 25, 224–244.

Dabbish, L.A., Wagstrom, P., Sarma, A. & Herbsleb, J.D. Coordination in innovative design and engineering: observations from a lunar robotics project. Proceedings of the 16th ACM international conference on Supporting group work, 2010. ACM, 225–234.

Eastman, C., Teicholz, P., Sacks, R. & Liston, K. 2008. BIM Handbook: A Guide to Building Information Modeling for Owners, Managers, Designers, Engineers, and Contractors, Hoboken, New Jersey, USA, John Wiley & Sons Inc.

Holzer, D. BIM for procurement - procuring for BIM. *In:* Crawford, R.H. & Stephan, A., eds. Living and Learning: Research for a Better Built Environment: 49th International Conference of the Architectural Science Association 2015, 2015 Melbourne, Australia. The Architectural Science Association and The University of Melbourne, 237–246.

Howard, R. & Björk, B.-C. 2008. Building information modelling–Experts' views on standardisation and industry deployment. *Advanced Engineering Informatics,* 22, 271–280.

Kassem, M., Succar, B. & Dawood, N. 2015. Building Information Modeling: analyzing noteworthy publications of eight countries using a knowledge content taxonomy. *In:* Issa, R. & Olbina, S. (eds.) *Building Information Modeling: Applications and practices in the AEC industry.* University of Florida: ASCE Press.

Loke, T.K. The Potential of Lifecycle Approach Through 5D. *In:* Rickers, U. & Krolitzki, S., eds. Proceedings of the 2012 Lake Constance 5D-Conference, 2012 Dusseldolf, Germany. VDI Verlag, 63–77.

Malone, T.W. 1987. Modeling coordination in organizations and markets. *Management science,* 33, 1317–1332.

Malone, T.W. & Crowston, K. 1994. The interdisciplinary study of coordination. *ACM Computing Surveys (CSUR),* 26, 87–119.

Malone, T.W. & Smith, S.A. 1988. Modeling the performance of organizational structures. *Operations Research,* 36, 421–436.

March, J.G. & Simon, H.A. 1958. *Organizations*, John Wiley and Sons, Inc.

Masterman, J. 1992. An introduction to building procurement systems, Spon.

Miettinen, R. & Paavola, S. 2014. Beyond the BIM utopia: Approaches to the development and implementation of building information modeling. *Automation in construction,* 43, 84–91.

Olson, E.M., Walker Jr, O.C. & Ruekert, R.W. 1995. Organizing for effective new product development: The moderating role of product innovativeness. *The Journal of Marketing*, 48–62.

Sebastian, R. BIM in different methods of project delivery. *In:* Zarli, A., Fiès, B., Egbu, C., Khalfan, M.M.A. & Amor, R., eds. Proceedings of the CIB W78-W102 Conference: Computer, Knowledge, Building, 2011a Sophia Antipolis, France. 800–809.

Sebastian, R. 2011b. Changing roles of the clients, architects and contractors through BIM. *Engineering, Construction and Architectural Management,* 18, 176–187.

Stank, T.P., Daugherty, P.J. & Gustin, C.M. 1994. Organizational structure: influence on logistics integration, costs, and information system performance. *The International Journal of Logistics Management,* 5, 41–52.

Turner Alan, E. 1997. Building Procurement. Macmillan Press Ltd.

Uher, T.E. & Davenport, P. 2009. *Fundamentals of building contract management*, UNSW Press.

Williamson, O.E. 1975. *Markets and hierarchies,* New York, Free Press.

Winch, G.M. 2002. *Managing construction projects,* Oxford, Blackwell Science Ltd.

Information & knowledge management (IV)—construction

eWork and eBusiness in Architecture, Engineering and Construction – Christodoulou & Scherer (Eds)

Construction information framework—the role of classification systems

P. Mêda
Construction Institute—CONSTRUCT-GEQUALTEC, Faculty of Engineering, Porto University, Porto, Portugal

E. Hjelseth
Department of Civil Engineering and Energy Technology, Faculty of Technology, Art and Design, Oslo and Akershus University College of Applied Sciences, Oslo, Norway

H. Sousa
CONSTRUCT-GEQUALTEC, Faculty of Engineering, Porto University, Porto, Portugal

ABSTRACT: This paper focuses on understanding the role of Construction Information Classification Systems (CICS). Information Classification Systems (ICS) that range several economic activities including construction are also explored in order to find touch points and define future trends. This starts with an overview of classification systems, framed with ISO 12006-2:2015. Two different realities, Norway and Portugal, were studied following ISO framework, providing a vision of the developments and applicability of both types of classification systems. Needs for further development have been discussed in a relation to the new possibilities of BIM. The outcome of this study is a list of proposals for further developments. Classification systems represent information. Information coherency is fundamental towards the objectives of efficiency on the industry. The use of BIM can enforce the development of a new generation of CICS.

1 INTRODUCTION

The built environment is the most remarkable physical product of human society (NATSPEC, 2008). It is constituted by a sum of construction processes that come from ancient times until present days. During time, construction evolved gaining more and more requirements, reaching bigger achievements and more complex productive processes. Nowadays the Architects, Engineers, Contractors, Operators and Owners (AECOO) supply-chain is composed by several agents that on a multistage process develop a sequence of different tasks that leading to the built object. During this productive process different disciplines work together following requirements in terms of regulations, products and technologies, demanding broad knowledge and strong collaborative effort. On the basis of all is the ability to communicate. Even the most rudimentary project relies on the participants being able to communicate and find relevant information at the appropriate time (NATSPEC, 2008).

The human communication can be distinguished into non-verbal, oral and written (Hollermann, 2012). Mutual understanding efficiency on all these forms of communication is vital for stakeholder's interaction and construction process development. With the introduction of information technologies, a new agent with very specific requirements is working along the process; the computer.

CICS play since long time an important role in construction. The development of these tools followed the industry needs, most related with specific problems and output requirements. Due to this, they had generally narrow scope, few or a single table and low complexity. Their applicability was constrained. In terms of format, they were paper based, setting structures that should be transposed for specific projects.

Other systems were developed with different and broader targets (wide range of economic activities), namely the categorization of economic activities and products.

With the introduction of higher information requirements on the construction process, broader and more complex CICS were needed. To frame these demands ISO standards were developed. Different countries and institutions adopted more or less these tools. The way the guidelines were followed was also heterogeneous.

In parallel, there are other systems, the ICS, not construction exclusive and cut-crossing different economic activities. They play similar roles (statistics, e-commerce, trading) and in many situations their applicability is mandatory, setting influence over construction.

The influence of ICS and the national/international CICS adoption was explored for two distinct realities, Norway and Portugal. The analysis was framed with the new ISO version in order to evaluate the situation for the different topics/tables, the state of development and foresee future trends and strategies, both at national and multi-national level.

2 OVERVIEW AND EVOLUTION

2.1 *Introduction*

This part explores different dimensions. A brief vision on CICS evolution is framed with ISO standards development. Regarding ICS's are explored some cases where higher "presence" on construction is observed. This overview allows an improved vision for the AECOO sector over the role of CICS and the "exploitation" of ICS in those systems.

2.2 *Overview on CICS and ISO evolution*

The ISO 12006-2:2015 is the latest result from the work developed by the ISO/TC 59 SC 13 committee. This group was gathered to work on common assumptions for CICS development. The first agreement on the subject was produced in 1994, the ISO/TR 14177. Later, the evolution of this document led to the publication, in 2001, of a new family of ISO standards (Parts 2 and 3). Part 2, had the scope of setting the framework for the development of built environment classification systems. As mentioned, systems for construction in application with releases that vary in time from the end of the 90's until the beginning of the 70's. Most had narrow visions of the construction and were developed with the objective of solving specific problems. Some of the more relevant examples, due to their extensive application are:

- the North American Uniformat II—a single table system, developed for building construction and with the aim of providing references for economic analysis and building management (Mêda, 2014a). It is centered on a Construction Elements vision.
- the CI/SfB, developed by Royal Institute of British Architects (RIBA), is composed by five tables (Race, 2013) (Yoon, 1995). It is geared for construction products and building elements. It meant to provide a satisfactory mean for structuring sets of detailed design drawings, working drawings and specifications.

These are just examples of an extensive list of successful and unsuccessful CICS developed to solve specific situations. In brief, it is possible to state that the main concern of this generation of CICS was the organization of information for the design stage and mainly geared for building construction projects.

This was a trend that the 2001 ISO version tried to counteract, by setting the built environment classification as a target. The first approach to this vision was set even before, in 1997. Following the lessons learned from previous developments and the considerations set on the 1994 technical report the British Uniclass was the first broad scope CICS to be published. Yet, its dimension resulted from the harmonization and update of several contents from other narrower systems that were already in use (Sousa, 2008) (RIBA, 1997). The first version of Uniclass was published with fifteen tables. The main scope in terms of types of constructions was still buildings, but it also integrated some contents for civil engineering works.

The North American Omniclass was the main system released after the standard publication. Among many others, followed its guidelines (Omniclass, 2013). This system can be considered quite similar to Uniclass in terms of topics and structure. Notwithstanding it has a broader scope. It is also supported by narrower systems as the above mentioned Uniformat II or the Masterformat.

With the development of other standards, the increasing requirements of construction life cycle thinking as well as the adoption of new technologies by the construction industry, ISO 12006-2:2001 was lacking in terms of scope and accommodation to new trends, such as Building Information Modelling (BIM) methodologies. The standard does not aims to provide a complete operational system. The objective is delivering common basis for easier understandings, guidelines towards homogenization and table titles. These elements constitute the key assumptions for the development of a new generation of CICS, with broader scope, digital philosophy and framed with the industry sustainability and efficiency requirements. This fact is relevant as many countries are foreseeing needs in terms of harmonization. One aspect mentioned on ISO is that mapping between national and international classifications should be fairly straightforward, as they are likely to differ in their detail due to differences in terms of construction culture and legislation (ISO, 2015).

2.3 *ICS that cut-cross the construction industry*

Other activities use similar conventions to foster mutual understandings. On this paper they are assumed as ICS. Detailed study of these systems is not intended. Yet, there are few that given their characteristics, relationship with construction and even influence on the supply-chain, worth to be mentioned.

The identified situations are mandatory systems within the European Union; adopted by the Statistical Office of the European Union—EUROSTAT or United Nations conventions.

2.3.1 *ICS—European Union*

Three systems play an important role on the EU economic activity survey (procurement, regulation and trading), crossing their scope with the construction industry. These are the Common Procurement Vocabulary—CPV; the Combined Nomenclature, Common Customs Tariff and Integrated Tariff of the European Union—Taric and the Product areas table from the Regulation (EU) Nº 305/2011 – Construction Products Regulation—CPR. Their scope and extension are very different as the purposes meet specific needs as follows:

- CPV – is the mandatory reference to be used on the publication of tender notices/EU public procurement. It is a single table and ranges products, services and works (OJEU, 2004) (BMEL, 2012) (CPV, 2008);
- Taric – is the regulation that supports goods clearance on the EU. It also provides means for collecting, exchanging and publishing data on EU external trade statistics (OJEU, 2014) (Taric, 2016);
- CPR table – sets product areas for the development of harmonized product standards (OJEU, 2011).

There is no difference in terms of development strategy between other areas and construction. These systems were set to attend specific needs and can be framed on the same assumptions of the first generation of CICS.

2.3.2 *ICS—Eurostat*

The Classification of Types of Construction—CC is one of the systems that countries within the "Eurostat family" must comply in what regards statistical data. CC is designed to serve different purposes such as statistics on construction activities, construction reports or building and housing censuses. CC is also structured to be used for the whole construction life cycle, namely changes in use, transactions, renovations, demolition (EUROSTAT, 1998). It has its roots on the provisional version of the Central Product Classification from the United Nations.

2.3.3 *ICS—United Nations*

The United Nations Standard Products and Services Code—UNSPSC is a hierarchical convention used to classify products and services. It has the purpose of enabling electronic commerce and providing the foundation for spend analysis. It was first developed in 1998 by Dun and Bradstreet as the proprietary code set called the SPSC. On the same year it was merged with the United Nations. Nowadays it is managed under contract by GS1 (CWA, 2010). Without being mandatory, this classification, given its structure, versioning dynamic and the release on worldwide used products, namely Enterprise Resource Planning—ERP software, it is gaining importance and achieving high levels of use (Mêda, 2014b).

2.4 *Classification vs code list*

A classification is a set of discrete, exhaustive and mutually exclusive observations, which can be assigned to one or more variables to be measured in the collation and/or presentation of data. The terms "classification" and "nomenclature" are often used interchangeably, despite the definition of a "nomenclature" being narrower than the "classification" (SDMX, 2016). In a classification tabs are often identified by a code and always by a designation. A classification may consist of several levels (hierarchical).

A code list is a predefined list from which some statistical concepts (coded concepts) take their values (SDMX, 2016). Each category in a code list has a unique code and a unique designation. There is no requirement for such classification, that the categories must be exhaustive and mutually exclusive.

2.5 *Importance of classification systems*

The previous mentioned objectives towards IDDS are grounded on the ability of setting effective and efficient communications at human and machine level.

Regarding the first communication type, terms, definitions and ways to assist interpretations will always be needed between the different construction agents involved. Laws, standards and guidelines will continue to provide this information on a more or less organized way.

Classification systems can provide the needed structures to organize this information, within the industry and on its borders. The role of the "new generation" of classification systems will be to aggregate and work the information from the industry and from its boundaries, developing structures that foster the needed mutual understanding, providing outputs for different intended visions. The harmonization of terms between tables constitutes here a major challenge.

Uniclass 2015 represents one of the latest developments (still undergoing). It is compliant with the new version of ISO, it is being drawn in order to support and be compatible with BIM methodologies. The main concern regarding harmonization of terms can be translated through the following:

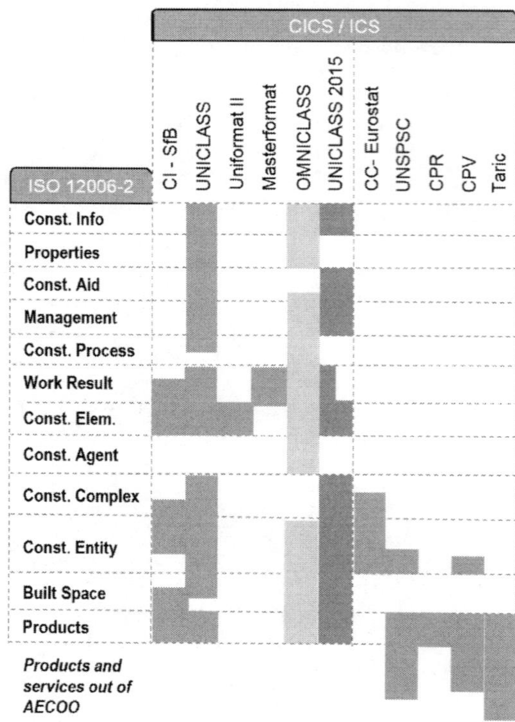

Figure 1. Overview of classification systems and framework within ISO 12006-2:2015 standard. Grey represents the topics that are covered. The analysis is just in terms of topic and does not differentiates, at the moment, scopes, ex: building construction VS built environment.

"Within the BIM toolkit a database of synonyms to make it as easy as possible to find the required classification using standards industry terminology" (Delany, 2015).

From the systems presented and following with the new version of ISO, Figure 1 can be drawn.

3 CASE STUDIES

3.1 *Cased studies from Norway and Portugal*

The research focused on a global vision as presented and used two countries as case study. The two countries Norway and Portugal, represent very different realities in terms of CICS definition and adoption, as well as distinct strategies in what regards implementation of new technologies and innovation on the AECOO industry.

3.2 *Classification systems in Norway*

In Norway most of the construction information is set on standards; Norwegian Standards.

A preliminary study (Bakkmoen, 2009) report commissioned by Standard Norway, shows the diversity of classifications systems in Norway and internationally. The report calls for the author a more comprehensive classification system in Norway, while he emphasizes the need for linking the national tables up against the one used in the Nordic countries and internationally. The report is basically the background for the creation of Standard Norway committee for classification in AECOO sector.

As an example, is the widely used standard specification system NS 3420 – Collection of all standards within the NS 3420 series—Specification texts for building, construction and installations that comes from the mid 70's. Specifically in terms of CICS, there are several standards that date from the early 80's and that have been revised and amended in order to meet new needs, extend the scope and attend to ISO revisions. As stated, these were focused on singular issues, narrow scope and specific problems. Examples of these standards, among others, are the NS 3451 – Table of building elements; NS 3455 – Table for building functions; NS 3457 Part 3: Classification of construction works—building types and Part 4: Classification of construction works—spatial functions. From lessons learned and following the trends, these standards have been updated with the main concern of getting similar terminology to promote interoperation between tables. As it is possible to observe, in terms of built environment they are geared for buildings, as other documents from specific public owners give answer to other construction types such as railways, roads, water pipe system, etc. In terms of construction stages they are set for the design stage. This situation suites the vision of the first generation of CICS.

From the operation and maintenance point of view it was developed by Statsbyggs the Tverrfaglig Merkesystem (TFM), a multidisciplinary labeling system that describes how building components and technical installations in the construction industry should be identified, systematized and labeled. The labeling process uses information from the above mentioned standards as well as information produced during the construction stage.

There is an initiative ongoing, coordinated by Standard Norway, with the scope of revise, harmonize and promote different use among different Norwegian actors of the above mentioned standards. At the same time, close follow up is being made on international systems. The main reason for this is the growing adoption of BIM methodologies by the industry.

Norway is a country outside the EU and therefore some of the previously mentioned ICS are not mandatory. Notwithstanding, it belongs to the EUROSTAT family and due to it the CC is followed.

NS 8360 standardizes type coding and classification of objects, connecting properties and values to the IFC model. It is the first BIM standard developed to take full advantage of building information models (NS 8360, 2015).

This standard will support automatic recognition of object types, information about object types and object instances between different IFC compliant applications, helping to increase efficiency and quality through the use of building information models. This will provide easier access to information, reduce errors, increase reusability of data and reduce duplication.

NS 8360 is appropriate for software vendors of BuildingSmart compatible software for modeling and analysis of BIM models, suppliers of BIM objects and requirements to BIM deliverables. Persons responsible for structuring BIM projects and object libraries. Use cases where tools based on this standard are recommended: coordination model (transparency model and collision control); assembly instructions; CO2 footprint; Life Cycle Costs (LCC); production management; progress planning; electrical sizing; indoor climate requirements for air quality and lighting; system for unique marking system and product coding.

Figure 2 presents a resume of the classification systems in application in Norway and their framework with the ISO.

ISO 12006-2:2015 / Norway	
Const. Info	
Properties	
Const. Aid	
Management	FM 3456
Const. Process	ISO 22263; NS 3454; AIN / RIF
Work Result	NS 3420
Const. Elem.	NS 3451
Const. Agent	*de facto*
Const. Complex	CC - Eurostat
Const. Entity	NS 3457; NWI Object Library; CC - Eurostat
Built Space	NS 3940; NWI; NS 3455
Products	TFM

Figure 2. Overview of classification systems in Norway and framework within ISO 12006-2:2015 standard. (based on Bakkmoen, 2013).

3.3 *Classification systems in Portugal*

The Portuguese construction and its organization follows since long time ago the French construction matrix. Therefore, most of the documents, standards or guidelines were translations or adaptations of that information. The main initiatives that launched the construction information systematization were set by the National Laboratory of Civil Engineering—LNEC that produced three main documents/guidelines geared for building design. The first focused on design organization, i.e. documents to be delivered, information templates, drawings scale and relations between them (LNEC, 1997).

With the influence of other countries guidelines, as the UK, it was produced a guideline with rules of measurement and bill of quantities organization. The last document established a breakdown system for construction elements setting unitary price scenarios for different construction technologies and products. These initiatives date from early 70's and constituted an effort on achieving to a similar organization for buildings. Other public owners, geared for civil engineering works such as roads, railways, worked on specific documents. These had different types of developments from CICS tables geared specific construction types to standard bill of quantities or complete construction specifications.

ISO 12006-2:2015 / Portugal	
Const. Info	
Properties	
Const. Aid	ProNIC
Management	
Const. Process	ISO 22263; D.L. n.º 18/2008
Work Result	ProNIC
Const. Elem.	ProNIC (not directly)
Const. Agent	D.L. n.º 18/2008; Lei n.º 31/2009
Const. Complex	CC - Eurostat
Const. Entity	CC – Eurostat; Portaria n.º 701-H/2008
Built Space	
Products	CPV; EU Combined Nomenclature; EU Regulation n.º 305/2011; APCMC Group of materials

Figure 3. Overview of classification systems in Portugal and framework within ISO 12006-2:2015 standard.

On the beginning of the 00's it was developed a project with the scope of producing an integrated breakdown structure for buildings and roads with complete technical specifications, standard bill of quantities, rules of measurement and cost scenarios. This project, ProNIC—Construction Information Standardization Protocol, was delivered in 2008 and had a broad application on a school refurbishment program launched in 2009 (Mêda, 2014a) (Sousa, 2012).

ProNIC is now being extended with the ambition of achieving to a broad scope, namely other types of civil engineering works and new organizations of information (definition and harmonization of new CICS tables).

At the same time and resulting from the transposition of the EU Directives, the Public Procurement Code, Decreto-Lei n.º 18/2088 was updated. One of the major recent innovations for the industry was brought by this new legal framework and it was the adoption of electronic procurement. This effort raised the bar in terms of construction process organization and technologies to support large amounts of electronic documents. Paper was highly reduced. The installed processes led to an accelerated adoption of one previously mentioned system, the CPV.

This legal framework implied requirements that were defined in additional diplomas:

- the definition of the agents, their intervention and the design documents was set on Portaria 701-H/2008;
- the roles and responsibilities of the construction agents were framed on Lei n.º 31/2009.

In what regards construction products, the manufacturers association uses since long time ago a list of groups of products, that is the APCMC Group of materials. This user friendly classification was recently mapped with CPV and Taric.

Portugal belongs to the EU and therefore it follows EU databases including EUROSTAT.

4 DISCUSSION

4.1 *Classification systems and IDDS research targets*

It is imperative for construction to gain a new breath in terms of efficiency, innovation and industrial leadership. As stated, given the broad and diverse supply-chain and outputs, wide strategies must be undertaken in order to fulfill the ambitious objectives. Terms as collaboration, interoperability, new processes and tools are mentioned on different construction strategic documents.

The CICS's cut-cross these terms as they provide the support and the base for the mutual understanding. The IDDS roadmap sets four priority research targets classified from near-term to long-term. The CICS development fits on targets one and two, providing results for the fulfillment of the topics as presented in Figure 4.

As mentioned, the CICS development followed national strategies with insight of other realities. Nowadays, the information technologies, the communications and the task groups provide new means for accessing and share knowledge between countries.

Notwithstanding, it is not plausible the adoption of a single system for the world, due to local differences (ISO mentions it and fosters the work of singularities on low levels of the breakdown structures) and the amount of systems that exist and will continue to interact with the AECOO industry (CICS and ICS). Mapping processes should be straightforward.

4.2 *Role in BIM*

There is no contradiction between classification systems and object-oriented way of working (Bakkmoen, 2009) or BIM methodology. From the machine communication point of view, for some situations the classification systems might lack in terms of precision that is, the necessary identification in order to make the information machine-readable (Bakkmoen, 2009). New solutions as the one implemented on Uniclass 2015 as well as mapping processes and cross-reference tools must be straightforward.

Classification is a way of information modelling where "Generalization ßà Specialization" is a widely used concept, to support colleting into general classes with limited of joint properties, instead of handling everything as unique elements without any requirement to which (information) element should be include. The last aspect is conditioning the development of standardized solutions with/for multiple users.

In this perspective is the openBIM data schema Industry Foundation Classes (IFC). IFC has information fields to support classification systems. On

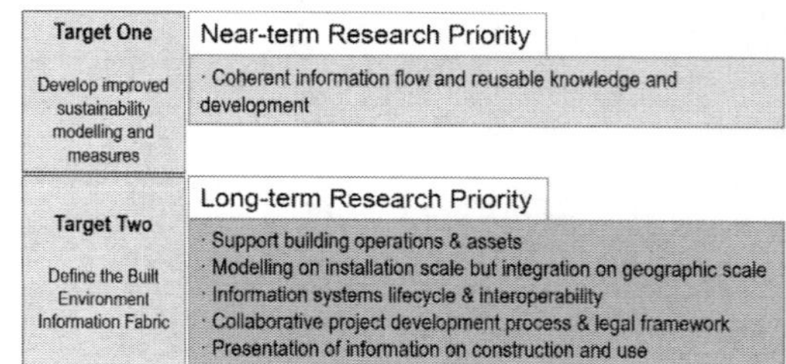

Figure 4. Identification of the IDDS Priority Research targets where the CICS developments give contributions. Adapted from (IDDS, 2013).

the other hand, the International Framework for Dictionaries (IFD) Library can play an essential role by setting a common reference for direct mapping between tables and providing the needed link between human understandable classification systems and machine readable information (IFD, 2016).

BIM enables structured information—which easily can be classified for relevant purposes. An example of this focus can be illustrated by the use of the Solibri Model Checker software. Just after importing an IFC model—a pop-up window about Classification appears. This functionality enables classification to be made easily, supported by filtering functions and color based visualization.

4.3 *Lessons learned*

Attending to the industry needs, the term CICS was assumed for structures that could provide organized information for specific activities, namely on preliminary stages of the construction process, as cost analysis or design specifications. Following the present framework (ISO tables) these contents fit in the Construction Elements and Work Results tables.

These structures gain different levels of importance in different countries; ex: Norway—standards; Portugal—guidelines. The scenario apart of the type of documents, is not as different as it was predictable. In terms of topics there is similarity. The most significant differences that can be pointed result from the timings and scope of the individual strategies for construction innovation. Portugal pushed the industry through the adoption of e-procurement while Norway is working on the adoption of BIM methodology.

Institutions not directly related with the AECOO sector worked other structures for their specific needs, resulting ICS that provide organized information for statistics and trade control, as construction types or products and services. These contents fit on the Construction Complex, Construction Entity and Products tables.

There are other topics where the effort on drawing organized structures is not so visible. Notwithstanding, the relevant information for the construction process can be found in distinct documents from legal diplomas until technical guidelines. Constitute examples of these situations contents that fit on the Construction Process, Construction Agents, Built Space and Information tables.

The vision from the two realities combined with the identification of relevant CICS and ICS and framed with the new ISO tables, can be resumed on the schema from Figure 5.

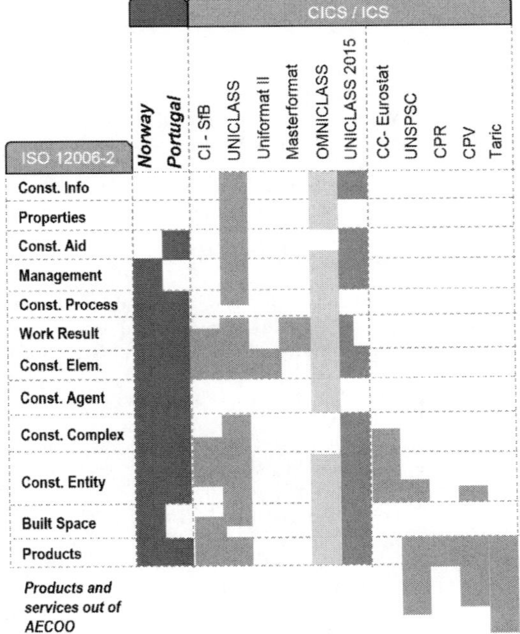

Figure 5. Overview of topics organized according with ISO 12006-2:2015 standard and their cover on the case studies (Norway/Portugal) and by different relevant classification systems.

4.4 *Need for further development*

The present developments in terms of information classification systems place different challenges at global, national and multi-national levels. Considering the outputs from the study it is possible to state that:

- Further developments on the Products, Construction Entity and Construction Complex tables should consider the contents from the already developed and in use ICS's. Additionally, mapping processes between these ICS should be straightforward. In terms of construction products future trends should be followed as Environmental Product Declarations (EPD) might be in short term mandatory. This information combined with Product Documentation Templates (PDT) can set the bridge for software solutions.
- Topics covered by Management, Construction Process, Construction Agent, Built Space and Construction Aid tables are spread on different documents from ISO standards to legal diplomas or guidelines. The collection and organization of this information is found to be the basis for the development of these topics. The Built Space table plays an important role when working with object-oriented software (definition of spaces).
- The Construction Elements table is one of the most relevant when working with object-

oriented software. It has a close relation with Work Results table and as presented, these constitute the core of the "old generation" of CICS. Deal lots of information with different visions and produced in different times constitutes here the major challenge. Guidance from newer CICS can be useful, namely on the upper levels, giving space for differences (accepted by the standard) on the lower ones;
- The remaining tables play an important role providing detailed level of information for specific issues (example: budgeting), complementing the visions of the previous mentioned.

5 CONCLUSIONS

Classification based standards are numerous and generally well implemented in the AECOO industry.

Classification systems are tools that provide an essential support for the accomplishment of the construction strategies targets. Innovation, efficiency and collaboration imply human to human (oral, written, other) and human-machine-human communication. Consistency and mutual understanding are key aspects for the effectiveness of the information exchanges across the construction process. Notwithstanding, even if CICS are well implemented, they should not be regarded as static solutions.

The literature review developed during this study identified, according with the new ISO 12006-2 version framework, the topics/main concerns of the relevant CICS and ICS.

Using that framework as reference it is possible to state that:

- ICS's should be considered on future CICS developments, namely on topics related with Products, Construction Entity and Construction Complex;
- For the fulfillment of the near-term IDDS targets, mapping processes between these ICS's and in use CICS should be straightforward;
- The Construction Element table is found to be one of the more relevant when working with object-oriented software;
- In what relates to Products and, by inherence, Construction Element, the Environmental Product Declarations (EPD) and Product Documentation Templates (PDT) can set the bridge for software solutions;
- Many topics from the framework are not defined on *de facto* CICS. From the case studies evidences, this relevant information is set on diverse documents that go from Standards to guidelines. Organization effort must be performed in order to set structures similar to those already developed for other tables.

Digitalization is an enabling factor that has to be explored. Digitalization plays an essential role and the exchange of objects should include information about classification. The Norwegian NS8360 is an example of a BIM standard which off classification. Further studies will focus on how applicable are the standards for implementation into digital software.

A new generation of CICS must therefore be developed following a wide multi-faceted vision geared for supporting the construction process and new technologies adoption; working together with other systems providing useful information for other activities and tasks that interact with the construction supply-chain; providing reports with practical and strategic information for the countries global economy and on a multi-national context.

REFERENCES

Bakkmoen, 2009. *Classification of building and civil engineering types for the AEC industry. Standards Norway*.

Bakkmoen, 2013. *Workshop om klassifikasjon i BAE sektoren, Oslo, Norway*.

BMEL, 2012. Bundesverband Materialwirtschaft, Einkauf und Logistik. *Final Report: Review of the Functioning of the CPV Codes/System*. European Commission, Germany.

CPV, 2008. Common Procurement Vocabulary—CPV. http://www.euro-tenders.com/ (ac. 2013–07–10).

CWA, 2010. European Committee for Standardization. *CWA 16138:2010 – Classification and catalogue systems used in electronic public and private procurement*. European Committee for Standardization, Brussels, 2010

Delany, S. (2015). https://toolkit.thenbs.com/articles/classification#classificationtables.

EUROSTAT, 1998. Classification of Types of Constructions. European Commission, Luxemburg. http://ec.europa.eu/eurostat/ramon/nomenclatures/index.cfm?TaretUrl=LST_NOM_DTL&StrNom=CC_1998&StrLanguagCode=EN&IntPcKey=&StrLayoutCode=HIERARCHIC.

Hollermann, S., 2012. *BIM—a challenge for communication between parties involved in construction*, ECPPM 2012—9th European Conference on Product and Process Modelling, Reykjavik—Iceland, 25–27th July 2012, 833–838.

IDDS. 2013. Research Roadmap Report CIB Integrated Design and Delivery Solutions (IDDS), Editors: Owen, R., Amor, A., Dickinson, J., Matthjis, P., and Kiviniemi, A. CIB Publication 370, *International Council for Building*. ISBN 978–90–6363–072–0. http://site.cibworld.nl/dl/publications/pub_370.pdf.

IFD, 2016. International Framework for Dictionaries, http://www.ifd-library.org/.

ISO, 2015. ISO 12006-2:2015 Building construction —Organization of information about construction works—Part 2: Framework for classification, International Organization Standardization, Oslo.

LNEC, 1997. Fonseca, M. Santos. *Curso sobre Regras de Medição na Construção*, LNEC, Lisboa.

Mêda, P., 2014a. *Integrated Construction Organization—Contributions to the Portuguese Framework*. Master Thesis in Civil Engineering, Faculty of Engineering of the University of Porto, Portugal.

Mêda, P., 2014b. Mêda, P., Sousa, H. *Information Consistency on Construction—Case study of correlation between Classification Systems for Construction Types*. 10th European Conference on Product & Process Modelling, 17–19 September, Vienna, Austria.

NATSPEC, 2008. *NATSPEC TECHreport—Information classification systems and the Australian construction industry*. http://www.natspec.com.au/images/PDF/TR02_Information_classification_systems.pdf (access 2016–02–05).

NS 8360, 2015. BIM-objekter, BIM objects—Naming, type encoding and properties for BIM objects and object libraries for construction works–Newsletter, https://www.standard.no/nyheter/nyhetsarkiv/bygg-anlegg-og-eiendom/2015/ns-83602015-bim-objekter/.

OJEU, 2004. Official Journal of the European Union. Directive 2004/17/EC of the European Parliament and of the Council, 31 March 2004.

OJEU, 2011. Official Journal of the European Union. Regulation (EU) n.º 305/2011 of the European Parliament and of the Council, of 9 march 2011. http://eur-lex.europa.eu/LexUriServ/LexUriServ.do?uri = OJ:L:2011:088:0005:0043:EN:PDF.

OJEU, 2014. Official Journal of the European Union. Regulation (EU) n.º 1101/2014, 31 October 2014. http://eur-lex.europa.eu/legal-content/pt/TXT/PDF/?uri = OJ:L:2014:312:FULL&from = pt.

Omniclass, 2013. A Strategy for Classifying the Built Environment. http://www.omniclass.org. (ac. 2013–07–10).

Race, S., 2013. *BIM demystified (2nd edition)*, RIBA Publishing, United Kingdom.

RIBA, 1997. Uniclass—Unified Classification for the Construction Industry, 1st edition, RIBA publications, UK.

SDMX (2016) Statistical Data and Metadata eXchange, https://sdmx.org/.

Sousa, H., Moreira, J. Mêda, P. 2008. *ProNIC© and the evolution of the Construction Information Classification Systems*, in: GEQUALTEC (Ed.), GESCON 2008 – International Forum on Construction Management, FEUP, Porto, 35–47.

Sousa, H., Mêda, P. 2012. Collaborative Construction based on Work Breakdown Structures, *ECPPM 2012 – 9th European Conference on Product and Process Modelling*, Reykjavik—Iceland, 25–27th July 2012, 839–845.

Taric, 2016. Taric consultation. http://ec.europa.eu/taxation_customs/dds2/taric/taric_consultation.jsp?Lang = en&Expand = true&SimDate = 20160413.

TFM, 2011. Tverrfaglig merkesystem (TFM), Prosjekteringsanvisning PA 0802. http://www.statsbygg.no/Files/publikasjoner/prosjekteringsanvisninger/0_Generelle/PA_0802-TFM.pdf.

Yoon, C.S., 1995. A chronological study on the U.K. building classification system since CI/SfB. *Korean Soc. of Arch. Eng.*, Seoul, 79–88.

eWork and eBusiness in Architecture, Engineering and Construction – Christodoulou & Scherer (Eds)
 ISBN 978-1-138-03280-4

A semantic web approach to efficient building product data procurement

N. Ghiassi, M. Taheri, U. Pont & A. Mahdavi
Department of Building Physics and Building Ecology, TU Wien, Vienna, Austria

ABSTRACT: Multiplicity of views and lack of a common ontological understanding of a building and its components among stakeholders involved in the building delivery and operation processes, result in an information gap between the requirements of various tasks and applications and the available building product representations. On the other hand, due to the extent and the dispersion of the available data, extraction of useful information in the right format has become a cumbersome and error-prone process. BAU-Web is an ongoing research effort aimed at exploring the potential of Semantic Web Technologies towards facilitating the utilization of web-based building product data. The present contribution describes the data-related challenges of the AEC (Architecture-Engineering-Construction) domain using the example of a loadbearing wall component, introduces the framework suggested by the BAU-Web project to address these challenges, and reports on the current state of the project.

1 INTRODUCTION

Coming from a variety of disciplines and occupational orientations, stakeholders and actors collaborating in the building industry have different perspectives and views on building as the final product of the collaboration. This multiplicity of views and lack of a common ontological understanding of a building and its components among stakeholders result in an information gap between the requirements of various tasks and applications (involved in the building delivery and operation processes) and the available building product representations. In other words, informational expectations of data clients (architects, auditors and performance evaluators, engineers, construction managers, constructors, building operators, etc.) with regard to building products are not always fully accounted for by data providers (manufacturers, third party product evaluators, and resellers). Past research has shown that difficulties in data procurement can go so far as to hinder the implementation of evaluation and analysis procedures for effective design decisions (Mahdavi and El-Bellahy 2005).

In this context, BAU-Web is an ongoing research effort aimed at exploring the potential of Semantic Web Technologies (Berners-Lee et al. 2001) towards facilitating the utilization of web-based building product data. The project targets the implementation of an appropriate ontology for a sample domain. It introduces a framework for the automated data extraction from multiple web-based sources, as well as the semi-automatic provision of user-requested data in the requested format.

2 BACKGROUND

2.1 *Semantic web, linked data and ontologies*

The Semantic Web vision, first introduced by Berners Lee, views the Web not just as an information space accessible to human users, but as a Web of well-defined and interlinked data that can be used by machines for automation and integration purposes. It involves augmenting the currently available web-based data with semantic information that render the data accessible and comprehensible for machine users.

"The term ontology is borrowed from philosophy, where an ontology is a systematic account of Existence "(Gruber 1993). In the field of knowledge-based systems, an ontology is defined as an explicit specification of a conceptualization, "an abstract, simplified view of the world that we wish to represent for some purpose" (same). An *ontology* defines a common vocabulary for stakeholders who need to share information in a domain. It includes machine-interpretable definitions of basic concepts in the domain and relations among them. Domain ontologies are the foundations of a Semantic Web, since they provide clear guidelines as to how to represent objects, their pertaining properties, and interrelations on the web. Once a domain ontology is agreed on and adopted by data providers, data retrieval, analysis, and integration can be automated based on this common language.

2.2 *Building product data representation, present practices*

The European Construction Products Regulation (CPR 2011) enforces the application of harmonized European standards on building products intended for the European market. These standards provide a system of harmonized technical specifications of building products, including all performance characteristics required by regulations in member states, with regard to mechanical resistance and stability of the products, fire safety, hygiene and health aspects, safety and accessibility of products in use, energy economy, and sustainability. These standards provide a solid basis for a building product ontology at the European level. Although the provided specifications suffice for a variety of AEC related inquiries, they do not support application domains for which a national standard has so far not been enforced (e.g., life cycle analysis, visual assessments).

Tools and methods developed for the assessment of various building performance indicators offer an insight to the informational requirements of various applications, and thus help enrich the representation schema offered by the CPR.

Various authorities and stakeholders in the AEC domain have developed schemas for the representation and distribution of building related data. In common standard building information models, including general purpose models such as IFC (BuildingSmart 2016), and specialized models such as gbXML (gbXML 2016), building products are represented through sets of semantic properties.

GbXML is intended for very few application domains (e.g., energy assessment, visualization, basic cost evaluations). It offers limited possibilities for the representation of building product features and is not extendable. IFC has a more elaborate and extendable schema. It has foreseen properties required for a wide range of material categories and applications. However, a systematic approach towards organization of the various product properties is lacking. The current representation of material properties seems to be the result of an ad-hock approach towards inclusion of the required property sets, rather than a well-structured ontological specification. This lack of a systematic approach renders the current schema unsuitable as the backbone of an ontology-based data repository.

2.3 *Utilization of ontologies in the building domain*

A first attempt towards utilization of ontologies to facilitate the automated retrieval and processing of web-based building product data was made in the SEMERGY project (Mahdavi et al. 2012a,b, Pont 2014, Pont et al. 2015). The project was concerned with the development of an optimization environment for informed building product selection. The developed ontological product representation supported the automated rule-based generation of component construction alternatives (as candidate solutions of the optimization process) and catered for the informational requirements of three application domains: Dynamic performance simulation, environmental footprint analysis based on the OI3 index (IBO 2011), and cost assessment (which constituted the optimization criteria).

Building on the SEMERGY experience, the present effort addresses the challenge of web-based building product data retrieval and utilization in a more general and comprehensive approach.

3 METHOD

3.1 *General approach*

The proposed framework (Fig. 1) relies on the development and wide-spread adoption of an extensive ontology for the representation and enhancement of web-based building product data. This ontology is to serve as a common vocabulary among data providers and data clients, to facilitate inquiry and delivery of building product data. BAU-Web is focused on the demonstration of the potential of a Semantic Web approach towards automated building data procurement.

The project does not concern itself with the data provision and maintenance processes. In this framework, data provision routines are consistent with the common practice in the AEC domain, where each data provider is responsible for the maintenance and quality control of their own online data repositories. However, these scattered repositories are expected to release data in machine readable format, as well as, adhere to the central building product ontology in representation of data.

The BAU-Web server hosts a building data repository structured according to the developed ontology. Specialized software agents are tasked with the automated extraction of relevant building product data from various websites. Thanks to a common ontological understanding of a building product, data pertaining to the same product can be collected from multiple sources and combined in the BAU-Web repository to generate complete product data profiles (e.g., technical properties from the manufacturer websites, price and availability information from reseller websites, etc.). Provided that the data distribution modalities of the online sources are not modified, these agents can periodically search these sources for updates and modifications.

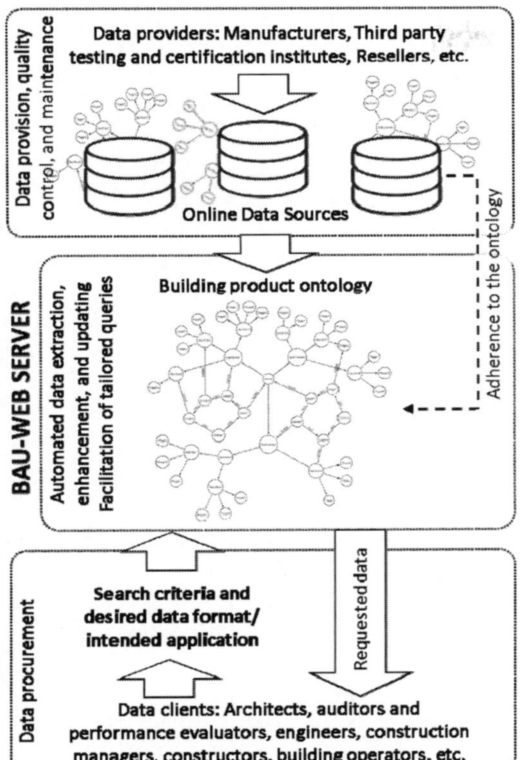

Figure 1. Schematic illustration of the proposed approach.

Data clients communicate their informational requirements (in terms of content and format) to the BAU-Web server; the server implements the relevant queries and issues the appropriate response, thereby relieving the client of the necessity to browse multiple web resources to find the required data.

3.2 *Study domain*

Considering the wide range of building products and application domains involved in the building construction process, in order to further restrict the bounds of the present project, a study domain involving a single building component and a limited number of building product categories and applications is defined. This simplification facilitated a more detailed and comprehensive study of the problem. The developed methods, of course, can be later applied to other product classes and application domains.

3.2.1 *Building product domains*

The selected case study is a conventional load-bearing masonry wall, which involves some or all of the following building product groups: Plaster, Brick, and Thermal insulation products.

3.2.2 *Application domains*

From the disciplines and processes involved in the life cycle of a building, the following application domains were selected to be addressed by the project:

i) Thermal, ii) Visual, and iii) Acoustical performance assessments, iv) Construction management, and v) Life Cycle Analysis (i.e. material preparation phase, building construction phase, building operation phase, demolition phase, waste treatment and recycling).

3.3 *Use-case scenarios and corresponding approaches*

BAU-Web intends to facilitate the utilization of web-based building product data with respect to two distinct use-case scenarios:

- Selection of the appropriate building products with regard to a set of physical attributes or functional characteristics (Product-oriented data retrieval).
- Retrieval of data on a known building product to fulfill the informational requirements of various tasks and applications (Task-oriented data retrieval).

The above objectives are distinguishable with respect to their data requirements.

3.3.1 *Scenario 1: Product-oriented data retrieval*

The objective of this scenario is to enable searching the wide-spread Web sources for building products that are suited for implementation in a certain building component. In the process of product selection, building products are viewed not in isolation but in relation to the building context, their role in the construction component, and their adjacent elements. A human user can easily determine whether a certain product is appropriate for use in a specific location or not based on visual clues, product characteristics, intuition, and domain knowledge. To automate this process, however, this decision-making ability has to be expressed in machine-processable form. In other words, a method has to be developed to allow the machine to determine the suitability of a certain product for a specific implementation. In this regard the following steps are necessary:

1. Identification of rules, preferences, and principles governing the building product selection processes.
2. Expressing the above principles in terms of search criteria.
3. Enhancing building product data with the informational requirements of the method developed in step 1.

A similar challenge was encountered in the SEMERGY project, in which an automated configuration of building components using available building product data was required. This was addressed through the development of layered construction templates for various building components. These templates incorporated rules according to which the suitability of products for a certain layer could be assessed. This general template-supported strategy has been adopted for the purpose of the present work. An alternative, more sophisticated approach would be to attempt at developing a formal language (Gladkii 1973) for the automatic generation of valid sequences of building products for various building components. We intend to investigate the potential of the latter approach in future studies.

3.3.2 *Scenario 2: Task-oriented data retrieval*

The second scenario aims at providing the user with pertinent information on a product according to the intended application. It involves the following steps:

1. Identification of the informational requirements of the intended application in terms of content and format.
2. Procurement of the necessary data on the building product.
3. Organization of the collected data in the desired format.

The challenges with respect to the above steps include finding the relevant data in the scattered domain of web-based sources, and creating an application-appropriate representation of the collected data. The former challenge is addressed through the BAU-Web data repository and its ability to create complete data profiles for various products through multiple sources. To address the second challenge, task oriented data templates were developed, which incorporated rules as to what product information is relevant to each task and in which format this information is to be supplied. Figure 2 schematically demonstrates the data retrieval process in response to the above scenarios.

By definition, the common building product ontology should provide a blue print for the representation of building product data (including various properties and characteristics of products and their associations) such that the resulting representation is sufficient to fulfill the informational requirements of the defined use-case scenarios within the bounds of the aforementioned case study. Compliance of the ontology with the pertinent standards and guidelines facilitates its wide-spread and effortless adoption by the data providers.

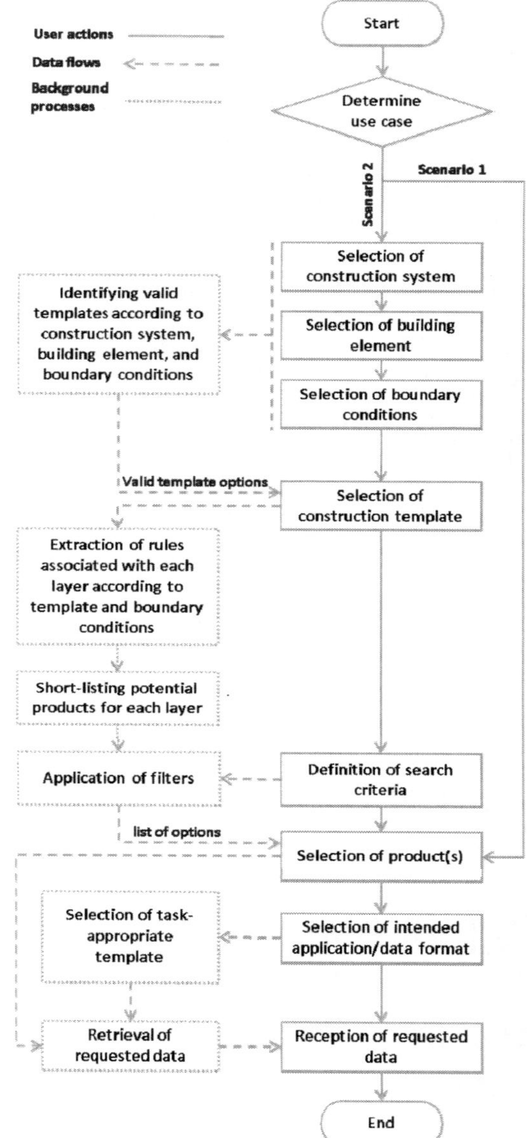

Figure 2. Data retrieval process.

The following chapter provides an overview of the preliminary results of the project towards the development of an appropriate building product ontology. Towards this end, the development of construction templates and task-oriented templates are discussed with illustrative examples. Inference-based data enrichment modalities are also briefly explained.

4 RESULTS AND DISCUSSION

4.1 *Construction templates*

A construction template is defined for a specific building component assuming a distinct role in a certain construction method. In this case the development of a template for a loadbearing wall in a massive masonry construction was intended. A template is a layered structure, incorporating constraints, requirements, and rules that define the characteristics of building products suited for implementation in each layer. Each template has a fixed number of layers (place holders for various building products), and two boundary conditions. The number and sequence of constituting layers and boundary conditions have implications for the products which can be used in each position (e.g., core insulation versus internal insulation). Each layer assumes a functional role in the intended component (e.g., loadbearing, finishing, insulation), and may have other specifications such as primary material class and format. Determining how general or specific the definition of a template should be is not a trivial task. On the one hand, the definition must insure the exclusion of illogical combinations. On the other hand, if made too specific, a very large number of templates are required to cover the common construction practices. Figure 3 displays a template developed for an external load bearing masonry wall. Note that this template represents only one of the many plausible compositions of the three considered product groups to form a load bearing wall component.

As stated before, the purpose of a template is to provide clear guidelines as to which building products can assume which layer positions. The following rules, deduced from the above template apply to products appropriate for each layer:

1. Layer 1: Products that can assume the finishing function in a wall, are of plaster material category and can be adjacent to clay products of block format.

Template ID		EW01			
Building element		Wall			
Construction system		Loadbearing masonry construction			
Boundary conditions		Internal space, "dry"	Outside air		
Description		External clay brick wall, with outside insulation and plaster finishing on both sides			
Layer requirements	Layer id	1	2	3	4
	Function	Finishing	Load bearing	Thermal Insulation	Finishing
	Material category	Plaster	Clay	Multiple categories	Plaster
	Adjacency i-1	Internal space, "dry"	Layer 1	Layer 2	Layer 3
	Adjacency i+1	Layer 2	Layer 3	Layer 4	Outside air

Figure 3. An instance of a construction template.

2. Layer 2: Products that can assume the loadbearing function in a wall and are formed in blocks (masonry) and entail clay as the constitutive material.
3. Layer 3: Products that can assume the external thermal insulation function in a wall and are stable in form, and can be covered by plaster.
4. Layer 4: Products that can assume the finishing function in a wall, are of plaster material class, can be exposed to outside air and can be applied on top of the insulation layer.

The informational requirements of the logical selection process should of course be reflected in the building product data repository. Accordingly, a number of property classes are required to sufficiently describe a building product for implementation of this logic. These property classes are displayed in Figure 4.

Note that the proposed approach can generate a very large number of valid solutions. There are thus appropriate methods and techniques are needed to limit and structure the solution spaces. Toward this end, we currently investigate two distinct possibilities. The first approach involves the user-based specification of the sequence in which concrete instances of template layers are identified. The second approach involves the concurrent generation of multiple realizations of the template. This set is subsequently structured via a ranking process that can involve multiple evaluation criteria pertaining to—for example—overall thermal performance, environmental impact, and cost.

4.2 *Task-oriented templates*

Similar to the construction templates, task-oriented templates provide the necessary rules and guidelines to address the second use-case scenario.

For a selected task or application domain, these templates determine the scope and format of the required building product information. This information should of course be included in the data representation schema (ontology). In order to define task-oriented templates, various standards and tools involved in the considered application domains were investigated and reverse-engineered for their informational requirements. Figure 5 displays a task-oriented template developed for Life Cycle Analysis. Similar templates have been defined for other application domains.

4.3 *Logical inferences and data enrichment*

BAU-Web is not aimed at addressing the data provision issue. Rather, it entrusts the data providers with the task. However, certain essential building

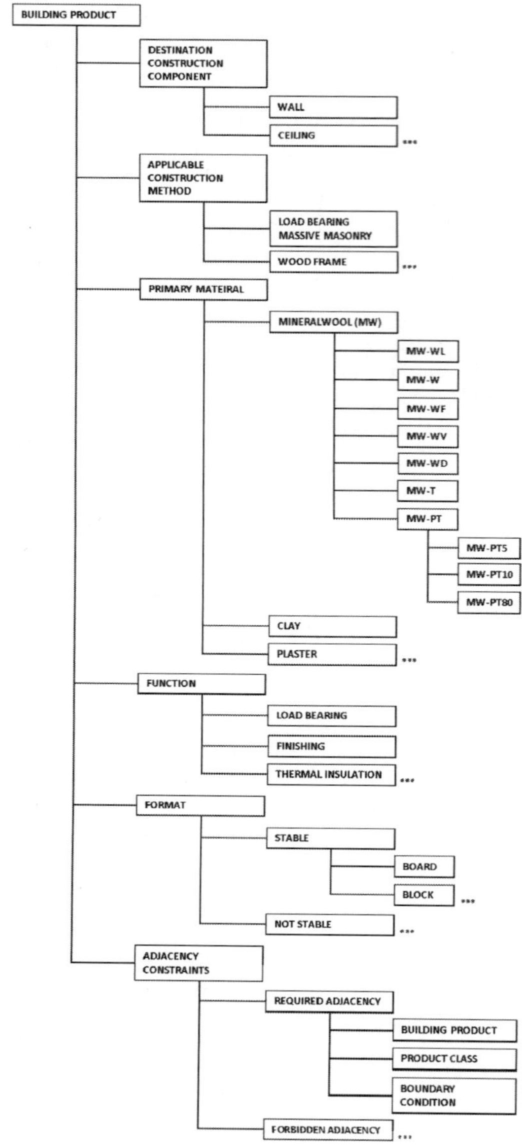

Figure 4. Building product properties required for implementation-oriented data retrieval.

product information (e.g., function, format, and adjacency constraints in certain cases) may be logically inferable based on the product descriptions provided by the manufacturers and third party evaluators. Some of these logical inference rules can be defined based on common knowledge and intuition. For instance all building products classified as Brick, have a stable block format. Other rules can be defined according to domain knowledge contained

Template ID		LCA_CradleToGate	
Source(s)		EN 15804:2012	
Output		Text	
Required Properties		Variable	Unit
	Environmental impact	Abiotic depletion potential for non fossil resources	kg Sb equiv.
		Abiotic depletion potential for fossil resources	MJ, net calorific value
		Acidification potential of soil and water	kg SO2 equiv.
		Depletion potential of the stratospheric ozone layer	kg CFC 11 equiv.
		Global warming potential	kg CO2 equiv.
		Eutrophication potential	kg PO4 equiv.
		Formation potential of tropospheric ozone	kg Ethene equiv.
	Resource use	Use of renewable primary energy excluding renewable primary energy resources used as raw materials	MJ, net calorific value
		Use of renewable primary energy resources used as raw materials	MJ, net calorific value
		Total use of renewable primary energy resources	MJ, net calorific value
		Use of non-renewable primary energy excluding renewable primary energy resources used as raw materials	MJ, net calorific value
		Use of non-renewable primary energy resources used as raw materials	MJ, net calorific value
		Total use of non-renewable primary energy resources	MJ, net calorific value
		Use of secondary material	kg
		Use of renewable secondary fuels	MJ, net calorific value
		Use of non-renewable secondary fuels	MJ, net calorific value
		Net use of fresh water	m3
	Waste generation	Hazardous waste disposed	kg
		Non-hazardous waste disposed	kg
		Radioactive waste disposed	kg

Figure 5. An instance of a task-oriented template.

in standards. For instance, the harmonized standards pertaining to factory made thermal insulation products, ÖNORM B 6000 (2014), provide rules (according to material category) as to whether an insulation product can be used adjacent to a plaster façade finishing or not. Such logical inference rules can be used to enrich the available data in accordance with the requirements of the common ontology.

In addition to the required data, the BAU-Web server can augment building products with additional semantic information, which enables user-friendly and application-oriented query processes. For instance, products can be ranked or classified according to specific properties (e.g., thermal performance indicators), offering more intuitive search possibilities (e.g., searching for environment friendly insulation material appropriate for application on external wall).

5 CONCLUSION AND FUTURE RESEARCH

The BAU-Web project addresses the challenge of building product data retrieval from scattered

Web-based data sources. Based on the Semantic Web vision, the project aims to develop a common data ontology, which serves as a schema for the provision and distribution of building product data. This data ontology is the core component of a data extraction, organization, and provision framework. The proposed method relies on the development of construction and task oriented templates to address two specific use-case scenarios with regard to data utilization: product-oriented data retrieval and task-oriented data retrieval. The developed templates define the scope and format of the necessary building product information (with regard to the limited study domain of a load bearing wall component).

Once combined, these templates provide a representation of building products, which caters for the requirements of the defined study domain. Future research efforts include the development of the common ontology for the study domain and demonstrative implementation of the data retrieval modalities. Moreover, alternative approaches (possibly based on formal languages) for the automated generation of valid sequences of building products for various building components need to be explored.

Last but not least, future work needs to resolve the aforementioned problem of the multiplicity of the generated valid solutions. Toward this end, options such as user-based sequential layer identification and multi-criteria ranking methods represent promising venues.

ACKNOWLEDGEMENTS

This research was conducted in the framework of the BAU-WEB Project. This project is funded by the Austrian Research Promotion Agency (FFG), grant No. 848583. The authors thank Rainer Bräuer, Melisa Covic, Alexandra Heiderer, Razvan-Gabriel Stanica, Cristina Tamas, Dawid Wolosiuk, and Andreas Wurm, for their valuable contribution.

REFERENCES

Berners-Lee, T., Hendler, J. & Lassila, O. 2001. The Semantic Web—A new form of Web content that is meaningful to computers will unleash a revolution of new possibilities. Scientific American, 2001. Available via http://www.scientificamerican.com/article/the-semantic-web/ (last accessed in March 2016).

BuildingSmart. 2016. http://www.buildingsmart.org/ (last accessed in March 2016).

CPR. 2011. European Construction Products Regulation. http://eur-lex.europa.eu/legal-content (last accessed in March 2016) gbXML 2016. http://www.gbxml.org/ (last accessed in March 2016).

Gladkii, A.V. 1973. Formal languages. Encyclopaedia of mathematics. Formal language. Encyclopedia of Mathematics. http://www.encyclopediaofmath.org/index.php?title=Formal_language&oldid=37513 (last accessed in March 2016).

Gruber, T.R. 1993. A translation approach to portable ontology specifications. Knowledge Acquisition 5(2). P. 199–220.

IBO. 2011. OI3-Indicatior. IBO-Guidelines to calculating the OI3 indicators for buildings. Österreichisches Institut für Bauen und Ökologie GmbH. Available via http://www.ibo.at/documents/OI3-LeitfadenV22_06_2011_english.pdf (last accessed in March 2016).

Mahdavi, A. & El-Bellahy, S. 2005. Effort and Effectiveness considerations in computational design evaluation: a case study. Building and Environment 40 (2005), P. 1651–1664.

Mahdavi, A., Pont, U., Shayeganfar, F., Ghiassi, N., Anjomshoaa, A., Fenz, S., Heurix, J., Neubauer, T. & Tjoa, A.M. 2012a. Exploring the utility of semantic web technology in building performance simulation. In: BauSIM 2012—Gebäudesimulation auf den Größenskalen Bauteil, Raum, Gebäude, Stadtquartier, C. Nytsch-Geusen et al. (ed.); Eigenverlag. wissenschaftliches Kommittee der IBPSA Germany-Austria.

Mahdavi, A., Pont, U., Shayeganfar, F., Ghiassi, N., Anjomshoaa, A., Fenz, S., Heurix, J., Neubauer, T. & Tjoa, A.M. 2012b. SEMERGY: Semantic web technology support for comprehensive building design assessment. In: eWork and eBusiness in Architecture, Engineering and Construction, G. Gudnason, R. Scherer (ed.); Taylor & Francis. P. 363–370.

ÖNORM B 6000. 2014. Factory made materials for thermal and/or acoustic insulation in building construction—Types, application and minimum requirements. Österreichisches Normungsinstitut. Vienna. Austria.

Pont, U. 2014. A comprehensive approach to web-enabled, optimization-based decision support in building design and retrofit. Dissertation, Vienna University of Technology. DOI 10.13140/RG.2.1.3115.9521.

Pont, U., Ghiassi, N., Fenz, S., Heurix, J. & Mahdavi, A. 2015. SEMERGY: Application of Semantic Web Technologies in Performance-Guided Building Design Optimization. Journal of Information Technology in Construction, 20 (2015), Special Issue ECPPM 2014, P. 107–120.

eWork and eBusiness in Architecture, Engineering and Construction – Christodoulou & Scherer (Eds)
© 2016 Taylor & Francis Group, ISBN 978-1-138-03280-4

BIM adoption for on-site reinforcement works—a work system view

A. Figueres-Munoz & C. Merschbrock
Oslo University College, Oslo, Norway

ABSTRACT: Building Information Modeling (BIM) technology can support the Architecture, Engineering and Construction (AEC) industry throughout all stages of project delivery. Nowadays, the quality and maturity of BIM models allow for their use beyond design. However, many firms struggle adopting BIM for supporting the construction phase of a project. The organizational challenges that BIM adoption entails are reported as one of the main obstacles for BIM aiding construction. In this article, we explore a successful case of BIM adoption by a Norwegian contractor. The industrial setting of the study consists of a major construction project where sophisticated models of the reinforcement were used to support on-site works. The theoretical approach used to structure the findings is the Work System Theory. The wide scope of this theory enables to complement prior knowledge about the adoption process. Our work shows that using BIM can further simplify the reinforced concrete supply chain.

1 INTRODUCTION

Building Information Modeling (BIM) has over the past few years become the preferred tool in building design (McGraw Hill, 2012). Thus, in many of today's building projects, virtual, high quality BIM models are the main outcome of the design phase. These resulting models can in turn support the construction, fabrication, and procurement activities through which the building is realized (Eastman et al., 2011). However, BIM technology diffusion in the Architecture, Engineering and Construction (AEC) industry is still mainly limited to design activities (Merschbrock & Munkvold, 2014). A barrier for the participation of contractors in innovative BIM practices is their late involvement in projects following the traditional Design-Bid-Build delivery method (Dossick & Neff, 2009, Eastman et al., 2011, Merschbrock & Munkvold, 2014). The transition from design to construction frequently represents the start of a decay in data continuity hindering BIM from being applied downstream in the construction supply chain (Goedert & Meadati, 2008). As a result, construction activities rely currently on an intensive use of paper drawings (Merschbrock, 2012, Sacks et al., 2010).

Extending BIM diffusion beyond design could benefit contractors in constructability analysis, quantity takeoffs, or simplify verification and guidance of construction (Eastman et al., 2011). Consequently, several contractors have undertaken pilot projects to adopt BIM tools for on-site activities (Davies & Harty, 2013). Succeeding in BIM adoption requires overcoming organizational and contractual challenges (Eastman et al., 2011, Davies & Harty, 2013, Dossick & Neff, 2009, Merschbrock & Nordahl-Rolfsen, 2016). Moreover, contractors' BIM adoption is difficult since almost every process and business relationship have to undergo some change (Eastman et al., 2011). This quote illustrates that BIM adoption entails "technochange" (Markus, 2004). Techno-change can be defined as a technology-driven organizational change encompassing a broad spectrum of implications within the whole organization (Harison & Boonstra, 2009). In other words, contractors need to create new work systems that suit BIM (Alter, 2013).

The usefulness of Information and Communication Technology (ICT) for supporting the placement of reinforcing bars has long been advocated in literature (Bernold & Salim, 1993). This is due to the large amounts of rework and numerous requests for information from the fabricators and builders to the designers (Navon et al., 2000, Tommelein & Ballard, 2005). BIM technology has become mature enough to support reinforcement works, and practitioners perceive it as useful (Merschbrock & Nordahl-Rolfsen, 2016). Thus, BIM can today change how information is created and exchanged throughout the whole reinforcement supply chain (Aram et al., 2013).

In this article, we contribute to research and practice by exploring the BIM adoption process and its implications for the work system of reinforcement placement. The research questions guiding our inquiry is: *How does BIM change on-site construction work systems?*

To answer the question, we conducted a case study in a major construction project ongoing in Oslo, namely, a new airport terminal. In this project, detailed BIM models of reinforcement were the only source of information used to design and mount reinforcement. To explore contractor's BIM adoption, a number of interviews were conducted. Interviewees included steel fixers, structural engineers and site engineers. The theory guiding our analysis is Alter's Work System Theory (Alter, 2013).

Following this introduction, we present the work system theory chosen for our analysis. Next, we provide a detailed description of the airport terminal case. Then, we present our findings based on the work system theory constructs. This is followed by a discussion of these findings and an overall conclusion of our work.

2 THEORY

A work system exists when technology, information and other resources are used to produce a specific outcome for a customer (Alter, 2013). Creating products and services based on new ICT entails building a new work system (Alter, 2015). This is the case for BIM adoption in the AEC industry. Construction firms struggle with establishing work systems facilitating the new technology (Merschbrock, 2012, Dossick & Neff, 2011, Dossick & Neff, 2013). Work System Theory (WST) explains the implementation of organizational systems. WST could help understanding how BIM implementation for on-site operations can be made more efficient (Alter, 2013).

The theory's explanatory power goes beyond the "technology-as-system" paradigm in Information Systems (IS) research. The core of this theory consists of two components, namely, the work system framework (Fig. 1) and the work systems life cycle (Fig. 2). The former defines the elements of which a work system is made-up. Organizational strategies, environment, and infrastructure represent the larger background, which influences and is influenced by the work system. The system itself consists of participants, and technology that uses information to perform processes and activities. These processes and activities create products/services for a customer. The arrows between two elements represent the need for alignment of those.

The work system life cycle, depicted in Figure 2, summarizes how work systems change over time. It suggests that work systems evolve through planned change (represented by arrows linking boxes) and unplanned change (represented by inward-facing arrows in boxes). The model assumes that new work-systems are initiated and funded by management. Developing a work system requires new procedures, software development, acquisition,

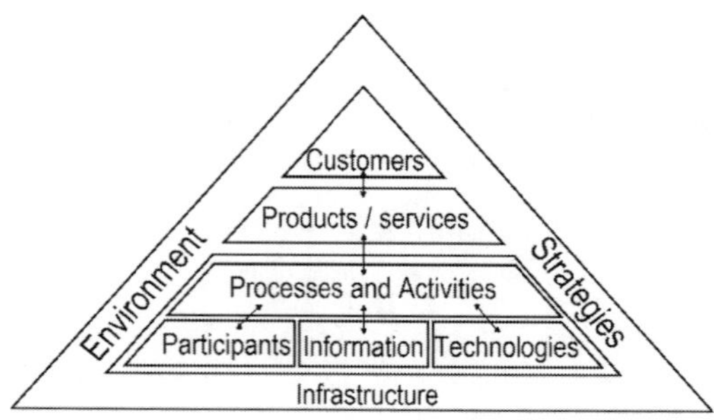

Figure 1. Work Systems Life Cycle (WSLC) model. Adapted from Alter (2013).

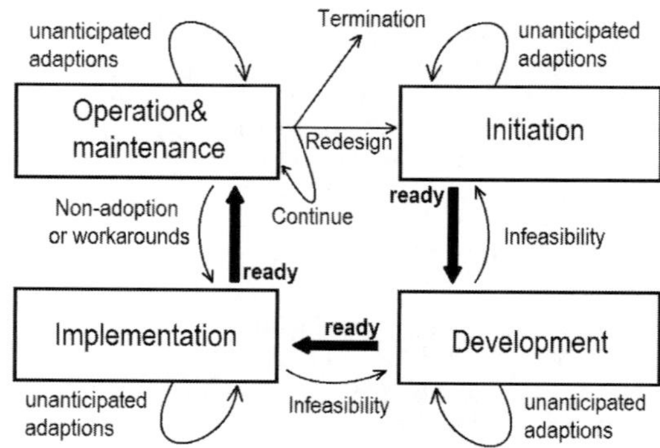

Figure 2. Work Systems Life Cycle. Adapted from Alter (2013).

and training. Implementing the work system in an organization requires operational change. Moreover, the operation and maintenance of the new work system would need to be planned. In all stages of this itinerary, unplanned change can transform the elements of the new work system.

The Work System Framework (WSF) and the Work System Life Cycle (WSLC) are instruments for understanding IT adoption. Alter (2015) suggests a three staged assessment process: (1) defining the new work system based on the WSF, (2) analyzing the adoption process based on the WSLC, and (3) interpreting the post-adoption situation. Studies informed by WST can complement knowledge about IT adoption beyond what has been explained by framewoks like the Technology Acceptance Model (Grover & Lyytinen, 2015).

3 METHOD

A case study was deemed appropriate for exploring the practice-based problems subject to our study. Case studies bring about an understanding of the actors involved and the context of the action (Benbasat et al., 1987). An important strength of case studies are their suitability to explore organizational innovations (ibid.). We considered a case study a good fit for studying ICT adoption for on-site practices. We conducted a case study of the construction of a new departure hall and pier at Oslo airport. A more detailed description of the project can be found elsewhere (Merschbrock & Nordahl-Rolfsen, 2016).

Due to the high complexity of the project, the client demanded the use of Building Information Modelling systems for design and construction. Furthermore, an open file format was used for collaboration in design. During the construction one of the contractors used BIM technology to support reinforcement works. We explore a pilot project of BIM use for the construction of the new pier's foundations. Steel fixers working on-site were able

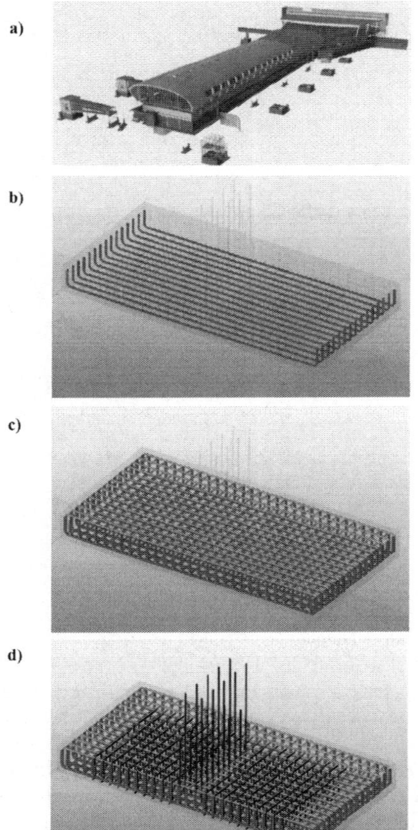

Figure 3. Screenshots from case study's BIM models. Figure 3a: overall view of the new pier. Figures 3b, 3c, 3d: stages of the suggested mounting sequence for reinforcement. Adapted from (Merschbrock & Nordahl-Rolfsen 2016).

Table 1. Interviews conducted in the case study.

Interviewee	Role
Client #1*	Structural engineer/ BIM manager
Contractor #1	Site manager
Contractor #2	Managing engineer
Sub-contractor #1	Superintendent
Sub-contractor #2, #3	Steel fixers
Steel supplier #1	Engineer

*Client #1 was part of the design team before he started working as BIM manager for the client.

to access the virtual models of the reinforcement and were provided with a collection of screenshots suggesting the sequence for their operations. Examples are shown in Figure 3.

The data collection consisted of a series of semi-structured interviews with the client's BIM manager, contractor's site engineers, sub-contractor steel fixers, their superintendent, and a rebar supplier. Thus, the interviews involved participants from across the concrete reinforcement supply chain. Interviews were conducted in September 2014. The interviews were voice recorded, transcribed and coded according to the core concepts of the work system theory. All participants gave their informed consent before starting up the interviews. Table 1 presents an overview of the interviewees; their names have been anonymized for confidentiality purposes.

4 FINDINGS

In what follows, we analyze the BIM adoption in the context of the chosen case study. This part of the article is structured as follows. First, we present a description of the unfolded BIM based work system for reinforcement works. This is presented based on the constructs of the work systems framework shown in Figure 1. Second, we report on the BIM adoption process. We present the initiation, development, and implementation phases of the new work system following Alter's work systems life cycle shown in Figure 2.

4.1 *The new BIM based work system*

Substituting 2D paper drawings with BIM models as information source for construction entailed changing established work processes not only for the contractor, but also for the client and subcontractor.

The resulting work processes that unfolded in the case study are depicted in Figure 4 as a workflow. This workflow includes input and output

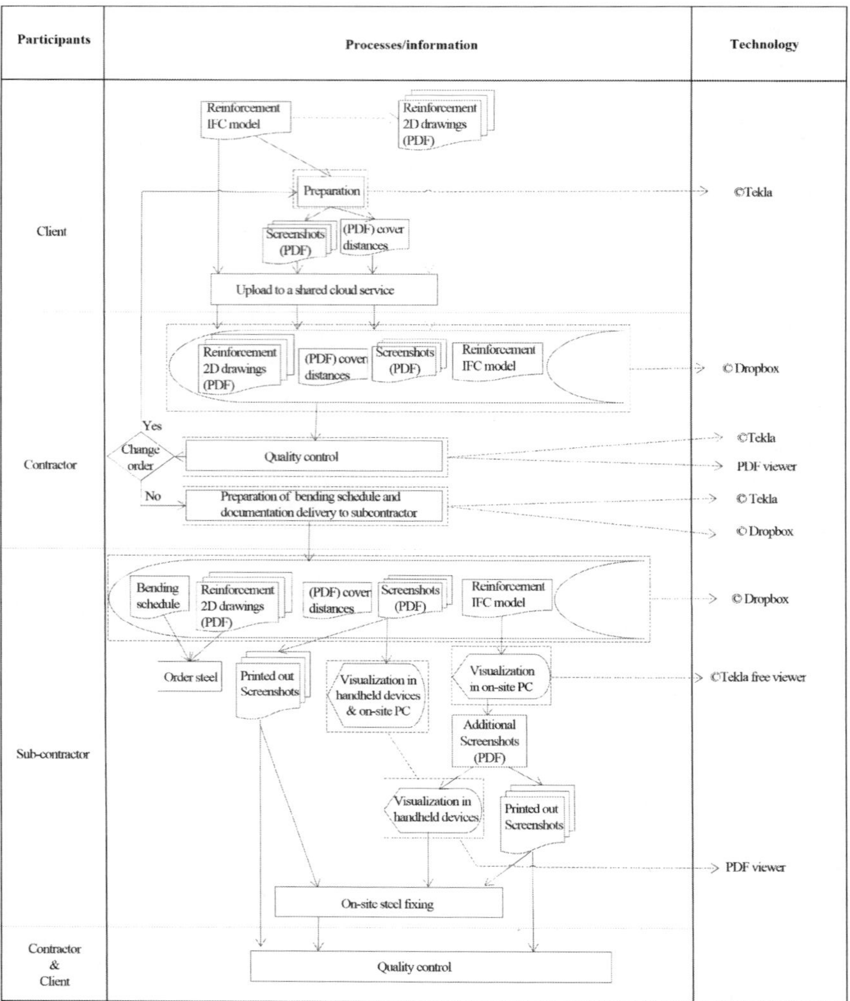

Figure 4. BIM based work system for reinforcement works.

information, participants involved and technology used in each step of the process. The agreement between client and contractor was to build the reinforcement based on BIM models. 2D drawings were nevertheless produced, as it was a contractual requirement. However, they were not deemed necessary for on-site activities. The following quotes illustrate this: "*[...] then we put the drawings in a cabinet, locked it, hung the key on the wall and they could use them if they needed*" (client#1), "*We had drawings, but we put them away, we didn't need them because we had everything on it [the model]*" (sub-contractor #1).

The reinforcement IFC model created during the design phase was further processed by the client to prepare a set of files to hand over to the main contractor. The client and main contractor firms used a cloud system to share information. Uploaded files by the client included IFC models of the foundations, PDF files with screenshots of the projected solutions and a suggested sequence to mount reinforcement frameworks, plus a document with cover distances for the different elements. 2D reinforcement drawings were also uploaded to the cloud system. The screenshots showed both the geometrical distribution of reinforcement families and the steel quality and position number of each rebar. The main contractor quality-checked these documents and if accepted, proceeded to prepare the bending schedule from the BIM model using Tekla. For the sub-contractor, only PDF viewers were necessary at the building site to access the required information. Nevertheless, the sub-contractor's superintendent allowed for the steel fixers to open the IFC models with a free Tekla viewer, so they could take their own screenshots, save them as PDF's and open them later in their own smartphones and tablets. In the words of one of the steel fixers: "*[...] we all fiddled around with it [the model], on the PC we had in [the site office]*" (Sub-contractor #3). Construction workers did not receive any formal training. However, they perceived the software as intuitive, and all were able to find what they needed. Even those with low IT literacy: "*it was easy [using a BIM tool]. Even for me, who is not very computer skilled*" (Sub-contractor #1).

Once the on-site works were done, the sub-contractor delivered "as-built" documentation back to the contractor. The "as-built" documents were copies of screenshots, signed by the superintendent to express compliance with the projected solution. According to the contractor, the accomplished level of quality achieved in the pilot project was as high as it would have been by using drawings. Furthermore, all interviewees at the contractor and sub-contractor firms stated that on-site works won in efficiency by using 3D visualization. Following quotes show this perceived gain in efficiency: "*[BIM] makes our work easier, we save a lot of time*" (sub-contractor #2), "*[Without the model] you need to look up for everything in the drawings and bending schedule*" (sub-contractor #2), "*[finding the drawings and discussing about them] can take up to 15 minutes*" (sub-contractor #3.

A potential participant that was left out of the BIM based work system was the steel supplier. Disagreements that interviewees did not disclose, prevented the steel supplier firm from being able to deliver reinforcement steel from the model. Thus, the 2D drawings were used by the sub-contractor to order the required materials.

4.2 *Life cycle of the new work system*

In this section, we present how the work system was initiated, developed, and implemented. The operation phase of a new work system is what happens after the initial pilot studies. This is why the operation and maintenance aspects of the life cycle have been left aside in our study.

The initiation of the work system was triggered by the client's construction management. The BIM manager stated that the idea of using BIM aiding reinforcement works was "*spontaneous, structural engineers went so far [in design] that we could deliver good quality models. So good quality that we could use them equally to drawings*". Then he started working in what he called "*a stunt in the middle of a hectic everyday*" to "*push the contractor into using the model*". The main contractor company was positive to the idea, but insisted on the contractual drawings requirement. Thus, they decided to print and store the drawings, but work based on BIM models downstream as long as possible. Reflecting on their own behavior, the site manager explained that, "*everything new is scary, you just distance yourself from it*"; while the managing engineers conceded that, "*you cannot stop it [BIM] either*".

The further development of the work system consisted in defining the scope of the implementation and detailing work processes. The agreement between the client and the contractor entailed that the later had to enhance IT capability, and support reinforcement works on a set of foundations with open format BIM models. The reason to use open formats was the intention to further handle all models generated in design and construction for facility management, and "*[basing design in one vendor's format] would have made the project files dead after 5 or 10 years [...] while the airport is to be used for the next 100*" (Client #1). IT training decided to be carried out by a super user from the client, and other one from the contractor's main office. At this stage, the contractor firm took contact with a steel supplier to explore the opportunities to order reinforcement from the model. The intentions were to cooperate to order the reinforcement direct from the model, but "*it happen something internal [in the contractor firm] that I did not understand. The meeting we were supposed to have was cancelled. I found out why later, but then we were out the process. We delivered the steel, but we did not partake in the process they had to reinforce without drawings*" (Steel supplier #1). No one of the interviewees disclosed the nature of the problems resulting in the exclusion of the steel supplier from the BIM based work system, but an early involvement was claimed to be the only requirement to include this participant in it. Finally, a sub-contractor that agreed to work based on the model instead the drawings was assigned the construction of the foundations.

The implementation of the work system can be considered as a pilot project, as the scope of the products to be developed by the new work system was limited to a set of foundations. Once client and contractor agreed the terms of their collaboration, the IT infrastructure was set in place and the engineers training started. Before on-site works started, the contractor sent a first draft of the screenshots to the sub-contractor, who complained about the quality: "*when we got [the screenshots], they were in black and white, and we saw the problem immediately. You should put colors on [rebar families]*" (Sub-contractor #1). This instance of unplanned change was handled by the contractor and resent the screenshots in color. Furthermore, client's BIM manager showed how to use a free Tekla viewer sub-contractor's superintendent. Steels workers were initially not supposed to use the model or get any training, but the superintendent showed it to them. They grasped quickly how to use the model viewer by themselves, took new screenshots and saved them in their smartphones and tablets. This can be seen as a second instance of unplanned change to the developed work system.

The overall conclusion of the implementation phase is that both the contractor and sub-contractor accepted the technology adoption, as they expressed in the interviews "*I have missed the tekla system afterwards*" (Sub-contractor #1), "*when you start working with it and realize how useful it is, it great!*"(Contractor #1). However, some challenges were pointed out that should be addressed before an eventual full implementation. Fist, the sub-contractor superintendent recognized he had a hard time with the language, as the model viewer was in English. In his own words "*my English is a little bit rusty, it [the software] would had been better in Norwegian*" (sub-contractor #1). Moreover, an economic analysis of the adoption was an important lack in the pilot project, as recognized by the clients BIM manager: "*there was no measure [of the costs of BIM adoption], but it is obvious that it costs a lot of money*". Two important factors this eventual analysis pointed out by participants are that working based on the project may (1) result in fewer errors during construction, and (2) allow saving money by not producing drawings.

5 DISCUSSION

We have presented a case of BIM adoption aiding on-site construction activities. We have reported on the adoption process and its result. The findings of our work show that adopting BIM for on-site activities is a process that entails organizational change for several participants of the project. This is in line with previous research (Davies & Harty, 2013). Moreover, BIM adoption was an owner driven process, triggered by a BIM "enthusiast" and happened at project level, as in similar case studies (Bråthen & Moum, 2015). What became apparent in our study is how organizational change can affect several participants of a project through both planned and unplanned manner. The result of the adoption process can be seen a new work system consisting of a new set of work processes that change how participants use information and technology.

We find that WST is a useful tool for explaining organizational change. Moreover, we argue that applying this theoretical framework to technology adoption can complement other approaches frequently used in literature. In this article, WST allowed exploring how and why an IT tool is adopted, a reported blind spot of the Technology Acceptance Model (Grover & Lyytinen, 2015).

Although we studied a system already implemented in practice, in our view WST could also be used to design future BIM based practice. This way, practitioners could manage early stage BIM implementations better and possibly some of the currently experienced issues in BIM adoption could be overcome.

Based on our findings we think that BIM technology is mature enough to allow working based on 3D models through the whole reinforcement steel supply, from design to on-site operations. Thus, using BIM tools can make 2D detail drawings superfluous. This could further simplify the suggested model for information creation and exchange in the reinforced concrete supply chain as proposed by Aram et al. (Aram et al., 2013). Since our work is limited to one case study, further research is needed to validate our findings and provide an exhaustive account on BIM adoption for on-site construction activities. Moreover, future work is required to establish the implications of BIM diffusion beyond design.

6 CONCLUSION

This article has presented a case study of BIM adoption by a contractor firm to support reinforcement works during the construction of a major building project. Based on the Work System Theory, we describe the new work processes, involved participants and technology used. Moreover, we analyze how adoption process emerged and evolved at project level.

Our contribution is twofold. First, we explore a new perspective on technology adoption by using the Work System Theory to explicate how and why it happens. Second, we point out how adopting BIM can affect the reinforced concrete supply chain.

REFERENCES

Alter, S. 2013. Work system theory: overview of core concepts, extensions, and challenges for the future. *Journal of the Association for Information Systems*, 72.

Alter, S. 2015. A Work System Perspective on Adoption Entities, Adoption Processes, and Post-Adoption Compliance and Noncompliance.

Aram, S., Eastman, C. & Sacks, R. 2013. Requirements for BIM platforms in the concrete reinforcement supply chain. *Automation in Construction,* 35, 1–17.

Benbasat, I., Goldstein, D.K. & Mead, M. 1987. The case research strategy in studies of information systems. *MIS quarterly*, 369–386.

Bernold, L.E. & Salim, M. 1993. Placement-oriented design and delivery of concrete reinforcement. *Journal of construction engineering and management,* 119, 323–335.

Bråthen, K. & Moum, A. Bridging the gap: taking BIM to the construction site. CIB W78 Conference, 2015 Eindhoven, The Netherlands. 79–88.

Davies, R. & Harty, C. 2013. Implementing 'Site BIM': a case study of ICT innovation on a large hospital project. *Automation in Construction,* 30, 15–24.

Dossick, C.S. & Neff, G. 2009. Organizational divisions in BIM-enabled commercial construction. *Journal of construction engineering and management,* 136, 459–467.

Dossick, C.S. & Neff, G. 2011. Messy talk and clean technology: communication, problem-solving and collaboration using Building Information Modelling. *The Engineering Project Organization Journal,* 1, 83–93.

Dossick, C.S. & Neff, G. 2013. Constructing Teams: Adapting Practices And Routines For Collaboration Through BIM.

Eastman, C., Teicholz, P., Sacks, R. & Liston, K. 2011. BIM handbook: A guide to building information modeling for owners, managers, designers, engineers and contractors, John Wiley & Sons.

Goedert, J.D. & Meadati, P. 2008. Integrating construction process documentation into building information modeling. *Journal of construction engineering and management,* 134, 509–516.

Grover, V. & Lyytinen, K. 2015. New State of Play in Information Systems Research: The Push to the Edges. *Mis Quarterly,* 39, 271–296.

Harison, E. & Boonstra, A. 2009. Essential competencies for technochange management: Towards an assessment model. *International Journal of Information Management,* 29, 283–294.

Markus, M.L. 2004. Technochange management: using IT to drive organizational change. *Journal of Information technology,* 19, 4–20.

McGraw Hill 2012. The Business Value of BIM in North America: Multi-Year Trend Analysis and User Ratings (2007–2012). Bedford, MA: McGraw-Hill Construction.

Merschbrock, C. 2012. Unorchestrated symphony: The case of inter–organizational collaboration in digital construction design. ITcon.

Merschbrock, C. & Munkvold, B.E. Succeeding with building information modeling: a case study of BIM diffusion in a healthcare construction project. System Sciences (HICSS), 2014 47th Hawaii International Conference on, 2014. IEEE, 3959–3968.

Merschbrock, C. & Nordahl-Rolfsen, C. 2016. BIM Technology acceptance among reinforcement workers–The case of Oslo Airport's terminal 2. *Journal of Information Technology in Construction (ITcon),* 21, 1–12.

Navon, R., Shapira, A. & Shechori, Y. 2000. Automated rebar constructability diagnosis. *Journal of construction engineering and management,* 126, 389–397.

Sacks, R., Radosavljevic, M. & Barak, R. 2010. Requirements for building information modeling based lean production management systems for construction. *Automation in construction,* 19, 641–655.

Tommelein, I.D. & Ballard, G. Restructuring the rebar supply system. Proc. Constr. Research Congress, 2005.

eWork and eBusiness in Architecture, Engineering and Construction – Christodoulou & Scherer (Eds)
© 2016 Taylor & Francis Group, ISBN 978-1-138-03280-4

Classification of detection states in construction progress monitoring

A. Braun & A. Borrmann
Chair of Computational Modeling and Simulation, Leonhard Obermeyer Center, Technical University of Munich, Munich, Germany

S. Tuttas & U. Stilla
Photogrammetry & Remote Sensing, Leonhard Obermeyer Center, Technical University of Munich, Munich, Germany

ABSTRACT: The research conducted in this publication focusses on automated progress monitoring. The recording of the current as-built state of a construction site is achieved by photogrammetric methods (e.g. UAVs) and compared to an as-planned (4D) BIM model. To visualize the detected elements and evaluate their respective detection rate, a schema for the classification of each individual element is presented. It compares the as-built ground truth with the actually detected elements and thus facilitates a quick and easy interpretation of the current construction state. Temporal data from construction schedules is added to further complete the provided information through visualization. Additionally, the classification helps to identify possible lacks of detection algorithms. New parameters for detection algorithms can be applied and the results are immediately visible with an easily understandable color scheme.

1 INTRODUCTION

In construction, progress supervision and monitoring is still a mostly analog and manual task. To prove that all work has been rendered as defined per contract, all performed tasks have to be monitored and documented. The demand for a complete and detailed monitoring technique rises for large construction sites where the complete construction area becomes too large to monitor by hand efficiently, and the amount of subcontractors rises. Main contractors that control their subcontractors' work, need to keep an overview of the current construction state. Regulatory problems add up on the requirement to keep track of the current status on the site.

The ongoing digitalization and the establishment of Building Information Modeling (BIM) technologies in the planning of construction projects can facilitate the use of digital methods in the construction phase. In an ideal implementation of the BIM idea, all semantic data on materials, construction methods, and even the process schedule are information consistent and connected. Therefore, it is possible to make statements about cost and the estimated project finalization. Possible deviations from the schedule can be detected and succeeding tasks are easy to identify.

This technological advancement allows new methods in construction monitoring. As described in (Braun et al., 2015) the authors propose a system for automated progress monitoring using photogrammetric point clouds. The main idea is to use common camera equipment on construction sites to capture the current construction state ("as-built") by taking pictures of all building elements (Tuttas et al., 2016). When sufficient images from different points of view are available, a 3D point cloud can be produced with the help of photogrammetric methods. This point cloud represents one particular timestamp of the construction progress and is then compared to the geometry of the BIM ("as-planned"). Details on the generation of the point clouds and the comparison algorithms can be found in (Tuttas et al., 2014, 2015).

For the visualization of the comparison results, of the detected elements and in order to verify the used algorithms, all gathered data are stored in a database that is accessible to the *progressTrack Viewer*. This tool is developed in the frame of this research project. It displays all building element information and in addition the process data. The detected elements are highlighted for an easy identification. Figure 2 shows one of the construction sites that were used for case studies during this research. The building mainly consists of in-situ concrete elements that were cast with formwork on site. In the screenshot, depicted in this figure, one specific acquisition date is selected and all detected elements are highlighted by means of a dedicated color. Green coloring represents elements that have been built and are detected and confirmed through the point cloud. All yellow elements are built but were not confirmed through the point cloud.

There are several reasons, why some of those elements were not detected:

One of the main reasons is occlusions. During construction, large amounts of temporary structures like scaffoldings, construction tools, and construction machinery obstruct the view on the element surfaces. Limited acquisition positions further reduce the visible surfaces and hence the overall quality of the generated point clouds. As introduced in (Huhnt, 2005), technological dependencies can help to formalize the schedule sequence. A precedence relationship graph (PRG) can hold this information and help to identify the described occluded elements (Braun et al. 2015b).

Another reason for weak detection rates is building elements which are currently under construction. As those elements count towards the overall progress, they must not be missed and play a crucial role in defining the exact position in the current process. Challenging parts are in general all construction methods, whose temporary geometry differs largely from the final element geometry. This accounts e.g. for reinforcements or formwork. On the one hand, formwork may obstruct the view of the element and thus make it impossible to be detected. On the other hand, the plane surface of a formwork for a slab might be detected as the surface of the slab itself and thus lead to false positives (Turkan et al., 2014).

Due to these challenges, detection algorithms need enhancements and require thorough testing. For the respective field studies, the ground truth is required to validate calculated results. Concerning construction monitoring, different construction sites are required and in addition, data recording should be tested under varying conditions to get different use cases for robust results. In the scope of this research, various case studies have been conducted for thorough testing results. Special focus lies on the points, that are counted as relevant for the individual surfaces of construction elements. Thus, thresholds need to be defined to filter unneeded and false points. Since these thresholds depend on various conditions like acquisition accuracy or construction methods, parameter studies are required for valid results.

This necessitates classification schemes to categorize all possible detection states.

2 RELATED WORK

Process monitoring has become a heavily researched topic recently. Capturing the as-built construction status is mainly achieved by laser scanners or cameras using photogrammetric methods. For the comparison with the as-planned state (BIM 3D geometry), three methods are currently established

i. Comparison of the as-built point clouds with points from the transformed as-planned geometry. These methods compare point clouds that are acquired by laser scanners (Bosché, 2010; Turkan, 2012) or photogrammetric methods and derived point clouds from as-planned surfaces (Kim et al., 2013). This is mainly done by Iterative Closest Point (ICP) algorithms.
ii. Feature detection in the acquired images from the as-built state. Using feature detection algorithms like SIFT, construction elements are directly identified from the acquired images. The derivation of the progress is based on a Bayesian approach, while learning the thresholds based on Support-Vector-Machines (SVM) (Golparvar-Fard et al., 2011).
iii. Matching the as-built point cloud onto the as-planned geometry surfaces. This approach matches relevant points from the point cloud directly onto triangulated surfaces of the as-planned model (Tuttas et al., 2015).

Monitoring with laser scanners or cameras proofed to be helpful according to these studies. It is possible to identify individual elements for each use case. In fact, most publications focus on identifying one particular element. All of these approaches lack the possibility to clearly identify building elements under construction. A first attempt to solve the problem of elements under construction is published by (Han et al., 2015). However, Han focusses on visibility issues. E.g. when an anchor bolt for a column is invisible as it is embedded into the concrete, it still must be present since the column on top of it requires the anchor bolt for structural reasons.

Process planning is often executed independently from architectural and structural design. Current research follows the concept of automation in the area of construction scheduling. Binding process information and the underlying building information model provides additional information that can be used in the context of progress monitoring.

Tauscher (2011) describes a method that allows automating the generation of the scheduling process at least partly. He chooses an object-oriented approach to categorizing each component according to its properties. Accordingly, each component is assigned to a process. Subsequently, important properties of components are compared with a process database to group them accordingly and assign the corresponding tasks to each object. Suitable properties for the detection of similarities are for example the element thickness or the construction material. With this method, a "semi—intelligent" support for process planning is implemented.

In (Huhnt, 2005) a mathematical formalism is introduced that is based on the quantity theory of the determination of technological dependencies as a basis for automated construction progress scheduling. In (Enge, 2010) a branch-and-bound algorithm is introduced to determine optimal decompositions of planning and construction processes into design information and process information.

Another important aspect for the as-planned vs. as-built comparison is dependencies. Technological dependencies show, which element is depending on another element, meaning, that it cannot be built after the first element is finished. These dependencies can be stored in so-called precedence relationships (Wu et al., 2010). A solution to store these dependencies in graphs is shown in (Szczesny et al., 2012).

However, the visualization and classification of the detection states have not been addressed in detail so far. (Golparvar-fard et al., 2009) have introduced an approach where a color scale for elements ahead of schedule (dark green) over elements on schedule (light green) to elements behind schedule (red color) is shown. This scale requires all elements to be visible and detected. As discussed before, this condition is not met at all times during monitoring.

3 CLASSIFICATION METHOD

During the comparison of the as-built and the as-planned state, different detection states occur. As introduced, a classification scheme is required to exactly visualize all possible detection states. In this case, the temporal factor is addressed. The generated point cloud for the as-built vs. as-planned comparison represents the actual situation on site at a certain time. However, this point cloud might not be perfect and have holes or low densities in several spots. For this reason, the detection algorithms might not identify all elements present. In order to correctly handle those elements, each element is considered and categorized independently.

As shown in Table 1, each building element is categorized in three different states:

i. **As-planned:** whether it should be built at a given time or not, according to the process plan. The main idea behind as-built vs. as-planned comparisons is to detect if there are any deviations on the site compared to this state.
ii. **As-built:** whether it is actually present or not. This state represents the ground truth and is not available during the as-built vs. as-planned comparison under real conditions. Corresponding data is gathered manually to prove scientific methods and refine them.
iii. **Detected:** whether it is detected by the detection algorithms or not. This state should equal the "as-built" state (ground truth). However, under real conditions it can only be approximated.

Derived from these three states with each two options (yes/no), the shown 3 × 8 matrix (2^3) shows all possible combinations.

These cases can occur due to different reasons. In general, those cases are desirable, where the ground truth and the detected elements align since this proves a correctly working algorithm.

3.1 *Description of the classification cases*

The mentioned cases have different reasons and importance that shall be discussed hereinafter.

3.1.1 *Case 1*

The considered element is not planned and also not built and detected. This applies to all elements which are installed at a later time.

3.1.2 *Case 2*

The element should be present according to the construction schedule, however, it is not yet built and also not detected. This case usually takes effect during a delay of the schedule.

3.1.3 *Case 3*

This object is built and should also be built according to the process plan. However, it has not been detected during the as-built vs. as-planned comparison.

Possible reasons are the already mentioned occlusions, low measuring accuracy or holes in the point cloud. Too few observation points or a smooth surface are other reasons for a bad point cloud quality. Additionally, low construction quality with too high variations from the as-planned model could indicate a not detected element.

3.1.4 *Case 4*

In this case, the component is present, however, been neither planned nor recognized. This can occur when the component was built earlier than in the construction process defined and in addition it is obscured or otherwise not recognized.

Table 1. Detection state cases for the as-built vs. as-planned comparison.

#	As-planned (process)	As-built (ground truth)	Detected (comparison)
1	–	–	–
2	X	–	–
3	X	X	–
4	–	X	–
5	–	X	X
6	X	X	X
7	X	–	X
8	–	–	X

However, this cannot be detected using the given boundary conditions. The case arises logically from the existing categories. A recognition of this case is only possible in the context of the research scenarios where the ground truth of the element is known.

3.1.5 *Case 5*

Case 5 is identical to Case 4 with the exception that in this case the element has been confirmed by the comparison. Again, the component has been built too early.

3.1.6 *Case 6*

Similar to Case 1 all categories are identical—the component has been built at the specified time and is also recognized.

3.1.7 *Case 7*

This case is critical. The component is planned and has also been recognized, but it is actually not built. This case is unusual and could be caused for example by an element which is currently under construction which will be completed with a slight delay. The geometry is confirmed by a sufficient amount of points; however, it is not yet completely finished.

On the other hand, this case can also be caused by errors in matching algorithms or weakly defined thresholds.

3.1.8 *Case 8*

This case is a similar situation as case 7, however, could have been caused by temporary structures or construction facilities.

3.2 *Description and color scheme*

The cases listed can all theoretically occur during matching with point clouds. However, based on previously evaluated results, the critical cases where a false positive occurs, meaning geometry is recognized without any actual geometry present, did not occur. Table 2 summarizes the reasons and descriptions for the classification states and corresponding problems. It should be stressed, that the noted problems solely focus on problems regarding the detection itself and not on problems for the construction site (e.g. delayed processes).

In order to visualize the defined classifications, a color scheme is introduced as depicted in Figure 1. The scheme emphasizes on the critical cases (7 and 8), marking them in red colors.

#	as-planned [process planned]	built [ground truth]	detected [PC to Geometry]	Color
1	-	-	-	
2	x	-	-	
3	x	x	-	
4	-	x	-	
5	-	x	x	
6	x	x	x	
7	x	-	x	
8	-	-	x	

Figure 1. Color scheme for the classification of detected elements.

Table 2. Description of the classifications.

#	Description	Problem
1	Element not yet built	–
2	Process delayed	–
3	Occlusions, low point cloud quality	No detection
4	Early finish, occlusions	No detection
5	Early finish	–
6	Element built	–
7	Element under construction, slight delay	False positive
8	Temporary structure detected	False positive

4 IMPLEMENTATION

During the scope of the conducted research project, three test sites have been monitored. The introduced classification scheme is applied to validate its meaningfulness.

4.1 *Prototype implementation*

The results of this research project necessitated a BIM viewer that implements all gathered data and provides detailed element information, tailored to the needs of progress monitoring.

Thus, the *progressTrack Viewer* was developed. The current development stage is depicted in Figure 2. It is based on a WPF framework and written in C#.

Building information models can be imported using the xBIM toolkit (Lockley, 2015). A Gantt diagram is used to display and bind construction elements to their respective processes. The software is connected to a database server that stores all semantic information in one place. The detection results from the implemented algorithms are stored there, too.

During this research, various parameter studies were conducted and the results can be viewed in this viewer. For a detailed evaluation of critical areas, the point cloud can be laid over the as-planned geometry. This feature makes it easier to understand, why certain elements were detected and others were not considered in an algorithm.

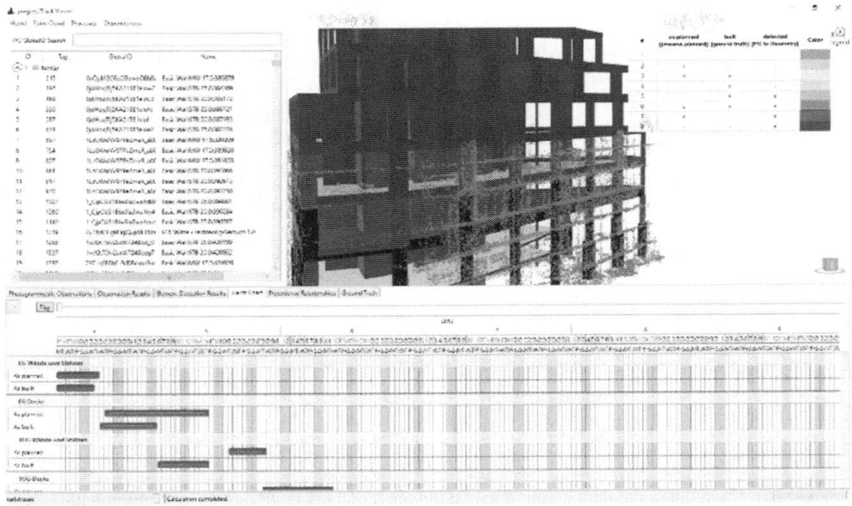

Figure 2. *progressTrack Viewer*: detected elements for construction site "A". Yellow elements are built, but not detected, green elements are built and detected.

4.2 *Model requirements*

For a valuable and precise as-built vs. as-planned comparison, the model itself needs to fulfill requirements regarding the detailing of all construction elements. In the scope of digital element representation, a detailed schema has been developed to classify the detailing of construction elements: the Level Of Detail (LOD).

According to this schema, a BIM necessitates at least LOD 300 for accurate construction site monitoring. The LOD states *"The Model Element is graphically represented within the Model as a specific system, object or assembly in terms of quantity, size, shape, location, and orientation. Non-graphic information may also be attached to the Model Element."* (The American Institute of Architects, 2013). Since the exact position, shape and measurements are required, this LOD seems accurate for the desired purpose.

Furthermore, Building Smart defined so called Model View Definitions (MVD) that describe the content of a BIM regarding the included elements and exchange requirements in the AEC industry (BuildingSmart, 2016). The general exchange definitions are labelled "Coordination View" (CV) or as per the newly defined IFC4 "Reference View" (RV).

Those views include all modelled elements with details and constructive parts like reinforcements. This view is required for the as-built vs. as-planned comparison.

Another important aspect regarding model quality are measurement rules for element boundaries. According to german standards, general construction requires an accuracy of around 1 cm for 1 meter of element length up to 3 cm for 30 meters of element length (DIN, 2013). The point cloud accuracy varies depending on several influences like lighting and is around 1–2 cm. Therefore, this approach is too inaccurate for exact quality measurements, however it is well suited for the as-planned vs. as-built comparison.

4.3 *Color scheme implementation*

Depicted in Figure 3, the results for the finished test site "C" are visualized. As discussed, the façade has very good coverage rates while the inner elements were not detected (*case 3*: planned, built, but not detected). The results are immediately understandable and thus, research results can be visualized in a suitable way.

4.4 *Analysis of false positives*

Figure 4 a) depicts a small sample of the color scheme. In this case, the as-built vs. as-planned comparison has been calculated with too weak thresholds, resulting in a higher error rate. Marked in light blue, the points that led to this false positive are visible.

They result from rebars, already installed on site, that reached up to the next level and thus led to a matching between the corresponding surface of the column in place. Figure 4 b) shows the problematic case. While the walls and columns one floor beneath the falsely detected element are marked as built / in progress correctly, only a few rebars reach up on the next level.

4.5 *Further refinement*

As shown in this example, the classification provides useful information for the user of the system, however not all data can be processed correctly. Especially, reasons for undetected elements under construction and large areas, obscured by scaffoldings or other constraints are hard to identify with the help of this classification.

To further detail and refine the classification, additional categories are defined:

i. **Under construction**: Elements under construction usually require additional elements like formwork or reinforcement, that is installed during the creation of the element. Those construction support elements can be identified with lower thresholds during the as-planned vs. as-built comparison.
ii. **Derived**: As discussed in (Braun et al., 2014), elements can be ordered according to their dependencies in relation to other elements. This leads to a precedence relationship graph, representing the technological dependencies of all elements of a building. To conclude, an element can be derived as "built", when following, depending elements are already marked as built. Thus, enabling algorithms to use this additional information to identify more elements during the as-planned vs. as-built comparison.

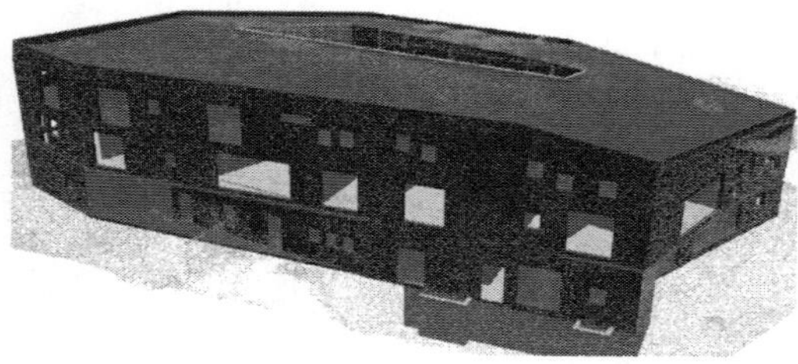

Figure 3. Case Study „Haus für Kinder" showing the developed color scheme. While the façade has been detected mostly correctly, inner elements could not be detected.

iii. **Previously detected**: Elements that have been detected in a previous time step may be occluded at a later stage due to added floors and slabs over the elements itself. However, previously detected elements do not vanish and already gathered information from previous steps can be used to improve detection rates even further.

This additional information, available through the consequent use of a BIM, bring great benefits to the construction progress monitoring methods and supply data to automate these systems even further.

5 DISCUSSION AND FUTURE WORK

This research focuses on the validation possibilities of as-built vs. as-planned element comparison in the scope of progress monitoring. A building information model provides all geometric and semantic data, representing the exact as-planned model of the construction site. The as-built state for individual time steps is generated by photogrammetric means, resulting in a three-dimensional point cloud. As mentioned, this point cloud often lacks quality due to e.g. occlusions or insufficient acquisition possibilities.

For this reason, the comparison of the as-planned geometry and the as-built point cloud returns incorrect or insufficient results that require attention during the development of algorithms and the design of suitable acquisition methods.

The presented classification scheme helps to get a quick and comprehensive overall impression of the current construction state of a building. Additionally, it proves helpful to identify falsely detected elements and visualize them accordingly. This is an essential part of the implementation and testing of new algorithms for the as-built vs. as-planned comparison.

Future improvements will include the use of even more available information through the BIM, like detailed process data and additionally, deeper research on the impact of detected elements, that are detected with a deviation to the current process plan.

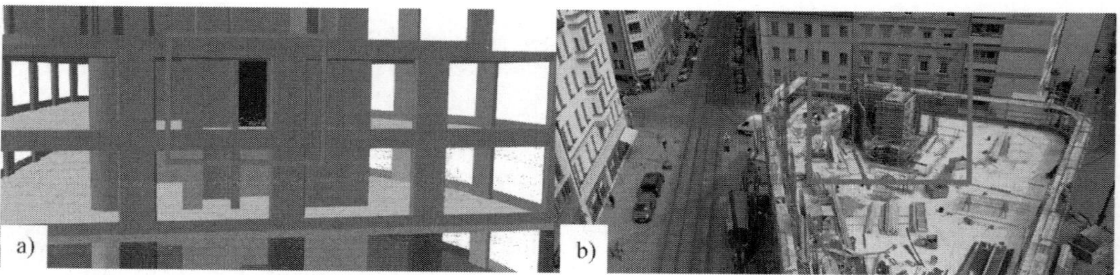

Figure 4. a) Too low thresholds may lead to wrong detection of a column (false positive) b) Picture of the current construction progress, resulting in falsely detected elements.

ACKNOWLEDGEMENTS

This work is supported by the German Research Foundation (DFG) under grants STI 545/6-1 and BO 3575/4-1. We like to thank Leitner GmbH & Co Bauunternehmung KG and Kuehn Malvezzi Architects (Test Site A), Staatliches Bauamt München, Baugesellschaft Brunner + Co and BKL (Baukran Logistik GmbH) (Test Site B) as well as Baureferat H5, Landeshauptstadt München, Baugesellschaft Mickan mbH & Co KG, h4a Architekten, Wenzel + Wenzel and Stadtvermessungsamt München (Test Site C) for their support during the case studies.

We thank the Leibniz Supercomputing Centre (LRZ) of the Bavarian Academy of Sciences and Humanities (BAdW) for the support and provisioning of computing infrastructure essential to this publication.

REFERENCES

Bosché, F., 2010. Automated recognition of 3D CAD model objects in laser scans and calculation of as-built dimensions for dimensional compliance control in construction. Adv. Eng. Informatics 24, 107–118. doi:10.1016/j.aei.2009.08.006.

Braun, A., Tuttas, S., Borrmann, A., Stilla, U., 2015. A concept for automated construction progress monitoring using BIM-based geometric constraints and photogrammetric point clouds. ITcon 20, 68–79.

Braun, A., Tuttas, S., Stilla, U., Borrmann, A., 2014. Towards automated construction progress monitoring using BIM-based point cloud processing, in: eWork and eBusiness in Architecture, Engineering and Construction: ECPPM 2014.

BuildingSmart, 2016. Model View Definitions [WWW Document]. URL http://www.buildingsmart-tech.org/specifications/mvd-overview.

DIN, 2013. DIN 18202 - Tolerances in building construction—Buildings. Baurechtliche Blätter.

Enge, F., 2010. Muster in Prozessen der Bauablaufplanung.

Golparvar-fard, M., Pena-Mora, F., Savarese, S., 2009. D4 AR—a 4 dimensional augmented reality model for automation construction progress monitoring data collection, processing and communication. J. Inf. Technol. Constr. 14, 129–153.

Golparvar-Fard, M., Pena-Mora, F., Savarese, S., Golparvar-Fard, M., Pena-Mora, F., Savarese, S., 2011. Monitoring changes of 3D building elements from unordered photo collections. Comput. Vis. Work. (ICCV Work. 2011 IEEE Int. Conf. 249–256. doi:10.1109/ICCVW.2011.6130250.

Han, K.K., Cline, D., Golparvar-Fard, M., 2015. Formalized knowledge of construction sequencing for visual monitoring of work-in-progress via incomplete point clouds and low-LoD 4D BIMs. Adv. Eng. Informatics. doi:10.1016/j.aei.2015.10.006.

Huhnt, W., 2005. Generating sequences of construction tasks, in: Conference on Information Technology in Construction.

Kim, C., Son, H., Kim, C., 2013. Fully automated registration of 3D data to a 3D CAD model for project progress monitoring. Autom. Constr. 35, 587–594. doi:10.1016/j.autcon.2013.01.005.

Lockley, S., 2015. xBIM.

Szczesny, K., Hamm, M., König, M., 2012. Adjusted recombination operator for simulation-based construction schedule optimization, in: Proceedings of the 2012 Winter Simulation Conference. pp. 1219–1229.

Tauscher, E., 2011. Vom Bauwerksinformationsmodell zur Terminplanung—Ein Modell zur Generierung von Bauablaufplänen.

The American Institute of Architects, 2013. AIA Document G202.

Turkan, Y., 2012. Automated Construction Progress Tracking using 3D Sensing Technologies. University of Waterloo.

Turkan, Y., Bosché, F., Haas, C.T., Haas, R., 2014. Tracking of secondary and temporary objects in structural concrete work. Constr. Innov. Information, Process. Manag. 14, 145–167. doi:10.1108/CI-12-2012-0063.

Tuttas, S., Braun, A., Borrmann, A., Stilla, U., 2016. Evaluation of acquisition strategies for image-based construction site monitoring, in: The International Archives of the Photogrammetry, Remote Sensing and Spatial Information Sciences. ISPRS Congress, Prague, Czech Republic.

Tuttas, S., Braun, A., Borrmann, A., Stilla, U., 2015. Validation of Bim Components By Photogrammetric Point Clouds for Construction Site Monitoring. ISPRS Ann. Photogramm. Remote Sens. Spat. Inf. Sci. II-3/W4, 231–237. doi:10.5194/isprsannals-II-3-W4-231-2015.

Tuttas, S., Braun, A., Borrmann, A., Stilla, U., 2014. Comparison of photogrammetric point clouds with BIM building elements for construction progress monitoring, in: The International Archives of the Photogrammetry, Remote Sensing and Spatial Information Sciences, Volume XL-3.

Wu, I.C., Borrmann, A., Beißert, U., König, M., Rank, E., 2010. Bridge construction schedule generation with pattern-based construction methods and constraint-based simulation. Adv. Eng. Informatics 24, 379–388. doi:10.1016/j.aei.2010.07.002.

eWork and eBusiness in Architecture, Engineering and Construction – Christodoulou & Scherer (Eds)

Introducing process mining for AECFM: Three experimental case studies

S. van Schaijk
Stam en De Koning, Eindhoven, The Netherlands

L.A.H.M. van Berlo
Netherlands Organization for Applied Scientific Research TNO, Delft, The Netherlands

ABSTRACT: The research field of process mining is relatively new and not been applied often in the Architecture, Engineering, Construction and Facility Management industry (AECFM). Process mining uses databases of existing IT systems to gain major insights in processes. Currently the AECFM industry increasingly adapts IT systems within all phases of the process. This creates the possibility to use process mining techniques to gain insight in the processes of construction projects. This paper introduces process mining by presenting three experimental case studies which are conducted in order to study the applicability of process mining in the AECFM. Studies are done in the design-, build—and operational phase. The study has proven to provide useful insight and potential applications. The research method that was used does not allow generalisation of the conclusions for the whole industry. Additional research is needed to study the potential of integrating different data sources from several phases.

Keywords: data mining, process mining, BIM, systems engineering, facility management

1 INTRODUCTION

During the 'data explosion' from last decades the capabilities of information systems expanded rapidly. As a result the digital universe and the physical universe are becoming more and more aligned. The growth of the technological possibilities with RFID (Radio Frequency Identification), GPS (Global Positioning System), Intelligent Imaging Camera systems, and sensor networks like the Internet of Things will stimulate further alignment of the digital and physical universe. Even in the construction industry, which generally is perceived as 'old fashioned', these new technologies are implemented. This data expension development makes it possible to record physical events as 'event logs' and analyse them digitally.

Process mining is all about exploiting event data in a meaningful way in order to provide insights, identify bottlenecks, anticipate problems, record policy violations, recommend countermeasures, and streamline processes. (van der Aalst, 2011)

Process mining uses databases of existing IT systems to gain major insights in processes. Currently the AECFM adapts IT systems within all phases more and more. Which gives the possibility to use process mining techniques to gain insight in the processes of construction projects.

In general these IT systems are not used to extract event logs, but are used to run daily activities. Thereby companies do not realize that they are creating valuable data.

Currently a lot of construction companies trying to optimize their processes with help of Building Information Modelling (BIM) and LEAN approaches. Companies claim to be LEAN. Where they try to optimize processes in order to increase customer satisfaction. One of the fundamentals of LEAN is knowing what you are doing and have insight into your own processes (Sayer & Williams, 2013). Business Process Management (BPM) and Process Intelligence (PI) approaches are used to monitor and control LEAN process flows.

Interviews with practitioners conducted by Quirijnen & van Schaijk, (2013) revealed that these monitoring systems mainly depend on human written notes. It was noticed that despite the importance, the current practices of monitoring systems are still-time-consuming, costly and prone to errors. Thereby monitoring is rarely be done and the data is nearly used.

Given (a) the interest in monitoring, observation of construction projects and the interest in construction process models, (b) the limited quality of current monitoring and observation tools and the unrealistic hand-made models, and

(c) the possibilities to autonomous create, store, and extract event data from IT systems, it seems legit to study the value of applying process mining in the construction. Therefore this research will explore the possibilities of improving construction projects with help of process mining.

This paper starts with elaborating the research goals and methods. There after process mining and event logs are explained more detailed. Followed by a description of the case studies which are conducted. The papers finishes with a conclusion, discussion and recommendations for further work.

2 RESEARCH

2.1 *Research goals*

The main goal of this research is to improve construction projects by providing insight in the construction process. This study explores if process mining techniques are suitable for providing insights which lead to more efficient processes in the AECFM.

This study targets to introduce process mining in the AECFM. Thereby it explores especially the applications and usability of process mining. In addition it gives insight in opinions about process mining of people working in the industry. But it does not give details about process mining statistics or algorithms. When one is interested in more details it is recommended to read the report BIM based process mining by van Schaijk (2016).

2.2 *Method*

Three experimental case studies are conducted in order to research out the applicability of process mining in the AECFM. The cases are executed in the design-, construction- and operational phase. Thereby a large part of the construction process is covered and a good overview can be created of the applicability of process mining.

Firstly a study is done within the design phase of a construction project. A systems engineering database of a civil project of a large contractor is analysed with process mining techniques.

Secondly a study is done within the construction phase of a project. As-planned- and as-built Building Information Models are used to create a process-oriented data warehouse which is analysed with process mining techniques.

Lastly a study is done within the operational phase of a building. This experimental study explored if facility management data is suitable for analysing processes around building elements with process mining techniques.

3 BACKGROUND

3.1 *Process mining*

Process mining provides an approach to gain insight and improve processes in a lot application domains. The goal of process mining is to gain event data, extract process-related information and discover a process model. Most organizations detect process problems based on fiction rather than facts. Van der Aalst (2011) describes process mining as an "*emerging discipline providing comprehensive sets of tools to provide fact-based insights and to support process improvements*" (p. 7). In comparison with other data mining techniques, like for example Business Process Management (BPM) and Business Intelligence (BI), process mining provides a full understanding of as-is to end-to-end processes. BI focuses on dashboards and reporting rather than clear process insights. BPM heavily relies on the experts which are modelling the to-be process and not help organizations to understand the actual as-is processes. (van der Aalst, 2011)

Little research is done in studying the potential of process mining in construction projects. A small study is executed by Terlouw and Mulder (2014) who explored the potential of process mining with the ISO standard for communication VISI. They used the events extracted from VISI archives in order to gain insight into communication processes in civil projects. They gained insights within three projects from a social interaction perspective and concluded that the potential of this data analyzing approach in the construction industry is high.

When looking at other industries several process mining studies have been done. Applications of process mining can be found in various economic sectors and industries like healthcare, governments, banking and insurance, educational instances, retail, transportation, cloud computing, capital goods industry. (van der Aalst, 2011)

The idea of process mining is not new. The roots of the research field can be found a half century ago. (Nerode, 1958) already presented an approach to synthesize finite-state machines from traces, (Petri, 1962) introduced the first modelling language capturing concurrency and (Gold, 1967) first explored different notions of learnability. While data mining gained more attention during the nineties little attention was given to process related mining. Since the first survey on process mining in 2003 (van der Aalst et al., 2003) a lot of progress in the research field has been made. Several techniques have been developed and various tools have come into existence. A comprehensive overview of the state-of-the-art in process mining is given in the book of Wil van der Aalst "Process

Mining Discovery, Conformance and Enhancement of Business Processes".

3.2 *Event logs*

The starting point of process mining is typically a raw data source. A raw data source may be every file, for example an Excel spreadsheet, transaction log, or a database. Often the necessary data is scattered over different files or data sources. For cross-organizational mining those sources may be distributed over multiple organisations. The raw data source has to be converted into event logs in order to be suitable for process mining analytics. The availability of high quality event logs is essential in order to enable process mining (van der Aalst, 2011).

3.2.1 *Event log structure*

Using Figure 1 as a reference, the following is assumed regarding event logs (van der Aalst, 2011):

- A process consists of cases.
- A case consists of events such that each event related to precisely one case.
- Events within a case are ordered.
- Events can have attributes. Examples of attributes are activity, time, costs, and resource.

The bare minimum to enable process mining requires a Case ID and an Activity log:

1. Case ID: A case identifier, also called process instance ID, is necessary to distinguish different executions of the same process. What precisely the case ID is depends on the domain of the process. For example in a hospital this would be the patient ID. In the case of construction process mining with BIM the case ID would be the Global Unique Identifier (GUID) of the building element when one is interested in this specific process. When someone is interested in the generic process of a project he could use the unique building ID as a case ID.
2. Activity: There should be names for different process steps or status changes that were performed in the process. If you have only one entry (one row) for each process instance, then your data is not detailed enough. The data needs to be on the transactional level (you should have access to the history of each case) and should not be aggregated to the case level. In the case of construction process mining of a building element there should be a history of this element.

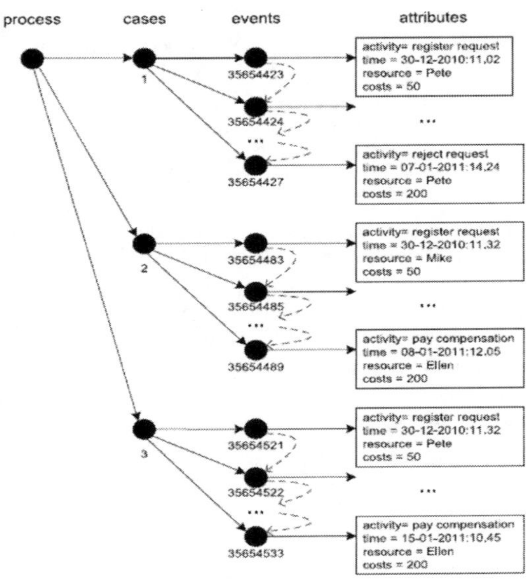

Figure 1. Event log structure (van der Aalst, 2011).

To analyze performance related properties additional attributes are useful in the event log like:

3. Timestamp: At least one timestamp is needed to bring events in the right order. This time stamp is also needed to identify delays between activities and bottleneck identification.
4. Resources: For example the person or the company who executed the activity.
5. ...More attributes can be added...

However the structure and information within the event logs may be the same, event logs can be described with different data formats. Detailed information about those formats can be read in (van der Aalst, 2011, Chapter 4), but the most occurring ones are CSV and XES (eXtensible Event Stream).

3.3 *Process mining analytics*

Mature process mining tools are available and offer advanced analytics. In this study process mining tools Disco (https://fluxicon.com/disco/) and MyInvenio (https://www.my-invenio.com/) are mainly used for analytics. However scientific researchers may prefer using ProM (http://www.promtools.org/) which gives an extensible framework where one can contribute by developing plug-ins.

Event logs enable three types of process mining (van der Aalst, 2011). The first one is *discovery* where an event log is taken as input and a process model (i.e. a Petri net or BPMN model) is generated as output. In addition it is also possible to discover resource-related models, such as social networks, when the log contains information about resources.

The second type of mining is *conformance*. This method compares reality with an existing process model using event logs of the process. In addition conformance checking can be used to detect, locate and explain deviations, and to measure the value of these deviations.

The last type of process mining is *enhancement*. This type focuses on extending or improving existing processes using event logs. Where conformance checking measures alignment, enhancement focusses on repairing, changing or extending the model. A model can be extended by adding performance data. Doing this enables one to show for example bottlenecks, service levels, throughput times, and frequencies.

4 CASE STUDIES

4.1 *Construction design mining*

Techniques that improved performance in other industries are more often adopted in the construction industry such as Systems Engineering (SE). Interviews with SE experts conducted within this study have indicated that SE IT systems supports with managing their project and helps to prove that the clients specifications are realized. However it is not known if those systems support an efficient process. Practioners indicated that a lot of people are involved in such projects who all work in parts of the IT systems but nobody has a clear overview of the total process. This case study explores the possibilities of discovering parts of the design process with process mining techniques. By use a large civil project the potential of process mining within the design phase of construction projects is discovered.

A database of the project is exported from SE tool Relatics (http://www.relatics.com/). It took some time to transform the data into event logs. But thereafter valuable analytics could be done with process mining tools Disco and MyInvenio

An as-desired process model of the project was made by the process engineers of the contractor. It is compared to the process model discovered with process mining. It seemed that both models did not match at all. In addition several process steps occurred repeatedly after each other. Some process-loops occurred more than 500 times, which was much more than expected.

Real bottlenecks are found in document flows, process variants are discovered, and social networks are exposed which proved that some people did not use the SE tool much. With this information improvements in the design process can be managed by the process engineers.

Due to this experimental case study the process engineers realized that the organization does not control the field of information—and data modelling enough. It is concluded that it is valuable to use process mining to give continuously feedback to the project managers at contractors. Process mining gives unique visualisations which enables refreshing insights. In addition process mining analytics did realize that some IT systems are used to store information, but those systems do not automatically facilitate an efficient process.

4.2 *Construction site process mining*

This case study targeted to get insight in the deviation of the as-planned and as-built process on construction sites. Since hardly any as-built data was available the researchers conducted an on-site research and collected the data.

It is seen that BIM based planning tools can be used to make as-planned models including planned tasks linked to building elements. Several monitoring technologies are available to capture the progress and planning deviation on construction sites. In this case drones flew repeatedly over the site. The pictures created with the drones are used to generate as-built point clouds models, which are compared to as-planned BIM models. The deviations of the as-planned model and as-built process where measured. By use of BIMserver (Beetz & Berlo, 2010) the as-planned model is translated into event logs. The as-built data is merged with those logs resulting in as-built event logs.

By use of the Plan-Capture-Analyse-Reuse workflow and several BIM and process mining tools, continuously learning loops for construction companies where realized. As part of this workflow the Planning consult software tool is developed. By use of this tool it is proven that event logs from previous projects can be reused in order to advice construction planners and identify risks in early phases of new to build construction projects. Thereby it serves as a basis for continuously learning and enables construction project time decrease.

It is noticed that with process mining tools bottlenecks from the event logs can be discovered. This case study proved that it is possible to gain insight in planning deviation, social networks (see Figure 2) and bottlenecks (see Figure 3) during the construction phase. In addition the technologies and methods proposed in this research can be implemented in current processes of construction companies. They will be useful and enable to reuse all captured data. Therefore it will form the basis for knowledge reassurance and fact based improving.

Based on the case study it can be said that the insights gained with the proposed methods have positive influence on managing construction projects.

4.3 *Facility management mining*

This experimental study explored if facility management data is suitable for analyzing processes around building elements with process mining techniques. By use of a case study at a hospital the potential of combining process mining with maintenance data is discovered. It can be concluded that with some data transformation maintenance data is suitable for process mining. Moreover the facility managers were surprised by the visualization techniques and they gained clear insight in the error handling process. As a result they discovered problematic building elements and odd processes. In addition as a result of this analysis the facility managers were surprised about the amount of money which was spend on to short (and unrealistic) jobs and they are going to monitor those errors for next months to figure out how this maybe can save them money. A notifiable quote was mentioned by one of the facility managers, she said "*We can probably save more money with investing in data analytics, than with firing our own people*".

The facility managers where definitely interested in using those kind of analytics more in the future. However this case study just explored the topic of facility management based process mining and has proven some useful applications. More (case-) studies should be elaborated to indicate the potential of this topic. In addition it would be useful to study the potential of integrating different data sources of other phases, for example the design—or construction—phase, in building elements lifecycle in order to gain insight in the process on a longer time span.

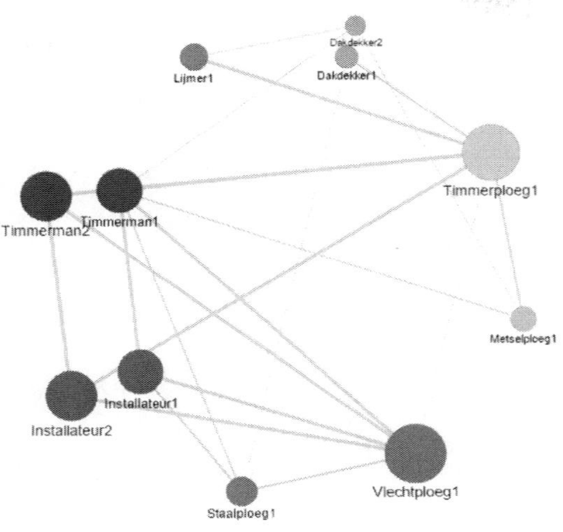

Figure 2. Social network visualizing the links between different people of the project. More lines means a bigger role in the project.

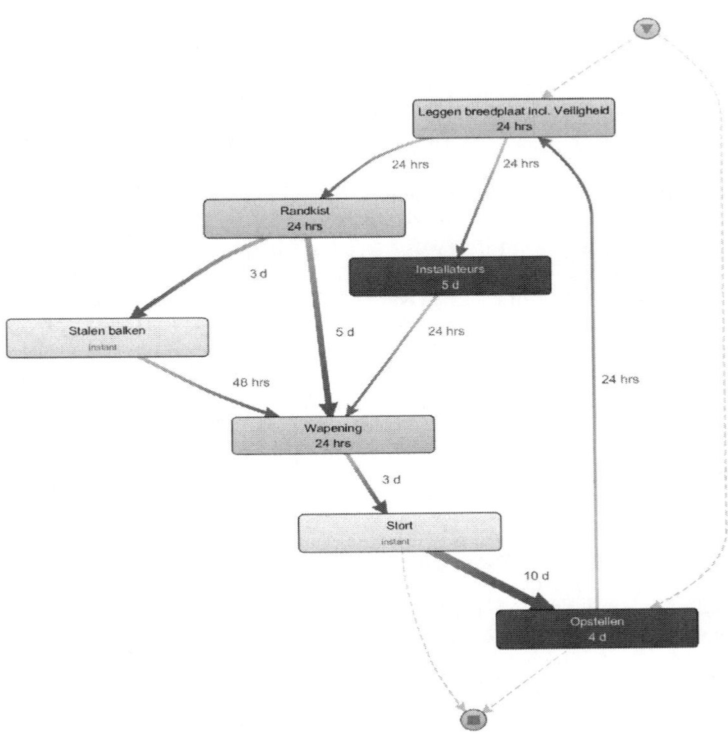

Figure 3. Process model of concrete floor elements with mean durations and statistics.

5 CONCLUSION

Companies in the AECFM do not always have clear sight on their processes (van Berlo et al, 2012). A lot of those processes are managed with IT systems. Especially in the plan—and maintain phase of buildings several tools are already applied. As a result of the experimental case studies within this research it can be concluded that some of these tools do not have functionality to give clear insight in overall processes. However the data which is in those tools can be extracted and analysed with process mining approaches.

Applying process mining can have different reasons. Organizational analytics, process variants, bottlenecks, planning deviation or conformance checking. Those applications are seen as valuable and have potential to help the construction industry understand and actually improve his processes.

It is seen that data preparation can be challenging. The used IT systems are not able to export clear event logs. This can cause a threshold to start with process mining experiments. However since event logs have a straightforward structure the researchers managed to transfer the data from the IT systems into event logs. In addition when one wants to do analytics on a more regular basis it is possible to automate the process of event log generation.

6 DISCUSSION AND FURTHER WORK

The research method that was used (experimental use cases) does not allow the conclusions to be applied to the whole industry. Additional research is needed to study the potential of integrating different data sources from several phases

More specific it is indicated that it can be useful to study the possibility of the integration of event log databases from different phases of construction projects. Meaning that event logs from a design phase, construction phase and operational phase could be merged in order to give unique insight in the lifecycle of a construction object. However, in contradiction to the plan and operate phase, event logs from the construction phase are difficult to find or generate (van Berlo & Natrop, 2015).

Interviews with domain experts revealed that several companies are interested in process mining services. When complex projects occur where the overall picture is difficult to manage, then process mining applications will be most promising. Since data is often available and existing algorithms are suitable such services are directly applicable (van Berlo & Bomhof, 2014), however companies have to be convinced of the power of those analytics. Therefore it is recommendable for further work to quantify the value of such analytics.

Event logs is a data structure which enables useful analytics, but storage of this data may be difficult. A possible solution as indicated in (van der Aalst, 2011) is one where companies can create a *process-oriented data warehouse* containing information about relevant events happened in the company. Possibilities for companies in the AECFM industry such as contractors can be studied as well.

Researchers which are interested in this subject can start with exploring the report and dataset (containing BIMs, Event logs, Point clouds and more) which is published along by van Schaijk (2016) and is freely downloadable.

REFERENCES

Beetz, Jakob, Berlo, LAHM van. "BIMserver. org–An open source IFC model server." Proceedings of the CIB W78 conference. 2010.

Gold, E.M. (1967). Language identification in the limit. *Information and Control.* http://doi.org/10.1016/S0019-9958(67)91165-5.

Nerode, A. (1958). Linear automaton transformations. *Proceedings of the American Mathematical Society, 9*(4), 541–541. http://doi.org/10.1090/S0002-9939-1958-0135681-9.

Petri, C.A. (1962). Kommunikation mit Automaten. *Fakultät Für Mathematik Und Physik.*

Quirijnen, R., & van Schaijk, S. (2013). *Meten = Weten.* Avans Hogeschool, Tilburg.

Sayer, N.J., & Williams, B. (2013). *Lean voor Dummies* (second edi). Amsterdam: Pearson Benelux bv.

Terlouw, L., & Mulder, H. (2014). Schatgraven in bouwprojecten. *Informatie, 1408*, 32–29.

van Berlo, L.A.H.M., Dijkmans, T.J.A., Hendriks, H., Spekkink, D., & Pel, W. (2012). BIM quickscan: benchmark of BIM performance in the Netherlands.

Van Berlo, L.A.H.M., & Bomhof, F. (2014). Creating the Dutch national BIM levels of development. American Society of Civil Engineers (ASCE).

van Berlo, L.A.H.M., and Mathijs Natrop. BIM on the construction site: providing hidden information on task specific drawings. Department of Computer Science, 2015.

van der Aalst, W.M.P. (2011). *Process Mining: Discovery, Conformance and Enhancement of Business Processes. Media* (Vol. 136). http://doi.org/10.1007/978-3-642-19345-3.

van der Aalst, W.M.P., Van Dongen, B.F., Herbst, J., Maruster, L., Schimm, G., & Weijters, a. J.M.M. (2003). Workflow mining: A survey of issues and approaches. *Data and Knowledge Engineering, 47*(2), 237–267. http://doi.org/10.1016/S0169–023X(03)00066–1.

van Schaijk, S. (2016). *BIM based process mining.* Eindhoven university of technology. Retrieved from http://www.slideshare.net/StijnvanSchaijk/bim-based-process-mining-master-thesis-presentation.

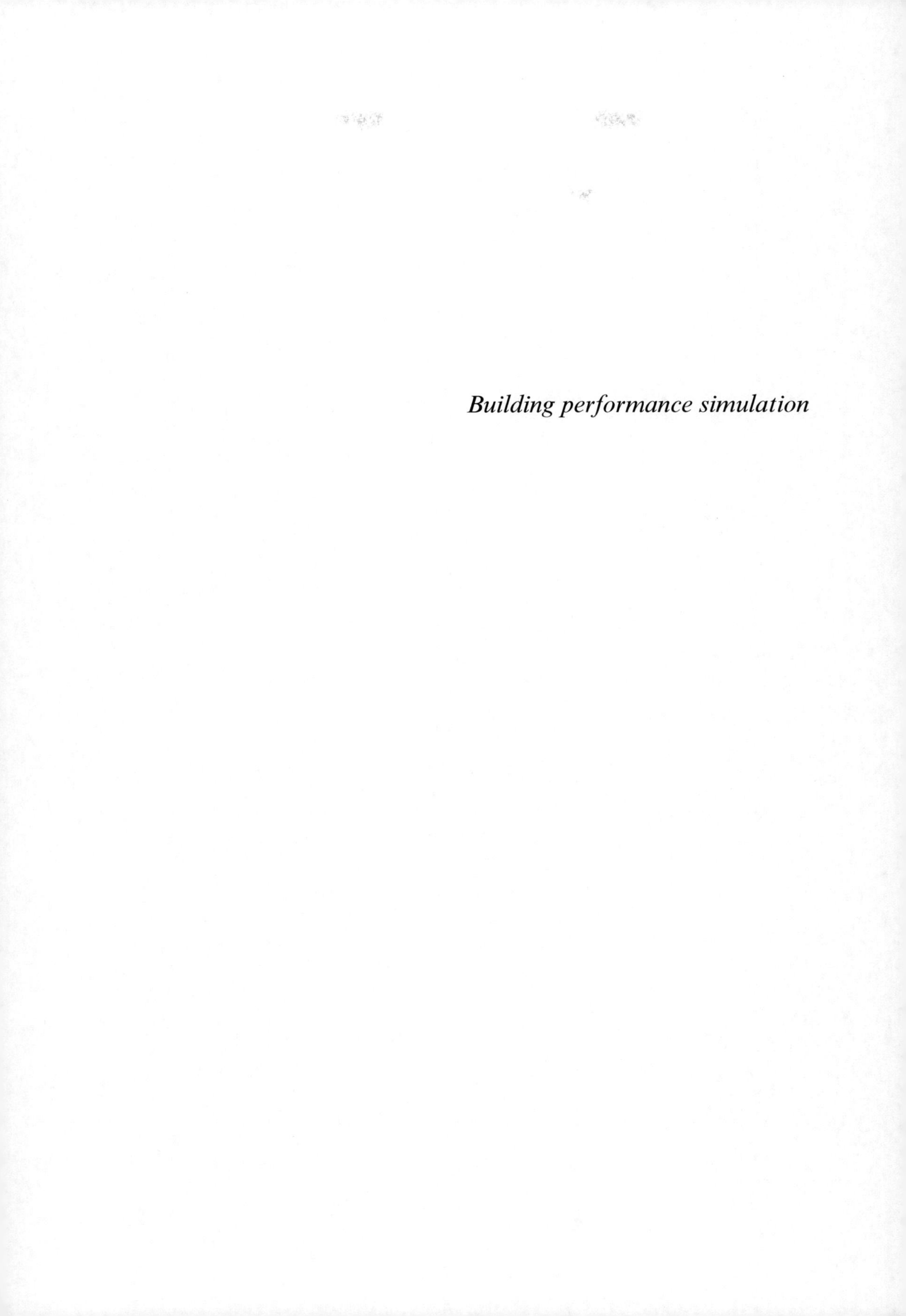

Building performance simulation

eWork and eBusiness in Architecture, Engineering and Construction – Christodoulou & Scherer (Eds)
© 2016 Taylor & Francis Group, ISBN 978-1-138-03280-4

Method of obtaining environmental impact data during the project development process through BIM tool use

Micheline Helen Cot Marcos
UNICURITIBA University, Curitiba, Paraná, Brazil

Erica Yukiko Yoshioka
University of São Paulo (USP), São Paulo, Brazil

ABSTRACT: The construction sector accounts for a significant portion of natural resources consumption, including energy, CO_2 emissions, water and building materials. The environment, economy and society should be considered in an integrated manner in the construction industry to meet society's expectations while reducing environmental impacts. New construction technologies combined with new projective technologies can contribute to the improvement of the built environment in the area of energy efficiency, thermal performance and environmental impacts. This research aims: to develop a method of obtaining informations of environmental impacts during the project development process, through the use of a modeling tool of Information Construction ("Building Information Modeling" - BIM), to assist in making decisions about building system that provides lower environmental impact. The use of BIM tools in building design process contributes to improving the quality of buildings, their thermal performance and the choice of building materials and systems that provide reduced environmental impacts. In the design phase, the BIM technologies provide the professional the possibility of conceiving a parameterized model, where you can view the volumes, check the impacts of sunlight, quantify and qualify the material applied with comfort variable settings and environmental impacts. For this research, the embedded CO_2 and energy in building materials were analyzed. For validation of the research, the method adopted was the case study which analyzed two building systems: steel frame and masonry, applied in the same condominium with twenty houses, which are in the design phase. From the literature review were extracted energy and CO_2 informations built the main building materials used in both building systems. Then these data were entered into a BIM tool. First were analyzed and the results obtained from twenty houses masonry then carried out the same process for a single house, and finally, a wall. This study is repeated for the steel frame system. As a result is obtained data of environmental impacts, assisting the professional in the choice of building materials and construction systems with lower environmental impact.

1 INTRODUCTION

1.1 *Hypothesis*

This article analyzes the energy and CO_2 embodied into building materials and unites these environmental analyzes in every building material used. Furthermore it provides alternative choices to projective changes before the completion of the project by using a BIM (Building Information Modelling) tool.

1.2 *Justification*

The global concern in making environmentally conscious buildings with their insertion in space comes from the fact that buildings consume more than half of all energy used in developed countries, which produce more than half of the greenhouse gases to the environment (ROAF, FUENTES & THOMAS, 2006). This consumption comes from the extraction of raw materials, through manufacturing of building materials, buildings erection, use and demolition. Currently are discussed, worldwide, actions to minimize environmental pollution and promote the reduction of greenhouse gas emissions that cause the greenhouse effect and resulting in global warming.

The Brazilian construction industry has great share of contribution in the current situation, identified as one of the sectors of the economy that generates greater impact on the natural environment. Therefore, this research is positioned in obtaining data from environmental impacts facing the energy consumption and CO_2 embodied into building materials, so that at the time of the project, it is possible to know the data and choose the construction materials that provides lower environmental impact, as use of a tool with BIM technology.

1.3 *Building Information Modeling tools (BIM) in project development to reduce environmental impacts*

The computational tools to support the building design process are currently indispensable for the construction industry, but with BIM technologies are still little explored by industry professionals. BIM tools occupy a small portion of the software to design market in South America (Marcos, 2015).

Performance, environmental sustainability of buildings and BIM (Building Information Modeling) are matters of evidence in the buildings sector and tools with BIM technology has been touted as a solution to the agents that enable the integrated work, easy identification of incompatibilities between the various designs, environmental simulation of the building, quantitative extraction, among others.

New construction technologies combined with new projective technologies contribute to the improvement of the built environment in the area of energy efficiency, thermal performance and environmental impacts (FREIRE & AMORIM, 2011). As described Agopyan & John (2011), design decisions and construction materials specification and its components, directly affect the consumption of natural energy resources. Regarding the consumption of these resources, the construction industry is positioning itself and becoming more and more relevant decisions. Building Information Modeling (BIM) is focused on the development process, use and transfer of digital information model of a building project to improve the design, construction and operation of a project. The commission "National Building Information Modeling Standards" (NBIMS, 2007), defines BIM as an easy and functional way to digital representation. BIM is a shared knowledge resource for information and form a reliable basis for decisions during the design process. The basic premise of BIM is to facilitate the collaboration of different actors (stakeholders) in different stages of the life cycle, to insert, extract, update or modify the information.

The Building Information Modeling tools represents a new generation of computational tools that manage object-oriented building the information in the project life cycle. It's a new way to be explored by professionals working in the area of Architecture, Engineering and Construction towards collaboration, interoperability and reuse of information. This approach aims to competitiveness and continuous improvement in the project development process (MARCOS, 2015). Therefore, a good design process conducted with the aid of appropriate information technology tools, is a key item for the quality of the buildings construction processes.

In the design phase defines the main guidelines of the projects in the construction industry. It has direct influence on the cost, time and production methods. However, there is a great distance between research in information technology applied to the construction and the methods actually practiced by the industry in daily life, both in Brazil and in other countries. The construction industry has a conservative character, which makes it difficult to incorporate the advances and keeps it technologically lagging behind the other industry segments. BIM requires specific domain for the users ability to modeled objects show intelligent behavior, imposing a practice that goes beyond the operational issues of labor. The language is seen not only in its descriptive aspect, but as a form of action directed to a mutual orientation with knowledge generation and exchange of information to improve the final quality of the project. In this sense, this paper aims to present the results of research through the use of a BIM tool.

1.4 *Product's modeling using BIM tools*

The product modeling in construction is the process for managing all information produced and used at different stages of the life cycle of the building, through a model which represents the physical and functional characteristics. New applications can be developed by mapping interoperability and to eliminate costly and time-consuming practice of creating multiple design process routines to get the same result (NIBIS, 2007).

According to Scheer & Ayres (2007) and Marcos (2015), the three-dimensional visualization added by use of a 3D CAD (Computer Aided Design) design increases the amount of information. However, 3D CADs have the same characteristic fragmentation information of geometric CADs, making it difficult to produce structured information, which normally constitute the core of the documentation of a project (plans, sections and elevations). On BIM tools, product modeling includes the concept of "virtual building": a set of parametric objects representing the building in a virtual environment. This set of objects are automatically extracted from the representations, documentation, quantitative reports, material and physical specifications. This is possible because in the BIM tools structure the model as databases containing information of each parametric object and from centralized access to them are held complex processing and the generation of structured documentation automatically. The level of information may be controlled according to need, from the design layout to constructive or performance analysis detailing. Parametric objects can also be direct references to products developed by manufacturers such as windows, prefabricated parts, accessories, etc. These objects and their updates can be obtained directly via the Internet and in the future automatically adjust their behavior to aspects of the project. For an example, objects representing structural parts, which automatically configures itself according to the types defined spans and supports. In Figure 1, it is shown as a BIM tool facilitates the design view.

Software with BIM technologies should be considered an evolution of the design process, in view of the new possibilities for visualization and processing of information. Among the main advantages to have better coordination of the constructive elements and their interference, reduction of working hours, increased productivity, better quality designs and detailing, centralized control of content and versions of project documents.

1.5 *BIM tool's applying a case study*

For the application of BIM tool in order to measure environmental impacts, two construction systems were used in order to obtain a comparison result at the end: ceramic brick and steel frame. Analyses were performed in a wall designed in ArchiCad software, detailed in both construction systems. Regarding environmental impacts were analyzed the embedded CO_2 and energy embodied in building materials for the two construction systems.

In this research we used the steel frame" system with light steel structure, so it can be so called "light steel frame".The system "Light Steel Framing" is a constructive system of rational design, whose main feature a structure formed by profiles formed of cold galvanized steel which are used for composing the structural and non-structural panels, secondary beams, floor joists, roof scissors and other components (Figures 2). Being an industrialized system, enables construction dry with speed of execution, so the system "Light Steel Framing" can be known by self-supporting dry construction system.

In the process using BIM, the virtual models may be understood as databases where they are either stored geometric data, such as textual each building element used in the design. The combination of these data enables automatic extraction of documents such as plans, sections, perspectives, or quantitative. The attention of the designer is therefore aimed primarily at design solutions and not to technical drawings, which are largely generated automatically by the tool. The advantages of using modeling go far beyond the creation of electronic models and streamlining the design specification documentation production process. As in metalworking industries, manufacturing and aerospace, model three-dimensional view lets you check the inadequacies and inconsistencies instantly, helping us intuitively decision processes in all stages of the project.

Another important point is the consolidation of information that make up the project. Once you use a unified database for all the information content, the changes in a document (e.g., a floor plan of the architectural design), spread to the other documents involved automatically, thus ensuring agility in updates, modifications and reliable access to information.

In this study, information of the embodied energy and CO_2 for each building material, are aggregated to the ArchiCad databases. Other data can be provided to the tool and thus get more results in the design process.

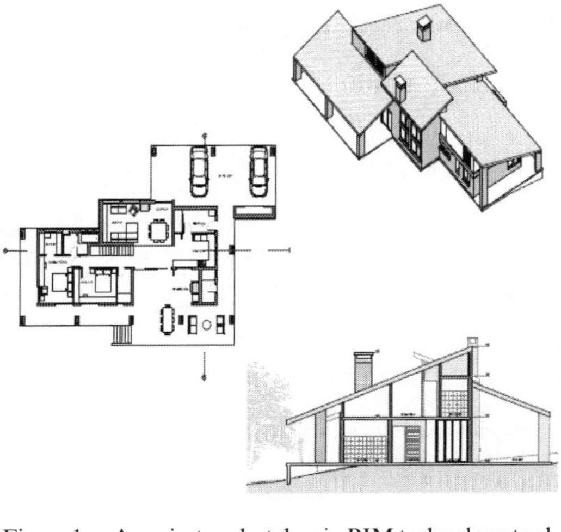

Figure 1. A project undertaken in BIM technology tool. Virtual model, low and cutting plant. Source: Scheer; Ayres (2007).

Figure 2. Residence structure in light steel frame with lock on cement board. House built for low incoming housing, at Curitiba—Paraná—Brazil. Source: Marcos (2015).

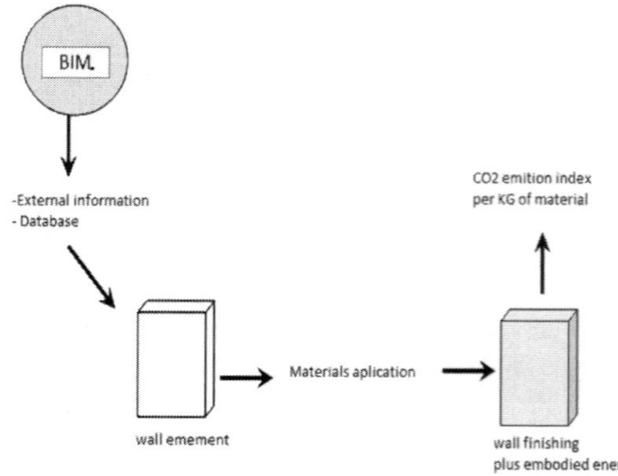

Figure 3. Information process in the BIM platform. Source: Marcos (2015).

In Figure 3 the process is outlined below with ArchiCad information. First element (in this case wall) is drawn, then apply the materials and finishes on this wall. The database is supplied with automatic association of tools that indicate the embedded carbon parameters in kilograms, generating a list of results.

In this way, results were obtained for the two building systems: ceramic brick and steel frame, with respect to corporate energy and CO2 embodied. By using this method, you can get data from various environmental impacts and analyze them at the time of project design as a way to assist in the choice of building system with less environmental impact.

2 RESULTS AND CONCLUSION

The wall analyzed in this research is part of the residence, as floorplan and is highlighted in blue.

Next, in Table 1, there are carbon values and embodied energy into a wall made by ceramic brick.

And, in Table 2, there are carbon values and embodied energy into the wall but made by steel frame.

With this, it's observed that, for the construction of a single wall made by steel frame, there is a consumption of 44% embedded CO_2 and 56% if the same wall is constructed of brick. And for the construction of a single wall made by steel frame will be a consumption of 63% of embodied energy and 37% of the same wall is constructed of brick.

The energy analysis, using the BIM system can be performed for the entire project or only a single element as shown in this study. The use of this methodology for project development helps the professional in the choice of building materials and therefore the most environmentally efficient building systems at the time of completion of the project. This construction professionals way can contribute to reducing environmental impacts at the time of completion of the project, thus being able to make choices of different building typologies according to the results of the analysis.

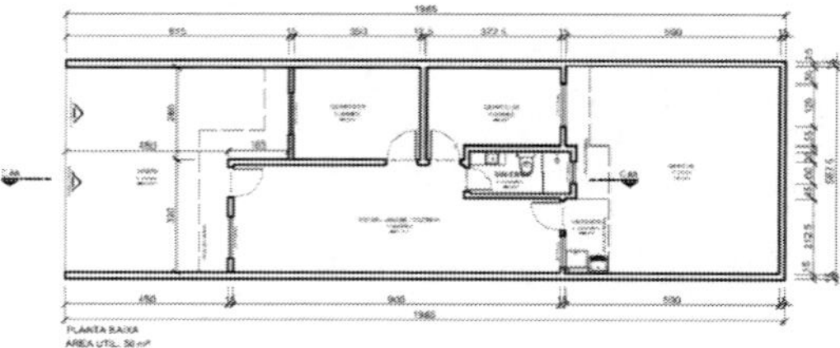

Figure 4. Wall analyzed in this research (blue). Source: Marcos (2015).

Table 1.

	Embodied CO_2	Embodied energy
Material	Kg	Kg
Roughcast	16,31	215,43
Brick – 9 cm	319,24	1641,49
Total	335,55	1856, 91

Table 2.

	Embodied CO_2	Embodied energy
Material	Kg	Kg
Structural steel	52,97	746,11
Dry wall	114	838,45
Cementitious board	96,55	1609,13
Total	263,52	3193,69

REFERENCES

Agopyan Vagan & Wanderley John (Editora Blucher) 2011. O desafio da sustentabilidade na construção civil. São Paulo, Brazil.

Freire, M. & Amorim, A. 2011. A abordagem BIM como contribuição para a eficiência energética no ambiente construiído. *Encontro de Tecnologia da Informação na Construção (TIC).2011*. Salvador: Brazil.

Marcos, Micheline (Novas Edições Acadêmicas) 2015. *Sustentabilidade na Arquitetura e o uso de ferramenta BIM*. Curitiba, Brazil.

Roaf, S. & Tomas, S. (Bookman) 2006. *Ecohouse – a casa ambientalmente sustentável*. Porto Alegre – Brazil.

Sergio S. & Ayres, C. 2007. Diferentes abordagens do uso do CAD no processo de projeto arquitetônico. *Workshop brasileiro de gestão do processo de projeto na construção de edifícios*. Curitiba: Brazil.

eWork and eBusiness in Architecture, Engineering and Construction – Christodoulou & Scherer (Eds)
© 2016 Taylor & Francis Group, ISBN 978-1-138-03280-4

A novel approach to building performance optimization via iterative operations on attribute clusters of designs options

A. Mahdavi, H. Shirdel & F. Tahmasebi
Department of Building Physics and Building Ecology, TU Wien, Vienna, Austria

ABSTRACT: The present paper explores the potential of a novel approach toward iterative global optimization of locally optimized attribute clusters of building design solutions. Thereby, clusters of design space attributes that are comprehensible to typical building designers as a compound yet coherent aspects of a design are made subject to multiple passes of local simulation-assisted optimizations. Hence, instead of allocating an individual dimension to each and every variable of a complex design within the context of a single-pass global optimization campaign, multiple iterative optimization steps target coherent clusters of such attributes and pursue those until the overall design meets the expected performance (or until further performance improvement is not forthcoming). The implementation of the proposed approach employs a number of existing—and freely accessible—computational applications, including an optimization software coupled to an energy simulation tool. We illustrate and document the performance of the current implementation of the proposed approach via an optimization case study. Thereby, different system operation options (i.e., sequential versus random cycling between attribute clusters) are demonstrated.

1 INTRODUCTION

In the last decades, a number of developments have made global optimization of large multi-dimensional design option spaces possible. Such developments include the increase in computing power, emergence of sophisticated optimization algorithms, and new techniques for the derivation of computationally efficient meta-models. Along with their promise, such developments also involve a number of potential drawbacks. For one thing, meta-models occasionally fail to capture the behavior of "non-conventional" and complex designs. Another critical problem pertains to the potentially opaque nature of large-scale global optimization exercises, which make them less amenable to provision of intuitively graspable support in the—typically iterative—design process. In this context, the present paper explores the potential of a novel approach toward iterative global optimization of locally optimized attribute clusters of building design solutions. Thereby, clusters of design space attributes that are comprehensible to typical building designers as a compound yet coherent aspects of a design are made subject to multiple passes of local simulation-assisted optimizations. Hence, instead of allocating an individual dimension to each and every variable of a complex design within the context of a single-pass global optimization campaign, multiple iterative optimization steps target coherent clusters of such attributes and pursue those until the overall design meets the expected performance (or until further performance improvement is not forthcoming). In our initial experimentation with this approach, we consciously do not aim at achieving a guaranteed optimum once the iteration is terminated. Rather, the distance of the emergent solutions to the theoretical global optimum are measured so as to evaluate their acceptability in the context the proposed method's benefits (i.e., use of original simulation models instead of meta-models as well as iterative, transparent, and intuitive navigation of the design space).

The implementation of the proposed approach employs a number of existing—and freely accessible—computational applications, including the optimization software GenOpt (Wetter 2001) coupled to the energy simulation tool EnergyPlus (Crawley et al. 2001). Moreover, the implementation targets scalability and flexibility: Specifically, users are to be provided with degrees of freedom in view of the selection of the clusters to be optimized. Moreover, additional clusters can be defined and variable sets in each cluster can be manipulated, while still achieving convergence within reasonable temporal horizons. We illustrate and document the performance of the current implementation of the proposed approach via an optimization case study. Thereby, different system operation options (e.g., random cycling between attribute clusters versus predefined sequences) are demonstrated.

The results thus far are highly encouraging: The proposed method delivers optimized solutions that are virtually indistinguishable from those of a reference one-shot global optimization run. But the results are not only obtained faster, but also via a transparent, traceable, and designer-friendly process.

2 BACKGROUND

High performance buildings are more likely to result from an effective performance-based design process. Such a process can benefit from computational support involving both building performance simulation and optimization techniques. As such, the application of numerical optimization in the design process is not a recent phenomenon. In fact research and development efforts in this domain have a track record over multiple decades, resulting in advanced in building informatics and mathematical optimization methods (Nguyen et al. 2014). However, currently bi-directional interfaces between optimization and building performance simulation tools that could automate the design alternative-evaluation loop are still under development (Attia 2012). Moreover, managing large number of independent variables in the optimization process still represents a challenge (Wetter 2011). The ideas presented in this paper have thus the potential to further advance the integration of coupled simulation and optimization environments in the building design process.

3 APPROACH

3.1 *Model*

The base model for the experiments is a modified version of a standard ASHRAE small office building retrieved from Commercial Prototype Building Models supported by the U.S. Department of Energy (DOE 2016). The office's dimensions are approximately 27.7 m (length), 18.5 m (width) and 3.1 m (height). Five thermal zones have been defined, namely one core zone and four perimeter zones (see Figure 1). The assumed location of the building is Vienna, Austria. It has 20 similarly sized (1.8 by 1.8 m) windows (6 windows on south and north facades and 4 windows on east and west facades). Each window has a 0.5 deep overhang and 2 lateral 0.5 m deep fins (see Figure 2). The opaque part of the exterior walls is modelled in terms of two thermally distinct (inner and outer) layers.

The office is assumed to be occupied by a total of 31 inhabitants (9 in the core zone, 7 each in the south and north zones, and 4 each in the east and west zones). Daylighting control sensor points are assigned to perimeter zones (half-way on the middle axis between the front and back walls, 0.8 m above the floor). Each window is assumed to have an exterior blind (movable shading device). Natural ventilation at the perimeter zones is controlled based on the outside air temperature.

3.2 *Definition of the clusters*

In this paper, to illustrate the proposed approach, we categorized the optimization variables in terms of four clusters. Three clusters entail design parameters and one pertains to control variables.

Façade Geometry (FG)
This cluster includes windows dimensions as well as overhangs and fins depth (see Figure 3). For each façade, dimensions of the windows and the depth of overhangs/fins vary simultaneously. As a whole, this cluster has 12 variables.

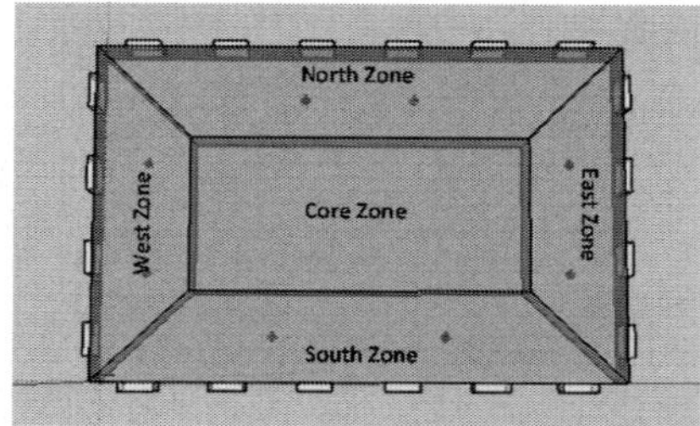

Figure 1. Schematic plan of the building model.

Figure 2. View of the building model.

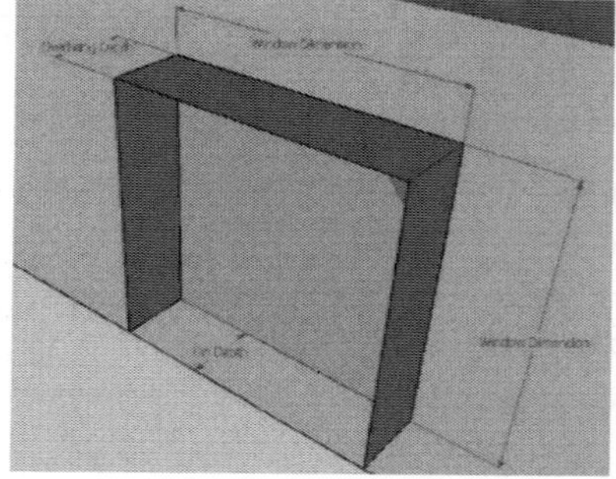

Figure 3. Variables of the cluster FG (Facade Geometry).

Material Properties (MP)
The façade is modelled in terms of opaque and transparent (glazing) components. The opaque part consists of an outer layer (with thermal conductivity as the pertinent variable) and an inner layer (with density as the pertinent variable). The pertinent variables of the glazing are assumed to be the thermal transmittance (U-value) and the visible transmittance. Note that for the purposes of the present demonstration, the glazing g-value is obtained as a function of its visible transmittance. Overall, this cluster has four variables.

Blind (movable shading device) Properties (BP)
This cluster comprises the pertinent variables the movable blind, namely Slat Width, Slat Angle, and Slat Solar Reflectance (Figure 4). Slat separation is obtained as a function of the slat width. Given the independence of the four facades in view of blind operation, this cluster entails 12 variables.

Control (CO)
Our final cluster is concerned with operation of blinds and windows openings during the year. For blinds, the variable subjected to optimization is a threshold value for incident solar irradiance, above which the blinds are closed. With regard to ventilation, a base mechanical ventilation system is assumed that delivers a prescribed fresh air supply rate of 7 l/s per person. However, given appropriate conditions (expressed in terms of an outdoor ambient temperature band) an additional magnitude of fresh air flow via window operation is supplied. The variables subject to optimization are in this case threshold values for outdoor temperature below which and above which windows are closed. This cluster includes thus 12 variables (4 for blind and 8 for window operation).

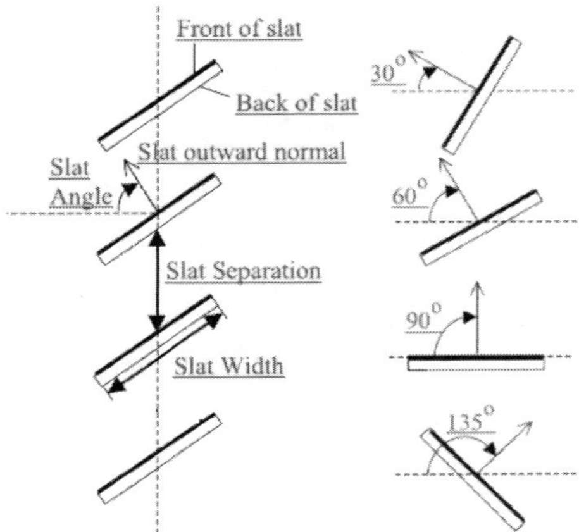

Figure 4. Exterior blind variables based on EnergyPlus documentation (DOE 2015).

3.3 *Tools and platform*

To integrate the simulation and optimization tools and automate the procedure, we developed a Java-based platform. The energy analysis and thermal load simulation program EnergyPlus was coupled with GenOpt, an optimization program for the minimization of a cost function that is computed using an external simulation program. GenOpt has a library with local and global multi-dimensional and one-dimensional optimization algorithms, as well as routines to support parametric runs. In our case, the optimization algorithm GPSPSOCCHJ was used (a Hybrid Generalized Pattern Search Algorithm with Particle Swarm Optimization Algorithm for Continuous and Discrete Variables).

To test the system, we implemented two approaches, one involving random cycling between attribute clusters (Figure 5) and one involving predefined sequences (Figure 6).

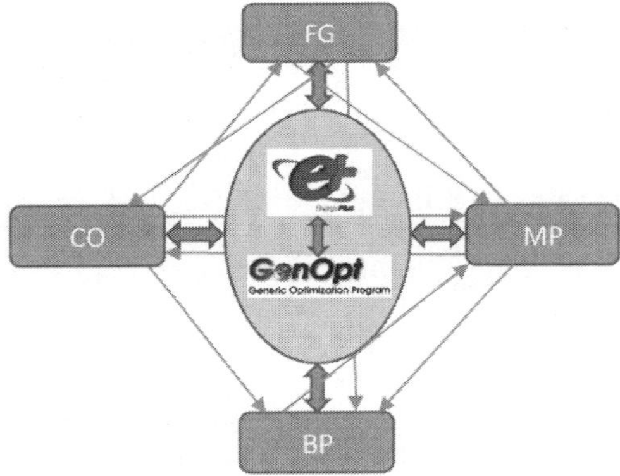

Figure 5. Illustration of random cycling between attribute clusters.

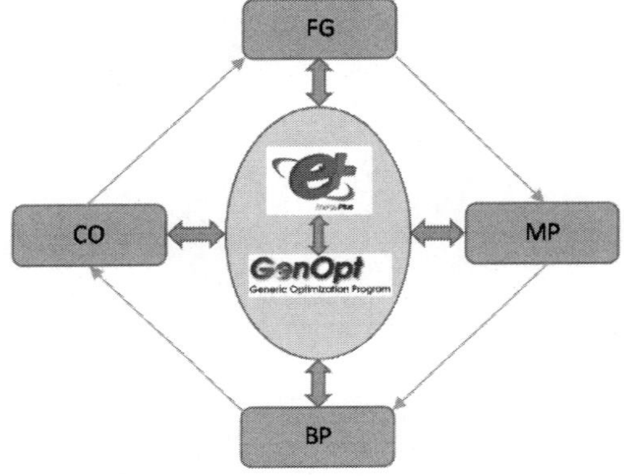

Figure 6. Illustration of predefined sequence between attribute clusters.

3.4 *Performance indicators and cost function*

The selected performance indicators in the present experiment—and the basis for the definition of the cost function U—include the building's annual heating (H), cooling (C), and lighting (L) energy demands. To establish the cost function, we assumed the building's heating systems uses natural gas and has a 85% efficiency. The cooling system was assumed to be electrically driven with a COP of 3.5. Furthermore, we assumed that, for the same energy content, the electricity price is 1.8 times the price of natural gas. This results in the following formulation for the cost function (U). Thereby, H, C, and L are in units of J and U in units of MWh.

$$U = ((H/0.85) + 1.8(C/3.5) + 1.8(L)) / (3.6 \times 10^9)$$

4 RESULTS

4.1 *Random cycling between attribute clusters*

This section includes results (variable values and their evolution) from optimization via random iterations between the aforementioned attribute clusters. Thereby abbreviations BM and GO refer to "Base Model" and "Global Optimization" respectively.

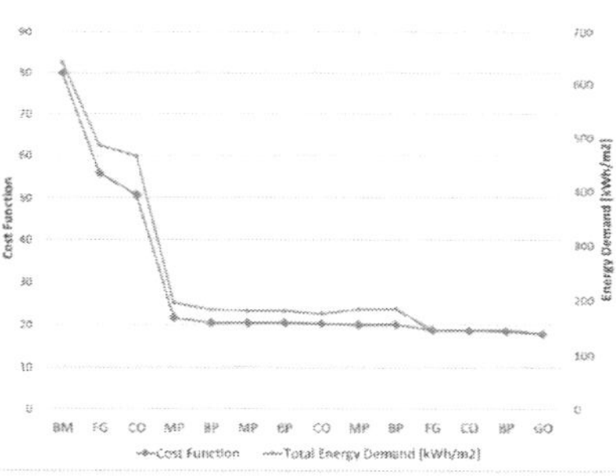

Figure 7. Cost function versus total energy demand (random iteration of the clusters).

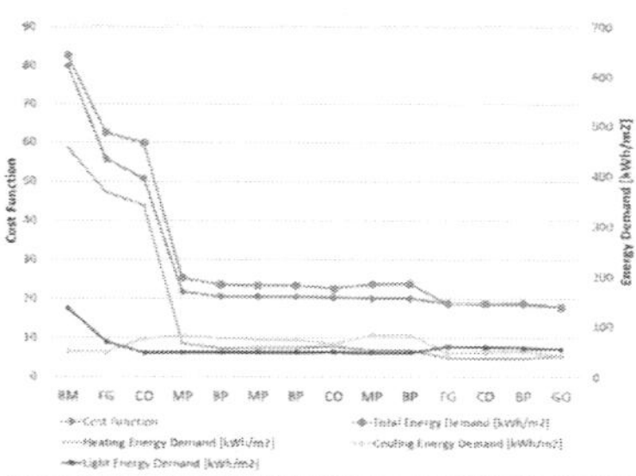

Figure 8. Energy demand evolution (random iteration of the clusters).

Figure 7 illustrates the evolution of the cost function together with the total energy demand in the course of 13 random iterations between attribute clusters. Figure 8 shows the energy demand (heating, lighting, cooling, and total) evolution for the same iterations. Figure 9 provides further details regarding the energy demand at the zone level.

4.2 *Predefined sequence of the attribute clusters*

This section includes results from optimization via predefined cycling between the aforementioned attribute clusters. Figure 10 illustrates the evolution

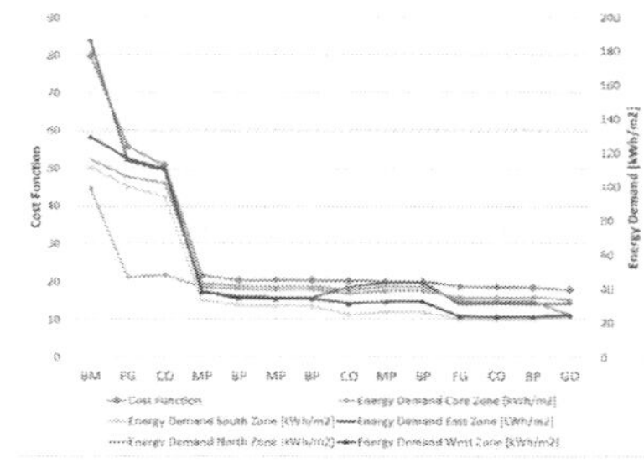

Figure 9. Zones energy demand evolution (random iteration of the clusters).

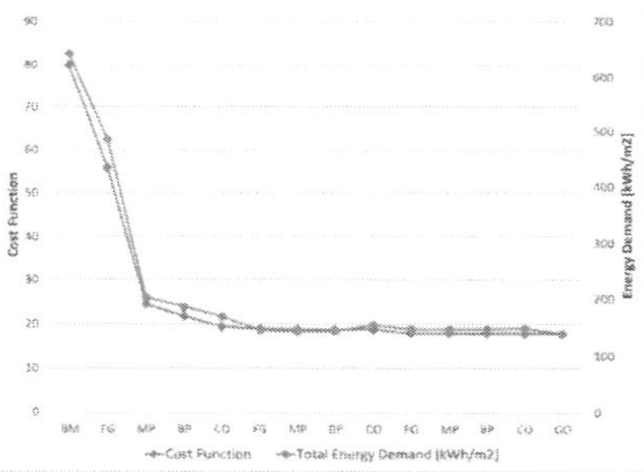

Figure 10. Cost function versus total energy demand (predefined sequences iteration of the clusters).

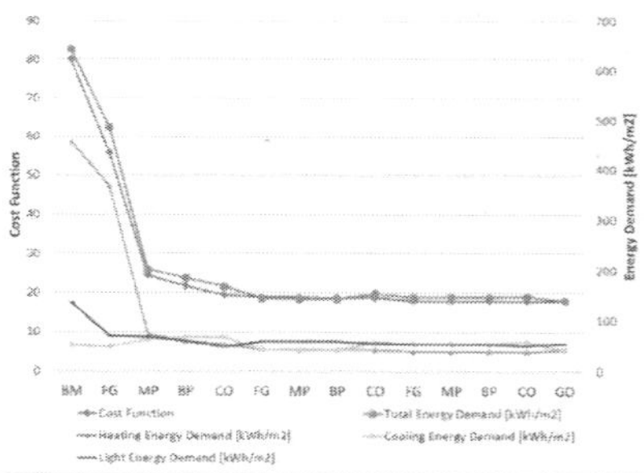

Figure 11. Energy demand evolution (predefined sequences iteration of the clusters).

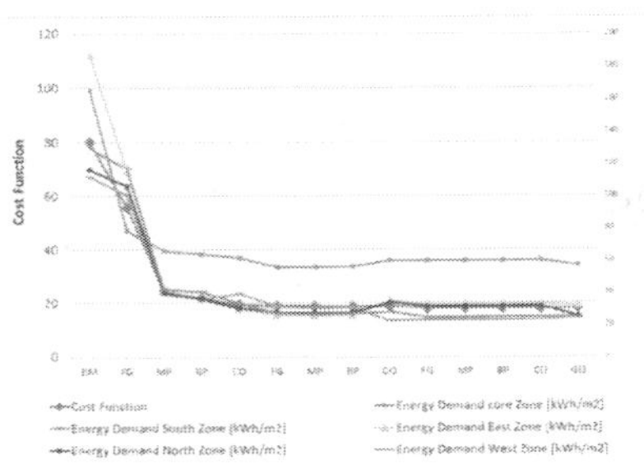

Figure 12. Zones energy demand evolution (predefined sequences iteration of the clusters).

of the cost function together with the total energy demand in the course of 13 iterations between attribute clusters. Figure 11 shows the energy demand (heating, lighting, cooling, and total) evolution for the same iterations. Figure 12 provides further details regarding the energy demand at the zone level.

5 DISCUSSION

Both examined procedures (random and predefined cycling between clusters) display the rapid convergence toward optima, as indicated by decrease in the value of energy use indicators as well as the cost function (see Figures 7 to 12). The bulk of optimization-based design improvement is in fact achieved during the first five iterations. Note that after some 10 iterations, no significant change in variable values is observed. Hence, in the present demonstration, only the first 12 iterations are illustrated. The process was terminated after 13 iterations.

As such, the end values of the individual design variables are not necessarily identical with those in the global optimization scenario. However, after the 13th iteration, the values of the performance indicators (as well as the value of the cost function) are virtually indistinguishable from those obtained in the global optimization scenario. Interestingly, global optimization requires roughly twice as much computational time as the proposed intra-cluster cycling approach.

6 CONCLUSION

We introduced a novel approach toward iterative global optimization of locally optimized attribute clusters of building design solutions. If well-structured, such clusters of design space attributes can be easily comprehensible to typical building designers as a compound yet coherent aspects of a design (e.g., building enclosure, building materials, building geometry, building systems, and control systems). Thus grouped, clusters can be made subject to multiple passes of local simulation-assisted optimizations, instead of a single-pass black-box global optimization step. We provided a proof of concept of the proposed approach via a prototypical implementation using existing simulation and optimization tools.

While not explicitly covered in this contribution, the implementation provides the users with degrees of freedom in view of the selection of the clusters to be optimized. Moreover, additional clusters can be defined and variable sets in each cluster can be manipulated, while still achieving convergence within reasonable temporal horizons.

We illustrated the performance of the current implementation of the proposed approach via an optimization case study, which contained different system operation options (e.g., random cycling between attribute clusters versus predefined sequences).

The results thus far are highly encouraging: The proposed method delivers optimized solutions that are—as far as the values of the energy performance indicators and the associate cost function are concerned—virtually indistinguishable from those of a reference one-shot global optimization run. However, in our approach the results are not only obtained faster and more efficiently, but also via a transparent, traceable, and designer-friendly process.

REFERENCES

Attia S. 2012. Computational optimization Zero Energy Building design: Interviews with 28 international experts. International Energy Agency (IEA) Task 40: Towards Net Zero Energy Buildings Subtask B.

Crawley D.B., Lawrie L.K., Winkelmann F.C., Buhl W.F., Huang Y.J., Pedersen C.O., Strand R.K., Liesen R.J., Fisher D.E., Witte M.J. & Glazer J. 2001. EnergyPlus: creating a new-generation building energy simulation program, Energy and Buildings 33 (2001) 319–331. doi:10.1016/S0378–7788(00)00114–6.

DOE 2015. Input Output Reference. University of Illinois, university of California, Lawrence Berkeley national laboratory, US department of energy.

DOE. 2016. Commercial Reference Buildings: http://energy.gov/eere/buildings/commercial-reference-buildings.

Nguyen, A. Reiter, S. & Rigo, P. 2014. A review on simulation-based optimization methods applied to building performance analysis, Applied Energy 113 (2014) 1043–1058. doi:10.1016/j.apenergy.2013.08.061.

Wetter M. 2001. GenOpt—A generic optimization program, Proceedings of the 7th IBPSA Conference, volume I, pages 601–608. Rio de Janeiro.

Wetter M. 2011. GenOpt, Generic optimization program—User manual, version 3.1.0. Technical report. Lawrence Berkeley National Laboratory.

eWork and eBusiness in Architecture, Engineering and Construction – Christodoulou & Scherer (Eds)
© 2016 Taylor & Francis Group, ISBN 978-1-138-03280-4

BIM-based building design platform—local environmental effects on building energy performances

D. Da Silva, P. Corralles & P. Tournier
CSTB—DEE, Champs-Sur-Marne, France

M. Cherepanova
ENGIE—CRIGEN, Seine-Saint-Denis, France

ABSTRACT: For complex building projects, energy engineers, economists and other domain experts are usually integrated in the design team, to take part in the decision-making process starting from the beginning of the project. Within the current design practices, the iterative design process is difficult to implement, mainly because of the data exchange between the designers and the evaluators is far from optimized thus causing unacceptable delays. We present a BIM-based, on-the-cloud, collaborative building design software platform. This platform will account for all physical phenomena at the building level, while also taking into account external, neighborhood level influences. Within this platform, the paper presents the physical models developed to take into account local environmental effects on building energy performances. Several models are presented and applied to two building types (residential and office building). The results allow drawing some impact quantification of these new parameters in building energy consumption. The paper exposes the application of this platform as a tool to derive optimal building design under different constraints and gives perspectives for further works and developments.

1 INTRODUCTION

Recent urbanization has put the emphasis on the energy and environmental quality of urban buildings. For complex building projects, energy engineers, economists and other domain experts are usually integrated in the design team, to take part in the decision-making process starting from the beginning of the project. Within the current design practices, the iterative design process is difficult to implement, mainly because of the data exchange between the designers and the evaluators is far from optimized thus causing unacceptable delays. These problems and the neighborhood design approach and the BIM development lead us to review building design.

The goal is to design, develop, and demonstrate a BIM-based, on-the-cloud, collaborative building design software platform that enables the user to access, modify and test solutions under different constraints. The optimization of the building design is then based in different segments, i.e. energy, environmental, acoustic, renewable energy and economics.

The energy consumption patterns of buildings located in dense city centres are highly dependent on the surrounding urban neighbourhood, compared to the low density, suburban/rural regions, where the building energy consumption patterns are similar to an isolated building energy consumption patterns. The design of sustainable neighbourhoods and restoring communities require a full scale modelling of urban neighbourhoods that is not consistently replicable with the current computational approaches.

In this paper, we deal with the question "How to take into account the local environmental effects on building energy performances?". The building design is often focused in the building itself, discarding the occupants behavior, surrounding environment and its microclimatic context.

One example is the use of an average meteorological weather file from an airport or from a rural area next to the city where the building is.

The building interacts with the environment through the exterior temperature, solar radiation, wind, etc. The effects of these parameters can have a high impact in the building consumption (Francisco Sanchez de la Flor, 2004).

The simulation of the neighborhood-level parameters is made by taking, in one hand, in statistical and stochastic modeling for occupant behavior and, on the other hand, measures-derived models to simulate the building surrounding environment. The tools developed must be adapted to the different building design actors constrains, i.e. low number of inputs, fast-computing time and be compatible with the existing software.

In detail, this paper presents an accurate and simple method to take into account local air temperature and wind variations, the impact of albedo and masks for energy evaluation. These environmental parameters impact are analyzed by taking into account two examples.

2 BIM-BASED, ON-THE-CLOUD, COLLABORATIVE BUILDING DESIGN SOFTWARE PLATFORM

Within the current design practices, the iterative design process is difficult to implement, mainly because of the data exchange between the designers and the evaluators (i.e. based on 2D or 3D drawings and text-based specification documents) is far from optimized thus causing unacceptable delays. The development of a platform that account for all physical phenomena at the building level, while also taking into account external, neighbourhood level influences could solve these problems. The HOLISTEEC project[1] is developing this platform. The design of this platform will rely on actual, field feedback and related business models/processes, while enabling building design and construction practitioners to take their practices one step forward, for enhanced flexibility, effectiveness, and competitiveness.

In more detail, this platform is a holistic multi-physical simulation engine, able to capture and assess in an integrated way building performances in various dimensions at building and neighborhood level: energy, environment, acoustics, and lighting.

Each design phase can be seen as an iterative loop over three distinct activities, as shown in the picture below:

- design activities, creating more detailed information concerning the future building and the way it will be constructed,
- simulation activities, updating the user defined objectives forecast values based on the available project information,
- project evaluation activities in which the user defined objective checklist is used as a criterion to decide whether to accept the current design and to move on to the next project phase, or on the contrary, to launch a new round of design/simulation activities.

The innovation is that the shared information model (BIM+NIM—Neighbourhood Information Model) shall reflect at all time the status of the design, allow for updates as soon decisions are taken and for instant assessment of these decisions.

[1]European Projet—Holistic and optimized life-cycle integrated support for energy-efficient building design and construction.

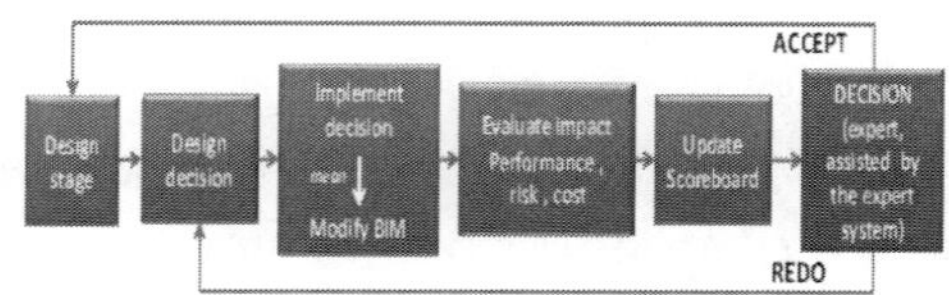

Figure 1. Schematic description of the early feedback loop HOLISTEEC approach.

This innovative design workflow will help fulfilling critical requirements from building design stakeholders, and actually support them in making the right choices at important decision points.

The present study represents one of the objects treated in this platform. A focus is made on how the local environmental effects influence the building energy performances. Next, we present the different analysed parameters.

3 BUILDING SURROUNDING ENVIRONMENT

With the aim of including neighbourhood environment parameters into the thermal simulation of buildings, we developed simple and outside-the-simulation-engine modules, so that these modules could be used with different building software's. Next we present the developed modules.

3.1 *Wind speed in urban canyon*

The wind flow inside canyons is driven and determined by the interaction of the flow field above buildings and the uniqueness of local effects as topography, building geometry and dimensions, streets, traffic and other local features.

The general idea of this module is to modify the wind conditions in order to take into account the canyon effect in building energy simulation.

The urban canyon module developed is based in works of Georgakis and Santamouris (Georgakis & Santamouris, 2008).

The computational methodology to calculate the wind speed in canyons is presented in the flow-chart below as a function of the wind speed and canyon geometry.

For building energy simulations the energy impact of wind speeds lower than 4 m/s can be neglected. Thus, in this module the highly complex models ("Part B of this paper" in the flow-chart) were not used.

The next figure shows an example of the canyon wind effect on the wind speed.

3.2 *Urban heat island effect*

Urbanization produces higher air temperatures in cities than in the undeveloped rural surround-

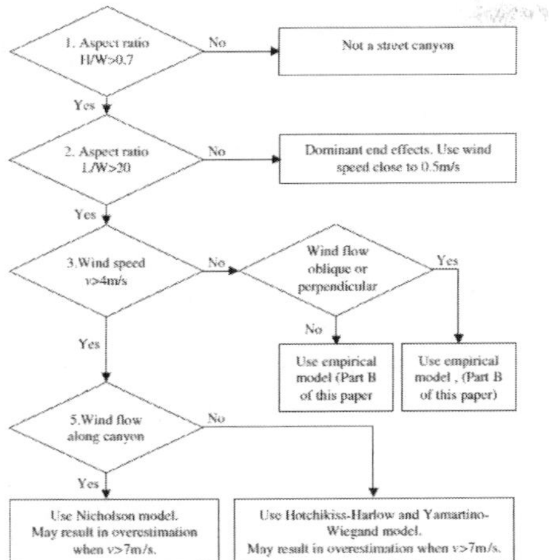

Figure 2. Flow-chart of the algorithm for estimating wind speed inside street.

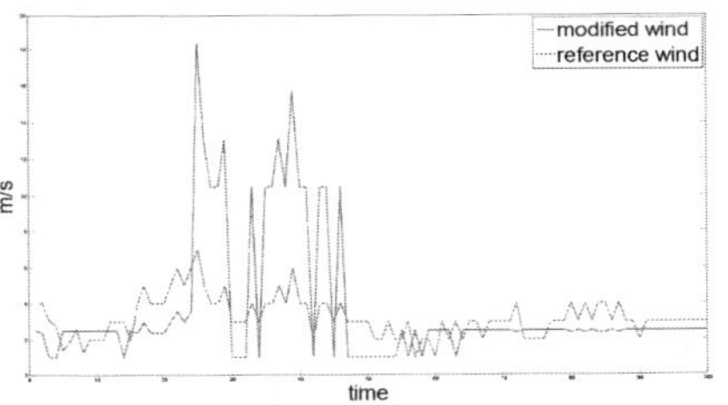

Figure 3. Wind speed with and without a canyon effect.

ings. This phenomenon, known as the Urban Heat Island (UHI) effect, is usually more intense under cloudless sky and light wind conditions and is mainly caused by the geometric and construction differences between the urban and rural surfaces and the anthropogenic heat released in the urban environment. The UHI effect can increase cooling energy consumption of buildings but can also reduce the heating energy consumption in winter.

As the major part of the meteorological data is obtained in rural areas, in order to take into account the UHI effect, a module was developed that modifies the external temperature of the meteorological file.

A large number of scientific articles cover this subject. However the majority of these articles deal with the effect of the building materials in the heat/cooling consumption or with urban canopy models for atmospheric models. The urban heat island physical models are very complex due to the large number of parameters needed.

Thus, the developed Urban Heat Island (UHI) module is based in field measures from the works of Oke, Kolokotroni and Giridharan (Giridharan & Kolokotroni, 2009) (Kolokotroni & Giridharan, 2008) (OKE, 1982).The UHI module does not take into account the thermal characteristics of the city like pedestrian pavements, tiles, road pavements to calculate an accurate albedo. The presence of green spaces is not taken into account in this study.

The works of Kolokotroni and Giridharan and Oke show that the UHI effet is correlated with the daily solar radiation, wind velocity and building position in the city (core, semi-urban and urban). The scheme presented below show the module procedure to modify the external temperature measured in a rural space next to the city.

The radiation indices ("ind_rad") are derived from the mean global horizontal radiation.

If the average wind velocity is lower than 5 m/s the urban heat island are considered maximum. If the values for the wind velocity are higher the urban heat island effect is considered to be ¼.

The values of the building height are derived from OKE study. The coefficients for building height ("*ind_height*") are:

- Average Building height < 50m—ind_height =1
- 50 < Average Building height <100m—ind_height =0.5
- Average Building height > 100m—ind_height =0

The measured values used to determine the UHI indicators and the daily temperature profiles are after the London city data from city centre to urban zones (Table 2). As no other measures where available, in this module we make the hypothesis that for other big cities the values are close to this ones.

An example of the temperatures between rural and city center, when the UHI module is used, is presented in Figure 5.

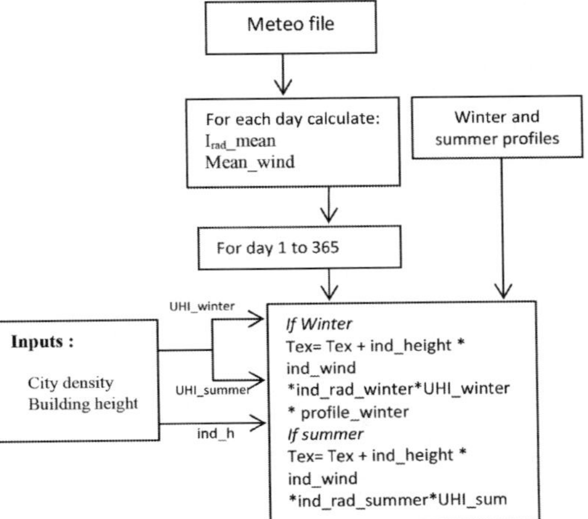

Figure 4. Module calculation scheme.

3.3 *Occupants behavior simulation*

The occupant behavior has a high impact on building the energy consumption. However, it is very difficult to predict the occupancy and the related activities. In order to take a more profound approach, we used a stochastic model 'Qiriel-Croniq' based on French national studies (Ansanay-Alex, Abdelouadoud, & Schetelat, 2016). This model alows the simulation of the occupants profiles and their activities (household equipements heat losses). This model was only used in the simulations of residential buildings as the office buildings section is under developpement.

Table 1. Sky indices used in the module derived from (Giridharan & Kolokotroni, 2009) (Kolokotroni & Giridharan, 2008).

Ind_rad	Below 300 W/m² – sky covered.	Between 300 and 500 W/m² – sky partiatly covered	Superior to 500 W/m² – clear sky
Winter	0.4	0.5	1
summer	0.5	0.8	1

Table 2. Urban heat island intensity values.

	UHI_winter	UHI_summer
Rural	0	0
Urban	1.2	1
Semi-urban	2	2.5
City center (core)	3	3.5

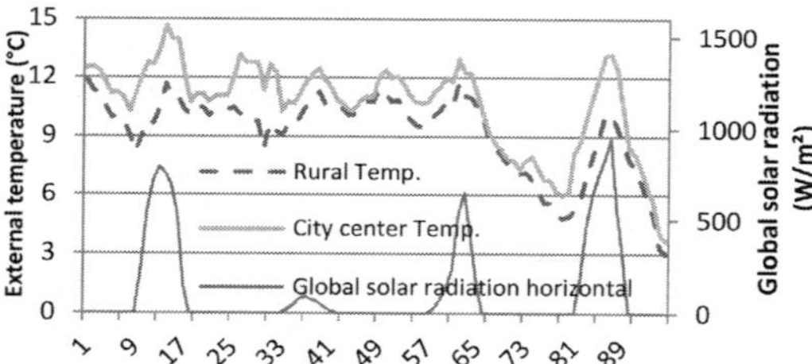

Figure 5. Difference between rural and city center when using the UHI module.

3.4 *Other building model parameters for urban environment simulation*

Some other parameters do not need the development of a complementary module as the building energy software's already take them into account. The list of the other parameters is presented below:

- the building height (low building = 10 m to very high building = 80 m): just the building height parameter is changed (the building geometry remains the same) to determine the models sensitivity to this parameter.
- the azimuthal masks (in order to simulate surrounding buildings masks): The buildings are simulated taking into account that the surrounding buildings have the same height and they are separated by a road. Note: The simulated buildings already present several shading devices.
- the exterior surfaces albedo (from albedo = 0.05 to 0.5).

4 SELECTED STUDY CASES

The objective of this study was to perform an analysis of different building model parameters to determine their significance on building energy consumption. All the calculations were made with the energy simulation software COMETh (Da Silva, et al., 2016).

Two different buildings (Table 3) have been selected to test the different modules and to perform a sensitivity analysis; a collective residential one and an office building, both are low-energy buildings (according to the French building reg-

Table 3. Resume of building characteristics.

	Residential building	Office building
Area (m²)	2116	8755
Leaking air flow rate	1 m³/h/m² (under 4 Pa)	1.7 m³/h/m² (under 4 Pa)
Ventilation system	Single-flow mechanical ventilation by extraction (1595 m³/h).	Double-flow mechanical ventilation (14925 m³/h) and the a single-flow mechanical ventilation (1500 m³/h).
Heating/cooling system	Collective condensing gas boiler with 95% efficiency (PCI) No cooling	Condensing gas boiler with efficiency of 96% / heat pump (cooling only) with EER coefficient of 2.8.
Total energy consumption (heat, cooling, light, ventilation)	55.42 kWh/m²	78.26 kWh/m²

Table 4. Bad and good cases for the two building types.

	Residential building	Office building
"Bad" case	Heating: conventional boiler	Heating: conventional boiler Cooling: EER = 2
"Good" case	Heating: high efficient (97%) boiler (condensing)	Heating: high efficient (97%) boiler (condensing) Cooling: EER = 3.5

ulation—RT2012). The meteorological file corresponds to the climate zone—H1a (in a rural area near to Paris).

The above system package is defined as "base case". We have defined 2 additional cases (Table 4) for these buildings a "bad" case and a "good" case. These cases will allow seeing how the building responds to the different environmental conditions.

5 RESULTS

The maximum individual impact of each variable (maximum variation of the results without signal) in building energy consumption is presented below.

The impact of the canyon effect and of the relative height of the building is low for the two buildings because these parameters will principally affect the heating and cooling consumptions and it is dependent of the infiltration rate. The consumptions and the infiltration rates are low (Table 5) as these buildings are very efficient ones.

The impact of the infiltration rate is very important. For example, if the infiltration rate of the residential building is set to 4 $m3/h.m^2$, the impact of the canyon effect is 14.5% of the total building consumption. For buildings where the major consumption posts are related with heating and cooling, the canyon effect can have an even higher impact.

The other parameters albedo and masks will also impact the lightning calculation so this phenomenon is reduced.

For the residential building the most sensible parameters are the UHI (with −7.97%) and building height (with −4.71%). As it is an efficient building the values of the UHI impact show that the heating is highly modified.

The results for the office building show that only the UHI effect (with 8.25%) and the albedo (with 7.50%) have a significant effect in building energy consumption. The albedo impact is due to

Table 5. Heating and cooling consumptions.

	Heating consumption	Cooling consumption
Residential (infiltration = 1 $m^3/h.m^2$ under 4 pa)	24.3 kWh/m^2	NA
Office (infiltration = 1 $m^3/h.m^2$ under 4 pa)	7.1 kWh/m^2	12.2 kWh/m^2

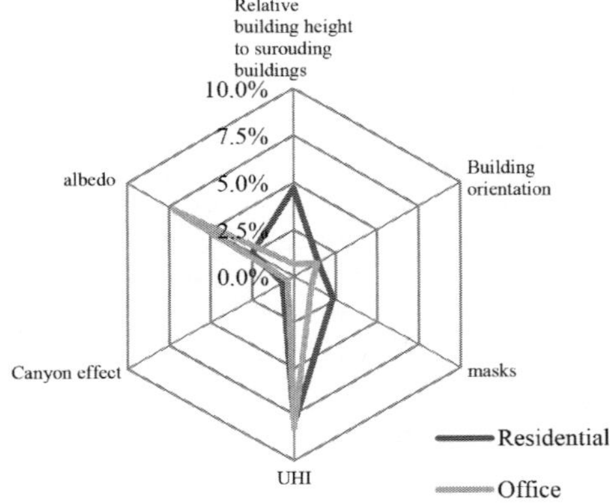

Figure 6. Maximum impact on building consumption per building surrounding parameter.

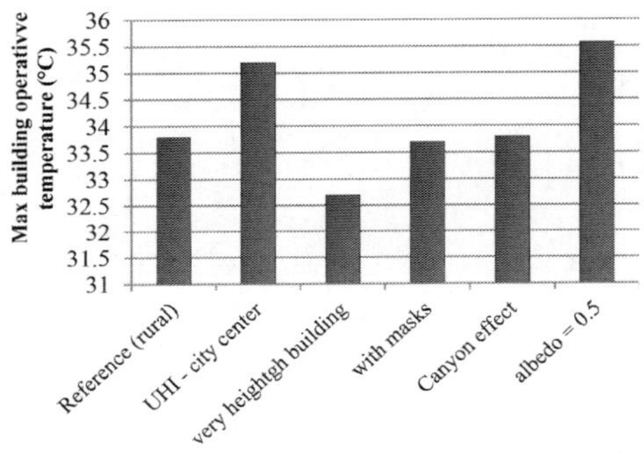

Figure 7. Maximum building operative temperatures for the different configurations.

high consumption of lightning of the building. The UHI effect is even higher than the residential building because of the cooling consumption.

As the residential building is not equipped with a cooling system, we analyzed the maximal reached operative temperature in the building.

The maximum operative temperature can be modified of −1.5 to +1.8 °C for the different simulated configurations.

6 DISCUSSION

The impact of shading by other buildings is small. However the tested buildings present several shading devices. As the tested masks do not shade the sun during summer (high solar altitude angles), the effects in summer comfort and energy consumption (cooling in the office building) are low. The BIM-Based platform should be able to take into account the surroundings geometry.

In the presented case the building orientation and design were fixed. Despite, the low effect of building orientation, this variable should be retained, as during the building design the window position can influence the building consumption.

Regarding the canyon effect, its impact in energy consumption is very low for these two buildings. However further tests should be made especially for building with different infiltration rates and high heating and cooling consumptions.

The impact of the UHI, albedo and building height are related. The surrounding albedo will disturb the UHI effect. If the building height is very different from surrounding buildings the UHI will be lower. The UHI model allows a simple quantification of the building position in the energy consumption. The developed model is based on a measure campaign done for the city of London (UK). Thus, to ensure the reliability of this model, it should be compared with more detailed models and measures in different conditions to validate / improve the current model and enable a correct building design evaluation.

Examples of the integration of these simulation modules to the BIM-based platform should allow deriving optimal building design in a neighborhood environment.

7 CONCLUSION AND PERSPECTIVES

This paper presents a new design platform that allows a multi-domain building design optimization, where during all the design status updates to the project are taken into account directly. The platform includes different domains of study (energy, economics, environment…). The paper develops a part of this platform which concerns the influence of the environmental effects on the building energy performance.

Several modules where developed to represent the building environment and applied to two buildings. The impact of the building surrounding environment is majorly due to the UHI effect and to the albedo of the surrounding. The canyon effect does not have a high impact for these two buildings due to low infiltration rate. However, a special attention should be made to this phenomenon for buildings with high infiltration rates.

The results show that the impact of each parameter is very different for the two building types. The results also show, briefly, that the impacts on summer comfort can also be non-negligible.

Due to the platform structure, these modules should be very flexible and simple. However, to correctly estimate the energy consumption, validation/improvement of the current modules should be done.

An example of the application of this platform will allow deriving optimal building design for different project objectives (economical and/or energy and/or environmental…).

ACKNOWLEDGEMENT

This study was partially funded by European Union Seventh Framework Programme FP7 2007-2013 under Grant Agreement n. EeB.NMP.2013-5-609138. The contents of this publication are the sole responsibility of the authors and don't necessarily reflect the views of the European Commission".

REFERENCES

Ansanay-Alex, G., Abdelouadoud, Y. & Schetelat, P., 2016. Statistical and Stochastic modelling of French Households and their energy consumption activities. *Clima 2016,* p. 9.

Da Silva, D. et al., 2016. Evaluation et perspectives du modèle thermique de COMETh, le cœur de calcul de la réglementation thermique des bâtiments neufs. *IBPSA.*

Francisco Sanchez de la Flor, S.A.D., 2004. Modelling microclimate in urban environments and assessing its influence on the performance of surrounding buildings. *Energy and Buildings,* Volume 36, pp. 403–413.

Georgakis, C. & Santamouris, M., 2008. On the estimation of wind speed in urban canyons for ventilation. Volume 43, pp. 1668–1682.

Giridharan, R. & Kolokotroni, M., 2009. Urban heat island characteristics in London during Winter. Volume 83, pp. 1668–1682.

Kolokotroni, M. & Giridharan, R., 2008. Urban heat island intensity in London: An investigation of the impact of physical characteristics on changes in outdoor air temperature during summer. Volume 82, pp. 986–998.

OKE, T., 1982. The energetic basis of the urban heat island. 108(455).

Xiaoshan Yang, L.Z.M.B.Q.M., 2012. An integrated simulation method for building energy performance assessment in urban environments. *Energy and Buildings,* pp. 243–251.

eWork and eBusiness in Architecture, Engineering and Construction – Christodoulou & Scherer (Eds)

Intelligent emergency exit signage system framework for real-time emergency evacuation guidance

J. Zhang & R.R.A. Issa
Rinker School of Construction Management, University of Florida, Gainesville, USA

ABSTRACT: Light-based single-function emergency exit signs may lead occupants to a fire-blocked route or a crowded exit resulting in a delayed evacuation and even more serious situation during an emergency evacuation. This paper will propose a real-time emergency evacuation sign guidance system which guides occupants to safe and fast evacuation routes considering fire source location and development trends, and human traffic flow trends in real time. The proposed system includes a fire detection sensor network system monitoring fire source and propagation, an occupant sensor network system evaluating the existing occupant traffic flow, and intelligent exit sign systems showing fire-free and crowd-free route directions. All fire detection sensor units communicate with each other to broadcast the danger location information. The occupant sensor network system detects the traffic flow coming in and going out for each direction, and detects and predicts crowd areas based on the traffic flow trends. The intelligent exit sign systems generate an evacuation strategy in response to the signals from the fire detection sensor network and occupant sensor network, and show the recommended directions to evacuate and the directions not suitable to evacuate. If crowded situation occurs, the exit sign systems also will show the directions to the fastest evacuation routes. The proposed system will generate first-hand evacuation sign guidance based on real-time emergency situations to achieve safe and fast evacuation. The installation of the proposed system will improve building evacuation performance and reduce injuries and fatalities.

1 INTRODUCTION

It is a key aspect of the design and operation of any building or facility to provide for life safety during fire emergency events. Although annual civilian deaths in structure fires has been dramatically reduced since 1997, statistics reports show little improvement from 2002 to 2013: about 3000 civilian deaths per year (Karter 2013). In addition, the enormous indirect costs associated with fires including temporary lodging, lost business, medical expenses, psychological damage, and others, are much higher. These statistics provide a stark illustration of the need to further improve fire escape even after the many fire protection and fire engineering measures have been implement in buildings from various aspects, e.g., materials, fire resistant structures, fire extinguisher facilities and so on.

Given the loss of lives and properties in fire disasters, the study of evacuation has become extremely necessary and important. Inefficient evacuation will result in a large number of casualties if fire occurs in a large place. Fire exit signs form one of the most important parts of any evacuation guidance system, they are traditionally designed to correctly mark the most efficient escape routes. Unambiguous guidance of the relevant fire exit signs will assist evacuation in times of emergency.

For example, for the evacuation of buildings typically floor-plans and escape signs attached to the walls show the nearest available building exits to be used when an emergency situation arises. After an incident occurs, the expectation is that the occupants look at the evacuation plan (or escape signs) and reach the nearest exit. However, the

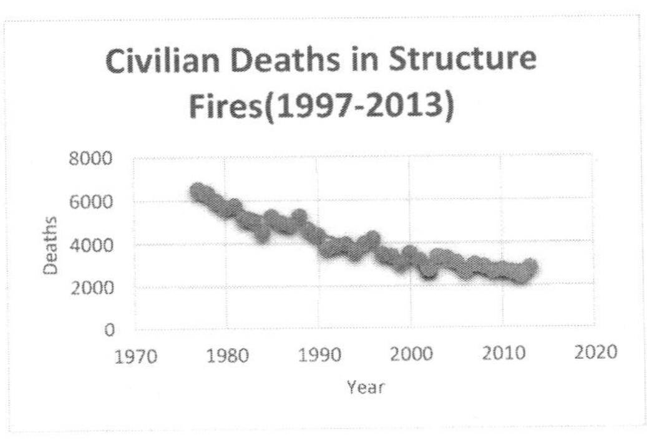

Figure 1. Civilian deaths annually in structure fires from 1997–2013.

actual effectiveness of any pre-set evacuation plan can be limited by several issues: the impossibility for occasional visitors (e.g., customers) to know (or look at) the evacuation plan; the unpredictability of human behavior in panic conditions; the lack of information about the type of emergency, the occupant traffic congestion situation inside the building. Current evacuation signage guidance systems suffer from rigid guidance to the closest exits regardless of the existence of fire blockage or serious traffic congestion. Moreover, they may not respond to ad hoc hazard development. with large buildings and huge complexes. There is a need to improve the building signage guidance systems to direct people to the safest way to reach an exit. Prior state of the art systems typically suggested only the shortest escape path, which may actually get people closer to the most dangerous or hazardous locations in the populated area. However, the safest escape route can dynamically change according to people locations and hazard locations. Multiple parameters may influence the calculation of the safest exit path. Examples of such parameters are: location of hazards, number of people to evacuate and their travel speed, wayfinding strategy. For this purpose, sensors distributed in or around the building can help in gaining a better understanding of the real-time situation and allow intelligent evacuation.

Light-based single-function emergency exit signs may lead occupants to a fire-blocked route or a crowded exit resulting in a delayed evacuation and even more serious situations during an emergency evacuation. This paper will propose a real-time emergency evacuation sign guidance system which guides occupants to safe and fast evacuation routes considering fire source location and development trends, and human traffic flow trends in real time. The proposed system includes a fire detection sensor network system monitoring fire source and propagation, an occupant sensor network system evaluating the existing occupant traffic flow, and intelligent exit sign systems showing fire-free and crowd-free route directions. All fire detection sensor units communicate with each other to broadcast the danger location information. The occupant sensor network system detects the traffic flow coming in and going out for each direction, and detects and predicts crowded areas based on the traffic flow trends. The intelligent exit sign systems generate an evacuation strategy in response to the signals from the fire detection sensor network and occupant sensor network, and show the recommended pathways to use for evacuation and those not suitable for evacuation. If a crowded situation occurs, the exit sign systems also will show the directions to the fastest evacuation routes. The proposed system will generate first-hand evacuation sign guidance based on real-time emergency situations to achieve safe and fast evacuation. The installation of the proposed system will improve building evacuation performance and reduce injuries and fatalities.

2 RELATED WORK

In this section, related work of path way finding to guide occupants to exits during emergency evacuation is examined.

Many studies demonstrate how the use of efficient wayfinding systems could significantly decrease the occupants' egress time (Jensen 1998; Jeon and Hong 2009; Vilar et al., 2014, Occhialin et al. 2016). They represent the easiest and most effective way to assist occupants. The traditional exit signage systems guide occupants to the closest exits during the evacuation process especially in particular environmental conditions (e.g. darkness, smoke) or when people are not familiar with the building (Benthorn and Frantzich 1996). Wayfinding signs must provide an immediate identification of escape routes and exits and guarantee a quick comprehension of all that information they make available in different environmental situations. Following these guidelines, the needed time for the best evacuation paths choice should decrease and the evacuation process in terms of motion speeds should also accelerate (Wong and Lo 2007). Safe condition signs and exit signage systems include: reflective signs (BSI 2000), photo-luminescent (PLM) signs (DIN 2002; Tonikian et al., 2006; UNI 2004), electrically-illuminated signs (BSI 2000), interactive wayfinding systems (Ran et al. 2014), and acoustic wayfinding systems (Nilsson and Frantzich 2010). The related effectiveness is influenced by the pedestrians' perception in relation to the signs position and environmental conditions (Kobes et al. 2010a). The accurate design of evacuation facilities layout (including plan positions and elevation from the ground) cannot exclude investigations on their use by the pedestrians. Experiments about identification distance (Wong and Lo 2007; Tuomisaari 1997) and influence on motion in terms of total evacuation time and speed are performed so as to define the wayfinding systems effectiveness (Jeon et al.2011; Proulx and Bénichou 2010). In particular, when signs (mainly: low placed exit signs) are clearly visible (e.g. electrically illuminated, PLM) in smoke conditions, people statistically tend to follow their directional indications and to use the suggested shortest evacuation paths (Kobes et al. 2010b; Kobes et al. 2010c). Studies on evacuation behavior attempted to understand the wayfinding in virtual game environments (Zhang and

Issa 2015). The most robust wayfinding systems is composed of PLM signs (Proulx and Bénichou 2010, Proulx et al. 2000; Kyle and Creek 1999) because of:

- no power supply (no interventions on building structure and layout),
- easy installation and uninstallation,
- low level of maintenance,
- high efficiency regardless of light conditions.

Requirements of PLM signs are also defined by regulations (DIN 2002; UNI 2004). Existing buildings scenarios have been tested and the results showed that PLM signs are able to guide the occupants towards the correct exit and to increase their evacuation speed in both corridors and stairs (Proulx and Bénichou 2010). Wayfinding systems are mainly divided in continuous (at least 1 directional sign per 5 m (~16 ft.) of path) and punctual (mainly placed at intersections and exits) applications (ISO 2004; Paulsen 1994). Applications on existing buildings are generally based on punctual systems. Tests with smoke conditions simulation are also performed (Jeon et al. 2011; Jeon and Hong 2009). Mainly, speeds sensibly increase (about +15% when using continuous PLM on the floor in respect to punctual guidance light) along the horizontal path, while the evacuation time decreases (up to about 65%) because people avoid wandering and more rapidly find the right direction (Jeon and Hong 2009). The positive effect on speed is more effective when directional arrows are closely placed (Jeon and Hong 2009). In staircase motion, despite high pedestrian densities, noticed speeds are higher than the predicted ones (up to about +15%) (Proulx et al. 2000). Finally, questionnaire responses qualitatively underlined that the evacuated people assessed a good environment visibility perception of the system in relation to lighting (70% of people defined as "acceptable" or "good" the PLM scenario without emergency lighting) (Proulx et al. 2000). Researchers has tempted to develop innovative emergency signage systems. Galea et al. (2016) developed an intelligent active dynamic signage system (IADSS) which could show the directions which are not shut down if needed. Instead of signage guidance systems, other researchers have attempted to develop indoor navigation systems using mobile devices (Inoue et al. 2008; Renaudin et al. 2007; Chittaro et al. 2008). Barnes et al. (2008) presented a distributed algorithm to direct occupants to exit using wireless sensor networks. However, intelligent emergency signage guidance system to clearly direct fast safe escape path has not been developed with consideration of real-time fire propagation and occupants' traffic congestion information.

3 PROPOSED SYSTEM

The proposed system consists of exit signage panel, fire and smoke detection sensor, and human traffic monitor sensor which are integrated as one single node to build up a graph network. If there exists a path for two nodes, an edge exists between these two nodes. If any node detects a serious hazard (fire or smoke), or is damaged during the event, all its edges will be disconnected.

3.1 *Exit signage unit description*

The proposed intelligent exit signage system consists of exit signage units as nodes in a graph network. Each unit consists of an exit sign panel with direction arrows, fire and smoke detection sensor, and occupant traffic monitoring sensor as shown in Figure 2.

Fire and smoke detection sensors detect the fire events and fire propagation and send warning information to the neighboring units. The occupant traffic monitoring sensors count the number of passing occupants and measure their speed. The sensing capability of each hazard sensor is not explored in this study. The operation of these devices is not discussed in this paper. To facilitate the operation of the algorithm, the data from the hazard and navigation graphs must be distributed throughout the network. The direction arrows on the exit signage panels have three colors to show the evacuee different information according to real-time traffic information and hazard location. The idea of color signals in the proposed exit signage unit is inspired by the general traffic signal system as shown in Figure 3.

3.2 *Network deployment*

Exit signs panels integrated with hazard sensors and monitor sensors are to be deployed throughout a building to maximize coverage of hazard detection. Information communication between neighboring units is required to form a real-time network. The neighboring units are defined as

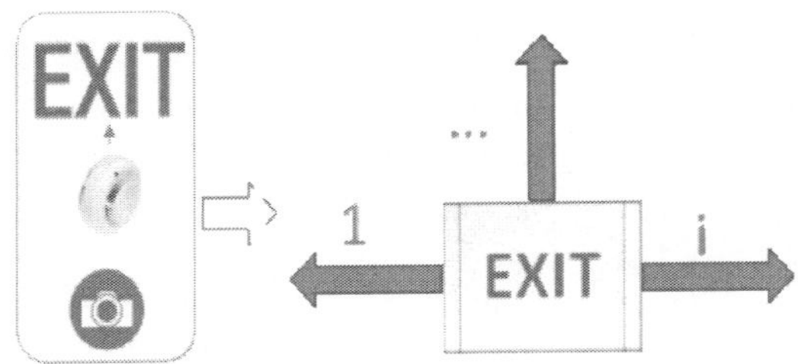

Figure 2. Node elements and description.

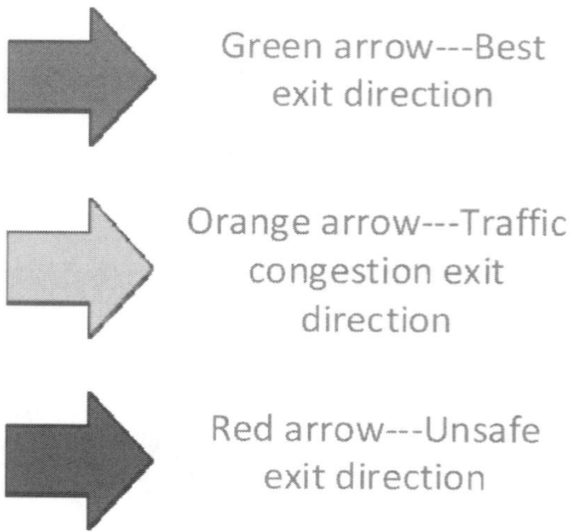

Figure 3. Direction arrow guidance with colors.

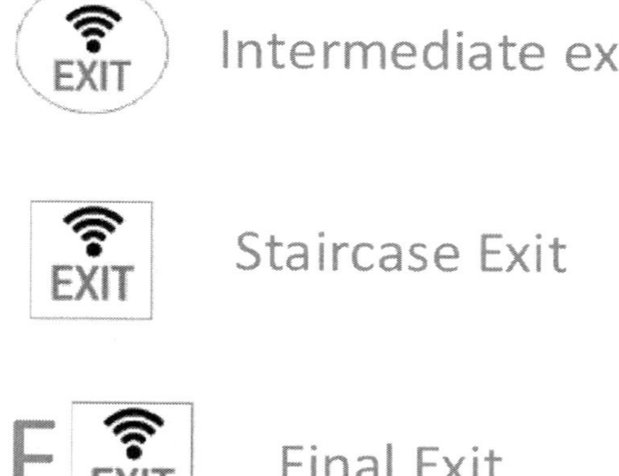

Figure 4. Node type in the graph network.

the units can be visually reached from each other without blocking building elements. The proposed framework uses the 3D building layout to construct a graph G in which different types of nodes indicate the location distribution difference, as illustrated in Figure 4. Final exits will be considered as the evacuation targets.

Each link between two neighbor nodes denotes the evacuation section from one exit signage position to another position in the building. To determine the evacuation time from node i to node j, the following steps need to be considered:

a. Average travel/walk time, Tt(i,j)

   Tt(i,j) = d(i,j)/v

   d(i,j): the distance between two exit signage position;
   v: occupant's walking speed, obtained from the occupant traffic monitoring sensors, or use the historical data from previous observation.
b. Queuing time if any, Tq(i,j)

   Tq(i,j) = N(i,j)/F(i,j)

   N(i,j): Number of occupants from i to j;
   F(i,j): traffic flow
c. Update of the queuing time to the graph network.

Since Tt(i,j) gives the total time to walk from node i to node j if there was no queuing, the additional of Tt(i,j) and Tq(i,j) would give a larger evacuation time than needed for the occupant to reach the exit. The network updates the Ad hoc queuing times, naming them "real time to queue" to consider the occupant traffic congestion, which may overwhelm time delay even it is on the shortest path.

For each unit in the network, the algorithm to perform real-time information communication with its neighbors is shown in Figure 6.

3.3 *Fastest (least cost) path finding*

Instead of using the shortest path algorithm, the proposed framework uses the fastest path finding algorithms to consider the real-time traffic congestion and danger fire propagation locations. The framework starts to work from the fire emergency alarm, then a dynamic graph with exit signage units will be built up according to the unsafe fire locations. The information communication then starts from the final exits in the building, the initial evacuation times for all final exits are set to

Figure 5. Evacuation time components.

```
FOR each node
  FOR each direction
    Send information
      IF it alarms
      Disconnect all its horizon-
tal edges, show red arrows
      END IF
    Receive information
      IF this neighbor alarms
       Disconnect this edge, show
red arrow to this direction.
      ELSE IF it's the best path
          Show green arrow to this
neighbor
      ELSE
       Show orange arrow as warn-
ing of delay
      END IF
  END FOR
END FOR
```

Figure 6. Information communication algorithms.

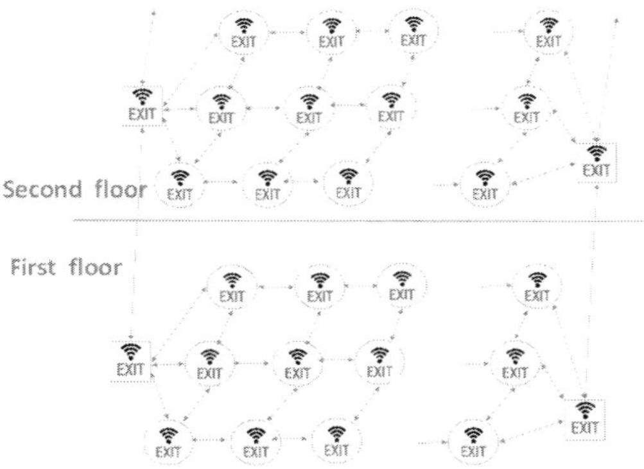

Figure 7. Exit signage network framework.

be 0, then the travel time and queuing time to the other nodes from bottom floors to top floors are updated. The staircase exit nodes only communicate with the neighboring staircases and the nodes on the same floor as illustrated in Figure 6. Each node will store the minimum evacuation time to the final exits, and show the corresponding directions to the best evacuation path, and show the warning signal if any direction is not safe to evacuate to or the evacuee has to delay their escape.

Figure 7 shows the floor of a building and the corresponding graph. The vertices of the graph correspond to locations where people can congregate (e.g. rooms, corridors, doorways or hallways). A link between two vertices of the graph represents a path that can be followed by the evacuees.

4 CONCLUSION AND FUTURE RESEARCH

The proposed exit signage guidance system that can be used during an emergency situation inside a building to support occupant decision-making to find the fast safe escape path. The system is able to function in real-time, adapt to the changes of the hazard situation and occupant traffic situations, and provide reliable suggestions to the evacuees regarding the direction of the best available exit.

By following the directions provided by the exit signage guidance system, occupants can evacuate the building using the best available paths while avoiding the hazardous areas and congested escape paths.

Optimization of exit signage unit deployment in the building need further study to make it compliant with current building standards as well as to reduce the evacuation time. The current framework focus on normal human evacuation, Assistance for disabled occupants with reduced travel speed, or occupants with difficulties for color signal recognition should be considered in future research.

REFERENCES

Barnes, M., Leather, H. & Arvind, D.K. 2007. Emergency evacuation using wireless sensor networks. In *Local Computer Networks, 2007. LCN 2007. 32nd IEEE Conference on* (pp. 851–857). IEEE.

Benthorn, L. & Frantzich, H. 1996. Fire alarm in a public building: How do people evaluate information and choose evacuation exit?. *LUTVDG/TVBB-3082-SE.*

British Standards Institution 2000. BS 5499–4:2000 - Safety signs, including fire safety signs, Code of practice for escape route signing.

Chittaro, L. & Nadalutti, D. 2008. Presenting evacuation instructions on mobile devices by means of location-aware 3D virtual environments. In *Proceedings of the 10th international conference on Human computer interaction with mobile devices and services* (pp. 395–398). ACM.

DIN, DIN 67510, Photoluminescent pigments and products.

D'Orazio, M., Longhi, S., Olivetti, P. & Bernardini, G. 2015. Design and experimental evaluation of an interactive system for pre-movement time reduction in case of fire. *Automation in Construction, 52*, 16–28.

Fahy, R.F. & Proulx, G. 2001. Toward creating a database on delay times to start evacuation and walking speeds for use in evacuation modeling.

Galea, E. R., Xie, H., Lawrence, P. 2016. *Intelligent Active Dynamic Signage System: Bringing the Humble Emergency Exit Sign into the 21st Century*, SFPE EUROPE MAGAZINE-ISSUE 3.

Gwynne, S.M.V. 2007. Optimizing fire alarm notification for high risk groups research project. *The Fire Protection Research Foundation, NFPA, Quincy, MA*.

Inoue, Y., Sashima, A., Ikeda, T. and Kurumatani, K. 2008. Indoor emergency evacuation service on autonomous navigation system using mobile phone. In *Universal Communication, 2008. ISUC'08. Second International Symposium on* (pp. 79–85). IEEE.

ISO, (2004). ISO 16069 Graphical symbols—Safety signs—Safety way guidance systems (SWGS).

Jensen, G. 1998. Wayfinding in heavy smoke: decisive factors and safety products. Fire safety 98. *Fire Protection in Nuclear Installations, Mumbai, India*.

Jeon, G.Y. & Hong, W.H. 2009. An experimental study on how phosphorescent guidance equipment influences on evacuation in impaired visibility. *Journal of Loss Prevention in the Process Industries*, *22*(6), 934–942.

Jeon, G.Y., Kim, J.Y., Hong, W.H. & Augenbroe, G. 2011. Evacuation performance of individuals in different visibility conditions. *Building and Environment*, *46*(5), 1094–1103.

Kobes, M., Helsloot, I., De Vries, B. & Post, J.G. 2010. Building safety and human behaviour in fire: A literature review. *Fire Safety Journal*, *45*(1), 1–11.

Kobes, M., Helsloot, I., de Vries, B. & Post, J. 2010. Exit choice, (pre-) movement time and (pre-) evacuation behaviour in hotel fire evacuation—Behavioural analysis and validation of the use of serious gaming in experimental research. *Procedia Engineering*, *3*, 37–51.

Kobes, M., Helsloot, I., de Vries, B., Post, J.G., Oberijé, N. & Groenewegen, K. 2010. Way finding during fire evacuation; an analysis of unannounced fire drills in a hotel at night. *Building and Environment*, *45*(3), 537–548.

Kyle, B. & Creak, J. 1999. *Assessment of photoluminescent material during office occupant evacuation*. Institute for Research in Construction.

Leslie, J. 2001. *A behavioural solution to the learned irrelevance of emergency exit signage*. In: Human Behaviour in Fires: Proceedings of the 2nd International Symposium in Fire. Interscience Communications, London, 23–33.

Nilsson, D. & Frantzich, H. 2010. Design of Voice Alarms—the Benefit of Mentioning Fire and the Use of a Synthetic Voice. In *Pedestrian and Evacuation Dynamics 2008* (pp. 135–144). Springer Berlin Heidelberg.

Occhialini, M., Bernardini, G., Ferracuti, F., Iarlori, S., D'Orazio, M. & Longhi, S. 2016. Fire exit signs: The use of neurological activity analysis for quantitative evaluations on their perceptiveness in a virtual environment. *Fire Safety Journal*, *82*, 63–75.

Paulsen, T.R.U.L.S. 1994. The Effect of Escape Route Information on Mobility and Way Finding Under Smoke Logged Conditions. *Fire Safety Science*, *4*, 693–704.

Purser, D.A. & Bensilum, M. 2001. Quantification of behaviour for engineering design standards and escape time calculations. *Safety Science*, *38*(2), 157–182.

Proulx, G. & Bénichou, N. 2010. Photoluminescent stairway installation for evacuation in office buildings. *Fire technology*, *46*(3), 471–495.

Proulx, G. 2002. Movement of people: the evacuation timing SFPE Handbook of Fire Protection Engineering, 342–366.

Proulx, G., Kyle, B. & Creak, J. 2000. Effectiveness of a photoluminescent wayguidance system. *Fire Technology*, *36*(4), 236–248.

Ran, H., Sun, L. & Gao, X. 2014. Influences of intelligent evacuation guidance system on crowd evacuation in building fire. *Automation in Construction*, *41*, 78–82.

Italian Organization for Standardization (UNI), (2004) UNI 7543:2004 - Safety colours and safety signs.

Renaudin, V., Yalak, O., Tomé, P. & Merminod, B. 2007. Indoor navigation of emergency agents. *European Journal of Navigation*, *5*(3), 36–45.

Tonikian, R., Proulx, G., Bénichou, N. & Reid, I. 2006. Literature review on photoluminescent material used as a safety wayguidance system. *PLM V6-2*.

Tuomisaari, M. 1997. *Visibility of exit signs and low-location lighting in smoky conditions*. VTT, Technical Research Centre of Finland.

Vilar, E., Rebelo, F., Noriega, P., Duarte, E. & Mayhorn, C.B. 2014. Effects of competing environmental variables and signage on route-choices in simulated everyday and emergency wayfinding situations. *Ergonomics*, *57*(4), 511–524.

Wong, L.T. & Lo, K.C. 2007. Experimental study on visibility of exit signs in buildings. *Building and Environment*, *42*(4), 1836–1842.

Xie, H., Filippidis, L., Gwynne, S., Galea E.R., Blackshields, D. & Lawrence, P.J. 2007. *Signage legibility distances as a function of observation angle,* Journal of Fire Protection Engineering, 17(1), 41–64.

Xudong, C., Heping, Z., Qiyuan, X., Yong, Z., Hongjiang, Z. & Chenjie, Z. 2009. Study of announced evacuation drill from a retail store. *Building and Environment*, *44*(5), 864–870.

Zhang, J. & Issa, R. R. 2015. "Collecting Fire Evacuation Performance Data using BIM-based Immersive Serious Games for Performance-based Fire Safety Design." Proceedings of the 2015 International Workshop on Computing in Civil Engineering, ASCE, Austin TX.

eWork and eBusiness in Architecture, Engineering and Construction – Christodoulou & Scherer (Eds)

Using BIM to support simulation of compliant building evacuation

J. Dimyadi & R. Amor
University of Auckland, Auckland, New Zealand

M. Spearpoint
University of Canterbury, Christchurch, New Zealand

ABSTRACT: Buildings must have a means of evacuating their occupants at times of emergency such as in the event of a fire. Such a provision typically needs to conform to applicable regulatory requirements. For simple buildings, this is achieved by incorporating a set of prescriptive requirements into the design. For more complex buildings, engineering analysis and simulation using advanced computational tools are often necessary to demonstrate conformance. This analysis and simulation process can be laborious if the required geometric and occupant data must be manually gathered from paper-based design information and relevant regulatory publications. BIM can provide an effective sharing of building information for the simulation stage and the output from the simulation can be used for compliance audit. In this paper, we develop a process of sharing BIM data with a probabilistic network evacuation simulation tool and use the output from the tool to inform a computer-aided compliance audit framework.

Keywords: BIM, building evacuation, compliance audit, simulation

1 INTRODUCTION

1.1 *Conventional compliant design practice*

Public buildings are typically required by law to have means of evacuation for the occupants in case of emergency such as in the event of a fire. The compliance with such a provision is usually subject to audit at the pre-construction design approval stage. However, compliance audit may also take place anytime in a building's life-cycle post-construction. For example, checking to ensure the intended occupancy and means of evacuation remain compliant with the applicable requirements.

One common method of compliance is to incorporate prescribed regulatory requirements directly into the design. Buildings with complex geometry and space configuration, however, may fall outside the scope of prescriptive requirements. In this case, a specific engineering design complying with some performance-based criteria would be necessary.

Compliant design audit has been a manual undertaking in the industry and involves checking paper-based drawings and specifications against a huge volume of regulatory documents and standards, which is laborious, error-prone and costly. Building Information Modelling (BIM) has gained popularity in recent years as a means of exchanging building design information digitally. However, regulatory requirements are still subject to human interpretation as they are written in the natural language not readily processable by computers. Recent research has endeavoured to represent regulatory knowledge in a Regulatory Knowledge Model (RKM) that is both human and machine readable (Dimyadi et al. 2016).

1.2 *Compliant fire evacuation design*

One primary objective of a building design is the occupant safety, which is generally governed by regulatory requirements. Prescriptive regulations typically specify the minimum number of escape routes that must be provided for a given type of occupancy, and stipulate thresholds for width, height, and length of escape routes that must not be exceeded. For simple building evacuation scenarios, demonstrating that these parameters have been incorporated as features of the building design is usually adequate for compliance. In more complex scenarios, however, the actual time needed for occupants to reach a safe place must be justified by computations to show that fire effects such as thermal radiation, low visibility and toxicity levels due to smoke would not impede the occupants' ability to evacuate. The tenability of escape routes can be evaluated by computations or simulations taking into account the type and amount of combustible materials, ventilation, and the fire dynamics in the enclosure.

A performance-based compliant fire evacuation design approach typically involves calculating the Required Safe Egress Time (RSET) for a particular escape route. It can then be demonstrated that it is within the allowable safety margin when compared to the Available Safe Egress Time (ASET), which is the time taken for the escape route to become untenable for evacuations based on specified thresholds. RSET is determined using calculations or simulations involving crowd dynamics including human behavior in fire events.

This paper develops a process of exchanging BIM and RKM data for compliant fire evacuation design, particularly for calculating RSET in the context of the New Zealand regulatory environment, although the approach is adaptable to regulatory frameworks in other parts of the world. The compliance document referred to in the example is the C/VM2 Verification Method: Framework for Fire Safety Design (MBIE 2014).

1.3 *EvacuatioNZ fire evacuation model*

There are many fire evacuation computational models available today with varying degrees of complexity and sophistication (Kuligowski & Peacock 2005). One such model is *EvacuatioNZ*, which is a risk-based coarse network evacuation model developed at the University of Canterbury in New Zealand that can be used to calculate building evacuation times such as RSET and other measures. The model is constructed of nodes connected by arcs or paths and is most suited for conducting a large number of simulations rapidly where many of the characteristics are represented by statistical distributions (Spearpoint & Xiang 2011). In *EvacuatioNZ*, spaces are represented as nodes where occupants can move from one node to another (Spearpoint 2009). Each node has properties such as length and width, or the floor area of the space it represents. Each path connecting two nodes has characteristics of evacuation components such as doors or staircases, as well as the distance between the nodes they connect. A "safe" node represents the final exit where occupants are considered safe (Figure 7).

EvacuatioNZ incorporates the Monte Carlo approach to produce probability distributions of evacuation times collected from repeated simulations of a specified scenario. It can simulate a given scenario for a specified duration or until all agents have reached a "safe" node. The simulation time steps are user-specified and typically in the order of 0.5 to 1 second intervals. The agent movement and queuing mechanics employed by *EvacuatioNZ* are based on mathematical equations published in the SFPE Handbook of Fire Protection Engineering (DiNenno 2002) and the Fire Engineering Design Guide (Spearpoint 2008).

1.4 *The building model*

An exemplar four-storey office building, which is one of nine types of building originally developed by the regulatory authority in New Zealand to evaluate the C/VM2 document, is used in this paper to illustrate the information exchange process. A Building Information Model (BIM) has been constructed in Autodesk's Revit for this exercise based on the information obtained from the original floor plans (Tan 2011). The open standard IFC (Industry Foundation Classes) version of the building model has been deposited into an online open repository (Amor & Dimyadi 2010) and freely available for access at http://openifcmodel.cs.auckland.ac.nz/

This is a relatively simple building with an internal stairwell that connects all the levels providing a single means of escape for the occupants to the final exit on the ground level. Each level is an open plan office measuring 26.5 m long, 12 m wide, and 3.2 m high containing the stairwell that is preceded by an enclosed corridor (Figure 1, Figure 2, and Figure 3). In this building model, doors are all 0.95 m wide and the staircase has 330 mm deep tread and 165 mm high riser and is 1.0 m wide throughout.

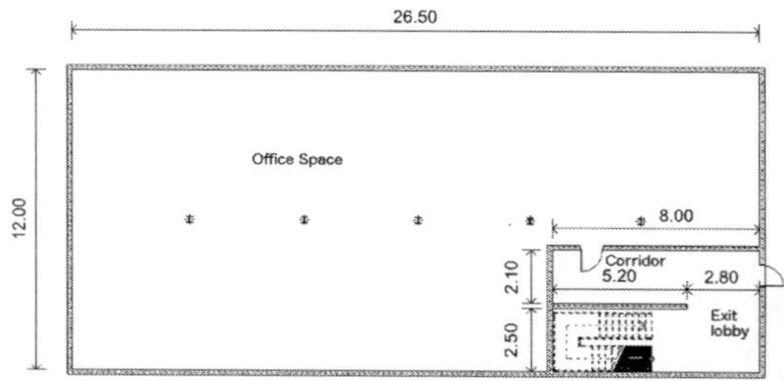

Figure 1. Ground Level floor plan of exemplar building.

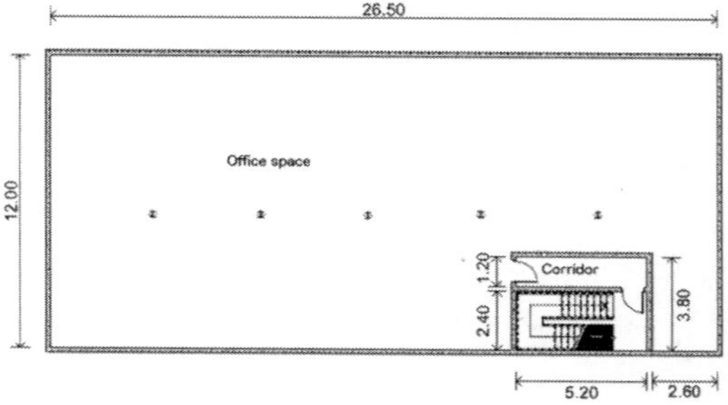

Figure 2. Typical floor plan of Levels 1 to 3 of exemplar building.

Figure 3. The exemplar building model in a BIM authoring environment.

This paper describes the process of generating the simulation input file for *EvacuatioNZ* using the BIM data of an exemplar building model. It also describes how the simulation output can be used to check for compliance against criteria set out in the RKM.

2 ACCESSING BUILDING INFORMATION

2.1 *Building Information Modelling (BIM)*

The ISO standard BIM (ISO 16739 2013) has gained popularity in recent years for sharing building design information among project stakeholders. A common method of exchanging BIM data is using the open standard Industry Foundation Classes (IFC) specification (ISO 16739 2013). However, accessing the right information for certain downstream applications may often be a challenge. This is because BIM is a highly complex model and necessarily so to capture every aspect of the building throughout its entire life-cycle. Furthermore, depending on the model's Level of Details (LODs), application specific information may not be available at the time of design and would need to be supplemented by a separate process.

To facilitate targeted and efficient information access for specific applications, a concept of BIM subset known as the Model View Definition (MVD) may be used (Hietanen 2006). The type of information to be included in each MVD can be specified using the open standard information delivery manual (IDM) (ISO 29481-1 2010). Every MVD is subject to certification before being accepted as a standard by buildingSMART International, which is the worldwide authority overseeing the creation and adoption of open international standards for the built environment.

An open software framework that is gaining popularity in managing and sharing the IFC data is the open source *BIMserver* (Beetz & van Berlo 2010). XML serializer plugins can be written and used with the *BIMserver* to extract sets of data using predefined schema to produce an MVD for a specific application.

Figure 4 is a model view of the ground level of the exemplar building showing the space, stairs and door objects as visualised using a software client of the *BIMserver*. The main office space is shown highlighted in green.

2.2 *Fire Compliance Model (FCM)*

An interim Fire Compliance Model (FCM) MVD schema has been defined in recent research to represent a BIM subset containing information specific to compliant fire safety design (Dimyadi et al. 2014). This FCM was used in a case study to illustrate the data exchange process in a computer-aided compliance audit framework. FCM is expressed in XML to facilitate data exchange with a number of fire engineering design software applications.

The minimum type of information required for evacuation-related fire design would generally include the geometry of spaces, particularly escape routes, the location of exit doors, and their dimensions, as well as the type of activities in each space. Spatial geometry information is typically expected to be available from BIM as it is an essential architectural design component. However, depending on the LOD of the model made available for design, certain information such as the space activity types may be missing and would need to be supplemented by a separate process. FCM allows one to identify what supplementary information is to be incorporated, either manually or via a software user interface, as long as they are predefined in the schema.

For a network model such as *EvacuatioNZ*, determining the absolute placement of objects in the model is not the primary objective. However, a few essential properties are needed, such as the floor areas, relative placement of space objects, and widths of doors with their locations relative to each space. Some of the objects and properties

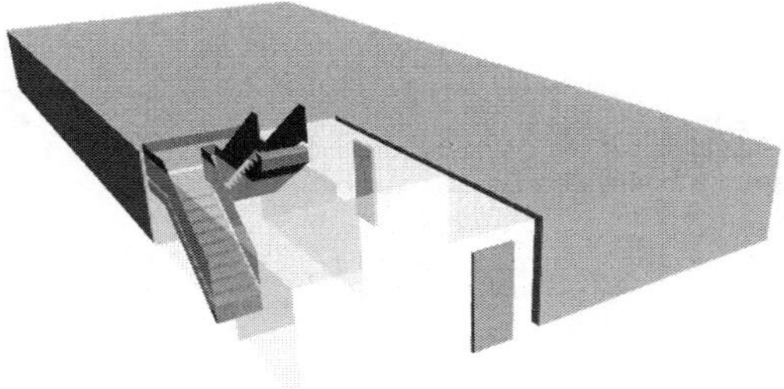

Figure 4. BIMserver rendered view of space, stairs, and door objects on the ground level.

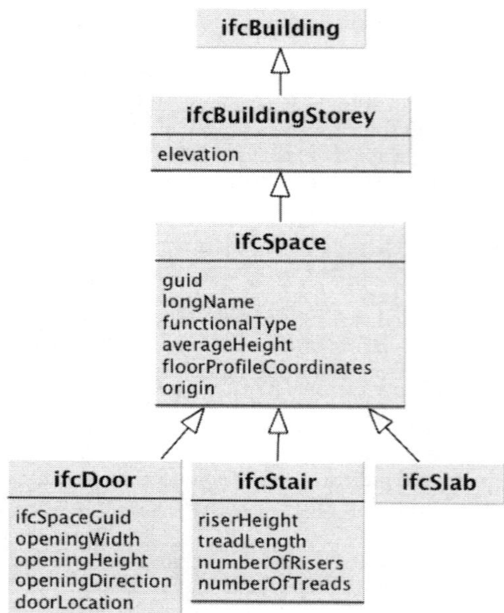

Figure 5. Class diagram of FCM showing some of the objects and properties.

defined in the FCM schema are given in the class diagram shown in Figure 5.

Not all object properties need to be represented directly in the FCM as some of these may be derived from another property. For example, the floor area and perimeter length of each space can be calculated from the *floorProfileCoordinates* of the space object. Additionally, critical traverse distances to the exit doors on each space may also be derived from the coordinates of the space object.

The geometry of stairs is an important input parameter for fire evacuation design and therefore must be defined accordingly in the FCM. Properties such as the *riserHeight*, *treadLength*, *numberOfRisers*, and *numberOfTreads* represent essential input data for fire evacuation simulations. In the case where these information is missing from the FCM, it is a common practice to assume a range of values and use them as variables in different simulation scenarios as part of the design process.

3 ACCESSING REGULATORY KNOWLEDGE

3.1 *Regulatory knowledge representation*

There have been several international efforts in recent years to standardise computable representations of regulatory knowledge. For example, the work by OASIS (Organisation for the Advancement of Structured Information Standards) in standardising the *Akoma Ntoso* (Architecture for Knowledge-Oriented Management of African Normative Texts using Open Standards and Ontologies) schema into *LegalDocML* (OASIS 2016). BuildingSMART International also took a similar initiative with their recent work in the Regulatory Room (BuildingSMART 2016).

3.2 *Regulator Knowledge Model (RKM)*

RKM has recently been developed as an interim method of representing regulatory knowledge that is both human and machine readable (Dimyadi et al. 2016). It is a digital version of a paper-based regulatory document designed to closely mimic the structure of the document it represents. This maintains user familiarity, and makes it easier to navigate the model for accessing the information.

The RKM representing the C/VM2 document is used in this paper to illustrate the compliant design audit aspect of the work.

4 COMPUTER-AIDED COMPLIANCE AUDIT

4.1 *Compliant Design Procedures (CDPs)*

The building engineering design process generally involves tasks that need to satisfy a set of requirements be it regulatory or the client's brief. The procedural nature of these compliant design tasks lend itself to automation.

A recent research investigated the use of BPMN (Business Process Model and Notation) workflow model to allow building designers or engineers to formally describe compliant design procedures for automated execution (Dimyadi et al. 2016). Each CDP workflow has script tasks with embedded instructions to query data from various sources or evaluate a given set of mathematical expressions. A high-level domain-specific query language was developed for that purpose. A dedicated process engine was also developed and incorporated into a compliance audit framework (ARCABIM) for processing these BPMN-compliant CDPs.

4.2 *ARCABIM framework*

ARCABIM takes FCM, RKMs, and CDPs as input components that can be maintained independently by the user or the domain experts instead of the system programmer.

ARCABIM supports performance-based design practice by allowing supplementary human input to be provided by the user. It also allows interfacing with simulation tools such as EvacuatioNZ. This is potentially beneficial as it can generate input data for these simulation tools and compare a specific output of these tools against criteria set out in RKMs to establish compliance.

5 FIRE EVACUATION SIMULATION

5.1 *Simulation input data*

EvacuatioNZ uses XML as its native input file format. It can also parse a basic subset of GraphML (Graph Markup Language) (Eiglsperger et al. 2013) that is supported by the *yEd* graph editing software application (yWorks 2016). *yEd* can be used to construct a network model representation of the evacuation scheme graphically and embed the required input parameters.

There are five primary groups of simulation input settings in *EvacuatioNZ*, as follows:

1. *Exit Behaviour*, sets up exit strategy definitions that includes following designated exit routes or finding the shortest path to a "safe" node.
2. *Agent Type*, defines occupant characteristics such as age group and walking speed as either static values or following a particular type of probability distribution. *Agent Types* are given one or more Exit Behaviour definitions.
3. *Populate*, defines groups of occupants with their characteristics inherited from the *Agent Type* definitions and their distribution within the building.
4. *Scenario*, sets up the number of simulations to execute and specifies the physical location of various input and output files.
5. *Simulation*, sets up the duration and time step of a particular simulation.

Each of the above groups of simulation setting can be defined in a separate XML document. Alternatively, they can all be combined into a single GraphML input file. Apart from these simulation settings, each network path must have a *ConnectionType* defining the type of connection it represents, such as a door or stair. Door properties such as doorWidth and specificFlow as well as tread and riser lengths of a stair object must also be defined.

5.2 *Modelling fire evacuation scenarios*

Typically, a number of fire scenarios would be evaluated in a fire evacuation design. For this exercise, one particular fire scenario is assumed where a fire starts in a normally unoccupied store room on the ground level within the stairwell, which has the potential of hindering egress (Figure 6). So, the objective of the exercise is to assess whether occupants are able

Figure 6. View of the store room adjacent the stairwell.

Table 1. Spaces, floor areas, and occupant load.

Space	Floor area (m^2)	Occupant load
Level 3 office	295.42	29
Level 3 corridor	6.24	–
Level 2 office	295.42	29
Level 2 corridor	6.42	–
Level 1 office	295.42	29
Level 1 corridor	6.42	–
Ground level office	276.92	27
Ground level corridor	10.92	–
Ground level store room	2.77	–
Ground level exit lobby	13.44	–

to evacuate the building before the fire breaks out of the store room. This is a scenario known as the "Unknown Threat" in the C/VM2 document.

The potential occupant load in each space is determined using the occupant density that relates to the activities in the space as prescribed by C/VM2. For an office occupancy, this is typically 10 m^2 per person. Table 1 shows all spaces in the building and the occupant load expected in each occupied space based on the prescribed occupant density. Spaces that have intermittent use such as corridors are not required to be occupied.

For simulation, it is assumed that occupants are randomly distributed in each space and there is an automatic building fire alarm system to alert occupants of a fire. The C/VM2 document prescribes a response time of 60 seconds before people begin their evacuation where fire occurs remotely from the occupied space. On activation of the fire alarm, occupants are expected to evacuate from the office space on each level by passing through a doorway into the enclosed corridor before entering the stairwell. Occupants from each level would merge with occupants descending from the upper levels and continue down towards the final exit on the ground level.

The evacuation scheme in this scenario can be represented graphically in a node network diagram using the *yEd* graph editor (Figure 7).

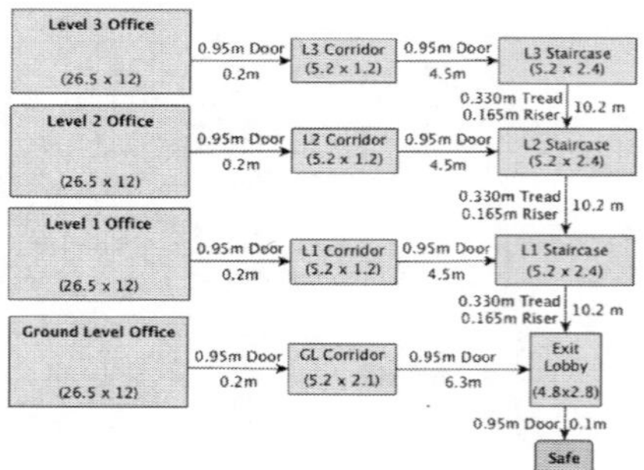

Figure 7. Node network representation of the evacuation scheme.

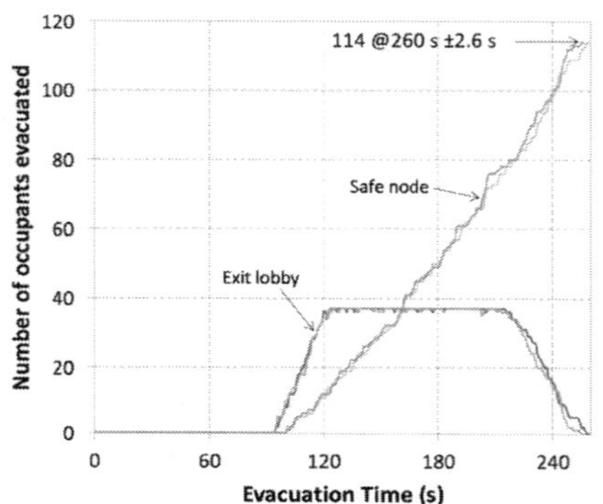

Figure 8. Cumulative graphs of number of occupants reaching the Exit lobby on the ground level and the final exit (safe node).

5.3 *Simulation run and result*

EvacuatioNZ is a command-line tool that can parse a GraphML input file and use the specified data to conduct the simulation. The output is a series of HTML and CSV files containing information such as the number of occupants evacuated in each time step. This can be used to plot a graph of cumulative occupants evacuated for each space for the entire duration of the simulation (Figure 8).

Running ten simulations for the particular case has produced an average RSET of 260 seconds with 2.6 seconds standard deviation to evacuate the 114 agents.

6 GENERATING SIMULATION INPUT DATA

6.1 *Sharing BIM data with EvacuatioNZ*

The FCM of the exemplar building model contains the geometry of each space, which is expressed in terms of its floor's profile coordinates. These coordinates can be used to calculate the space's floor area and perimeter length, which can be used to populate the *length* and *width* parameters of each node. More importantly, they can be used to establish the likely evacuation paths, i.e. the sequence of nodes in an assumed evacuation scheme, as follows:

1. For a building storey with a single means of escape, an escape route would start at the most remote location from the point of exit on that level.
2. Where there are multiple exit points that connect to spaces on multiple levels, the situation becomes more complex. One option is to automatically assume that an independent escape route exists for each exit point regardless of whether the start of the escape route is near an exit point. This would be applicable in a scenario where one exit may be blocked due to the location of a fire or other reasons.
3. Alternatively, a network diagram rendition of the building in GraphML can be generated as a provisional layout template for editing in the *yEd* graph editor. The template would have spaces represented as nodes and doors represented as paths, all with their respective properties embedded in them. A designer can then use this template in *yEd* to develop evacuation schemes for different simulations.

Distances between the centroid of nodes can also be derived from each space's profile coordinates. The geometry of the stairs may be available from the FCM. Otherwise, the tread and riser dimensions can be treated as variables with assumed values. The length travelled in the stairs can be determined knowing the elevations of each level and the assumed tread and riser dimensions.

6.2 *Sharing RKM data with EvacuatioNZ*

Parameters such as occupant densities and pre-evacuation time are given in C/VM2, which are represented and accessible from the RKM. The occupant density value can be used to calculate the expected occupant load in spaces intended as office occupancy. Space activity types are currently indicated in both FCM and RKM using the OmniClass standard space function classifications system (CSI 2012). For example, an office space is classified as "13-55 11 00".

The calculated occupant load for each space is then used by the "*Populate*" input specification for *EvacuatioNZ*.

6.3 *Generating input file for EvacuatioNZ*

ARCABIM can use data extracted from FCM and RKM, as well as some assumed values for object sizes and placement to generate a GraphML file representing the evacuation scheme that can be visualised and adjusted as necessary in the *yEd* graph editing environment.

To ensure correctness of representation, every aspect of the generated input file should be checked in *yEd* before being used with *EvacuatioNZ* for simulations.

6.4 *CDP modelling*

C/VM2 also prescribes the method of checking if RSET of an escape route would be less than that for the adjacent store room to sustain a complete burnout. This can be described in a CDP workflow (Figure 9).

The task to evaluate RSET checks for the presence of an input file for *EvacuatioNZ* in a specified folder. If missing, then it will trigger the process of generating the input file.

The burnout time of a space can be calculated using the method prescribed by C/VM2, which can be described in a CDP workflow (Figure 10).

One method of calculating the burnout time is using the "Time Equivalence" formula as described in C/VM2. The CDP workflow (Figure 10) can either specify all the mathematical equations and the steps required to obtain all the parameters or, alternatively, extract those equations directly from RKM. The execution of this CDP workflow in ARCABIM produces a burnout time for the store room of 212 minutes (Figure 11).

6.5 *Sharing EvacuatioNZ data with ARCABIM*

By specifying the path of the *EvacuatioNZ* output files as part of the instructions in the CDP workflow (Figure 10), ARCABIM would then be able to extract the RSET value as calculated by *EvacuatioNZ*.

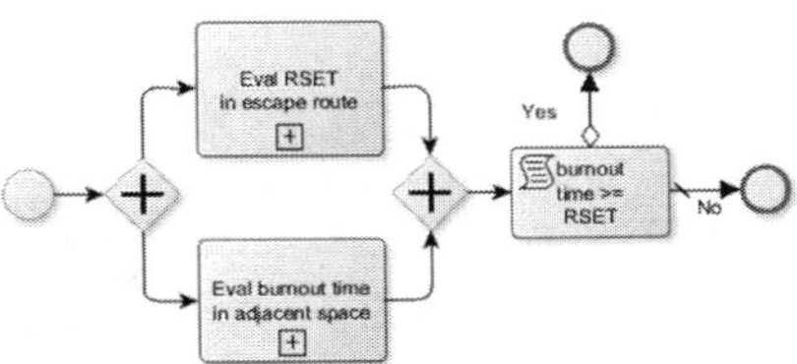

Figure 9. CDP workflow to check if RSET would be less than the burnout time of an adjacent space.

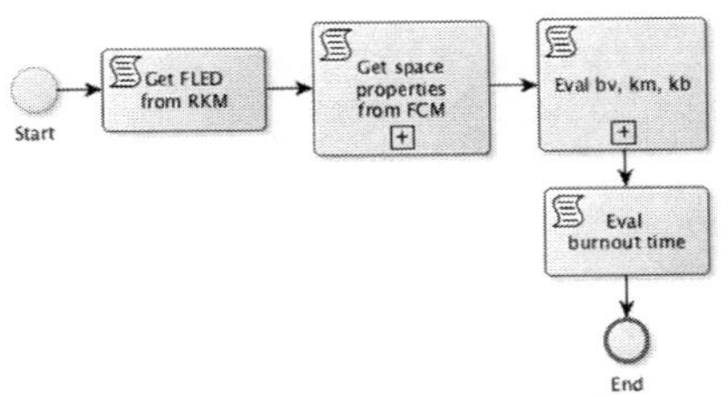

Figure 10. CDP workflow to calculate the burnout time.

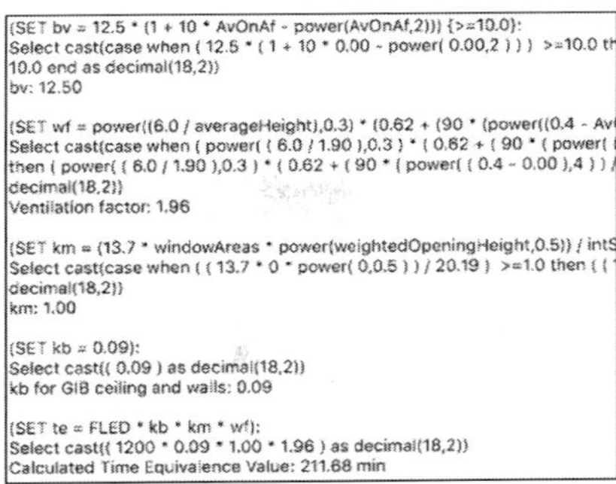

```
(SET bv = 12.5 * (1 + 10 * AvOnAf - power(AvOnAf,2))) {>=10.0}:
Select cast(case when ( 12.5 * ( 1 + 10 * 0.00 - power( 0.00,2 ) ) )  >=10.0 th
10.0 end as decimal(18,2))
bv: 12.50

(SET wf = power((6.0 / averageHeight),0.3) * (0.62 + (90 * (power((0.4 - Av
Select cast(case when ( power( ( 6.0 / 1.90 ),0.3 ) * ( 0.62 + ( 90 * ( power(
then ( power( ( 6.0 / 1.90 ),0.3 ) * ( 0.62 + ( 90 * ( power( ( 0.4 - 0.00 ),4 ) ) /
decimal(18,2))
Ventilation factor: 1.96

(SET km = (13.7 * windowAreas * power(weightedOpeningHeight,0.5)) / intS
Select cast(case when ( ( 13.7 * 0 * power( 0,0.5 ) ) / 20.19 )  >=1.0 then ( (
decimal(18,2))
km: 1.00

(SET kb = 0.09):
Select cast(( 0.09 ) as decimal(18,2))
kb for GIB ceiling and walls: 0.09

(SET te = FLED * kb * km * wf):
Select cast(( 1200 * 0.09 * 1.00 * 1.96 ) as decimal(18,2))
Calculated Time Equivalence Value: 211.68 min
```

Figure 11. The execution intermediate output in ARCABIM.

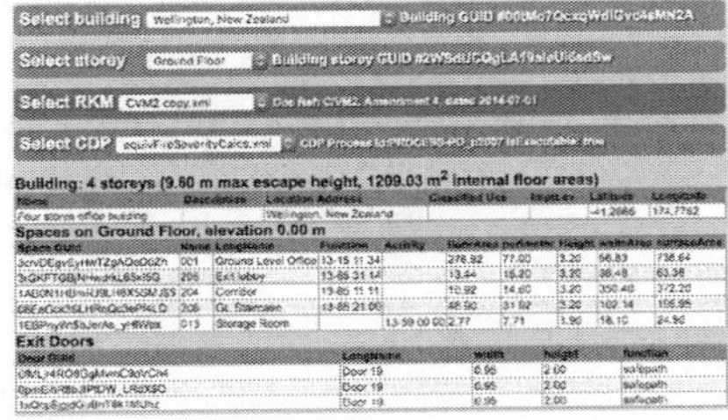

Figure 12. Screenshot of ARCABIM during an audit session.

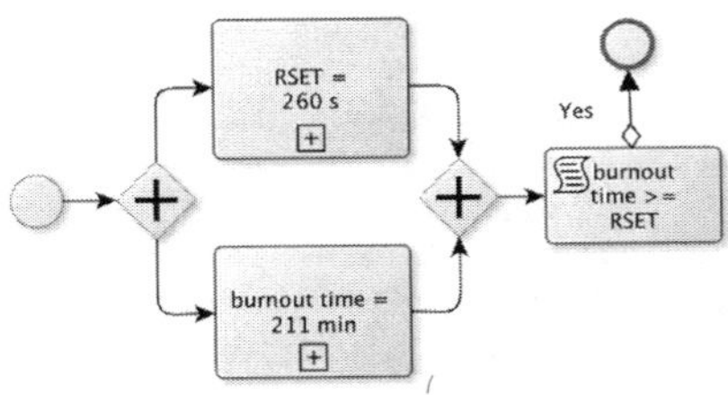

Figure 13. A successful outcome of the compliance audit.

6.6 *Compliance audit process*

The compliance audit process would start with importing the FCM into ARCABIM. As the audit is conducted one building storey at a time, the user would choose the building storey, then select the RKMs to audit against and the CDPs to guide the audit process (Figure 12).

As part of the compliance audit process, the RSET value shared by *EvacuatioNZ* is compared with the calculated burnout time of the store room to establish compliance.

In this case, as the simulated RSET value is only 260 seconds, occupants can clearly evacuate the building before the fire in the store room breaks out. Therefore, the outcome of the compliance audit is successful (Figure 13).

7 SUMMARY AND CONCLUSION

This paper has described a method of sharing BIM data with *EvacuatioNZ*, a risk-based fire evacuation simulation tool. The data exchange process has been illustrated using an exemplar building model. The paper has described how the input file for *EvacuatioNZ* can be automatically generated and how the output from the simulation can be used to check for compliance in a computer-aided compliance audit framework.

For simple buildings, particularly one with single means of escape such as that illustrated in this paper, the process of automatically constructing the likely escape routes is relatively straightforward. For a more complex layout with multiple evacuation options, the user can request a network diagram rendition of the building for further editing in *yEd*. All nodes and paths in the template would have inherited the properties of their respective spaces, doors, and stairs. This makes it relatively straightforward for the user to rearrange the nodes and paths to suit any evacuation scheme for simulation.

Taking advantage of the available data from BIM or FCM and RKM to generate an input file for simulation would save considerable time, particularly if the required information would otherwise need to be measured from paper drawings. However, a possible challenge is that the required information, such as space activity types, may not always be available from the BIM model. Therefore, there needs to be a set of pre-defined default or assumed values to use as part of the input data generation process. Additionally, there needs to be a separate process where supplementary data can be provided manually, if necessary.

In any case, the system generated input file should always be considered provisional and is subject to review and verification by the designer for correctness and appropriateness before being used for simulations.

One area worth investigating further is identifying the scope of information that needs to be included in the FCM to enable sharing of BIM data with other types of fire simulations.

REFERENCES

Amor, R. & Dimyadi, J., 2010. An Open Repository of IFC Data Models and Analyses to Support Interoperability Deployment. In *Proceedings of CIB W78 Conference*. Cairo, Egypt, Egypt, pp. 16–18.

Beetz, J. & van Berlo, L., 2010. bimserver.org—An Open Source IFC Model Server. In *Proceedings of CIB W78 International Conference*. Cairo, Egypt, pp. 1–8.

BuildingSMART, 2016. buildingSMART Regulatory Room. Available at: www.buildingsmart.org/standards/standards-organization/rooms/regulatory-room [Accessed April 9, 2016].

CSI, 2012. *OmniClass—A Strategy for Classifying the Built Environment Table 13 – Spaces by Function*, Construction Classification System.

Dimyadi, J. et al., 2014. Computer-aided Compliance Audit to Support Performance-based Fire Engineering Design. In *Proceedings of 10th International Conference on Performance-based Codes and Fire Safety Design Methods*. Gold Coast, Queensland.

Dimyadi, J. et al., 2016. Computerizing Regulatory Knowledge for Building Engineering Design. *Journal of Computing in Civil Engineering*, C4016001, pp.1–13.

DiNenno, P.J. ed., 2002. *SFPE Handbook of Fire Protection Engineering* 3rd ed., National Fire Protection Association Inc.

Eiglsperger, M. et al., 2013. Graph Markup Language (GraphML). *Handbook of Graph Drawing and Visualization*, pp.517–541.

Hietanen, J., 2006. IFC model view definition format. *International Alliance for Interoperability*, pp.1–29. Available at: http://www.secondschool.net/one/IAI_IFC_framework.pdf.

ISO 16739, 2013. *ISO 16739:2013 Industry Foundation Classes (IFC) for data sharing in the construction and facility management industries*, Geneva, Switzerland.

ISO 29481-1, 2010. *ISO 29481-1: 2010 Building information modelling—Information delivery manual*, British Standards.

Kuligowski, E.D. & Peacock, R.D., 2005. *A Review of Building Evacuation Models (Technical Note 1471)*, Washington, D.C., USA.

MBIE, 2014. *C/VM2 Verification Method: Framework for Fire Safety Design For New Zealand Building Code Clauses C1-C6 Protection from Fire* 4th ed., Wellington, New Zealand: The Ministry of Business, Innovation and Employment.

OASIS, 2016. OASIS LegalDocumentML (LegalDocML) TC. *OASIS*. Available at: https://www.oasis-open.org/committees/tc_home.php?wg_abbrev=legaldocml [Accessed April 10, 2016].

Spearpoint, M. ed., 2008. *Fire Engineering Design Guide* 3rd ed., Christchurch, New Zealand: New Zealand Centre for Advanced Engineering.

Spearpoint, M. & Xiang, X., 2011. Calculating Evacuation Times from Lecture Theatre Type Rooms Using a Network Model. In *Fire Safety Science*. Maryland, USA, pp. 599–612.

Spearpoint, M.J., 2009. Comparative Verification Exercises on a Probabilistic Network Model for Building Evacuation. *Journal of Fire Sciences*, 27(5), pp.409–430.

Tan, Y.K., 2011. *Evacuation Timing Computations Using Different Evacuation Models*. Christchurch, New Zealand: University of Canterbury, New Zealand.

yWorks, 2016. yEd Graph Editor. *yWorks GmbH Germany*. Available at: https://www.yworks.com/products/yed [Accessed April 7, 2016].

BIM implementation and deployment (II)— human resources and economics

eWork and eBusiness in Architecture, Engineering and Construction – Christodoulou & Scherer (Eds)
 ISBN 978-1-138-03280-4

Human-resources optimization & re-adaptation modelling in enterprises

S. Zikos, S. Rogotis, S. Krinidis, D. Ioannidis & D. Tzovaras
Information Technologies Institute, Centre for Research and Technology Hellas, Thermi-Thessaloniki, Greece

ABSTRACT: Optimization of Human Resources (HR) and re-adaptation are of vital importance to enterprises in order to keep the workload balanced and maintain high performance levels when unexpected events or exceptions occur, such as an arrival of a new unscheduled task. In this paper, a novel HR optimization & re-adaptation model for enterprises is introduced. The model integrates different entities such as employees, processes, work schedules, resources, and location information. The heterogeneous information is translated to a common vocabulary in order to be utilized for assigning tasks to human resources automatically without the need of supervision. Conditional Random Fields (CRFs) probabilistic models are trained, so as to learn the already applied task assignment patterns, and their output is taken into account in the decision process. The HR optimization toolkit, which comprises the models and the HR optimization tool, has been tested with real data acquired from an industrial environment achieving favourable results towards HR assignment.

1 INTRODUCTION

Resource management is crucial for the performance of an enterprise. Efficient management of resources in enterprises, and especially in dynamic environments such as shop floors, is a great challenge as various factors have to be taken into account in order to deal with possible unexpected events. Examples of such real-time events include machine failures, arrival of urgent jobs, due date changes, changes in job processing time, etc. (Varela & Riberio 2014). The goal is to achieve the optimal management of the resources according to the current condition and based on the preferred policies.

The two main types of resources that have to be managed in enterprises are machinery resources and human resources. These resource types have different characteristics. Machines are subject to unexpected failures and their performance is characterized by low variability. On the contrary, the performance of an employee when executing a task depends on various parameters such as the experience level, skills, time of the day etc. Furthermore, workload balancing among employees is vital to ensure fairness and employee convenience.

A resource assignment algorithm deals with the assignment of resources by selecting the most suitable from the ones available and by taking into account resource constraints. On the other hand, a scheduling algorithm assigns resources to tasks while setting the execution starting time of tasks as well. The present paper focuses on the first problem category. Depending on the execution phase, resource allocation can be static or dynamic. In the former case a set of tasks and a set of resources are available at design time and the allocation of resources is decided before the execution starts. Contrariwise, dynamic resource allocation is performed when a new task arrives at run time.

Due to the high interest in task scheduling and resource assignment, there are several studies in the literature proposing various approaches to deal with the above-mentioned problems in the context of human resources. References of recent work on personnel assignment and human resource scheduling are provided in the following two paragraphs, respectively.

Wibisono et al. (2015) propose an on-the-fly dynamic human resource allocation method in Business Process Management Systems which is based on Naïve Bayes model. The results showed that their approach outperformed other rule-based methods applied, such as Random and Order-based. A mechanism in which the dynamic resource allocation optimization problem is modelled as Markov decision processes is proposed by Huang et al. (2011). The mechanism observes its environment to learn appropriate policies which optimize resource allocation in business process execution. An alternative approach presented by Liu et al. (2008) applies a machine learning algorithm (decision tree, Naïve Bayes, SVM) to the workflow event log to learn various kinds of activities that each employee undertakes. When staff assignment is needed, the classifiers suggest a suitable actor to undertake the specified activities. The approach is characterized by the authors as semi-automated because the workflow initiator needs to make the final decision. Another interesting solution for the human resource assignment problem is proposed by Cabanillas et al. (2013), where the authors specify and utilize a metamodel to define preferences. In that work the human resources are prioritized according to preferences, based on a ranking mechanism.

Recent work on the NP-hard complex resource scheduling problem includes approaches based on multi-agent systems and heuristics. More specifically, Skobelev et al. (2013) have proposed an adaptive scheduling method for resource management in manufacturing workshops, which is based on a multi-agent system. The system's model includes agent classes such as Order, Worker, Machine and other. Agnetis et al. (2014) propose two heuristics (STA-OMSB, MSB-DOS) that solve job shop scheduling problems in handicraft production, taking into account also the skills of each employee. The case study that was examined revealed that the MSB-DOS method yielded close-to-optimal solutions in reasonable time. Lastly, Bouzidi-Hassini et al. (2015) discuss a new approach to integrate the scheduling of production and maintenance operations. The approach takes into account human resources availability and is based on multi-agent systems for modelling the production workshop. However, the allocation of activities to the best human resource based on its competence is not considered.

This paper presents a novel model for Human Resource optimization in enterprises. Various static and dynamic parameters of the involved entities such as tasks and employees are incorporated and combined. The model definition allows a monitoring component to assess the operational status of the enterprise in real-time and respond to changes regarding the workload. By using the defined model, multiple criteria such as the workload and task priority can be taken into account for assigning human resources to tasks automatically. Furthermore, (a) Conditional Random Fields probabilistic models (Sutton & McCallum 2006) are trained and employed in the resource selection process and (b) the use of thermal cameras for extracting the location of employees is examined. The components of the system and its architecture are also presented.

The remainder of the paper is organized as follows: Section 2 presents the HR model and the HR assignment method applied. Section 3 describes the architecture of the HR optimization toolkit and section 4 presents case study results. Lastly, conclusions and future directions are summarized in section 5.

2 PROBLEM DESCRIPTION & MODELLING

2.1 *Problem description*

The primary goal is to build a HR model for use in enterprises. The model has to be as generic as possible and be extendable with additional parameters or entities. The latter is important since operations and requirements in an enterprise may change over time. The defined model is utilized by an HR optimization component which helps to automate and optimize the HR assignment process in static or dynamic resource allocation, and to provide the manager with visualized information about the progress of task execution.

The manual assignment of urgent tasks to employees by the manager or the supervisor can be time-consuming. Thus, in order to deal with this issue, the primary objective is to automatically select the most suitable employee(s) to perform an arriving task, by taking into account multiple criteria in real-time. The HR model includes all the necessary information used as input for making the final decision, such as the current work schedule per employee and other information about the current state.

2.2 *HR model description*

The HR model is a bridge between the HR strategy and other key areas of HR management, such as processes, actors, etc. The model adopts entities and the vocabulary of Business To Manufacturing Markup Language (B2MML). B2MML (MESA XML Committee 2013) consists of a set of schemas written using the XML Schema language (XSD) that implement the data models in the ISA-95 standard. Furthermore, in addition to the B2MML schemas, new XSD schemas have been defined regarding the format of sensor information, measurements and localization.

BIMs (Building Information Models) are digital representations of physical and functional attributes of infrastructures such as houses, factories etc. (Azhar 2011, Volk et al. 2014). A Building Information Model (BIM) which contains the locations of critical assets and equipment is utilized by the HR model. BIM information is extracted and stored in gbXML format. In this way, the layout of the facility along with the positions of critical equipment and employees can be visualized and presented to the manager.

The two main entities employed in the HR assignment modelling are the human resources (employees) and the tasks. A task T can be atomic or composite. A composite task CT is composed of $k>1$ atomic tasks (T_i, $i = 2, 3, \ldots, k$). Composite tasks or processes are common in enterprises operations and are usually represented by directed graphs, where each node is an atomic task. The execution order of tasks in a process has to be defined in addition to the number of employees required per each trade. To this end, additional properties for each task belonging to a *CT* are defined, such as the ID of the *CT*, flags showing whether the task is the first or the final in the *CT*, and a set of the IDs of the preceding tasks. The definition of these properties allows us to model different types of workflows, such as workflows that include only sequential tasks or workflows that include tasks that can be executed in parallel. The properties of tasks that are utilized in the assignment process are listed below:

- ID
- Employee trade required
- Number of employees required
- Estimated duration in minutes
- Scheduled starting time
- Priority weight
- Location
- Assets required

The specification of task information utilized by the HR model, such as the qualifications of employees required, is the result of job analysis (Morgeson et al. 2004). Job analysis provides information to enterprises which helps to determine the best fit employees for specific tasks. Prototypes of the common tasks occurring in the enterprise are built and are part of the model. Each task prototype defines the static parameters of the task and can be identified by its unique ID number. The availability of the characteristics of common tasks facilitates the automated creation of new tasks in the system when emergency events, such as critical malfunctions, are detected. Each task usually requires an employee with specific trade e.g. a technician and therefore it cannot be performed by an employee of a different trade. We also assume that an estimated duration of each task in minutes is also known. It can be the average execution time of past instances of the task. For each already scheduled task, it is expected the starting time of the task to have been defined. For new urgent tasks we consider that the scheduled starting time is also defined (as the task starts shortly after the arrival). Priority weight denotes the level of importance of the task and is a real number in [0…1]. The approximate location of the task is the name of the area where the task is performed. Placement information is retrieved from the BIM. Assets (e.g. machines, spares) required by a task are considered as constraints in HR optimization.

The static and dynamic properties of employees are listed below:

Table 1. Main parameters of the HR model.

Parameter	Employee	Task
ID	Yes	Yes
Trade	Yes	Yes
Location	Yes	Yes
Work schedule	Yes	–
Estim. duration	–	Yes
Starting time	–	Yes
Priority weight	–	Yes

- ID
- Hierarchy scope
- Trade
- Experience level
- Shift starting/ending time
- Current Position as X,Y coordinates
- Work schedule
- Running task

The trade of the employee denotes his/her expertise. An employee can have more than one trade. The experience level is also defined per each employee (Trainee, Novice, Experienced). The HR model supports preemption. As a human resource can be regarded as a processing unit which processes tasks, preemption means that a task being performed by an employee can be temporarily interrupted and resumed at a later time. This property allows an employee to execute a new task of higher priority shortly after its arrival. In this work we assume that all the employees working at each time instance share common starting and ending shift times. Table 1 below summarizes the most important parameters utilized in the proposed HR model.

2.3 *Towards HR localization using thermal cameras*

In order to support the real-time localization of employees, a method for calibrating thermal cameras against the BIM, which is used as reference point, has been developed and evaluated. Thermal cameras are utilized offering notable advantages in respect to other imaging systems. They have gained popularity over the past few years, since they yield good results under adverse illumination conditions. This renders them particular popular for both indoor/outdoor monitoring applications. Moreover, their performance is not greatly influenced by shadow effects or other physical limitations that might be encountered in uncontrolled enterprise environments due to the presence of smoke, gas, increased humidity, etc. For thermal camera's geometric calibration, a calibration pattern was manufactured and prepared, exhibiting high-thermal contrast. A ray-tracing approach against the shop floor's geometry, obtained through the BIM, is performed in order to produce a detailed 3D thermal model. Each surface can be projected on a parallel plane (orthographic projection), resulting in images with geometric scale and no distortion, called orthothermograms. The proposed approach enables an end-to-end camera pose estimation and that allows spatially accurate motion detection and indoor monitoring for employee localization within the shop floor (Fig. 1). Information extracted from the thermal cameras can be fused with RFID or another similar technology in order to identify with improved accuracy an employee's position within the inspected area.

2.4 *HR assignment method*

The proposed dynamic HR assignment method, which utilizes the HR model presented in the previous sections, ranks the candidate employees and then selects the required employee(s) for a task. The higher the ranking of an employee, the lower the assignment cost is. One of the factors that affect the assignment cost of employees is derived from the output of Conditional Random Fields (CRFs) probabilistic models. The probabilistic models incorporate the enterprise's practices as they are built using historical task assignment examples. However, since these past task assignments may not be the optimal, additional criteria are also considered for ranking the candidate employees. A CRF model is created per each employee. The main advantage of a CRF when compared to a Hidden Markov Model (HMM) is the ability to include more complex non-independent features of the observations. The probability of producing a state sequence y given an observation sequence $\times$ can be computed by Equation 1 below:

$$p(y|x) = \frac{1}{Z(x)} exp\left\{\sum_{k=1}^{K} \lambda_k f_k\left(y_t, y_{t-1}, x_t\right)\right\} \quad (1)$$

Figure 1. 3D thermal model of the inspected area.

where f_k = feature function; λ_k = learned weight associated with feature f_k; and Z(x) = the normalization function shown in Equation 2.

$$Z(x) = \sum_y \exp\left\{\sum_{k=1}^{K} \lambda_k f_k\left(y_t, y_{t-1}, x_t\right)\right\} \quad (2)$$

At each time step, the model takes as input a feature vector (observation) and infers the hidden state. The estimation produced at each time step is the model's hidden state and in our case it can be either 0 (not recommended for assignment) or 1 (candidate for assignment). The feature vector constructed at each time step comprises the following features, normalized to [0...1]:

1. the type (ID) of the arriving task to be assigned
2. the employee's conformance to the trade required by the task
3. the employee's remaining workload level

The use of the task type in employees models favours the selection of employees with frequent past assignments of the particular task. The particular behaviour is desirable since these employees tend to be more familiar with the task resulting in high execution performance. The second feature denotes if the employee is able to perform the task. In case the trade does not match with the trade required, its value is 0. Regarding the employee's remaining workload, five different levels are possible (None, Low, Medium, High, Max). It is computed based on the estimated durations of the employee's scheduled tasks.

A training phase is required in advance in order to create each CRF model. A sequence of observations along with the actual ground truth label per each observation must be provided as input for learning the model parameters per each employee. The training data per model are created concurrently for all employees by using historical assignment events in the form: < *Time, TaskID, EmployeeID* >.

The total selection cost for each employee E is computed as follows:

$$E_{totalcost} = (1 - W) \times E_{cap\_cost} + W \times E_{adapt\_cost} \quad (3)$$

where W = weight factor set to 0.6, E_{cap_cost} denotes the capability cost and E_{adapt_cost} denotes the adaptation cost. They are computed respectively by the two equations below

$$E_{cap\_cost} = W_1 \times E_{model\_cost} + W_2 \times E_{rem\_wload} + W_3 \times E_{dist} \quad (4)$$

where $W_1 + W_2 + W_3 = 1$.

$$E_{adapt\_cost} = \left(E_{overlap} + E_{weight\_comp} + E_{task\_inter}\right)/3 \quad (5)$$

Each CRF model's current belief probability (bp) of being in state 1, which is a real value in [0…1], is retrieved. E_{model_cost} is derived from the belief probability (bp) of the employee's CRF probabilistic model and is equal to 1-bp.

E_{rem_wload} denotes the normalized remaining workload of the employee until the end of the shift. It is computed by dividing the total duration of remaining work in minutes by the total minutes left until the end of the shift. Remaining workload is an important parameter and considered in both the probabilistic model and the cost function. Only considering employee capabilities during task assignment may result in most capable employees being assigned a heavier workload (Shen et al. 2003).

The current location of employees can be taken into account for selecting the employee who is closer to the location of the task in order to improve responsiveness. The normalized distance cost factor (E_{dist}) is the Euclidean distance between the task location and the employee position divided by the maximum possible distance that can be observed based on the architectural plan derived from the BIM model.

$$E_{dist} = \frac{\sqrt{(E\_x - T\_x)^2 + (E\_y - T\_y)^2}}{\max\_distance} \quad (6)$$

The cost factors in the employee capability equation are not of the same weight. Instead, the weight values depend on the selected assignment policy mode, which can be either 'Performance' or 'Balancing'. Thus, when operating under the 'Performance' policy $W_1 = 0.5 > W_2 = 0.3$ and when operating under the 'Balancing' policy $W_1 = 0.3 < W_2 = 0.5$. The policy can be selected by the manager.

In order to compute the adaptation cost, a check is performed for determining whether the arriving task overlaps the employee's running task or a scheduled task. If this is not the case (the employee is available for the whole specific time period), then $E_{adapt\_cost} = 0$, otherwise the list of overlapping tasks is returned. $E_{overlap}$ is defined as the ratio the arriving task is overlapped (based on its scheduled starting time and estimated duration) by the employee's running and scheduled tasks. E_{weight_comp} cost factor takes into account the priority weights of the tasks. The priority weight of the arriving task is compared with the weights of the overlapped scheduled tasks. $E_{weight\_comp} = 1$ in case the arriving task's weight is lower than the weight of at least one overlapped scheduled task.

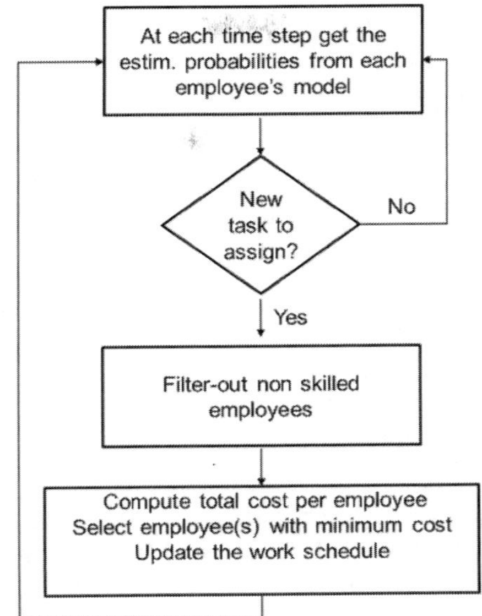

Figure 2. Steps for HR assignment.

The task interruption cost (E_{task_inter}) equals to the elapsed execution time of the currently running task divided by its estimated duration. This factor implements the concept of introducing cost proportionate to a running task's remaining execution time. The motivation behind this factor is to declare that it is undesirable to interrupt a task with short remaining execution time.

In case of insertion of an unexpected task into the work schedule, the execution of an already scheduled task may not be able to be started on time because of employee unavailability due to time overlap. Therefore, a check is performed in order to examine if the affected task can be postponed by advancing the scheduled starting time without introducing new overlaps (re-adapt the employee's work schedule). If this is not the case, a strategy has to be defined for determining the remaining tasks that have to be rescheduled.

The steps of the HR assignment method are presented schematically in Figure 2.

3 SYSTEM ARCHITECTURE

The HR optimization toolkit comprises the models, the localization component and the HR optimization component. All information is available through APIs and a shared common vocabulary is defined. The middleware is responsible for the distribution of information. Furthermore, all information of the HR model is stored in its database.

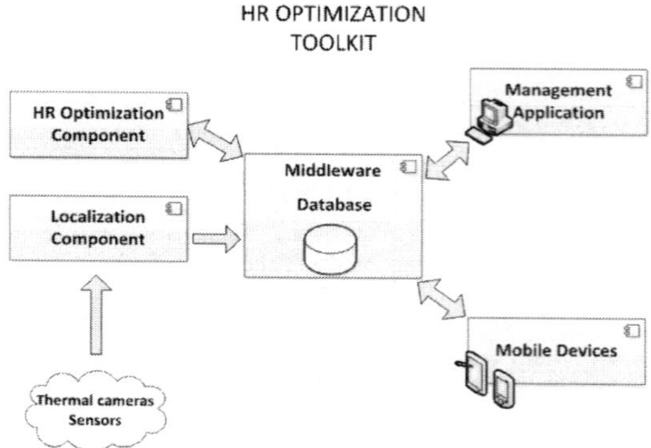

Figure 3. System architecture.

The localization component is responsible for the extraction of the positions of employees in the enterprise premises. As such, the location of each employee at each time instance is sent to the HR model for storing and distribution.

The HR optimization component executes the HR assignment in case a new task has to be performed. Moreover, it analyses the work schedules and monitors the execution progress of tasks. Thus, possible exceptions that can affect both the current work executed and the scheduled future work can be detected. The supervisors and the manager can be notified about an exception in order to take appropriate actions.

The management application is used by the manager and runs on a desktop PC. The manager can review the current progress of tasks and access information such as the completed tasks and the availability of each employee. New tasks can be also added to the system through the management application.

Each employee is able to view the daily work schedule and his scheduled tasks by using a mobile device. Upon an assignment of a new task, the employee is notified accordingly. Moreover, the employee declares the actual starting time and ending time of each task he/she performs. The system architecture is depicted in Figure 3.

4 CASE STUDY

In order to test the functionality of the HR optimization toolkit, real static and dynamic data were acquired from an industrial shop floor for chemical processes. Information about different tasks, information about employees and historical data were embodied in the HR model. The HR resource assignment algorithm was run in offline mode under different initial state cases.

The specific shop floor involves employees of different trades such as process operators, maintenance supervisors and technicians. Automation, Electrical and Process technicians are the main distinct technician trades employed. An excerpt from the definition of employee static information in XML format is illustrated in Figure 4. The scenarios that will be presented concern the assignment of Automation technicians. More specifically, the case study is about a maintenance composite task (replacement of a heating resistance) of high priority with estimated duration of about 150 minutes that has to be assigned and executed as soon as possible. One Automation technician is required (along with employees of other trades) and must be selected from the two candidates Automation technicians (ID1: 14-NM, ID2:18-PK) who are present. The scheduled starting time of the maintenance task is 10:06 a.m.

4.1 *Experimental results*

In the first scenario, which is the most straightforward, there is not any scheduled or running task during the time interval required by the arriving task for neither of the two candidate employees. Therefore, both technicians are available, no task overlap exists and the employee adaptation cost $E_{adapt\_cost} = 0$ in both cases. As it is illustrated in Figure 5, the technician with ID 14-NM has two

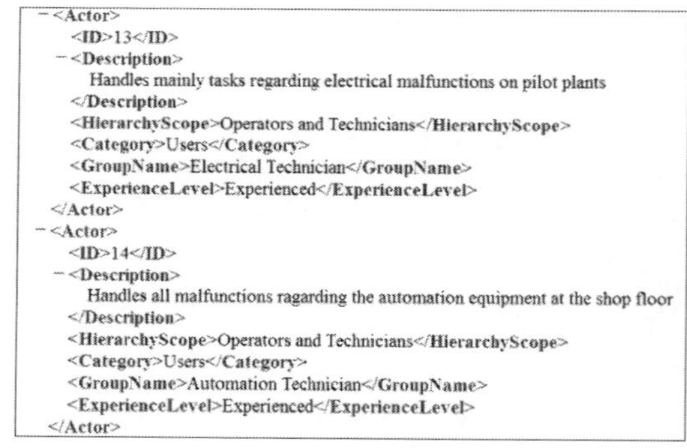

```
- <Actor>
    <ID>13</ID>
  - <Description>
      Handles mainly tasks regarding electrical malfunctions on pilot plants
    </Description>
    <HierarchyScope>Operators and Technicians</HierarchyScope>
    <Category>Users</Category>
    <GroupName>Electrical Technician</GroupName>
    <ExperienceLevel>Experienced</ExperienceLevel>
  </Actor>
- <Actor>
    <ID>14</ID>
  - <Description>
      Handles all malfunctions ragarding the automation equipment at the shop floor
    </Description>
    <HierarchyScope>Operators and Technicians</HierarchyScope>
    <Category>Users</Category>
    <GroupName>Automation Technician</GroupName>
    <ExperienceLevel>Experienced</ExperienceLevel>
  </Actor>
```

Figure 4. Static information about employees (actors) in XML format.

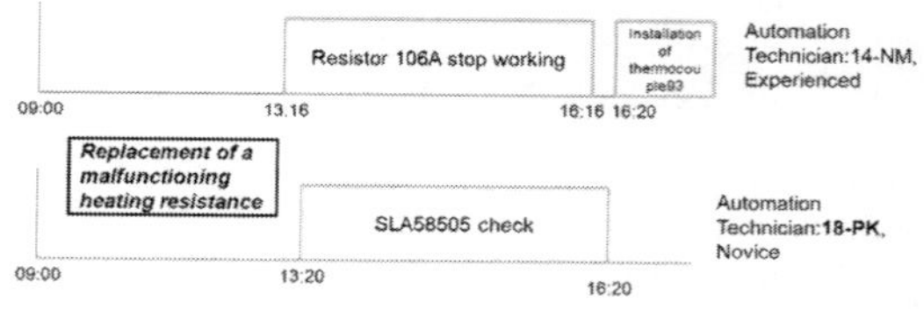

Figure 5. Assigning a technician to an arriving task according to the first scenario.

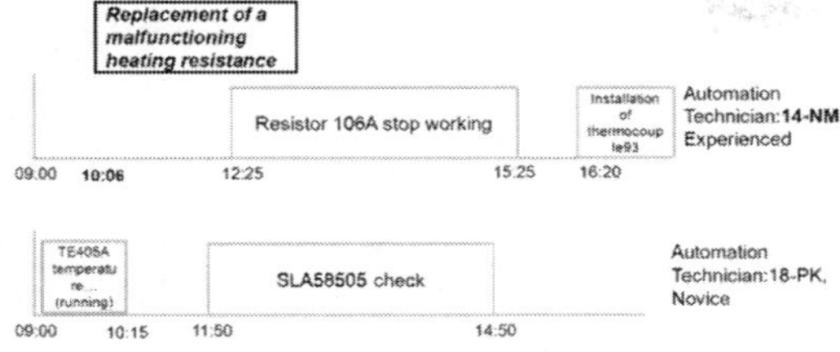

Figure 6. Assigning a technician to an arriving task according to the second scenario.

scheduled tasks in his work list: the first one has an estimated duration of 3 hours while the second one lasts 30 minutes. On the contrary, only a 3-hour task is scheduled to be performed by the technician with ID 18-PK. Eventually the latter Automation Technician was selected for the task, yielding the lower cost, which was 0.167 as opposed to 0.182 for 14-NM technician, due to the lighter remaining workload.

In a second scenario employing the same technicians, which is illustrated in Figure 6, the already assigned task of each employee overlaps with the arriving high priority maintenance task. Moreover, the technician with ID 18-PK is performing another task which has started at 09:15 when the high priority task arrives. Thus, the adaptation cost E_{adapt_cost} of technician 18-PK is much higher (0.533 versus 0.034) due to the longer overlap ratio and the need to interrupt the currently running task which is about to finish soon. The total cost E_{total_cost} calculated for technician 14-NM is 0.203, which is lower than the cost that corresponds to technician 18-PK (0.491). As a result, technician 14-NM is selected in this case even though his remaining workload is higher. We have to note that in both cases tested, the estimations produced by the CRF models were almost the same for both technicians (1 versus 0.99) and as a result the respective cost factors were negligible.

In both the aforementioned scenarios, the applied assignment policy for calculating E_{cap_cost} was set to 'Balancing'. We also have to note that distance computation was not utilized in the case study since employee location information was not available. The installation of the equipment required for the localization of employees is a work in progress.

5 CONCLUSIONS AND FUTURE WORK

This paper presented a feature-rich HR model which allows the optimization of human resources in enterprises. The modelling approach that was followed is described along with the main entities and parameters that are taken into account. Furthermore, a novel multi-criteria task assignment method which attempts to assign an arriving task to the most appropriate employee(s) in enterprises was presented. The task assignment method takes into consideration diverse information such as the remaining workload, task overlapping and estimations provided by CRF probabilistic models trained with historical task assignments in order to rank the candidate employees based on suitability. Initial evaluation showed that effective and even optimal decisions can be made according to the selected assignment policy.

The advantage of the proposed HR model is its ability to encapsulate the entities and information required for allowing the automated decision making and minimizing human intervention when HR management-related actions are needed. However, the estimations produced by the probabilistic models are not available in case a new type of task which has not been seen before has to be assigned. Furthermore, the ranking of employees based on proximity to the task requires localization hardware to be installed. We believe that since advanced sensing solutions are becoming more widespread and IT infrastructure is gaining a key role in enterprises' shop floors, automated real-time monitoring and evaluation of the work, operations and production process towards decision support is feasible. Since such functionality could improve convenience and reduce response times when unexpected events occur, we believe it is worthwhile for an enterprise to invest in such a solution.

As a future work, we intend to extend the proposed method to support task scheduling by extracting as output the scheduled starting time of tasks as well. We also plan to enhance the CRF probabilistic models by adding extra features, and utilize the information about the current location of employees into the HR assignment method.

ACKNOWLEDGEMENT

This work was partially supported by the EU funded H2020 – IA – 636302 – SatisFactory project.

REFERENCES

Agnetis, A., Murgia, G. & Sbrilli, S. 2014. A job shop scheduling problem with human operators in handicraft production. *International Journal of Production Research* 52(13): 3820–3831.

Azhar, S. 2011. Building Information Modeling (BIM): Trends, benefits, risks, and challenges for the AEC industry. *Leadership and Management in Engineering* 11(3): 241–252.

Bouzidi-Hassini, S., Tayeb, F.B.S., Marmier, F. & Rabahi, M. 2015. Considering human resource constraints for real joint production and maintenance schedules. *Computers & Industrial Engineering* 90: 197–211.

Cabanillas, C., García, J.M., Resinas, M., Ruiz, D., Mendling, J. & Ruiz-Cortés, A. 2013. Priority-based human resource allocation in business processes. In S. Basu, C. Pautasso, L. Zhang & X. Fu (eds), *Service-Oriented Computing*: 374–388. Berlin: Springer Berlin Heidelberg.

Huang, Z., van der Aalst, W.M., Lu, X. & Duan, H. 2011. Reinforcement learning based resource allocation in business process management. *Data & Knowledge Engineering* 70(1): 127–145.

Liu, Y., Wang, J., Yang, Y. & Sun, J. 2008. A semi-automatic approach for workflow staff assignment. *Computers in Industry* 59(5): 463–476.

MESA XML Committee, B2MML V0600. mesa.org. 2013. http://www.mesa.org/en/B2MML.asp Web. 10 Apr. 2016.

Morgeson, F.P., Delaney-Klinger, K., Mayfield, M.S., Ferrara, P. & Campion, M.A. 2004. Self-presentation processes in job analysis: a field experiment investigating inflation in abilities, tasks, and competencies. *Journal of Applied Psychology* 89(4): 674–686.

Shen, M., Tzeng, G.H. & Liu, D.R. 2003. Multi-criteria task assignment in workflow management systems. In IEEE (ed.), *System Sciences; Proc. intern. conf.*, Hawaii, 6–9 January 2003. IEEE.

Skobelev, P., Kolbova, E., Kazanskaia, D., Shepilov, Y., Tsarev, A., Shpilevoy, V. & Shishov, A. 2013. Multi-agent system smart factory for real-time workshop management in aircraft jet engines production. *Intelligent Manufacturing Systems* 11(1): 204–209.

Sutton, C. & McCallum, A. 2006. An introduction to conditional random fields for relational learning. In L. Getoor & B. Taskar (eds), *Introduction to statistical relational learning*: 93–128. Cambridge, MA: MIT press.

Varela, M.L.R. & Ribeiro, R.A. 2014. Distributed manufacturing scheduling based on a dynamic multi-criteria decision model. In L.A. Zadeh, A.M. Abbasov, R.R. Yager, S.N. Shahbazova, & M.Z. Reformat (eds.), *Recent Developments and New Directions in Soft Computing*: 81–93. Springer International Publishing.

Volk, R., Stengel, J. & Schultmann, F. 2014. Building Information Modeling (BIM) for existing buildings—Literature review and future needs. *Automation in construction* 38: 109–127.

Wibisono, A., Nisafani, A.S., Bae, H. & Park, Y.J. 2015. On-the-fly performance-aware human resource allocation in the business process management systems environment using naïve Bayes. *Asia Pacific Business Process Management*: 70–80. Springer International Publishing.

eWork and eBusiness in Architecture, Engineering and Construction – Christodoulou & Scherer (Eds)
© 2016 Taylor & Francis Group, ISBN 978-1-138-03280-4

A review of resource based view in the construction industry: A BIM case as a strategic resource

O. Geylani
Graduate School of Science and Engineering and Technology, Construction Sciences, Istanbul Technical University, Istanbul, Turkey

A. Dikbas
Dean of the Faculty of Fine Arts, Design and Architecture, Istanbul Medipol University, Istanbul, Turkey

ABSTRACT: Resource Based View (RBV) as an economy based strategic management theory, became an important management strategy for project management executives during the last decade. On the other hand, Building Information Modelling (BIM) is a valuable resource that projects and other outputs produced by its software leads the firm to differ in the market and may be considered as a strategic resource for the construction firms. This research will present the findings that are gathered from a case study in one of the Turkish light steel prefabricated construction firm, which has a new step with the Building Information Modelling system implementation including its tools and capabilities. This paper aims to test the Resource Based View in Turkish construction industry, while applying the BIM as a strategic resource within the environment of project management. Moreover, future research will take a closer look at the outcomes of the case study of this research, such benefits of the BIM as a strategic resource may be revised for other research studies, the outputs of the case study as a time study, has additional benefits for the construction firms such as; re-organization of the organizational structure of the firm, team development of the BIM execution plan, human resource management, the management strategy development for the firm.

Keywords: Resource Based View, Project Management, Building Information Modelling, strategic resource, construction industry

1 INTRODUCTION

Resource Based View is one of the key research subjects that has been studied during the last decade by the scholars since 1950s. On the other hand, Resource Based View (RBV) as an economy based strategic management theory became an important management strategy for Project management executives. However, the construction industry had various theoretical and empirical case studies in different research fields, RBV has a limited background within especially in the construction industry.

Moreover, RBV related studies include IT industry and its outputs which may take into account as strategic resources. Hence, as a strategic resource with the interface of RBV, PM and construction industry, Building Information Modelling (BIM) and the effects of its implementation, is targeted during this research. Hence, this research will present the findings that is gathered from a case study in one of the Turkish light steel prefabricated construction firm, which has a new step in the BIM system implementation, its tools and capabilities.

So, this paper aims to test the RBV in Turkish construction industry, while applying the BIM as a strategic resource within the environment of PM.

2 LITERATURE REVIEW

Today, if we ask why do some firms persistently outperform others? This will be the key question for RBV; which shift the focus on firms, resources to understand the performance of a firm.

In the early 1950s, the business study was generally centered on the study of individual business people and firms (Barney and Clark, 2007) and the late 1950s, RBV phenomenon came about and formed after the work of Edith Penrose (1959) with the contribution of other researchers (Selznick, 1957; Chandler, 1962). Conforming to Penrose (1959), a firm's growth was related to the resources that defined as human or non-human resources.

Moreover, the resources had the impact on the firms' strengths and weaknesses.

Later, in 1980s the business history included the terms of strategic management and the description of the growth and success of the firms (Chandler, 1984). While Michael Porter concentrated on a firm's product market position and its competitiveness in the market, he defined firm's resources as opportunities and threats to explain the theory of competitive advantage, this definition was close to Penrose's thought. Hence, as a further study, Porter regarded a firm's performance was linked to the external environment of the firm which refer to the macro economy, industry, market place, trade, which was explained by the industry view perspective (Porter, 1996).

Moreover, the attention to the competitive advantage was advanced by the work of Wernerfelt (1984) that he realized the competition among the firms linked to the resources which would support the firm in order to gain competitive advantage. Almost, at the same time period, Wernerfelt and Rumelt published papers about RBV in 1984. On one hand, Wernerfelt (1984) emphasized performance of the firms could be analyzed regarding the resources of a firm. On the other hand, according to Rumelt (1984) worked on the ability of firms to use the resources and a kind of strategic theory could explain the question; why firms exits? He, especially, took attention on two of the subjects; the rent generating and appropriating characteristics of firms (Barney and Clark, 2007).

Later on, these previous studies; Barney (1986a) concentrated on the resources of a firm and emphasized that the superior firm performance was related to the characteristics of the resources; comparable to Wernerfelt (1984) their proposal could become a theory according to him. Moreover, the studies of Barney about RBV could serve more than the theories of competitive advantage, depending on the product market positions of firms.

As a consequence of the history of RBV; Wernerfelt (1984), Rumelt (1984), Barney (1986), Dierickx and Cool (1989); these four papers actually pointed out the main principles of RBV. Especially, do a research about persistent superior firm performance, that would be possible by doing an analysis about a firm's resources, this research also could turn into a theory.

Another consequence was that; in order to gain the sustained superior firm performance; a firm's resources should have some specific characteristics. Rumelt (1984) suggested two different term value and isolating mechanisms (Barney, 1986a) these resources should generate economic rents. More, the resources would be kind of a resource that help the firm gain and sustain superior performance (Barney and Clark, 2007).

Today, there is still a gap in the construction industry about RBV studies. The outcomes of strategic resources should be revised and the arguments of the case studies may become a basis for all construction firms which look for an appropriate strategy for their firms in order to gain and sustain competitive advantage.

3 RESOURCE CATEGORIZATION

A firm's bundle of resources is categorized by scholars and their managerial focus. Firstly, resources may have an effect on externally or internally on the forum. While strategic resources are related a firm's product market and its components are company vision, business portfolio, market opportunities, customer base and reputation in the market, market position (such as market leader, first mover, etc.) Moreover, the resources are divided into two categories; tangible resources that the financial resources, including cash flow, debt capacity etc. And intangible resources like human resources (individual) and organizational resources (organizational). In addition, Resource based view has also its own categorization which step by step leads to the competitive advantage.

3.1 *Valuable resources*

Having valuable resources is the first step for a firm in order to gain competitive advantage or sustained competitive advantage. Related to the valuable resources, a firm is able to maintain a strategy to keep and develop its accomplishment. Concerning to the SWOT model (Strengths-Weaknesses-Opportunities-Threats) if a firm's performance and its strategy increase opportunities and decreases threats than its resources have the potential to gain competitive advantage. Moreover, another mission of valuable resources is that; they increase the economic rent for the firm (Barney and Clark, 2007), as an example, customer interest will rise and the costs will decrease by these kind of resources.

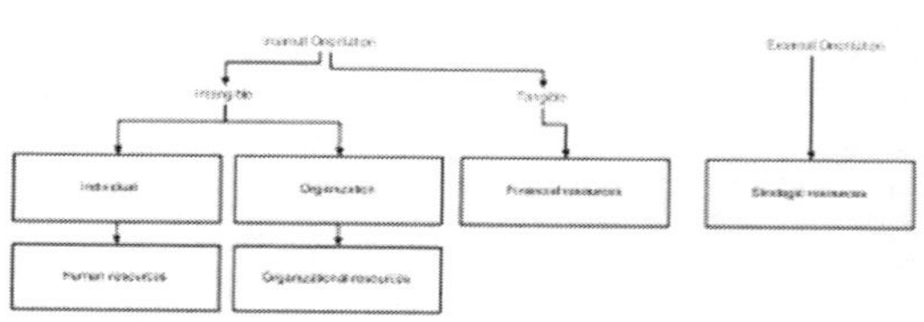

Figure 1. Resource Categorization (Lim, 2008).

Figure 2. VRIO framework.

Moreover, firms should also posses valuable characteristics to have the potential for sustained competitive advantage. The firm "strategy and the specific market environment" are the main criteria that a firm should measure the value of the resources, according to these criteria.

3.2 *Rare resources*

While generally every firm may have valuable resources, competitive advantage requires more economic value for a firm. Regarding an industry that many firms have the same kind of valuable resources at that point and a similar strategy if implemented, there will not be any positive change for the competitive advantage. In fact, the competitive advantage is possible if a firm has valuable and rare resources. Otherwise, the valuable resources which are common in every firm will create just the "competitive parity" which is only helpful for the firm's economic survival (Porter, 1980; Mc Kelvey, 1982).

However, firms may have a valuable resource or a bundle of resources, competitive advantage may exist in an industry if some of the firms generate "perfect competition dynamics" (Barney and Clark, 2007).

3.3 *Imperfectly imitable resources*

The competitive advantage of the firm is related to its valuable and rare resources.

More, sustained competitive advantage, is possible when other firms are not able to possess or copy the same resources. These sources are called imperfectly imitable resources (Lippman and Rumelt, 1982; Barney, 1986a,b).

Moreover, the imitation of the resources will be costly for other firms because of three reasons;

- The capability of having a resource for a firm is related to its sole "historical conditions".
- It is difficult to understand the reason and the network between the resources of a firm and its sustained competitive advantage.
- It is also difficult to understand the social reason that a resource could make a firm to get an advantage (Dierickx and Cool, 1989) so other firms to imitate that resource is socially complex (Barney and Clark, 2007).

3.4 *Organization*

The literature review includes two assumptions for the organization. One of them is that; valuable strategies may be implemented because of the potential of them which present for a firm. Second one is; some of the resources may become a reason for sustained competitive advantage and they are costly to imitate (Barney and Clark, 2007).

On the other hand, according to Barney (2001) firms should be organized to use the resources and capabilities efficiently, which means to use the economic potential of them.

However, strategy term and RBV has a connection there is a gap in empirical studies regarding strategy-RBV implementation. Moreover, the scholars focused on the reasons of sustained competitive advantage if the only resource was the ability of the firm to gain it. The ability of the firms was studied in detail (Teece, Pisano and Shuen, 1997) and according to their study; a firm may develop new capabilities, which later called "dynamic capabilities". So, an organization is an organized firm that is capable of implementing a firm's strategy to gain competitive advantage.

3.5 *The VRIO framework*

Valuable, rare, inimitable resources, organization and the heterogeneity of the resources, all of these criteria has a relation to making a firm to gain sustained competitive advantage. So, in order to understand the correspondence of a resource for sustained competitive advantage, a framework was developed and called VRIO framework. VRIO includes four key parameters and to analyze the resources.

The VRIO framework of the RBV, proposes several steps to gain sustained competitive advantage. These steps, of course, related to the resources, and the first step began with the valuable resources. If the firm posses valuable resources that means the firm has competitive parity, that the firm makes "normal profits" and nearly equal to its competitors. As well, more than valuable resources a firm may have rare resources also, at that time the firm has a "temporary competitive advantage" hence, any time a competitor may also gain this advantage. In addition to all, however a firm has valuable and rare resources, a firm may prevent competitors to gain the competitive advantage of its inimitable resources, on that occasion the firm has the competitive advantage. According to the VRIO framework, the organization should be capable of using the resources and the capabilities, granted that the firm will have sustained competitive advantage. Sustain competitive advantage as a situation means that the firm has over profit than its competitors and the competition is not easy to "understand or copy".

4 TESTING RESOURCE BASED THEORY

There is a link between a firm's resources and that its strategy is implemented. According to Barney (2001) the firm resources regarded as the tangible and intangible asset which are controlled by the firm. On the other hand, they should posses the "potential to generate economic value" described Porter (1991). The term "value" actually is related to the firm's strategy and the capabilities to use the resources appropriately (Amit & Shoemaker, 1993).

Scholars, emphasize that a firm could measure this potential of its resources by measuring the value which is created by firm strategies. Moreover, it is mentioned that some resources have no value, they could still have the potential during a strategy implementation.

Besides, the strategy examples could be counted as; cost leadership, product differentiation, vertical integration, flexibility, tacit collusion, strategic alliances, corporate diversification, mergers and acquisitions and other international strategies (Barney, 2001).

The literature review shows there had been little research about the specific firm resources and capabilities which has the potential to "create and implement" the previously mentioned firm strategies. Also, because of the diversity of firm resources Barney (2001) made a classification of firm resources that; these are financial, physical, human and organizational resources. The classification was assumed to be useful to understand the link between the resources and the firm specific strategies.

In order to examine the resources, capabilities and the firm strategy two of the sample examples given in the literature. First, in 1994 a study was made in pharmaceutical by Henderson and Cockburn. They worked on the effectiveness of new patentable drugs, comparing the different pharmaceutical firms. Here, the patent is a criteria for testing because it is a source of economic value of pharmaceutical firms (Mansfield, Schwartz and Wagner, 1981). Morever, the number of patented drugs have an effect on the revenue. This is explained by Henderson and Cockburn regarding a specific resource called "architectural competence". According to them, this specific leads some firms to have more patents compared to other firms. So, their study Showed that when this specific resource is used to "develop" new patentable drugs, it was potentially generated economic value.

Another important study was about IT industry, the IT function and the customer service function were examined in the study of Ray, Barney and Wuhanna (2004). One of the North American insurance company's customer service had an important role in insurance companies because it has information intensive function. The interface of IT and customer service is that IT is a helpful tool to connect the customer service professionals with customers' needs. The aim of the study was to measure "the level of cooperation between the IT and customer service functions". Actually, that relationship was regarded as "a socially complex resource" which would also have the potential to create economic value for the insurance firms, but it was also needed a strategy which was the usage of that resource during two "develop customer service-specific IT applications".

With a common sense of the previous study, they both examined the potential of a specific resource whether it creates economic value for the firm and they also have similar characteristics, that; these kind of studies generally "quantitative case studies". They focused on the relationship and tried to understand the value of the firm strategies in a small sample size of the firms in a specific industry which helped the authors "clearly" identify industry-specific resources and capabilities, the aim was to built industry specific measures of these resources".

Likely, in quantitative researches, they used quantitative methods, they made a correlation between a firm's economic performance and firm resources and "attributes" after examination of the relationship between the measures of firm resources and attributes (Barney and Clark, 2007).

4.1 *The integration of RBV and PM*

Resource Based View (RBV) as a strategic management theory took attention during the last decades, hence it supports a firm to gain and keep competitive advantage. Regarding, the integration of RBV and PM, it should be better concentrating on PM capabilities and resources and PM which has an effect on a firm's competitiveness.

During the literature review, it was understood that RBV started to be searched with PM methodology because PM experts began to concentrate on strategic resources, the RBV. The strategy of a firm leads the organization, to share out the resources and capabilities and also the ones that related to PM.

As it is mentioned above, from the beginning of a project, until its closing phase PM also uses its own tangible and intangible resources including PM processes, tools and techniques. As an example of tangible resources; the tools, templates, pm bodies of knowledge or guidelines. Here, the intangible resources are also critical, so these are the "know-how" and "know-what" of a firm for PM methodology.

A research was done by Judgev et al. (2007) to examine the relationship between key project management assets and these Project management process characteristics. The research presented findings also to distinguish the PM resources and their characteristics for VRIO framework.

Firstly, an organization which "invest in" PM methodologies, uses a guide book, a "guide line" or "checklists" during PM processes. As well, these may be taken into account as PM resources in order reach appropriate project outcomes. On one hand, they are valuable PM resources, but on the other hand, any organization may easily use them as a valuable resource during their projects.

Secondly, the technology of the firm, including "hardware" and "software"; are likewise used during PM processes. "Knowledge sharing" (flow) and "decision making" are made by the technology that supported by the firm's and used by the PM executives. Knowledge, as constituents of the projects required knowledge management systems and IT systems that are a kind of "physical tools and techniques". As it is seen, these PM related firm resources also are valuable for a firm but they are not "rare or unique".

In addition to all above, PM practitioners are educated and afterwards, use the PM knowledge bodies and the PM standards that are set by well-known institutions. The bodies of knowledge are again, a kind of valuable PM resources, but the findings showed that they may used by any firm, so they are not rare resources.

As an organization, working with immense projects, need to be more concentrated on and specialization of PM. So, that kind of organizations chose the way of establishing Project Management Offices (PMOs). The technology, methodology, templates, standards is typical PM resources that they may be valuable but every PMOs may have them (Judgev, et al., 2007).

RBV is an appropriate theory to use in this thesis research for following reasons;

- RBV has a rich 30 year history
- RBV addresses knowledge and process assets and this fits with the exploration of PM as a strategic resource (Judgev, 2003).
- RBV focus on strategic resources to gain and sustain competitive advantage, so this study also focus on strategic resources for the construction firms.
- RBV interested the "within" structure of the firm's the organization itself; so, this research analyzes the organizational structure of the construction firms.
- RBV both emphasizes tangible and intangible resources, including human resources and capabilities also intangible PM resources, moreover, this research considers Building Information Modelling (BIM) as a PM related strategic resource for the construction firms.

4.2 *Resource based view and BIM*

The BIM application research Project was emerged from the needs of one of the prefabricated Turkish construction firm. During the first meetings the founders of the company explained that they had Autodesk BIM Revit software on the personal' computers for three years. However, there had been an attempt to get the personnel learned the use of Revit, that attempt was unsuccessful because of the efficient use of the program. Moreover, other managers of the company were doubtful after the fail of the education session of BIM. After three years time, the company again searched for solutions in order to have organizational development and innovation.

The founders of the company have always been close to innovative solutions and asked for the executives' views for their problems. So, during the first meetings the problems are mainly explained under a list;

- The firm has many national and international ongoing projects, but the workers, including architects, engineers and site workers complained about the limited time; the bidding projects, the ordered projects and the application projects were eventuate concurrently. However, every project is important to the revenue of the company and client satisfaction, it was hard to concentrate on the projects even if they have the same type of projects.
- The founders of the firm were stated that the personnel could not turn every bidding projects into ordered projects, unfortunately that resulted in the lose of %50 new clients.
- The firm has 5 different brands, each brand has a manager, a design team including architects and engineers. When a new client asks for a new project to see the design and the cost, the design team uses AutoCAD to prepare 2D drawings, after the completion of the drawing phase the cost of the project was possible to calculate, but this is generally too late for the impatient clients.

4.3 *The decision making of BIM investment*

The review of the firm A requirements are quite similar to the BIM4M2 research that is the "BIM for manufacturers and manufacturing" survey conducted in 2014 and updated on 2015. The root causes of why the manufacturers invest in BIM listed as below, so the customer satisfaction, time saving outcomes and competitive advantage of the firm is some reason that why these firms apply BIM in their organizations.

- There is a demand of customers about perfect and correct 3D modelling of their prefabricated construction projects.
- Using Revit (BIM) the typical projects will be produced and saved in the firm's BIM database to save time when the demand of the customers arise, the price offers and the context of the project will be calculated with zero mistakes.
- Using BIM will help to outperform the rivals.
- BIM will lead the firm to become unique in the market (if other firms do not apply same solutions) more, the firm will gain competitive advantage.
- Policymakers support BIM in some countries, so if the firm has international projects it will be the easiest way to communicate by using BIM as remote partner of the projects.
- BIM will increase the efficiency before the project is constructed on the site, the project team will decrease the risks of mistakes/misunderstandings.

4.4 *Resource based view and BIM interface*

"BIM is an enabling technology with the potential for improving communication among business partners, improving the quality of information available for decision making, improving the quality of services delivered, reducing cycle time, and reducing cost at every stage in the life cycle of a building" (Smith and Tardif, 2009).

Regarding a developing construction firm, BIM serves more than a software, it serves a system which will support the firm's competitiveness in the marketplace. The reason to combine BIM and Resource Based View at a first glance is that they for related to management strategy of the firm and the resources.

For example, RBV supports the firms to gain and sustain competitive advantage and BIM offers a tool (Autodesk-Revit) to produce the projects that may be unique to firm, so the firm may become unique because of using the innovative software and the technology.

RBV distinguishes resources step by step, according to VRIO framework to gain sustained competitive advantage. Valuable resources are chosen according to the firm's goals, BIM on the other hand is a valuable resource that the projects and other outputs produced by BIM software leads the firm to differ in the market. If the other firms do not apply BIM, this is a successful step to competitive advantage. BIM also linked with firm's patented product data that is kept in database. The way of data protection with other softwares which worked with BIM tools is helpful for the firm to protect their rare intangible resources, such as product information or formulas.

5 THE CASE STUDY

5.1 *Data collection method*

Research methods of our research include, face to face meeting and case study method. Data was collected during the meetings within the firm. The firm describes their needs, strengths and weaknesses which is also important to understand their resources. Also, A group of 50 employers, including group managers, architects, engineers and other technical personnel had been asked about their needs during their working hours. Commonly, the first impression was about a need of miracle to heal the time and cost of the projects in order to be successful.

The meetings were lasted more than a month than the firm was asked to start an education session of BIM and its tools.

5.2 *The firm V*

The case study presents a two year BIM application project in a light steel prefabricated construction firm which was founded in 1990 Istanbul, Turkey. First, It is better to understand the history and characteristics of the firm; that it has 13 brands and 10 companies which produces product and services in individual, corporate and industrial "means".

The firm's growth has mostly been through geographic expansion that its production in the manufacturing facilities of four different provinces around Istanbul, Turkey (Dilovası, Sultanbeyli, Bilecik and Tuzla) spreading in 67.000 m2 closed and 151.000 m2 total area.

The exportation power of the firm was expanded %70 of its production in 2015. The application of the products (houses and other prefabricated facility buildings) nearly around 3 million m2 in Turkey and in more than 70 countries around the world.

The history of the company has many innovative and successful milestones, but in the year of 2015, they decided to invest in the development of the organizational structure and strategic management, especially they took consultation about how to value of human resources and capabilities.

The BIM system was set, the personnel were educated and the organizational structure reconfigured according to BIM executives, tangible resources (computers), intangible resources (BIM software and license) revised, the know-how of the years applied to BIM.

During our research, approximately in one year time there had been 90 times of visits to the firm for the BIM application phases. The outcomes of the phases may be listed as;

- 40 of the personnel of the firm had an education of BIM and also BIM Revit.
- The resources of the firm regarding especially human resources, organized again according to the BIM execution plan.
- The new capabilities of personnel are discovered.
- The performance of the personnel was tested by practice drawings with a time limitation.

6 CONCLUSION

This paper highlights the possibility of implementation of Resource Based View, Project Management and BIM topics. Theoretical background of the strategic management concept in the construction industry is better be developed by the case studies especially. This case study is a necessary step for construction firms which regard BIM as a strategic resource. Moreover, future research will take a closer look at the outcomes of the case study of this research, such benefits of the BIM as a strategic resource may be revised for other research studies, the outputs of the case study as a time study, has additional benefits for the construction firms such as; re-organization of the organizational structure of the firm, team development for BIM execution plan including human resource management, the development of the management strategy for the firm. However, this research is conducted in one of the Turkish construction firm, BIM system application, the outputs of the research will hopefully encourage the other construction firms to outperform their rivals.

REFERENCES

[1] Almarri, K. and Gardiner, P. (2014). "Application of Resource-Based View on Project Management Research: Supporters and Opponents". Social and Behavioral Sciences 119 (2014) 437–445.

[2] Barney, J.B. (1991). "Firm Resources and Sustained Competitive Advantage." Journal of Management 17: 99–120.

[3] Barney, J.B. (2001). "Resource-Based Theories of Competitive Advantage" Journal of Management 27: 643–650.

[4] Barney, J.B. (2001). "Resource-Based Theories of Competitive Advantage: A Ten-Year Retrospective on the Resource-Based View." Journal of Management 27: 643–650.

[5] Chandler, A. (1962). "Strategy and Structure": Chapters in the history of the American industrial enterprise. Cambridge, MA: MIT Press.

[6] Jugdev, K. and Mathur, G. (2006), "Project management elements as strategic assets: preliminary findings", Management Research News, Vol. 29 No. 10, pp. 604–17.

[7] Jugdev, K., et. al. (2007). "Project management assets and their relationship with the project management capability of the firm", International Journal of Project Management, Vol. 25 No. 6, pp. 560–8.

[8] Judgev, K. and Mathur, G. (2013). "Bridging Situated Learning Theory to the Resource-Based View of Project Management" International Journal of Managing Projects in Business 6(4): 633–653.

[9] Galbreath (2005). "Which resources matter the most to firm success? An exploratory study of resource-based theory." Technovation 25 (2005) 979–987.

[10] Killen, C., et al. (2011). "Advancing Project and Portfolio Management Research: Applying Strategic Management Theories". International Journal of Management 30(2012): 525–538.

[11] Lim, S.K. (2008). Dynamic Resource-Based View of Entrepreneurial Firm Growth. School of Graduate and Postdoctoral Studies: The University of Western Ontario London, Ontario, Canada.

[12] Lopez (2005). "Competitive advantage and strategy formulation: The key role of dynamic capabilities". Management Decision 43(5): 661–669.

[13] Mathur, G., et al. (2007). "Intangible Project Management Assets as Determinants of Competitive Advantage". Management Research News 30(7): 460–475.

[14] Newbert, S.L. (2007). "Empirical Research on the Resource-Based View of the Firm: An Assessment and Suggestions for Future Research" Strategic Management Journal 28(2): 121–146.

[15] Paiva, E.L. et. al. (2007). "Organizational knowledge and the manufacturing strategy process: A resource-based view analysis." Journal of Operations Management 26(2008): 115–132.

[16] Penrose, E. (1959). "The Theory of the Growth of the Firm". London: Basil Blackwell.

[17] Porter, M.E. (1980). "Competitive Strategy: Techniques for Analyzing Industries a nd Competitors". New York: The Free Press.

[18] Rodríguez, T.F.E. and V.P. Robaina (2006). "A review of outsourcing from the resource-based view of the firm." International Journal of Management Reviews 8(1): 49–70.

[19] Rumelt, R.P. 1984. "Towards A Strategic Theory of the Firm." In Competitive Strategic Management, ed. R.B. Lamb, pp 566–70. Englewood Cliffs, NJ: Prentice-Hall.

[20] Selznick, P. (1957). "Leadership and Administration". Harper & Row: New York.

[21] Schmidt, J. and T. Keil (2013). "What Makes a Resource Valuable? Identifying the Drivers of Firm-Idiosyncratic Resource Value." Academy of Management Review 38(2): 206–228.

[22] Wernerfelt, B. (2013). "Small forces and large firms: Foundations of the RBV." Strategic Management Journal 34(6): 635–643.

eWork and eBusiness in Architecture, Engineering and Construction – Christodoulou & Scherer (Eds)

Building information modeling in use: How to evaluate the return on investment?

A. Guerriero, S. Kubicki & S. Reiter
Luxembourg Institute of Science and Technology, Luxembourg

ABSTRACT: BIM (Building Information Modeling) is a new paradigm, which changes the way facilities are designed, built and managed. It is now recognized that implementing BIM in AEC (Architecture, Engineering and Construction) projects implies both organizational and technological challenges. The benefits of BIM are also well described in scientific and professional literature, in public and private BIM guidelines published by governments and owners, as well as in the marketing campaigns of software editors. However, implementing BIM internally in organizations or in AEC projects' teams remains a strategic investment. The value of BIM investments (i.e. Return on Investment) is not well defined nowadays, especially when it comes to projects' organizations, i.e. sharing of costs vs. sharing of benefits among the design/construction/facility management chain. The article proposes a project-level and structured method, enabling to assess the BIM ROI among project partners and projects' workflows implementing BIM (BIM uses). This method is then deployed on real construction projects and results are discussed.

1 INTRODUCTION

Building Information Modelling constitutes a lever of modernization of the construction industry that profoundly modifies the business practices in this sector. Eastman (Eastman et al., 2011) defines BIM as "*a digital representation of the building process to facilitate exchange and interoperability of information in digital format*". BIM is considered as a socio-technological system that implies for the construction sector not only a technological change, but also a process change. Indeed BIM modifies the way that architectural design artefacts are created, but also profoundly the collaborative process associated to the act of building structures and managing facilities and assets with virtual models. The Bews-Richards BIM Maturity Model (Bew and Richards, 2008) presents the different levels that an organization has to overpass in order to reach the BIM level 3 where the collaborative process is expected to be practically and technologically integrated. These changes resulting from the progressive sophistication of the BIM-related processes have a considerable financial impact that probably slows down the adoption of BIM. The question of the BIM ROI (Return on Investment) is therefore crucial to convince the construction sector to convert, as well as to strategically manage this change.

The article presents a method of project-based BIM ROI calculation relying on BIM uses. Then first application results are discussed. Finally prospects and further developments are developed.

2 RETURN ON INVESTMENT OF BIM

According to (NRC, 2012), measuring the ROI of BIM on construction projects has different finalities: 1) contributing to understand the impact on the organizational performance; 2) planning, monitoring, measuring and identifying the benefits (including future benefits); 3) taking into account the risks associated to the benefits; 4) collecting data that could be used for other IT-based investment analysis.

The following literature review presents various research works addressing BIM ROI topic that are representative of these various objectives.

2.1 *Existing literature on BIM ROI*

The ROI is classically defined as follows:

$$ROI = \frac{Gain - Cost}{Cost}$$

Even if one can see a wide interest in measuring the costs, benefits and Return on Investment (ROI) of BIM, only a few research outputs have been published on this topic:

- (Autodesk, 2007) suggested a method for evaluation of BIM ROI dedicated to the architecture and engineering offices. It is based on the loss of productivity during the BIM training period of time and the gain of productivity when employees become operational and efficient.
- In (Giel et al., 2010), the authors suggest an analysis of the ROI based on the gain resulting from the direct cost of subcontracting out panel shop drawings, the direct cost in preventable change orders and the indirect costs for preventable time overrun. Then, the authors propose a comparison between similar projects adopting or not BIM.
- (Qian, 2012) proposed a model, which is dedicated to the prediction of cash flow of new BIM implementers on an annual basis. It takes into account a lot of parameters such as the average salary of the different specialists, the savings due to productivity gains, annual maintenance costs of BIM-enabled workstation, etc.
- In (Lu et al., 2014), the authors suggest using time-effort distribution curves in order to evaluate the costs/benefits of BIM implementation. By comparison between a BIM and non-BIM project, the researchers highlight that the effort input in the design stage increases significantly but during the execution stage, the cost per square meter decreased significantly.
- (Poirier et al., 2015) measure the impact of BIM on labour productivity based on an action-research methodology.
- In (Teicholz, 2013), the author presents an analysis of BIM ROI from the owners' point of view. The gains and costs are segmented in two parts: initial elements (i.e. due to the implementation of the integrated system) and permanent elements (i.e. resulting from the use of the integrated system).
- In (Love et al., 2013) the research focuses on the asset owner point of view and proposes a benefit evaluation framework incorporating intangible benefits (e.g. improvement of customer service and satisfaction) and indirect costs (e.g. process re-design, staff turnover).
- Other reports or publications (McGraw-Hill, 2010, Kreider et al., 2010) present approaches based on the perceived impact of BIM in the professional practice, relying on surveys and questionnaires.

The analysis of the existing methods for calculating the ROI of BIM highlights that no unified nor global approach has been proposed so far. Moreover, in these research works, the measure of BIM ROI is global for a project or an organization but does not take into account the particular uses of BIM.

In Penn state's BIM project execution planning guide, twenty-five BIM uses have been identified and described (e.g. 3D coordination, design reviews) (Pennsylvania State University, 2011). Based on this research work (Kreider et al., 2010), we suggest a method for measuring the impact of the BIM uses on an AEC project. The method proposed in this article is an evaluation of the BIM ROI distinguishing the quantitative and qualitative metrics and decomposing the analysis according to both the stakeholders involved and the particular BIM uses implied in a given AEC project, all along its lifecycle. A prototype tool has been developed to support the method, using MS Excel and tested on real AEC projects.

3 AN ADAPTIVE ROI IDENTIFICATION METHOD

The aim of the proposed method is to provide a structured feedback about BIM costs and gains across a series of AEC projects. To achieve this goal, the method has first to be generic enough, in order to take into account the specificities of each project in terms of practitioners' teams, nature of building being designed and built, as well as regarding the types of exchanges conducted through BIM models (i.e. BIM uses). Second, the ROI identification cannot only target the productivity aspect of BIM. Indeed it is quite difficult to harmonize productivity value across very dissimilar projects' organization. Moreover, in a practical way, one can recognize that construction practitioners are usually not disposed to accounts' transparency.

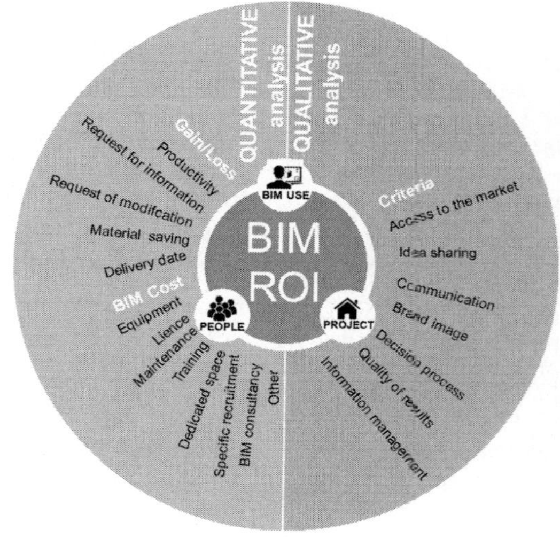

Figure 1. Metrics for BIM ROI assessment.

3.1 *BIM ROI metrics*

As the quantitative metrics are limited to interpret BIM ROI, the proposed method adds an assessment of qualitative metrics (See Figure 1). The model for quantitative analysis is principally based on the financial valorization of the gains/losses of productivity and the indirect gains as presented in (Giel et al., 2010). Regarding quantitative metrics, the *productivity* is usually evaluated through the "gains vs. costs" formula. Gains are basically constituted by the capability of an organization to gain time in completing tasks. A financial translation can be obtained by linking it with labor costs. The management of *Requests for Information* is usually a cumbersome process that can be facilitated by BIM, and the method suggests identifying if RFIs augment, are stable or diminish. The *Requests for Modification* are also considered as important metrics as BIM should better integrate the technical design disciplines and thus reduce the rework and modifications across the design tasks. Moreover *savings regarding materials and building products* are often announced in BIM literature and the method also considers it. Finally, the optimization of the design and construction processes thanks to BIM exchange processes is often claimed, therefore the method considers *earlier or later building delivery* as an essential quantitative metric. The second set of metrics considers the qualitative value of BIM. The proposed method considers first the interest of BIM for *accessing the construction market*, e.g. when tendering for public bids where BIM is explicitly required. The interest of BIM for *sharing ideas* across project's stakeholders is another important aspect. Its usages to enhance *communication* among stakeholders as well as with end-users are also considered. At the scale of an organization (company), BIM experience is often marketed to develop the *brand image*. *Decision-making processes* are also qualitatively impacted by BIM processes and BIM datasets, e.g. for owners decision-making. Then, the *quality of results* (both design and construction results) is a global expectation regarding digital modeling and BIM. Finally, the method assesses the overall *information management* as a qualitative feedback of BIM.

3.2 *BIM ROI assessment process*

The proposed method aims to be adaptive in order to enable the assessment of BIM ROI in heterogeneous AEC projects environments. This adaptation is possible thanks to the concepts of "BIM uses" and "BIM users". The assessment process is summarized in Figure 2, which also defines the main interaction with the toolset developed to assist the assessor (Figure 3). The first step is related the identification of the BIM uses for the project being assessed. Selecting the project's BIM uses among the whole list guarantees a good understanding of the implementation of IM across the processes. In the proposed method, the BIM uses can also be identified per project phase in order to refine the analysis (e.g. BIM model authoring can be considered at design, construction and/or management phases).

A "configurator" sheet assists the assessor in this preliminary task. The second step defines the roles and users of BIM (i.e. stakeholders both producing and consuming BIM data). A macro then generates sheets from a template in order to provide the assessor with an adapted set of fields. For example, regarding productivity, four sheets can be generated (programming, design, construction and facility management), comprising the specific BIM uses and involved actors selected. The assessor then fills in the various sections related to productivity (gain vs. loss of worktime per BIM use and per BIM user) and other quantitative gain/loss. Then the qualitative feedback is gathered per BIM use and per BIM user, on the basis of a Likert scale: 1 (strongly disagree) to 5 (strongly agree). A specific sheet is committed to identifying the costs of BIM that can be specifically charged to the project implementation: additional hardware and software (license) costs, staff training, specific recruitment or consultancy.

Finally, reports are produced to graphically visualize the Return on Investment: 1) Per actor: quantitative ROI and qualitative metrics compared, and 2) Per BIM use: quantitative visualisation (showing gains and/or losses per BIM use), and qualitative view.

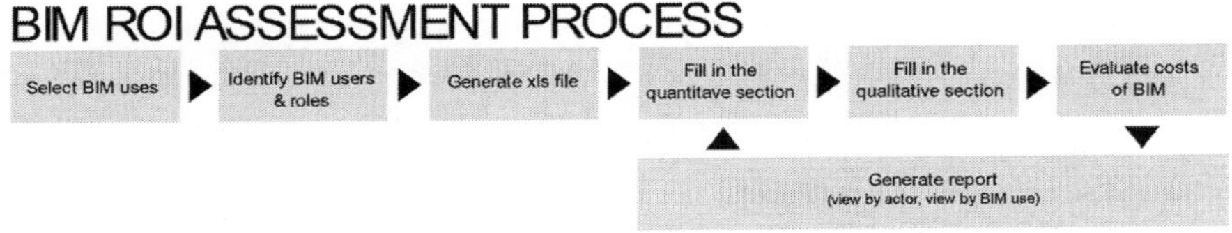

Figure 2. BIM ROI assessment process.

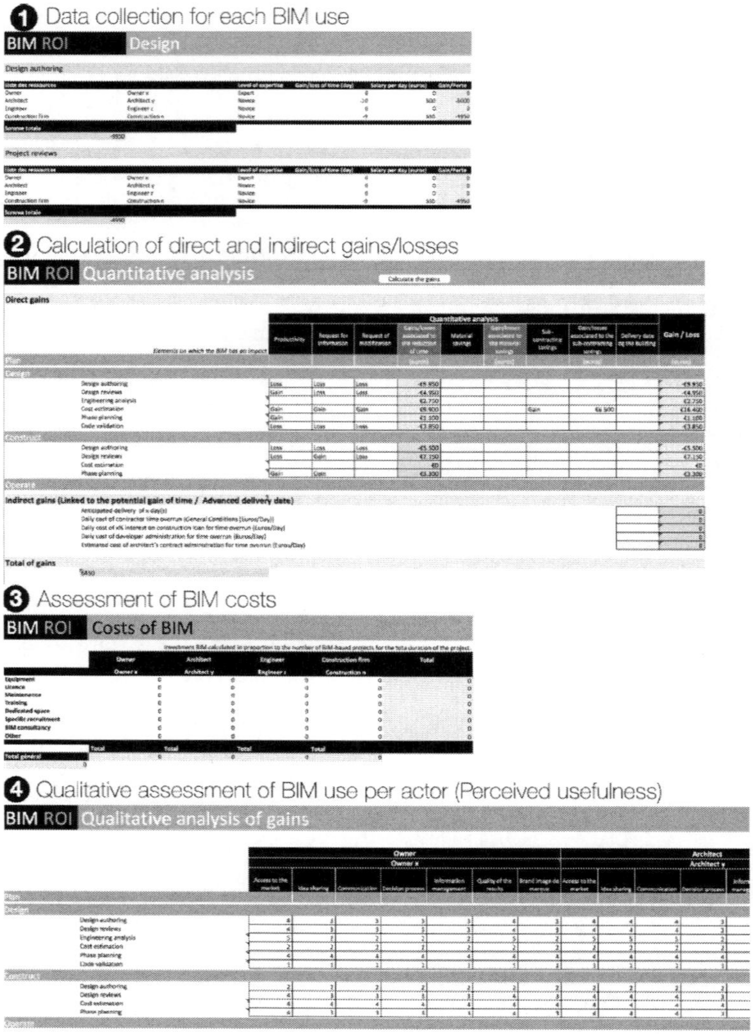

1 Data collection for each BIM use

BIM ROI Design

2 Calculation of direct and indirect gains/losses

BIM ROI Quantitative analysis

3 Assessment of BIM costs

BIM ROI Costs of BIM

4 Qualitative assessment of BIM use per actor (Perceived usefulness)

BIM ROI Qualitative analysis of gains

Figure 3. Interface of the xls tool.

4 FIRST DEPLOYMENT

The method for BIM ROI assessment has been deployed on two pilot projects.

Case study 1

The first project is a design and build project of 14 M€ that was constructed in France. The Building Information Model was developed on the construction firm's initiative, which aimed at developing its own BIM strategy. The BIM uses deployed in this project are the following (See definition in the Table 1):

- Design authoring (during the design and execution stage),
- Code validation (during the design stage),
- Design reviews (during the design and execution stage),
- Phase planning (during the design and execution stage),
- Cost estimation (during the design and execution stage).

The principal BIM users implied in this project were the architect, the structural engineer, and the construction firm. The owner was not an actor in this experimentation but was interested to observe the impact of the new exchange processes based on BIM. Due to the fact that the project was under construction when we began the analysis, it was

Table 1. BIM uses definition, extract from (PennState, 2011).

BIM use	Description
Design Authoring	A process in which 3D software is used to develop a Building Information Model based on criteria that is important to the translation of the building's design.
Code Validation	A process in which code validation software is utilized to check the model parameters against project specific codes. Code validation is currently in its infant stage of development within the U.S. and is not in widespread use. However, as model checking tools continue to develop, code compliance software with more codes, code validation should become more prevalent within the design industry.
Design Reviews	A process in which stakeholders view a 3D model and provide their feedbacks to validate multiple design aspects. These aspects include evaluating meeting the program, previewing space aesthetics and layout in a virtual environment, and setting criteria such as layout, sightlines, lighting, security, ergonomics, acoustics, textures and colors, etc.
Phase Planning (4D Modeling)	A process in which a 4D model (3D models with the added dimension of time) is utilized to effectively plan the phased occupancy in a renovation, retrofit, addition, or to show the construction sequence and space requirements on a building site.
Cost Estimation (Quantity Take-Off)	A process in which BIM can be used to assist in the generation of accurate quantity take-offs and cost estimates throughout the lifecycle of a project.
Engineering Analysis (Structural, Lighting, Energy, Mechanical, Other)	A process in which intelligent modeling software uses the BIM model to determine the most effective engineering method based on design specifications.

difficult to collect data, especially concerning the costs of BIM. For this operation, the costs of BIM were not filled in, only the gains/losses were assessed. Consequently it is not possible to obtain a ROI value. Nevertheless, it is interesting to analyze the partial collected data (See the Figure 4):

- The quantitative analysis by BIM use reveals that even if the project is an experimentation and the users are novice for the BIM uses characterizing the project, we can identify positive gains by using BIM:
  - o During the design stage, the design reviews, the cost estimation and the phase planning contribute to significant gains.
  - o During the execution stage, the design reviews (even if losses are identified during the design stage) and the phase planning appears as the most positive and valuable uses.
- The qualitative analysis highlights that during the design stage, the BIM use "Design Reviews" and "Design authoring" are the most interesting, and they cover largely all the facets of the qualitative analysis. The Design reviews are specifically appreciated for their capability to reinforce the sharing of ideas among stakeholders.

Finally in the quantitative analysis part as well as in the qualitative part, the construction firm appears as the user who beneficiates more largely of BIM in this operation. We can suppose that the effort in the design stage contributes largely to the strategy of zero defects, and rework of the contractor and sub-contractors on site.

It is important to note that the collected data are only partial, so the analysis should be reinforced by a more global data collection. The actors implied in the BIM ROI process were interested by the method structured by BIM use and actors, as the strategy of BIM implementation on this pilot project largely focused on the implementation of particular BIM use, and take place in a context where actors are "BIM use testers". Nevertheless, the method requires a large collection of data coming from all the stakeholders. This is incompatible with a data collection at the end of the project. The BIM ROI assessment process requires to be completed progressively all along the project's life cycle by all the BIM actors.

4.1 *Case study 2*

The case study 2 is a project of 21 M€ that is under construction in France. The Building Information Model was developed, as for the first case study, on the construction firm's initiative, which aimed at developing its own BIM strategy. The BIM uses deployed in this project are the following (See definition in the Table 1):

- Design authoring (during the design and execution stage),
- Code validation (during the design stage),
- Design reviews (during the design and execution stage),
- Phase planning (during the design and execution stage),
- Cost estimation (during the design and execution stage),
- Engineering analysis (during the design stage).

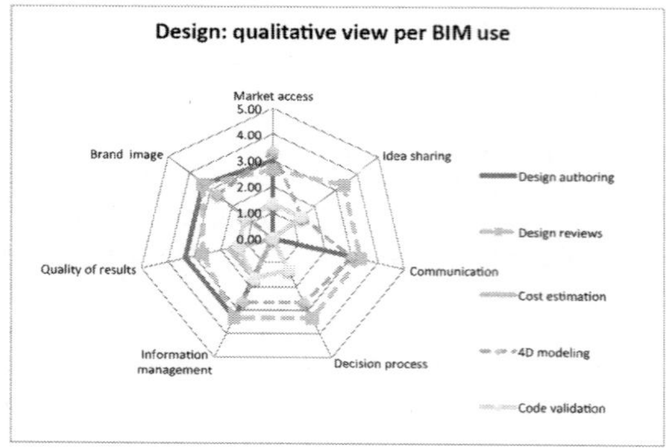

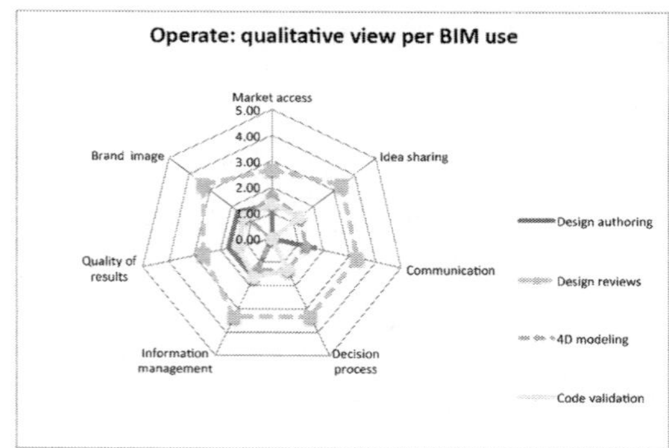

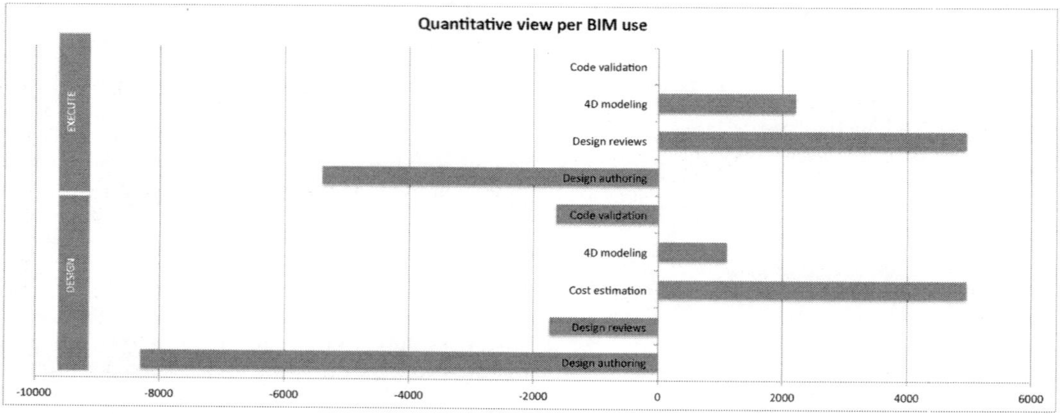

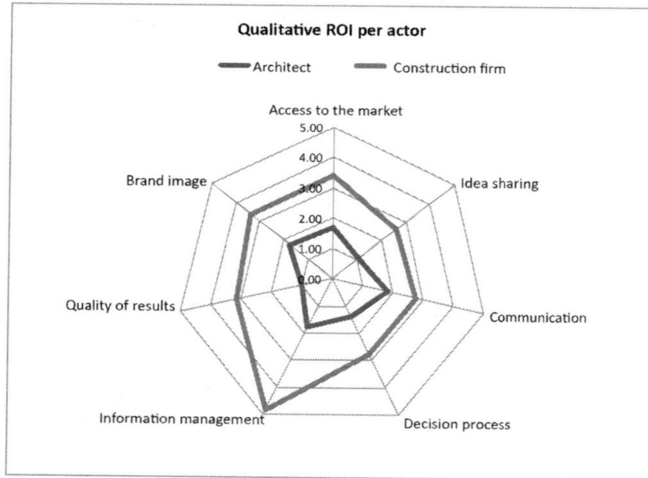

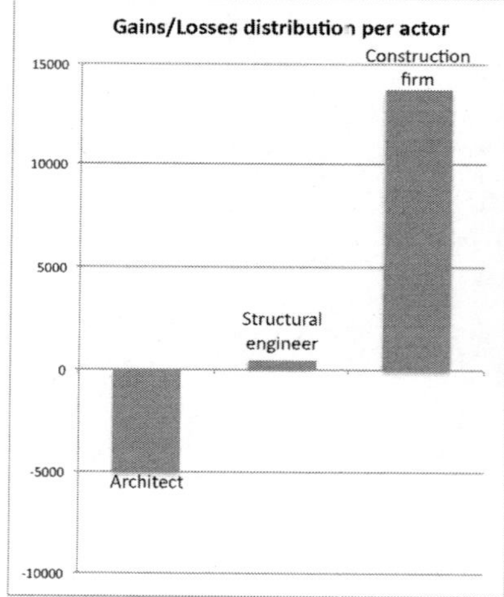

Figure 4. Case study 1: data analysis.

The BIM users implied in this project were the engineer, the architect and the construction firm.

It is important to note that this project is a pilot project for which we have only a few data; for example, the costs of BIM have not been filled in. Moreover, this operation is under construction and consequently it is not possible to make a final assessment. Nevertheless, the analysis of collected data highlight that (See the Figure 5):

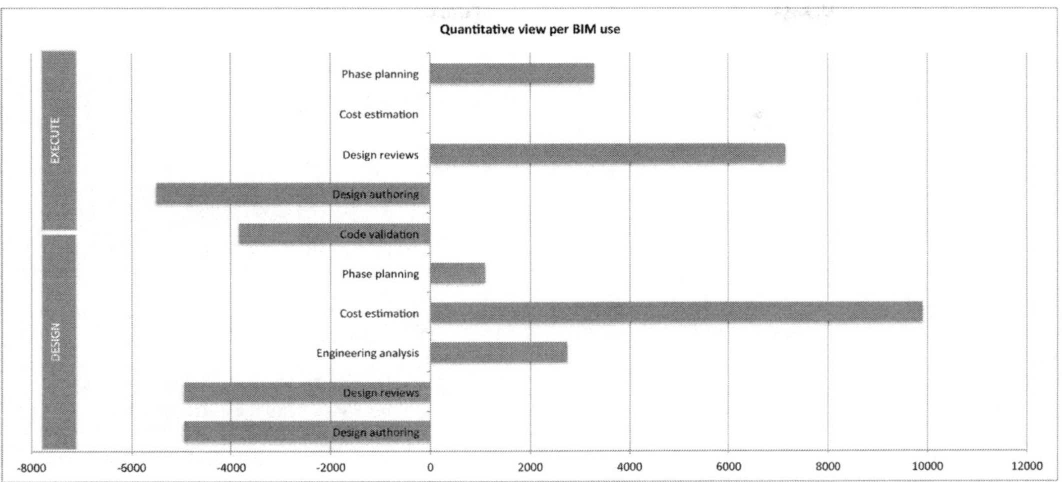

Figure 5. Case study 2: data analysis.

- During the design phase, the energy analysis, the phase planning and particularly the cost estimation, appear as the BIM uses having a positive impact.
- During the execution phase, the design reviews (contrary to the same BIM use during the design phase), and the phase planning are at the origin of positive gains.

The construction firm is the actor who has the best expertise level. This is also the first beneficiary of the BIM implementation on this operation.

These two case studies are not sufficient to validate the method but they constitute an interesting first step of deployment. The framework structured by BIM uses and BIM users includes the parameters for direct/indirect gains and the costs of the investment. The MS Excel tool supporting the method appears easy to use and the resulting reports constitute a good decision support tool during the BIM implementation on a project. The limit as for all the ROI methods is related to data collection, which is a time-consuming effort. The actors are not sufficiently organized to collect these types of data. The task is consequently hard, especially when the project is finalized. Therefore, a progressive data collection all along the project's life cycle should be recommended for all the BIM users.

5 PROSPECTS AND CONCLUSION

In this article, a method is proposed to assess the Return on Investment of BIM. The method has been partially applied on several case studies, and two detailed examples are provided in order to illustrate it.

The aim of this method is to assess projects in a repeatable manner, in order to envisage consolidating the results across several projects and to finally conclude on the added value of BIM regarding:

- its particular *uses*. Indeed, most of BIM guidelines consider BIM implementation through limited information exchanges workflow, involving a few stakeholders. This approach enables closed understanding of BIM exchanges when defining projects' BIM strategies, but often raises the question of the added value of each one of these processes. The method should then bring clarity to this. Beyond, the dependencies between each sub-process should be considered, as in most of the cases low-added value processes (e.g. BIM authoring, or applying classification) are required in order to achieve high-added value BIM uses (e.g. 4D modeling).
- and the *actors involved* in producing and consuming BIM-based data. Only a few actors are involved in each BIM use, some of them producing BIM datasets, some others consuming it for other purposes. It is important to identify clearly their workload, as well as to estimate if each one benefits from these efforts in his direct uses. For example, in some countries where the architectural mission is often limited to the design stage (no construction management), the added value of BIM is perceived only for downstream practitioners (i.e. contractors, facility managers) ((McGraw-Hill, 2012), p. 25).

The method has been applied on several projects in 2015; however gathering datasets for assessing ROI has often been difficult. The primary reason is that the authors were only able to apply it to on-going or finished AEC projects. In such cases, the principles for recording data are not clearly defined for all the projects partners, and therefore it is difficult to gather it and especially to compare it. But this is also due to the particular nature of such productivity-related information. Indeed, practitioners are usually not transparent and not disposed to make such sensible information public. The combination of qualitative and quantitative analyses is an option to address this issue. Public incentives, in the framework of BIM trials or experiments, could be another one.

In the results, the authors also recurrently observed a negative ROI at the project level. It should be mentioned that in all of the cases, the BIM maturity of stakeholders involved were quite low. Indeed, only a few (large) projects have been completed with BIM from the beginning to the end in Europe. Most common projects always have to face different maturity levels, and low experience of owners, designers and construction teams. Then, traditional tasks (such as design authoring, construction documentation) performed with BIM methods take more time (negative value). And they are often not counterbalanced by other tasks like simulation tasks, handling of design changes etc. which could bring value to the whole project balance. This last issue is interesting as it highlights the actual barriers to BIM implementation, opening prospects to the future cost effectiveness, accompanied with new services adding value, quality and time efficiency to architectural projects, design and construction teams and finally building owners.

ACKNOWLEDGEMENTS

This research is a part of the results of the BIMetric project funded by the PUCA (*Plan Urbanisme Construction Architecture)* France, for more details see http://bimetric.list.lu/. The authors thank the partners of the project coming from the LRA (Laboratoire de Recherche Architecturale, Toulouse, France) and the MAACC (Laboratoire de Modélisation pour l'Assistance à l'Activité Cognitive de la Conception, Paris, France) for the collaboration and rich discussions.

REFERENCES

Autodesk. 2007. BIM's Return on Investment. 1–5.

Bew, M. & Richards, M. 2008. Bew-Richards BIM Maturity Model.

Eastman, C., Teicholz, P., Sacks, R. & Liston, K. 2011. *BIM Handbook: A Guide to Building Information Modeling for Owners, Managers, Designers, Engineers and Contractors*, New Jersey, John Wiley & Sons Ltd.

Giel, B., Issa, R. R., A. & Olbina, S. Return on investment analysis of building information modeling in construction. Proceedings of the International Conference on Computing in Civil and Building Engineering, 2010/05/30 2010 Nottingham University. 1–6.

Kreider, R., Messner, J. & Dubler, C. 2010. Determining the Frequency and Impact of Applying BIM for Different Purposes on Projects. *Innovation in AEC Conference.* The Pennsylvania State University, University Park, PA.

Love, P. E. D., Simpson, I., Hill, A. & Standing, C. 2013. From justification to evaluation: Building information modeling for asset owners. *Automation in Construction,* 35, 208–216.

Lu, W., Fung, A., Peng, Y., Liang, C. & Rowlinson, S. 2014. Cost-benefit analysis of Building Information Modeling implementation in building projects through demystification of time-effort distribution curves. *Building and Environment,* 82, 317–327.

McGraw-Hill 2010. La valeur commerciale du BIM en Europe, Obtenir des résultats en matière de modélisation des données d'un bâtiment (BIM) au Royaume-Uni, en France et en Allemagne. *Rapport SmartMarket.*

McGraw-Hill. 2012. SmartMarket Report. 1–72.

NRC. 2012. NRC Publications Archives des publications du CNRC. 1–28.

Pennsylvania State University. 2011. BIM project execution planning guide version 2.1 released—May 2011.

Poirier, E. A., Staub-French, S. & Forgues, D. 2015. Measuring the impact of BIM on labor productivity in a small specialty contracting enterprise through action-research. *Automation in Construction,* 58, 74–84.

Qian, A. Y. 2012. Benefits and ROI of BIM for multi-disciplinary project management. National University of Singapore.

Teicholz, P. 2013. *BIM for facility managers*, John Wiley & Sons.

eWork and eBusiness in Architecture, Engineering and Construction – Christodoulou & Scherer (Eds)
© 2016 Taylor & Francis Group, ISBN 978-1-138-03280-4

A new training concept for implementation of 5D planning with regard to construction of large-scale projects

L. Herter, K. Silbe & J. Díaz
THM, Giessen, Germany

ABSTRACT: This paper describes the possible incorporation of a training concept in support of the nationwide introduction of Building Information Modelling (BIM) as from 2016. After analysing the requirements, standards and general conditions set by the *Bundesministerium für Verkehr und digitale Infrastruktur (BMVI)* [Federal Ministry of Transport and Digital Infrastructure] and their cooperation partners, including *Deutsche Bahn AG* [German Railway Corporation] (Germany's largest property owner and operator of the country's national railway infrastructure), the *Technische Hochschule Mittelhessen (THM)* [Mittelhessen University of Applied Sciences] in Giessen developed a concept for training BIM experts and specialists. The concept is intended for the coworkers of both the principal and the contractor, whose job it is to promote the implementation and development of the new work method BIM in the construction sector.

1 INTRODUCTION

1.1 *State of the art*

In the German construction industry, the majority of planning services are currently done in 2D and passed on in different data formats (Braun et al. 2015). For many years, most tendering has been done entirely in text form with specifically priced service items on a bill of quantities. A 3D planning and computer simulation is extremely rare. According to the *Honorarordnung für Architekten und Ingenieure (AHO 2013)* [Official Scale of Fees for Services by Architects], such simulations are regarded as "special services" and have to be dealt with separately. Due to the complexity of the whole planning process, it is not unusual with complicated and challenging projects that there are mistakes and collisions which do not become obvious until work is in progress. As a result, such projects suffer disruptions and get severely delayed, thus leading to considerable extra costs for the principal.

Such a development is unacceptable, especially in cases which are financed by public funds. A good example for this is the *Elbphilharmonie* [Elbe Philharmonic Hall] in Hamburg. In 2004, the construction costs were estimated at approx. 77 million euros. Currents estimates put the figure at 789 million euros (Loewenstein 2014). The new airport under construction in Berlin, German's capital, shows a similar trend. Here, construction costs were originally estimated at 1.9 billion euros, but have meanwhile risen to approx. 4.7 billion euros[1], with the final figure now forecast to be 5.4 billion euros[2]. All this clearly shows that in the face of increasing complexity regarding current and, above all, future construction projects, there is a need to adapt planning and implementation accordingly.

At present, there is insufficient understanding among the individual specialist disciplines in Germany with regard to "Building Information Modelling" (BIM). Too little is known about this method as a whole, about its advantages and disadvantages, and about the costs, risks and chances involved both before and during the construction process. Since many questions remain open and practical guidelines for its implementation are lacking, BIM is still not widely used in Germany.

Figure 1 shows the current degree of digitalisation in the respective industrial sectors in Germany. It is obvious that the construction industry with just 1.9 points has an extremely low level of digitalisation in comparison to other leading branches of the economy. For example, the IT and telecommunications sector is in first place with 3.2 points (Riemensperger

[1] Hartl, Robert: Flughafen Berlin [Online] – visited on 14th April 2016—http://www.flughafen-berlin-kosten.de/
[2] http://de.statista.com/statistik/daten/studie/245914/umfrage/kosten-des-flughafens-berlin-brandenburg/—visited on 14th April 2016.

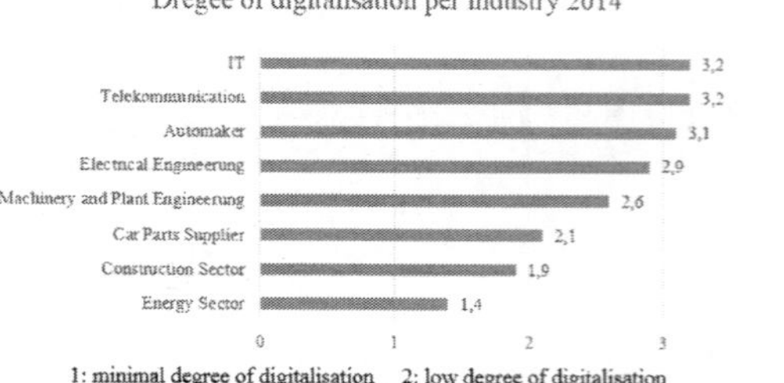

Figure 1. Degree of digitalisation per industry. Author's own illustration based on an Accenture strategy of 2015 (Riemensperger et al. 2015).

et al. 2015). It is essential in future that the German construction industry wakes up to the importance of digitalisation at all stages of a project.

1.2 *Legal regulations*

The BIM projects which have so far been initiated and carried out by principals in Europe are regarded as a success. This is mainly due to the fact that the virtual model was used throughout the respective project and all knowledge gained could be documented in detail. In several EU member states (Finland, Norway, Netherlands, Denmark, Great Britain) the use of BIM has been made mandatory for construction projects financed by public funds. In these countries, the use of BIM is regulated by existing guidelines which comprise prescribed procedures and working processes (Díaz 2014).

The German construction industry has no alternative but to adopt BIM as a working method if it wants to remain competitive on the international market. An EU guideline released by the European Parliament on 28th March 2014 (2014/24/EU) refers to the introduction of electronic data models for public works contracts. As from 2016, such data models will be required for all contracts involving projects valued higher than EU procurement thresholds. This regulation applies to all 28 member states and has to be implemented at national level. In many cases, unresolved legal questions are the reason for the delay in introducing BIM in the German construction sector (Eschenbruch 2014).

In Germany, there are still no laws governing how to work with a virtual model. Nor are there clear indications concerning the authority and responsibility of the stakeholders and the principal's representatives while the project is being executed. This applies to such issues as liability and discretionary powers.

2 REQUIREMENTS FOR IMPLEMENTATION

2.1 *BIM Stufenplan (Road Map for Digital Design and Construction)*

The Road Map for Digital Design and Construction (Bramann et al. 2015) introduced by the *Reformkommission Bau von Großprojekten* [Commission on Construction Reform for Major Projects] on behalf of the *Bundesministerium für Verkehr und digitale Infrastruktur (BMVI)* [Federal Ministry of Transport and Digital Infrastructure], which is regarded as the official schedule for the implementation of BIM, is primarily directed at principals from the public sector. One essential aim of the Road Map is to sensitise all stakeholders. Besides the demands made on the new method, the time periods for completing the different steps are important. The first stage up to 2017 constitutes the preparatory phase, which is to be used for clarifying different procedures, carrying out pilot projects, training staff, and developing standardisation measures. During the second stage, the extended pilot phase begins with the application of Performance Level 1. This stage should be completed in 2020. At the end of 2020, the third stage begins. This comprises implementation of Performance Level 1 as part of infrastructure construction and structural engineering. The requirements of Performance Level 1 comprise areas such as "data", "processes" and "qualifications". Concerning data, the amount of detail of the 3D model and the data exchange format are just a couple of points mentioned in the so-called *Auftraggeber-Informations-Anforderungen (AIA)* [Contractor's Information Requirements]. Concerning processes, it is important that the so-called *BIM-Abwicklungsplan (BAP)* [BIM Resolution Plan] contains regulations with regard to essential roles, functions, processes, interfaces and interactions as well as rules for fair and cordial collaboration. These rules comprise all the performance phases from preliminary draft to execution of the construction work (Bramann et al. 2015). The BIM qualification of the principal and of the contractor is also an essential element of the *Road Map for Digital Design and Construction* (Bramann et al. 2015). It is necessary to ensure that the contractor has the BIM competences to fulfil the contract in the appropriate manner while the principal must be in a position to demand such competences by setting the respective requirements. In this way, both parties can act in harmony and the BIM concept can be implemented effectively.

It is not only the initial steps which are an important criterion for the application of BIM. This method is also useful in optimising lifecycle costing of a given project. Due to the high degree of digitalisation and the use of a virtual model with individual digital elements, a great deal of information is available (e.g. manufacturer's name, material, estimated lifetime, costs, maintenance schedules). This information provides valuable insight into the whole-life cost of a building from original construction to final demolition. Thus the principal knows from the beginning which costs he will have to reckon with in terms of production, operation and maintenance. Whole-life cost is an extremely important aspect in public-private partnerships (PPP). Under the terms of such contracts, the private sector principal is not only committed to having the building constructed, but also to operating and maintaining it for several decades. Thus it is of advantage if the contractor can estimate the costs involved.

2.2 *Pilot projects in Germany*

Among the first projects which the *BMVI* successfully initiated with *Deutsche Bahn AG* [German Railway Corporation] on the basis of BIM were a number of minor construction measures at railway stations, e.g. modernisation of the station at Coppenbrügge to the extent of 400,000 euros (Ruehl 2014, Aschmann 2014) and modernisation of the interchange station in Werbig to the extent of approx. 1 million euros (Horstmann 2016). The successful completion of these reference projects encouraged *Deutsche Bahn AG* to undertake further measures to verify the initial results and to compile a catalogue of standards (Deutsche Bahn AG 2015). These standards represent a broad outline of the entire BIM process from inception to completion (including an as-built model) in the execution of small-scale and medium-scale projects.

The standards set up by *Deutschen Bahn AG* also influence the demands made on the planners. For example, the scope of services to be offered by external architects and construction engineers in the planning and supervision phases now has to include the BIM method. Traffic installation planners, surveyors and other specialists have to take into account the extra work involved due to the use of BIM and add this to the standard services which are currently listed on the *Honorarordnung für Architekten und Ingenieure* (AHO 2013) [Official Scale of Fees for Services by Architects]. Thus the principal is charged an appropriate fee for the services provided during the project phases. In addition, the already established Level of Detail (LoD) and Level of Information (LoI) stipulate clear standards concerning the nature of the virtual construction model in the planning, execution and completion phases for both internal and external planners.

Furthermore, the list of standard services now has to contain a coordinated database of construction elements for compiling the 3D model with the defined attributes. The quality and sustainability of the virtual construction model is a key factor of the new work method. For this purpose, there has to be a firm structure for compiling and maintaining the model in the different project phases (planning, construction, manufacturing, completion, delivery). In this context, *Deutsche Bahn AG* has issued a scope statement as a BIM standard for small and medium-sized railway stations which, among other things, regulates the modelling of a depictable object. This so-called "Modelling Manual" also describes the way individual elements are to be integrated in the BIM application, so that an automatic quantity calculation and an automatic bill of quantities can be created in a so-called *AVA-Programm (Ausschreibung-Vergabe-Abrechnung)* [Tendering—Awarding contracts—Invoicing (TIA)] (Deutsche Bahn AG 2015).

In order to standardise the planning process, *Deutsche Bahn AG* created their own database for

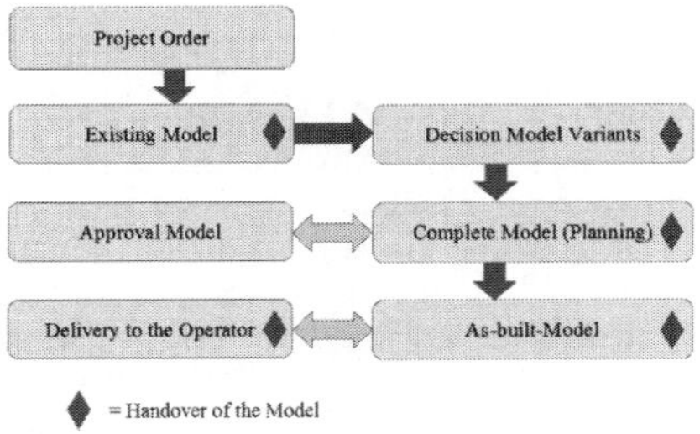

Figure 2. BIM process from project commissioning to delivery of as-built model. Author's own depiction based on BIM standards of *Deutsche Bahn* (Deutsche Bahn AG 2015).

Table 1. Level of detail of construction model per project phase. Author's own depiction based on BIM standards of *Deutsche Bahn* (Deutsche Bahn AG 2015).

Model type	Level of detail	Performance
as-built model (demolition)	LoD: 100 Lol: -	surveying
as-built model (interfaces)	LoD: 100 Lol: 100	surveying
model variants	LoD: 100 Lol: 100	planning / preliminary planning
complete model (Stage 1)	LoD: 200 Lol: 200	planning / design planning
complete model (Stage 2)	LoD: 300 Lol: 300	planning / execution planning
as-built model	LoD: 400 Lol: 400	documentation

structural elements and had the additional planning software *iceBIMrail* developed (for partially automated planning of railway stations) within the framework of an R&D contract as a plug-in for AUTODESK REVIT®.

A further aim is to incorporate all major projects which are in the planning phase and some which are already in the execution phase in the existing BIM spectrum. Such projects include the Rastatt railway tunnel (completion in 2018) to the extent of 450 million euros (BMVI 2014), the Filstal railway bridge (completion in 2018) to the extent of 53 million euros (Schuett 2015), and the new Petersdorf railway bridge (completion in 2018) to the extent of 35.5 million euros. A number of cooperation projects in other areas of infrastructure—e.g. Auenbach viaduct near Chemnitz partnered by the service provider *Deutsche Einheit Fernstraßenplanungs und—bau GmbH (DEGES)*—serve to gain more experience and develop a holistic work method.

2.3 *BIM implementation by the principal*

An important task for the principal is to provide adequate induction training to those members of staff who will be working with the new BIM method. This group includes not only architects and engineers involved in the early performance phases (design and planning), but also those responsible at a later stage (tendering, awarding contracts) and during construction. The special training is based on an analysis of the standards set up by the *BMVI* (Puestow et al. 2015) and laid down in the *Road Map for Digital Design and Construction* (Bramann et al. 2015) as well as the BIM standards of *Deutschen Bahn AG* (Deutsche Bahn AG 2015), who function as a cooperation partner in the field of BIM application in pilot projects in public-sector construction. Based on these conditions, the *Technische Hochschule Mittelhessen (THM)* [Mittelhessen University of Applied Sciences] used its wide experience in BIM to develop a concept for the further education and certification of staff members in both the private and public sector. Another group that requires training in BIM are those working in planning offices and other companies in the construction sector who are engaged by the principal. Such a qualification has to cover a range of methods related to model-oriented exchange of information and partner-based project management.

In addition, all course participants have to develop sufficient skills for handling data models and coordinating them with customary software. Such persons who are competent in BIM management and BIM coordination (i.e. those employed by the principals and the companies they commission) will subsequently implement BIM in pilot projects and establish its use in the German construction sector.

3 TRAINING CONCEPT

Based on the standards of the *BMVI* represented by the *Reformkommission Bau von Großprojekten* (Puestow et al. 2015) and the *Road Map for Digital Design and Construction* (Bramann et al. 2015) as well as the previous experience of *Deutsche Bahn AG* in the form of BIM standards for small and medium-sized railway stations, this paper will present the main components of a training concept which was developed at the THM for the further education of staff members working with BIM. This concept comprises modules relating to both the planning phase and the execution phase.

3.1 *Basics of digital planning and construction*

The introductory module "Basics of Digital Planning and Construction" gives a general outline of 5D/BIM methodology and elaborates on the federal initiative supporting the introduction of BIM in the German construction sector. Participants learn the basic meaning and terminology of Building Information Modelling. With this background, it is then possible to respond to specific tasks, to understand standards, to transfer know-how to processes in 5D planning and, last but not least, to document all findings in a comprehensible manner.

A further important course content are the advantages which result from the introduction of BIM. These include changes in the planning phase, higher expectations relating to planning quality, more efficient use of relevant databases, better communication between planning and construction teams, and easier collaboration between partners. Such a positive development will lead to a more fluid workflow for all parties involved. The knowledge gained from the various BIM-based pilot projects already initiated by the *BMVI* will provide an overview of the revised roles in the planning and execution of construction projects and the need to introduce BIM as a work method.

3.2 *3D and information modelling*

The module on "3D and Information Modelling" deals with the procedure involved in the sustainable modelling of a virtual building and its elements. Special importance is attached to information regarding virtual construction elements which can be used at a later stage in different BIM-compatible software applications. In addition, the module deals with standards recommended by the software producer and those laid down by the *BMVI* (Puestow et al. 2015) and the *Road Map for Digital Design and Construction* (Bramann et al. 2015) with regard to the way databases are maintained and supplemented in order to ensure efficient and uniform use of data. This applies to both the Level of Detail (LoD) and the Level of Information (LoI). Depending on the stage of planning, the data can be integrated into structural engineering software (e.g. Tekla), into software for calculating energy requirements (e.g. IDA ICE), or into project-controlling software (e.g. RIB iTWO® 5D). This requires software interfaces such as Industry Foundation Classes (ifc) or Construction Process Integration (cpi). Depending on the application, it is important to configure certain data contents of the construction model via virtual components and interfaces and to coordinate the data models of the individual planning disciplines.

3.3 *Integration of 5D planning*

In the module "Integration of 5D Planning", participants learn different ways of structuring and executing an integrated planning process which ensures smooth collaboration among the involved parties. The contents are first taught in theory and then applied in a practical context.

It is a basic requirement in today's world to consider the entire lifecycle of a modern building and to carry out any changes which are necessary during the planning phase. Such processes are explained in this module. Planning has become a holistic exercise, involving specialists from many disciplines who should be integrated in the procedures as early as possible so as to avoid changes at a later stage resulting from miscalculations in planning. Figure 3 depicts how cost-intensive a design change can be during the ongoing planning process.

3.4 *Planning coordination*

In the module "Planning Coordination", participants become familiar with concepts for executing cooperative BIM projects, organising the underlying digital information flow between different specialist disciplines, and ensuring planning quality through automatic processes.

Among other things, the module aims to teach the necessary practice-oriented methods, explain the technologies with the help of examples, and

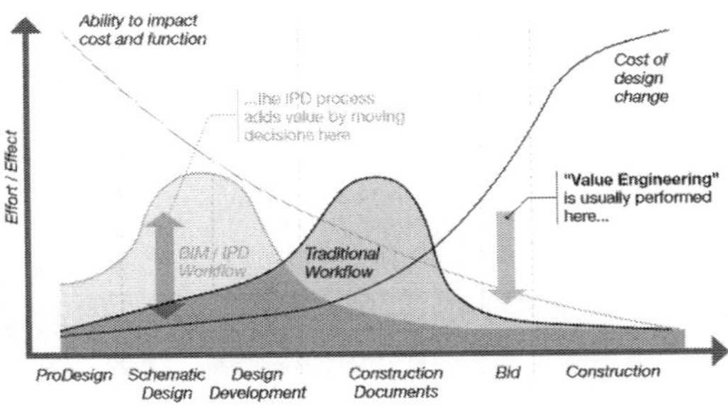

Figure 3. MacLeamy Curve (Chidambaram et al. 2011).

train future BIM experts by using suitable exercises. Participants are shown how to reflect on project planning, manage information through digital exchange in a cooperative BIM process, and use applications for model checking and clash detection. In this context, participants gain experience with the communication and processing platform of the BIM server that determines which parties have right of access and authorisation to edit with regard to the virtual construction model.

3.5 *5D Cost planning & TIA*

The module "Cost planning & TIA" covers the essential aspects of any construction project, including tendering, awarding of contracts and invoicing. The participants become familiar with model-based cost planning (5D) with the help of a digital building model. A sample project is used to put the theoretical BIM-specific topics into a practical context. Among other things, this involves linking the data derived from cost-planning and tendering to the virtual elements contained in the 3D model.

The next step is to calculate the value added for the day-to-day work with the new 5D work method in the form of automatic processes with RIB iTWO® 5D. In addition to providing a more or less automated performance description with a bill of quantities, the BIM application gives insight into the expected costs during the preliminary planning and design planning phases. These costs can be carried forward and updated throughout the subsequent service phases (i.e. awarding contracts and execution), so that it is possible to make comparisons and thus improve project controlling.

Similarly, various procurement processes can be illustrated. Due to the fact that the bill of quantities is compiled automatically, tenders and contract awards can be generated for each individual phase based on the model. These costs are also part of project controlling.

3.6 *5D Scheduling and Controlling*

In the module "Scheduling and Controlling", participants acquire in-depth knowledge about the execution of construction work with the help of a sample project. The following points are important: assigning processes to a virtual construction model; linking individual processes; presenting construction procedures; tracking individual stages of completion in the execution of the construction work; linking and mapping control measures. The construction work is displayed in a simulation by means of Gantt charts, so that discrepancies can be detected and visualised at an early stage. In this module, participants will gain more insight into scheduling and controlling and be able to weigh up the opportunities and threats involved in a given project.

3.7 *Additional modules*

Besides the elementary modules regarding BIM application, the THM also offers specialised modules in the areas of "BIM Management", "BIM Coordination", and "BIM-related Project Management".

In order to bundle the large volume of existing data, a module entitled "Digital Information on the Construction Site" has been created. Participants learn how to deal with the data which arise during the execution of a project and which now have to be integrated into the virtual construction model. Furthermore, the data will be needed for calculating the final cost of the building measure or perhaps for other purposes at a later date.

Another module with the title "Change Management" deals with restructuring procedures in companies and reshaping work processes. The new developments require a new way of thinking on the part of the principal and the contractor, so that they can work together as partners. All these ideas have to be made clear to both sides before a changeover from conventional practice to BIM application can be achieved.

The question of remuneration for services provided by architects and engineers is dealt with separately in the module "Fees". This sets out an appropriate payment scale for those involved in the planning and execution of a BIM project.

When a construction project is being planned, it is important that the general public are informed of its purpose and usefulness. Such issues are included in the module "Public Relations Activities Based on Models". Digital simulations creating virtual accessibility are designed to arouse the interest and raise the level of acceptance of a construction project among the inhabitants.

Last but not least, the module entitled "Copyright" looks in detail at the situation not only of architects and engineers, but also of compilers of databanks. An important issue is the legal background and the ownership rights in relation to the data which has been created.

4 CONCLUSION

Due to the introduction of BIM, a considerable degree of restructuring will take place. Together with their cooperation partners, the *BMVI* has done a lot to facilitate this development in Germany. The contribution made by the *THM* is to offer a training concept which closes the gap between theory and practice. In order to do this, the *THM* was able to draw on its considerable experience in the field, firstly on the software level and secondly in the construction sector.

The *THM* offers a training concept which fulfils both the requirements of the principal and those of the contractor. In line with the *Road Map for Digital Design and Construction* for the changeover to BIM, the concept first of all strives to sensitise, qualify and certify all those who are involved in the construction process. It is the declared aim of the *BMVI* that a holistic approach to construction work, especially major projects, should be introduced by 2020. As this date comes closer, there is a growing need for such a training concept.

REFERENCES

Aschmann, M. 2014. Alles für 400.000 Euro. Coppenbrügge.

Ausschuss der Verbände und Kammern der Ingenieure und Architekten für die Honorarordnung e.V. (AHO) 2013. Honorarordnung für Architekten und Ingenieure. Berlin.

Braun, S., Rieck, A. & Koehler-Hammer, C. 2015. Ergebnisse der BIM-Studie für Planer und Ausführenden >>Digitale Planungs—und Fertigungsmethoden<<: 11, 18.

Bramann, H. & May, I. 2015. *Road Map for Digital Design and Construction*. Berlin: Federal Ministry of Transport and Digital Infrastructure.

Chidambaram, S., Palanisamy, L., Leong, N.K., Wee, T.K., Leong, T.K. & Kwang, T.K. 2011. Maximising BIM Integration with the Project Team. *The BIM Issue*: 7.

Deutsche Bahn AG, 2015. BIM-Vorgaben für kleine und mittlere Verkehrsstationen. Berlin: DB Station & Service AG.

Díaz, J. 2014. Wann kann Building Information Modeling in der deutschen Bauwirtschaft eingesetzt werden. *RKW-Kompetenzzentrum, Informationen/Bau-Rationalisierung 1/2014*: 4–6.

Eschenbruch, K., Malkwitz, A., Gruener, J., Poloceck, A. & Karl, C.K. 2014, Maßnahmenkatalog zur Nutzung von BIM in der öffentlichen Bauverwaltung unter Berücksichtigung der rechtlichen ordnungspolitischen Raumbedingung: 107–112.

Federal Ministry of Transport and Digital Infrastructure 2014. *BIM-Pilotprojekt Rastatter Tunnel*. Berlin: Federal Ministry of Transport and Digital Infrastructure.

Federal Ministry of Transport and Digital Infrastructure 2014. *BIM-Pilotprojekt Brücke Petersdorfer See*. Berlin: Federal Ministry of Transport and Digital Infrastructure.

Federal Ministry of Transport and Digital Infrastructure 2014. *BIM-Pilotprojekt Südverbund Chemnitz*. Berlin: Federal Ministry of Transport and Digital Infrastructure.

Horstmann, W. 2016. Einführung der BIM-Methodik bei der DB Station&Service: 4.

Loewenstein, H.-H. 2014. Bauen in Deutschland—ein ökonomisches Desaster. *BRZ-Mittelstands-Forum 2014*: 9.

Puestow, M., May, I. & Peitsch, D. 2015. *Reformkomission Bau von Großprojekten*. Berlin: Federal Ministry of Transport and Digital Infrastructure.

Riemensperger, F., Hagemeier, W., Pfannes, P, Wahrendorf, M. & Feldmann, M. 2015. *Digitalisierungsstrategien der deutschen Top500*. Berlin: Accenture.

Ruehl, T. 2014. BIM im Verkehrswegebau—die Anwendung von BIM-Methoden bei standardisierten Verkehrsstationen. *BRZ-Mittelstands-Forum 2014*: 9–10.

Schuett, B. 2015. BIM in der Ausführungsplanung und Bauausführung—Pilotprojekt Eisenbahnüberführung Filstal. DB Projekt Stuttgart-Ulm GmbH.

eWork and eBusiness in Architecture, Engineering and Construction – Christodoulou & Scherer (Eds)

Combining BIM models and data with game technology to improve the decision making process: 'PlayConstruct'

H. Jeffrey
Skanska, Maple Cross, UK

ABSTRACT: Using BIM models and associated data, to enable visualisations of how the built environment will be constructed, appear and be used is increasingly well established. The potential to combine these visualisations and the associated data with game technology to illustrate, by 'gamification', the consequence of a decision is being explored in this research project. For example, the components and products that could all potentially meet the brief but might have variable impact on the environment, programme and procurement, durability as well as cost could be selected but not necessarily in the optimum combinations, depending on the drivers. By creating a game 'template' which is called 'Playconstruct', these varying parameters can be installed then selected in game play and compared to achieve the optimum solution for the project and the reasoning captured and illustrated. The benefits should be the engagement of parties interested in the design and construction process as well as the owners and end users. There is also potential for the principle to be beneficial in the broader education and training sectors. For example, illustrating to students the consequences of decisions in a fun and interactive way. Also where training methods can be enhanced by the game experience and scenarios created that are only limited by the imagination of the game designers. This has the potential to be able to create simulations to illustrate circumstances and consequences to be avoided, such as damaging underground services during excavations, or reinforcing good practice. There are many such instances where unnecessary cost and injury occur, despite no doubt rigorous training and toolbox talks etc. that the complementary use of Playconstruct simulations might reduce. The game template can be adapted to suit almost any process in any sector. The illustration of the consequences of decisions, both good and bad, in real time which previously may not have become apparent for some time is likely to find many applications. The technology will bring some of the time served traditional training methods right up to date and have the power to elicit during the currency of the game play a response from what could be far in the future.

Keywords: Gamification, decisions, consequences, process.

1 INTRODUCTION

This paper summarises the activity undertaken in an Innovate UK funded project involving Unit 9, an electronic game designer, Skanska UK and BRE together with the rationale for applying the principles of gamification to the serious business of construction process improvement. The original title of the funded project was BIMCity and this evolved into the more user friendly 'PlayConstruct' as the project developed but are currently interchangeable.

Oh and Hock (2001) summarised simulation "*as a powerful educational tool that has been used successfully in many different settings, such as flight simulation, military training and hardware design. Learning via simulation provides significant educational benefits: valuable experience is accumulated without the potentially dramatic consequences in case of failure, Unknown situations can be introduced and practiced, experiences can be repeated, alternatives can be explored and a general freedom of experimentation and play is promoted in the training exercise. Simulation has the benefit of showing and teaching users cause and effect in a practical manner. If they make a wrong decision the undesired effects and consequences will become clear and is quicker, cheaper and potentially safer than on the job learning*".

The early aspiration for the project outcome was for the BIMCity simulator to be a non-linear synthetic training environment that allows learners/players/designers to rehearse different scenarios, tasks, problems or activities in advance of state-of-art building design putting into practice these techniques. The project intended to develop a gaming platform that would support the development of games and applications that will improve the efficacy of the construction process by encouraging designers and contractors to explore different approaches to design and construction. It would enable 'zero prototyping' of the design of the built environment and the process needed to build it. The platform will be compatible with existing BIM file formats and will provide a link between the complex technologies around BIM. It will assist digital companies, building product manufacturers, and contractors to successfully develop and manage commercial BIM games.

By applying 'gamification' to construction design, BIMCity/PlayConstruct will enable users to experiment with different approaches to meeting clients' needs whilst optimising key performance indicators far more effectively than is permitted by current technology. BIMCity/PlayConstruct will utilise data from 'Building Information Models' (BIM)—developed for a specific building or structure—to allow users to explore different approaches to design and construction via a range of games. Building on past experience of designs and building it will enable each construction project to be turned into a game, providing an immersive experience for the user in optimising a building at the design stage. Games will be developed for different stages of the design and construction process. The player will receive a score based on a number of 'key performance indicators' (KPIs; e.g. cost, time & environmental impact) linked to the stage of the design and construction process being addressed in the specific application.

During the progress of the project it became apparent that the gamification process was more likely to lend itself to assisting in the understanding and resolution of specific issues rather than as a general design tool. Examples identified as relevant to the industry now are:

- The consequences of and the compromises made, during design and procurement/selection decisions in relation to cost, durability and environmental impact.
- The future impact of retrospectively changing a design decision made at the early stage of a project, usually with input from those who are not involved with the later change.
- Underground service strikes that occur with alarming regularity and have potential far reaching impact, despite, training, toolbox talks, method statements etc.

Once it was identified that these specific elements could be incorporated in gameplay scenarios it became apparent that the developing BIMCity/ PlayConstruct game platform could be use-fully applied to simulation training in areas of Environmental, Health and Safety, operational activity as well as client engagement with process understanding. Additionally which we believe will be of great significance going forward, is that this technological interaction with BIM models will be useful across the education sector and provide an alternative image of what can be done in construction that is more appealing than current perceptions, to young people. This may help address the growing skills shortage in the industry and the lack of appeal to new entrants to it.

2 GAMIFICATION DEFINED

Nick Pelling coined the term "gamification" in 2002, meaning "applying game-like accelerated user interface design to make electronic transactions both enjoyable and fast."

'*Gamification is the concept of applying game mechanics and game design techniques to engage and motivate people to achieve their goals. Gamification taps into the basic desires and needs of the users impulses which revolve around the idea of status and achievement*' (Badgeville).

The Wikipedia 2016 definition is: "*Gamification is the application of typical elements of game playing, such as point scoring, competition with others, rules of play, to other areas of activity. In the case of online marketing techniques, to encourage engagement with a product or service*".

Hinchcliffe and Alter (2012) describe gamification "*as a set of techniques that generally revolve around engaging and motivating individuals and groups to perform specific actions, but technology is a primary enabler. Gamification doesn't actually involve playing a game at all. Rather it uses the principles of gaming to make an experience more fun and engaging. Gamification is the use of game mechanics and rewards in a real world, non-game setting in order to compel users to behave and engage in a specific way*".

When people hear gamification, they envision games created for a business purpose. Gamification is not about creating something new. It is about amplifying the effect of an existing, core experience by applying the motivational techniques that make games so engaging. When you gamify high-value interactions with customers, employees, and partners, you achieve greater understanding and empathy with the process, drive more sales, get stronger collaboration, better ROI, deeper loyalty and higher customer satisfaction.

3 GAME MECHANICS

Gamification is built upon 10 primary game mechanics, that motivate and engage users, and may use any combination of these techniques to accomplish business goals. These game mechanics are:

- **Fast Feedback**: Immediate feedback or response to actions.
- **Transparency:** Show users exactly where they stand on the metrics that matter.
- **Goals:** Challenges give users a purpose for interaction, and educate users about what is valued and possible within the experience.
- **Badges**: Evidence of accomplishments.
- **Levelling Up**: Levels indicate long-term or sustained achievement. Used to identify status within a community and to unlock new missions, badges, activities, and rewards.
- **Onboarding**: Video games train you how to play as you play – users learn by doing. Simple missions help new users become engaged immediately as they master basic tasks, rather than being stumped by an unfamiliar interface or a detailed manual.
- **Competition**: Encourage competition with time-based, individualized leaderboards.
- **Collaboration:** Show team members how they are contributing to the group's success and what the consequence of their actions and decisions may be.
- **Community:** Sharing participant achievements creates energy in the community by making people aware of what others are doing.
- **Points:** Tangible, measurable evidence of accomplishments and the basis for a record to prove to external parties (Bunchball 2016).

4 WHY GAMIFY?

The goals of gamefication generally fall into the following categories and provide the motivation for engagement:

- Improve engagement by customers, workers, and the marketplace
- Enable personal development and growth
- Encourage competition toward achieving business goals
- Foster collaboration for shared outcomes
- Improve productivity in core business activities

The rationale for playing games is grounded in good learning theory. We learn and retain information faster when playing. There are many words now being written, especially in marketing circles, about the "gamification" of Business Process Management and process improvement. However, there appears to be little consensus on what it might be and what it might mean (McGregor, Bloor 2011).

As it stands today, the process of gamification tends to fork early into two main approaches. The first applies elements of gameplay to existing enterprise applications or processes. The second conceives an entirely new experience from the ground up, intertwining game mechanics with the application itself. Either way, the motivational aspects remain the most important element for success (Hinchcliffe and Alter 2012).

"*The problem is that a lot of businesses are not really thinking about the incentives and the motivational aspect behind why you would gamify anything. If you try to put game mechanics into a process without looking at the incentives and the motivational patterns around it, your effort is going to fail.*" says Ari Lightman, director of Carnegie Mellon University's CIO Institute. Companies can apply gaming techniques to marketing campaigns, product development efforts, sales activities, or any other business or non business process. To provide the desired effect—and for gamification technologies to be useful and the outcomes achieved—the goals of the game elements being applied must be connected in some well-defined and meaningful way to the business activity. The key is to embed the gamification technology into the process of getting work done, and connect that to the desired business goals. The application of game play is usually to in-duce some kind of behavioural change to improve BMP. According to Hinchcliffe and Alter 2012, in order to facilitate this game-based design technologies can enable the following behavioural drivers, adapted to specific audiences or scenarios. User identity is a primary point of integration between technology and existing enterprise applications and IT infrastructure. Today's user directories generally are not game-ready. Gamification tools and plat-forms typically augment user directories, so user in-formation related to the gaming activity is tracked, stored, and made available to gamified applications.

- **Reward:** Rewards are discrete benefits to a participant in the gaming environment.
- **Status:** Status can be a way to motivate game participants who want to improve their reputations. Generally, status is most effective in social or collaborative games, where other participants can perceive mutual status.
- **Achievement:** Achievements are milestones a participant accrues by meeting objectives or attaining goals. These milestones demonstrate progress, guide desired behaviour, and psychologically reinforce involvement. They are often tracked in a participant's user profile through points, levels, badges, virtual goods, and other marks of accomplishment.
- **Self-expression:** Looking at participants as more than just cogs in a gaming environment is often required to sustain long-term involvement. It

can also be essential for encouraging activity that has meaningful and useful outcomes in terms of the contributions to the gaming activity itself.

- **Competition:** All humans love challenges, yet many people are often the most motivated when they compete with each other. Gaming environments that pit participants against each other using leaderboards, status, achievements, and other techniques can help increase both initial participation and sustained use longer term.
- **Altruism:** Some game situations benefit from having participants reward each other.

5 AIMS, GAME DESIGN AND PSYCHOLOGICAL ENGAGEMENT

While gamification should align with business objectives and support a particular business process, the effort must also attempt to get inside the participants' heads. One of the more delicate aspects of game-based design is the relationship among the user experience, the gaming technology, and the business process. While some vendors provide a default user interface, such as a leaderboard or a reputation score wired into an existing social network, many leave the exercise up to the implementer, assuming that it must be situated appropriately for the local environment by those who know it best.

The inclusion of game mechanics and game dynamics in business applications will vary depending on the technologies selected and the business requirements. However, the resulting gamified solution typically delivers the following capabilities in some formal or informal way:

- Feedback: Provides visual, social, and psychological feedback mechanisms inside existing digital user experiences to encourage sought-after behaviour by end users.
- Analytics: Connects the gamification feedback mechanisms to the relevant big data analytics sources that measure and depict progress against desired business outcomes.
- Business intelligence: Supplies useful and relevant business intelligence, usually in near real time, on user behaviour.
- Management and administration tools: Consists of a set of management and administration tools to adjust performance targets, gaming objectives, business rules, and other in-game parameters.
- Ready integration: Offers a simple, easy-to-use set of lightweight technology components that allow ready integration into existing enterprise applications.
- APIs: Provides a more formal and sophisticated set of structured Application Programming Interfaces (APIs) that allows deeper integration of the gamification platform into line of business systems. While lightweight integration is useful for basic gamification, getting to higher-impact results can require deeper integration of the business process with gaming technology.
- Underlying platform: Consists of the underlying gamification platform itself, which provides various capabilities to the enterprise applications that need it. The platform includes the user experience, robust systems integration features, a gamification engine, a reporting system, administration tools, programming language libraries etc.

6 BENEFITS OF GAMIFICATION

Gamification techniques strive to leverage people's natural desires and the behavioural drivers outlined above. Businesses can use gamefication to drive desired user behaviours that are advantageous to their brand. One common technique is to increase engagement by rewarding users who accomplish desired tasks, using the game mechanics described previously. Gamification is a very practical technique to take advantage of big data. It is a powerful tool for motivating better performance, driving business results, and generating a competitive advantage. By capturing and analysing the big data on behaviours, businesses can create a more engaging experience that motivates employees and users. By providing extensive insight into user behaviour, big data can indicate what activities, content types, and frequencies are yielding the best results. It also allows companies to adapt to various user behaviours and motivations. In order to get the most value out of gamification it is important to go beyond the data and try to understand not only how the users are be-having but also to ask 'why', and also come up with creative ideas to improve the system. Gamification includes a number of psychological concepts, especially regarding motivation, behaviour, and personality. Deep fluency and understanding of these concepts is one of the most important keys to proper gamification implementation.

- Improves employee confidence.
- Offers immediate and applicable feedback.
- Improves knowledge retention.
- Employees can master the art of practice makes perfect.
- Mistakes become invaluable training opportunities.

- Gives employees the chance to try out new performance behaviours.
- Reduces training time and cost (itystudio 2016).

The benefits should be the engagement of parties interested in the design and construction process as well as the owners and end users. There is also potential for the principle to be beneficial in the broader education and training sectors. For example, illustrating to students the consequences of decisions in a fun and interactive way. Also where training methods can be enhanced by the game experience and scenarios created that are only limited by the imagination of the game designers. This has the potential to be able to create simulations to illustrate circum-stances and consequences to be avoided, such as damaging underground services during excavations, or reinforcing good practice. There are many such instances where unnecessary cost and injury occur, despite no doubt rigorous training and toolbox talks etc. that the complementary use of Playconstruct simulations might reduce.

7 CONNECTING THE GAME PLAY TO PROCESS AND GAME DEVELOPMENT

There are a set of questions to pose when considering the game play. For example how will operatives make decisions and judgements then analyse different types of information under conditions of risk and uncertainty? How will they respond to customers, apply regulatory procedures and solve problems when under pressure? To start to address these questions it is necessary to identify the personnel that will make errors in high risk situations and the behaviours that cause high risk performance errors. The game should then be able to help deliver targeted performance improvement and drive actions based upon unique performance insights (Caspian)

The process for the development of the prototype games can be summarised as follows:

- Define design needs. Based on assessments of current inefficiencies in building design and construction process identify steps where gaming could be beneficial and identify priority applications
- Identify and review potential gaming approaches. Based on latest state-of-the-art and experience from other gaming applications identify approaches to introduce gaming techniques to construction process modelling
- Develop BIMCity scenarios. This task will develop application and business scenarios for BIMCity/PlayConstruct based on the outcomes of the above

The work involved identifying the relevant target stages of construction to apply the simulation scenarios. Then formulating the appropriate high level gaming approaches to gamefication. This would use the Unity gaming platform which would assimilate where possible and applicable the data and information from existing construction BIM models from real projects supplied by Skanska. Extensive research was undertaken and two suggested subjects for prototype development were identified. The research aimed to produce insight starting from the scenarios guidelines and it focused the best practice for the design of the game Prototype. The next step was to develop the game platform. This was to enable a template to be created so that future users of the game could put in their own criteria to create their own gameplay for their specific circumstances. The intention was always to avoid trying to create a 'one size fits all' scenario. Experiences from across Skanska were elicited throughout the course of the project. Initially by describing the intent and asking for specific issues of relevance to each part of the business. The implementation and testing of the core modules, as well as the identification and development of key components, needed to move ahead with the most promising leads. One of these turned out to be from the environmental department who were still using paper plans and a description of 'what if scenarios'. This method would have been recognizable one hundred years ago. This together with the realisation that the education sector might be very interested led to the development of the Classroom Designer (www.playconstruct.co.uk/demo/classroom).

As the prototype developed with interaction, and the player can decide initially where the school is located based on the advantages and disadvantages of the site. Figure 1 shows a still from the classroom prototype using an actual BIM model from a Skanska school project in Bristol.

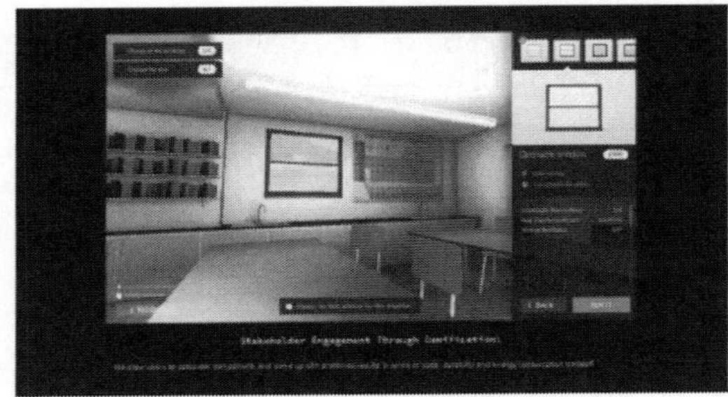

Figure 1. Interior of classroom game showing window selection and impact of choice.

Playing the classroom game gives an introduction to the complexities of compromise and the residual impact of those decisions in terms of cost, convenience and environmental impact for example. One element that might be beneficial to look into as a natural progression from the school scenario of window change in the prototype to the cumulative impact on the environment.

- The small change of one element in one building can be cumulated in many elements in one building to many buildings to show the impact on the city.
- Some intermediate steps from window to city can be shown but also the marker put down to show a link to green and enviro-friendly procurement, such as doors from a sustain-able timber source.
- This will illustrate the effects from element to buildings to cities and thus from national to global effects. This is of great potential interest to the education sector.

The other area of significance was provided by the utilities sector in the form of 'underground cable strikes'. There are approximately 500 per year in one single company, with an average consequential impact of around £2,000 per event. If this is extrapolated to all the other services damaged during excavations and by other utility companies throughout the world this amounts to an extremely important is-sue. Of even more significance strikes can also lead to injury and death. Figure 2 shows the amount of subterranean services in a typical urban dig and the ease with which they are damaged and the impact on a hand tool from a high voltage cable strike. Figure 3 is a still from the 'strike dodger' prototype game (Strike Dodger: www.playconstruct.co.uk/demo/strikedodger).

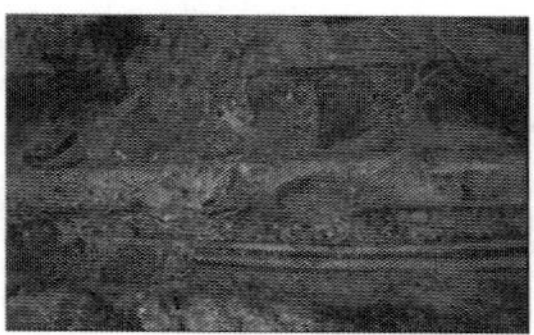
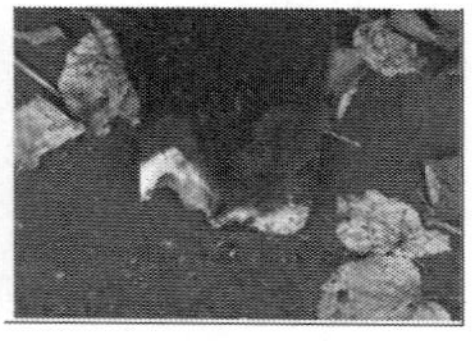

Figure 2. Damaged HV cable and 'melted shovel'.

Figure 3. scenario incorporating photo reality of actual site.

8 CONCLUSIONS

The game template can be adapted to suit almost any process in any sector. The illustration of the consequences of decisions, both good and bad, in real time which previously may not have become apparent for some time is likely to find many applications. The technology will bring some of the time served traditional training methods right up to date and have the power to elicit during the currency of the game play a response from what could be far in the future.

In the short term the prototypes created to date could be developed into an interactive tool to help reduce the incidence of cable strikes by Utilities and bring the environmental induction presentation up to date and interactive with the latest technology. Game technology can bring additional dimensions to simulations, training and specific problem solving that is only possible using those techniques. Given the example that all the traditional training methods are given yet there are more than 500 cable strikes per year in Utilities alone, the simulations are only limited by the imagination of the game designer. This principle can be applied to many areas of the company to improve the construction process and thus reduce costs and improve efficiency. The immediate and potential uses of PlayConstruct are as follows:

1. As a decision tool benefitting from gaming to validate and understand design or construction decisions and their future consequences.
2. As a training aid, working through prescribed generic scenarios. Example of environmental training Decisions resulting in different out-comes. This would give far better engagement for the participant & the resulting scores would give good metrics on under-standing / engagement. This would gain from a 'gaming' approach, especially if there are clear visual consequences & there is a linear work stream.
3. Themes to generate for future review could fall under the following headings:

   - Change management control
   - Capex/Opex rationale and protection of optimised design during development
   - Education both within and for schools but also the broader appeal of the industry
   - Client engagement
   - Management of expectations

The PlayConstruct prototypes are still under development and it is understood that the effect and impact of the game play is only restricted by the imagination of the game designer. Not only is there great scope in the education field for illustrating future consequences of decisions and actions now: Eliciting a 'response from the future'. There is also scope to 'shock and awe' people, hopefully into remembering things they have encountered in the simulation and gameplay that they would not normally retain from traditional training methods. This we believe connects with the psychology of gameplay and human nature in that there is a natural tendency to forget risk and consequence. Familiarity with processes and no actual experience of accident or incident leads to a natural complacency. We believe this does not only apply to the construction industry. At critical moments in operations the human mind may be pre-occupied with a recent argument, for example, or money trouble with, or even an immediate distraction in the mind or by the mobile phone. We believe gameplay can bring the consequences of this inattention in a very forceful and memorable way that will go some way to reducing the number of incidents and process failures. A short demo video exists at: www.playconstruct.co.uk.

REFERENCES

Advantages Of Online Training Simulations In Employee... www.itystudio.com/.../7-advantages-of-online-training-simulations-in-e...http://webcache.googleusercontent.com/search?q=cache:eAhGM9IDUdgJ:www.itystudio.com/en/7-advantages-of-online-training-simulations-in-employee-performance/+&cd = 4&hl = en &ct = clnk&gl = uk.

Business Simulation Training for Your Employees www.designingdigitally.com/blog/2015/11/business simulation-training.

Caspian.co.uk/ http://webcache.googleusercontent.com/search?q=cache:O4lCzQwQ0gEJ:caspian.co.uk/+&cd =15&hl=en&ct=clnk&gl=uk.

Felicia, P. 2011. Education adapting game technology to support individual and organizational learning. In proc: Software Process Improvement and Practice.

Handbook of Research on Improving Learning and Motivation... https://books.google.co.uk/books?isbn=1609604962.

How to improve the customer and employee experience www.pwc.com/.../technology.../feature-gaming-technology-improve-bus http://webcache.googleusercontent.com/search?q=cache:bDhvhwqph8YJ:www.pwc.com/us/en/technology-forecast/2012/issue3/features/feature-gaming-technology-improve-business.html+&cd=9&hl=en&ct=clnk&gl=ukwww.ics.uci.edu/~andre/papers/C11.pdf http://webcache.googleusercontent.com/search?q=cache:Pxm8FyUqM0AJ:www.ics.uci.edu/~andre/papers/C11.pdf+&cd=10&hl = en&ct = clnk&gl = uk.

Oh, E. & Hoek, A. V.-D. 2001. Adapting game technology to support individual and organisational learning. In proc: SEKE, 2001.

The Game of Process Improvement | Bloor www.bloor-research.com/analysis/game-process/ http://webcache.googleusercontent.com/search?q=cache:CfwTLYoJyuIJ:www.bloorresearch.com/analysis/game-process-improvement/+&cd=1&hl=en&ct=clnk&gl=uk.

What is Gamification? How Does Gamification Work ... www.bunchball.com/gamification

What is Gamification? | Gamification.org-Badgeville https://badgeville.com/wiki/Gamification.

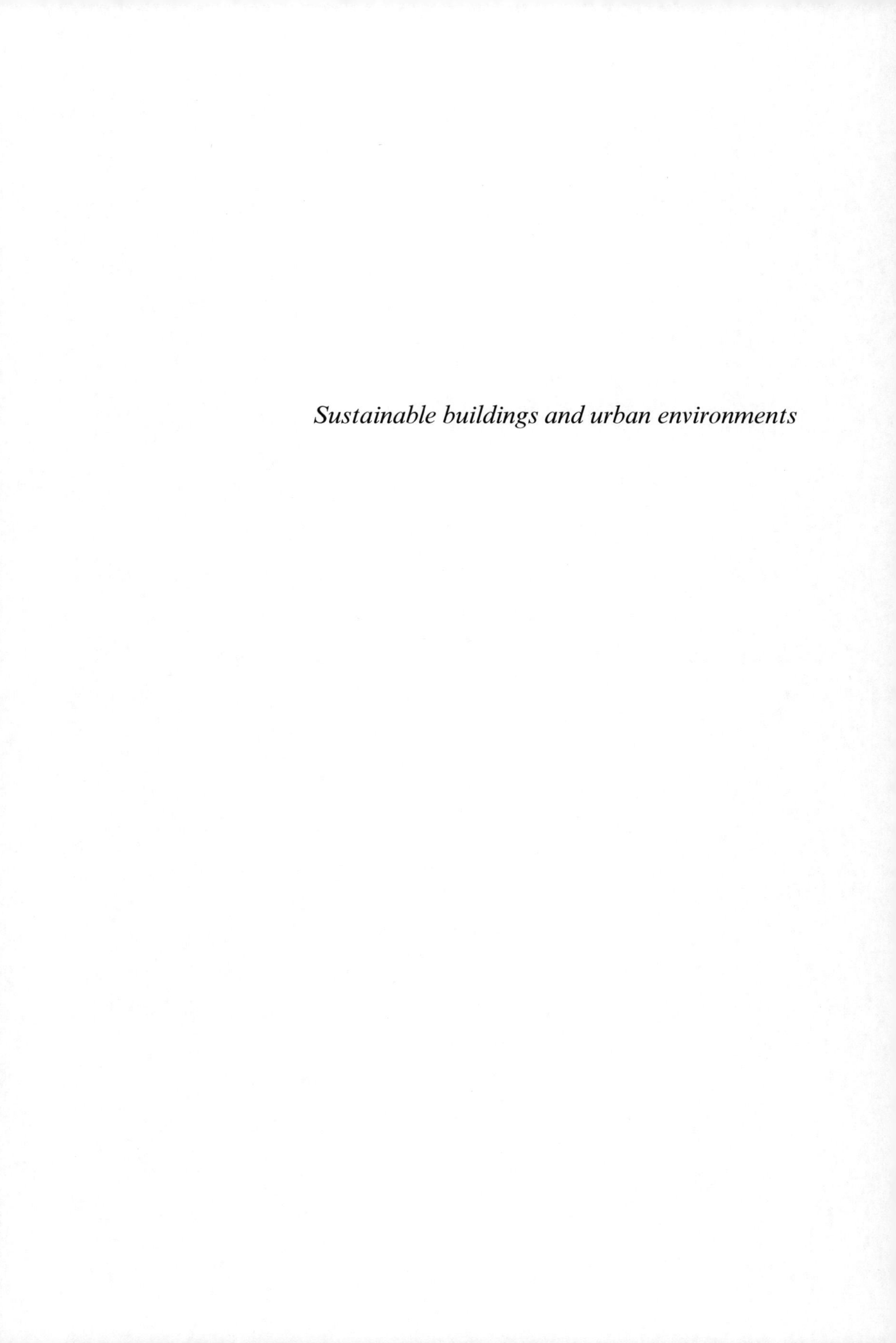

Sustainable buildings and urban environments

eWork and eBusiness in Architecture, Engineering and Construction – Christodoulou & Scherer (Eds)
© 2016 Taylor & Francis Group, ISBN 978-1-138-03280-4

Energy savings and maintenance optimization through the implementation of GESTENSIS energy management system

M. Macarulla, M. Casals, M. Gangolells & B. Tejedor
Group of Construction Research and Innovation (GRIC), Department of Project and Construction Engineering, Universitat Politècnica de Catalunya, Barcelona, Spain

ABSTRACT: Buildings in Europe are responsible of 40% of the final energy consumption. With the aim to accomplish the European 20/20/20 targets, buildings play an important role. Moreover, public buildings should exemplify the best practices in terms of energy efficiency. In this context this paper describes an energy management system for buildings developed in the GESTENSIS project. The system is composed by 5 modules that give tools to building managers to optimize the energy consumption. The building managers have a set of metrics to understand how their building is working and to know if the building performance is decreasing. The system also supports building managers in their daily building operation, optimizing the different types of spaces (corridors, lecture rooms and PC rooms). The system also helps building managers to carry out maintenance activities. It is expected to reduce 15.26% of the gas consumption and 37.59% of the electricity consumption.

1 INTRODUCTION

In developed countries the energy consumed by buildings is around 20–40% of total energy use (Pérez-Lombard et al. 2008). In Europe, the contribution from buildings towards energy consumption is nearly 40% of final energy consumption and 36% of the greenhouse emissions (EC, 2013). Buildings consume energy in their whole life cycle, but the 80–90% of their lifecycle energy use is consumed during the operation stage (Ramesh et al. 2010).

At a European level, the '20/20/20' commitments agreed under the EU Climate Change and Energy Package set three targets for 2020: a minimum 20% reduction in GHG emissions based on 1990 levels; 20% of final energy consumption to be produced by renewable energy resources; and 20% reduction in primary energy use compared with projected levels to be achieved by improving energy efficiency. In addition, the energy performance of buildings directive (EU 2010) requires all new public buildings to be nearly zero-energy by the end of 2018, and all new buildings must be nearly zero-energy by 2018.

The introduction of Energy Management Systems (EMS) in buildings will contribute to achieve European common objectives, but also will generate the opportunity to create new services for buildings. EMS enables to control and manage loads and optimize the building energy consumption, but also provides a tremendous amount of data that can be used for different usages.

This paper is aimed at describing an energy management system for buildings developed in the GESTENSIS project. First, it is described which was the starting point of the GESTENSIS project and it is identified the previous research limitations. Second, the pilot of the GESTENSIS project is presented. Third, the GESTENSIS system is described. Finally, it is presented the expected savings and benefits in the maintenance management due to the use of the GESTENSIS system.

2 GESTENSIS PROJECT

GESTENSIS project is a follow up project of the ENCOURAGE project aimed at rationalizing energy usage in building by implementing a smart energy grid based on intelligent scheduling of energy consuming appliances, renewable energy production, and inter-building energy trading (Albano 2012).

The architecture of the ENCOURAGE project was divided into 4 modules (Albano 2012):

- Device Management module provides access to, and control of, devices.
- Supervisory Control implements strategies to orchestrate the operation of different subsystems in a cell.
- Energy Brokerage and Business Intelligence modules supports inter-building energy exchange in the Macrocell, and supports the building energy

managers deciding when to "produce, store, buy, sell, use" energy.

- The Middleware module is an event processing system that takes the data from the building network and processes it as a stream of events.

The GESTENSIS project aims to broaden the functionalities of the Energy Brokerage and Business Intelligence Module in order to develop an energy management system for buildings. The purpose of the GESTENSIS energy management system (GESTENSIS for here on) is not only to achieve energy savings, but also to offer new services to building managers with the aim to optimize the building maintenance.

The GESTENSIS project used the results previously obtained in ENCOURAGE to develop new strategies to reduce the energy consumption in academic buildings.

Previous research and experiments carried out in academic buildings were the use of Key Performance Indicators (KPIs) to characterize and optimize the use of energy in buildings. These KPIs were used to develop strategies based on user awareness to achieve 20% of energy reduction. The proposed user awareness was done at two levels: building managers and university campus users.

The developed KPIs were displayed to the building managers using a web platform. The aim was to give tools to building managers to characterize their buildings and determine strategies to reduce the energy consumption of the building (Macarulla et al. 2014a). The KPIs also were used to create messages to be sent to the Twitter platform in order to develop user awareness strategies. The aim of this strategy was to promote good practices in the use of public buildings and reduce the energy consumption (Macarulla & Albano 2014b).

In terms of the limitations of the adopted strategies, two potential limitations were identified. The first one was the difficulty to reach all users, because the Twitter platform was not used by the whole building users. The second one was the students' engagement as they did not perceive the problem (Le Guilly 2016).

3 PILOT DESCRIPTION

The GESTENSIS pilot is the Escola Tècnica Superior d'Enginyeries Industrial i Aeronàutica de Terrassa (ETSEIAT), an academic building with a floor area of 11.600 m^2. The building has 3 floors mainly devoted to academic uses and is used by 2,600 students and 240 lecturers and administrative staff. The selected areas for the experiments are located in the whole building. The demonstrator involve the 4 main corridors, 5 lecture rooms and 1 PC room. All the above mentioned rooms are located on the third floor.

Each corridor is equipped with sensors of temperature, humidity and CO_2 concentration (THC), and one power analyzer per line of lights. The lecturer rooms are equipped with THC sensors, light sensors, and power analyser.

The management of the HVAC is limited to the building boiler management, because the HVAC system does not enable a room oriented control.

4 THE GESTENSIS SYSTEM

The development of energy management systems should follow the principles of interoperability, scalability, and creation of new market opportunities. The development of GESTENSIS followed the aforementioned principals and used the architecture of the previous ENCOURAGE project.

GESTENSIS was divided into 5 main modules or components: acquisition and storage module, visualization module, diagnosis module, prediction module and control module. This section describes each component of GESTENSIS.

4.1 *Acquisition and storage module*

Buildings usually have different systems for its operation. For example the monitoring system, the control system or the access control system. To ensure an optimal building operation all systems implemented in a building should be interoperable.

The acquisition and storage module allows the platform to obtain data from sensors or other energy management systems. Right now GESTENSIS is able to exchange information with Powerstudio platform and Concordia platform.

4.2 *Visualization module*

The visualization module enables building managers to see a set of Key Performance Indicators (KPIs) to check the performance of the building. In addition, building managers can use the KPIs to help identifying and prioritizing which building areas are suitable to improve in terms of energy efficiency.

The KPIs implemented are those reported by Macarulla et al. (2014a), adding 2 new KPIs: the level of CO_2 concentration and the percentage of time that each room is in a comfort standard (Table 1).

To determine the standard of comfort the Spanish legislation was used. The Spanish regulation about installations (RITE 2007) indicates that the comfort levels are different during the summer and

Table 1. Strategic objectives, Key Performance Questions (KPQs) and KPIs for the GESTENSIS project. Adapted from Macarulla et al. (2014a).

Strategic objective	KPQ	KPI
OBJ-A-EFF	Are we increasing energy efficiency of the building/macro cell?	Energy savings
		Energy savings rate
	Are we increasing energy efficiency of the appliances?	Stand-by power
		Hours of stand-by
		Appliances consumption per category
	Where can we improve energy optimization?	Consumption/m^2
		Consumption/occupant
		Consumption/DD
		Consumption/(occupant·m^2·DD)
		Optimizable part of consumption
		Appliances time of use
	Are the consumption patterns consistent?	Consumption per hour type
		Consumption per day type
	Are we increasing overall sustainability?	CO_2 savings
OBJ-D-COMF	Is the user comfort being compromised?	Internal temperature variability
		Internal humidity variability
		CO_2 concentration level
		% of time in comfort levels
OBJ-E-ECO	Are we decreasing the total costs?	Costs savings
		Costs per kWh consumed

DD: Degree day.

Table 2. Indoor air conditions established by RITE.

	Temperature (°C)		Humidity (%)	
	T_{refh}	T_{refl}	H_{refh}	H_{refl}
Summer	23	25	45	60
Winter	21	23	40	50

during the winter. The comfort regions are determined by a range of temperature and humidity. Table 2 presents the aforementioned levels of temperature and humidity for summer and winter.

Another relevant aspect to determine the comfort level is the CO_2 concentration. Different studies demonstrated the relationship between the levels of CO_2 concentration and the impacts on learning outcomes. Those studies concluded that limit for CO_2 concentration in indoor air is 1.000 ppm (Toftum 2015).

The KPI "% of time in comfort levels" is calculated accounting the time when the following expressions are true:

- Summer: **IF** $T_{refh\_summer} > T_{room} > T_{refl\_summer}$ **AND** $H_{refh\_summer} > H_{room} > H_{refl\_summer}$ **AND** $CO2_{room} < CO2_{ref}$
- Winter: **IF** $T_{refh\_winter} > T_{room} > T_{refl\_winter}$ **AND** $H_{refh\_winter} > H_{room} > H_{refl\_winter}$ **AND** $CO2_{room} < CO2_{ref}$

During the ENCOURAGE project the need of a cheap and reliable methodology to determine the occupation of a room was arisen (Macarulla 2014a). In the GESTENSIS project CO2 sensors are used to estimate the occupation of a room. An algorithm was developed to estimate the building occupancy with the CO2 sensors. In addition the algorithm enables to know which is the current usage of each space.

4.3 *Diagnosis module*

The diagnosis module uses the information stored in the acquisition and storage module to identify malfunctions in the installations. Two main functionalities are implemented in this module: determination of boiler performance and determination of burnt out bulbs.

The objective of this module is to provide information to the building maintenance service to support their activities.

Generally, maintenance can be either preventive or corrective (Motawa & Almarshad 2013). Preventive maintenance reduces the operating costs and catastrophic breakdown risk (Sheu et al.

2015). On the other hand, corrective maintenance is a reactive action in response to a cause of failure or break down (Motawa & Almarshad 2013).

Using the abovementioned definitions, the determination of boiler performance is a tool to carry out preventive maintenance. Building managers are able to know in real time which is the status of the boiler and detect poor boiler's performance. With this information, building managers are able to plan maintenance activities in order to optimize the boiler energy consumption and avoid boiler break downs. On the other hand, the burnt out bulbs functionality helps to identify burnt bulbs in the corridors. The corridor bulbs are not a critical system, and their performance do not decrease enough to carry out a preventive maintenance. For this reason this functionality enables practitioners to know when the failure occurs and can solve the problem before any user complains about the burned bulb.

4.4 *Prediction module*

The prediction module enables building managers to plan out the energy usage over time, and optimize the electricity bill.

The prediction module also enables building managers to know which should be the energy consumption of the building each day. In this way, building managers can detect if the building is losing performance or the users are not carrying out good practices in terms of energy consumption.

4.5 *Control module*

The last module is the control module. Four main algorithms are implemented in GESTENSIS: controlling the boiler, controlling corridor lights, controlling loads in lecturer rooms, and controlling loads in PC rooms.

4.5.1 *Controlling the boiler*

Building managers have the challenge to optimize boilers use. One of the biggest problematic that building managers face is to determine when the boiler has to be switched on to ensure building thermal comfort when users arrive to the building.

The developed algorithm determines each day at what time it has to be turned on the boiler. The algorithm is based on neural network methods. This algorithm optimize the boiler consumption and the interior building thermal comfort.

4.5.2 *Corridor lights*

The lighting system benefits from the real time building occupancy provided by the occupancy algorithm, and the information of light sensors. With this two inputs the system can regulate the level of light in the corridors.

In the GESTENSIS demonstrator corridors have spaces used as a studying areas for the students. Depending if the studying areas are empty or not the level of light in corridors should be higher or lower. When the studying areas are used the level of light in this areas has to be higher than when the corridor is only used as a passageway. In addition, to optimize the use of energy when students use the studying areas they have to be grouped. In this way the lighting level could be reduced in the areas that are not used.

An algorithm decides each quarter of hour depending on the occupancy estimation and the level of exterior light which lines of lights should be opened.

4.5.3 *Controlling loads in lecturer rooms*

In the demonstrator students have free access to the lecturer rooms either when the students have lessons or not. This produce that lecturer rooms are used as a studying areas, and the students do not use the specific areas for this purpose. As a consequence the use of spaces are not optimized and is spend more energy than it is really needed. In addition when students left the lecture rooms lots of times they forget to turn off the lights. Another problem is that sometimes the teacher's computer or the projector are not turned on during hours without use.

The proposed system keeps user awareness messages and introduces the automatization in this type of rooms. The lights and plugs only can be used in the lecture rooms if a key is introduced in the room's electrical panel. In this way if someone needs to use this space, he needs a key that should remain introduced all the time that the space is used. When the user finish their activity in the room, he takes the key and left the room. To avoid that in the teacher exchange the lights remains turned on, when a user leave can turn off the lights or not. If he do not turn off the lights after 10 minutes the lights are turned off automatically.

With the proposed system two benefits should be achieved. The first one is the optimization in the use of spaces. The studying areas will be opened progressively depending on the demand. The students will not be able to use the room that they want, they will use the available rooms. In the other hand, lights, computers and projectors from the lecture rooms will remain turned off when the rooms are not used.

4.5.4 *Controlling loads in PC rooms*

PC rooms have a high energy consumption when the room is not used due to the computers standby energy consumption (Macarulla 2012a). GESTENSIS implemented an algorithm to cut the electricity in the room when it was not used.

The initial idea was to implement the same solution implemented in the lecture rooms explained in the previous sections. However, when PC are not used for teaching purposes, this rooms are used as a studying rooms. During the project an exploratory experiment will be carried out in order to test if it is possible to optimize the use of this rooms. The idea is to use the occupancy algorithm to determine the occupancy, if the occupancy is low and the people can be located in another studying room, the system will start the closing room protocol. With that experiment is expected to reduce the energy consumption due to an optimization of the building space usage.

5 POTENTIAL SAVINGS AND BENEFITS

GESTENSIS is not fully implemented in the demonstrator. Currently the system is capturing the information of all sensors, but the diagnosis and control module are partially implemented.

The monitoring results show that the building exceeds the thermal comfort levels during winter. During the working hours the building exceeds the 23°C, punctually reaching 25°C in the midday hours (from 11 am to 14 pm). During March, April, May, October and November at 8 am (when the building starts its activity) the temperature inside the building reaches 23°C. On Mondays of January, February and December (the coldest months) the temperature at 8 am in some cases is lower than 18°C. As a consequence, the building managers receive complaints due to the building thermal comfort is not achieved. These inefficiencies are produced due to the building boiler is switched on every day at same time during the whole winter.

The proposed system will turn on the boiler at different time each day depending on the needs. The aim is to ensure the building thermal comfort at 8 am.

The expected results are estimated using the results of a previous experience in another building, where the same algorithm to control boilers was implemented. In that building during the coldest months the gas energy savings were 9%; the rest of the months with heating demand the reduction was 15%. In addition, an extra 3% of reduction for every degree that the thermostat is set back is added. Table 3 shows the expected gas savings for each month. The expected reduction of gas demand for the whole year is 15.26%.

Savings are not the unique expected benefit due to the boiler management. It is expected that complains due to thermal comfort will be reduced because the system will ensure that at 8 am the building thermal comfort is not compromised.

Table 3. Expected gas savings per month.

	Baseline (MWh)	Expected Savings (%)	Expected Savings (%)
January	138.76	12	16.65
February	103.39	12	12.41
March	61.78	18	11.12
April	25.08	21	5.27
May	2.10	21	0.44
June	0.00	0	0.00
July	0.00	0	0.00
August	0.00	0	0.00
September	0.00	0	0.00
October	0.04	21	0.01
November	39.97	21	8.18
December	88.28	18	15.89
Total	458.39	15.26	69.97

At the present, corridors are managed by the security service. When the building opens, the security service turns on the lights. When the building closes, the security service turns off the lights. The expected electrical savings due to the optimization of the corridor lights are estimated based on an occupancy study and a determination of the lighting needs. During periods without lessons electrical savings can be near 50%. During periods with lessons, the light requirements are higher and the estimated electrical savings are limited to 35%.

Similar study was developed for the lecture room. The study revealed that, on average, lights remains turned on 1.5 hours per day in each room when these rooms are not being used. The study also revealed that the projectors remains turned on 0.5 hours per day without use. In addition, it was found that the standby energy consumption of this type of rooms is 50 W. The expected electrical savings due to controlling loads in lecturer rooms are 3%.

Finally, in case of the PC room the controlling loads function will cut electricity during building closing periods. The expected electrical savings due to this measure are 35%. On the whole, the GESTENSIS control approach is expected to achieve 37.59% of electrical savings.

It is also expected that GESTENSIS could help building managers with the maintenance activities. The proposed system should help to identify preventive maintenance. In this way, the life of the involved equipment should increase. Moreover, the building managers will have information of the poor performance of the installations. As a consequence, preventive maintenance should be carried out in order to ensure that the equipment is operating in optimal conditions.

Another expected benefit is the reduction in the maintenance rounds to detect burn bulbs in corridors. Currently, the maintenance team carries out rounds periodically to detect burned bulbs. The proposed system will detect burned bulbs and the maintenance team will be informed. In this way complains related with this issue could be reduced.

6 CONCLUSIONS

The GESTENSIS project has developed a modular energy management system capable to provide tools to optimize the energy consumption and building maintenance activities in academic buildings.

The system was designed based on interoperability, scalability, and creation of new market opportunities. The system aims to obtain data from different sources and produce relevant information for the building managers.

The system provides information to the building manager about the building performance. In this way the building manager can identify action to optimize the energy consumption. In addition the prediction module enables the building manager to optimize the energy bill. The system also assist the building managers in the optimization of the building operation. The system controls the boilers predicting the time needed to warm the building each day and setting a different hour to turn on the boiler each day. In addition, the system controls the corridor lights depending on the needs of each moment.

GESTENSIS also optimizes the energy use in lecture rooms and PC rooms. Ensuring that when those type of rooms are not used any light, computer or projector is turned on.

Finally the system helps building managers to carry out maintenance activities. The system helps to identify preventive maintenance actions and corrective maintenance actions.

The expected savings due to the implementation of the system are an electricity reduction of 15.26%, and a gas reduction of 37.59%. Despite the system is not fully implemented and the tests are not completely finished, the initial results and previous experiences suggest that the expected savings could be achieved. The results of this project can help to design or retrofit academic buildings to be more energy efficient and optimize the maintenance of this kind of buildings.

ACKNOWLEDGEMENTS

The GESTENSIS project has been funded by Ministerio de Industria, Energía y Turismo from Spain. Authors also wish to acknowledge ESVALL Project S.A., GNARUM and ADVANTICSYS for their support and funding.

REFERENCES

Albano, M., Ferreira, L.L., Le Guilly, T., Ramiro, M., Faria, E., Pérez, L., Ferreira, R., Gaylard, E., Jorquera, D., Roarke, E., Lux, D., Scalari, S., Sørensen, S.M., Gangolells, M., Pinho, L.M., & Skou, A. 2013. The ENCOURAGE ICT architecture for heterogeneous smart grids. In *IEEE Eurocon Conference (Eurocon 2013)*, Zagreb, Croatia, 1–4 July 2013: 1383–1390.

Directive 2010/31/EU of the European parliament and of the council of 19 May 2010 on the energy performance of buildings, 2010 OJ L 153/13.

EC, 2013. Financial support for energy efficiency in buildings, Report from the Commision to the Eurpean parliament and the Council. [Accessed on 19 April 2016]. Available at: <https://ec.europa.eu/energy/sites/ener/files/documents/report_financing_ee_buildings_com_2013_225_en.pdf>.

Macarulla, M., Casals, M., Gangolells, M. & Forcada, N. 2014a. Reducing energy consumption in public buildings through user awareness. *European Conference on Product and Process Modelling. "eWork and eBusiness in Architecture, Engineering and Construction: ECPPM 2014"*. Vienna, Austria, 17–19 September 2014, 637–642.

Macarulla, M. & Albano, M. 2014b. Smarter grid through collective intelligence: user awareness for enhanced performance. *Socialines Technologijos (Journal of Social Technologies)*, 4(2): 292–305.

Ministerio de la presidencia, 2007. Reglamento de Instalaciones

Térmicas en los Edificios (RITE), 1027/2007.

Motawa, I., & Almarshad, A. 2013. A knowledge-based BIM system for building maintenance. *Automation in Construction*, 29, 173–182.

Le Guilly, T., Skou, A., Olsen, P., Madsen, P.P., Albano, M., Ferreira, L., Pinho, L.M., Casals, M., Macarulla, M., Gangolells, M. & Pedersen, K. 2016. ENCOURAGEing results on ICT for energy efficient buildings. Emerging Technologies and Factory Automation. Berlin, Germany, 6–9 September 2016. Submitted.

Pérez.Lomabard, L., Ortiz, J. & Pout, C. 2008. A review on building energy consumption information. *Energy and Buildings*, 40(3): 394–398.

Ramesh, T., Prakash, R. & Shukla, K.K. 2010. Life cycle energy analysis of buildings: An overview. *Energy and Buildings*, 42(10): 1592–1600.

Sheu, S., Chang, C., Chen, Y. & Zhang, Z.G. 2015. Optimal preventive maintenance and repair policies for multi-state Systems. *Reliability Engineering & System Safety*, 140: 78–87.

Toftum, J., Kjeldsen, B.U., Wargocki, P., Menå, H.R., Hansen, E.M.N. & Clausen, G. 2015 Association between classroom ventilation mode and learning outcome in Danish schools, *Building and Environment*, 92: 494–503.

eWork and eBusiness in Architecture, Engineering and Construction – Christodoulou & Scherer (Eds)
© 2016 Taylor & Francis Group, ISBN 978-1-138-03280-4

Responsiveness based material—[a] passive shading control system

M.J. de Oliveira
DINÂMIA'CET-IUL, Instituto Universitário de Lisboa (ISCTE-IUL), Lisboa, Portugal

V. Rato
ISTAR-IUL, Instituto Universitário de Lisboa (ISCTE-IUL), Lisboa, Portugal

C. Leitão
Pratt Institute of Design GAUD. Brooklyn, NY, USA
School of Architecture, Rensselaer Polytechnic Institute, Troy, NY, USA

ABSTRACT: During the last decades Architecture has been looking to its basic principles, finding in nature a natural and obvious inspiration. Materials and environment have been playing an important and essential role in this process. Recovering the ideals of the 1950's intellectually movement *Performative Turn,* performance-oriented design finds its fundaments on the understanding that architecture unfold their performative capacity by absorbing the complexity conditions and processes. Following this premise, architecture and environment are simultaneously set at a spatial, material and temporal level.

The following article has the goal to describe a methodology to find the material and environmental driven parameters to be considered in the design and construction of a passive shading system. This research aims to develop a universal parametric definition, based on cork material and environmental essential and determinant driven parameters that could enable us to design a totally personalized passive shading system to any location and time.

1 INTRODUCTION

Throughout the nineteenth century, human constructions relied on available/raw materials inertia for thermal regulation—buildings had thick walls and small narrow windows, enabling us to sustain the heat in the interior of the spaces during the winter, and protecting us from the intensive heat during the summer. Narrow windows helped us controlling the ventilation, minimizing/optimizing thermal behavior in the interior of the spaces according with uses and needs. In the twentieth century, advancements in insulation assemblies and composites of materials create possibilities for floor/ceiling walls with narrow steel frames—and mass production of these assemblies enables the pervasive reliance by humans of artificial air ventilation in the interior of the buildings and structures.

At this point of the 21st century, architecture and design find concepts and development, through layers of complementary information including those concerning constraints and criteria for the design of buildings envelopes—often constituted of assemblies of interrelated/responsive parts among each other with and with surrounding environment. Often in these building envelopes, form, material and structure are expressed and worked as distinct components from the same body, working independently from each other. However in nature there is no such distinction. There is no natural body or system, where structure is independent from material, or where material works independently from form, or even form independent from structure. In nature, bodies emerged as interrelated systems, with no assemblies or parts. The natural body is at once structure, form and material.

1.1 *Related work*

In recent years, new strategies for design and techniques for making materials and large constructions have emerged, based on biological models of the processes by which natural material forms are produced. Biological organisms have evolved multiple variations of form that should not be thought of as separate from their structure and material properties. Structural and informational forces collaborate to create what is ultimately a material, and then what we call form—these processes are very difficult to completely analyze and final natural bodies nearly impossible to be reproduced through the same growth methodology they were 'produced'. The self-organization of biological material systems is a dynamic process that occurs over time and produces the capacity for inducing change in the order and structure of a system, modifying its behavior and performance (Kauffman, 1993).

Passive systems—such as those observed in forms for feedback in nature between structural forces and form—create new questions and potentials to rethink architectural systems and bodies.

One example of this is the shading system Bloom, developed by the Dosu Studio 2012. Bloom is a bimetal structure installation that reacts to heat, generating increasing or decreasing openings in a pattern, enabling a shading system to adapt to its environment. The keygen was a crucial part of understanding the material, its behavior conditions and characteristics, enabling a clearly development of the pattern behavior and performance.

Other iconic experience, in 2013, is the 'The Hygroscopic Envelop Prototype' from Achim Mengues based on a structural responsive and hygroscopic system. This system consists of a structural surface composed of several regions sensitive to local humidity concentration moving and adapting to climate change. More than finding architectural surfaces as solutions Menges "form follows performance" strategy mixes appearance and organization of patterned skins and structures in nature, enabling to explore materials behaviors and effects—biomimetics and biomimicry (Kolarevic and Klinger, 2008).

1.2 *Purpose and project*

In order to design a periodic shading control structure, mutable and (re)adaptable to a specific external environment and context, a customized definition is produced, parametrically reconfigurable, allowing for a total adaptation to external inputs and environmental data. The main focus of this research is the knowledge of material properties and performative responsiveness within the scope of some specific morphological conditions and environmental exposure. While keeping performative and responsive issues as digital layers, all the data is considered in the digital and virtual plane, being tested during the design process. This methodology aims to contribute to solve site-specific conditions and constraints.

2 OBJECTIVES

2.1 *General scope*

This paper presents an ongoing research focused in develop a passive shading control system, using parametric methodologies based material and environmental knowledge. The results are informed by design decisions based on contextual factors and consequently we are interested in stablish a parameter-driven methodology.

2.2 *Main goal*

The main goal defined, for this investigation stage, was to develop a parametric environmental system that could balance a multi-functioning driver parameter. The fundamental hypothesis supporting this system find its basis in the natural behavior of the applied materials (Cork and Metal), as well as in its internal and external exchange of data and environmental inputs. Inspired by natural physical elements, the target is to develop a parametric definition that expresses a shading control system that respond and adapts to a pre-determined environment, with specific characteristics, functions and to its inhabitants occupancy.

3 METHODOLOGY

The methodology (being) used to develop the passive shading system encompass three stages.

Stage 01: Material driven-parameters—Aims to find and describe through several material experiences and physical essays the most relevant and dominant material parameters that could influence the system.

The Stage 02: System—Form, structure, material—aims to point at two possible paths to achieve the shading system, explaining pattern decisions and some assembly considerations. The Stage 03: Integration design/Environmental analysis exposes the most relevant design parameters and design tools that are being used during the form-finding process.

3.1 *Stage 01: Material driven-parameters*

Conducting a material base investigation it's essential to understand the material characteristics and properties.

Cork it's a natural anisotropic material. 100% renewable, 100% organic, with excellent acoustic and thermal behavior. Cork is used and applied in several types of product and contexts—since insulation construction panels to umbrellas textiles. There are mainly three types of "cork": (a) The first one is the cork as a raw material—mainly used in the stoppers industry. (b) The second type of cork products, still 100% natural are expanded agglomerate—produced from the waste of all types of cork transformations. In this case the cork is granulated, heated and compacted in an autoclave at 350° degrees. This process inputs more than 3X the grain size and releases the suberin from the cork cells enabling the natural aggregation of the grains and the compaction of the material. (c) Cork agglomerated possibly the most used and known sub cork product. Contrary to the expanded agglomerate in the cork agglomerate the grain are not expanded, and the product conserve the natural cork color. There are several densities defined by the grain size and quantity. Grains could be mixed with resins, pigments, rubber and other aggregates.

The selected products for our research pilot prototypes were the expanded agglomerate cork—MD (Medium Density) and HD (High Density) – and agglomerate cork samples. Three different types of grain between 0,5 mm to 25 mm, and with a reference of self-weight of 300 kg to 470 kg/m^3 – all high density.

3.1.1 *Cork design boundaries*

Knowing the anisotropic and cellular characteristics of the cork raw material, we could easily identify and understand several cork behavior and properties, such as lightness, compressibility, flexibility, expansion and so on. In a way to potentiate distortion and higher flexion cork cell microstructures were brought to board scale.

Four patterns were developed based on cork microcellular observation (Figure 2). Using as a primary structure the radial section, also known as the 'honeycomb' (Figure 1) parametric patterns were defined using as primary inputs the characteristics of the material: (1) Structural integrity and average number of the grain size (in the several boards) conditioning the scale of the patterns; (2) pattern equilibrium, properties and equilibrium based cork radial section.

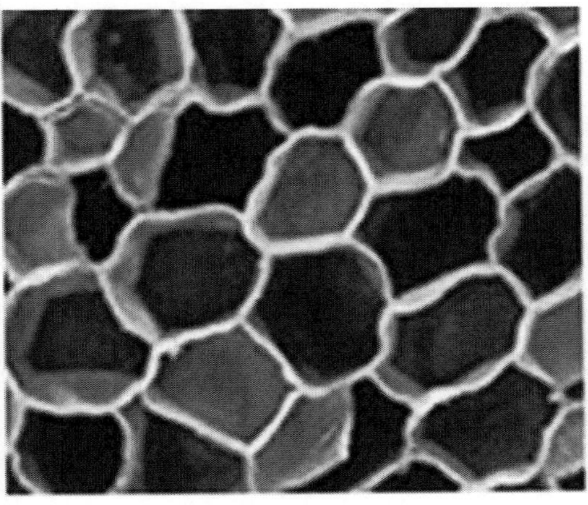

Figure 1. Cork radial section—cellular microstructure. (Silva et al.).

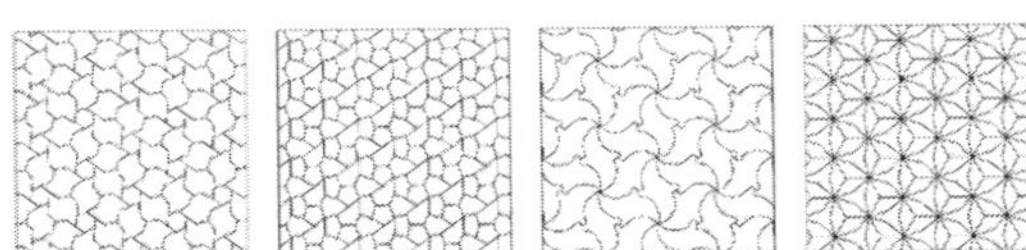

Figure 2. Parametric patterns based cork radial section.

3.1.2 *Cork samples*

Four patterns in five different types of cork products were prototyped. 500 × 500 × 20 mm boards, ½ depth cut, Ø 3 mm milling tool, were the common parameters to the five types of cork boards and to the four patterns (Figures 3–4).

In order to obtain some specific input to design of the main shading system some distortion and flexion testes were conduct in order to establish geometrical boundaries, assembly potential conditions and hypothetical board's organization.

3.1.3 *Flexion*

The flexion test was conduct with the following tools: (1) Two idlers with predetermined diameter (related with the material thickness), working as two support points, apart with a pre-establish distance related with the length of the material sample; (2) A semi cylindrical axe.

The goal was to analyze the behavior of the sample, simulating a bi-supported beam by a bending test at three points (Figure 5).

3.1.4 *Distortion*

This test was conducted with the following tools: (1) Two idlers with predetermined diameter (related

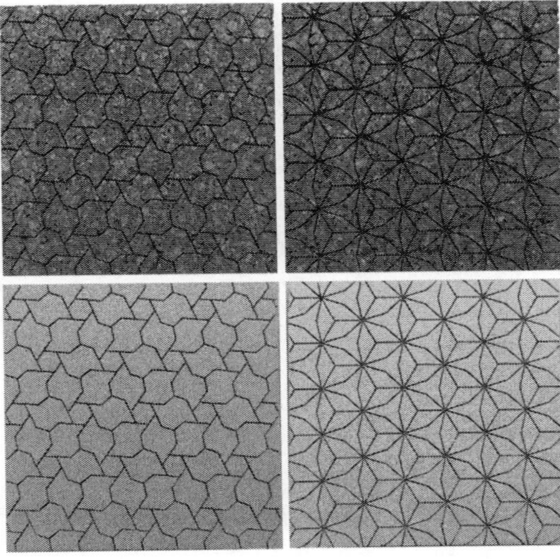

Figure 3. Two different patterns (Escher on the left and Earth on the right, two different types of agglomerated cork—expanded agglomerated cork (top samples), composite agglomerated cork (down samples).

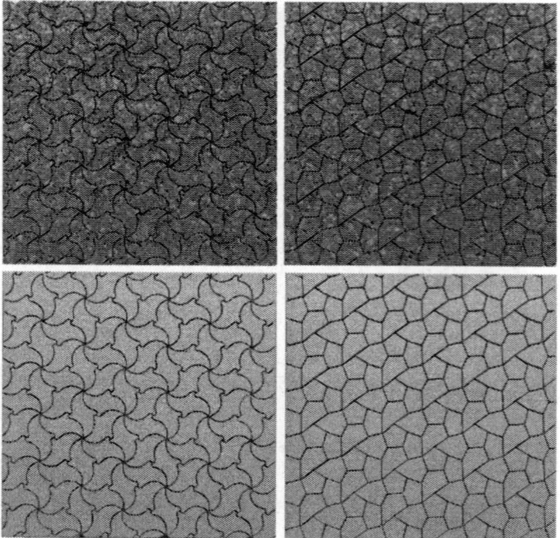

Figure 4. Two different patterns—Tile on the left and Islamic on the right—two different types of agglomerated cork—expanded agglomerated cork (top samples), composite agglomerated cork (down samples).

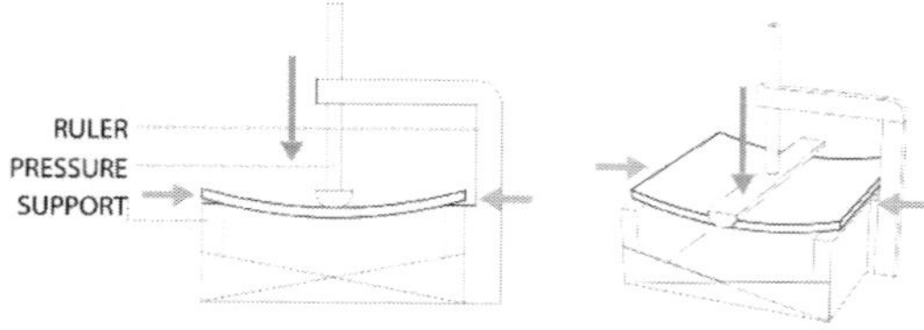

Figure 5. Flexion test representation.

with the material thickness), working as two diagonal support points, apart with a pre-establish distance related with the diagonal length of the material sample; (2) A semi cylindrical sand container axe hung to the two free diagonal points.

The goal was to analyze extract values from the curvature behavior of the sample, simulating a bi-supported beam by a bi-bending test at two points (Figure 6).

3.2 *Stage 02: System—form, structure, material*

With input from the material tests, parameters were generated and several boundaries and values were determined as design drivers. From the material tests two possibilities emerged as separated logics with the same fundaments: (a) explore the pattern potential as a self-material assembly, and work the boards as plan corps or (b) explore the pattern potential at two levels, at the bending board possibility and as self-assembly boards. At this point we know that the bending potential only could exist

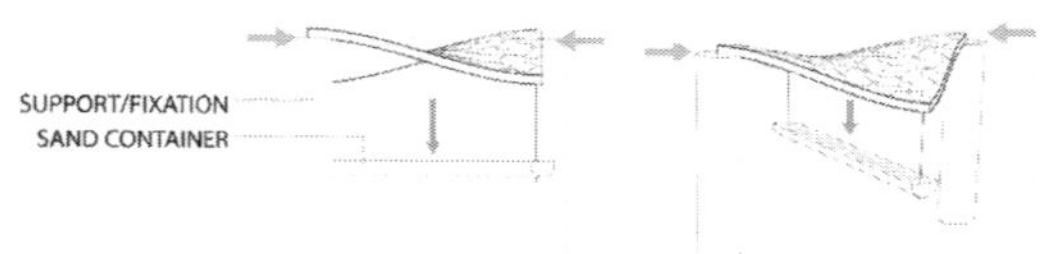

Figure 6. Distortion test representation.

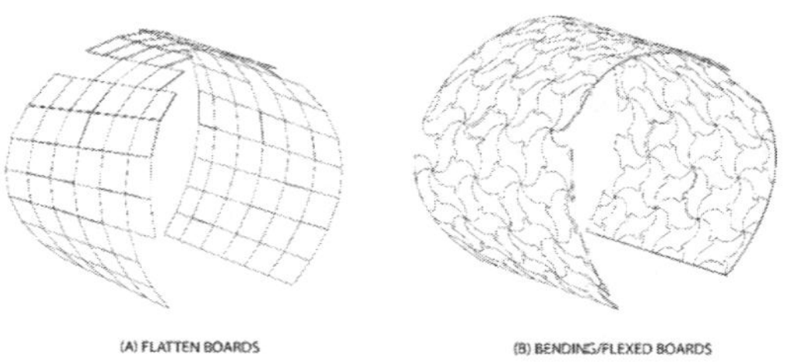

Figure 7. Shading system—structural essays and pattern application.

adding a malleable but resistant material, such as metal (Figure 7).

In both possible solutions, the assembly process only considers the material properties, its weight, friction, elasticity, flexibility, traction and resistance. The form of the system is intimately related with the chosen solution. On the one hand the system could work as diamond defined by its limits, uniform and consistent, reflecting a secure compromise between the different elements of the organism. Or, on the other hand, the system could be unbound at edges, working as a two face structure (two materials), self-supported and assembled, exposing its full strength as a material, as a structure and form—a full integrated system. This latter system is being simulated in a Rhino/GH (Grasshopper) workflow. Material properties and characteristics are being loaded through Geco and structure is being tested through Kangaroo and Karamba (GH plug-ins).

3.3 *Stage 03: Integration design/Environmental analysis*

Towards material tests and shading system structural simulation, environmental analysis is being integrated in the workflow. All the processes run simultaneously such that the results of one interaction, potentiate and improve aspects of the other interactions and forms. At this stage of the research, the work considers climatic analysis, based on weather file information using LadyBug (GH plug-in) for connections between the available data, geometry and materials. It integrates confort and measurable indicators such as *Ther-*

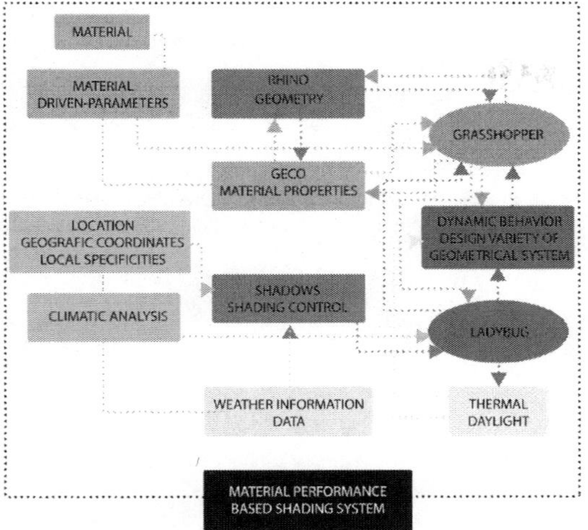

Figure 8. Diagram of the process, resources order and connections.

mal (PET—Physiological Equivalent Temperature) and *Daylight* (adding Honeybee GH plug-in). The parameters are: Thermal—influence the heat balance of the human body—temperature, shadow, radiation, air direction and velocity, etc. Daylight—illuminance and luminance values.

Geometric reconfigurations are being made so that the system is able to react to those climate conditions represented through thermal and daylight comfort. The parametrization of the process defines the dynamic movements of the elements. The predominance of and sustainability of the system will always reinforce the design driver criteria that mostly constrains its geometric parameters (Figure 8).

4 RESEARCH INSIGHTS

At this point, the research work is still open to the two possibilities announced in 3.2. However other factors have been added such as the importance of keeping the system restricted to one material, and therefore keeping the production capacity tie to one unique type of industry. This will avoid additional costs and enable a faster production and control of the entire process—from creation, fabrication, transportation to local assembly.

Another important conclusion has been extracted from the cork samples physical essays. The durability of the system could be an important factor for its commercialization and potential application. Being made (in its majority) by cork it's an assumed seasonal/temporary system. However the system should contemplate the possibility of being renewable or even prevent the possible substitution of parts during its life cycle time-lapse.

5 FUTURE WORK

Additional aspects of research and testing can bring further potential towards the development of this system. Other unmentioned material variables could determine decision making criteria for the proposed possibilities. Such variables as material thickness, density, the cut depth, the diameter of the milling tool, the section of the milling tool—these are all parameters that are determining to the form-finding process.

Other following steps are related with a more exhaustive structure of the parametric definition. The connection between geometry, design driver, environmental data—through plugins—can find different methodological choices which are informed from a logic process that prioritize some parameters and solutions according to the initial inputs of the target environment and user.

Towards this objective, the development of this work includes the creation of a pattern for a specific physical location. Variables of a specific shading system will be created and a virtual essay will be prototyped to scale 1:1. The prototype will be tested in laboratorial conditions.

ACKNOWLEDGMENTS

The authors would like to thanks to the Amorim Group. A special thanks to Dr. Carlos Manuel, Engº Lopes Infante, and Dr. José Andrade from the Amorim Insulation Industries and to Dr. Marina Rodrigues from the Amorim Cork Composites Industries.

REFERENCES

Burry, J., Salim, F., Sharaidin, K., Burry, M., & Nielsen, S.A., 2013. Understanding Heat Transfer Performance for Designing Better Façades. Acadia, pp. 71–78.

De Oliveira, M. & Rato, V.M., 2014. CORK'EWS From microstructural composition into macrostructural performance. 2nd International Conference of biodigital architecture & genetics, pp.320–330.

De Oliveira, M.J. & Moreira Rato, V., 2015. From Morphogenetic Data To Performa—Tive Behaviour. Emerging Experiences in the PAst, Present and Future of Digital Architecture. CAADRIA 2015, 20th International Conference of the Association for Computer-Aided Architectural Design Research in Asia, pp.765–774.

Hensel, M., Menges, A., & Weinstock, M. 2010. Emergent Technologies and Design. Towards a biological paradigm for architecture. USA and Canada; Routledge.

Kauffman, S., 1993. The Origins of Order: Self-Organization and Selection in Evolution. Oxford University Press.

Kolarevic, B., & Klinger, K., 2008. Manufacturing Material effects. Rethinking Design and Making in Architecture. USA and Canada; Routledge.

Oxman, R., 2008. Performance-based Design : Current Practices and Research Issues. International Journal of Architectural Computing, 06(01), pp.1–17.

Pearce, P., 1980. Structure in Nature is a Strategy for Design. The MIT Press; Reprint edition (June 16, 1980).

Silva, S.P., Sabino, M.A., Fernandes, E.M., Correio, V.M., Boesel, L.F. & Reis, R.L., 2005. Cork: properties, capabilities and applications, International Materials Reviews, 50 (6) (2005), 345–365.

Schumacher, P., 2009. Parametricism: A new global style for architecture and urban design. Architectural Design, 79(4), pp.14–23.

eWork and eBusiness in Architecture, Engineering and Construction – Christodoulou & Scherer (Eds)
© 2016 Taylor & Francis Group, ISBN 978-1-138-03280-4

Promoting energy users' behavioural change in social housing through a serious game

M. Casals, M. Gangolells & M. Macarulla
Group of Construction Research and Innovation, Department of Project and Construction Engineering, Technical University of Catalonia, Barcelona, Spain

A. Fuertes, R. Jones & S. Pahl
Plymouth University, Plymouth, Devon, UK

M. Ruiz
Fremen Corp., Troyes, France

ABSTRACT: Housing represents about 29% of the total energy consumption in Europe and contributes with around 20% of emissions (European Commission 2013). Social housing represents about 12% of the total European housing stock and therefore is a significant target for energy efficiency measures by governments of EU member states. This paper is aimed at exploring how an innovative serious game could contribute to energy consumption and carbon emissions reduction in social housing by increasing the social tenants' understanding and engagement in energy efficiency. The proposed solution is being developed under the auspices of the EnerGAware project (Energy Game for Awareness of energy efficiency in social housing communities), funded by the European Commission under the Horizon 2020 programme.

1 INTRODUCTION

The housing sector is one of the priority areas in Europe with regard to energy efficiency – not only because it consumes a great amount of energy, but also because it remains greatly inefficient. Housing represents about 29% of the total energy consumption in Europe and contributes with around 20% of emissions (European Commission 2013). Social housing represents about 12% of the total European housing stock and therefore is a significant target for energy efficiency measures by governments of EU member states. The proportion of social housing does however vary significantly between countries; the Netherlands has the highest share of social housing in Europe, accounting for 32% of the total housing stock, followed by Austria (23%) and Denmark (19%). The UK (18%), Sweden (18%), France (17%) and Finland (16%) also have a relatively large social housing sector (CECODHAS 2011). Increased cost of fuel, the liberalisation of energy markets and decreased levels of welfare provision in Europe since the 1970s, has also resulted in an increasing number of households living in social housing that cannot afford the energy bills. In 2011, 9.8% of households in the EU could not afford to heat their home adequately, whilst 8.8% of households were in arrears on their utility bills (Thomson & Snell 2013).

One of the ways of addressing this challenge is though social tenants' behavioural change. This paper is aimed at exploring how an innovative serious game could contribute to energy consumption and carbon emissions reduction in social housing by increasing the social tenants' understanding and engagement in energy efficiency. The proposed solution is being developed under the auspices of the EnerGAware project (Energy Game for Awareness of energy efficiency in social housing communities), funded by the European Commission under the Horizon 2020 programme.

2 THE ENERGAWARE SERIOUS GAME

Serious games are a simulation environment, based on social interaction and scenario experimentation, designed to highlight, although virtually, potential realistic outcomes. Therefore, the main objective of serious games is to change human behaviour through education and training.

The EnerGAware project aims to bridge the gap in people's understanding and awareness of energy consumption by developing a serious game linked to the real energy consumption of the users' homes. The following subsection describes the process of eliciting specific user, building and game requirements necessary to design the EnerGAware integrated serious game and metering system solution. Subsection 2.2 introduces the concept behind the EnerGAware serious game whereas subsection 2.3 describes the game mechanics. The methodology used to assess the impact of the EnerGAware serious game in terms of energy saving and peak demand but also in terms of perceived physical comfort, self-reported energy consumption behaviours and awareness, energy knowledge, social media activity, IT-literacy and socio-economic status and health is also described.

2.1 *Eliciting requirements*

A comprehensive identification and analysis of the specific user, building and game requirements that are necessary to design the EnerGAware integrated serious game and metering system solution was carried out. Requirements were defined using a range of different datasets and methods including (1) literature review; (2) a large-scale, city-wide survey, undertaken in Plymouth, UK, during 2015, which was administered to all the 2,772 social houses managed by the social housing provider partaking in the EnerGAware project; (3) three game-play scenarios focus groups undertaken with social housing tenants in Plymouth, UK, during 2015 and (4) asocial housing building stock database gathered and managed by project partner DCH (Building Stock Condition Database).

2.1.1 *User requirements*

Results suggested that the EnerGAware serious game virtual world should be based on a domestic environment (e.g. virtual home), so as to help the players to relate to. Results revealed the existence of a large group of older people, high presence of retired people and a large group with low educational level, suggesting that the EnerGAware game should put special attention when designing the visual aspects of the game to those requirements derived from human aging process and novice users. In relation to the didactic approach of the game, the game should adapt to different learning levels and provide clear and easy to understand goals. Regarding the educational content, the game should allow users to learn how to balance the energy consumption, comfort and financial cost of a house; gain knowledge on how much energy is used by the typical end-uses existing in a domestic environment, poor practices of use that might increase the energy consumption, as well as the most efficient ways to use them to save energy. The game should also help the player to assess the potential energy savings from different behaviour actions and energy-efficient changes to the virtual house. From the game functionalities point of view, the link to social media platforms to enhance communication and information sharing amongst players was found relevant.

2.1.2 *Building requirements*

The most common building characteristics, building envelope, building services and controls and renewable energy generation were analysed and transformed into the 'typical' social dwelling which was used to influence the design the virtual home contained in the EnerGAware serious game. Data related to the energy metering and monitoring systems existing in social homes (e.g. smart meters, end-use metering, etc.), internet availability and coverage was used to design the energy metering and data communication infrastructure.

2.1.3 *Game requirements*

Results validated that a significant part of the social tenants have a good IT-literacy, Internet and social networks habits, and experience in playing video games. Therefore, the results suggest that the online serious game approach adopted for the EnerGAware serious game should not be a barrier for the targeted audience. Both the focus group and the Social Housing Survey results suggested that the EnerGAware serious game should be an energy management game (home management, resources management) focused on a virtual house customization game. Regarding the graphical aspect and the setting of the EnerGAware serious game, the results of the Social Housing Survey suggested that this is not a major criterion of game choice for the targeted players. However, the focus groups concluded that a pseudo-realistic game setting would be better than a fantasy world (or sci-fi, or cartoon) and better than a fully-realistic simulation. A tactile tablet was identified as the most suitable IT device (both technically and cost-effectively) for the deployment of the EnerGAware serious game.

2.2 *The EnerGAware concept*

The serious game will allow users to design their own virtual home using a simple drag-and-drop interface. The users will have an initial limited financial budget (in-game currency) available to construct their house and choose domestic appliances, lighting and furniture. Users will be able to earn in-game rewards to upgrade their home and buy more in-game objects by improving the energy efficiency of the house (e.g. increase insulation) or change the game characters energy efficiency behavior.

A building energy consumption simulation engine (Figure 1) will calculate the current energy consumption of the virtual house and provide the potential options for improving its energy efficiency. The options provided will demonstrate the potential energy savings from (1) upgrading buildings' envelopes (i.e. no insulation vs. cavity wall insulation or solid wall insulation, etc.); (2) replacing the existing domestic appliances and lighting (i.e. incandescent bulbs vs. CFL bulbs or LED bulbs, etc.); and (3) user behaviour change (i.e. reducing heating and cooling temperatures or durations; leaving appliances in standby mode, etc.). Users will be able to click on appliances and HVAC devices in the game

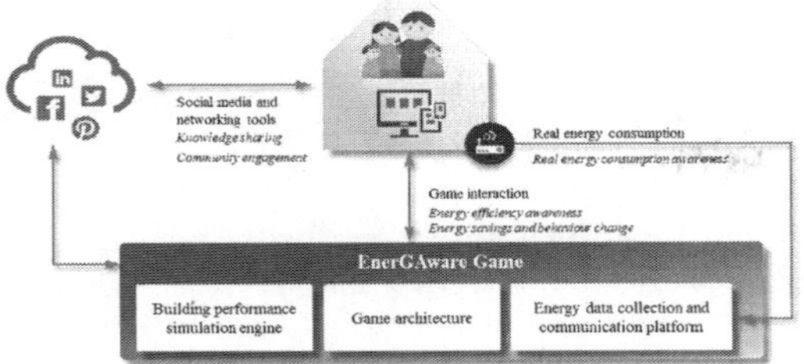

Figure 1. EnerGAware serious game schematic.

and receive feedback about the energy consumption in different modes (e.g. active, standby, off; set-point temperatures; heating/cooling periods).

The successful balancing of energy consumption, comfort and financial cost will lead to the user generating extra in-game rewards which can then be reinvested in the home. The user will need to decide whether to invest in low cost options providing low energy savings (i.e. replacing plasma by LED TV, installing draft excluders, etc.), or high cost options providing high energy savings (i.e. solid wall insulation, solar photovoltaic panels, etc.) with the latter taking more time to save up for. In addition, the user will be able to play a series of missions related to energy use integrated into the main EnerGAware serious game. These will contribute to increase energy awareness and will provide the user with an opportunity to earn further in-game rewards. The EnerGAware serious game will also be connected to the actual energy consumption (smart meter data) of the house in which the user lives (Figure 1). This connection will have three purposes; firstly, real world energy savings will translate into in-game rewards; secondly, the user will be able to view their current and historic energy consumption of their homes' through the serious game interface; and thirdly, to validate the energy savings achieved in the real world from playing the serious game.

Finally and as shown in Figure 1, the EnerGAware serious game will be embedded in wider social media and networking tools (e.g. Facebook, Twitter, etc.). In the simplest form, these links with social media and networking tools will be used to disseminate the game and enable users to enter or re-enter the serious game from potentially anywhere in the world, thus reducing barriers to participation and encourage large scale uptake beyond the project's lifetime. The social media features will also provide users a platform to share data of their achievements, compete with each other, give energy advice, as well as, join together to form virtual energy communities.

2.3 *The EnerGAware game mechanics*

As shown in Figure 2, the cat is the main character of the game and the only one that can be controlled by the player. The human characters living in the virtual house are non-player characters controlled by the computer. They have non-energy efficient behaviours, which the cat will try to address. Neighbors may request the player's help in several kinds of situations (e.g. advices to choose an energy provider or actions such as turning off all the lights). Within the house customisation mode, the player is able to create his/her dream house. An editor function allows the player to buy appliances, furniture, decoration items and energy efficiency upgrades (e.g. wall insulation) in a realistic environment. A mission mode provides knowledge about energy efficiency and educates the player about right energy management behaviours. Missions take place in neighbours' houses with a fixed geometry that substantially ease energy consumption simulations. The mission mode also shows ideas about how the player's house could evolve.

Figure 2. Screenshot of the EnerGAware prototype.

The main gameplay loop starts with a daily pool of energy points. The player has an operational house with a global energy consumption and she/he will have to save energy points to complete energy efficiency objectives. It will allow him/her to unlock game content, mainly new items and upgrades for already owened items (e.g. appliances, insulation). New items might be more efficient, i.e. consume less energy points (e.g. a more energy efficient fridge), or just be a smarter version (e.g. a bigger TV).

If the player chooses the more energy efficient item, this will impact favourably the global energy consumption. Upgrades and new items will increase global happiness which will in turn increase daily money income. Money is then used to buy upgrades and new items.

The delicate balance between energy consumption and occupants' happiness is an important part of the gameplay. Reducing the happiness level too much to save energy points will decrease humans' productivity and as a consequence they will earn less money. The goal is to make the players understand the need to invest in better equipment, smart connected devices and insulation to reduce their need in energy points without decreasing the happiness level of their humans.

2.4 *Methodology to evaluate the impact of the EnerGAware serious game*

The game will be tested in a social housing pilot of 100 homes in Plymouth (United Kingdom) by studying the actual real-world energy savings achieved from playing the game, as well as households' reactions, feedback and improved energy literacy, while exploring the wider community benefits of implementing such a game.

In accordance to the International Performance Measurement and Verification Protocol (IPMVP) (EVO 2012), pre-post comparison will be applied in the EnerGAware project. As shown in Figure 3, the effect of the serious game will be estimated through comparison of the energy consumption, peak demand and other indicators before (baseline evaluation), after the intervention starts (mid-term evaluation) and at the end of the intervention (final evaluation).

In addition, a control group approach will be also implemented following the recommendation in the European ICT PSP Methodology for calculating energy savings in buildings (BECA 2012). An advantage of the control group approach is that the data are collected over the same time period for both the experimental and control group; therefore external influences (e.g. energy price changes, longer school holidays, sports events, etc.) which could have an effect on the measured dependent variables are also considered (Figure 3).

In the EnerGAware project, 50 households will be assigned to the experimental group and 50 households will be assigned to the control group. Social housing tenants in the experimental group will play the EnerGAware serious game and have their energy (gas and electricity) consumption monitored. The social housing tenants in the control group will have only their energy consumption monitored. Households in both groups will complete a series of additional surveys, at the end of the mid-term evaluation and final evaluation periods, regarding their self-reported energy consumption behaviours and awareness, energy knowledge, social media activity and IT-literacy as well as perceived physical comfort and socio-economic status and health.

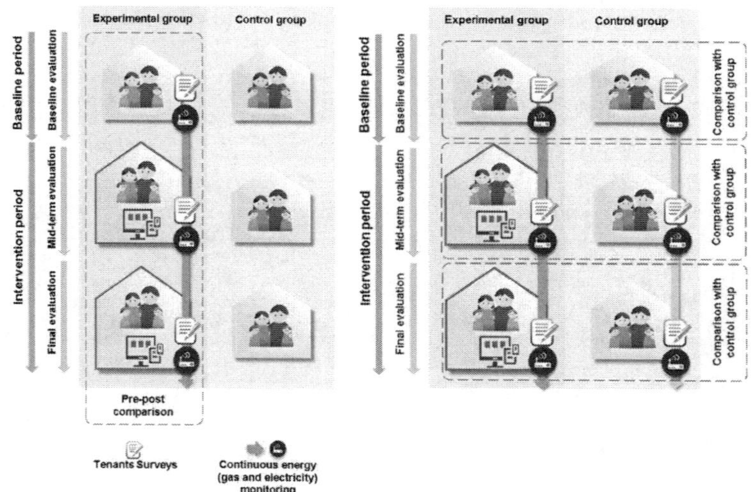

Figure 3. Pre-post comparisons (left) and comparison with control group (right).

Households in the experimental group will also complete a survey about the usability, usefulness, attractiveness and interaction level of the EnerGAware serious game.

3 CONCLUSIONS

This paper has presented the EnerGAware project aimed to reduce energy consumption and carbon emissions in a sample of European social housing by changing the energy efficiency behaviour of the social tenants through the implementation of a serious game linked to the real energy use of the participants' homes.

Regarding energy efficiency in social housing, past research initiatives have mainly focused on displaying real-time energy consumption data and optimising energy management (generation and usage) using ICT systems. The EnerGAware project will significantly advance the current state-of-the-art by developing and testing the effect of providing social tenants with a serious game, that is both linked to social media and networking tools (e.g. Facebook, Twitter, etc.) and to the actual energy consumption (smart meter data) of the house in which the game user lives.

In relation to serious gaming, the EnerGAware serious game will be the first of its kind: combining real-time energy consumption of the home in which the game player lives, with energy saving feedback and rewards (i.e. unlocking features and content in the serious game). The EnerGAware serious game will step beyond existing e-learning solutions and games as it is designed to appeal to the new generation of digital natives and to trigger them to stay engaged, play, absorb and learn. The serious game will also be linked to social media and networking tools, to address the communication and social routines of people in the new digital age.

Regarding Internet of Things (IoT), current games regarding energy consumption either use simulated data, or data collected once. To improve the realism of the game, and the educative impact on the final user, cyber-physical systems will collect data from the user's house. Another technical challenge that will be tackled by the project is the interaction with existing energy monitoring devices. Embracing the paradigm of a real IoT can lead to standardized and interoperable installations, which will result accessible to a broader public.

In relation to behaviour change psychology, the project will deliver principles and insights that transcend the transient nature of current ICT-based solutions for energy efficiency: understanding the cognitive and social processes that determine people's decision making and behaviours. This will provide a basis for further development of ICT-based solutions that can be used to make energy usage visible and enable people to achieve energy savings.

Regarding social media and networking, the project will explore whether social networking sites may be able to play a role in helping to support behavioural change, both structurally and by shaping beliefs and culture. Social media and networks will be exploited to engage a wider range of people, beyond the social housing pilots, with the EnerGAware serious game. We will evaluate which types of energy information and data people are willing to share (and which attract most attention and debate in their social network) and examine how people use these to discuss and reduce energy use.

ACKNOWLEDGEMENTS

This paper is based on the EnerGAware project (Energy Game for Awareness of energy efficiency in social housing communities), contract no. 649673. The research project is supported by the EU H2020 programme (H2020-EE-2014-2-RIA) and is being carried out jointly by the Universitat Politecnica de Catalunya (Spain), the University of Plymouth (United Kingdom), the Instituto Superior de Engenharia do Porto (Portugal), Fremen Corp (France), Advantic Sistemas y Servicios (Spain), DCH Devon and Cornwall Housing (United Kingdom) and EDF Energy R&D UK Centre Limited (United Kingdom).

REFERENCES

BECA Balanced European Conservation Approach 2012. *The ICT PSP Methodology for Energy Saving Measurement: A common deliverable from projects of ICT for sustainable growth in the residential sector, Version 3.*

CECODHAS Housing Europe Observatory 2011. *Housing Europe Review 2012: The nuts and bolts of European social housing systems.* Building and Social Housing Foundation (BSHF). Available at: <http://tinyurl.com/mtorf8l>. Accessed on 27 March 2016.

EVO Efficiency Valuation Organisation 2012. *International performance measurement and verification protocol. Concepts and options for determining energy and water savings, vol. 1, Technical Report.* Available at: <http://www.coned.com/energyefficiency/PDF/EVO%20-%20IPMVP%202012.pdf>. Accessed on 2 October 2015.

EnerGAware 2015. *Energy Game for Awareness of energy efficiency in social housing communities.* EU funded project. Contract number: 649673. Available at: <http://energaware.eu/>. Accessed on 26 March 2016.

European Commission 2013. *EU energy in figures: Statistical Pocketbook 2013.* Luxembourg: Publications Office of the European Union. Available at: < http://ec.europa.eu/energy/sites/ener/files/documents/2013_pocketbook.pdf>. Accessed on 15 January 2016.

Thomson, H. & Snell, C. 2013. *Energy Poverty in the EU: Policy Brief.* Available at: <https://www.openaire.eu/search/publication?articleId=datacite____::08e3ab33b677e1c3bc8df10c55245ead>. Accessed on 15 January 2016.

eWork and eBusiness in Architecture, Engineering and Construction – Christodoulou & Scherer (Eds)
© 2016 Taylor & Francis Group, ISBN 978-1-138-03280-4

Total life cycle and near real time environmental assessment approach: An application to district and urban environment?

C. Kuster, Y. Rezgui, J.-L. Hippolyte & M. Mourshed
BRE Trust Centre for Sustainable Engineering, School of Engineering, The Parade, Cardiff University, Cardiff, UK

ABSTRACT: This paper presents the ongoing research on the development of a total life cycle and near real time environmental assessment. Over 30 widely used sustainability assessment tools have been reviewed. This is done through investigating different characteristics of each tools namely: locality, scale, life cycle stage implementation, chosen criteria, chosen weighting system; and the definition of common patterns and issues. Additionally, forecasting models applied for power and energy consumption have been reviewed with around 120 applications spread into 46 papers. Forecasting applications are compared in order to define particular pattern. The methodology of the future application is presented in this paper as well. The methodology and technical approach used in the new neighbourhood environmental assessment framework is described. Cardiff Urban Sustainability Platform (CUSP), base for the development of this project, is presented with a brief overview of its main features and its ongoing application on the city of Ebbw Vale.

1 INTRODUCTION

During the past decades, many efforts and studies have been done toward the good practices in the design and operation of a neighbourhood (Gil & Duarte 2013; Ameen et al. 2015). Frameworks developed by companies or governmental institutions are numerous (Gil & Duarte 2013; Criterion Planners 2014). Internationally distributed, they constitute a good base on the definition of sustainability criteria within a neighbourhood or an extended urban area. However, there is no consensus on the definition of these criteria and their relative importance in a project (Sharifi & Murayama 2013; Sullivan et al. 2014). More, these frameworks are often if not always static methods, fixed in space and time (Kyrkou et al. 2011; Sharifi & Murayama 2013; Sullivan et al. 2014). There is a lack of models that could assess and promote good practices across the entire neighbourhood life cycle or that could adapt themselves to different localities. The growing interests on ICTs can help to improve this matter (Sein & Harindranath 2004). Indeed, there are no doubt that these two brownfield are linked in some extent since new technologies and efficient monitoring processes lead to a better understanding of the environmental key issues and those in real time (Hilty et al. 2006).

One decade has passed since the introduction of the first Neighbourhood Sustainability Assessments (Sharifi & Murayama 2013; Sharifi & Murayama 2014). Various research have been evaluating the performance and effectivity of the existing tools. Overall, these studies address a common problem among neighbourhood sustainability assessment concerning weighting, criteria selection and lack of systematic approach.

The main purpose of this research is to improve the temporal and local adaptability of actual neighbourhood sustainability assessments by integrating ICT and artificial intelligences in the scope.

This paper begins with an introduction of background of sustainability assessment practices and their importance in the current context, especially at the urban scale. Section 2 provides a brief overview of the main features of neighbourhood sustainability assessments and their operation, and opens the discussion on possible improvement by critically reviewing in depth various existing frameworks. Section 3 introduces forecasting models and their importance in assessment in order to help decision making. A critical review of different models found in the literature is done and define the ideal models to integrate in the future framework. Finally, Section 4 presents the methodology used for the development of the framework.

2 NEIGHBOURHOOD SUSTAINABILITY ASSESSMENT

2.1 *Frameworks analysis*

The analysis of the different frameworks focuses on the definition of the themes, criteria and indicators across them. The frameworks studied have been taken from "*A Global Survey of Urban*

Sustainability Rating Tools" (Criterion Planners 2014), report written by Criterion Planner, an urban and regional planning firm that operates a global registry of urban sustainability rating tools as a non-profit public service.

Overall, 61 current tools distributed in 21 countries have been reviewed. It appeared that 32 frameworks were irrelevant for the study because of a lack of information in the literature or because they were too specific to a field and did not take the overall urban system in consideration. Thus, it remains 29 tools that fit the interest of the study and that have been review in depth.

2.2 *Sustainability assessment scheme*

The themes, criteria and indicators have been collected from the frameworks and compared. Sharifi in his paper (Sharifi & Murayama 2013) describes themes as 'broad topics of concern to sustainability', for instance, climate and resources or urban design. Munier defines criteria as "parameters used to evaluate the contribution of a project to meet the required objective" (Nolberto Munier 2004). Criteria define aspects that need to be considered in order to achieve the sustainability of a project, for instance, water quality or GHG. Indicators, the last level of the assessment scheme, are defined as qualitative or quantitative measurable elements, (Hamedani & Huber 2011; Sharifi & Murayama 2013). They are targets for which credits are awarded if they are met.

Among the frameworks, the number of themes addressed varies from 4 to 10. The high majority of the frameworks use field such as "energy", "transport" or "health" to define their themes. In some frameworks, themes are defined by the sustainability dimensions namely, "environmental", "social" and "economic"; such as in CASBEE for cities, Gold standard Cities or DGNB-UD. Moreover, some aspects can be found as themes in some tools and as criteria in others. For instance "Transportation "is a theme in ELITE cities or BERDE for Clustered Residential Development and a criteria in STAR Communities or BREEAM Communities. Thus, there are disparities in the hierarchy of the assessment features. The number of criteria present within each theme varies from 1 to 20, exceeding 20 in really few cases (e.g. CEEQUAL). These variations mainly depend on how narrow is the theme addressed and on voluntary emphasis toward some aspects. Relatively few frameworks present the indicators used to assess sustainability. There is a clear lack of transparency on this matter. About the framework that do present the indicators used, a remarkable variation of their number can be noticed. Indeed, the number of indicators varies from 1 (e.g. "innovation" into BREEAM Communities) to over 100 (e.g. "social aspects" into CASBEE for cities). This large range of indicators brings real interrogations on the number of indicators that need to be used. Does the efficiency of an assessment depends on the number of its indicators? On this matter, Ameen states that an increased number of criteria (and thus of indicators) can reduce overlap between sustainability dimensions and themes, making it clearer for the different stakeholders (Ameen et al. 2015). Finally, it is noticeable that similar theme are addressed by different terms, for instance "Site" in Global Sustainability Assessment System for Districts covers the same aspects than "Smart location and linkage" in LEED-ND. These differences of terminology brings confusion and complexity in a field still in development.

These observations show a real lack of consensus on the definition of sustainability. This issue is supported by many studies(Orova & Reith 2013; Sharifi & Murayama 2013; Ameen et al. 2015) (Raed Fawzi Mohammed Ameen, 2015; Sharifi, 2013; Milanda Orova, 2013). There is a need for standardisation of the main features of neighbourhood assessment tools. This standardisation should define a clear hierarchy between the themes, criteria and indicators, address which themes and/or criteria are mandatory and establish a common terminology without affecting the flexibility of the assessment. Another issue is the lack of transparency on the frameworks themselves and on their structure which make the understanding of the method more difficult for the stakeholders.

Nevertheless, some common patterns can be found across the different frameworks. 8 main themes have been identified by gathering common ideas, concepts and terms, namely "Resources and Climate", "Land use and ecology", "Urban Design", "Health and well being", "Equity and diversity", "Governance", "Innovation", "Resilient economy".

2.3 *Sustainability dimensions*

Sustainable development and its definition are still controversial topics and nowadays, there are still no real consensus on this matter. Nevertheless, one generally well accepted feature is the 3 pillars or dimensions of sustainability: environmental dimension, economic dimension and social dimension.

In the scope of this study, it is interesting to analyse the repartition of the 3 dimensions of the sustainability. Figure 1 is a sample of the methodology used to visualized the themes repartition. A colour code corresponding to the previously defined themes has been applied. When an actual theme fits into a previously-defined theme, it holds the colour of this theme. In some cases, themes

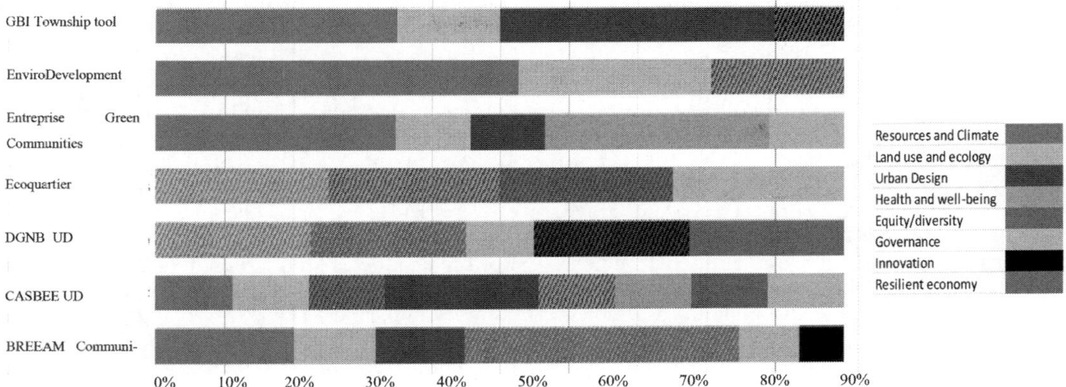

Figure 1. Sample of the themes distribution.

present in the frameworks address aspects that can fits into more than one defined theme. In this case, the theme holds several colours of the themes it fits in. For this reason, the following figures do not show a precise repartition in matter of percentage. However, it still gives a good overview of it. This methodology has been applied to the 29 tools studied. Overall, the theme "Resources and climate" is the most consider theme weighting around 30% in overall average, followed by "Land use and ecology" and "Urban Design" with around 18%, then "Health and well-being" counting for approximatively 13%, "Governance" 10% and finally "Resilient Economy", "Innovation" and "Equity/Diversity" with respectively around 6%, 5% and 3%. There is a clear emphasis toward "Resources and climate", "Land use and ecology" and "Urban Design" which are themes closely related to the environmental dimension. The study highlight an unbalanced consideration regarding the different dimensions of the sustainability. A good emphasis on environmental performance is observed and in some extent, social issue through the urban design. These observations follow the outcomes of many studies (Murgante et al. 2011; Sharifi & Murayama 2013; Sullivan et al. 2014; Ameen et al. 2015).

2.4 *Weighting system and adaptability*

There are great disparities in the weighting system across the frameworks. By nature, the weighting system consists of assigning a relative importance between different aspects. It is a really subjective process and therefore controversial. These noticeable differences toward the weighting system come partially from the fact that stakeholders have different interests, focus on distinct aspects. For example, BREEAM Communities is more focus on environmental concerns while LEED-ND seek more the "new urbanism" (site location and connectivity) (Sullivan et al. 2014). These differences are also closely related to the adaptability with the local characteristics (Retzlaff 2009).

The differences in the weighting system raise an essential issue within the sustainability assessment in general which is the issue of adaptability. Indeed, in a high majority of the cases, neighbourhood sustainability assessment is suitable for its original country but miss flexibility in order to perform in another environment (Haapio 2012). On the 29 studied frameworks, only 2 (EcoDistricts Protocol and STAR Community) integrate a consulting stage in order to define unique weight related to each projects. BREEAM communities address this issue by introducing an "International Bespoke" process to determine if the tool meet the requirements of international projects and allowed changes following the locality (BRE Global Limited About 2012). In its standard(CASBEE 2015), CASBEE-Cities specifies than it "can be customized (localized) to reflect local context" without explaining the process to do so. In any cases, when local adaptability is addressed, there is a lack of automatisms and flexibility.

The issue of adaptability is not restricted to the locality only, but also to temporal changes. The next decades will observe significant changes in the global environment and thus environmental requirements. Therefore, there is a clear need to track temporal changes within the neighbourhood life cycle. On the 29 frameworks, 24 are applied at the design stage of the project, providing guidance for the development or assessing the quality of the future neighbourhood. 4 are applied at the operation phase, assessing the quality of already existing neighbourhood or cities. Only 1 (EcoDistricts Protocol) covers the operation of an existing district and the design stage for retrofitting development and none are applied at the post-occupancy stage. The first reason is a matter of scale, many frameworks has developed tools for assessing buildings in operation, e.g. CASBEE with CASBEE for existing buildings, BREEAM with BREEAM In-use, LEED with LEED building operation and Maintenance, but these schemes have not been yet extended to a bigger scale. The second reason is the goals of these frameworks. They are destined, for the majority of them, to the professional of the construction and/or of the built environment. However, various organisation could be interested by the implementation of sustainability assessment for neighbourhood in operation such as governmental institution or energy and water companies. Thus, there is a need for the development of assessment that could be applicable at the design, operation and possibly post-occupancy stage.

3 FORECASTING MODELS

3.1 *Overall description*

The study presents a comparative analysis of forecasting models used in the energy and electricity forecasts. In his paper presenting the result of the M3 competition (the 3rd competition of a series of competition intended to evaluate and compare the accuracy of different forecasting methods, respectively in 1982, 1993 and 2000), Makridakis states that "simple methods developed by practicing forecasters do as well, or in many cases better, than sophisticated ones"(Makridakis & Hibon 2000). Which mean that there are no evidences that complex models recently developed will do better than "simple" ones. There is therefore a need to identify which model fit the best a particular situation.

The need for forecast varies following the cases. The setting of a model is subject to numerous changes: the available data used as inputs, the time term wanted, the time resolution, the scale (from a handful of domestic appliances to a whole country consumption). All these characteristics affect positively or negatively the performance of a forecasting model and thus have been investigated in this comparative study.

3.2 *Studies features*

Overall, 120 applications within 46 papers have been studied. From the 120 cases studied, 18 different models have been identified. It appeared than 4 of the 120 cases were irrelevant because of a lack of information on the model or the use of models that does not fit the interest of the study (e.g. physical based model). Moreover, the models AR, MA and their declination can be seen as variations of the Box-Jenkins model. Therefore they have been grouped into this label. Figure 2 shows the distribution of the different forecasting models through the papers. The regression model (often multiple regressions or multivariate regressions) are the most spread, present in 19 papers, 39.13%, followed by the Artificial Neural Network (ANN) present in 16 papers (34.78%). Box-Jenkins models represent 30.43% of the overall models with 13 papers listed. In a lesser proportion, SVM and Bottom up models follow, present in 6 and 5 papers respectively (13.04% and 10.87%). The others models are singularities. The relatively high amount of regression, ANN and Box-Jenkins models is due to the fact that they are often seen as a reference models in order to do comparison. This reinforces their status as the dominant models in the field. SVM and bottom up models are used in a lesser extent but there is a clear interest around these models supported by an increasing number of studies.

It is interesting to underline that in 53.45% of the cases, a data pre-process has been done. Three main kind of pre analysis can be observed: smoothing, study of significance and decomposition. Some mathematical tools can be used in this purpose such as Principal Component Analysis (PCA), Pearson Correlation (PCC), P-value, Analysis of Variance (ANOVA), Kernel Density Estimation (KDE), Canonical Correspondence Analysis (CCA).

From one minute ahead to several decades, the use of forecasts is numerous. The papers reviewed reflect this plurality and the terms have been classified on the following system: "Very short Term" (< 1h), Short term (1h to several days), "Mid term" (1 month to a season), "Long term" (1 to several years).

The Table 1 gives the distribution of the terms through the papers and study cases. With respectively 56.52% and 41.30%, the long-term and short term prediction represent the actual needs in mat-

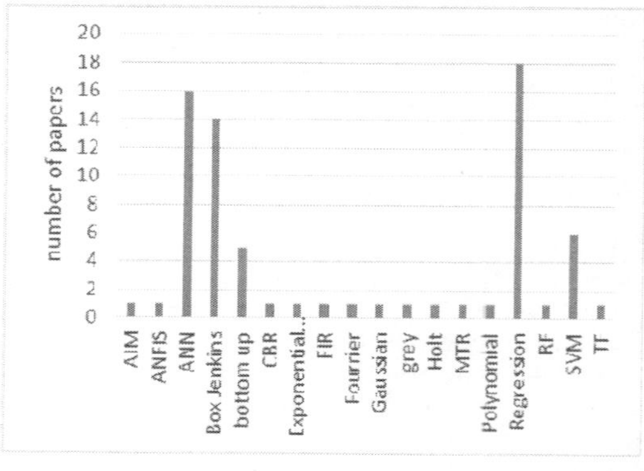

Figure 2. Model distribution through the reviewed papers.

Table 1. Term distribution through the reviewed papers and studies.

Term	Number of studies	Number of papers	Distribution percentage
Very Short-Term	5	1	2.17%
Short-term	44	19	41.30%
Mid-term	6	2	4.35%
Long-Term	65	26	56.52%

ter of energy forecasting in building. There is a lack of attention on very short term and mid-term prediction.

The forecasting models can be implemented with a large range of inputs. Electricity patterns, incomes, occupancy, electricity price, temperature, building size, rainfall, dwelling type, GDP, population are just few example of the various possible inputs of a forecasting model. For the study, inputs have been grouped into 5 different categories: "Energy" (dependant variable), "Socio-economic", "Meteorological", "Building" (data concerning the built or the occupancy), "Time index" (data that specifies the time and date).

The Table 2 present the input distribution through the papers and cases. The energy past pattern with 76.09% is the most used input. The remaining 24% are used to set a model or to train it but not necessarily introduced as an input. Moreover, if the use of the building characteristics is slightly preferred with 54.35%, there is no real preference in the use of meteorological and socio-economic data, both are present in 41.30% of the papers. Finally, the time index data are used in a lesser proportion with 26.09% of papers found.

Overall, the use of the independent inputs depends mainly of the meaning we want to give to our model, putting forward socio-economic correlation or meteorological correlation, at a building or country scale. For big scale forecasts (entire country and long time horizon) socio-economic inputs are preferred which explains their large representation across the papers. Time index is often introduced to increase the accuracy of model.

3.3 *Results*

Figure 3 shows the distribution of the forecasting term within the different models. The majority of the regression models are used for a long-term prediction, 1 year or more. Only 4 out of 20 configuration are against short term prediction. Conversely, the ANN is mainly used for a short term prediction with 10 papers studying this configuration. In a lesser extent, the use of ANN for a long term prediction follows with 5 papers. Likewise, the box-Jenkins and SVM models have been

Table 2. Input distribution through the reviewed papers and studies.

Input	Number of studies	Number of paper	Distribution percentage
Socio-economic	51	19	41.30%
Meteo	41	19	41.30%
Energy	87	35	76.09%
Building	50	25	54.35%
Time index	34	12	26.09%

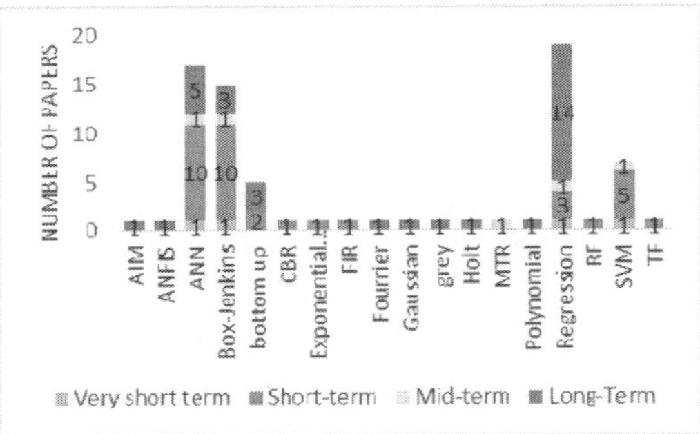

Figure 3. Term distribution through the models.

mainly applied for a short term prediction with respectively 9 and 5 different references considering this configuration. The bottom up model seems to be equally used for short and long term forecast. The singularity of the other models does not allowed any interpretations about the way they are applied.

Figure 4 shows the distribution of the independent inputs for the different models through the reviewed papers. The regression models have been mainly set up with socio-economic inputs. This observation agrees with the statement done in section 3.3 that socio-economic inputs are preferred for large scale and long-term forecasts. Building features, occupancy and meteorological data are the most used across the papers with around the half of the studies using them. The ANN presents the largest range of independent variables. This shows the flexibility of this model toward the data introduced as inputs. In the case of the ANN and SVM, the relatively high amount of time index data used reflects a need in order to increase their accuracy. In the case of the Box-Jenkins model, the introduction of exogenous inputs often aims to improve the accuracy of the time series analysis. The configuration the most widely used across the papers is the use of the past energy pattern as dependent variable and meteorological and building features as independent variables (13.04% of them). However, in many cases, the only use of energy time series is sufficient (28.26% of them).

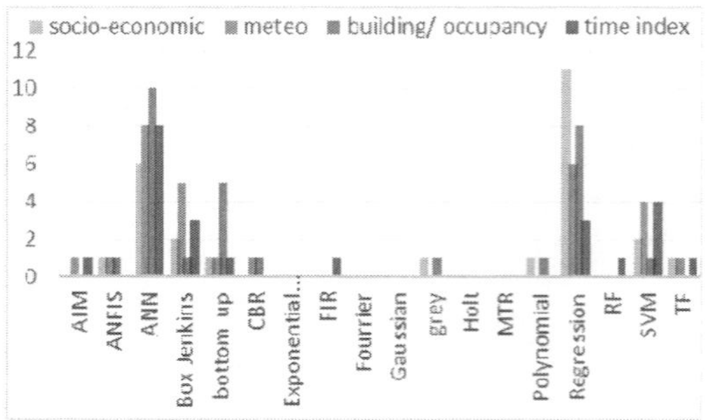

Figure 4. Independent inputs distribution through the models.

3.4 *Forecasting models in sustainability assessment*

In their report (Coplák & Raksanyi 2003), Coplak and Raksanyi state that the main purpose of the assessment tools was to play the role of decision support tool (Sharifi & Murayama 2013). Thus, sustainability assessment can be seen as more than simple proof of environmental, social and economic quality. It should provide guidance, help decision and assess the relevance and quality of those.. In this scope, forecasting models and simulations play an essential role. Forecasting models should be use in order to assess the consistence and performance of a strategy and thus, help the different actor to take the best actions. The relevance of strategies can be considered as a criteria of sustainability.

4 METHODOLOGY AND TECHNICAL APPROACH

4.1 *Definition of the framework*

The first step in the development of the framework is the definition of its main components. The definition of a relevant framework will be done by the DELPHI approach. The DELPHI approach seeks to find a consensus on a specific issue by conducting an expert consultation. The Figure 5 shows the DELPHI method applied in the study in order to define the original framework.

First of all, the review of the different sustainability assessment frameworks (see section 2) will enable a first definition of the themes, criteria and indicators that must be addressed as well as their relative importance in a project. Then, a panel of expert will be selected among different domains of expertise, local and/or international, professionals and/or academics. Once these panel of experts selected, the study proceeds to a consultation via a report or a questionnaire. The document should contain a detail description of the future approach considered for such tool and of the issues addressed (e.g. local adaptability, district life time application, user in the scope etc). After the expert consultation, the project members gather all the comments, recommendations, and issues and redefine the framework.The process is repeated until a consensus is found.

4.2 *Implementation of the framework*

The framework implementation has to be contextualised with the recent technological advances. In the current context, the increasing interests on Information and Communication Technologies (ICTs) is very valuable (Rezgui et al., 2011). ICT are meant to change the field by enabling the collection of a high amount of accurate data in a quick and efficient way. Moreover, this increasing amount of data must be coupled with efficient and relevant processing. The development of artificial intelligences and semantic systems aims to give significant meaning to the data. They allow a better understanding of a system and help a quick and consistent decision making.

The scheme shown in Figure 6 presents the main features of the future platform. The data are collected from various sources such as meters, survey or statistical data.

They are implemented into the ontology and other data processing. The ontology aims to give meaning to the different elements of the built environment and to define their possible inter-connection across domains and scales. It gives a picture of the overall concepts and theirs influences rather than simply considering them individually. The data processing include clusterization, aggregation and disaggregation, scenario prediction and optimisation methods. These elements are underlying a 3D interface visible by the user. The 3D interface gives a good representation of the urban environment, enable a user-friendly navigation and provides the labelling of various components. Finally, dashboard will display the main outcomes such as the key performance indicators based on the framework definition, the real time information, scenario predictions, alerts, recommendations, reports etc.

4.3 *Cardiff Urban Sustainability Platform (CUSP)*

The Cardiff Urban Sustainability Platform (CUSP) is currently being developed at Cardiff University. It is an immersive decision support tool built to deliver Cardiff University powerful urban analytics. CUSP enable interactive monitoring and inform decision making through a web interface (see Figure 7). Currently based on the city of Ebbw

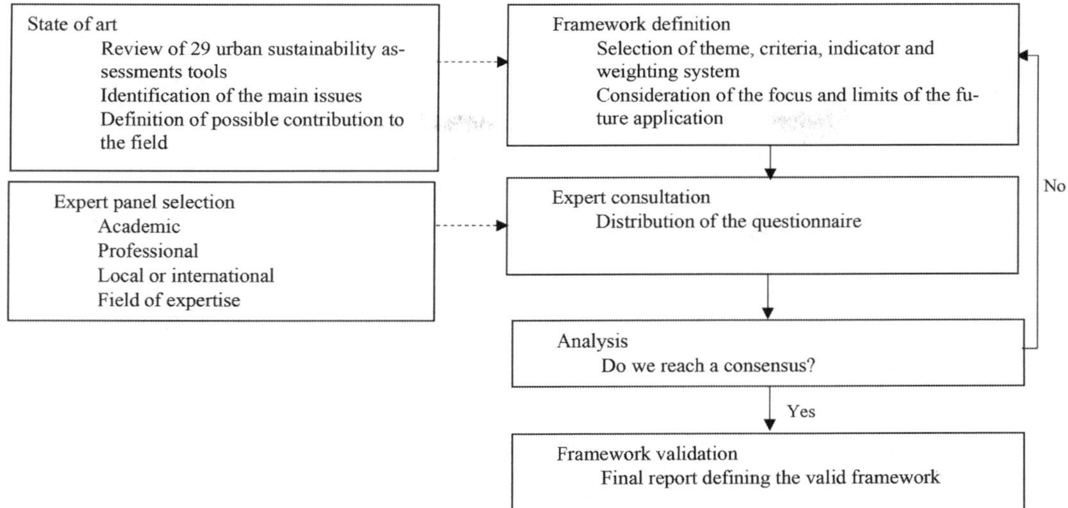

Figure 5. DELPHI approach.

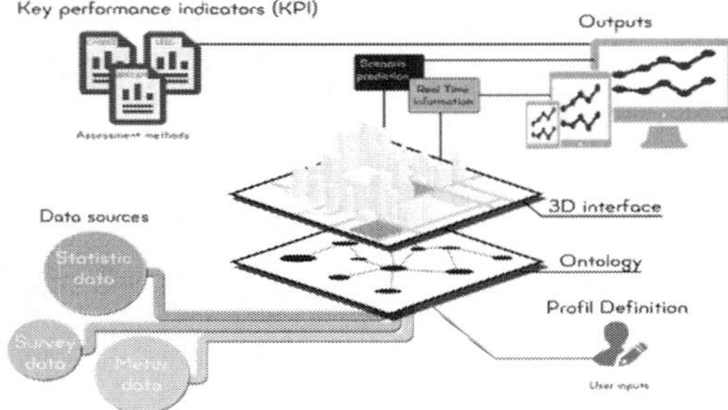

Figure 6. Platform scheme.

Figure 7. CUSP interface on Ebbw Vale case study.

Vale (UK), CUSP provide a 3D information model of the site and its buildings.

IFC being widely used as a standard for data exchange in the building information modelling, it has been chosen as the scheme of the information model. The production of 3D IFC models is helped by the use of 3D point data collection conducted via laser scanning. Point clouds are collected and exported into Autodesk Revit where building information models are created by tracing over them.The IFC models holds semantic information essential in the global semantic-based approach.

3D building information models are then exported into Unity Game Engine. These 3D models are enriched with semantic information across domains and scales.

Some dashboards provide the direct outcomes of the energy and water domains by clicking on a specific building. In the energy dashboard, the operational schedule for each energy production unit, key energy performance indicators and day-ahead predicted demand are displayed. In the water dashboard, alerts to warn of existing or predicted issues such as water quality, flood risk or network leakages are displayed. This provides useful insight into the management of the network and allows the manager to take decisions.

Finally, CUSP allows an physical-based simulation of the district-heating network in order to identify potential issues and inefficiencies.

CUSP has been chosen as a base for the development of the tool. In the frame of the study, the project team will especially look at the definition of the KPIs, the design of the dashboard and the way KPI are displayed, the forecasting models, and the artificial intelligence mechanisms enabling local and temporal adaptability.

4.4 *Validation of the framework*

Last stage of the project, the validation consists of analysing the outcomes of the framework.

The implementation of the framework on Ebbw Vale as well as on Cardiff University Campus will allow the comparison and highlight the possible adaptability issues. Significant improvements and issues will be identified and reported with a special emphasis on the contribution of such tool on the field of sustainable urban environment assessment and monitoring. Afterward, the future stages of development will be discussed. The reception of such tool by different parties (i.e. tenants, experts and governmental institution) should be investigated as well. A consultation would allow to understand how people perceive the tool and might use it.

5 CONCLUSION

The literature review of a large number of neighbourhood sustainability assessment has permitted the definition of several issues. First of all, there is a lack of consensus on the theme, criteria and indicator that need to be addressed as well as a clearly unbalanced consideration of the different dimensions of the sustainability. Secondly, there are no mechanisms for local and temporal adaptability. Moreover, the high majority of the framework are design to be apply for the design and development of an urban area and relatively few frameworks considers the evaluation of the operational quality. Last but not least, there is a low consideration of the tenant's opinion in the development of such assessments. The development of a total life cycle and near-real time environmental assessment based on expert consultation, ICTs and artificial intelligences aims to solve these issues. The expert consultation ensures the development of a consensus-based framework while the ICTs and artificial intelligences ensure an improved adaptability of the model, a better assessment of the KPIs, the use of the tool across the life-cycle and the involvement of every parties.

REFERENCES

Ameen, R.F.M., Mourshed, M. & Li, H., 2015. A critical review of environmental assessment tools for sustainable urban design. *Environmental Impact Assessment Review*, 55, pp.110–125..

BRE Global Limited About, 2012. *BREEAM Communities. Technical Manual*,

CASBEE, 2015. *CASBEE for Cities. Pilot version for worldwise use*,

Coplák, J. & Raksanyi, P., 2003. *Planning sustainable settlements*,

Criterion Planners, 2014. *A Global Survey of Urban Sustainability Rating Tools*.

Gil, J. & Duarte, J.P., 2013. Tools for evaluating the sustainability of urban design: a review. *Urban Design and Planning*, 166(DP6), pp.311–325.

Haapio, A., 2012. Towards sustainable urban communities. *Environmental Impact Assessment Review*, 32(1), pp.165–169.

Hamedani, a. Z. & Huber, F., 2011. A comparative study of "DGNB" certificate system in urban sustainability. *WIT Transactions on Ecology and the Environment 155, pp. 121–132*, 155(August), pp.121–13.

Hilty, L.M. et al., 2006. The relevance of information and communication technologies for environmental sustainability—A prospective simulation study. *Environmental Modelling and Software*, 21(11), pp.1618–1629.

Kyrkou, D. et al., 2011. Urban sustainability assessment systems How appropriate are global sustainability assessment systems? *27th International Conference on Passive and Low Energy Architecture*, (July), pp.2–7.

Makridakis, S. & Hibon, M., 2000. The M3-Competition: results, conclusions and implications. *International Journal of Forecasting*, 16(4), pp.451–476.

Murgante, B., Borruso, G. & Lapucci, A., 2011. Sustainable Development: Concepts and Methods for Its Application in Urban and Environmental Planning., pp.1–15.

Nolberto Munier, 2004. *Multicriteria Environmental Assessment*, Dordrecht: Kluwer Academic Publishers. Available at: http://link.springer.com/10.1007/1-4020-2090-2.

Orova, M. & Reith, A., 2013. Comparison and evaluation of neighbourhood sustainability assessment systems. *PLEA 2013: Sustainable Architecture for a Renewable Future*, (September).

Retzlaff, R.C., 2009. The use of Leed in Planning and Development Regulation: An Exploratory Analysis. *Journal of Planning Education and Research*, pp.1–11.

Rezgui Y, Boddy S, Wetherill M, Cooper G, Past, present and future of information and knowledge sharing in the construction industry: Towards semantic service-based e-construction? Computer-Aided Design, 43 (5) (2011) 502–515. ISSN 0010-4485. 10.1016/j.cad.2009.06.005.

Sein, M.K. & Harindranath, G., 2004. Conceptualizing the ICT Artifact: Toward Understanding the Role of ICT in National Development. *The Information Society*, 20(1), pp.15–24.

Sharifi, A. & Murayama, A., 2013. A critical review of seven selected neighborhood sustainability assessment tools. *Environmental Impact Assessment Review*, 38, pp.73–87.

Sharifi, A. & Murayama, A., 2014. Neighborhood sustainability assessment in action: Cross-evaluation of three assessment systems and their cases from the US, the UK, and Japan. *Building and Environment*, 72, pp.243–258.

Sullivan, L., Rydin, Y. & Buchanan, C., 2014. Neighbourhood Sustainability Frameworks—A Literature Review, (May), p.22.

Information & knowledge management (V)—infrastructure

eWork and eBusiness in Architecture, Engineering and Construction – Christodoulou & Scherer (Eds)
© 2016 Taylor & Francis Group, ISBN 978-1-138-03280-4

Detecting, classifying and rating roadway pavement anomalies using smartphones

C. Kyriakou, S.E. Christodoulou & L. Dimitriou
Department of Civil and Environmental Engineering, University of Cyprus, Nicosia, Cyprus

ABSTRACT: Pavements are major roadway infrastructure assets, and pavement maintenance to the preferred level of serviceability comprises one of the most challenging problems faced by civil and transportation engineers. Presented herein is a study on the utilization of low-cost technology for the data collection and classification of roadway pavement anomalies, by using sensors from smartphones and from automobiles' On-Board Diagnostic (OBD-II) devices while vehicles are in movement. The smartphone-based data collection is com-plimented with artificial neural network techniques, various algorithms and classification models for the clas-sification of detected roadway anomalies. The proposed system architecture and methodology utilize nine metrics in the analysis, are checked against three types of roadway anomalies, and are validated against hun-dreds of roadway runs (relating to several thousands of data points) with an accuracy rate of about 90%. The study's results confirm the value of smartphone sensors in the low-cost (and eventually crowd-sourced) detec-tion of roadway anomalies.

Keywords: roadway pavement condition assessment, smartphones

1 INTRODUCTION

In recent years, the area of interest for transportation authorities, researchers and practitioners has shifted from the construction of new roads to the management of existing ones (Panagopoupou & Chassiakos 2012). One of the most significant indicators for road quality is the pavement surface condition, which is identified by the anomalies in the pavement surface that have an effect on the ride quality of a vehicle. Road anomalies can cause unpleasant driving, increase fuel consumption, damage vehicles and in many cases be the reason for traffic accidents, injuries and/or fatalities. Pavement surface can deteriorate in time from causes related to location, materials used, traffic, weather, etc. Identifying road anomalies related to transverse defects, longitudinal defects, potholes and cracking on the pavement surface can collaborate to surveying road condition quality. Pavement surface condition monitoring systems could raise the value of road surface, protect vehicles from damage as a result of bad roads and improve traffic safety.

As an outcome of the fast and powerful development of smartphones in current years, connected vehicle technology has obtained noteworthy consideration within the transportation, infrastructure, and automotive industries. This paper examines the use of smartphones and connected vehicle applications, in the interest of improving the condition evaluation and management of roadway pavement surfaces, and by extension it investigates the possibility that connected vehicle data may contribute to monitoring the condition of transport infrastructure. Modern smartphones can be utilized to capture vehicle sensor data without integration with built-in vehicle systems, since they come with a range of built-in sensors, such as accelerometer, gyroscope and GPS sensors. Further, vehicle system (Controller Area Network, CAN, bus) data can be collected through On-Board Diagnosis (OBD) Bluetooth connectors (ELM 327 Bluetooth Car Diagnostic Scanner) to a smartphone (e.g. the DashCommand™ software application). This combination of hardware and software components enables the monitoring of, among others, forward, lateral and vertical acceleration, vehicle roll and pitch, GPS latitude and longitude, GPS vehicle speed, engine RPM and current acceleration based on the last two speed readings.

The vision for roadway anomaly detection by use of smartphone technology is set in parallel with the premise that such technology can be utilized for crowd-sourced data collection and analysis in GIS-based Pavement Management Systems (PMS) (Fig. 1). Up-to-date and future primary datasets will be different in form commencing traditional pavement surface condition data, and in order to over-

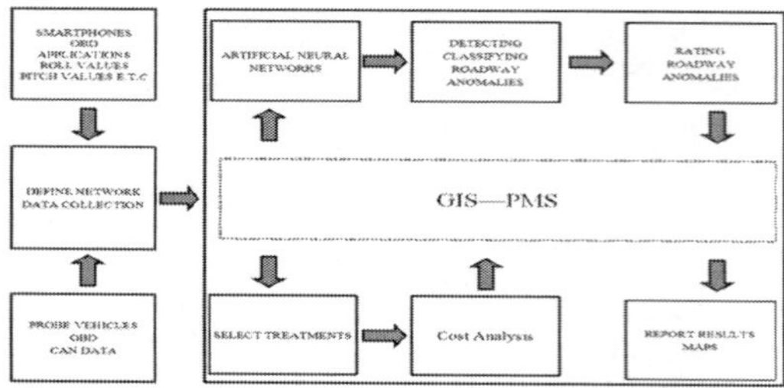

Figure 1. GIS—PMS.

take this barrier, it is prudent that crowd-sourced data be collected from a statistically significant number of probe-vehicles (Dennis et al. 2014). A number of vehicles collecting this data could be used in order to highlight rough pavement and potholes within a roadway network. Further, crowd-sourced data would create a geocoded event at points where the car develops abnormal behaviour. Populating a database with crowd-sourced event points from a number of connected vehicles will allow engineers and PMS program managers to identify where vehicles are experiencing rough riding conditions.

Research on road anomaly detection and classification utilizing smartphones, artificial neural networks, various algorithms and classification models. Further to this short introduction, a literature review section provides a brief outline of existing work concerned to roadway anomaly detection utilizing PMS and pothole detection using smartphones. The section on methodology setup section outlines the developed data collection structure and methods, while the results and discussion section exhibits the methods and tools used to classify the data and experiment results. The paper concludes with key findings and an outline of future research directions.

2 LITERATURE REVIEW

The American Association of Highway Transportation Officials (AASHTO, 1993) states that the "…function of a PMS is to improve the efficiency of decision making, expand its scope, provide feedback on the consequences of decisions, facilitate the coordination of activities within the agency, and ensure the consistency of decisions made at different management levels within the same organization." Unquestionably, the investment in a PMS is worthwhile as it provides the tools an agency demands for the balanced resource allocation, ideal use of funds, pavement treatment selection, pavement treatment cost reductions and enhanced credibility with stakeholders (Washington Department of Transportation, 1994).

It is essential to understand the benefits and related cost of any expenses in pavement management before starting the process (Khattak et al. 2008). In developed countries, PMS are specialized platforms equipped with expensive equipment built-in specialized Pavement Evaluation Vehicles (Seraj et al. 2014). The costs related to a PMS include software (purchase and installation), data collection, database setting up and system maintenance, as well as, updates, consultant services, employee training, personnel time and actual expenditures on the pavement and rehabilitation (AASHTO, 1990).

The first step in developing a pavement management procedure is to define the roadway network. The second step in pavement management design process is to verify the survey methodology for collecting the distress data. At present a PMS utilizes surveys, such as the Pavement Surface Evaluation and Rating (PASER) rating system, which is built on a series of photographs and descriptions for each of the individual rating categories (Walker et al. 2002). A rater uses this series of photographs to evaluate the overall condition of an individual pavement surface segment (Wolters et al. 2011). After verifying the survey methodology for collecting the distress data, an agency must opt between the two main approaches of collecting road surface condition data, i.e. manual or automated data collection (McQueen & Timm 2005). In manual surveys, pavement surface condition data are collected from a moving vehicle (windshield surveys) or by "walking" the pavement. Automated surveys are performed using vehicles being fitted with specialized camera that collect images and other sensing devices that collect sensor data relevant to the pavement being under examination (AASHTO, 2006). The third step concerns the prediction of the pavement surface condition. Pavement network conditions can be evaluated using either average deterioration rates or prediction models via statistical modelling like regression analysis. Furthermore, some systems use probabilistic type models which have mostly been based on Markovian theory (Wolters et al. 2011). The fourth step of PMS design is to select the suitable treatments for the pavement network. The recommended treatments are arranged using cyclical treatment selection, ranking or benefit/cost/analysis. The fifth step refers to the development of reports comprising the results obtain from the previous steps. The sixth step is to select pavement management tool such as a pavement management software, customized spreadsheets and/or GIS software. The selection of a pavement management tool being used varies according to the requirements of the agency and user needs. The final step is keeping the process updated. Notably, pavement management is a dynamic process that requires regular updates (Washington Department of Transportation, 1994).

De Zoysa et al. (2007) proposed a public transport system called "BusNet" in order to monitor environmental pollution and pavement surface condition by adding acceleration sensors boards to the system. BusNet implements a sensor network, placed on top of public buses. The acceleration sensors identify potholes through changes in the vertical acceleration and determine the car speed modification using the horizontal acceleration. Erikson et al. (2008) used seven taxis running in the Boston area and developed a mobile sensor system called "Pothole Patrol". Each taxi needed a computer running the Linux operating system, a WiFi card for transmitting collected data, an external GPS and a 3-axis accelerometer. In general, there are three main problems concerning the above systems. First, there is a large number of events (such as doors being knocked, unexpected swerves) and road anomalies (such as road expansion joints) that are difficult to distinguish from potholes. The second problem is that the systems cannot identify between a pothole that merits fixing and a bump in the road. The third problem is that the values reported by the sensors depend on a car's speed and how the sensors are mounted on the car.

Tai et al. (2010) used a mobile phone with triaxial accelerometer to collect acceleration data while riding a motorcycle. Strazdins et al. (2011) proposed a method requiring an Android smartphone with GPS, 3-axis accelerometer and a communication channel (cellular or Wifi). The system consists of two application components, one for the Android device and one for a data server. Seraj et al. (2014) proposed a system that detects road anomalies using mobile equipped with inertial accelerometers and gyroscopes sensors. They applied a method to remove the effects of speed, slopes and drifts from sensor signals. For future work they aim to apply this method for road anomalies detection in participatory sensing, using clustering by geo-coordinates. Alessandroni et al. (2014) described a system which included a combination of a custom mobile application and a georeferenced database system. The roughness values computed and stored into a backend geographic information system enable visualization of road conditions. This proposed approach introduced an integrated system for monitoring applications in a scalable, crowd-sourcing collaborative sensing environment. Mohamed et al. (2015) suggested the gyroscope around gravity rotation as the primary indicator for road anomalies, in addition to the accelerometer sensor, in order to avoid false-positive indications; especially when there is a sudden stop or sudden change in motion acceleration. The above systems, despite hardware differences in terms of GPS accuracy and accelerometer sampling rate and noise, they show that pothole detection is possible. The 2014 Mercedes-Benz S-Class used a Light-Detection-and-Ranging (lidar) scanner to measure pavement roughness as a component of an active suspension system. Recently, the Jaguar Land Rover automaker announced that is examining a new connected vehicle technology which permits a vehicle to point dangerous potholes in the road and then allocate this data in real time with other vehicles and road authorities (Nick O'Donnell, 2015).

3 METHODOLOGICAL SETUP

As aforementioned, the work presented herein investigates the utilization of smartphone technology for the detection of roadway anomalies and for their classification. Vehicle and smartphone data can be selected by both state-operated fleet vehicles and privately owned vehicles, and smartphones operated by the general public. Vehicle data are collected by sensors already installed on typical vehicles and smartphones. Further, the smartphones can perform data screening, fusion, and transmission of collected data for further data processing, analysis, or storage. The technology required to implement crowd-sourced pavement condition monitoring from smartphones is already established and has been proven workable. The power of crowd-sourced data is that large data sets, which are collected through multiple data sources, negate the limitations in generalizability of data collected from a single data source. Even though multiple vehicles might provide conflicting data relating to pavement condition, the total effect and 'knowledge' inherent in the data provides an accurate model of the roadway condition in relation to how an average user experiences the pavement condition.

The study focuses on three types of common roadway anomalies (transverse depressions Figure 2a; longitudinal depressions, Figure 2b; and potholes or manholes, Figure 2c), which it examines first individually and then in tandem. Data on these types of roadway anomalies is collected in-situ by use of a car fitted with a smartphone (mounted on the car's windshield) with its GPS, accelerometers and gyroscopes activated, and with an On-Board Diagnosis (OBD-II) reader connected to it. Further to the sensors, the smartphone had its video camera working as well for recording the routes travelled and consequently for visually validating the existence of roadway anomalies (as detected by the data analysis and ANN classification). The smartphone was also fitted with the *DashCommand* application for recording (and exporting) sensor readings, date/time stamps, and GPS locations of taken data. Vehicle system (CAN) data can be relayed through the OBD-II reader to the smart device and then transmitted for processing or storage via digital cellular connection or other means.

The collected case-study data are of high spatial resolution (at intervals of 0.1 seconds) and pertain to both uni-dimensional (e.g. X, Y, Z accelerations, speed, etc.) and two-dimensional indicators (e.g. the smartphone's roll and pitch values, Figure 3a, which can be related to the traveling host car's roll and pitch values). In essence, the roll relates to a car's acceleration difference between its left and right front wheels, while the pitch relates to a car's acceleration difference between its front and rear wheels. In tandem, roll and pitch point out in what manner the host car is off balance, sideways and front/back

The field investigation included the collection of sensor data for a total of nine parameters (as shown in Table 1), which were also presumed to influence the accuracy of detection and classification of the roadway anomalies examined. Most of the nine factors examined were varied during the field investigation (e.g. speed, acceleration, engine RPM) and some were left constant (e.g. type of car/engine and fuel type). Two Variables (VAR_3, VAR_4) were calculated posterior the data collection, as they relate to the difference in roll and pitch, respectively, between two data points (0.1 seconds in difference between the two data points).

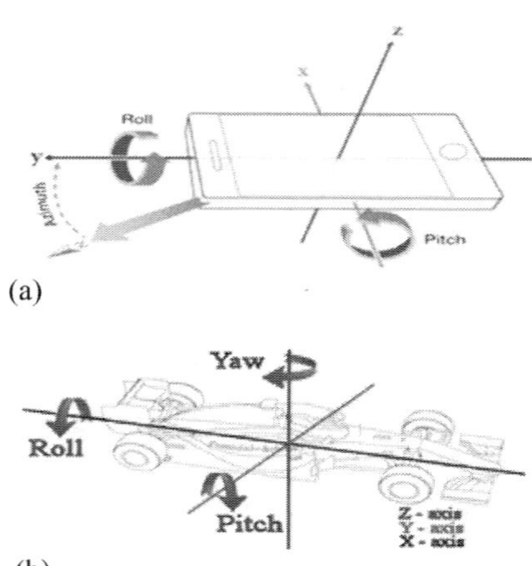

Figure 3. Smartphone's (a) roll and pitch directions (Physics Forums, 2015) and (b) relation to car's wheels' differential (White-Smoke, 2010).

Table 1. Data collected and variables used for classifying roadway anomalies.

Variable	Variable Name	Variable Description
VAR _1	Aux.Accel. Forward (Gs)	Forward Acceleration
VAR _2	Aux.Accel. Lateral (Gs)	Lateral Acceleration
VAR _3	Roll 2 – Roll 1(Â°)	Numerical Difference Between Two Successive Roll Values
VAR _4	Pitch 2 – Pitch 1(Â°)	Numerical Difference Between Two Successive Pitch Values
VAR _5	Aux.Gps. Latitude (Â°)	GPS Latitude
VAR _6	Aux.Gps. Longitude (Â°)	GPS Longitude
VAR _7	Aux.Gps.Speed (Km/Hr)	GPS Vehicle Speed
VAR _8	Sae.Rpm (rpm)	Engine RPM
VAR _9	Calc.Acceleration (m/s²)	Current acceleration based on the last two speed readings

Figure 2. Roadway anomaly types examined for detection and classification: (a) transverse defect/anomaly; (b) longitudinal defect/anomaly; (c) potholes/manholes.

The field investigation included the collection of sensor data for a total of nine parameters (as shown in Table 1), which were also presumed to influence the accuracy of detection and classification of the roadway anomalies examined. Most of the nine factors examined were varied during the field investigation (e.g. speed, acceleration, engine RPM) and some were left constant (e.g. type of car/engine and fuel type). Two variables (VAR_3, VAR_4) were calculated posterior the data collection, as they relate to the difference in roll and pitch, respectively, between two data points (0.1 seconds in difference between the two data points).

4 RESULTS AND DISCUSSION

A first look at the collected raw data brings to light the complexity of the problem, as variables such as the X/Y/Z acceleration and vehicular speed thought to point out roadway anomalies are not as conclusive as originally thought. A plot of acceleration in one direction over time, such as the vertical acceleration (Figure 3), helps demonstrate the lack of any pattern in the taken readings. The vertical variability is random, and even at a point of known roadway anomaly the variability in the acceleration is not definite of the existence of the anomaly. A related situation displays itself for various vehicular speeds (20, 40, 60, 80, 100 km/hr).

The situation can be improved, should one weigh in the vehicular roll and pitch values over time (Figure 4a, 4b), but even the aforementioned indicators fail to safely point out locations of roadway anomalies. Despite the fact these plots indicate areas of suspicion (highs and lows, away from the running average values), they are not reliable indicators. Further, the plots fail to provide information on other running parameters which could affect the accuracy of the data (e.g. the vehicle's speed at the time of data sensing). The difference in values between subsequent locations (Figure 4c, 4d) appears to be a better predictor (points of high peaks indicate a roadway anomaly), but still that is not fail-proof in the absence of other complimentary parameters.

4.1 *Ann clasification*

For that reason, the datasets are then fed into an artificial neural network (ANN) consisting of 9 inputs (I1, I2, …, I9), 8 hidden neurons (H1, H2, …, H10) and 2 outputs (I1, I2). The ANN's architecture (shown in Figure 6) was implemented in MATLAB™. The ANN inputs are the parameters listed in Table 1, while the outputs are binary in nature ('0' for no defect, '1' for defect) and they are used to classify data readings into classes of roadway anomalies.

The ANN is first trained for each case of roadway anomaly (as given by Figure 1), and then trained with all three roadway anomalies in tandem (ClassType 0, ClassType 1, ClassType 2, ClassType 3). 'ClassType 0' refers to the no-defect case, 'ClassType 1' refers to transverse defect/anomaly (Figure 1a), 'ClassType 2' refers to longitudinal defect/anomaly (Figure 1b) and 'ClassType 3' refers to potholes/manholes (Figure 1c). Each case 70% of the data is used for training, 15% for testing and 15% for validating the ANN, with a synopsis of the obtained classification results shown in (Table 2).

The intent was to first train the ANN to detect each defect in isolation of the others, and then

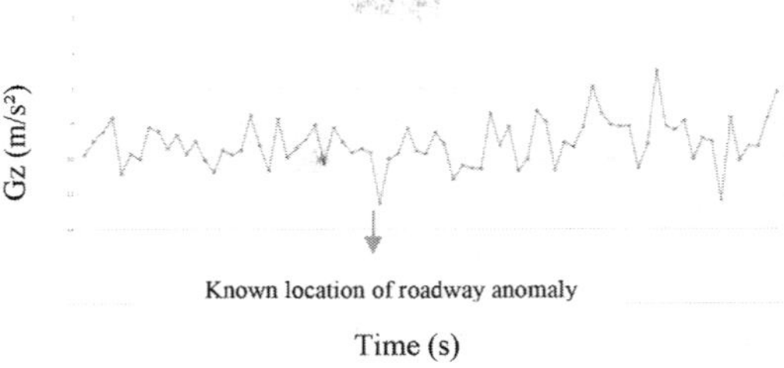

Figure 4. Vertical acceleration over time.

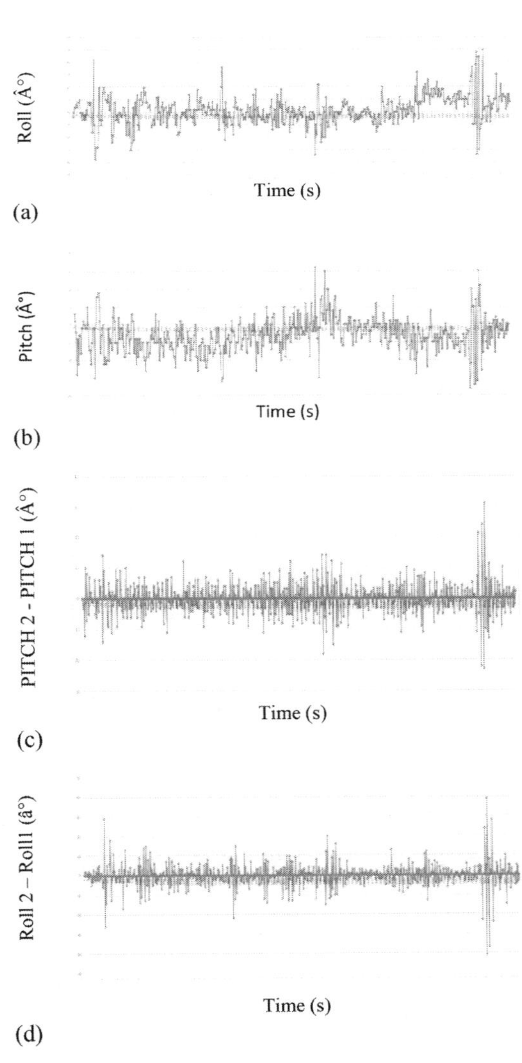

Figure 5. Raw and processed sensor data: (a) roll, over time; (b) pitch, over time; (c) point-to-point roll variation, over time; (d) point-to-point pitch variation, over time.

train the ANN to distinguish defects between the three defect classes in examination. As can be seen the ANN classification arrives at a high degree of accuracy, not only when examining for the existence of a specific roadway anomaly but also when examining for all roadway anomalies at once (last case in Table 2). This high degree of accuracy is also evident when the ANN results are examined closer, by means of the produced ANN confusion matrices for training, testing and validating the classification of defects.

The ANN confusion matrices enable us to investigate the numbers (and percentages) of not only the accurate classifications (i.e. perfect matches between target and output classes) but also of erroneous and of false-positive classifications. The horizontal axis in each confusion matrix indicates the target class (what the ANN classification should result to) and the vertical axis indicates the output class (what the ANN classification actually results to). As a backdrop, let us note that a good classifier yields a confusion matrix that will look dominantly diagonal, and that all off-diagonal elements on the confusion matrix represent misclassified data.

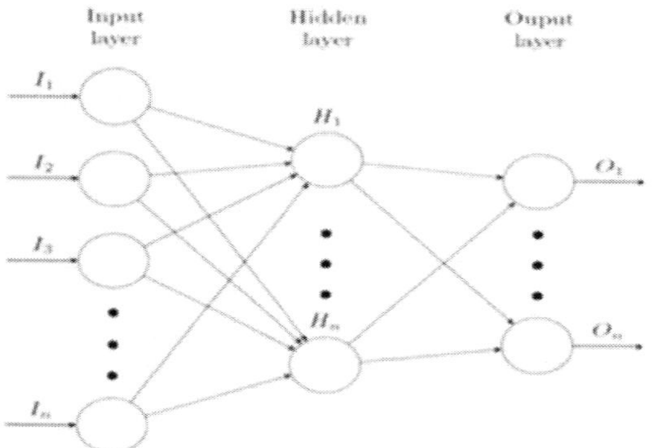

Figure 6. ANN model schematic architecture.

In essence, the ANN classifier detects and precisely categorizes the three roadway anomalies (target classes '2', '3' and '4') while also distinguishing the 'no defect' condition (target class '1'), thus separating normal and abnormal roadway pavement conditions (Tables 3–6).

4.2 *Classification models*

In addition, classification can be performed with supervised machine learning in which an algorithm "learns" to classify new observations from examples of labeled data. In order to classify and validate the data, supervised learning algorithms were used for multiclass problems. A test was performed between various algorithms to train and cross validate classification models for binary or multiclass problems. A training was performed to search for the best classification model type, including decision trees using various classifiers, discriminant analysis, support vector machines, logistic regression, nearest neighbors, and ensemble classification. After cross validating multiple models, compare their cross-validation errors side-by-side, the best model was selected (bagged trees). Supervised machine learning was done by supplying a known set of input data (observations or examples) and known responses to the data (i.e., labels or classes). The aforementioned datasets were fed into classification models consisting of 9 observations and 4 responses. The classification models observations are the parameters listed in Table 1, while the responses are '1' for no defect, '2' for transverse defects, '3' for longitudinal defects and '4' for potholes. The data used to train a model that generates predictions for the response to new data. The classification models architecture (shown in Figure 7) was implemented in MATLAB™.

Bagged trees use Breiman's 'random forest' algorithm. 'Random forests are a combination of tree predictors such that each tree depends on the values of a random vector sampled independ-

Table 2. Ann training and validation statistics for the various roadway anomaly cases examined.

Refers to	Classes	Train. Accuracy	Valid. Accuracy	Test. Accuracy	Over. Accuracy	Total Accuracy
Figure 1a	0:Type_0	100%	100%	100%	100%	99.8%
	1:Type_1	100%	96%	100%	99%	
Figure 1b	0:Type_0	84%	83%	79%	83%	91%
	1:Type_2	95%	98%	94%	95%	
Figure 1c	0:Type_0	100%	100%	100%	100%	100%
	1:Type_3	100%	100%	100%	100%	
Figure 1a,1b,1c	0:Type_0	100%	97.6%	100%	99.7%	99.8%
	1:Type_1	99.1%	100%	100%	99.3%	
	2:Type_2	100%	100%	100%	100%	
	3:Type_3	100%	100%	100%	100%	

Table 3. Training confusion matrix.

Target class						
1	2	3	4	Overall		
201	0	0	0	99.5%	1	Output class
23.9%	0.0%	0.0%	0.0%	0.5%		
0	110	0	0	100.0%	2	
0.0%	13.0%	0.0%	0.0%	0.0%		
0	0	223	0	100.0%	3	
0.0%	0.0%	26.5%	0.0%	0.0%		
0	0	0	305	100.0%	4	
0.0%	0.0%	0.0%	36.3%	0.0%		
100%	99.1%	100.0%	100.0%	99.9%	Overall	
0.0%	0.9%	0.0%	0.0%	0.1%		

Class 1: No defect
Class 2: Transverse defect/anomaly (Fig. 1a)
Class 3: Longitudinal defect/anomaly (Fig. 1b)
Class 4: Potholes/manholes (Fig. 1c).

Table 4. Validation confusion matrix.

Target class						
1	2	3	4	Overall		
40	0	0	0	100.0%	1	Output class
22.2%	0.0%	0.0%	0.0%	0.0%		
1	22	0	0	95.7%	2	
0.6%	12.2%	0.0%	0.0%	4.3%		
0	0	55	0	100.0%	3	
0.0%	0.0%	30.6%	0.0%	0.0%		
0	0	0	62	100.0%	4	
0.0%	0.0%	0.0%	34.4%	0.0%		
97.6%	100.0%	100.0%	100.0%	99.4%	Overall	
2.4%	0.0%	0.0%	0.0%	0.6%		

Class 1: No defect
Class 2: Transverse defect/anomaly (Fig. 1a)
Class 3: Longitudinal defect/anomaly (Fig. 1b)
Class 4: Potholes/manholes (Fig. 1c).

Table 5. Test confusion matrix.

Target Class						
1	2	3	4	Overall		
48	0	0	0	100.0%	1	Output class
26.7%	0.0%	0.0%	0.0%	0.0%		
0	20	0	0	100.0%	2	
0.0%	11.1%	0.0%	0.0%	0.0%		
0	0	51	0	100.0%	3	
0.0%	0.0%	28.3%	0.0%	0.0%		
0	0	0	61	100.0%	4	
0.0%	0.0%	0.0%	33.9%	0.0%		
100.0%	100.0%	100.0%	100.0%	100.0%	Overall	
0.0%	0.0%	0.0%	0.0%	0.0%		

Class 1: No defect
Class 2: Transverse defect/anomaly (Fig. 1a)
Class 3: Longitudinal defect/anomaly (Fig. 1b)
Class 4: Potholes/manholes (Fig. 1c).

Table 6. All confusion matrix.

Target class						
1	2	3	4	Overall		
289	1	0	0	99.7%	1	Output class
24.1%	0.1%	0.0%	0.0%	0.3%		
1	152	0	0	99.3%	2	
0.1%	12.7%	0.0%	0.0%	0.7%		
0	0	329	0	100.0%	3	
0.0%	0.0%	27.4%	0.0%	0.0%		
0	0	0	428	100.0%	4	
0.0%	0.0%	0.0%	35.7%	0.0%		
99.7%	99.3%	100.0%	100.0%	99.8%	Overall	
0.3%	0.7%	0.0%	0.0%	0.2%		

Class 1: No defect
Class 2: Transverse defect/anomaly (Fig. 1a)
Class 3: Longitudinal defect/anomaly (Fig. 1b)
Class 4: Potholes/manholes (Fig. 1c).

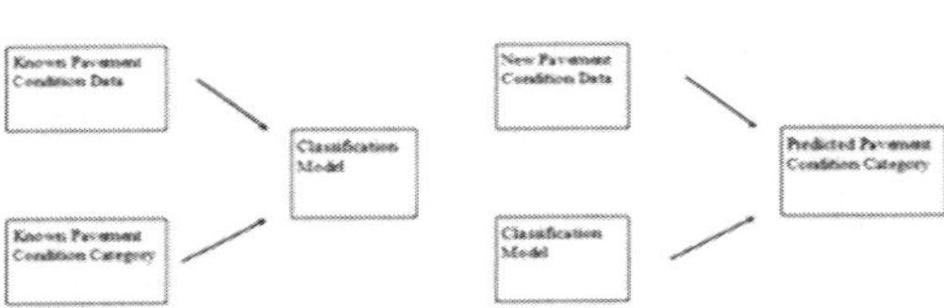

Figure 7. Classification models schematic architecture.

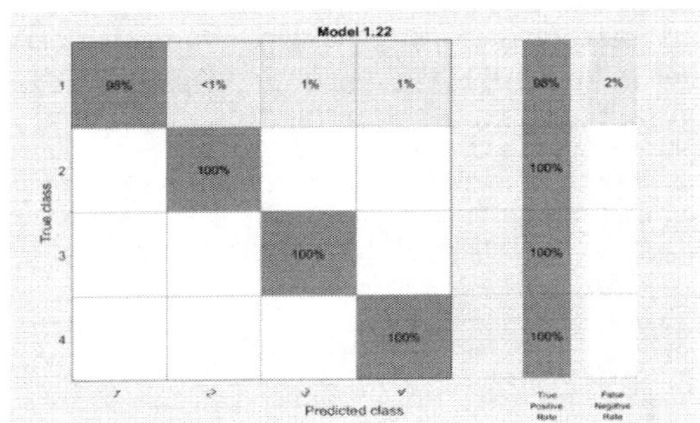

Figure 8. Bagged Trees confusion matrices.

ently and with the same distribution for all trees in the forest. The generalization error for forests converges a.s. to a limit as the number of trees in the forest becomes large. The generalization error of a forest of tree classifiers depends on the strength of the individual trees in the forest and the correlation between them' (Breiman 2001).

As shown in (Figure 8), the ANN classification rightly classifies roadway anomalies (diagonal elements for each of the confusion matrices) with approximately absolute accuracy. In essence, the classification model (bagged trees)

Table 7. Bagged Trees training, validation and prediction statistics for the various roadway anomaly cases examined.

Classifier Type	Accuracy	(%)
Bagged Trees	Train	99.8
Bagged Trees	Validation	99.65
Bagged Trees	Prediction	100

distinguishes and accurately categorizes the three roadway anomalies (target classes '2', '3' and '4') while also differentiating the 'no defect' condition (target class '1'), hence separating normal and abnormal roadway pavement conditions.

Each case the data is used for training, for validating and for predicting the classification model (bagged trees), with a synopsis of the obtained classification results shown in (Table 7).

5 CONCLUSIONS AND FUTURE WORK

Transportation agencies can improve the condition and operation of their transportation networks by implementing a Pavement Management System (PMS) that utilizes vehicle-based data collection, OBD connections and decision support software. The popularity of smartphone technology in vehicles provides an opportunity to efficiently collect vehicle data and process it by use of connected and distributed systems. Even though connected vehicle data is not likely to directly provide us with traditional assessment metrics (such as IRI and PCI), new metrics might supplement and eventually supplant traditional metrics. The paper presented a study on the utilization of smartphones for the detection of roadway anomalies and on the utilization of artificial neural networks and classification models for the classification of such defects. The applied methodology is instantly available, low-cost and precise, and can be utilized in crowd-sourced applications leading to roadway assessment and pavement management systems. The presented study documents the detection and classification of three types of roadway anomalies, exhibiting accuracy levels higher than 90%. The proposed methodology is currently field-tested with larger datasets and a higher number of roadway defect types, with links created to GIS mapping and database management systems for the use of the proposed methodology in PMS.

REFERENCES

AASHTO 2006. *Asset Management Data Collection Guide, Task Force 45 Report.* Washington, DC: American Association of State Highway and Transportation Officials.

AASHTO 1993. *Guide for Design of Pavements Structures.* Washington, D.C: American Association of State Highway and Transportation Officials.

AASHTO 1990. *Guidelines for Pavement Management Systems.* Washington, D.C: American Association of State Highway and Transportation Officials.

Alessandroni, G., Klopfenstein, L., Delpriori, S., Dromedari, M., Luchetti, G., Paolini, B., Seraghiti, A., Lattanzi, E., Freschi, V. & Carini, A. 2014. SmartRoadSense: Collaborative Road Surface Condition Monitoring. *Proc.of UBICOMM-2014.IARIA.*

Breiman, L. 2001. Random forests. *Machine Learning,* 45(1), pp. 5–32.

De Zoysa, K., Keppitiyagama, C., Seneviratne, G.P. & Shihan, W. 2007. A public transport system based sensor network for road surface condition monitoring, *Proceedings of the 2007 workshop on Networked systems for developing regions* 2007, ACM, pp. 9.

Khattak, J., Baladi, Y., Zhang, Z. & Ismail, S. 2008. *A Review of the Pavement Management System of the State of Louisiana—Phase I.* Washington D.C: Transportation Research Board.

Mcqueen, J.M. & Timm, D.H. 2005. Part 2: Pavement Monitoring, Evaluation, and Data Storage: Statistical Analysis of Automated Versus Manual Pavement Condition Surveys. *Transportation Research Record: Journal of the Transportation Research Board,* 1940(1), pp. 53–62.

Mohamed, A., Fouad, M., Elhariri, E., El-Bendary, N., Zawbaa, H.M., Tahoun, M. & Hassanien, A.E., 2015. RoadMonitor: An intelligent road surface condition monitoring system. *Intelligent Systems' 2014.* Springer, pp. 377–387.

Panagopoupou, I.M. & Chassiakos, P.A., 2012. An Optimization Model For Pavement Maintenace Planning And Resource Allocation. *Transportation Research Circular,* Number E-C136.

Seraj, F., Zwaag, B.J., Dilo, A., Luarasi, T. & Havinga, P. 2014. RoADS: A road pavement monitoring system for anomaly detection using smart phones.

Strazdins, G., Mednis, A., Kanonirs, G., Zviedris, R. & Selavo, L. 2011. Towards vehicular sensor networks with android smartphones for road surface monitoring, *2nd international workshop on networks of cooperating objects, Chicago, USA* 2011.

Tai, Y., Chan, C. & Hsu, J.Y. 2010. Automatic road anomaly detection using smart mobile device, *conference on technologies and applications of artificial intelligence, Hsinchu, Taiwan* 2010.

Walker, D., Entine, L. & Kummer, S. 2002. *Pavement Surface Evaluation and Rating: PASER manual.* Madison WI: University of Wisconsin, Transportation Information Center.

Washington Department of Transportation 1994. A Guide for Local Agency Pavement Managers, Washington State Department of Transportation, Trans Aid Service Center. The Northwest Technology Transfer Center.

Wolters, A., Zimmerman, K., Schattler, K. & Rietgraf, A. 2011. *Implementing Pavement Management Systems For Local Agencies.* ICT-R27-87. Urbana, IL 61801: Illinois Center for Transportation.

eWork and eBusiness in Architecture, Engineering and Construction – Christodoulou & Scherer (Eds)
© 2016 Taylor & Francis Group, ISBN 978-1-138-03280-4

Patch defects detection for pavement assessment, using smartphones and support vector machines

G.M. Hadjidemetriou & S.E. Christodoulou
University of Cyprus, Nicosia, Cyprus

ABSTRACT: The condition evaluation of roadway transport networks is conducted to provide decision support for appropriate maintenance activities, preventing the possibility of detrimental effects. The costly, time-consuming and subjective current pavement assessment methods lead to the requirement for automation of the underlying process. Presented herein is an automated methodology for pavement patches detection; a process which is crucial for pavement surface evaluation and rating. Support Vector Machine (SVM) Classification is utilized, whilst the possibility of collecting pavement frames from smartphones, positioned insides of cars is examined. The SVM is trained and tested by feature vectors generated from the histogram and two texture descriptors of non-overlapped square blocks, which constitute an image. The outcome is the indication of the frames that include patches and the image blocks which are characterized as parts of patches.

1 INTRODUCTION

Paved roads are one of the great man-made creations that directly affect humans' life prosperity. The infrastructure system of roads serve mobility and contribute to well-being, productivity and areas development (Cook 2011). There have currently been approximately 4,192,874 km of paved roads, surfaced with concrete, asphalt or composite in the United States (Federal Highway Administration 2011). Their overall valuation score has been nearby to failing, rated as 'D' (ASCE 2013), while action is needed for upgrading. Roads pavement condition is greatly related with the safety of vehicle passengers. Furthermore, a disruption, due to degradation, in roadway networks might have extensive effects not only on travelers, but also on the society at large. Possible broad impacts could be the loss of reachability to areas of interest, the re-routing of vehicular traffic, the reduction of productivity caused by re-routings and the entropy loads in the roadway network system.

The prevention of the possibility of disruption could achieved through maintenance activities or pavement replacement. Condition valuation of roadway transport networks is conducted in order to facilitate the 'repair or replace' decision. In case of the decision for repair, condition assessment is crucial for choosing, designing and applying the appropriate maintenance activity. The significance of efficient condition assessment of lifeline systems, including roadway transport networks, has received increasing attention, in recent years. The knowledge of the condition status and the provided level of service by the road are the initial requirement for planning decision support systems (NAMS Group 2006). The field of interest for experts and researchers has recently shifted from the construction of novel roads to the management of the existing roads (Panagopoulou & Chassiakos 2012). This change reinforces the importance of efficient and accurate pavement condition evaluation as the appropriate maintenance program is chosen based on the assessment data.

The stages of raw data collection, distresses identification and distresses evaluation compose the most common current pavement assessment methodologies. Although each of the three phases can be conducted either manually or automated, a number of significant issues are identified in both cases. The manual data collection (first stage) possesses safety threats to the inspectors, while the manual pavement defects identification and evaluation (second and third stages) are tedious, time-consuming and expensive because of the vast amount of data; as well as subjective due the perception of different raters (Gavilán et al. 2011, Bianchini et al. 2010). Regarding automated methods, dedicated vehicles can be used for data collection and specialized software for distresses detection and assessment. However, the overwhelming majority of transportation authorities are not able to buy or use them for the inspection of their entire network. The aforementioned boundaries lead to the challenge for the development of a reliable and cost-effective automated technique which detects, locates and quantifies roadway defects.

A range of individual roadway defects determine pavement condition rating. The rating systems indicate the provided service level by roadway as well as the need for improvement of the current situation. Pavement patches detection and quantification is critical because it is one of the pavement distress types, whose extend and condition determines road pavement condition ratings (Wisconsin Transportation Information Center 2013). The basic types of pavement distresses are longitudinal, transverse and alligator cracks, potholes, patches, rutting and depressions (UKPMS 2005). Although patching is a pavement maintenance activity, it is also considered as a pavement defect, as it does not reinstate the pavement to its initial condition. A patch can be defined as a portion of pavement surface that has been detached and replaced, or for which extra material has been added to the original road pavement surface. Patching is one of the most extensive and expensive pavement maintenance actions, conducting by transportation authorities to repair potholes, cracking, rutting, failed patches or utility cuts.

The development of a novel algorithm for automated detection of pavement patches in road surface images is in progress. The proposed methodology utilizes Support Vector Machine (SVM) Classification and it is based on the differences between the color and the texture of the patch area and the surrounding non-patch area. The previously presented patch detection algorithm (Hadjidemetriou et al. 2015, Hadjidemetriou et al. 2016) uses images collected by a generic dash camera position on the rear of a passenger vehicle. In this paper, the focus is on the improvement of the existing algorithm and the test of the method using images collected from smartphones, positioned inside a car. The possibility of the combination of the proposed patch detection algorithm with other techniques, which use smartphone sensors, is examined to propose in the near future an integrated low cost and crowd-sourced pavement condition assessment system. In the following section, the current practice, the state of research, the previous work in patch detection using SVM and the research objective are presented. Then, the proposed methodology is analyzed, followed by the experimental implementation. The Matlab™ programming language has been used for the algorithm development. The algorithm was tested using 200 pairs of pavement images. Each pair consists of an image collected from a smartphone positioned inside a car and an image captured from a generic dash camera positioned on the rear of the vehicle for the same part of a road, to investigate whether the novel idea of using smartphones for pavement evaluation is as accurate as the previously proposed methodology. The pavement surface videos, wherefrom images were extracted, were recorded on roads of Nicosia, Cyprus. The paper last section presents the conclusions and future work.

2 BACKGROUND

2.1 *State of practice*

Condition assessment of pavements facilitates national and local transportation authorities to decide whether there is a need for maintenance or replacement actions. In case of maintenance, condition evaluation data assist the choice, plan and design of the applicable repair action. The current practice in pavement condition valuation is normally separated into the following main stages: 1) data collection, 2) defects detection and 3) defects condition evaluation.

The stage of raw data collection is performed automated or manually. The former is the most effective approach and it is conducted by inspection vehicles, such as the Automatic Road Analyzer (ARAN) platform, which is able to perform a range of roadway data collection tasks. These dedicated vehicles collect data in different forms with the aid of cameras, Global Positioning Systems, high frequency laser scanners, road profilers and accelerometers (ARAN 2014, D. for T.U. DfT 2011, Fugro Roadware 2010). Laser scanners are used to measure the transverse and longitudinal road profile, presenting the elevation differences (ARAN 2014, Fugro Roadware 2010). Cameras collect data, used from the inspectors to visually assess pavement condition. In spite of the inspection vehicles efficiency and their capability of operating at highway speeds (up to 100 km/h), some countries (e.g. Cyprus) are not able to purchase them, while some others (e.g UK, USA) are able to afford only a few of them, due to their extremely high purchase and operational costs. Hence, their utilization is limited to the principal road network (e.g. 9.6% of the UK entire network), while frequent inspections are not feasible (D. for T.U. DfT 2013). Consequently, labor intensive manual surveys are conducted to inspect a remarkable percentage of road networks. Inspectors walk along the pavement or use passenger vehicles, which move at low speeds of 8–15 mph, to manually collect data (SDDOT 2011). The manual collection of data could be considered as time consuming, whereas it poses safety threats for the inspectors. In countries where the inspection of the entire road network is not affordable, inspectors collect data only for the principal network or for parts of the entire network that users complain about their condition.

The second and third stages of the process include the detection and evaluation of distresses as well as the calculation of the Road Condition

Indicator (RCI) (D. for T.U. DfT 2011) or the pavement roughness (MnDOT 2013) to realize whether a specific part of a road needs further detailed investigation. The detection and assessment of distresses, are tedious and time-consuming processes, since they require the use of trained raters for the analysis of massive amounts of collected data. The raters view and examine collected digital road videos manually, sitting in front of monitors, to identify distressed areas and evaluate their status, based on their experience and pavement condition manuals, which consist of guidelines and criteria (MnDOT 2006). The direction from which the inspectors view the road pavement, as well as the appearance of the road surface which depends on the weather conditions, in combination with raters' subjectivity, directly affect pavement ratings (Gavilán et al. 2011, Bianchini et al. 2010). Another challenge worth mentioning is the need for significant volume of storage due to the amounts of the collected data. The aforementioned major drawbacks of current pavement condition assessment practices lead to the need for automation of the process. Based on this requirement, a number of companies developed software, which is supplementary to the provided equipment by the same firms. The data collected from the previously mentioned dedicated vehicles are post-processed to assess all defects simultaneously or individually (Sy et al. 2008). Nevertheless, this adds to the already high cost of specialized vehicles.

2.2 *State of research*

Overcoming the limitation of the current practices in pavement defects detection has attracted the interest of many researchers in recent years. They have been trying to automate the process of the detection, classification and measurement of different types of pavement distresses. Some of their research studies are based on computer-vision techniques since images or videos are used as data. A noteworthy proportion of the most remarkable research studies concern with cracks, raveling, potholes and patches.

The performance of six different segmentation algorithms was evaluated by Tsai et al. (2010), associated with cracks, concluding that the dynamic optimization based method exceeds the performance of the other approaches. A variety of methods focused on crack identification and assessment (Huang & Xu 2006, Sy et al. 2008), crack depth calculation (Amarasiri et al. 2009) and cracks classification (Sun et al. 2009, Ying & Salari 2010). The defect of raveling was the research subject for Tsai and Wang (2015), who tried to detect raveling utilizing emerging 3D technology and macrotexture analysis; as well as for Mathavan et al. (2014), who quantified raveling from a combination of two- and three-dimensional images. Koch and Brilakis (2011) focused on automated detection of potholes in images, proposing a camera-based method. Their method is based on pothole texture and shape. They subsequently improved the detection reliability of the method to achieve effective counting of potholes, using video data (Koch et al. 2012). Yu and Salari (2011) used laser imaging to identify potholes, whilst Mednis et al. (2011) utilized accelerometers and smartphones for automated detection of potholes.

Pavement patching, the subject of the current study, has drawn the attention of Yao et al. (2008), who proposed an algorithm that detects patches amongst punch-outs and cracks. Cafiso et al. (2006) presented a method which differentiates cracks from pothole/patch area. Nonetheless, it is not able to distinguish a patch from a pothole. Zhou et al. (2006) proposed an approach for defects recognition, isolation and assessment on road images. However, this method main limitation is the inability for differentiating patches borders from cracks. A patch detection method in images of asphalt pavement was presented by Radopoulou et al. (2013), taking into consideration criteria for patch width, length and area. Nevertheless, this approach performs high accuracy detection results under certain circumstances. It requires the existence of only one patch in each picture that occupies 10%–40% of the image and its entire area is included in the image. Radopoulou and Brilakis (2014, 2015) upgraded their previous method by comparing the texture of the patch area with the texture of the non-patch surrounding area and by using a vision-tracking algorithm to identify patches on videos data. However, their approach is limited to detecting and tracing patch defects in subsequent frames; and the patch should still occupy a predetermined proportion of the frame (2.5%–65%).

Some research studies have investigated the possibility of pavements defects detection, using smartphones and crowd-sourced applications which might lead in the near future to integrated pavement management systems. A system which includes the integration of a custom mobile application and a georeferenced database system was described by Alessandroni et al. (2014). Smartphones equipped with GPS, gyroscope and inertial accelerometer were used by Seraj et al. (2016) to monitor pavement surface. Kyriakou et al. (2016) presented a study which explores the possibility of roadway anomalies detection using smartphones sensors automobiles' On-Board Diagnostic (OBD-II) devices while passengers' vehicles are in movement. The Jaguar Land Rover automaker has recently publicized that it has been exploring a novel connected vehicle technology that allows a vehicle to identify the position of dangerous potholes and then share this information with other vehicles and road authorities in rea time (O'Donnell & McConomy 2015).

2.3 *Previous work in the use of support vector machine classification for automated patch detection*

In our previous work we proposed an algorithm which automatically detects patches in pavement frames (Hadjidemetriou et al. 2015). The previously proposed method uses three SVMs to classify each of the image blocks of 20 × 20 pixels as part of a "patch" or "no-patch", three times, while the final blocks classification is based on voting. The first SVM is trained from two intensity histograms for each frame; one for the pixels of patches areas and one for the pixels non-patches pixels. The second and third SVM are trained, taking data from blocks within the image; the segments are 20 × 20 pixels in width and height. The second SVM feature vector is generated from the local intensity histogram of each block, whilst the third one from the Discrete Cosine Transform (DCT). Despite the promising results of this method, it uses a small data set of 50 images; it is slow and computationally inefficient because of the three support vector machines.

At a later stage of our research, only one SVM is trained based on a feature vector which takes information from blocks within the image, whose size is again 20 × 20 pixels (Hadjidemetriou et al. 2016). The feature vector is produced by the local intensity histogram of the block and two different texture descriptors, named two-dimensional Discrete Cosine Transform (DCT) and Gray-Level Co-occurrence Matrix (GLCM). The performance of the algorithm is slightly improved and the computational time is significantly lower, compared to the previous algorithm. However, the method does not distinguish the frames which include patches from the frames that do not have any patches. Moreover, it needs a camera positioned on the rear of a vehicle or a parking camera; a low-cost requirement which could easily be implemented by transportation authorities. Nonetheless, a crowd-sourced method which could be applied by all road networks users could be explored.

2.4 *Research objective*

The aforementioned major limitations of the current pavement assessment methods give rise to the necessity of data reduction, model simplification and automation in pavement condition assessment. An intelligent and automated decision support system for 'repair-or-replace' strategies could prevent roadway network disruptions, optimize roadway maintenance, and decrease the environmental footprint of traffic movements (gas emissions) caused by unconditioned roadway pavements. The overarching objective of the current research is the development of an automated method for the detection, location and quantification of road pavement defects, in a most reliable and cost-effective way, improving on existing processes.

Pavement patches detection and quantification is critical because it is one of the pavement defects types, whose extend and condition affects road pavement condition ratings, while the rating systems indicate the provided service level and the need for improvement of the current situation. The objective of the current paper is the proposal of a crowd-sourced method that automatically detects pavement patches on pavement surface frames, utilizing smartphones and custom computer-vision algorithms.

3 METHODOLOGY

SVM classification is utilized to facilitate the automated detection of patches in pavement frames. It is a supervised machine learning method, since the algorithm is trained by a set of examples. SVM could be defined as a discriminative classifier, linked with learning algorithms, which recognize patterns, after taking labelled training data. It is divided into two main stages: (i) SVM training and (ii) SVM testing. It is efficiently used for two-group classification problems (Cortes & Vapnik 1995), such as the presented problem, where the algorithm separates the testing data into "patch" and "no-patch" categories.

Figure 1 presents the basic steps of the proposed algorithm for both training and testing stages. The first phase of the algorithm (SVM training) starts with the transformation of images into grey-scale, without losing any crucial data since the colors of the pavement are mainly shades of grey. The image processing of grey-scale images has lower computation time and complexity, compared to RGB images. The described novel methodology uses only one SVM, which is trained based on the ground truth and a feature vector. The ground truth is entered in the algorithm to provide information regarding the pixels of each frame which are part of a pavement patch. The feature vector is generated and subsequently the SVM is trained, extracting information from blocks within the frame, whose size is 20 × 20 pixels in width and height. One should also note that in order to assist the SVM to be able to distinguish "patch" from "no-patch" areas, blocks which includes weighty proportions of both patch and non-patch areas (the patch area is more than 5% and less than 95% of the block) are not used for the training of the SVM. The

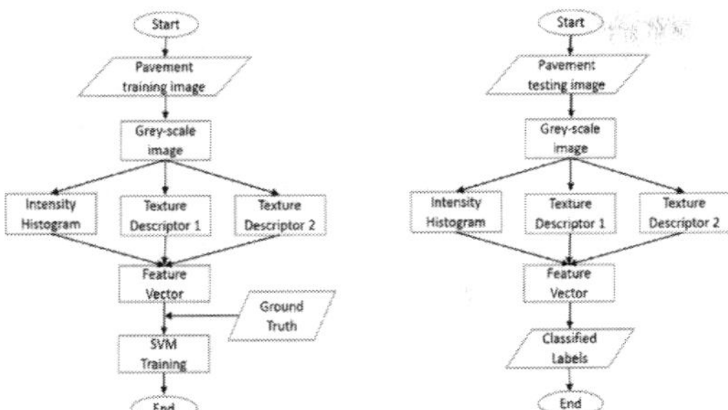

Figure 1. Training and testing phases of the proposed patch detection algorithm.

feature vector of every square block is formed by the local intensity histogram and two texture descriptors, termed two-dimensional Discrete Cosine Transform (DCT) and Gray-Level Co-occurrence Matrix (GLCM). DCT, which is efficiently used for pattern recognition purposes, states a finite amount of data points in terms of a weighted sum of cosine functions oscillating at diverse frequencies (Ahmed et al. 1974). The statistical system of GLCM considers the spatial relationship of pixels (Haralick et al. 1973). The functions of GLCM describe the texture of a frame by generating a matrix which contains the calculated frequency of the occurrence of pixel pairs with definite values and in a specific spatial relationship. Our algorithm extracts data from this matrix to estimate and use the statistical measures of contrast, correlation, energy and homogeneity.

The second phase of the presented algorithm is comprised primarily of the SVM testing, following a similar path with the phase of the SVM training. Testing frames are transformed into gray-scale and divided into square areas of 20×20 pixels. The local intensity histogram and the two texture descriptors form a feature vector for each block. SVM uses this feature vector to classifies every frame block in "patch" or "no-patch". Finally, a predefined number of connected "patch" blocks indicates the presence of a patch in an image. Consequently, the algorithm discriminates between the images which include parts of patches and the frames which do not contain any patches parts.

4 EXPERIMENTAL IMPLEMENTATION

A field study was conducted, by positioning a smartphone inside a car and a generic dash camera on the rear of a passenger vehicle to record pavement videos from the road networks of Nicosia, Cyprus. The two different cameras recorded videos of the same pavement areas so that the comparison of the patch detection performance of the two methods being feasible. The videos were recorded under sunny or cloudy weather conditions, during daytime and transformed into pavement frames that were processed by the proposed algorithm. The pixels resolution of the pictures was 640×480 and the proposed method was performed using the programming language of Matlab™.

The metrics of accuracy (1), precision (2) and recall (3) were utilized in evaluating the performance of the automated patch detection method. The performance assessment was conducted for blocks classification ("patch" or "no-patch") and for images classification ("including patches" or "not including patches"). They are calculated by using True Positives (TP), which is the number of blocks or pictures that were correctly identified as part of a patch or including a patch, respectively; False Positives (FP), which are the blocks or frames incorrectly recognized as parts of a patch or including a patch; True Negatives (TN), which are the cells or pictures that correctly labelled as belonging to non-patch areas or as not including patches; and False Negatives (FN), which are blocks or frames that were wrongly not recognized as belonging to a patch or including a patch. Accuracy characterizes the average correctness of the detection system. High precision ratio designates that the blocks or frames, identified as parts of a patch or including patches respectively, correspond to actual patches. High proportion of recall indicates that the majority of the patch areas were correctly identified.

$$Accuracy = \frac{TP + TN}{TP + TN + FP + FN} \tag{1}$$

$$Precision = \frac{TP}{TP + FP} \tag{2}$$

$$Recall = \frac{TP}{TP + FN} \tag{3}$$

The testing of the proposed method was implemented by using 200 pairs of frames. Each pair contains a picture collected from the camera positioned on the rear of the car and a frame collected from the smartphone positioned inside the passenger vehicle. Both of the images were collected from the same part of the road. 5-fold cross-validation was utilized for the valuation of the system performance. The algorithm run five times and in every phase the training set included 160 images, with testing set consisted of the remaining 40 frames. Rotating the set of 40 images for each run thereby tests all of the 200 pictures.

100 of the 200 pairs of images included parts of at least one patch. For each of the pairs that include parts of patches the performance of the blocks classification ("patch" or "no-patch") has also been estimated. The blocks classification performance of

the rest of the 100 images was not calculated since they include only negative ("no-patch") blocks. Table 1 represents the blocks and images classification for images collected from the generic dash camera and for images from the smartphone camera. Figure 2 depicts the way that patch areas are identified by the algorithm (red color for positives and blue color for negatives), including examples of blocks which are characterized as TP, FP, TN or FN.

Figures 3 and 4 show the way accuracy, precision and recall are changed by increasing the number of testing images. The number of testing frames and algorithm performance are directly correlated as the rise of the former causes the improvement of the latter. This procedure will be continued until the growth of the number of pictures will worsen or will not cause noteworthy enhancement on the method performance.

Table 1. Overall performance of the proposed algorithm.

Generic Dash Camera			
Blocks Classification		**Images Classification**	
Accuracy	82.85 %	Accuracy	82.5 %
Precision	65.61 %	Precision	77.78 %
Recall	91.97 %	Recall	91 %
Smartphone Camera			
Blocks Classification		**Images Classification**	
Accuracy	80.46 %	Accuracy	80 %
Precision	63.84 %	Precision	75.42 %
Recall	89.35 %	Recall	89 %

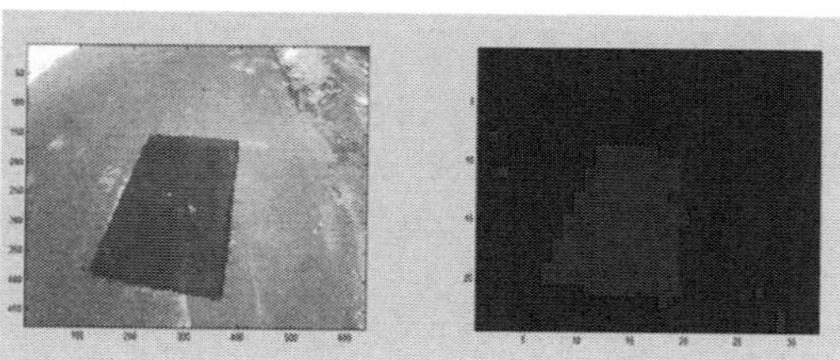

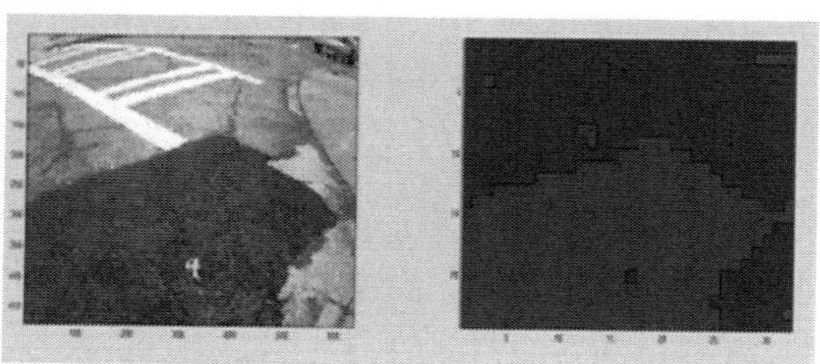

Figure 2. Examples of processed images by the proposed algorithm.

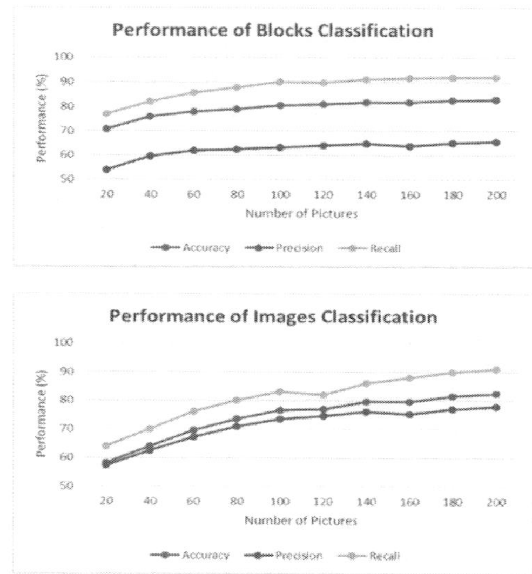

Figure 3. Correlation between the proposed method performance and the number of testing frames collected from the generic dash camera.

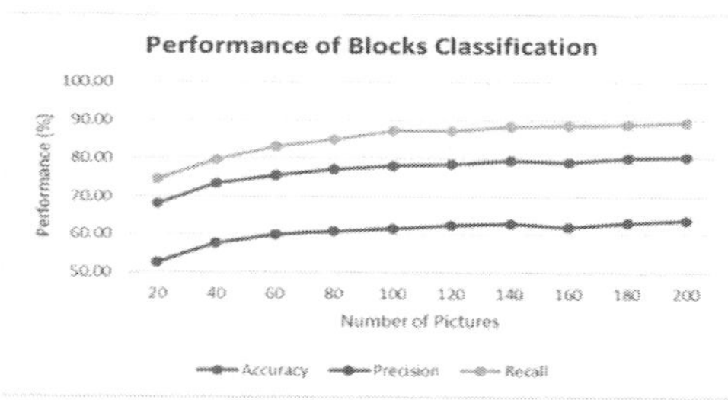

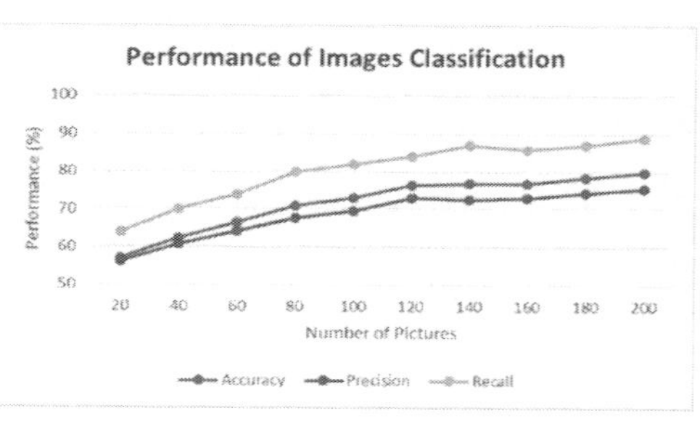

Figure 4. Correlation between the proposed method performance and the number of testing frames collected from the smartphone camera.

5 CONCLUSIONS

The current techniques of pavement evaluation, used by transportation authorities, are characterized by numerous major limitations, indicating the need for automation of the processes, utilizing low cost techniques. Presented herein is the automated detection of pavement patches using support vector machines, which is a fast classification method. The method, for the first time, classifies images as "including patches" and "not including patches", based on the neighboring 20×20 pixels blocks that can be labelled as "patch" or "no-patch" using the already developed algorithm by the authors. Another novel element investigated in the current paper worth mentioning is the utilization of smartphone cameras positioned inside of passengers' vehicles to collect pavement videos. Additionally, the data set has been increased and 200 pairs of frames were used to evaluate the methodology performance.

The performance results are slightly better when a camera is positioned on the rear of car, compared to the placement of a smartphone camera inside a vehicle. This is explained by the fact that videos collected from the smartphone, which is inside the vehicle, are affected by the cleanness of the window. Nevertheless, the results are still promising, indicating the contribution of the current paper since the combination of the proposed patch detection algorithm with other techniques that also use smartphone sensors, might lead to an integrated low cost and crowd-sourced pavement condition assessment system. The presented method is still characterized by some strong advantages such as the identification of multiple patches in a single image or the detection of proportions of patches when their entire area is not included in the image. However, the presented system has room for improvement and work is currently under way to further develop the classification accuracy. Future work includes the measurement of the pavement patch area, the modification of the presented algorithm to a video-based detection method, the data set increase to achieve better training of the SVM and the investigation of alternative ways to collect the pavement videos such as drones or motorcycle helmet cameras.

REFERENCES

Ahmed, N., Natarajan, T. & Rao, K.R. 1974. Discrete cosine transform. IEEE transactions on Computers, (1): 90–93.

Alessandroni, G., L. Klopfenstein, S. Delpriori, M. Dromedari, G. Luchetti, B. Paolini, A. Seraghiti, E. Lattanzi, V. Freschi, & A. Carini 2014. SmartRoadSense: Collaborative Road Surface Condition Monitoring. Proc. of UBICOMM–2014.

Amarasiri, S., Gunaratne, M. & Sarkar, S., 2009. Modeling of crack depths in digital images of concrete pavements using optical reflection properties. Journal of Transportation Engineering, 136(6): 489–499.

ASCE 2013. 2013 Report card for America's infrastructure. New York.

Bianchini, A., Bandini, P. & Smith, D.W. 2010. Interrater reliability of manual pavement distress evaluations. Journal of Transportation Engineering, 136(2): 165–172.

Cafiso, S., Graziano, A.D. & Battiato, S. 2006. Evaluation of pavement surface distress using digital image collection and analysis. In Seventh International Congress on Advances in Civil Engineering.

Cook A. 2011. A fresh start for the Strategic Road Network: Managing our roads better to drive economic growth, boost innovation and give road users more for their money. https://www.gov.uk/government/uploads/system/uploads/attachment_data/file/4378/strategic-road-network.pdf (accessed April 05, 2016).

Cortes, C. & Vapnik, V. 1995. Support-vector networks. Mach Learn, 20(3): 273–297.

Federal Highway Administration 2011. 'Highway Statistics 2011'.

Gavilán, M., Balcones, D., Marcos, O., Llorca, D.F., Sotelo, M.A., Parra, I., Ocaña, M., Aliseda, P., Yarza, P. and Amírola, A. 2011. Adaptive road crack detection system by pavement classification. Sensors, 11(10): 9628–9657.

Hadjidemetriou G., Christodoulou S. and Vela P. 2016. Automated Detection of Pavement Patches utilizing Support Vector Machine Classification. 18th Mediterranean Electrotechnical Conference—MELECON, Limassol, Cyprus.

Hadjidemetriou G., Serrano M., Vela P. & Christodoulou S. 2015. Patch Defects in Images using Support Vector Machines. Proceedings in the Fifteenth International Conference on Civil, Structural and Environmental Engineering Computing. Prague, Czech Republic.

Haralick, R.M., Shanmugam, K. & Dinstein, I.H. 1973. Textural features for image classification. Systems, Man and Cybernetics, IEEE Transactions on, (6): 610–621.

Huang, Y. & Xu, B. 2006. Automatic inspection of pavement cracking distress. Journal of Electronic Imaging, 15(1): 013017–013017.

Koch, C. & Brilakis, I., 2011. Pothole detection in asphalt pavement images. Advanced Engineering Informatics, 25(3): 507–515.

Koch, C., Jog, G.M. & Brilakis, I. 2012. Automated pothole distress assessment using asphalt pavement video data. Journal of Computing in Civil Engineering, 27(4): 370–378.

Kyriakou, C., Christodoulou, S. & Dimitriou, L. 2015. Road Anomaly Detection and Classification using Smartphones and Artificial Neural Networks. In the 95th Transportation Research Board Annual Meeting, Washington D.C., U.S.A.

Mathavan, S., Rahman, M., Stonecliffe-Jones, M. & Kamal, K. 2014. Pavement Raveling Detection and Measurement from Synchronized Intensity and Range Images. Transportation Research Record: Journal of the Transportation Research Board, (2457): 3–11.

Mednis, A., Strazdins, G., Zviedris, R., Kanonirs, G. & Selavo, L. 2011. Real time pothole detection using android smartphones with accelerometers. In Distributed Computing in Sensor Systems and Workshops (DCOSS), 2011 IEEE International Conference.

NAMS Group 2006. International infrastructure management manual. National Asset Management Steering Group.

O'Donnell, N., K. McConomy 2015. Jaquar Land Rover Announces Technology Research Project to Detect, Predict and Share Data on Potholes. http://Newsroom.Jaguarlandrover.Com/En-in/Jlr-Corp/News/2015/06/jlr_pothole_alert_research_100615/ (accessed June 15, 2015).

Panagopoulou, M.I. & Chassiakos, A.P. 2012. Optimization model for pavement maintenance planning and resource allocation. Transportation Research Circular, Number E-C163, Maintenance Management: 25–38.

Radopoulou S.C. & Brilakis I. 2014. Improving patch defect detection using vision tracking on video data, in the 21st EG-ICE Workshop on Intelligent Computing in Engineering, Cardiff, UK.

Radopoulou, S.C. & Brilakis, I. 2015. Patch detection for pavement assessment. Automation in Construction, 53: 95–104.

Radopoulou, S.C., Jog, G.M. & Brilakis, I. 2013. Patch Distress Detection in Asphalt Pavement Images. In 30th International Symposium on Automation and Robotics in Construction (ISARC 2013): 1572–1580.

Seraj, F., van der Zwaag, B.J., Dilo, A., Luarasi, T. & Havinga, P., 2016. RoADS: A road pavement monitoring system for anomaly detection using smart phones. In Big Data Analytics in the Social and Ubiquitous Context: 5th International Workshop on Modeling Social Media, MSM 2014, 5th International Workshop on Mining Ubiquitous and Social Environments, MUSE 2014, and First International Workshop on Machine Learning for Urban Sensor Data, SenseML 2014, Revised Selected Papers, 128–146. Springer International Publishing.

Sun, Y., Salari, E. & Chou, E. 2009. Automated pavement distress detection using advanced image processing techniques. In Electro/Information Technology. IEEE International Conference: 373–377.

Sy, N.T., Avila, M., Begot, S. & Bardet, J.C. 2008. Detection of defects in road surface by a vision system. In Electrotechnical Conference, 2008. MELECON 2008. The 14th IEEE Mediterranean: 847–851.

Tsai, Y.C. & Wang, Z. 2015. Development of an Asphalt Pavement Raveling Detection Algorithm Using Emerging 3D Laser Technology and Macrotexture Analysis. IDEA Program Transportation Research Board The National Academies.

Tsai, Y.C., Kaul, V. & Mersereau, R.M. 2009. Critical assessment of pavement distress segmentation methods. Journal of transportation engineering, 136(1): 11–19.

UKPMS 2005. The UKPMS user manual http://www.pcis.org.uk/index.php?p = 6/8/0/list,0,61 (accessed July 11, 2015).

Wisconsin Transportation Information Center 2013. Asphalt Pavement Surface Evaluation and Rating Manual, University of Wisconsin Madison.

Yao, X., Yao, M. & Xu, B. 2008. Automated detection and identification of area-based distress in concrete pavements. In Seventh International Conference on Managing Pavement Assets.

Ying, L. & Salari, E. 2010. Beamlet Transform-Based Technique for Pavement Crack Detection and Classification. Computer-Aided Civil and Infrastructure Engineering, 25(8): 572–580.

Yu, X. & Salari, E. 2011, May. Pavement pothole detection and severity measurement using laser imaging. In Electro/Information Technology (EIT), 2011 IEEE International Conference.

Zhou, J., Huang, P.S. & Chiang, F.P. 2006. Wavelet-based pavement distress detection and evaluation. Optical Engineering, 45(2): 027007–027007.

eWork and eBusiness in Architecture, Engineering and Construction – Christodoulou & Scherer (Eds)
© 2016 Taylor & Francis Group, ISBN 978-1-138-03280-4

Comparing diurnal patterns of domestic water consumption: An international study

J. Terlet, T.H. Beach & Y. Rezgui
Cardiff School of Engineering, Cardiff, UK

G. Bulteau
Centre Scientifique et Technique du Batiment, Nantes, France

ABSTRACT: The increasing variability of water supply and demand makes the creation of improved sustainable water management systems crucial. Optimized decision making tools are required to better manage urban water resources. This paper presents the implementation of the WISDOM project aiming at (a) increasing user awareness and modifying behaviours, (b) encouraging water conservation and (c) reducing peak-period of water distribution loads. By collecting near real-time data, smart metering encourages behaviour changes and informs water companies about their customers' consumption while feedback provided through in-home displays educates people about their usage. Collecting disaggregated data about domestic water use also allows the profiling of behaviours that need to be targeted to promote conservation. In the context of the WISDOM project, these two devices were implemented within households in Cardiff (UK) allowing the collection of disaggregated data. These new technologies will optimize the peak-period management of water while encouraging water conservation and behavioural change.

1 INTRODUCTION

Current trends in population growth, urbanization and climate change increase the demand for water worldwide. Water users are key players in the water value chain and are at the heart of water management. Research is necessary to better understand users behaviours and habits in order to determine water consuming attitudes and promote behavioural change. In Australia and in the United-States (USA) where water scarcity is an important issue, studies have been conducted in order to encourage behavioural change. However, in Europe, water scarcity is not deemed such an important issue and literature on this subject is thus less developed. More specifically, only a few studies focus on disaggregated data about hourly diurnal patterns of water consumption indoor and outdoor in Europe. This information is crucial to determine users' behaviours and activities in order to increase water conservation. Additionally, information about hourly patterns of water consumption allows the redistribution of domestic water usage to reduce peak demand on the water network. This paper aims at (a) reviewing current studies on disaggregation of water usage and its benefits all around the world, (b) highlighting the advantages and the necessity of collecting such data in Europe and (c) introducing the work done in the context of the WISDOM project[1] that will allow the collection of disaggregated water usage data in Cardiff (UK), through the implementation of smart meters and surveys.

2 OVERVIEW OF CURRENT WATER USAGE DATA COLLECTED AROUND THE WORLD

Collecting information about diurnal patterns of water consumption is crucial in order to implement efficient restriction regimes, water tariffs and improved demand management initiatives (Cole & Stewart, 2013). Identifying these diurnal patterns helps determining daily water consumption trends among users belonging to different socio-demographic groups and living in different climatic regions (Cole & Stewart, 2013). Data related to water usage is available in many countries and gives an overview of water consumption patterns worldwide.

When it comes to water consumption, there are important discrepancies in the daily amount of water used per person around the world. According to the 2006 Human Development Report, between 1998 and 2002, the USA and Australia were found to be the most water consuming countries with

[1]This research is funded by the European Union's Seventh Framework Programme for research, technological development and demonstration under grant agreement no. 619795.

respectively 575 litres used per person per day on average for the former and 490 litres for the latter (United Nations Development Program, 2006). Due to growing water scarcity, countries like these countries have now adopted measures to reduce their overall water consumption. In 2016, after having experienced a few years of severe drought, water consumption per capita and per day in California reached its lowest per person rate since 2014 and dropped to about 230 litres (California Water Boards, 2015, 2016). In the city of Santa Fe in New Mexico, water usage per capita per day has been reduced by more than 50 percent since 1995 after the implementation of a water conservation program but still amounted to 359 litres in 2014 (City of Santa Fe, 2016). In Australia, in 2009, daily water usage per capita amounted to 203 to 222 litres (Cahill & Lund, 2011). In Melbourne where management programmes have been launched to reduce water usage, the daily consumption per capita dropped from 247 litres to 149 litres between 2000 to 2012 (Gan & Redhead, 2013; Rathnayaka et al., 2014). As mentioned in section 4 of this paper, research and studies conducted in these two countries has developed innovative techniques to encourage water conservation and behavioural change (Beal & Stewart, 2013; Froehlich et al., 2012; Larson et al., 2012; Willis et al., 2010). Between 1980 and 2006, a large majority of the empirical studies related to water demand management (72%) came from the USA and Australia (Romano et al., 2014; Worthington & Hoffman, 2008).

In Europe, daily water consumption has also been reduced. In 2002, domestic water usage in Italy amounted to about 206 litres per capita per day but dropped to 170 litres in 2009 (Romano et al., 2014). Likewise, the launch of diverse awareness campaigns in Spain has led to a reduction of water usage from 183 litres per capita per day in 2001 to 141 litres in 2010 (Gutierrez-escolar et al., 2014). In France, daily water consumption has been only slightly reduced since 2001 and amounts to about 150 litres per capita, just like in the United Kingdom where the same amount of water per person per day has been maintained over the years (Energy Saving Trust, 2013; Ministere de l'Environnement de l'Energie et de la Mer, 2011; United Nations Development Program, 2006).

In Asia, Japan daily domestic water use in 2000 amounted to 322 litres per person but decreased and reached 297 litres per capita in 2010 (Organisation for Economic Co-Operation and Development, 2015). Urban domestic water consumption in China has also decreased and dropped from 230 litres per person per day in 1997 to 131 litres in 2008 (Zhang et al., 2012).

In the early 2000s, in South America, Mexican water usage amounted to about 360 litres per person and per day with an average of 297 litres in metropolitan areas (Tortajada & Castelan, 2003; United Nations Development Program, 2006; WWF, 2011). During the same period, a few African countries such as Ghana, Nigeria, Kenya and Asian countries such as Bangladesh and Cambodia cross the water poverty threshold with less than 50 litres per person and per day (United Nations Development Program, 2006).

Table 1. Percentages of water used by specific activities in daily household water usage.

	Showers	Flushing toilets	Washing purposes*
Countries	Percentage	Percentage	Percentage
Australia	29.38	17.15	21.64
China	27	–	–
Eastern Africa	38.67	–	36.22
France	39	–	28
UK	25	22	13
USA	16.8	26.7	21.7
Spain	32.73	--	10.33
Venezuela	–	36	14

*Washing machines, dishwashers, hand washing.

Nowadays, the use of water per capita in developing countries however tends to increase due to the development of industrial activity that leads to higher standards of living (Dray et al., 2008; UNESCO, 2014). Identifying the most water-consuming appliances within households across the world allows the identification of cultural differences and similarities in domestic water usage. According to diverse studies, the domestic activities that appear to use most water within households worldwide are activities related to hygiene purposes such showering, washing and flushing toilets as shown in Table 1 (Albiol & Agulló, 2014; Blanco et al., 2014; Carragher et al., 2012; Centre d'Information Sur l'Eau, 2011; Energy Saving Trust, 2013; Lu & Smout, 2008; Mayer et al., 1999).

While daily water consumption data per appliances is available in a lot of different countries around the world, it is more difficult to get information about hourly diurnal patterns of water consumption. This data is however useful in order to understand people's habits and daily use of water but also to offer better water services and initiatives to encourage water conservation.

3 METHODS AND TECHNIQUES FOR COLLECTING DISAGGREGATED DATA RELATED TO DOMESTIC WATER USAGE

Obtaining detailed information about the way individuals use water allows the forecasting of future water demand with a higher accuracy (Fontdecaba et al., 2013). Disaggregation of domestic water consumption of end users can help identifying the average and peak end-use water consumption volumes and thus inform the planning process of water services (Beal & Stewart, 2013; Fontdecaba et al., 2013).

Smart metering in the water domain aims at collecting and communicating up-to-date water usage reading on a near real-time or real-time basis (Cole & Stewart, 2013). By recording hourly intervals of consumption, it also allows the identification of the degree of outdoor water usage within a water network (Cole & Stewart, 2013). Disaggregated water data can be collected by installing pressure sensors on deployment sites or by implementing ground truth sensors on individual fixtures within households (Froehlich, et al., 2011). It gives indications on current demand flow rates that can help configuring network distribution models and urban water planning (Cole & Stewart, 2013). Flow trace analysis as proposed by Dziegielewski et al. (1992) is a commonly used technique to identify water usage events (Larson et al., 2012). It analyses the "aggregate flow-trace patterns off of an inline water meter to determine the source of the water usage event" relying on the fact that domestic water usage activities usually use consistently the same amount of water i.e. flushing a specific toilet requires the same volume and flow of water over time (Larson et al., 2012). Using this tool, residential end uses can be disaggregated and quantified "from a flow data set recorded" from an household's water meter (DeOreo et al., 2011). Trace wizard is a computer software used in several studies worldwide to disaggregate the data in a repository of end water uses, such as toilets, showers, washing machines and garden watering (Beal & Stewart, 2013; DeOreo et al., 2011; Gato-Trinidad et al., 2011; Mead, 2008; Willis et al., 2011).

Nevertheless, obtaining disaggregated data and identifying specific water-consuming activities can be limited by the inability of discerning several simultaneous events and by the low flow of external consumption (Beal & Stewart, 2013; Wilkes et al., 2005). Indeed, when two water events overlap, the performance of flow trace analysis can drop from 83% to only 24% of correctly categorized water events (Lyons et al., 2011). To overcome these limits, high resolution meters, involving experienced analysts, can be used (Beal & Stewart, 2013). However, contrary to the use of smart meters for billing purposes, using very high resolution data requires more readings per hour and reduces the battery lifetime of water meters. These devices thus need to be replaced and maintained more often. In order to obtain precise end-use disaggregation and to verify the identification of user behaviours by the flow trace software, the use of smart meters can instead be coupled with the use of surveys. For instance, in the context of their study, Beal & Stewart (2013) conducted a water audit of household water use behaviours and asked participants to complete a water use diary in addition to the implementation of a flow trace software within households.

4 DISAGGREGATED DATA AND HOURLY PATTERNS OF WATER CONSUMPTION: EXAMPLES OF AUSTRALIA AND THE USA

Water consumption varies depending on many factors such as the climate of a country, water users behaviours, consumer socio-demographics and household water efficiency (Beal & Stewart, 2013). Warmer temperatures are often associated with an increased water consumption (Beal & Stewart, 2013). Thus, in warm countries affected by water shortages such as Australia and the USA, studies have been conducted to analyse diurnal patterns of domestic water consumption using hourly disaggregation (Beal & Stewart, 2013; Cole & Stewart, 2013; Gato-Trinidad et al., 2011; Mayer et al., 1999). Collecting longitudinal hourly data over a long period of time can increase the understanding of the factors influencing peak demand (Cole & Stewart, 2013; Stewart et al., 2010). It gives indications to water suppliers and consumers about the times at which water is used, how much and how often (Gato-Trinidad et al., 2011).

Domestic water consumption varies on a daily basis (Beal & Stewart, 2013). It includes an average peak day hour and peak days on an annual basis (Beal & Stewart, 2013). Peak hour demand can be defined as the "peak hourly demand a system will be called on to deliver" (Cole & Stewart, 2013). Peak days on the other hand are the "maximum demand in any one day of the year" (Cole & Stewart, 2013). Peak hour demand is affected by outdoor water usage while peak days involve both indoor and outdoor water usage. Water consumption is embedded in people's daily life and follows routine (Cole & Stewart, 2013). It is therefore possible to identify a morning and an afternoon peak of water consumption including series of water consumption habits that are usually repeated (Cole & Stewart, 2013; Kappel & Grechenig, 2009). In their study conducted in Hervey Bay, Australia, Cole & Stewart (2013) found that early morning water consumption occurring between 12am and 5am usually includes consumption volumes less than 10 litres. This peak consumption can be associated with singular or multiple activities like hand washing, toilet flush or cooking (Cole & Stewart, 2013). However, when water consumption reaches 100 litres or more, the identification of such activities becomes harder (Cole & Stewart, 2013). This study found that peak consumption occurred between 8 and 9 am and 6 and 7pm (Cole & Stewart, 2013).

The morning peak, which occurs an hour later in summer months, usually includes the use of showers and toilets, followed by the use of clothes washers while the evening peak mainly involves the use of dishwasher (DeOreo et al., 2011; Gato-Trinidad et al., 2011). In the USA, indoor peak consumption occurs around 7am with a peak at 9am while evening consumption reaches an evening peak between 6 and 10pm (DeOreo et al., 2011; Mayer et al., 1999). Concerning outdoor use, it increases from 5am to 10am and reaches its evening peak between 6pm and 9pm (DeOreo et al., 2011; Mayer et al., 1999). Additionally, domestic water consumption during work days and consumption during weekends appear to be similar (Fontdecaba et al., 2013).

5 BENEFITS OF COLLECTING HOURLY DISAGGREGATED DATA

In the energy domain, hourly adaptive pricing is considered an efficient way of providing incentives to achieve sustainable energy usage (Doostizadeh & Ghamesi, 2012). Collecting hourly data can be useful when implementing pricing and non-pricing measures restricting water usage. Using an "hourly inclining block tariff" with a "peak hour charge" when a certain level of water consumption is reached on an hourly basis can reduce peak demand (Cole & Stewart, 2013). Collecting hourly data also informs about the changes occurring in water consumption habits and patterns and evaluates the success of pricing and non-pricing methods (Cole & Stewart, 2013).

Up-to date peak demand data is necessary to adapt the water supply networks to usage patterns (Beal & Stewart, 2013; Froehlich et al., 2011). Hourly patterns of water usage can inform water suppliers of the needs and habits of consumers and help them tailor their services to the demand (Cole & Stewart, 2013). Likewise, it allows the identification of behaviours that offer the most potential for conservation (Deoreo et al., 1996). That way, it helps improving the design of conservation measures and evaluating the effectiveness of such measures (Deoreo et al., 1996; Larson et al., 2012; Mead, 2008). Collecting disaggregated data can increase people's awareness of their water consumption by enabling eco-feedback and sustainability applications (Froehlich et al., 2011). Eco-feedback can encourage water conservation by making the link between an household's activities and the use of resources (Froehlich et al., 2012). Displaying hourly disaggregated data by appliances can also affect the social dynamics of a household by giving information about domestic habitudinal activities (Froehlich et al., 2012). In addition to just allowing the visualization of water consumption, eco-feedback reports household activities (Froehlich et al., 2012). Consumers can thus receive bills detailing their hourly water consumption during a certain period of time (Cole & Stewart, 2013).

6 THE WISDOM PROJECT: HOURLY PATTERNS OF WATER CONSUMPTION IN EUROPE

When it comes to identifying the factors affecting water consumption, there is a need for "country and location specific research" as cultural, environmental and behavioural conditions can vary depending on the country people live in (Rathnayaka et al., 2014). Contrary to European countries, the large amount of daily water consumption per capita in the USA and in Australia creates more feasible opportunities for water savings in those countries. When in Australian urban areas, residential water use includes a "significant outdoor component" amounting to 29% of households' overall water consumption, in Europe, outdoor water consumption represents only 3% of average household water usage (Benito et al., 2009; Cole & Stewart, 2013; Gato-Trinidad et al., 2011). Due to the averagely low consumption of water in European households, reducing water consumption in those countries appears challenging. In the United-Kingdom, daily consumption per capita amounts to only 142 litres (Energy Saving Trust, 2013). Reducing water consumption thus implies changing indoor consumption and water activities that are often considered as "essential" to individuals (Willis et al., 2011). Moreover, severe droughts happening in countries affected by water scarcity have encouraged consumer education, an increased usage of water saving devices and the implementation of restrictions to domestic water usage (Cole & Stewart, 2013). Individuals living in areas affected by drought are also more likely to be concerned about water conservation as personally experiencing water shortages can increase their willingness to change their behaviours in order to save water (Gilbertson et al., 2011). However, due to different climatic conditions in Europe, Europeans might not be as aware of the importance of water saving as the Australian population. Thus, there is a need to find ways to educate people about an issue that they might not feel concerned about at first in order to promote more sustainable water uses.

The WISDOM Project is conducted in Cardiff (UK) and in La Spezia (Italy) and aims at developing and testing an intelligent Information and Communication Technologies (ICTs) system that enables "just in time" monitoring of the water value chain from water abstraction to discharge, in order to optimize the management of water resources. The socio-technical dimension of the water value is a key aspect of the WISDOM project. To this end, the influences of water customers' usage characteristics and how they interact with the water network will be considered into the overall view of the water value chain. The WISDOM project intends to (a) increase user awareness and modify behaviours concerning the use of water, (b) achieve quantifiable reduction of water consumption, and (c) reduce peak-period of water distribution loads by using ICTs.

Prior to the deployment of this instrumentation, two online questionnaires have been sent within Cardiff.

The first questionnaire contained ten questions related to water usage and was included in a bigger survey i.e. Ask Cardiff Survey, that is conducted every year within the city. The 2000 responses collected gave us an overview of current household water usages, as reported by the participants. We found that the main reason motivating people to save water was to "help the environment (34.52%) followed by "reduce bills" (31.43%) and "reduce wastage"(28.80%). Participants were then asked if they agreed to take part in another survey for the WISDOM project. A more detailed questionnaire containing about 40 questions was then sent to the 800 people who had volunteered in order to learn more about their environmental knowledge and beliefs, their water consumption behaviours, their household's characteristics and the water-saving devices installed within their home. The 200 responses collected informed us about people's current perceptions of their water usage. A large majority of participants appeared to consider water conservation as an important issue (92.42%) and would use water-saving devices if they were provided for them (93.43%). However, only 39.39% would actually invest in these devices. Population in Cardiff can thus be considered as following expected trends when it comes to environmental issues. People indeed want to be known as saving water and as being eco-friendly persons, but our survey has shown that this does not always translate into action.

The selection of participative households then started with Welsh Water, the only water supplier in Cardiff, sending a letter to every metered house in the area of this city. People who were interested in taking part in the WISDOM study contacted Welsh Water and were visited for a water audit. After that, they could make their final decision to volunteer. The trial phase of this project involves about 50 selected households divided in three different groups with two different types of houses. Type 1 houses will only have smart meters installed whereas type 2 houses will have smart meters, electricity monitoring and in-home processing implemented. In order to evaluate the impact of the user oriented feedback system tested, a control sample of households will have smart meters installed within their properties without any further intervention. Group 1 will have smart meters installed and will be able to view their consumption on their mobile phone. However, this group will not have access to disaggregated data related to their water usage by appliances. Smart-meters and in-home displays will also be implemented within houses of participants from Group 2. These participants will be able to easily access information and receive feedback about their daily water consumption on a mobile application containing approximately the same content as the tablet. More importantly, this group of participants will receive disaggregated data and will be able to know when and where the water is used. Flow trace analysis will be used to collect this data at regular and frequent intervals. Disaggregated data will be provided through the interface of the in-home display and the applications and will take the form of a graph allowing participants to identify the most water-consuming appliances. By comparing Group 1 and Group 2, access to disaggregated data and its impact on people's consumption will be evaluated.

Following the study made by Beal & Stewart (2013), it appeared useful to collect accurate disaggregated data from both smart meters and a survey. Disaggregated data will thus be obtained using smart meters and by conducting a survey with the participants using a mobile application mid-phase through the trial. The aim is to obtain direct information about hourly patterns of water consumption and to identify water consuming activities and appliances. This information will also help determining the behavioural lifestyles that impact water usage both positively and negatively and the habits that need to be targeted to encourage water savings. Using disaggregated data will also allow initiatives and measures taken by authorities and water companies to be tailored to individuals' consumption and habits in order to efficiently reduce their consumption. Likewise, water suppliers will be able to adapt their services to their customers' needs and design efficient strategies to achieve better demand management (Harou et al., 2014). Collecting information about hourly domestic water usage will help redistributing people's daily consumption to reduce the peak demand on the water network. Consumers will be more aware of the times at which their household uses more water and of the appliances that consume the most allowing them to achieve larger water savings by targeting wasteful practices (Larson et al., 2012).

7 CONCLUSION

Growing water scarcity makes it necessary to adapt services and to encourage sustainable use of water resources in urban areas. Disaggregated data collected by smart meters can provide information about hourly patterns of domestic water consumption and is thus necessary to identify user behaviours that need to be targeted in order to promote conservation. While many studies using disaggregated data have been conducted in developed countries affected by water scarcity such as Australia and the USA, research in Europe is less developed. By using smart meters, in-home displays and surveys to collect and display disaggregated data, the European WISDOM Project will provide information about hourly domestic water usage in Europe. This will allow the choice of adequate strategies to promote water conservation and behavioural change while giving water companies the possibility to tailor their services to their customers' needs. Additionally, results from the WISDOM pilots will be used to identify how the project can be replicated throughout all countries and differing European climatic areas.

REFERENCES

Albiol, C., & Agulló, F. (2014). *La reducción del consumo de agua en España: causas y tendencias. Aquae Papers.*

Beal, C.D., & Stewart, R. a. (2013). Identifying Residential Water End Uses Underpinning Peak Day and Peak Hour Demand. *Journal of Water Resources Planning and Management*, 1–10. doi:10.1061/(ASCE)WR.1943–5452.0000357.

Benito, P., Mudgal, S., Dias, D., Jean-Baptiste, V., Kong, M.A., Inman, D., & Muro, M. (2009). *Water Efficiency Standards.*

Blanco, H.A., Williams, M.L. De, Ana C. Velezmoro, & Aguilar, V.H. (2014). Consumo De Agua En Actividades Domesticas. Caso de Estudio: Estudientes de La Asignatura Saneamiento Ambiental de La UCV. *Revista de La Faculta de Ingenierie U.V.C, 29*, 51–55.

Cahill, R., & Lund, J. (2011). Residential Water Conservation in Australia and California, (November), 1–14.

California Water Boards. (2015). *Urban Water Conservation Drops From 22 Percent to Near 9 Percent in January.*

California Water Boards. (2016). *January 2016 Statewide Conservation Data.*

Carragher, B.J., Stewart, R. a., & Beal, C.D. (2012). Quantifying the influence of residential water appliance efficiency on average day diurnal demand patterns at an end use level: A precursor to optimised water service infrastructure planning. *Resources, Conservation and Recycling, 62*, 81–90. doi:10.1016/j.resconrec.2012.02.008

Centre d'Information Sur l'Eau. (2011). *Utiliser—Le Bon Usage De L'Eau.*

City of Santa Fe. (2016). Water Conservation. Retrieved March 7, 2016, from http://www.santafenm.gov/water_conservation.

Cole, G., & Stewart, R.A. (2013). Smart meter enabled disaggregation of urban peak water demand: Precursor to effective urban water planning. *Urban Water Journal, 10*(3), 174–194. doi:10.1080/1573062X.2012.716446.

Deoreo, W.B., Heaney, J.P., & Mayer, P.W. (1996). Flow Trace Analysis to Assess Water Use. *Journal American Water Works Association*, (January), 79–90.

DeOreo, W.B., Mayer, P.W., Martien, L., Hayden, M., Funk, A., Kramer-, M., … Heberger, M. (2011). *California Single Family Water Use Efficiency Study. Aquacraft, Inc. Water Engineering and Management.* Retrieved from http://www.aquacraft.com/sites/default/files/pub/DeOreo-(2011)-California-Single-Family-Water-Use-Efficiency-Study.pdf

Doostizadeh, M., & Ghamesi, H. (2012). A Day-ahead Electricity Pricing Model Based on Smart Metering and Demand-side Management. *Energy, 46*(1), 221–230.

Dray, D.M., Samuelson, A., Zepf, M., & Kejriwal, A. (2008). *The Essentials of Investing in the Water Sector; version 2.0.*

Dziegielewski, B., Opitz, E., Kiefer, J., & Baumann, D. (1992). *Evaluation of Urban Water Conservation Programs: A Procedures Manual.*

Energy Saving Trust. (2013). At Home with Water, (At Home With Water).

Fontdecaba, S., Sánchez-Espigares, J. a., Marco-Almagro, L., Tort-Martorell, X., Cabrespina, F., & Zubelzu, J. (2013). An Approach to Disaggregating Total Household Water Consumption into Major End-Uses. *Water Resources Management, 27*(7), 2155–2177. doi:10.1007/s11269–013–0281–8.

Froehlich, J., Larson, E., Saba, E., Campbell, T., & Atlas, L. (2011). A Longitudinal Study of Pressure Sensing to Infer Real-World Water Usage Events in the Home. *Pervasive Computing, 6696*, 50–69. doi:10.1007/978–3-642–21726–5_4.

Froehlich, J., Patel, S., Landay, J. a., Findlater, L., Ostergren, M., Ramanathan, S., … Bai, M. (2012). The design and evaluation of prototype eco-feedback displays for fixture-level water usage data. *Proceedings of the 2012 ACM Annual Conference on Human Factors in Computing Systems—CHI '12*, 2367. doi:10.1145/2207676.2208397.

Gan, K., & Redhead, M. (2013). *Melbourne Residential Water Use Studies.* Retrieved from https://www.yvw.com.au/yvw/groups/public/documents/document/yvw1004065.pdf.

Gato-Trinidad, S., Jayasuriya, N., & Roberts, P. (2011). Understanding urban residential end uses of water. *Water Science & Technology, 64*(1), 36–42. doi:10.2166/wst.2011.436.

Gilbertson, M., Hurlimann, A., & Dolnicar, S. (2011). Does water context influence behaviour and attitudes to water conservation ? *Australasian Journal of Environmental Management, 18:1*(February), 47–60. doi:10.1080/14486563.2011.566160.

Gutierrez-escolar, A., Castillo-martinez, A., Gomez-pulido, J.M., & Garcia-lopez, E. (2014). A New System for Households in Spain to Evaluate and Reduce Their Water Consumption. *Water, 6*, 181–195. doi:10.3390/w6010181.

Harou, J.J., Garrone, P., Rizzoli, A.E., Maziotis, A., Castelletti, A., & Fraternali, P. (2014). Smart metering, water pricing and social media to stimulate residential water efficiency : opportunities for the SmartH2O project. In *16th Conference on Water Distribution System Analysis* (Vol. 00).

Kappel, K., & Grechenig, T. (2009). "Show-me": Water Consumption at a glance to promote Water Conservation in the Shower. In *Proceedings of the 4th International Conference on Persuasive Technology*.

Larson, E., Froehlich, J., Campbell, T., Haggerty, C., Atlas, L., & Fogarty, J. (2012). Disaggregated water sensing from a single, pressure-based sensor : An extended analysis of HydroSense using staged experiments. *Pervasive and Mobile Computing*, *8*(1), 82–102. doi:10.1016/j.pmcj.2010.08.008.

Lu, T., & Smout, I. (2008). Domestic Water Consumption : A field study in Harbin, China. In *33rd WEDC International Conference, Ghana* (pp. 1–4).

Lyons, K., Hightower, J., & Huang, E.M. (2011). Pervasive Computing. In *9th International Conference on Pervasive Computing* (p. 370). San Francisco, USA.

Mayer, P.W., Deoreo, W.B., Opitz, E.M., Kiefer, J.C., Davis, W.Y., Dziegielewski, B., & Nelson, J.O. (1999). Residential End Uses of Water. *Awwarf*, 310. Retrieved from http://books.google.com/books?hl=en&lr=&id=cHKl-eV-Q5MC&pgis=1.

Mead, N. (2008). *Investigation of Domestic Water End Use*. University of Southern Queensland.

Ministere de l'Environnement de l'Energie et de la Mer. (2011). Eau Potable: La Consommation. Retrieved March 4, 2016, from http://www.statistiques.developpement-durable.gouv.fr/lessentiel/ar/306/305/eau-potable-consommation.html.

Organisation for Economic Co-Operation and Development (OECD). (2015). *Water and Cities: Ensuring Sustainable Futures*. IWA Publishing.

Rathnayaka, K., Maheepala, S., Nawarathna, B., George, B., Malano, H., Arora, M., & Roberts, P. (2014). Factors affecting the variability of household water use in Melbourne, Australia. *Resources, Conservation and Recycling*, *92*, 85–94. doi:10.1016/j.resconrec.2014.08.012.

Romano, G., Salvati, N., & Guerrini, A. (2014). Estimating the Determinants of Residential Water Demand in Italy. *Water*, *6*, 2929–2945. doi:10.3390/w6102929.

Stewart, R. a., Willis, R., Giurco, D., Panuwatwanich, K., & Capati, G. (2010). Web-based knowledge management system: linking smart metering to the future of urban water planning. *Australian Planner*, *47*(2), 66–74. doi:10.1080/07293681003767769.

Tortajada, C., & Castelan, E. (2003). Water Management for a Megacity : Mexico City Metropolitan Area. *Ambio - A Journal of the Human Environment*, *32*(2), 124–129. doi:10.1639/0044-7447(2003)032.

UNESCO. (2014). *The United Nations World Water Development Report – N° 5 – 2014: Water and Energy – vol. 1; Facing the challenges—vol. 2*.

United Nations Development Program. (2006). *Human Development Report 2006 Beyong Scarcity: Power, poverty and the global water crises. Journal of Government Information* (Vol. 28). doi:10.1016/S1352-0237(02)00387-8.

Wilkes, C.R., Mason, A.D., & Hern, S.C. (2005). Probability Distributions for Showering and Bathing Water-Use Behavior for Various U. S. Subpopulations. *Risk Analysis*, *25*(2), 317–337.

Willis, R.M., Stewart, R. a., Panuwatwanich, K., Jones, S., & Kyriakides, A. (2010). Alarming visual display monitors affecting shower end use water and energy conservation in Australian residential households. *Resources, Conservation and Recycling*, *54*(12), 1117–1127. doi:10.1016/j.resconrec.2010.03.004.

Willis, R.M., Stewart, R.A., Panuwatwanich, K., Williams, P.R., & Hollingsworth, A.L. (2011). Quantifying the influence of environmental and water conservation attitudes on household end use water consumption. *Journal of Environmental Management*, *92*(8), 1996–2009. doi:10.1016/j.jenvman.2011.03.023.

Worthington, A.C., & Hoffman, M. (2008). An Empirical Survey of Residential Water Demand Modelling. *Journal of Economic Surveys*, *22*(5), 842–871.

WWF. (2011). *Big Cities. Big Water. Big Challenges. Water in an Urbanizing World*.

Zhang, Q., Kobayashi, Y., Alipalo, M.H., & Zheng, Y. (2012). *Drying Up—What to do about droughts in the People's Republic of China With a Case Study from Guiyang Municipality, Guizhou Province*.

eWork and eBusiness in Architecture, Engineering and Construction – Christodoulou & Scherer (Eds)

Waterloss detection in streaming water meter data using wavelet change-point anomaly detection

S.E. Christodoulou, E. Kourti, A. Agathokleous & C. Christodoulou
Nireas International Water Research Center, Department of Civil and Environmental Engineering, University of Cyprus, Nicosia, Cyprus

ABSTRACT: Described herein is a method for the detection of pipe breaks and water losses in urban water distribution networks, by use of a wavelet change-point anomaly detection algorithm and streaming water consumption data from an urban locale. The wavelet change-point method utilizes the Continuous Wavelet Transform (CWT) of signals to analyze how the frequency content of a signal changes over time, and wavelet coherence to reveal time-varying frequency content common in multiple signals. The method also utilizes streaming water consumption data from consumers ('automatic meter reading' devices, AMR) and from District Meter Areas (DMA), to acquire inherent knowledge of water consumption at normal conditions at house and area-wide levels, and to make inferences about water consumption under abnormal conditions. This temporal anomaly detection is then georeferenced and used for spatial anomaly detection, producing 'heatmap' representations of the areas in the city with high probability of waterloss incidents.

1 INTRODUCTION

1.1 *Introduction*

Water Distribution Networks (WDNs) are degrading worldwide with damages manifesting as pipe-failure incidents and leading to significant levels of non-revenue water (typically in the range 20%–30%). While data on pipe-failures can, post-event, be analyzed and provide useful knowledge on the causalities of the failures, records of these incidents are usually hard to produce, are inconsistent and poorly recorded, and most often only span over short periods of time. Of outmost importance is the need for timely detection of any anomalies in the WDN operations and in network behavior, and the correlation of such anomalies to waterloss.

The paper proposes an efficient solution for the timely identification of anomalies in WDN by use of a wavelet anomaly detection algorithm and streaming water consumption data from consumers and from District Meter Areas (DMA), in the form of 'Automatic Meter Reading' devices (AMR). The wavelet anomaly detection approach is then coupled with spatio-temporal analysis to arrive at spatial Decision Support Systems (DSS) that automate the process of waterloss detection, reduce water loss and increase the efficiency in the management WDN, and eventually positively contribute to the sustainability of such networks.

1.2 *System architecture*

The system developed is an integrated near-real-time remote monitoring system of water distribution networks which provides real time performance monitoring (Figure 1).

The proposed methodology (originally reported in Gagatsis et al. 2012) is an integrated monitoring solution in the form of application software integrating a relational database system, Automatic Metered Reading (AMR), Geographic Information Systems technology (GIS) and spatio-temporal clustering techniques. AMR readings are stored in frequent intervals, and used to identify leakages in house connections and district metered areas in real time.

It should be noted, though, that the architecture shown in Figure 1 relies heavily on two key components: (1) AMR, for real-time consumption readings and (2) leakage detection algorithms.

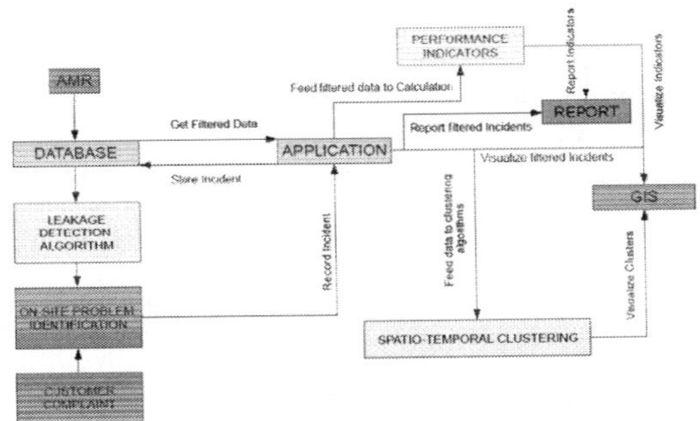

Figure 1. Schematic of the pipe-failure and water leak monitoring and visualization solution proposed (Gagatsis et al. 2012).

2 STATE OF KNOWLEDGE ON WATERLOSS DETECTION

Anomalies in WDN operations usually manifest themselves in the form of pipe-failure incidents (pipe breaks) and in the form of water loss. While pipe-failures are not a perfect indicator of WDNs, they are amongst the primary symptoms of network degradation and one of the few sources of information on the condition of WDNs. However, recording pipe-failure incidents is a relatively recent practice. Furthermore, knowledge of the type of pipes and their condition over an area is often unknown prior to a pipe-failure incident which requires repair.

Past efforts had led to the identification of several risk factors influencing the fragility of WDNs and devising decision support systems for sustainably managing them. For example, Christodoulou & Deligianni (2009) used data mining techniques and neurofuzzy systems in order to identify pipe-failure data clusters and convert these into decision support rules for waterloss management. Subsequently, Christodoulou & Agathokleous (2012) studied the effects of intermittent water supply on the vulnerability of WDNs, reporting that intermittent water supply policies resulted in increased stresses in the network they were applied to that eventually lead to an abnormally high number of pipe-failure incidents and higher volumes of water losses. The spatial distribution of pipe-failure incidents was studied by Gagatsis et al. (2012) who, first used spatial analysis techniques and a spatial clustering method to study the extend of pipe-failure clustering in the Limassol WDN and, then, transformed the deduced knowledge into a decision support system.

To date, pipe-failure incident records are usually the result of customer complaints (e.g. low pressure, no water, high bills etc) which are then investigated by Water Board technicians. However, all of these sources of information often require a certain amount of time before the source of a leak is discovered thus resulting in significant water losses until the damage is repaired. In order to resolve this problem, the use of Automatic Meter Reading (AMR) is being advocated and is currently implemented widely.

Automatic meter readers are devices embedded in house connection meters which can provide readings in frequent intervals. These readings are transferred to a monitoring server. By examining unusual rises or drops in water consumption, water board technicians can be made aware of potential leaks and investigate them, thus resulting in a faster, more efficient leak detection method. Furthermore, the use of AMR can provide near real-time information on the water consumption within a District Metered Area (DMA) and compare it with the total water input in that DMA. A DMA is an area of the WDN where water originates and is distributed to the network from a single point of entry and where the network's water pressure is approximately uniform. Large differences between periodic readings indicate large leaks within the network.

In principle, the use of AMR devices permits Water Boards to detect water leaks before they are reported by customers. This is particularly useful in cases when customers are away on vacation or in cases when the leak is not detectable by the customers. The AMR devices send consumption information in frequent intervals to the central database. This data is then analyzed for detecting abnormal increases in water consumption readings sent from the AMR units. When such an abnormal increase is detected an alert is sent to the WDN operators of possible WDN malfunction, and then the WDN operators are tasked with linking the event with a possible water leakage.

This process is in many aspects subjective (in terms of how the anomaly thresholds are set) and manual (thus time consuming) with regards to the analysis of indicators (outliers) of water consumption anomalies. Further, the architecture shown in Figure 1 relies heavily on the wide use of AMR devices in the WDN (one for each house connection) and thus it is not cost-efficient.

A solution could be the selective location of AMR devices (or sensors) within the WDN, for which several sensor placement optimization techniques exist (Ostfeld et al. 2008, Christodoulou et al. 2013, Christodoulou 2015). The problem, though, of analyzing the sensed operational data of a WDN remains, in terms of processing time-related data and sifting through this data for possible anomalies in the WDN's behavior.

3 WAVELET ANOMALY DETECTION

3.1 *Wavelets*

A wavelet is a wave-like oscillation with amplitude beginning at zero, increasing, and then decreasing back to zero. As a mathematical tool, wavelets can be used to extract information from many different kinds of data, and can be used for detecting anomalies in such data. Thus, they are particularly useful in analyzing time-series of patterned data (such as the subject water flow data from a WDN).

In data mining, anomaly detection (or outlier detection) is the identification of items, events or observations which do not conform to an expected pattern or other items in a dataset, and typically the anomalous items relate to some form of a problem in the system examined.

In the context of water loss in WDN, detected anomalies may often be: (1) either unexpected drops in water consumption, indicating normal or abnormal (e.g. water theft) usage, or (2) bursts in activity (e.g. increased water flow) that they may relate to leakage events in the network. This pattern does not adhere to the common statistical definition of an outlier as a rare object, and many outlier detection methods (in particular unsupervised methods) will fail on such data, unless it has been aggregated appropriately.

A practical step-by-step guide to wavelet analysis is given, with examples taken from time series of the El Niño–Southern Oscillation (ENSO), in the work by Torrence & Compo (2010). The guide includes a comparison to the windowed Fourier transform, the choice of an appropriate wavelet basis function, edge effects due to finite-length time series, and statistical significance tests for wavelet power spectra.

With regards to the application of wavelets to water distribution networks and leakage detection notable is the work by Srirangarajan et al. (2013), in which a technique was reported for detecting and locating transient pipe burst events in water distribution systems. Their method uses multiscale wavelet analysis of recorded high rate pressure data to detect transient events, while the used wavelet coefficients and Lipschitz exponents provide additional information about the nature of the signal feature detected and can be used for feature classification. Further, a local search method is proposed to estimate the arrival time of the pressure transient associated with a pipe burst event, and a graph-based localization algorithm is proposed to determine the actual location (or source) of the pipe burst.

3.2 *Case study: Anomaly detection*

Let us first examine the efficiency of the wavelet change-point detection algorithm using a time series of duration 7,000 hours (January-October), depicting the water consumption of a 3-person household following the consumer profile of Figure 2. The average total daily water consumption of the household is presumed to be 419 L (based on records from a local Water Board), distributed hourly across the day as per Figure 2.

The hourly consumption profile of Figure 2 considers both the indoor and outdoor water consumption, and is typical of a 3-person household. Water consumption starts in the early morning hours (around 06:00), increases peaking up at around 10:00, then drops until the early afternoon hours (16:00), peaking up again in the late afternoon and early evening hours (18:00–21:00), before dying down at night (21:00–06:00). It should be also noted that the household's daily water consumption does not, at any point in time, zero out. This is in agreement with the 'Minimum Night Flow (MNF)' concept, commonly used in WDN operations. MNF is a common method used to evaluate water loss in a water network, and refers to the water volume flowing through the network even when all true water demand (by consumers) is zero (typically in the time band of 02:00–04:30).

Further to the profiled daily water consumption, a week-long drop in the household's consumption during July is recorded ($t \in [5050, 5100]$ hours).

The water consumption's time series (Figure 3) is first processed macroscopically (the signal duration is set to 7000 hours) to identify the time periods of concern, and then microscopically to zoom in on possible consumption anomalies. The wavelet parameters used in the macroscopic analysis are $\alpha = 0.0$, $n = 50$ and $k = 10$. As aforementioned, for $\alpha = 0.0$ the divergence is not bounded, and the model even though is less accurate it is good for visualization purposes (thus for macroscopic analysis). The value of parameter n affects the resolution (time interval) of the analysis and the value of k affects the immediacy (time shift) of the anomaly detection.

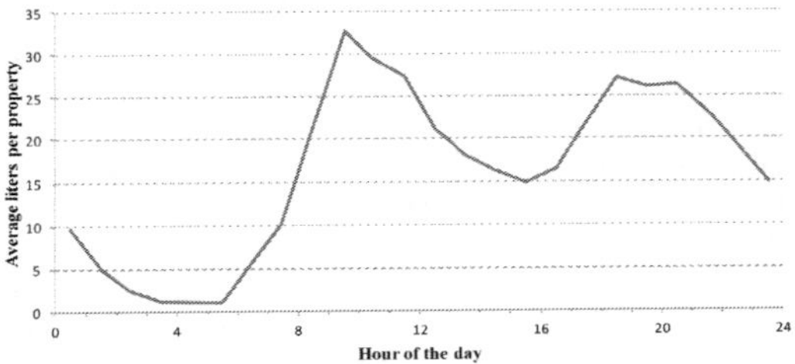

Figure 2. Hourly water consumption profile.

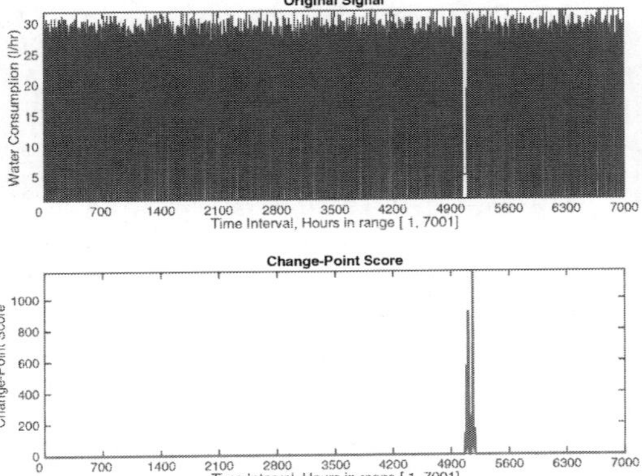

Figure 3. Water consumption signal and change-point score (Scenario 1, Hours 1–7000).

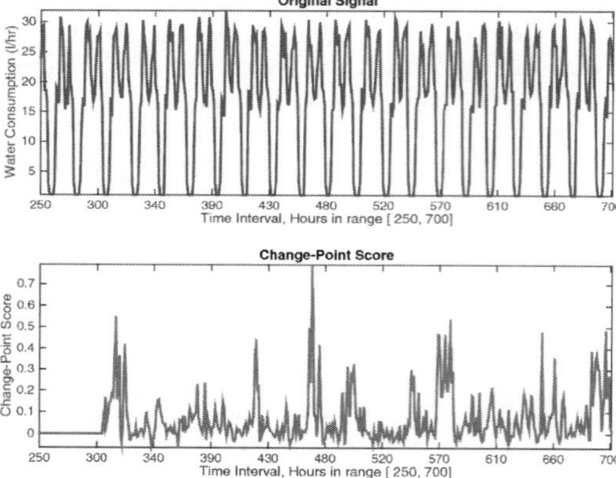

Figure 4. Water consumption signal and change-point score (Scenario 1, Hours 250–700).

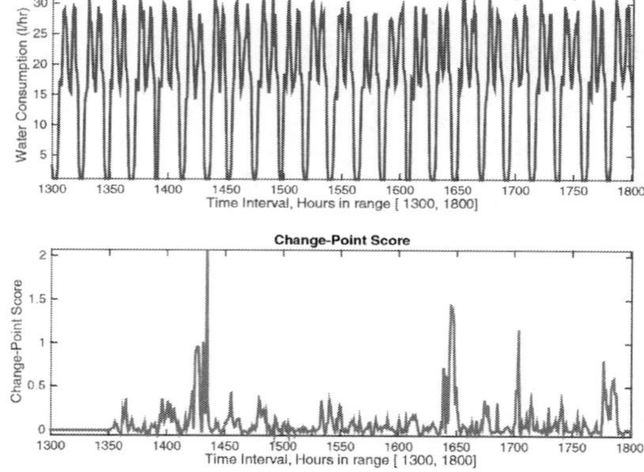

Figure 5. Water consumption signal and change-point score (Scenario 1, Hours 1300–1800).

The results of the macroscopic analysis are depicted in Figure 3. The water consumption signal oscillates over time, with a gap shown in the region of $t \in [5050, 5100]$, as expected. This gap is picked up by the wavelet change-point detection algorithm, as an anomaly in the water consumption, with an anomaly score that seems to outweigh any other anomalies in the time-series.

In order to get a closer look at the underlying time-series, a microscopic analysis is performed at other time intervals of concern. The wavelet parameters used in the microscopic analysis are now $\alpha = 0.0$, $n = 25$ and $k = 1$, striving for a finer resolution and a smaller time shift in the analysis and prediction. The oscillating signal patterns in the underlying time series are now clearer, and once the region in analysis excludes the region of high anomaly (i.e. the summer period of $t \in [5050, 5100]$), the anomaly scores of the other regions are not dominated by the region of high anomaly score.

For example, a review of time periods $t \in [250, 700]$ and $t \in [1300, 1800]$ indicates varying small-scale anomalies in the observed water consumption (Figures 4 and 5) but with smaller anomaly scores compared to the anomaly stemming from the reduction in consumption during the summer vacation period ($t \in [5050, 5100]$, Figure 3).

These anomalies are hard to detect in the original time-series (consumption signal), and easier to see in the wavelet change-point detection score chart.

3.3 *Case study: Waterloss detection*

We now turn our attention to the suitability of the wavelet change-point detection algorithm for waterloss detection, re-examining the case-study time series assuming a water leak is presented at $t \in [300, 500]$ hours. The water leak is presumed to be 2 L/hr for the first 100 hours and then 5 L/hr for the next 100 hours. As in the first case-study, the water consumption signal is first analyzed macroscopically and then microscopically, using the sets of wavelet parameters indicated before. Further, in order to avoid the dominating influence of the week-long water consumption drop during the time interval of $t \in [5050, 5100]$, the analysis only considers the first 5000 hours of the water consumption signal.

The macroscopic analysis (Figure 6) shows the increase in water consumption during the water leak period ($t \in [200, 400]$) and identifies this as an anomaly in the general water consumption signal. Further, it scores this anomaly high compared to other signal anomalies, indicating the high probability that the detected deviation in water consumption is highly unusual. As aforementioned, the time shift in detection shown in the anomaly score chart (Figure 6) stems from the wavelet parameters used in the analysis.

A microscopic analysis (Figure 7) confirms the findings of the macroscopic review (Figure 6). A spike in the change-point score is first recorded at approximately $t = 400$ hours, followed by a bigger spike at approximately $t = 520$ hours, highlighting the anomaly in the water consumption pattern. The spikes in the anomaly score coincide with the two phases of recorded water leak (2 L/hr in range $t \in [300, 400]$ and 5 L/hr in range $t \in [400, 500]$). As with the previous case, the time shift in the anomaly score spikes are due to the wavelet parameters used in the analysis.

As with the previous case study, the anomaly score for the waterloss period ($t \in [300, 500]$ hours, Figure 7) outweighs anomaly scores in other periods of the water consumption time series (such as the ones shown in Figure 8), indicating that the detected anomaly is not attributed to 'signal noise' (i.e. small variations in water consumption patterns).

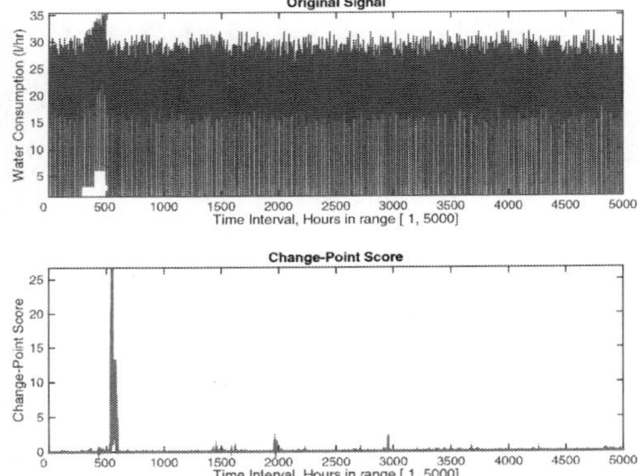

Figure 6. Water consumption signal and change-point score (Scenario 2, Hours 1–5000).

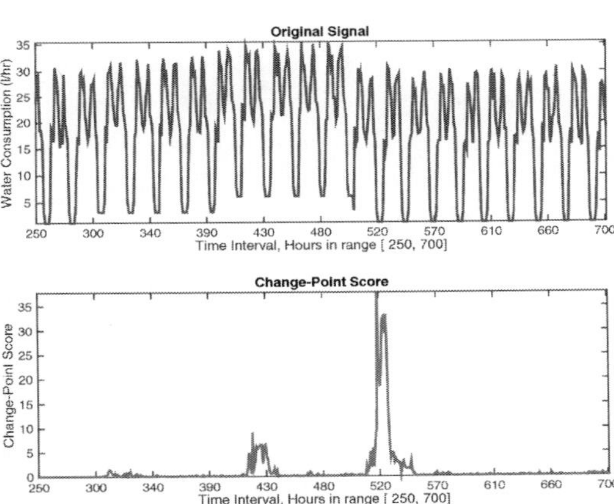

Figure 7. Water consumption signal and change-point score (Scenario 2, Hours 250–700).

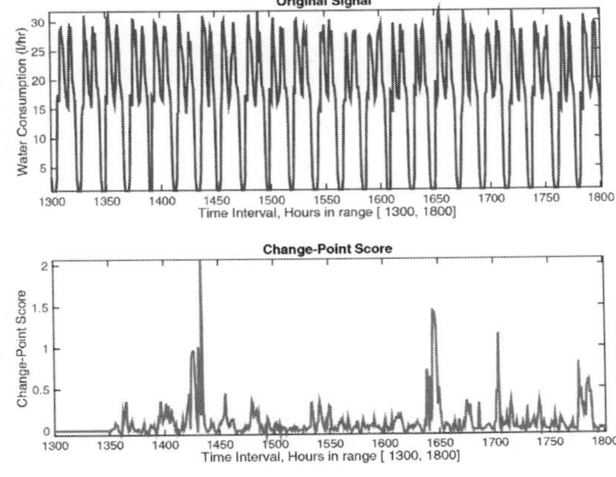

Figure 8. Water consumption signal and change-point score (Scenario 2, Hours 1300–1800).

4 CONCLUSIONS

The work presented herein evaluates the utilization of a wavelet change-point detection algorithm for the detection of anomalies in water consumption time series, and its applicability to waterloss detection in water distribution networks. The wavelet change-point detection method was implemented on a hourly water consumption signal, of 7000-hours duration, successfully detecting unusual reductions and increases in the water consumption patterns, and classifying them as anomalies. The first water consumption anomaly type was related to a discontinuity in the signal (a break in the consumer's water consumption patterns), whereas the second type related to an unusual increase in the signal (a waterloss incident).

Unlike traditional approaches that monitor the average current periodic consumption and compare this average to the one from a corresponding past period (e.g. this month's average vs. last month's average, or the average from the same month a year ago), the wavelet change-point detection approach is dynamic and sensitive to any change in the time series in study, Further, the method dynamically assigns anomaly scores to the detected changes in the signal, thus easing waterloss detection.

Ongoing research work on wavelet change-point detection will address spatial mapping and heat-map generation for the identification of areas of concern in a distribution network, as well as the blending of Automatic Meter Reading (AMR) technologies and water auditing at a District Meter Area (DMA) level, for the dynamic (and near-real-time) detection of water leaks in an urban water distribution network.

REFERENCES

Christodoulou, S. 2015. Smarting up water distribution networks with an entropy-based optimal sensor placement Strategy. *Journal of Smart Cities* 1(1): 47–58.

Christodoulou, S. & Agathokleous, A. 2012. A study on the effects of intermittent water supply on the vulnerability of urban water distribution networks. *Journal of Water Science and Technology* 12(4): 523–530.

Christodoulou, S. & Deligianni, A. 2009. A neurofuzzy decision framework for the management of water distribution networks. *Water Resources and Management* 24(1): 139–156.

Christodoulou, S., Gagatsis, A., Xanthos, S., Kranioti, S., Agathokleous, A. & Fragiadakis, M. 2013. Entropy-based sensor placement optimization for waterloss detection in water distribution networks. *Water Resources Management* 27(13): 4443–4468.

Gagatsis, A., Kranioti, S., Christodoulou, S., Agathokleous, A. & Xanthos, S. 2012. An integrated software solution for identifying, monitoring and visualizing water leak incidents in Water Distribution Networks.

Proc., IWA's intern. conf. on new developments in IT & water, Amsterdam, Netherlands, 4–6 November 2012.

Ostfeld, A., Uber, J.G., Salomons, E., Berry, J.W., Hart, W.E., Phillips, C.A., Watson, J-P., Dorini, G., Jonkergouw, P., Kapelan, Z, di Pierro, F., Khu, S-T., Savic, D., Eliades, D., Polycarpou, M., Ghimire, S.R., Barkdoll, B.D., Gueli, R., Huang, J.J., McBean, E.A., James, W., Krause, A., Leskovec, J., Isovitsch, S., Xu, J., Guestrin, C., VanBriesen, J., Small, M., Fischbeck, P., Preis, A., Propato, M., Piller, O., Trachtman, G.B.,Wu & Z.Y.,Walski, T. 2008, The battle of the water sensor networks (BWSN): A design challenge for engineers and algorithms. *Water Resources Planning and Management* 134: 556–569.

Srirangarajan, S., Allen, M., Preis, A., Iqbal, M., Lim, H.B., & Whittle, A.J. 2013. Wavelet-based burst event detection and localization in water distribution systems. Journal of Signal Processing Systems 72(1): 1–16.

Torrence, C. & Compo, G.P., 1998. A practical guide to wavelet analysis. *Bulletin of the American Meteorological Society* 79(1): 61–78.

Construction/risk management, regulatory and legal aspects

eWork and eBusiness in Architecture, Engineering and Construction – Christodoulou & Scherer (Eds)
© 2016 Taylor & Francis Group, ISBN 978-1-138-03280-4

Managing constructability on a construction stage: BIM methods

M. Tauriainen & J. Helminen
Lujatalo Oy, Espoo, Finland
Aalto University, Espoo, Finland

J. Puttonen
Aalto University, Espoo, Finland

ABSTRACT: Constructability has emerged as an important tool for improving construction performance, productivity and quality. Recently, it has been suggested that constructability can be promoted through the use of Building Information Modeling (BIM). This research focused on BIM based methods used on the building site in order to improve constructability at the construction stage. An empirical study was carried out by using three different research methods; interviews, action and literature research. The research group helped personnel in charge of construction works solving daily issues and to search optimum solutions for constructability with BIM-based methods. Solutions were tested with the personnel. Constructability development was winded around 4D schedule including visualization, 3D site layout and modeling of temporary support and structures. With BIM-based 4D schedule the constructability of structures could be analyzed, and it was easier to control and instruct workers, plan tasks beforehand, make procurements and plan an assembly sequence than using conventional construction planning methods.

1 INTRODUCTION

1.1 *BIM at the construction stage*

On construction sites, the use of building information models is still restricted. The traditional way of planning and manage construction works and the lack of suitable modeling software for contractors, the lack of the knowledge to use modeling software, compatibility issues between software and investment costs in software are mentioned as a hindrance to the adaptation of BIM on sites (Mäki & Kerosuo 2015; Eastmann et al. 2011). The use of BIM requires changes in the accustomed way of action. Furthermore, few studies concentrate on the use of BIM on site. (Mäki & Kerosuo 2015).

In Finland, a few site engineers and project managers are able to use and manage BIM-based working at the construction stage. The use of BIM depends on individual tasks, roles and responsibilities and unfortunately often clear instructions about the use of BIM are missing and so site managers are free to manage use cases into to suit best their own tasks. (Mäki & Kerosuo 2015). The BIM in leading companies is under a limited use in BIM-based construction planning and the visualization in 3D or 4D as 3D site layout planning, 4D scheduling, static 3D visualization, dynamic 4D simulation of planned construction workflow and site status on a specific date and automated BIM-based safety planning. Clash detection using the combined model and the visual examination of BIM is widely used. (Sulankivi et al. 2014).

The site layout is an important part of the planning of construction. The site layout promotes factors significant for constructability as the use of site area, tidiness and the safety on site (Sulankivi et al. 2009). The site layout can also be used to reduce the cost of building material handling on site and to improve productivity (Srinath et al. 2015). BIM-based site layout illustrates risk areas, distances and space dimensions. The range of cranes or danger zones of lifting and driving ways reserved for emergency vehicles can be visualized for example. (COBIM 2012). The site layout changed during the construction stage, because the space needed for temporary material storages and building activities changes. So the site layout should be dynamic and project specific. The planning of layout based on the idea of the type and number of temporary spaces needed, space requirements and time period needed on site. (Srinath et al. 2015).

Recently, the use of 4D schedule was researched by Zhou et al. (2014). In the building process, the use of 4D schedule makes the efficient detection of design errors possible. 4D schedule shows the building process from a wider perspective and makes the simulation of the order of activities possible at a very exact level. The study showed that with the help of the simulation of 4D schedule project personnel got the better understanding of the dependency between structures, chosen tools and building methods. The use of 4D schedule helped making decisions, site planning and scheduling resources.

1.2 *Constructability*

Since the 1970s, research concerning buildability and constructability evolved from the studies focusing on methods increasing cost efficiency and quality in the construction industry. Constructability aims to optimize the use of construction knowledge and experience in project planning, design, engineering, procurement and construction to achieve overall objectives. Accordingly, constructability is an approach that links the design and construction processes. (Tauriainen 2015).

BIM simulates the construction project in a virtual environment. This allows contractors and designer to make adjustments to constructability in BIMs, and to practice construction before actualizing it on a building site. BIM provides the opportunity to develop more systematic, detailed and concrete methods for evaluating and improving constructability. However, only a limited number of researches have attempted to examine constructability using BIM as an intensifying technology. (Tauriainen et al. 2014).

2 RESEARCH OBJECTIVES AND METHODOLOGY

2.1 *Objectives*

The objective of this research was to clarify how constructability could be improved at the construction stage using BIM as an intensifying technology. As a working hypothesis suitable BIM based solutions could be 3D site layout, 4D schedule and the visualization of assembly schedules.

2.2 *Methodology*

An empirical study in three cases was carried out by using three different research methods; interviews, action- and literature research. The basic information of cases is presented in Table 1. The progress of the study is shown in Figure 1.

The first author worked as a BIM specialist in the back office of a construction company and the second author worked as a BIM developer and a facilitator on the construction site of a school building project making his master's thesis at the same time. The interviews were executed as semi-structured interviews centralizing in two main themes; constructability and BIM. In the interviews, several constructability issues were identified.

In case 3, the action research part was executed both in the back office and on site. The researchers helped personnel in charge of construction works solving daily issues and to search optimum solutions for constructability with BIM-based methods.

Table 1. The basic information of the case projects.

Project	Case 1	Case 2	Case 3
Type of research	Interview	Interview	Action research
Number of interviews	3	1	3
Construction type	Renovation building	New building	New building
Type of project	Management contract	Contract of general charge and charge of planning	Divided contract of general charge
Available models	Structural, architectural and HVAC	Structural and HVAC	Structural, architectural and HVAC

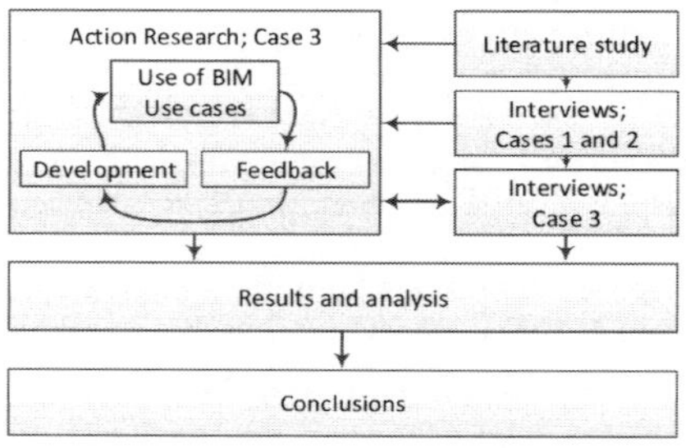

Figure 1. The progress of the study.

Constructability issues were visualized and simulated using BIM with the site personnel. In the discussions, the personnel gave important feedback. The feedback was collected both in informal and formal interviews. Use cases of BIM were revisited and new solutions were developed by the researchers. Afterwards the suitability of BIM solutions was again tested with the personnel.

Results and conclusions were simultaneously led next to an empirical study. The utility of the study was confirmed and verified by the relevant literature during the study.

3 RESULTS AND ANALYSIS

3.1 *Constructability issues, the useful use of BIM and means to intensify the use of BIM in the construction stage*

In the interviews and during the action research, several constructability issues were perceived. The perception of issues could be wrong so a root cause analysis was performed by the researchers and 26 root causes were traced in total. Root causes are presented in Table 2. Root causes were classified to the constructability factor areas and 20 from 26 causes were identified mainly (13) or partially (7) to the factors of design solutions and BIMs, drawings and specifications. As the factors of building economy and building process, 6 (5 mainly 1 partially) causes were identified; and 7 (1 mainly 6 partially) causes were identified to belonging the factors of assemblies, construction works and health and safety.

In the interviews, as the useful use of BIM in the construction stage was mentioned:

- 4D schedule,
- 3D site layout,

Table 2. Identified problems in the case projects divided in the categories of constructability factors and BIM methods.

Categorizations / Root causes of the identified issues	Constructability factor areas								In Cases	BIM methods											
	I		II		III					i			ii			iii					
	Building economy	Building process	Design solutions	BIMs, drawings and specifications	Standardisation and prefabrication	Assemblies	Construction works	Health and safety		Design control	4D schedule	3D site layout	Modeling of temporary structures	Visualization	Information search	Revision managament	Model updating	Clash detection	Validity of models	Accuracy of models
1. Defining and classifying acquisitions time schedule.	x	x							3		x									
2. Delayed HVAC design changes.		x							1	x										
3. Seldom update of BIMs.		x							1								x			
4. Defective control of the progress of construction works.		x							3		x									
5. Communication issues with disturbed stakeholders.		x							3		x									
6. Difficulties in precast structure design.		x	x						2	x										
7. Low floor height.			x						1					x						
8. Defective preliminary study of existing structures.			x						1										x	
9. Defective measurement of height levels of existing structures.			x						1										x	
10. Inadequate design of temporary MEP systems.			x	x					3					x						
11. Imperfect serviceability design of MEP systems.			x	x					2					x						
12. Exceptional assembly schedule of hollow core slabs.			x			x			2		x									
13. Mixed (precast, cast in place and steel) frame structures.			x			x	x		3		x									
14. Inaequate site layout design.			x				x		3			x								
15. The lack of information of needed site arrengements.			x				x		3			x								
16. Challenging costruction works of cast in place structures and temporary supports.			x				x		3		x		x							
17. Difficult soil conditions.			x				x		3			x								
18. Depencies of structures are difficult to detect from 2D-drawings.				x					1,2,3					x						
19. Large number of the drawings.				x					1,2,3						x					
20. Inadequate integration of BIMs.				x					1,2									x		
21. Inaccuracy of details and MEP systems in BIMs.				x					1,2									x		x
22. Lack of architectural BIM.				x					2	x									x	
23. Design of complementary and support structures.				x					2				x							
24. Inaccuracy of electrical model.				x					2									x		x
25. Comparison of new and old drawings is complicated and laborious.				x					3					x				x		
26. Inadequate planning of fall protection.								x	3			x	x							

- visualization of structures that are difficult to build,
- the planning of health and safety on site,
- the order of prefabricated elements using BIM as a source of logistic information,
- mobile devices on site.

Means to intensify the use of BIM in the construction stage were proposed:

- practical training and education to use BIM on site,
- user-oriented BIM software on site,
- fast local area network and Internet connections to and on site,
- proper IT support on site,
- the unrestricted availability of BIMs for all contractors on site.

3.2 *3D site layout*

In the case 3, a 3D site layout was modeled (see Figure 2c) at the beginning of the construction stage under the control of general superintendent of the building. The timing of modeling was late and as site manager later confirmed the modeling of 3D site layout should be scheduled from the cost estimation and tendering stage to the preliminary construction stage. In spite of the late timing of modeling the 3D site layout was considered as a useful tool in the planning and visualizing the site arrangements, site barracks, logistic areas and access roads. The 3D site layout was modeled using SketchUp Pro. The model was saved in ifc-file format and integrated with architectural, structural and building services BIMs in Solibri Model Checker, so the 3D site layout was easy to visualize for subcontractors, foremen and construction workers.

3.3 *Visualizing the reach range of a rotary tower crane*

The reach range of a rotary tower crane was visualized using structural BIM (Figure 2b). After the preliminary selection of the tower crane the maximum lifting capacities to different reach range were clarified using capacity curves. The lifting

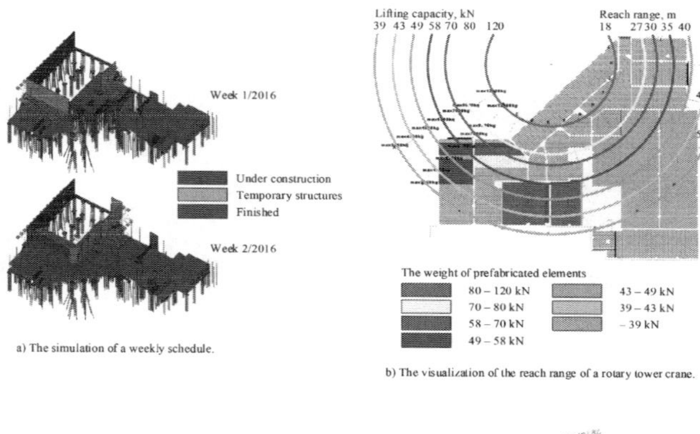

a) The simulation of a weekly schedule.

b) The visualization of the reach range of a rotary tower crane.

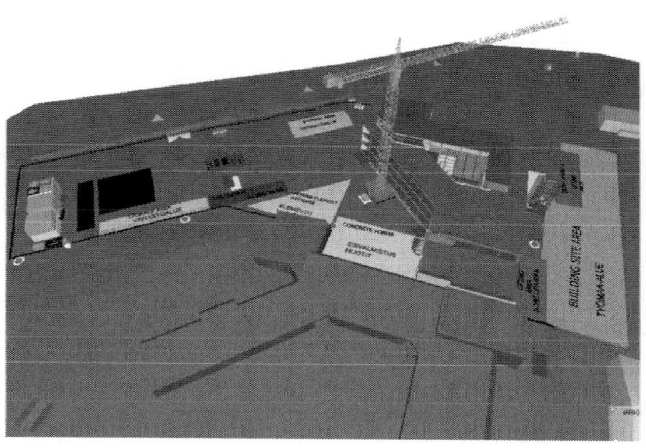

c) The visualization of site layout.

Figure 2. The simulation of weekly schedule (a), the visualization of reach range of a rotary tower crane (b) and site layout BIM (c).

capacities were visualized as cylinders in BIM. In the BIM, prefabricated elements were categorized and visualized in different colors easily in groups according to the weights of elements. After that, the visual inspection of elements in relation to each reach ranges revealed the elements outside the lifting capacity of the selected rotary tower crane.

The examination of the lifting capacity of the crane made a quick and efficient observation possible from what the elements can be raised by crane and what not. The examinations of the lifting capacity could be used to choose as optimal size and location as possible and to estimate the need of other lifters. This facilitated the acquisition and planning of the lifting equipment needed and the optimum cost of lifting could be analyzed. Site manager kept important that the examinations of the lifting capacity of the crane should already be examined at the cost estimation and tendering stage. He also hoped that the capacity examination of the crane would be made automatic with the BIM so that, those elements which the crane is not able to raise would become of different colors automatically.

3.4 *The 4D schedule of structural frame structures*

In the case 3, a structural frame system comprised a mixed structural system of steel frames, prefabricated and in-situ concrete structures. The constructability of frame structures was assessed by site manager and general superintendent to be difficult and the visualization of 4D schedule was assumed to reveal several constructability issues. Superintendent with site engineer planned the general schedule using scheduling software typically in the Finnish construction company. Durations of activities were calculated based on the resource standards of construction works. The researchers prepared the first version a 4D schedule and visualizations based on the general schedule of footing and frame works at the construction stage before the construction of footings and structural frame system began.

Tekla Structures Construction Modeling software was used for 4D scheduling and visualization. At first, activities and the as-planned beginning and ending dates were copied to the modeling software and sub-activities were added by the researchers. Model objects (building elements) in the structural model were organized in locations (phases and floors) and afterwards the objects were linked to the sub activities. At second, several scaffolding and studs were modeled and linked to the activities by the researchers. Falling protection safe fences and handrails were also modeled and visualized. Thirdly, the visualization focused on the structures which constructability was assessed to be difficult. In the building, there were several tall in-situ concrete walls and heavy precast beam structures. Special attention was paid for the visualization of formwork and supporting structures, the order of iron mounting and casting. The lifting of heavy beams and lifting places of a truck crane were carefully planned using the model. At fourth, the construction works between new and old building was tested and simulated before the actual construction work began because the old building was used at the same time. The order of new construction works, demolition works of old structures and the building of temporary protective wall was planned and simulated step by step. The simulation was presented for acceptance to the client's personnel. At fifth, actual beginning and actual finishing dates of activities were added to the schedule. The control of activities could be arranged and the visualization of performed activities late, on time or before as-planned schedule was easily simulate.

The 4D schedule were visualized on the daily bases and on weekly bases (Figure 2a). The retiming of activities and sub-activities were scheduled according to the observations done by site manager, general superintendent and foremen. On site, the 3D views based on weekly 4D schedules were printed and hung on the walls of the meeting room for foremen and construction workers.

With the 4D schedule the progress of construction works were effectively managed and controlled and that facilitated staying on schedule. Site manager weighted in the feedback interview that 4D schedule views concretized the content of works more effectively than the traditional 2D bar or the line of balance schedules. It was possible to show foremen and construction workers clearly with the model views what activities in a given week will be made and where they locate. This eliminates misunderstandings and helped workers to understand the wholeness and the constructability of structures. Site engineer considered important the fact that from the model views it could be seen how the structures are connected to other structures. The geometry, location and surrounding structures were seen and an understanding was obtained what it should look like when the structure will be ready. Superintendent mentioned that 4D schedule helped the procurement of building elements and materials. Using BIMs and 4D scheduling it was possible to see necessary elements and materials and divide them into orders.

3.5 *Categorization of BIM methods with the constructability factor areas*

To analyze the connection between constructability issues and the used BIM methods at the construction stage a categorization of BIM methods would be needed. In the literature, several BIM methods were presented without any specific categorization and due to it the researchers decided

Table 3. The categories of BIM methods.

(i) Management, optimization and control of construction	
	1. Design control
	2. 4D schedule
	3. 3D site layout
(ii) BIM-based working	4. Modeling of temporary structures
	5. Visualization
	6. Information search
(iii) Reliability of BIM	7. Revision management
	8. Clash detection
	9. Model updating
	10. Validity of models
	11. Accuracy of models

Table 4. Constructability factor areas (Tauriainen 2015).

(I) Economy and management	
	1. Building economy
	2. Building process
(II) Design solutions and visualization of design output	
	3. Design solutions
	4. BIMs, drawings and specifications
(III) Construction and production	
	5. Standardization and prefabrication
	6. Assemblies
	7. Construction works
	8. Health and safety

to develop and propose their own categorization. They grouped several BIM methods in three categories as expressed in Table 3.

The content of constructability and factor areas are grouped in three main categories as expressed in the Table 4.

The combination of categorization of constructability factor areas and BIM methods with the results of the analysis of constructability issues are presented in Table 2.

4 CONCLUSION

The objective of the research was to clarify how constructability could be improved at the construction stage using BIM as an intensifying technology. Altogether 26 root causes in six constructability factor areas were classified. The root causes leading to constructability issues were focused in six constructability factor areas.

In the research, a new categorization of BIM methods was proposed. Constructability issues related to the building process, design solutions and construction works can be solved effectively using BIM methods as 3D site layout, 4D schedule, and design control. This represents 50% of the identified root causes leading to identified root causes of the perceived constructability issues. Constructability issues related to the content of BIMs, drawings and specifications could be identified effectively using methods as visualization, clash detection, controlling the accuracy and validity of BIM models. This represents 54% of identified root cases. Using BIM methods as 4D schedule, visualizations, clash detection and 3D site layout 73% of all root causes of constructability issues could be identified.

Even though the number of cases, interviews and identified root causes of the constructability issues were minor the value of BIM improving constructability in the construction stage and the working hypothesis could not be contradicted in this research.

REFERENCES

Common Bim requirements (COBIM) 2012. *Series 13: Use of models in construction.* [Referenced 30.4.2016]. Http://www.en.buildingsmart.kotisivukone.com/3.

Eastman, C., Teicholz, P., Sacks, R. & Liston, K. 2011. *BIM Handbook: A Guide to Building Information Modeling for Owners, Managers, Designers, Engineers, and Con-tractors.* 2nd edition. Hoboken, New Jersey, USA: John Wiley & Sons.

Mäki, T. & Kerosuo, H. 2015. Site managers daily work and the use of building information modelling in construction site management. *Construction Management and Economics.* 33(3): 163–175.

Srinath, S.K. & Jack C.P. 2015. A BIM-based automated site layout planning framework for congested construction sites. *Automation in Construction.* 59: 24–37.

Sulankivi, K., Mäkelä, T. & Kiviniemi, M. *Tietomalli ja työmaan turvallisuus.* Tutkimusraportti, VTT. 2009.

Sulankivi, K., Tauriainen M., Kiviniemi M. 2014. Safety aspect in constructability analysis with BIM. *In the proceedings of CIB W099 International Conference: Achieving Sustainable Construction Health and Safety:* 586–596. Lund University, Sweden, June 2014.

Tauriainen, M. 2015. *The content of constructability. Preliminaries for co-operational development of constructability in Finland.* Licentiate's Thesis. Department of Civil and Structural Engineering, Aalto University. Espoo, Finland.

Tauriainen, M., Mero, A-K., Lemström, A., Puttonen, J. & Saari, A. 2014. The development of constructability using BIM as an intensifying technology. In Gudnason & Scherer (eds.), *eWork and eBusiness in Architecture, Engineering and Construction:* 713–716. London: Taylor & Francis Group.

Zhou, Y., Ding, L., Wang, X., Truijens, M. & Lou, H. 2014. Applicability of 4D modeling for resource allocation in mega liquefied natural gas plant construction. *Automation in construction.* 50: 50–63.

eWork and eBusiness in Architecture, Engineering and Construction – Christodoulou & Scherer (Eds)

Integrating BIM and agent-based modelling for construction operational optimization—a LBS approach

F.L. Rossini, G. Novembri, A. Fioravanti & C. Insola
Department of Civil, Building and Environmental Engineering, Sapienza—University of Rome, Rome, Italy

ABSTRACT: The need to manage complexity, and the current necessity of interventions on existing buildings to provide project and construction methodologies and tools capable to support them in a proficient way. With the scope to define in advance the places occupied by workers to accomplish a task, is defined a methodology and a related tool to integrate Building Information Modeling (BIM) with an Agent-Based simulation of workers. The goal is to know as early as possible: where it is possible to work in a productive and safe way; how it is possible to be more efficient placing in the same working space different working phases; when it is possible to allow the continuity of building operations.

1 INTRODUCTION

The industry of A/E/C presents a huge of complexity that in building construction sites has multifaceted aspects due to the variety of new material, building solutions and the need to respect pressing timetables and narrow budgets. This ever-growing complexity, furthermore, is less manageable when the building is a construction site where construction workers and users are present at the same time, ensuring the continuity of building services.

Moreover, it is difficult foresee behavior of users and workers by the usual Building Codes or rules of thumb, expressed in handbooks—even if digital—as we were in the XIX century. Now, with the introduction and the progressive spreading of the engineering approach, it is time to deal with new and old complexities.

During the last decades, the development of ICT allowed the simplification of building process management, thanks to the automation of reasoning tasks, for instance 'clash detections' and 'rule checking' (Solihin & Eastman, 2015).

An important contribution to the correct understanding of phenomena that affect the quality of architecture was the development of Information and Communication Technologies (ITC) and Computer Science.

Evidently, these provided designers useful tools able to predict the effects of their choices and, consequently, to avoid as possible mistakes and misunderstandings, which are important complementary causes of construction delay, unexpected costs and possible injuries during the working-phases.

The paper's aim is to describe a use of the Artificial Intelligence (AI) technique of agents to simulate the project of building design and organizational choices in order to prevent risks like overlaying of activities, wasted time, under-used spaces and resources and, consequently, improve productivity of construction sites allowing, where required, the partial continuity of building use.

Nowadays Information Modeling methodologies have shown great potential in A/E/C field, contributing to support designers (architects, urban planners, engineers, etc.) in managing the complexity of large quantity of information, allowing also the automatic identification of conflicts, mainly geometrical (Singh & al., 2011). At the same time, they have not shown clear capabilities in associating building entities and their assigned resources with their construction methods, required materials, execution time and generated interferences.

The support that these tools provide in architectural design, and the need to move toward the simplification of the overall project organization model construction from the initial stages of design, could suggest appropriate choices of production techniques, optimizing required construction time and mitigating risks.

However, BIM and integration of project management tool is difficult to realize by means of PERT and CPM techniques; it requires, conversely, an explicit representation of management and assessment for:

- Working-team behaviours on the construction site;
- Space required for the execution of the work to be carried out;
- Number and type of the resources involved.

Thus, techniques based on Location-Based Structure (LBS) are more suitable to develop such a different approach (Kenley & Seppanen, 2009).

Location is very important in AEC as it is linked to main characteristic of a building whether used space for construction-related activities. In this paper is described how automatically define a single "location" called <Room> in a BIM environment via a specific application. Term 'location' means the space required by a working-team to reach its goal.

2 STATE OF THE ART

2.1 *BIM (Building Information Modeling) in construction management field*

There is a multitude of different definitions of BIM, coined by the first generation involved in this field (Eastman & al., 2011) but, for the purpose of this paper, the building SMART alliance definition (2012) describes BIM as a 'digital representation of physical and functional characteristics of a facility. As such, it serves as a shared knowledge resource for information about a facility forming a reliable basis for decisions during its lifecycle from inception onward. A basic premise of BIM is collaboration by different stakeholders at different lifecycle phases of a facility to insert, extract, update or modify information in the BIM to support and reflect the roles of actors involved. This defines how this methodology acquire essential information to manage a construction site. In fact, interventions on existing buildings have as first requirement knowing the starting-point situation.

Thus, a BIM model can contain appropriate information for construction like materials present in the existing constructions, potential risks (like asbestos) or details can improve the level of the working-area conditions awareness.

However, uncertainty in existing building data cannot be avoided unless a complete field inspection is undertaken, and there is still a risk of human error. Three approaches may be used to manage the uncertainty of data in a BIM model: (1) verification, (2) acceptance or (3) avoidance. In the first, key data not known or certain field-verified and field-measured; this is the costliest way but has the lowest risk. In the second, a rating is tagged for each element. These are determined individually—for example if a subset of equipment was field-located and 50% were found exactly as indicated on the available drawings, they would be 50% certain—and only elements with a minimum certainty are modeled. Third, data below this certainty level (which may be 99%+ for some organizations) is omitted from the model. This final option is the least expensive but may severely limit the model functionality (McArthur, 2015).

Finally, BIM is a promising digital methodology to collect and represent information, but lacks technique, methods and related tools capable to predict the feasibility, time and costs related to a construction process. To exceed this gap, in this research is proposed to link Agent-Based simulation to BIM environment, via an application that, in existing modeled in BIM, specify the space needed to complete a working activity.

2.2 *ABM (Agent-Based Modeling) in architectural construction sites*

Building, is a complex activity. Nevertheless, this complexity is the sum of simple issues, which are solved often by small worker-teams or, in many cases, by single worker. For this reason, agent-based modeling is very near to the real phenomenon, and can model accurately the interaction among them and between agents and the context given, in this case, by the related BIM model.

Furthermore, in ABM the ontological correspondence between the computer agents in the model and in the real world actors makes it easy and evident to represent actors and the environment and their relationship (Gilbert, 2008), working also on different levels of abstraction, starting from the lower level 'reactive', to the higher level 'proactive' (Novembri & al., 2015).

The project construction management is a realm very near to the ABM method capabilities, because this field involves cross-disciplinary problems like social and human aspects, and both spatial and temporal interactions among different participating teams (Liang, 2012). On the other hand, current BIM and Construction Management tools provide embedded agents-inference engine: these tools are able to represent the working phase duration (4D) or costs (5D) only if the designer sets data following his implicit knowledge. Indeed, the method here described allows to solve a part of the general problem. This application in effect provide managers to know how much space is needed to workers to reach their goals by automatic reasoning given by agents' inference.

3 METHODOLOGY

3.1 *Modelling agents to interact with building construction issues: brief program framework*

To describe agent-based program operation and to give an actual example of it, a small Windows Presentation Foundation (WPF) in C# language with more updated pattern of parallelization with the aim of enhancing concurrency among agents. In this prototype the thread elimination is only a visual omission, because the thread continues computing operation in background.

Effectively, in every application, the <CancellationToken> management is fundamental to ensure a fast agent execution, avoid CPU overloading for computation which, following this procedure, is discarded.

More precisely, when a dimensional input given in a BIM environment is defined, starts the interaction among several agents, characterized by different rules, behavior and, substantially, goals (Castelfranchi & Falcone, 1998).

Here, the aim is the optimization of spaces where the working-phases are located during the global operation of the building, avoiding inhibitions of its social function.

Agents are located in a workspace that have specific requirement and constraints: when all agents find a satisfying condition, the solution is given.

Specifically, every agent geometrically modifies this working-space and, when the modifications required by agents are not in collision among them, the final boundary is designed.

Since these processes happens in parallel mode, the conclusion of a method stops other agents modifying the state of starting instance, until the next iterative cycle.

Every execution, therefore, produces similar but different results; it is not possible, effectively, to control the thread priority and execution speed because memory access is exclusively random: we can associate this selection process to the spermatozoa journey toward the ovum.

Thus, the agent that modifies the space occupied by the workers is the first that ends the entire optimization process in one of the possible ways that satisfies all other agents requirements. Obviously, an agent intervening first in the first cycle, will not arrive necessary first also in the second one, because the whole cycle is random based; this randomness is addressed exclusively to the prior interest to solve the problem in the most effective way, without privileges assigned to agents.

3.2 *Modelling agents to interact with building construction issues: The first-programming phase*

With the sake to allow a coherent graphic representation of the algorithm, the application was encapsulated in a WPF project. The base-class is the one that describes the space vertex, or rather the descriptive base-elements and allow to the method to calculate the Euclidean parameters.

The Room-class constructor, conversely, has the task to foresee the other two missing vertices (the other two vertices of the rectangle assigned in the previous method) necessary to represent the process. The <Draw> method, allows to design the instance in the Canvas, previously defined. <SetAsLast()> highlights the instance border, while <SetAsBase()> refreshes the current state. The <Area> and <FormFactor> properties define the surface dimensions and the related form-factor, set up as an important parameter to give working area reliable dimensions.

```
using System;
using System.Collections.Generic;
using System.Windows;
using System.Windows.Controls;
using System.Windows.Media;
using System.Windows.Shapes;

namespace TestRoomAgents
{
   public class Room
   {
      public List<Vertex> Vertexes { get; private set; }
      public Polyline Drawing { get; private set; }
      public Room(Vertex topleft, Vertex bottomright)
      {
         Vertex topRight = new Vertex(bottomright.X, topleft.Y);
         Vertex bottomLeft = new Vertex(topleft.X, bottomright.Y);
         Vertexes = new List<Vertex>() { topleft, topRight, bottomright,
bottomLeft };
      }

      public void Draw(Canvas canvas, TextBlock info)
      {
         Random rnd = new Random();
         Drawing = new Polyline();
         Drawing.Stroke = System.Windows.Media.Brushes.SlateGray;
         Drawing.StrokeThickness = 2;
         Drawing.Fill   =   new   SolidColorBrush(Color.FromArgb(100,
(byte)rnd.Next(0, 250), (byte)rnd.Next(0, 250), (byte)rnd.Next(0, 250)));
         Drawing.FillRule = FillRule.EvenOdd;
         PointCollection plc = new PointCollection();
         Drawing.Points = plc;
         canvas.Children.Add(Drawing);
         int i = 1;
         foreach (var vertex in Vertexes)
         {
            TextBlock letterVisual = new TextBlock();
            letterVisual.Text = i.ToString();
            letterVisual.FontSize = 14;
            canvas.Children.Add(letterVisual);
            letterVisual.Foreground = new SolidColorBrush(Colors.Black);
            letterVisual.SetValue(Canvas.TopProperty, vertex.Y + 2);
            letterVisual.SetValue(Canvas.LeftProperty, vertex.X + 2);
            plc.Add(new Point(vertex.X, vertex.Y));
            i++;
         }
         plc.Add(new Point(Vertexes[0].X, Vertexes[1].Y));
         Drawing.MouseEnter += (s, e) =>
         {
            Drawing.StrokeThickness = 6;
            info.Text = "Area: " + this.Area.ToString("n") + "; Forma: "
+ this.FormFactor.ToString("n");
         };
         Drawing.MouseLeave += (s, e) =>
         {
            Drawing.StrokeThickness = 2;
            info.Text = "Passa sopra una forma per informazioni...";
         };
      }

      public double Area
      {
         get
         {
            var l1 = Math.Abs(Vertexes[1].X) - Math.Abs(Vertexes[0].X);
            var l2 = Math.Abs(Vertexes[2].Y) - Math.Abs(Vertexes[1].Y);
            var val = l1*l2;
            return val;
         }
      }

      public double FormFactor
      {
         get
         {
            double latoOriz = Vertexes[0].DistanceTo(Vertexes[1]);
            double latoVert = Vertexes[0].DistanceTo(Vertexes[3]);
            if (latoOriz > latoVert) return latoOriz / latoVert;
            else return latoVert / latoOriz;
         }
      }

      public void SetAsLast()
      {
         if (Drawing != null)
         {
            Drawing.Stroke = new SolidColorBrush(Colors.Red);
            Drawing.StrokeThickness = 3;
         }
      }

      public void SetAsBase()
      {
         Drawing.Stroke = new SolidColorBrush(Colors.Gray);
         Drawing.StrokeThickness = 2;
      }
   }
}
```

To finish the first phase, we have to set up the WPF control, in which several application will visualized.

```
<Window x:Class="TestRoomAgents.MainWindow"
        xmlns="http://schemas.microsoft.com/winfx/2006/xaml/presentation"
        xmlns:x="http://schemas.microsoft.com/winfx/2006/xaml"
        xmlns:d="http://schemas.microsoft.com/expression/blend/2008"
        xmlns:mc="http://schemas.openxmlformats.org/markup-
compatibility/2006"
        xmlns:local="clr-namespace:TestRoomAgents"
        mc:Ignorable="d"
        Title="MainWindow" Height="500" Width="800">
    <Grid>
      <Grid.ColumnDefinitions>
         <ColumnDefinition Width="6*" />
         <ColumnDefinition Width="4*" />
      </Grid.ColumnDefinitions>
      <Canvas Grid.Column="0" x:Name="canvas"
MouseDown="canvas_MouseDown" Background="AntiqueWhite" IsEnabled="True" />
      <Grid Grid.Column="1" Margin="10">
         <Grid.RowDefinitions>
            <RowDefinition Height="Auto" />
            <RowDefinition Height="*" />
            <RowDefinition Height="Auto" />
         </Grid.RowDefinitions>
         <TextBlock Grid.Row="0" TextWrapping="Wrap">
            Per iniziare, fare click con il mouse nel settore a sinistra
per disegnare il primo vertice della stanza da letto.
            Disegnare quindi il vertice opposto facendo click. Quando
saranno stati disegni entrambi i vertici, gli agenti verranno avviati.
         </TextBlock>
         <ScrollViewer Grid.Row="1" Margin="0 10 0 0">
            <StackPanel x:Name="stackpanel"></StackPanel>
         </ScrollViewer>
         <Border Background="WhiteSmoke" Grid.Row="2">
            <TextBlock x:Name="Info" FontStyle="Italic">passa sopra una
forma per saperne di più...</TextBlock>
         </Border>
      </Grid>
   </Grid>
</Window>
```

The proposed framework show the principal graphical interface of the application. The pink surface, the Canvas, is the area where the <Room> is located.

3.3 *Agents execution logic*

To describe the base-structure of an agent we defined an interface that, when implemented, allows to agents to define a name, a block-function and a worker-function.

In our case, we have not implemented the block-functions, and the worker function requires access in a Room (the working-area) and the output of a new Room.

Finally, the program produce an object and requires the restitution of a new instance of the same object type.

```
using System;
namespace TestRoomAgents.Agents
{
   public interface IAgent
   {
      string Name { get; }
      void AbortRequired();
      Func<Room, Room> WorkingFunc { get; }
   }
}
```

For these agents, the implementation required is <nullable> type. The <null> type, in effect, will be used in agreement of current norms to warn the agent about a completed work, or rather in a compatible state. In this way, we optimize the program interface. For example, the implementation between two agents is:

```
using System;
using System.Threading;
namespace TestRoomAgents.Agents
{
   class AgentArea : IAgent
   {
      const double MAXAREA = 10000;
      const double MINAREA = 2000;

      public string Name { get; private set; }
      public Func<Room, Room> WorkingFunc { get; private set; }
      public AgentArea()
      {
         Name = "AREA";
         WorkingFunc = (Room startingRoom) =>
         {
            Thread.Sleep(150);
            double area = startingRoom.Area;
            if (area > MINAREA && area < MAXAREA) return null;

            Vertex topleft = startingRoom.Vertexes[0];
            Vertex bottomright = startingRoom.Vertexes[2];
            double factorLato1 = 0.15 * top-
left.DistanceTo(startingRoom.Vertexes[1]);
            double factorLato2 = 0.15 * top-
left.DistanceTo(startingRoom.Vertexes[3]);
            Room returnedRoom;
            if (area > MAXAREA)
            {
               Vertex newBottomRight = new Vertex(bottomright.X - fac-
torLato1, bottomright.Y - factorLato2);
               returnedRoom = new Room(topleft, newBottomRight);
            }
            else
            {
               Vertex newBottomRight = new Vertex(bottomright.X + fac-
torLato1, bottomright.Y + factorLato2);
               returnedRoom = new Room(topleft, newBottomRight);
            }
            return returnedRoom;
         };
      }

      public void AbortRequired()
      {

      }
   }
}

using System;
using System.Collections.Generic;
using System.Linq;
using System.Text;
using System.Threading;
using System.Threading.Tasks;

namespace TestRoomAgents.Agents
{
   class FormFactorAgent : IAgent
   {
      const double MAXFACTOR = 1.2;
      const double MINFACTOR = 0.8;
      public string Name { get; private set; }
      public Func<Room, Room> WorkingFunc { get; private set; }
      public FormFactorAgent()
      {
         Name = "FORMFACTOR";
         WorkingFunc = (Room startingRoom) =>
         {
            Thread.Sleep(150);
            Vertex topleft = startingRoom.Vertexes[0];
            Vertex bottomright = startingRoom.Vertexes[2];
            double latoOriz = top-
left.DistanceTo(startingRoom.Vertexes[1]);
            double latoVert = top-
left.DistanceTo(startingRoom.Vertexes[3]);
            double formfactor = latoOriz / latoVert;
            if(formfactor > MINFACTOR && formfactor < MAXFACTOR) { re-
turn null; }
            Room returnedRoom;
            if (latoOriz > latoVert)
            {
               double factorOriz = latoOriz * 0.15;
               double factorVert = latoVert * 0.15;
               Vertex newBottomRight = new Vertex(formfactor > MAXFAC-
TOR ? bottomright.X - factorOriz : bottomright.X + factorOriz, bottom-
right.Y);
               returnedRoom = new Room(topleft, newBottomRight);
            }
            else
            {
               double factorOriz = latoOriz * 0.15;
               double factorVert = latoVert * 0.15;
               Vertex newBottomRight = new Vertex(bottomright.X, form-
factor > 1/MAXFACTOR ? bottomright.Y + factorVert : bottomright.Y - fac-
torVert);
               returnedRoom = new Room(topleft, newBottomRight);
            }
            return returnedRoom;
         };
      }

      public void AbortRequired()
      {

      }
   }
}
```

3.3 *Agents management and parallelization*

To manage the parallelization process, then the sending and reception of data, we used Task Parallel Library (TPL) dataflow, which extends the namespace. <System.Thread> with several functions, turned toward data-oriented programming: a detailed documentation about this topic is provided by the Microsoft web-library, while the whole package is available on Nuget, because it is not included in the .NET framework (Microsoft, 2016).

The Scheduler class manage the access to four public functions that can be used by the User Interface (UI) with the aim to receive notices about the agents' state.

```
using System;
using System.Collections.Generic;
using System.Linq;
using System.Reflection;
using System.Runtime.CompilerServices;
using System.Text;
using System.Threading;
using System.Threading.Tasks;
using System.Threading.Tasks.Dataflow;

namespace TestRoomAgents.Agents
{
    public static class Scheduler
    {
        public static event EventHandler<Room> AgentIstanced;
        public static event EventHandler<ReactEventArgs> AgentReacted;
        public static event EventHandler<IAgent> AgentReactionAborted;
        public static event EventHandler<IAgent> AgentPosted;

        private static CancellationTokenSource ThreadDestroyer { get; set;
}
        private static List<IAgent> Agents { get; set; }
        private static Room WorkingRoom { get; set; }
        private static Dictionary<IAgent, TransformBlock<Room, Room>>
AgentsDictionary { get; set; }
        private static Dictionary<IAgent, RectStatus> AgentLastRe-
turnedStatus { get; set; }
        private static volatile IAgent SyncronizedAgent;
    ...
```

<SyncronizedAgent> will memorize information about agents, which has the synchronization priority because entered the process firstly.

<AgentLastReturnedStatus> contains information about the output of single agents, <WorkingRoom> is the currently synchronized <Room>.

<AgentsDictionary> contains the <TransformBlock> obtained by <Reflection> process.

The definition of auxiliary class is:

```
public class ReactEventArgs
{
    public virtual object Result { get; private set; }
    public RectStatus Status { get; private set; }
    public ReactEventArgs(object Result, RectStatus Status)
    {
        this.Result = Result;
        this.Status = Status;
    }
}

public enum RectStatus
{
    Wrong,
    Ok,
    OkConfirmed
}
```

To allow an effective operation of the agents' systems, initialization function is required and, essentially, upload the starting room—in this case the predefined working area—and, through the reflection mode, produces all the agents defined.

```
...
        public static void Init(Room room)
        {
            WorkingRoom = room;
            Agents = new List<IAgent>();
            AgentLastReturnedStatus = new Dictionary<IAgent, RectStatus>();
            var agentsTypes = from t in Assem-
bly.GetExecutingAssembly().GetTypes() where
t.GetInterfaces().Contains(typeof(IAgent)) &&
t.GetConstructor(Type.EmptyTypes) != null select t;
            foreach (var agentType in agentsTypes)
            {
                IAgent agent = Activator.CreateInstance(agentType) as
IAgent;
                AgentLastReturnedStatus.Add(agent, RectStatus.Wrong);
                Agents.Add(agent);
                if (AgentIstanced != null) { AgentIstanced(agent, room); }
            }
        }
...
```

The agent broadcast has to run in a synchronous way, and includes a sort of logical re-verification of results, fundamental in this experiment.

```
...
        [MethodImpl(MethodImplOptions.Synchronized)]
        public static void Brodcast(IAgent skipAgent = null)
        {
            SyncronizedAgent = null;
            ThreadDestroyer = new CancellationTokenSource();
            AgentsDictionary = new Dictionary<IAgent, TransformBlock<Room,
Room>>();
            lock (AgentsDictionary)
            {
                foreach (var agent in Agents)
                {
                    if (skipAgent != agent)
                    {
                        TransformBlock<Room, Room> runningAgent = new Trans-
formBlock<Room, Room>(agent.WorkingFunc, new ExecutionDataflowBlockOptions {
CancellationToken = ThreadDestroyer.Token });
                        AgentsDictionary.Add(agent, runningAgent);
                        if (AgentPosted != null) AgentPosted(null, agent);
                    }
                }

                Parallel.ForEach(AgentsDictionary, currentAgent =>
                {
                    PostAndReceive(currentAgent.Key, currentAgent.Value);
                });
            }

        }
...
```

<PostAndReceive> is the key-function: its task is data checking.

```
        private static void PostAndReceive(IAgent agent, Transform-
Block<Room, Room> worker)
        {
            worker.Post(WorkingRoom);
            Task<Room> receiver = worker.ReceiveAsync();
            receiver.ContinueWith((Task<Room> t) =>
            {
                if (SyncronizedAgent == null)
                {
                    SyncronizedAgent = agent;
                    Room newRoom = t.Result;
                    if (newRoom != null)
                    {
                        WorkingRoom = newRoom;
                        AgentLastReturnedStatus[agent] = RectStatus.Wrong;
                    }
                    else
                    {
                        AgentLastReturnedStatus[agent] = AgentLastRe-
turnedStatus[agent] == RectStatus.Ok ? RectStatus.OkConfirmed : RectSta-
tus.Ok;
                    }

                    if (AgentReacted != null) AgentReacted(agent, new Re-
actEventArgs(WorkingRoom, AgentLastReturnedStatus[agent]));
                    if (AgentLastReturnedStatus[agent] != RectSta-
tus.OkConfirmed) { PostAndReceive(agent, worker); }
                    SyncronizedAgent = null;
                }
                else
                {
                    AbortAgents();
                }
            });
        }
```

As we can see, the implementation is rather simple, and guides to the replacement of the previous <Room> instance, to a new one. The validate state of <Room> is given by the return of the working-function with <null> value.

The complete version of the Scheduler class is:

```
using System;
using System.Collections.Generic;
using System.Linq;
using System.Reflection;
using System.Runtime.CompilerServices;
using System.Text;
using System.Threading;
using System.Threading.Tasks;
using System.Threading.Tasks.Dataflow;

namespace TestRoomAgents.Agents
{
    public static class Scheduler
    {
        public static event EventHandler<Room> AgentIstanced;
        public static event EventHandler<ReactEventArgs> AgentReacted;
        public static event EventHandler<IAgent> AgentReactionAborted;
        public static event EventHandler<IAgent> AgentPosted;

        private static CancellationTokenSource ThreadDestroyer { get; set;
}
        private static List<IAgent> Agents { get; set; }
        private static Room WorkingRoom { get; set; }
        private static Dictionary<IAgent, TransformBlock<Room, Room>>
AgentsDictionary { get; set; }
        private static Dictionary<IAgent, RectStatus> AgentLastRe-
turnedStatus { get; set; }
        private static volatile IAgent SyncronizedAgent;
        public static void Init(Room room)
        {
            WorkingRoom = room;
            Agents = new List<IAgent>();
            AgentLastReturnedStatus = new Dictionary<IAgent, RectStatus>();
            var agentsTypes = from t in Assem-
bly.GetExecutingAssembly().GetTypes() where
t.GetInterfaces().Contains(typeof(IAgent)) &&
t.GetConstructor(Type.EmptyTypes) != null select t;
            foreach (var agentType in agentsTypes)
            {
                IAgent agent = Activator.CreateInstance(agentType) as
IAgent;
                AgentLastReturnedStatus.Add(agent, RectStatus.Wrong);
                Agents.Add(agent);
                if (AgentIstanced != null) { AgentIstanced(agent, room); }
            }
        }

        [MethodImpl(MethodImplOptions.Synchronized)]
        public static void Brodcast(IAgent skipAgent = null)
        {
            SyncronizedAgent = null;
            ThreadDestroyer = new CancellationTokenSource();
            AgentsDictionary = new Dictionary<IAgent, TransformBlock<Room,
Room>>();
            lock (AgentsDictionary)
            {
                foreach (var agent in Agents)
                {
                    if (skipAgent != agent)
                    {
                        TransformBlock<Room, Room> runningAgent = new Trans-
formBlock<Room, Room>(agent.WorkingFunc, new ExecutionDataflowBlockOptions {
CancellationToken = ThreadDestroyer.Token });
                        AgentsDictionary.Add(agent, runningAgent);
                        if (AgentPosted != null) AgentPosted(null, agent);
                    }
                }

                Parallel.ForEach(AgentsDictionary, currentAgent =>
                {
                    PostAndReceive(currentAgent.Key, currentAgent.Value);
                });
            }
        }

        private static void AbortAgents()
        {
            foreach (var Agent in AgentsDictionary)
            {
                IAgent subAgent = Agent.Key;
                if (SyncronizedAgent != null && subAgent != SyncronizedA-
gent)
                {
                    subAgent.AbortRequired();
                    if (AgentReactionAborted != null) AgentReactionAbort-
ed(new object(), subAgent);
                    PostAndReceive(subAgent, Agent.Value);
                    if (AgentPosted != null) AgentPosted(new object(), subA-
gent);
                }
            }
        }
    private static void PostAndReceive(IAgent agent, Transform-
Block<Room, Room> worker)
    {
        worker.Post(WorkingRoom);
        Task<Room> receiver = worker.ReceiveAsync();
        receiver.ContinueWith((Task<Room> t) =>
        {
            if (SyncronizedAgent == null)
            {
                SyncronizedAgent = agent;
                Room newRoom = t.Result;
                if (newRoom != null)
                {
                    WorkingRoom = newRoom;
                    AgentLastReturnedStatus[agent] = RectStatus.Wrong;
                }
                else
                {

                    AgentLastReturnedStatus[agent] = AgentLastRe-
turnedStatus[agent] == RectStatus.Ok ? RectStatus.OkConfirmed : RectSta-
tus.Ok;
                }

                if (AgentReacted != null) AgentReacted(agent, new Re-
actEventArgs(WorkingRoom, AgentLastReturnedStatus[agent]));
                if (AgentLastReturnedStatus[agent] != RectSta-
tus.OkConfirmed) { PostAndReceive(agent, worker); }
                SyncronizedAgent = null;
            }
            else
            {
                AbortAgents();
            }
        });
    }

    public class ReactEventArgs
    {
        public virtual object Result { get; private set; }
        public RectStatus Status { get; private set; }
        public ReactEventArgs(object Result, RectStatus Status)
        {
            this.Result = Result;
            this.Status = Status;
        }
    }

    public enum RectStatus
    {
        Wrong,
        Ok,
        OkConfirmed
    }
  }

}
```

3.4 *Enrollment of WPF events*

The graphic interface file is to completed with the <CoreBehind> capable to connect itself with the agent-based system, in addition to the definition of areas. The code task is simply to enroll the several statistic events involved in the <Scheduler> and provide a graphical output to be visualized in the stockpanel.

```
using System.Collections.Generic;
using System.Linq;
using System.Windows;
using System.Windows.Controls;
using System.Windows.Input;
using System.Windows.Media;
using System.Windows.Shapes;
using TestRoomAgents.Agents;

namespace TestRoomAgents
{
    public partial class MainWindow : Window
    {
        Room _lastRoom;
        Room lastRoom
        {
            get { return _lastRoom; }
            set
            {
                if(_lastRoom != null) _lastRoom.SetAsBase();
                _lastRoom = value;
                _lastRoom.SetAsLast();
            }
        }
        List<Vertex> drawnVertexes;
        public MainWindow()
        {
            InitializeComponent();
            drawnVertexes = new List<Vertex>();
            //NOTIFICO A VIDEO LA "MESSA IN ONDA" DI UN NUOVO AGENTE
            Scheduler.AgentIstanced += (object agent, Room startingRoom) =>
            {
                stackpanel.Dispatcher.Invoke(() => {
                    TextBlock t = new TextBlock();
                    t.Text = string.Format("[{0}] è stato avviato. L'area
iniziale è: {1:n} - Il fattore di forma è: {2:n}", (agent as IAgent).Name,
startingRoom.Area, startingRoom.FormFactor);
                    t.Foreground = new SolidColorBrush(Colors.Gray);
                    stackpanel.Children.Add(t);
                });
            };
```

```
            Scheduler.AgentPosted += (object sender, IAgent agent) =>
            {
                stackpanel.Dispatcher.Invoke(() => {
                    TextBlock t = new TextBlock();
                    t.Text = string.Format("[{0}] è stato avviato", (agent
as IAgent).Name);
                    t.Foreground = new SolidColorBrush(Colors.Orange);
                    stackpanel.Children.Add(t);

                });
            };

            Scheduler.AgentReactionAborted += (object sender, IAgent agent)
=>
            {
                stackpanel.Dispatcher.Invoke(() => {
                    TextBlock t = new TextBlock();
                    t.Text = string.Format("[{0}] è stato fermato", (agent
as IAgent).Name);
                    t.Foreground = new SolidColorBrush(Colors.Red);
                    stackpanel.Children.Add(t);
                });
            };

            Scheduler.AgentReacted += (object agent, Sched-
uler.ReactEventArgs react) =>
            {
                stackpanel.Dispatcher.Invoke(() => {
                    TextBlock t = new TextBlock();
                    t.Text = string.Format("[{0}] ha migliorato lo stato.
Area è: {1:n} - Fattore di forma è: {2:n}", (agent as IAgent).Name, (re-
act.Result as Room).Area, (react.Result as Room).FormFactor);
                    t.TextWrapping = TextWrapping.Wrap;
                    t.Padding = new Thickness(3);
                    stackpanel.Children.Add(t);

                    if(react.Status != Scheduler.RectStatus.Wrong)
                    {
                        TextBlock r = new TextBlock();
                        r.Text = string.Format(react.Status == Sched-
uler.RectStatus.Ok ? "[{0}] ha risolto il problema." : "[{0}] ha confermato
il risultato", (agent as IAgent).Name);
                        r.Foreground = new SolidColorBrush(Colors.Green);
                        r.FontWeight = FontWeights.Bold;
                        stackpanel.Children.Add(r);
                    } else
                    {
                        (react.Result as Room).Draw(canvas, Info);
                        lastRoom = (react.Result as Room);
                    }
                });
            };
        }

        /// <summary>
        /// Viene invocato dopo il disegno del secondo punto del quadrato.
Avvia tutti gli agenti disponibili.
        /// </summary>
        public void StartAgents(Room startingRoom)
        {
            Scheduler.Init(startingRoom);
            Scheduler.Brodcast();
        }

        #region DrawingModule
        /// <summary>
        /// Disegno un quadrato partendo da punti. Rinizio al disegno del
terzo punto.
        /// Dopo che è stato inviato il secondo punto, avvio il programma
agenti
        /// </summary>
        /// <param name="sender"></param>
        /// <param name="e"></param>
        private void canvas_MouseDown(object sender, MouseButtonEventArgs
e)
        {
            if(e.RightButton == MouseButtonState.Pressed) { return; }
            if (drawnVertexes.Count == 2) {
                drawnVertexes = new List<Vertex>();
                canvas.Children.Clear();
                stackpanel.Children.Clear();
            }

            Point newVertex = e.GetPosition(canvas);
            drawnVertexes.Add(new Vertex(newVertex.X, newVertex.Y));

            Ellipse vertexVisual = new Ellipse();
            vertexVisual.Width = 20;
            vertexVisual.Height = 20;
            vertexVisual.Stroke = new SolidColorBrush(Colors.DarkGray);
            vertexVisual.StrokeThickness = 2;
            vertexVisual.SetValue(Canvas.TopProperty, -vertexVisual.Height
/ 2 + newVertex.Y);
            vertexVisual.SetValue(Canvas.LeftProperty, vertexVisual.Width /
2 + newVertex.X);
            canvas.Children.Add(vertexVisual);

            if (drawnVertexes.Count == 2)
            {
                var topLeft = new Vertex(drawnVertexes.Min(x => x.X), drawn-
Vertexes.Min(y => y.Y));
                var bottomRight = new Vertex(drawnVertexes.Max(x => x.X),
drawnVertexes.Max(y => y.Y));

                Room room = new Room(topLeft, bottomRight);
                room.Draw(canvas, Info);
                StartAgents(room);
            }
        }

        /// <summary>
        /// Fornisco informazioni extra al click del tasto destro (mi serve
per debug)
        /// </summary>
        /// <param name="e"></param>
        protected override void OnMouseRightButtonDown(MouseButtonEventArgs
e)
        {

            if(e.OriginalSource is Polyline)
            {
                Polyline p = e.OriginalSource as Polyline;
                StackPanel infoStackPanel = new StackPanel();
                infoStackPanel.Background = p.Fill;
                infoStackPanel.Children.Add(new TextBlock { Text = "Informa-
zioni per la 'room' selezionata:", FontWeight = FontWeights.Bold });
                int j = 1;
                for(int i = 0; i<= 3; i++)
                {
                    Point current = p.Points[i];
                    Point next = p.Points[i + 1];
                    TextBox t = new TextBox();
                    int jn = j + 1;
                    if (jn == 5) { jn = 1; }
                    t.Text = string.Format("[{0}-{1}] lungo: {2}",
j.ToString(), jn.ToString(), (new Vertex(current.X, cur-
rent.Y)).DistanceTo(new Vertex(next.X, next.Y)));
                    j++;
                    infoStackPanel.Children.Add(t);
                }
                stackpanel.Children.Add(infoStackPanel);
            }

            e.Handled = true;
        }
        #endregion
    }
```

3.4 *Linking ABM application and BIM*

The application developed is conceived to link the geometrical result of the iteration among intelligent agents with the "BIM-world" via the API (Advanced Programming Interface). The dataset provided by the BIM, in this case represented by the instances "surface", define the agents' interaction: they react to events generated every time the surface analyzed is modified. In summary, instances are classified through their <Unique_ID> and their geometrical data are imported into the Canvas (Fig. 1); Here, the program identifies geometrical values and starts its computation, until the optimized working surface is obtained.

Furthermore, when an object is modified, the system updates data in order to avoid inconsistency within the BIM model. However, the interaction mode varies as a function of the type of interaction. Any request made via the BIM interface is immediately forwarded to the combined ABM application, and the BIM awaits the completion of the task: in this case, a synchronous mode of interaction is established. Otherwise, when the ABM application needs to interact with the BIM world, a request will be visulizaed in the alert palette, waiting for the project manager validation.

4 RESULTS

The result of this process is depicted in Figure 1. The green rectangle is the surface inherited by the BIM environment, while the red-bounded one is the space needed to achieve the construction goal by a specific worker-team: this show us where the risky activity is located and consequently allow actors to more appropriate evaluate the risky conditions.

5 DISCUSSION AND CONCLUSION

5.1 *Beyond the real: Mixed reality as a new dimension of construction site*

The described methodology allows project managers to use a system able to support them in deciding where, when and what to do with the awareness of coded risks and the results of simulations performed. In this study, the forecast is limited to the spatial aspect; in future works the focus will be shifted to the 4D / 5D aspects, determining in advance spaces, time and the expected resources to accomplish a defined task.

Another step for this research will be the moving of these computational simulations to the visual environment of Augmented Reality (AR) tool, that can be used by managers to improve the awareness of operations, and by workers as a class, to prepare them to solve working-task related problems.

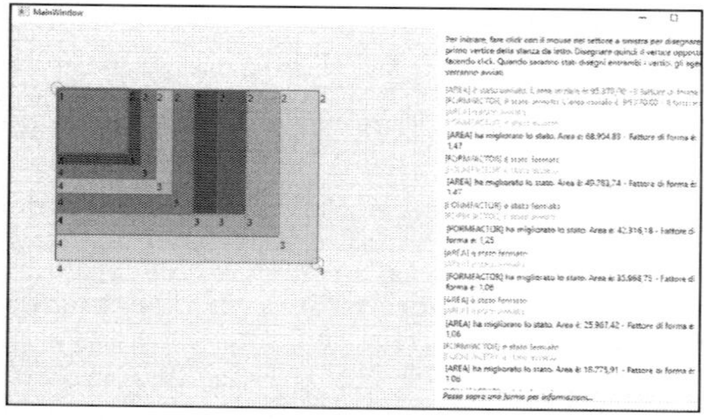

Figure 1. ABM program, the result of an iteration. On the left, the Canvas containing the optimized space. On the right side, the code (programmed by C. Insola).

5.2 *Summary: Pseudo-code process structure*

[When] a Revit Transaction concluded (and Revit is idling):

1. Load all the installed agents as 'Agents';
2. Start a parallel execution of every Agent;
3. Wait for Agents action (the program is in a suspended state).

   [When] An Agent starts the thread:

1. Agent asks the agent manager for a collection of observed Revit items (a 'Collection');
2. The Agents manager reacts to the request and checks if the required items are changed.
   a. If not:
      i. the agent manager kills the invoker agent;
      ii. Agent manager stores a "current valid configuration" status message for the invoker agent;
      iii. *If yes*, the changed items are sent to the agent and a status message is awaited.
3. The agent checks if the Collection has changed and the new items configuration matches its own valid range.

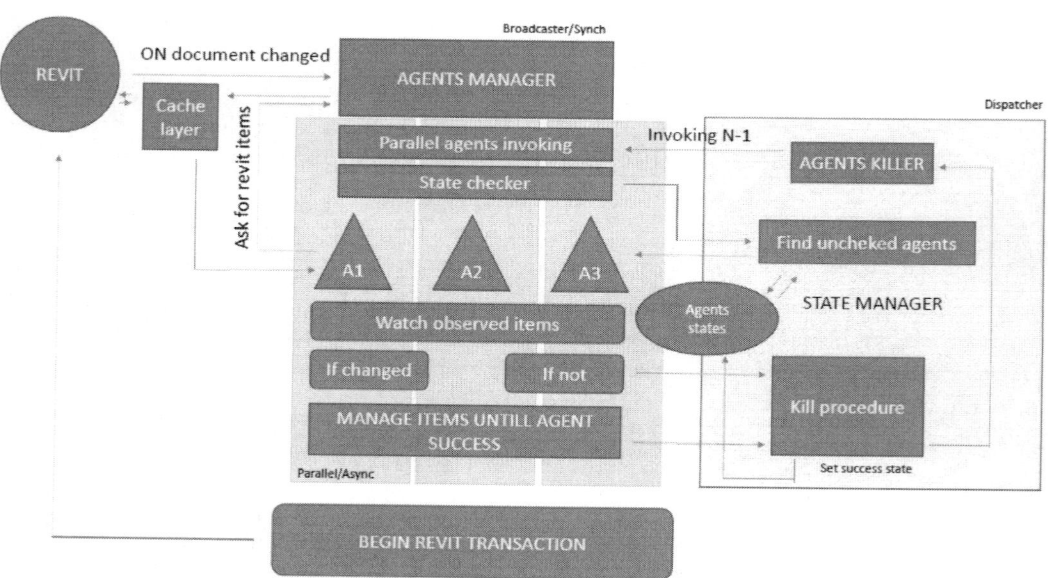

Figure 2. the synthesis about the process proposed. Linking BIM object to Agent-based simulation.

4. *If* the new configuration is valid, then the agent sends a 'current valid configuration' status response to the agent manager, then kills itself.
5. *If* the new configuration is invalid:
   a. The agents start to execute one or more internal algorithms to get a valid configuration.
   b. *When* a new configuration has been found, the new collection of Revit items is sent to the agent manager
   a. The agent manager recalls all other agents
      i. *If* they were performing some tasks, they are interrupted to allow attention to be paid to the new deal
      ii. The agent manager sends the new Collection proposed as valid by the agent in point 5 to all the agents
      iii. Points 3 to 5 are performed until a universal 'current valid configuration' status is sent by all agents
   b. *When* all agents have sent a "current valid configuration" the agent manager starts a Revit transaction
   e. The changed collection of Revit items is suggested to the Revit users
   f. The agent manager suspends itself and waits for a Revit document change.

REFERENCES

Building Smart, international home of BIM <http://www.buildingsmart.org/> last access 13/04/2016.

Castelfranchi, C., Falcone, R., 1998, Toward a theory of delegation for Agents-Based systems, Robotics and Autonomous systems, 24 (24), pp. 141–157.

Gilbert, N., 2008, Agent-Based Models. Sage Publications, Los Angeles, p. 14.

Eastman, C. Teicholz, P., Sacks, R., & Liston, K. (2011). BIM handbook: a guide to building information modeling for owners, managers, designers, engineers, and contractors. Wiley.

Kenley, R, Seppänen, O, 2009, Location-based management of construction projects: part of a new typology for project scheduling methodologies in Proceedings of the 2009 Simulation Conference, Eds. Rossetti M, D, Hill, R, R, Johansson, A, Dunkin A and Ingalls, RG. Austin, TX, USA, December 13–16, pp 2563–2570;

Liang, C., 2012, Agent-based modeling in urban and architectural research: A brief literature review. Frontiers of architectural Research vol. 1 (2012) pp. 166–177.

McArthur, J.J., 2015, A building information management (BIM) framework and supporting case study for existing building operations, maintenance and sustainability, in Procedia Engineering 118 (2015), pp. 1104–1111.

Novembri, A., Fioravanti, A., Rossini, F.L., 2015, Geometria qualitativa nel BIM-world: Generazione della Location Breakdown Structure (LBS) per un processo di costruzione sostenibile, in Environmental sustainability, circular economy and building production. Maggioli Editore, pp. 502–521.

Singh, V, Ning, G, Xiangyu, W, 2011, A theoretical framework of a BIM-based multi-disciplinary collaboration platform in Automation in construction 20 (2011) pp. 134–144

Solihin, W., Eastman, C., 2015, Classification of rules for automated BIM rule checking development, in Automation in construction, Vol. 53 (2015), pp. 69–82.

Websites:

Laiserin J. in LaiserinLetter

http://www.laiserin.com/index.php.

eWork and eBusiness in Architecture, Engineering and Construction – Christodoulou & Scherer (Eds)
 ISBN 978-1-138-03280-4

Topological robustness and vulnerability assessment of water distribution networks

A. Agathokleous, C. Christodoulou & S.E. Christodoulou
Nireas International Water Research Center, Department of Civil and Environmental Engineering, University of Cyprus, Nicosia, Cyprus

ABSTRACT: Being able to assess the reliability of the network against different hazards helps water distribution agencies prioritize their interventions and ensure a minimum reliability level of the network. Research to-date has helped identify a number of potential time-invariant and time-dependent risk factors contributing to pipe fragility and network reliability. Among them are factors such as a pipe's age, diameter, material and number of previous breaks, as well as the network's topology, operating pressure and water flow. In terms of introducing a network's topology to its risk level, recent work has highlighted the importance of a network's connectivity to its reliability and the need for robust appraisal methods of network connectivity metrics. The work described herein discusses such a method based on a network's 'betweenness centrality' index and demonstrates its importance using a case-study Water Distribution Network (WDN) under both normal and abnormal operating conditions. The proposed method is also coupled with spatial mapping to indicate areas of concern in the network, and with a decision support system to assist in prioritizing actions to improve on the network's robustness and resilience.

1 INTRODUCTION

One of the biggest challenges faced by city agencies is the management of their WDN, especially in light of the increasing network complexity and of the increasing rate of pipe deterioration. As a result, it is becoming increasingly important to being able to assess the network reliability and resilience to catastrophic events by use of automated process models. Such models should take into consideration both time-invariant and time-dependent risk factors, with topological reliability being an early key indicator.

The management system of a well-organized WDN includes Decision Support Systems (DSS) to assess the pipeline system condition and the extent of damage due to pipe failures. The development of these tools requires installation within the network of monitoring devices and sensors, hydraulic simulation models, extended database, Geographical Information Systems (GIS) and mathematical models that simulate the behavior of the WDN through time (Figure 1).

The development of an integrated tool for condition assessment of WDNs is an expensive, time consuming and multi-faceted procedure which, despite its complexity, most times fails to reflect in great detail the actual real-time behavior of network operation. Further, the development, operations and maintenance of the underlying hydraulic and mathematical tools require specialized knowledge, not typically found within the water agencies, and analysis of real-time WDN operational data that hinder the applicability of the models.

The aforementioned difficulties give rise to the need for an automated early-stage appraisal system of a WDN's vulnerability, based on the WDN's topological characteristics, which could then be easily amended with additional information for a more holistic vulnerability model. Such information are the hydraulic characteristics, the previ-

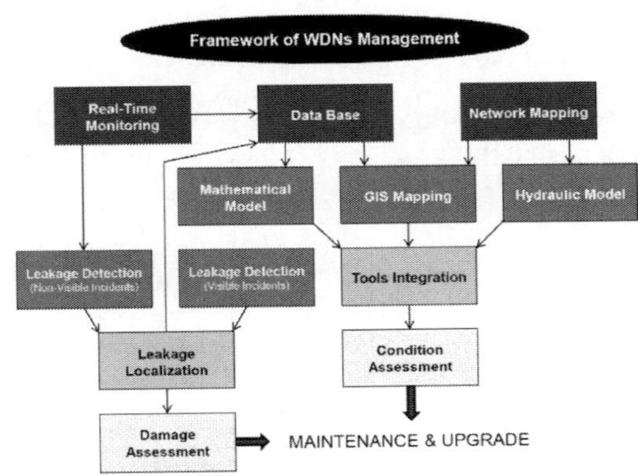

Figure 1. Schematic framework of a WDN management system.

ous pipe breaks and past network performance under normal and abnormal operating conditions, and indicative models can be found in the works by Fragiadakis et al. (2013), Christodoulou & Fragiadakis (2014) and Fragiadakis et al. (2016).

The goal is the development of a model that: (1) it is cost efficient; (2) its use does not require hydraulic and mathematical knowledge and advanced software skills; (3) it does not require operational data beyond the topography of the WDN; (4) its upgrading and updating procedures are not painful and time consuming; (5) it can lead to reliable conclusions within a short time; (6) its capabilities are not limited to the risk and network condition assessment, but it can also be used as a management tool for designing the operation mode of a WDN.

This paper proposes a model based on the principle that the sections of WDN which are hubs and support large segments of the pipeline system, are subject to stresses whose size is of greater intensity and frequency relative to other network sections, thus being more vulnerable to failure. The importance of each node, relative to the other nodes, with regards to its contribution to the provided network service is computed factoring in the topology of the WDN as described by a betweenness centrality metric.

Finally, for validating the applicability and accuracy of the proposed model, the deduced by the model spatial risk levels (heatmaps) are compared to corresponding maps created by the use of actual WDN failure data (pipe bursts).

1.1 *Case-study network*

The case-study network utilized in the analysis is a sub-network (a 'District Metered Area', DMA) from the water distribution of the city of Nicosia (Cyprus). The studied DMA (DMA6) is divided into 4 sub-district metered areas (Sub-DMAs) and it is remotely monitored through a Supervisory Control and Data Acquisition (SCADA) system across the DMA, which continually collects and transmits operational information. All leakage incident data reports since 2003 had been maintained in a specially designed database. The data utilized for verification / validation purposes of the proposed model, covers a time period of approximately eight years (01/01/2003 to 31/12/2010) and includes 548 incidents associated with water mains (WM).

2 STATE OF KNOWLEDGE

2.1 *Condition assessment models for WDNs*

To date, several mathematical models simulating the operations and condition of WDNs have been developed. Initial research efforts were based on the development of a single-objective mathematical model expressing the failure pattern or system reliability, while subsequent studies expanded the work by presenting multi-objective failure models. More recent efforts focused on the development of models simulating WDN behavior. Finally some recent research activities focused on abnormal operating conditions due to exogenous factors that affect the condition of the network.

Kleiner and Rajani (2001) provide an overview of the mathematical models related to structural deterioration of the WDNs, with the different models grouped into classes according to the governing equations and the types of data that are required for the implementation of each model. In addition, critiques and comparisons of the various models have been carried out. Mutikanga et al. (2013) provide a review on the current tools and methodologies applied to assess, monitor, and control losses in WDNs. The main aim was the identification of the tools and methods that have been applied. The range of the available management systems varies, from simple managerial tools such as performance indicators to highly sophisticated optimization methods such as evolutionary algorithms.

2.2 *Betweenness centrality*

The centrality metric is the most widely used indicator for identifying important nodes of a network based on topological characteristics. The importance of a node refers to different meanings, depending on the network type and the examined case study. Centrality is given in terms of a real-valued function for each network node (vertices of a graph), and the resulting values provide a ranking which identifies the most important nodes (Bonacich 1987, Borgatti, 2005). The ease in calculation of the centrality metric relates to the complexity of the network, which depends on the network size (number of vertices and edges), the connectivity of vertices (number of edges linked on each vertex, directed / undirected edges and connected / disconnected vertices) and the weights assigned to the edges (Opsahl et al. 2010).

Betweenness centrality is an indicator for identifying the vertices of a network that have high contribution to the transfer of items within the network, and it is equal to the degree of which a vertex falls on the shortest path between the other vertices (Freeman 1977). Mathematically, the betweenness centrality of a vertex, u, is defined by (Anthonisse 1971):

$$C_B(u) = \sum_{s \neq u \neq t} \frac{\sigma_{st}(u)}{\sigma_{st}} \tag{1}$$

where σ_{st} is the total number of shortest paths from vertex s to vertex t and $\sigma_{st}(u)$ is the number of those paths that pass through u.

The complexity of real-life networks has prompted researchers to develop various algorithms to efficiently calculate the betweenness centrality of such networks. The research works are numerous and the range of the developed algorithms is large because of the networks' diversity. An example that highlights the need of providing different algorithms for calculating betweenness centrality is the work presented by Brandes (2008), which discusses several variants of betweenness centrality.

3 ANALYSIS, RESULTS AND DISCUSSION

3.1 *Pipeline system of DMA6*

The WDN in study is first analyzed topologically, with betweenness centrality metrics computed for every node in the network and geographically mapped. The result of the analysis is then spatially mapped using a heatmap, showing the levels of the betweenness centrality index across the case-study DMA (Figure 2).

The areas of high betweenness centrality values (shown in red) indicate the nodes through which a high number of origin-to-destination paths passes, when considering only the topology of the network. As expected, betweenness centrality numbers are lower on the periphery of the network, and increase towards its center.

3.2 *Vulnerability heatmap of Continuous Water Supply (CWS)*

The WDN in study is then analyzed under normal (continuous water flow) operating conditions (data period before March 2008). For this, the Continuous Water Supply (CWS) mode, the betweenness centrality metrics are computed in conjunction with the underlying hydraulic model for the DMA. That is, the hydraulic model and the deduced nodal pressures and pipe water flows in essence dictate the origin-to-destination paths for every calculation and thus affect the betweenness centrality computation. As with the previous case, the result of the analysis is then spatially mapped by use of a heatmap, showing the levels of the betweenness centrality index across the case-study DMA (Figure 3).

A comparison between Figures 2 and 3 shows the difference in the spatial allocation of the nodal betweenness centrality indices and of the resulting network vulnerability, when the hydraulic behavior of the DMA under continuous water supply operations is considered.

Now the betweenness centrality values are higher along major water mains, as the deduced

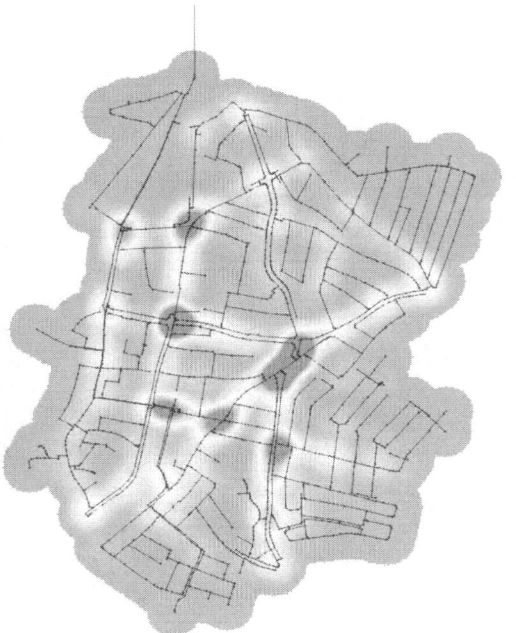

Figure 2. Betweenness centrality heatmap, for network topology.

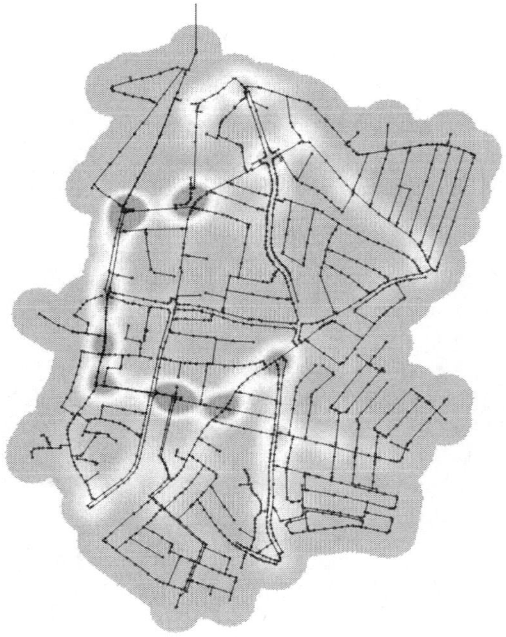

Figure 3. Betweenness centrality heatmap, for Continuous Water Supply (CWS) operations.

water flows at the nodes along these pipes create more water pathways passing through these nodes. Noteworthy is also the fact that the change in network vulnerability (Figure 3) is actually obtained by opening/closing 9 valves (nodes) in the original network (Figure 2). This change in vulnerability indicates the effects of hydraulic pressure on the network performance.

3.3 *Vulnerability heatmap of Intermittent Water Supply (IWS)*

When in abnormal operating conditions (such as intermittent water supply operations), the network in study exhibits a different behavior, with the betweenness centrality indices varying compared to the ones deduced for the continuous water supply operations (Figure 3). Under IWS operations (data period after March 2008), the DMA in study is subdivided into four subDMAs (Figure 4), each receiving water for 12 hours every 48 hours. During this operation mode the betweenness centrality indices change, depending on which subDMA is active (Figures 5–8) and how the varying water flow affects the water pathways and the origin-destination pairs in the network.

As with the CWS case (Figure 3), the betweenness centrality metrics are computed in conjunction with the underlying hydraulic model for the subDMA, and the deduced water flows in the piping network dictate the origin-to-destination paths in the network, thus affecting the computation of the nodal betweenness centrality indices.

3.4 *Spatial distribution of the failure incidents*

The results of shifting in the betweenness centrality indices as a result of the change in the water

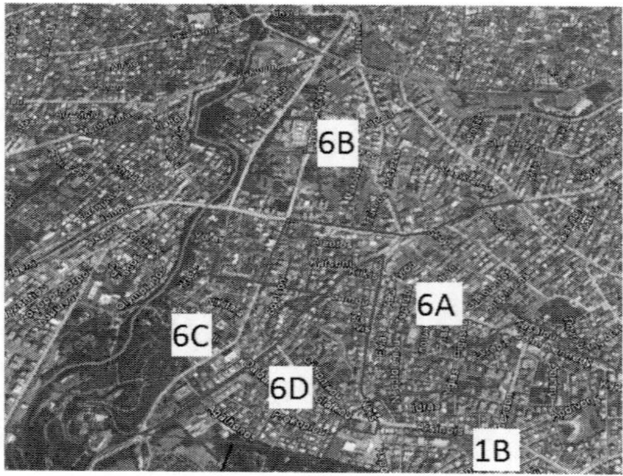

Figure 4. The four subDMAs examined under IWS operations.

Figure 5. Betweenness centrality heatmap (subDMA 6 A), for Intermittent Water Supply (IWS) operations.

Figure 6. Betweenness centrality heatmap (subDMA 6B), for Intermittent Water Supply (IWS) operations.

Figure 7. Betweenness centrality heatmap (subDMA 6C), for Intermittent Water Supply (IWS) operations.

Figure 8. Betweenness centrality heatmap (subDMA 6D), for Intermittent Water Supply (IWS) operations.

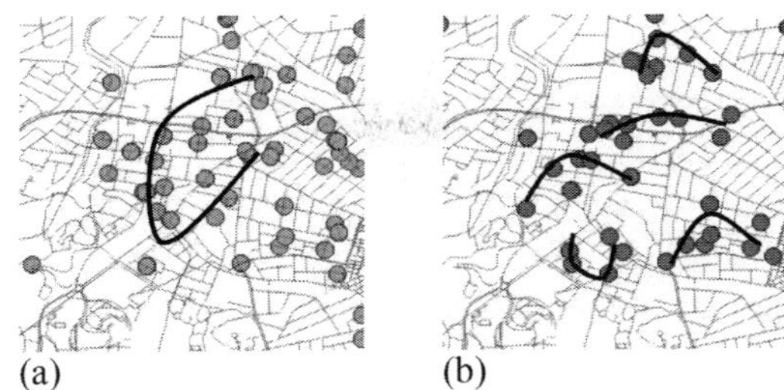

Figure 9. Failure incidents (pipe breaks) in studied WDN (a) during CWS operations (before March 2008); (b) during IWS operations (after March 2008).

supply operations (from CWS to IWS) can be seen in Figures 9a and 9b, depicting the failure incidents in the WDN in study during the time periods of CWS and IWS, respectively.

As can be seen in Figure 9, the spatial distribution of the failure incidents changes as the mode of operations changes, and the center of gravity shifts as the failure incidents cluster around regions of high betweenness centrality indices.

This spatial shift and the observed clustering can be attributed to the 'reorganization' of the WDN, as evidenced by the betweenness centrality heatmaps, stemming from the changes in water flow/pressure across the network.

4 CONCLUSIONS

The work presented herein provides evidence of the links between the behaviour of a WDN under varying operating conditions, the betweenness centrality indices of the network's nodes, and the network's vulnerability. Given these links, WDN operators can forecast a WDN's behaviour under several scenarios and plan for them, optimizing the behaviour of the WDN and minimizing its vulnerability against endogenous and exogenous threats, without the need for dynamic hydraulic models. Furth research work is currently under way to link the observed fragility of a WDN with its betweenness centrality characteristics, using a larger dataset (higher number of incidents, and longer time periods).

ACKNOWLEDGMENT

The work presented herein is part of research initiatives funded by the University of Cyprus's "Post-Doctoral Researchers" program. Special thanks are also extended to Nicosia's Water Board for providing operational data on its network.

REFERENCES

Anthonisse, J. 1971. The rush in a directed graph. Technical Report BN 9/71, Stichting Mathematisch Centrum Amsterdam, Netherlands.

Bonacich, P. 1987. Power and centrality: A family of measures. *American Journal of Sociology* 92(5): 1170–1182.

Borgatti, S.P. 2005. Centrality and network flow. *Social Networks* 27(1): 55–71.

Brandes, U. 2008. On variants of shortest-path betweenness centrality and their generic computation. *Social Networks* 30(2): 136–145.

Christodoulou, S. & Fragiadakis, M. 2014. Vulnerability assessment of water distribution networks considering performance data. *Infrastructure Systems* 10.1061/(ASCE) IS.1943–555X.0000224, 04014040.

Fragiadakis, M., Christodoulou, S.E. & Vamvatsikos, D. 2013. Reliability assessment of urban water distribution networks under seismic loads. *Water Resources Management* 27(10): 3739–3764.

Fragiadakis, M., Xanthos, S., Eliades, D.G., Gagatsis, A. & Christodoulou, S.E., 2016. Graph-based hydraulic vulnerability assessment of water distribution networks. Lecture Notes in Computer Science, 8985: 81–87.

Freeman, L.C. 1977. A set of measures of centrality based on betweenness. *Sociometry* 40(1): 35–41.

Kleiner, Y. & Rajani, B. 2001. Comprehensive review of structural deterioration of water mains: Statistical models. *Urban Water* 3(3): 131–150.

Mutikanga, H.E., Sharma, S.K. & Vairavamoorthy, K. 2013. Methods and tools for managing losses in water distribution Systems. *Journal of Water Resources Planning and Management* 139(2): 166–174.

Opsahl, T., Agneessens, F. & Skvoretz, J. 2010. Node centrality in weighted networks: Generalizing degree and shortest paths. *Social Networks* 32: 245–251.

eWork and eBusiness in Architecture, Engineering and Construction – Christodoulou & Scherer (Eds)
© 2016 Taylor & Francis Group, ISBN 978-1-138-03280-4

Contractual and legal issues for building information modelling in Turkey

Z. Sözen & A. Dikbaş
Istanbul Medipol University, Istanbul, Turkey

ABSTRACT: BIM is being used with increased frequency by Turkish contractors on building and infrastructure projects in Turkey. Although there is agreement among Project stakeholders that BIM is a valuable tool in the achievement of project goals and the management of projects, a full fledged implementation of BIM is not yet achieved due to several issues. The present paper shall attempt to identify a range of contractual and legal issues concerning the implementation of BIM in Turkey. The contractual and legal issues that present barries to the full implementation of BIM may be categorized under the following headings: the legal framework, liability and responsibility, confidentiality and traditional approaches to damages. The legal framework in Turkey lends more support to a fragmented structure within the construction industry, delineating the rights and responsibilities of the parties clearly. Current intellectual property law may not be adequate with the enhanced use of BIM in a contract. Furthermore traditional approaches to contractual liability and responsibility favour clear cut allocations of risk between the designers, contractors, subcontractors and owners. BIM allows the sharing of information between the project parties and authorisation of access to otherwise confidential data. The present tendency of the parties is to protect commercial information and to preserve commercial competitiveness. Finally the traditional approach to contractual damages is another factor stifling the widespread use of BIM in construction projects.

1 INTRODUCTION

Building Information Modelling (BIM) is being used with increased frequency by Turkish contractors on building and infrastructure projects in Turkey. Although there is agreement among project stakeholders that BIM is a valuable tool in the achievement of Project goals and the management of projects, a full fledged implementation of BIM is not yet achieved due to several issues.

BIM can add value in all project delivery methods, but it is particularly aligned with an Integrated Project Delivery (IPD) contract. There has been rapid progress in the use of Building Information Modelling (BIM) around the world, especially in the last decade. The Turkish Construction industry has not yet fully implemented or made use of BIM but thereis a strong drive from the private sector and municipalities to adopt BIM in construction schemes.

This paper shall not address the advantages of BIM over the traditional methods, as this has been undertaken by many researchers. Rather it will attempt to identify the barriers to the implementation of BIM in the Turkish construction industry.

Construction has played a significant role in the economic development of Turkey since the 70's, both domestically and overseas, currently accounting for 5.9% of GDP. The share of GDP attributable to construction is estimated around 30%, with the impact of the construction sector on other sectors of the economy taken into consideration according to a report by the European International Contractors (FIEC Construction Activity in Europe, 2015).

In addition, Turkish contractors are rapidly integrating into the global construction sector. According to the ENR magazine's annual report, Turkey took the second place after China in terms of the number of contracting companies building the largest volume of projects around the world in the last seven years. The ENR Sourcebook is an annual publication that ranks international contractors and design firms based on contracting revenues generated abroad.

ENR's "The World's Top 250 International Contractors" list published in August 2015, based on 2014 figures, included 43 Turkish contractors. (ENR, 2015, The 2015 The Top 250 International Contractors 101–200).

The implications of this is that Turkish contractors shall have to adopt innovative methods to sustain their competitiveness in the global construction arena.

A few words about the method employed in this paper are due here:

Our observational work was undertaken in two forms—as expert witnesses and as claim consultants to Turkish contractor firms over a period of 20 years.

The author's experiences as an expert witnesses and claim consultant in the construction sector indicate that contractual relationships in the Turkish construction sector are characterized by:

- A lack of trust in the other party
- Specific mistrust in the contractor
- Adversarial attitudes
- Confrontational atmosphere
- Confidentiality concerns

The method used approximates participant observation in the sense that the authors have shared and observed the activities of the parties.

2 BARRIERS

2.1 *Legal barriers*

The existing legal system in Turkey is not well equipped to deal with regulating a multi-party, collaborative contractual relationship. Instead, it is heavily biased in favour of a two-party relationship, whereas BIM basically assumes the early and sustainable relationships between employers, lead designers, design consultants, contractors and subcontractors. The rationale behind this assumption is that these different parties may make appropriate contributions or corrections to design and construction.

As such the legal framework in Turkey lends more support to a fragmented structure within the construction industry, delineating the rights and responsibilities of the parties clearly. Turkey is a civil law jurisdiction The Law of Obligations was borrowed from the Swiss Civil Code in 1926 after the foundation of the Turkish Republic.

Private contractual relations are governed by the Turkish Code of Obligations numbered 6098 ("TCO"), entered into force as of July 1st, 2012. Construction contracts are regulated in the Articles of 471–486 of Turkish Law of Obligations (Türk Borçlar Yasası, 2012).

On the other hand, Public Procurement in Turkey is governed by the Public Procurement Law no. 4734(67) and the Public Procurement Contracts Law no. 4735 (68) both dated 2002. These Laws govern all public procurement, with a few exceptions (Kamu İhale Kurumu 2002 a/b).

The Law on Intellectual and Artistic Works No. 5846 ("Law No. 5846") regulates intellectual property rights. (Fikir ve Sanat Eserleri Kanunu, 2001) Within the scope of this law, a work is any kind of intellectual and artistic product bearing the characteristics of its owner and which is considered a work of science and literature, music, fine arts or cinema. Architectural and construction documents are therefore treated as "works of science and literature, music, fine arts or cinema" and the architect/engineer as the "author", who owns the intellectual property thereof.

The Law of Obligations focuses mainly contractors' liabilities in detail. These are namely:

- Duty of care and loyalty
- To perform the work personally or cause the work to be performed by a third party under its control.
- To provide all equipment necessary for the performance of the work unless agreed otherwise.
- Obligation to Commence and Carry out the Work
- Defects Liability

The Law does nor place an equivalent emphasis on the Employer's liabilities. The Employer's main responsibility is payment and reasonable examination of the work for patent defects.

On the other hand selection of contractors under the Public Procurement Law is based on the financial proposal by selected bidders after biddability review according to defined pre-qualification criteria. Liability for public welfare offences is a major determinant of contractual behaviour for the public employer in Turkey and this liability is cost-sensitive. Public authorities exhibit special sensitivity to cost during the evaluation process. Regulatory offences thus correspond to an incredibly diverse and complex series of activities. Negligence may suffice and penalties may be serious.

The public authority is particularly sensitive to cost claims filed by the contractor, while granting extension of time is not a major issue. Cost claims in compensable delay cases can be a serious cause of concern for public employers.

"In current practice, most of Turkish publicclients avoid forming non-price criteria for bid evaluation procedure in spite of the authorization given them in Article 40. The reason for avoidance is the probability of "subjective evaluation of contractors and bids as to nonprice" criteria that may not be translated into numerical impact data. Subjective and arbitrary treatments of publicclients may harm public benefits." (Yılmaz & Ergönül, 2011, p. 479).

Cost, therefore is a lead driver both in bid evaluation and contract administration. What naturally flows from this is the under pricing of bids, eventually leading to higher costs and claims.

In addition to the foregoing, it should be noted that the public procurement system in Turkey assumes a design-bid-build delivery system.

The contract types defined by the law no 4735 are restricted to the following:

a. turn key lump-sum contracts over the total tender price for the entire work proposed by the tenderer on the basis of application projects and site lists thereof.
b. in procurement of goods or services; lump-sum contracts over the total tender price proposed by the tenderer for the entire work, the detailed specifications and quantities of which are predetermined by the contracting entity.
c. Unit price contracts over the total price calculated by multiplying the quantity for each work of item specified in the schedule prepared by the contracting entity, with unit prices proposed by the tenderer for each corresponding work of item,on the basis of, preliminary or final projects and site lists thereof along with unit price definitions in procurement of works whereas on the basis of detailed specifications of the work involved in procurement of goods or services (Kamu İhale Kurumu, 2002b).

In addition to the foregoing, it should be noted that the public procurement system in Turkey assumes a design-bid-build delivery system.

The standard force majeure provision is restricted to the events listed under Article 10. These are

- Natural disasters.
- Legal strikes.
- Epidemics.
- Announcement of partial or general mobilization.
- Other similar circumstances that may be determined by the Authority where necessary.

The Contracting authority may accept any of the above circumstances as force majeure, provided that the event does not result from the contractor's default, it constitutes an obstacle to the fulfillment of contractual obligations, the contractor was not able to remove such obstacle, the contractor has notified the contracting authority in writing within twenty days as from the date of the force majeure, and it has been documented / certified by competent authorities (Kamu İhale Kurumu, 2002b).

Whether "other similar circumstances" are qualified as force majeure is entirely at the dicretion of the Contracting authority and the decision usually leads to delays and disputes.

In addition, the "General Specifications for Public Works" contains a no damages for delay provision for employer delays (T.C. Çevre ve Şehircilik Bakanlığı, 2011).

On the other hand FIDIC international forms and FIDIC based World Bank contracts are widely used in infrastructure projects in Turkey. However, the Contracting Authority generally extends the contractual liability of the contractor by particular conditions.This issue will be discussed in more detail under 2.2. *Traditional approaches to contractual liability and responsibility.*

To summarize, the existing legal framework is neither equipped to handle early involvement of different parties, whereas BIM derives advantages from fornt loading, engaging stakeholders at an early stage of the Project.

2.2 *Traditional approaches to contractual liability and responsibility*

Furthermore traditional approaches to contractual liability and responsibility favour clear cut allocations of risk between the designers, contractors, subcontractors and owners. Literature on contractual relationships in Turkey tends to focus on the contractors' liabilities and duties rather than the employers', let alone on collaborative approaches.

Traditionally both private and public contracts are unilateral. Employers (owners and contractors vis a vis subcontractors) usually dictate the terms of the contract, minimizing employer risks while imposing onerous terms on the contractor or the subcontractor. These contracts are characterized by exclusions of damages and disclaimers for the owner and limitations of remedies for contractors. In fact, most contracts in use specifically attempt to remove all risk from the employer and transfer it to the contractor. Generally the contractor or the subcontractor is not involved in contract drafting. One sided termination clauses, "pay when paid" clauses and "no damages for delay" clauses have become regular inclusions to standard contracts and subcontracts.

In 2006, Onur et al have studied risk allocation in public procurement projects in Turkey thorugh a survey administered to 18 construction firms (Onur et al, 2006). The findings of the study indicate that contractors traditionally carry a larger proportion of risk as compared to employers.

In a study conducted in 2010, Onur has compared the risks undertaken by Turkish contractors underTurkish General Conditions of Construction (GCC) which is mandatory for public construction projects in Turkey with contractors' risks under FIDIC Red Book General Conditions of Contract (Onur, 2010). The findings of this later study indicate that risks undertaken by contractor firms under Turkish General Conditions of Construction are significantly higher than the risks undertaken under FIDIC Red Book.

As noted earlier, FIDIC standard forms and World Bank contracts, though used extensively in Turkey, are generally amended to offer extra protection for the Contracting Authority by deleting and/or replacing standard contract clauses, shifthing the risk to the contractor.

In a recent dispute between a Municipality and a Turkish contractor under a World Bank contract, the contractor was interrupted two times. The first interruption was due to an expropriation problem. The pipe line consequently had to be re-designed. The second interruption was due to archeological findings and the pipe line had to be re-designed once more. The Contracting Authority rejected overhead claims and re-design expenses in these compensable delay situations because of amended contract clauses.

Another common dispute source is expropriation. Lack of possession of / access to the Site is one of the main causes of delays/disruptions to public infrastructure projects in Turkey.This is a significant problem for the infrastructure contractor. The Contracting authority usually sets the commencement date as the date of hand over, thus evading overhead claims. Alternatively, the Contracting authority may give the Contractor access to the Site gradually in sections while requiring the Contractor to adjust its baseline programme. The Contractor also waives rights to claim costs in respect of the gradual handing over of the Site.

From a risk allocation perspective, fundamental changes would be required under an IPD/BIM model including mutual waivers of damages and claims.

2.3 *Confidentiality barriers*

BIM allows the sharing of information between the Project parties and authorisation of access to otherwise confidential data. The present tendency of the private and public parties in Turkey is to protect commercial information and to preserve commercial competitiveness. The operation of BIM on a project may allow parties to access information which is otherwise confidential.

Lack of confidence in the other party is a major barrier to information sharing. Differences in corporate cultures and frames of reference may be the root causes of this particular resistance. Neither the employers nor the contractors are interested in developing partnering relationships. And precisely for these reasons, parties are suspicious of the other parties.

The main principle in public tenders under the new Procurement Laws is achievement of transparency through open tender procedures. However, commercial confidentiality during the performance of the contract remains a serious concern for public procurement (Sözen, Z., 2016).

Another factor is duplicate documentation. Difficulties associated with obtaining building permissions account for this. The parties basically intend to comply with existing codes, while reserving the potential to make alterations to a scheme after it has been granted planning approval. The result is duplicate documentation, one of which is not official. It is impossible to disclose non-official documentation to other parties.

One of the assumptions of BIM is that all information should be contained in one place, allowing the parties to gain access to information simultaneously and concurrently. Duplicate projects and duplicate information are incompatible with the BIM model.

2.4 *Traditional approaches to contractual damages*

Finally the traditional approach to contractual damages is another factor stifling the widespread use of BIM in construction projects. These contracts are characterized by exclusions of damages, disclaimers for the owner and limitations of remedies for contractors. IPD aims to minimize disputes and claims by removing sanctions.

At present, the traditional Turkish contract is laden with damage provisions and boilerplate clauses. For employers and contractors a contract without sanctions would seem unimaginable.

Dikmen, in her survey of 51 Turkish contractors has tried to assess the perceptions of the construction professionals in Turkey on the concept of partnering (Dikmen et al, 2008). The survey results indicated that experience of the potential partner in similar projects, image of the company, relations of the company with clients, and financial, technical, and managerial capability were perceived to be the foremost factors in partnering relations. The findings implicate that the contractors are basically seeking elements of mutual trust in the formation of long term relationships.

Kog and Yaman report the findings of a meta-classification and analysis based on a literature survey into the relative importance of contractor selection criteria (Kog & Yaman, 2014) The results lend support to Dikmen's findings with financial standing, management capability, technical ability, reputation, health and ability.

However, long term business alliances are not uncommon in the Turkish construction sector. This type of a relationship is characterized by patterns of interorganizational exchange, long term recurrent exchanges interfirm collaborations (Sözen & Kayahan, 2011, 131). Sözen and Kayahan have studied the length of the relationship between main and specialist trade contractors, control methods and the number of rival specialty contractors in the Turkish construction industry, drawing on the data provided by a study of 27 specialist trade contractors. The results demonstrated that the length of the relationship between main and specialty contractors is inversely related to the number of rival specialist trade contractors, but directly related to the flexibility of control exercised by the main contractor.

The implications of these findings are that flexibility in control can be a cause and consequence of long term relationships. Predictability of partner behavior and trust seem to replace formal contractual governance.

Furthermore, the number of rivals within the scope of a single project was inversely related to the formation of dyadic long term relationships by disrupting the relative balance between two firms. Intense rivalry among specialty contractors may render dyadic relationships less attractive.

Supporting evidence was provided earlier by Küçük (1995) and Kayahan (1996). These two studies reported the tendencies of subcontractors to establish and maintain long term and enduring relationships with main contractors There is also data supporting the decrease of control problems between main and subcontractors with increasing informality of control methods (Kayahan, 1996).

In addition, early involvement in a project may significantly reduce the risks of delay and consequently alleviate the age old delay damages issues. Clevenger and Khan (Clevenger & Khan, 2014) report the findings of two case studies, where design to fabrication models yielded benefits in avoiding delay damages.

3 CONCLUSIONS

A common goal for the stakeholders of a project is a debatable issue. In the traditional model, parties enter the project at different stages, creating friction and interface coordination issues.

The present paper is by no means a full analysis of the contractual and legal issues for Building Information Modelling in Turkey. The paper, as noted earlier is based on the authors' long term associations with the contractors and employers in the Turkish construction industry in their capacity as expert witnesses and claim consultants.

The present paper does recognise that the lack of a compatible legal framework, traditional approaches to contractual liability and responsibility, confidentiality concerns and traditional approaches to damages have inhibited widespread adoption of BIM in private and public sector projects.

It also recognizes that in the absence of mutual goals, collaboration and integration between parties from different specialist areas requires significant effort.

It is therefore by no means easy to immediately replace the existing framework and traditional approaches with new processes that are capable of providing a better environment for BIM, such as IPD.

It seems more reasonable to revise the existing framework and to modify traditional adversarial approaches gradually. What is probably required is a specialised construction law for Turkey, with a more balanced risk distribution and with a flexibility for a multi-party system.

The authors believe that for the more basic forms of BIM, the Turkish legal framework can be used with appropriate amendments made to the contracts

Amendments in the public procurement system are also necessary, allowing for design build procurement.

Risk shifting practices of employers and Contracting authorities is not only incompatible with the basic BIM/IPD philosophy but also may lead to higher tender prices. Contractors systematically add a premium to their bids to account for added risks.

An additional issue is higher project costs generated by cost and time overruns and contractor replacement due to termination practices.

This paper also concludes that organizational cultures of the parties to a construction contract need to be similar and non-adversarial for the adoption of BIM. An implication of this is the need to investigate the inherent organizational cultures of the parties in moving from an adversarial to a non-adversarial relationship.

And training is another crucial ingredient in the adoption of new and innovative technologies that require a new mind-set. Curriculau changes in architecture-engineering faculties may help develop BIM awareness.

REFERENCES

Clevenger, C. & Khan, R. 2014. Impact of BIM-Enabled Design-to-Fabrication on Building Delivery. Pract. Period. Struct. Des. Constr., 10.1061/(ASCE) SC.1943–5576.0000176, 122–128.

Dikmen, I., Birgonul M.T., Ozorhon, B. & Eren, K. 2008. Critical success factors for partnering in the Turkish construction industry, 24th Annual ARCOM (Association of Researchers in Construction Management) Conference, 1013–1022, Cardiff, UK.

Kamu İhale Kurumu 2002a. 4734 Sayılı Kamu İhale Kanunu.

Kamu İhale Kurumu 2002b. 4735 Sayılı Kamu İhale Sözleşmeleri Kanunu.

Kayahan, O. 1996. Control of subcontractors by main contractors from the subcontractors' point of view, Master's dissertation, Istanbul Technical University.

Küçük, M.A. 1995. İnşaat Sektöründe Alt Yüklenici davranışları, Master' s dissertation, Istanbul Technical University.

Kog, F. & Yaman H. 2014. A Multi Agent Systems based Contractor Pre-qualification Model "A Meta Classification and Analysis of Contractor Selection and Prequalification". *Procedia Engineering*, 85(12/2014):302–310.

Onur, L.O, Mürsel, E. & Baykan, U. 2006. Yapim İşleri Genel Şartnamesi'nde İşveren ile Yüklenicinin Sorumluluk Paylaşiminin Proje Maliyetine Etkisi,Teknik Bilimler Meslek Yüksekokulu, 5(3):2006.

Onur, L.O. 2010 Quantitative Comparison of Responsibilities' and Risks' Distribution Between Turkish General Conditions of Construction and FIDIC Red Book General Conditions, e-Journal of New World Sciences Academy 2010, 5(2), Article Number: 1A0069.

Sözen, Z. 2015, FIDIC Genel Koşullarından Örneklerle İnşaat Sözleşmelerinin Yönetimi, Legal, Istanbul. Türk Borçlar Yasası, 2012.

Sözen, Z. & Kayahan, O. 2001. Correlates of the length of the relationship between main and specialist trade contractors in the construction industry. *Construction Management and Economics*, 19(2):131–133.

TBMM 2001 Türkiye Cumhuriyeti Fikir ve Sanat Eserleri Kanunu.

T.C. Çevre ve Şehircilik Bakanlığı, 2011.

Yilmaz, A. & Ergönül, S. 2011. Selection of Contractors for Middle-Sized Projects in Turkey. *Gazi University Journal of Science*, 24(3), 477–485.

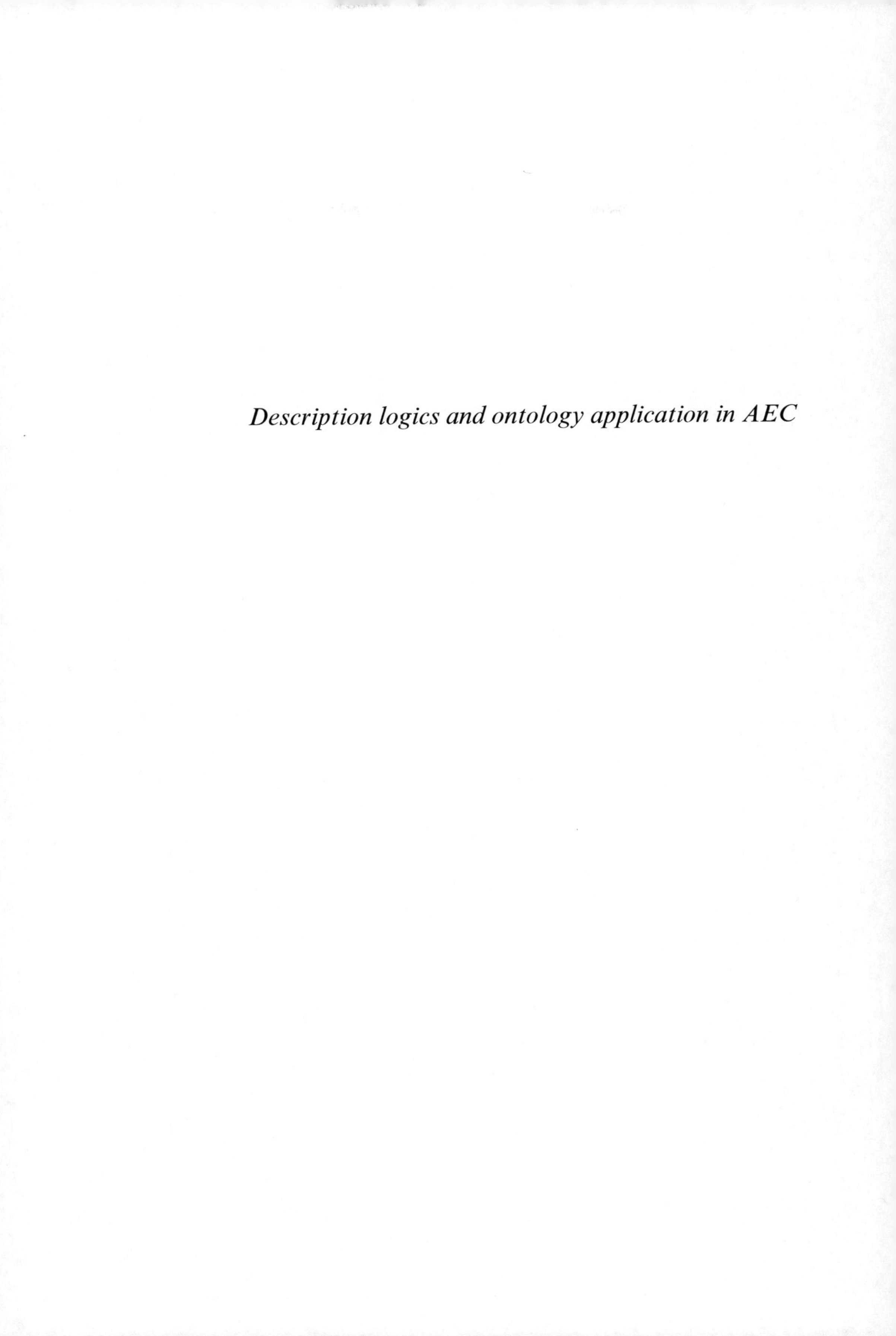

Description logics and ontology application in AEC

eWork and eBusiness in Architecture, Engineering and Construction – Christodoulou & Scherer (Eds)
© 2016 Taylor & Francis Group, ISBN 978-1-138-03280-4

Structured building monitoring: Ontologies and platform

A. Mahdavi, S. Glawischnig, M. Schuss, F. Tahmasebi & A. Heiderer
Department of Building Physics and Building Ecology, TU Wien, Vienna, Austria

ABSTRACT: Building data monitoring can provide performance feedback for operational optimisation of existing facilities and improve future designs. It can support energy and performance contracting, smart load balancing, model-predictive building systems control, and preventive building maintenance. However, a closer look at the current practice suggest that the commonly deployed technical infrastructures are not mature enough and their hardware resilience and software interoperability are in need of improvement. To address these issues, we first introduce in this paper an ontology for the representation and incorporation of multiple layers of data in pertinent computational applications such as building performance simulation tools and building automation systems. We then address common data processing requirements and exemplify a number of typical queries that building monitoring data repositories must support. Finally, we describe a specific technical platform for the structured collection, storage, processing, and multi-user exchange of monitored data.

Keywords: buildings, inhabitants, behaviour, building monitoring, building controls, performance simulation

1 INTRODUCTION

Systematic and continuous scanning of buildings' operational states can offer multiple benefits. It can provide performance feedback for operational optimisation of existing facilities and improve future designs. It can support energy and performance contracting, smart load balancing, model-predictive building systems control, and preventive building maintenance. Last but not least, systematically monitored high-resolution and high-fidelity data can advance the state of knowledge in a wide range of domains in building science, including building integrity, building automation, indoor environment, and human factors. This potential is not unknown to the relevant professional community. Accordingly, there are numerous instances of installed building monitoring systems. However, a closer look at the current practice suggest that the commonly deployed technical infrastructures are not mature enough and their hardware resilience and software interoperability may be argued to remain wanting.

Aside from this general background, the present contribution has a specific motivational grounding as it emerged in a response to requirements formulated within an ongoing Annex of the International Energy Agency Annex 66 (IEA 2016) pertaining to the computational representation of building inhabitants in view of their presence and actions in buildings. Thereby, it was found necessary to address the paucity of richly structured approaches to the collection, storage, sharing, and analyses of monitored data relevant to the inhabitants' presence/activities in (and impact on) the buildings. It is important to understand that this circumstance requires efforts far beyond the mere inclusion of classes of real and virtual sensors and meters in building information systems. We thus first introduce in this paper an ontology for the representation and incorporation of multiple layers of data in pertinent computational applications such as building performance simulation tools and building automation systems. Such data layers include, aside from those directly relevant to detection of people's presence, movement, and (control-oriented) actions in buildings, a number of external and internal boundary conditions (i.e., prevailing weather conditions, indoor environmental circumstances) as well as relevant states of buildings' devices and systems. We then proceed to address common data processing requirements and exemplify thereby a number of typical queries that building monitoring data repositories need to support. Finally, we describe a specific technical platform for the structured collection, storage, processing, and multi-user exchange of monitored data.

2 TOWARD AN ONTOLOGY FOR BUILDING MONITORING DATA

2.1 *General monitoring data categories*

Developmental efforts pertaining to representations of inhabitants in building information models require multiple layers of empirical information. Broadly speaking, five general information categories may be distinguished, namely *i)* inhabitants, *ii)* indoor environmental conditions, *iii)* outdoor environmental conditions, *iv)* systems, equipment, and devices, and *v)* energy flows.

2.1.1 *Inhabitants*

Time series of inhabitants' location (presence, movement) in building is obviously a prerequisite of all relevant modelling efforts. Likewise, inhabitants' actions (specifically, operation of indoor environmental control devices as well as equipment and appliances) are central to formulation of predictive routines. Such actions must be either directly monitored, or—in case of exclusively user-driven device and equipment actuators—extracted from corresponding device/equipment state change data. Moreover, depending on the building systems' type and configuration, inhabitants may have the possibility to control pertinent set-points for heating, cooling, ventilation, lighting, etc. Hence, inhabitant-driven changes in the values of such set-points (e.g. via operating a thermostat) need to be registered. Depending on the resolution and coverage of intended predictive models, additional data concerning inhabitants may be required. This includes inhabitants' state data (clothing, activity, physiology) as well as perceptual and attitudinal information (e.g., subjective evaluation of indoor environmental conditions). Note that, aside from technical feasibility issues, data collection campaigns addressing inhabitants may be also considerably constrained due to privacy issues.

2.1.2 *Indoor environment*

Causal theories of inhabitants' control-oriented behaviour typically involve one or more indoor environmental parameters as independent variables (e.g., ambient air temperature, illuminance levels). Collection of such data is thus an indispensable precondition of any serious model development effort. High-resolution spatial and temporal data from multiple domains (hygro-thermal, visual, acoustical, air quality) would be obviously most preferable. Practical and economic considerations may, however, limit the scope and coverage of respective monitoring campaigns and associated infrastructures.

2.1.3 *Outdoor environment*

Behavioural models frequently require information regarding outdoor conditions. For instance, prediction of adaptive actions (such as operation of windows) may need information concerning the prevailing outdoor conditions (e.g., temperature, solar radiation, precipitation, sound levels). While standard weather stations can provide a good part of required data, special modelling efforts may require additional sensory equipment.

2.1.4 *Systems and devices*

As alluded to earlier, prediction of inhabitants' interactions with buildings' control systems and devices represents one of the main use cases of behavioural models. The knowledge of the state of the relevant devices (windows, luminaires, shades, etc.) and the corresponding actuators is thus mandatory. Depending on the sophistication level of a building, monitored data might include state information from devices that can be controlled: *i)* only automatically (or centrally); *ii)* only by inhabitants; *iii)* both automatically and by inhabitants (e.g., via user override of automated control routines).

Note that only in case *ii* above there is an unambiguous relationship between occupants' actions and devices' states. As mentioned previously, inhabitants may have the possibility to manipulate the values of the control parameters of the buildings' environmental systems. For example, set-point temperatures for room heating and cooling can be adjusted via thermostats. Thus inhabitants' system-related actions include also adjusting the control parameter values, which must be continuously monitored.

In case of device states and control set-points, changes of values in successive observations denote control events or actions. In other words, such events/actions are implicitly present in the monitored state data, as they can be extracted from time series data of the state of the devices and associated control set-points. However, events/actions should be ideally accompanied with respective causes (e.g., human initiators or agents). Different formalisms could be used to address this matter in the ontology. One simple approach would be to assign an agent or actor ID to every monitored device or set-point state at time t_i, if it displays a change with respect to the previous observation at time t_{i-1}. Main instances of such actors are of course human agents or control software.

2.1.5 *Energy*

Inhabitants' presence and actions not only influence indoor environmental conditions, but have also major implications for the energy performance of buildings and their systems. This implies the critical importance of continuous energy use monitoring. Ideally, such monitoring should be conducted via a high-resolution sub-metering infrastructure, such that different energy sources and different spatial zones could be differentiated.

2.2 *Date points and their attributes*

2.2.1 *Variables*

Sensor, meters, and other data sources (e.g., simulation-powered virtual sensors, human agents) in the above five categories generate streams of information (values of corresponding variables) subject to monitoring, storage, and processing. Variables have values and various attributes.

2.2.2 *Values*

By definition, variables pertain to properties that are subject to change. This is reflected in the changing values of the variables. Observational data are typically in the category of measured (quantitative) data. The can be expressed in numeric order. Measured values of scalar nature (such as ambient air temperature) have a magnitude. Those of vector nature (such as air flow velocity or sound intensity) have a magnitude and a direction.

While most measured variables in building monitoring have values that can be expressed in terms of real numbers, there are variables whose values are not quantitative in nature. A class of such variables may be characterised as nominal data. For instance, user feed-back regarding thermal comfort conditions may be expressed in terms of classifications and categories. Another class of data, referred to as ranked data, can serve the characterisation of variables whose values display a certain order. For instance, successive positions of a valve represent a ranked order. Applied to actuators, device states may be captured in terms of a variable with a Boolean variable. Both nominal and ranked data can be made subject to quantitative operations when, for instance, variable values are mapped into a set of ordinal numbers and treated subsequently—e.g., via statistical operations.

2.2.3 *Attributes*

Spatial and temporal attributes (or extensions) can be assigned to variable values. A variable value may have a spatial attribute in that it may be assigned to a one-dimensional point, a two-dimensional plane (represented, for instance, in terms of a polygon), or a three-dimensional volume (represented, for instance, in terms of a polyhedron). In case individual sensor readings of different points are aggregated (e.g., in the course of post-processing) for a plane or a volume, the mode of aggregation (e.g., arithmetic averaging) would be of interest. A variable value usually has also a temporal attribute. Specifically, the point in time when the reading occurs must be recorded in terms of a time stamp. Moreover, sensor readings may be assigned to a discrete time interval (sampling interval).

2.2.4 *Spatial destination of data points*

Both sensors and their respective sensing target need to be specified and mapped to the building's spatial structure (topology). The spatial hierarchy of a building can be captured in 3-D representational framework such as a BIM (Building Information Modelling) environment. Depending on the embedded spatial hierarchy, sensors and their sensing targets can be associated with entities such as whole building, floors, sections, rooms, zones, workstations, etc. The spatial association of data sources and building's constitutive geometric and functional units is essential for a seamless representation of monitoring hardware, networks, topology, and architecture in a BIM environment.

2.2.5 *Specification of data sources*

The nature of the source of monitored data (sensors, meters, or human agents) must be specifically documented. This includes technical specifications pertaining to items such as measuring range, precision, and accuracy. Given the potentially extensive and heterogeneous nature of the corresponding information, the ontology would be typically including a reference to an external document. Sensors, meters, and other sources of data must be of course specified in view of their position and the respective topological reference to the location in the building model.

2.3 *The ontology matrix*

A compact representation of the categories of the aforementioned monitored variables together with a generic description of their values and sources is given in Table 1.

3 DATA PROCESSING AND TYPICAL QUERIES

The preparation of monitored data can involve very different data processing paths and options. The necessary steps of the related processing routines are strongly dependent on the specific attributes and behaviour of the data collection sequence (see Figure 1) containing the sensor, the signal convertor, pre-processing, and data storage.

Generally, data processing could be separated into two main categories, one for periodic data and the other for event triggered or event related data. The result for most typical data processing routines consists of periodic data streams with fixed intervals, whereby the time stamps are synchronized. In case periodically measured data are exported into building simulation tools, discrete values are generated in a manner to represent the averaged raw data value of the preceding interval period.

3.1 *Periodic raw data*

Periodic data is provided by systems that store measurements at regular time intervals based on an internal cycle timer. Corresponding typical

Table 1. A basic ontology of major categories of monitored data in buildings.

			Categories of monitored variables				
			Inhabitants	Indoor conditions	Outdoor conditions	Systems & Devices	Energy flows
		Monitored variable	*Presence*	*Air temperature*	*Global irradiance*	*Window state*	*Electricity use*
Value	Measured (quantitative)	Scalar component					
		Vector component					
	Nominal, ordinal, Boolean, …						
		Unit					
	Spatial attribute	Point					
		Plane					
		Volume					
		Topological reference					
	Temporal attribute	Time stamp					
		Sampling interval					
Actor		ID					
Sensor (data source)		Specification					
		Position					
		Topological reference					
		Notes					

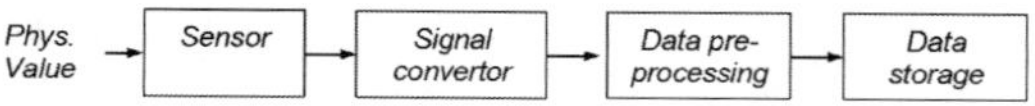

Figure 1. Data measurement, pre-processing, and storage.

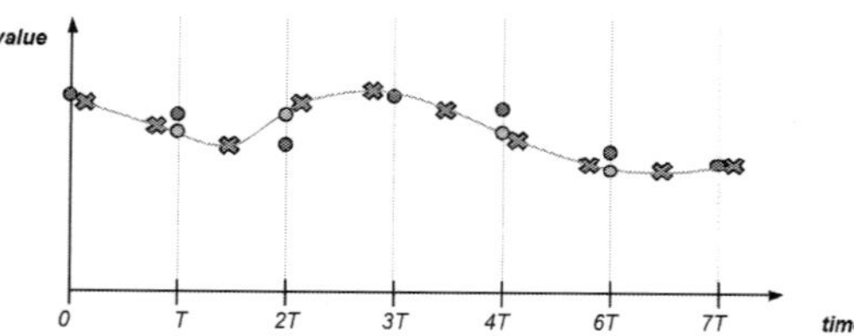

Figure 2. Illustration of measured (x), snap shots or instantaneous data (green), and generated periodic data (red).

systems are Building Management Systems and measurement systems or data loggers. The interval is usually defined by an internal setup value of such systems. A cycle timer triggers the execution of an internal polling algorithm and the data storage routine. Data processing for this type of data is mainly a simple averaging or an interpolation of the raw data as illustrated in Figure 2. The red dots show the periodically generated data as classic weighted averages of the raw data in the preceding interval. The green dots represent generated periodic values for the exact periodic timestamp based on interpolations. As mentioned earlier, for most cases where the data is used for building simulation, the first method is used.

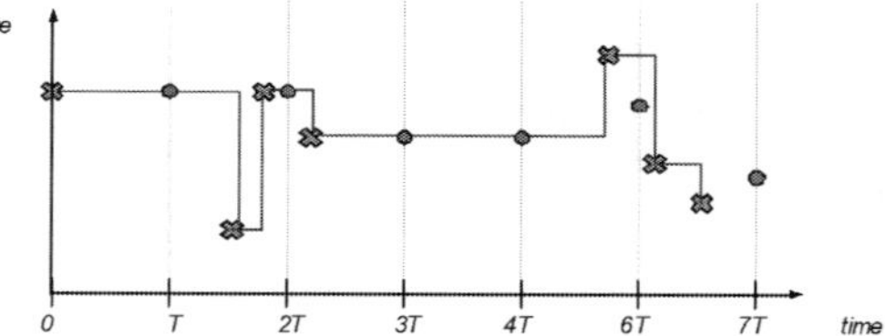

Figure 3. Illustration of event based data (x) and generated periodic data (red).

3.2 *Event related raw data*

Data monitoring systems that are triggered by events such as the detection of a movement, the opening of a door or window, activation of devices, alarms or warnings tend to store the raw data with corresponding—typically irregular—timestamps. Usually this data has to be post-processed to generate periodic synchronized data for the subsequent analysis, evaluation or export into other applications (e.g., building simulation tools). Figure 3 shows a typical trend of event based raw data together with an example of generated periodic data from a data processing algorithm. It is very common that the event based data is only stored when a change occurs. The periodic data generation process has to work in terms of a "sample and hold" process and repeats the last value as long as no new one is recorded. If more than one value was measured during an interval, different post processing options may be relevant. For instance, periodic instantaneous data may be generated using the last recorded value at each interval (red dots in Figure 3). However, in certain use cases (e.g., building energy simulation), multiple measurements within an interval are aggregated (for instance via temporally weighted averaging) and assigned as the periodic interval value.

3.3 *Interval data from BMS integrated sensors or data loggers*

Raw data from BMS logging routines usually contain time stamps that are not synchronized, hence a data processing with an interpolation and subsequent weighted averaging is necessary. An interpolation of values is always needed when the polling interval is in a similar range or smaller as the needed periodic data. Likewise, data extrapolation may be necessary to account for a gap in measurements (see Figure 4).

3.4 *Generation of occupancy schedules from PIR-motion raw sensor data*

PIR Motion sensors usually report a change of state (i.e. from occupied to vacant or vice versa). Internally, after reporting the occupied state, these sensors delay switching back to the vacancy state with an internal timer. This value could be a fixed or variable depending to the specific product. This internal function is necessary to avoid unreasonably rapid state fluctuation. Depending on the time interval of the desired periodic data, even with this sensor-integrated filtering, the raw data may contain multiple fluctuations within each interval. For the generation of typical binary occupancy schedules the raw data has to be processed as follows (see Figure 6):

1. Generation of weighted average values for each interval.
2. A sample-hold pattern to generate values for the intervals with no measurements.
3. Generation of periodic binary occupancy data using a threshold value. The designation of this threshold value requires some experience with the sensor's behaviour (latency) and is necessary to distinguish between transitional events (e.g., a co-worker passing by a non-occupied workstation in an open-plan office) and actual occupancy.

Event related data from sensors reporting not binary values has to be processed in a similar way as illustrated before but without the last step.

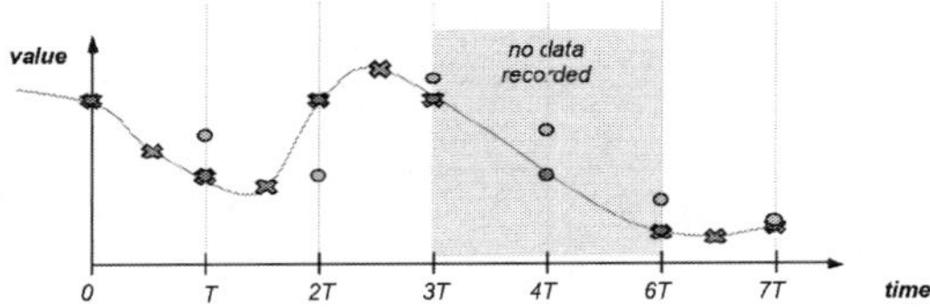

Figure 4. Data pre-processing for periodic interval BMS data.

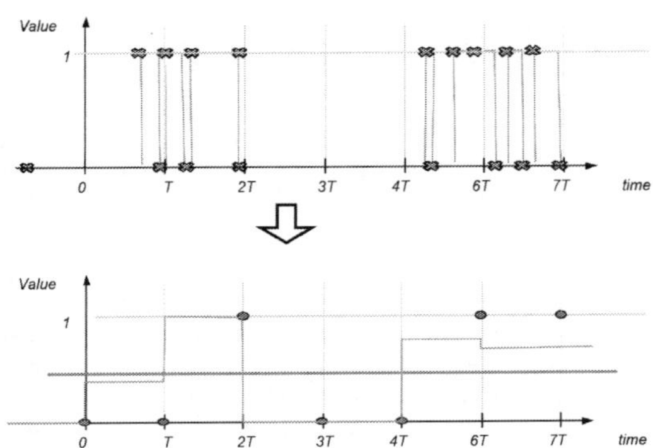

Figure 5. Data pre-processing for event related occupancy raw data from PIR sensors.

4 PROTOTYPICAL IMPLEMENTATION—MONITORING SYSTEM TOOLKIT (MOST)

To support the introduced variety of data sources and queries, MOST (Zach & Mahdavi 2012, MOST 2012) defines and implements a variety of software modules to cover data handling, (pre-) processing and access. MOST intends to provide a platform- and vendor independent toolkit for building data. Thus, the toolkit consists of open source components that are utilized to access third party (also proprietary) building information systems. The following chapter introduces the (i) software architecture of MOST (data storage, processing and access), (ii) the modules, which the toolkit consists of, (iii) the approach to accessing data sources that have previously been defined, and (iv) use cases for virtual data points.

4.1 *Toolkit architecture*

The proposed toolkit is designed to offer high-performance and scalable building monitoring functionality for a large variety of buildings. Monitoring dynamic data offers certain challenges. At times, high frequency data flows must be monitored without slowing down the system's overall performance. To accomplish this within one application, stateless core components were designed, which allow adding new module instances during runtime. These additional instances can be removed during "cool down" periods to free resources that might be

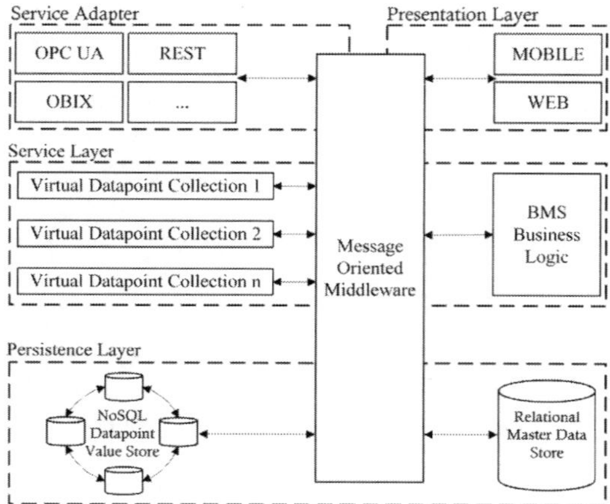

Figure 6. Illustration of the system architecture of the proposed urban monitoring framework (Glawischnig et al. 2014a).

required by other tasks. Such flexibility requires a central distribution mechanism that routes incoming requests between modules as well as different physical servers that might be located in different buildings.

As it can be seen in Figure 6, this functionality is realized by a Message Oriented Middleware (MOM), which establishes connections between the persistence layer, the service layer, the presentation layer, and the service adapters. As the proposed toolkit's core modules are written in Java, the Java Message Service (JMS) API is used. Every instance of redundant components listens to the same queue. The transferred messages can only be handled by one instance (the instance that initially took the message from the queue). Datapoints define topics that allow observers to listen to changes in the monitored values. In the proposed case, the datapoint topics are identified via the owner instance's unique name and the prefix "OBSRC_": OBSRV_ < dp_name>. For instance, the activity observation of the specific occupancy sensor "con5" (e.g. contact sensor mounted at a door) would be established via the topic ID *OBSRV_con5*. Accordingly, all entities (e.g. actors/agents) define their own queues that can be observed by specific monitoring modules. MOM thus facilitated communication between distributed redundant components.

Beside this centralized data communication mechanism, it can be seen that the application is divided into four layers. The service adapter offers access to the data collection via standardized industry protocol implementations such as (OPC UA, oBIX, REST). These services can be used to develop custom client applications. In the present

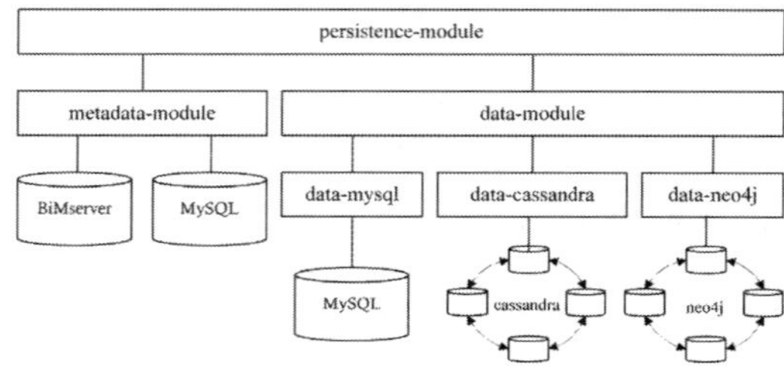

Figure 7. Persistence layer: specific module implementation.

Table 2. MOST currently implemented modules.

MOST-Module	Description
Connector (most-connector)	Driver for a data-source (Sensor/ Actor) to MOST
MySQL (most-mysql)	Handles database access for meta-data like Datapoints and ones.
Neo4 j (most-neo4 j)	Handles database requests for atapoint measurements stored in Neo4 j.
Cassandra (most-casandra)	Handles database requests for datapoint measurements stored in Cassandra.
Calibration (most-calibration)	Calibrates a building simulation model (currently EnergyPlus only) in an automated and periodic manner.
Virtual Datapoint (most-vdp)	Contains implementations of Datapoint interface to provide common access for not directly measureable data.
MOST Server (most-server)	Routes requests between different MOST modules. Only required for distributed deployments
REST (most-rest)	Exposes MOST data as RESTful web service.
OPC UA (most-opcua)	Exposes MOST data through OPC Unified Architecture.
Obix (most-obix)	Exposes MOST data through oBIX.
Web (most-web)	Provides an out of the box web application to query, visualize and export monitored data.

case, two client applications, namely a web client and a mobile client are implemented. The BMS logic (data analysis, data processing routines) as well as the virtual datapoints are implemented within the service layer.

Data handling is a vital point when developing scaling applications. In the proposed system, highly dynamic sensor data (regardless the actual sensor type) is stored in suitable data stores. As it can be seen in Figure 7, data distribution is handled by the MOST-persistence-module. This module merges configuration data with the actual datapoint values. Incoming requests are divided into two categories: (i) static configuration—metadata and (ii) actual data. Metadata and security sensitive data, such as user information and sensor configuration is stored in a relational database. The metadata-module handles configuration requests and accesses the necessary databases. The actual implementation can vary, the current implementation uses MySQL. Sensor data is stored in a suitable data store (e.g. neo4 j, Cassandra).

4.2 *Module overview*

Table 2 shows an overview of the current MOST module implementations.

4.3 *Data access*

Data access to various sensors, fieldbus technologies and building management systems is implemented in the most-connector module (Zach et al. 2012). Currently support for the following technologies is implemented:

i. JDBC compatible data sources and therefore most relational databases (MySQL, Microsoft SQL, Oracle, etc.). As most BMS implement this drivers, access to common systems is supported.
ii. OPC DA data sources, which enables access to most sensor and fieldbus technologies (KNX, BACnet, LonWorks, M-Bus, etc.).
iii. EnOcean USB 300 supporting low cost monitoring with a Raspberry Pi and energy harvesting sensors (Zach et al. 2014).

Beside the common APIs, future work could focus on support for IoTSyS (Jung et al. 2012), which allows heterogeneous access via Obix to several building automation systems such as BACnet, KNX, ZigBee, DALI, and Lon Works.

As mentioned in the previous sections, data access can be granted via various standardized industry protocol implementations. Based on the previous work by Hofstätter (2012), opcua4 j, the module OPC Unified Architecture (UA) server was implemented. Zonal information is used for the OPC UA address space. A zone connects to its datapoints with a hasComponent reference. Attributes such as a datapoint' s unit are represented as OPC UA properties. This module enables data access for all processing applications supporting the OPC UA interface.

The module oBIX provides data access based on the Open Building Information eXchange standard (oBIX 2014). This standard provides data with the default contracts *Points*, *Alarm* and *History*. To support the proposed pre-processed queries a new contract was developed.

4.4 *Pre-processing*

Standardized data definitions are used to exchange data and to calculate data adjustments. This way, data processing and data processing algorithms can easily be added at runtime. As discussed previously, certain data sources require a specific query logic. To ensure a customizable, reusable data processing process, the *most-preproc* library was introduced. As it can be seen in Figure 8, certain algorithms, such as data comparison and data validation can be used in a stand-alone library. Moreover, the *preproc* library includes algorithms to calculate periodic data.

4.5 *Virtual sensor implementation*

The aforementioned virtual sensors represent a vital concept in the proposed monitoring structure. Virtual sensors can provide data regarding phenomena that cannot be directly measured with physical sensors (e.g. average temperature across multiple zones). Currently, MOST includes the following virtual sensor implementations in the *most-vdp* module, which can easily be extended as the modules can be redundantly deployed:

- A virtual datapoint, which takes the mean surface temperature of a radiator and calculates its heating power based on radiator specifications and the (applicable) room temperature.
- A virtual datapoint wrapping the MOST domain specific language (*most-DSL*) implemented in Scala. It enables users to weave datapoint's values into mathematical expressions where particular values are evaluated at runtime based on the requested timeframe for evaluation. An expression computing the average temperature in °C of two datapoints *"tem1"* and *"tem2"* would be written as follows:

dp("tem1") + dp("tem2")) / 2

Integrating *most-DSL* as a VDP allows nesting an arbitrary graph of *most-DSL* expressions, whereas loops are not allowed. Assuming the last expression would be accessible as the VDP *"avgTem"*, we could build a new VDP converting the result to °F:

dp("avgTem") * 1.8 + 32

- A prototype of a simulation based virtual datapoint is currently in development. Using the simulation tool EnergyPlus (EnergyPlus 2016) it calculates data, which is not directly measured (e.g. thermal comfort). A calibrated simulation model is used to provide reliable simulated results.

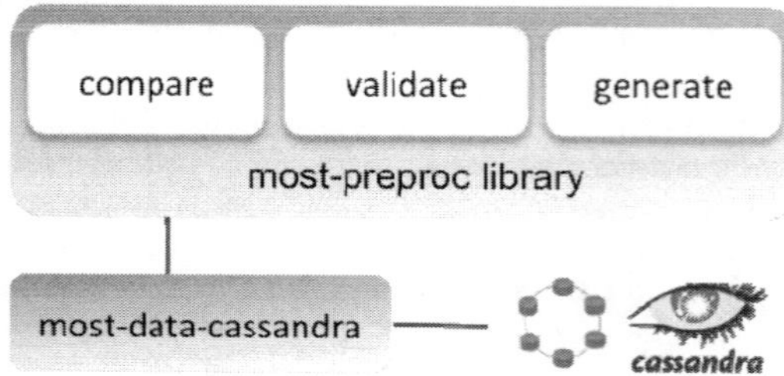

Figure 8. Database- pre-processing library communication (Glawischnig et al. 2014b).

5 CONCLUSION

In this paper, we highlighted the multi-faceted potential of building monitoring in supporting the building design, delivery, and operation process. However, we argued that the current practice in this area in not optimal, as there is al lack of richly structured approaches to the collection, storage, sharing, and analyses of monitored data, particularly as relevant to the inhabitants' presence/activities in (and their impact on) the buildings. To address this circumstance, we first introduced an ontology for the representation and incorporation of multiple layers of data (inhabitants' presence, control actions, indoor climate, outdoor conditions, device states) in pertinent computational applications such as building performance simulation tools and building automation systems. We then enumerated a number of common data processing requirements and typical as relevant to building monitoring data repositories. To provide a concrete instance of a modular and scalable monitoring system architecture, we described a specific technical platform for the structured collection, storage, processing, and multi-user exchange of monitored data.

REFERENCES

EnergyPlus 2016. EnergyPlus Engineering Reference. http://apps1.eere.energy.gov/buildings/energyplus/. accessed: 20.02.2016.

Glawischnig S., Hofstätter H. & Mahdavi A. 2014a. A Distributed Generic Data Structure for Urban Level Building Data Monitoring. In *Information and Communication Technology*, Vol. 8407.

Glawischnig S., Hofstätter H. & Mahdavi A. 2014b. Applications of Web-Technologies in Building Information Systems: the case of the Monitoring System Toolkit (MOST). In *2nd ICAUD – Proceedings.*

Hofstätter H., 2012. opcua4 j - open source implementation of an opc ua server in java. Available: https://code.google.com/p/opcua4 j/.

IEA 2016. http://www.annex66.org/. accessed: 23.03.2016.

Jung M., Chelkal J., Schober J., Kastner W., Zhou L. & Nam G., 2014. IoTSyS: an integration middleware for the Internet of Things. In *Proceedings of the 4th international conference on the internet of things.*

MOST. 2012. Monitoring System Toolkit, http://most.bpi.tuwien.ac.at. accessed: 10.03.2016.

oBIX 2016. Open Building Information Xchange. http://www.obix.org. accessed: 20.01.2016.

Zach R. & Mahdavi A. 2012. MOST-designing a vendor and technology independent toolkit for building monitoring, data preprocessing, and visualization. In *Proceedings – First International Conference on Architecture and Urban Design.* Vol. 1.

Zach R., Hofstätter H. & Mahdavi A. 2014. A distributed and scalable approach to building monitoring. In *Proceedings of the 10th European Conference on Product and Process Modelling. pp 231–236.*

eWork and eBusiness in Architecture, Engineering and Construction – Christodoulou & Scherer (Eds)
© 2016 Taylor & Francis Group, ISBN 978-1-138-03280-4

SemCat: Publishing and accessing building product information as linked data

G. Gudnason
Innovation Center Iceland, Reykjavik, Iceland

P. Pauwels
Department of Architecture and Urban Planning, Ghent University, Ghent, Belgium

ABSTRACT: Building product information is still mostly distributed, heterogeneous, unstructured, with inconsistent semantics and communicated using low technology means. Over the last decade, there has been little progress towards change. Architecture, Engineering and Construction (AEC) practitioners still face challenges when searching and acquiring necessary information and re-using it in their ICT tools. Whilst manufacturer information is more focused on trade and business processes, the demand for precision technical data by designers and engineers is continually growing with more mature and complex standards and ICT tools used in energy performance analysis, Life-Cycle Cost (LCC), Life-Cycle Assessment (LCA) and BIM processes. The majority of product manufacturers are small, regional companies, many with limited ICT capabilities and resources to keep up with this growing demand. In this paper we report on an enhanced approach to provide standard structured building product information with consistent semantics across multidisciplinary processes. The approach is based on Linked Data technologies.

1 INTRODUCTION

1.1 *Sharing private building product data*

The building industry is predominantly comprised of small companies (95% with less than 20 employees) that operate in dynamic, temporary, multi-organizational value chains that make it difficult to bond long lasting permanent relationships. A significant part in this dynamic value chain relies on information about building products, which represents an estimated 40% of the total costs. Digital building product information is one of the key drivers for digital building success and as such a valuable asset for manufacturers as well as architects and contractors. The following question has always been key in this context: *how do we provide information about building products that can be transparently shared, exchanged and re-used in different domains across the building life-cycle?*

However, this question is always obscured by a seemingly *conflicting commercial interest* towards maintaining a competitive position in the market as a product manufacturer or a supplier. Namely, manufacturers, especially Small and Medium Enterprises (SMEs), are not economically motivated to openly share detailed building product information if there are many overlapping and even conflicting industry specifications with incompatible semantics that means high investment costs for manufacturers to comply with. Also, an important aspect for manufacturers is maintaining a competitive edge over their competitors: manufacturers need to be able to maintain their distinctive market values and differentiate product characteristics also in terms of digital building product information. Furthermore, manufacturer information has traditionally been more focused on trade and business processes (e.g. production and inventory management, ordering, sales and after sales and logistics). This is changing, as a stronger response is required to the growing demand for precision technical data by designers and engineers, who use increasingly mature and complex standards and ICT tools for energy performance analyses, Life-Cycle Costing (LCC), Life-Cycle Assessment (LCA) and other Building Information Modelling (BIM) processes.

1.2 *Increasingly semantic digital means for sharing building product data*

With the advent of the BIM in combination with the World Wide Web (WWW), building product information publication transitioned to online web pages, web portals and other sorts of online directories, but mostly replicating previous paper-based media. The paper-based and online web presentations have always to a large extent aimed at human-readability. Recent developments aimed at making information available in an object-oriented or at least structured form (CSV, XML, JSON) to support digital building and BIM processes. This has made a significant disruption in this traditional publication process of building information.

Also the building product representations follow this evolution, starting from isometries and 2D paper-based drawings to the delivery of 3D object models in various file formats, including major proprietary file formats (RVT, DWG, DGN, and so forth) as well as neutral data formats (IFC). These are generic parametric BIM objects representing building elements and products (e.g. windows, doors, sanitary equipment, HVAC elements, floor coverings) that are made available in BIM object libraries in relation to specific design software. Building product manufactures are driven by industry to similarly provide BIM objects and climb on board the BIM train.

Advanced BIM object technologies offered by BIMobject™ and the NBS National BIM Library have made a substantial impact in servicing product manufactures to this extent. These repositories typically contain predefined BIM building blocks associated with documentation, e.g. specifications and product descriptions. This product documentation commonly contains the textual description of the product models, including their functions, dimensions, materials, performances, and so forth.

1.3 *A need for standard Product Data Templates (PDTs)*

There are, however, many considerations that manufactures must take into account when deciding on creating BIM objects for their products. For example; does a BIM object add value to my customers, for which products in my product range do I need to have BIM objects, how much detail will I need to specify, which software formats do I need to support, how do I handle different professional and process information requirements and, last but not least, will there be a return on my investment and subsequent increase in sales. Given that manufacturers are motivated to embrace BIM, high investment and operational cost in creating and maintaining BIM objects is still a major barrier for a majority of midsized and small manufacturers.

An alternative, more economical and practical solution for manufacturers to deliver BIM data has been around for a decade now based on standard Product Data Templates (PDTs) specifying different product types. The Specifier's Properties information exchange (SPie) is an open standard for the exchange of building product, equipment, and material information that defines over 1200 PDTs for individual products and assembled products (East et al. 2011). The key to the PDT approach is having standard structured building product data that enables sharing and exchanging it between different tools used in different stages of the project life-cycle. The template model and exchange format are based on open industry standards such as ISO 16739 for model and objects properties, ISO 10303 Format, ifcXML and FM handover MVD. This approach of starting from open industry standards and standard structured PDTs allows manufactured building product information to be available across BIM processes, from building conception to operation.

As a second example, the UK counterpart to the above approach was initiated in an industry effort (http://www.constructionproducts.org.uk) and implemented by NBS as part of the NBS BIM Toolkit (https://toolkit.thenbs.com), to allow manufactures to meet the UK Government BIM level 2 requirements. The NBS BIM Toolkit currently contains over 5000 PDTs for various product ranges (Chapman I. 2015). The SPie and NBS initiatives ultimately focus on handover of as-built data in Cobie for operational stages. They vary in detail, however. SPie sets its focus on project-related product data for operational purposes whereas the NBS PDT also contains more detailed general data usable in other BIM processes.

The coBuilder goBIM (http://gobim.com), a commercial platform, as a third example, also adds value in their PDT structure by supporting manufacturers in delivering EU mandated Declaration of Performance (DoP) and voluntary, but recommended Environment Performance Declaration (EPD), besides delivering general product properties, IFC and Cobie data.

Different from BIM objects, PDTs are specified and developed in harmony by industry specialists with varying relevant backgrounds, e.g. professional associations, trade association, industry standard bodies, and specifier and product manufacturer organizations. The idea of this joint effort is to agree on a common shared view in specifying product types, including properties, naming, units etc. and thus guaranteeing a common set of properties and attributes in accordance with harmonized, national and industry standards and guidelines.

If such a standard set of PDTs can be devised (left in Figure 1), product manufacturers are able to use their in-house PDFs and product databases, while also making them available using the agreed templates (center in Figure 1). Moreover, the templates can be used to import and export the product data supplied by the manufacturers to diverse tools in the building life-cycle (right in Figure 1). Hence, the level of standardization involved in

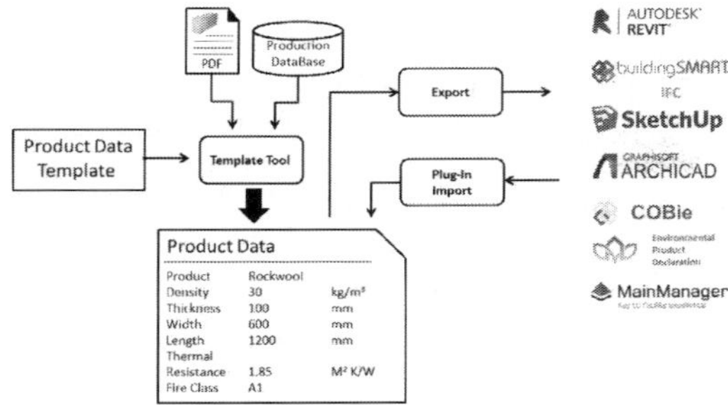

Figure 1. Product Data Template open BIM support.

PDT development will greatly enhance product data interoperability by providing the right information exchange for use by various tools either by exporting the information in a computer-readable format or through a tool specific plug-in.

Whilst the PDT approach offers undisputable benefits and added value across the construction value chain, their marketing is ultimately based on the same business models that have existed for decades where information providers collect product information in web portals from manufacturers, usually for a fee, to allow designers to search, analyze, compare and reuse the information.

The SemCat approach we introduce in this paper provides opportunities for development of new and innovative business models in the supply chain based on Linked (Open) Data technologies and standard semantics to, for example, supporting supply chain networks in linking product data with other relevant data sets, creating customized data sets for specific stakeholders and providing suppliers with detailed and structured product data, and so forth. This approach is presented in Section 2, including the way in which semantics can be published and accessed for building products across multidisciplinary processes. We hereby present a sample use case, indicating industrial set up and relevance. In Section 3, we give an indication of related work in terms of already existing building product ontologies, and Section 4 outlines how such ontology can be used from within a set of building product templates. The last section then outlines discussions, conclusions and possible future steps.

2 BUILDING PRODUCT INFORMATION AS LINKED DATA: A SAMPLE USE CASE

2.1 *Why linked data?*

Linked data technologies are meant for representing any information that would be available in a global web of data in a human- and computer-readable format. As a result, linked data technologies allow by design to represent nearly any kind of data as Resource Description Framework (RDF) graphs, including product manufacturer data. By taking this step, the data would become available as RDF data, supplementing other AEC-related data that can also be made available as RDF graphs, including regulations (Beach et al. 2015) and building data (Pauwels et al. 2016). As a result, the data can more easily be found on the Web, so that it can be used by a diverse type of services, including semantic (federated) search algorithms, reasoning and rule-checking systems, building performance analysis services, and so forth. Hence, these technologies can make product manufacturer data more easily available for AEC practitioners (Costa & Pauwels 2015; Gao et al. 2015). This would in turn increase visibility and market reach for the product manufacturers and make their data available industry-wide.

Making building product information available as linked data is not just a matter of applying technology. There are quite a number of barriers and approaches in making building product information available as linked data. One of the key issues is deciding on the ontology that is to be followed. Every domain and, in fact, every user relies on his own terms, concepts and relationships (see also Beach et al., 2015). The ontologies used by product manufacturers will thus notably differ from the ontologies used by AEC practitioners. This is crucial to the idea of a PDT as displayed in Fig. 1, as it allows communication between the native sources of the manufacturer (top in Fig. 1) and the in-house tools of the AEC practitioners (right in Fig. 1). A system needs to be devised that makes the building product data on the one hand interpretable by AEC practitioners, but, on the other hand, leaves enough room to the product manufacturers to still remain in control of the data they publish. This is why PDTs have a significant role, as they specify in a bottom-up industrial-driven manner how the data should be structured in order to make them semantically exchangeable.

2.2 *The proposed business case: SemCat*

Our work in terms of a business case focuses on the needs of small manufacturing companies and provides an Excel-based data management tool for structured building product information. The aim is not to develop yet another specification for building PDTs, but to develop a user friendly tool, SemCat, to assist manufacturers in supplying structured data about their products using existing product data templates such as NBS PDTs or SPie. Maintaining dozens or hundreds of Excel data sheets containing the product data structured according to different PDTs, as with the current set-up, is thus one of the challenges for manufacturers that we wish to address. By using the appropriate web technologies, agreed PDTs, and intuitive (for manufacturers: Excel) interfaces, we aim to address it with the proposed tool.

In addition, when product data is available in Excel documents, it still needs to be published so we can be sure it will reach prospective customers. The tool aims to simplify this process as well. It namely aims at enabling manufacturers to manage various needed PDTs, as well as the product information in product catalogues, along with other business data in a single source. As such, the tool adds business value for manufacturers in directly delivering product information for different business processes.

2.3 *The use case tool: SemCat*

Figure 2 shows the proposed modular software configuration of the SemCat tool. The tool is currently available in an Excel interface. It consists of two parts, 1) an Excel template which defines the company and product catalogue structure and various predefined data sets such as classification tables, properties and units (top of Fig. 2), and 2) an Excel plug-in, which provides the user interface and SemCat functionalities (center of Fig. 2).

The SemCat Excel Plug-in has four components: the Template Manager, the Catalogue Manager, the Export Template Engine, and the Catalogue Registration. The Template Manager allows 1) development of product data templates using internally defined property sets and their export in Excel format (left in Fig. 2) and 2) import of existing product data templates e.g. NBS and SPie formats using plugin data translators (right in Fig. 2). The Catalogue Manager allows end users to select PDTs and fill in relevant product data in a user friendly interface with various built-in measures such as a multilingual data entry. The two main other features, the Export Template Engine and the Catalogue Registration, facilitate end users in exporting the product catalogue data to different data formats and in notifying product data archives of the catalogue and version updates respectively (bottom in Fig. 2).

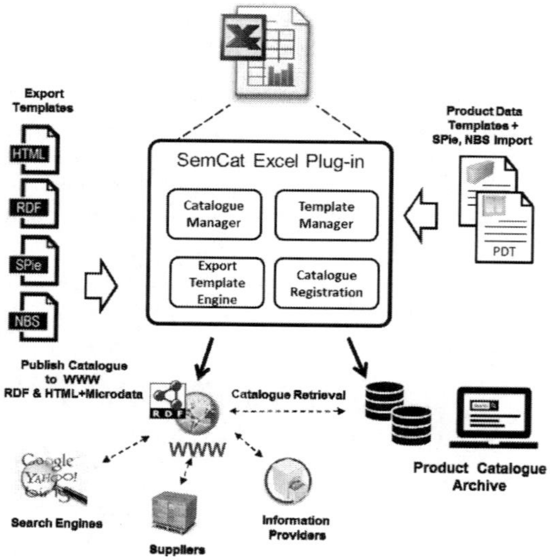

Figure 2. Overview of the SemCat tool.

The catalogue export facilities are based on user customizable external scripting templates, loaded at run-time that can facilitate practically any type of data format export including the NBS and SPie Excel product data sheet format. Most important here is the possibility of making the product data available in RDF graphs, thereby allowing the targeted Linked Data functionalities. Pre-installed scripts allow export of company and catalogue data in two ways 1) as a product catalogue represented in RDF and 2) as product web pages with embedded semantic mark-up in Microdata/RDFa for company and product data. Both are meant to be included in the manufacturer website or as HTML snippets in existing HTML product pages.

Through this approach we hope to provide open standard multilingual semantic product data by using industry established PDTs. This allows:

1. search engines to index at property level and thus facilitate searching products that match the designed product specification using more meaningful semantic search approaches,
2. information providers with product catalogues based on a standard ontology i.e. schema.org to facilitate innovative business ventures in linking various construction data with manufactured product data,
3. suppliers and product distributers' access to detailed structured product and business data e.g. packaging and shipping,
4. supporting BIM object development with external detailed product data, generic objects or manufacturer specific, through links to the actual product data by their unique web URI or class of products by classification code thus supporting data requirements of non BIM tools e.g. an energy simulation tool,
5. extending the product data by linking to company specific and market data (e.g. suppliers and identifying location, regions, countries where the product is eligible for shipment etc.),
6. and finally making the product data part of the global web of data.

3 (BUILDING) PRODUCT ONTOLOGIES

If we wish to make (building) product information available as linked data (bottom in Fig. 2), we need one or more ontologies that allow structuring the information so that it is useful for the stakeholders using the information down the value chain (architects, contractors, facility managers), while maintaining usability by the product manufacturers. We propose to represent these ontologies in the Web Ontology Language (OWL), so that they can easily be used to generate RDF data and RDFa or microdata. These ontologies are at a minimum to be used when exporting the RDF data using the SemCat tool, but they could eventually also be used for storing and publishing the native product data. As usability will mainly be guaranteed by the availability of an intuitive user interface for product manufacturers and building product templates (see Section 4), we will here primarily focus on available ontologies for representing the actual building product information.

There are a couple of strategies and ontologies that can be used to capture building product data. The choice in opting for one or the other strategy depends a lot on what the data will afterwards be used for. This is closely related to information exchange and software architecture diagrams as the one in Figure 2. In the considered use case (Section 2), we relied on the GoodRelations ontology (Hepp 2008). This is a powerful e-Business ontology for publishing information on commercial products and services in a way friendly to search engines, mobile applications and browser extensions. The ontology is industry-neutral, meaning that it can describe products and services in commercial airline companies as well as construction companies.

The main counterargument against GoodRelations would be that it aims primarily at the commercial end user interface (cost of products and services) and less at the actual product data (e.g. energy performance features, material data, structural strength, durability aspects), which is precisely what would be of value in the construction industry. An existing example that relies on GoodRelations is the BauDataWeb portal (http://semantic.eurobau.com), which lists building materials European-wide.

If more precise and trustworthy technical information is to be stored, so that decision support can be provided to various stakeholders (clients, architects, contractors, engineers, facility managers) throughout the entire building life cycle, one will need an ontology that supports representing and providing such more trustworthy, technically useful and certified data. In this case, it might be an option to relate to an ontology that is centrally managed and provides central data quality certification. To some extent, the buildingSMART Data Dictionary (bSDD) aims to provide such a service. However, the way in which the bSDD is currently built, it provides at best a solid terminology that can be used to tag external object models in order to make them semantically more meaningful and allow classification for example. As an example, it would be possible to model building products in any authoring tool (BIM tools, but also spreadsheet tables and textual documents) and declare links to the specific terms in the bSDD, so that the authored building products become more well-defined. The downside of this approach is that the semantic enrichment mainly occurs through a tagging approach, leaving the actual data (BIM objects, spreadsheet data, textual documents) semantically less meaningful and diverse.

Third, it would be possible to represent building products using the IFC data model (possibly linked with the bSDD) and as such build a database of 3D building products that are directly linked to relevant product data (cfr. Fig. 2). From the IFC models, it is possible to directly generate 3D models with relevant product data included in BIM authoring tools like Revit, ArchiCAD, Tekla and so forth. In this case, the ifcOWL ontology (Pauwels & Terkaj 2016) would form a good place to start modelling building product data.

A fourth and last alternative comes close to the first and third approach presented in this section, and involves creating one or more new dedicated ontologies that focuses specifically on the kind of data it wants to capture. Such an ontology thus has a specific scope, similar to the GoodRelations ontology that scopes on capturing market data (prices, availability, contact data, etc.), and similar to the ifcOWL ontology that scopes the description of building data in a component-based manner (as opposed to space-based ontologies). Instead of opting for an industry-wide building data ontology, like ifcOWL, however, one could set the scope of an ontology also to the manufacturer side in construction industry, or even more narrow, to the building products offered by individual manufacturers. For example, individual window glazing companies, precast concrete suppliers, or roof tile resellers could develop their own distinct in-house ontologies to capture their data (which is likely close to the already available in-house structures and databases). The downside of this approach is that anyone outside of the scope of the ontology has difficulties in finding and reusing the information. This downside is real and important, as this is already one of the main challenges in information search and retrieval of building product models that are offered in all sorts of formats (RVT, DGN, IFC, …), but which are only searchable via their (limited) metadata (Gao et al. 2015).

To conclude, the approach that we wish to promote here is the usage of dedicated ontologies, namely those targeting the representation of technical building product data that can be used throughout the building life-cycle by various stakeholders, but with a large enough scope. There exist a number of candidate structures and ontologies. If considering IFC to be inappropriate because of its focus on buildings, an alternative example is ETIM (Electro-Technical Information Model-http://www.etim-international.com/), which aims at structuring features of electro-technical products.

4 BUILDING PRODUCT TEMPLATES

As outlined in Section 1, several existing schemes already define standard product data templates. Each of these has different emphasis, structure, and detail and each targets different end usages. The SemCat tool has its own internal structure for product data templates based on a collection of predefined product property sets (e.g. thermal, hygrothermal, environmental, structural, sustainable, service life, packaging, production etc.). All in all, about 200 defined properties are available that use the UN/CEFACT Common Code for specifying the unit of measure.

Although SPie and NBS PDTs can be modelled using the built-in data and template structure, with many property terms in common, it is not practical as each of the template schemes potentially uses different terminology and naming conventions for properties as well as in expressing units of measure. Supporting many different template schemes using the internal template model could result in complex and unsupportable mapping arrangements or substantial overlapping in property terms. The SemCat tool therefore imports NBS and SPie PDTs preserving their original setup and vocabulary for properties and units without any type of mapping.

The route taken for implementing the SemCat PDT structure focused on minimizing data entry by end users and flexibility, allowing manufacturers to add proprietary data attributes as an extension to templates. In Figure 3, for example, a typical template structure is shown for a Rockwool insulation product. First of all, all templates belong to a classification group, in this case the OmniClass™ table 23 Products. Secondly, the template structure divides product properties into two models, a master product and a variant product. The relationship between the two is simple. The master product specifies properties that are shared among all the product variants whereas; the variant product contains unique properties that belong only to the variant product. To illustrate this, insulation is produced in different thicknesses, but in same density. Density therefore becomes a property of the master product model, whereas thickness and thermal resistance vary based on thickness for each product article and therefore become properties of the variant product model, as shown in Figure 3. Thirdly, properties in a template can be *metadata*, for descriptive purposes (e.g. version information, template type), *fixed*, meaning that the value for the property is fixed for all instances of the product template, *mandatory*, meaning that its value is required as a minimum, *recommended*, meaning that the manufacturer is expected to enter a value for this property and/or *optional* allowing manufacturers the option to enter data for such properties.

Users can define or import as many product data templates as they need for the product catalogue. For example, a manufacturer can create templates for internally maintaining product properties required for generation of EPDs and import

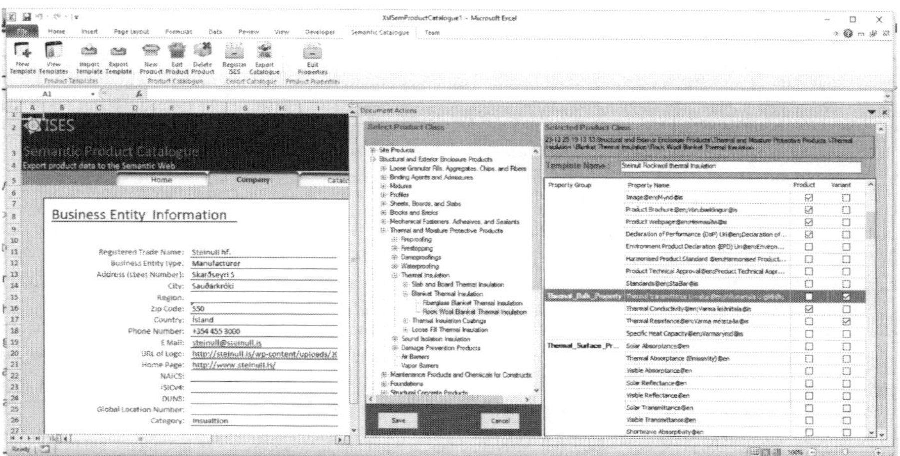

Figure 3. SemCat template manager prototype.

templates for NBS PDTs and/or SPie for the same product in one or more catalogues to support format requirements of different end users. Ideally, product data templates should be developed and distributed by individual product manufacturer associations using expertise in each product field (e.g. brick manufactures, glazing manufacturers etc.). If this is not the case, the SemCat tool offers manufacturers the ability to specify PDTs based on standardized structure and data sets (e.g. vocabulary and terminology) that can make an impact for re-usability and interoperability in the short term.

Whilst PDTs are specifically directed at manufactured products, generic products as used in many engineering analysis tools can similarly be specified using the tool just like for manufactured products. The only difference is that the generic products are linked to a specific generic manufacturer. This will enable semantic data tools to querying the linked data for generic products in the same way as manufactured data.

5 BUILDING PRODUCT CATALOGUES (AS SEMANTIC WEB DATA AND LINKED DATA)

The SemCat user interface is multilingual. In the catalogue tab the user sets the default catalogue language as shown in Figure 4. This will display all text properties in the chosen language. Entering product data is quite straightforward, as the end user selects the product type template from the list of available templates and the tool builds the user interface accordingly (Figure 4). Furthermore, the tool provides many features in the user interface to assist the end user in entering property values, for example multilingual text fields, selection lists for enumerations etc. In accordance with the explained linked data approach and business model, other product relevant information can be linked to the product data as shown in the lower tab area as an example accessory and spare parts, consumables, similar products and older and newer models.

The main objective of the tool is to make the standard product data usable by different tools across multidisciplinary AEC processes. The catalogue export facility, as noted earlier in Section 2, uses script based template files, which essentially are documents whose script elements collectively dictate how the template processing engine will process the catalogue data to generate the desired output format (e.g. RDF, gbXML, ifcXML, ISO 10303 part 21, HTML, JSON, and Excel). The set-up allows users to adopt and modify existing template files and create customized templates for processing the catalogue data and thereby providing support for different exports. The tool supports several data exports up-front focusing on semantic web formats through preinstalled script templates such as:

1. Template that generates a HTML web page for each product in the catalogue and the company data with embedded semantic mark-up in Microdata or RDFa format. These will produce either a complete web page or a HTML snippet that can be included in existing web pages. The template can be modified by the manufacturer and adopted to specific requirements of the product web site with relatively small effort. Furthermore, a skeleton sitemap (sitemap.xml) is automatically created for the generated web pages, thus facilitating search engine optimization and web site indexation.

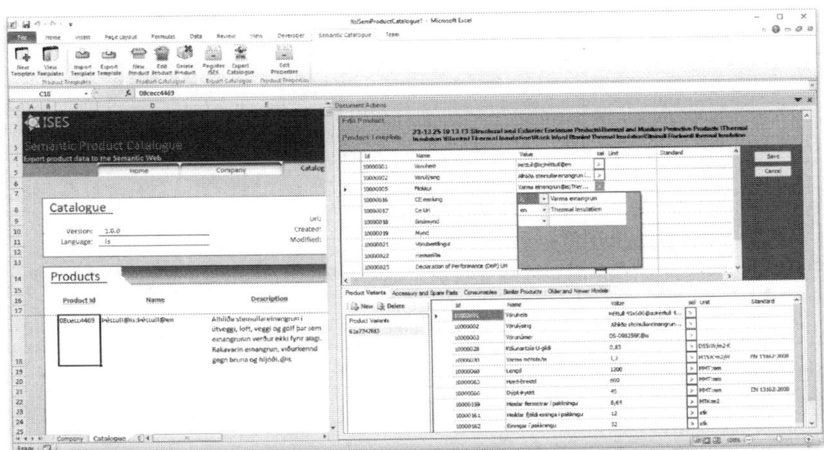

Figure 4. The SemCat user interface is multilingual (Icelandic in this case) and provides an intuitive user interface to add and manage product data.

2. Template that generates an ontology for the company and catalogue data expressed as an RDF graph. The generated product ontology hosted on the manufacturer website will become part of the growing Web of Data in the AEC sector.
3. Template for generation of the classification ontology. The current classification table in the tool is the OmniclassTM table 23 products, however, any other classification system tables can be used for this purpose e.g. UniClass, eCl@ss. The method used for this ontology transformation is described in (Hepp & de Bruijn 2007)
4. Template for generation of the properties ontology. The schema.org (GoodRelation) lacks the definition of properties needed for detailed description of building products, but provides guidelines as how to extend the ontology to include various properties (Radinger et al. 2013). Whilst this has more limited capabilities compared to data modelling schemas in the AEC industry, for example, the IFC schema (ISO 16739) and as required for modelling properties with complex values such as material moisture transport properties. It nevertheless provides a solid starting point for producing standard product descriptions and facilitating structured product data exchange in the AEC domain.

6 CONCLUSIONS

In this experiment we have examined the feasibility of using a simple, easy to use tool to develop standard generic and manufactured product data using linked data technologies in a way that the product data sets do not need to be developed or hosted by commercial information providers at substantial cost to manufacturers, but is manageable within the manufacturing organization using relatively limited IT resources and knowhow.

Publishing building product information as Linked Data largely depends on the scope of the ontology selected. The GoodRelations ontology, as discussed earlier, has a number of drawbacks for publishing building product information as linked data, especially when expressing the numerous and complex properties used in the life-cycle of buildings. There is also no concept for a product catalogue in the ontology, which is necessary when modelling product catalogues, but metadata such as for versioning, responsibility information are needed to effectively use and manage the product catalogues. Whilst these drawbacks exist, they have, to a point, been previously addressed in (Radinger 2013) and enhanced using PDTs as illustrated in this paper. Nevertheless, the development of a dedicated ontology for manufactured building products remains an open issue.

The SemCat tool is still in its development stage with several limitations in the current design that will need to be addressed in following versions. To name a few: synchronization of existing product catalogues with new versions of PDTs will need to be implemented; the development of a more consistent method to structure and naming property sets in the properties table is required; and further alignment with SPie and NBS PDTs needs to be implemented. Formal definitions of enumerations for enumerated properties need to be considered and support for product DoP and EDP publication would greatly add value to manufacturers and facilitate further acceptance by manufacturers.

The industry established Product Data Template approach has made a considerable contribution towards making product information available as standard structured data across BIM processes. The enhanced approach as presented in this paper, further builds on this work to publish product data as Linked Data allowing building product information to be part of the growing Web of Data in the AEC industry.

REFERENCES

Beach, T.H., Rezgui, Y., Li, H., & Kasim, T. 2015. A rule-based semantic approach for automated regulatory compliance in the construction sector. Expert Systems with Applications 42: 5219–5231. DOI: 10.1016/j.eswa.2015.02.029.

Chapman, I. 2015 PRODUCT DATA TEMPLATES FOR MANUFACTURERS, https://toolkit.thenbs.com/articles/pdts

Costa, G. & Pauwels, P. 2015. Building product suggestions for a BIM model based on rule sets and a semantic reasoning engine. Proceedings of the 32nd International CIB W78 conference, 98–107.

East, B., McKay, D., Bogen, C., & Kalin, M. 2011. Developing Common Product Property Sets (SPie). Computing in Civil Engineering (2011): pp. 421–429.

Gao, G., Liu, Y.-S., Wang, M., Gu, M., & Yong, J.-H. 2015. A query expansion method for retrieving online BIM resources based on Industry Foundation Classes. Automation in Construction 56: 14–25. DOI: 10.1016/j.autcon.2015.04.006.

Hepp, M. & de Bruijn, J. 2007. GenTax: A Generic Methodology for Deriving OWL and RDF-S Ontologies from Hieratical classifications, Thesauri, and Inconsistent Taxonomies. In the proceedings of the 4th European Semantic Web Conference (ESWC 2007), p. 129–144 Innsbruck, Austria

Hepp, M. 2008. GoodRelations: An Ontology for Describing Products and Services Offers on the Web In A. Gangemi and J. Euzenat (Eds.): EKAW 2008, LNCS 5268, pp. 329–346, 2008. © Springer-Verlag Berlin Heidelberg 2008

Pauwels, P. & Terkaj, W. 2016. EXPRESS to OWL for construction industry: Towards a rec-ommendable and usable ifcOWL ontology. Automation in Construction 63: 100–133. DOI: 10.1016/j.autcon.2015.12.003.

Radinger, A. Rodriguez-Castro, B. Soltz, A. & Hepp, M. 2013. BauDataWeb: The Austrian Building and Construction Materials Market as Linked Data. In proceedings of the ISEM'13, September 04—06 2013, Graz, Austria.

eWork and eBusiness in Architecture, Engineering and Construction – Christodoulou & Scherer (Eds)
© 2016 Taylor & Francis Group, ISBN 978-1-138-03280-4

Automatic ontology-based green building design parameter variation and evaluation in thermal energy building performance analyses

K. Baumgärtel & R.J. Scherer
Institute of Construction Informatics, Technische Universität Dresden, Dresden, Germany

ABSTRACT: Thermal energy analyses based on Building Information Models (BIMs) are becoming more and more practicable in architecture, engineering and construction. This enables detailed studies about the building energy behaviour with predefined energy-relevant parameters. Although this is an absolute advantage there are also some problems regarding the daily work of energy experts. The simulation configurations and executions cost much time and the pre-processing can be very erroneous due to design modelling problems, wrong material assignments etc. To allow assignments from external data like product catalogues or climate information to BIM data, the energy-extended BIM (eeBIM) framework was developed as multimodel concept for energy simulations. This multimodel was extended by ontologies to allow semantic enrichments and constraints for checking the model quality of inter-linked models. It can be used as an input data set for thermal energy performance analyses. An energy performance platform, called Virtual Energy Laboratory (VEL), integrates different energy tools and data management functions to allow complex thermal energy simulations based on a BIM and additional energy-relevant data. This paper shows how an optimized Green Building Design (GBD) can automatically be derived from a building information model using semantic technologies and highly-scalable processing methods based on an ontology-controlled workflow in the VEL.

1 INTRODUCTION

The search of an optimal green building design is a huge task for architects and energy planners. Buildings have to be designed in a resource-efficient way following ecological principles and a minimized impact on its environment (Sinha 2009; Kibert 2012). Various requirements of building owners, national institutions and physical limitations have to be considered. In the Virtual Energy Laboratory a sensitivity analysis regarding thermal energy helps to identify Green Building Design Parameters (GBDPs) which have a big impact on the building indoor climate.

The way to an optimized GBD is complex and mostly cyclic. Starting from a design, simulation variants with appropriate GBDP, representing constructions with different material layers and occupancy data, have to be configured by an energy expert so that given target Key Performance Indicators (KPIs) like the maximum heating energy consumption will be fulfilled.

An ontology-controlled energy simulation workflow which uses a RDF data graph and basic services integrated in the VEL will be explained in this paper to allow an enhanced and simplified simulation target definition setup. This enables the automatic configuration of multiple thermal energy simulations and the generation of multiple simulation models to compare different green building designs alternatives after the simulation runs with the calculated KPIs. The concept of an automatic simulation configuration will be presented together with its workflows, data integration and semantic descriptions, rules and queries based on ontologies.

In section 2 the semantic model concept of the VEL is introduced. It follows the multimodel approach of the energy-extended BIM which was defined in (Katranuschkov et al. 2011) and (Scherer et al. 2011). The eeBIM serves as central model framework bringing together information from different domains like BIM, cost models, energy models and other data models by using a link model to formulate entity relationships between them. This multimodel approach is implemented in the VEL based on ontologies. The laboratory integrates simulation tools that use the eeBIM to calculate KPIs. The KPIs of different simulation variants can be compared with each other within the VEL. The variant management is done via a Simulation Matrix where the parameters are defined.

With those semantic models a sensitivity analysis can be executed. Section 3 presents the sensitivity analysis provided by the VEL based on ontologies, constraints and rules. The analysis is split different phases where all necessary data is processed step-wise starting with a user-defined configuration.

Section 4 presents an extension of that workflow to allow an automatic optimization of the GBD following user requirements and target KPIs.

Finally, section 5 provides an evaluation of the concept and gives a summary of this paper.

2 SEMANTIC MODEL MANAGEMENT IN THE VIRTUAL ENERGY LABORATORY

2.1 *Virtual Energy Laboratory*

The Virtual Energy Laboratory is a platform for managing BIM data and thermal building energy performance analyses. It consists of different functionalities for:

- Importing BIM data like Industrial Foundation Classes (IFC) files and non-BIM data files like construction data with material arrangements, occupancy data and Test Reference Years (TRYs) as climate data;
- Homogenising different data sets with semantic technologies using ontologies;
- Visualising building information and allowing assignments of energy-relevant data to BIM data to build an energy extended BIM;
- Checking of BIM and eeBIM model quality to exclude semantic problems;
- Transforming and enriching level of details of the building information model;
- Testing building information against user requirements;
- Configuring and starting multiple thermal energy simulations;
- Creating Key Performance Indicators (KPIs) based on simulation results and visualise them;
- Selecting optimal thermal energy simulation variants or reassign other GBDP to create new eeBIM variants and start new simulations (providing simulation cycles).

The modules providing the functionality are described in other papers (Baumgärtel et al. 2011; Baumgärtel et al. 2012; Baumgärtel 2013) in detail.

2.2 *From BIM to eeBIM*

There is a complex BIM management in the VEL starting with an analysis of a given IFC file (currently IFC2 × 3) and a semantic check against energy-relevant exchange requirements. Such exchange requirements are for example the check for defined rooms and space boundaries between a room and its bounding elements.

If the check succeeds, the IFC file is transformed to a representation in Web Ontology Language (OWL) (World Wide Web Consortium 2004) using IfcOWL from (Beetz et al. 2009). The idea is to have all building information in an easily extendible data format like OWL so that other non-IFC data can be easily inter-linked with it. This multimodel integration is described in (Baumgärtel et al. 2014). In this concept the term extended BIM (*eBIM*) is used when it is based on IfcOWL and semantic enrichments and simplifications helps to identify energy concepts like façade elements and to set room types based on a room book.

Using eBIM allows the application of SPARQL (Prud'hommeaux & Seaborne 2008) for querying and semantic rules for inferring new knowledge. Together with the GBDP which are also integrated in additional ontologies based on OWL the eeBIM can be created. This is done by the user when assigning parameter for the climate like the TRY or constructions to building elements. The eeBIM is the highest level of detail and is absolutely needed before a thermal energy simulation can be started. Mostly, the information models like IFC, climate data, occupancy data etc. are provided in heterogeneous data formats. Therefore, a homogeneous data format using OWL and RDF simplifies the linkage of different models. Ontology rules and conditions guarantee that all information needed is available and the VEL is responsible for automatically checking and informing the user about potential problems.

The eeBIM is expressed through RDF triples and forms the overall RDF graph containing all necessary information to run thermal energy simulations as well as the evaluation of results after the simulation runs with Key Performance Indicators (KPIs). With the eeBIM we also allow the validation of a Green Building Design against energy regulations like the European Directives 2010/31/EU and 2012/27 (Baumgärtel et al. 2014).

2.3 *Simulation matrix*

In the sensitivity analysis in the VEL the user focuses on changing Green Building Design Parameters of building entities while the geometry data should be changed when an optimization of materials or occupancy data still not suffices. Therefore, most of the simulation input data is not changed when generating multiple GBD variants.

The building geometry and the building entities like walls, windows, rooms etc. expressed in the eBIM as RDF graph are the base conditions. Together with the GBDP assignments to the building entities they represent one eeBIM variant. The changes to the base condition are structured in a *simulation matrix*. The simulation matrix is a XML-based file which can be transferred together with the eBIM to the simulation program to start multiple thermal energy simulations (Katranuschkov et al. 2015). It consists of *Targets*, *Variables*, *Assignment Groups* and *Combinations*.

Targets are spatial structures of the building like the building itself, one or more storeys or several rooms which specify the simulation scopes. With that definition it is possible to restrict the simulations on parts of the building.

Variables consist of a list of GBDP variations like constructions with different thermal transmittance values or occupancy data.

Assignment Groups are either groups of building elements (e.g. façade elements) or spatial structures (e.g. rooms) and connections to Variables. Such links can vary from 1:1 up to m:n bindings.

Combinations are links between Targets, Variables and Assignment Groups to define each simulation variant. Thus each combination defines exactly which parameter variations are to be applied in each specified simulation task. In addition, there are predefined cross-product and parallel single-value combinations which are useful shortcuts for the sensitivity analysis steps.

3 ONTOLOGY-CONTROLLED SENSITIVITY ANALYSIS WORKFLOW

Figure 1 presents the ontology-controlled workflow in the VEL (Baumgärtel et al. 2015) where semantic checks guide the user through the assignment of appropriate materials or room usage scenarios. Starting with a *setup phase* a user specifies room requirements which are used in the *pre-processing phase* for preparing the *simulation phase* and in the *post-processing phase* for evaluating the results. Below each phase is described in detail:

3.1 *Setup phase*

In the setup phase a user specifies requirements for each room like GDBP ranges, e.g. maximum thermal transmittance values (u-value) for windows, and KPI target values, e.g. maximum heating energy consumption, in a graphical user interface. The GDBP requirements can be selected out of a requirement catalogue, as described in (Baumgärtel et al. 2014), or can be created by using various input fields and are transformed into SPARQL *queries*, while the KPI target values are translated to ontology *rules*. For example, by applying

```
PREFIX BIMOnto: http://openeebim.bau.tu-dresden.de/dev/ontology/BIMOnto#
PREFIX eeBIM: http://openeebim.bau.tu-dresden.de/dev/ontology/eeBIMOnto#
SELECT ?c
WHERE {
    ?c  a   BIMOnto:Construction;
        eeBIM:type  ?type;
        eeBIM:uValue    ?uv.
    FILTER (regex(?type,"window","i") && ?uv < 1.3)
}
```

Listing 1: SPARQL query for retrieving a list with constructions with thermal transmittance value less than 1.3 W/m²K.

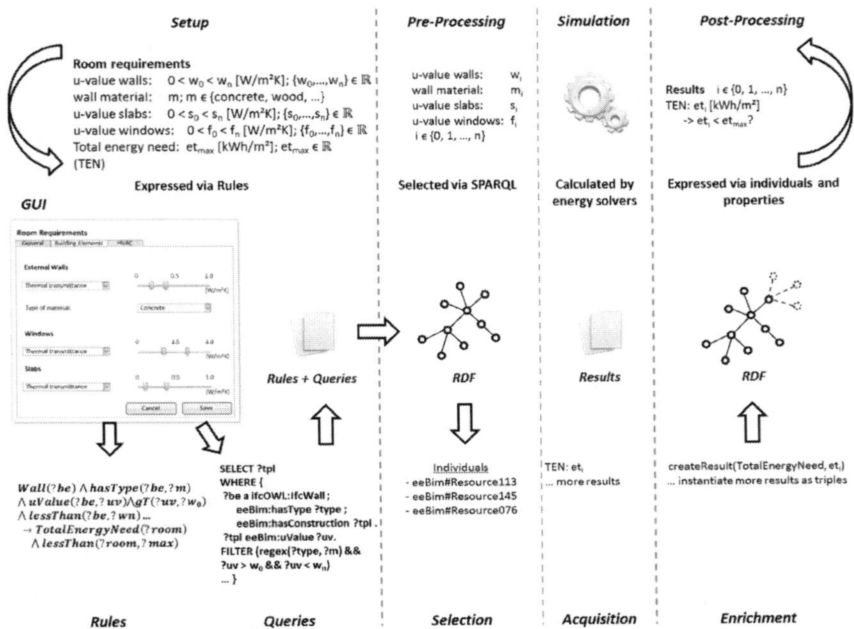

Figure 1. Ontology-controlled simulation workflow.

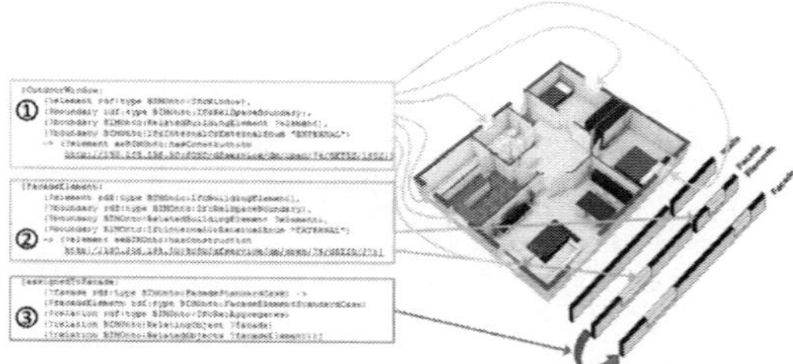

Figure 2. Ontology-based GBDP assignment.

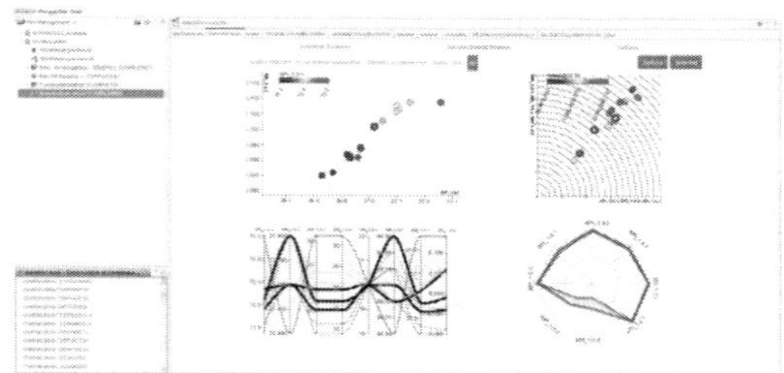

Figure 3. Multi KPI analysis view in the VEL.

the requirements of a passive house, the user can define thermal insulations or parameter of the HVAC system. These definitions have an impact on the *selection* of GBDP, like constructions that have a thermal transmittance value within a user-defined range or suitable occupancy data in the pre-processing phase. There are currently GBDPs for outdoor wall and indoor wall, slab, door and window constructions and occupancy for offices, kitchen etc. provided in the VEL.

3.2 *Pre-processing phase*

In the pre-processing phase the queries from the setup are applied to find all suitable GBDPs. Such an example SPARQL query is presented in listing 1. The user defined that there is an overall u-value of 1.3 W/m²K for window types. The result is a list of possible glazings which fits to this maximum u-value. The user can choose one glazing for all windows which is linked to building entities in the RDF graph via the owl:ObjectProperty *hasConstruction* (the eeBIM ontology can be found under the URL http://openeebim.bau.tu-dresden.de/dev/ontology/) and is shown in Figure 2. The links between the BIM and non-BIM energy relevant data forms an energy extended BIM for the simulations.

3.3 *Simulation phase*

In the simulation phase the eeBIM variants are consumed and transformed to simulation models which are used from the simulation tool to start the processing of new output data. Therefore, this phase is used for acquisition of how specific GBDPs have impacts on different KPIs in a given GBD variant. The tool included in the VEL is called Therakles (Bauklimatik Dresden 2016) and can be used perfectly for starting multiple single-zone simulations parallel. It produces several simulation results, like heating and cooling energy consumptions.

3.4 *Post-processing phase*

After the simulations are completed, all results are collected and computed to KPIs. The rules from the setup phase are used to check the KPIs against user-defined target-KPIs. If computed KPIs of specific variants don't fulfil target KPIs, the variant is marked as not suitable under the given circumstances. If the test succeeds, the results are created in the RDF graph as enrichment of the eeBIM. If all tests fail or the user wants to reassign other GBDP he can start a new cycle beginning in the setup by changing requirements or in the pre-processing phase by changing GBDPs.

Hundreds of simulation variants and it computed KPIs can be viewed in the VEL in the Multi KPI analysis view (see Figure 3). With that view it is possible for the user to filter out variants, which don't fulfil the GBD requirements. All information that is used in a variant is presented so that the user can decide manually which GBD variant is the optimal one.

4 GREEN BUILDING DESIGN VARIATION METHOD

In the VEL we also allow the automatic green building design optimization of a BIM based on the user setup and the ontology-controlled sensitivity analysis workflow. The aim is to find all possible GBD variations which fulfil all user requirements in a fast and comfortable way. The basic conditions are:

- A highly scalable and flexible computation method for creating multiple parallel thermal energy simulations;
- Less user efforts by configuring hundreds or thousands of simulations;
- Automatic check against GBD requirements and target values;
- Concise presentations of best-suitable green building design variants;
- Easy extendible concept to cover complex requirements and to concretize optimizations results.

Therefore, we decided to adapt Google's MapReduce method for creating multiple variations and filtering irrelevant variants. Google's search engine use MapReduce for the fast processing of search requests (Dean & Ghemawat 2008) in a cloud environment. In the *Map* phase input data is distributed to multiple processes and intermediate results are computed out of it. After completion of all processes, the intermediate results are used to produce output data in the *Reduce* phase. Between both phases there is a *Shuffle* phase which distributes the data on the nodes of the Reduce phase.

In the foreground of the MapReduce method the focus is on automatic parallelization and distribution of complex calculations by minimizing the computation time. Therefore, this concept is also suitable in the virtual energy laboratory to determine optimal GBD variants. We use an adapted method with five steps (see Figure 4):

4.1 *Splitting*

In the Splitting phase the BIM is divided in building entity types *ET*. Each $et \in ET$ represents an IFC type which is important for the GBD like IfcWall (semantically categorized in indoor and outdoor walls) for walls, IfcSlab for slabs, IfcWindow for windows or IfcSpace for rooms (see equation 1). The entity types are tightly coupled to GBDP which can be assigned to them. Using all of those entity types the assignment of GBDPs leads to multiple different green building design variants.

$$ET = [Room, Outdoor\,Wall, Indoor\,Wall\ldots] \quad (1)$$

4.2 *Mapping*

The Mapping phase is characterized by the definition of key/value pairs whereat each building entity type *et* is the key and a set *m* of suitable GBDPs is the value. An assigned GBDP is appropriate if it fits to the requirements type (occupancy, construction etc.) and it provides values within the given range (equation 2). The function *map* uses the GBDP requirements from the simulation configuration in the setup phase and assigns all GBDPs

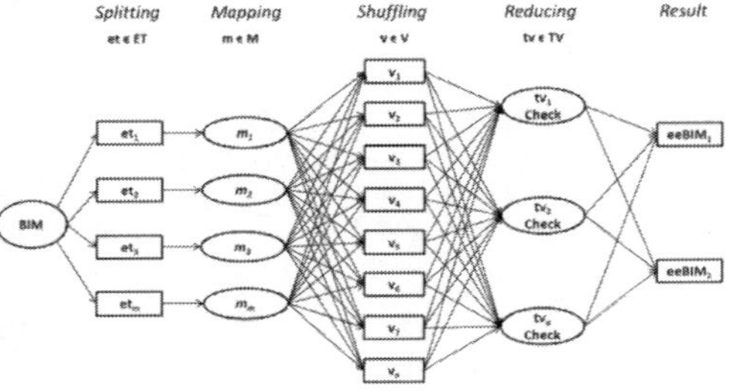

Figure 4. Green building design parameter variation following the MapReduce algorithm.

which fulfil the requirements to the corresponding *et* (equation 3).

For example, et_1 are outdoor walls where the requirement is that they have a maximum thermal transmittance value (u-value) of 0.24 W/m²K, while et_2 are all rooms in a building where a specific room book formulates the requirements of a sanitary room, a kitchen and multiple offices with occupation times 8 a.m. till 8 p.m. from Monday till Friday. m_1 consists of all constructions ($GBDP_1$... $GBDP_n$) out of the provided catalogues which can be used for outdoor walls with a u-value equal or below 0.24 W/m²K. The set m_2 are the room types with the specified occupation times defined in the room book. If the tests succeed, the retrieved GBDPs are added as variables to the simulation matrix.

$$m = \{\mathrm{GBDP}_1 \ldots \mathrm{GBDP}_n | \text{fulfil req.values}\} \quad (2)$$

$$map\colon ET \to M \quad (3)$$

4.3 *Shuffling*

In the Shuffle phase all possible green building design variants (*V*) are created in the simulation matrix by using all result sets *M* for all building entity types *ET*. A variant $v \in V$ consists of all building entity types and one GBDP $g(x)$ per each type. The function *f* maps one building entity type to one parameter of the corresponding GBDP type (see equation 4). The number of all GBD variants results from equation 5. For example, there are five building entity types and the numbers of appropriate GBDPs are $m_1 = 10$, $m_2 = 2$, $m_3 = 5$, $m_4 = 5$, $m_5 = 1$. The total number of GBD variants is $10 \times 2 \times 5 \times 5 \times 1 = 500$. After that, all of those possible variants are added to the simulation matrix as assignment groups.

$$f : et \mapsto g(x) \mid g(x) \epsilon m \quad (4)$$

$$V = ET \times M \quad (5)$$

4.4 *Reducing*

In the Reduce phase the simulation matrix is dissolved to prepare the thermal energy simulations. Each combination presents one simulation and is computed parallel. Each simulation run produces results. The results are calculated to different KPIs like total, heating and cooling energy need. With the user-defined target KPIs target value (*TV*) checks are executed. For each $tv \in TV$ specified in the setup phase the corresponding KPIs of all GBD variants are compared and poor results will be filtered out automatically (equation 6).

$$tv = \begin{cases} \text{succeeds, if KPI meet target value} \\ \text{fails, otherwise} \end{cases} \quad (6)$$

4.5 *Result*

Finally, the results which pass all checks are the green building design variants which fulfil all requirements. They are presented in the multi KPI analysis view (Figure 3) to the user so that they can be compared in detail and to decide on the final design. The aim is to only provide a best-of all variant out of hundreds of variants for the user. This accelerates and facilitates his work significantly.

5 EVALUATION AND CONCLUSIONS

The ontology-controlled simulation workflow helps to automatically check the green building design by comparing requirements with used parameters and specified target with calculated key performance indicator values. The virtual energy laboratory guides the user so that design problems will not appear and problematic assignments to building elements can be excluded.

Starting from a building design given as BIM a step-wise enrichment and conversion of model information with the mapping of an IFC file to its corresponding RDF graph is part of the VEL. This helps to identify façade elements and heated rooms and leads to a more detailed green building design study. The eeBIM represents the assignments of GBDPs to the building information model and we get the possibility to specify constraints and semantic rules so that requirements can be formulated and simulation input data can be validated.

With the use of the adapted MapReduce method an automatic variation of all appropriate parameters and the complete definition of green building design variants can be executed in the laboratory. This is done in a highly flexible and scalable way so that key performance indicators of hundreds of simulations can be calculated and optimal variations of the result set based on target key performance values are selected in parallel.

The user can pick variants that he is most interested in and can compare the input and output data of those simulations in the multi KPI analysis view of the VEL and decide on the final green building design.

ACKNOWLEDGMENTS

We kindly acknowledge the support of the European Commission to the project eeEmbedded, Grant Agreement No. 609349, http://eeembedded.eu.

REFERENCES

Bauklimatik Dresden, 2016. THERAKLES. Available at: http://bauklimatik-dresden.de/therakles/index.php [Accessed April 18, 2016].

Baumgärtel, K., 2013. Multimodellintegration in einem virtuellen Labor. In *Proceedings Forum Bauinformatik 2013*. Munich, Germany, pp. 25–39.

Baumgärtel, K., Guruz, R., Katranuschkov, P. & Scherer, R.J., 2011. Use Cases, Challenges and Software Architecture of a Virtual Laboratory for Life Cycle Building Energy Management. In P. Cunningham & M. Cunningham, eds. *Proceedings eChallenges e–2011*. Florence, Italy.

Baumgärtel, K., Kadolsky, M. & Scherer, R.J., 2014. An ontology framework for improving building energy performance by utilizing energy saving regulations. In B. Martens, A. Mahdavi, & R. Scherer, eds. *eWork and eBusiness in Architecture, Engineering and Construction: ECPPM 2014*. London, England: Taylor & Francis.

Baumgärtel, K., Katranuschkov, P. & Scherer, R.J., 2012. Design and Software Architecture of a Cloud-based Virtual Energy Laboratory for Energy-Efficient Design and Life Cycle Simulation. In G. Gudnason & R.J. Scherer, eds. *eWork and eBusiness in Architecture, Engineering and Construction: ECPPM 2012*. Reykjavik, Iceland.

Baumgärtel, K., Katranuschkov, P. & Scherer, R.J., 2015. Ontology-controlled Energy Simulation Workflow. In *Sustainable Places 2015*. Savona, Italy: Sigma Orionis, pp. 28–36.

Beetz, J., van Leeuwen, J. & de Vries, B., 2009. IfcOWL: A case of transforming EXPRESS schemas into ontologies. *Artificial Intelligence for Engineering Design, Analysis and Manufacturing*, 23(01), p. 89.

Dean, J. & Ghemawat, S., 2008. MapReduce : Simplified Data Processing on Large Clusters. *Communications of the ACM*, 51(1), pp. 1–13.

Katranuschkov, P., Guruz, R., Liebich, T. & Bort, B., 2011. Requirements and Gap Analaysis for BIM Extension to an Energy Efficient BIM Framework. *2nd Workshop on eeBuildings Data Models (CIBW078-W102)*.Sophia Antipolis, France.

Katranuschkov, P., Scherer, R. & Hoch, R., 2015. Optimizing Energy-Efficient Building Design Using BIM.

Kibert, C.J., 2012. Sustainable Construction: Green Building Design and Delivery. John Wiley & Sons, 2012.

Prud'hommeaux, E. & Seaborne, A., 2008. SPARQL Query Language for RDF. W3C Recommendation 15 January 2008. *World Wide Web Consortium*.

Scherer, R.J., Grunewald, J. & Baumgärtel, K., 2011. A Framework Approach for eeBIM and Heterogeneous eeAnalysis Data Models. In *2nd Workshop of eeBuilding Data Models*. Sophia Antipolis, France.

Sinha, R., 2009. Green Building: A Step Towards Sustainable Architecture. *ICFAI Journal of Infrastructure*, 7(2), pp. 91–102.

World Wide Web Consortium, 2004. OWL Web Ontology Language. Available at: https://www.w3.org/TR/owl-features/ [Accessed April 21, 2016].

eWork and eBusiness in Architecture, Engineering and Construction – Christodoulou & Scherer (Eds)
© 2016 Taylor & Francis Group, ISBN 978-1-138-03280-4

A comprehensive ontologies-based framework to support the retrofitting design of energy-efficient districts

G. Costa & Á. Sicilia
ARC, La Salle Engineering and Architecture, Ramon Llull University, Barcelona, Spain

G.N. Lilis
Department of Production Engineering and Management, Technical University of Crete, Chania, Greece

D.V. Rovas
Institute for Environmental Design and Engineering, The Bartlett School of Environment, Energy and Resources, University College London, London, UK

J. Izkara
Construction Unit, Tecnalia Research and Innovation, Bizkaia, Spain

ABSTRACT: As part of the Europe 2020strategy, one of the challenges for the European construction sector is to reduce the energy footprint and CO_2 emissions from new and renovated buildings. This interest is also fostered at a district scale with new technological solutions being developed to achieve more efficient designs. In response to this challenge, a web-based platform for district energy-efficient retrofitting design projects has been proposed in the context of OptEEmAL research project. In order to provide data integration and interoperability between BIM/GIS models and energy simulation tools through this platform, a District Data Model (DDM) has been devised. In this model, fields for urban sustainable regeneration (energy, social, environment, comfort, urban morphology and economic) are related to existing ontological models based on the CityGML and IFC schemas. This paper discusses how the semantic representation from IFC and CityGML files with different levels of detail can be integrated to obtain consistent description of the district in the proposed district data model.

1 INTRODUCTION

Energy efficiency is a fundamental part of the Europe 2020 strategy to reduce the energy footprint and the CO_2 emissions of new and renovated buildings. To achieve this aim, the optimized, integrated and systemic design of energy efficient retrofitted buildings is also required at district scale. To be more efficient in this design, new technological solutions are needed to carry out different types of simulations through comprehensive tools which assess the impact of different actions through a holistic approach. In this approach, these tools should be able to integrate the different data required to perform these simulations including GIS, BIM, energy, economic, weather, monitoring, social, targets, etc.

One way to assess the impact on the energy efficiency of buildings is by applying energy conservation measures to the project in order to obtain a comparative evaluation of the results. On the basis of this approach an "Optimised Energy Efficient Design Platform has been devised for refurbishment at district level" in the context of the OptEEmAL project[1]. This paper presents the project's vision, lines and general approach proposed to support the retrofitting design in the simulation terms described above.

The platform aims to support decision-making in retrofitting projects at a district level. There are several overreaching objectives that have to be set as part of the retrofitting decision-making process. Such objectives include the improvement of the quality of living of the inhabitants, the use of spaces, the economic dynamism of the areas, or the reduction of poverty. These objectives are not only related to energy consumption, but also to social and economic aspects given that cost-effective solutions and more sustainable developments are also common objectives. Based on these factors, different actions can be taken in the definition of a district retrofitting plan. However, this often requires stakeholders to make complex decisions through the use of different tools whose objective parameters have to be brought to light and investigated through specific technical studies. A holistic approach, solutions able

[1]https://www.opteemal-project.eu/

to integrate stakeholders, tools and all the relevant information about retrofitting in a single framework seem to be missing in the plans and strategies to overcome these drawbacks and to better support the design of district retrofitting.

The management and conservation of urban districts requires an approach that considers each of the buildings and other city elements as a part of an environment that should be conserved, updated and showcased. This approach requires the integration of Geographic Information Systems (GIS) and Building Information Models (BIM), while at the same time bearing in mind the particular nature of urban districts (Döllner & Hagedorn 2007).

However, when the retrofitting is addressed at district scale the complexity of decision making grows disproportionately due to: (1) great number of factors to be considered (e.g., economic, social, technical), (2) the interactions between them, and (3) the number of stakeholders involved in the decision-making process. Consequently, the need arises for an interactive and user-friendly decision support tool to enable an analysis of the impact of the building energy oriented retrofitting project on the sustainability of the urban district in a holistic way and also to facilitate the necessary communication mechanisms that can forge agreement between the multiple stakeholders that are involved in this process (Romero et al. 2014).

The paper is structured as follows: Section 2 introduces the OptEEmAL platform as a candidate solution addressing the need to apply energy efficient district retrofitting actions at district level. An ontology-based approach to facilitate district data integration and promote interoperability with multiple simulation tools is introduced as the key component of this platform. The three essential components of this platform are: 1. The Energy Conservation Measures Catalogue, 2. The District Data Model, and 3. The Simulation Manager. These components are introduced in Section 2. The role of ontologies to facilitate the integration of different data models and tools is reviewed in Section 3 in order to provide the theoretical base to explain how the DDM provides the intertwining of standard data models (e.g., CityGML, IFC) with ontologies in domains related with sustainable regeneration (energy, social, environment, comfort, urban morphology and economic), which is described in Section 4. Conclusions and discussion on ongoing research in the OptEEmAL project is provided in the last section of this paper.

2 OPTEEMAL PROJECT

2.1 *Context and purpose*

To respond to the need for innovative design tools for refurbishment at building and district level, a development of a platform has been proposed in the context of the OptEEmAL project with the aim of delivering an optimized, integrated and systematic design for building and district-level retrofitting projects. Based on given initial district conditions, the platform provides the necessary information for simulation tools according to multiple candidate Energy Conservation Measures (ECM). This information includes buildings, urban areas, weather, sensors, etc., and project constrains (costs, barriers, targets, etc.). Through a comparison of the respective simulation results, the platform identifies the most suitable energy conservation measure, which achieves the desired reduction of the district energy demand and consumption under certain constraints. By selecting the best case, the design is updated according to the measures implemented to obtain an enhanced version which can be returned to the BIM authoring tools.

District and building scales have been addressed separately but both are very much connected in the process of energy efficient retrofitting of districts. Strategic decisions—such as the prioritization of areas to retrofit and the implementation of district heating—are taken at district scale, while executive decisions are mainly addressed at building level. Urban scale and the influence and restrictions imposed by urban environment in the building retrofitting should be taken into consideration in the initial stages of the retrofitting process (e.g., feasibility studies and conceptual design). In the early stages of an energy retrofitting process, administrations and managers are mainly involved and the level of detail of the information required is low. In the later stages of the process (e.g., design or implementation of the interventions) a detailed description of building components and their characteristics are required. The main stakeholders in these steps are architects and constructors, and the decisions to be taken are focused on the building level more than on the component level. In this context, it is critical to identify solutions which can cover the need for the connection between the strategic scale (urban) and the executive scale (building), through the definition of a common, multi-scale and interoperable data model which contains the geometric and semantic information required for the management and decision making in the district retrofitting.

2.2 *Platform components*

The OptEEmAL platform provides services for current situation diagnosis, retrofitting scenarios generation, evaluation, optimization and data export, and comprises three main components: (1) an integrated ontology-based District Data

Model (DDM) that is connected to a (2) catalogue of Energy Conservation Measures (ECMs) and which is accessed by (3) a simulation manager that is responsible for generating the required simulation models for each tool (Figure 1).

The DDM is the central component of the platform which has been conceived as a comprehensive ontologies-based framework for district information representation. This enables the intertwining of standardized data models (e.g., CityGML, IFC) with ontologies in domains related to sustainable regeneration (energy, social, environment, comfort, urban morphology and economic). The DDM provides a semantically integrated data model (including information about the geometry, materials, equipment, and indicators, at the building and urban scales) that the platform needs to carry out retrofitting processes.

Energy Conservation Measures (ECM) both at the building and district level, are contained in an ECM catalogue. These measures contain key information to generate applicable scenarios, but also to overcome the existing barriers in the district and they should comply with user objectives in terms of efficiency improvement, cost constraints, financial schemes, etc. The catalogue includes a wide range of measures to reduce the district energy demand and consumption through passive, active, local Renewable Energy Sources (RES) integration and control strategies measures.

According to the data available, the simulation manager generates inputs for the simulation tools. The corresponding tool (e.g., EnergyPlus) is then invoked to perform the computation. By post-processing the outputs generated by simulation tools, the District Performance Indicators (DPIs) are computed. In short, the simulation manager

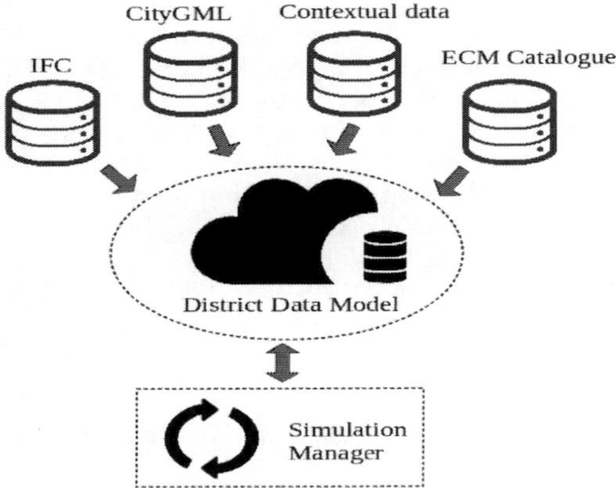

Figure 1. Overview of the architecture platform outlined in OptEEmAL project.

calculates the DPIs according to the baseline scenario to diagnose the current status of the district, and the retrofitting scenarios generated by applying different energy conservation measures. Once the DPIs are calculated according to candidate optimized designs, they are compared with those calculated in the baseline design in order to assess the level of improvement the energy performance or other relevant indicators.

2.3 *Performance evaluation*

DPIs are defined in order to assess the impact of the measures contained in the ECM catalogue using different performance criteria, which include energy demand, consumption, cost and others. In this way the selected scenarios not only conform to a number of different performance constraints but they also achieve the desired performance values. Some of the DPIs require complex calculations while others are obtained by simple operations using other DPIs. DPIs requiring complex operations are evaluated through building simulations. In order to evaluate these DPIs in an automated manner and assess different retrofitting scenarios in a relatively short time, an automated simulation model generation process is established, using data across different time and space scales.

2.4 *Data requirements*

The required data of the OptEEmAL platform can be classified using space and time criteria. Using a spatial differentiation these data can be distinguished into building-level BIM data and district data. BIM data contain information regarding each individual building that includes geometry, construction materials, and building services. District data are related to building groups and include geometric description of multiple building envelopes, and district level systems, serving multiple buildings, such as district heating.

Using time criteria, the required data can be classified into (1) static data, which remain unchanged during simulation executions such as building geometric data, and construction material data, and (2) dynamic data which change during simulation executions such as the operation schedules of building devices.

Managing the above plethora of different data appears to be a challenging issue for the automation of the simulation model generation process performed by the simulation manager. This automation involves the careful selection of subsets of the above data sets based on the requirements and the characteristics of each individual simulation. Two general simulation cases can be distinguished: 1. District-scale simulations performed by CitySim where the district data are required, and 2. Building-scale simulations performed by EnergyPlus where both district and building-scale data must be taken into account. To cover both simulation types IFC, CityGML and Contextual data are required, as described next.

2.5 *Input data sources: IFC, CityGML and contextual data*

Some of the data required for the population of DDM are obtained from instantiated data structures that follow the widely used IFC and CityGML schemas and some—not contained in the previous schemas—are inserted manually, characterized as contextual data. IFC is a popular data schema adopted by the AEC industry as a standard (ISO 16739 2013) which can be used to describe a variety of building entities including architectural, structural and mechanical components. In the context of the present work, data from the architectural section of IFC are required to describe the geometry of the buildings and the properties of building materials and information from the mechanical part will be used to describe building systems.

District-level data are collected from CityGML data structures. These include mostly geometric data of the building envelopes and also the geometric representation of other urban elements (e.g., green areas, roads, city furniture). These other urban element data, play an indirect role in building simulations as shading surfaces and they do not form part of any retrofitting scenario. These data will either augment each individual building simulation model to improve shading calculations, by including additional neighbor shading objects, or will provide the necessary geometric input of district-scale simulation models suitable for CitySim.

IFC and CityGML structures do not provide all the data needed for simulation model generation. Missing data appear in both the building and district contexts and are characterized as contextual data. These include: weather data, operation schedules of devices and inhabitants, simulation parameters, energy prices and building typologies (Vimmr et al. 2013).

Each individual data component should pass three checking stages, namely: correctness, completeness and consistency, before being inserted in the DDM. During the correction stage the inserted data are checked for compliance to certain correctness rules. Towards this direction, error detection and correction mechanisms, such as the ones developed for the geometric data of IFC files (Lilis et al. 2015), can be used in order to guarantee data correctness in DDM. Similarly, during the completeness stage, the data inserted in the DDM are checked against certain completeness rules. For example, completeness rules are defined by the minimum data requirements for simulation model generation. In the case of any of these requirements not being satisfied, a completeness error is reported. Finally, at the consistency stage, the inserted data structures are checked for compatibility to other existing DDM data structures. For example, an inserted IFC BIM model should be correctly placed (location/orientation) with respect to an existing three dimensional CityGML geometric context. Any inconsistency is also reported for correction.

3 USE OF ONTOLOGIES TO INTEGRATE DATA MODELS AND TOOLS

3.1 *Implementation scenarios for data integration*

Several European initiatives and guidelines consider data models of vital importance for the improvement of the energy performance of building information (e.g., EeB[2], ECTP[3]) and try to establish a common geospatial information infrastructure at European Level (e.g., INSPIRE[4]), reaching the conclusion that a better understanding of the urban system is necessary in order to achieve sustainable development goals for cities (e.g., EPIC[5], SEMCITY[6]). Accurate 3D urban models are an important tool for a better understanding of urban systems and thus for sustainable urban development. The solution, based on 3D digital models, has grown in importance over recent years as it offers complete support which is easily updated, allowing information storage and visualization on an urban scale (Mao & Ban 2011).

There are international standards for the management of data related with construction processes based on BIM (Building Information Modelling) and GIS (Geospatial Information System). At a building level, data models based on XML facilitate the validation and exchange between Computer-Aided Design applications or energy assessment tools (gbXML, Architecture Engineering and Construction XML, Building Information Model XML, Industry Foundation Classes XML, etc.). On an urban or city scale GML and KML are the most used data formats for 3D representation. However, both can store geometry but nei-

[2]http://ec.europa.eu/research/industrial_technologies/energy-efficient-buildings_en.html

[3]http://www.ectp.org

[4]http://inspire.ec.europa.eu

[5]http://www.semcity.net/cms

[6]http://www.epic-cities.eu

ther are designed to store semantic information. On a building scale, the problems are similar, CAD tools used to work with lines and polygons and the semantic information available was almost zero. A significant step forward was made towards the semantic enrichment of the building scale model when the BIM concept appeared, and Industry Foundation Classes (IFC) embodied the international open standard for BIM (Succar 2009). A multi-scale data model that integrates both scales is CityGML. The aim of the development of CityGML was to reach a common definition and understanding of the basic entities, attributes, and relations within a 3D city model (Kolbe 2009). This is especially important, since it allows the reuse of the same data in different fields of application (Gröger & Plümer 2012). It has been designed to store both types of information and permits the storage of 3D information, considering both urban scale and building level. CityGML is a standard widely used in Europe. Most of the German cities have a CityGML model, at least at its lowest level of detail (LoD1), some of them with highest level. For example Berlin has one of the most advanced CityGML models).

Combining domain specific information with data models for the construction sector is a task which is being addressed through different approaches. Urban energy tool developers at city level (e.g., CitySim, NEST) have developed their own tailor made urban information models. The extension of existing standard data models for construction sector (e.g., IFC, CityGML) in order to complete them with domain specific information has been mainly addressed through the use of extension mechanisms for each of the models. For example, the Application Domain Extensions (ADE) mechanism has been developed to extend CityGML to other domains. This way, it is possible the storage of relevant domain specific data in a common open city data model which can be used to perform domain-specific simulations. However, the interoperability of the data model is reduced when the extension is implemented and tailored to the data model.

An alternative approach to centralized standard data models is the integration of data using semantic web technologies. This way, data from different existing models can be integrated, for example, to carry out holistic analysis. Multiple domains, scales, and levels of detail have to be modelled, for example, in urban renovation projects (Crapo et al. 2011, Sicilia et al., 2014).

3.2 *Semantic-based interoperability*

Semantic interoperability solutions are based on providing a shared understanding of the meaning associated to the data from different sources and domains in order to facilitate the exchange across networked information systems. The meaning can be provided through ontologies and by making the semantics of data explicit through the use of formal languages. Ontologies specified in OWL (Ontology Web Language) allow the specification of a description-logic-based formal structure to RDF graphs. If an ontology in OWL is instantiated in an RDF graph, generic queries and reasoning engines are able to easily reuse the data of this graph.

Semantic-based solutions can be applied to integrate data from different data models including IFC, CityGML and data from other sources such as cadastre, climate, consumption of buildings, and others. Semantic-based interoperability using Semantic Web technologies is a reasonable technological solution to integrate data from multiple heterogeneous sources and to ensure the communication between the integrated data and an open set of tools. The use of semantic technologies to enhance IFC and CityGML has already been explored in some research works (Laat & Berlo 2011, Amirebrahimi et al. 2015, and others).

3.3 *Data integration and transformation process in OptEEmAL*

The goal of the data integration and transformation process in the OptEEmAL platform is to populate the District Data Model with different input data provided by an end-user (Figure 2). In the first step of the process, the information from input data models is transformed into semantic data models by means of ontologies which define the particular domain of the input data. Then, in between the semantic data models and the simulation tools,

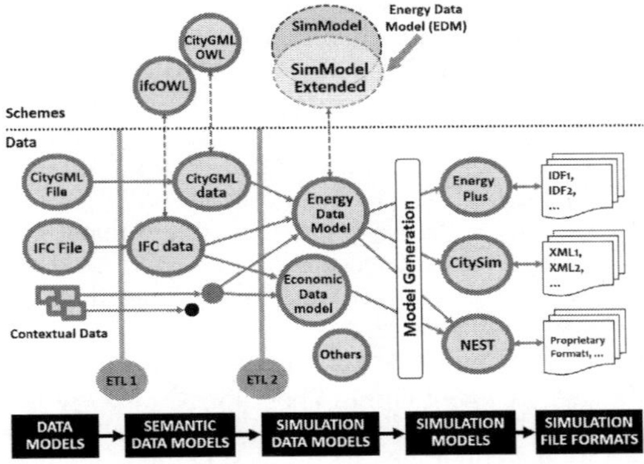

Figure 2. Overview of the data integration and transformation process in OptEEmAL platform. In this figure the process is exemplified for the case of EDM, and EnergyPlus, CitySim and NEST tools.

there are the simulation data models which are ontology-based models to represent a simulation domain such as energy and economic. The simulation data models are generic enough and are representative in order to feed different simulation tools. This way, different simulation models are derived from the simulation data models as specific inputs generated for each simulation tool. For example, the input simulation model for EnergyPlus is an Input Data Files (IDF). For NEST (Neighborhood Evaluation for Sustainable Territories)[7], which is a tool for Life Cycle Assessment (LCA) calculation, the input data file is a proprietary format. For CitySIM[8], which is an urban performance simulation engine, the input data file is an XML based on an application-specific schema.

In the DDM, the semantic data models and simulation data models must be represented by ontologies that define the particular domain of the models. This is needed (1) to carry out the transformation of input data to a specific semantic domain, and (2) to provide a representation in RDF format in order to facilitate their querying through SPARQL language. Queries in this language enable data retrieval to generate simulation models in a flexible way.

Since ontologies are required to represent the input data (IFC, CityGML and contextual data) in RDF in order to facilitate their integration into simulation data models such as the EDM (Energy Data Model).

Different prototypes of ontologies in these domains have been developed in the last decade in order to provide their representation as semantic data. For example, Katranuschkov et al. (2003) created an ontological framework as part of an extensible and open architecture to access data in IFC format. More recently, and as a result of additional research work (Schevers & Drogemuller 2005, Beetz 2009, Pauwels & Terkaj 2016), IFC is now available as an ontology (ifcOWL) with the support of the BuildingSMART. The ifcOWL ontology enables extensions towards other structured data sets using semantic web technologies (buildingSMART 2015). Regarding CityGML, Métral et al. (2010) presented various approaches based on the use of ontologies to improve the interoperability between 3D urban models. This demonstrated that ontologies can overcome semantic limitations in CityGML data models.

In the energy domain, SimModel is the prime example of simulation data model (O'Donnell 2011). It is an XML-based data scheme designed to support building-scale simulation models which has an OWL version (Pauwels, 2014). SimModel does not contain district-related data structures. Therefore, to integrate district data an extension can be developed. Given that simulation data models—such as SimModel—are represented by means of ontologies, they can be easily extended.

The data integration and transformation process in OptEEmAL is based on three steps:

- ETL1: between data models and semantic data models. This is a transformation from raw data sources stored in CSV files, relational databases, XML, Json, etc., to RDF. In the Semantic Web community exists several technological solutions to deal with this kind of sources such as relational-to-RDF translators (e.g., morph-RDB) and mapping languages (e.g., R2RML).
- ETL2: between semantic data models and simulation data models. This is a transformation from a RDF graph to another RDF graph with a different structure defined by simulation data models (e.g., SimModel).
- Model generation: between simulation data models to simulation models (data inputs for simulation tools). This transformation has to be created ad-hoc for each simulation tool using SPARQL queries.

4 DISTRICT DATA MODEL (DDM)

4.1 *Data integration process to generate simulation models*

The approach adopted in the OptEEmAL platform in order to assess the performance of district retrofitting scenarios, through the calculation of different types of DPIs, is divided into four stages: (1) input data quality checking, (2) transformation of input data to OWL/RDF input data (structured according to semantic data models: CityGML OWL, IFC OWL, Contextual Data OWL and ECM data OWL), (3) conversion of OWL/RDF input data to OWL/RDF simulation data models (structured according to semantic data models: SimModel Extended OWL and Other Simulation Domains OWL), and (4) generation of simulation models, as illustrated in Figure 3. This section shows the role played by the DDM in this process to ensure the interoperability between simulation data models and tools, and shows its relation with the rest of the previously introduced platform components.

The process starts with the entry of project data by users of the platform (a CityGML model and different IFC models (noted as IFC_b), but also other data such as socio-economic, sensors monitoring, energy prices and weather). Input data are checked in this first part of the process to verify their correctness, completeness and consistency,

[7]http://www.nobatek-nest.com
[8]http://citysim.epfl.ch

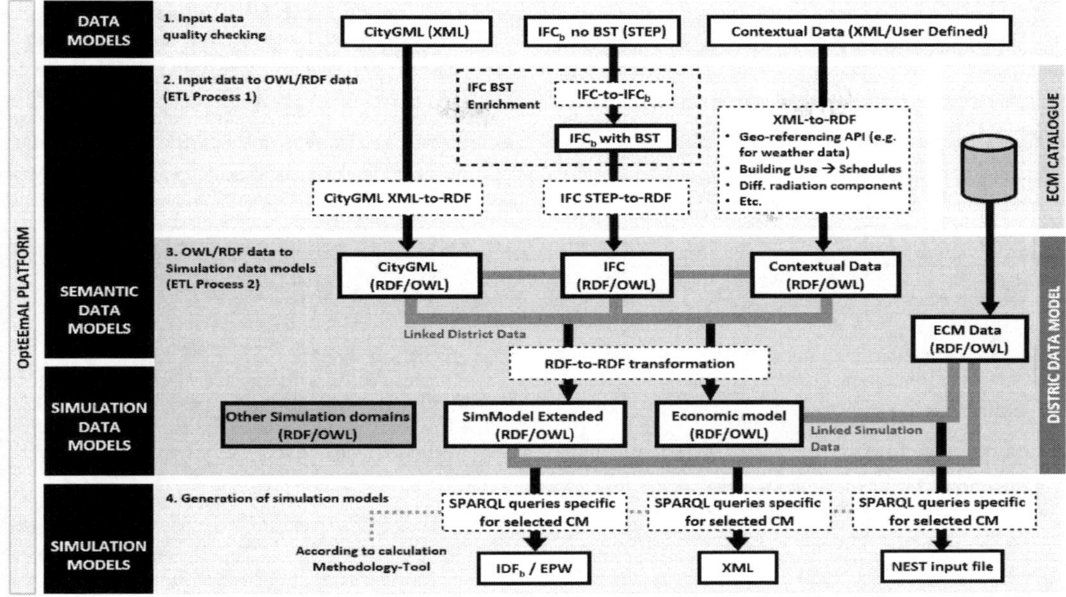

Figure 3. Data flow of the process to generate simulation models.

as described in Section 2.5. Since the information provided in the IFC files cannot be used directly as inputs to energy simulation programs because they require further processing related to the generation of the second-level space boundary geometric topology, a converter is used to carry out this transformation of these IFC files into a Boundary Surface Topology (BST) (Lilis et al. 2016).

In a second step, the input data from CityGML and IFC files are transformed into data described in OWL and RDF languages, according to ontologies that correspond with versions of these standards, and ontologies from other domains related with sustainable regeneration (energy, social, environment, comfort, urban morphology and economic). Through a mapping between ontologies it is possible to perform ETL (Extract, Transform and Load) processes in which the data defined in each of the input domains are transformed to different simulation data models. When the data are defined in the domain of a simulation data model, measures from the ECM Catalogue can be applied as new data aggregated, for example, as new properties of elements and materials. Information described in the simulation data models (e.g., EDM) is queried through SPARQL queries to generate the simulation models in the last step of this process. This models are generated according to each specific tool (EnergyPlus, SimCity, NEST, etc.).

The two following sections show the way in which simulation models are generated in more detail, and how the different levels of detail of the input data provided can be used to create models achieving different levels of accuracy for the DPI calculation.

4.2 *Simulation data models*

The simulation programs which will be used for DPI evaluation (EnergyPlus, CitySim, NEST) require different forms of input data which populate different simulation data models. Furthermore, depending on the data availability, different calculation methodologies can be established using the same simulation tool. As a result, multiple simulation data models can be formed, using different queries as part of the ETL process mentioned earlier, depending on the selected simulation tool and calculation methodology (as highlighted in Figure 3, step 3).

In EnergyPlus input data must be structured using different classes defined, depending on the version of the program, in an input data dictionary file (*.idd file). Based on the data class descriptions contained in the IDD file, a single input data file (*.idf file) will be generated by the simulation manager for each selected retrofitting scenario and calculation methodology.

Similarly, in CitySim, inputs are structured in an XML file format according to a predefined XSD schema. Again here, this XML file will be populated based on the selected scenario and calculation methodology.

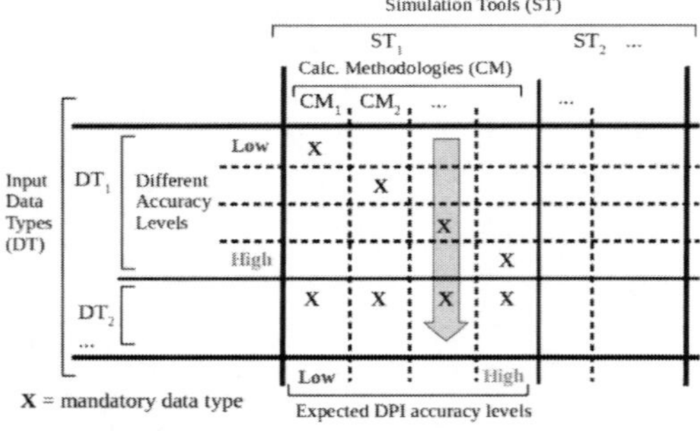

Figure 4. Expected DPI accuracy variation depended on the selected calculation methodology.

Finally, as a future line of work, an interface to integrate the NEST tool with the OptEEmAL platform will be developed.

4.3 *Calculation methodologies*

Sophisticated simulation tools such as EnergyPlus, accept different representations in multiple input Data Types (DTs). For example, the building construction, input data type, may have two descriptions, a detailed multi-layer description where the properties of each material layer are defined, or an equivalent single layer description. Such variability in the accuracy of the description of the input DTs generates multiple simulation execution possibilities, defined as Calculation Methodologies (CMs), for every Simulation Tool (ST). The expected accuracy of the DPI evaluation of each CM, varies as well, as illustrated in Figure 4. Consequently, for each refurbishment scenario and simulation tool, a finite number of Calculation Methodologies (CM) can be established, which depends on the number of possible combinations of the input data type descriptions. Each CM is defined by a unique combination of input data type descriptions, highlighted by the solid arrow in Figure 4.

5 CONCLUSIONS

This paper has introduced a District Data Model devised for the OptEEmAL platform as a framework to support retrofitting designs in districts. Its implementation as an ontology-based approach has been discussed in this paper as a possible solution to the challenge of integrating the different design information (BIM, GIS and contextual data) needed to perform simulations using different tools.

The use of simulation data models in the context of the DDM is presented as a plausible approach to solving the interoperability issues between data models (e.g., IFC, CityGML) and different simulation tools. In the case of the energy domain, SimModel is a simulation data model which is tailored towards the energy-related data required by most popular energy simulation tools. In other domains, such as the economic one, it is still not clear that there is such representative model.

One of the conclusions that can be highlighted in the research outlined in this paper is that even with greater flexibility to generate the simulation models from the data models to perform simulations for each tools, ad-hoc adapters still need to be developed to provide this interoperability.

The next steps in this research are to further develop the ontology-based solution adopted in the DDM in order to carry out fully-automated energy simulations with EnergyPlus and CitySIM.

ACKNOWLEDGEMENTS

Part of the work presented in this paper is based on research conducted within the project "Optimised Energy Efficient Design Platform for Refurbishment at District Level", which has received funding from the European Union Horizon 2020 Framework Programme (H2020/2014–2020) under grant agreement n° 680676.

REFERENCES

Amirebrahimi, S., Rajabifard, A., Mendis, P., & Ngo, T. 2015. A Data Model for Integrating GIS and BIM for Assessment and 3D Visualisation of Flood Damage to Building. *Locate*, 15, 10–12.

Beetz, J., Van Leeuwen, J., & De Vries, B. 2009. IfcOWL: A case of transforming EXPRESS schemas into ontologies. *Artificial Intelligence for Engineering Design, Analysis and Manufacturing*, 23(01), 89–101.

Bontas, E. P., Mochol, M., & Tolksdorf, R. 2005. Case studies on ontology reuse. In *Proceedings of the IKNOW05 International Conference on Knowledge Management* (Vol. 74).

BuildingSMART. (2015). IFC4 Release Summary. http://www.buildingsmart-tech.org/specifications/ifc-releases/ifc4-release.

Crapo, Andrew, Ray Piasecki, and Xiaofeng Wang. 2011. The smart grid as a semantically enabled internet of things. In Implementing Interoperability Advancing Smart Grid Standards, Architecture and Community: Information Interoperability Track, Grid-Interop Forum (Vol. 2011).

Döllner, J., & Hagedorn, B. 2007. Integrating urban GIS, CAD, and BIM data by service based virtual 3D city models. R. et al. (Ed.), *Urban and Regional Data Management-Annual*, 157–160.

Gröger, G., & Plümer, L. 2012. CityGML–Interoperable semantic 3D city models. *ISPRS Journal of Photogrammetry and Remote Sensing*, 71, 12–33.

Gröger G, Kolbe T H, Nagel C, Häfele K H. 2012. OGC City Geography Markup Language (CityGML) Encoding Standard Version 2.0. http://www.opengis.net/spec/citygml/2.0.

ISO 16739. 2013 Industry Foundation Classes (IFC) for data sharing in the construction and facility management industries.

Kolbe, T. H. 2009. Representing and exchanging 3D city models with CityGML. In *3D geo-information sciences* (pp. 15–31). Springer Berlin Heidelberg.

Laat, R., and Berlo, L. 2011. Integration of BIM and GIS: The Development of the CityGML GeoBIM Extension. In *Advances in 3D Geo-Information Sciences*, Berlin, Germany, 211–225.

Lilis, G. N., Giannakis, G., & Rovas, D. 2015. Detection and Semi-automatic correction of geometric inaccuracies in IFC files. IBSPA Conference, Hyderabad, India.

Lilis, G. N., Giannakis, G.I., Katsigarakis, K., Rovas, D.V., Costa, G., Sicilia, A., & Garcia, M.A. 2016. Simulation model generation combining IFC and CityGML data. In *Proceedings of the 11th European Conference on Product & Process Modelling*. Limassol, Cyprus.

Mao, B., & Ban, Y. 2011. Online Visualization of 3D City Model Using CityGML and X3DOM. *Cartographica: The International Journal for Geographic Information and Geovisualization*, 46(2), 109–114.

Métral, C., Billen, R., Cutting-Decelle, A. F., & Van Ruymbeke, M. 2010. Ontology-based approaches for improving the interoperability between 3D urban models. *Journal of Information Technology in Construction*, 15, 169–184.

O'Donnell, J., See, R., Rose, C., Maile, T., Bazjanac, V., & Haves, P. 2011. SimModel: A domain data model for whole building energy simulation. *Proceedings of Building Simulation 2011: 12th Conference of International Building Performance Simulation Association*, Sydney.

Paslaru, E. 2005. Using context information to improve ontology reuse. In Doctoral Workshop at the 17th Conference on Advanced Information Systems Engineering CAiSE (Vol. 5).

Pauwels, P., Corry, E., & O'Donnell, J. 2014. Representing SimModel in the web ontology language. In *American Society of Civil Engineers* (pp. 2271–2278).

Pauwels, P., & Terkaj, W. (2016). EXPRESS to OWL for construction industry: towards a recommendable and usable ifcOWL ontology. *Automation in Construction*, 63, 100–133.

Romero, A.; Egusquiza, A.; Izkara, J.L. 2014. Integrated decision support tool in energy retrofitting projects for sustainable urban districts. In *proceedings of the SB14 World Sustainable Buildings*. Barcelona. October 2014.

Schevers, H., & Drogemuller, R. 2005. Converting the industry foundation classes to the web ontology language. In *Semantics, Knowledge and Grid, 2005. SKG'05. First International Conference on* (pp. 73–73). IEEE

Sicilia, A., Madrazo, L., & Pleguezuelos J. 2014. Integrating multiple data sources, domains and tools in urban energy models using semantic technologies. In *Proceedings of the 10th European Conference on Product & Process Modelling*. Vienna, Austria, pp. 837–844.

Succar, B. 2009. Building information modelling framework: A research and delivery foundation for industry stakeholders. *Automation in Construction*, 18(3), 357–375.

Vimmr, T., Loga, T., Diefenbach, N., Brita, S., & Lucie, B. 2013. Tabula—Residential Building Typologies. In *12th European Countries - Good practice example from the Czech Republic*. CESB13 Prague, Czech Republic.

eWork and eBusiness in Architecture, Engineering and Construction – Christodoulou & Scherer (Eds)
© 2016 Taylor & Francis Group, ISBN 978-1-138-03280-4

Author index